U0360944

郭柏灵院士

郭柏灵论文集

第十三卷

Selected Papers of Guo Boling

Volume 13

SCIENCE PRESS
Beijing

Responsible Editor: Chen Yuzhuo

Published by Science Press
16 Donghuangchenggen North Street
Beijing 100717, P. R. China

Printed in Beijing

ISBN 978-7-03-059560-7

序

今年是恩师郭柏灵院士 70 寿辰，华南理工大学出版社决定出版《郭柏灵论文集》。郭老师的弟子，也就是我的师兄弟，推举我为文集作序。这使我深感荣幸。我于 1985 年考入北京应用物理与计算数学研究所，师从郭柏灵院士和周毓麟院士。研究生毕业后我留在研究所工作，继续跟随郭老师学习和研究偏微分方程理论。老师严谨的治学作风和对后学的精心培养与殷切期望，给我留下了深刻的印象，同时老师在科研上的刻苦精神也一直深深地印在我的脑海中。

郭老师 1936 年出于福建省龙岩市新罗区龙门镇，1953 年从福建省龙岩市第一中学考入复旦大学数学系，毕业后留校工作。1963 年，郭老师服从祖国的需要，从复旦大学调入北京应用物理与计算数学研究所，从事核武器研制中有关的数学、流体力学问题及其数值方法研究和数值计算工作。他全力以赴地做好了这项工作，为我国核武器的发展做出了积极的贡献。1978 年改革开放以后，他又在非线性发展方程数学理论及其数值方法领域开展研究工作，现为该所研究员、博士生导师，中国科学院院士。迄今他共发表学术论文 300 余篇、专著 9 部，1987 年获国家自然科学奖三等奖，1994 年和 1998 年两度获得国防科工委科技进步奖一等奖，为我国的国防建设与人才培养做出了巨大贡献。

郭老师的研究方向涉及数学的多个领域，其中包括非线性发展方程的数学理论及其数值解、孤立子理论、无穷维动力系统等，其研究工作的主要特点是紧密联系数学物理中提出的各种重要问题。他对力学及物理学等应用学科中出现的许多重要的非线性发展方程进行了系统深入的研究，其中对 Landau-Lifshitz 方程和 Benjamin-Ono 方程的大初值的整体可解性、解的唯一性、正则性、渐近行为以及爆破现象等建立了系统而深刻的数学理论。在无穷维动力系统方面，郭老师研究了一批重要的无穷维动力系统，建立了有关整体吸引子、惯性流形和近似惯性流形的存在性和分形维数精细估计等理论，提出了一种证明强紧吸引子的新方法，并利用离散化等方法进行理论分析和数值计算，展示了吸引子的结构和图像。下面我从这几个方面介绍郭老师的一些学术成就。

Landau-Lifshitz 方程 (又称铁磁链方程) 由于其结构的复杂性，特别是强耦合性和不带阻尼时的强退化性，在 20 世纪 80 年代之前国内外几乎没有从数学上进行理论研究的成果出现。最先进行研究的，当属周毓麟院士和郭老师，他们在 1982 年到 1986 年间，采用 Leray-Schauder 不动点定理、离散方法、Galerkin 方法证明了从一维到高维的各种边值问题整体弱解的存在性，比国外在 1992 年才出现的同类结果早了将近 10 年。

20 世纪 90 年代初期，周毓麟、郭柏灵和谭绍滨，郭柏灵和共敏纯得到了两个在国内外至今影响很大的经典结果。第一，通过差分方法结合粘性消去法，利用十分巧妙的先验估计，证明了一维 Landau-Lifshitz 方程光滑解的存在唯一性，对于一维问题给出了完整的答案，解决了长期悬而未决的难题。第二，系统分析了带阻尼的二维 Landau-Lifshitz 方程弱解的奇性，发现了 Landau-Lifshitz 方程与调和映照热流的联系，其弱解具有与调和映照热流完全相同的奇性。现在，国内外这方面的文章基本上都引用这个结果。调和映照的 Landau-Lifshitz 流的概念，即是源于此项结果。

20 世纪 90 年代中期，郭老师对于 Landau-Lifshitz 方程的长时间性态、Landau-Lifshitz 方程耦合 Maxwell 方程的弱解及光滑解的存在性问题进行了深入的研究，得到了一系列的成果。铁磁链方程的退化性以及缺少相应的线性化方程解的表达式，对研究解的长时间性态带来很大困难。郭老师的一系列成果克服了这些困难，证明了近似惯性流形的存在性、吸引子的存在性，给出了其 Hausdorff 和分形维数的上、下界的精细估计。此外，我们知道，与调和映照热流比较，高维铁磁链方程的研究至今还很不完善。其中最重要的是部分正则性问题，其难点在于单调不等式不成立，导致能量衰减估计方面的困难。另外一个是 Blow-up 解的存在性问题，至今没有解决；而对于调和映照热流来说，这样的问题的研究是比较成熟的。

对于高维问题，20 世纪 90 年代后期至今，郭老师和陈韵梅、丁时进、韩永前、杨士山一道，得出了许多成果，大大地推动了该领域的研究进程。第一，证明了二维问题的能量有限弱解的几乎光滑性及唯一性，这个结果类似于 Freire 关于调和映照热流的结果。第二，得到了高维 Landau-Lifshitz 方程初边值问题的奇点集合的 Hausdorff 维数和测度的估计。第三，得到了三维 Landau-Lifshitz-Maxwell 方程的奇点集合的 Hausdorff 维数和测度的估计。第四，得到了一些高维轴对称问题的整体光滑解和奇性解的精确表达式。郭老师还开创了一些新的研究领域。例如，关于一维非均匀铁磁链方程光滑解的存在唯一性结果后来被其他数学家引用并推广到一般流形上。第五，率先讨论了可压缩铁磁链方程测度值解的存在性。最近，在 Landau-Lifshitz 方程耦合非线性 Maxwell 方程与面，也取得了许多新的进展。

多年来，郭老师还对一大批非线性发展方程解的整体存在唯一性、有限时刻的爆破性、解的渐近性态等开展了广泛而深入的研究，受到国内外同行的广泛关注。研究的模型源于数学物理、水波、流体力学、相变等领域，如含磁场的 Schrödinger 方程组、Zakharov 方程、Schrödinger-Boussinesq 方程组、Schrödinger-KdV 方程组、长短波方程组、Maxwell 方程组、Davey-Stewartson 方程组、Klein-Gordon-Schrödinger

方程组、波动方程、广义 KdV 方程、Kadomtsev-Petviashvili (KP) 方程、Benjamin-Ono 方程、Newton-Boussinesq 方程、Cahn-Hilliard 方程、Ginzburg-Landau 方程等。其中不少耦合方程组都是郭老师得到了第一个结果，开创了研究的先河，对国内外同行的研究产生了深远的影响。

郭老师在无穷维动力系统方面也开展了广泛的研究，取得了丰硕的成果。对耗散非线性发展方程所决定的无穷维动力系统，研究了整体吸引子的存在性、分形维数估计、惯性流形、近似惯性流形、指数吸引子等问题。特别是在研究无界域上耗散非线性发展方程的强紧整体吸引子存在性时所提出的化弱紧吸引子成为强紧吸引子的重要方法和技巧，颇受同行关注并广为利用。对五次非线性 Ginzburg-Landau 方程，郭老师利用空间离散化方法将无限维问题化为有限维问题，证明了该问题离散吸引子的存在性，并考虑五次 Ginzburg-Landau 方程的定态解、慢周期解、异宿轨道等的结构。利用有限维动力系统的理论和方法，结合数值计算得到具体的分形维数 (不超过 4) 和结构以及走向混沌、湍流的具体过程和图像，这是一种寻求整体吸引子细微结构新的探索和尝试，对其他方程的研究也是富有启发性的。1999 年以来，郭老师集中于近可积耗散的和 Hamilton 无穷维动力系统的结构性研究，利用孤立子理论、奇异摄动理论、Fenichel 纤维理论和无穷维 Melnikov 函数，对于具有小耗散的三次到五次非线性 Schrödinger 方程，证明了同宿轨道的不变性，并在有限维截断下证明了 Smale 马蹄的存在性，目前，正把这一方法应用于具小扰动的 Hamilton 系统的研究上。他对于非牛顿流无穷维动力系统也进行了系统深入的研究，建立了有关的数学理论，并把有关结果写成了专著。以上这些工作得到国际同行们的高度评价，被称为“有重大的国际影响”“对无穷维动力系统理论有重要持久的贡献”。最近，郭老师及其合作者又证明了具耗散的 KdV 方程 L^2 整体吸引子的存在性，该结果也是引人注目的。

郭老师不仅自己辛勤地搞科研，还尽心尽力培养了大批的研究生 (硕士生、博士生、博士后)，据不完全统计，有 40 多人。他根据每个人不同的学习基础和特点，给予启发式的具体指导，其中的不少人已成为了该领域的学科带头人，有些人虽然开始时基础较差，经过培养，也得到了很大提高，成为了该方向的业务骨干。

《郭柏灵论文集》按照郭老师在不同时期所从事的研究领域，分成多卷出版。文集中所搜集的都是郭老师正式发表过的学术成果。把这些成果整理成集出版，不仅系统地反映了他的科研成就，更重要的是对于从事这方面学习、研究的学者无疑大有裨益。这本文集的出版得到了多方面的帮助与支持，特别要感谢华南理工大学校长李元元教授、华南理工大学出版社范家巧社长和华南理工大学数学科学院吴敏院长的支持。还要特别感谢华南理工大学的李用声教授、华南师范大学的丁时进教授、北京应用物理与计算数学研究所的苗长兴研究员等人在论文的搜集、选择与校对等工作中付出了辛勤的劳动。感谢华南理工大学出版社的编辑对文集的精心编排工作。

谭绍滨

2005 年 8 月于厦门大学

致　谢

《郭柏灵论文集》于 2006 年开始由华南理工大学出版社连续出版了十一卷，论文的收集截止到 2012 年。从第十二卷开始《郭柏灵论文集》将由科学出版社出版发行，从该卷起将收集郭柏灵院士于 2013 年以来先后发表的科研成果。在出版过程中科学出版社的同志，特别是陈玉琢老师进行了精心策划和细致编排，付出了辛勤的劳动。再次谨对他们的帮助表示衷心的感谢。

谭绍滨

2018 年 3 月于厦门大学

Contents

2014 年

2015 年

H^2-regularity Random Attractors of Stochastic Non-Newtonian Fluids with Multiplicative Noise*

Guo Chunxiao (郭春晓), Guo Boling (郭柏灵) and Yang Hui (杨慧)

Abstract In this paper, the authors study the long time behavior of solutions to stochastic non-Newtonian fluids in a two-dimensional bounded domain, and prove the existence of H^2-regularity random attractor.

Keywords random attractors; non-Newtonian fluid; multiplicative noise; Stratonovich process

1 Introduction

In this paper, we consider the following stochastic incompressible non-Newtonian fluids with multiplicative noise, and assume that $D \subset \mathbf{R}^2$ is a two-dimensional bounded smooth open domain,

$$\mathrm{d}u + \big(u \cdot \nabla u - \nabla \cdot \tau(e(u)) + \nabla \pi\big)\mathrm{d}t = g(x)\mathrm{d}t + \sum_{j=1}^{m} b_j u \circ \mathrm{d}\omega_j(t)\,, \quad x \in D,\ t > 0\,, \tag{1.1}$$

$$u(x,0) = u_0(x)\,, \quad x \in D\,, \tag{1.2}$$

$$\nabla \cdot u(x,t) = 0, \tag{1.3}$$

subject to the boundary conditions

$$u(x,t) = 0,\ , \tau_{ijl}\kappa_j\kappa_l = 0,\ \ x \in \partial D, \tag{1.4}$$

where $\circ$ denotes the Stratonovich sense in the stochastic term, and $\omega_j(t), 1 \leqslant j \leqslant m$ are mutually independent two-sided Wiener processes, $b_j \in \mathbf{R}, 1 \leqslant j \leqslant m$ are given. The unknown vector function u denotes the velocity of the fluids, g is the external

* Appl. Math. Mech., 2014, 5(1): 105–116. DOI:doi.org/10.1007/s10483-014-1776-7.

body force vector, and the scalar function π represents the pressure, $\tau_{ij}(e(u))$ is a symmetric stress tensor. Many fluid materials, such as liquid foams, molten plastics, paints etc., whose viscous stress tensors are represented by the form

$$\tau_{ij}(e(u)) = 2\mu_0(\epsilon + |e(u)|^2)^{\frac{p-2}{2}} e_{ij}(u) - 2\mu_1 \Delta e_{ij}(u), \quad \epsilon > 0, \quad i,j = 1,2, \tag{1.5}$$

$$e_{ij}(u) = \frac{1}{2}\left(\frac{\partial u_i}{\partial x_j} + \frac{\partial u_j}{\partial x_i}\right), \quad |e(u)|^2 = \sum_{i,j=1}^{2} |e_{ij}(u)|^2,$$

and $\tau_{ijl} = 2\mu_1 \dfrac{\partial e_{ij}}{\partial x_l}(i,j,l = 1,2)$, $\kappa = (\kappa_1, \kappa_2)$ denotes the exterior unit normal to the boundary ∂D. The first condition represents the usual no-slip condition associated with a viscous fluid, while the second one expresses the fact that the first moments of the traction vanish on ∂D, it is a direct consequence of the principle of virtual work.

When $p = 2$, $\mu_1 = b_j = 0$, equation (1.1) is a deterministic Navier-Stokes equation. $\mu_0 = \mu_1 = b_j = 0$, it is Euler equation, they both are Newtonian fluids. The fluid is called shear thinning, while $1 < p < 2$, and is shear thickening, while $p > 2$.

There are many works concerning the deterministic non-Newtonian fluids, involving the questions of existence, uniqueness and regularity of the solution, the existence of attractor and manifold (see [1,2,3,10,11,18,20]).

In the past decades, for stochastic Navier-Stokes equation, Burgers equation, etc., there have been much work and interesting results related to their existence, uniqueness and attractors, for these topics and the progresses in these fields (see [4,5,6,7,8,9,16]). After that, there are a series of papers to investigate stochastic non-Newtonian fluids. Some important results have been obtained, such as [12,14,15,19], and so on. Recently, Zhao et al. [19] proved the existence of global random attractor for two-dimensional stochastic non-Newtonian fluids with multiplicative noise in the case of $1 < p < 2$, and Guo [13] extended the result to the case of $2 < p < 3$. The appearance of operator $\nabla \cdot (\Delta e)$ makes the non-Newtonian fluids have a higher regularity than Navier-Stokes equation. Many authors have obtained the H^2-regularity attractors for deterministic system (see [3,17,20]). We naturally want to know whether the similar result is also true for non-newtonian fluid with multiplicative noise. It is to this goal that the present paper is addressed. In this paper, we prove that there exist H^2-regularity random attractors for stochastic non-Newtonian fluids with multiplicative noise in the case of $1 < p < 2$.

Crauel, Debussche and Flandoli (see[4,5]) present a general theory to study the random attractors. The main general result on random attractors relies heavily on the existence of a random compact attracting set. In this paper, we will apply the theory to prove the existence of H^2-regularity random attractors for two-dimensional stochastic non-Newtonian fluids in the case of $1 < p < 2$. Firstly, we make use of the Stratonovich transform to change the stochastic equation to a deterministic equation with random parameter; Secondly, we obtain the existence of bounded absorbing sets by some estimates of solution in space V; Thirdly, we use the compact embedding of Sobolev space to obtain the existence of a H^2-compact random set.

The paper is organized as follows. In section 2, we recall some definitions and already known results concerning random attractors; In section 3, we develop all the results needed to prove the existence of random attractors in space V with $1 < p < 2$, $g \in H^1$. In section 4, we establish the existence of a compact random attractor in V by compactness of Sobolev embedding.

Remark 1.1 *After making the Stratonovich transform, the larger p, the more difficult to obtain some norm estimates of v. Especially, in the case of $p > 2$, we need to restrict the lower bound of p in order to obtain the existence of H^2-compact absorbing set. We would discuss this problem in a forthcoming paper by elaborate estimates.*

2 Preliminaries

We introduce some functional spaces and notations for the convenience of the following contents.

$L^q(D)$-the Lebesgue space with norm $||\cdot||_{L^q}$, especially, $||\cdot||_{L^2} = ||\cdot||$.

$H^\sigma(D)$-the Sobolev space $\{u \in L^2(D),\ \ D^k u \in L^2(D),\ \ k \leqslant \sigma\}$, and $||\cdot||_{H^\sigma} = ||\cdot||_\sigma$.

Define a space of smooth functions

$$\mathscr{V} = \{u \in C_0^\infty(\overline{D}) : \nabla \cdot u = 0\},$$

H=the closure of $\mathscr{V}$ in $L^2(D)$ with norm $||\cdot||$, $(\cdot,\cdot)$ denotes the inner product in H.

$H_0^1(D)$ = the closure of $\mathscr{V}$ in $H^1(D)$ with norm $||\cdot||_1$.

V = the closure of $\mathscr{V}$ in $H^2(D)$ with norm $||\cdot||_2$, and V' is the dual space of V.

$\mathscr{C}(I,X)$-the space of continuous functions from the interval I to X. By simple computation, we can conclude the results $\nabla \cdot e(u) = \dfrac{1}{2}\Delta u$, and $\nabla \cdot (\Delta e(u)) = \dfrac{1}{2}\Delta^2 u$.

Thus, $2\mu_1\nabla\cdot(\Delta e(u))=\mu_1\Delta^2 u$.

For notational simplicity, C is a generic constant, and may assume various values from line to line throughout this paper.

We introduce the linear operator A as follows: consider the positive definite V-elliptic symmetric bilinear form $a(\cdot,\cdot):V\times V\to R$ given by

$$a(u,v)=\int_D \Delta u\,\Delta v\mathrm{d}x\,,\quad (u,\ v\in V).$$

As a consequence of the Lax-Milgram lemma, we obtain an isometry $A\in\mathscr{L}(V,V')$,

$$\langle Au,v\rangle_{V'\times V}=a(u,v)=\langle f,v\rangle_{V'\times V}\,,\quad \forall v\in V,$$

where V' is the dual space of V, and the domain of A is

$$D(A)=\{u\in V\ :a(u,v)=(f,v),\ \ f\in H\subset V',\ \ \forall v\in V\}.$$

In fact $A=P\Delta^2$, P is the projection from $L^2(D)$ to H.

Define the trilinear form b on $H_0^1(D)\times H_0^1(D)\times H_0^1(D)$ given by

$$b(u,v,\psi)=\int_D u_i\frac{\partial v_j}{\partial x_i}\psi_j\mathrm{d}x\,,\quad u,v,\psi\in H_0^1(D)\,.$$

Next, define a bilinear map B on $H_0^1(D)\times H_0^1(D)$ by

$$(B(u,u),\psi)=b(u,u,\psi),\ \ u,\psi\in H_0^1(D).$$

Define the map $N(u)$ on $H_0^1(D)$ as follows

$$(N(u),\psi)=\int_D\gamma(u)e_{ij}(u)e_{ij}(\omega)\mathrm{d}x,\quad u,\omega\in H_0^1(D)\,,$$

where $\gamma(u)=2\mu_0(\epsilon+|e(u)|^2)^{\frac{p-2}{2}}$.

From these preparation, equations (1.1)–(1.4) can be translated into the following abstract problems in H :

$$\mathrm{d}u+[\mu_1 Au+N(u)+B(u,u)]\mathrm{d}t=g\mathrm{d}t+\sum_{j=1}^m b_j u\circ\mathrm{d}\omega_j(t),\ \ t>s\,,\tag{2.1}$$

$$u(s)=u_s\,,\ s\in\mathbf{R}\,.\tag{2.2}$$

We next recall some definitions and results concerning the random attractors, which can be found in [4,5]. Let (X,d) be a complete separable metric space and $(\Omega,\mathscr{F},\mathbb{P})$ be a complete probability space. We will consider a family of mappings $S(t,s;\omega):X\to X,\ \ -\infty<s\leqslant t<\infty$, parameterized by $\omega\in\Omega$ in the following contents.

Definition 2.1 *Given $t \in R$ and $\omega \in \Omega$, $K(t,\omega) \subset X$ is an attracting set, if for all bounded sets $B \subset X$*

$$d(S(t,s;\omega)B, K(t,\omega)) \to 0, \quad s \to -\infty,$$

where $d(A,B)$ is the semidistance defined by

$$d(A,B) = \sup_{x \in A} \inf_{y \in B} d(x,y).$$

Definition 2.2 *A family $A(\omega)$, $\omega \in \Omega$ of closed subsets of X is measurable, if for all $x \in X$, the mapping $\omega \mapsto d(x, A(\omega))$ is measurable.*

Definition 2.3 *Let $\{\theta_t : \Omega \to \Omega,\ t \in \mathbf{R}\}$ be a family of measure preserving transformation of $(\Omega, \mathscr{F}, \mathbb{P})$ such that $\theta_0 = id_\Omega$ and $\theta_{t+s} = \theta_t \circ \theta_s$ for all $t, s \in \mathbf{R}$. Here we assume θ_t is ergodic under $\mathbb{P}$. Especially, for all $s < t \in \mathbf{R}$, and $x \in X$,*

$$S(t,s;\omega)x = S(t-s,0;\theta_s\omega)x, \qquad \mathbb{P}-a.e..$$

Definition 2.4 *Define the random omega limit set of a bounded set $B \subset X$ at time t as*

$$A(B,t,\omega) = \bigcap_{T<t} \overline{\bigcup_{s<T} S(t,s;\omega)B}.$$

Definition 2.5 *Let $S(t,s;\omega)_{t \geqslant s, \omega \in \Omega}$ be a stochastic dynamical system, $A(t,\omega)$ is stochastic set satisfying the following conditions:*

(1) *It is the minimal closed set such that for $t \in \mathbf{R}$, $B \subset X$,*

$$d(S(t,s;\omega)B, A(t,\omega)) \to 0, \quad s \to -\infty,$$

which implies $A(t,\omega)$ attracts B (B is a deterministic set).

(2) *$A(t,\omega)$ is the largest compact measurable set, which is invariant in sense that*

$$S(t,s;\omega)\,A(\theta_s\omega) = A(\theta_t\omega), \quad s \leqslant t.$$

Then $A(t,\omega)$ is said to be the random attractor.

Theorem 2.1 (see [4]) *Let $S(t,s;\omega)_{t \geqslant s, \omega \in \Omega}$ be a stochastic dynamical system satisfying the following conditions*

(1) *$S(t,r;\omega)S(r,s;\omega)x = S(t,s;\omega)x$, for all $s \leqslant r \leqslant t$ and $x \in X$,*

(2) *$S(t,s;\omega)$ is continuous in X, for all $s \leqslant t$,*

(3) *for all $s < t$ and $x \in X$, the mapping*

$$\omega \mapsto S(t,s;\omega)x$$

is measurable from $(\Omega, \mathscr{F})$ *to* $(X, \mathscr{B}(X))$,

(4) *for all* t , $x \in X$ *and* $\mathbb{P}$-*a.e.* ω, *the mapping*

$$s \mapsto S(t, s; \omega)x$$

is right continuous at any point.

Assume that there exists a group θ_t, $t \in \mathbf{R}$, *of measure preserving mappings such that*

$$S(t, s; \omega)x = S(t - s, 0; \theta_s\omega)x \ , \qquad \mathbb{P} - a.e. \ s < t, x \in X \tag{2.3}$$

holds and for $\mathbb{P}$-*a.e.* ω *,there exists a compact attracting set* $K(\omega)$ *at time 0, for* $\mathbb{P}$-*a.e.* $\omega \in \Omega$ *, we set* $\Lambda(\omega) = \overline{\bigcup_{B \subset X} A(B, \omega)}$ *, where the union is taken over all the bounded subsets of* X *and* $A(B, \omega)$ *is given by*

$$A(B, \omega) = A(B, 0, \omega) = \bigcap_{T<0} \overline{\bigcup_{s<T} S(0, s; \omega)B},$$

then $\Lambda(\omega)$ *is a random attractor.*

Remark 2.1 *For* (2.1)–(2.2), *let* $\Omega = \{\omega \in C(R, l^2) | \, \omega(0) = 0\}$, *with* $\mathbb{P}$ *being the product measure of two Wiener measures on the negative and the positive time parts of* Ω, *then* $(\beta_1(t, \omega), \beta_2(t, \omega), ..., \beta_k(t, \omega), ...) = \omega(t)$. *In this case, time shift* θ_t *is defined as*

$$(\theta_t\omega)(s) = \omega(t + s) - \omega(t) \ , \quad s, t \in \mathbf{R},$$

so the condition (2.3) *is satisfied.*

3 Some a priori estimate

In this section, we introduce an auxiliary stratonovich process which enables us to change the stochastic equation to an evolution equation depending on a random parameter. Considering the process $\eta(t) = \mathrm{e}^{-\sum_{j=1}^{m} b_j \omega_j(t)}$ satisfies the Stratonovich equation

$$\mathrm{d}\eta(t) = -\sum_{j=1}^{m} b_j \eta(t) \circ \mathrm{d}\omega_j(t). \tag{3.1}$$

Set $v(t) = \eta(t)u(t)$ satisfy the following equation

$$\frac{\mathrm{d}v}{\mathrm{d}t} + 2\mu_1 Av + \eta B(u, u) + \eta N(u) = \eta g, \tag{3.2}$$

$$v(x, s) = v_s = \eta(s)u_s(x), \ \ x \in D \, , \ s \in R, \tag{3.3}$$

Similarly (see [9,10]), we can use the Galerkin method to prove that the following results hold for $\mathbb{P}$-a.e. $\omega \in \Omega$:

- For $g \in H$, $v_s \in H$, $s < T \in \mathbf{R}$, there exists a unique weak solution to (3.2)–(3.3) satisfying $v \in \mathscr{C}(s,T;H)\bigcap L^2(s,T;V)$ with $v(s) = v_s$.
- For $g \in H^1$, $v_s \in V$, $s < T \in \mathbf{R}$, there exists a unique weak solution to (3.2)–(3.3) satisfying $v \in \mathscr{C}(s,T;V)\bigcap L^2(s,T;D(A))$ with $v(s) = v_s$.

We define the stochastic dynamical system $(S(t,s;\omega))_{t\geqslant s,\omega\in\Omega}$ by

$$S(t,s;\omega)u_s = u(t,\omega;s,u_s) = \eta^{-1}(t,\omega)v(t,\omega;s,\eta(s,\omega)u_s).$$

It can be easily checked that the assumptions (1)–(4) are satisfied in Theorem 2.1. In the following, we will prove the existence of a compact attracting set $K(\omega)$ at time 0 in V. First, we try to obtain some estimates in H, $H_0^1(D)$ and V; Second, we use the compactness of the embedding to prove the existence of compact random attractors.

Lemma 3.1 *Let $1 < p < 2, g \in H$, there exist random radius $r_1(\omega), r_2(\omega)$, such that $\forall\rho > 0$, there exists $\overline{s}(\omega) \leqslant -1$, such that for all $s \leqslant \overline{s}(\omega)$, and for all $u_s \in H$, with $\|u_s\| \leqslant \rho$, the solution of equation (3.2)–(3.3) with $v_s = \eta(s)u_s$ satisfies the following inequalities*

$$\|v(t,\omega;s,\eta(s,\omega)u_s)\|^2 \leqslant r_1^2(\omega),\ t\in[-1,0]\ \mathbb{P}-a.e.,$$

$$\int_{-1}^{0}\|v(t)\|_2^2\mathrm{d}t \leqslant r_2^2(\omega),$$

where

$$r_1^2(\omega) = \mathrm{e}^{\mu_1\lambda^2}\left(1+\frac{2\|g\|^2}{\mu_1\lambda^2}\int_{-\infty}^{0}\mathrm{e}^{\mu_1\lambda^2\sigma}\eta^2(\sigma)\mathrm{d}\sigma\right),$$

$$r_2^2(\omega) = \frac{r_1^2(\omega)}{\mu_1} + \frac{2\|g\|^2}{\mu_1^2\lambda^2}\int_{-1}^{0}\eta^2(t)\mathrm{d}t.$$

Proof Taking the inner product of equation (3.2) with v in H, and noticing that $b(u,u,v) = 0$, we obtain

$$\frac{1}{2}\frac{\mathrm{d}}{\mathrm{d}t}\|v\|^2 + \mu_1\|\Delta v\|^2 + 2\mu_0\eta\int_D(\epsilon+|e(u)|^2)^{\frac{p-2}{2}}e(u)e(v)\mathrm{d}x = (\eta g, v). \tag{3.4}$$

Let $I = 2\mu_0\eta\int_D(\epsilon+|e(u)|^2)^{\frac{p-2}{2}}e(u)e(v)\mathrm{d}x$, applying the condition $v = \eta u$, namely, $e(v) = \eta e(u)$, thus $I \geqslant 0$.

We drop the term involving μ_0 in (3.4), then deduce

$$\frac{1}{2}\frac{\mathrm{d}}{\mathrm{d}t}\|v\|^2 + \mu_1\|\Delta v\|^2 \leqslant \frac{\eta^2\|g\|^2}{\mu_1\lambda^2} + \frac{\mu_1\lambda^2\|v\|^2}{4}, \tag{3.5}$$

where we have used the ϵ-Young inequality, and λ is a constant satisfying the inequality $\|\nabla v\|^2 \geqslant \lambda\|v\|^2$, furthermore,

$$\frac{1}{2}\frac{\mathrm{d}}{\mathrm{d}t}\|v\|^2+\frac{\mu_1\|\Delta v\|^2}{2}+\frac{\mu_1\lambda^2\|v\|^2}{2}\leqslant\frac{\eta^2\|g\|^2}{\mu_1\lambda^2}+\frac{\mu_1\lambda^2\|v\|^2}{4}. \tag{3.6}$$

Obviously,

$$\frac{\mathrm{d}}{\mathrm{d}t}\|v\|^2+\mu_1\|v\|_2^2+\frac{\mu_1\lambda^2\|v\|^2}{2}\leqslant\frac{2\eta^2\|g\|^2}{\mu_1\lambda^2}. \tag{3.7}$$

For $s\leqslant -1$, and $-1\leqslant t\leqslant 0$, by Gronwall lemma on the interval $[s,t]$, we can deduce

$$\begin{aligned}\|v(t)\|^2 &\leqslant \mathrm{e}^{-\mu_1\lambda^2(t-s)}\|\eta(s)u(s)\|^2+\int_s^t \mathrm{e}^{-\mu_1\lambda^2(t-\sigma)}\frac{2\|g\|^2}{\mu_1\lambda^2}\eta^2(\sigma)\mathrm{d}\sigma\\ &\leqslant \mathrm{e}^{-\mu_1\lambda^2 t}\left(\mathrm{e}^{\mu_1\lambda^2 s}\eta^2(s)\|u_s\|^2+\frac{2\|g\|^2}{\mu_1\lambda^2}\int_s^t \mathrm{e}^{\mu_1\lambda^2\sigma}\eta^2(\sigma)\mathrm{d}\sigma\right)\\ &\leqslant \mathrm{e}^{\mu_1\lambda^2}\left(\mathrm{e}^{\mu_1\lambda^2 s}\eta^2(s)\|u_s\|^2+\frac{2\|g\|^2}{\mu_1\lambda^2}\int_{-\infty}^0 \mathrm{e}^{\mu_1\lambda^2\sigma}\eta^2(\sigma)\mathrm{d}\sigma\right).\end{aligned} \tag{3.8}$$

On the other hand, by standard argument

$$\lim_{t\to-\infty}\frac{1}{t}\sum_{j=1}^m b_j\omega_j(t)=0,\quad \mathbb{P}-a.s..$$

It follows that $\sigma\mapsto \mathrm{e}^{\mu_1\lambda^2\sigma}\eta^2(\sigma)$ is pathwise integrable over $(-\infty,0]$,

$$\lim_{s\to-\infty}\mathrm{e}^{\mu_1\lambda^2 s}\eta^2(s)=0\quad \mathbb{P}-a.s..$$

Let $r_1^2(\omega)=\mathrm{e}^{\mu_1\lambda^2}\left(1+\frac{2\|g\|^2}{\mu_1\lambda^2}\int_{-\infty}^0 \mathrm{e}^{\mu_1\lambda^2\sigma}\eta^2(\sigma)\mathrm{d}\sigma\right)$, which is finite $\mathbb{P}-a.s.$ by the above conditions. Given $\rho>0$, there exists $\overline{s}(\omega)$ such that $\mathrm{e}^{\mu_1\lambda^2}\eta^2(s)\rho^2\leqslant 1$, for all $s\leqslant\overline{s}(\omega)$, it follows that

$$\|v(t,\omega;s,\eta(s,\omega)u_s)\|^2\leqslant r_1^2(\omega),\ t\in[-1,0].$$

Integrating (3.7) with respect to t from $[-1,0]$, we get

$$\|v(0)\|^2+\mu_1\int_{-1}^0\|v(t)\|_2^2\mathrm{d}t\leqslant\frac{2\|g\|^2}{\mu_1\lambda^2}\int_{-1}^0\eta^2(t)\mathrm{d}t+\|v(-1)\|^2, \tag{3.9}$$

we can easily deduce that

$$\begin{aligned}\int_{-1}^0\|v(t)\|_2^2\mathrm{d}t &\leqslant \frac{r_1^2(\omega)}{\mu_1}+\frac{2\|g\|^2}{\mu_1^2\lambda^2}\int_{-1}^0\eta^2(t)\mathrm{d}t\\ &\doteq r_2^2(\omega).\end{aligned}$$

Lemma 3.2 *Let $1 < p < 2$, $g \in H$, there exist random radius $r_3(\omega), r_4(\omega)$, such that $\forall \rho > 0$, there exists $\overline{s}(\omega) \leqslant -1$, such that for all $s \leqslant \overline{s}(\omega)$, and for all $u_s \in H$, with $\|u_s\| \leqslant \rho$, the solution of equation (3.2)–(3.3) with $v_s = \eta(s)u_s$ satisfies the inequalities*

$$\|v(t,\omega; s, \eta(s,\omega)u_s)\|_2^2 \leqslant r_3^2(\omega), \quad t \in [-1, 0], \mathbb{P} - a.e..$$

$$\int_{-1}^{0} \|v(t)\|_4^2 \mathrm{d}t \leqslant r_4^2(\omega).$$

Proof Taking the inner product of equation (3.2) with $\Delta^2 v$ in H, we can get

$$\frac{1}{2}\frac{\mathrm{d}}{\mathrm{d}t}\|\Delta v\|^2 + \mu_1\|\Delta^2 v\|^2 + \eta b(u, u, \Delta^2 v)$$
$$+ 2\mu_0\eta \int_D (\epsilon + |e(u)|^2)^{\frac{p-2}{2}} e(u)e(\Delta^2 v)\mathrm{d}x = (\eta g, \Delta^2 v). \tag{3.10}$$

Obviously, applying the ϵ-Young inequality, we obtain

$$|(\eta g, \Delta^2 v)| \leqslant \frac{\mu_1\|\Delta^2 v\|^2}{4} + \frac{\eta^2\|g\|^2}{\mu_1}. \tag{3.11}$$

By Hölder inequality, Gagliardo-Nirenberg inequality and ϵ-Young inequality, we get

$$\begin{aligned}
|\eta b(u, u, \Delta^2 v)| &\leqslant C_1\|\eta u\|_{L^4}\|\nabla u\|_{L^4}\|\Delta^2 v\| \\
&\leqslant \frac{\mu_1\|\Delta^2 v\|^2}{8} + \frac{2C_1^2}{\mu_1}\|\eta u\|_{L^4}^2\|\nabla u\|_{L^4}^2 \\
&\leqslant \frac{\mu_1\|\Delta^2 v\|^2}{8} + \frac{2C_1^2 C_2}{\mu_1}\|\eta u\|\|\eta\nabla u\|\|\nabla u\|\|\Delta u\| \\
&\leqslant \frac{\mu_1\|\Delta^2 v\|^2}{8} + \frac{2C_1^2 C_2 C_3}{\mu_1}\|u\|\|u\|_2\|v\|_2\|\Delta^2 v\| \\
&\leqslant \frac{\mu_1\|\Delta^2 v\|^2}{4} + \frac{C}{\mu_1^2}\|u\|^2\|u\|_2^2\|v\|_2^2,
\end{aligned} \tag{3.12}$$

where $C = 8C_1^4 C_2^2 C_3^2$. Similar to the derivation in [20], we can obtain

$$|2\mu_0\eta \int_D (\epsilon + |e(u)|^2)^{\frac{p-2}{2}} e(u)e(\Delta^2 v)\mathrm{d}x| \leqslant \frac{\mu_1\|\Delta^2 v\|^2}{4} + C\|v\|_2^2. \tag{3.13}$$

Combining the above estimates, we can deduce

$$\frac{1}{2}\frac{\mathrm{d}}{\mathrm{d}t}\|\Delta v\|^2 + \frac{\mu_1\|\Delta^2 v\|^2}{4} \leqslant \left(\frac{C}{\mu_1^2}\|u\|^2\|u\|_2^2 + C\right)\|v\|_2^2 + \frac{\eta^2\|g\|^2}{\mu_1}. \tag{3.14}$$

It is also that

$$\frac{\mathrm{d}}{\mathrm{d}t}\|\Delta v\|^2 + \frac{\mu_1\|\Delta^2 v\|^2}{2} \leqslant \left(\frac{2C}{\mu_1^2}\|u\|^2\|u\|_2^2 + 2C\right)\|v\|_2^2 + \frac{2\eta^2\|g\|^2}{\mu_1}. \tag{3.15}$$

Let $L(t)=\dfrac{2C}{\mu_1^2}\|u\|^2\,\|u\|_2^2+2C$, $M(t)=\dfrac{2\eta^2\|g\|^2}{\mu_1}$, then

$$\frac{\mathrm{d}}{\mathrm{d}t}\|\Delta v\|^2\leqslant L(t)\|v\|_2^2+M(t). \tag{3.16}$$

For $-1\leqslant\xi\leqslant t\leqslant 0$, using Gronwall inequality on the interval $[\xi,t]$, we obtain

$$\begin{aligned}\|v(t)\|_2^2&\leqslant\|v(\xi)\|_2^2\exp\left(\int_\xi^t L(\sigma)\mathrm{d}\sigma\right)+\int_\xi^t M(\sigma)\exp\left(\int_\sigma^t L(\tau)\mathrm{d}\tau\right)\mathrm{d}\sigma\\&\leqslant(\|v(\xi)\|_2^2+\int_{-1}^0 M(\sigma)\mathrm{d}\sigma)\exp\left(\int_{-1}^0 L(\sigma)\mathrm{d}\sigma\right).\end{aligned}$$

Integrating above inequality with respect to ξ over $[-1,0]$, then we get

$$\|v(t)\|_2^2\leqslant\left(\int_{-1}^0\|v(\xi)\|_2^2\mathrm{d}\xi+\int_{-1}^0 M(\sigma)\mathrm{d}\sigma\right)\exp\left(\int_{-1}^0 L(\sigma)\mathrm{d}\sigma\right). \tag{3.17}$$

To obtain the boundedness of $\|v(t)\|_2$, next, we estimate the right hand side of inequality (3.17) one by one.

$$\begin{aligned}\int_{-1}^0 L(\sigma)\mathrm{d}\sigma&=\int_{-1}^0(\frac{2C}{\mu_1^2}\|u(\sigma)\|^2\,\|u(\sigma)\|_2^2+2C)\mathrm{d}\sigma\\&\leqslant 2C+\frac{2C}{\mu_1^2}\int_{-1}^0\eta^{-4}(\sigma)\|v(\sigma)\|^2\,\|v(\sigma)\|_2^2\mathrm{d}\sigma\\&\leqslant 2C+\frac{2C}{\mu_1^2}\sup_{-1\leqslant\sigma\leqslant 0}\eta^{-4}(\sigma)\cdot\sup_{-1\leqslant\sigma\leqslant 0}\|v(\sigma)\|^2\int_{-1}^0\|v(\sigma)\|_2^2\mathrm{d}\sigma\\&\leqslant 2C+\frac{2Cr_1^2(\omega)r_2^2(\omega)}{\mu_1^2}\cdot\sup_{-1\leqslant\sigma\leqslant 0}\eta^{-4}(\sigma),\end{aligned} \tag{3.18}$$

where we have used the conclusion of Lemma 3.1.

$$\int_{-1}^0 M(\sigma)\mathrm{d}\sigma=\int_{-1}^0\frac{2\eta^2(\sigma)\|g\|^2}{\mu_1}\mathrm{d}\sigma=\frac{2\|g\|^2}{\mu_1}\int_{-1}^0\eta^2(\sigma)\mathrm{d}\sigma,$$

by the property of Stratonovich process $\eta(t)$ in Lemma 3.1, we deduce that both $\int_{-1}^0 L(\sigma)\mathrm{d}\sigma$ and $\int_{-1}^0 M(\sigma)\mathrm{d}\sigma$ are bounded. Combining these estimates, we can conclude that

$$\|v(t)\|_2^2\leqslant r_3^2(\omega),\ \ \forall t\in[-1,0].$$

Integrating (3.15) with respect to t over $[-1,0]$, then

$$\begin{aligned}\|\Delta v(0)\|^2+\frac{\mu_1}{2}\int_{-1}^{0}\|\Delta^2 v(t)\|^2\mathrm{d}t \leqslant& \int_{-1}^{0}\left(\frac{2C}{\mu_1^2}\|u\|^2\|u\|_2^2+2C\right)\|v(t)\|_2^2\mathrm{d}t\\&+\int_{-1}^{0}\frac{2\eta^2(t)\|g\|^2}{\mu_1}\mathrm{d}t+\|\Delta v(-1)\|^2\\ \leqslant& \int_{-1}^{0}\left(\frac{2C}{\mu_1^2}\|v(t)\|^2\|v(t)\|_2^2\eta^{-4}(t)+2C\right)\|v(t)\|_2^2\mathrm{d}t\\&+\frac{2\|g\|^2}{\mu_1}\int_{-1}^{0}\eta^2(t)\mathrm{d}t+\|\Delta v(-1)\|^2. \qquad (3.19)\end{aligned}$$

In other words, $\displaystyle\int_{-1}^{0}\|v(t)\|_4^2\mathrm{d}t\leqslant r_4^2(\omega)$.

Actually, from this Lemma we get

$$\begin{aligned}\|u(t,\omega;s,u_s)\|_2^2 &= \|\eta^{-1}(t,\omega)v(t,\omega;s,\eta(s,\omega)u_s)\|_2^2\\&\leqslant \sup_{-1\leqslant t\leqslant 0}\frac{1}{\eta^2(t,\omega)}\|v(t,\omega;s,\eta(s,\omega)u_s)\|_2^2\\&\leqslant r_3^2(\omega)\sup_{-1\leqslant t\leqslant 0}\frac{1}{\eta^2(t,\omega)}, \qquad (3.20)\end{aligned}$$

and from above results, $r_3^2(\omega)\sup_{-1\leqslant t\leqslant 0}\dfrac{1}{\eta^2(t,\omega)}$ is bounded. This Lemma implies the existence of bounded absorbing set.

4 Existence of random attractors in V

In this section, we establish the existence of a compact random attractor in V by compactness of Sobolev embedding. First, we need the following Lemma.

Lemma 4.1 *Let $1<p<2$, $g\in H^1$, there exists random radius $r_4(\omega)$, such that $\forall\,\rho>0$, there exists $\overline{s}(\omega)\leqslant -1$, such that for all $s\leqslant\overline{s}(\omega)$, and for all $u_s\in H$, with $\|u_s\|\leqslant\rho$, the solution of equation* (3.2)–(3.3) *with $v_s=\eta(s)u_s$ satisfies the inequality*

$$\|v(0)\|_3^2\leqslant r_7^2(\omega).$$

Proof Taking the inner product of (3.2) with $-A^{\frac{1}{2}}v_t$ in H,

$$\begin{aligned}(v_t,-A^{\frac{1}{2}}v_t)+\mu_1(Av,-A^{\frac{1}{2}}v_t)=&-\eta(N(u),-A^{\frac{1}{2}}v_t)+(\eta g(x),-A^{\frac{1}{2}}v_t)\\&-\eta b(u,u,-A^{\frac{1}{2}}v_t), \qquad (4.1)\end{aligned}$$

the first two terms in the left hand side of (4.1) have the following estimates,

$$(v_t,-A^{\frac{1}{2}}v_t)=(A^{\frac{1}{4}}v_t,A^{\frac{1}{4}}v_t)=\|A^{\frac{1}{4}}v_t\|^2\ ,$$

$$\mu_1(Av,-A^{\frac{1}{2}}v_t)=\mu_1\frac{1}{2}\frac{\mathrm{d}}{\mathrm{d}t}\|v\|_3^2.$$

We can obtain easily

$$|(\eta g,-A^{\frac{1}{2}}v_t)|\leqslant\frac{1}{8}\|A^{\frac{1}{4}}v_t\|^2+2\eta^2\|g\|_1^2.$$

Now, we estimate the nonlinear terms in the equation (4.1).

$$\begin{aligned}|\eta b(u,u,-A^{\frac{1}{2}}v_t)|&\leqslant\eta\int_D[\,|\nabla u||\nabla u|+|u||D^2u|\,]\,|A^{\frac{1}{4}}v_t|\mathrm{d}x\\&\leqslant\eta[\,\|\nabla u\|_{L^\infty}\|\nabla u\|+\|u\|_{L^\infty}\|D^2u\|\,]\|A^{\frac{1}{4}}v_t\|\ ,\end{aligned}$$

where we have used the Hölder inequality. Furthermore, we can obtain

$$\begin{aligned}&|\eta b(u,u,-A^{\frac{1}{2}}v_t)|\\&\leqslant\eta(\,\|u\|_3\|\nabla u\|+\|u\|_2^2\,)\,\|A^{\frac{1}{4}}v_t\|\\&=[\eta^{-1}(\|v\|_3\|v\|_1+\|v\|_2^2)]\,\|A^{\frac{1}{4}}v_t\|\\&\leqslant\frac{1}{4}\|A^{\frac{1}{4}}v_t\|^2+(\eta^{-2}\|v\|_1^2)\|v\|_3^2+\frac{1}{4}\|A^{\frac{1}{4}}v_t\|^2+\eta^{-2}\|v\|_2^4\ ,\end{aligned}$$

where we have used the Sobolev embedding $H^2(D)\hookrightarrow L^\infty(D)$, ε-Young inequality, and the condition $v=\eta u$.

For other nonlinear term, using the Hölder inequality, then

$$\eta|(N(u),-A^{\frac{1}{2}}v_t)|\leqslant2\mu_0\eta\left\{\int_D\left[\frac{\partial^2}{\partial x_i\partial x_k}((\epsilon+|e(u)|^2)^{\frac{p-2}{2}}e_{ij}(u))\right]^2\mathrm{d}x\right\}^{\frac{1}{2}}\|A^{\frac{1}{4}}v_t\|\ .$$

Let

$$\mathcal{J}=\left\{\int_D\left[\frac{\partial^2}{\partial x_i\partial x_k}((\epsilon+|e(u)|^2)^{\frac{p-2}{2}}e_{ij}(u))\right]^2\mathrm{d}x\right\}^{\frac{1}{2}}.$$

The estimate of $\mathcal{J}$ is similar to [14], by the integration by parts, the assumed condition $1<p<2$, sobolev embedding, and some technological computation, then we can get

$$\mathcal{J}\leqslant C\|u\|_3+C\|u\|_2\|u\|_3^2.$$

Thus

$$\begin{aligned}\eta|(N(u),-A^{\frac{1}{2}}v_t)| &\leqslant 2\mu_0\eta[C\|u\|_3+C\|u\|_2\|u\|_3^2]\|A^{\frac{1}{4}}v_t\|\\&\leqslant [2\mu_0C\|v\|_3+2\mu_0C\eta^{-2}\|v\|_2\|v\|_3^2]\|A^{\frac{1}{4}}v_t\|\\&\leqslant \frac{1}{8}\|A^{\frac{1}{4}}v_t\|^2+8\mu_0^2C^2\|v\|_3^2+\frac{1}{8}\|A^{\frac{1}{4}}v_t\|^2+8\mu_0^2C^2\eta^{-4}\|v\|_2^2\|v\|_3^4,\end{aligned}\tag{4.2}$$

where we have used ε-Young inequality, and the condition $v=\eta u$.

Putting above inequalities together, we obtain

$$\begin{aligned}\frac{1}{2}\frac{\mathrm{d}}{\mathrm{d}t}(\mu_1\|v\|_3^2)+\frac{1}{8}\|A^{\frac{1}{4}}v_t\|^2 \leqslant&(\eta^{-2}\|v\|_1^2+8\mu_0^2C^2)\|v\|_3^2+\eta^{-2}\|v\|_2^4\\&+2\eta^2\|g\|_1^2+(8\mu_0^2C^2\eta^{-4}\|v\|_2^2\|v\|_3^2)\|v\|_3^2,\end{aligned}\tag{4.3}$$

changing the inequality suitably to use the Gronwall inequality, then

$$\begin{aligned}\frac{\mathrm{d}}{\mathrm{d}t}(\mu_1\|v\|_3^2)+\frac{1}{4}\|A^{\frac{1}{4}}v_t\|^2 \leqslant&\frac{2(\eta^{-2}\|v\|_1^2+8\mu_0^2C^2+8\mu_0^2C^2\eta^{-4}\|v\|_2^2\|v\|_3^2)}{\mu_1}(\mu_1\|v\|_3^2)\\&+2\eta^{-2}\|v\|_2^4+4\eta^2\|g\|_1^2.\end{aligned}\tag{4.4}$$

Let

$$X(t)=\mu_1\|v\|_3^2,$$

$$Y(t)=\frac{2(\eta^{-2}\|v\|_1^2+8\mu_0^2C^2+8\mu_0^2C^2\eta^{-4}\|v\|_2^2\|v\|_3^2)}{\mu_1},$$

$$Z(t)=2\eta^{-2}\|v\|_2^4+4\eta^2\|g\|_1^2,$$

Thus,

$$\frac{\mathrm{d}}{\mathrm{d}t}X(t)\leqslant Y(t)X(t)+Z(t).$$

For $-1\leqslant\zeta\leqslant\xi\leqslant 0$, by Gronwall inequality on the interval $[\zeta,\xi]$, we obtain

$$\begin{aligned}X(\xi)&\leqslant X(\zeta)\exp\left(\int_\zeta^\xi Y(t)\mathrm{d}t\right)+\int_\zeta^\xi Z(t)\exp\left(\int_t^\xi Y(\tau)\mathrm{d}\tau\right)\mathrm{d}t\\&\leqslant \left(X(\zeta)+\int_{-1}^0 Z(t)\mathrm{d}t\right)\exp\left(\int_{-1}^0 Y(t)\mathrm{d}t\right).\end{aligned}$$

Also integrating above inequality with respect to ζ over $[-1,0]$, then

$$X(\xi)\leqslant\left(\int_{-1}^0 X(\zeta)\mathrm{d}\zeta+\int_{-1}^0 Z(t)\mathrm{d}t\right)\exp\left(\int_{-1}^0 Y(t)\mathrm{d}t\right).\tag{4.5}$$

Next, we prove the right hand side of (4.5) is bounded.

$$\int_{-1}^{0} X(\zeta)\mathrm{d}\zeta = \int_{-1}^{0} \mu_1 \|v(\zeta)\|_3^2 \mathrm{d}\zeta \leqslant \mu_1 \int_{-1}^{0} \|v(\zeta)\|_4^2 \mathrm{d}\zeta \leqslant \mu_1 r_4^2(\omega), \tag{4.6}$$

where we have used the conclusion of Lemma 3.2.

Note that

$$\begin{aligned}
&\int_{-1}^{0} Y(t)\mathrm{d}t \\
=& \int_{-1}^{0} \frac{2(\eta^{-2}(t)\|v(t)\|_1^2 + 8\mu_0^2 C^2 + 8\mu_0^2 C^2 \eta^{-4}(t)\|v(t)\|_2^2 \|v(t)\|_3^2)}{\mu_1}\mathrm{d}t \\
\leqslant& \int_{-1}^{0} \frac{2\|v(t)\|_2^2}{\mu_1}\eta^{-2}(t)\mathrm{d}t + \frac{16\mu_0^2 C^2}{\mu_1} + \int_{-1}^{0} \frac{16\mu_0^2 C^2 \|v(t)\|_2^2 \|v(t)\|_3^2}{\mu_1}\eta^{-4}(t)\mathrm{d}t \\
\leqslant& \frac{2r_3^2(\omega)}{\mu_1}\int_{-1}^{0} \eta^{-2}(t)\mathrm{d}t + \frac{16\mu_0^2 C^2}{\mu_1} + \frac{16\mu_0^2 C^2 r_3^2(\omega)}{\mu_1}\int_{-1}^{0} \|v(t)\|_3^2 \eta^{-4}(t)\mathrm{d}t \\
\leqslant& \frac{2r_3^2(\omega)}{\mu_1}\int_{-1}^{0} \eta^{-2}(t)\mathrm{d}t + \frac{16\mu_0^2 C^2}{\mu_1} + \frac{16\mu_0^2 C^2 r_3^2(\omega)}{\mu_1}\sup_{-1\leqslant t\leqslant 0} \eta^{-4}(t)\int_{-1}^{0} \|v(t)\|_4^2 \mathrm{d}t \\
\leqslant& \frac{2r_3^2(\omega)}{\mu_1}\int_{-1}^{0} \eta^{-2}(t)\mathrm{d}t + \frac{16\mu_0^2 C^2}{\mu_1} + \frac{16\mu_0^2 C^2 r_3^2(\omega) r_4^2(\omega)}{\mu_1}\sup_{-1\leqslant t\leqslant 0} \eta^{-4}(t) \\
\doteq& r_5^2(\omega),
\end{aligned}$$

where we have used the property of Stratonovich process $\eta(t)$ and the conclusion of Lemma 3.2.

Similarly,

$$\begin{aligned}
\int_{-1}^{0} Z(t)\mathrm{d}t &= \int_{-1}^{0} (2\eta^{-2}(t)\|v(t)\|_2^4 + 4\eta^2(t)\|g\|_1^2)\mathrm{d}t \\
&\leqslant 2r_3^4(\omega)\int_{-1}^{0} \eta^{-2}(t)\mathrm{d}t + 4\|g\|_1^2 \int_{-1}^{0} \eta^2(t)\mathrm{d}t \\
&\doteq r_6^2(\omega).
\end{aligned}$$

From (4.5) and above inequalities, we obtain

$$\|v(\xi)\|_3^2 = X(\xi) \leqslant \frac{1}{\mu_1}(\mu_1 r_4^2(\omega) + r_6^2(\omega))\mathrm{e}^{r_5^2(\omega)}, \qquad \xi \in [-1, 0]. \tag{4.7}$$

Now we put $r_7^2(\omega) \doteq \frac{1}{\mu_1}(\mu_1 r_4^2(\omega) + r_6^2(\omega))\mathrm{e}^{r_5^2(\omega)}$, especially, for $\xi = 0$,

$$\|u(0)\|_3^2 = \|v(0)\|_3^2 \leqslant r_7^2(\omega),$$

Theorem 4.1 *Let* $1 < p < 2$, $g \in H^1$, $u_s \in H$, *there exist random attractors for the stochastic non-Newtonian with multiplicative noise* (2.1)–(2.2) *in* V.

Proof Let $K(\omega)$ be the ball in $H_0^3(D)$ of radius $r_7(\omega)$, we have proved that for any B bounded in V, there exists $\overline{s}(\omega)$ such that for $s \leqslant \overline{s}(\omega)$,

$$S(0, s; \omega)B \subset K(\omega)\ \mathbb{P} - a.e..$$

This clearly implies that $K(\omega)$ is an attracting set at time $t = 0$. Since it is compact in V, and Theorem 2.1 applies.

Theorem 4.1 implies that the asymptotic smoothing effect of the fluids in the sense that solution becomes eventually more regular than the initial data.

Acknowledgements This paper is supported by NNSF of China under grant number:11126160,11201475.

References

[1] Bellout H, Bloom F, Nečas J. Young measure-valued solutions for non-Newtonian incompressible viscous fluids [J]. Commun. Part. Differ. Equ., 1994, 19: 1763–1803.

[2] Bloom F. Attractors of non-newtonian fluids [J]. J. Dyn. Diff. Eqs., 1995, 7(1): 109–140.

[3] Bloom F, Hao W. Regularization of a non-Newtonian system in an unbounded channel:existence of a maximal compact attractor [J]. Nonl. Anal. TMA, 2001, 43: 743–766.

[4] Crauel H, Debussche A, Flandoli F. Random attractors [J]. J. Dyn. Diff. Equ., 1997, 9: 307–341.

[5] Crauel H, Flandoli F. Attractors for random dynamical systems [J]. Prob. Th. Rel. Fields, 1994, 100: 365–393.

[6] Da Prato G, Debussche A, Temam R. Stochastic Burgers' equation [J]. NoDEA, 1994, 1: 389–402.

[7] de Bouard A, DebusscheA. On the stochastic korteweg-de vries equation [J]. J. Functional Analysis, 1998, 154: 215–251.

[8] de Bouard A, Debussche A. A stochastic nonlinear Schrödinger equation with multiplicative noise [J]. Comm. Math. Phys., 1999, 205: 161–181.

[9] Da Prato G, Zabczyk J. Stochastic Equations in Infinite Dimensions [M]. Cambridge: Cambridge University Press, 1992.

[10] Guo B, Lin G, Shang Y. Non-Newtonian Fluids Dynamical Systems [M]. Beijing, National Defense Industry Press, 2006.

[11] Guo B, Guo C. The convergence of non-Newtonian fuids to Navier-Stokes equations [J]. J. Math. Anal. Appl., 2009, 357: 468–478.

[12] Guo B, Guo C, Zhang J. Martingale and stationary solutions for stochastic non- Newtonian fluids [J]. Diff. Int. Equ., 2010, 23: 303–326.

[13] Guo C, Guo B. Remark on random attractor for a two dimensional incompressible non-Newtonian fluid with multiplicative noise [J]. Commun. Math. Sci., 2012, 10: 821–833.

[14] Guo B, Guo C, Han Y. Random attractors of stochastic non-Newtonian fluid [J]. Acta Math. Appl. Sin., English Series, 2012, 28: 165–180.

[15] Li J, Huang J. Dynamics of 2D Stochastic non-Newtonian fluids driven by fractional Brownian motion [J]. Appl. Math. Mech., English Edition, 2013, 34(2) :189–208.

[16] Krylov N V, Rozovsikii B L. Stochastic evolution equations [J]. J. Soviet Math., 1981, 16: 1233–1277.

[17] Zhao C, Li Y. H^2-compact attractor for a non-Newtonian system in two-dimensional unbounded domains [J]. Nonlinear Analysis, 2004, 56: 1091–1103.

[18] Zhao C, Li Y. A note on the asymptotic smoothing effect of solutions to a non-Newtonian system in 2-D unbounded domains [J]. Nonlinear Analysis, 2005, 60: 475–483.

[19] Zhao C, Li Y, Zhou S. Random attractor for a two-dimensional incompressible non-Newtonian fluid with multiplicative noise [J]. Acta Mathematica Scientia, 2011, 31: 567–575.

[20] Zhao C, Zhou S. Pullback attractors for a non-autonomous incompressible non-Newtonian fluid [J]. J. Diff. Equ., 2007, 238: 394–425.

Global Existence of Smooth Solutions to the k-ε Model Equations for Turbulent Flows*

Bian Dongfen (边东芬) and Guo Boling (郭柏灵)

Abstract In this paper we are concerned with the global existence of smooth solutions to the k-ε model equations for turbulent flows in $\mathbb{R}^3$. The global well-posedness is proved under the condition that the initial data are close to the standard equilibrium state in H^3-framework. The proof relies on energy estimates about velocity, temperature, turbulent kinetic energy and rate of viscous dissipation. We use several new techniques to overcome the difficulties from the product of two functions and higher order norms. This is the first result concerning k-ε model equations.
Keywords Turbulent flow equations; compressible flows; k-ε model equations; classical solution; global existence

1 Introduction

All flows encountered in engineering practice, both simple ones such as two-dimensional jets, wakes, pipe flows and flat plate boundary layers and more complicated three-dimensional ones, become unstable above a certain Reynolds number. At low Reynolds numbers flows are laminar. At high Reynolds numbers flows are observed to become turbulent. Turbulence stands out as a prototype of multi-scale phenomenon that occurs in nature. It involves wide ranges of spatial and temporal scales and this makes it very difficult to study analytically and prohibitively expensive to simulate computationally. Up to now, there are not any general theory suitable for the turbulent flows. Many, if not most, flows of engineering significance are turbulent, so the turbulent flow regime is not just of theoretical interest. Fluid engineers need access to viable tools capable of representing the effects of turbulence [2, 7]. Hence, research about the above system (1.1) is initial and very important.

* Commun. Math. Sci., 2014, 4(12): 707 – 721.

We consider in this work the k-ε model equations for turbulent flows on $\mathbb{R}^3$,

$$\begin{cases} \rho_t + \operatorname{div}(\rho u) = 0, \\ (\rho u)_t + \operatorname{div}(\rho u \otimes u) - \Delta u - \nabla \operatorname{div} u + \nabla p = -\dfrac{2}{3}\nabla(\rho k), \\ (\rho h)_t + \operatorname{div}(\rho u h) - \Delta h = \dfrac{Dp}{Dt} + S_k, \\ (\rho k)_t + \operatorname{div}(\rho u k) - \Delta k = G - \rho\varepsilon, \\ (\rho\varepsilon)_t + \operatorname{div}(\rho u \varepsilon) - \Delta\varepsilon = \dfrac{C_1 G\varepsilon}{k} - \dfrac{C_2\rho\varepsilon^2}{k}, \\ (\rho, u, h, k, \varepsilon)(x,t)|_{t=0} = (\rho_0(x), u_0(x), h_0(x), k_0(x), \varepsilon_0(x)), \end{cases} \tag{1.1}$$

with

$$S_k = \left[\mu\left(\frac{\partial u^i}{\partial x_j} + \frac{\partial u^j}{\partial x_i}\right) - \frac{2}{3}\delta_{ij}\mu\frac{\partial u^k}{\partial x_k}\right]\frac{\partial u^i}{\partial x_j} + \frac{\mu_t}{\rho^2}\frac{\partial p}{\partial x_j}\frac{\partial \rho}{\partial x_j},$$

$$G = \frac{\partial u^i}{\partial x_j}\left[\mu_e\left(\frac{\partial u^i}{\partial x_j} + \frac{\partial u^j}{\partial x_i}\right) - \frac{2}{3}\delta_{ij}\left(\rho k + \mu_e\frac{\partial u^k}{\partial x_k}\right)\right],$$

where $\delta_{ij} = 0$ if $i \neq j$, $\delta_{ij} = 1$ if $i = j$, and μ, μ_t, C_1, C_2 are four positive constants satisfying $\mu + \mu_t = \mu_e$.

The system (1.1) is formed by combining effect of turbulence on time-averaged Navier-Stokes equations with the k-ε model equations. Here ρ, u, h, k and ε denote the density, velocity, total enthalpy, turbulent kinetic energy and rate of viscous dissipation, respectively. The pressure p is a smooth function of ρ. In this paper, without loss of generality, we have renormalized some constants to be one.

This paper is devoted to the study of the global existence of smooth solutions for the system (1.1) under suitable assumptions. We mainly apply the standard energy method in [1, 3, 4, 6] to prove the global well-posedness for the k-ε model equations (1.1). Our result is expressed in the following.

Theorem 1.1 *Assume that the initial data are close enough to the constant state* $(\bar\rho, 0, 0, \bar k, 0)$, *i.e. there exists a constant* δ_0 *such that if*

$$\|(\rho_0 - \bar\rho, u_0, h_0, k_0 - \bar k, \varepsilon_0)\|_{H^3(\mathbb{R}^3)} \leqslant \delta_0, \tag{1.2}$$

then the system (1.1) admits a unique smooth solution $(\rho, u, h, k, \varepsilon)$ *such that for any* $t \in [0, \infty)$,

$$\begin{aligned} &\|(\rho - \bar\rho, u, h, k - \bar k, \varepsilon)\|_{H^3}^2 + \int_0^t \|\nabla\rho\|_{H^2}^2 + \|(\nabla u, \nabla h, \nabla k, \nabla\varepsilon)\|_{H^3}^2 ds \\ &\leqslant C\|(\rho_0 - \bar\rho, u_0, h_0, k_0 - \bar k, \varepsilon_0)\|_{H^3}^2, \end{aligned}$$

where C *is a positive constant.*

Remark 1.2 *The existence of the local solution for* (1.1) *can be obtained from the standard method based on the Banach theorem and contractivity of the operator defined by the linearization of the problem on a small time interval (see also* [4,5,8]*). Hence, we omit the local existence part for simplicity. The global existence of smooth solutions will be proved by extending the local solutions with respect to time based on a-priori global estimates.*

Remark 1.3 *Although our proofs are in spirit of those for Navier-Stokes and MHD equations* [1,3,4,6], *we shall derive several new estimates arising from the presence of the total enthalpy, turbulent kinetic energy and rate of viscous dissipation and overcome the difficulties from the product of two functions and higher order norms.*

Notation 1.4 *Throughout the paper, C stands for a general constant, and may change from line to line. The norm $\|(A,B)\|_X$ is equivalent to $\|A\|_X + \|B\|_X$. The notation $L^p(\mathbb{R}^3)$, $1 \leqslant p \leqslant \infty$, stands for the usual Lebesgue spaces on $\mathbb{R}^3$ and $\|\cdot\|_p$ denotes its L^p norm.*

The rest of this paper is organized as follows. In Section we establish Proposition 2.1 and some a-priori estimates used in the proof of Theorem 1.1. In Section we complete the proof of Theorem 1.1.

2 A priori estimates

In this section we establish a-priori estimates of the solutions. we assume that $(v,u,\theta,k,\varepsilon)$ is a smooth solution to (1.1) on the time interval $(0,T)$ with $T>0$. We shall establish the following proposition.

Proposition 2.1 *There exists a constant $\delta \ll 1$ such that if*

$$\sup_{0\leqslant t\leqslant T} \|(\rho-\bar{\rho},u,h,k-\bar{k},\varepsilon)\|_{H^3} \leqslant \delta, \tag{2.3}$$

then for any $t\in[0,T]$, there exists a constant $C_1>1$ such that it holds

$$\begin{aligned}&\|(\rho-\bar{\rho},u,h,k-\bar{k},\varepsilon)\|_{H^3}^2 + \int_0^t \|\nabla\rho\|_{H^2}^2 + \|\nabla u,\nabla h,\nabla k,\nabla\varepsilon\|_{H^3}^2 ds\\ &\leqslant C_1\|(\rho_0-\bar{\rho},u_0,h_0,k_0-\bar{k},\varepsilon_0)\|_{H^3}^2.\end{aligned} \tag{2.4}$$

Proof First, let $\rho=a+\bar{\rho}$, $k=m+\bar{k}$, $f'(\rho)=\dfrac{p'(\rho)}{\rho}$, we rewrite the system (1.1) as follows:

$$\begin{cases} a_t + \operatorname{div}((a+\bar{\rho})u) = 0, \\ u_t + u\cdot\nabla u - \dfrac{1}{a+\bar{\rho}}(\Delta u + \nabla \operatorname{div} u) + \nabla[f(a+\bar{\rho}) - f(\bar{\rho})] \\ \quad = -\dfrac{2}{3(a+\bar{\rho})}\nabla((a+\bar{\rho})(m+\bar{k})), \\ h_t + u\cdot\nabla h - \dfrac{1}{a+\bar{\rho}}\Delta h = -f'(a+\bar{\rho})(a+\bar{\rho})\operatorname{div} u + \dfrac{1}{a+\bar{\rho}}S_k, \\ m_t + u\cdot\nabla m - \dfrac{1}{a+\bar{\rho}}\Delta m = \dfrac{1}{a+\bar{\rho}}G - \varepsilon, \\ \varepsilon_t + u\cdot\nabla\varepsilon - \dfrac{1}{a+\bar{\rho}}\Delta\varepsilon = \dfrac{C_1 G\varepsilon}{(a+\bar{\rho})(m+\bar{k})} - \dfrac{C_2\varepsilon^2}{m+\bar{k}}, \\ (a,u,h,m,\varepsilon)(x,t)|_{t=0} = (a_0(x), u_0(x), h_0(x), m_0(x), \varepsilon_0(x)), \end{cases} \tag{2.5}$$

with

$$S_k = \left[\mu\left(\frac{\partial u^i}{\partial x_j} + \frac{\partial u^j}{\partial x_i}\right) - \frac{2}{3}\delta_{ij}\mu\frac{\partial u^k}{\partial x_k}\right]\frac{\partial u^i}{\partial x_j} + \frac{\mu_t}{(a+\bar{\rho})^2}\frac{\partial p}{\partial x_j}\frac{\partial a}{\partial x_j},$$

$$G = \frac{\partial u^i}{\partial x_j}\left[\mu_e\left(\frac{\partial u^i}{\partial x_j} + \frac{\partial u^j}{\partial x_i}\right) - \frac{2}{3}\delta_{ij}\left((a+\bar{\rho})(m+\bar{k}) + \mu_e\frac{\partial u^k}{\partial x_k}\right)\right].$$

From a priori assumption (2.3) and the Sobolev inequality together with the first equation of (2.5), we have

$$\sup_{x\in\mathbb{R}}|(a, a_t, \nabla a, u, \nabla u, h, \nabla h, m, \nabla m, \varepsilon, \nabla\varepsilon)| \leqslant C\|(a,u,h,m,\varepsilon)\|_{H^3} \leqslant C\delta. \tag{2.6}$$

Moreover,

$$\frac{\bar{\rho}}{2} \leqslant \rho = a + \bar{\rho} \leqslant 2\bar{\rho}, \quad \frac{\bar{k}}{2} \leqslant k = m + \bar{k} \leqslant 2\bar{k}, \tag{2.7}$$

and

$$0 < \frac{1}{C_0} \leqslant f'(\rho) \leqslant C_0 < \infty, \quad |f^{(n)}(\rho)| \leqslant C_0 \text{ for any positive integer } n, \tag{2.8}$$

with C_0 a positive constant.

In what follows, we will always use the smallness assumption of δ and (2.6)–(2.8). We divide the a priori estimates into three steps.

Step 1 L^2-norms of u, h, m, ε

Multiplying the second equation of (2.5) by u and integrating over $\mathbb{R}^3$, one can deduce that

$$\begin{aligned}
&\frac{1}{2}\frac{d}{dt}\|u\|_2^2+\frac{1}{a+\bar{\rho}}(\|\nabla u\|_2^2+\|\text{div}u\|_2^2)\\
=&-\int_{\mathbb{R}^3}u\cdot\nabla u\cdot u\text{d}x+\int_{\mathbb{R}^3}(f(a+\bar{\rho})-f(\bar{\rho}))\text{div}u\text{d}x\\
&+\int_{\mathbb{R}^3}\frac{1}{(a+\bar{\rho})^2}[(\nabla a\otimes u):\nabla u+\text{div}u(\nabla a\cdot u)]\text{d}x\\
&-\int_{\mathbb{R}^3}\frac{2}{3(a+\bar{\rho})}\nabla((a+\bar{\rho})(m+\bar{k}))\cdot u\text{d}x.
\end{aligned}\tag{2.9}$$

With the help of Hölder's inequality, $\dot{W}^{1,2}(\mathbb{R}^3)\hookrightarrow L^6(\mathbb{R}^3)$ and (2.6), we estimate the right-hand side of (2.9) as

$$-\int_{\mathbb{R}^3}u\cdot\nabla u\cdot u\text{d}x\leqslant C\|u\|_6\|\nabla u\|_2\|u\|_3\leqslant C\delta\|\nabla u\|_2^2,\tag{2.10}$$

$$\begin{aligned}
&\int_{\mathbb{R}^3}\frac{1}{(a+\bar{\rho})^2}[(\nabla a\otimes u):\nabla u+\text{div}u(\nabla a\cdot u)]\text{d}x\\
&\leqslant C\|\nabla a\|_2\|\nabla u\|_2\|u\|_{L^\infty}\leqslant C\delta\|(\nabla a,\nabla u)\|_2^2,
\end{aligned}\tag{2.11}$$

$$\begin{aligned}
&-\int_{\mathbb{R}^3}\frac{2}{3(a+\bar{\rho})}\nabla((a+\bar{\rho})(m+\bar{k}))\cdot u\text{d}x\\
=&-\int_{\mathbb{R}^3}\frac{2(m+\bar{k})}{3(a+\bar{\rho})}\nabla a\cdot u-\frac{2}{3(a+\bar{\rho})}(a+\bar{\rho})\nabla m\cdot u\text{d}x\\
\leqslant&\, C\delta\|(\nabla a,\nabla m)\|_2^2,
\end{aligned}\tag{2.12}$$

$$\begin{aligned}
&\int_{\mathbb{R}^3}(f(a+\bar{\rho})-f(\bar{\rho}))\text{div}u\text{d}x\\
&=-\int_{\mathbb{R}^3}[f(a+\bar{\rho})-f(\bar{\rho})]\left(\frac{a_t+u\cdot\nabla a}{a+\bar{\rho}}\right)\text{d}x\\
&=-\frac{\text{d}}{\text{d}t}\int_{\mathbb{R}^3}F(a)\text{d}x-\int_{\mathbb{R}^3}\frac{f'(\bar{\rho}+\theta a)}{a+\bar{\rho}}au\cdot\nabla a\text{d}x\\
&\leqslant-\frac{\text{d}}{\text{d}t}\int_{\mathbb{R}^3}F(a)\text{d}x+C\delta\|\nabla a\|_2^2,
\end{aligned}\tag{2.13}$$

with $\theta\in(0,1)$ and $F(a)$ is defined as

$$F(a)=\int_0^a\frac{f(s+\bar{\rho})-f(\bar{\rho})}{s+\bar{\rho}}\text{d}s.\tag{2.14}$$

Combining (2.10)–(2.13) with (2.9), we get

$$\frac{1}{2}\frac{d}{dt}\left[\|u\|_2^2+\int_{\mathbb{R}^3}F(a)\text{d}x\right]+\frac{1}{a+\bar{\rho}}(\|\nabla u\|_2^2+\|\text{div}u\|_2^2)\leqslant C\delta\|(\nabla a,\nabla u,\nabla m)\|_2^2.\tag{2.15}$$

Multiplying the energy equation, governing equation for turbulent kinetic energy k and ε-equation of (2.5) by h, m and ε respectively, and integrating them over the whole space $\mathbb{R}^3$, similarly we can get that

$$
\begin{aligned}
&\frac{1}{2}\frac{d}{dt}\|h\|_2^2+\frac{1}{a+\bar{\rho}}\|\nabla h\|_2^2\\
=&\int_{\mathbb{R}^3}\frac{1}{(a+\bar{\rho})^2}\nabla a\cdot\nabla h\cdot h\mathrm{d}x-\int_{\mathbb{R}^3}f'(a+\bar{\rho})(a+\bar{\rho})\mathrm{div}uh\mathrm{d}x\\
&+\int_{\mathbb{R}^3}\frac{1}{a+\bar{\rho}}S_k\cdot h\mathrm{d}x-\int_{\mathbb{R}^3}u\cdot\nabla h\cdot h\mathrm{d}x,
\end{aligned}\tag{2.16}
$$

$$
\begin{aligned}
&\frac{1}{2}\frac{d}{dt}\|m\|_2^2+\frac{1}{a+\bar{\rho}}\|\nabla m\|_2^2\\
=&\int_{\mathbb{R}^3}\frac{1}{(a+\bar{\rho})^2}\nabla m\cdot\nabla a\cdot m\mathrm{d}x-\int_{\mathbb{R}^3}\varepsilon\cdot m\mathrm{d}x+\int_{\mathbb{R}^3}\frac{1}{a+\bar{\rho}}G\cdot m\mathrm{d}x\\
&-\int_{\mathbb{R}^3}u\cdot\nabla m\cdot m\mathrm{d}x,
\end{aligned}\tag{2.17}
$$

$$
\begin{aligned}
&\frac{1}{2}\frac{d}{dt}\|\varepsilon\|_2^2+\frac{1}{a+\bar{\rho}}\|\nabla\varepsilon\|_2^2\\
=&\int_{\mathbb{R}^3}\frac{1}{(a+\bar{\rho})^2}\nabla\varepsilon\cdot\nabla a\cdot\varepsilon\mathrm{d}x+\int_{\mathbb{R}^3}\frac{C_1G\varepsilon^2}{(a+\bar{\rho})(m+\bar{k})}\mathrm{d}x\\
&-\int_{\mathbb{R}^3}\frac{C_2\varepsilon^3}{m+\bar{k}}\mathrm{d}x-\int_{\mathbb{R}^3}u\cdot\nabla\varepsilon\cdot\varepsilon\mathrm{d}x.
\end{aligned}\tag{2.18}
$$

A direct computation gives that

$$\int_{\mathbb{R}^3}\frac{1}{(a+\bar{\rho})^2}\nabla a\cdot\nabla h\cdot h\mathrm{d}x\leqslant C\|\nabla a\|_2\|\nabla h\|_2\|h\|_{L^\infty}\leqslant C\delta\|(\nabla a,\nabla h)\|_2^2,\tag{2.19}$$

$$-\int_{\mathbb{R}^3}f'(a+\bar{\rho})(a+\bar{\rho})\mathrm{div}uh\mathrm{d}x\leqslant C\|\nabla a\|_2\|\nabla u\|_2\|h\|_3\leqslant C\delta\|(\nabla a,\nabla u)\|_2^2,\tag{2.20}$$

$$
\begin{aligned}
\int_{\mathbb{R}^3}\frac{1}{a+\bar{\rho}}S_k\cdot h\mathrm{d}x=&\int_{\mathbb{R}^3}\frac{1}{a+\bar{\rho}}\left\{\left[\mu(\frac{\partial u^i}{\partial x_j}+\frac{\partial u^j}{\partial x_i})-\frac{2}{3}\delta_{ij}\mu\frac{\partial u^k}{\partial x_k}\right]\frac{\partial u^i}{\partial x_j}\right.\\
&\left.+\frac{\mu_t}{(a+\bar{\rho})^2}\frac{\partial p}{\partial x_j}\frac{\partial a}{\partial x_j}\right\}\cdot h\mathrm{d}x\\
\leqslant&\, C\delta\|(\nabla a,\nabla u)\|_2^2,
\end{aligned}\tag{2.21}
$$

$$-\int_{\mathbb{R}^3}u\cdot\nabla h\cdot h\mathrm{d}x\leqslant C\|u\|_3\|\nabla h\|_2\|h\|_6\leqslant C\delta\|\nabla h\|_2^2,\tag{2.22}$$

$$\int_{\mathbb{R}^3}\frac{1}{(a+\bar{\rho})^2}\nabla m\cdot\nabla a\cdot m\mathrm{d}x\leqslant C\|\nabla m\|_2\|\nabla a\|_2\|m\|_{L^\infty}\leqslant C\delta\|(\nabla a,\nabla m)\|_2^2,\tag{2.23}$$

$$-\int_{\mathbb{R}^3}\varepsilon\cdot m\mathrm{d}x\leqslant C\|\varepsilon\|_2\|m\|_6\|m+\bar{k}\|_6\|m+\bar{k}\|_6\leqslant C\delta\|\nabla m\|_2^2,\tag{2.24}$$

$$\begin{aligned}\int_{\mathbb{R}^3}\frac{1}{a+\bar{\rho}}G\cdot m\mathrm{d}x&=\int_{\mathbb{R}^3}\frac{1}{a+\bar{\rho}}\frac{\partial u^i}{\partial x_j}[\mu_e(\frac{\partial u^i}{\partial x_j}+\frac{\partial u^j}{\partial x_i})-\frac{2}{3}\delta_{ij}((a+\bar{\rho})(m+\bar{k})\\&\quad+\mu_e\frac{\partial u^k}{\partial x_k})]\cdot m\mathrm{d}x\\&\leqslant C\delta\|(\nabla u,\nabla m)\|_2^2,\end{aligned}\tag{2.25}$$

$$-\int_{\mathbb{R}^3}u\cdot\nabla m\cdot m\mathrm{d}x\leqslant C\|u\|_3\|\nabla m\|_2\|m\|_6\leqslant C\delta\|\nabla m\|_2^2,\tag{2.26}$$

$$\int_{\mathbb{R}^3}\frac{1}{(a+\bar{\rho})^2}\nabla\varepsilon\cdot\nabla a\cdot\varepsilon\mathrm{d}x\leqslant C\|\nabla\varepsilon\|_2\|\nabla a\|_2\|\varepsilon\|_{L^\infty}\leqslant C\delta\|(\nabla a,\nabla\varepsilon)\|_2^2,\tag{2.27}$$

$$\begin{aligned}\int_{\mathbb{R}^3}\frac{C_1G\varepsilon^2}{(a+\bar{\rho})(m+\bar{k})}\mathrm{d}x&=\int_{\mathbb{R}^3}\frac{C_1\varepsilon^2}{(a+\bar{\rho})(m+\bar{k})}\frac{\partial u^i}{\partial x_j}\left[\mu_e\left(\frac{\partial u^i}{\partial x_j}+\frac{\partial u^j}{\partial x_i}\right)\right.\\&\quad\left.-\frac{2}{3}\delta_{ij}\left((a+\bar{\rho})(m+\bar{k})+\mu_e\frac{\partial u^k}{\partial x_k}\right)\right]\mathrm{d}x\\&\leqslant C\delta\|(\nabla u,\nabla\varepsilon)\|_2^2,\end{aligned}\tag{2.28}$$

$$-\int_{\mathbb{R}^3}\frac{C_2\varepsilon^3}{m+\bar{k}}\mathrm{d}x\leqslant C\|\varepsilon\|_2\|\varepsilon\|_6\|\varepsilon\|_6\|m+\bar{k}\|_6\leqslant C\delta\|\nabla\varepsilon\|_2^2,\tag{2.29}$$

$$-\int_{\mathbb{R}^3}u\cdot\nabla\varepsilon\cdot\varepsilon\mathrm{d}x\leqslant C\|u\|_3\|\nabla\varepsilon\|_2\|\varepsilon\|_6\leqslant C\delta\|\nabla\varepsilon\|_2^2.\tag{2.30}$$

The estimates (2.19)–(2.30) together with (2.16)–(2.18) imply

$$\frac{1}{2}\frac{\mathrm{d}}{\mathrm{d}t}\|h\|_2^2+\frac{1}{a+\bar{\rho}}\|\nabla h\|_2^2\leqslant C\delta\|(\nabla a,\nabla u)\|_2^2,\tag{2.31}$$

$$\frac{1}{2}\frac{\mathrm{d}}{\mathrm{d}t}\|m\|_2^2+\frac{1}{a+\bar{\rho}}\|\nabla m\|_2^2\leqslant C\delta\|(\nabla a,\nabla u)\|_2^2,\tag{2.32}$$

$$\frac{1}{2}\frac{\mathrm{d}}{\mathrm{d}t}\|\varepsilon\|_2^2+\frac{1}{a+\bar{\rho}}\|\nabla\varepsilon\|_2^2\leqslant C\delta\|(\nabla a,\nabla u)\|_2^2.\tag{2.33}$$

Step 2 L^2-norms of $\nabla^3u,\nabla^3h,\nabla^3m,\nabla^3\varepsilon$

Applying the differential operator ∂_{lmn} to the momentum equation of (2.5), then multiplying it by $\partial_{lmn}u$, and integrating over $\mathbb{R}^3$, one gets

$$\begin{aligned}&\frac{1}{2}\frac{\mathrm{d}}{\mathrm{d}t}\|\partial_{lmn}u\|_2^2\\&=\int_{\mathbb{R}^3}\partial_{lmn}\left(\frac{1}{a+\bar{\rho}}\Delta u\right)\partial_{lmn}u\mathrm{d}x+\int_{\mathbb{R}^3}\partial_{lmn}\left(\frac{1}{a+\bar{\rho}}\nabla\mathrm{div}u\right)\partial_{lmn}u\mathrm{d}x\\&\quad-\int_{\mathbb{R}^3}\partial_{lmn}(u\cdot\nabla u)\cdot\partial_{lmn}u\mathrm{d}x+\int_{\mathbb{R}^3}\partial_{lmn}[f(a+\bar{\rho})-f(\bar{\rho})]\cdot\partial_{lmn}\mathrm{div}u\mathrm{d}x\\&\quad-\frac{2}{3}\int_{\mathbb{R}^3}\partial_{lmn}\left[\frac{1}{a+\bar{\rho}}\nabla((a+\bar{\rho})(m+\bar{k}))\right]\cdot\partial_{lmn}u\mathrm{d}x.\end{aligned}\tag{2.34}$$

The first term on the right-hand side of (2.34) can be estimated as

$$
\begin{aligned}
&\int_{\mathbb{R}^3} \partial_{lmn}\left(\frac{1}{a+\bar{\rho}}\Delta u\right)\partial_{lmn}u\mathrm{d}x \\
=&-\int_{\mathbb{R}^3}\frac{1}{a+\bar{\rho}}|\partial_{lmn}\nabla u|^2\mathrm{d}x+\int_{\mathbb{R}^3}\frac{1}{(a+\bar{\rho})^2}\nabla\partial_{lmn}u:(\nabla a\otimes\partial_{lmn}u)\mathrm{d}x \\
&+\int_{\mathbb{R}^3}\left[-\frac{6}{(a+\bar{\rho})^4}\partial_l a\partial_m a\partial_n a\Delta u+\frac{2}{(a+\bar{\rho})^3}(\partial_{lm}a\partial_n a\Delta u+\partial_{ln}a\partial_m a\Delta u\right. \\
&+\partial_{mn}a\partial_l a\Delta u+\partial_l a\partial_m a\partial_n\Delta u+\partial_l a\partial_n a\partial_m\Delta u+\partial_n a\partial_m a\partial_l\Delta u) \\
&-\frac{1}{(a+\bar{\rho})^2}(\partial_{lmn}a\Delta u+\partial_{lm}a\partial_n\Delta u+\partial_{ln}a\partial_m\Delta u+\partial_{mn}a\partial_l\Delta u \\
&\left.+\partial_l a\partial_{mn}\Delta u+\partial_m a\partial_{ln}\Delta u+\partial_n a\partial_{lm}\Delta u)\right]\cdot\partial_{lmn}u\mathrm{d}x \\
\leqslant& C\delta\|(\nabla^2u,\nabla^3u,\nabla^4u,\nabla^2a,\nabla^3a)\|_2^2-\int_{\mathbb{R}^3}\frac{1}{a+\bar{\rho}}|\partial_{lmn}\nabla u|^2\mathrm{d}x.
\end{aligned}
\tag{2.35}
$$

Similarly, we can estimate the second and third terms as

$$
\begin{aligned}
&\int_{\mathbb{R}^3}\partial_{lmn}\left(\frac{1}{a+\bar{\rho}}\nabla\mathrm{div}u\right)\partial_{lmn}u\mathrm{d}x \\
\leqslant& C\delta\|(\nabla^2u,\nabla^3u,\nabla^4u,\nabla^2a,\nabla^3a)\|_2^2-\int_{\mathbb{R}^3}\frac{1}{a+\bar{\rho}}|\partial_{lmn}\mathrm{div}u|^2\mathrm{d}x,
\end{aligned}
\tag{2.36}
$$

$$
-\int_{\mathbb{R}^3}\partial_{lmn}(u\cdot\nabla u)\cdot\partial_{lmn}u\mathrm{d}x\leqslant C\delta\|\nabla^3u\|_2^2. \tag{2.37}
$$

Now, let's estimate the fourth term on the right-hand side of (2.34) as

$$
\begin{aligned}
&\int_{\mathbb{R}^3}\partial_{lmn}[f(a+\bar{\rho})-f(\bar{\rho})]\cdot\partial_{lmn}\mathrm{div}u\mathrm{d}x \\
=&-\int_{\mathbb{R}^3}\partial_{lmn}[f(a+\bar{\rho})-f(\bar{\rho})]\partial_{lmn}\left(\frac{a_t+u\cdot\nabla a}{a+\bar{\rho}}\right)\mathrm{d}x \\
&-\int_{\mathbb{R}^3}[f'''(\rho)\partial_l a\partial_m a\partial_n a+f''(\rho)\partial_{lm}a\partial_n a+f''(\rho)\partial_{ln}a\partial_m a \\
&+f''(\rho)\partial_{mn}a\partial_l a+f'(\rho)\partial_{lmn}a]\times\left[-\frac{6}{(a+\bar{\rho})^4}\partial_l a\partial_m a\partial_n a(a_t+u\cdot\nabla a)\right. \\
&+\frac{2}{(a+\bar{\rho})^3}(\partial_{lm}a\partial_n aa_t+\partial_{ln}a\partial_m aa_t+\partial_{mn}a\partial_l aa_t+\partial_l a\partial_m a\partial_n a_t \\
&+\partial_l a\partial_n a\partial_m a_t+\partial_n a\partial_m a\partial_l a_t+u\cdot\nabla a\partial_{lm}a\partial_n a+u\cdot\nabla a\partial_{ln}a\partial_m a \\
&+u\cdot\nabla a\partial_{mn}a\partial_l a+\partial_l u\cdot\nabla a\partial_m a\partial_n a+\partial_m u\cdot\nabla a\partial_l a\partial_n a+\partial_n u\cdot\nabla a\partial_l a\partial_m a \\
&+u\cdot\partial_l\nabla a\partial_m a\partial_n a+u\cdot\partial_m\nabla a\partial_l a\partial_n a+u\cdot\partial_n\nabla a\partial_l a\partial_m a) \\
&-\frac{1}{(a+\bar{\rho})^2}(a_t\partial_{lmn}a+\partial_l a_t\partial_{mn}a+\partial_m a_t\partial_{ln}a+\partial_n a_t\partial_{lm}a+\partial_{lm}a_t\partial_n a
\end{aligned}
$$

$$
\begin{aligned}
&+ \partial_{ln}a_t\partial_m a + \partial_{mn}a_t\partial_l a + u\cdot\nabla a\partial_{lmn}a + \partial_l u\cdot\nabla a\partial_{mn}a\\
&+ \partial_m u\cdot\nabla a\partial_{ln}a + \partial_n u\cdot\nabla a\partial_{lm}a + u\cdot\nabla\partial_l a\partial_{mn}a + u\cdot\nabla\partial_m a\partial_{ln}a\\
&+ u\cdot\nabla\partial_n a\partial_{lm}a + \partial_{lm}u\cdot\nabla a\partial_n a + \partial_{ln}u\cdot\nabla a\partial_m a + \partial_{mn}u\cdot\nabla a\partial_l a\\
&+ u\cdot\nabla\partial_{lm}a\partial_n a + u\cdot\nabla\partial_{ln}a\partial_m a + u\cdot\nabla\partial_{mn}a\partial_l a + \partial_l u\cdot\nabla\partial_m a\partial_n a\\
&+ \partial_l u\cdot\nabla\partial_n a\partial_m a + \partial_m u\cdot\nabla\partial_l a\partial_n a + \partial_m u\cdot\nabla\partial_n a\partial_l a + \partial_n u\cdot\nabla\partial_l a\partial_m a\\
&+ \partial_n u\cdot\nabla\partial_m a\partial_l a) + \frac{1}{a+\bar\rho}(\partial_{lmn}a_t + \partial_{lmn}u\cdot\nabla a + \partial_{lm}u\cdot\nabla\partial_n a\\
&+ \partial_{ln}u\cdot\nabla\partial_m a + \partial_{mn}u\cdot\nabla\partial_l a + \partial_l u\cdot\nabla\partial_{mn}a + \partial_m u\cdot\nabla\partial_{ln}a\\
&\left.+ \partial_n u\cdot\nabla\partial_{lm}a + u\cdot\nabla\partial_{lmn}a)\right]\mathrm{d}x\\
\leqslant{}& C\delta\|(\nabla a,\nabla^2 a,\nabla a_t,\nabla^3 a,\nabla^2 a_t,\nabla^2 u,\nabla^3 u)\|_2^2 - \int_{\mathbb{R}^3}[f'''(\rho)\partial_l a\partial_m a\partial_n a\\
&+ f''(\rho)\partial_{lm}a\partial_n a + f''(\rho)\partial_{ln}a\partial_m a + f''(\rho)\partial_{mn}a\partial_l a + f'(\rho)\partial_{lmn}a]\\
&\times\left[\frac{1}{a+\bar\rho}(\partial_{lmn}a_t + u\cdot\nabla\partial_{lmn}a)\right]\mathrm{d}x. \qquad (2.38)
\end{aligned}
$$

Since

$$
\begin{aligned}
-\int_{\mathbb{R}^3}\frac{1}{a+\bar\rho}f'''(\rho)\partial_l a\partial_m a\partial_n a\partial_{lmn}a_t\mathrm{d}x &= \int_{\mathbb{R}^3}\partial_l\left(\frac{1}{a+\bar\rho}f'''(\rho)\partial_l a\partial_m a\partial_n a\right)\partial_{mn}a_t\mathrm{d}x\\
&\leqslant C\delta\|(\nabla a,\nabla^2 a,\nabla^2 a_t)\|_2^2,\\
-\int_{\mathbb{R}^3}\frac{1}{a+\bar\rho}f''(\rho)\partial_{lm}a\partial_n a\partial_{lmn}a_t\mathrm{d}x &= \int_{\mathbb{R}^3}\partial_l\left(\frac{1}{a+\bar\rho}f''(\rho)\partial_{lm}a\partial_n a\right)\partial_{mn}a_t\mathrm{d}x\\
&\leqslant C\delta\|(\nabla^2 a,\nabla^2 a_t,\nabla^3 a)\|_2^2,
\end{aligned}
$$

$$
\begin{aligned}
&-\int_{\mathbb{R}^3}\frac{1}{a+\bar\rho}f'''(\rho)\partial_l a\partial_m a\partial_n a u\cdot\nabla\partial_{lmn}a\mathrm{d}x\\
={}&\int_{\mathbb{R}^3}\partial_j\left(\frac{1}{a+\bar\rho}f'''(\rho)\partial_l a\partial_m a\partial_n a u^j\right)\partial_{lmn}a\mathrm{d}x\\
\leqslant{}& C\delta\|(\nabla^2 a,\nabla^3 a)\|_2^2,
\end{aligned}
$$

and

$$
\begin{aligned}
-\int_{\mathbb{R}^3}\frac{1}{a+\bar\rho}f''(\rho)\partial_{lm}a\partial_n a u\cdot\nabla\partial_{lmn}a\mathrm{d}x &= \int_{\mathbb{R}^3}\partial_j\left(\frac{1}{a+\bar\rho}f''(\rho)\partial_{lm}a\partial_n a u^j\right)\partial_{lmn}a\mathrm{d}x\\
&\leqslant C\delta\|\nabla^3 a\|_2^2,
\end{aligned}
$$

we obtain

$$
\begin{aligned}
&-\int_{\mathbb{R}^3}[f'''(\rho)\partial_l a\partial_m a\partial_n a + f''(\rho)\partial_{lm}a\partial_n a + f''(\rho)\partial_{ln}a\partial_m a + f''(\rho)\partial_{mn}a\partial_l a]\\
&\times\left[\frac{1}{a+\bar\rho}(\partial_{lmn}a_t + u\cdot\nabla\partial_{lmn}a)\right]\\
\leqslant{}& C\delta|(\nabla a,\nabla^2 a,\nabla^2 a_t,\nabla^3 a)\|_2^2. \qquad (2.39)
\end{aligned}
$$

Finally,

$$\begin{aligned}
&-\int_{\mathbb{R}^3}\frac{f'(\rho)}{a+\bar\rho}\partial_{lmn}a(\partial_{lmn}a_t+u\cdot\nabla\partial_{lmn}a)\mathrm{d}x\\
=&-\frac{1}{2}\int_{\mathbb{R}^3}\frac{f'(\rho)}{a+\bar\rho}(\partial_{lmn}a)^2_t\mathrm{d}x+\frac{1}{2}\int_{\mathbb{R}^3}\frac{f'(\rho)}{a+\bar\rho}(\partial_{lmn}a)^2\mathrm{div}u+\frac{f''(\rho)}{a+\bar\rho}(\partial_{lmn}a)^2u\cdot\nabla a\\
&-\frac{f'(\rho)}{(a+\bar\rho)^2}(\partial_{lmn}a)^2u\cdot\nabla a\mathrm{d}x\\
\leqslant&-\frac{1}{2}\frac{\mathrm{d}}{\mathrm{d}t}\int_{\mathbb{R}^3}\frac{f'(\rho)}{a+\bar\rho}(\partial_{lmn}a)^2\mathrm{d}x+\frac{1}{2}\int_{\mathbb{R}^3}\left[\frac{f'(\rho)}{a+\bar\rho}\right]_t(\partial_{lmn}a)^2\mathrm{d}x+C\delta\|\nabla^3a\|_2^2\\
\leqslant&-\frac{1}{2}\frac{\mathrm{d}}{\mathrm{d}t}\int_{\mathbb{R}^3}\frac{f'(\rho)}{a+\bar\rho}(\partial_{lmn}a)^2\mathrm{d}x+C\delta\|\nabla^3a\|_2^2,
\end{aligned}$$

together with (2.39), thus (2.38) can be replaced by

$$\begin{aligned}
&\int_{\mathbb{R}^3}\partial_{lmn}[f(a+\bar\rho)-f(\bar\rho)]\cdot\partial_{lmn}\mathrm{div}u\mathrm{d}x\\
\leqslant&\, C\delta\|(\nabla a,\nabla a_t,\nabla^2a,\nabla^2a_t,\nabla^3a,\nabla^2u,\nabla^3u)\|_2^2\\
&-\frac{1}{2}\frac{\mathrm{d}}{\mathrm{d}t}\int_{\mathbb{R}^3}\frac{f'(\rho)}{a+\bar\rho}(\partial_{lmn}a)^2\mathrm{d}x.
\end{aligned}\tag{2.40}$$

The last term on the right-hand side of (2.34) can be estimated as

$$\begin{aligned}
&-\frac{2}{3}\int_{\mathbb{R}^3}\partial_{lmn}[\frac{1}{a+\bar\rho}\nabla((a+\bar\rho)(m+\bar k))]\cdot\partial_{lmn}u\mathrm{d}x\\
=&\frac{2}{3}\int_{\mathbb{R}^3}\partial_{mn}[\frac{m+\bar k}{a+\bar\rho}\nabla a+\nabla m]\cdot\partial_{llmn}u\mathrm{d}x\\
=&\frac{2}{3}\int_{\mathbb{R}^3}\Bigg[\frac{2(m+\bar k)}{(a+\bar\rho)^3}\partial_ma\partial_na\nabla a-\frac{1}{(a+\bar\rho)^2}((m+\bar k)\partial_{mn}a\nabla a+\partial_ma\partial_nm\nabla a\\
&+\partial_na\partial_mm\nabla a+(m+\bar k)\partial_ma\partial_n\nabla a+(m+\bar k)\partial_na\partial_m\nabla a)+\frac{1}{a+\bar\rho}(\partial_{mn}m\nabla a\\
&+\partial_mm\nabla\partial_na+\partial_nm\nabla\partial_ma+(m+\bar k)\nabla\partial_{mn}a)+\partial_{mn}\nabla m\Bigg]\cdot\partial_{llmn}u\mathrm{d}x\\
\leqslant&\, C\delta\|(\nabla a,\nabla^2a,\nabla^2m,\nabla^3m,\nabla^4m,\nabla^4u)\|_2^2+C\int_{\mathbb{R}^3}(\nabla\partial_{mn}a)^2\mathrm{d}x,
\end{aligned}$$

which together with (2.35)–(2.37) and (2.40) gives

$$\begin{aligned}
&\frac{1}{2}\frac{\mathrm{d}}{\mathrm{d}t}[\|\partial_{lmn}u\|_2^2+\int_{\mathbb{R}^3}\frac{f'(\rho)}{a+\bar\rho}(\partial_{lmn}a)^2\mathrm{d}x]+\int_{\mathbb{R}^3}\frac{1}{a+\bar\rho}|\partial_{lmn}\nabla u|^2\mathrm{d}x\\
&+\int_{\mathbb{R}^3}\frac{1}{a+\bar\rho}|\partial_{lmn}\mathrm{div}u|^2\mathrm{d}x\\
\leqslant&\, C\delta\|(\nabla a,\nabla a_t,\nabla^2a,\nabla^2a_t,\nabla^3a,\nabla^2u,\nabla^3u,\nabla^2m,\nabla^3m,\nabla^4m)\|_2^2\\
&+C\int_{\mathbb{R}^3}(\nabla\partial_{mn}a)^2\mathrm{d}x.
\end{aligned}\tag{2.41}$$

Applying ∂_{lmn} to the energy equation, k-equation and ε-equation of (2.5), multiplying the resulting equations by $\partial_{lmn}h$, $\partial_{lmn}m$ and $\partial_{lmn}\varepsilon$ respectively, then integrating them over the whole space $\mathbb{R}^3$, one can prove that

$$\begin{aligned}&\frac{1}{2}\frac{\mathrm{d}}{\mathrm{d}t}\|\partial_{lmn}h\|_2^2\\&=\int_{\mathbb{R}^3}\partial_{lmn}\left(\frac{1}{a+\bar{\rho}}\Delta h\right)\cdot\partial_{lmn}h\mathrm{d}x+\int_{\mathbb{R}^3}\partial_{lmn}\left(\frac{S_k}{a+\bar{\rho}}\right)\cdot\partial_{lmn}h\mathrm{d}x\\&\quad-\int_{\mathbb{R}^3}\partial_{lmn}(f'(a+\bar{\rho})(a+\bar{\rho})\mathrm{div}u)\cdot\partial_{lmn}h\mathrm{d}x-\int_{\mathbb{R}^3}\partial_{lmn}(u\cdot\nabla h)\cdot\partial_{lmn}h\mathrm{d}x,\end{aligned}\tag{2.42}$$

$$\begin{aligned}&\frac{1}{2}\frac{\mathrm{d}}{\mathrm{d}t}\|\partial_{lmn}m\|_2^2\\&=\int_{\mathbb{R}^3}\partial_{lmn}(\frac{1}{a+\bar{\rho}}\Delta m)\cdot\partial_{lmn}m\mathrm{d}x-\int_{\mathbb{R}^3}\partial_{lmn}\varepsilon\cdot\partial_{lmn}m\mathrm{d}x\\&\quad+\int_{\mathbb{R}^3}\partial_{lmn}(\frac{G}{a+\bar{\rho}})\cdot\partial_{lmn}m\mathrm{d}x-\int_{\mathbb{R}^3}\partial_{lmn}(u\cdot\nabla m)\cdot\partial_{lmn}m\mathrm{d}x,\end{aligned}\tag{2.43}$$

$$\begin{aligned}&\frac{1}{2}\frac{\mathrm{d}}{\mathrm{d}t}\|\partial_{lmn}\varepsilon\|_2^2\\&=\int_{\mathbb{R}^3}\partial_{lmn}\left(\frac{1}{(a+\bar{\rho})^2}\Delta\varepsilon\right)\cdot\partial_{lmn}\varepsilon\mathrm{d}x+\int_{\mathbb{R}^3}\partial_{lmn}\left(\frac{C_1G\varepsilon}{(a+\bar{\rho})(m+\bar{k})}\right)\cdot\partial_{lmn}\varepsilon\mathrm{d}x\\&\quad-\int_{\mathbb{R}^3}\partial_{lmn}\left(\frac{C_2\varepsilon^2}{m+\bar{k}}\right)\cdot\partial_{lmn}\varepsilon\mathrm{d}x-\int_{\mathbb{R}^3}\partial_{lmn}(u\cdot\nabla\varepsilon)\cdot\partial_{lmn}\varepsilon\mathrm{d}x.\end{aligned}\tag{2.44}$$

As same as the estimate (2.35), we can deduce

$$\begin{aligned}&\int_{\mathbb{R}^3}\partial_{lmn}\left(\frac{1}{a+\bar{\rho}}\Delta h\right)\partial_{lmn}h\mathrm{d}x\\&\leqslant C\delta\|(\nabla^2h,\nabla^3h,\nabla^4h,\nabla^2a,\nabla^3a)\|_2^2-\int_{\mathbb{R}^3}\frac{1}{a+\bar{\rho}}|\partial_{lmn}\nabla h|^2\mathrm{d}x,\end{aligned}\tag{2.45}$$

$$\begin{aligned}&\int_{\mathbb{R}^3}\partial_{lmn}\left(\frac{1}{a+\bar{\rho}}\Delta m\right)\partial_{lmn}m\mathrm{d}x\\&\leqslant C\delta\|(\nabla^2m,\nabla^3m,\nabla^4m,\nabla^2a,\nabla^3a)\|_2^2-\int_{\mathbb{R}^3}\frac{1}{a+\bar{\rho}}|\partial_{lmn}\nabla m|^2\mathrm{d}x,\end{aligned}\tag{2.46}$$

$$\begin{aligned}&\int_{\mathbb{R}^3}\partial_{lmn}\left(\frac{1}{a+\bar{\rho}}\Delta\varepsilon\right)\partial_{lmn}\varepsilon\mathrm{d}x\\&\leqslant C\delta\|(\nabla^2\varepsilon,\nabla^3\varepsilon,\nabla^4\varepsilon,\nabla^2a,\nabla^3a)\|_2^2-\int_{\mathbb{R}^3}\frac{1}{a+\bar{\rho}}|\partial_{lmn}\nabla\varepsilon|^2\mathrm{d}x.\end{aligned}\tag{2.47}$$

Again, from the Hölder inequality, (2.6)–(2.8), $\dot{W}^{1,2}(\mathbb{R}^3)\hookrightarrow L^6(\mathbb{R}^3)$ and $H^2(\mathbb{R}^3)\hookrightarrow$

$L^\infty(\mathbb{R}^3)$, we get

$$
\begin{aligned}
&\int_{\mathbb{R}^3} \partial_{lmn}\left(\frac{S_k}{a+\bar{\rho}}\right)\cdot\partial_{lmn}h\mathrm{d}x\\
=&\int_{\mathbb{R}^3}\partial_{lmn}\left\{\frac{1}{a+\bar{\rho}}\left[\left(\mu\left(\frac{\partial u^i}{\partial x_j}+\frac{\partial u^j}{\partial x_i}\right)-\frac{2}{3}\delta_{ij}\mu\frac{\partial u^k}{\partial x_k}\right)\frac{\partial u^i}{\partial x_j}\right.\right.\\
&\left.\left.+\frac{\mu_t}{(a+\bar{\rho})^2}\frac{\partial p}{\partial x_j}\frac{\partial a}{\partial x_j}\right]\right\}\cdot\partial_{lmn}h\mathrm{d}x\\
\leqslant& C\delta\|(\nabla^3 a,\nabla^3 u,\nabla^4 u,\nabla^3 h,\nabla^4 h)\|_2^2,
\end{aligned}
$$

$$
\begin{aligned}
&\int_{\mathbb{R}^3} \partial_{lmn}\left(\frac{G}{a+\bar{\rho}}\right)\cdot\partial_{lmn}m\mathrm{d}x\\
=&\int_{\mathbb{R}^3}\partial_{lmn}\left\{\frac{1}{a+\bar{\rho}}\frac{\partial u^i}{\partial x_j}\left[\mu_e\left(\frac{\partial u^i}{\partial x_j}+\frac{\partial u^j}{\partial x_i}\right)\right.\right.\\
&\left.\left.-\frac{2}{3}\delta_{ij}\left((a+\bar{\rho})(m+\bar{k})+\mu_e\frac{\partial u^k}{\partial x_k}\right)\right]\right\}\cdot\partial_{lmn}m\mathrm{d}x\\
\leqslant& C\delta\|(\nabla^2 a,\nabla^3 a,\nabla^3 m,\nabla^4 m,\nabla^3 u,\nabla^4 u)\|_2^2,
\end{aligned}
$$

$$
\begin{aligned}
&\int_{\mathbb{R}^3} \partial_{lmn}\left(\frac{C_1 G\varepsilon}{(a+\bar{\rho})(m+\bar{k})}\right)\cdot\partial_{lmn}\varepsilon\mathrm{d}x\\
=&\int_{\mathbb{R}^3}\partial_{lmn}\left\{\frac{C_1\varepsilon}{(a+\bar{\rho})(m+\bar{k})}\frac{\partial u^i}{\partial x_j}\left[\mu_e\left(\frac{\partial u^i}{\partial x_j}+\frac{\partial u^j}{\partial x_i}\right)\right.\right.\\
&\left.\left.-\frac{2}{3}\delta_{ij}\left((a+\bar{\rho})(m+\bar{k})+\mu_e\frac{\partial u^k}{\partial x_k}\right)\right]\right\}\cdot\partial_{lmn}\varepsilon\mathrm{d}x\\
\leqslant& C\delta\|(\nabla^3 a,\nabla^3 u,\nabla^4 u,\nabla^3 m,\nabla^3\varepsilon)\|_2^2,
\end{aligned}
$$

$$
\begin{aligned}
&-\int_{\mathbb{R}^3}\partial_{lmn}(f'(a+\bar{\rho})(a+\bar{\rho})\mathrm{div}u)\cdot\partial_{lmn}h\mathrm{d}x\\
\leqslant& C\delta\|(\nabla^3 a,\nabla^3 u,\nabla^4 u,\nabla^3 h,\nabla^4 h)\|_2^2,\\
&-\int_{\mathbb{R}^3}\partial_{lmn}(u\cdot\nabla h)\cdot\partial_{lmn}h\mathrm{d}x\leqslant C\delta\|(\nabla^3 u,\nabla^3 h,\nabla^4 h)\|_2^2,\\
&-\int_{\mathbb{R}^3}\partial_{lmn}\varepsilon\cdot\partial_{lmn}m\mathrm{d}x\leqslant C\delta\|(\nabla^3\varepsilon,\nabla^3 m,\nabla^4 m)\|_2^2,\\
&-\int_{\mathbb{R}^3}\partial_{lmn}(u\cdot\nabla m)\cdot\partial_{lmn}m\mathrm{d}x\leqslant C\delta\|(\nabla^3 m,\nabla^4 m,\nabla^3 u)\|_2^2,\\
&-\int_{\mathbb{R}^3}\partial_{lmn}\left(\frac{C_2\varepsilon^2}{m+\bar{k}}\right)\cdot\partial_{lmn}\varepsilon\mathrm{d}x\leqslant C\delta\|(\nabla^3 m,\nabla^3\varepsilon)\|_2^2,\\
&-\int_{\mathbb{R}^3}\partial_{lmn}(u\cdot\nabla\varepsilon)\cdot\partial_{lmn}\varepsilon\mathrm{d}x\leqslant C\|(\nabla^3\varepsilon,\nabla^4\varepsilon,\nabla^3 u)\|_2^2.
\end{aligned}
$$

Incorporating the above estimates and (2.42)–(2.44) yields that

$$
\begin{aligned}
&\frac{1}{2}\frac{\mathrm{d}}{\mathrm{d}t}\|\partial_{lmn}h\|_2^2+\int_{\mathbb{R}^3}\frac{1}{a+\bar{\rho}}|\partial_{lmn}\nabla h|^2\mathrm{d}x\\
&\leqslant C\delta\|(\nabla^2a,\nabla^3a,\nabla^2h,\nabla^3h,\nabla^3u,\nabla^4u)\|_2^2,
\end{aligned}
\tag{2.48}
$$

$$
\begin{aligned}
&\frac{1}{2}\frac{\mathrm{d}}{\mathrm{d}t}\|\partial_{lmn}m\|_2^2+\int_{\mathbb{R}^3}\frac{1}{a+\bar{\rho}}|\partial_{lmn}\nabla m|^2\mathrm{d}x\\
&\leqslant C\delta\|(\nabla^2a,\nabla^3a,\nabla^2m,\nabla^3m,\nabla^3\varepsilon,\nabla^3u,\nabla^4u)\|_2^2,
\end{aligned}
\tag{2.49}
$$

$$
\begin{aligned}
&\frac{1}{2}\frac{\mathrm{d}}{\mathrm{d}t}\|\partial_{lmn}\varepsilon\|_2^2+\int_{\mathbb{R}^3}\frac{1}{a+\bar{\rho}}|\partial_{lmn}\nabla \varepsilon|^2\mathrm{d}x\\
&\leqslant C\delta\|(\nabla^2a,\nabla^3a,\nabla^3m,\nabla^2\varepsilon,\nabla^3\varepsilon,\nabla^3u,\nabla^4u)\|_2^2.
\end{aligned}
\tag{2.50}
$$

Step 3 L^2-norms of $\nabla a,\nabla^3a$

We first estimate for ∇a. For this purpose, we calculate as

$$
\begin{aligned}
&\int_{\mathbb{R}^3}\left[\frac{1}{2}|\nabla a|^2+\frac{(a+\bar{\rho})^2}{2}\nabla a\cdot u\right]_t\\
=&\int_{\mathbb{R}^3}\nabla a\cdot\nabla a_t+(a+\bar{\rho})a_t\nabla a\cdot u+\frac{(a+\bar{\rho})^2}{2}\nabla a_t\cdot u+\frac{(a+\bar{\rho})^2}{2}\nabla a\cdot u_t\mathrm{d}x.
\end{aligned}
\tag{2.51}
$$

The first term of the right-hand side of (2.51) can be estimated as follows:

$$
\begin{aligned}
\int_{\mathbb{R}^3}\nabla a\cdot\nabla a_t\mathrm{d}x&=-\int_{\mathbb{R}^3}\nabla a\cdot\nabla\mathrm{div}[(a+\bar{\rho})u]\mathrm{d}x\\
&=-\int_{\mathbb{R}^3}\nabla a\cdot(\nabla^2a\cdot u)+\nabla a\cdot(\nabla a\cdot\nabla u)\\
&\quad+(a+\bar{\rho})\nabla a\cdot\nabla\mathrm{div}u+|\nabla a|^2\mathrm{div}u\mathrm{d}x\\
&\leqslant C\delta\|\nabla a\|_2^2-\int_{\mathbb{R}^3}(a+\bar{\rho})\nabla a\cdot\nabla\mathrm{div}u\mathrm{d}x.
\end{aligned}
\tag{2.52}
$$

Also, we estimate the rest three terms as

$$
\int_{\mathbb{R}^3}(a+\bar{\rho})a_t\nabla a\cdot u\mathrm{d}x\leqslant C\delta\|(a_t,\nabla a)\|_2^2\leqslant C\delta\|(\nabla a,\nabla u)\|_2^2,
\tag{2.53}
$$

$$
\int_{\mathbb{R}^3}\frac{(a+\bar{\rho})^2}{2}\nabla a_t\cdot u\mathrm{d}x\leqslant C\delta\|(\nabla a,\nabla u)\|_2^2,
\tag{2.54}
$$

$$
\begin{aligned}
&\int_{\mathbb{R}^3}\frac{(a+\bar{\rho})^2}{2}\nabla a\cdot u_t\mathrm{d}x\\
=&\int_{\mathbb{R}^3}\frac{(a+\bar{\rho})^2}{2}\nabla a\cdot\Big(-u\cdot\nabla u+\frac{1}{a+\bar{\rho}}\Delta u+\frac{1}{a+\bar{\rho}}\nabla\mathrm{div}u
\end{aligned}
$$

$$
\begin{aligned}
& -\nabla[f(a+\bar\rho)-f(\bar\rho)]-\frac{2}{3(a+\bar\rho)}\nabla((a+\bar\rho)(m+\bar k))\Big)\mathrm{d}x \\
\leqslant\; & C\delta\|(\nabla a,\nabla u)\|_2^2-\int_{\mathbb{R}^3}\frac{(a+\bar\rho)(m+\bar k)}{3}|\nabla a|^2\mathrm{d}x \\
& -\int_{\mathbb{R}^3}\frac{(a+\bar\rho)^2}{2}f'(\rho)|\nabla a|^2\mathrm{d}x+\int_{\mathbb{R}^3}\frac{a+\bar\rho}{2}\nabla a\cdot(\Delta u+\nabla\mathrm{div}u)\mathrm{d}x. \qquad (2.55)
\end{aligned}
$$

Since

$$
\begin{aligned}
& \int_{\mathbb{R}^3}\frac{a+\bar\rho}{2}\nabla a\cdot(\Delta u-\nabla\mathrm{div}u)\mathrm{d}x=\int_{\mathbb{R}^3}\frac{a+\bar\rho}{2}\partial_i a\cdot(\partial_{jj}u^i-\partial_i\partial_j u^j)\mathrm{d}x \\
=\; & \int_{\mathbb{R}^3}-\frac{1}{2}\partial_i a\partial_j a\partial_j u^i-\frac{a+\bar\rho}{2}\partial_{ij}a\partial_j u^i+\frac{1}{2}\partial_i a\partial_j a\partial_i u^j+\frac{a+\bar\rho}{2}\partial_{ij}a\partial_i u^j\mathrm{d}x=0,
\end{aligned}
$$

together with (2.51)–(2.55) and (2.7)–(2.8), we show that

$$
\int_{\mathbb{R}^3}\left[\frac{1}{2}|\nabla a|^2+\frac{(a+\bar\rho)^2}{2}\nabla a\cdot u\right]_t+C\|\nabla a\|_2^2\leqslant C\delta\|\nabla u\|_2^2. \qquad (2.56)
$$

Now, we turn to estimate for $\nabla^3 a$. Almost parallel to the inequality (2.51), we get

$$
\begin{aligned}
& \int_{\mathbb{R}^3}\left[\frac{1}{2}|\partial_{lmn}a|^2+\frac{(a+\bar\rho)^2}{2}\partial_{lmn}a\partial_{lm}u^n\right]_t\mathrm{d}x \\
=\; & \int_{\mathbb{R}^3}-\partial_{lmn}a\partial_{lmn}\mathrm{div}[(a+\bar\rho)u]+(a+\bar\rho)a_t\partial_{lmn}a\partial_{lm}u^n \\
& -\frac{(a+\bar\rho)^2}{2}\partial_{lmn}\mathrm{div}[(a+\bar\rho)u]\partial_{lm}u^n+\frac{(a+\bar\rho)^2}{2}\partial_{lmn}a\partial_{lm}u_t^n\mathrm{d}x. \qquad (2.57)
\end{aligned}
$$

We estimate the right-hand side of (2.57) as follows:

$$
\begin{aligned}
& \int_{\mathbb{R}^3}-\partial_{lmn}a\partial_{lmn}\mathrm{div}[(a+\bar\rho)u]\mathrm{d}x \\
=\; & \int_{\mathbb{R}^3}-\partial_{lmn}a[u\cdot\nabla\partial_{lmn}a+\partial_l u\cdot\nabla\partial_{mn}a+\partial_m u\cdot\nabla\partial_{ln}a+\partial_n u\cdot\nabla\partial_{lm}a \\
& +\partial_{lm}u\cdot\nabla\partial_n a+\partial_{ln}u\cdot\nabla\partial_m a+\partial_{mn}u\cdot\nabla\partial_l a+\partial_{lmn}u\cdot\nabla a+\partial_{lmn}a\mathrm{div}u \\
& +\partial_{lm}a\partial_n\mathrm{div}u+\partial_{ln}a\partial_m\mathrm{div}u+\partial_{mn}a\partial_l\mathrm{div}u+\partial_l a\partial_{mn}\mathrm{div}u+\partial_m a\partial_{ln}\mathrm{div}u \\
& +\partial_n a\partial_{lm}\mathrm{div}u+(a+\bar\rho)\partial_{lmn}\mathrm{div}u]\mathrm{d}x \\
\leqslant\; & C\delta\|(\nabla^3 a,\nabla^3 u)\|_2^2-\int_{\mathbb{R}^3}(a+\bar\rho)\partial_{lmn}a\partial_{lmn}\mathrm{div}u\mathrm{d}x, \qquad (2.58)
\end{aligned}
$$

$$
\int_{\mathbb{R}^3}(a+\bar\rho)a_t\partial_{lmn}a\partial_{lm}u^n\mathrm{d}x\leqslant C\delta\|(\nabla^3 a,\nabla^2 u)\|_2^2, \qquad (2.59)
$$

$$
\begin{aligned}
&-\int_{\mathbb{R}^3}\frac{(a+\bar\rho)^2}{2}\partial_{lmn}\mathrm{div}[(a+\bar\rho)u]\partial_{lm}u^n\mathrm{d}x\\
=&-\int_{\mathbb{R}^3}\frac{(a+\bar\rho)^2}{2}\partial_{lm}u^n[\partial_{lmn}(u\cdot\nabla a)+\partial_{lmn}a\mathrm{div}u+\partial_{lm}a\partial_n\mathrm{div}u\\
&+\partial_{ln}a\partial_m\mathrm{div}u+\partial_{mn}a\partial_l\mathrm{div}u+\partial_l a\partial_{mn}\mathrm{div}u+\partial_m a\partial_{ln}\mathrm{div}u+\partial_n a\partial_{lm}\mathrm{div}u\\
&+(a+\bar\rho)\partial_{lmn}\mathrm{div}u]\mathrm{d}x\\
\leqslant& C\delta\|(\nabla^2a,\nabla^3a,\nabla^2u,\nabla^3u,\nabla^4u)\|_2^2,
\end{aligned}\tag{2.60}
$$

$$
\begin{aligned}
&\int_{\mathbb{R}^3}\frac{(a+\bar\rho)^2}{2}\partial_{lmn}a\partial_{lm}u^n_t\mathrm{d}x\\
=&\int_{\mathbb{R}^3}\frac{(a+\bar\rho)^2}{2}\partial_{lmn}a\partial_{lm}\Big[-u^i\partial_iu^n+\frac{1}{a+\bar\rho}\partial_{ii}u^n+\frac{1}{a+\bar\rho}\partial_{in}u^i\\
&-\partial_n(f(a+\bar\rho)-f(\bar\rho))-\frac{2}{3(a+\bar\rho)}\partial_n((a+\bar\rho)(m+\bar k))\Big]\mathrm{d}x\\
\leqslant& C\delta\|(\nabla^2a,\nabla^3a,\nabla^2u,\nabla^3u,\nabla^3m,\nabla^4m)\|_2^2\\
&+\int_{\mathbb{R}^3}\frac{a+\bar\rho}{2}\partial_{lmn}a(\partial_{lmii}u^n+\partial_{lmni}u^i)\mathrm{d}x-\int_{\mathbb{R}^3}\frac{(a+\bar\rho)^2}{2}f'(\rho)(\partial_{lmn}a)^2\mathrm{d}x\\
&-\int_{\mathbb{R}^3}\frac{(a+\bar\rho)(m+\bar k)}{3}(\partial_{lmn}a)^2\mathrm{d}x.
\end{aligned}\tag{2.61}
$$

Noting that

$$
\begin{aligned}
&-\int_{\mathbb{R}}(a+\bar\rho)\partial_{lmn}a\partial_{lmn}\mathrm{div}u\mathrm{d}x+\int_{\mathbb{R}^3}\frac{a+\bar\rho}{2}\partial_{lmn}a(\partial_{lmii}u^n+\partial_{lmni}u^i)\mathrm{d}x\\
=&\int_{\mathbb{R}^3}\frac{a+\bar\rho}{2}\partial_{lmn}a(\partial_{lmii}u^n-\partial_{lmni}u^i)\mathrm{d}x\\
=&-\frac{1}{2}\int_{\mathbb{R}^3}\partial_ia\partial_{lmn}a\partial_{lmi}u^n+(a+\bar\rho)\partial_{lmni}a\partial_{lmi}u^n-\partial_ia\partial_{lmn}a\partial_{lmn}u^i\\
&-(a+\bar\rho)\partial_{lmni}a\partial_{lmn}u^i\mathrm{d}x\\
=&-\frac{1}{2}\int_{\mathbb{R}^3}\partial_ia\partial_{lmn}a\partial_{lmi}u^n-\partial_ia\partial_{lmn}a\partial_{lmn}u^i\mathrm{d}x\\
\leqslant& C\delta\|(\nabla^3a,\nabla^3u)\|_2^2,
\end{aligned}
$$

based on the estimates (2.57)–(2.61), with the help of (2.7)–(2.8) and the interpolation inequality, one obtains

$$
\begin{aligned}
&\int_{\mathbb{R}^3}\Big[\frac{1}{2}|\partial_{lmn}a|^2+\frac{(a+\bar\rho)^2}{2}\partial_{lmn}a\partial_{lm}u^n\Big]_t\mathrm{d}x+C\|\nabla^3a\|_2^2\\
\leqslant& C\delta\|(\nabla^2a,\nabla^2u,\nabla^3u,\nabla^4u,\nabla^3m,\nabla^4m)\|_2^2.
\end{aligned}\tag{2.62}
$$

Step 4 Conclusion

Consequently, multiplying (2.41) by a appropriate small constants α, together with (2.15), (2.31)–(2.33), (2.48)–(2.50), (2.56) and (2.62), we have

$$\begin{aligned}&\frac{\mathrm{d}}{\mathrm{d}t}\Big\{\|(u,h,\nabla^3 h,m,\nabla^3 m,\varepsilon,\nabla^3\varepsilon\|_2^2+\int_{\mathbb{R}^3}F(a)\mathrm{d}x+\alpha[\|\nabla^3 u\|_2^2\\&+\int_{\mathbb{R}^3}\frac{f'(\rho)}{a+\bar\rho}(\partial_{lmn}a)^2\mathrm{d}x]+\int_{\mathbb{R}^3}\Big[\frac12|\nabla a|^2+\frac{(a+\bar\rho)^2}{2}\nabla a\cdot u\\&+\frac12|\partial_{lmn}a|^2+\frac{(a+\bar\rho)^2}{2}\partial_{lmn}a\partial_{lm}u^n\Big]\mathrm{d}x\Big\}\\&+C(\alpha)\|\nabla a,\nabla^3 a,\nabla u,\nabla^4 u,\nabla h,\nabla^4 h,\nabla m,\nabla^4 m,\nabla\varepsilon,\nabla^4\varepsilon\|_2^2\\ \leqslant{}& 0,\end{aligned}\tag{2.63}$$

where we have used the fact that

$$|\nabla^i a_t|\leqslant C\sum_{k=1}^{i+1}(|\nabla^k a|+|\nabla^k u|),\quad i=1,2.$$

Integrating the inequality (2.63), from (2.6)–(2.8), (2.14) and the smallness of α, we can finish the proof of Proposition 2.1. □

3 Proof of global existence

We will finish the proof of Theorem 1.1 in this section. First, let's state the local existence. Since it can be proved in a standard way as that in [4, 8], we omit the proof.

Proposition 3.2 *Under the assumption of the Theorem* 1.1, *then there exists a constant* $T>0$ *such that the system* (2.5) *admits a unique smooth solution* (ρ,u,h,k,ε) *which satisfies that there exists a constant* $C_2>1$ *such that for any* $t\in[0,T]$,

$$\begin{aligned}&\|(\rho-\bar\rho,u,h,k-\bar k,\varepsilon)\|_{H^3}^2+\int_0^t(\|\nabla\rho\|_{H^2}^2+\|(\nabla u,\nabla h,\nabla k,\nabla\varepsilon)\|_{H^3}^2)\mathrm{d}s\\ \leqslant{}& C_2\|(\rho_0-\bar\rho,u_0,h_0,k_0-\bar k,\varepsilon_0)\|_{H^3}^2.\end{aligned}\tag{3.64}$$

In the following, by a continued argument, combining the local existence and the a priori estimates proposition, we will prove the global existence of smooth solutions.

First, suppose

$$E_0=\|(\rho_0-\bar\rho,u_0,h_0,k_0-\bar k,\varepsilon_0)\|_{H^3}<\min(\delta/\sqrt{C_2},\delta/\sqrt{C_1C_2}),\tag{3.65}$$

where δ is defined in Proposition 2.1. Since the initial data satisfy $E_0 < \delta/\sqrt{C_2}$, then by Proposition 3.2, there exists a constant $T^* > 0$ such that there exists a unique solution on $[0, T^*]$ satisfying

$$E_1 := \sup_{0\leqslant t\leqslant T^*} \|(\rho - \bar{\rho}, u, h, k - \bar{k}, \varepsilon)\|_{H^3} \leqslant \sqrt{C_2}E_0. \tag{3.66}$$

Therefore, using the inequality $E_0 < \delta/\sqrt{C_1C_2}$, from Proposition 2.1, we have

$$E_1 \leqslant \sqrt{C_1}E_0 < \delta/\sqrt{C_2}. \tag{3.67}$$

Notice that T^* depends only on E_0. Starting from T^*, then the initial problem (2.5) with initial data $(\rho, u, h, k, \varepsilon)(T^*)$ still has a unique solution on $[T^*, 2T^*]$, and from Proposition 3.2, we get

$$\sup_{T^*\leqslant t\leqslant 2T^*} \|(\rho - \bar{\rho}, u, h, k - \bar{k}, \varepsilon)\|_{H^3} \leqslant \sqrt{C_2}E_1 \leqslant \sqrt{C_1C_2}E_0 \leqslant \delta.$$

Again from Proposition 2.1, one can deduce

$$E_2 = \sup_{0\leqslant t\leqslant 2T^*} \|(\rho - \bar{\rho}, u, h, k - \bar{k}, \varepsilon)\|_{H^3} \leqslant \sqrt{C_1}E_0 < \delta/\sqrt{C_2}.$$

Repeating the procedure for $0 \leqslant t \leqslant NT^*$, $N = 1, 2, 3, \cdots$, we can extend the local solution to infinity as far as the initial data are small enough such that $E_0 \leqslant \min(\delta/\sqrt{C_2}, \delta/\sqrt{C_1C_2})$. Thus the proof of Theorem 1.1 is complete.

References

[1] Chen Q and Tan Z. *Global existence and convergence rates of smooth solutions for the compressible magnetohydrodynamical equations*, Nonl. Anal., 72, 4438–4451, 2010.

[2] Launder B E and Spalding D B. *Mathematical Models of Turbulence*, Academic Press, London and New York, 1972.

[3] Matsumura A and Nishita T. *The initial value problem for the equations of motion of compressible viscous and heat conductive fluids*, Proc. Japan Acad. Ser. A, 55, 337–342, 1979.

[4] Matsumura A and Nishita T. *The initial value problem for the equations of motion of viscous and heat conductive gses*, J. Math. Kyoto Univ., 20(1), 67–104, 1980.

[5] Nash J. *Le problème de cauchy pour les équations différentielles d'un fluide général*, Bull. Soc. Math. France, 90, 487–497, 1962.

[6] Pu X K and Guo B L. *Global existence and convergence rates of smooth solutions for the full compressible magnetohydrodynamical equations*, Z. Angew Math. Phys., 2012, in press.

[7] Versteeg H K and Malalasekera W. *An troduction to computational fluid dynamics, The finite volume method*, Longman Scientific and Technical, London and New York, 1995.
[8] Volpert A I and Hudjaev S I. *On the cauchy probelm for composite systems of nonlinear differential equations*, Math. USSR-Sb, 16, 517–544, 1972.

Global Random Attractors for the Stochastic Dissipative Zakharov Equations*

Guo Yanfeng (郭艳凤), Guo Boling (郭柏灵) and Li Donglong (李栋龙)

Abstract The stochastic dissipative Zakharov equations with white noise are mainly investigated. The global random attractors endowed with usual topology for the stochastic dissipative Zakharov equations are obtained in the sense of usual norm. The method is to transform the stochastic equations into the corresponding partial differential equations with random coefficients by Ornstein–Uhlenbeck process. The crucial compactness of the global random attractors will be obtained by decomposition of solutions.

Keywords Stochastic dissipative Zakharov equations; Global random attractor; Ornstein-Uhlenbeck process; Compactness

1 Introduction

In recent decades, there has been an increasing interest in the investigation of the counterpart for some classical plasma physics phenomena. It is well known that in the theory of laser plasma interaction, the most general useful model for the Langmuir turbulence is described by Zakharov equations, which consist of one equation for the envelope of the Langmuir wave of the high–frequency electric field and another for the ion–acoustic wave density perturbation[17].

$$\mathrm{i}E_t + \Delta E - nE = 0, \tag{1.1}$$

$$n_{tt} - \Delta n - \Delta |E|^2 = 0, \tag{1.2}$$

where $E : D \times \mathbf{R}^+ \to \mathbf{C}$ is the envelope of the high frequency electric field and $n : D \times \mathbf{R}^+ \to \mathbf{R}$ is the plasma density measured from its equilibrium value. Zakharov equations play an important role in the turbulence theory for plasma waves

* Acta Math. Appl. Sin.–E, 2014, 30(2): 289–304. DOI: 10.1007./s10255-014-0288-9.

and resemble closely to nonlinear Schrödinger equation. Due to the importance of them, Zakharov equations draw greatly attention of many mathematicians and physicists, and have been quite extensively studied theoretically and numerically. Many properties of solutions for (1.1)–(1.2) in one dimension to three dimension have been studied in numerical and theoretical proofs, see [1,8,10,12,13]. In order to get better qualitative properties, it is necessary to include damping effects or effects of the loss of energy. Many are interested in the dissipative Zakharov equations of the form

$$\mathrm{i}E_t + \Delta E - nE + \mathrm{i}\gamma E = f, \tag{1.3}$$

$$n_{tt} + \alpha n_t - \Delta n - \Delta|E|^2 = g, \tag{1.4}$$

which have been extensively investigated for the asymptotic behaviors of the solution for (1.3)–(1.4) in one dimension, see [5,6] .

Recently, the importance of taking random effects into account in modeling, analyzing, simulating and predicting complex phenomena has been widely recognized in geophysical and climate dynamics, materials science, chemistry, biology and other areas, see [4,11]. Stochastic partial differential equations (SPDEs) are appropriate mathematical models for complex systems under random influences or noise. Usually, the noise can be regarded as a simple approximation of turbulence in fluids. As an example using of stochastic forces, the stochastic Zakharov equations can be deduced when stochastic forces are considered in Zakharov equations[7]. For the stochastic dissipative Zakharov equations, the global random attractors only endowed with weak topology have been obtained on a regular bounded domain D in Itô sense instead of usual topology in [9].

However, in this paper, we will study the global random attractors endowed with usual topology. We know that when stochastic forces are considered in every equation of stochastic Zakharov equations there exist some difficulties to obtain the global random attractors endowed with usual topology. Here the stochastic force only in (1.4) will be considered, that is, the following coupled stochastic dissipative Zakharov equations on a bounded domain D

$$n_{tt} + \alpha n_t - \Delta n - \Delta|E|^2 = \sum_{j=1}^{m} \phi_j(x)\frac{\mathrm{d}\omega_j(t)}{\mathrm{d}t}, \tag{1.5}$$

$$\mathrm{i}E_t + \Delta E - nE + \mathrm{i}\gamma E = 0, \tag{1.6}$$

where ϕ_j are time–independent functions, and ω_j are independent two–sided Brownian motions. We endow with the initial conditions $n(0) = n_0(x), n_t(0) = n_1(x), E(0) = E_0(x), x \in D$ and Dirichlet boundary conditions $n|_{\partial D} = 0,\ E|_{\partial D} = 0$ for (1.5)–(1.6). The random dynamical system associated with (1.5)–(1.6) has been obtained[5,9]. The some difficulties can be overcome, and the new results of global random attractors endowed with usual topology will be obtained. It is crucial to obtain the compactness of the random attractors endowed with the usual topology by transforming the stochastic equations into partial differential equations with random coefficients[2,3,6,15], but not in the weak topology.

The paper is organized as follows. In Section 2, some definitions and known results for random attractors, and the main results are given. In Section 3, on the basis of transformation of the Ornstein–Uhlenbeck (O–U) process, a series of time uniform priori estimates in different spaces are established. In Section 4, *a* priori estimates for the decomposition of the solutions will be given, which is the key step for the compactness. In Section 5, proof of the main results, the existence of a global random attractors endowed with usual topology, will be given. In the whole paper, the letter C or c is generic positive constant independent of T which may change their values from terms to terms.

2 Preliminary and main results

The considered domain is a bounded open set $D = [0, L]$ in $\mathbf{R}$. The usual functions spaces $L^2(D), H^m(D), H_0^1(D), H^{-m}(D)$ are used. The scalar product, norm on $L^2(D)$ or p-norm on $L^p(D)$ $(p \geqslant 1)$ will be denoted by $(u, v) = \int_D u(x)\overline{v(x)}\,\mathrm{d}x,\ \|u\| = \left(\int_D |u|^2\,\mathrm{d}x\right)^{\frac{1}{2}}$ or $\|u\|_{L^p} = \left(\int_D |u|^p\,\mathrm{d}x\right)^{\frac{1}{p}}$, respectively. The norm in $H^s(D)$ (s is a nonnegative integer) is denoted by $\|u\|_{H^s} = \left(\sum_{0\leqslant|\alpha|\leqslant s} \|D^\alpha u\|_{L^2(D)}^2\right)^{\frac{1}{2}}$ .We define $H^{-s}(D)$ is the dual space of $H^s(D)$.

Now, some basic concepts related to random attractors for random dynamical system are referred to [2,3]. Let $(X, \|\cdot\|_X)$ be a separable Hilbert space with Borel σ-algebra $\mathscr{B}(X)$ endowed with the distance d, and let $(\Omega, \mathcal{F}, \mathbf{P})$ be a probability space. We also denote the mappings $S(t, s; \omega) : X \to X, -\infty < s \leqslant t < \infty$ with explicit dependence on ω. In most application there exists a group θ_t, $t \in \mathbf{R}$, of measure preserving transformations of $(\Omega, \mathcal{F}, \mathbf{P})$ with the property

$$S(t, s; \omega)x = S(t - s, 0; \theta_s\omega)x, \quad s < t, \quad x \in X, \quad P-\text{a.e.}.$$

In applications to stochastic evolution equations driven by white noise, $\omega(t)$ is the two–sided Wiener space $C_0(\mathbf{R}; X)$ of continuous functions with values in a Banach space X, and equal to 0 at $t = 0$. In this case θ_t is defined as

$$(\theta_t\omega)(s) = \omega(t+s) - \omega(t), \quad s, t \in \mathbf{R}.$$

Definition 2.1 *Let $t \in \mathbf{R}$ and $\omega \in \Omega$. A random dynamical system (RDS) with time t on a metric, complete and separable space (X, d) with Borel σ-algebra $\mathscr{B}$ over $\{\theta_t\}$ on $(\Omega, \mathcal{F}, \mathbf{P})$ is a measurable map*

$$S(t, s; \omega) : X \to X, \quad -\infty < s \leqslant t < \infty$$

such that $S(0, 0; \omega) = id$ and $S(t, 0; \omega) = S(t, s; \omega)S(s, 0; \omega)$ for all $t, s \in \mathbf{R}$ and for all $\omega \in \Omega$.

A RDS is said to be continuous or differentiable if $S(t, s; \omega) : X \to X, -\infty < s \leqslant t < \infty$ is continuous or differentiable, respectively, for all $t \in \mathbf{R}$ outside a P-nullset.

Definition 2.2 *Given $t \in \mathbf{R}$ and $\omega \in \Omega$, $K(t, \omega) \subset X$ is called an attracting set, if for all bounded sets $B \subset X$*

$$d(S(t, s; \omega)B, K(t, \omega)) \to 0, \quad s \to -\infty,$$

where $d(A, B)$ is the semidistance defined by

$$d(A, B) = \sup_{x\in A} \inf_{y\in B} d(x, y).$$

Definition 2.3 *A family $A(\omega)(\omega \in \Omega)$ of the closed subsets of X is measurable, if for all $x \in X$, the mapping $\omega \mapsto d(A(\omega), x)$ is measurable.*

Definition 2.4 *Define the random omega limit set of a bounded set $B \subset X$ at time t as*

$$A(B, t; \omega) = \bigcap_{T<t} \overline{\bigcup_{s<T} S(t, s; \omega)B}.$$

Definition 2.5 *Let $S(t, s; \omega)_{t\geqslant s, \omega\in\Omega}$ be a random dynamical system, $A(\omega)$ is a random set satisfying the following conditions*

(1) *It is the minimal closed set such that for $t \in \mathbf{R}, B \subset X$,*

$$d(S(t, s; \omega)B, A(\omega)) \to 0, \quad s \to -\infty,$$

then we called $A(\omega)$ attractors B (B is a deterministic set).

(2) *$A(\omega)$ is the largest compact measurable set, which is invariant in sense that*

$$S(t,s;\omega)A(\theta_s\omega)=A(\theta_t\omega),\quad s\leqslant t.$$

Then $A(\omega)$ is said to be the random attractors.

Theorem 2.1 (see [2,3]) *Let $S(t,s;\omega)_{t\geqslant s,\omega\in\Omega}$ be a random dynamical system satisfying the following conditions*

(i) *$S(t,r;\omega)S(r,s;\omega)x=S(t,s;\omega)x$ for all $s\leqslant r\leqslant t$ and $x\in X$,*

(ii) *$S(t,s;\omega)$ is continuous in X, for all $s\leqslant t$,*

(iii) *for all $s<t$ and $x\in X$, the mapping $\omega\mapsto S(t,s;\omega)x$ is measurable from $(\Omega,\mathcal{F})$ to $(X,\mathscr{B}(X))$,*

(iv) *for all $t\in \boldsymbol{R}, x\in X$ and P-a.e. ω, the mapping $s\mapsto S(t,s;\omega)x$ is right continuous at any point.*

Assume that there exists a group θ_t, $t\in\boldsymbol{R}$ of measure preserving mappings such that

$$S(t,s;\omega)x=S(t-s,0;\theta_s\omega)x,\quad P-a.e.$$

holds and for P-a.e., there exists a compact attracting set $K(\omega)$ at time 0, for P-a.e. $\omega\in\Omega$, we set $\Lambda(\omega)=\overline{\bigcup_{B\subset X}A(B,\omega)}$, where the union is taken over all the bounded subsets of X and $A(B,\omega)$ is given by

$$A(B,0;\omega)=\bigcap_{T<0}\overline{\bigcup_{s<T}S(0,s;\omega)B},$$

then $\Lambda(\omega)$ is random attractors.

3 *A* time uniform priori estimates

First for $j=1,\cdots,m$, let W_j be the solution of $\mathrm{d}W_{j,t}+\beta W_{j,t}\mathrm{d}t=d\omega_j$. Set $W=\sum_{j=1}^m\phi_j(x)W_j(t)$, $N=n-W$. Therefore we have

$$W_{tt}+\beta W_t=\sum_{j=1}^m\phi_j(x)\frac{\mathrm{d}\omega_j(t)}{\mathrm{d}t}.$$

We know that the regularity of $W_{j,t}$ and W_j should coincide with the regularity of ϕ_j in [2,3]. Then we can see that (1.5) and (1.6) become

$$N_{tt}+\alpha N_t-\Delta N-\Delta|E|^2=(\beta-\alpha)W_t+\Delta W, \tag{3.1}$$

$$\mathrm{i}E_t+\Delta E-(N+W)E+\mathrm{i}\gamma E=0, \tag{3.2}$$

$$E(0)=E_0(x),N(0)=n_0(x)-W(0),N_t(0)=n_1(x)-W_t(0). \tag{3.3}$$

As in [14], for the nonlinear wave equations, we set $M = N_t + \varepsilon N$, where $\varepsilon > 0$ should be enough small and chosen later. So we have three unknown variables which satisfy the following equations

$$M = N_t + \varepsilon N, \tag{3.4}$$

$$M_t + (\alpha - \varepsilon)M - \varepsilon(\alpha - \varepsilon)N - \Delta N - \Delta|E|^2 = (\beta - \alpha)W_t + \Delta W, \tag{3.5}$$

$$\mathrm{i}E_t + \Delta E - (N + W)E + \mathrm{i}\gamma E = 0, \tag{3.6}$$

where $\alpha > 0, \gamma > 0, \varepsilon > 0$. The corresponding initial conditions are

$$M(0) = N_t(0) + \varepsilon N(0), N(0) = n_0 - W(0), E(0) = E_0.$$

For equations (3.4)–(3.6), we also define the product spaces V_0, V_1 and V_2 as follows

$$V_0 = H^{-1}(D) \times L^2(D) \times H_0^1(D),$$
$$V_1 = L^2(D) \times H_0^1(D) \times (H^2(D) \cap H_0^1(D)),$$
$$V_2 = H_0^1(D) \times (H^2(D) \cap H_0^1(D)) \times (H^3(D) \cap H_0^1(D)).$$

Endow each V_i with the usual norm $\|\cdot\|_{V_i}$ and we have $V_2 \subset V_1 \subset V_0$ with compact embeddings. In the following, we give a priori estimates for the equations (3.4)–(3.6) on different spaces.

3.1 A priori estimates in $V_0 = H^{-1}(D) \times L^2(D) \times H_0^1(D)$

Proposition 3.1 *If $E(s) \in L^2(D)$ and $\|E(s)\| \leqslant R$, then for all $0 \geqslant t \geqslant s$, we have $\|E(t)\| \leqslant C$, where C is independent of t and ω.*

Proof Taking the inner product of (3.6) with E over D and considering the imaginary part, gets

$$\frac{\mathrm{d}}{\mathrm{d}t}\|E\|^2 + 2\gamma\|E\|^2 = 0. \tag{3.7}$$

Then integrating from s to t for $0 \geqslant t \geqslant s$, one obtains

$$\|E(t)\|^2 \leqslant \mathrm{e}^{2\gamma(s-t)}\|E(s)\|^2 \leqslant \|E(s)\|^2 \leqslant C, \quad 0 \geqslant t \geqslant s, \tag{3.8}$$

where C is independent of t and ω.

Proposition 3.2 *Let $\phi_j(x) \in H^1(D)$ and $(M(t), N(t), E(t))$ be the solution of (3.4)–(3.6). For any given $R > 0$ and $(M(s), N(s), E(s)) \in V_0$ satisfying*

$$\|(M(s), N(s), E(s))\|_{V_0} \leqslant R,$$

there exist a random radius $r_0(\omega)$ and a $s_1(\omega) \leqslant -1$, such that for all $s \leqslant s_1(\omega)$, the following inequality holds P-a.e.

$$\|(M(0), N(0), E(0))\|_{V_0}^2 \leqslant r_0(\omega).$$

Moreover for all $0 \geqslant t \geqslant s$, we also have

$$\begin{aligned} &\|M(t)\|_{H^{-1}}^2 + \|N(t)\|^2 + \|\nabla E(t)\|^2 \\ &\leqslant ce^{\eta(s-t)}H_0(s) + c\int_s^t e^{\eta(\tau-t)}g_1(\tau)\,\mathrm{d}\tau + cg_2(t) := f_1(t), \end{aligned}$$

where $g_1(\tau)$, $g_2(t)$ are defined below.

Proof Taking the inner product of (3.6) with $\gamma E + E_t$ over D and considering the real part, gets

$$\frac{1}{2}\frac{\mathrm{d}}{\mathrm{d}t}\|\nabla E\|^2 + \gamma\|\nabla E\|^2 + \mathrm{Re}\int_D (N+W)E(\bar{E}_t + \gamma\bar{E})\,\mathrm{d}x = 0. \tag{3.9}$$

Taking the inner product of (3.5) with $(-\Delta)^{-1}M$ over D, one gets

$$\begin{aligned} &\frac{1}{2}\frac{\mathrm{d}}{\mathrm{d}t}(\|M\|_{H^{-1}}^2 + \|N\|^2) + (\alpha-\varepsilon)\|M\|_{H^{-1}}^2 - \varepsilon(\alpha-\varepsilon)\int_D N(-\Delta)^{-1}M\,\mathrm{d}x \\ &+ \varepsilon\|N\|^2 + \int_D N_t|E|^2\,\mathrm{d}x + \varepsilon\int_D N|E|^2\,\mathrm{d}x \\ = &(\beta-\alpha)\int_D (W_t + \Delta W)(-\Delta)^{-1}M\,\mathrm{d}x. \end{aligned} \tag{3.10}$$

Notice that

$$\frac{\mathrm{d}}{\mathrm{d}t}\int_D (N+W)|E|^2\,\mathrm{d}x = \int_D (N_t+W_t)|E|^2\,\mathrm{d}x + 2\mathrm{Re}\int_D (N+W)E\bar{E}_t\,\mathrm{d}x.$$

Then taking $\varepsilon \leqslant \min\left\{\frac{\alpha}{4}, \frac{2\lambda_1}{\alpha}\right\}$, where λ_1 is the first eigenvalue of the operator Δ, from (3.9) and (3.10), one gets

$$\begin{aligned} &\frac{\mathrm{d}}{\mathrm{d}t}\left(\|\nabla E\|^2 + \frac{1}{2}(\|M\|_{H^{-1}}^2 + \|N\|^2) + \int_D (N+W)|E|^2\,\mathrm{d}x\right) + 2\gamma\|\nabla E\|^2 \\ &+ \frac{\varepsilon}{2}(\|M\|_{H^{-1}}^2 + \|N\|^2) \leqslant -2\gamma\int_D (N+W)|E|^2\,\mathrm{d}x + \int_D W_t|E|^2\,\mathrm{d}x \\ &- \varepsilon\int_D N|E|^2\,\mathrm{d}x + (\beta-\alpha)\int_D (W_t+\Delta W)(-\Delta)^{-1}M\,\mathrm{d}x. \end{aligned} \tag{3.11}$$

Choosing

$$\eta = \min\left\{\gamma, \frac{\varepsilon}{2}\right\}$$

and setting

$$H_0(t)=\|\nabla E\|^2+\frac{1}{2}(\|M\|_{H^{-1}}^2+\|N\|^2)+\int_D(N+W)|E|^2\,\mathrm{d}x,$$

then (3.11) can be changed into

$$\begin{aligned}&\frac{\mathrm{d}}{\mathrm{d}t}H_0(t)+\eta H_0(t)\leqslant(\eta-2\gamma)\int_D(N+W)|E|^2\,\mathrm{d}x+\int_D W_t|E|^2\,\mathrm{d}x\\&-\varepsilon\int_D N|E|^2\,\mathrm{d}x+(\beta-\alpha)\int_D(W_t+\Delta W)(-\Delta)^{-1}M\,\mathrm{d}x.\end{aligned}\tag{3.12}$$

We now estimate each term on the right hand side of (3.12). Using the Hölder inequality, Gagliardo–Nirenberg inequality and Young's inequality, they have

$$\begin{aligned}&(\eta-2\gamma)\int_D(N+W)|E|^2\,\mathrm{d}x\leqslant\frac{\gamma}{4}\|\nabla E\|^2+\frac{\varepsilon}{8}\|N\|^2+c(\|E\|^6+\|W\|^2),\\&\int_D W_t|E|^2\,\mathrm{d}x\leqslant\|W_t\|\|E\|^{\frac{3}{2}}\|\nabla E\|^{\frac{1}{2}}\leqslant\frac{\gamma}{4}\|\nabla E\|^2+c(\|E\|^6+\|W_t\|^2),\\&-\varepsilon\int_D N|E|^2\,\mathrm{d}x\leqslant\frac{\gamma}{4}\|\nabla E\|^2+\frac{\varepsilon}{8}\|N\|^2+c\|E\|^6,\\&(\beta-\alpha)\int_D(W_t+\Delta W)(-\Delta)^{-1}M\,\mathrm{d}x\leqslant\frac{\varepsilon}{8}\|M\|_{H^{-1}}^2+c(\|W\|_{H^1}^2+\|W_t\|_{H^{-1}}^2).\end{aligned}$$

Instituting above inequalities into (3.12) yields

$$\frac{\mathrm{d}}{\mathrm{d}t}H_0(t)+\eta H_0(t)\leqslant cg_1(t),\tag{3.13}$$

where $g_1(t)=\|E\|^6+\|W\|^2+\|W_t\|^2+\|W\|_{H^1}^2+\|W_t\|_{H^{-1}}^2$ grows at most polynomial as $t\to-\infty$. Then for (3.13) integrating from s to t with respect to t for $0\geqslant t\geqslant s$, one can obtain

$$H_0(t)\leqslant \mathrm{e}^{\eta(s-t)}H_0(s)+c\int_s^t \mathrm{e}^{\eta(\tau-t)}g_1(\tau)\,\mathrm{d}\tau.\tag{3.14}$$

From the first of above inequalities, one gets

$$\begin{aligned}&\frac{1}{2}\|M(t)\|_{H^{-1}}^2+\frac{1}{4}\|N(t)\|^2+\frac{1}{2}\|\nabla E(t)\|^2\leqslant H_0(t)+c(\|E(t)\|^6+\|W(t)\|^2)\\&\leqslant \mathrm{e}^{\eta(s-t)}H_0(s)+c\int_s^t \mathrm{e}^{\eta(\tau-t)}g_1(\tau)\,\mathrm{d}\tau+g_2(t),\quad 0\geqslant t\geqslant s,\end{aligned}$$

where $g_2(t)$ grows at most polynomial as $t\to-\infty$. So there is

$$\begin{aligned}&\|M(t)\|_{H^{-1}}^2+\|N(t)\|^2+\|\nabla E(t)\|^2\leqslant cH_0(t)+c(\|E(t)\|^6+\|W(t)\|^2)\\&\leqslant c\mathrm{e}^{\eta(s-t)}H_0(s)+c\int_s^t \mathrm{e}^{\eta(\tau-t)}g_1(\tau)\,\mathrm{d}\tau+cg_2(t):=f_1(t),\quad 0\geqslant t\geqslant s.\end{aligned}\tag{3.15}$$

When $t=0$, we can deduce

$$\begin{aligned}&\|M(0)\|_{H^{-1}}^2+\|N(0)\|^2+\|\nabla E(0)\|^2\\&\leqslant c\mathrm{e}^{\eta s}H_0(s)+c\int_s^0 \mathrm{e}^{\eta\tau}g_1(\tau)\,\mathrm{d}\tau+cg_2(0)\leqslant r_0(\omega).\end{aligned}\tag{3.16}$$

The proof is complete.

3.2 A priori estimates in $V_1=L^2(D)\times H_0^1(D)\times(H^2(D)\cap H_0^1(D))$

Proposition 3.3 *Let $\phi_j(x)\in H^2(D)$ and $(M(t),N(t),E(t))$ be the solution of (3.4)–(3.6). For any given $R>0$ and $(M(s),N(s),E(s))\in V_1$ satisfying*

$$\|(M(s),N(s),E(s))\|_{V_1}\leqslant R,$$

there exist a random radius $r_1(\omega)$ and a $s_2(\omega)\leqslant -1$, such that for all $s\leqslant s_2(\omega)$, the following inequality holds P-a.e.

$$\|(M(0),N(0),E(0))\|_{V_1}^2\leqslant r_1(\omega).$$

Moreover for all $0\geqslant t\geqslant s$, we also have

$$\begin{aligned}&\|M(t)\|^2+\|\nabla N(t)\|^2+\|\Delta E(t)\|^2\\&\leqslant ce^{\eta(s-t)}a_2(s)+c\int_s^t e^{\eta(\sigma-t)}g_6(\sigma)\,\mathrm{d}\sigma+g_7(t):=f_4(t),\end{aligned}\tag{3.17}$$

where $a_2(s),g_6(\tau)$, $g_7(t)$ are defined below.

Proof Taking the inner product of (3.6) with $\gamma\Delta E+\Delta E_t$ over D and considering the real part, one gets

$$\frac{1}{2}\frac{\mathrm{d}}{\mathrm{d}t}\|\Delta E\|^2+\gamma\|\Delta E\|^2-\mathrm{Re}\int_D(N+W)E(\Delta\bar{E}_t+\gamma\Delta\bar{E})\,\mathrm{d}x=0.\tag{3.18}$$

Taking the inner product of (3.5) with M over D, one gets

$$\begin{aligned}&\frac{1}{2}\frac{\mathrm{d}}{\mathrm{d}t}(\|M\|^2+\|\nabla N\|^2)+(\alpha-\varepsilon)\|M\|^2-\varepsilon(\alpha-\varepsilon)\int_D NM\,\mathrm{d}x+\varepsilon\|\nabla N\|^2\\&+\int_D(N_t+\varepsilon N)\Delta|E|^2\,\mathrm{d}x=\int_D((\beta-\alpha)W_t+\Delta W)M\,\mathrm{d}x.\end{aligned}\tag{3.19}$$

Notice that

$$\begin{aligned}&2\frac{\mathrm{d}}{\mathrm{d}t}\mathrm{Re}\int_D(N+W)E\Delta\bar{E}\,\mathrm{d}x\\&=\int_D(N_t+W_t)\Delta|E|^2\,\mathrm{d}x-\int_D(N_t+W_t)|\nabla E|^2\,\mathrm{d}x\\&\quad+2\mathrm{Re}\int_D(N+W)E_t\Delta\bar{E}\,\mathrm{d}x+2\mathrm{Re}\int_D(N+W)E\Delta\bar{E}_t\,\mathrm{d}x.\end{aligned}$$

Taking ε as former, then from (3.18) and (3.19), one gets

$$\begin{aligned}
&\frac{\mathrm{d}}{\mathrm{d}t}\left(\frac{1}{2}(\|M\|^2+\|\nabla N\|^2)+\|\Delta E\|^2-2\mathrm{Re}\int_D(N+W)E\Delta\bar{E}\,\mathrm{d}x\right)\\
&+2\gamma\|\Delta E\|^2+\frac{\varepsilon}{2}(\|M\|^2+\|\nabla N\|^2)\\
\leqslant\ &2\int_D(N_t+W_t)|\nabla E|^2\,\mathrm{d}x-2\mathrm{Re}\int(N+W)E_t\Delta\bar{E}\,\mathrm{d}x\\
&+2\gamma\mathrm{Re}\int_D(N+W)E\Delta\bar{E}\,\mathrm{d}x-\int W_t\Delta|E|^2\,\mathrm{d}x\\
&+\varepsilon\int_D N\Delta|E|^2\,\mathrm{d}x+\int_D((\beta-\alpha)W_t+\Delta W)M\,\mathrm{d}x. \qquad (3.20)
\end{aligned}$$

Choosing η as former and setting

$$H_1(t)=\frac{1}{2}(\|M\|^2+\|\nabla N\|^2)+\|\Delta E\|^2-2\mathrm{Re}\int_D(N+W)E\Delta\bar{E}\,\mathrm{d}x,$$

then (3.20) can be changed into

$$\begin{aligned}
&\frac{\mathrm{d}}{\mathrm{d}t}H_1(t)+\eta H_1(t)\\
\leqslant\ &2\int_D(N_t+W_t)|\nabla E|^2\,\mathrm{d}x-2\mathrm{Re}\int_D(N+W)E_t\Delta\bar{E}\,\mathrm{d}x-\int_D W_t\Delta|E|^2\,\mathrm{d}x\\
&+(2\eta+2\gamma)\mathrm{Re}\int_D(N+W)E\Delta\bar{E}\,\mathrm{d}x+\varepsilon\int_D N\Delta|E|^2\,\mathrm{d}x\\
&+\int_D((\beta-\alpha)W_t+\Delta W)M\,\mathrm{d}x. \qquad (3.21)
\end{aligned}$$

By the same arguments as before, one gets

$$\frac{\mathrm{d}}{\mathrm{d}t}H_1(t)+\eta H_1(t)\leqslant c(f_1^6(t)+g_3(t)), \qquad (3.22)$$

where $g_3(t)=\|W_t\|^4+\|W\|_{H^2}^{12}+1$ grows at most polynomial as $t\to-\infty$. Therefore for (3.22), integrating from s to t with respect to t for $0\geqslant t\geqslant s$, one obtains

$$H_1(t)\leqslant \mathrm{e}^{\eta(s-t)}H_1(s)+c\int_s^t \mathrm{e}^{\eta(\tau-t)}(f_1^6(\tau)+g_3(\tau))\,\mathrm{d}\tau. \qquad (3.23)$$

However, one knows

$$\begin{aligned}
f_1^6(\tau)&=\left(c\mathrm{e}^{\eta(s-\tau)}H_0(s)+c\int_s^\tau \mathrm{e}^{\eta(\sigma-\tau)}g_1(\sigma)\,\mathrm{d}\sigma+cg_2(\tau)\right)^6\\
&\leqslant c\mathrm{e}^{6\eta(s-\tau)}H_0^6(s)+c\left(\int_s^\tau \mathrm{e}^{\eta(\sigma-\tau)}g_1(\sigma)\,\mathrm{d}\sigma\right)^6+cg_2^6(\tau). \qquad (3.24)
\end{aligned}$$

On one hand, it is easy to see

$$c\int_s^t \mathrm{e}^{\eta(\tau-t)}\mathrm{e}^{6\eta(s-\tau)}H_0^6(s)\,\mathrm{d}\tau = c\mathrm{e}^{-\eta t}\mathrm{e}^{6\eta s}H_0^6(s)\int_s^t \mathrm{e}^{-5\eta\tau}\,\mathrm{d}t \leqslant cH_0^6(s)\mathrm{e}^{\eta(s-t)}. \quad (3.25)$$

On the other hand, using the Hölder inequality, one gets

$$\begin{aligned}
&c\int_s^t \mathrm{e}^{\eta(\tau-t)}\left(\int_s^\tau \mathrm{e}^{\eta(\sigma-\tau)}g_1(\sigma)\,\mathrm{d}\sigma\right)^6 \mathrm{d}\tau\\
\leqslant\, & c\int_s^t \mathrm{e}^{\eta(\tau-t)}\left(\int_s^\tau g_1^6(\sigma)\mathrm{e}^{3\eta(\sigma-\tau)}\,\mathrm{d}\sigma\right)\left(\int_s^\tau \mathrm{e}^{\frac{3\eta}{5}(\sigma-\tau)}\,\mathrm{d}\sigma\right)^5 \mathrm{d}\tau\\
\leqslant\, & c\int_s^t \left(\int_\sigma^t \mathrm{e}^{\eta(\tau-t)}g_1^6(\sigma)\mathrm{e}^{3\eta(\sigma-\tau)}\,\mathrm{d}\tau\right)\mathrm{d}\sigma \leqslant c\int_s^t \mathrm{e}^{\eta(\sigma-t)}g_1^6(\sigma)\,\mathrm{d}\sigma. \qquad (3.26)
\end{aligned}$$

Then from (3.23) one can obtain

$$\begin{aligned}
H_1(t) &\leqslant c\mathrm{e}^{\eta(s-t)}(H_1(s)+H_0^6(s)) + c\int_s^t \mathrm{e}^{\eta(\sigma-t)}(g_1^6(\sigma)+g_2^6(\sigma)+g_3(\sigma))\,\mathrm{d}\sigma\\
&:= c\mathrm{e}^{\eta(s-t)}a_1(s) + c\int_s^t \mathrm{e}^{\eta(\sigma-t)}g_4(\sigma)\,\mathrm{d}\sigma := g_5(t),\quad 0\geqslant t\geqslant s, \qquad (3.27)
\end{aligned}$$

where $g_4(\sigma)$ grows at most polynomial as $\sigma\to-\infty$. Then one has

$$\begin{aligned}
&\|M(t)\|^2+\|\nabla N(t)\|^2+\|\Delta E(t)\|^2 \leqslant cH_1(t)+cf_1^2(t)+c\|W(t)\|^4\\
\leqslant\, & cg_5(t)+cf_1^2(t)+c\|W(t)\|^4 := f_3(t),\quad 0\geqslant t\geqslant s. \qquad (3.28)
\end{aligned}$$

Then one can obtain

$$\begin{aligned}
&f_3(t)\\
\leqslant\, & c\mathrm{e}^{\eta(s-t)}(a_1(s)+H_0^2(s)) + c\int_s^t \mathrm{e}^{\eta(\sigma-t)}(g_1^2(\sigma)+g_4(\sigma))\,\mathrm{d}\sigma + cg_2^2(t)+c\|W(t)\|^4\\
:=\, & c\mathrm{e}^{\eta(s-t)}a_2(s) + c\int_s^t \mathrm{e}^{\eta(\sigma-t)}g_6(\sigma)\,\mathrm{d}\sigma + g_7(t) := f_4(t),
\end{aligned}$$

that is

$$\|M(t)\|^2+\|\nabla N(t)\|^2+\|\Delta E(t)\|^2 \leqslant f_4(t),\quad 0\geqslant t\geqslant s, \qquad (3.29)$$

where $g_6(\sigma)$ and $g_7(\sigma)$ grow at most polynomial as $\sigma\to-\infty$. When $t=0$, from (3.28) one can obtain

$$\|M(0)\|^2+\|\nabla N(0)\|^2+\|\Delta E(0)\|^2 \leqslant f_4(0) \leqslant r_1(\omega). \qquad (3.30)$$

Then the proof is complete.

3.3 A priori estimates in $V_2 = H_0^1(D) \times (H^2(D) \cap H_0^1(D)) \times (H^3(D) \cap H_0^1(D))$

Proposition 3.4 *Let $\phi_j(x) \in H^3(D)$ and $(M(t), N(t), E(t))$ be the solution of* (3.4)–(3.6). *For any given $R > 0$ and $(M(s), N(s), E(s)) \in V_2$ satisfying*

$$\|(M(s), N(s), E(s))\|_{V_2} \leqslant R,$$

there exist a random radius $r_2(\omega)$ and a $s_3(\omega) \leqslant -1$, such that for all $s \leqslant s_3(\omega)$, the following inequality holds P-a.e.

$$\|(M(0), N(0), E(0))\|_{V_2}^2 \leqslant r_2(\omega).$$

Moreover for all $0 \geqslant t \geqslant s$, we also have

$$\begin{aligned}
&\|\nabla M(t)\|^2 + \|\Delta N(t)\|^2 + \|\nabla \Delta E(t)\|^2 \\
&\leqslant ce^{\eta(s-t)} a_4(s) + c\int_s^t e^{\eta(\sigma-t)} g_{10}(\sigma)\,\mathrm{d}\sigma + g_{11}(t) := f_5(t),
\end{aligned}$$

where $a_4(s), g_{10}(\sigma)$, $g_{11}(\sigma)$ are defined below.

Proof Taking the inner product of (3.6) with $\gamma\Delta^2 E + \Delta^2 E_t$ over D and considering the real part, one gets

$$\frac{1}{2}\frac{\mathrm{d}}{\mathrm{d}t}\|\nabla\Delta E\|^2 + \gamma\|\nabla\Delta E\|^2 + \mathrm{Re}\int_D (N+W)E(\Delta^2\bar{E}_t + \gamma\Delta^2\bar{E})\,\mathrm{d}x = 0. \tag{3.31}$$

Taking the inner product of (3.5) with $-\Delta M$ over D, one gets

$$\begin{aligned}
&\frac{1}{2}\frac{\mathrm{d}}{\mathrm{d}t}(\|\nabla M\|^2 + \|\Delta N\|^2) + (\alpha-\varepsilon)\|\nabla M\|^2 - \varepsilon(\alpha-\varepsilon)\int_D \nabla N\nabla M\,\mathrm{d}x + \varepsilon\|\Delta N\|^2 \\
&+\int_D \Delta(N_t + \varepsilon N)\Delta|E|^2\,\mathrm{d}x = -\int_D ((\beta-\alpha)W_t + \Delta W)\Delta M\,\mathrm{d}x. \qquad (3.32)
\end{aligned}$$

Notice that

$$\begin{aligned}
&\frac{\mathrm{d}}{\mathrm{d}t}\mathrm{Re}\int_D (N+W)E\Delta^2\bar{E}\,\mathrm{d}x \\
=\ &\mathrm{Re}\int_D (N_t+W_t)E\Delta^2\bar{E}\,\mathrm{d}x + \mathrm{Re}\int_D (N+W)E_t\Delta^2\bar{E}\,\mathrm{d}x \\
&+\mathrm{Re}\int_D (N+W)E\Delta^2\bar{E}_t\,\mathrm{d}x \\
=\ &\frac{1}{2}\int_D \Delta|E|^2\Delta N_t\,\mathrm{d}x + \frac{1}{2}\int_D \Delta|E|^2\Delta W_t\,\mathrm{d}x - 2\int_D |\nabla E|^2\Delta(N_t+W_t)\,\mathrm{d}x \\
&+\mathrm{Re}\int_D (N_t+W_t)\Delta E\Delta\bar{E}\,\mathrm{d}x + 2\mathrm{Re}\int_D (\nabla N_t + \nabla W_t)\nabla E\Delta\bar{E}\,\mathrm{d}x \\
&+\mathrm{Re}\int_D (N+W)E_t\Delta^2\bar{E}\,\mathrm{d}x + \mathrm{Re}\int_D (N+W)E\Delta^2\bar{E}_t\,\mathrm{d}x.
\end{aligned}$$

Then taking ε and η as former, from (3.31) $\times$ 2 + (3.32) one gets

$$\begin{aligned}
&\frac{\mathrm{d}}{\mathrm{d}t}H_2(t)+\eta H_2(t)\\
&\leqslant\int_D\Delta|E|^2\Delta W_t\,\mathrm{d}x-4\int_D|\nabla E|^2\Delta(N_t+W_t)\,\mathrm{d}x+2\mathrm{Re}\int_D(N_t+W_t)\Delta E\Delta\bar{E}\,\mathrm{d}x\\
&+4\mathrm{Re}\int_D(\nabla N_t+\nabla W_t)\nabla E\Delta\bar{E}\,\mathrm{d}x+2\mathrm{Re}\int_D(N+W)E_t\Delta^2\bar{E}\,\mathrm{d}x\\
&-\varepsilon\int_D\Delta|E|^2\Delta N\,\mathrm{d}x+2(\eta-\gamma)\mathrm{Re}\int_D(N+W)E\Delta^2\bar{E}\,\mathrm{d}x\\
&-\int_D((\beta-\alpha)W_t+\Delta W)\Delta M\,\mathrm{d}x,
\end{aligned}\tag{3.33}$$

where $H_2(t)=\frac{1}{2}(\|\nabla M\|^2+\|\Delta N\|^2)+\|\nabla\Delta E\|^2+2\mathrm{Re}\int_D(N+W)E\Delta^2\bar{E}\,\mathrm{d}x$.

Similar to the former estimates, using the Hölder inequality, Gagliardo-Nirenberg inequality and Young's inequality, and since $E_t=\mathrm{i}(\Delta E-(N+W)E+\mathrm{i}\gamma\nabla E)$, the terms of the right hand side of (3.33) can be estimated. Then from (3.33) one can obtain

$$\frac{\mathrm{d}}{\mathrm{d}t}H_2(t)+\eta H_2(t)\leqslant c(f_4^4(t)+g_8(t)),\tag{3.34}$$

where $g_8(t)=\|W_t\|_{H^2}^2+\|\nabla W\|^8+\|\nabla\Delta W\|^2+1$ grows at most polynomial as $t\to-\infty$. Therefore from (3.34), integrating from s to t with respect to t for $0\geqslant t\geqslant s$, and using the same method as former, one can obtain

$$H_2(t)\leqslant c\mathrm{e}^{\eta(s-t)}a_3(s)+c\int_s^t\mathrm{e}^{\eta(\sigma-t)}g_9(\sigma)\,\mathrm{d}\sigma:=g_{10}(t),\quad 0\geqslant t\geqslant s,\tag{3.35}$$

where $g_9(\sigma)$ grows at most polynomial as $\sigma\to-\infty$. Moreover since

$$2\mathrm{Re}\int_D(N+W)E\Delta^2\bar{E}\,\mathrm{d}x\leqslant\frac{1}{2}\|\nabla\Delta E\|^2+c(\|\Delta E\|^4+\|\nabla N\|^4+\|\nabla W\|^4),$$

one gets

$$\begin{aligned}
&\|\nabla M(t)\|^2+\|\Delta N(t)\|^2)+\|\nabla\Delta E(t)\|^2\leqslant g_{10}(t)+cf_4^2(t)+\|\nabla W(t)\|^4\\
&\leqslant c\mathrm{e}^{\eta(s-t)}a_4(s)+c\int_s^t\mathrm{e}^{\eta(\sigma-t)}g_{11}(\sigma)\,\mathrm{d}\sigma+g_{12}(t):=f_5(t),\quad 0\geqslant t\geqslant s,
\end{aligned}\tag{3.36}$$

where $g_{11}(\sigma)$ and $g_{12}(\sigma)$ grow at most polynomial as $\sigma\to-\infty$. Then when $t=0$, from (3.36) one can obtain

$$\|\nabla M(0)\|^2+\|\Delta N(0)\|^2+\|\nabla\Delta E(0)\|^2\leqslant f_5(0)\leqslant r_2(\omega).\tag{3.37}$$

Then the proof is complete.

4 Decomposition of the solution

For fixed $K \in \mathbf{N}$, we denote by $P_K = P$ the orthogonal projector in L^2 onto the space spanned by the first K eigenvectors of $-\Delta$, and set $Q_K = Q = I - P_K$. In addition, let $E = P_K E + Q_K E$ and $y = P_K E$. Now we split Q_K of E as $Q_K E = Z + \chi$ and N as $N = p + q$, where (p, Z) is solution of

$$\mathrm{i}Z_t + \Delta Z - Q_K((N+W)(y+Z)) + \mathrm{i}\gamma Z = 0, \tag{4.1}$$

$$p_{tt} + \alpha p_t - \Delta p = \Delta|y+Z|^2 + (\beta - \alpha)W_t + \Delta W, \tag{4.2}$$

$$p(s) = 0, p_t(s) = 0, Z(s) = 0, \tag{4.3}$$

and (q, χ) is solution of

$$\mathrm{i}\chi_t + \Delta\chi - Q_K((N+W)\chi) + \mathrm{i}\gamma\chi = 0, \tag{4.4}$$

$$q_{tt} + \alpha q_t - \Delta q = \Delta|\chi|^2 + 2\mathrm{Re}(v\bar{\lambda}), \tag{4.5}$$

$$q(s) = N(0), q_t(s) = N_t(0), \chi(x, s) = E_0, \tag{4.6}$$

where $v = y + Z$.

In this section, we consider *a* priori estimates for the equations (4.1)–(4.3). By the usual method, we can rewrite the equations (4.1)–(4.3) as

$$\mathrm{i}Z_t + \Delta Z - Q_K((N+W)(y+Z)) + \mathrm{i}\gamma Z = 0, \tag{4.7}$$

$$p_1(t) = p_t + \varepsilon p, \tag{4.8}$$

$$p_{1t} + (\alpha - \varepsilon)p_1(t) - \varepsilon(\alpha - \varepsilon)p - \Delta p = \Delta|y+Z|^2 + (\beta - \alpha)W_t + \Delta W, \tag{4.9}$$

$$p_1(s) = 0, p(s) = 0, Z(s) = 0, \tag{4.10}$$

where ε can be chose as before. By the results former, one knows that

$$\|(M(t), N(t), E(t))\|_{V_1} \leqslant f_4(t)$$

when $0 \geqslant t \geqslant s$.

Proposition 4.1 *Let $\phi_j(x) \in H^3(D)$ and (p_1, p, Z) be the solution of* (4.7)–(4.10)*. Then there exist random a radius $r_5(\omega)$ and a $s_4(\omega) \leqslant -1$, such that for all $s \leqslant s_4(\omega)$ the following inequality holds P-a.e.*

$$\|(p_1(0), p(0), Z(0))\|_{V_0}^2 \leqslant r_5(\omega).$$

Moreover we also have

$$\|\nabla Z(t)\|^2 \leqslant c\int_s^t e^{\eta(\sigma-t)}h_2(\sigma)\,\mathrm{d}\sigma := h_3(t),\quad 0\geqslant t\geqslant s,$$

and

$$\|p_1(t)\|_{H^{-1}}^2 + \|p(t)\|^2 \leqslant c\int_s^t e^{\eta(\sigma-t)}h_4(\sigma)\,\mathrm{d}\sigma := h_5(t),\quad 0\geqslant t\geqslant s,$$

where $h_2(\sigma)$, $h_4(\sigma)$ grow at most polynomial as $\sigma\to-\infty$.

Proof Taking the inner product of (4.7) with Z over D, and considering the imaginary part, one gets

$$\frac{\mathrm{d}}{\mathrm{d}t}\|Z\|^2 + 2\gamma\|Z\|^2 = 2\mathrm{Im}(Q_K((N+W)(y+Z)),Z). \tag{4.11}$$

Since $Q_K = Q_K^*$, according to the former results one can have

$$\begin{aligned}&2\mathrm{Im}(Q_K((N+W)(y+Z)),Z) = 2\mathrm{Im}\int_D (p+W)y\bar{Z}\,\mathrm{d}x\\ &\leqslant \gamma\|Z\|^2 + c(\|\nabla E\|^2 + \|N\|^4 + \|W\|^4) \leqslant \gamma\|Z\|^2 + c(f_1(t) + f_1^2(t) + \|W(t)\|^4).\end{aligned}$$

Then from (4.11) one can obtain

$$\frac{\mathrm{d}}{\mathrm{d}t}\|Z\|^2 + \eta\|Z\|^2 \leqslant c(f_1(t) + f_1^2(t) + \|W(t)\|^4). \tag{4.12}$$

Integrating from s to t for $0\geqslant t\geqslant s$, one obtains

$$\begin{aligned}\|Z(t)\|^2 &\leqslant \mathrm{e}^{\eta(s-t)}\|Z(s)\|^2 + c\int_s^t \mathrm{e}^{\eta(\sigma-t)}(f_1(\sigma) + f_1^2(\sigma) + \|W(\sigma)\|^4)\,\mathrm{d}\sigma\\ &\leqslant c\int_s^t \mathrm{e}^{\eta(\sigma-t)}g_{13}(\sigma)\,\mathrm{d}\sigma := h_0(t),\quad 0\geqslant t\geqslant s,\end{aligned} \tag{4.13}$$

where $g_{13}(\sigma)$ grows at most polynomial as $\sigma\to-\infty$. When $t=0$, one can deduce

$$\begin{aligned}\|Z(0)\|^2 &\leqslant c\int_s^0 \mathrm{e}^{\eta\sigma}(f_1(\sigma) + f_1^2(\sigma) + \|W(\sigma)\|^4)\,\mathrm{d}\sigma\\ &\leqslant c\int_s^0 \mathrm{e}^{\eta\sigma}g_{13}(\sigma)\,\mathrm{d}\sigma \leqslant r_3(\omega).\end{aligned}$$

Moreover taking the inner product of (4.7) with $\gamma Z + Z_t$ over D, and taking the real part, one gets

$$\begin{aligned}&\frac{\mathrm{d}}{\mathrm{d}t}\|\nabla Z\|^2 + 2\gamma\|\nabla Z\|^2\\ &= -2\gamma\mathrm{Re}((N+W)(y+Z),Z) - 2\mathrm{Re}((N+W)(y+Z),Z_t).\end{aligned} \tag{4.14}$$

Now we consider the right hand side of (4.14). Since

$$Z_t = \mathrm{i}(\Delta Z - Q_K((N+W)(y+Z)) + \mathrm{i}\gamma Z),$$

from the second term of the right hand side of (4.14) one can obtain

$$\begin{aligned}&-2\gamma\mathrm{Re}((N+W)(y+Z),Z) - 2\mathrm{Re}((N+W)(y+Z),Z_t)\\&= -2\mathrm{Re}\int_D \mathrm{i}(N+W)\overline{(y+Z)}(\Delta Z - Q_K((N+W)(y+Z)))\,\mathrm{d}x.\end{aligned}\tag{4.15}$$

On the one hand, the first term of the right hand side of (4.15) is estimated as follows

$$\begin{aligned}&-2\mathrm{Re}\int_D \mathrm{i}(N+W)\overline{(y+Z)}\Delta Z\,\mathrm{d}x\\&= -2\mathrm{Im}\int_D (\nabla(N+W)\overline{(y+Z)} + (N+W)\nabla\overline{(y+Z)})\nabla Z\,\mathrm{d}x\\&\leqslant \frac{\gamma}{8}\|\nabla Z\|^2 + c(f_4^2(t) + f_4^4(t) + h_0^2(t) + \|\nabla W\|^4 + \|W\|^4).\end{aligned}$$

On the other hand, one knows that the second term of the right hand side of (4.15)

$$2\mathrm{Re}\int_D \mathrm{i}(N+W)\overline{(y+Z)}Q_K((N+W)(y+Z))\,\mathrm{d}x = 0.$$

Therefore from (4.14) one can obtain

$$\frac{\mathrm{d}}{\mathrm{d}t}\|\nabla Z\|^2 + \eta\|\nabla Z\|^2 \leqslant c(f_4^2(t) + f_4^4(t) + h_0^2(t) + \|\nabla W(t)\|^4 + \|W(t)\|^4).$$

Integrating from s to t with respect to t, and using the same method in the former, one can have

$$\begin{aligned}&\|\nabla Z(t)\|^2\\&\leqslant c\int_s^t \mathrm{e}^{\eta(\sigma-t)}(f_4^2(\sigma) + f_4^4(\sigma) + h_0^2(\sigma) + \|\nabla W(\sigma)\|^4 + \|W(\sigma)\|^4)\,\mathrm{d}\sigma\\&\leqslant c\int_s^t \mathrm{e}^{\eta(\sigma-t)}g_{14}(\sigma)\,\mathrm{d}\sigma := h_1(t),\quad 0\geqslant t\geqslant s,\end{aligned}\tag{4.16}$$

where $g_{14}(\sigma)$ grows at most polynomial as $\sigma\to-\infty$. Moreover when $t=0$, one has

$$\|\nabla Z(0)\|^2 \leqslant c\int_s^0 \mathrm{e}^{\eta\sigma}g_{14}(\sigma)\,\mathrm{d}\sigma \leqslant r_3(\omega).$$

Now taking the inner product of (4.9) with $(-\Delta)^{-1}p_1(t)$ over D, and taking η as former, one can obtain

$$\begin{aligned}&\frac{\mathrm{d}}{\mathrm{d}t}H_5(t) + \eta H_5(t) + \frac{\varepsilon}{4}\left(\|p_1\|_{H^{-1}}^2 + \|p\|^2\right)\\&\leqslant (\Delta|y+Z|^2, (-\Delta)^{-1}p_1(t)) + ((\beta-\alpha)W_t + \Delta W, (-\Delta)^{-1}p_1(t)).\end{aligned}\tag{4.17}$$

where $H_5(t)=\dfrac{1}{2}(\|p_1\|_{H^{-1}}^2+\|p\|^2)$. Then from (4.17), one can obtain

$$\frac{\mathrm{d}}{\mathrm{d}t}H_5(t)+\eta H_5(t)\leqslant c(f_1^2(t)+h_1^2(t)+\|W_t+\Delta W\|_{H^1}^2).$$

From the above inequality, one can obtain

$$H_5(t)\leqslant c\int_s^t \mathrm{e}^{\eta(\sigma-t)}g_{15}(\sigma)\,\mathrm{d}\sigma:=h_2(t),\quad 0\geqslant t\geqslant s. \tag{4.18}$$

where $g_{15}(\sigma)$ grows at most as $\sigma\to-\infty$. Moreover when $t=0$, one gets

$$H_5(0)\leqslant c\int_s^0 \mathrm{e}^{\eta\sigma}g_{15}(\sigma)\,\mathrm{d}\sigma\leqslant r_4(\omega). \tag{4.19}$$

Taking $r_5(\omega)=\max\{r_3(\omega),r_4(\omega)\}$, then the results of the theorem are obtained. Then the proof is complete.

Similarly to the discussion of the Proposition 4.1, one can obtain the following propositions.

Proposition 4.2 *Let $\phi_j(x)\in H^3(D)$ and (p_1,p,Z) be the solution of (4.7)–(4.10). Then there exist a random radius $r_8(\omega)$ and a $s_5(\omega)\leqslant -1$, such that for all $s\leqslant s_5(\omega)$ the following inequality holds P-a.e.*

$$\|(p_1(0),p(0),Z(0))\|_{V_1}^2\leqslant r_8(\omega).$$

Moreover for all $0\geqslant t\geqslant s$, we also have

$$\|\Delta Z(t)\|^2\leqslant c\int_s^t e^{\eta(\sigma-t)}g_{16}(\sigma)\,\mathrm{d}\sigma+g_{17}(t):=h_3(t),$$

$$\|p_1(t)\|^2+\|\nabla p(t)\|^2\leqslant c\int_s^t e^{\eta(\sigma-t)}g_{18}(\sigma)\,\mathrm{d}\sigma:=h_4(t),$$

where $g_{16}(\sigma)$, $g_{17}(\sigma)$ and $g_{18}(\sigma)$ grow at most polynomial as $\sigma\to-\infty$.

Proposition 4.3 *Let $\phi_j(x)\in H^3(D)$ and (p_1,p,Z) be the solution of (4.7)–(4.10). Then there exist a random radius $r_9(\omega)$ and a $s_6(\omega)\leqslant -1$, such that for all $s\leqslant s_6(\omega)$ the following inequality holds P-a.e.*

$$\|(p_1(0),p(0),Z(0))\|_{V_2}^2\leqslant r_9(\omega). \tag{4.20}$$

Now *a* priori estimates for the equations (4.4)–(4.6) are investigated. By usual method the equations (4.4)–(4.6) can be rewritten as

$$\mathrm{i}\chi_t + \Delta\chi - Q_K((N+W)\chi) + \mathrm{i}\gamma\chi = 0, \tag{4.21}$$

$$q_{1t} = q_t + \varepsilon q, \tag{4.22}$$

$$q_{1t} + (\alpha-\varepsilon)q_1 - \varepsilon(\alpha-\varepsilon)q + \Delta q = \Delta|\chi|^2 + 2\mathrm{Re}(v\bar{\lambda}), \tag{4.23}$$

$$q(s) = N(s) + W(s), q_1(s) = N_1(s) + W_t(s), \chi(s) = E_0, \tag{4.24}$$

where $v = y + Z$. We consider the proposition of solution $(q_1(t), q(t), \chi(t))$ of (4.21)–(4.24).

Proposition 4.4 *Let $\phi_j(x) \in H^3(D)$ and $(q_1(t), q(t), \chi(t))$ be the solution of (4.21)–(4.24). For any given $R > 0$ and $(q_1(s), q(s), \chi(s)) \in V_1$ satisfying*

$$\|(q_1(s), q(s), \chi(s))\|_{V_1} \leqslant R$$

and for arbitrary ε, there exists a $s_{10}(\omega) \leqslant -1$, such that for all $s \leqslant s_{10}(\omega)$ the following inequality holds P-a.e.

$$\|(q_1(0), q(0), \chi(0))\|_{V_1} \leqslant \varepsilon.$$

Proof On one hand, taking the inner product of (4.21) with χ and taking the imagine part, one can obtain

$$\frac{\mathrm{d}}{\mathrm{d}t}\|\chi\|^2 + 2\gamma\|\chi\|^2 = 0.$$

Then integrating on the both sides of above equality, one can obtain

$$\|\chi(t)\|^2 = \mathrm{e}^{2\gamma(s-t)}\|\chi(s)\|^2 \leqslant \mathrm{e}^{2\eta(s-t)}\|\chi(s)\|^2, \quad 0 \geqslant t \geqslant s.$$

Moreover, when $t = 0$, one has $\|\chi(0)\|^2 \leqslant \mathrm{e}^{2\eta s}\|\chi(s)\|^2 \to 0$ as $s \to -\infty$. In addition, taking the inner product of (4.21) with $\gamma\chi + \chi_t$ and taking the real part, one can obtain

$$\begin{aligned} &\frac{\mathrm{d}}{\mathrm{d}t}\|\nabla\chi\|^2 + 2\gamma\|\nabla\chi\|^2 \\ &+ 2\gamma\mathrm{Re}\int_D (N+W)|\chi|^2\,\mathrm{d}x + 2\mathrm{Re}\int_D (N+W)\chi\bar{\chi}_t\,\mathrm{d}x = 0. \end{aligned} \tag{4.25}$$

Notice that

$$2\gamma\mathrm{Re}\int_D (N+W)|\chi|^2\,\mathrm{d}x \leqslant \frac{\gamma}{4}\|\nabla\chi\|^2 + c(\|N\|^2 + \|W\|^2 + 1)\|\chi\|^2.$$

Since $\chi_t = \mathrm{i}(\Delta\chi - Q_K((N+W)\chi) + \mathrm{i}\gamma\chi)$, one gets

$$\begin{aligned}&2\mathrm{Re}\int_D \mathrm{i}(N+W)\bar{\chi}(\Delta\chi - Q_K((N+W)\chi) + \mathrm{i}\gamma\chi)\,\mathrm{d}x\\ =\;&2\mathrm{Re}\int_D \mathrm{i}(N+W)\bar{\chi}(\Delta\chi + \mathrm{i}\gamma\chi)\,\mathrm{d}x.\end{aligned}$$

Now we estimate the terms of the right hand side of above equality as follows

$$\begin{aligned}&2\gamma\mathrm{Re}\int_D (N+W)|\chi|^2\,\mathrm{d}x \leqslant \frac{\gamma}{4}\|\nabla\chi\|^2 + c(\|N\|^2 + \|W\|^2 + 1)\|\chi\|^2;\\ &2\mathrm{Re}\int_D \mathrm{i}(N+W)\bar{\chi}\Delta\chi\,\mathrm{d}x = 2\mathrm{Im}\int_D \nabla(N+W)\bar{\chi}\nabla\chi\,\mathrm{d}x\\ &\qquad\leqslant \frac{\gamma}{4}\|\nabla\chi\|^2 + c(\|\nabla N\|^4 + \|\nabla W\|^4)\|\chi\|^2.\end{aligned}$$

Then instituting above inequalities into (4.25) one can obtain

$$\frac{\mathrm{d}}{\mathrm{d}t}\|\nabla\chi\|^2 + \eta\|\nabla\chi\|^2 \leqslant c(\|\nabla N\|^4 + \|\nabla W\|^4 + 1)\|\chi\|^2. \tag{4.26}$$

Integrating (4.26) from s to t with respect to t, one gets

$$\begin{aligned}&\|\nabla\chi(t)\|^2\\ \leqslant\;&\mathrm{e}^{\eta(s-t)}\|\nabla\chi(s)\|^2 + c\int_s^t \mathrm{e}^{\eta(\sigma-t)}(\|\nabla N(\sigma)\|^4 + \|\nabla W(\sigma)\|^4 + 1)\|\chi(\sigma)\|^2\,\mathrm{d}\sigma\\ \leqslant\;&\mathrm{e}^{\eta(s-t)}\|\nabla\chi(s)\|^2 + c\int_s^t \mathrm{e}^{\eta(\sigma-t)}(f_4^4(\sigma) + \|\nabla W(\sigma)\|^2 + 1)\|\chi(\sigma)\|^2\,\mathrm{d}\sigma.\end{aligned} \tag{4.27}$$

As the discussion of the former, from (4.27), one can obtain

$$\begin{aligned}&\|\nabla\chi(t)\|^2 \leqslant \mathrm{e}^{\eta(s-t)}\|\nabla\chi(s)\|^2\\ &+ c\mathrm{e}^{\eta(s-t)}\|\chi(s)\|^2\int_s^t \mathrm{e}^{\eta(s-\sigma)}\left(\mathrm{e}^{4\eta(s-\sigma)}a_2^4(s) + g_6^4(\sigma) + g_7^4(\sigma) + \|\nabla W(\sigma)\|^2 + 1\right)\mathrm{d}\sigma\\ &:= h_5(t),\quad 0 \geqslant t \geqslant s.\end{aligned} \tag{4.28}$$

Moreover when $t=0$, one can see

$$\|\nabla\chi(0)\|^2 \to 0,\quad s\to-\infty. \tag{4.29}$$

Taking the inner product of (4.23) with $(-\Delta)^{-1}q_1(t)$ over D, one can obtain

$$\begin{aligned}&\|q_1(t)\|_{H^{-1}}^2 + \|q(t)\|^2 \leqslant \mathrm{e}^{\eta(s-t)}(\|q_1(s)\|_{H^{-1}}^2 + \|q(s)\|^2) +\\ &+ c\mathrm{e}^{\eta(s-t)}\int_s^t \mathrm{e}^{\eta(s-\sigma)}g_{19}(\sigma)\,\mathrm{d}\sigma := h_6(t),\quad 0\geqslant t\geqslant s,\end{aligned} \tag{4.30}$$

where $g_{19}(\sigma)$ grows at most polynomial as $\tau \to -\infty$. Moreover when $t = 0$, one can see $\|q_1(0)\|^2_{H^{-1}} + \|q(0)\|^2 \to 0$ as $s \to -\infty$.

By the same method and similar estimates, one can obtain $\|\Delta\chi(0)\|^2 \to 0$ and $\|q_1(0)\|^2 + \|\nabla q(0)\|^2 \to 0$ as $s \to -\infty$. Then the proof is complete.

5 Proof of main result Theorem 5.1

It is well known that there exists a continuous random dynamical system associated with (1.5)–(1.6) in [5,9]. By the transformation of Ornstein–Uhlenbeck process, it can be seen that the research of the existence of a global compact random attractor of equation (1.5)–(1.6) is equivalence with corresponding investigation of equations (3.1)–(3.3). On the basis of the above series of *a* priori estimates, it is well known that the crucial step is to obtain the compactness of the global compact random attractor of equations (3.1)–(3.3). Then the main result has been given as following.

Theorem 5.1 *The random dynamical system associated with stochastic dissipative Zakharov equations* (1.5)–(1.6) *with noise has a global compact random attractor in* V_1 *endowed with usual topology in the sense of Theorem* 2.1.

Remark 5.2 *Notice that the main results of a global compact random attractor in* V_1 *endowed with usual topology in the sense of Theorem* 2.1 *not only coincide with the results in the sense of the weak topology in* [9], *but also obtain the new results.*

Proof By the transformation of Ornstein–Uhlenbeck process, we can see that the existence of a global compact random attractor of equations (1.5)–(1.6) is equivalence with corresponding investigation of equations (3.1)–(3.3). It is well known that there exists a continuous random dynamical system by the usual discussion in [5,9]. Therefore we now consider the existence of a global compact random attractor of equations (3.1)–(3.3) in the sense of Theorem 2.1.

Let $K(\omega)$ be the ball in V_1 of radius $r_8^{\frac{1}{2}}(\omega)$. Since $E = y + Z + \chi$, $N = p + q$, by Proposition 4.2, we can obtain that $K(\omega)$ is an attracting set at time 0. Moreover, according to Proposition 3.1–3.4 and Proposition 4.1–4.4, we can know that $K(\omega)$ is a compact subset of V_1. This shows that the random dynamical system associated with stochastic dissipative Zakharov equations (1.5)–(1.6) with noise has a global compact random attractor in V_1 in the sense of Theorem 2.1. Then the proof is complete.

We known that the random attractor endowed with weak topology instead of usual topology had been only obtained in V_2, which has been investigated in Itô sense in [9]. Although here considering the stochastic force only on (1.4), we can prove the existence of a global compact random attractor with respect to usual or strong topology in V_1 following the approach of [6], which not only coincides with the random attractors in the sense of the weak topology in V_2 in [9], but also obtain the new better result. However, in Theorem 5.1, the compact property of the random attractors is the key result, which is obtained by the decomposition of solutions of stochastic dissipative Zakharov equations (1.5)–(1.6). Here the decomposition is different from the deterministic equations in [6], which is constructed by the new decomposition for our aim.

So for our main result, there is important significance in mathematical theory and physical application for the global random attractor endowed with usual topology.

Acknowledgements The authors would like to thank the referee's valuable suggestions. This paper is supported by National Natural Science Foundation of China (No. 11061003, No. 11301097), Guangxi Natural Science Foundation (No. 2013GXNSFAA019001), and Guangxi Education Institution Scientific Research Item (No. 2013YB170).

References

[1] Bejenaru I, Herr S, Holmer J, Tataru D. On the 2d Zakharov equations with L^2 Schrödinger data[J]. Nonlinearity, 2009, 22: 1063–1089.

[2] Crauel H, Flandoli F. Attractors for random dynamical systems[J]. Probab. Theory Rel., 1994, 100: 365–393.

[3] Crauel H, Debussche A, Flandoli F. Random attractors[J]. J. Dyn. Differ. Equ., 1997, 9: 307–341.

[4] E W, Li X, Vanden–Eijnden E. Some recent progress in multiscale modeling, Multiscale modeling and simulation[M], Lect. Notes in Computer Science Engineering, Springer, Berlin, 2004.

[5] Flahaut I. Attractors for the dissipative Zakharov equations[J]. Nonlinear Anal. TMA. , 1991, 16(7): 599–633.

[6] Goubet O, Moise I. Attractors for dissipative Zakharov equations[J]. Nonlinear Anal. TMA. , 1998, 31(7): 823–847.

[7] Guio P, Forme F. Zakharov simulations of Langmuir turbulence: Effects on the ion–acoustic waves in incoherent scattering[J]. Phys. Plasmas, 2006, 13: 122902–10.

[8] Guo B. On the IBVP for some more extensive Zakharov equations[J]. J. Math. , 1987, 7(3): 269–275.

[9] Guo B, Lv Y, Yang X. Dynamics of Stochastic Zakharov Equations[J]. J. Math. Phys., 2009, 50: 052703.

[10] Guo Y, Dai Z, Li D. Explicit Heteroclinic Tube Solutions for the Zakharov System with Periodic Boundary[J]. Chin. J. Phys., 2008, 46(5): 570–577.

[11] Imkeller P, Monahan A H. Conceptual stochastic climate models[J]. Stoch. Dynam., 2002, 2: 311–326.

[12] Li Y. On the initial boundary value problems for two dimensional systems of Zakharov equations and of complex–Schrödinger–real–Boussinesq equations[J]. J. Partial Differential Equations, 1992, 5(2): 81–93.

[13] Masselin V. A result on the blow–up rate for the Zakharov equations in dimension 3[J]. SIAM J. Math. Anal., 2001, 33(2): 440–447.

[14] Temam R. Infinite–dimensional Dynamical Systems in Mechanics and Physics[M]. Springer-Verlag, 1988.

[15] Wang B. Random attractors for the stochastic Benjamin–Bona–Mahony equation on unbounded domains[J]. J. Diff. Equ., 2009, 246: 2506–2537.

[16] Wang B. Random attractors for the stochastic FitzHugh–Nagumo equations on unbounded domains[J]. Nonlinear Anal. TMA., 2009, 71(7–8): 2811–2828.

[17] Zakharov V E. Collapse of Langmuir waves[J]. Sov. Phys. JETP, 1972, 35: 908–914.

Global Smooth Solutions of the Generalized KS-CGL Equations for Flames Governed by a Sequential Reaction*

Guo Changhong (郭昌洪), Fang Shaomei (房少梅) and Guo Boling (郭柏灵)

Abstract In this paper, we investigate the periodic initial value problem and Cauchy problem of the generalized Kuramoto-Sivashinsky-complex Ginzburg-Landau (GKS-CGL) equations for flames governed by a sequential reaction. We prove the global existence and uniqueness of solutions to these two problems in various spatial dimensions via delicate a priori estimates, the Galerkin method and so-called continuity method.

Keywords global existence; generalized KS-CGL equations; sequential reaction; a prior estimates; Galerkin method

1 Introduction

In this paper, we consider the global existence and uniqueness of solutions for the following coupled generalized Kuramoto-Sivashinsky-complex Ginzburg-Landau (GKS-CGL) equations for flames

$$P_t = \xi P + (1+i\mu)\Delta P - (1+i\nu)|P|^2 P - \nabla P \nabla Q - r_1 P\Delta Q - g r_2 P \Delta^2 Q, \quad (1.1)$$

$$Q_t = -\Delta Q - g\Delta^2 Q + \delta\Delta^3 Q - \frac{1}{2}|\nabla Q|^2 - \eta |P|^2, \quad (1.2)$$

with the periodic initial conditions

$$P(x+Le_i, t) = P(x,t),\ Q(x+Le_i,t) = Q(x,t),\ x\in\Omega,\ t\geqslant 0, \quad (1.3)$$

$$P(x,0) = P_0(x), \quad Q(x,0) = Q_0(x), \quad x\in\Omega. \quad (1.4)$$

* Commun. Math. Sci., 2014, 12(8): 1457−1474. DOI: 10.4310/CMS.2014.v12.n8.a5.

or the initial conditions

$$P(x,0)=P_0(x),\quad Q(x,0)=Q_0(x),\quad x\in\mathbb{R}^n. \tag{1.5}$$

The spatial domain Ω is a bounded domain in n-dimensional real Euclidean space $\mathbb{R}^n(n=1,2)$, and the time $t\geqslant 0$. The complex function $P(x,t)$ is the rescaled amplitude of the flame oscillations, the real function $Q(x,t)$ is the deformation of the first front, and both of them are L-periodic. The coefficient $\xi=\pm 1$. The Landau coefficients μ,ν and the coupling coefficient $\eta>0$ are real, while the parameters r_1 and r_2 are complex, $r_1=r_{1r}+ir_{1i}$, $r_2=r_{2r}+ir_{2i}$. The coefficient $g>0$ is proportional to the supercriticality of the oscillatory mode. $\delta>0$ is a constant, $L>0$ is the period and e_i is the standard coordinate vector.

If we take $\delta=0$ in Eq.(1.2), the coupled GKS-CGL equations (1.1)(1.2) are reduced to the classical KS-CGL equations[1], which describe the nonlinear interaction between the monotonic and oscillatory modes of instability of the two uniformly propagating flame fronts in a sequential reaction. Specifically, they describe both the long-wave evolution of the oscillatory mode near the oscillatory instability threshold, and the evolution of the monotonic mode. For the background of the uniformly propagating premixed flame fronts and the derivation of the KS-CGL model, we refer to [1,2,3,4] for details. If there were no coupling with the monotonic mode (terms with Q in Eq.(1.1)), then Eq.(1.1) would be the well-known CGL equation that usually describes the weakly nonlinear evolution of a long-scale instability[5]. And if taking $\delta=0$ and the coupling coefficient $\eta=0$ in Eq.(1.2), the Eq.(1.2) reduces to the well-known KS equation[6], which governs the flame front's spatio-temporal evolution and occurs monotonic instability. As can be seen, the coupled GKS-CGL equations (1.1)(1.2) can better describe the dynamical behavior for flames governed by a sequential reaction, since they generalize the KS equations, the CGL equations or the KS-CGL equations.

So far, the mathematical analysis and physical study about the CGL equation and KS equation have been done by many researchers. For example, the existence of global solutions and attractor for the CGL equation and KS equation are studied in [7,8,9,10,11]. For some other results, see [12,13,14,15] and reference therein. However, little progress has been obtained for the coupled KS-CGL equations since they were derived to describe the nonlinear evolution for flames by A. A. Golovin, et al[1]. They studied the traveling waves of the coupled equations numerically and continued to study the spiral waves[16], which exhibit new types of instabilities. Meanwhile, there is few work to consider mathematical analytical properties of the KS-CGL equations and generalized KS-CGL equations, even the existence and uniqueness of the solutions. In this paper, we are concerned with the global existence and uniqueness theory for the periodic initial value problem (1.1)–(1.4) and the Cauchy problem (1.1)(1.2)(1.5) via delicate a prior estimates and the Galerkin method. For the Cauchy problem (1.1)(1.2)(1.5), we suppose that $P(x,t), Q(x,t)$ and some of their derivatives with respect to x tend to zero as $|x| \to \infty$.

The rest of paper is organized as follows. In Section , we briefly give some notations and preliminaries. In Section , we will establish a prior estimates for the solutions of the periodic initial value problem (1.1)–(1.4). In Section , the existence and uniqueness of the global smooth solutions for the periodic initial value problem (1.1)–(1.4) are obtained via the Galerkin method and so-called continuity method. In Section 1, we employ the usual method of limiting process to obtian the solutions for the Cauchy problem (1.1)(1.2)(1.5).

2 Notations and Preliminaries

We shall use the following conventional notations throughout the paper. Let L^k_{per} and $H^k_{per}, k = 1, 2, \cdots$ denote the Hilbert and Sobolev spaces of L-periodic, complex-valued functions endowed with the usual L^2 inner product $(u, v) = \int_\Omega u(x)\overline{v}(x)\mathrm{d}x$ and norms

$$\|u\|_{L^2} = \sqrt{(u,u)}, \quad \|u\|_{H^k} = \left(\sum_{|\alpha|\leqslant k} \|D^\alpha u(x)\|\right)^{\frac{1}{2}}.$$

Here $\overline{v}$ denotes the complex conjugate of v. For brevity, we write $\|u\| = \|u\|_{L^2}$ and denote the L^p-norm by $\|u\|_p = \left(\int_\Omega |u|^p \mathrm{d}x\right)^{1/p}$. Without any ambiguity, we denote a generic positive constant by C which may vary from line to line.

In the following sections, we frequently use following inequalities.

Lemma 2.1 (Young's inequality with ε)[17] *Let $a > 0,\ b > 0,\ 1 < p, q < \infty$, $\dfrac{1}{p} + \dfrac{1}{q} = 1$. Then*

$$ab \leqslant \varepsilon a^p + C(\varepsilon) b^q,$$

for $C(\varepsilon) = (\varepsilon p)^{-q/p} q^{-1}$.

Lemma 2.2 (Gagliardo-Nirenberg inequality)[18] *Let Ω be a bounded domain with $\partial\Omega$ in C^m, and let u be any function in $W^{m,r}(\Omega)\cap L^q(\Omega)$, $1\leqslant q,r\leqslant\infty$. For any integer j, $0\leqslant j<m$, and for any number a in the interval $j/m\leqslant a\leqslant 1$, set*

$$\frac{1}{p}=\frac{j}{n}+a\left(\frac{1}{r}-\frac{m}{n}\right)+(1-a)\frac{1}{q}.$$

If $m-j-n/r$ is not a nonnegative integer, then

$$\|D^j u\|_{L^p}\leqslant C\|u\|_{W^{m,r}}^a\|u\|_{L^q}^{1-a}. \tag{2.1}$$

If $m-j-n/r$ is a nonnegative integer, then (2.1) holds for $a=j/m$. The constant C depends only on Ω,r,q,j,a.

In the sequel, we will use the following inequalities as the specific cases of the Gagliardo-Nirenberg inequality:

$$\|D^j u\|_{L^\infty}\leqslant C\|u\|_{H^m}^a\|u\|^{1-a},\quad ma=j+n/2, \tag{2.2}$$

$$\|D^j u\|_{L^2}\leqslant C\|u\|_{H^m}^a\|u\|^{1-a},\quad ma=j, \tag{2.3}$$

$$\|D^j u\|_{L^4}\leqslant C\|u\|_{H^m}^a\|u\|^{1-a},\quad ma=j+n/4. \tag{2.4}$$

3 A priori estimates

In this section, we derive some a priori estimates for the solutions of the problem (1.1)–(1.4). Firstly we have

Lemma 3.1 *Assume $P_0(x)\in L^2_{per}(\Omega)$, $Q_0(x)\in H^1_{per}(\Omega)$, and suppose that $2\delta-g^2r_{2r}^2>0$. Then for the solutions of the problem (1.1)–(1.4), we have*

$$\|P\|^2\leqslant e^{K_1t}(\|P_0\|^2+\|\nabla Q_0\|^2),\quad \|\nabla Q\|^2\leqslant e^{K_1t}(\|P_0\|^2+\|\nabla Q_0\|^2), \tag{3.1}$$

where K_1 is a positive constant.

Proof Firstly we differentiate Eq.(1.2) with respect to x once and set

$$W=\nabla Q. \tag{3.2}$$

Then Eqs.(1.1) and (1.2) can be rewritten as

$$P_t=\xi P+(1+i\mu)\Delta P-(1+i\nu)|P|^2P-\nabla PW-r_1P\nabla W-gr_2P\nabla\Delta W, \tag{3.3}$$

$$W_t = -\Delta W - g\Delta^2 W + \delta\Delta^3 W - W\nabla W - \eta\nabla(|P|^2). \tag{3.4}$$

Multiplying (3.3) by $\overline{P}$, integrating with respect to x over Ω and taking the real part, we obtain

$$\begin{aligned}\frac{1}{2}\frac{\mathrm{d}}{\mathrm{d}t}\|P\|^2 =& \xi\|P\|^2 - \|\nabla P\|^2 - \int_\Omega |P|^4 \mathrm{d}x - \mathrm{Re}\int_\Omega \nabla P\overline{P}W \mathrm{d}x \\ &- r_{1r}\int_\Omega |P|^2\nabla W \mathrm{d}x - gr_{2r}\int_\Omega |P|^2\nabla\Delta W \mathrm{d}x,\end{aligned} \tag{3.5}$$

where

$$-\mathrm{Re}\int_\Omega \nabla P\overline{P}W \mathrm{d}x = \frac{1}{2}\int_\Omega |P|^2\nabla W \mathrm{d}x. \tag{3.6}$$

On the other hand, multiplying (3.4) by W and integrating over Ω, we have

$$\begin{aligned}\frac{1}{2}\frac{\mathrm{d}}{\mathrm{d}t}\|W\|^2 =& \|\nabla W\|^2 - g\|\Delta W\|^2 - \delta\|\nabla\Delta W\|^2 \\ &- \int_\Omega W^2\nabla W \mathrm{d}x - \eta\int_\Omega \nabla(|P|^2)W \mathrm{d}x,\end{aligned} \tag{3.7}$$

where

$$\int_\Omega W^2\nabla W \mathrm{d}x = \sum_{i=1}^n \int_\Omega \frac{\partial W}{\partial x_i}W^2 \mathrm{d}x = \frac{1}{3}\sum_{i=1}^n \int_\Omega \frac{\partial}{\partial x_i}(W^3)\mathrm{d}x = 0, \tag{3.8}$$

and

$$-\eta\int_\Omega \nabla\left(|P|^2\right)W \mathrm{d}x = \eta\int_\Omega |P|^2\nabla W \mathrm{d}x. \tag{3.9}$$

Adding (3.5) and (3.7) together, and noticing (3.6), (3.8) and (3.9), there holds

$$\begin{aligned}\frac{\mathrm{d}}{\mathrm{d}t}\left(\|P\|^2 + \|W\|^2\right) =& 2\xi\|P\|^2 - 2\|\nabla P\|^2 - 2\int_\Omega |P|^4 \mathrm{d}x + 2\|\nabla W\|^2 - 2g\|\Delta W\|^2 \\ &- 2\delta\|\nabla\Delta W\|^2 + (1 + 2\eta - 2r_{1r})\int_\Omega |P|^2\nabla W \mathrm{d}x \\ &- 2gr_{2r}\int_\Omega |P|^2\nabla\Delta W \mathrm{d}x.\end{aligned} \tag{3.10}$$

According to the Gagliardo-Nirenberg inequality (2.3), we have

$$
\begin{aligned}
&2\|\nabla W\|^2+(1+2\eta-2r_{1r})\int_\Omega |P|^2\nabla W\mathrm{d}x\\
&\leqslant 2\|\nabla W\|^2+|1+2\eta-2r_{1r}|\left(\int_\Omega |P|^4\mathrm{d}x\right)^{\frac{1}{2}}\|\nabla W\|\\
&\leqslant \varepsilon_1\int_\Omega |P|^4\mathrm{d}x+C\|\nabla W\|^2\\
&\leqslant \varepsilon_1\int_\Omega |P|^4\mathrm{d}x+C\|W\|_{H^3}^{\frac{2}{3}}\|W\|^{\frac{4}{3}}\\
&\leqslant \varepsilon_1\int_\Omega |P|^4\mathrm{d}x+\frac{\varepsilon_2}{2}\|\nabla\Delta W\|^2+C_1\|W\|^2, \qquad (3.11)
\end{aligned}
$$

and

$$
\left|-2gr_{2r}\int_\Omega |P|^2\nabla\Delta W\mathrm{d}x\right|\leqslant\frac{\varepsilon_2}{2}\|\nabla\Delta W\|^2+\frac{2g^2r_{2r}^2}{\varepsilon_2}\int_\Omega |P|^4\mathrm{d}x. \qquad (3.12)
$$

Combining (3.10)–(3.12) and noticing $|\xi|=1$, we have

$$
\begin{aligned}
\frac{\mathrm{d}}{\mathrm{d}t}\left(\|P\|^2+\|W\|^2\right)\leqslant{}& 2\|P\|^2+C_1\|W\|^2-2\|\nabla P\|^2-2g\|\Delta W\|^2\\
&-(2\delta-\varepsilon_2)\|\nabla\Delta W\|^2-\left(2-\frac{2g^2r_{2r}^2}{\varepsilon_2}-\varepsilon_1\right)\int_\Omega |P|^4\mathrm{d}x. \qquad (3.13)
\end{aligned}
$$

Under the condition $2\delta-g^2r_{2r}^2>0$, we can choose ε_2 such that $0<g^2r_{2r}^2<\varepsilon_2\leqslant 2\delta$ and choose ε_1 to be sufficiently small such that $2-\dfrac{2g^2r_{2r}^2}{\varepsilon_2}-\varepsilon_1>0$. Then we have

$$
\begin{aligned}
&\frac{\mathrm{d}}{\mathrm{d}t}\left(\|P\|^2+\|W\|^2\right)+2\|\nabla P\|^2+2g\|\Delta W\|^2\\
&\quad+(2\delta-\varepsilon_2)\|\nabla\Delta W\|^2+\left(2-\frac{2g^2r_{2r}^2}{\varepsilon_2}-\varepsilon_1\right)\int_\Omega |P|^4\mathrm{d}x\\
&\leqslant 2\|P\|^2+C_1\|W\|^2\\
&\leqslant K_1(\|P\|^2+\|W\|^2), \qquad (3.14)
\end{aligned}
$$

where $K_1=\max(2,C_1)$. By the Gronwall's inequality, we have

$$
\|P\|^2+\|W\|^2\leqslant e^{K_1t}(\|P_0\|^2+\|W_0\|^2), \qquad (3.15)
$$

where K_1 is a positive constant. Combining the transformation (3.2), one can complete the proof of Lemma 3.1.

Lemma 3.2 *Assume $P_0(x)\in H_{per}^1(\Omega), Q_0(x)\in H_{per}^3(\Omega)$, and $2\delta-g^2r_{2r}^2>0$. If $n=2$, we also suppose $|\nu|\leqslant\sqrt{3}$. Then for the solutions of the problem* (1.1)–(1.4), *we have*

$$
\|\nabla P\|^2\leqslant e^{K_2t}(\|\nabla P_0\|^2+\|\Delta Q_0\|^2+\|\nabla\Delta Q_0\|^2+Ct), \qquad (3.16)
$$

$$\|\Delta Q\|^2+\|\nabla\Delta Q\|^2\leqslant e^{K_2t}(\|\nabla P_0\|^2+\|\Delta Q_0\|^2+\|\nabla\Delta Q_0\|^2+Ct), \tag{3.17}$$

where K_2 and C are positive constants.

Proof Similar to the first step in Lemma 3.1, we use the transformed equations. Multiplying (3.3) by $(-\Delta\overline{P})$, integrating with respect to x over Ω and taking the real part, we get

$$\begin{aligned}\frac{1}{2}\frac{\mathrm{d}}{\mathrm{d}t}\|\nabla P\|^2=&\,\xi\|\nabla P\|^2-\|\Delta P\|^2+\mathrm{Re}\int_\Omega(1+i\nu)|P|^2P\Delta\overline{P}\mathrm{d}x\\&+\mathrm{Re}\int_\Omega\nabla P\Delta\overline{P}W\mathrm{d}x+\mathrm{Re}\int_\Omega r_1P\Delta\overline{P}\nabla W\mathrm{d}x\\&+\mathrm{Re}\int_\Omega gr_2P\Delta\overline{P}\nabla\Delta W\mathrm{d}x.\end{aligned} \tag{3.18}$$

Taking the inner product of (3.4) with $(-\Delta W)$ and Δ^2W over Ω respectively, then

$$\begin{aligned}\frac{1}{2}\frac{\mathrm{d}}{\mathrm{d}t}\|\nabla W\|^2=&\,\|\Delta W\|^2-g\|\nabla\Delta W\|^2-\delta\|\Delta^2W\|^2\\&+\int_\Omega W\nabla W\Delta W\mathrm{d}x+\eta\int_\Omega\nabla(|P|^2)\Delta W\mathrm{d}x,\end{aligned} \tag{3.19}$$

and

$$\begin{aligned}\frac{1}{2}\frac{\mathrm{d}}{\mathrm{d}t}\|\Delta W\|^2=&\,\|\nabla\Delta W\|^2-g\|\Delta^2W\|^2-\delta\|\nabla\Delta^2W\|^2\\&-\int_\Omega W\nabla W\Delta^2W\mathrm{d}x-\eta\int_\Omega\nabla(|P|^2)\Delta^2W\mathrm{d}x.\end{aligned} \tag{3.20}$$

Adding (3.18), (3.19) and (3.20) together yields that

$$\begin{aligned}&\frac{\mathrm{d}}{\mathrm{d}t}(\|\nabla P\|^2+\|\nabla W\|^2+\|\Delta W\|^2)\\&=2\xi\|\nabla P\|^2-2\|\Delta P\|^2+2\|\Delta W\|^2-2(g-1)\|\nabla\Delta W\|^2\\&\quad-2(\delta+g)\|\Delta^2W\|^2-2\delta\|\nabla\Delta^2W\|^2\\&\quad+2\mathrm{Re}\int_\Omega(1+i\nu)|P|^2P\Delta\overline{P}\mathrm{d}x+2\mathrm{Re}\int_\Omega\nabla P\Delta\overline{P}W\mathrm{d}x\\&\quad+2\mathrm{Re}\int_\Omega r_1P\Delta\overline{P}\nabla W\mathrm{d}x+2\mathrm{Re}\int_\Omega gr_2P\Delta\overline{P}\nabla\Delta W\mathrm{d}x\\&\quad+2\int_\Omega W\nabla W\Delta W\mathrm{d}x+2\eta\int_\Omega\nabla(|P|^2)\Delta W\mathrm{d}x\\&\quad-2\int_\Omega W\nabla W\Delta^2W\mathrm{d}x-2\eta\int_\Omega\nabla(|P|^2)\Delta^2W\mathrm{d}x.\end{aligned} \tag{3.21}$$

Now we need to majorize the right hand side of (3.21). Notice that when the spatial dimension $n=1$, it is easy to find that

$$\begin{aligned}\left|2\mathrm{Re}\int_\Omega(1+i\nu)|P|^2P\overline{P}_{xx}\mathrm{d}x\right| &\leqslant 2|1+i\nu|\|P\|_{L^8}^2\|P\|_{L^4}\|P_{xx}\| \\ &\leqslant C\|P_{xx}\|^{\frac{3}{8}}\|P\|^{\frac{13}{8}}\|P_{xx}\|^{\frac{1}{8}}\|P\|^{\frac{7}{8}}\|P_{xx}\| \\ &\leqslant \frac{1}{2}\|P_{xx}\|^2+C\|P\|^{10} \\ &\leqslant \frac{1}{2}\|P_{xx}\|^2+C. \end{aligned}\tag{3.22}$$

While $n=2$, we handle this term as follows

$$\begin{aligned}&2\mathrm{Re}\int_\Omega(1+i\nu)|P|^2P\Delta\overline{P}\mathrm{d}x \\ =&-2\mathrm{Re}\int_\Omega(1+i\nu)(|P|^2|\nabla P|^2+P\nabla\overline{P}\nabla(|P|^2))\mathrm{d}x \\ =&-2\int_\Omega|P|^2|\nabla P|^2\mathrm{d}x-\int_\Omega|\nabla(|P|^2)|^2\mathrm{d}x+\nu\int_\Omega\nabla(|P|^2|)\cdot i(\overline{P}\nabla P-P\nabla\overline{P})\mathrm{d}x \\ =&-\frac{1}{2}\int_\Omega(3|\nabla(|P|^2)|^2-2\nu\nabla(|P|^2)\cdot i(\overline{P}\nabla P-P\nabla\overline{P})+|\overline{P}\nabla P-P\nabla\overline{P}|^2)\mathrm{d}x.\end{aligned}\tag{3.23}$$

We observe that the integrand in the last term in (3.23) is a quadratic form, and under the condition of $|\nu|\leqslant\sqrt{3}$, the matrix $\begin{pmatrix}3 & -\nu\\ -\nu & 1\end{pmatrix}$ is nonnegative definite, which implies the integrand is nonnegative. Thus we have

$$2\mathrm{Re}\int_\Omega(1+i\nu)|P|^2P\Delta\overline{P}\mathrm{d}x<0\leqslant\frac{1}{2}\|\Delta P\|^2.\tag{3.24}$$

Meanwhile, according to the Young's inequality with ε and Gagliardo-Nirenberg inequality (2.2), (2.3) and (2.4), we obtain the following estimates

$$\begin{aligned}&2\|\Delta W\|^2-2(g-1)\|\nabla\Delta W\|^2+\left|2\mathrm{Re}\int_\Omega\nabla P\Delta\overline{P}W\mathrm{d}x\right| \\ &+\left|2\mathrm{Re}\int_\Omega r_1P\Delta\overline{P}\nabla W\mathrm{d}x\right|+\left|2\mathrm{Re}\int_\Omega gr_2P\Delta\overline{P}\nabla\Delta W\mathrm{d}x\right|+\left|2\int_\Omega W\nabla W\Delta W\mathrm{d}x\right| \\ \leqslant&2\|\Delta W\|^2+2|g-1|\|\nabla\Delta W\|^2+2\|W\|_\infty\|\nabla P\|\|\Delta P\| \\ &+2|r_1|\|P\|\|\Delta P\|\|\nabla W\|_\infty+2g|r_2|\|P\|\|\Delta P\|\|\nabla\Delta W\|_\infty+2\|W\|\|\nabla W\|_\infty\|\Delta W\| \\ \leqslant&2\|\Delta W\|^2+2|g-1|\|\nabla\Delta W\|^2+C\|W\|_{H^4}^{\frac{n}{8}}\|W\|^{\frac{8-n}{8}}\|P\|_{H^2}^{\frac{1}{2}}\|P\|^{\frac{1}{2}}\|\Delta P\|\end{aligned}$$

$$
\begin{aligned}
&+\frac{1}{6}\|\Delta P\|^2+C\|W\|_{H^3}^{\frac{2+n}{3}}\|W\|^{\frac{4-n}{3}}+\frac{1}{6}\|\Delta P\|^2\\
&+C\|\nabla\Delta W\|_\infty^2+C\|W\|_{H^4}^{\frac{2+n}{8}}\|W\|^{\frac{6-n}{8}}\|\Delta W\|\\
\leqslant&\, 2\|\Delta W\|^2+C\|W\|_{H^4}^{\frac{3}{2}}\|W\|^{\frac{1}{2}}+C\|W\|_{H^4}^{\frac{n}{8}}\|W\|^{\frac{8-n}{8}}\|P\|_{H^2}^{\frac{1}{2}}\|P\|^{\frac{1}{2}}\|\Delta P\|\\
&+\frac{1}{6}\|\Delta P\|^2+C\|W\|_{H^3}^{\frac{2+n}{3}}\|W\|^{\frac{4-n}{3}}+\frac{1}{6}\|\Delta P\|^2+\frac{1}{6}\|\Delta P\|^2\\
&+C\|W\|_{H^5}^{\frac{6+n}{5}}\|W\|^{\frac{4-n}{5}}+C\|W\|_{H^4}^{\frac{2+n}{8}}\|W\|^{\frac{6-n}{8}}\|\Delta W\|\\
\leqslant&\,\frac{1}{2}\|\Delta P\|^2+\frac{\delta+g}{2}\|\Delta^2 W\|^2+\frac{\delta}{2}\|\nabla\Delta^2 W\|^2+C_2\|\nabla P\|^2\\
&+C_3\|\nabla W\|^2+C_4\|\Delta W\|^2+C.
\end{aligned}\tag{3.25}
$$

Similar to compute (3.25), we also get

$$
\begin{aligned}
\left|2\eta\int_\Omega\nabla(|P|^2)\Delta W\mathrm{d}x\right|=&\left|2\eta\int_\Omega|P|^2\nabla\Delta W\mathrm{d}x\right|\\
\leqslant&\, 2\eta\|\nabla\Delta W\|\|P\|_{L^4}^2\\
\leqslant&\, C\|W\|_{H^4}^{\frac{3}{2}}\|W\|^{\frac{1}{2}}\|P\|_{H^1}^{\frac{n}{2}}\|P\|^{\frac{4-n}{2}}\\
\leqslant&\,\frac{\delta+g}{2}\|\Delta^2W\|^2+C_5\|\nabla P\|^2+C_6\|\nabla W\|^2\\
&+C_7\|\Delta W\|^2+C,
\end{aligned}\tag{3.26}
$$

and

$$
\begin{aligned}
&\left|-2\int_\Omega W\nabla W\Delta^2W\mathrm{d}x-2\eta\int_\Omega\nabla(|P|^2)\Delta^2W\mathrm{d}x\right|\\
\leqslant&\left|\int_\Omega W^2\nabla\Delta^2W\mathrm{d}x\right|+2\eta\left|\int_\Omega|P|^2\nabla\Delta^2W\mathrm{d}x\right|\\
\leqslant&\,\frac{\delta}{2}\|\nabla\Delta^2W\|^2+C(\|W\|_{L^4}^4+\|P\|_{L^4}^4)\\
\leqslant&\,\frac{\delta}{2}\|\nabla\Delta^2W\|^2+C(\|W\|_{H^1}^n\|W\|^{4-n}+\|P\|_{H^1}^n\|P\|^{4-n})\\
\leqslant&\,\frac{\delta}{2}\|\nabla\Delta^2W\|^2+C_8\|\nabla P\|^2+C_9\|\nabla W\|^2+C.
\end{aligned}\tag{3.27}
$$

Substituting (3.22)–(3.27) into (3.21) and noticing $|\xi|=1$, we have

$$
\begin{aligned}
&\frac{d}{dt}(\|\nabla P\|^2+\|\nabla W\|^2+\|\Delta W\|^2)+\|\Delta P\|^2\\
&+(\delta+g)\|\Delta^2W\|^2+\delta\|\nabla\Delta^2W\|^2\\
\leqslant&\,(2+C_2+C_5+C_8)\|\nabla P\|^2+(C_3+C_6+C_9)\|\nabla W\|^2\\
&+(C_4+C_7)\|\Delta W\|^2+C\\
\leqslant&\, K_2(\|\nabla P\|^2+\|\nabla W\|^2+\|\Delta W\|^2)+C,
\end{aligned}\tag{3.28}
$$

where $K_2 = \max(2 + C_2 + C_5 + C_8, C_3 + C_6 + C_9, C_4 + C_7)$ is a positive constant. Applying the Gronwall's inequality, we have

$$\|\nabla P\|^2 + \|\nabla W\|^2 + \|\Delta W\|^2 \leqslant e^{K_2 t}(\|\nabla P_0\|^2 + \|\nabla W_0\|^2 + \|\Delta W_0\|^2 + Ct). \tag{3.29}$$

This completes the proof of Lemma 3.2 with the transformation (3.2).

Corollary 3.1 *Under the conditions of Lemma* 3.2, *we have the following estimates*

$$\|P\|_{H^1_{per}} \leqslant C, \quad \|\nabla Q\|_\infty \leqslant C, \tag{3.30}$$

where C are positive constants.

Proof From Lemma 3.1, Lemma 3.2 and the Gagliardo-Nirenberg inequality (2.2), we have $\|\nabla Q\|_\infty \leqslant C\|\nabla Q\|_{H^2}^{\frac{n}{4}}\|\nabla Q\|^{\frac{4-n}{4}} \leqslant C$, which concludes (3.30).

Lemma 3.3 *Assume that $P_0(x) \in H^2_{per}(\Omega), Q_0(x) \in H^4_{per}(\Omega)$, and under the conditions of Lemma* 3.2. *Then for the solutions of the problem* (1.1)–(1.4), *we have estimates*

$$\|\Delta P\|^2 + \|\Delta^2 Q\|^2 \leqslant e^{K_3 t}(\|\Delta P_0\|^2 + \|\Delta^2 Q_0\|^2 + Ct), \tag{3.31}$$

where K_3 and C are positive constants.

Proof After making the transformation (3.2), we take inner product of (3.3) with $\Delta^2\overline{P}$ over Ω and take the real part, we obtain

$$\begin{aligned}\frac{1}{2}\frac{\mathrm{d}}{\mathrm{d}t}\|\Delta P\|^2 = {}& \xi\|\Delta P\|^2 - \|\nabla\Delta P\|^2 - \mathrm{Re}\int_\Omega (1+i\nu)|P|^2 P\Delta^2\overline{P}\mathrm{d}x \\ & - \mathrm{Re}\int_\Omega \nabla P\Delta^2\overline{P}W\mathrm{d}x - \mathrm{Re}\int_\Omega r_1 P\Delta^2\overline{P}\nabla W\mathrm{d}x \\ & - \mathrm{Re}\int_\Omega gr_2 P\Delta^2\overline{P}\nabla\Delta W\mathrm{d}x.\end{aligned} \tag{3.32}$$

Multiplying (3.4) by $(-\Delta^3 W)$ and integrating with respect to x over Ω, then

$$\begin{aligned}\frac{1}{2}\frac{\mathrm{d}}{\mathrm{d}t}\|\nabla\Delta W\|^2 = {}& \|\Delta^2 W\|^2 - g\|\nabla\Delta^2 W\|^2 - \delta\|\Delta^3 W\|^2 \\ & + \int_\Omega W\nabla W\Delta^3 W\mathrm{d}x + \eta\int_\Omega \nabla(|P|^2)\Delta^3 W\mathrm{d}x.\end{aligned} \tag{3.33}$$

Adding the above two equalities arrives at

$$\begin{aligned}
&\frac{\mathrm{d}}{\mathrm{d}t}(\|\Delta P\|^2+\|\nabla\Delta W\|^2)\\
&=2\xi\|\Delta P\|^2-2\|\nabla\Delta P\|^2+2\|\Delta^2 W\|^2-2g\|\nabla\Delta^2 W\|^2-2\delta\|\Delta^3 W\|^2\\
&\quad-2\mathrm{Re}\int_\Omega(1+i\nu)|P|^2P\Delta^2\overline{P}\mathrm{d}x-2\mathrm{Re}\int_\Omega\nabla P\Delta^2\overline{P}W\mathrm{d}x\\
&\quad-2\mathrm{Re}\int_\Omega r_1P\Delta^2\overline{P}\nabla W\mathrm{d}x-2\mathrm{Re}\int_\Omega gr_2P\Delta^2\overline{P}\nabla\Delta W\mathrm{d}x\\
&\quad+2\int_\Omega W\nabla W\Delta^3 W\mathrm{d}x+2\eta\int_\Omega\nabla(|P|^2)\Delta^3W\mathrm{d}x.
\end{aligned}\tag{3.34}$$

Firstly according to the Gagliardo-Nirenberg inequality, Lemma 3.1 and Lemma 3.2, we have

$$\begin{aligned}
&\left|-2\mathrm{Re}\int_\Omega(1+i\nu)|P|^2P\Delta^2\overline{P}\mathrm{d}x\right|\\
&=\left|2\mathrm{Re}\int_\Omega(1+i\nu)(|P|^2\nabla P\nabla\Delta\overline{P}+(P\nabla\overline{P}+\overline{P}\nabla P)P\nabla\Delta\overline{P})\mathrm{d}x\right|\\
&\leqslant 6|1+i\nu|\|P\|_\infty^2\|\nabla P\|\|\nabla\Delta P\|\\
&\leqslant\frac{1}{3}\|\nabla\Delta P\|^2+C\|P\|_{H^2}^{n}\|P\|^{4-n}\|\nabla P\|^2\\
&\leqslant\frac{1}{3}\|\nabla\Delta P\|^2+C_{10}\|\Delta P\|^2+C,
\end{aligned}\tag{3.35}$$

and

$$\begin{aligned}
\left|-2\mathrm{Re}\int_\Omega\nabla P\Delta^2\overline{P}W\mathrm{d}x\right|&=\left|2\mathrm{Re}\int_\Omega\nabla P\nabla\Delta\overline{P}\nabla W\mathrm{d}x+2\mathrm{Re}\int_\Omega\Delta P\nabla\Delta\overline{P}W\mathrm{d}x\right|\\
&\leqslant 2\|\nabla W\|_\infty\|\nabla P\|\|\nabla\Delta P\|+2\|W\|_\infty\|\Delta P\|\|\nabla\Delta P\|\\
&\leqslant\frac{1}{3}\|\nabla\Delta P\|^2+C\|\nabla W\|_{H^2}^{\frac{n}{2}}\|\nabla W\|^{\frac{4-n}{2}}+C_{11}\|\Delta P\|^2\\
&\leqslant\frac{1}{3}\|\nabla\Delta P\|^2+C_{11}\|\Delta P\|^2+C_{12}\|\nabla\Delta W\|^2+C.
\end{aligned}\tag{3.36}$$

In the same way, we can handle these terms as follows

$$\begin{aligned}
&\left|2\|\Delta^2W\|^2-2\mathrm{Re}\int_\Omega r_1P\Delta^2\overline{P}\nabla W\mathrm{d}x-2\mathrm{Re}\int_\Omega gr_2P\Delta^2\overline{P}\nabla\Delta W\mathrm{d}x\right|\\
&\leqslant 2\|\Delta^2W\|^2+\left|2\mathrm{Re}\int_\Omega r_1(P\nabla\Delta\overline{P}\Delta W+\nabla P\nabla\Delta\overline{P}\nabla W)\mathrm{d}x\right|\\
&\quad+\left|2\mathrm{Re}\int_\Omega gr_2(P\nabla\Delta\overline{P}\Delta^2W+\nabla P\nabla\Delta\overline{P}\nabla\Delta W)\mathrm{d}x\right|
\end{aligned}$$

$$
\begin{aligned}
&\leqslant 2\|\Delta^2 W\|^2 + 2|r_1|\|P\|\|\nabla\Delta P\|\|\Delta W\|_\infty \\
&\quad +2|r_1|\|\nabla P\|\|\nabla\Delta P\|\|\nabla W\|_\infty + 2g|r_2|\|P\|\|\nabla\Delta P\|\|\Delta^2 W\|_\infty \\
&\quad +2g|r_2|\|\nabla P\|\|\nabla\Delta P\|\|\nabla\Delta W\|_\infty \\
&\leqslant 2\|\Delta^2 W\|^2 + \frac{1}{3}\|\nabla\Delta P\|^2 + C\|\nabla W\|_{H^3}^{\frac{2+n}{3}}\|\nabla W\|^{\frac{4-n}{3}} \\
&\quad +C\|\nabla W\|_{H^2}^{\frac{n}{2}}\|\nabla W\|^{\frac{4-n}{2}} + C\|\nabla W\|_{H^5}^{\frac{6+n}{5}}\|\nabla W\|^{\frac{4-n}{5}} \\
&\quad +C\|\nabla W\|_{H^4}^{\frac{4+n}{4}}\|\nabla W\|^{\frac{4-n}{4}} \\
&\leqslant \frac{1}{3}\|\nabla\Delta P\|^2 + g\|\nabla\Delta^2 W\|^2 + \frac{\delta}{4}\|\Delta^3 W\|^2 + 4\|\Delta^2 W\|^2 \\
&\quad +C\|\nabla\Delta W\|^2 + C \\
&\leqslant \frac{1}{3}\|\nabla\Delta P\|^2 + g\|\nabla\Delta^2 W\|^2 + \frac{\delta}{4}\|\Delta^3 W\|^2 + C\|\nabla W\|_{H^4}^{\frac{3}{2}}\|\nabla W\|^{\frac{1}{2}} \\
&\quad +C\|\nabla\Delta W\|^2 + C \\
&\leqslant \frac{1}{3}\|\nabla\Delta P\|^2 + g\|\nabla\Delta^2 W\|^2 + \frac{\delta}{2}\|\Delta^3 W\|^2 + C_{13}\|\nabla\Delta W\|^2 + C.
\end{aligned}
\tag{3.37}
$$

For the last two terms, we can also have that

$$
\begin{aligned}
&\left|2\int_\Omega W\nabla W\Delta^3 W dx + 2\eta\int_\Omega \nabla(|P|^2)\Delta^3 W \mathrm{d}x\right| \\
&\leqslant 2\|W\|_{L^\infty}\|\nabla W\|\|\Delta^3 W\| + 4\eta\|P\|_{L^\infty}\|\nabla P\|\|\Delta^3 W\| \\
&\leqslant \frac{\delta}{2}\|\Delta^3 W\|^2 + C\|P\|_{H^2}^{\frac{n}{2}}\|P\|^{\frac{4-n}{2}} \\
&\leqslant \frac{\delta}{2}\|\Delta^3 W\|^2 + C_{14}\|\Delta P\|^2 + C.
\end{aligned}
\tag{3.38}
$$

Then combining (3.34)–(3.38) and noticing $|\xi| = 1$, there holds

$$
\begin{aligned}
&\frac{\mathrm{d}}{\mathrm{d}t}(\|\Delta P\|^2 + \|\nabla\Delta W\|^2) + \|\nabla\Delta P\|^2 + g\|\nabla\Delta^2 W\|^2 + \delta\|\Delta^3 W\|^2 \\
&\leqslant (2 + C_{10} + C_{11} + C_{14})\|\Delta P\|^2 + (C_{12} + C_{13})\|\nabla\Delta W\|^2 + C \\
&\leqslant K_3(\|\Delta P\|^2 + \|\nabla\Delta W\|^2) + C,
\end{aligned}
\tag{3.39}
$$

where $K_3 = \max(2 + C_{10} + C_{11} + C_{14}, C_{12} + C_{13})$ is a positive constant. Applying the Gronwall's inequality, we have

$$
\|\Delta P\|^2 + \|\nabla\Delta W\|^2 \leqslant e^{K_3 t}(\|\Delta P_0\|^2 + \|\nabla\Delta W_0\|^2 + Ct), \tag{3.40}
$$

where K_3 and C are positive constants. Noticing the transformation (3.2), the proof of Lemma 3.3 is completed.

Corollary 3.2 *Under the conditions of Lemma 3.3, we have the following estimates*

$$\|P\|_{L^\infty} \leqslant C, \quad \|\Delta Q\|_{L^\infty} \leqslant C, \tag{3.41}$$

where C are positive constants.

Proof Based on the results of Lemma 3.1–Lemma 3.3 and the Gagliardo-Nirenberg inequality, one can obtain this corollary easily.

Lemma 3.4 *Under the conditions of Lemma 3.3, then for the solutions of the problem (1.1)–(1.4), we have*

$$\|P\|_{H^2_{per}} \leqslant C, \quad \|Q\|_{H^4_{per}} \leqslant C, \tag{3.42}$$

where C are positive constants.

Proof From the estimates in Lemma 3.1-Lemma 3.3, we see $\|P\|_{H^2_{per}} \leqslant C$. However, we need to estimate $\|Q\|$ for $\|Q\|_{H^4_{per}}$. Considering the original Eq.(1.2), multiplying this equation by Q and integrating over Ω, we have

$$\frac{1}{2}\frac{\mathrm{d}}{\mathrm{d}t}\|Q\|^2 = \|\nabla Q\|^2 - g\|\Delta Q\|^2 - \delta\|\nabla\Delta Q\|^2 - \frac{1}{2}\int_\Omega |\nabla Q|^2 Q\mathrm{d}x - \eta\int_\Omega |P|^2 Q\mathrm{d}x. \tag{3.43}$$

By the previous lemmas and the corollary (3.30), we can get

$$\begin{aligned}
&\left|\|\nabla Q\|^2 - \frac{1}{2}\int_\Omega |\nabla Q|^2 Q\mathrm{d}x - \eta\int_\Omega |P|^2 Q\mathrm{d}x\right| \\
&\leqslant C\|Q\|_{H^2}\|Q\| + \frac{1}{2}\|\nabla Q\|_{L^\infty}\|\nabla Q\|\|Q\| + \eta\|P\|_{L^\infty}\|P\|\|Q\| \\
&\leqslant C\|Q\|^2 + C.
\end{aligned} \tag{3.44}$$

Substituting (3.44) into (3.43), we find that

$$\frac{\mathrm{d}}{\mathrm{d}t}\|Q\|^2 \leqslant 2C\|Q\|^2 + C. \tag{3.45}$$

By Gronwall's inequality and Lemma 3.1-Lemma 3.3, we get $\|Q\|_{H^4_{per}} \leqslant C$.

Lemma 3.5 *Assume that $P_0(x) \in H^3_{per}(\Omega), Q_0(x) \in H^5_{per}(\Omega)$, and under the conditions of Lemma 3.3. Then for the solutions of the problem (1.1)–(1.4), we have the following estimates*

$$\|\nabla\Delta P\|^2 + \|\nabla\Delta^2 Q\|^2 \quad \leqslant e^{K_4 t}(\|\nabla\Delta P_0\|^2 + \|\nabla\Delta^2 Q_0\|^2 + Ct), \tag{3.46}$$

where K_4 and C are positive constants.

Proof Using the transformed equations (3.3)(3.4) as before. Taking inner product of (3.3) with $(-\Delta^3\overline{P})$ over Ω and taking the real part, we can obtain

$$\begin{aligned}\frac{1}{2}\frac{\mathrm{d}}{\mathrm{d}t}\|\nabla\Delta P\|^2=&\ \xi\|\nabla\Delta P\|^2-\|\Delta^2P\|^2+\mathrm{Re}\int_\Omega(1+i\nu)|P|^2P\Delta^3\overline{P}\mathrm{d}x\\&+\mathrm{Re}\int_\Omega\nabla P\Delta^3\overline{P}W\mathrm{d}x+\mathrm{Re}\int_\Omega r_1P\Delta^3\overline{P}\nabla W\mathrm{d}x\\&+\mathrm{Re}\int_\Omega gr_2P\Delta^3\overline{P}\nabla\Delta W\mathrm{d}x.\end{aligned}\tag{3.47}$$

Taking the inner product of (3.4) with Δ^4W over Ω, then we get

$$\begin{aligned}\frac{1}{2}\frac{\mathrm{d}}{\mathrm{d}t}\|\Delta^2W\|^2=&\ \|\nabla\Delta^2W\|^2-g\|\Delta^3W\|^2-\delta\|\nabla\Delta^3W\|^2\\&-\int_\Omega W\nabla W\Delta^4W\mathrm{d}x-\eta\int_\Omega\nabla(|P|^2)\Delta^4W\mathrm{d}x.\end{aligned}\tag{3.48}$$

Adding the above two equalities gives

$$\begin{aligned}\frac{\mathrm{d}}{\mathrm{d}t}(\|\nabla\Delta P\|^2+\|\Delta^2W\|^2)=&\ 2\xi\|\nabla\Delta P\|^2-2\|\Delta^2P\|^2+2\|\nabla\Delta^2W\|^2\\&-2g\|\Delta^3W\|^2-2\delta\|\nabla\Delta^3W\|^2\\&+2\mathrm{Re}\int_\Omega(1+i\nu)|P|^2P\Delta^3\overline{P}\mathrm{d}x\\&+2\mathrm{Re}\int_\Omega\nabla P\Delta^3\overline{P}W\mathrm{d}x+2\mathrm{Re}\int_\Omega r_1P\Delta^3\overline{P}\nabla W\mathrm{d}x\\&+2\mathrm{Re}\int_\Omega gr_2P\Delta^3\overline{P}\nabla\Delta W\mathrm{d}x\\&-2\int_\Omega W\nabla W\Delta^4W\mathrm{d}x-2\eta\int_\Omega\nabla(|P|^2)\Delta^4W\mathrm{d}x.\end{aligned}\tag{3.49}$$

Now using the estimates of previous lemmas and corollaries, we can majorize the right hand of (3.49) as follows. Firstly, we have

$$\begin{aligned}&\left|2\mathrm{Re}\int_\Omega(1+i\nu)|P|^2P\Delta^3\overline{P}\mathrm{d}x\right|\\=&\left|2\mathrm{Re}\int_\Omega(1+i\nu)(2(\nabla P)^2\overline{P}+2|P|^2\Delta P+4P|\nabla P|^2+P^2\Delta\overline{P})\Delta^2\overline{P}\mathrm{d}x\right|\\\leqslant&\ 2|1+i\nu|(6\|\nabla P\|_\infty^2\|P\|+3\|P\|_\infty^2\|\Delta P\|)\|\Delta^2P\|\\\leqslant&\ \frac{1}{4}\|\Delta^2P\|^2+C\|P\|_{H^3}^{\frac{2+n}{3}}\|P\|^{\frac{4-n}{3}}+C\\\leqslant&\ \frac{1}{4}\|\Delta^2P\|^2+C_{14}\|\nabla\Delta P\|^2+C,\end{aligned}\tag{3.50}$$

$$
\begin{aligned}
&\left|2\mathrm{Re}\int_\Omega \nabla P\Delta^3\overline{P}W\mathrm{d}x\right| \\
&=\left|2\mathrm{Re}\int_\Omega (\Delta W\nabla P+2\nabla W\Delta P+W\nabla\Delta P)\Delta^2\overline{P}\mathrm{d}x\right| \\
&\leqslant 2(\|\Delta W\|\|\nabla P\|_\infty+2\|\nabla W\|_\infty\|\Delta P\|+\|W\|_\infty\|\nabla\Delta P\|)\|\Delta^2P\| \\
&\leqslant \frac{1}{4}\|\Delta^2P\|^2+C\|\nabla P\|_{H^2}^{\frac{n}{2}}\|\nabla P\|^{\frac{4-n}{2}}+C\|\nabla\Delta P\|^2+C \\
&\leqslant \frac{1}{4}\|\Delta^2P\|^2+C_{15}\|\nabla\Delta P\|^2+C,
\end{aligned}
\tag{3.51}
$$

and

$$
\begin{aligned}
&\left|2\mathrm{Re}\int_\Omega r_1P\Delta^3\overline{P}\nabla W\mathrm{d}x\right| \\
&=\left|2\mathrm{Re}\int_\Omega (\Delta P\nabla W+2\nabla P\Delta W+P\nabla\Delta W)\Delta^2\overline{P}\mathrm{d}x\right| \\
&\leqslant 2(\|\nabla W\|_\infty\|\Delta P\|+2\|\Delta W\|\|\nabla P\|_\infty+\|P\|_\infty\|\nabla\Delta W\|)\|\Delta^2P\| \\
&\leqslant \frac{1}{4}\|\Delta^2P\|^2+C\|\nabla P\|_{H^2}^{\frac{n}{2}}\|\nabla P\|^{\frac{4-n}{2}}+C \\
&\leqslant \frac{1}{4}\|\Delta^2P\|^2+C_{16}\|\nabla\Delta P\|^2+C.
\end{aligned}
\tag{3.52}
$$

Similarly, using Gagliardo-Nirenberg inequality and previous estimates to obtain that

$$
\begin{aligned}
&\left|2\|\nabla\Delta^2W\|^2+2\mathrm{Re}\int_\Omega gr_2P\Delta^3\overline{P}\nabla\Delta W\mathrm{d}x\right| \\
&\leqslant 2\|\nabla\Delta^2W\|^2+\left|2\mathrm{Re}\int_\Omega gr_2(\nabla\Delta^2WP+2\Delta^2W\nabla P+\nabla\Delta W\Delta P)\Delta^2\overline{P}\mathrm{d}x\right| \\
&\leqslant 2\|\nabla\Delta^2W\|^2+2g|r_2|(\|P\|_\infty\|\nabla\Delta^2W\|+2\|\nabla P\|\|\Delta^2W\|_\infty)\|\Delta^2P\| \\
&\quad+2g|r_2|\|\Delta P\|\|\nabla\Delta W\|_\infty\|\Delta^2P\| \\
&\leqslant \frac{1}{4}\|\Delta^2P\|^2+2\|\nabla\Delta^2W\|^2+C\|W\|_{H^6}^{\frac{5}{3}}\|W\|^{\frac{2}{3}} \\
&\quad+C\|\Delta W\|_{H^4}^{\frac{4+n}{3}}\|\Delta W\|^{\frac{4-n}{4}}+C\|\nabla W\|_{H^4}^{\frac{4+n}{3}}\|\nabla W\|^{\frac{4-n}{4}} \\
&\leqslant \frac{1}{4}\|\Delta^2P\|^2+\frac{g}{2}\|\Delta^3W\|^2+4\|\nabla\Delta^2W\|^2+C\|\Delta^2W\|^2+C \\
&\leqslant \frac{1}{4}\|\Delta^2P\|^2+g\|\Delta^3W\|^2+C_{17}\|\Delta^2W\|^2+C.
\end{aligned}
\tag{3.53}
$$

For the last two terms, we get

$$
\begin{aligned}
&\left|-2\int_\Omega W\nabla W\Delta^4 W dx - 2\eta\int_\Omega \nabla(|P|^2)\Delta^4 W \mathrm{d}x\right| \\
\leqslant &\left|2\int_\Omega ((\nabla W)^2 + W\Delta W)\nabla\Delta^3 W dx\right| \\
&+ 2\eta\left|\int_\Omega (\Delta P\overline{P} + 2|\nabla P|^2 + P\Delta\overline{P})\nabla\Delta^3 W \mathrm{d}x\right| \\
\leqslant &\, 2(\|\nabla W\|_\infty \|\nabla W\| + \|W\|_\infty \|\Delta W\|)\|\nabla\Delta^3 W\| \\
&+ 2\eta(\|\Delta P\|\|P\|_\infty + 2\|\nabla P\|_\infty\|\nabla P\| + \|P\|_\infty\|\Delta P\|)\|\nabla\Delta^3 W\| \\
\leqslant &\, \delta\|\nabla\Delta^3 W\|^2 + C_{18}\|\nabla\Delta P\|^2 + C.
\end{aligned} \tag{3.54}
$$

Substituting (3.50)–(3.54) into (3.49) yields that

$$
\begin{aligned}
&\frac{\mathrm{d}}{\mathrm{d}t}(\|\nabla\Delta P\|^2 + \|\Delta^2 W\|^2) + \|\Delta^2 P\|^2 + g\|\Delta^3 W\|^2 + \delta\|\nabla\Delta^3 W\|^2 \\
\leqslant &\, (2 + C_{14} + C_{15} + C_{16} + C_{18})\|\nabla\Delta P\|^2 + C_{17}\|\Delta^2 W\|^2 + C \\
\leqslant &\, K_4(\|\nabla\Delta P\|^2 + \|\Delta^2 W\|^2) + C,
\end{aligned} \tag{3.55}
$$

where $K_4 = \max(2 + C_{14} + C_{15} + C_{16} + C_{18}, C_{17})$ is a positive constant. Applying Gronwall's inequality and transformation (3.2) completes the proof of Lemma 3.5.

Corollary 3.3 *Under the conditions of Lemma* 3.5, *we have the following estimates*

$$
\|\nabla P\|_{L^\infty} \leqslant C, \quad \|\nabla\Delta Q\|_{L^\infty} \leqslant C, \tag{3.56}
$$

for the solutions of the problem (1.1)–(1.4), *where* C *are positive constants.*

Lemma 3.6 *Under the conditions of Lemma* 3.1*–Lemma* 3.5, *we have the following estimates*

$$
\|P_t\|^2 + \|Q_t\|^2 + \|\nabla Q_t\|^2 \leqslant C, \tag{3.57}
$$

for the solutions of the problem (1.1)–(1.4), *where* C *is a positive constant.*

Proof We differentiate Eqs.(1.1) and (1.2) with respect to t once, take the inner product of the resulting equations with $\overline{P}_t$ and $(Q_t - \Delta Q_t)$ respectively, and take the real parts, we obtain

$$
\begin{aligned}
\frac{1}{2}\frac{\mathrm{d}}{\mathrm{d}t}\|P_t\|^2 = &\,\xi\|P_t\|^2 - \|\nabla P_t\|^2 - \mathrm{Re}\int_\Omega (1 + i\nu)(2|P|^2|P_t|^2 + P^2\overline{P}_t^2)\mathrm{d}x \\
&- \mathrm{Re}\int_\Omega (\nabla P_t\nabla Q + \nabla P\nabla Q_t)\overline{P}_t \mathrm{d}x \\
&- \mathrm{Re}\int_\Omega r_1(|P_t|^2\Delta Q + P\Delta Q_t\overline{P}_t)\mathrm{d}x \\
&- \mathrm{Re}\int_\Omega g r_2(|P_t|^2\Delta^2 Q + P\Delta^2 Q_t\overline{P}_t)\mathrm{d}x,
\end{aligned} \tag{3.58}
$$

and

$$\begin{aligned}\frac{1}{2}\frac{\mathrm{d}}{\mathrm{d}t}(\|Q_t\|^2+\|\nabla Q_t\|^2)=&\|\nabla Q_t\|^2-(g-1)\|\Delta Q_t\|^2-(\delta+g)\|\nabla\Delta Q_t\|^2\\&-\delta\|\Delta^2 Q_t\|^2-\int_\Omega \nabla Q\nabla Q_t(Q_t-\Delta Q_t)\mathrm{d}x\\&-\eta\int_\Omega(P_t\overline{P}+P\overline{P}_t)(Q_t-\Delta Q_t)\mathrm{d}x,\end{aligned}\tag{3.59}$$

Adding (3.58) and (3.59) together, noticing $|\xi|=1$ and using the estimates in Lemma 3.1-Lemma 3.5, there yields

$$\begin{aligned}&\frac{\mathrm{d}}{\mathrm{d}t}(\|P_t\|^2+\|Q_t\|^2+\|\nabla Q_t\|^2)\\&=2\xi\|P_t\|^2-\|\nabla P_t\|^2+2\|\nabla Q_t\|^2-2(g-1)\|\Delta Q_t\|^2-2(\delta+g)\|\nabla\Delta Q_t\|^2\\&\quad-2\delta\|\Delta^2Q_t\|^2-2\mathrm{Re}\int_\Omega(1+i\nu)(2|P|^2|P_t|^2+P^2\overline{P}_t^2)\mathrm{d}x\\&\quad-2\mathrm{Re}\int_\Omega(\nabla P_t\nabla Q+\nabla P\nabla Q_t)\overline{P}_t\mathrm{d}x\\&\quad-2\mathrm{Re}\int_\Omega r_1(|P_t|^2\Delta Q+P\Delta Q_t\overline{P}_t)\mathrm{d}x\\&\quad-2\mathrm{Re}\int_\Omega gr_2(|P_t|^2\Delta^2Q+P\Delta^2Q_t\overline{P}_t)\mathrm{d}x\\&\quad-2\int_\Omega\nabla Q\nabla Q_t(Q_t-\Delta Q_t)dx-2\eta\int_\Omega(P_t\overline{P}+P\overline{P}_t)(Q_t-\Delta Q_t)\mathrm{d}x\\&\leqslant 2\|P_t\|^2-\|\nabla P_t\|^2+2\|\nabla Q_t\|^2+2|g-1|\|\Delta Q_t\|^2-2\delta\|\nabla\Delta Q_t\|^2\\&\quad-2\delta\|\Delta^2Q_t\|^2+6|1+i\nu|\|P\|_\infty^2\|P_t\|^2\\&\quad+2(\|\nabla Q\|_\infty\|\nabla P_t\|+\|\nabla P\|_\infty\|\nabla Q_t\|)\|P_t\|\\&\quad+2r_{1r}\|\Delta Q\|_\infty\|P_t\|^2+2|r_1|\|P\|_\infty\|\Delta Q_t\|\|P_t\|\\&\quad+4g|r_2|\|\nabla\Delta Q\|_\infty\|P_t\|\|\nabla P_t\|+2g|r_2|\|\nabla P\|_\infty\|\nabla\Delta Q_t\|\|P_t\|\\&\quad+2g|r_2|\|P\|_\infty\|\nabla\Delta Q_t\|\|\nabla P_t\|\\&\quad+2\|\nabla Q\|_\infty\|Q_t\|(\|\nabla Q_t\|+\|\Delta Q_t\|)+4\eta\|P\|_\infty\|P_t\|(\|\nabla Q_t\|+\|\Delta Q_t\|)\\&\leqslant C(\|P_t\|^2+\|Q_t\|^2+\|\nabla Q_t\|^2),\end{aligned}\tag{3.60}$$

where we apply the Young's inequality with ε and the Gagliardo-Nirenberg inequality repeatedly. Thus, Gronwall's inequality yields the estimates of Lemma 3.6.

Generally based on the results of the previous lemmas and the mathematical deduction, we have the following lemma for problem (1.1)–(1.4).

Lemma 3.7 *Assume that* $P_0(x) \in H^k(\Omega), Q_0(x) \in H^{k+2}(\Omega)$ $(k \geqslant 3)$, *and* $2\delta - g^2r_{2r}^2 > 0$. *If* $n = 2$, *we also suppose* $|\nu| \leqslant \sqrt{3}$. *Then for the solutions of the problem* (1.1)–(1.4), *we have the following estimates*

$$\left\|\nabla^k P\right\|^2 + \left\|\nabla^{k+2}Q\right\|^2 \leqslant C, \tag{3.61}$$

Furthermore, there also holds

$$\left\|\nabla^{k-3} P_t\right\|^2 + \left\|\nabla^{k-2}Q_t\right\|^2 \leqslant C, \tag{3.62}$$

where the positive constant C *depends on* $\left\|\nabla^k P_0\right\|$ *and* $\left\|\nabla^{k+2}Q_0\right\|$ *and independent of the period* L.

4 The local solutions and global solutions

In this section, we will obtain the existence and uniqueness of the local solutions and global solutions for the periodic initial value problem (1.1)–(1.4). Firstly, we adopt the Galerkin method to construct the approximate solutions for the problem (1.1)–(1.4). Let $\omega_j(x)(j = 1, 2, \cdots)$ be the unit eigenfunctions satisfying the equation

$$\Delta\omega_j + \lambda_j\omega_j = 0, \quad j = 1, 2, \cdots, \quad \omega_j \in H_0^1(\Omega) \cap L^4(\Omega), \tag{4.1}$$

with periodicity $\omega_j(x) = \omega_j(x + Le_i)(i = 1, 2)$ and $\lambda_j(j = 1, 2, \cdots)$ is the corresponding eigenvalues different from each other $\{\omega_j(x)\}$ consists of the orthogonal base in $L^2(\Omega)$. Thus the approximate solutions can be written as

$$P_m(x, t) = \sum_{j=1}^{m} \alpha_{jm}(t)\omega_j(x), \quad Q_m(x, t) = \sum_{j=1}^{m} \beta_{jm}(t)\omega_j(x). \tag{4.2}$$

According to the Galerkin method, these undetermined coefficients $\alpha_{jm}(t)$ and $\beta_{jm}(t)$ have to satisfy the following initial value problem of a system of the ordinary differential equations

$$\begin{aligned}(P_{mt}, \omega_j) = {} & \xi(P_m, \omega_j) - (1 + i\mu)(\nabla P_m, \nabla\omega_j) - (1 + i\nu)(|P_m|^2 P_m, \omega_j) \\ & - (\nabla P_m \nabla Q_m, \omega_j) - r_1(P_m \Delta Q_m, \omega_j) - gr_2(P_m \Delta^2 Q_m, \omega_j),\end{aligned} \tag{4.3}$$

$$\begin{aligned}(Q_{mt}, \omega_j) = {} & (\nabla Q_m, \nabla\omega_j) - g(\Delta Q_m, \Delta\omega_j) - \delta(\nabla\Delta Q_m, \nabla\Delta\omega_j) \\ & - \frac{1}{2}(|\nabla Q_m|^2, \omega_j) - \eta(|P_m|^2, \omega_j),\end{aligned} \tag{4.4}$$

with initial conditions

$$P_m(x,0) = P_{0m}(x), \quad Q_m(x,0) = Q_{0m}(x), \tag{4.5}$$

where $0 \leqslant t \leqslant T$ and $j = 1, 2, \cdots, m$.

We assume that

$$P_{0m}(x) \xrightarrow{H^3_{per}(\Omega)} P_0(x), \quad Q_{0m}(x) \xrightarrow{H^5_{per}(\Omega)} Q_0(x), \quad m \to \infty. \tag{4.6}$$

Similar to complete the proof of Lemma 3.1–Lemma 3.6, we can establish the estimates of the solutions of the problem (1.1)–(1.4) which are uniform for m. By using the compact principle, we can prove

Theorem 4.1 (Local existence) *Assume that $P_0(x) \in H^3_{per}(\Omega)$, $Q_0(x) \in H^5_{per}(\Omega)$, and $2\delta - g^2 r^2_{2r} > 0$. If $n = 2$, we also suppose $|\nu| \leqslant \sqrt{3}$. Then the periodic initial value problem* (1.1)–(1.4) *possesses the periodic local solutions $P(x,t)$ and $Q(x,t)$, which satisfy*

$$P(x,t) \in L^\infty(0,t_0; H^3_{per}(\Omega)), \quad P_t(x,t) \in L^\infty(0,t_0; L^2_{per}(\Omega)),$$

$$Q(x,t) \in L^\infty(0,t_0; H^5_{per}(\Omega)), \quad Q_t(x,t) \in L^\infty(0,t_0; H^1_{per}(\Omega)),$$

where t_0 depends on $\|P_0(x)\|_{H^3_{per}}$ and $\|Q_0(x)\|_{H^5_{per}}$.

Theorem 4.2 (Global existence) *Under the conditions of Theorem* 4.1*. Then there exists global solutions $P(x,t)$ and $Q(x,t)$, which satisfy*

$$P(x,t) \in L^\infty(0,T; H^3_{per}(\Omega)), \quad P_t(x,t) \in L^\infty(0,T; L^2_{per}(\Omega)),$$

$$Q(x,t) \in L^\infty(0,T; H^5_{per}(\Omega)), \quad Q_t(x,t) \in L^\infty(0,T; H^1_{per}(\Omega)),$$

for the periodic initial value problem (1.1)–(1.4).

Proof From Theorem 4.1 we know that the local solutions for the problem (1.1)–(1.4) exist and t_0 depends on $\|P_0(x)\|_{H^3_{per}}$ and $\|Q_0(x)\|_{H^5_{per}}$. According to the priori estimates in Section and by the so-called continuity method, we can obtain the global solutions for the problem (1.1)–(1.4) easily.

Theorem 4.3 (Uniqueness for global solutions) *Under the conditions of Theorem* 4.2*, the global solutions $P(x,t)$ and $Q(x,t)$ of the periodic initial value problem* (1.1)–(1.4) *are unique.*

Proof Suppose that $P_1(x,t), Q_1(x,t)$ and $P_2(x,t), Q_2(x,t)$ are two solutions of problem (1.1)–(1.4), then the differences $P = P_1(x,t) - P_2(x,t)$, $Q(x,t) = Q_1(x,t) - Q_2(x,t)$ will satisfy

$$P_t = \xi P + (1+i\mu)\Delta P - (1+i\nu)(|P_1|^2 P_1 - |P_2|^2 P_2) - (\nabla P_1 \nabla Q_1 - \nabla P_2 \nabla Q_2)$$

$$-r_1(P_1\Delta Q_1 - P_2 \Delta Q_2) - g r_2 (P_1 \Delta^2 Q_1 - P_2 \Delta^2 Q_2), \tag{4.7}$$

$$Q_t = -\Delta Q - g\Delta^2 Q + \delta \Delta^3 Q - \frac{1}{2}\left(|\nabla Q_1|^2 - |\nabla Q_2|^2\right) - \eta(|P_1|^2 - |P_2|^2), \tag{4.8}$$

$$P(x+Le_i, t) = P(x,t), \quad Q(x+Le_i,t) = Q(x,t), \tag{4.9}$$

$$P(x,0) = 0, \quad Q(x,0) = 0. \tag{4.10}$$

Taking the inner product of (4.7) with $\overline{P}$ and taking the real parts, taking inner product of (4.8) with $-\Delta Q$ over Ω, then adding these two equations together, we obtain

$$\begin{aligned}
&\frac{\mathrm{d}}{\mathrm{d}t}(\|P\|^2 + \|\nabla Q\|^2) \\
&= 2\|P\|^2 - 2\|\nabla P\|^2 + 2\|\Delta Q\|^2 - 2g\|\nabla \Delta Q\|^2 - 2\delta\|\Delta^2 Q\|^2 \\
&\quad - \mathrm{Re}\int_\Omega 2(1+i\nu)(|P_1|^2 P_1 - |P_2|^2 P_2)\overline{P}\mathrm{d}x \\
&\quad - 2\mathrm{Re}\int_\Omega (\nabla P_1 \nabla Q_1 - \nabla P_2 \nabla Q_2)\overline{P}\mathrm{d}x \\
&\quad - 2\mathrm{Re}\int_\Omega r_1(P_1 \Delta Q_1 - P_2 \Delta Q_2)\overline{P}\mathrm{d}x \\
&\quad - 2\mathrm{Re}\int_\Omega g r_2 (P_1 \Delta^2 Q_1 - P_2 \Delta^2 Q_2)\overline{P}\mathrm{d}x \\
&\quad + \int_\Omega \left(|\nabla Q_1|^2 - |\nabla Q_2|^2\right)\Delta Q dx + 2\eta \int_\Omega (|P_1|^2 - |P_2|^2)\Delta Q \mathrm{d}x \\
&\leqslant C(\|P\|^2 + \|\nabla Q\|^2),
\end{aligned} \tag{4.11}$$

where we majorize the right-hand side of (4.11) with Young's inequality with ε and the Gagliardo-Nirenberg inequality, since $P_1(x,t)$, $Q_1(x,t)$ and $P_2(x,t)$, $Q_2(x,t)$ are the solutions of the problem (1.1)–(1.4) satisfying the estimates in Lemma 3.1–Lemma 3.6

By the Gronwall's inequality and noticing the conditions (4.10), we can complete the proof of the Theorem 4.3.

More generally, we have the following existence and uniqueness theorems of the global smooth solutions from Lemma 3.7.

Theorem 4.4 (Existence and uniqueness for global smooth solutions) *Suppose that* $P_0(x) \in H^k_{per}(\Omega)$, $Q_0(x) \in H^{k+2}_{per}(\Omega)(k \geqslant 3)$ *and* $2\delta - g^2 r_{2r}^2 > 0$. *If* $n = 2$, *we also assume* $|\nu| \leqslant \sqrt{3}$. *Then there exists unique global smooth solutions* $P(x,t)$ *and* $Q(x,t)$, *which satisfy*

$$P(x,t) \in L^\infty(0,T; H^k_{per}(\Omega)), \quad P_t(x,t) \in L^\infty(0,T; H^{k-3}_{per}(\Omega)),$$

$$Q(x,t) \in L^\infty(0,T; H^{k+2}_{per}(\Omega)), \quad Q_t(x,t) \in L^\infty(0,T; H^{k-2}_{per}(\Omega)),$$

for the periodic initial value problem (1.1)–(1.4).

5 Cauchy problem

In previous sections, we studied the existence and uniqueness of the global smooth solutions for the periodic initial value problem (1.1)–(1.4). In this section, we will discuss the Cauchy problem (1.1)(1.2)(1.5) in the infinity domain $\Omega_T = \{(x,t)|x \in \mathbb{R}^n, 0 \leqslant t \leqslant T\}(n = 1,2)$. Since we have supposed that $P(x,t), Q(x,t)$ and some of their derivatives with respect to x tend to zero as $|x| \to \infty$, then the a priori estimates in Section also hold for the solutions of the problem (1.1)(1.2)(1.5) . Furthermore, the a priori estimates are bounded and independent of the period L of the domain Ω, thus we can choose sequence $L_k (k \to \infty, L_k \to \infty)$ and obtain global existence in $[0, T_k]$. Then we can employ the usual method of limiting process for $L_k \to \infty (k \to \infty)$ which is also so-called the diagonal selection to obtain the solutions of Cauchy problem. The global existence and uniqueness theorems for the Cauchy problem (1.1)(1.2)(1.5) which are parallel to Theorem 4.2, Theorem 4.3 and Theorem 4.4 can be stated as follows.

Theorem 5.1 *Suppose that* $2\delta - g^2 r_{2r}^2 > 0$, *and if* $n = 2$, *we also assume* $|\nu| \leqslant \sqrt{3}$. *If* $P_0(x) \in H^3(\mathbb{R}^n)$, $Q_0(x) \in H^5(\mathbb{R}^n)$. *Then there exists global solutions*

$$P(x,t) \in L^\infty(0,T; H^3(\mathbb{R}^n)), \quad P_t(x,t) \in L^\infty(0,T; L^2(\mathbb{R}^n)),$$

$$Q(x,t) \in L^\infty(0,T; H^5(\mathbb{R}^n)), \quad Q_t(x,t) \in L^\infty(0,T; H^1(\mathbb{R}^n)),$$

for the Cauchy problem (1.1)(1.2)(1.5).

Furthermore, if $P_0(x) \in H^k(\mathbb{R}^n)$, $Q_0(x) \in H^{k+2}(\mathbb{R}^n)(k \geqslant 4)$. *Then there exists unique global smooth solutions* $P(x,t), Q(x,t)$ *for the Cauchy problem* (1.1)(1.2)(1.5), *which satisfy*

$$P(x,t) \in L^\infty(0,T;H^k(\mathbb{R}^n)), \quad P_t(x,t) \in L^\infty(0,T;H^{k-3}(\mathbb{R}^n)),$$

$$Q(x,t) \in L^\infty(0,T;H^{k+2}(\mathbb{R}^n)), \quad Q_t(x,t) \in L^\infty(0,T;H^{k-2}(\mathbb{R}^n)).$$

References

[1] Golovin A A, Matkowsky B J, Bayliss A, Nepomnyashchy A A. Coupled KS–CGL and coupled Burgers-CGL equations for flames governed by a sequential reaction [J]. Phys. D, 1999, 129: 253–298.

[2] Williams F A. Combustion Theory [M]. Benjamin Cummings, Menlo Park, 1985.

[3] Peláez J, Liñán A. Structure and stability of flames with two sequential reactions [J]. SIAM J. Appl. Math., 1985, 45: 503–522.

[4] Peláez J. Stability of premixed flames with two thin reaction layers [J]. SIAM J. Appl. Math., 1987, 47(4): 781–799.

[5] Nepomnyashchy A A. Order parameter equations for long wavelength instabilities [J]. Phys. D, 1995, 86: 90–95.

[6] Sivashinsky G I. Nonlinear analysis of hydrodynamic instability in laminar flames I. Derivation of basic equations [J]. Acta Astro., 1977, 4(11-12): 1177–1206.

[7] Li D, Guo B, Liu X. Existence of global solution for complex Ginzburg Landau equation in three dimensions [J]. Appl. Math. J. Chinese Univ. Ser. A, 2004, 19(4): 409–416.

[8] Li D, Guo B, Liu X. Regularity of the attractor for 3-D complex Ginzburg-Landau equation [J]. Acta Math. Appl. Sin. Engl. Ser., 2011, 27(2): 289–302.

[9] Ghidaglia J M, Héron B. Dimension of the attractor associated to the Ginzburg-Landau equation [J]. Phys. D, 1987, 28(3): 282–304.

[10] Guo B. The existence and nonexistence of a global smooth solution for the initial value problem of generalized Kuramoto-Sivashinsky type equations [J]. J. Math. Res. Exposition, 1991, 11(1): 57–70.

[11] Guo B. The nonlinear Galerkin methods for the generalized Kuramoto-Sivashinsky type equations [J]. Adv. Math., 1993, 22(2): 182–184.

[12] Postlethwaite C M, Silber M. Spatial and temporal feedback control of traveling wave solutions of the two-dimensional complex Ginzburg-Landau equation [J]. Phys. D, 2007, 236(1): 65–74.

[13] Sherratt J A, Smith M J, Rademacher J D M. Patterns of sources and sinks in the complex Ginzburg-Landau equation with zero linear dispersion [J]. SIAM J. Appl. Dyn. Syst., 2010, 9(3): 883–918.

[14] Doroninl G G, Larkin N A. Kuramoto-Sivashinsky model for a dusty medium [J]. Math. Methods Appl. Sci., 2003, 26(3): 179–192.

[15] MacKenzie T, Roberts A J, Accurately model the Kuramoto-Sivashinsky dynamics with holistic discretization [J]. SIAM J. Appl. Dynam. Sys., 2006, 5(3): 365–402.

[16] Golovin A A, Nepomnyashchy A A, Matkowsky B J. Traveling and spiral waves for sequential flames with translation symmetry: coupled CGL-Burgers equations [J]. Phys. D, 2001, 160: 1–28.

[17] Evans L C. Partial differential equations, Graduate studies in mathematics [M]. Providence, Rhode Island, American Mathematical Society, 1988.

[18] Friedman A. Partial differential equations [M]. Holt, Reinhart and Winston, New York, 1969.

Blow-up of Smooth Solutions to the Isentropic Compressible MHD Equations*

Bian Dongfen (边东芬) and Guo Boling (郭柏灵)

Abstract This article studies the blow-up of smooth solutions to the isentropic compressible MHD equations (1.1) in $\mathbb{R}^N$, $2 \leqslant N \leqslant 3$. We obtain that under physical conditions, if the initial density ρ_0 and magnetic H_0 have compact support, then any smooth radially symmetric solutions (ρ, u, H) will blow up in finite time.

Keywords compressible MHD equations; isentropic fluids; blow-up

1 Introduction

We consider in this paper the compressible MHD equations on $\mathbb{R}^N$, $2 \leqslant N \leqslant 3$,

$$\begin{cases} \partial_t \rho + \operatorname{div}(\rho u) = 0, \\ \partial_t(\rho u) + \operatorname{div}(\rho u \otimes u - H \otimes H) + \nabla \left(p + \dfrac{1}{2}|H|^2 \right) - \mu \Delta u - (\mu + \lambda) \nabla \operatorname{div} u = 0, \\ \partial_t H - \operatorname{curl}(u \times H) = 0, \\ \operatorname{div} H = 0, \\ (\rho, u, H)_{|t=0} = (\rho_0, u_0, H_0), \end{cases} \tag{1.1}$$

which describes the motion of electrically conducting media in the presence of a magnetic field. Here ρ, u, H and p denote the density, velocity field, magnetic field and pressure respectively. The pressure p is a function of ρ and satisfies $p(\rho) = a\rho^\gamma$ where $a > 0$ is a constant, $\gamma > 1$ is the adiabatic exponent. We denote by λ and μ the two viscosity coefficients of the fluid, which are assumed to satisfy the physical restrictions: $\mu > 0$ and $\lambda + \dfrac{2\mu}{N} \geqslant 0$.

For smooth initial data such that the density ρ_0 is bounded away from zero (i.e. $0 < \underline{\rho} \leqslant \rho_0(x) \leqslant M$), the existence and uniqueness of local classical solutions to

* Appl. Anal., 2013, DOI: http://dx.doi.org/10.1080/00036811.2013.766324.

the compressible Navier-Stokes equations has been known for a long time (see the pioneering work of J.Nash [33] or the paper of N.Itaya [28]). Global well-posedness to the compressible Navier-Stokes equations has been proved by Matsumura and Nishida [31] for smooth data close to equilibrium. The reader may refer to [10, 12-17, 21, 29] for more recent advances on the subject. Concerning the global well-posedness of weak solutions to the compressible Navier-Stokes equations for the large initial data, readers refer to [6,7,30,32], and refer to [8,10,38] and references therein for the viscous shallow water equations.

Recently, Danchin has obtained several important well-posedness in the critical spaces for the compressible Navier-Stokes equations [12-16]. Chen-Miao-Zhang [10] has also proved the local well-posedness in $\dot{B}^1_{2,1} \times (\dot{B}^0_{2,1})^2$ for the viscous shallow water equations and for the compressible Navier-Stokes equations with density dependent viscosities in the critical Besov spaces $\dot{B}^{\frac{N}{p}}_{p,1}$ [9]. Bian-Yuan has obtained local well-posedness in the critical Besov spaces [2] and super critical Besov spaces [3] for the compressible MHD equations. Bian-Guo has showed local well-posedness in the critical Besov spaces [1] for the full compressible MHD equations.

However, the regularity and uniqueness of weak solutions to compressible flows remains open. About blow-up criterions for the strong (smooth) solutions to the compressible Navier-Stokes equations, readers can refer to [5, 19, 20, 22-27, 36]. Also, there are many results about regularity of incompressible Navier-Stokes equations ([4,35,37,42,43]) and incompressible MHD equations ([39,40,44,45]). In the absence of heat conduction, it was first proved by Z. P. Xin that any non-zero smooth solution to the Cauchy problem of the non-barotropic compressible Navier-Stokes system with initially compact supported density would blow up in finite time (see [41]). This result was generalized to the cases for the non-barotropic compressible Navier-Stokes system with heat conduction ([11]) and for non-compact but rapidly decreasing at far field initial densities ([34]). Recently, Du-Li-Zhang [18] prove that any smooth spherically symmetric solutions to the compressible Navier-Stokes equations for two-dimensional isothermal fluids will blow up in finite time as long as the initial densities are compactly supported. In this paper we prove that under physical conditions, if the initial density ρ_0 and magnetic H_0 have compact support, then any smooth radially symmetric solutions (ρ, u, H) to the isentropic compressible MHD equations (1.1) on $\mathbb{R}^N$, $2 \leqslant N \leqslant 3$, will blow up in finite time. Our main result is as follows:

Theorem 1.1 *Suppose that initial density ρ_0 and magnetic H_0 have compact support, if the initial datum are radially symmetric, i.e. $\rho = \rho(|x|, t)$, $u = \frac{x}{|x|}\bar{u}(|x|, t)$, $H = \frac{x}{|x|}\bar{H}(|x|, t)$, then the smooth solution $(\rho, u, H) \in C^1([0, T], W^{m,1}(\mathbb{R}^N))$ $(m > 2,\ 2 \leqslant N \leqslant 3)$ to system (1.1) with nontrivial initial density and magnetic will blow up in finite time.*

Remark 1.1 *When $H = 0$, system (1.1) reduces to the compressible Navier-Stokes equations. If we choose $N = 2$, then we obtain the result as in [18] for the isothermal compressible MHD equations. Moreover, our result also holds for the N-dimension $(2 < N \leqslant 3)$ isentropic compressible Navier-Stokes equations.*

Remark 1.2 *In order to overcome the difficulty arising from the strong coupling and interaction between the magnetic field and the fluid variables, we need energy estimates for the magnetic field. Also, we apply the idea used in the incompressible MHD equations to deal with the nonlinear terms.*

Remark 1.3 *Our method works for the isothermal case. The differences consists in that, to estimate $\int_{\mathbb{R}^N} u \cdot \nabla \rho dx$, we need multiply the density equation of* (1.1) *by* $\ln(\rho + \varepsilon)$.

Throughout the paper, C stands for a "harmless" constant and may be different from line to line. We denote $\|f\|_p$ by $\|f\|_{L^p}$.

2 Proof of Theorem 1.1

In this section, we will follow four steps to show the proof of Theorem 1.1.

Step 1 The density and magnetic keep support all the time.

Since the initial density ρ_0 and magnetic H_0 have compact support, there exists two constants $R,\ R' > 0$ such that

$$
\begin{aligned}
&\mathrm{supp}\rho_0(x) \cap \mathrm{supp}H_0(x) \subseteq B_R, \\
&\mathrm{supp}\rho_0(x) \cup \mathrm{supp}H_0(x) \subseteq B_{R'},
\end{aligned} \tag{2.1}
$$

where $B_R := \{x \in \mathbb{R}^N \| x| \leqslant R\}$ and $B_{R'} := \{x \in \mathbb{R}^N \| x| \leqslant R'\}$.

Denote by $x(t; \bar{x})$ the particle path starting from $\bar{x}$, i.e. $x(t; \bar{x})$ satisfies the following equation

$$
\begin{cases}
\dfrac{\mathrm{d}x}{\mathrm{d}t} = u(x, t), \\
x(0) = \bar{x},
\end{cases} \tag{2.2}
$$

here u is the velocity.

Denote by Ω_t the closed region that is the image of B_R under the flow map (2.2),

$$\Omega_t := \{(x,t)|x = x(t;\bar{x}), \bar{x} \in B_R\}. \tag{2.3}$$

Note that on the particle path $x = x(t;\bar{x})$, the density density ρ and magnetic H satisfy homogeneous ordinary differential equations. Hence, if ρ and H are zero at some position $\bar{x}$, then on the particle path starting from $\bar{x}$, ρ and H will be zero all the time. By (2.1), we get

$$\rho = H \equiv 0 \quad \text{in} \quad \Omega_t^c, \tag{2.4}$$

which together with the momentum equation of (1.1), one has

$$\mu\Delta u + (\mu + \lambda)\nabla \mathrm{div} u = 0 \quad \text{in} \quad \Omega_t^c. \tag{2.5}$$

Because $u = \dfrac{x}{|x|}\bar{u}(|x|,t)$, where $\bar{u}$ is a radially symmetric function, it follows from (2.5) that

$$\bar{u}_{rr} + (N-1)\left(\frac{\bar{u}}{r}\right)_r = 0 \quad \text{in} \quad \Omega_t^c.$$

From the condition $u(\cdot,t) \in W^{m,1}(\mathbb{R}^N)$, we know that

$$\bar{u}_r + (N-1)\frac{\bar{u}}{r} = 0 \quad \text{in} \quad \Omega_t^c.$$

One can easily compute that

$$\bar{u} \equiv C(t)r^{-(N-1)}, \quad \text{where } C(t) \text{ is a constant that only depends on } t,$$

is the general solution to this equation. Since $u \in C([0,T], L^1(\mathbb{R}^N))$, therefore, we obtain for a.e. $t \in [0,T]$,

$$\int_{\mathbb{R}^N} |u(x,t)|\mathrm{d}x = \int_{\omega_N}\int_0^{+\infty} |\bar{u}(x,t)|r^{N-1}\mathrm{d}r\mathrm{d}s < +\infty,$$

which shows $\bar{u}(r,t) \equiv 0$ in Ω_t^c. Thus $\Omega_t \equiv \Omega_0 \subseteq B_{R'}$.

Step 2 Derivation of the contraction inequality.

After multiplying the momentum equation of (1.1) by weight x, then integrating the resulting equation on the whole space $\mathbb{R}^N$, it holds that

$$\begin{aligned}
&\int_{\mathbb{R}^N} \rho u(x,t)\cdot x\mathrm{d}x - \int_{\mathbb{R}^N} \rho_0 u_0 \cdot x\mathrm{d}x \\
=&\int_0^t\int_{\mathbb{R}^N} (\rho|u|^2 - |H|^2)\mathrm{d}x\mathrm{d}\tau + \int_0^t\int_{\mathbb{R}^N}\left(Na\rho^\gamma + \frac{1}{2}N|H|^2\right)\mathrm{d}x\mathrm{d}\tau.
\end{aligned}$$

From the fact that $\operatorname{supp}\rho_0(x) \cup \operatorname{supp}H_0(x) \subseteq B_{R'}$, we then obtain

$$\begin{aligned}&\int_{B_{R'}} \rho u(x,t)\cdot x\mathrm{d}x - \int_{B_{R'}} \rho_0 u_0 \cdot x\mathrm{d}x\\&= \int_0^t \int_{B_{R'}} \left(\rho|u|^2 + \frac{N-2}{2}|H|^2\right) \mathrm{d}x\mathrm{d}\tau + \int_0^t \int_{B_{R'}} Na\rho^\gamma \mathrm{d}x\mathrm{d}\tau.\end{aligned}$$

Note that $\gamma > 1$, by Sobolev inequality we can get $L^\gamma(B_{R'}) \hookrightarrow L^1(B_{R'})$. Hence, it follows from $N \geqslant 2$ that

$$\begin{aligned}&\int_{B_{R'}} \rho u(x,t)\cdot x\mathrm{d}x - \int_{B_{R'}} \rho_0 u_0 \cdot x\mathrm{d}x\\&\geqslant \int_0^t \int_{B_{R'}} \rho|u|^2 \mathrm{d}x\mathrm{d}\tau + \int_0^t \int_{B_{R'}} Na\rho \mathrm{d}x\mathrm{d}\tau \geqslant Nam_0 t,\end{aligned} \tag{2.6}$$

where we use $\int_{\mathbb{R}^N} \rho \mathrm{d}x = \int_{\mathbb{R}^N} \rho_0 \mathrm{d}x := m_0$.

Next we show $\int_{B_{R'}} \rho u(x,t)\cdot x\mathrm{d}x - \int_{B_{R'}} \rho_0 u_0 \cdot x\mathrm{d}x$ is bounded, which will give the contradiction with the induction hypothesis that the MHD equations have a global solution. To get the boundness of the term $\int_{B_{R'}} \rho u(x,t)\cdot x\mathrm{d}x - \int_{B_{R'}} \rho_0 u_0 \cdot x\mathrm{d}x$, we need to prove the following energy conservation.

Step 3 Energy conservation.

After multiplying the momentum equation of (1.1) by u and using the equation of density, one gets

$$\begin{aligned}&\frac{\mathrm{d}}{\mathrm{d}t}\int_{\mathbb{R}^N} \left(\frac{1}{2}\rho|u|^2 + \frac{a}{\gamma-1}\rho^\gamma\right) \mathrm{d}x + \int_{\mathbb{R}^N} (\mu|\nabla u|^2 + (\lambda+\mu)|\mathrm{div}u|^2)\mathrm{d}x\\&= \int_{\mathbb{R}^N} \left(\mathrm{div}(H\otimes H) - \nabla\left(\frac{1}{2}|H|^2\right)\right)\cdot u\mathrm{d}x.\end{aligned} \tag{2.7}$$

The term on the right-hand side of (2.7) can be rewritten as

$$\int_{\mathbb{R}^N} \left(\mathrm{div}(H\otimes H) - \nabla\left(\frac{1}{2}|H|^2\right)\right)\cdot u\mathrm{d}x = -\int_{\mathbb{R}^N} \left(H^\top \nabla u H + \frac{1}{2}\nabla(|H|^2)\cdot u\right) \mathrm{d}x.$$

Hence, (2.7) becomes

$$\begin{aligned}&\frac{\mathrm{d}}{\mathrm{d}t}\int_{\mathbb{R}^N} \left(\frac{1}{2}\rho|u|^2 + \frac{a}{\gamma-1}\rho^\gamma\right) \mathrm{d}x + \int_{\mathbb{R}^N} (\mu|\nabla u|^2 + (\lambda+\mu)|\mathrm{div}u|^2)\mathrm{d}x\\&= -\int_{\mathbb{R}^N} \left(H^\top \nabla u H + \frac{1}{2}\nabla(|H|^2)\cdot u\right) \mathrm{d}x.\end{aligned} \tag{2.8}$$

Multiplying the third equation of (1.1) by H, integrating over $\mathbb{R}^N$ and using the condition $H \in \mathcal{C}^1([0,T], L^1(\mathbb{R}^N))$, one has

$$\frac{1}{2}\frac{\mathrm{d}}{\mathrm{d}t}\int_{\mathbb{R}^N}|H|^2\mathrm{d}x - \int_{\mathbb{R}^N}(\mathrm{curl}(u\times H))\cdot H\mathrm{d}x = 0. \tag{2.9}$$

A direct calculation shows that

$$\int_{\mathbb{R}^N}\mathrm{curl}(u\times H)\cdot H\mathrm{d}x = \int_{\mathbb{R}^N}\left(H^\top\nabla uH + \frac{1}{2}\nabla(|H|^2)\cdot u\right)\mathrm{d}x.$$

Thus, (2.9) implies that

$$\frac{1}{2}\frac{\mathrm{d}}{\mathrm{d}t}\int_{\mathbb{R}^N}|H|^2\mathrm{d}x = \int_{\mathbb{R}^N}\left(H^\top\nabla uH + \frac{1}{2}\nabla(|H|^2)\cdot u\right)\mathrm{d}x. \tag{2.10}$$

Combining (2.8) with (2.10), we get

$$\frac{\mathrm{d}}{\mathrm{d}t}\int_{\mathbb{R}^N}\left(\frac{1}{2}\rho|u|^2 + \frac{a}{\gamma-1}\rho^\gamma + \frac{1}{2}|H|^2\right)\mathrm{d}x + \int_{\mathbb{R}^N}(\mu|\nabla u|^2 + (\lambda+\mu)|\mathrm{div}u|^2)\mathrm{d}x = 0.$$

Integrating in time from 0 to T, we have

$$\begin{aligned}&\int_{\mathbb{R}^N}\left(\frac{1}{2}\rho|u|^2 + \frac{a}{\gamma-1}\rho^\gamma + \frac{1}{2}|H|^2\right)(\cdot,T)\mathrm{d}x + \int_0^T\int_{\mathbb{R}^N}(\mu|\nabla u|^2 + (\lambda+\mu)|\mathrm{div}u|^2)\mathrm{d}x\mathrm{d}\tau\\ &= \int_{\mathbb{R}^N}\left(\frac{1}{2}\rho_0|u_0|^2 + \frac{a}{\gamma-1}\rho_0^\gamma + \frac{1}{2}|H_0|^2\right)\mathrm{d}x.\end{aligned}$$

Subsequently, we obtain

$$\int_0^T\int_{B_{R'}}|\nabla u|^2\mathrm{d}x\mathrm{d}\tau \leqslant C. \tag{2.11}$$

Then using Hölder inequality, Newton-Leibniz formula, $N \leqslant 3$ and $|\nabla_x u|^2 = \frac{|\bar{u}|^2}{r^2} + |\partial_r u|^2$, it can be derived as

$$\begin{aligned}\left(\int_{B_{R'}}|\rho u\cdot x|\mathrm{d}x\right)^2 &\leqslant m_0^2\|u\cdot x\|_{L_x^\infty}^2 = m_0^2\|r\bar{u}(r,t)\|_{L_x^\infty}^2\\ &\leqslant \left(\int_0^{R'}(|\bar{u}| + |r\bar{u}_r|)\mathrm{d}r\right)^2\\ &\leqslant R'\int_0^{R'}(|\bar{u}(r,t)|^2 + |r\bar{u}_r|^2)\mathrm{d}r\end{aligned}$$

$$
\begin{aligned}
&\leqslant R'^{4-N}\left(\int_0^{R'}\frac{|\bar{u}(r,t)|^2}{r^2}\cdot r^{N-1}\mathrm{d}r+\int_0^{R'}|\bar{u}_r|^2\cdot r^{N-1}\mathrm{d}r\right)\\
&\leqslant \frac{R'^{4-N}}{N\alpha(N)}\int_{B_{R'}}|\nabla_x u|^2\mathrm{d}x\\
&\leqslant C\int_{B_{R'}}|\nabla u|^2\mathrm{d}x.
\end{aligned}
\tag{2.12}
$$

Step 4 Derivation of blow-up.

In view of (2.6), (2.11), (2.12) and Cauchy-Schwartz inequality, we get

$$
\begin{aligned}
\frac{N^2a^2m_0^2T^3}{3}&=\int_0^T N^2a^2m_0^2t^2\mathrm{d}t\\
&\leqslant\int_0^T\left(\int_{B_{R'}}\rho u\cdot x\mathrm{d}x-\int_{B_{R'}}\rho_0u_0\cdot x\mathrm{d}x\right)^2\mathrm{d}t\\
&\leqslant 2\int_0^T\left(\int_{B_{R'}}\rho u\cdot x\mathrm{d}x\right)^2\mathrm{d}t+2\int_0^T\left(\int_{B_{R'}}\rho_0u_0\cdot x\mathrm{d}x\right)^2\mathrm{d}t\\
&\leqslant C+CT.
\end{aligned}
$$

Hence, we obtain that the smooth solutions to (1.1) will blow up in finite time, which concludes the proof of Theorem1.1.

Acknowledgements The research was partially supported by the National Natural Science Foundation of China (No.11071057, No.11271052). The authors would like to thank Prof. C. Miao and Prof. Q. Chen for their value discussions.

References

[1] Bian D F and Guo B L. Well-posedness in critical spaces for the full compressible MHD equations, *Acta Math. Sci.*, in press. (2012)

[2] Bian D F and Yuan B Q. Local well-posedness in critical spaces for compressible MHD equations, (2010), 1–30, submitted.

[3] Bian D F and Yuan B Q. Well-posedness in super critical Besov spaces for compressible MHD equations, to appear in *Int. J. Dynamical System and Differential Equations*, 3(3)(2011), 383–399.

[4] Bian D F and Yuan B Q. Regularity of Weak Solutions to the Generalized Navier-Stokes Equations, *Acta Math. Sci.*, 31A(6)(2011), 1601–1609.

[5] Bian D F and Yuan B Q. Extension criterion for local solutions to the 3-D compressible Navier-Stokes Equations (in Chinese), (2011), 1–11, submitted.

[6] Bresch D, Desjardins B. Existence of global weak solutiona for a 2D viscous shallow water equations and convergence to the quasi-geostrophic model, *Commun. Math. Sci.*, 238(2003), 211–223.

[7] Bresch D, Desjardins B and Chi-Kun Lin, On some compressible fluid models: Korteweg, lubrication, and shallow water systems, *Comm. Partial Differential Equations*, 28(2003), 843–868.

[8] Bresch D, Desjardins B and Métivier G. Recent mathematical results and open problems about shallow water systems, Analysis and simulation of fluid dynamics, 15–31, *Adv. Math.Fluid Mech.*, Birkhäuser, Basel, 2007.

[9] Chen Q L, Miao C X, Zhang Z F. Well-posedness in critical spaces for compressible Navier-Stokes equations with density dependent viscosities, *Rev. Mat. Iberoamericana*, 26(2010), 915–946.

[10] Chen Q L, Miao C X, Zhang Z F. On the well-posedness for the viscous shallow water equations, *SIAM J. Math. Anal.*, 40(2008), 443–474.

[11] Cho Y, Jin B J. Blow-up of viscous heat-conducting compressible flows, *J. Math. Anal. Appl.*, 320(2)(2006), 819–826.

[12] Danchin R. Local theory in critical spaces for compressible viscous and heat-conductive gases, *Comm. Partial Differential Equations*, 26(2001), 1183–1233.

[13] Danchin R. Global existence in critical spaces for flows of compressible viscous and heat-conductive gases, *Arch. Rational Mech. Anal*, 160(2001), 1–39.

[14] Danchin R. Global existence in critical spaces for compressible Navier-Stokes equations, *Invent. Math. Anal.*, 141(2000), 579–614.

[15] Danchin R. On the uniqueness in critical spaces for compressible Navier-Stokes equations, *Nonlinear Differrential Equations Appl.*, 12(2005), 111–128.

[16] Danchin R. Well-posedness in critical spaces for barotropic viscous fluids with truly not constant density, *Comm. Partial Differential Equations*, 32(2007), 1373–1397.

[17] Danchin R. Density-dependent incompressible viscous fluids in critical spaces, *Proc. Roy. Soc. Edinburgh Sect.A*, 133(2003), 1311–1334.

[18] Du D P, Li J Y and Zhang K J. Blowup of smooth solutions to the Navier-Stokes equations for compressible isothermal fluids, (2011), 1–5, preprint.

[19] Fan J S, Jiang S. Blow-up criterion for the Navier-Stokes equations of compressible fluids, *J. Hyper. Diff. Equa.*, 5(1)(2008), 167–185.

[20] Fan J S, Jiang S, Ou Y B. A blow-up criterion for the compressible viscous heat-conductive flows, *Ann. I. H. Poincare*, 27(1)(2010), 337–350.

[21] Hoff D and Zumbrun K. Multi-dimensional diffusion waves for the Navier-Stokes equations of compressible flow, *Indiana Univ. Math. J.*, 44(1995), 603–676.

[22] Huang X D. Some results on blowup of solutions to the compressible Navier-Stokes equations, Ph. D Thesis. The Chinese University of Hong Kong, (2009).

[23] Huang X D, Li J. A blow-up criterion for the compressible Navier-Stokes equations in

the absence of vacuum, *Methods Appl. Anal.*, in press. (2010)

[24] Huang X D, Li J, Xin Z P. Blow-up criterion for the compressible flows with vacuum states, preprint, *Commun. Math. Phys.*, 301(2011), 23–35.

[25] Huang X D, Xin Z P. A blow-up criterion for classical solutions to the compressible Navier-Stokes equations, *Sci. in China*, 53(3)(2010), 671–686.

[26] Huang X D, Li J, Xin Z P. Serrin Type Criterion for the three-dimensional viscous compressible flows, *SIAM J. Math. Anal.*, 43(4)(2011), 1872–1886.

[27] Huang X D, Xin Z P. A blow-up criterion for the compressible Navier-Stokes equations, *arXiv*: 0902.2606v1.

[28] Itaya N. On initial value problem of the motion of compressible viscous fluid, especially on the problem of uniqueness, *J. Math. Kyoto Univ.*, 16(1976), 413–427.

[29] Kobayashi T and Shibata Y. Dacay estimates of solutions for the equations of motion of compressible viscous and heat-conductive gases in an exterior domain in $\mathbb{R}^3$, *Comm. Math. Phys.*, 200(1999), 621–660.

[30] Lion P L. Mathematics Topic in Fluid Mechanics, **vol. 2**, Oxford Lecture Series in Mathematics and its Applications, Clarendon Press, Oxford, (1998).

[31] Matsumura A and Nishida T. The initial value problem for the equations of motion of viscous and heat-conductive gases, *J. Math. Kyoto Univ.*, 20(1980), 67–104.

[32] Mellet A, Vasseur A. On the barotropic compressible Navier-Stokes equations, *Comm. Partial Differential Equations*, 32(2007), 431–452.

[33] Nash J. Le problème de Cauchy pour les équations différentielles d'un fluide général, *Bull. Soc. Math. France*, 90(1962), 487–497.

[34] Rozanova O. Blow up of smooth solutions to the compressible Navier-Stokes equations with the data highly decreasing at infinity, *J. Differential Equations*, 245(2008), 1762–1774.

[35] Serrin J. On the interior regularity of weak solutions of the Navier-stokes equations, *Arch. Ration. Mech. Anal.*, 9(1962), 187–195.

[36] Sun Y Z, Wang C, Zhang Z F. A Beale-Kato-Majda blow-up criterion for the 3-D compressible Navier-Stokes equations, *J. Math. Pures Appl.*, 95(1)(2011), 36–47.

[37] Beirào da Veiga H, A new regularity class for the Navier-Stokes equations in $\mathbb{R}^n$, *Chinese Ann. Math. Ser. B*, 16(1995), 407–412.

[38] Wang W K and Xu C J. The Cauchy problem for viscous shallow water equations, *Rev. Mat. Iberoamericana*, 21(2005), 1–24.

[39] Wu J H. Generalized MHD equations, *J. Differemtial Equations*, 195(2003), 284–312.

[40] Wu J H. Global regularity for a class of generalized magnetohydrodynamic equations, *J. Math. Fluid Mech.*, doi 10.1007/s00021-009-0017-y.

[41] Xin Z P. Blow up of smooth solutions to the compressible Navier-Stokes equation with compact density, *Commun. Pure Appl. Math.*, 51(1998), 229–240.

[42] Zhou Y. A new regularity criterion for weak solutions to the Navier-Stokes equations,

J. Math. Pures Appl., 84(2005), 1496–1514.

[43] Zhou Y. Milan Pokorný, On a regularity criterion for the Navier-Stokes equations involving gradient of one velocity component, *J. Math. Phys.*, 50, 123514 (2009).

[44] Zhou Y. Regularity criteria for the generalized viscous MHD equations, *Ann. I. H. Poincaré-AN*, 24(2007), 491–505.

[45] Zhou Y. A new regularity criterion for weak solutions to the viscous MHD equations in terms of the vorticity field, *Nonlinear Anal.*, 72(2010), 3643–3648.

Random Attractors of Stochastic Non-Newtonian Fluids on Unbounded Domain*

Guo Chunxiao (郭春晓), Guo Boling (郭柏灵) and Guo Yanfeng (郭艳凤)

Abstract We consider the stochastic non-Newtonian fluids defined on a two-dimensional Poincaré unbounded domain, and prove that it generates an asymptotically compact random dynamical system. Then, we establish the existence of random attractor for the corresponding random dynamical system. Random attractor is invariant and attracts every pullback tempered random set.

Keywords random attractor; non-Newtonian fluids; energy method; asymptotic compactness

1 Introduction

Let $D \subset \mathbf{R}^2$ be an open set, which is a channel, and suppose that D satisfies the Poincaré inequality, i.e., there exists a constant $\lambda_1 > 0$ such that

$$\lambda_1 \int_D \varphi^2 \mathrm{d}x \leqslant \int_D |\nabla \varphi|^2 \mathrm{d}x.$$

We consider the following stochastic incompressible non-Newtonian fluids in two-dimensional unbounded domain D,

$$\mathrm{d}u + \big(u \cdot \nabla u - \nabla \cdot \tau(e(u)) + \nabla \pi\big)\mathrm{d}t = f(x)\mathrm{d}t + \sum_{j=1}^{m} h_j(x)\, \mathrm{d}\omega_j(t), \quad x \in D,\ t > 0, \tag{1.1}$$

$$u(x, 0) = u_0(x), \quad x \in D, \tag{1.2}$$

$$\nabla \cdot u(x, t) = 0, \tag{1.3}$$

$$u = 0, \tau_{ijk} n_j n_k = 0, \ x \in \partial D. \tag{1.4}$$

* Stochastics and Dynamics, 2014,14(01):1350008. DOI:doi.org/10.1142/S0219493713500081.

The unknown vector function u denotes the velocity of the fluids, vector function $f(x)$ is the time-independent external body force, we assume $f(x) \in H$ (which will be specified in the following notation), and the scalar function π represents the pressure. Furthermore, due to the divergence free condition, we know

$$(\nabla\pi, u) = \int_D \nabla \cdot (u\pi)dx = 0.$$

The symmetric stress tensor of the fluid is denoted by $\tau_{ij}(e(u))$ which has the following form:

$$\tau_{ij}(e(u)) = 2\mu_0(\epsilon + |e(u)|^2)^{\frac{p-2}{2}} e_{ij}(u) - 2\mu_1\Delta e_{ij}(u), \quad \epsilon > 0, \quad i,j = 1,2, \tag{1.5}$$

$$e_{ij}(u) = \frac{1}{2}\left(\frac{\partial u_i}{\partial x_j} + \frac{\partial u_j}{\partial x_i}\right), \quad |e(u)|^2 = \sum_{i,j=1}^{2} |e_{ij}(u)|^2.$$

Obviously, the relation between $\tau_{ij}(e(u))$ and $e(u)$ is nonlinear, the fluid is called to be non-Newtonian in this case. For example, liquid foams and polymeric fluids tend to be non-Newtonian fluids. And $h_j(x)$ $(j = 1, 2, \cdots, m)$ are smooth enough functions defined on $\mathbf{R}^2$, $\{\omega_j(t)\}_{j=1}^m$ are independent two-sided real-valued Wiener processes adapted to a filtration $(\mathscr{F}_t)_{t\in\mathbf{R}}$ on a fixed probability space $(\Omega, \mathscr{F}, \mathbb{P})$. Define the time shift by

$$\theta_t\omega(\cdot) = \omega(\cdot + t) - \omega(t), \quad \omega \in \Omega, \, t \in \mathbf{R}.$$

Then $(\Omega, \mathscr{F}, \mathbb{P}, \theta_t)$ is a metric dynamical system.

For the convenience of the following contents, first, we set some notations.

- $L^p(D)$ - the Lebesgue space with norm $|\cdot|_{L^p}$, particularly, $|\cdot|_{L^2} = |\cdot|$.
- $H^\sigma(D)$ - the Sobolev space $\{u \in L^2(D),\ D^k u \in L^2(D),\ k \leqslant \sigma\}$, $|\cdot|_{H^\sigma} = |\cdot|_\sigma$.
- $\mathscr{C}(I, X)$ - the space of continuous functions from the interval I to X.
- Denote

$$\mathscr{V} = \{u \in C_0^\infty(D),\ \nabla \cdot u = 0,\ u = 0,\ x \in \partial D\},$$

- H=the closure of $\mathscr{V}$ in $L^2(D)$ with norm $|\cdot|$, $(\cdot,\cdot)$–the inner product in H.
- $H_0^1(D)$= the closure of $\mathscr{V}$ in $H^1(D)$ with norm $|\cdot|_1$.
- V= the closure of $\mathscr{V}$ in $H^2(D)$ with norm $|\cdot|_2$.

Obviously, when $p = 2$, $\mu_1 = 0$, and $h_j(x) = 0$, Eq. (1.1) is a deterministic equation and reduces into Navier-Stokes equation. $\mu_0 = \mu_1 = 0$, it is Euler equation, they both are Newtonian fluids. In this paper, we will concentrate our attention on the case $1 < p < 2$.

Before describing our work, we recall some facts about the theory on the deterministic non-Newtonian fluids. Many papers have been devoted to the well-posedness of solution, the existence of attractor and manifold for deterministic non-Newtonian fluids, see [4, 5, 6, 15, 21, 20]. In fact, the deterministic system model usually neglects the impact of many small perturbations, and stochastic equation can conform to physical phenomena better. Many authors have proposed the study of stochastic equations, and there has been much work related to stochastic equations and dynamical systems. For important equations, such as the Navier-Stokes equation, Burgers equation, Schrödinger equation etc., there have been much work and interesting results related to their existence, uniqueness and attractors, for these topics and the progresses in these fields, see [8, 9, 10, 11, 12, 13]. From then on, there are also a series of papers [14, 16, 19] to study stochastic non-Newtonian fluids. But there are little papers about stochastic non-Newtonian fluids in unbounded domain.

Random attractor is a compact invariant set depending on chance and moves with time, attracting any orbit starting from $-\infty$. The main result on random attractors relies heavily on the existence of a compact random attracting set, see Crauel *et al.* [9, 10]. However, because Sobolev embeddings are no longer compact in the case of unbounded domain, the unboundedness of domain induces a major difficulty for proving the existence of an attractor. In the deterministic case, this difficulty was solved by different methods, the authors [6] proved the existence of the global attractor by using weighted space method. Ball [1] overcame this difficulty by the energy equation approach. Recently, these methods have been also generalized to a stochastic framework. It is worth mentioning that, in the case of lattice systems defined on the entire integer set, the existence of a random attractor was proved in [2]. Some authors proved the existence of a random attractor by a tail estimate method, see Bates *et al.* [3], and Wang *et al.* [17, 18] for more details. In particular, in [7] Brzeźniak *et al.* proved the existence of random attractors to a 2D stochastic Navier-Stokes equations in an unbounded domain with additive white noise. The key of the approach in order to obtain asymptotic compactness of the RDS is energy type estimates. Along this clue, we want to know whether there exists random attractor for stochastic non-Newtonian fluid defined on unbounded domain. This is the main subject that we will develop in this work. In this paper, we use the ideas in [7] to show the existence of the random attractors in H. The new difficulty comes from the more complicated nonlinear terms in the equation due to the lack of compactness of Sobolev embeddings. First, we establish the existence of absorbing set by some uniform estimates of solution, then prove that the RDS is asymptotically compact by obtaining the convergence of these nonlinear terms. In fact, this is the main result of Theorem 4.1.

Equations (1.1)–(1.4) can be translated into the following abstract problems in H:

$$du+[\mu_1 Au+\mu_0 A_p u+B(u,u)]dt=f(x)dt+\sum_{j=1}^{m}h_j(x)\,d\omega_j(t)\,,\quad x\in D,\ t>s\,, \tag{1.6}$$

$$u(x,s)=u_s(x)\,,\ s\in\mathbf{R},\ \ x\in D\,, \tag{1.7}$$

where $A=P\Delta^2$, and P is the projection from $L^2(D)$ to H,

$$B(u,u)=u\cdot\nabla u,\quad (A_p u)_i=-\frac{\partial}{\partial x_j}\left([\epsilon+|e(u)|^2]^{\frac{p-2}{2}}e_{ij}(u)\right).$$

Define

$$(B(u,v),\omega)=b(u,v,\omega)=\int_D u_i\frac{\partial v_j}{\partial x_i}\omega_j dx\,,\quad u,v,\omega\in H^1(D)\,,$$

$$(A_p(u),\omega)=\int_D\gamma(u)e_{ij}(u)e_{ij}(\omega)dx,\quad u,\omega\in H^1(D)\,,$$

where $\gamma(u)=(\epsilon+|e(u)|^2)^{\frac{p-2}{2}}$.

The paper is organized as follows. In Section 2, we recall some definitions and already known results concerning random attractors; In Section 3, we prove the existence of continuous random dynamical system (RDS); In Section 4, we derive some uniform estimates on the solutions and prove the existence of a bounded random absorbing set; In Section 5, we show that the RDS is asymptotically compact, and prove the existence of random attractors.

2 Preliminaries

We introduce the linear operator A as follows: consider the positive definite V-elliptic symmetric bilinear form $a(\cdot,\cdot):V\times V\to R$ given by

$$a(u,v)=\int_D\Delta u\,\Delta v dx\,,\quad (u,\,v\in V).$$

As a consequence of the Lax-Milgram lemma, we obtain an isometry $A\in\mathscr{L}(V,V')$,

$$\langle Au,v\rangle_{V'\times V}=a(u,v)=\langle f,v\rangle_{V'\times V}\,,\quad\forall v\in V,$$

where V' is the dual space of V, and the domain of A is

$$D(A) = \{u \in V : a(u,v) = (f,v),\ f \in H \subset V',\ \forall v \in V\}.$$

In fact $A = P\Delta^2$, and P is the projection from $L^2(D)$ to H.

According to Rellich Theorem, A^{-1} is compact in H, then

$$A\phi_n = \lambda_n \phi_n\ ,\quad \phi_n \in D(A), \tag{2.1}$$

where $\{\phi_n\}_{n=1}^{\infty}$ are the eigenfunctions and also are basis of V, $\lambda_n > 0, \lambda_n \to \infty$, when $n \to \infty$.

We next recall some definitions and results concerning the random attractors for stochastic dynamical systems, which can be found in [9,10]. Let (X,d) be a separable metric space and let $(\Omega, \mathscr{F}, \mathbb{P})$ be a complete probability space.

Definition 2.1 *Let $\{\theta_t : \Omega \to \Omega,\ t \in \mathbf{R}\}$ be a family of measure preserving transformation of $(\Omega, \mathscr{F}, \mathbb{P})$ such that $\theta_0 = id_\Omega$ and $\theta_{t+s} = \theta_t \circ \theta_s$ for all $t, s \in \mathbf{R}$. Here we assume θ_t is ergodic under $\mathbb{P}$.*

Definition 2.2 *A continuous random dynamical system (RDS) on X over a metric dynamical system $(\Omega, \mathscr{F}, \mathbb{P}, (\theta_t)_{t\in\mathbf{R}})$ is a mapping*

$$\varphi : \mathbf{R}_+ \times \Omega \times X \ni (t,\omega,x) \mapsto \varphi(t,\omega,x) \in X$$

such that for all $\omega \in \Omega$, $s,t \in \mathbf{R}_+$, and $x \in X$,

(1) $\varphi(0,\omega,x) = x$;

(2) $\varphi(t+s;\omega)x = \varphi(t,\theta_s\omega) \circ \varphi(s,\omega)x$;

(3) $\varphi(t,\omega,\cdot) : X \to X$ *is continuous for all* $t \in \mathbf{R}_+$.

For the sake of simplicity of notation, RDS φ will be denoted as RDS φ.

Definition 2.3 *Define the random omega limit set of a bounded set B as*

$$\Omega(B,\omega) = \Omega_B(\omega) = \bigcap_{T\geqslant 0} \overline{\bigcup_{t\geqslant T} \varphi(t,\theta_{-t}\omega)B(\theta_{-t}\omega)}.$$

Definition 2.4 *A set $K(t,\omega) \subset X$ is said to be attracting, if for all $\omega \in \Omega$, bounded random sets $B(\omega) \subset X$,*

$$d(\varphi(t,\theta_{-t}\omega, B(\theta_{-t}\omega)), K(t,\omega)) \to 0, \quad t \to \infty,$$

where $d(A,B)$ is the semidistance defined by

$$d(A,B) = \sup_{x\in A} \inf_{y\in B} d(x,y).$$

Definition 2.5 *A set $K(t,\omega)\subset X$ is said to be absorbing, if for all $\omega\in\Omega$, bounded random sets $B(\omega)\subset X$,*

$$\varphi(t,\theta_{-t}\omega,B(\theta_{-t}\omega))\subset K(t,\omega)),\quad t\geqslant t_B(\omega).$$

Definition 2.6 *We say that a RDS φ defined on a separable Banach space X is asymptotically compact if for all $\omega\in\Omega$, for any sequence t_n, such that $t_n\to\infty$ and any bounded X-valued sequence $\{x_n\}_n$, the set $\{\varphi(t_n,\theta_{-t_n}\omega)x_n: n\in N\}$ is relatively compact in X.*

Definition 2.7 *A stochastic set $A(t,\omega)$ is said to be a random attractor, if the following conditions hold:*

(1) *It is the minimal closed set such that for $t\in\mathbf{R}$, $B\subset X$,*

$$d(\varphi(t,\theta_{-t}\omega,B(\theta_{-t}\omega)),A(t,\omega))\to 0,\quad t\to\infty,$$

which implies $A(t,\omega)$ attracts random set B.

(2) *$A(t,\omega)$ is the largest compact measurable set, which is $\varphi-$invariant in sense that*

$$\varphi(t,\theta_{-t}\omega)B(\theta_{-t}\omega)=B(\omega),\quad t\geqslant 0.$$

Theorem 2.1 *Let $(\Omega,\mathscr{F},\mathbb{P},\theta_t)$ be a metric Dynamical System, $\mathfrak{D}$ is a nonempty class of closed and bounded random sets on X, and φ a continuous RDS on X over $(\Omega,\mathscr{F},\mathbb{P},\theta_t)$. Suppose that $\{K(\omega)\}_{\omega\in\Omega}$ is a closed random absorbing set for φ in $\mathfrak{D}$ and φ is $\mathfrak{D}$-asymptotically compact in X. Then φ has a unique $\mathfrak{D}$-random attractor $\{A(\omega)\}_{\omega\in\Omega}$ which is given by*

$$A(\omega)=\bigcap_{T\geqslant 0}\overline{\bigcup_{t\geqslant T}\varphi(t,\theta_{-t}\omega)B(\theta_{-t}\omega)}.$$

3 Existence of continuous RDS

In this section, we show that there is a continuous random dynamical system (RDS) generated by the stochastic non-Newtonian fluid with additive noise. Thus, we give the Ornstein-Uhlenbeck process as follows.

For any $\alpha>0$, $z_j(t)=\displaystyle\int_{-\infty}^{t}\mathrm{e}^{-\alpha(t-s)}\mathrm{d}w_j(s)$, $(j=1,\cdots,m,)$ is an Ornstein-Uhlenbeck process satisfying the equation

$$\mathrm{d}z_j+\alpha z_j\mathrm{d}t=\mathrm{d}w_j(t),\tag{3.1}$$

where the trajectories of z_j are $\mathbb{P}$-a.e. continuous.

Let $Z(t) = \sum_{j=1}^{m} h_j(x) z_j(t)$, we have

$$\mathrm{d}Z + \alpha Z \mathrm{d}t = \sum_{j=1}^{m} h_j(x) \mathrm{d}w_j(t). \tag{3.2}$$

Next, we need to convert the stochastic equation with a random additive term into a deterministic equation with a random parameter. To study (1.6), it is usual to translate the unknown $v = u - Z$ (Z is a stationary process solving the problem (3.2)) and obtain the following equation for the new variable v, see [9, 10],

$$\begin{aligned} &\frac{\mathrm{d}v}{\mathrm{d}t} + \mu_1 A v + \mu_0 A_p(v+Z) + B(v+Z, v+Z) \\ &\quad = \alpha Z + f(x), \quad x \in D, \quad t > s, \end{aligned} \tag{3.3}$$

$$v(s,\omega) = v_s = u_s - Z(s,\omega), \quad x \in D, \quad s \in \mathbf{R}. \tag{3.4}$$

Our previous results yield the existence and the uniqueness of solutions to problem (3.3)–(3.4) in bounded domain, as well as its continuous dependence on the data. For stochastic dynamical system, our results are similar to the results of [8], we can prove that the following results hold for $\mathbb{P}$-a.e. $\omega \in \Omega$,

- for $f \in H$, $h_i \in H^2$, $(i = 1, 2, \cdots, m)$, $v_s \in H$, $s < T \in \mathbf{R}$, there exists a unique weak solution to (3.3)–(3.4) satisfying $v \in \mathscr{C}(s,T;H) \bigcap L^2(s,T;V)$ with $v(s) = v_s$.
- for $f \in H$, $h_i \in H^2$, $(i = 1, 2, \cdots, m)$, $v_s \in V$, $s < T \in \mathbf{R}$, there exists a unique weak solution to (3.3)–(3.4) satisfying $v \in \mathscr{C}(s,T;V) \bigcap L^2(s,T;D(A))$ with $v(s) = v_s$.

Moreover, for $u_0 \in H$, $\omega \in \Omega$, $t \geqslant s$, if we define

$$u(t,s;\omega)u_0 = \varphi(t-s;\theta_s\omega)u_0 = v(t,s;\omega,u_0 - Z(s,\omega)) + Z(t,\omega). \tag{3.5}$$

Then the process $u(t)$, $t \geqslant s$ is a solution to problem (1.1)–(1.4).

Definition 3.1 *We define a map* $\varphi : \mathbf{R} \times \Omega \times H \mapsto H$ *by*

$$(t,\omega,u_0) \mapsto v(t,s;\omega,u_0 - Z(s,\omega)) + Z(t,\omega). \tag{3.6}$$

Then φ satisfies conditions (1)–(3) of Definition 2.2. Therefore, φ is a continuous RDS on H. In the next two sections, we establish uniform estimates for the solutions and prove the existence of random attractors for RDS φ.

4 Energy estimates and existence of absorbing sets in H

In this section, we derive some uniform estimates on the solutions of (3.3)–(3.4) with the purpose of proving the existence of a bounded random absorbing set.

Lemma 4.1 *Suppose that $v(t,\omega;s,u_s-Z(s,\omega))$ is the solution of (3.3)–(3.4), with $h_i\in H^2$, $(i=1,2,...m)$, $f\in H$, $u_s\in H$. Then, for $t\geqslant s$, and $\mathbb{P}$-a.e. $\omega\in\Omega$,*

$$
\begin{aligned}
|v(t)|^2\leqslant&|v(s)|^2\exp\left(-\frac{\mu_1\lambda_1(t-s)}{4}+\frac{16C^2}{\mu_1\lambda_1}\int_s^t|Z(\sigma)|_1^2\mathrm{d}\sigma\right)\\
&+2\int_s^t g(\sigma)\exp\left(-\frac{\mu_1\lambda_1(t-\sigma)}{4}+\frac{16C^2}{\mu_1\lambda_1}\int_\sigma^t|Z(\tau)|_1^2\mathrm{d}\tau\right)\mathrm{d}\sigma, \qquad (4.1)
\end{aligned}
$$

where $g(t)=\dfrac{2\alpha^2|Z|^2}{\mu_1\lambda_1}+\dfrac{8C^2}{\mu_1\lambda_1}|Z|_1^2|Z|^2+\dfrac{\mu_1\lambda_1}{8}|Z|_1^2+\dfrac{\mu_0^2\epsilon^{p-2}}{\mu_1}|Z|_1^2+\dfrac{2|f|^2}{\mu_1\lambda_1}$.

And the following energy equality holds for $t\geqslant s$, and $\mathbb{P}$-a.e. $\omega\in\Omega$,

$$
\begin{aligned}
|v(t)|^2=&\,\mathrm{e}^{-\mu_1\lambda_1^2(t-s)}|v(s)|^2+2\int_s^t\mathrm{e}^{-\mu_1\lambda_1^2(t-\tau)}\left(-\left[\mu_1|\Delta v(\tau)|^2-\frac{\mu_1\lambda_1^2}{2}|v(\tau)|^2\right]\right)\mathrm{d}\tau\\
&-2\mu_0\int_s^t\mathrm{e}^{-\mu_1\lambda_1^2(t-\tau)}\int_D[\epsilon+|e(v+Z)|^2]^{\frac{p-2}{2}}e_{ij}(v+Z)e_{ij}(v)\mathrm{d}x\mathrm{d}\tau\\
&-2\int_s^t\mathrm{e}^{-\mu_1\lambda_1^2(t-\tau)}b(v+Z,v+Z,v)\mathrm{d}\tau\\
&+2\int_s^t\mathrm{e}^{-\mu_1\lambda_1^2(t-\tau)}(f(x)+\alpha Z,v(\tau))\mathrm{d}\tau. \qquad (4.2)
\end{aligned}
$$

Proof Taking the inner product of (3.3) with v in H,

$$
\begin{aligned}
&\frac{1}{2}\frac{\mathrm{d}}{\mathrm{d}t}|v|^2+\mu_1|\Delta v|^2+\mu_0\int_D[\epsilon+|e(v+Z)|^2]^{\frac{p-2}{2}}e_{ij}(v+Z)e_{ij}(v)\mathrm{d}x\\
&=(f(x),v)+\alpha(Z,v)-b(v+Z,v+Z,v), \qquad (4.3)
\end{aligned}
$$

$$
\begin{aligned}
\frac{1}{2}\frac{\mathrm{d}}{\mathrm{d}t}|v|^2+\mu_1|\Delta v|^2\leqslant&\,\mu_0\left|\int_D[\epsilon+|e(v+Z)|^2]^{\frac{p-2}{2}}e_{ij}(Z)e_{ij}(v)\mathrm{d}x\right|+\alpha|(Z,v)|\\
&+|(f(x),v)|+|b(v+Z,v+Z,v)|, \qquad (4.4)
\end{aligned}
$$

where the term $\displaystyle\int_D[\epsilon+|e(v+Z)|^2]^{\frac{p-2}{2}}e_{ij}(v)e_{ij}(v)dx$ is positive, so we drop it in the left-hand side of above inequality.

By Hölder inequality and ϵ-Young inequality, we can easily obtain

$$
|(f(x),v)|\leqslant|f(x)||v|\leqslant|f(x)||v|_1\leqslant\frac{\mu_1\lambda_1|v|_1^2}{8}+\frac{2|f(x)|^2}{\mu_1\lambda_1},
$$

$$\alpha|(Z,v)| \leqslant \alpha|Z||v| \leqslant \alpha|Z||v|_1 \leqslant \frac{\mu_1\lambda_1|v|_1^2}{8}+\frac{2\alpha^2|Z|^2}{\mu_1\lambda_1},$$

where λ_1 is the constant satisfying Poincaré inequality.

$$\begin{aligned}|b(v+Z,v+Z,v)| &= |b(v+Z,Z,v+Z)| \\ &\leqslant |v+Z|_{L^4}|v+Z|_{L^4}|Z|_1 \\ &\leqslant C|v+Z||v+Z|_1|Z|_1 \\ &\leqslant \frac{4C^2}{\mu_1\lambda_1}|Z|_1^2|v+Z|^2+\frac{\mu_1\lambda_1}{16}|v+Z|_1^2 \\ &\leqslant \frac{8C^2|Z|_1^2}{\mu_1\lambda_1}|v|^2+\frac{\mu_1\lambda_1}{8}|v|_1^2+\frac{8C^2}{\mu_1\lambda_1}|Z|_1^2|Z|^2+\frac{\mu_1\lambda_1}{8}|Z|_1^2,\end{aligned}$$

where the first equality is owing to the divergence free condition, the second inequality is owing to the Gagliardo-Nirenberg inequality

$$|u|_{L^4} \leqslant C|u|^{\frac{1}{2}}|\nabla u|^{\frac{1}{2}},$$

and the third inequality is ϵ-Young inequality.

Thanks to $1<p<2$,

$$\mu_0\left|\int_D[\epsilon+|e(v+Z)|^2]^{\frac{p-2}{2}}e_{ij}(Z)e_{ij}(v)\mathrm{d}x\right| \leqslant \mu_0\epsilon^{\frac{p-2}{2}}|v|_1|Z|_1 \leqslant \frac{\mu_1}{4}|v|_2^2+\frac{\mu_0^2\epsilon^{p-2}}{\mu_1}|Z|_1^2.$$

Putting above estimates together, we can obtain the following inequality:

$$\begin{aligned}&\frac{1}{2}\frac{\mathrm{d}}{\mathrm{d}t}|v|^2+\frac{\mu_1}{4}|v|_2^2+\left(\frac{\mu_1\lambda_1}{8}-\frac{8C^2|Z|_1^2}{\mu_1\lambda_1}\right)|v|^2 \\ &\leqslant \frac{2\alpha^2|Z|^2}{\mu_1\lambda_1}+\frac{8C^2}{\mu_1\lambda_1}|Z|_1^2|Z|^2+\frac{\mu_1\lambda_1}{8}|Z|_1^2+\frac{\mu_0^2\epsilon^{p-2}}{\mu_1}|Z|_1^2+\frac{2|f|^2}{\mu_1\lambda_1}.\end{aligned} \tag{4.5}$$

Let $g=\frac{2\alpha^2|Z|^2}{\mu_1\lambda_1}+\frac{8C^2}{\mu_1\lambda_1}|Z|_1^2|Z|^2+\frac{\mu_1\lambda_1}{8}|Z|_1^2+\frac{\mu_0^2\epsilon^{p-2}}{\mu_1}|Z|_1^2+\frac{2|f|^2}{\mu_1\lambda_1}$, it follows that

$$\frac{\mathrm{d}}{\mathrm{d}t}|v|^2+\frac{\mu_1}{2}|v|_2^2+\left(\frac{\mu_1\lambda_1}{4}-\frac{16C^2|Z|_1^2}{\mu_1\lambda_1}\right)|v|^2 \leqslant 2g. \tag{4.6}$$

By Gronwall inequality, for $s \leqslant -1$, and $t \in [-1,0]$, we deduce

$$\begin{aligned}|v(t)|^2 \leqslant & |v(s)|^2\exp\left(-\int_s^t\left(\frac{\mu_1\lambda_1}{4}-\frac{16C^2|Z(\sigma)|_1^2}{\mu_1\lambda_1}\right)\mathrm{d}\sigma\right) \\ &+2\int_s^t g(\sigma)\exp\left(-\int_\sigma^t\left(\frac{\mu_1\lambda_1}{4}-\frac{16C^2|Z(\tau)|_1^2}{\mu_1\lambda_1}\right)\mathrm{d}\tau\right)\mathrm{d}\sigma \\ = & |v(s)|^2\exp\left(-\frac{\mu_1\lambda_1(t-s)}{4}+\int_s^t\frac{16C^2|Z(\sigma)|_1^2}{\mu_1\lambda_1}\mathrm{d}\sigma\right) \\ &+2\int_s^t g(\sigma)\exp\left(-\frac{\mu_1\lambda_1(t-\sigma)}{4}+\int_\sigma^t\frac{16C^2|Z(\tau)|_1^2}{\mu_1\lambda_1}\mathrm{d}\tau\right)\mathrm{d}\sigma.\end{aligned}$$

In (4.3), by adding and subtracting $\frac{\mu_1\lambda_1^2|v(t)|^2}{2}$, then integrating the energy equation, it is immediate that for all $s \leqslant t$, and $v_s \in H$,

$$
\begin{aligned}
|v(t)|^2 = & \mathrm{e}^{-\mu_1\lambda_1^2(t-s)}|v(s)|^2 + 2\int_s^t \mathrm{e}^{-\mu_1\lambda_1^2(t-\tau)}\left(-\left[\mu_1|\Delta v(\tau)|^2 - \frac{\mu_1\lambda_1^2}{2}|v(\tau)|^2\right]\right)\mathrm{d}\tau \\
& - 2\mu_0\int_s^t \mathrm{e}^{-\mu_1\lambda_1^2(t-\tau)}\int_D [\epsilon + |e(v+Z)|^2]^{\frac{p-2}{2}} e_{ij}(v+Z)e_{ij}(v)\mathrm{d}x\mathrm{d}\tau \\
& - 2\int_s^t \mathrm{e}^{-\mu_1\lambda_1^2(t-\tau)} b(v+Z, v+Z, v)\mathrm{d}\tau \\
& + 2\int_s^t \mathrm{e}^{-\mu_1\lambda_1^2(t-\tau)}(f(x) + \alpha Z, v(\tau))\mathrm{d}\tau.
\end{aligned}
$$

Lemma 4.2 *Under the above assumptions in Lemma 4.1, for each ω,*

$$
\lim_{t\to-\infty} |Z(\omega)(t)|^2 \exp\left(\frac{\mu_1\lambda_1 t}{4} + \frac{16C^2}{\mu_1\lambda_1}\int_t^0 |Z(s)|_1^2\mathrm{d}s\right) = 0.
$$

Lemma 4.3 *Under the above assumptions in Lemma 4.1, for each ω,*

$$
\int_{-\infty}^0 \exp\left(\frac{\mu_1\lambda_1 t}{4} + \frac{16C^2}{\mu_1\lambda_1}\int_t^0 |Z(s)|_1^2\mathrm{d}s\right)\mathrm{d}t < \infty.
$$

Lemma 4.4 *Under the above assumptions in Lemma 4.1, for each ω,*

$$
\begin{aligned}
&\int_{-\infty}^0 (1 + |Z(t)|^2 + |Z(t)|_1^2 + |Z(t)|^2|Z(t)|_1^2) \\
&\quad \times \exp\left(\frac{\mu_1\lambda_1 t}{4} + \frac{16C^2}{\mu_1\lambda_1}\int_t^0 |Z(s)|_1^2\mathrm{d}s\right)\mathrm{d}t < \infty.
\end{aligned}
$$

We refer the readers to [7] for more details about these lemmas.

Remark 4.1 *From Lemma 4.3 and Lemma 4.4, we can deduce that*

$$
\int_{-\infty}^0 g(t)\exp\left(\frac{\mu_1\lambda_1 t}{4} + \frac{16C^2}{\mu_1\lambda_1}\int_t^0 |Z(s)|_1^2\mathrm{d}s\right)\mathrm{d}t < \infty.
$$

Before obtaining the existence of random absorbing set, we give the following definition, which is necessary to prove the existence of random attractor.

Definition 4.1 *A function $r: \Omega \to (0,\infty)$ belongs to the class $\mathfrak{R}$ if and only if*

$$
\limsup_{t\to\infty} r^2(\theta_{-t}\omega)\exp\left(-\frac{\mu_1\lambda_1 t}{4} + \frac{16C^2}{\mu_1\lambda_1}\int_{-t}^0 |Z(\sigma)|_1^2\mathrm{d}\sigma\right) = 0.
$$

From now on, we denote by $\mathfrak{QR}$ the class of all closed and bounded random sets Q on H such that the function $\Omega \ni \omega \mapsto r(Q(\omega))$ belongs to the class $\mathfrak{R}$. Next, we prove the existence of bounded absorbing set.

Let Q be a random set from the class $\mathfrak{QR}$, and let $r_Q(\omega)$ be the radius of $Q(\omega)$, $r_Q(\omega) := \sup\{|x|_H : x \in Q(\omega)\}$, $\omega \in \Omega$.

Proposition 4.1 *The following functions $r_i(\omega)$, $i = 1, 2, 3$ belong to class $\mathfrak{R}$.*

$$r_1^2(\omega) := |Z(0)|^2,$$

$$r_2^2(\omega) := \sup_{t \leqslant 0} |Z(\omega)(t)|^2 \exp\left(\frac{\mu_1\lambda_1 t}{4} + \frac{16C^2}{\mu_1\lambda_1}\int_t^0 |Z(s)|_1^2 ds\right),$$

$$r_3^2(\omega) := \int_{-\infty}^0 g(t) \exp\left(\frac{\mu_1\lambda_1 t}{4} + \frac{16C^2}{\mu_1\lambda_1}\int_t^0 |Z(s)|_1^2 ds\right) dt.$$

The next theorem is the main result of this paper.

Theorem 4.1 *Considering the RDS φ on H generated by the 2D stochastic non-Newtonian equation with additive noise, then the following properties hold:*

(1) *there exists a $\mathfrak{QR}$-absorbing set $B \in \mathfrak{QR}$,*

(2) *the random dynamical system is $\mathfrak{QR}$-asymptotically compact.*

Thus the family A of sets defined by $A(\omega) = \Omega_B(\omega)$ for all $\omega \in \Omega$, is the minimal $\mathfrak{QR}$-attractor.

Obviously, the $\mathfrak{D}$ class in Theorem 2.1 is taken as $\mathfrak{QR}$-the collection of all random subsets of H. If the conditions (1) and (2) are satisfied, we can apply Theorem 2.1 to obtain the conclusion of Theorem 4.1. The proof is divided into two main parts: the existence of absorbing set which is shown in the rest of this section, and the other part is the asymptotic compactness for RDS φ which will be dealt with in Sec. 5.

Lemma 4.5 *Assume that for each random set Q belonging to $\mathfrak{QR}$, there exists a random set $K(\omega)$ belonging to $\mathfrak{QR}$ such that $K(\omega)$ absorbs Q.*

Proof Let Q be a random set from the class $\mathfrak{QR}$. Let $r_Q(\omega)$ be the radius of $Q(\omega)$,

$$r_Q(\omega) := \sup\{|x| : x \in Q(\omega)\}, \quad \omega \in \Omega.$$

Let v be the solution of (3.3)–(3.4) on time interval $[s, \infty)$ with the initial condition $v(s) = u_s - Z(s)$. Applying (4.1) with $t = 0, s \leqslant 0$, then

$$|v(0)|^2 \leqslant 2(|v_s|^2+|Z(s)|^2)\exp\left(\frac{\mu_1\lambda_1 s}{4}+\frac{16C^2}{\mu_1\lambda_1}\int_s^0|Z(\sigma)|_1^2\mathrm{d}\sigma\right)$$
$$+2\int_s^0 g(\sigma)\exp\left(\frac{\mu_1\lambda_1\sigma}{4}+\frac{16C^2}{\mu_1\lambda_1}\int_\sigma^0|Z(\tau)|_1^2\mathrm{d}\tau\right)\mathrm{d}\sigma\,, \tag{4.7}$$

for $\omega\in\Omega$, set

$$r_{11}^2(\omega)\leqslant 2+2|Z(s)|^2\exp\left(\frac{\mu_1\lambda_1 s}{4}+\frac{16C^2}{\mu_1\lambda_1}\int_s^0|Z(\sigma)|_1^2\mathrm{d}\sigma\right)$$
$$+2\int_s^0 g(\sigma)\exp\left(\frac{\mu_1\lambda_1\sigma}{4}+\frac{16C^2}{\mu_1\lambda_1}\int_\sigma^0|Z(\tau)|_1^2\mathrm{d}\tau\right)\mathrm{d}\sigma\,, \tag{4.8}$$
$$r_{12}(\omega)=|Z(0)|.$$

Because both $r_{11}(\omega)$ and $r_{12}(\omega)$ belong to the class $\mathfrak{R}$, $r_{13}=r_{11}+r_{12}$ belongs to $\mathfrak{R}$ as well. Furthermore, the random set $K(\omega)$ defined by $K(\omega)=\{u\in H,\,|u|\leqslant r_{13}\}$ belongs to the family $\mathfrak{D}\mathfrak{R}$. Next, we need to prove that random set $K(\omega)$ absorbs random set $Q(\omega)$. Since $r_Q(\omega)\in\mathfrak{R}$, there exists $t_Q(\omega)\leqslant 0$, such that

$$r_Q^2(\theta_{-t}\omega)\exp\left(-\frac{\mu_1\lambda_1 t}{4}+\frac{16C^2}{\mu_1\lambda_1}\int_{-t}^0|Z(\sigma)|_1^2\right)\mathrm{d}\sigma\leqslant 1,\ \ for\ t\geqslant t_Q(\omega).$$

Hence from (4.7), we obtain $|v(0,s;\omega,u_s-Z(s))|\leqslant r_{11}(\omega)$. Furthermore,

$$|u(0,\omega;s,u_s)|\leqslant|v(0,s;\omega,u_s-Z(s))|+|Z(0)|\leqslant r_{11}(\omega)+r_{12}(\omega)\leqslant r_{13}(\omega).$$

In other words, $u(0,\omega;s,u_s)\in K(\omega)$, for all $s\geqslant t_Q(\omega)$. This Lemma gives the existence of random bounded absorbing sets $K(\omega)$.

Acknowledgement This paper is supported by NNSF of China under grant numbers 11126160, 11201475, 11101423 and 11061003; and also by the Fundamental Research Funds for the Central Universities No. 2010QS04 and the Guangxi Natural Science Foundation No. 2013GXNSFAA019001.

References

[1] Ball J M. Global attractors for damped semilinear wave equations [J]. Discrete Contin. Dyn. Syst., 2004, 10: 31–52.

[2] Bates P W, Lisei H, Lu K. Attractors for stochastic lattice dynamical systems [J]. Stochastic and Dynamics, 2006, 6:1–21.

[3] Bates P W, Lu K, Wang B. Random attractors for stochastic reaction-diffusion equations on unbounded domains [J]. J. Diff. Eq., 2009, 246: 845–869.

[4] Bellout H, Bloom F, Nečas J. Young measure-valued solutions for non-Newtonian incompressible viscous fluids [J]. Commun. PDE, 1994, 19: 1763–1803.

[5] Bloom F. Attractors of non-newtonian fluids [J]. J. Dyn. Diff. Eqs., 1995, 7: 109–140.

[6] Bloom F, Hao W. Regularization of a non-Newtonian system in an unbounded channel:existence of a maximal compact attractor [J]. Nonl. Anal. TMA, 2001, 43: 743–766.

[7] Brzeźniak Z, Caraballo T, Langa J A, Li Y, Lukaszewicz G, and Real J. Random attractors for stochastic 2D-Navier-Stokes equations in some unbounded domains [J]. J. Diff. Equ., 2013, 255: 3897–3919.

[8] Brzeźniak Z, Li Y. Asymptotic compactness and absorbing sets for 2D stochastic Navier-Stokes equations on some unbounded domains [J]. Trans. Amer. Math. Soc., 2006, 358: 5587–5629.

[9] Crauel H, Debussche A, Flandoli F. Random attractors [J]. J. Dyn. Diff. Equ., 1997, 9: 307–341.

[10] Crauel H, Flandoli F. Attractors for random dynamical systems [J]. Prob. Th. Rel. Fields, 1994, 100: 365-393.

[11] Da Prato G, Debussche A, Temam R. Stochastic Burgers' equation [J]. NoDEA, 1994, 1: 389–402.

[12] Da Prato G, Zabczyk J. Stochastic Equations in Infinite Dimensions [M]. Cambridge: Cambridge University Press, 1992.

[13] de Bouard A, Debussche A. The stochastic nonlinear Schrödinger equation in H^1 [J]. Stochastic Analysis and Applications, 2003, 21: 97–126.

[14] Guo B, Guo C, Zhang J. Martingale and stationary solutions for stochastic non-Newtonian fluids [J]. Differential and Integral Equations, 2010, 23: 303–326.

[15] Guo B, Lin G, Shang Y. Non-Newtonian Fluids Dynamical Systems [M]. Beijing, National Defense Industry Press, 2006.

[16] Guo C, Guo B, Han Y. Random attractors of Stochastic non-Newtonian fluid [J]. Acta Math. Appl. Sinica (English Series), 2012, 28: 165–180.

[17] Wang B. Attractors for reaction-diffusion equations in unbounded domains [J]. Phys. D, 1999, 128: 41–52.

[18] Wang B, Gao X. Random attractors for wave equations on unbounded domains, Disc. Cont. Dyn. Sys., 2009, 2009: 800–809.

[19] Zhao C, Li Y, Zhou S. Random attractor for a two-dimensional incompressible non-Newtonian fluid with multiplicative noise [J]. Acta Mathematica Scientia, 2011, 31: 567–575.

[20] Zhao C, Li Y. A note on the asymptotic smoothing effect of solutions to a non-Newtonian system in 2-D unbounded domains [J]. Nonlinear Analysis, 2005, 60: 475–483.

[21] Zhao C, Zhou S. Pullback attractors for a non-autonomous incompressible non-Newtonian fluid [J]. J. Diff. Equ., 2007, 238: 394–425.

Long Time Behavior of Solutions to Coupled Burgers-Complex Ginzburg-Landau (Burgers-CGL) Equations for Flames Governed by Sequential Reaction*

Guo Changhong (郭昌洪), Fang Shaomei (房少梅) and Guo Boling (郭柏灵)

Abstract This paper studies the existence and long time behavior of the solutions for the coupled Burgers-complex Ginzburg-Landau(Burgers-CGL) equations, which were derived from the nonlinear evolution of the coupled long-scale oscillatory and monotonic instabilities of a uniformly propagating combustion wave governed by a sequential chemical reaction, having two flame fronts corresponding to two reaction zones with a finite separation distance between them. We firstly show the existence of the global solutions for this coupled equations via subtle transforms, delicate a priori estimates and so-called continuity method, then prove the existence of the global attractor and establish the estimates of the upper bounds of Hausdorff and fractal dimensions for the attractor.

Keywords coupled Burgers-complex Ginzburg-Landau(Burgers-CGL) equations; global solution; global attractor; Hausdorff and fractal dimension

1 Introduction

As we know, the uniformly propagating planar premixed flame fronts can become unstable leading to kinds of flame structures, such as pulsating flames, spinning and spiral flames, flames with traveling or standing waves on the flame fronts[1]. Premixed gaseous flames generally involve many reactants and many reactions, and the

* Appl. Math. Mech. -Engl. Ed., 2014, 35(4): 515 – 534. DOI: 10.1007/s10483-014-1809-7.

simplest model is the case of a single limiting reactant and called one-stage chemical reaction. In premixed gaseous flames with a one-stage chemical reaction, there are two kinds of diffusional thermal instability, the first one is called monotonic instability, whose spatio-temporal evolution is governed by the Kuramoto–Sivashinsky (KS) equation[2], the other one is relative to oscillatory instability which occurs with a nonzero wavenumber and frequency at the instability threshold, and its nonlinear evolution is described by complex Ginzburg-Landau(CGL) equations[3].

However, the actual chemical kinetics governing the structure of flames can be quite complex and the effect of complex chemical kinetics on the dynamics of propagating flame fronts is governed by the two-stage sequential reaction, which leads to the occurrence of two flame fronts with various regimes of propagation[4,5]. One of the possible regimes of propagation is that of two planar uniformly propagating fronts with a constant separation distance between them. A linear stability analysis of this regime for the sequential reaction problem showed that, as in the case of a one-stage reaction, the flame fronts can become unstable with respect to either monotonic or oscillatory instabilities[6]. The nonlinear interaction between the monotonic and oscillatory modes of instability of the two uniformly propagating flame fronts can be governed by the following coupled KS-CGL equations[7]:

$$P_t + \nabla P \cdot \nabla Q = \xi P + (1+iu)\nabla^2 P - (1+iv)|P|^2 P - (r_1 \nabla^2 Q + r_2 g \nabla^4 Q)P, \quad (1.1)$$

$$Q_t = -m\nabla^2 Q - g\nabla^4 Q - \frac{1}{2}|\nabla Q|^2 - \omega |P|^2. \quad (1.2)$$

The system of coupled equations (1.1) and (1.2) describe the interaction of the evolving monotonic and oscillatory modes of instability of the uniform flame fronts. Specifically, when $m > 0$ and $\xi = 1$, both monotonic and oscillatory modes are excited. In the case of $m > 0$ and $\xi = -1$, the oscillatory mode is excited and the monotonic mode is damped. However when the monotonic mode is excited and the oscillatory mode is damped, there is $m < 0$, and $g = 0$, and the interaction between the excited oscillatory mode and the damped monotonic mode will be described by a complex Ginzburg-Landau equation for P coupled to a Burgers equation for Q, the system of coupled KS-CGL equations (1.1) and (1.2) are simplified to the following Burgers-CGL equations[7]:

$$P_t + \nabla P \cdot \nabla Q = \xi P + (1+iu)\nabla^2 P - (1+iv)|P|^2 P - r_1 \nabla^2 Q P, \quad (1.3)$$

$$Q_t = |m|\nabla^2 Q - \frac{1}{2}|\nabla Q|^2 - \omega |P|^2, \quad (1.4)$$

where $P(x,t)$ is the rescaled complex amplitude of the flame oscillations, $Q(x,t)$ is the deformation of the first front. u, v are the Landau coefficients, r_1 and ω are the coupling coefficients. The parameters r_1 is complex, $r_1 = r_{1r} + ir_{1i}$, and all other parameters are real.

If there were no coupling with the monotonic mode (terms with Q in (1.3)), the (1.3) would be the well-known complex Ginzburg-Landau equation that usually describes the weakly nonlinear evolution of a long-scale instability[8]. And if the coupling coefficient ω is zero in (1.4), then the equation (1.4) would be the well-known Burgers equation[9]. There are so many works concerning the CGL equation[10,11,12,13,14] and Burgers equation[15,16,17] physically and mathematically.

However, there is few work to consider the existence, uniqueness, and the long time behavior of the solutions for the coupled Burgers-CGL equations (1.3)(1.4) mathematically. In this paper, we will consider the following periodic initial value problem for the Burgers-CGL equations in one-dimensional version:

$$P_t = \xi P + (1+iu)P_{xx} - (1+iv)|P|^2P - P_xQ_x - r_1PQ_{xx} + f(x), (x,t) \in \Omega \times \mathbb{R}^+, \quad (1.5)$$

$$Q_t = |m|Q_{xx} - \frac{1}{2}|Q_x|^2 - \omega|P|^2 + g(x), \quad (x,t) \in \Omega \times \mathbb{R}^+, \quad (1.6)$$

$$P(x+2L,t) = P(x,t), \quad Q(x+2L,t) = Q(x,t), \quad (x,t) \in \Omega \times \mathbb{R}^+, \quad (1.7)$$

$$P(x,0) = P_0(x), \quad Q(x,0) = Q_0(x), \quad x \in \Omega. \quad (1.8)$$

where $\Omega = [-L, L]$ and $2L$ is the period. $\xi > 0, \omega > 0$, $f(x), g(x)$ are given real functions.

In what following, we will study the existence and uniqueness of the global smooth solutions for the periodic initial value problem (1.5)–(1.8), and then we will prove the existence of the global attractor and establish the estimates of the upper bounds of Hausdorff and fractal dimensions for the attractor.

The rest of paper is organized as follows. In Section , we briefly give some notations and preliminaries. In Section , we will establish a prior estimates for the solutions of the periodic initial value problem (1.5)–(1.8) and obtain the existence for the solutions. In Section , the global attractor $\mathcal{A} \subset H^2(\Omega) \times H^3(\Omega)$ for the periodic initial value problem are obtained. In Section , we establish the estimates of the upper bounds of Hausdorff and fractal dimensions for the attractor. In the last Section , we make some conclusions.

2 Notations and preliminaries

Throughout this paper, we denote the spaces of complex valued functions and real valued functions by the same symbols, and denote by $\|\cdot\|$ the norm of $L^2(\Omega)$ with the usual inner product $(\cdot,\cdot)$, $\|\cdot\|_{L^p}$ denotes the norm of $L^p(\Omega)$ for $1 \leqslant p \leqslant \infty (\|\cdot\| = \|\cdot\|_2)$. $W^{k,p}(\Omega)$ is the usual k-th order Sobolev space with its norm $\|u\|_{W^{k,p}(\Omega)}$. When $p=2$, we write $\|\cdot\|_{H^k} = \|\cdot\|_{W^{k,2}(\Omega)}$. Generally, $\|\cdot\|_X$ denotes the norm of any Banach space X. $L^\infty(0,T;X)$ denotes the space of functions $u(x,t)$ which belong to X as functions of x for every $t(0\leqslant t\leqslant T)$. Without any ambiguity, we denote a generic positive constant by C which may vary from line to line, and $C_i, E_i, k_i, \kappa_i, (i=1,2,\cdots)$ are positive constants depending on the known values.

In the following sections, we will use the following inequalities.

Lemma 2.1 (Gagliardo-Nirenberg inequality)[18] *Let Ω be a bounded domain with $\partial\Omega$ in C^m, and let u be any function in $W^{m,r}(\Omega)\cap L^q(\Omega)$, $1\leqslant q,r\leqslant\infty$. For any integer j, $0\leqslant j<m$, and for any number a in the interval $j/m\leqslant a\leqslant 1$, set*

$$\frac{1}{p} = \frac{j}{n} + a\left(\frac{1}{r} - \frac{m}{n}\right) + (1-a)\frac{1}{q}.$$

If $m-j-n/r$ is not a nonnegative integer, then

$$\|D^j u\|_{L^p} \leqslant C\|u\|_{W^{m,r}}^a \|u\|_{L^q}^{1-a}. \tag{2.9}$$

If $m-j-n/r$ is a nonnegative integer, then (2.9) holds for $a=j/m$. The constant C depends only on Ω, r, q, j, a.

Lemma 2.2 (Generalization of the Sobolev-Lieb-Thirring inequality)[19] *Assume that $\Omega\subset R^n$ is of class C^{2m} or that it enjoys the prolong property, and we denote by $|\Omega|$ the volume of Ω. Let $\varphi_j, 1\leqslant j\leqslant N$, be a finite family of $H^m(\Omega)$ which is orthonormal in $L^2(\Omega)$ and set, for almost every $x\in\Omega$,*

$$\rho(x) = \sum_{j=1}^N |\varphi_j(x)|^2.$$

Then for every p satisfying

$$\max\left(1, \frac{n}{2m}\right) < p \leqslant 1 + \frac{n}{2m},$$

there exists a constant κ_1 such that

$$\left(\int_\Omega \rho(x)^{p/(p-1)}\mathrm{d}x\right)^{2m(p-1)/n} \leqslant \frac{\kappa_1}{|\Omega|^{2m/n}}\int_\Omega \rho(x)\mathrm{d}x + \kappa_1\sum_{j=1}^N\int_\Omega |D^m\varphi_j|^2\mathrm{d}x.$$

The constant κ_1 depends on m, n, p; it also depends on the shape of Ω but not on its size. It is independent of the family φ_j and of N.

Especially, when $p = 1 + \dfrac{n}{2m}$ and $m = 1$, there is

$$\sum_{j=1}^{N} \|\nabla \varphi_j\|^2 \geqslant \kappa_2 |\Omega|^{-\frac{2}{n}} N^{\frac{n+2}{n}} - \kappa_3 |\Omega|^{-\frac{2}{n}} N, \qquad (\kappa_2 > 0,\ \kappa_3 > 0), \tag{2.10}$$

and when $m = 2$

$$\sum_{j=1}^{N} \|\Delta \varphi_j\|^2 \geqslant \kappa_2 |\Omega|^{-\frac{4}{n}} N^{\frac{n+4}{n}} - \kappa_3 |\Omega|^{-\frac{4}{n}} N. \qquad (\kappa_2 > 0,\ \kappa_3 > 0), \tag{2.11}$$

where κ_2, κ_3 are positive constants only depending on κ_1.

3 The existence of solutions

In this section, we will obtain the existence and uniqueness of the solutions for the periodic initial value problem (1.5)–(1.8). Firstly, we establish some a prior estimates for the solutions.

Lemma 3.1 *Assume that $P_0(x) \in L^2(\Omega), Q_0(x) \in H^1(\Omega),\ f(x) \in L^2(\Omega), g(x) \in L^2(\Omega)$, and suppose that $2r_{1r} - 1 > 0$. Then for the solutions of the problem (1.5)–(1.8), we have*

$$\|P\|^2 \leqslant E_0, \quad \|Q_x\|^2 \leqslant E_1, \quad \forall\, 0 \leqslant t \leqslant T, \tag{3.12}$$

where the constant E_0, E_1 depend on $\|P(0)\|$ and $\|Q_x(0)\|$.

Proof Firstly, we differentiate equation (1.6) with respect to x once and set

$$W = Q_x, \tag{3.13}$$

then the equations (1.5) and (1.6) are changed into

$$P_t = \xi P + (1 + iu) P_{xx} - (1 + iv)|P|^2 P - W P_x - r_1 W_x P + f, \tag{3.14}$$

$$W_t = |m| W_{xx} - W W_x - \omega(|P|^2)_x + g_x. \tag{3.15}$$

Then multiplying (3.14) by $\overline{P}$, integrating with respect to x over Ω and taking the real part, we can get

$$\begin{aligned} \frac{1}{2}\frac{\mathrm{d}}{\mathrm{d}t}\|P\|^2 = \xi \|P\|^2 - \|P_x\|^2 - \int_\Omega |P|^4 \mathrm{d}x - \mathrm{Re}\int_\Omega W P_x \overline{P} \mathrm{d}x \\ - \mathrm{Re}\int_\Omega r_1 W_x |P|^2 \mathrm{d}x + \mathrm{Re}\int_\Omega f \overline{P} \mathrm{d}x, \end{aligned} \tag{3.16}$$

where

$$-\text{Re}\int_\Omega WP_x\overline{P}\mathrm{d}x=-\text{Re}\left(\frac{1}{2}W|P|^2|_{-L}^{L}-\frac{1}{2}\int_\Omega|P|^2W_x\mathrm{d}x\right)=\frac{1}{2}\int_\Omega|P|^2W_x\mathrm{d}x.\quad(3.17)$$

From (3.16) and (3.17), there is

$$\frac{\mathrm{d}}{\mathrm{d}t}\|P\|^2=2\xi\|P\|^2-2\|P_x\|^2-2\int_\Omega|P|^4\mathrm{d}x+(1-2r_{1r})\int_\Omega|P|^2W_x\mathrm{d}x+2\text{Re}\int_\Omega f\overline{P}\mathrm{d}x.\quad(3.18)$$

On the other hand, multiplying the equation (3.15) by W and integrating over Ω, we have

$$\frac{1}{2}\frac{\mathrm{d}}{\mathrm{d}t}\|W\|^2=-|m|\|W_x\|^2-\int_\Omega W^2W_x\mathrm{d}x-\omega\int_\Omega\left(|P|^2\right)_xW\mathrm{d}x+\int_\Omega g_xW\mathrm{d}x,\quad(3.19)$$

where

$$-\int_\Omega W^2W_x\mathrm{d}x=-\frac{1}{3}W^3|_{-L}^{L}=0,\quad(3.20)$$

$$-\omega\int_\Omega\left(|P|^2\right)_xW\mathrm{d}x=-\omega(|P|^2W)|_{-L}^{L}+\omega\int_\Omega|P|^2W_x\mathrm{d}x=\omega\int_\Omega|P|^2W_x\mathrm{d}x.\quad(3.21)$$

Then from (3.19), (3.20) and (3.21), there holds

$$\frac{\mathrm{d}}{\mathrm{d}t}\|W\|^2=-2|m|\|W_x\|^2+2\omega\int_\Omega|P|^2W_x\mathrm{d}x+2\int_\Omega g_xW\mathrm{d}x.\quad(3.22)$$

Multiplying (3.18) by 2ω and (3.22) by $(2r_{1r}-1)$ respectively, and adding theses two equations together, we can get

$$\begin{aligned}\frac{\mathrm{d}}{\mathrm{d}t}\left(2\omega\|P\|^2+(2r_{1r}-1)\|W\|^2\right)=&\,4\omega\xi\|P\|^2-4\omega\|P_x\|^2-4\omega\int_\Omega|P|^4\mathrm{d}x\\&-2|m|(2r_{1r}-1)\|W_x\|^2+4\omega\text{Re}\int_\Omega f\overline{P}\mathrm{d}x\\&+2(2r_{1r}-1)\int_\Omega g_xW\mathrm{d}x.\end{aligned}\quad(3.23)$$

Noting the condition $2r_{1r}-1>0$, there holds

$$\begin{aligned}\left|4\omega\xi\|P\|^2+4\omega\text{Re}\int_\Omega f\overline{P}\mathrm{d}x\right|&\leqslant 4\omega\xi\|P\|^2+4\omega\|f\|\|P\|\\&\leqslant(4\omega\xi+2\omega)\|P\|^2+2\omega\|f\|^2\\&\leqslant 2\omega\int_\Omega|P|^4\mathrm{d}x+2\omega\|f\|^2+C(|\Omega|),\end{aligned}\quad(3.24)$$

and

$$\left|2(2r_{1r}-1)\int_\Omega g_xW\mathrm{d}x\right|\leqslant 2(2r_{1r}-1)\|g\|\|W_x\|\leqslant|m|(2r_{1r}-1)\|W_x\|^2+C\|g\|^2.\quad(3.25)$$

Combining (3.23)–(3.25) together yields

$$\frac{\mathrm{d}}{\mathrm{d}t}(2\omega\|P\|^2+(2r_{1r}-1)\|W\|^2)+2\omega\int_\Omega|P|^4\mathrm{d}x+|m|(2r_{1r}-1)\|W_x\|^2$$
$$\leqslant 2\omega\|f\|^2+C\|g\|^2+C(|\Omega|)=M_0. \tag{3.26}$$

From Hölder's inequality, there is $2\omega\|P\|^2\leqslant 2\omega\int_\Omega|P|^4\mathrm{d}x+C(|\Omega|)$. And since $\int_\Omega W\mathrm{d}x=\int_{-L}^{L}Q_x\mathrm{d}x=Q|_{-L}^{L}=0$, thus applying the Poincaré's inequality, there holds $\|W\|^2\leqslant C_W\|W_x\|^2$, where C_W is a positive constant. Therefore we have

$$\frac{\mathrm{d}}{\mathrm{d}t}(2\omega\|P\|^2+(2r_{1r}-1)\|W\|^2)+2\omega\|P\|^2+\frac{|m|}{C_W}(2r_{1r}-1)\|W\|^2\leqslant M_0, \tag{3.27}$$

which implies

$$\frac{\mathrm{d}}{\mathrm{d}t}(2\omega\|P\|^2+(2r_{1r}-1)\|W\|^2)+K_1(2\omega\|P\|^2+(2r_{1r}-1)\|W\|^2)\leqslant M_0. \tag{3.28}$$

where $K_1=\min(1,|m|/C_W)$.

By the Gronwall's inequality, we infer that

$$2\omega\|P\|^2+(2r_{1r}-1)\|W\|^2\leqslant(2\omega\|P(0)\|^2+(2r_{1r}-1)\|W(0)\|^2)e^{-K_1t}+\frac{M_0}{K_1}. \tag{3.29}$$

And therefore,

$$\|P\|^2\leqslant E_0,\quad \|W\|^2\leqslant E_1,\quad \forall\, 0\leqslant t\leqslant T. \tag{3.30}$$

Combining the transformation (3.13), the proof of Lemma 3.1 is completed.

Lemma 3.2 *Under the conditions of Lemma* 3.1, *and assume that* $P_0(x)\in H^1(\Omega)$, $Q_0(x)\in H^2(\Omega)$, $f(x)\in L^2(\Omega)$, $g(x)\in H^1(\Omega)$. *Then for the solutions of the problem* (1.5)–(1.8), *we have*

$$\|P_x\|^2+\|Q_{xx}\|^2\leqslant E_2,\quad \forall\, 0\leqslant t\leqslant T, \tag{3.31}$$

where the constant E_2 *depends on* $\|P_x(0)\|$ *and* $\|Q_{xx}(0)\|$.

Proof Similarly to the first step in the proving the Lemma 3.1, we can make use of the transformed equations. Firstly, multiplying (3.14) by $(-\overline{P}_{xx})$, integrating with respect to x over Ω and taking the real part, we can get

$$\frac{1}{2}\frac{\mathrm{d}}{\mathrm{d}t}\|P_x\|^2=\xi\|P_x\|^2-\|P_{xx}\|^2+\mathrm{Re}\int_\Omega(1+iv)|P|^2P\overline{P}_{xx}\mathrm{d}x$$
$$+\mathrm{Re}\int_\Omega WP_x\overline{P}_{xx}\mathrm{d}x+\mathrm{Re}\int_\Omega r_1W_xP\overline{P}_{xx}\mathrm{d}x-\mathrm{Re}\int_\Omega f\overline{P}_{xx}\mathrm{d}x, \tag{3.32}$$

where

$$\mathrm{Re}\int_\Omega r_1 W_x P\overline{P}_{xx}\mathrm{d}x = -\mathrm{Re}\int_\Omega r_1 W_{xx} P\overline{P_x}\mathrm{d}x - \mathrm{Re}\int_\Omega r_1 W_x |P_x|^2\mathrm{d}x. \tag{3.33}$$

Secondly, we take the inner product of (3.15) with $(-W_{xx})$ over Ω, then we can obtain

$$\frac{1}{2}\frac{\mathrm{d}}{\mathrm{d}t}\|W_x\|^2 = -|m|\|W_{xx}\|^2 + \int_\Omega W W_x W_{xx}\mathrm{d}x + \omega\int_\Omega \left(|P|^2\right)_x W_{xx}\mathrm{d}x - \int_\Omega g_x W_{xx}\mathrm{d}x. \tag{3.34}$$

Combining (3.32)–(3.34), there is

$$\begin{aligned}
&\frac{\mathrm{d}}{\mathrm{d}t}\left(\|P_x\|^2 + \|W_x\|\right)\\
&= 2\xi\|P_x\|^2 - 2\|P_{xx}\|^2 - 2|m|\|W_{xx}\|^2 + 2\mathrm{Re}\int_\Omega (1+iv)|P|^2 P\overline{P}_{xx}\mathrm{d}x\\
&\quad + 2\mathrm{Re}\int_\Omega W P_x\overline{P}_{xx}\mathrm{d}x - 2\mathrm{Re}\int_\Omega r_1 W_x|P_x|^2\mathrm{d}x - 2\mathrm{Re}\int_\Omega r_1 W_{xx} P\overline{P}_x\mathrm{d}x\\
&\quad + 2\omega\int_\Omega \left(|P|^2\right)_x W_{xx}\mathrm{d}x + 2\int_\Omega W W_x W_{xx}\mathrm{d}x\\
&\quad - 2\mathrm{Re}\int_\Omega f\overline{P}_{xx}\mathrm{d}x - 2\int_\Omega g_x W_{xx}\mathrm{d}x.
\end{aligned} \tag{3.35}$$

According to the Gagliardo-Nirenberg inequality and Lemma 3.1, we have

$$\|P_x\|^2 \leqslant C\|P_{xx}\| + C, \tag{3.36}$$

or

$$\|P_x\|^2 \leqslant \varepsilon_1\|P_{xx}\|^2 + C. \tag{3.37}$$

Similarly, we have

$$\|W_x\|^2 \leqslant C\|W_{xx}\| + C, \quad \text{or} \quad \|W_x\|^2 \leqslant \varepsilon_2\|W_{xx}\|^2 + C, \tag{3.38}$$

where ε_1 and ε_2 are positive constants. Thus we choose $\varepsilon_1 = 1/12\xi$ in (3.37), we can obtain

$$2\xi\|P_x\|^2 \leqslant \frac{1}{6}\|P_{xx}\|^2 + C. \tag{3.39}$$

Similarly, from the Gagliardo-Nirenberg inequality, the Lemma 3.1 and (3.37), there holds

$$\begin{aligned}\left|2\mathrm{Re}\int_\Omega(1+iv)|P|^2P\overline{P}_{xx}\mathrm{d}x\right| &\leqslant 2|1+iv|\|P\|_{L^8}^2\|P\|_{L^4}\|P_{xx}\|\\ &\leqslant C\|P\|_{H^2}^{\frac{3}{8}}\|P\|^{\frac{13}{8}}\|P\|_{H^2}^{\frac{1}{8}}\|P\|^{\frac{7}{8}}\|P_{xx}\|\\ &\leqslant C((\|P\|^2+\|P_x\|^2+\|P_{xx}\|^2)^{\frac{1}{2}})^{\frac{1}{2}}\|P_{xx}\|\\ &\leqslant \frac{1}{6}\|P_{xx}\|^2+C, \end{aligned}\tag{3.40}$$

and

$$\begin{aligned}\left|2\mathrm{Re}\int_\Omega WP_x\overline{P}_{xx}\mathrm{d}x\right| &\leqslant 2\|P_x\|_{L^\infty}\|W\|\|P_{xx}\|\\ &\leqslant C((\|P\|^2+\|P_x\|^2+\|P_{xx}\|^2)^{\frac{1}{2}})^{\frac{3}{4}}\|P\|^{\frac{1}{4}}\|P_{xx}\|\\ &\leqslant \frac{1}{6}\|P_{xx}\|^2+C. \end{aligned}\tag{3.41}$$

On the other hand, from (3.36), we can also have

$$\begin{aligned}\left|-2\mathrm{Re}\int_\Omega r_1W_x|P_x|^2\mathrm{d}x\right| &\leqslant 2|r_1|\|W_x\|_{L^\infty}\|P_x\|^2\\ &\leqslant C((\|W\|^2+\|W_x\|^2+\|W_{xx}\|^2)^{\frac{1}{2}})^{\frac{3}{4}}\|W\|^{\frac{1}{4}}(C\|P_{xx}\|+C)\\ &\leqslant \frac{|m|}{4}\|W_{xx}\|^2+\frac{1}{6}\|P_{xx}\|^2+C. \end{aligned}\tag{3.42}$$

and

$$\begin{aligned}&\left|-2\mathrm{Re}\int_\Omega r_1W_{xx}P\overline{P}_x\mathrm{d}x+2\omega\int_\Omega\left(|P|^2\right)_xW_{xx}\mathrm{d}x\right|\\ &\leqslant (2|r_1|+4\omega)\|P\|_{L^\infty}\|P_x\|\|W_{xx}\|\\ &\leqslant \frac{|m|}{4}\|W_{xx}\|^2+\frac{1}{6}\|P_{xx}\|^2+C. \end{aligned}\tag{3.43}$$

For the next term, we can compute that in the same way

$$\begin{aligned}\left|2\int_\Omega WW_xW_{xx}\mathrm{d}x\right| &\leqslant 2\|W_x\|_{L^\infty}\|W\|\|W_{xx}\|\\ &\leqslant C((\|W\|^2+\|W_x\|^2+\|W_{xx}\|^2)^{\frac{1}{2}})^{\frac{3}{4}}\|W\|^{\frac{1}{4}}\|W_{xx}\|\\ &\leqslant \frac{|m|}{4}\|W_{xx}\|^2+C. \end{aligned}\tag{3.44}$$

For the last two terms, we apply Hölder's inequality to obtain

$$\begin{aligned}&\left|-2\mathrm{Re}\int_\Omega f\overline{P}_{xx}\mathrm{d}x-2\int_\Omega g_xW_{xx}\mathrm{d}x\right|\\ &\leqslant 2\|f\|\|P_{xx}\|+2\|g_x\|\|W_{xx}\|\\ &\leqslant \frac{1}{6}\|P_{xx}\|^2+\frac{|m|}{4}\|W_{xx}\|^2+6\|f\|^2+\frac{4}{|m|}\|g_x\|^2. \end{aligned}\tag{3.45}$$

Substituting (3.39)–(3.45) into (3.35), there holds

$$\frac{\mathrm{d}}{\mathrm{d}t}\left(\|P_x\|^2+\|W_x\|^2\right)+\|P_{xx}\|^2+|m|\|W_{xx}\|^2\leqslant 6\|f\|^2+\frac{4}{|m|}\|g_x\|^2+C=M_1. \quad (3.46)$$

From (3.37) and (3.38) again, we find

$$\|P_x\|^2\leqslant\frac{1}{2}\|P_{xx}\|^2+C,(\varepsilon_1=1/2),\ \|W_x\|^2\leqslant\frac{|m|}{2}\|W_{xx}\|^2+C,(\varepsilon_2=|m|/2). \quad (3.47)$$

Thus combining the transformation (3.13), there holds

$$\frac{\mathrm{d}}{\mathrm{d}t}\left(\|P_x\|^2+\|Q_{xx}\|^2\right)+\left(\|P_x\|^2+\|Q_{xx}\|^2\right)+\frac{1}{2}\|P_{xx}\|^2+\frac{|m|}{2}\|Q_{xxx}\|^2\leqslant M_1. \quad (3.48)$$

By the Gronwall's inequality, we arrive at

$$\|P_x\|^2+\|Q_{xx}\|^2\leqslant E_2,\quad \forall\, 0\leqslant t\leqslant T, \quad (3.49)$$

where the constant E_2 depends on $\|P_x(0)\|$ and $\|Q_{xx}(0)\|$.

Furthermore, combining (3.48) and (3.49), we can also obtain that

$$\int_0^t\|P_{xx}\|^2\mathrm{d}s\leqslant C,\quad \int_0^t\|Q_{xxx}\|^2\mathrm{d}s\leqslant C, \quad (3.50)$$

where C is a positive constant. The proof of Lemma 3.2 is completed.

Corollary 3.1 *Under the conditions of Lemma* 3.1 *and Lemma* 3.2, *the solutions of the problem* (1.5)–(1.8) *satisfy*

$$\|P\|_{L^\infty}\leqslant C,\quad \|Q_x\|_{L^\infty}\leqslant C, \quad (3.51)$$

where C are positive constants.

Lemma 3.3 *Under the conditions of Lemma* 3.1 *and Lemma* 3.2, *then for the solutions of the problem* (1.5)–(1.8), *we have*

$$\|Q\|^2\leqslant C,\quad \|Q\|_{L^\infty}\leqslant C,\quad \forall\, 0\leqslant t\leqslant T,$$

where C are positive constants.

Proof Considering the equation (1.6), we multiply this equation by Q and integrate over Ω, then we have

$$\frac{1}{2}\frac{\mathrm{d}}{\mathrm{d}t}\|Q\|^2=-|m|\|Q_x\|^2-\frac{1}{2}\int_\Omega|Q_x|^2Q\mathrm{d}x-\omega\int_\Omega|P|^2Q\mathrm{d}x+\int_\Omega gQ\mathrm{d}x. \quad (3.52)$$

By the previous lemmas and the corollary (3.51), we can get

$$\left|-\frac{1}{2}\int_\Omega |Q_x|^2 Q \mathrm{d}x\right| \leqslant \frac{1}{2}\|Q_x\|_{L^\infty}\|Q_x\|\|Q\| \leqslant |m|\|Q_x\|^2 + C_1\|Q\|^2, \tag{3.53}$$

and

$$\left|-\omega\int_\Omega |P|^2 Q \mathrm{d}x + \int_\Omega gQ\mathrm{d}x\right| \leqslant (\omega\|P\|_{L^\infty}\|P\| + \|g\|)\|Q\| \leqslant C_2\|Q\|^2 + C. \tag{3.54}$$

Substituting (3.53) and (3.54) into (3.52), we find that

$$\frac{\mathrm{d}}{\mathrm{d}t}\|Q\|^2 \leqslant 2(C_1 + C_2)\|Q\|^2 + C. \tag{3.55}$$

By Gronwall's inequality and Agmon's inequality, we can complete the proof of Lemma 3.3.

Lemma 3.4 *Under the conditions of Lemma 3.2 , and assume that* $P_0(x) \in H^2(\Omega)$, $Q_0(x) \in H^3(\Omega)$, $f(x) \in H^1(\Omega)$, $g(x) \in H^2(\Omega)$. *Then we have the following estimates:*

$$\|P_{xx}\|^2 + \|Q_{xxx}\|^2 \leqslant E_3, \quad \forall\, 0 \leqslant t \leqslant T, \tag{3.56}$$

where the constant E_3 *depends on* $\|P_{xx}(0)\|$ *and* $\|Q_{xxx}(0)\|$.

Proof In the same way, taking inner product of (3.14) with $\overline{P}_{xxxx}$ and taking the real part. And multiplying (3.15) by W_{xxxx} and integrating with respect to x, then adding together yields

$$\begin{aligned}\frac{\mathrm{d}}{\mathrm{d}t}\left(\|P_{xx}\|^2 + \|W_{xx}\|^2\right) =\ & 2\xi\|P_{xx}\|^2 - 2\|P_{xxx}\|^2 - 2|m|\|W_{xxx}\|^2 \\ & - 2\mathrm{Re}\int_\Omega (1+iv)|P|^2 P\overline{P}_{xxxx}\mathrm{d}x \\ & - 2\mathrm{Re}\int_\Omega WP_x\overline{P}_{xxxx}\mathrm{d}x - 2\mathrm{Re}\int_\Omega r_1 W_x P\overline{P}_{xxxx}\mathrm{d}x \\ & - 2\int_\Omega WW_xW_{xxxx}\mathrm{d}x - 2\omega\int_\Omega (|P|^2)_x W_{xxxx}\mathrm{d}x \\ & + 2\mathrm{Re}\int_\Omega f\overline{P}_{xxxx}\mathrm{d}x + 2\int_\Omega g_x W_{xxxx}\mathrm{d}x.\end{aligned} \tag{3.57}$$

According to the Gagliardo-Nirenberg inequality, Lemma 3.1 and Lemma 3.2, we have

$$\|P_{xx}\|^2 \leqslant \varepsilon_3\|P_{xxx}\|^2 + C. \tag{3.58}$$

Similarly, we have

$$\|W_{xx}\|^2 \leqslant \varepsilon_4\|W_{xxx}\|^2 + C, \tag{3.59}$$

where ε_3 and ε_4 are positive constants. Thus we choose $\varepsilon_3 = 1/12\xi$ in (3.58), we can obtain

$$2\xi\|P_{xx}\|^2 \leqslant \frac{1}{6}\|P_{xxx}\|^2 + C. \tag{3.60}$$

According above lemmas and the Corollary 3.1, there is

$$\begin{aligned}\left|-2\mathrm{Re}\int_\Omega (1+iv)|P|^2P\overline{P}_{xxxx}\mathrm{d}x\right| &= \left|2\mathrm{Re}\int_\Omega (1+iv)(2PP_x\overline{P}+P^2\overline{P_x})\overline{P}_{xxx}\mathrm{d}x\right| \\ &\leqslant \frac{1}{6}\|P_{xxx}\|^2 + C(6|1+iv|)^2\|P\|_{L^\infty}^4\|P_x\|^2 \\ &\leqslant \frac{1}{6}\|P_{xxx}\|^2 + C. \end{aligned} \tag{3.61}$$

From the Lemma 3.2 and Corollary 3.1, we know that $\|W_x\| = \|Q_{xx}\| \leqslant C$ and $\|W\|_{L^\infty} = \|Q_x\|_{L^\infty} \leqslant C$, then according to the Gagliardo-Nirenberg inequality, we have

$$\begin{aligned}\left|-2\mathrm{Re}\int_\Omega WP_x\overline{P}_{xxxx}\mathrm{d}x\right| &= \left|2\mathrm{Re}\int_\Omega W_xP_x\overline{P}_{xxx}\mathrm{d}x + 2\mathrm{Re}\int_\Omega WP_{xx}\overline{P}_{xxx}\mathrm{d}x\right| \\ &\leqslant 2\|P_x\|_{L^\infty}\|W_x\|\|P_{xxx}\| + 2\|W\|_{L^\infty}\|P_{xx}\|\|P_{xxx}\| \\ &\leqslant C(C+\varepsilon_3^{\frac{3}{8}}\|P_{xxx}\|^{\frac{3}{4}})\|P_{xxx}\| + C(\varepsilon_3\|P_{xxx}\|^{\frac{1}{2}}+C)\|P_{xxx}\| \\ &\leqslant \frac{1}{6}\|P_{xxx}\|^2 + C, \end{aligned} \tag{3.62}$$

and from (3.59)

$$\begin{aligned}&\left|-2\mathrm{Re}\int_\Omega r_1W_xP\overline{P}_{xxxx}\mathrm{d}x\right| \\ &= \left|2\mathrm{Re}\int_\Omega r_1W_{xx}P\overline{P}_{xxx}\mathrm{d}x + 2\mathrm{Re}\int_\Omega r_1W_xP_x\overline{P}_{xxx}\mathrm{d}x\right| \\ &\leqslant 2|r_1|\|P\|_{L^\infty}\|W_{xx}\|\|P_{xxx}\| + 2|r_1|\|P_x\|_{L^\infty}\|W_x\|\|P_{xxx}\| \\ &\leqslant C\|W_{xx}\|\|P_{xxx}\| + C(\|P\|^{\frac{3}{4}} + \|P_x\|^{\frac{3}{4}} + \|P_{xx}\|^{\frac{3}{4}})\|P_{xxx}\| \\ &\leqslant C(\varepsilon_4^{\frac{1}{2}}\|W_{xxx}\|^{\frac{1}{2}}+C)\|P_{xxx}\| + C(C+\varepsilon_3^{\frac{3}{8}}\|P_{xxx}\|^{\frac{3}{4}}))\|P_{xxx}\| \\ &\leqslant \frac{1}{6}\|P_{xxx}\|^2 + \frac{|m|}{4}\|W_{xxx}\|^2 + C. \end{aligned} \tag{3.63}$$

Similarly to getting (3.63), we can also obtain that

$$
\begin{aligned}
&\left|-2\int_{\Omega} WW_xW_{xxxx}\mathrm{d}x\right| \\
=&\left|2\int_{\Omega} W_x^2W_{xxx}\mathrm{d}x+2\int_{\Omega} WW_{xx}W_{xxx}\mathrm{d}x\right| \\
\leqslant& 2\|W_x\|_{L^\infty}\|W_x\|\|W_{xxx}\|+2\|W\|_{L^\infty}\|W_{xx}\|\|W_{xxx}\| \\
\leqslant& C(C+\varepsilon_4^{\frac{3}{8}}\|W_{xxx}\|^{\frac{3}{4}})\|W_{xxx}\|+C(\varepsilon_4^{\frac{1}{2}}\|W_{xxx}\|^{\frac{1}{2}}+C)\|W_{xxx}\| \\
\leqslant& \frac{|m|}{4}\|W_{xxx}\|^2+C.
\end{aligned}
\tag{3.64}
$$

And for the next term, we have

$$
\begin{aligned}
&\left|-2\omega\int_{\Omega}(|P|^2)_xW_{xxxx}\mathrm{d}x\right| \\
=&\left|2\omega\int_{\Omega}(P_{xx}\overline{P}+2P_x\overline{P_x}+P\overline{P}_{xx})W_{xxx}\mathrm{d}x\right| \\
\leqslant& 4\omega\|P\|_{L^\infty}\|P_{xx}\|\|W_{xxx}\|+4\omega\|P_x\|_{L^\infty}\|P_x\|\|W_{xxx}\| \\
\leqslant& C(\varepsilon_3^{\frac{1}{2}}\|P_{xxx}\|^{\frac{1}{2}}+C)\|W_{xxx}\|+C(C+\varepsilon_3^{\frac{3}{8}}\|P_{xxx}\|^{\frac{3}{4}})\|W_{xxx}\| \\
\leqslant& \frac{1}{6}\|P_{xxx}\|^2+\frac{|m|}{4}\|W_{xxx}\|^2+C.
\end{aligned}
\tag{3.65}
$$

At last, we can compute the last two terms by Hölder's inequality as

$$
\begin{aligned}
&\left|-2\mathrm{Re}\int_{\Omega} f\overline{P}_{xxxx}\mathrm{d}x-2\int_{\Omega} g_xW_{xxxx}\mathrm{d}x\right| \\
\leqslant& 2\|f_x\|\|P_{xxx}\|+2\|g_{xx}\|\|W_{xxx}\| \\
\leqslant& \frac{1}{6}\|P_{xxx}\|^2+\frac{|m|}{4}\|W_{xxx}\|^2+6\|f_x\|^2+\frac{4}{|m|}\|g_{xx}\|^2.
\end{aligned}
\tag{3.66}
$$

Then combining (3.57)–(3.66), there holds

$$
\frac{\mathrm{d}}{\mathrm{d}t}\left(\|P_{xx}\|^2+\|W_{xx}\|^2\right)+\|P_{xxx}\|^2+|m|\|W_{xxx}\|^2\leqslant 6\|f_x\|^2+\frac{4}{|m|}\|g_{xx}\|^2+C=M_2.
\tag{3.67}
$$

From (3.58) and (3.59) again, we find

$$
\|P_{xx}\|^2\leqslant\frac{1}{2}\|P_{xxx}\|^2+C,\ (\varepsilon_3=1/2),\ \|W_{xx}\|^2\leqslant\frac{|m|}{2}\|W_{xxx}\|^2+C,\ (\varepsilon_4=|m|/2).
\tag{3.68}
$$

Thus by virtue of the transformation (3.13), there holds

$$
\frac{\mathrm{d}}{\mathrm{d}t}\left(\|P_{xx}\|^2+\|Q_{xxx}\|^2\right)+\left(\|P_{xx}\|^2+\|Q_{xxx}\|^2\right)+\frac{1}{2}\|P_{xxx}\|^2+\frac{|m|}{2}\|Q_{xxxx}\|^2\leqslant M_2.
\tag{3.69}
$$

According to the Gronwall's inequality, we can obtain

$$\|P_{xx}\|^2 + \|Q_{xxx}\|^2 \leqslant E_3, \quad \forall\, 0 \leqslant t \leqslant T, \tag{3.70}$$

where the constant E_3 depends on $\|P_{xx}(0)\|$ and $\|Q_{xxx}(0)\|$.

Furthermore, combining (3.69) and (3.70), we can also obtain that

$$\int_0^t \|P_{xxx}\|^2 \mathrm{d}s \leqslant C, \quad \int_0^t \|Q_{xxxx}\|^2 \mathrm{d}s \leqslant C, \tag{3.71}$$

where C is a positive constant. Thus, the proof of Lemma 3.4 is completed.

Corollary 3.2 *Under the conditions of Lemma* 3.1 *and Lemma* 3.4, *the solutions of the problem* (1.5)–(1.8) *satisfy*

$$\|P_x\|_{L^\infty} \leqslant C, \quad \|Q_{xx}\|_{L^\infty} \leqslant C,$$

where C are positive constants.

From the results of Lemma 3.1 – Lemma 3.4, we find that

$$\|P\|_{H^2}^2 \leqslant C, \quad \|Q\|_{H^3}^2 \leqslant C, \quad \forall\, 0 \leqslant t \leqslant T, \tag{3.72}$$

where C depend only on $||P_0||_{H^2}, ||Q_0||_{H^3}$ and T. And thus, by the standard methods we can extend the local solutions $P(x,t), Q(x,t)$ of the problem (1.5)–(1.8) to global solutions, that is, we have

Theorem 3.1 *Suppose that $P_0(x) \in H^2(\Omega)$, $Q_0(x) \in H^3(\Omega)$, $f(x) \in H^1(\Omega)$, $g(x) \in H^2(\Omega)$ and suppose that $2r_{1r} - 1 > 0$. Then the problem (1.5)–(1.8) have unique solutions $P(x,t), Q(x,t)$ satisfying*

$$P(x,t) \in L^\infty(0,T;H^2(\Omega)), \quad P_t(x,t) \in L^\infty(0,T;L^2(\Omega)),$$

$$Q(x,t) \in L^\infty(0,T;H^3(\Omega)), \quad Q_t(x,t) \in L^\infty(0,T;H^1(\Omega)).$$

4 Existence of the global attractor

In this section, we construct the global attractor for the problem (1.5)–(1.8). We first note that by Theorem 3.1, there exists a dynamical system $S(t)(t \geqslant 0)$ which maps $H^2(\Omega) \times H^3(\Omega)$ to $H^2(\Omega) \times H^3(\Omega)$ such that $S(t)(u_0, Q_0) = (P(t), Q(t))$, the solutions of problem (1.5)–(1.8).

In order to obtain the existence of the global attractor associated with semigroup $S(t)$, we need uniform a priori estimates in time. From (3.29), we see that when $\|P_0\| \leqslant R$ and $\|Q_0\|_{H^1} \leqslant R$, there is

$$\begin{aligned} \|P(t)\|^2 &\leqslant \frac{1}{2\omega}(2\omega\|P(0)\|^2 + (2r_{1r}-1)\|Q_x(0)\|^2)e^{-K_1 t} + \frac{M_0}{2\omega K_1} \\ &\leqslant \frac{R^2}{2\omega}(2\omega + 2r_{1r} - 1)e^{-K_1 t} + \frac{M_0}{2\omega K_1} \\ &\leqslant \frac{M_0}{\omega K_1}, \quad (\forall\, t \geqslant t_1), \end{aligned} \tag{4.73}$$

$$\begin{aligned} \|Q_x(t)\|^2 &\leqslant \frac{1}{2r_{1r}-1}(2\omega\|P(0)\|^2 + (2r_{1r}-1)\|Q_x(0)\|^2)e^{-K_1 t} + \frac{M_0}{(2r_{1r}-1)K_1} \\ &\leqslant \frac{R^2}{2r_{1r}-1}(2\omega + 2r_{1r} - 1)e^{-K_1 t} + \frac{M_0}{(2r_{1r}-1)K_1} \\ &\leqslant \frac{2M_0}{(2r_{1r}-1)K_1}, \quad (\forall\, t \geqslant t_1), \end{aligned} \tag{4.74}$$

where $t_1 = \dfrac{1}{K_1} \ln \dfrac{R^2 K_1(2\omega + 2r_{1r} - 1)}{M_0}$ depends on the data (f, g, R) when $\|P_0\| \leqslant R$, $\|Q_0\|_{H^1} \leqslant R$.

Applying (4.73), (4.74) and repeating the procedure of Lemma 3.2, we can derive that

$$\frac{\mathrm{d}}{\mathrm{d}t}(\|P_x\|^2 + \|Q_{xx}\|^2) + (\|P_x\|^2 + \|Q_{xx}\|^2) + \frac{1}{2}\|P_{xx}\|^2 + \frac{|m|}{2}\|Q_{xxx}\|^2 \leqslant C_1, \forall\, t \geqslant t_1, \tag{4.75}$$

where C_1 is a constant depending on the data (f, g) and the known parameters.

By the Gronwall's inequality and (4.75), we find that when $t \geqslant t_1$, there is

$$\begin{aligned} \|P_x(t)\|^2 + \|Q_{xx}(t)\|^2 &\leqslant (\|P_x(t_1)\|^2 + \|Q_{xx}(t_1)\|^2)e^{-(t-t_1)} + C_1 \\ &\leqslant E_2 e^{-(t-t_1)} + C_1 \quad \text{(By Lemma 3.2)} \\ &\leqslant 2C_1 \quad (\forall\, t \geqslant t_2), \end{aligned} \tag{4.76}$$

where $t_2 = \max\left(t_1, t_1 + \ln \dfrac{E_2}{C_1}\right)$ depends on the data (f, g, R) when $\|P_0\|_{H^1} \leqslant R$, $\|Q_0\|_{H^2} \leqslant R$.

Applying (4.73), (4.76) and repeating the procedure of the proof of Lemma 3.4, we obtain that

$$\begin{aligned} &\frac{\mathrm{d}}{\mathrm{d}t}(\|P_{xx}\|^2 + \|Q_{xxx}\|^2) + (\|P_{xx}\|^2 + \|Q_{xxx}\|^2) \\ &+ \frac{1}{2}\|P_{xxx}\|^2 + \frac{|m|}{2}\|Q_{xxxx}\|^2 \leqslant C_2, \forall t \geqslant t_2, \end{aligned} \tag{4.77}$$

where C_2 depends on the data (f,g) and the known parameters.

Similarly to (4.76), by the Gronwall's inequality and Lemma 3.4, we can derive that

$$\begin{aligned}\|P_{xx}(t)\|^2+\|Q_{xxx}(t)\|^2&\leqslant(\|P_{xx}(t_2)\|^2+\|Q_{xxx}(t_2)\|^2)e^{-(t-t_2)}+C_2\\&\leqslant E_3e^{-(t-t_2)}+C_2\\&\leqslant 2C_2\quad(\forall\, t\geqslant t_3),\end{aligned}\tag{4.78}$$

where $t_3=\max\left(t_2,t_2+\ln\dfrac{E_3}{C_2}\right)$ depends on the data (f,g,R) when $\|P_0\|_{H^2}\leqslant R$, $\|Q_0\|_{H^3}\leqslant R$.

We observe that (4.73), (4.74), (4.76) and (4.78) imply that

$$\|P(t)\|_{H^2}\leqslant K,\quad \|Q(t)\|_{H^3}\leqslant K,\quad \forall\, t\geqslant t_3,\tag{4.79}$$

where K ia s positive constant.

In what follows, we are going to show that the semigroup $S(t):H^2(\Omega)\times H^3(\Omega)\to H^2(\Omega)\times H^3(\Omega)$ is compact for large t. That is

Lemma 4.1 *Assume that the conditions of Theorem* 3.1 *hold, and suppose that* $f\in H^2(\Omega),g\in H^3(\Omega)$. *Then for the solutions of the problem* (1.5)–(1.8), *we have*

$$\|P_{xxx}\|^2\leqslant E_4\quad \|Q_{xxxx}\|^2\leqslant E_4,\quad \forall\, t\geqslant t_3,\tag{4.80}$$

where the constant E_4 *depends on* $\|f_{xx}\|$, $\|g_{xxx}\|$ *and* R *when* $\|P_0\|_{H^2}\leqslant R$, $\|Q_0\|_{H^3}\leqslant R$.

Proof Firstly by virtue of the transformed equations(3.14)(3.15), we take inner product of (3.14) with $-\overline{P}_{xxxxxx}$ over Ω, take the real part and we can obtain

$$\begin{aligned}\frac{1}{2}\frac{\mathrm{d}}{\mathrm{d}t}\|P_{xxx}\|^2=&\ \xi\|P_{xxx}\|^2-\|P_{xxxx}\|^2+\mathrm{Re}\int_\Omega(1+iv)|P|^2P\overline{P}_{xxxxxx}\mathrm{d}x\\&+\mathrm{Re}\int_\Omega WP_x\overline{P}_{xxxxxx}\mathrm{d}x+\mathrm{Re}\int_\Omega r_1W_xP\overline{P}_{xxxxxx}\mathrm{d}x\\&-\mathrm{Re}\int_\Omega f\overline{P}_{xxxxxx}\mathrm{d}x.\end{aligned}\tag{4.81}$$

Secondly, multiplying (3.15) by $-W_{xxxxxx}$ and integrating with respect to x over Ω, then we can get

$$\begin{aligned}\frac{1}{2}\frac{\mathrm{d}}{\mathrm{d}t}\|W_{xxx}\|^2=&-|m|\|W_{xxxx}\|^2+\int_\Omega WW_xW_{xxxxxx}\mathrm{d}x\\&+\omega\int_\Omega(|P|^2)_xW_{xxxxxx}\mathrm{d}x-\int_\Omega g_xW_{xxxxxx}\mathrm{d}x.\end{aligned}\tag{4.82}$$

Adding the (4.81) and (4.82) together yields

$$\frac{\mathrm{d}}{\mathrm{d}t}(\|P_{xxx}\|^2+\|W_{xxx}\|^2)=2\xi\|P_{xxx}\|^2-2\|P_{xxxx}\|^2-2|m|\|W_{xxxx}\|^2$$
$$+2\mathrm{Re}\int_\Omega(1+iv)|P|^2P\overline{P}_{xxxxxx}\mathrm{d}x+2\mathrm{Re}\int_\Omega WP_x\overline{P}_{xxxxxx}\mathrm{d}x$$
$$+2\mathrm{Re}\int_\Omega r_1W_xP\overline{P}_{xxxxxx}\mathrm{d}x+2\int_\Omega WW_xW_{xxxxxx}\mathrm{d}x$$
$$+2\omega\int_\Omega(|P|^2)_xW_{xxxxxx}\mathrm{d}x-2\mathrm{Re}\int_\Omega f\overline{P}_{xxxxxx}\mathrm{d}x-2\int_\Omega g_xW_{xxxxxx}\mathrm{d}x. \tag{4.83}$$

According to the previous lemmas and the corollaries, we have

$$\left|2\mathrm{Re}\int_\Omega(1+iv)|P|^2P\overline{P}_{xxxxxx}\mathrm{d}x\right|=\left|2\mathrm{Re}\int_\Omega(1+iv)(|P|^2P)_{xx}\overline{P_{xxxx}}\mathrm{d}x\right|$$
$$\leqslant\frac{1}{2}\|P_{xxxx}\|^2+C, \tag{4.84}$$

and

$$\left|2\mathrm{Re}\int_\Omega WP_x\overline{P}_{xxxxxx}\mathrm{d}x\right|=\left|2\mathrm{Re}\int_\Omega(WP_x)_{xx}\overline{P}_{xxxx}\mathrm{d}x\right|$$
$$\leqslant\frac{1}{2}\|P_{xxxx}\|^2+C_3\|P_{xxx}\|^2+C. \tag{4.85}$$

Similar to getting (4.85), we can obtain

$$\left|2\mathrm{Re}\int_\Omega r_1W_xP\overline{P}_{xxxxxx}\mathrm{d}x\right|\leqslant\frac{1}{2}\|P_{xxxx}\|^2+C_4\|W_{xxx}\|^2+C. \tag{4.86}$$

On the other hand, for the next two terms, we have

$$\left|2\int_\Omega WW_xW_{xxxxxx}\mathrm{d}x\right|=\left|2\int_\Omega(WW_x)_{xx}W_{xxxx}\mathrm{d}x\right|$$
$$\leqslant\frac{2|m|}{3}\|W_{xxxx}\|^2+C_5\|W_{xxx}\|^2+C, \tag{4.87}$$

and

$$\left|2\omega\int_\Omega(|P|^2)_xW_{xxxxxx}\mathrm{d}x\right|=\left|2\omega\int_\Omega(|P|^2)^{(n)}W_{xxxx}\mathrm{d}x\right|$$
$$\leqslant\frac{2|m|}{3}\|W_{xxxx}\|^2+C_6\|P_{xxx}\|^2+C. \tag{4.88}$$

Finally by the Hölder's inequality, we have

$$\begin{aligned}
&\left|-2\mathrm{Re}\int_\Omega f\overline{P}_{xxxxxx}\mathrm{d}x-2\int_\Omega g_x W_{xxxxxx}\mathrm{d}x\right|\\
=&\left|-2\mathrm{Re}\int_\Omega f_{xx}\overline{P}_{xxxxxx}\mathrm{d}x-2\int_\Omega g_{xxx} W_{xxxxxx}\mathrm{d}x\right|\\
\leqslant& 2\|f_{xx}\|\|P_{xxxx}\|+2\|g_{xxx}\|\|W_{xxxx}\|\\
\leqslant& \frac{1}{2}\|P_{xxxx}\|^2+\frac{2|m|}{3}\|W_{xxxx}\|^2+2\|f_{xx}\|^2+\frac{3}{4|m|}\|g_{xxx}\|^2.
\end{aligned}\tag{4.89}$$

Combining (4.86)–(4.89), we have

$$\begin{aligned}
\frac{\mathrm{d}}{\mathrm{d}t}(\|P_{xxx}\|^2+\|W_{xxx}\|^2)\leqslant&(2\xi+C_3+C_6)\|P_{xxx}\|^2+(C_4+C_5)\|W_{xxx}\|^2\\
&+2\|f_{xx}\|^2+\frac{3}{4|m|}\|g_{xxx}\|^2+C\\
\leqslant& C_7(\|P_{xxx}\|^2+\|W_{xxx}\|^2)+C_8,
\end{aligned}\tag{4.90}$$

where $C_7=\max(2\xi+C_3+C_6,C_4+C_5)$ is a constant.

Remembering the transformation (3.13), this implies that

$$\frac{\mathrm{d}}{\mathrm{d}t}(\|P_{xxx}\|^2+\|Q_{xxxx}\|^2)\leqslant C_7(\|P_{xxx}\|^2+\|Q_{xxxx}\|^2)+C_8.\tag{4.91}$$

On the other hand, integrating (4.77) between t and $t+1$, and applying (4.78), we find that

$$\int_t^{t+1}(\|P_{xxx}\|^2+\|Q_{xxxx}\|^2)dt\leqslant C_9,\quad \forall\, t\geqslant t_3.\tag{4.92}$$

And thus, by the uniform Gronwall lemma, it follows from (4.91) (4.92) that

$$\|P_{xxx}\|^2+\|Q_{xxxx}\|^2\leqslant E_4,\tag{4.93}$$

which concludes the proof of Lemma 4.1.

In order to prove the existence of global attractor of problem (1.5)–(1.8), we need the following result:

Theorem 4.1 [10] *We assume that H is a metric space and that the nonlinear operator $S(t)$ of H into itself for $t\geqslant 0$ satisfied*

$$S(t+s)=S(t)\cdot S(s),\quad \forall\, s,t\geqslant 0,\quad S(0)=I,\quad (\textit{Identity in } H).$$

And also $S(t)$ are continuous and uniformly compact for large t. That means for every bounded set B_0, there exists t_0, which may depend on B_0 that $\bigcup_{t\geqslant t_0} S(t)B_0$ is relatively *compact in H. We also assume that there exists an open set U and a bounded set B of U such that B is absorbing in U.*

Then the ω-limit set of B: $\mathcal{A} = \omega(B) = \bigcap_{s\geqslant 0}\overline{\bigcup_{t\geqslant s} S(t)B}$ is a compact attractor, which attracts the bounded set of U. It is the maximal bounded attractor in U.

Theorem 4.2 *Assume that the conditions of Theorem 3.1 hold, and suppose that $f \in H^2(\Omega), g \in H^3(\Omega)$. Then there exists a global attractor $\mathcal{A} \subset H^2(\Omega) \times H^3(\Omega)$ for the periodic initial problem* (1.5)–(1.8), *i.e., there is a set $\mathcal{A}$ such that*

(1) $S_t\mathcal{A} = \mathcal{A}$, $t \in R^+$;

(2) $\lim\limits_{t\to\infty} dist(S(t)B, \mathcal{A}) = 0$, *for any bounded set $B \subset H^2(\Omega) \times H^3(\Omega)$, where*

$$dist(X, Y) = \sup_{x\in X}\inf_{y\in Y} \|x - y\|_E,$$

and $S(t)(P_0, Q_0)$ is a semigroup operator generated by the problem (1.5)–(1.8).

Proof On account of the result of Theorem 4.1, we will prove this theorem by checking the conditions in the Theorem 4.1.

We observer that (4.79) shows that the ball

$$B = \{(P, Q) \in H^2(\Omega) \times H^3(\Omega) : \|P(t)\|_{H^2} \leqslant K,\ \|Q(t)\|_{H^3} \leqslant K\}$$

is an absorbing set of $S(t)$ in $H^2(\Omega) \times H^3(\Omega)$. In addition, Lemma 4.1 implies the dynamical system $S(t)$ is uniformly compact for large t. Thus, according to the Theorem 4.1, we can conclude that the ω-limit set of B: $\mathcal{A} = \omega(B) = \bigcap_{s\geqslant 0}\overline{\bigcup_{t\geqslant s} S(t)B}$ is a compact attractor on $H^2(\Omega) \times H^3(\Omega)$, where the closure is taken in $H^2(\Omega) \times H^3(\Omega)$.

This completes the proof of Theorem 4.2.

5 Dimension of the global attractor

In this section, we prove that the global attractor $\mathcal{A}$ as a compact subset of $H^2(\Omega) \times H^3(\Omega)$ has finite Hausdorff dimension and fractal dimension. To this end, we consider the first variation equation of (1.5)–(1.8):

$$\begin{aligned} U_t \ &= \xi U + (1 + iu)U_{xx} - (1 + iv)(2|P|^2U + P^2\overline{U}) - Q_xU_x - P_xV_x \\ &\quad -r_1Q_{xx}U - r_1PV_{xx}, \end{aligned} \tag{5.94}$$

$$V_t = |m|V_{xx} - Q_xV_x - \omega P\overline{U} - \omega\overline{P}U, \tag{5.95}$$

with the initial value

$$U(x, 0) = U_0(x), \quad V(x, 0) = V_0(x), \tag{5.96}$$

and the periodic boundary conditions

$$U(x+2L,t)=U(x,t),\quad V(x+2L,t)=V(x,t), \tag{5.97}$$

where $(U_0,V_0)\in H^2(\Omega)\times H^3(\Omega)$, $(P(t),Q(t))=S(t)(P_0,Q_0)$ is the solution of (1.5)–(1.8) with $(P_0,Q_0)\in\mathcal{A}$.

Given $(P_0,Q_0)\in\mathcal{A}$, and the solution $S(t)(P_0,Q_0)\in H^2(\Omega)\times H^3(\Omega)$, by standard methods we can easily prove that for any $(U_0,V_0)\in H^2(\Omega)\times H^3(\Omega)$, the linear initial boundary value problem (5.94)–(5.97) possesses a unique solution

$$U(x,t)\in L^2(0,T;H^2(\Omega)),\quad U_t(x,t)\in L^2(0,T;H^2(\Omega)),$$

$$V(x,t)\in L^2(0,T;H^3(\Omega)),\quad V_t(x,t)\in L^2(0,T;H^3(\Omega)),$$

and thus $(U(t),V(t))\in C(0,T;H^2(\Omega)\times H^3(\Omega))$.

Let $Y_0=(P_0,Q_0)$, $Z_0=(U_0,V_0)$, denote by $Y(t)=S(t)Y_0$, $Z(t)=(U(t),V(t))=DS(t)Z_0$ the solution of (5.94)–(5.97). We now show $S(t)$ is uniformly differentiable on the bounded set of $H^2(\Omega)\times H^3(\Omega)$. More precisely, we have

Lemma 5.1 *Let R and T be two positive numbers. Then there exists a constant C depending on R and T such that for every Y_0,Z_0,t satisfying $\|Y_0\|_{H^2\times H^3}\leqslant R$, $\|Y_0+Z_0\|_{H^2\times H^3}\leqslant R$, $0\leqslant t\leqslant T$, we have*

$$\|S(t)(Y_0+Z_0)-S(t)Y_0-DS(t)Z_0\|_{H^2\times H^3}\leqslant C\|Z_0\|_{H^2\times H^3}.$$

The proof of this result, which is lengthy but classical, is omitted here. For the techniques used, we refer readers to [19] and reference therein. Note that Lemma 5.1 shows that $S(t)$ is uniformly differentiable on $\mathcal{A}$.

Now we study the transformation of m-dimensional volumes in $H^2(\Omega)\times H^3(\Omega)$ by the operator $DS(t)Y_0, Y_0\in\mathcal{A}$. Denote by $Z_1(t),Z_2(t),\cdots,Z_N(t)$ the solution of linear equations (5.94)–(5.97) corresponding respective to initial data $Z_1(0)=\xi_1,Z_2(0)=\xi_2,\cdots,Z_N(0)=\xi_N$, here $\xi_j\in\overline{H}=L^2(\Omega)\times H^1(\Omega),j=1,2,\cdots,N$. By the simple computation, we can deduce that

$$\begin{aligned}|Z_1\wedge Z_2\wedge\cdots\wedge Z_N|_{\wedge^N\overline{H}}=&|\xi_1\wedge\xi_2\wedge\cdots\wedge\xi_N|_{\wedge^N\overline{H}}\\&\cdot\exp\left(\int_0^t \mathrm{Re}Tr(L(Y(t))\circ Q_N(\tau))d\tau\right),\end{aligned} \tag{5.98}$$

where $L(Y(t))=L(S(t)Y_0)$ is a linear map $L(t,Y_0)\xi=Z(t)$. $\wedge$ denotes the exterior product, Tr the trace of the operator, and $Q_N(t)$ the orthogonal projection of space

$\overline{H}$ to the spanning subspace generated by $Z_1(t), Z_2(t), \cdots, Z_N(t)$. Therefore, we can then estimate

$$\omega_N(L(t,Y_0)) = \sup_{\substack{\xi_i\in\overline{H}\\|\xi_i|\leqslant 1\\i=1,\cdots,N}} |Z_1\wedge Z_2\wedge\cdots\wedge Z_N|_{\wedge^N\overline{H}}, \tag{5.99}$$

$$\omega_N(L(t,Y_0)) \leqslant \sup_{\substack{\xi_i\in\overline{H}\\|\xi_i|\leqslant 1\\i=1,\cdots,N}} \exp\left(\int_0^t \mathrm{Re}Tr(L(t,Y_0)\circ Q_N(\tau))d\tau\right). \tag{5.100}$$

At a given time τ, let $\Phi_j = (\phi_j,\psi_j)^{\mathrm{T}}$ be orthonormal basis of $\overline{H}$ such that $Q_N(\tau)\overline{H} = \mathrm{Span}[Z_1(\tau), Z_2(\tau), \cdots, Z_N(\tau)]$. Thus we have

$$\begin{aligned}\mathrm{Re}Tr(L(t,Y_0)\circ Q_N(\tau)) &= \sum_{j=1}^{\infty}\mathrm{Re}(L(t,Y_0)\circ Q_N(\tau),\Phi_j(\tau),\Phi_j(\tau))\\ &= \sum_{j=1}^{N}\mathrm{Re}(L(t,Y_0)\Phi_j(\tau),\Phi_j(\tau))_{\overline{H}},\end{aligned} \tag{5.101}$$

where the scalar product $(Z,\Phi)_{\overline{H}} = \int_\Omega (U\phi_j + V_x\psi_{jx})dx$. Omitting temporarily the variable τ, we see that

$$\begin{aligned}&\mathrm{Re}(L(t,Y_0)\Phi_j(\tau),\Phi_j(\tau))_{\overline{H}}\\ =\;&\mathrm{Re}(\xi\phi_j + (1+iu)\phi_{jxx} - 2(1+iv)|P|^2\phi_j - (1+iv)P^2\overline{\phi}_j\\ &-Q_x\phi_{jx} - P_x\psi_{jx} - r_1Q_{xx}\phi_j - r_1P\psi_{jxx},\overline{\phi}_j)\\ &+\mathrm{Re}(|m|\psi_{jxxx} - Q_{xx}\psi_{jx} - Q_x\psi_{jxx} - \omega P_x\overline{\phi}_j\\ &-\omega P\overline{\phi}_{jx} - \omega\overline{P}_x\phi_j - \omega\overline{P}\phi_{jx},\psi_{jx})\\ =\;&\xi\|\phi_j\|^2 - \|\phi_{jx}\|^2 - |m|\|\psi_{jxx}\|^2\\ &-\mathrm{Re}\int_\Omega(1+iv)(2|P|^2\phi_j + P^2\overline{\phi}_j)\overline{\phi}_j\mathrm{d}x\\ &-\mathrm{Re}\int_\Omega Q_x\phi_{jx}\overline{\phi}_j\mathrm{d}x - \mathrm{Re}\int_\Omega P_x\psi_{jx}\overline{\phi}_j\mathrm{d}x\\ &-\mathrm{Re}\int_\Omega r_1Q_{xx}\phi_j\overline{\phi}_j\mathrm{d}x - \mathrm{Re}\int_\Omega r_1P\psi_{jxx}\overline{\phi}_j\mathrm{d}x\\ &-\mathrm{Re}\int_\Omega Q_{xx}\psi_{jx}\psi_{jx}\mathrm{d}x - \mathrm{Re}\int_\Omega Q_x\psi_{jxx}\psi_{jx}\mathrm{d}x\\ &-\omega\mathrm{Re}\int_\Omega P_x\overline{\phi}_j\psi_{jx}\mathrm{d}x - \omega\mathrm{Re}\int_\Omega P\overline{\phi}_{jx}\psi_{jx}\mathrm{d}x\\ &-\omega\mathrm{Re}\int_\Omega \overline{P}_x\phi_j\psi_{jx}\mathrm{d}x - \omega\mathrm{Re}\int_\Omega \overline{P}\phi_{jx}\psi_{jx}\mathrm{d}x.\end{aligned} \tag{5.102}$$

According the results in the previous section, we can estimate that

$$\left|-\mathrm{Re}\int_\Omega (1+iv)(2|P|^2\phi_j + P^2\overline{\phi}_j)\overline{\phi}_j dx\right| \leqslant 3|1+iv|\|P\|^2_{L^\infty}\|\phi_j\|^2 \leqslant k_1\|\phi_j\|^2, \quad (5.103)$$

$$\begin{aligned} &\left|-\mathrm{Re}\int_\Omega Q_x\phi_{jx}\overline{\phi}_j\mathrm{d}x - \mathrm{Re}\int_\Omega P_x\psi_{jx}\overline{\phi}_j\mathrm{d}x\right| \\ \leqslant\ & \|Q_x\|_{L^\infty}\|\phi_{jx}\|\|\phi_j\| + \|P_x\|_{L^\infty}\|\psi_{jx}\|\|\phi_j\| \\ \leqslant\ & \frac{1}{6}\|\phi_{jx}\|^2 + \|\psi_{jx}\|^2 + k_2\|\phi_j\|^2, \end{aligned} \quad (5.104)$$

and

$$\begin{aligned} &\left|-\mathrm{Re}\int_\Omega r_1Q_{xx}\phi_j\overline{\phi}_j\mathrm{d}x - \mathrm{Re}\int_\Omega r_1P\psi_{jxx}\overline{\phi}_j\mathrm{d}x\right| \\ \leqslant\ & |r_1|\|Q_{xx}\|_{L^\infty}\|\phi_j\|^2 + |r_1|\|P\|_{L^\infty}\|\psi_{jxx}\|\|\phi_j\| \\ \leqslant\ & \frac{|m|}{4}\|\psi_{jxx}\|^2 + k_3\|\phi_j\|^2. \end{aligned} \quad (5.105)$$

In the similar way, we can also obtain that

$$\begin{aligned} &\left|-\mathrm{Re}\int_\Omega Q_{xx}\psi_{jx}\psi_{jx}\mathrm{d}x - \mathrm{Re}\int_\Omega Q_x\psi_{jxx}\psi_{jx}\mathrm{d}x\right| \\ \leqslant\ & \|Q_{xx}\|_{L^\infty}\|\psi_{jx}\|^2 + \|Q_x\|_{L^\infty}\|\psi_{jxx}\|\|\psi_{jx}\| \\ \leqslant\ & \frac{|m|}{4}\|\psi_{jxx}\|^2 + k_4\|\psi_{jx}\|^2, \end{aligned} \quad (5.106)$$

and

$$\begin{aligned} &\left|-\omega\mathrm{Re}\int_\Omega P_x\overline{\phi}_j\psi_{jx}\mathrm{d}x - \omega\mathrm{Re}\int_\Omega P\overline{\phi}_{jx}\psi_{jx}\mathrm{d}x\right| \\ \leqslant\ & \omega\|P_x\|_{L^\infty}\|\phi_j\|\|\psi_{jx}\| + \omega\|P\|_{L^\infty}\|\phi_{jx}\|\|\psi_{jx}\| \\ \leqslant\ & \|\psi_{jx}\|^2 + k_5\|\phi_j\|^2 + \frac{1}{6}\|\phi_{jx}\|^2 + k_6\|\psi_{jx}\|^2 \\ \leqslant\ & \frac{1}{6}\|\phi_{jx}\|^2 + k_5\|\phi_j\|^2 + k_7\|\psi_{jx}\|^2. \end{aligned} \quad (5.107)$$

For the last two terms, there holds

$$\begin{aligned} &\left|-\omega\mathrm{Re}\int_\Omega \overline{P}_x\phi_j\psi_{jx}\mathrm{d}x - \omega\mathrm{Re}\int_\Omega \overline{P}\phi_{jx}\psi_{jx}\mathrm{d}x\right| \\ \leqslant\ & \omega\|P_x\|_{L^\infty}\|\phi_j\|\|\psi_{jx}\| + \omega\|P\|_{L^\infty}\|\phi_{jx}\|\|\psi_{jx}\| \\ \leqslant\ & \|\psi_{jx}\|^2 + k_8\|\phi_j\|^2 + \frac{1}{6}\|\phi_{jx}\|^2 + k_9\|\psi_{jx}\|^2 \\ \leqslant\ & \frac{1}{6}\|\phi_{jx}\|^2 + k_8\|\phi_j\|^2 + k_{10}\|\psi_{jx}\|^2. \end{aligned} \quad (5.108)$$

Combining (5.102)–(5.108) together yields that

$$\begin{aligned}\mathrm{Re}(L(t,Y_0)\Phi_j(\tau),\Phi_j(\tau))_{\overline{H}} \leqslant & -\frac{1}{2}\|\phi_{jx}\|^2-\frac{|m|}{2}\|\psi_{jxx}\|^2\\ &+(\xi+k_1+k_2+k_3+k_5+k_8)\|\phi_j\|^2,\\ &+(1+k_4+k_7+k_{10})\|\psi_{jx}\|^2\\ \leqslant & -\frac{|m|}{4}\|\psi_{jxx}\|^2-\frac{1}{2}\|\phi_{jx}\|^2+k_{11}\|\phi_j\|^2+k_{12}\|\psi_j\|^2,\end{aligned} \tag{5.109}$$

where k_{11}, k_{12} are two positive constants depending on the known values.

From (5.101) and (5.109), we know that

$$\begin{aligned}\mathrm{Re}Tr(L(t,Y_0)\circ Q_N(\tau)) \leqslant & -\frac{|m|}{4}\sum_{j=1}^N\|\psi_{jxx}\|^2-\frac{1}{2}\sum_{j=1}^N\|\phi_{jx}\|^2+k_{11}\sum_{j=1}^N\|\phi_j\|^2\\ &+k_{12}\sum_{j=1}^N\|\psi_j\|^2.\end{aligned} \tag{5.110}$$

From (2.10) and (2.11), the results of the generalization of the Sobolev–Lieb–Thirring inequality, we find that

$$\sum_{j=1}^N\|\phi_{jx}\|^2 \geqslant \frac{\kappa_2}{4L^2}N^3-\frac{\kappa_3}{4L^2}N, \quad \sum_{j=1}^N\|\psi_{jxx}\|^2 \geqslant \frac{\kappa_2}{16L^4}N^5-\frac{\kappa_3}{16L^4}N. \tag{5.111}$$

Substituting (5.111) into (5.110), we can get

$$\begin{aligned}&\mathrm{Re}Tr(L(t,Y_0)\circ Q_N(\tau))\\ \leqslant & -\frac{|m|}{4}(\frac{\kappa_2}{16L^4}N^5-\frac{\kappa_3}{16L^4}N)-\frac{1}{2}(\frac{\kappa_2}{4L^2}N^3-\frac{\kappa_3}{4L^2}N)+k_{11}N+k_{12}N\\ \leqslant & -\frac{\kappa_2|m|}{64L^4}N^5-\frac{\kappa_2}{8L^2}N^3+\left(\frac{\kappa_3|m|}{64L^4}+\frac{\kappa_3}{8L^2}+k_{11}+k_{12}\right)N\\ \leqslant & -\frac{\kappa_2|m|}{128L^4}N^5+\kappa_4,\end{aligned} \tag{5.112}$$

where we have handled the last two terms with the Young's inequality, and κ_4 is a positive constant.

Assuming that Y_0 belong to the global attractor $\mathcal{A}$, we can majorize the quantity

$q_N = \limsup_{t\to\infty} \sup_{\substack{\Phi_i \in H \\ |\Phi_i|\leqslant 1 \\ i=1,\cdots,N}} \left(\frac{1}{t}\int_0^t \mathrm{Re}Tr(L(t,Y_0)\circ Q_N(\tau))d\tau\right)$ as follows

$$q_N \leqslant -\frac{\kappa_2|m|}{128L^4}N^5 + \kappa_4. \tag{5.113}$$

We see that if N is defined by

$$N-1 < \left(\frac{256\kappa_4 L^4}{\kappa_2|m|}\right)^{\frac{1}{5}} \leqslant N, \tag{5.114}$$

then $q_N \leqslant -\kappa_4 < 0$.

By application of Proposition V.2.1 and Theorem V.3.3 in [19], we have thus proved the following theorem

Theorem 5.1 *Assume that the conditions of Theorem 4.2 hold, and and we consider the dynamical system associated with the periodic initial problem (1.5)–(1.8). We denote by $\mathcal{A}$ the corresponding global attractor whose existence is stated in Theorem 4.2, and let N be defined by (5.114).Then the Hausdorff dimension of $\mathcal{A}$ is less then or equal to N, and its fractal dimension is less then or equal to $2N$.*

6 Conclusions

The coupled Burgers-CGL equations were derived from the nonlinear evolution of the coupled long-scale oscillatory and monotonic instabilities of a uniformly propagating combustion wave governed by a sequential chemical reaction. Thus it is an interesting topic for mathematicians and other researchers. In this paper, we studied the periodic initial value problem (1.5)–(1.8) for these coupled equations mathematically. Firstly, the existence and uniqueness of the global smooth solutions are obtained by using some subtle transforms, delicate a priori estimates and so-called continuity method, which assure the well-posedness of the rescaled complex amplitude of the flame oscillations $P(x,t)$ and the deformation of the first front $Q(x,t)$ in the time $[0,T]$. Secondly, we consider the long time behavior for these solutions, which study the motions of $P(x,t)$ and $Q(x,t)$ as $t\to\infty$. We obtained the existence of the global attractor $\mathcal{A}\subset H^2(\Omega)\times H^3(\Omega)$ for problem (1.5)–(1.8) and established the estimates of the upper bounds of Hausdorff and fractal dimensions for the attractor. More precisely, the Hausdorff dimension of $\mathcal{A}$ is less then or equal to N, and its fractal dimension is less then or equal to $2N$, where N is defined in (5.114). For the physical significance, we know that the rescaled complex amplitude of the flame oscillations $P(x,t)$ and the deformation of the first front $Q(x,t)$ attract the bounded sets of $H^2(\Omega) \times H^3(\Omega)$ as $t \to \infty$ under the initial conditions $P_0(x)$ and $Q_0(x)$, and the dimensions of the bounded sets are bounded.

Acknowledgements This work was supported by the National Natural Science Foundation of China No.11271141.

References

[1] Williams F A. Combustion Theory [M]. Benjamin Cummings, Menlo Park, 1985.

[2] Sivashinsky G I. Nonlinear analysis of hydrodynamic instability in laminar flames I. Derivation of basic equations [J]. Acta Astro., 1977, 4(11-12): 1177–1206.

[3] Olagunju D O, Matkowsky B J. Coupled complex Ginzburg-Landau type equations in gaseous combustion [J]. Stability Appl. Ana. Continu. Media, 1992, 2: 31–58.

[4] Kapila A K, Ludford G S S. Two-step sequential reactions for large activation energies [J]. Combust. Flame, 1977, 29: 167–176.

[5] Margolis S B, Matkowsky B J. Steady and pulsating modes of sequential flame propagation [J]. Comb. Sci. Technol., 1982,27(5-6): 193–213.

[6] Peláez J. Stability of premixed flames with two thin reaction layers [J]. SIAM J. Appl. Math., 1987, 47(4): 781–799.

[7] Golovin A A, Matkowsky B J, Bayliss A, Nepomnyashchy A A. Coupled KS–CGL and coupled Burgers-CGL equations for flames governed by a sequential reaction [J]. Phys. D, 1999, 129: 253–298.

[8] Nepomnyashchy A A. Order parameter equations for long wavelength instabilities [J]. Phys. D, 1995, 86: 90–95.

[9] Burgers, J. M. A mathematical model illustrating the theory of turbulence. *Adv. Appl. Mech.*, 1948, 1: 171–199.

[10] Ghidaglia J M, Héron B. Dimension of the attractor associated to the Ginzburg-Landau equation [J]. Phys. D, 1987, 28(3): 282–304.

[11] Doering C R, Gibbon J D, Holm D D, Nicolaenko B. Low-dimensional behavior in the complex Ginzburg-Landau equation [J]. Nonlinearity, 1988, 1(2): 279–309.

[12] Yang L, Guo B, Xu H. Inhomogeneous initial boundary value problem for Ginzburg-Landau equations [J]. Appl. Math. Mech. -Engl. Ed., 2004, 25(4): 373–380.

[13] Li D, Guo B. Asymptotic behavior of 2D generalized stochastic Ginzburg-Landau equation with additive noise [J]. Appl. Math. Mech. -Engl. Ed., 2009, 30(8): 945–956.

[14] Li D, Guo B, Liu X. Regularity of the attractor for 3-D complex Ginzburg-Landau equation [J]. Acta Math. Appl. Sin. Engl. Ser., 2011, 27(2): 289–302.

[15] Hopf E. The partial differential equation $u_t + uu_x = \mu u_{xx}$ [J]. Comm. Pure Appl. Math., 1950, 3: 201–230.

[16] Zhu C, Wang R. Numerical solution of Burgers equation by cubic B-spline quasi-interpolation [J]. Appl. Math. Comput., 2009, 208: 260–272.

[17] Chidella S R, Yadav M K. Large time asymptotics for solutions to a nonhomogeneous Burgers equation [J]. Appl. Math. Mech. -Engl. Ed., 2010, 31(9): 1189–1196.

[18] Friedman A. Partial differential equations [M]. Holt, Reinhart and Winston, New York, 1969.

[19] Temam R. Infinite dimensional dynamical systems in mechanics and physics [M]. Springer-Verlag, New York. 1997.

High-order Rogue Waves in a Vector Nonlinear Schrödinger Equations*

Ling Liming (凌黎明), Guo Boling (郭柏灵) and Zhao Lichen (赵立臣)

Abstract We study the dynamics of high-order rogue waves (RWs) in two-component coupled nonlinear Schrödinger equations. We find that four fundamental rogue waves can emerge from second-order vector RWs in the coupled system, in contrast to the high-order ones in single-component systems. The distribution shape can be quadrilateral, triangle, and line structures by varying the proper initial excitations given by the exact analytical solutions. The distribution pattern for vector RWs is more abundant than that for scalar rogue waves. Possibilities to observe these new patterns for rogue waves are discussed for a nonlinear fiber.

Introduction Rogue wave (RW) is the name given by oceanographers to isolated large amplitude wave, which occurs more frequently than expected for normal, Gaussian distributed, statistical events [1−3]. It depicts a unique event that seems to appear from nowhere and disappear without a trace, and can appear in a variety of different contexts [4−7]. RW has been observed experimentally in nonlinear optics [8,9], water wave tank [10], and even in plasma system [11]. These experimental studies suggest that the rational solutions of related dynamics equations can be used to describe these RW phenomena [12,13]. Moreover, there are many different pattern structures for high-order RW [14−16], which can be understood as a nonlinear superposition of fundamental RW (the first order RW). Recently, many efforts were devoted to classify the hierarchy for each order RW [17,18], since the superpositions are nontrivial and admit only a fixed number of elementary RWs in each high order solution.

Recent studies were extended to RWs in multi-component coupled systems, since complex systems usually involve more than one component [19−21]. For the coupled system, the usual coupled effects are cross-phase modulation. The linear stability

* Phys. Rev. E, 2014, 89(4): 041201(R). Doi: 10.1103/PhysRevE.89.041201.

analysis on the coupled system indicates that the cross-phase modulation term can vary the instability regime characters [22−24]. Moreover, for scalar system, the velocity of the background field has no real effects on the pattern structure for RW, since the corresponding solutions can be correlated through Galileo transformation. But for coupled system, the relative velocity between different component fields has real physical effects, and can not be erased by any trivial transformation. Therefore, the extended studies on vector ones are nontrivial and meaningful. Recently, some novel patterns for RW were presented in the coupled systems, such as dark RWs [25,26], the interaction between RWs and other nonlinear waves [26−28], four-petaled flower structure [29,30] and so on. These studies indicate that there are much abundant pattern dynamics for RW in the multi-component coupled systems, which are quite distinctive from the ones in scalar systems.

Very recently, high-order RW [31,32] were excited successfully in a water wave tank. This suggests that high-order analytic RW solution is meaningful physically and can be realized experimentally [33]. However, as far as we know, the high-order vector RW has not been taken seriously until now. In [34], the authors consider the high-order RW solutions, which can be reduced into scalar ones, by modified DT method. As high-order scalar RWs are nontrivial superpositions of elementary RWs, the high-order vector ones could be nontrivial and possess more abundant dynamics characters. The knowledge about them would enrich our realization and understanding of RW's complex dynamics.

In this paper, we introduce a family of high-order rational solution in coupled nonlinear Schrödinger equations(CNLSE), which describe RW phenomena in multi-component system prototypically. We find that four fundamental RWs can emerge for the second-order vector RW in the coupled system, which is quite different from the scalar high-order ones for which it is impossible for four fundamental RWs to emerge. Moreover, six fundamental RWs can emerge on the distribution for the second-order vector RW.

The two-component coupled model We begin with the well known CNLSE in dimensionless form

$$\begin{aligned} &\mathrm{i}q_{1,t} + q_{1,xx} + g(|q_1|^2 + |q_2|^2)q_1 = 0, \\ &\mathrm{i}q_{2,t} + q_{2,xx} + g(|q_1|^2 + |q_2|^2)q_2 = 0, \end{aligned} \tag{1}$$

where g is the nonlinear coefficient. The CNLSE model can describe the dynamics of matter wave in quasi-1 dimensional two-component Bose-Einstein condensate [20], the evolution of optical fields in a two-mode or polarized nonlinear fiber [21], and even the vector financial system [35]. With $g=2$, Eq.(1) admits the following Lax pair:

$$\begin{aligned}\Phi_x&=(\mathrm{i}\lambda\Lambda+Q)\Phi,\\ \Phi_t&=(3\mathrm{i}\lambda^2\Lambda+3\lambda Q+\mathrm{i}\sigma_3(Q_x-Q^2))\Phi,\end{aligned}\tag{2}$$

where

$$Q=\begin{pmatrix}0&q_1&q_2\\-\bar{q}_1&0&0\\-\bar{q}_2&0&0\end{pmatrix},\quad \begin{aligned}\Lambda&=\mathrm{diag}(-2,1,1),\\ \sigma_3&=\mathrm{diag}(1,-1,-1),\end{aligned}$$

the symbol overbar represents complex conjugation. The compatibility condition $\Phi_{xt}=\Phi_{tx}$ gives the CNLSE (1).

The standard Darboux transformation (DT) [36] for linear system (2) is

$$\begin{aligned}&\Phi[1]=T\Phi,\ T=I+\frac{\bar{\lambda}_1-\lambda_1}{\lambda-\bar{\lambda}_1}\frac{\Phi_1\Phi_1^\dagger}{\Phi_1^\dagger\Phi_1},\\ &Q[1]=Q+\mathrm{i}(\bar{\lambda}_1-\lambda_1)[P_1,\Lambda],\end{aligned}\tag{3}$$

where Φ_1 is a special solution for system (1) at $\lambda=\lambda_1$, the symbol † represents the hermitian transpose. It is well known that the standard $N-$fold DT should be done with different spectral parameters, or there will be some singularity in the DT matrix. The generalized DT was presented to solve this problem in [15], which can be used to derive high-order RW conveniently by taking a special limit about the parameters λ_is.

The studies on the first-order vector RW in the ref. [28, 26], indicate that there should be some restriction conditions on the plane wave background fields ($\alpha_j\exp[\mathrm{i}(k_jx+w_jt)]$, $j=1,2$) to obtain the general RW solutions for CNLSE. The wave vector difference of background plane wave $|k_1-k_2|$ between the two components should satisfy certain relation with the background amplitudes α_j and nonlinear coefficient g, namely, $\alpha_1=\alpha_2=\alpha$ and $|k_1-k_2|=\sqrt{g/2}\,\alpha$ [26]. In ref. [28], they choose the seed solution

$$q_1=\alpha\exp\left[\mathrm{i}\left(\frac{1}{2}\alpha x+\frac{15}{4}\alpha^2t\right)\right],\ q_2=\alpha\exp\left[\mathrm{i}\left(-\frac{1}{2}\alpha x+\frac{15}{4}\alpha^2t\right)\right].$$

In fact, the parameter α can be re-scaled by scaling transformation. Thus we can consider a much simpler seed solution as the background where RW exist without losing generality

$$q_1=\exp\theta_1,\ q_2=\exp\theta_2,\tag{4}$$

where $\theta_1 = \left[\mathrm{i}\left(\frac{1}{2}x + \frac{15}{4}t\right)\right]$, $\theta_2 = \left[\mathrm{i}\left(-\frac{1}{2}x + \frac{15}{4}t\right)\right]$.

We have proved that high-order RW can be derived by taking a certain limit of the spectral parameter [15]. To take the limit conveniently, we set $\lambda_j = \frac{\sqrt{3}i}{2}(1+\epsilon_j^3)$, $j = 1, 2, \cdots, N$. Substituting seed solution (4) into equation (2), we can obtain the fundamental solution

$$\Phi_i(\lambda_j) = D\begin{bmatrix} \left[\mathrm{i}\left(\lambda_j + \frac{1}{2}\right) - \xi_i\right]\left[\mathrm{i}(\lambda_j - \frac{1}{2}) - \xi_i\right]\exp\omega_i \\ \left[\mathrm{i}(\lambda_j - \frac{1}{2}) - \xi_i\right]\exp\omega_i \\ \left[\mathrm{i}(\lambda_j + \frac{1}{2}) - \xi_i\right]\exp\omega_i \end{bmatrix}$$

where $i = 1, 2, 3$,

$$D = \mathrm{diag}\left(\exp\left(\frac{5it}{2}\right), \exp\left[-\frac{\mathrm{i}}{4}(2x+5t)\right], \exp\left[\frac{\mathrm{i}}{4}(2x-5t)\right]\right),$$

$$\omega_i = \xi_i x + \left(\mathrm{i}\xi_i^2 + 2\lambda_j\xi_i + 2\mathrm{i}\lambda_j^2 + \frac{3\mathrm{i}}{2}\right)t$$

and ξ_i satisfies the following cubic equation

$$\xi^3 - \left(\frac{9}{2}\epsilon_j^3 + \frac{9}{4}\epsilon_j^6\right)\xi - \frac{3}{2}\sqrt{3}\epsilon_j^3 - \frac{9}{4}\sqrt{3}\epsilon_j^6 - \frac{3}{4}\sqrt{3}\epsilon_j^9 = 0. \tag{5}$$

By the Taylor expansion of the fundamental solution form as done in [15], the generalized DT can be used to derive RW solution. However, Taylor expansions of the fundamental solution form are quite complicated which bring much complex calculation. We find this process can be simplified greatly by the following special solution form

$$\begin{aligned} \Psi_1(\lambda_j) &= \frac{1}{3}\left(\Phi_1(\lambda_j) + \Phi_2(\lambda_j) + \Phi_3(\lambda_j)\right), \\ \Psi_2(\lambda_j) &= \frac{\sqrt[3]{2}}{3\epsilon_j}\left(\Phi_1(\lambda_j) + \omega^2\Phi_2(\lambda_j) + \omega\Phi_3(\lambda_j)\right), \\ \Psi_3(\lambda_j) &= \frac{\sqrt[3]{4}}{3\epsilon_j^2}\left(\Phi_1(\lambda_j) + \omega\Phi_2(\lambda_j) + \omega^2\Phi_3(\lambda_j)\right), \end{aligned} \tag{6}$$

where $\omega = \exp[2\mathrm{i}\pi/3]$, which are also the solution of the Lax pair with the seed solutions (4). We can prove that

$$\Psi(\epsilon_j) = f\Psi_1 + g\Psi_2 + h\Psi_3 \tag{7}$$

where

$$f = f_1 + f_2\epsilon_j^3 + f_3\epsilon_j^6 + \cdots + f_N\epsilon_j^{3(N-1)},$$
$$g = g_1 + g_2\epsilon_j^3 + g_3\epsilon_j^6 + \cdots + g_N\epsilon_j^{3(N-1)},$$
$$h = h_1 + h_2\epsilon_j^3 + h_3\epsilon_j^6 + \cdots + h_N\epsilon_j^{3(N-1)},$$

and f_i,g_i,h_i are complex numbers, can be expanded around $\epsilon_j = 0$ with the following form

$$\Psi(\epsilon_j) = \Psi^{[1]} + \Psi^{[2]}\epsilon_j^3 + \Psi^{[3]}\epsilon_j^6 + \cdots + \Psi^{[N]}\epsilon_j^{3(N-1)} + O(\epsilon_j^{3N}).$$

To obtain the vector RW solution, we merely need to take limit $\epsilon_j \to 0$ [17]. After performing the generalized DT, we can present N-th order localized solution on the plane backgrounds with the same spectral parameter $\lambda_j = \frac{\sqrt{3}\mathrm{i}}{2}(1+\epsilon_j^3)$ as

$$q_1[N] = \exp\theta_1 \frac{\det(M_1)}{\det(M)},$$
$$q_2[N] = \exp\theta_2 \frac{\det(M_2)}{\det(M)}, \tag{8}$$

where

$$M_1 = M - 3\mathrm{i}Y_2^\dagger Y_1,\ M_2 = M - 3\mathrm{i}Y_3^\dagger Y_1,$$
$$X = \begin{bmatrix} X_1 \\ X_2 \\ X_3 \end{bmatrix} = \left[\Psi^{[1]},\ \Psi^{[2]},\ \cdots,\ \Psi^{[N]}\right], \tag{9}$$

$Y_1 = X_1 \exp\left(-\frac{5it}{2}\right)$, $Y_2 = X_2 \exp\left[\frac{\mathrm{i}}{4}(2x+5t)\right]$, $Y_3 = X_3\exp\left[\frac{\mathrm{i}}{4}(-2x+5t)\right]$, and $M = (M_{l,m})_{1\leqslant l,m\leqslant N}$. The $M_{l,m}$ can be derived by

$$\frac{\langle\Psi(\epsilon_j), \Psi(\epsilon_j)\rangle}{\lambda_j - \bar{\lambda}_j} = \sum_{l,m=1}^{+\infty,+\infty} M_{l,m}\epsilon_j^{3(m-1)}\bar{\epsilon}_j^{3(l-1)}.$$

The compact solution formula (8) can be used to derive $N-$th order RW solution. With $N = 1$, the first order vector RW can be derived directly, which agrees well with the ones in [28, 26]. We find the dynamics structure of high-order RW in the coupled system are much more abundant than the ones in scalar systems [14, 15, 16, 17, 18]. Even for the second-order RW, whose distributions possess many different structures, which are quite different from the second-order scalar ones. As an example, we exhibit the dynamics behavior of second-order vector RW solution. Since the expressions of high-order RW solution are quite complicated, we will present them elsewhere.

The dynamics of second-order vector rogue waves There are six free parameters in the generalized second-order RW solution, denoted by f_j, g_j and h_j $(j=1,2)$, which can be used to obtain different types or patterns for the rogue wave dynamics. We find that there are mainly two kinds of RW solutions which correspond to four fundamental RWs and six fundamental ones obtained by setting $f_1=0$ and $f_1\neq 0$ respectively.

Firstly, we discuss the second-order RW solution which possesses four fundamental RWs. The pattern is quite different from the ones in scalar NLSE system [17,18], for which it is impossible for four fundamental RWs to emerge on the temporal-spatial distribution plane. To obtain this kind of solution, we merely need to choose parameter $f_1=0$. We classify them by parameters f_2, g_1, g_2, h_1, h_2 whether or not are zeros. By this classification, there could be 2^5 kinds of different solution which correspond to different patterns on the temporal-spatial distribution. We find there are mainly three types of the patterns, such as quadrilateral, triangle, and line structures.

The explicit shape of the quadrilateral can be varied through the parameters. As an example, we show two cases for the quadrilateral structure in Fig. 1. The first case: the four RWs arrange with the "rhombus" structure (Fig. 1(a,b)). The spatial-temporal distribution are similar globally in the two components, but the RW with highest peak emerge at different time, it appear at time $t=-5$ for the component q_1; and at $t=5$ for the component q_2. The second case: the four RWs arrange with "rectangle" structure (Fig. 1(c,d)). It is seen that the peak values of two RWs on the right hand is much higher than the ones on the left hand in the component q_1. The character is inverse for the component q_2.

Varying the other parameters, we can observe the interaction between the four RWs. When two of them fuse into be a new RW, the three RWs can emerge with "triangle" structure on the temporal-spatial distribution (Fig. 2(a,b)). The structure of the triangle can be changed by vary the parameters. Especially, the three RWs can emerge in a "line" (Fig. 2(c,d)), which is perpendicular with the t axes. Namely, at a certain time, three or four RWs can emerge synchronously.

Secondly, we consider the second case of the second-order RW, which possesses six fundamental RWs. To obtain this kind of RW, we choose the parameter $f_1=1$. We find that the six fundamental RWs can constitute many different structures, such as "pentagon", "quadrilateral", "triangle", and "line" structures. As an example, we show the "pentagon" structure in Fig. 3(a,b). The pentagon structure can be varied too through changing the parameters. There is one RW in the internal region of the pentagon, and its location in the distribution plane can be varied too. This case is similar to the "pentagon" structure of the third order RW of scalar NLS equation [17,18]. The six RWs can be arranged with the "rectangle" structure through varying the parameters, such as the one in Fig. 3(c,d). The structure is similar to the one in Fig. 3(c,d). But there is a new RW inside the "rectangle", which is formed by the interaction of the other two fundamental RWs.

The six RWs can be also arranged with the "triangle" structure, shown in Fig. 4(a,b). In this case, there are two fundamental RWs and a new RW to form a triangle. The new RW is formed by the interaction between the other four fundamental RWs. Moreover, the RWs can be arranged with "line" structure too, shown in Fig. 4(c,d).

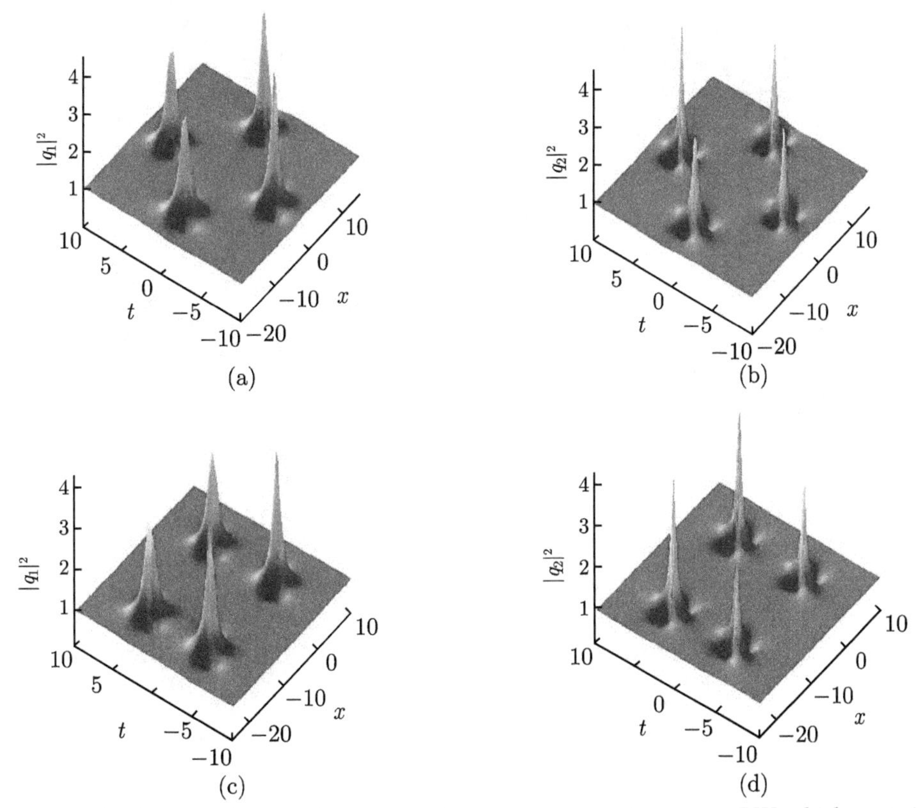

Figure 1 (a,b) The "rhombus" structure for the second-order vector RW which contains four fundamental ones. The parameters are $f_1 = 0, f_2 = 0, g_1 = 1, g_2 = 0, h_1 = 0, h_2 = 10000$, (c,d) The "rectangle" structure for the second-order vector RW which contains four fundamental ones. The parameters are $f_1 = 0, f_2 = 0, g_1 = 1, g_2 = 1000, h_1 = 10, h_2 = 0$.

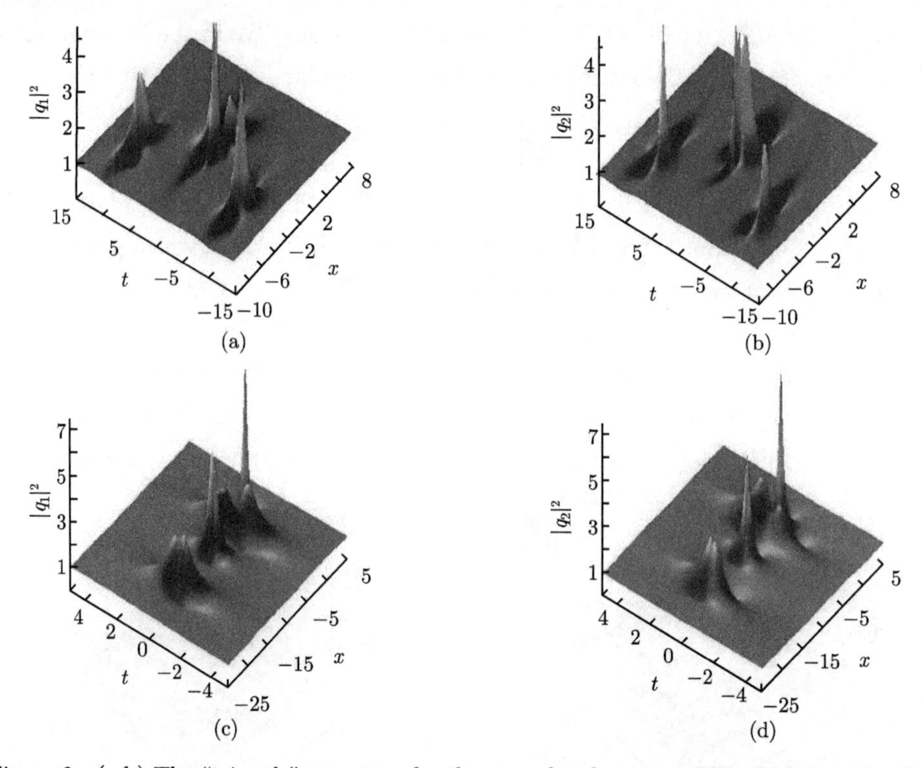

Figure 2 (a,b) The "triangle" structure for the second-order vector RW which contains four fundamental ones. The parameters are $f_1 = 0, f_2 = 100, g_1 = 1, g_2 = 0, h_1 = 0, h_2 = 0$, (c,d) The "line" structure for the second-order vector RW which contains four fundamental ones. The parameters are $f_1 = 0, f_2 = 0, g_1 = 1, g_2 = 0, h_1 = 10, h_2 = 0$.

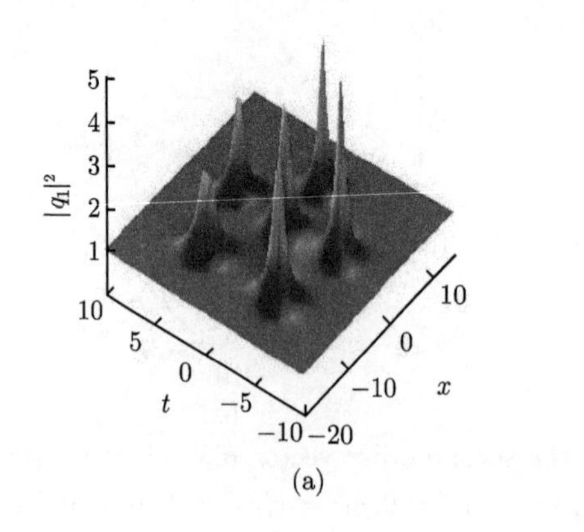

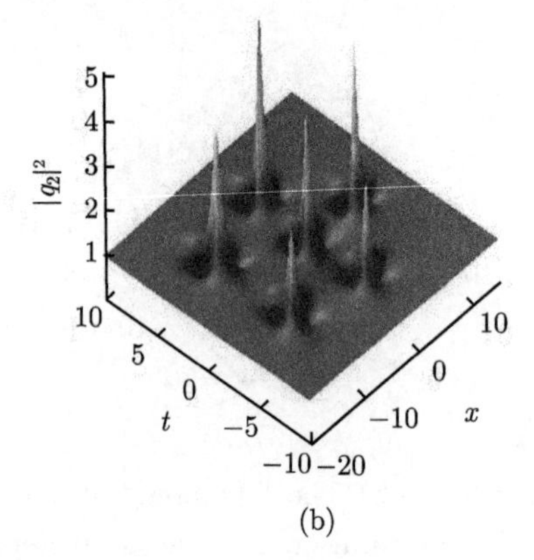

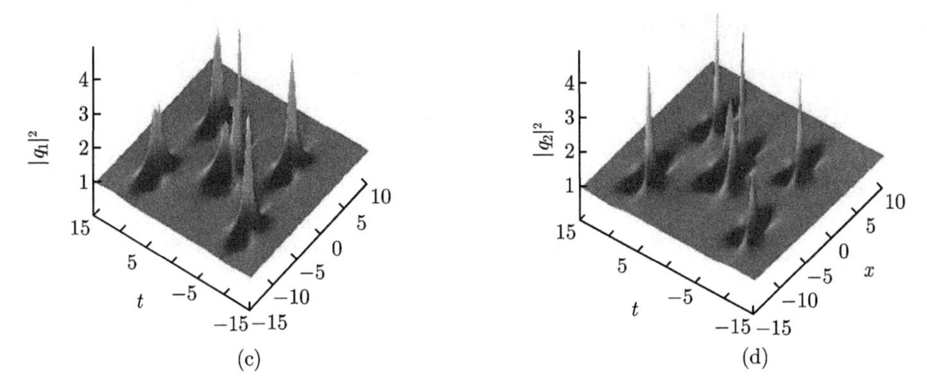

Figure 3 (a,b) The "pentagon" structure for the second-order vector RW which contains six fundamental ones. The parameters are $f_1 = 1, f_2 = 0, g_1 = 0, g_2 = 0, h_1 = 0, h_2 = 10000$, (c,d) The "rectangle" structure for the second-order vector RW which contains six fundamental ones. The parameters are $f_1 = 1, f_2 = 0, g_1 = 0, g_2 = 0, h_1 = 100, h_2 = 0$.

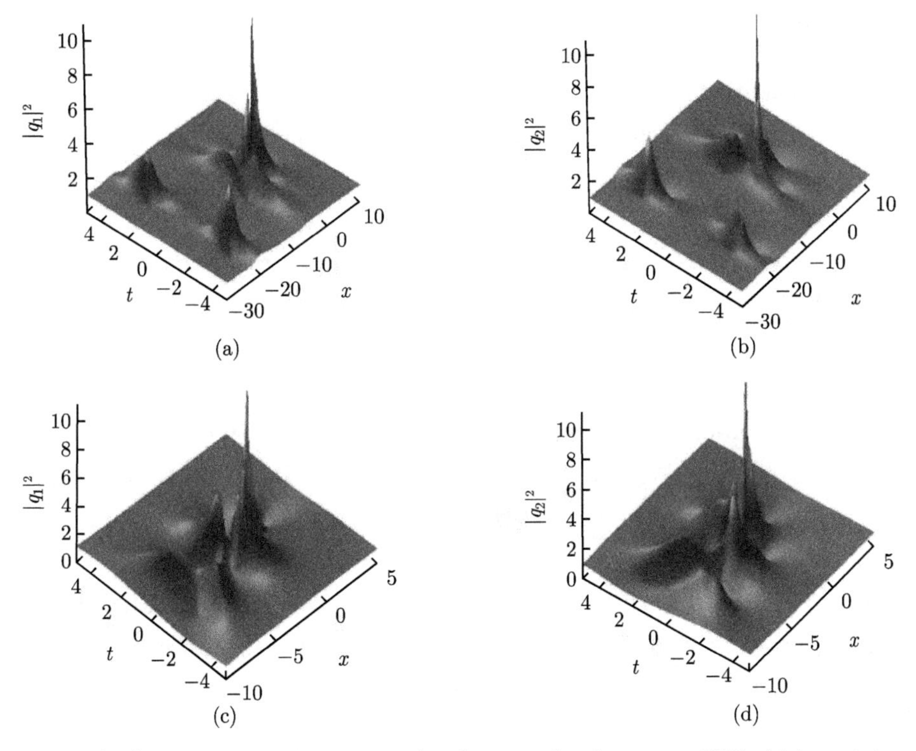

Figure 4 (a,b) The "triangle" structure for the second-order vector RW which contains six fundamental ones. The parameters are $f_1 = 1, f_2 = 0, g_1 = 10, g_2 = 0, h_1 = 0, h_2 = 0$. (c,d) The "line" structure for the second-order vector RW which contains six fundamental ones. The parameters are $f_1 = 1, f_2 = 0, g_1 = 0, g_2 = 0, h_1 = 0, h_2 = 0$.

There should be six fundamental RWs arranged in one line case, but it is very complicated to derive the case since the parameters are too many to be managed well. We just show a particular one of the "line" cases.

Possibilities to observe these vector rogue waves It is expected that these vector RW could be observed in two-mode nonlinear fibers [8,9]. As an example, we consider the case that the operation wavelength of each mode is nearly 1.55 μm, the GVD coefficients are -20 ps^{2}km^{-1} in the anomalous regime, and the Kerr coefficients are nearly 1.1 W^{-1}km^{-1}, corresponding to the self-focusing effect in the fiber [37]. The unit in x direction will be denoted as 0.23 ps, and the one in t will be denoted as 0.55 km. One can introduce two distinct modes to the nonlinear fiber operating in the anomalous GVD regime [38,39]. The spontaneous development of RW seeded from some perturbation should be on the continuous waves as the ones in [8,9]. The continuous wave background intensities in the two modes should be equal (assume to be 1 W), and the frequency difference between the two modes should be 0.23 ps^{-1} to observe the vector RW. We can manipulate the initial perturbation approaching the ideal initial condition to observe these vector RW patterns. The ideal initial condition including intensity and phase profiles can be given by the exact second-order vector RW solution with given parameters in the scaled units. Moreover, the initial intensity and phase profiles can be made through related modulation operator [9,37]. The vector RWs could be observed in the nonlinear fiber with approaching the ideal initial excitation form presented here [9].

Conclusions We present a generalized RW solution of CNLSE, which can be used to obtain arbitrary order vector RW. We find that there are mainly two kinds of rogue wave solutions for the second-order vector RW in CNLSE, which correspond to four fundamental RWs and six fundamental ones. Based on these results, we expect that there are much more abundant exotic patterns for RW in three or more component NLSE.

Acknowledgements This work is supported by National Natural Science Foundation of China (Contact No. 11271052) and Fundamental Research Funds for the Central Universities (Contact No. 2014ZB0034).

References

[1] Onorato M, Residori S, Bortolozzo U, Montina A, Arecchi F T. Rogue waves and their generating mechanisms in different physical contexts [J]. Phys. Rep., 2013, 528:

47–89.

[2] Kharif C, Pelinovsky E, Slunyaev A. Rogue Waves in the Ocean [M]. Heidelberg: Springer, 2009.

[3] Osborne A R. Nonlinear Ocean Waves and the Inverse Scattering Transform [M]. New York: Elsevier, 2010.

[4] Ruban V, Kodama Y, Ruderman M, et al., Rogue waves C towards a unifying concept?: Discussions and debates [J]. Eur. Phys. Journ. Special Topics, 2010, 185: 5–15.

[5] Akhmediev N, Pelinovsky E, Editorial C introductory remarks on "discussion & debate: Rogue waves-towards a unifying concept?" [J]. Eur. Phys. J. Special Topics, 2010, 185: 1.

[6] Kharif C, Pelinovsky E. Physical mechanisms of the rogue wave phenomenon [J]. Eur. J. Mech. B/Fluids, 2003, 22: 603.

[7] Pelinovsky E, Kharif C. Extreme Ocean Waves [M]. Berlin: Springer, 2008.

[8] Solli D R, Ropers C, Koonath P, Jalali B. Optical rogue waves[J]. Nature, 2007, 450: 06402.

[9] Kibler B, Fatome J, Finot C, Millot G, Dias F, Genty G, Akhmediev N, Dudley J M. The Peregrine soliton in nonlinear fibre optics [J]. Nature Phys., 2010, 6: 790.

[10] Chabchoub A, Hoffmann N P, Akhmediev N. Rogue wave observation in a water wave tank[J]. Phys. Rev. Lett. 2011, 106:204502.

[11] Bailung H, Sharma S K, Nakamura Y. Observation of Peregrine solitons in a multicomponent plasma with negative ions [J]. Phys. Rev. Lett., 2011, 107:255005.

[12] Osborne A R. The random and deterministic dynamics of 'rogue waves' in unidirectional, deep-water wave trains [J], Mar. Struct., 2001, 14: 275; Akhmediev N, Ankiewicz A, Taki M. Waves that appear from nowhere and disappear without a trace [J], Phys. Lett. A, 2009, 373: 675–678.

[13] Akhmediev N, Ankiewicz A, Soto-Crespo J M, Dudley J M. Rogue wave early warning through spectral measurements? [J]. Phys. Lett. A, 2011, 375: 541–544.

[14] Ohta Y, Yang J K. General high-order rogue waves and their dynamics in the nonlinear Schrödinger equation [J]. Proc. R. Soc. A, 2012, 468: 1716–1740.

[15] Guo B, Ling L, Liu Q P. Nonlinear Schrödinger equation: generalized Darboux transformation and rogue wave solutions [J]. Phys. Rev. E 2012, 85: 026607; Guo B, Ling L, Liu Q P. High-Order Solutions and Generalized Darboux Transformations of Derivative Nonlinear Schrödinger Equations [J]. Stud. Appl. Math., 2013, 130: 317–344.

[16] He J S, Zhang H R, Wang L H, Porsezian K, Fokas A S. Generating mechanism for higher-order rogue waves [J]. Phys. Rev. E, 2013, 87: 052914.

[17] Ling L, Zhao L-C. Simple determinant representation for rogue waves of the nonlinear Schrödinger equation [J]. Phys. Rev. E, 2013, 88: 043201.

[18] Kedziora D J, Ankiewicz A, Akhmediev N. Circular rogue wave clusters[J]. Phys. Rev.

E, 2013, 88: 013207.

[19] Becker C,Stellmer S, et al. Oscillations and interactions of dark and dark-bright solitons in Bose-Einstein condensates [J]. Nat. Phys., 2008, 4: 496 C 501.

[20] Hamner C,Chang J J, Engels P,Hoefer M A. Generation of dark-bright soliton trains in superfluid-superfluid counterflow [J]. Phys. Rev. Lett., 2011, 106: 065302.

[21] Tang D Y, Zhang H, Zhao L M, Wu X. Observation of high-order polarizatio $N-$locked vector solitons in a fiber laser [J]. Phys. Rev. Lett., 2008, 101: 153904.

[22] Wright O C, Forest M G. On the Bäcklund-gauge transformation and homoclinic orbits of a coupled nonlinear Schrödinger system [J]. Physica D, 2000, 141: 104 C 116; M.G. Forest, S.P. Sheu, and P.C. Wright, On the construction of orbits homoclinic to plane waves in integrable coupled nonlinear Schrödinger systems [J]. Phys. Lett. A, 2000, 266: 24–33.

[23] Forest M G, Wright O C. An integrable model for stable: unstable wave coupling phenomena [J]. Physica D, 2003, 178: 173–189.

[24] Chow K W, Wong K K Y, Lam K. Modulation instabilities in a system of four coupled, nonlinear Schrödinger equations [J]. Phys. Lett. A, 2008, 372: 4596–4600.

[25] Bludov Y V, Konotop V V, Akhmediev N. Vector rogue waves in binary mixtures of Bose-Einstein condensates [J]. Eur. Phys. J. Special Topics, 2010, 185: 169.

[26] Zhao L-C, Liu J. Localized nonlinear waves in a two-mode nonlinear fiber [J]. J. Opt. Soc. Am. B, 2012, 29: 3119–3127.

[27] Baronio F, Degasperis A, Conforti M, Wabnitz S. Solutions of the vector nonlinear Schrödinger equations: evidence for deterministic rogue waves [J]. Phys. Rev. Lett., 2012, 109: 044102.

[28] Guo B, Ling L. Rogue wave, breathers and bright-dark-rogue solutions for the coupled Schrödinger equations [J]. Chin. Phys. Lett., 2011, 28: 110202.

[29] Zhao L-C, Liu J. Rogue-wave solutions of a three-component coupled nonlinear Schrödinger equation [J]. Phys. Rev. E, 2013, 87: 013201.

[30] Baronio F, Conforti M, Degasperis A, Lombardo S. Rogue waves emerging from the resonant interaction of three waves [J]. Phys. Rev. Lett., 2013, 111: 114101.

[31] Chabchoub A, Akhmediev N. Observation of rogue wave triplets in water waves [J]. Phys. Lett. A, 2013, 377: 2590–2593.

[32] Chabchoub A, Hoffmann N, Onorato M, et al. Observation of a hierarchy of up to fifth-order rogue waves in a water tank [J]. Phys. Rev. E, 2012, 86: 056601.

[33] Erkintalo M, Hammani K, Kibler B, Finot C, Akhmediev N, Dudley J M, Genty G. Higher-order modulation instability in nonlinear fiber optics [J]. Phys. Rev. Lett., 2011, 107: 253901.

[34] Zhai B G, Zhang W G, Wang X L, Zhang H Q. Multi-rogue waves and rational solutions of the coupled nonlinear Schrödinger equations [J]. Nonlinear Analysis: Real World Applications, 2013, 14: 14–27.

[35] Yan Z Y. Vector financial rogue waves [J]. Phys. Lett. A, 2011, 375: 4274–4279.
[36] Matveev V B, Salle M A. Darboux transformations and solitons [M]. Berlin: Springer-Verlag, 1991.
[37] Dudley J M, Genty G, Dias F, Kibler B, Akhmediev N. Modulation instability, Akhmediev Breathers and continuous wave supercontinuum generation [J]. Opt. Express, 2009, 17: 21497–21508.
[38] Afanasyev V V, Kivshar Yu S, Konotop V V, Serkin V N. Dynamics of coupled dark and bright optical solitons [J]. Opt. Lett., 1989, 14: 805.
[39] Ueda T, Kath W L. Dynamics of coupled solitons in nonlinear optical fibers [J]. Phys. Rev. A, 1990, 42: 563.

Schrödinger Limit of Weakly Dissipative Stochastic Klein-Gordon-Schrödinger Equations and Large Deviations*

Guo Boling (郭柏灵), Lv Yan (吕艳) and Wang Wei (王伟)

Abstract This paper derives a Schrödinger approximation for weakly dissipative stochastic Klein-Gordon-Schrödinger equations with a singular perturbation and scaled small noises on a bounded domain. Detail uniform estimates are given to pass the limit as perturbation and noise disappear. Approximation in two different spaces are considered. Furthermore a large deviation principe of solutions is derived by weak convergence approach.

Keywords Klein-Gordon-Schrödinger equations; Yukawa coupling; large deviation principle; weak convergence

1 Introduction

In the model that describes the interaction of scalar nucleons interacting with neutral scalar mesons, the dynamics of these fields through Yukawa coupling is given by the following Klein-Gordon-Schrödinger equations (KGS)

$$\mathrm{i}\psi_t + \Delta\psi = -\phi\psi\,, \tag{1.1}$$

$$\frac{1}{c^2}\phi_{tt} - \Delta\phi + \mu^2\phi = |\psi|^2\,, \tag{1.2}$$

where ψ, ϕ represent a complex scalar nucleon field and a real meson field respectively, c, μ describe the speed and mass of the meson respectively. In physical point, $c \to \infty$ system (1.1)–(1.2) corresponds to the infinite limit of the velocity of propagation of disturbances in the Klein-Gordon equation, and so to an instantaneous response of the field ϕ to variations of the field ψ [3,25]. So an effective approximation of solution (ψ^ν, ϕ^ν) is desirable for large $c > 0$.

* Disc. Cont. Dyns. Syst. A, 2014, 34(7): 2795–2818. Doi:10.3934/dcds.2014.34.2795.

The equations (1.1)–(1.2), which are a conservative system, have been studied extensively. Most of the results concerned the existence of local and global solutions, blow-up and the initial boundary value problem for KGS [1, 10, 11, 13]. When the damping is considered we have the dissipative KGS

$$\mathrm{i}\psi_t + \Delta\psi + \mathrm{i}\alpha\psi + \phi\psi = f\,, \tag{1.3}$$

$$\nu\phi_{tt} + \beta\phi_t - \Delta\phi + \phi - |\psi|^2 = g\,, \tag{1.4}$$

where $\nu = 1/c^2, \alpha, \beta$ are positive constants, driving term f is complex. Here we assumed $\mu = 1$.

The dissipative mechanism in (1.3)–(1.4) is introduced by the terms $i\alpha\psi^\nu$ and $\beta\phi_t^\nu$. The long time behavior of (1.3)–(1.4) defined on bounded or unbounded domain is also studied by many authors. For system (1.3)–(1.4) in a bounded domain, Wahl [3] proved the existence of a weak global attractor in space $H^1 \times H^1 \times L^2$. Further Wang and Lange [29] proved that the weak global attractor is indeed the strong one. For (1.3)–(1.4) in unbounded domain Guo and Li [12] constructed a global attractor in space $H^2 \times H^2 \times H^1$ which attracts bounded sets in space $H^3 \times H^3 \times H^2$. Further Lu and Wang [18] obtain the global attractor in space $H^1 \times H^1 \times L^2$. Further Li and Guo [17] proved the global attractor in space $H^1 \times H^1 \times L^2$ coincides with the one in space $H^2 \times H^2 \times H^1$ which yields the smooth effect of the KGS system.

Recently the need to taking random effects into account in modeling, analyzing, simulating and predicting complex phenomena has been widely recognized in geophysical and climate dynamics, materials science, chemistry, biology and other areas [8, 14, e.g.]. Stochastic partial differential equations (SPDEs or stochastic PDEs) are appropriate mathematical models for complex systems under random influences or noise.

Here we consider the randomly forced dissipative KGS system on a bounded domain $D \subset \mathbf{R}^n$, $1 \leqslant n \leqslant 3$, with smooth boundary ∂D, which is described by the following Itô stochastic Klein-Gordon-Schrödinger (SKGS) equations

$$\mathrm{i}\psi_t^\nu + \Delta\psi^\nu + \mathrm{i}\nu\psi^\nu + \phi^\nu\psi^\nu = \sqrt{\nu}\dot{W}_1\,, \quad \text{on} \quad D \tag{1.5}$$

$$\nu\phi_{tt}^\nu + \nu^\alpha\phi_t^\nu - \Delta\phi^\nu + \phi^\nu - |\psi^\nu|^2 = \sqrt{\nu^{1+\alpha}}\dot{W}_2\,, \quad \text{on} \quad D \tag{1.6}$$

with zero Dirichlet boundary condition

$$\psi^\nu = 0\,, \quad \phi^\nu = 0\,, \quad \text{on} \quad \partial D \tag{1.7}$$

and initial condition

$$\psi^\nu(x,0) = \psi_0(x)\,, \quad \phi^\nu(x,0) = \phi_0(x)\,, \quad \phi_t^\nu(x,0) = \phi_1(x)\,. \tag{1.8}$$

Here $1 \geqslant \alpha \geqslant 0$ is a constant, $W = (W_1, W_2)$ is random force which are smooth in space and white in time, which is detailed in next section. As c is very large, that is ν is very small. We here are concerned with the limit of solution (ψ^ν, ϕ^ν) as $\nu \to 0$.

Notice that we need the strength of noise to be of $\mathcal{O}(\sqrt{\nu^{1+\alpha}})$ to balance singular terms $\nu\phi_{tt}^\nu$ and $\nu^\alpha\phi_t^\nu$. We also need the dissipation term in ψ^ν equation has same order as the damping term in ϕ^ν equation.

For $1 \geqslant \alpha > 0$, passing the limit $\nu \to 0$ in (1.5)–(1.6), we have the following Schrödinger–Yukawa equation (Theorem 4.1 and 4.2)

$$\mathrm{i}\psi_t + \Delta\psi + \phi\psi = 0\,, \quad \text{on} \quad D \tag{1.9}$$

$$-\Delta\phi + \phi - |\psi|^2 = 0\,, \quad \text{on} \quad D \tag{1.10}$$

$$\psi(0) = \psi_0\,, \phi(0) = \phi_0 \tag{1.11}$$

with zero Dirichlet boundary condition. Then the solution of (1.5)–(1.8), providing $\phi_0 = A^{-1}|\psi_0|^2$, converges to solution of the following Schrödinger equation (Theorem 4.3)

$$\mathrm{i}\psi_t + \Delta\psi + (A^{-1}|\psi|^2)\psi = 0\,, \quad t > 0 \quad \text{and} \quad \psi(x,0) = \psi_0(x) \quad \text{on} \quad D \tag{1.12}$$

where $A = -\Delta + 1$ on D with zero Dirichlet boundary condition. For $\alpha = 0$, we have the following limit equation

$$\mathrm{i}\psi_t + \Delta\psi + \phi\psi = 0\,, \quad \text{on} \quad D \tag{1.13}$$

$$\phi_t - \Delta\phi + \phi - |\psi|^2 = 0\,, \quad \text{on} \quad D \tag{1.14}$$

$$\psi(0) = \psi_0\,, \phi(0) = \phi_0 \tag{1.15}$$

with zero Dirichlet boundary condition.

Large deviation principle (LDP) is an effective tool to describe rare events of differential system driven by small noise [9, 24]. Lots of work have been devoted to derive LDP for SPDEs [5,6,15,23,26,27,30,31]. Recently developed weak convergence method, which is related Laplace principle, is a more simple and direct approach [4]. Here we apply the weak convergence method to derive large deviation principle for the processes (ψ^ν, ϕ^ν) as $\nu \to 0$ (Theorem 5.2) for the case $\alpha = 0$. In this approach we also have to deal with the localized integral estimates of time increments (Lemma 5.3), which is also used for large deviations of stochastic Boussinesq equations [6].

2 Preliminary

We give a mathematical framework and some results used frequently in the following sections.

By the transformation $\theta^\nu = \phi_t^\nu + \delta\phi^\nu$ with δ a small positive constant which will be specified later, the equations (1.5)–(1.6) become

$$\mathrm{i}\psi_t^\nu + \Delta\psi^\nu + \mathrm{i}\nu^\alpha\psi^\nu + \phi^\nu\psi^\nu = \sqrt{\nu^\alpha}\dot{W}_1\,, \tag{2.16}$$

$$\nu\theta_t^\nu + (\nu^\alpha - \nu\delta)\theta^\nu - \Delta\phi^\nu + (1 - \delta(\nu^\alpha - \nu\delta))\phi^\nu - |\psi^\nu|^2 = \sqrt{\nu^{1+\alpha}}\dot{W}_2 \tag{2.17}$$

with initial condition

$$\psi^\nu(x,0) = \psi_0(x)\,,\ \ \phi^\nu(x,0) = \phi_0(x)\,,\ \ \theta^\nu(x,0) = \theta_0(x)\,. \tag{2.18}$$

Here $\psi_0^\nu(x)$ is a complex valued function and $\phi_0^\nu(x)$, $\theta_0^\nu(x)$ are real valued functions. Denote by $L^p(D)$ the space consisting of p-square integrable complex valued functions defined on D with norm $\|\cdot\|_{L^p}$. For $p = 2$, the inner product on $H := L^2(D)$ is defined by

$$\langle u\,,v\rangle = \mathrm{Re}\int_{\mathbf{R}^n} u(x)\bar{v}(x)\,\mathrm{d}x\,, \quad u\,,v \in L^2(\mathbf{R}^n)\,,$$

and denote the norm by $\|\cdot\|_0$. We use $H^m = H^m(D)$, $m > 0$, to denote the usual Sobolev spaces and the norm is denoted by $\|\cdot\|_m$. For $m = 1$, we denote by $H_0^1 = H_0^1(D)$. Now we define spaces $\mathcal{E}_0$ as

$$\mathcal{E}_0 = H_0^1 \times H_0^1 \times H$$

endowing with the usual product norm $\|\cdot\|_{\mathcal{E}_0}$.

We are given a complete probability space $(\Omega, \mathcal{F}, \{\mathcal{F}_t\}_{t\geqslant 0}, \mathbb{P})$. For simplicity we assume stochastic term $W(t)$ is defined on $(\Omega, \mathcal{F}, \{\mathcal{F}_t\}_{t\geqslant 0}, \mathbb{P})$ in the following form

$$W(x,t) = (W_1(x,t), W_2(x,t)) = (q_1(x)B_1(t), q_2(x)B_2(t))$$

with $B_1(t)$ is a standard complex valued Wiener process, $B_2(t)$ is a real value Wiener process and $q_1(x)$, $q_2(x)$ are smooth enough real valued functions. Here B_1 and B_2 are mutually independent. For a general case, $W_1(x,t)$ and $W_2(x,t)$ are both H-valued

Q-Wiener process [24], with compact covariance operator Q_1 and Q_2, the extension of our result is immediate. In our case Q_1 and Q_2 are defined as

$$Q_1 f(x) = \int_D q_1(x) f(y) q_1(y)\,\mathrm{d}y, \quad Q_2 g(x) = \int_D q_2(x) g(y) q_2(y)\,\mathrm{d}y$$

for any f, $g \in H$.

In our approach we need the following lemma.

Lemma 2.1 *Let E, E_0 and E_1 be Banach spaces such that $E_1 \Subset E_0$, the interpolation space $(E_0, E_1)_{\theta,1} \subset E$ with $\theta \in (0,1)$ and $E \subset E_0$ with $\subset$ and $\Subset$ denoting continuous and compact embedding respectively. Suppose $p_0, p_1 \in [1,\infty]$ and $T > 0$, such that*

$$\begin{aligned} &\mathcal{V} \text{ is a bounded set in } L^{p_1}(0,T;E_1) \\ \text{and} \quad &\partial\mathcal{V} := \{\partial v : v \in \mathcal{V}\} \text{ is a bounded set in } L^{p_0}(0,T;E_0). \end{aligned}$$

Here ∂ denotes the distributional derivative. If $1-\theta > 1/p_\theta$ with

$$\frac{1}{p_\theta} = \frac{1-\theta}{p_0} + \frac{\theta}{p_1},$$

then $\mathcal{V}$ is relatively compact in $C(0,T;E)$.

Hereafter, we denote by C, $C(\cdots)$, C_T and $C_T(\cdots)$ any positive constants which may change from line to line. The subscript and $(\cdots)$ show what constants depend on.

3 Uniform estimates for solutions

First by an exactly same discussion in [19], we have

Theorem 3.1 *Assume $q_1 \in H_0^1$, $q_2 \in H$ and $(\psi_0, \phi_0, \phi_1) \in \mathcal{E}_0$, for any fixed $\nu > 0$ and $T > 0$ equations (1.5)–(1.8) has a unique solution in space $L^2(\Omega, C(0,T;\mathcal{E}_0))$.*

Let $(\psi^\nu, \phi^\nu, \theta^\nu)$ be any solution to (2.16)–(2.17). We give some estimates for $(\psi^\nu, \phi^\nu, \theta^\nu)$ in spaces $\mathcal{E}_0$. Most discussions are similar as those in [19], so here we do not give detail analysis. We just consider the case $n = 3$. For $n \leqslant 2$ the estimates can be obtained similarly by the same discussion.

We first have the following estimates.

Lemma 3.1 *Assume $\psi_0 \in H$, $q_1 \in H$. Then every solution of problem (2.16)–(2.18) satisfies $\psi^\nu \in L^\infty(\mathbf{R}; L^{2p}(\Omega; H)) \cap L^{2p}(\Omega, C(0,T;H))$ for any $T > 0$ and $p \geqslant 1$.*

Proof Consider the equation

$$\psi_t^\nu = \mathrm{i}\Delta\psi^\nu - \nu\psi^\nu + \mathrm{i}\phi^\nu\psi^\nu - \mathrm{i}\sqrt{\nu}\dot{W}_1, \tag{3.19}$$

the Itô's formula yields

$$\begin{aligned}&\|\psi^\nu(t)\|_0^2 - \|\psi_0\|_0^2\\ =& -2\nu\int_0^t \|\psi^\nu(s)\|_0^2\,\mathrm{d}s - \mathrm{i}2\sqrt{\nu}\int_0^t \mathrm{Re}\int_D \psi^\nu(s)\,\mathrm{d}W_1(s)\,\mathrm{d}x + \nu\|q_1\|_0^2 t.\end{aligned} \tag{3.20}$$

Then Gronwall lemma implies

$$\mathbb{E}\|\psi^\nu(t)\|_0^2 \leqslant \mathbb{E}\|\psi_0\|_0^2 + \|q_1\|_0^2, \quad t \geqslant 0.$$

Furthermore applying Itô's formula to $\|\psi^\nu\|_0^{2p}$ yields

$$\begin{aligned}\|\psi^\nu(t)\|_0^{2p} - \|\psi_0\|_0^{2p} \leqslant& p\int_0^t \|\psi^\nu(s)\|_0^{2p-2}(-\nu\|\psi^\nu(s)\|_0^2 + \nu\|q_1\|_0^2)\,\mathrm{d}s\\ &+ 2p\int_0^t \|\psi^\nu(s)\|_0^{2p-2}\langle\psi^\nu(s), -\mathrm{i}\sqrt{\nu}\,\mathrm{d}W_1(s)\rangle\\ &+ 2p(p-1)\nu\int_0^t \|\psi^\nu(s)\|_0^{2p}\|q_1\|_0^2\,\mathrm{d}s\\ \leqslant& -\frac{\nu}{2}\int_0^t \|\psi^\nu(s)\|_0^{2p}\,\mathrm{d}s + \nu C_{2p}\|q_1\|_0^{2p} t\\ &+ 2p\int_0^t \|\psi^\nu(s)\|_0^{2p-2}\langle\psi^\nu, -\mathrm{i}\sqrt{\nu}\,\mathrm{d}W_1(s)\rangle\end{aligned}$$

for some constant $C_{2p} > 0$. Still by Gronwall lemma we have

$$\mathbb{E}\|\psi^\nu(t)\|_0^{2p} \leqslant \mathbb{E}\|\psi_0\|_0^{2p} + C_{2p}\|q_1\|_0^{2p}, \quad t \geqslant 0. \tag{3.21}$$

Moreover for any $T > 0$

$$\sup_{0\leqslant t\leqslant T}\|\psi^\nu(t)\|_0^{2p} \leqslant \|\psi_0\|_0^{2p} + \nu C_p\|q_1\|_0^{2p} T + 2p\sup_{0\leqslant t\leqslant T}\int_0^t \|\psi^\nu(s)\|_0^{2p-2}\langle\psi^\nu, -\mathrm{i}\sqrt{\nu}\,\mathrm{d}W_1(s)\rangle.$$

By the maximal estimates for stochastic integral [24, Lemma 7.2] and estimate (3.21)

$$\mathbb{E}\sup_{0\leqslant t\leqslant T}\left|\int_0^t \|\psi^\nu(s)\|_0^{2p-2}\langle\psi^\nu(s), -\mathrm{i}\sqrt{\nu}\,\mathrm{d}W_1(s)\rangle\right|^2 \leqslant C_{2p}(\mathbb{E}\|\psi_0\|_0^{4p} + \|q_1\|_0^{4p})T.$$

Then we have

$$\mathbb{E}\sup_{0\leqslant t\leqslant T}\|\psi^\nu(t)\|_0^{2p} \leqslant \mathbb{E}\|\psi_0\|_0^{2p} + C_{2p}(\mathbb{E}\|\psi_0\|_0^{4p} + \|q_1\|_0^{4p})T.$$

The proof is complete.

Then we give the following estimates of solutions to (2.16)–(2.18) in $\mathcal{E}_0$.

Lemma 3.2 *Assume* $(\psi_0, \phi_0, \theta_0) \in \mathcal{E}_0$, $q_1 \in H_0^1$ *and* $q_2 \in H$. *Then for any* $T > 0$ *every solution of problem (2.16)-(2.18) satisfies* $(\psi^\nu, \phi^\nu, \theta^\nu) \in L^\infty(0, T; L^2(\Omega, \mathcal{E}_0))$.

Proof By Itô's formula, from (2.17) we have

$$\begin{aligned}
&\nu\|\theta^\nu(t)\|_0^2 - \nu\|\theta_0\|_0^2 \\
=&2\int_0^t \left[-(\nu^\alpha - \nu\delta)\|\theta^\nu\|_0^2 - \langle\nabla\phi^\nu, \nabla\theta^\nu\rangle - (1 - \delta(\nu^\alpha - \nu\delta))\langle\phi^\nu, \theta^\nu\rangle\right] \mathrm{d}s \\
&+ 2\int_0^t \langle|\psi^\nu|^2, \theta^\nu\rangle \,\mathrm{d}s + 2\sqrt{\nu^{1+\alpha}}\int_0^t \langle\theta^\nu(s), \mathrm{d}W_2(s)\rangle + \nu^\alpha\|q_2\|_0^2 t .
\end{aligned}$$

Then choosing $\delta = \nu^\alpha/2 > 0$, we have

$$\begin{aligned}
&\nu\|\theta^\nu(t)\|_0^2 - \nu\|\theta_0\|_0^2 + \|\phi^\nu(t)\|_1^2 - \|\phi_1\|_1^2 + \|\phi^\nu(t)\|_0^2 - \|\phi_0\|_0^2 \\
&+ \frac{\nu^\alpha}{4}\int_0^t \left[\nu\|\theta^\nu(s)\|_0^2 + \|\phi^\nu(s)\|_1^2 + \|\phi^\nu(s)\|_0^2\right] \mathrm{d}s \\
\leqslant& 2\int_0^t \langle|\psi^\nu(s)|^2, \theta^\nu(s)\rangle \,\mathrm{d}s + 2\sqrt{\nu^{1+\alpha}}\int_0^t \langle\theta^\nu, \mathrm{d}W_2(s)\rangle + \nu^\alpha\|q_2\|_0^2 t
\end{aligned}$$

for some constant $C > 0$ independent of ν . We treat the second term of the righthand of the above inequality. Applying Itô formula to $\|\psi^\nu(t)\|_1^2$ yields

$$\begin{aligned}
\|\psi^\nu(t)\|_1^2 - \|\psi_0\|_1^2 =& -2\nu\int_0^t \|\psi^\nu(s)\|_1^2 \,\mathrm{d}s + 2\int_0^t \langle\phi^\nu(s)\psi^\nu(s), \psi_t^\nu(s) + \nu\psi^\nu(s)\rangle \,\mathrm{d}s \\
&- 2\int_0^t \langle\psi_t^\nu(s) + \nu\psi^\nu(s), \sqrt{\nu}\,\mathrm{d}W_1(s)\rangle + \nu\|q_1\|_1^2 t .
\end{aligned}$$

Notice that, by exactly same discussion in [19]

$$\frac{\mathrm{d}}{\mathrm{d}t}\langle\phi^\nu, |\psi^\nu|^2\rangle = \langle\phi_t^\nu, |\psi^\nu|^2\rangle + \langle\phi^\nu, \psi_t^\nu\bar{\psi^\nu} + \psi^\nu\bar{\psi^\nu}_t\rangle + \nu\mathrm{Tr}\,(\phi^\nu q_1^2) , \tag{3.22}$$

and

$$2\langle\phi^\nu\psi^\nu, \psi_t^\nu + \nu\psi^\nu\rangle = \langle\phi^\nu, \psi_t^\nu\bar{\psi^\nu} + \psi\bar{\psi^\nu}_t\rangle + 2\nu\langle\phi^\nu, |\psi^\nu|^2\rangle . \tag{3.23}$$

Then introduce

$$H^\nu(t) = \nu\|\theta^\nu(t)\|_0^2 + \|\phi^\nu(t)\|_0^2 + \|\phi^\nu(t)\|_1^2 + 2\|\psi^\nu(t)\|_1^2 - 2\langle\phi^\nu(t), |\psi^\nu(t)|^2\rangle , \tag{3.24}$$

we have

$$\begin{aligned}
&H^\nu(t)-H^\nu(0)+\frac{\nu^\alpha}{8}\int_0^t H^\nu(s)\,\mathrm{d}s+\frac{\nu^\alpha}{8}\int_0^t[\nu\|\theta^\nu(s)\|_0^2+\|\phi^\nu(s)\|_1^2+\|\phi^\nu(s)\|_0^2]\,\mathrm{d}s\\
&+\frac{15\nu^\alpha}{4}\int_0^t\|\psi^\nu(s)\|_1^2-\frac{19\nu^\alpha}{4}\int_0^t\langle\phi^\nu(s),|\psi^\nu(s)|^2\rangle\,\mathrm{d}s\\
\leqslant&-2\nu^\alpha\int_0^t\mathrm{Tr}(\phi^\nu(s)q_1^2)\,\mathrm{d}s-4\sqrt{\nu^\alpha}\int_0^t\langle\psi_t^\nu(s)+\nu\psi^\nu(s),\mathrm{d}W_1(s)\rangle\\
&+2\sqrt{\nu^{1+\alpha}}\int_0^t\langle\theta^\nu(t),\mathrm{d}W_2(s)\rangle+2\nu\|q_1\|_1^2t+\nu^\alpha\|q_2\|_0^2t\,.
\end{aligned}$$

Notice that for some ν-independent constant $C>0$

$$\begin{aligned}
\frac{19\nu^\alpha}{4}\langle\phi^\nu,|\psi^\nu|^2\rangle&\leqslant\frac{\nu^\alpha}{16}\|\phi^\nu\|_1^2+\frac{\nu^\alpha}{16}\|\psi^\nu\|_1^2+C\nu^\alpha\|\psi^\nu\|_0^6\\
2\nu^\alpha\mathrm{Tr}(\phi^\nu q_1^2)&\leqslant\frac{\nu^\alpha}{32}\|\phi^\nu\|_1^2+C\nu^\alpha(\|q_1\|_1^2+\|q_1\|_0^6)\,.
\end{aligned}$$

Then we deduce that

$$\begin{aligned}
&H^\nu(t)-H^\nu(0)+\frac{\nu^\alpha}{8}\int_0^t H^\nu(s)\,\mathrm{d}s\\
\leqslant&C\nu^\alpha\int_0^t\|\psi^\nu(s)\|_0^6-4\sqrt{\nu^\alpha}\int_0^t\langle\psi_t^\nu(s)+\nu\psi^\nu(s),\mathrm{d}W_1(s)\rangle+\nu^\alpha\|q_2\|_0^2t\\
&+2\sqrt{\nu^{1+\alpha}}\int_0^t\langle\theta^\nu(t),\mathrm{d}W_2(s)\rangle+(2+C)\nu\|q_1\|_1^2t+C\nu\|q_1\|_0^6t\,.
\end{aligned}$$

By the Gronwall's lemma we have

$$\mathbb{E}H^\nu(t)\leqslant e^{-t\nu/4}\mathbb{E}H^\nu(0)+C(\mathbb{E}\|\psi_0\|_0^6,\|q_1\|_0^6,\|q_1\|_1^2,\|q_2\|_0^2) \tag{3.25}$$

for some constant $C(\mathbb{E}\|\psi_0\|_0^6,\|q_1\|_0^6,\|q_1\|_1^2,\|q_2\|_0^2)>0$. Furthermore, for some constant $C>0$ [19]

$$H^\nu(t)\geqslant\nu\|\theta^\nu\|_0^2+\|\phi^\nu\|_0^2+\frac{1}{2}\|\phi(t)\|_1^1+\|\psi^\nu(t)\|_1^2-C\|\psi^\nu(t)\|_0^6-2\|\psi^\nu(t)\|_0^2$$

we then have

$$\begin{aligned}
&\mathbb{E}\left[\nu\|\theta^\nu(t)\|_0^2+\|\phi^\nu(t)\|_1^2+\|\phi^\nu(t)\|_0^2+\|\psi^\nu(t)\|_1^2\right]\\
\leqslant&e^{-t\nu/4}\mathbb{E}H^\nu(0)+C(\mathbb{E}\|\psi_0\|_0^6,\|q_1\|_0^6,\|q_1\|_1^2,\|q_2\|_0^2)\,.
\end{aligned}$$

Now by exactly same discussion as that for SKGS equations [19], we also have for any $p>0$

$$\mathbb{E}\sup_{0\leqslant t\leqslant T}[H^\nu(t)]^p\leqslant\mathbb{E}[H^\nu(0)]^p+C_T(\mathbb{E}\|\psi_0\|_0^{6p},\|q_1\|_0^2,\|q_1\|_1^2,\|q_2\|_0^2) \tag{3.26}$$

for some constant $C_T(\mathbb{E}\|\psi_0\|_0^{6p}, \|q_1\|_0^2, \|q_1\|_1^2, \|q_2\|_0^2) > 0$.

Furthermore from (1.5)–(1.6) one easily has

$$\mathbb{E}\|\psi_t^\nu\|_{L^2(0,T;H^{-1})} \leqslant C_T(\mathbb{E}H_0^\nu(0), \mathbb{E}\|\psi_0\|_0^6, \|q_1\|_1^2, \|q_1\|_0^2, \|q_2\|_0^2)$$

and

$$\mathbb{E}\|\nu^{\alpha/2}\phi_t^\nu\|_{L^2(0,T;H)} \leqslant C_T(\mathbb{E}H_0^\nu(0), \mathbb{E}\|\psi_0\|_0^6, \|q_1\|_1^2, \|q_1\|_0^2, \|q_2\|_0^2).$$

Then by the above estimates and Lemma 2.1 we have the following result

Theorem 3.2 *Assume $q_1 \in H_0^1$ and $q_2 \in H$. Then for any $T > 0$ and any $(\psi_0, \phi_0, \phi_1) \in \mathcal{E}_0$, the solution $(\psi^\nu, \phi^\nu, \phi_t^\nu)$ to equations (1.5)–(1.8) satisfies*

$$\begin{aligned}&\mathbb{E}\sup_{0\leqslant t\leqslant T}\|(\psi^\nu(t), \phi^\nu(t), \sqrt{\nu}\phi_t^\nu(t))\|_{\mathcal{E}_0}^p + \mathbb{E}\|(\psi_t^\nu, \nu^{\alpha/2}\phi_t^\nu)\|_{L^2(0,T;H^{-1}\times H)}^2\\ &\leqslant C_T(\psi_0, \phi_0, \phi_1, q_1, q_2)\end{aligned}$$

for any $p \geqslant 2$ and some constant $C_T(\psi_0, \phi_0, \phi_1, q_1, q_2) > 0$. Moreover probability measures

$$\{\mathcal{D}((\psi^\nu, \nu^{\alpha/2}\phi^\nu))\}_{0<\nu\leqslant 1},$$

the distributions of $\{(\psi^\nu, \nu^{\alpha/2}\phi^\nu)\}_\nu$, is tight in space $C(0,T; H\times H)$.

4 Limit of the solutions

Now we are concerned with the limit of the solutions of the SKGS equations (1.5)–(1.8) as $\nu \to 0$.

First we have the following convergence result.

Theorem 4.1 *Assume $q_1 \in H_0^1$ and $q_2 \in H$, $(\psi_0, \phi_0, \phi_1) \in \mathcal{E}_0$. For $0 < \alpha \leqslant 1$ and any $T > 0$, ψ^ν converges strongly in probability to ψ in space $C(0,T;H)$ and ϕ^ν converges weakly in probability to ϕ in space $L^2(0,T;H_0^1)$ with (ψ,ϕ) solving equations (1.9)–(1.11). For $\alpha = 0$, (ψ^ν, ϕ^ν) converges strongly in probability to (ψ,ϕ) in space $C(0,T;H\times H)$ with (ψ,ϕ) solving (1.13)–(1.15).*

Proof For $0 < \alpha \leqslant 1$, by Theorem 3.2, for any $\kappa > 0$, there are bounded sets $K_{1,\kappa} \subset L^2(0,T;H_0^1)$, which is compact in $C(0,T;H)$, $K_{2,\kappa} \subset L^2(0,T;H_0^1)$ and $K_{3,\kappa} \subset L^2(0,T;H)$ such that

$$\mathbb{P}\{\psi^\nu \in K_{1,\kappa}, \phi^\nu \in K_{2,\kappa} \text{ and } \sqrt{\nu}\phi_t^\nu \in K_{3,\kappa}\} \geqslant 1-\kappa.$$

Next we restrict on the set of $\Omega_\kappa \subset \Omega$ as

$$\Omega_\kappa = \{\omega \in \Omega : \psi^\nu \in K_{1,\kappa}\ \phi^\nu \in K_{2,\kappa} \text{ and } \sqrt{\nu}\phi_t^\nu \in K_{3,\kappa}\}.$$

We show that for almost all $\omega \in \Omega_\kappa$, there is a unique $(\psi, \phi) \in C(0,T; H \times H)$, which solves (1.9)–(1.11), such that

$$\begin{aligned} &\psi^\nu \text{ converges strongly to } \psi \text{ in } C(0,T;H),\\ &\phi^\nu \text{ converges weakly to } \phi \text{ in } L^2(0,T;H_0^1) \end{aligned}$$

as $\nu \to 0$. Then by the arbitrariness of κ, we have the above convergence in probability.

By the compactness of $K_{1,\kappa}$ and the boundedness of $K_{2,\kappa}$, for any sequences $\{\psi^{\nu_n}\}$, $\{\phi^{\nu_n}\}$ there is a subsequence, which we still denote by $\{\psi^{\nu_n}\}, \{\phi^{\nu_n}\}$, convergent in space $C(0,T;H)$ strongly and $L^2(0,T;H_0^1)$ weakly respectively. Write the limit as ψ, ϕ respectively, then next we show (ψ,ϕ) solves (1.9)–(1.10). For simplicity we omit the subscript of ν_n. Choosing testing function $\rho \in C^\infty([0,T];C_0^\infty(D))$ and multiplying on both sides of (1.5)–(1.6), we have

$$i\langle\psi^\nu(t),\rho(t)\rangle - i\langle\psi_0,\rho(0)\rangle - \int_0^t \langle\nabla\psi^\nu(s),\nabla\rho(s)\rangle \mathrm{d}s + i\nu\int_0^t \langle\psi^\nu(s),\rho(s)\rangle\,\mathrm{d}s$$
$$+\int_0^t \langle\phi^\nu(s)\psi^\nu(s),\rho(s)\rangle\,\mathrm{d}s = \sqrt{\nu}\int_0^t \langle\rho(s),\mathrm{d}W_1(s)\rangle, \tag{4.27}$$

$$\begin{aligned}\nu\langle\phi_t^\nu(t),\rho(t)\rangle &+ \nu^\alpha\langle\phi^\nu(t),\rho(t)\rangle - \nu\langle\phi_1,\rho(0)\rangle - \nu^\alpha\langle\phi_0,\rho(0)\rangle\\ &+\int_0^t \langle\nabla\phi^\nu(s),\nabla\rho(s)\rangle\,\mathrm{d}s + \int_0^t\langle\phi^\nu(s),\rho(s)\rangle\,\mathrm{d}s\\ &-\int_0^t \langle|\psi^\nu(s)|^2,\rho(s)\rangle\,\mathrm{d}s = \sqrt{\nu^{1+\alpha}}\int_0^t\langle\rho(s),\mathrm{d}W_2(s)\rangle.\end{aligned} \tag{4.28}$$

Notice that

$$\nu^{1+\alpha}\mathbb{E}\left|\int_0^t \langle\rho(s),\mathrm{d}W_2(s)\rangle\right|^2 = \nu^{1+\alpha}\int_0^t\langle\rho(s),q_2\rangle\,\mathrm{d}s \to 0, \quad \text{as} \quad \nu\to 0$$

and by the choice of $K_{3,\kappa}$,

$$\nu\mathbb{E}\langle\phi_t^\nu(t),\rho(t)\rangle \leqslant \left[\mathbb{E}\nu^2\|\phi_t^\nu(t)\|_0^2\right]^{1/2}\|\rho(t)\|_0 \to 0, \quad \text{as} \quad \nu \to 0.$$

Furthermore by the choice of $K_{1,\kappa}$ and the compactly embedding $H^1 \subset L^4(D)$, $|\psi^\nu|^2$ is bounded in $L^2(0,T;H)$ and

$$|\psi^\nu(x,t) - \psi(x,t)| \to 0, \quad \text{as} \quad \nu\to 0$$

for almost all $(x,t) \in [0,t] \times D$. Then by Lions–Aubin Lemma, $|\psi^\nu|^2$ converges strongly to $|\psi|^2$ in $L^2(0,T;H)$. Then from (4.28), ϕ solves

$$-\Delta\phi + \phi = |\psi|^2 .$$

Now we derive the equation solved by ψ. Similar discussion one easily has

$$\nu\mathbb{E}\left|\int_0^t \langle\rho(s), \mathrm{d}W_1(s)\rangle\right|^2 = \nu^\alpha \int_0^t \langle\rho(s), q_1\rangle \,\mathrm{d}s \to 0, \quad \text{as} \quad \nu \to 0$$

and

$$\left|\nu\int_0^t \langle\psi^\nu(s), \rho(s)\rangle \,\mathrm{d}s\right|^2 \leqslant \nu^2 \int_0^t \|\psi^\nu(s)\|_0^2 \,\mathrm{d}s \int_0^t \|\rho(s)\|_0^2 \,\mathrm{d}s \to 0, \quad \text{as} \quad \nu \to 0 .$$

Also notice that

$$\int_0^t \langle\psi^\nu(s)\phi^\nu(s) - \psi(s)\phi(s), \rho(s)\rangle \,\mathrm{d}s = \int_0^t \langle(\psi^\nu(s) - \psi(s))\phi^\nu(s), \rho(s)\rangle \,\mathrm{d}s$$
$$+ \int_0^t \langle(\phi^\nu(s) - \phi(s))\psi(s), \rho(s)\rangle \,\mathrm{d}s \to 0, \quad \text{as} \quad \nu \to 0 .$$

Then we conclude that (ψ,ϕ) solves the limit equation (1.9)–(1.11). By the uniqueness of solution to equation (1.9)–(1.11) [16, 28], we complete the proof for $0 < \alpha \leqslant 1$.

For $\alpha = 0$, by Theorem 3.2, $K_{2,\kappa}$ can be chosen as a compact set in $C(0,T;H)$. Then similar discussion as above we have, from (4.28), (ψ,ϕ) solves (1.13)–(1.15). By the uniqueness of solution to equation (1.13)–(1.15). The proof is complete.

By a similar discussion for deterministic case [2], for $0 < \alpha \leqslant 1$ we show a strong convergence of ϕ^ν in space $L^2(0,T;H_0^1)$ for any $T > 0$. In fact we have the following result.

Theorem 4.2 *Assume $q_1 \in H_0^1$ and $q_2 \in H$, $(\psi_0,\phi_0,\phi_1) \in \mathcal{E}_0$. For $0 \leqslant \alpha \leqslant 1$ and any $T > 0$, (ψ^ν,ϕ^ν) converges strongly in probability to (ψ,ϕ) in space $L^2(0,T;H_0^1 \times H_0^1)$.*

Proof For any $\kappa > 0$, we still consider $\omega \in \Omega_\kappa$ which defined in the proof of Theorem 4.1.

First for $0 < \alpha \leqslant 1$, let

$$H_0^\nu(t) = \frac{1}{2}\left[\nu\|\phi_t^\nu(t)\|_0^2 + \|\phi^\nu(t)\|_1^2 + \|\phi^\nu(t)\|_0^2\right] + \|\psi^\nu(t)\|_1^2 - \langle\phi^\nu(t), |\psi^\nu(t)|^2\rangle$$

and

$$H_0(t) = \frac{1}{2}\left[\|\phi(t)\|_1^2 + \|\phi(t)\|_0^2\right] + \|\psi(t)\|_1^2 - \langle\phi(t), |\psi(t)|^2\rangle .$$

From equations (1.5)–(1.6) we have

$$
\begin{aligned}
&H_0^\nu(t) - H_0^\nu(0) \qquad (4.29)\\
=&-\nu^\alpha\int_0^t \|\phi_t^\nu(s)\|_0^2\,\mathrm{d}s - 2\nu\int_0^t \|\psi^\nu(s)\|_1^2\,\mathrm{d}s + \sqrt{\nu^{1+\alpha}}\int_0^t \langle\phi^\nu(s),\mathrm{d}W_2(s)\rangle\\
&+\nu^\alpha\|q_2\|_0^2\,t + \nu\|q_1\|_1^2\,t + 2\nu\int_0^t\langle\phi^\nu(s),|\psi^\nu(s)|^2\rangle\,\mathrm{d}s - \nu\int_0^t \mathrm{Tr}(\phi^\nu(s)q_1^2)\,\mathrm{d}s\\
&-2\sqrt{\nu}\int_0^t\langle\psi_t^\nu(s)+\nu\psi^\nu(s),\mathrm{d}W_1(s)\rangle\,.
\end{aligned}
$$

Notice that

$$
\begin{aligned}
&\frac{1}{2}\left[\|\phi^\nu(t)-\phi(t)\|_1^2 + \|\phi^\nu(t)-\phi(t)\|_0^2 + \nu\|\phi_t^\nu(t)\|_0^2\right] + \|\psi^\nu(t)-\psi(t)\|_1^2\\
=&H_0^\nu(t) - H_0(t) + \langle\phi^\nu(t),|\psi^\nu(t)|^2\rangle - \langle\phi(t),|\psi(t)|^2\rangle \qquad (4.30)\\
&-\mathrm{Re}\int_D\Big[\nabla(\phi^\nu(t)-\phi(t))\nabla\phi(t) + (\phi^\nu(t)-\phi(t))\phi(t)\\
&+2\nabla(\psi^\nu(t)-\psi(t))\nabla\bar{\psi}(t)\Big]\,\mathrm{d}x\,, \qquad (4.31)
\end{aligned}
$$

and by the conservation of the limit equation (1.9)–(1.10),

$$
H_0^\nu(t) - H_0(t) = H_0^\nu(t) - H_0(0) = H_0^\nu(t) - H_0^\nu(0)\,.
$$

Then from (4.29), we have

$$
\begin{aligned}
&\frac{1}{2}\left[\|\phi^\nu(t)-\phi(t)\|_1^2 + \|\phi^\nu(t)-\phi(t)\|_0^2 + \nu\|\phi_t^\nu(t)\|_0^2\right] + \|\psi^\nu(t)-\psi(t)\|_1^2\\
&+\nu^\alpha\int_0^t \|\phi_t^\nu(s)\|_0^2\,\mathrm{d}s\\
=&-2\nu^\alpha\int_0^t\|\psi^\nu(s)\|_1^2\,\mathrm{d}s + \sqrt{\nu}\int_0^t\langle\phi^\nu(s),\mathrm{d}W_2(s)\rangle + \nu^\alpha\|q_2\|_0^2\,t + \nu\|q_1\|_1^2\,t\\
&+2\nu^\alpha\int_0^t\langle\phi^\nu(s),|\psi^\nu(s)|^2\rangle\,\mathrm{d}s - \nu^\alpha\int_0^t\mathrm{Tr}(\phi^\nu(s)q_1^2)\,\mathrm{d}s\\
&-2\sqrt{\nu}\int_0^t\langle\psi_t^\nu(s)+\nu\psi^\nu(s),\mathrm{d}W_1(s)\rangle\\
&-\mathrm{Re}\int_D\left[\nabla(\phi^\nu(t)-\phi(t))\nabla\phi(t) + (\phi^\nu(t)-\phi(t))\phi(t) + 2\nabla(\psi^\nu(t)-\psi(t))\nabla\bar{\psi}(t)\right]\mathrm{d}x\\
&+\langle\phi^\nu(t),|\psi^\nu(t)|^2\rangle - \langle\phi(t),|\psi(t)|^2\rangle\,.
\end{aligned}
$$

By the weak convergence of $\phi^\nu\to\phi$ in space $L^2(0,T;H_0^1)$ and strong convergence of $|\psi^\nu(t)|^2\to|\psi(t)|^2$ in space $L^2(0,T;H)$, we have

$$\int_0^T \langle\phi^\nu(t), |\psi^\nu(t)|^2\rangle\,\mathrm{d}t - \int_0^T \langle\phi(t), |\psi(t)|^2\rangle\,\mathrm{d}t$$
$$= \int_0^T \langle\phi^\nu(t), |\psi^\nu(t)|^2 - |\psi(t)|^2\rangle\,\mathrm{d}t + \int_0^T \langle\phi^\nu(t) - \phi(t), |\psi(t)|^2\rangle\,\mathrm{d}t \to 0\,, \quad \nu \to 0\,.$$

Similarly by the weak convergence of ϕ^ν and ψ^ν in space $L^2(0,T;H_0^1)$ we have

$$\int_0^T \mathrm{Re}\int_D \big[\nabla(\phi^\nu(t) - \phi(t))\nabla\phi(t) + (\phi^\nu(t) - \phi(t))\phi(t)$$
$$+ 2\nabla(\psi^\nu(t) - \psi(t))\nabla\bar{\psi}(t)\big]\,\mathrm{d}x \to 0\,, \quad \nu \to 0\,.$$

Now for $\alpha = 0$. Let $\Phi^\nu(t) = \phi^\nu(t) - \phi(t)$, then

$$\Phi^\nu_t = \Delta\Phi^\nu - \Phi^\nu + |\psi^\nu|^2 - |\psi|^2 + \sqrt{\nu}\dot{W}_2 - \nu\phi^\nu_{tt}\,, \quad \Phi^\nu(0) = 0\,.$$

By the Itô formula we have

$$\|\Phi^\nu(t)\|_0^2 + 2\int_0^t \big[\|\Phi^\nu(s)\|_1^2 + \|\Phi^\nu(s)\|_0^2\big]\,\mathrm{d}s$$
$$=2\int_0^t \langle|\psi^\nu(s)|^2 - |\psi(s)|^2, \Phi^\nu(s)\rangle\,\mathrm{d}s + 2\sqrt{\nu}\int_0^t \langle\Phi^\nu(s), \mathrm{d}W_2(s)\rangle$$
$$- 2\nu\int_0^t \langle\phi^\nu_{tt}(s), \Phi^\nu(s)\rangle\,\mathrm{d}s + \nu\|q_2\|_0^2\,.$$

Noticing that $|\psi^\nu|^2$ strongly converges to $|\psi|^2$ in $L^2(0,T;H)$ and

$$\nu\int_0^t \langle\phi^\nu_{tt}, \Phi^\nu(s)\rangle\,\mathrm{d}s \tag{4.32}$$
$$=\nu\langle\phi^\nu_t(t), \phi^\nu(t)\rangle - \nu\langle\phi^\nu_t(0), \phi^\nu(0)\rangle - \nu\int_0^t \|\phi^\nu(s)\|_0^2\,\mathrm{d}s$$
$$- \nu\langle\phi^\nu_t(t), \phi(t)\rangle + \nu\langle\phi^\nu_t(0), \phi(0)\rangle + \nu\int_0^t \langle\phi^\nu_t(s), \phi_t(s)\rangle\,\mathrm{d}s$$

by the choice of Ω_κ, we have for almost all $\omega \in \Omega_\kappa$

$$\phi^\nu \to \phi \quad \text{strongly in} \quad L^2(0,T;H_0^1)\,.$$

Let $\Psi^\nu(t) = \psi^\nu(t) - \psi(t)$, then

$$\Psi^\nu_t = \mathrm{i}\Delta\Psi^\nu - \nu\psi^\nu + \mathrm{i}\psi^\nu\phi^\nu - \mathrm{i}\psi\phi - \mathrm{i}\sqrt{\nu}\dot{W}_1\,, \quad \Psi^\nu(0) = 0\,.$$

Multiplying Ψ^ν on both sides of the above equation and taking imaginary part yield

$$\int_0^t \|\Psi^\nu(s)\|_1^2\,\mathrm{d}s = \int_0^t \mathrm{Im}\int_D \big[\psi^\nu(s)\phi^\nu(s) - \psi(s)\phi(s)\big]\Psi^\nu(s)\,\mathrm{d}x\,\mathrm{d}s$$
$$- \nu\int_0^t \mathrm{Im}\int_D \Psi^\nu(s)\psi^\nu(s)\,\mathrm{d}x\,\mathrm{d}s - \sqrt{\nu}\int_0^t \mathrm{Im}\int_D \Psi^\nu(s)dW_1(s)\,\mathrm{d}x\,.$$

By the strong convergence of Ψ^ν in $C(0,T;H)$ and weak convergence of $\psi^\nu\phi^\nu$ to $\psi\phi$ in $L^2(0,T;H)$, we have

$$\int_0^t \|\Psi^\nu(s)\|_1^2\,\mathrm{d}s \to 0, \quad \nu \to 0.$$

The proof is complete.

Now assume

$$-\Delta\phi_0 + \phi_0 = |\psi_0|^2 \tag{4.33}$$

we show the Schrödinger type limit equation.

Theorem 4.3 *Assume $q_1 \in H_0^1$, $q_2 \in H$ and (4.33) holds. For $0 < \alpha \leqslant 1$, any $T > 0$ and $(\psi_0, \phi_0, \phi_1) \in \mathcal{E}_0$, ψ^ν converges to ψ in space $C(0,T;H)$ with ψ solving Schrödinger equation (1.12).*

Proof For any $\kappa > 0$, we consider $\omega \in \Omega_\kappa$ defined in the proof of Theorem 4.1. Rewrite (4.27) as

$$\begin{aligned}
&\mathrm{i}\langle\psi^\nu(t), \rho(t)\rangle - \mathrm{i}\langle\psi_0, \rho(0)\rangle - \int_0^t \langle\nabla\psi^\nu(s), \nabla\rho(s)\rangle\,\mathrm{d}s \\
&+ \int_0^t \langle\psi^\nu(s)A^{-1}|\psi^\nu|^2(s), \rho(s)\rangle\,\mathrm{d}s \\
=& \sqrt{\nu}\int_0^t \langle\rho(s), \mathrm{d}W_1(s)\rangle + \int_0^t \langle\psi^\nu(s)[\phi^\nu(s) - A^{-1}|\psi^\nu|^2], \rho(s)\rangle\,\mathrm{d}s \\
&+ \mathrm{i}\nu\int_0^t \langle\psi^\nu(s), \rho(s)\rangle\,\mathrm{d}s.
\end{aligned}$$

Noticing the choice of $K_{1,\kappa}$ and the smooth property of A^{-1}, we have

$$\int_0^t \langle\psi^\nu(s)A^{-1}|\psi^\nu|^2(s), \rho(s)\rangle\,\mathrm{d}s \to \int_0^t \langle\psi(s)A^{-1}|\psi(s)|^2, \rho(s)\rangle\,\mathrm{d}s$$

Next we consider the limit of term

$$D^\nu(t) = \phi^\nu(t) - A^{-1}|\psi^\nu(t)|^2.$$

We repeat the argument for Schrödinger approximation to deterministic Zakharov equations [25, Theorem 5]. For this define

$$R^\nu(t) = \int_0^t \int_0^s D^\nu(\tau)\,\mathrm{d}\tau\,\mathrm{d}s.$$

Then R^ν solves the following equation

$$\nu R_{tt}^\nu + \nu^\alpha R_t^\nu + AR^\nu = \nu^\alpha f^\nu(t), \quad R^\nu(0) = 0, R_t^\nu(0) = 0 \tag{4.34}$$

with

$$\nu^\alpha f^\nu(t) = \nu\phi_0 + \nu\phi_1 t + \nu^\alpha \phi_0 t - \nu^\alpha \int_0^t A^{-1}|\psi^\nu(s)|^2 \, ds + \sqrt{\nu^{1+\alpha}} \int_0^t W_2(s) \, ds .$$

By the estimates in Section and the maximal estimate of Wiener process, we have

$$\nu^\alpha \mathbb{E} \sup_{0 \leqslant t \leqslant T} \|f^\nu(t)\|_0^2 \leqslant \nu^\alpha C_T$$

for some constant C_T. From equation (4.34)

$$\frac{1}{2}\frac{d}{dt}\left[\nu\|R_t^\nu\|_0^2 + \|R^\nu\|_1^2\right] \leqslant -\frac{\nu^\alpha}{2}\|R_t^\nu(t)\|_0^2 + \frac{\nu^\alpha}{2}\|f^\nu(t)\|_0^2 .$$

Then we have

$$\mathbb{E} \sup_{0 \leqslant t \leqslant T} \|R^\nu(t)\|_1^2 \leqslant \nu^\alpha C_T . \tag{4.35}$$

Now for any $\varrho \in C_0^\infty([0,T] \times D)$, we have

$$\mathbb{E} \int_0^t \langle D^\nu(s), \varrho(s) \rangle \, ds = \mathbb{E} \int_0^t \langle R^\nu(s), \varrho_{tt}(s) \rangle \, ds \to 0, \quad \nu \to 0 .$$

By the density of $C_0^\infty([0,T] \times D)$ in $L^2(0,T;H)$, we have

$$\mathbb{E} \int_0^t \langle \psi^\nu(s)[\phi^\nu(s) - A^{-1}|\psi^\nu|^2], \rho(s) \rangle \, ds \to 0, \quad \nu \to 0 . \tag{4.36}$$

Finally by the uniqueness of solution to equation (1.12) (see e.g. [16,28]), we complete the proof.

5 Large deviations for SKGS equations

Next we consider the large deviations for the solution to SKGS equations (1.5)–(1.8) with $\alpha = 0$ by a weak convergence approach. For any $T > 0$, we build the large deviation principle for (ψ^ν, ϕ^ν) in space $C(0,T; H \times H) \cap L^2(0,T; H_0^1 \times H_0^1)$.

We recall some standard definitions and results on large deviations theory [24, e.g.]. Let $\{u^\nu\}$ be a family of random variables defined on a probability space $(\Omega, \mathcal{F}, \mathbb{P})$ and taking values in a Polish space E. The theory of large deviations is mainly concerned with events B for which probabilities $\mathbb{P}(u^\nu \in B)$ converge to zero exponentially quickly as the fast-time parameter $\nu \to 0$. The exponential decay rate of such probabilities is characterised by a "rate function".

Definition 5.1 *A function $I : E \to [0,\infty]$ is called a* rate function *if it is lower semicontinuous. A rate function I is called a* good rate function *if for each $M < \infty$ the level set $\{u \in E : I(u) \leqslant M\}$ is compact in E.*

Definition 5.2 *The sequence $\{u^\nu\}$ is said to satisfy the large deviation principle (LDP) on E with rate function I and speed $\beta(\nu) \to \infty$ as $\nu \to 0$, if the following two conditions hold:*

1. *there exists a large deviation upper bound, for each closed subset F of E,*

$$\limsup_{\nu\to 0} \frac{1}{\beta(\nu)} \log \mathbb{P}(u^\nu \in F) \leqslant -I(F) = -\inf_{u\in F} I(u);$$

2. *there exists a large deviation lower bound, for each open subset G of E,*

$$\liminf_{\nu\to 0} \frac{1}{\beta(\nu)} \log \mathbb{P}(u^\nu \in G) \geqslant -I(G) = -\inf_{u\in G} I(u).$$

The large deviation principle and the following Laplace principle are equivalent in a Polish space [7].

Definition 5.3 *The sequence $\{u^\nu\}$ is said to satisfy the Laplace principle on E with rate function I and speed $\beta(\nu) \to \infty$, as $\nu \to 0$, if for each real-valued, bounded continuous function h defined on E,*

$$\lim_{\nu\to 0} \frac{1}{\beta(\nu)} \log \mathbb{E} \exp\left[-\beta(\nu) h(u^\nu)\right] = -\inf_{y\in E} \{h(y) + I(y)\}.$$

Here we consider the LDP for solution processes to SPDEs driven by a Wiener process. We follow a weak convergence approach to show a Laplace principle. For this we give the following sufficient conditions for a stochastic process $\{u^\nu\}$ satisfying the Laplace principle.

Let $(U_1, \|\cdot\|_{U_1})$ be a separable Hilbert space and $\{W(t)\}_{t\geqslant 0}$ be a U_1 valued Q-Wiener process with respect to a complete filtered probability space $(\Omega, \mathcal{F}, \{\mathcal{F}_t\}_{t\geqslant 0}, \mathbb{P})$ and let $U = Q^{1/2}U_1$. Suppose $g^\nu : C([0,T]; U_1) \to E$ is a measurable map and $u^\nu(\cdot) = g^\nu(W(\cdot))$. Let

$$\mathcal{A} = \Big\{ h \mid h \text{ is a } U\text{-valued } \mathcal{F}_t\text{-predictable process such that} \int_0^T \|h(s)\|_U^2 \, \mathrm{d}s < \infty \text{ a.s.} \Big\},$$

and for $M > 0$

$$\mathcal{S}_M = \left\{ h \in L^2(0,T;U) \mid \int_0^T \|h(s)\|_U^2 \,\mathrm{d}s \leqslant M \right\}$$

which is a Polish space endowed with the weak topology. Define

$$\mathcal{A}_M = \{ h \in \mathcal{A} : h \in \mathcal{S}_M, \mathbb{P}\text{-a.s.} \} .$$

Now we give sufficient conditions for u^ν satisfying the Laplace principle.

Condition 5.1 *There exists a measurable map $g^0 : C(0,T;U_1) \to E$ and function $\beta(\nu)$ with $\beta(\nu) \to \infty$ as $\nu \to 0$, such that the following conditions hold.*

1. *Let $\{h^\nu : \nu > 0\} \subset \mathcal{A}_M$ for some $M < \infty$. If h^ν converges in distribution to h as $\mathcal{S}_M$-valued random elements, then*

$$g^\nu \left(W(\cdot) + \sqrt{\beta(\nu)} \int_0^\cdot h^\nu(s) \,\mathrm{d}s \right) \to g^0 \left(\int_0^\cdot h(s) \,\mathrm{d}s \right)$$

in distribution as parameter $\nu \to 0$.

2. *For every $M < \infty$, the set*

$$K_M = \left\{ g^0 \left(\int_0^\cdot h(s) \,\mathrm{d}s \right) \mid h \in \mathcal{S}_M \right\}$$

is a compact subset of E.

For each $f \in E$, and using the notation $\inf \emptyset = \infty$, define

$$I(f) = \inf_{\{h \in L^2(0,T;U) \mid f = g^0(\int_0^\cdot h(s)\,\mathrm{d}s)\}} \left\{ \frac{1}{2} \int_0^T \|h(s)\|_U^2 \,\mathrm{d}s \right\} . \tag{5.37}$$

Then Budhiraja et al. [4] proved the following theorem.

Theorem 5.1 *If $u^\nu = g^\nu(W(\cdot))$ satisfies Condition 5.1, then the family $\{u^\nu \mid \nu > 0\}$ satisfies the Laplace principle on E with rate function I given by (5.37) and speed $\beta(\nu)$.*

Next we consider large deviations for (ψ^ν, ϕ^ν). Let

$$B(t) = (B_1(t), B_2(t)) .$$

In this case $E=C(0,T;H\times H)\cap L^2(0,T;H_0^1\times H_0^1)$, $U=U_1=\mathbb{C}\times\mathbf{R}$ and $\beta(\nu)=\nu^{-1/2}$. Then by the well-posedness result of SKGS equations (1.5)–(1.8) and Yamada–Watanabe theorem in Banach space [21], there exists a Borel measurable function

$$g^\nu: C(0,T;U_1)\to E$$

such that $(\psi^\nu(t),\phi^\nu(t))=g^\nu(B(t))$ almost surely.

Now for $h=(h_1,h_2)\in\mathcal{A}_M$ with $(h_1(t),h_2(t))$ is U_1-valued process, the controlled analogue of SKGS equations are

$$\mathrm{i}\psi_t^{h,\nu}+\Delta\psi^{h,\nu}+\mathrm{i}\nu\psi^{h,\nu}+\phi^{h,\nu}\psi^{h,\nu}=\sqrt{\nu}\dot{W}_1+q_1h_1,\quad \text{on}\quad D \tag{5.38}$$

$$\nu\phi_{tt}^{h,\nu}+\phi_t^{h,\nu}-\Delta\phi^{h,\nu}+\phi^{h,\nu}-|\psi^{h,\nu}|^2=\sqrt{\nu}\dot{W}_2+q_2h_2,\quad \text{on}\quad D \tag{5.39}$$

with zero Dirichlet boundary condition and initial value $\psi^{h,\nu}(0)=\psi_0$, $\phi^{h,\nu}(0)=\phi_0$ and $\phi_t^{h,\nu}(0)=\phi_1$. Then we have

Lemma 5.1 *Assume $q_1\in H_0^1$, $q_2\in H$ and $(\psi_0,\phi_0,\phi_1)\in\mathcal{E}_0$, for any fixed $T>0$ and $h\in\mathcal{A}_M$, equations (5.38)–(5.39) has a unique solution in $L^2(\Omega,C(0,T;\mathcal{E}_0))$. Moreover for any $p\geqslant 1$*

$$\begin{aligned}&\mathbb{E}\sup_{0\leqslant t\leqslant T}\|(\psi^{h,\nu}(t),\phi^{h,\nu}(t),\sqrt{\nu}\phi_t^{h,\nu}(t))\|_{\mathcal{E}_0}^p+\mathbb{E}\|(\psi_t^{h,\nu},\phi_t^{h,\nu})\|_{L^2(0,T:H^{-1}\times H)}\\&\leqslant C_T(\|h\|_{L^2(0,T;U_1)},\|q_1\|_1,\|q_1\|_0,\|q_2\|_0)\end{aligned}$$

with some constant $C_T(\|h\|_{L^2(0,T;U_1)},\|q_1\|_1,\|q_1\|_0,\|q_2\|_0)>0$.

Proof The existence of solution is direct by applying the Girsanov argument [31, e.g.] and follow the approach to Theorem 3.1. The estimates of solution can be followed exactly by the approach in Section .

First from equation (5.38), by Itô's formula we have

$$\begin{aligned}&\|\psi^{h,\nu}(t)\|_0^2\\&=\|\psi_0\|_0^2-2\nu\int_0^t\|\psi^{h,\nu}(s)\|_0^2\,\mathrm{d}s+2\int_0^t\langle-\mathrm{i}q_1h_1(s),\psi^{h,\nu}(s)\rangle\,\mathrm{d}s+\nu\|q_1\|_0^2t\\&\quad-2\mathrm{i}\int_0^t\langle\psi^{h,\nu}(s),\sqrt{\nu}\mathrm{d}W_1(s)\rangle\end{aligned}\tag{5.40}$$

and

$$
\begin{aligned}
&\|\psi^{h,\nu}(t)\|_1^2-\|\psi_0\|_1^2\\
=&-2\nu\int_0^t\|\psi^{h,\nu}(s)\|_1^2\,\mathrm{d}s+2\int_0^t\langle\phi^{h,\nu}(s)\psi^{h,\nu}(s),\psi_t^{h,\nu}(s)+\nu\psi^{h,\nu}(s)\rangle\,\mathrm{d}s\\
&-2\int_0^t\left\langle\psi_t^{h,\nu}(s)+\nu\psi^{h,\nu}(s),\sqrt{\nu}\mathrm{d}W_1(s)\right\rangle+\nu\|q_1\|_1^2t\\
&+2\int_0^t\langle h_1(s)\nabla q_1,\nabla\psi^{h,\nu}(s)\rangle\,\mathrm{d}s-2\int_0^t\langle q_1h_1(s),i\psi^{h,\nu}(s)\phi^{h,\nu}(s)\rangle\,\mathrm{d}s\\
&+2\int_0^t\langle q_1h_1(s),i\sqrt{\nu}\mathrm{d}W_1(s)\rangle. \qquad (5.41)
\end{aligned}
$$

Then for any $T>0$, by similar discussion for ψ^ν, from (5.38) we have

$$
\mathbb{E}\sup_{0\leqslant t\leqslant T}\|\psi^{h,\nu}(t)\|_0^{2p}\leqslant C_{T,p}(\mathbb{E}\|h_1\|_{L^2(0,T)}^{4p},\|q_1\|_0^{4p})
$$

for some constant $C_{T,p}>0$.

Define $\theta^{h,\nu}(t)=\phi^{h,\nu}(t)-\delta\phi_t^{h,\nu}(t)$, with $\delta=1/2$, then from (5.39),

$$
\begin{aligned}
&\nu\theta_t^{h,\nu}+(1-\nu\delta)\theta^{h,\nu}-\Delta\phi^{h,\nu}+(1-\delta(1-\nu\delta))\phi^{h,\nu}-|\psi^{h,\nu}|^2\\
=&\sqrt{\nu}\dot{W}_2+q_2h_2
\end{aligned}
$$

and

$$
\begin{aligned}
&\nu\|\theta^{h,\nu}(t)\|_0^2-\nu\|\theta_0\|_0^2\\
=&2\int_0^t\left[-(1-\nu\delta)\|\theta^{h,\nu}\|_0^2-\langle\nabla\phi^{h,\nu},\nabla\theta^{h,\nu}\rangle-(1-\delta(1-\nu\delta))\langle\phi^{h,\nu},\theta^{h,\nu}\rangle\right]\mathrm{d}s\\
&+2\int_0^t\langle|\psi^{h,\nu}|^2,\theta^{h,\nu}\rangle\,\mathrm{d}s+\|q_2\|_0^2t+\int_0^t\langle q_2h_2(s),\theta^\nu(s)\rangle\,\mathrm{d}s\\
&+2\sqrt{\nu}\int_0^t\langle\theta^\nu(s),\mathrm{d}W_2(s)\rangle. \qquad (5.42)
\end{aligned}
$$

Now define

$$
\begin{aligned}
H^{h,\nu}(t)=&\nu\|\theta^{h,\nu}(t)\|_0^2+\|\phi^{h,\nu}(t)\|_0^2+\|\phi^{h,\nu}(t)\|_1^2+2\|\psi^{h,\nu}(t)\|_1^2\\
&-2\langle\phi^{h,\nu}(t),|\psi^{h,\nu}(t)|^2\rangle,
\end{aligned}
$$

then from (5.41), (5.42) and relations (3.22), (3.23), by same discussion in Section we still have

$$
\mathbb{E}\sup_{0\leqslant t\leqslant T}[H^{h,\nu}(t)]^p\leqslant C_T(\|q_1\|_0^2,\|q_1\|_1^2,\|q_2\|_0^2,\mathbb{E}\|h\|_{L^2(0,T;U_1)}^{6p}). \qquad (5.43)
$$

Moreover from (5.38)–(5.39) one can easily have

$$\mathbb{E}\|\psi_t^{h,\nu}\|_{L^2(0,T;H^{-1})} \leqslant C_T(\|q_1\|_1, \|q_1\|_0, \|q_2\|_0, \mathbb{E}\|h\|_{L^2(0,T;U_1)})$$

and

$$\mathbb{E}\|\phi_t^{h,\nu}\|_{L^2(0,T;H)} \leqslant C_T(\|q_1\|_1, \|q_1\|_0, \|q_2\|_0, \mathbb{E}\|h^\nu\|_{L^2(0,T;U_1)}) .$$

The proof is complete.

Then by the above lemma, $(\psi^{h,\nu}, \phi^{h,\nu})$ can be represented by

$$(\psi^{h,\nu}(t), \phi^{h,\nu}(t)) = g^\nu \left(B(t) + \nu^{-1/2} \int_0^t h(s)\, \mathrm{d}s \right) .$$

Furthermore for $h = (h_1, h_2) \in \mathcal{A}_M$ we have [22]

Lemma 5.2 *Assume $q_1 \in H_0^1$ and $q_2 \in H$. For any $T > 0$, the equations*

$$\mathrm{i}\psi_t^h + \Delta\psi^h + \phi^h\psi^h = q_1 h_1 , \quad on \quad D \tag{5.44}$$

$$\phi_t^h - \Delta\phi^h + \phi^h - |\psi^h|^2 = q_2 h_2 , \quad on \quad D , \tag{5.45}$$

with zero Dirichlet boundary condition, $\psi^h(0) = \psi_0$, $\phi^h(0) = \phi_1$ has a unique solution solution $(\psi^h, \phi^h) \in C(0,T; H \times H) \cap L^2(0,T; H_0^1 \times H_0^1)$.

Then define a measurable function $g^0 : C(0,T;U_1) \to E$ as

$$g^0(w(t)) = (\psi^h(t), \phi^h(t)), \quad \text{if} \quad w(t) = \int_0^t h(s)\, \mathrm{d}s \quad \text{for some} \quad h \in L^2(0,T;U)$$

and otherwise $g^0(w) = 0$. Then to show the large deviations for (ψ^ν, ϕ^ν) in space E we need to verify Condition 5.1. To show the convergence in space $L^2(0,T; H_0^1 \times H_0^1)$ we need a technique lemma, which is concerned with the time increment of the solution to the stochastic control problem. Similar results approach is also applied to study LDP for stochastic Boussinesq equations [6]. For any integer $k = 0, \dots, 2^n - 1$ and $s \in [kT2^{-n}, (k+1)T2^{-n})$, define $\underline{s}_n = kT2^{-n}$ and $\bar{s}_n = (k+1)T2^{-n}$. Given $T > 0$, $N > 0$, define for any $t \in [0,T]$,

$$B_N(t) = \left\{ \omega \in \Omega : \sup_{0 \leqslant s \leqslant t} \|(\psi^{h,\nu}(s), \phi^{h,\nu}(s))\|_{H_0^1 \times H_0^1}^2 + \|\phi_t^{h,\nu}\|_{L^2(0,T;H)}^2 < N \right\} . \tag{5.46}$$

Lemma 5.3 *For M, $N > 0$ and any $h \in \mathcal{A}_M$,*

$$\mathbb{E}\left[\chi_{B_N(T)} \int_0^T \|(\psi^{h,\nu}(s), \phi^{h,\nu}(s)) - (\psi^{h,\nu}(\bar{s}_n), \phi^{h,\nu}(\bar{s}_n))\|_{H \times H}^2\, \mathrm{d}s \right] \leqslant C2^{-n/2}$$

for some constant $C > 0$ which is independent of ν.

Proof For any $h \in \mathcal{A}_M$, by Itô formula and the definition of $B_N(t)$ we have

$$
\begin{aligned}
&\mathbb{E}\left[\chi_{B_N(T)}\int_0^T \|\psi^{h,\nu}(s)-\psi^{h,\nu}(\bar{s}_n)\|_0^2\,\mathrm{d}s\right]\\
=&2\mathbb{E}\left[\chi_{B_N(T)}\int_0^T \mathrm{d}s\int_s^{\bar{s}_n}\langle \mathrm{i}\nabla\psi^{h,\nu}(r),\nabla\psi^{h,\nu}(s)\rangle \mathrm{d}r\right]\\
&-2\mathbb{E}\left[\chi_{B_N(T)}\int_0^T \mathrm{d}s\int_s^{\bar{s}_n}\langle \nu\psi^{h,\nu}(r),\psi^{h,\nu}(r)-\psi^{h,\nu}(s)\rangle\,\mathrm{d}r\right]\\
&+2\mathbb{E}\left[\chi_{B_N(T)}\int_0^T \mathrm{d}s\int_s^{\bar{s}_n}\langle \mathrm{i}\phi^{h,\nu}(r)\psi^{h,\nu}(r),\psi^{h,\nu}(r)-\psi^{h,\nu}(s)\rangle\,\mathrm{d}r\right]\\
&-2\mathbb{E}\left[\chi_{B_N(T)}\int_0^T \mathrm{d}s\int_s^{\bar{s}_n}\langle \sqrt{\nu}\mathrm{d}W_1(r),\psi^{h,\nu}(r)-\psi^{h,\nu}(s)\rangle\right]\\
&+2\mathbb{E}\left[\chi_{B_N(T)}\int_0^T \mathrm{d}s\int_s^{\bar{s}_n}\langle h_1(r),\psi^{h,\nu}(r)-\psi^{h,\nu}(s)\rangle\,\mathrm{d}r\right]+\nu\|q_1\|_0^2T^2 2^{-n}\\
\leqslant& C(N,T,M)2^{-n}\\
&+2\mathbb{E}\left[\chi_{B_N(T)}\int_0^T \mathrm{d}s\int_s^{\bar{s}_n}\langle \mathrm{i}\phi^{h,\nu}(r)\psi^{h,\nu}(r),\psi^{h,\nu}(r)-\psi^{h,\nu}(s)\rangle\,\mathrm{d}r\right]\\
&-2\mathbb{E}\left[\chi_{B_N(T)}\int_0^T \mathrm{d}s\int_s^{\bar{s}_n}\langle \sqrt{\nu}\mathrm{d}W_1(r),\psi^{h,\nu}(r)-\psi^{h,\nu}(s)\rangle\right]\\
&+2\mathbb{E}\left[\chi_{B_N(T)}\int_0^T \mathrm{d}s\int_s^{\bar{s}_n}\langle h_1(r),\psi^{h,\nu}(r)-\psi^{h,\nu}(s)\rangle\,\mathrm{d}r\right].
\end{aligned}
$$

Notice that, by Gagliardo–Nirenberg inequality

$$
\begin{aligned}
&|\langle\phi^{h,\nu}(r)\psi^{h,\nu}(r),\psi^{h,\nu}(r)-\psi^{h,\nu}(s)\rangle|\\
\leqslant&\|\phi^{h,\nu}(r)\|_1\|\psi^{h,\nu}(r)\|_1^{1/2}\|\psi^{h,\nu}(r)\|_0^{1/2}\|\psi^{h,\nu}(r)-\psi^{h,\nu}(s)\|_0,
\end{aligned}
$$

then by the definition of $B_N(t)$ we have

$$
2\mathbb{E}\left[\chi_{B_N(T)}\int_0^T \mathrm{d}s\int_s^{\bar{s}_n}\langle \mathrm{i}\phi^{h,\nu}(r)\psi^{h,\nu}(r),\psi^{h,\nu}(r)-\psi^{h,\nu}(s)\rangle\,\mathrm{d}r\right]\leqslant C(N,T,M)2^{-n}.
$$

By Cauchy inequality, maximal estimates of stochastic integral [24, Lemma 7.2]and the definition of $B_N(t)$

$$\begin{aligned}
&\mathbb{E}\left[\chi_{B_N(T)}\int_0^T \mathrm{d}s\int_s^{\bar{s}_n}\langle\sqrt{\nu}\mathrm{d}W_1(r),\psi^{h,\nu}(r)-\psi^{h,\nu}(s)\rangle\right]\\
\leqslant&\int_0^T \mathrm{d}s\left[\mathbb{E}\left(\int_s^{\bar{s}_n}\langle\sqrt{\nu}\mathrm{d}W_1(r),\psi^{h,\nu}(r)-\psi^{h,\nu}(s)\rangle\chi_{B_N(t)}\right)^2\right]^{1/2}\\
\leqslant&\sqrt{\nu}\int_0^T \mathrm{d}s\left[\int_s^{\bar{s}_n}\|q_1\|_0^2\,\mathbb{E}\left[\|\psi^{h,\nu}(r)-\psi^{h,\nu}(s)\|_0^2\,\chi_{B_N(T)}\right]\mathrm{d}r\right]^{1/2}\\
\leqslant&C(N,T,M)2^{-n/2}.
\end{aligned}$$

Also by Cauchy inequality, Fubini's theorem and the definition of $B_N(t)$ we have

$$\begin{aligned}
&2\mathbb{E}\left[\chi_{B_N(T)}\int_0^T \mathrm{d}s\int_s^{\bar{s}_n}\langle h_1(r),\psi^{h,\nu}(r)-\psi^{h,\nu}(s)\rangle\,\mathrm{d}r\right]\\
\leqslant&2\mathbb{E}\left[\chi_{B_N(T)}\int_0^T \mathrm{d}s\int_s^{\bar{s}_n}\|h(r)\|_0\|\psi^{h,\nu}(r)-\psi^{h,\nu}(s)\|_0\,\mathrm{d}r\right]\\
\leqslant&2\mathbb{E}\left[\chi_{B_N(T)}\int_0^T \|h(r)\|_0\,\mathrm{d}r\int_r^{\bar{r}_n}\|\psi^{h,\nu}(s)-\psi^{h,\nu}(r)\|_0\,\mathrm{d}s\right]\\
\leqslant&C(N,T,M)2^{-n}.
\end{aligned}$$

Then by the above estimates we have

$$\mathbb{E}\left[\chi_{B_N(T)}\int_0^T \|\psi^{h,\nu}(s)-\psi^{h,\nu}(\bar{s}_n)\|_0^2\,\mathrm{d}s\right]\leqslant C2^{-n/2}.$$

Further by Itô formula we have

$$\begin{aligned}
&\mathbb{E}\left[\chi_{B_N(T)}\int_0^T \nu\|\theta^{h,\nu}(s)-\theta^{h,\nu}(\bar{s}_n)\|_0^2\,\mathrm{d}s\right]\\
&\quad+\mathbb{E}\left[\chi_{B_N(T)}\int_0^T \|\phi^{h,\nu}(s)-\phi^{h,\nu}(\bar{s}_n)\|_1^2\,\mathrm{d}s\right]\\
&\quad+\mathbb{E}\left[\chi_{B_N(T)}\int_0^T \|\phi^{h,\nu}(s)-\phi^{h,\nu}(\bar{s}_n)\|_0^2\,\mathrm{d}s\right]\\
=&2\mathbb{E}\left[\chi_{B_N(T)}\int_0^T \mathrm{d}s\int_s^{\bar{s}_n}\langle-(1-\nu\delta)\theta^{h,\nu}(r),\theta^{h,\nu}(r)-\theta^{h,\nu}(s)\rangle\,\mathrm{d}r\right]\\
&-2\delta\mathbb{E}\left[\chi_{B_N(T)}\int_0^T \mathrm{d}s\int_s^{\bar{s}_n}\langle\nabla\phi^{h,\nu}(r),\nabla\phi^{h,\nu}(r)-\nabla\phi^{h,\nu}(s)\rangle\,\mathrm{d}r\right]
\end{aligned}$$

$$
\begin{aligned}
&-2\delta\mathbb{E}\left[\chi_{B_N(T)}\int_0^T \mathrm{d}s\int_s^{\bar{s}_n}\langle\delta(1-\delta(1-\nu\delta))\phi^{h,\nu}(r),\phi^{h,\nu}(r)-\phi^{h,\nu}(s)\rangle\,\mathrm{d}r\right]\\
&+2\mathbb{E}\left[\chi_{B_N(T)}\int_0^T \mathrm{d}s\int_s^{\bar{s}_n}\langle|\psi^{h,\nu}(r)|^2,\theta^{h,\nu}(r)-\theta^{h,\nu}(s)\rangle\,\mathrm{d}r\right]\\
&+2\mathbb{E}\left[\chi_{B_N(T)}\int_0^T \mathrm{d}s\int_s^{\bar{s}_n}\langle\sqrt{\nu}\mathrm{d}W_2(r),\theta^{h,\nu}(r)-\theta^{h,\nu}(s)\rangle\right]\\
&+2\mathbb{E}\left[\chi_{B_N(T)}\int_0^T \mathrm{d}s\int_s^{\bar{s}_n}\langle h_2(r),\theta^{h,\nu}(r)-\theta^{h,\nu}(s)\rangle\,\mathrm{d}r\right]+\int_0^T\mathrm{d}s\int_s^{\bar{s}_n}\|q_2\|_0^2\,\mathrm{d}r\,.
\end{aligned}
$$

By Cauchy inequality, Fubini's theorem and the construction of $B_N(t)$ we have

$$
\begin{aligned}
&\mathbb{E}\left[\chi_{B_N(T)}\int_0^T \mathrm{d}s\int_s^{\bar{s}_n}\langle-(1-\nu\delta)\theta^{h,\nu}(r),\theta^{h,\nu}(r)-\theta^{h,\nu}(s)\rangle\,\mathrm{d}r\right]\\
\leqslant\ &\mathbb{E}\left[\chi_{B_N(T)}\int_0^T \mathrm{d}s\int_s^{\bar{s}_n}\|\theta^{h,\nu}(r)\|_0\|\theta^{h,\nu}(r)-\theta^{h,\nu}(s)\|_0\,\mathrm{d}r\right]\\
\leqslant\ &\mathbb{E}\left[\chi_{B_N(T)}\int_0^T \|\theta^{h,\nu}(r)\|_0\,\mathrm{d}r\int_r^{\bar{r}_n}\|\theta^{h,\nu}(r)-\theta^{h,\nu}(s)\|_0\,\mathrm{d}s\right]\\
\leqslant\ &\mathbb{E}\left\{\chi_{B_N(T)}\int_0^T \|\theta^{h,\nu}(r)\|_0\,\mathrm{d}r\left[\int_r^{\bar{r}_n}\mathrm{d}s\right]^{1/2}\left[\int_r^{\bar{r}_n}\|\theta^{h,\nu}(r)-\theta^{h,\nu}(s)\|_0^2\,\mathrm{d}s\right]^{1/2}\right\}\\
\leqslant\ &C(N,T,M)2^{-n/2}\,.
\end{aligned}
$$

Then similar discussions as above we also have

$$
\mathbb{E}\left[\chi_{B_N(T)}\int_0^T\|\phi^{h,\nu}(s)-\phi^{h,\nu}(\bar{s}_n)\|_0^2\,\mathrm{d}s\right]\leqslant C2^{-n/2}\,.
$$

The proof is complete.

Now we can draw the following weak convergence result.

Lemma 5.4 *Assume* $q_1\in H_0^1$, $q_2\in H$ *and* $(\psi_0,\phi_0,\phi_1)\in\mathcal{E}_0$. *Let* h^ν *converges in distribution to* h *as* $\mathcal{S}_M$*-valued random variables, then*

$$
g^\nu\left(B(\cdot)+\frac{1}{\sqrt{\nu}}\int_0^\cdot h^\nu(s)\,\mathrm{d}s\right)\to g^0\left(\int_0^\cdot h(s)\,\mathrm{d}s\right)
$$

in space E.

Proof As we just verify the convergence in distribution of $(\psi^{h^\nu,\nu},\phi^{h^\nu,\nu})$, by the Skorokhod representation theorem, we can construct processes, which we still denote by (h^ν, h, B), without changing distribution and h^ν converges almost surely to h in $\mathcal{S}_M$ with weak topology. Then $\int_0^t h^\nu(s)\,\mathrm{d}s$ converges weakly to $\int_0^t h(s)\,\mathrm{d}s$ in U for almost all $t \in [0,T]$.

Now let $\Phi^{h^\nu,\nu} = \phi^{h^\nu,\nu} - \phi^h$ and $\Psi^{h^\nu,\nu} = \psi^{h^\nu,\nu} - \psi^h$. Then

$$\Phi_t^{h^\nu,\nu} = \Delta\Phi^{h^\nu,\nu} - \Phi^{h^\nu,\nu} + |\psi^{h^\nu,\nu}|^2 - |\psi^h|^2 + \sqrt{\nu}\dot{W}_2 + q_2(h_2^\nu - h_2) - \nu\phi_{tt}^\nu$$

and

$$\mathrm{i}\Psi_t^{h^\nu,\nu} + \Delta\Psi^{h^\nu,\nu} + \mathrm{i}\nu\psi^{h^\nu,\nu} + \phi^{h^\nu,\nu}\psi^{h^\nu,\nu} - \phi^h\psi^h = \sqrt{\nu}\dot{W}_1 + q_1(h_1^\nu - h_1)$$

with $\Phi^{h^\nu,\nu}(0) = 0$ and $\Psi^{h^\nu,\nu}(0) = 0$. The similar discussion in Section , by Itô's formula we have

$$\begin{aligned}
&\|\Phi^{h^\nu,\nu}(t)\|_0^2 + 2\int_0^t \left[\|\Phi^{h^\nu,\nu}(s)\|_1^2 + \|\Phi^{h^\nu,\nu}(s)\|_0^2\right] \mathrm{d}s \qquad (5.47)\\
=&2\int_0^t \langle |\psi^{h^\nu,\nu}(s)|^2 - |\psi^h(s)|^2, \Phi^{h^\nu,\nu}(s)\rangle \,\mathrm{d}s + 2\sqrt{\nu}\int_0^t \langle \Phi^{h^\nu,\nu}(s), \mathrm{d}W_2(s)\rangle\\
&- 2\nu\int_0^t \langle \phi_{tt}^{h^\nu,\nu}(s), \Phi^{h^\nu,\nu}(s)\rangle \,\mathrm{d}s + \nu\|q_2\|_0^2 t + 2\int_0^t \langle q_2(h_2^\nu(s) - h_2(s)), \Phi^{h^\nu,\nu}(s)\rangle \,\mathrm{d}s,
\end{aligned}$$

$$\begin{aligned}
&\|\Psi^{h^\nu,\nu}(t)\|_0^2 \qquad (5.48)\\
=&2\int_0^t \mathrm{Re}\int_D \mathrm{i}[\psi^{h^\nu,\nu}(s)\phi^{h^\nu,\nu}(s) - \psi^h(s)\phi^h(s)]\Psi^{h^\nu,\nu}(s)\,\mathrm{d}x\,\mathrm{d}s\\
&- 2\nu\int_0^t \mathrm{Re}\int_D \psi^{h^\nu,\nu}(s)\Psi^{h^\nu,\nu}(s)\,\mathrm{d}x\,\mathrm{d}s - 2\sqrt{\nu}\int_0^t \mathrm{Re}\int_D \mathrm{i}\Psi^{h^\nu,\nu}(s)\mathrm{d}W_1(s)\,\mathrm{d}x\\
&- 2\int_0^t \mathrm{Re}\int_D \mathrm{i}q_1(h_1^\nu(s) - h_1(s))\Psi^{h^\nu,\nu}(s)\,\mathrm{d}x\,\mathrm{d}s + \nu\|q_1\|_0^2 t,
\end{aligned}$$

and multiplying $\Psi^{h^\nu,\nu}$ on both sides of the above equation and taking imaginary part yield

$$\begin{aligned}
&\int_0^t \|\Psi^{h^\nu,\nu}(s)\|_1^2\,\mathrm{d}s \qquad (5.49)\\
=&\int_0^t \mathrm{Im}\int_D \mathrm{i}[\psi^{h^\nu,\nu}(s)\phi^{h^\nu,\nu}(s) - \psi^h(s)\phi^h(s)]\Psi^{h^\nu,\nu}(s)\,\mathrm{d}x\,\mathrm{d}s\\
&- \nu\int_0^t \mathrm{Im}\int_D \Psi^{h^\nu,\nu}(s)\psi^{h^\nu,\nu}(s)\,\mathrm{d}x\mathrm{d}s - \sqrt{\nu}\int_0^t \mathrm{Im}\int_D \mathrm{i}\Psi^{h^\nu,\nu}(s)\mathrm{d}W_1(s)\,\mathrm{d}x\\
&- \int_0^t \mathrm{Im}\int_D \mathrm{i}q_1(h_1^\nu(s) - h_1(s))\Psi^{h^\nu,\nu}(s)\,\mathrm{d}x\,\mathrm{d}s.
\end{aligned}$$

Notice that, by using the compact embedding $H^1 \subset L^4(D)$

$$
\begin{aligned}
&\int_0^t \langle |\psi^{h^\nu,\nu}(s)|^2 - |\psi^h(s)|^2, \Phi^{h^\nu,\nu}(s)\rangle \,\mathrm{d}s \\
=&\int_0^t \langle \Psi^{h^\nu,\nu}(s)\bar{\psi}^{h^\nu,\nu}(s), \Phi^{h^\nu,\nu}(s)\rangle \,\mathrm{d}s + \int_0^t \langle \psi^h(s)\bar{\Psi}^{h^\nu,\nu}(s), \Phi^{h^\nu,\nu}(s)\rangle \,\mathrm{d}s \\
\leqslant&\int_0^t \|\Phi^{h^\nu,\nu}(s)\psi^{h^\nu,\nu}(s)\|_0 \|\Psi^{h^\nu,\nu}(s)\|_0 \,\mathrm{d}s + \int_0^t \|\Phi^{h^\nu,\nu}(s)\psi^h(s)\|_0 \|\Psi^{h^\nu,\nu}(s)\|_0 \,\mathrm{d}s \\
\leqslant&\int_0^t \|\Phi^{h^\nu,\nu}(s)\|_1 [\|\psi^{h^\nu,\nu}(s)\|_1 + \|\psi^h(s)\|_1] \|\Psi^{h^\nu,\nu}(s)\|_0 \,\mathrm{d}s \\
\leqslant&\frac{1}{2}\int_0^t \|\Phi^{h^\nu,\nu}(s)\|_1^2 \,\mathrm{d}s + \frac{1}{2}\int_0^t [\|\psi^{h^\nu,\nu}(s)\|_1^2 + \|\psi^h(s)\|_1^2] \,\|\Psi^{h^\nu,\nu}(s)\|_0^2 \,\mathrm{d}s
\end{aligned}
$$

and similarly

$$
\begin{aligned}
&\int_0^t \int_D [\psi^{h^\nu,\nu}(s)\phi^{h^\nu,\nu}(s) - \psi^h(s)\phi^h(s)]\Psi^{h^\nu,\nu}(s) \,\mathrm{d}x \,\mathrm{d}s \\
=&\int_0^t \int_D [\Psi^{h^\nu,\nu}(s)\phi^{h^\nu,\nu}(s) - \psi^h(s)\Phi^{h^\nu,\nu}(s)]\Psi^{h^\nu,\nu}(s) \,\mathrm{d}x \,\mathrm{d}s \\
\leqslant&\frac{1}{8}\int_0^t \|\Psi^{h^\nu,\nu}(s)\|_1^2 \,\mathrm{d}s + 2\int_0^t [\|\phi^{h^\nu,\nu}(s)\|_1^2 + \|\psi^h(s)\|_1^2][\|\Psi^{h^\nu,\nu}(s)\|_0^2 + \|\Phi^{h^\nu,\nu}(s)\|_0^2] \,\mathrm{d}s\,.
\end{aligned}
$$

Then we have

$$
\begin{aligned}
&\|\Phi^{h^\nu,\nu}(t)\|_0^2 + \|\Psi^{h^\nu,\nu}(t)\|_0^2 + \frac{1}{2}\int_0^t [\|\Phi^{h^\nu,\nu}(s)\|_1^2 + \|\Psi^{h^\nu,\nu}(s)\|_1^2] \,\mathrm{d}s \qquad (5.50)\\
\leqslant&\int_0^t [\|\phi^{h^\nu,\nu}(s)\|_1^2 + \|\psi^{h^\nu,\nu}(s)\|_1^2 + \|\phi^h(s)\|_1^2 + \|\psi^h(s)\|_1^2] [\|\Phi^{h^\nu,\nu}(s)\|_0^2 \\
&+ \|\Psi^{h^\nu,\nu}(s)\|_0^2] \,\mathrm{d}s + R_1^\nu(t) + R_2^\nu(t) + R_3^\nu(t)
\end{aligned}
$$

with

$$
\begin{aligned}
R_1^\nu(t) =& 2\int_0^t \langle q_2(h_2^\nu(s) - h_2(s)), \Phi^{h^\nu,\nu}(s)\rangle \,\mathrm{d}s \\
R_2^\nu(t) =& -2\int_0^t \mathrm{Re} \int_D \mathrm{i}q_1(h_1^\nu(s) - h_1(s))\Psi^{h^\nu,\nu}(s) \,\mathrm{d}x \,\mathrm{d}s \\
&- \int_0^t \mathrm{Im} \int_D iq_1(h_1^\nu(s) - h_1(s))\Psi^{h^\nu,\nu}(s) \,\mathrm{d}x \,\mathrm{d}s
\end{aligned}
$$

and

$$R_3^\nu(t)=2\sqrt{\nu}\int_0^t\langle\Phi^{h^\nu,\nu}(s),\mathrm{d}W_2(s)\rangle-2\nu\int_0^t\langle\phi_{tt}^{h^\nu,\nu}(s),\Phi^{h^\nu,\nu}(s)\rangle\,\mathrm{d}s+\nu\|q_2\|_0^2t+\nu\|q_1\|_0^2t$$
$$-2\nu\int_0^t\mathrm{Re}\int_D\psi^{h^\nu,\nu}(s)\Psi^{h^\nu,\nu}(s)\,\mathrm{d}x\,\mathrm{d}s-2\sqrt{\nu}\int_0^t\mathrm{Re}\int_D\mathrm{i}\Psi^{h^\nu,\nu}(s)\mathrm{d}W_1(s)\,\mathrm{d}x$$
$$-\nu\int_0^t\mathrm{Im}\int_D\Psi^{h^\nu,\nu}(s)\psi^{h^\nu,\nu}(s)\,\mathrm{d}x\,\mathrm{d}s-\sqrt{\nu}\int_0^t\mathrm{Im}\int_D\mathrm{i}\Psi^{h^\nu,\nu}(s)\mathrm{d}W_1(s)\,\mathrm{d}x\,.$$

By the definition of $B_N(t)$ (see (5.46)) and similar discussion as (4.32),

$$\mathbb{E}\left[\chi_{B_N(T)}\sup_{0\leqslant t\leqslant T}R_3^\nu(t)\right]\to0\,,\quad\nu\to0\,.$$

Now we treat $R_1^\nu(t)$. For any $k=0,\ldots,2^n$, set $t_k=kT2^{-n}$. Then

$$\sup_{0\leqslant t\leqslant T}|R_1^\nu(t)|\leqslant\sup_{0\leqslant t\leqslant T}\left|\int_0^t\langle q_2(h_2^\nu(s)-h_2(s)),\Phi^{h^\nu,\nu}(s)-\Phi^{h^\nu,\nu}(\bar{s}_n)\rangle\,\mathrm{d}s\right|$$
$$+\sum_{k=1}^{2^n}\left|\left\langle\int_{t_{k-1}}^{t_k}q_2(h_2^\nu(s)-h_2(s))\,\mathrm{d}s,\Phi^{h^\nu,\nu}(t_k)\right\rangle\right|$$
$$+\sup_{1\leqslant k\leqslant2^n}\sup_{t_{k-1}\leqslant t\leqslant t_k}\left|\left\langle\int_{t_{k-1}}^{t}q_2(h_2^\nu(s)-h_2(s))\,\mathrm{d}s,\Phi^{h^\nu,\nu}(t_k)\right\rangle\right|$$

By Lemma 5.3, and Cauchy–Schwartz inequality

$$\mathbb{E}\left[\chi_{B_N(T)}\sup_{0\leqslant t\leqslant T}\left|\int_0^t\langle q_2(h_2^\nu(s)-h_2(s)),\Phi^{h^\nu,\nu}(s)-\Phi^{h^\nu,\nu}(\bar{s}_n)\rangle\,\mathrm{d}s\right|\right]$$
$$\leqslant\left[\mathbb{E}\int_0^T\|q_2(h_2^\nu(s)-h_2(s))\|_0^2\,\mathrm{d}s\right]^{1/2}\left[\mathbb{E}\chi_{B_N(T)}\int_0^T\|\Phi^{h^\nu,\nu}(s)-\Phi^{h^\nu,\nu}(\bar{s}_n)\|_0^2\,\mathrm{d}s\right]^{1/2}$$
$$\leqslant C_12^{-n/4}$$

for some constant $C_1>0$. Similarly by Cauchy–Schwartz inequality

$$\mathbb{E}\left[\chi_{B_N(T)}\sup_{1\leqslant k\leqslant2^n}\sup_{t_{k-1}\leqslant t\leqslant t_k}\left|\left\langle\int_{t_{k-1}}^{t}q_2(h_2^\nu(s)-h_2(s))\,\mathrm{d}s,\Phi^{h^\nu,\nu}(t_k)\right\rangle\right|\right]$$
$$\leqslant2N\mathbb{E}\left[\sup_{1\leqslant k\leqslant2^n}\int_{t_{k-1}}^{t_k}\|q_2(h_2^\nu(s)-h_2(s))\|_0\,\mathrm{d}s\right]$$
$$\leqslant C_22^{-n/2}$$

for some constant $C_2>0$. By the weak convergence of h_2^ν to h_2 in $L^2(0,T)$, for any $0\leqslant a<b\leqslant T$, $\int_a^b(h_2^\nu(s)-h_2(s))\,\mathrm{d}s\to0$. Then by the dominated convergence

theorem, we have

$$\mathbb{E}\sum_{k=1}^{2^n}\left|\left\langle\int_{t_{k-1}}^{t_k} q_2(h_2^\nu(s)-h_2(s))\,\mathrm{d}s, \Phi^{h^\nu,\nu}(t_k)\right\rangle\right| \to 0\,, \quad \nu\to 0\,.$$

By the arbitrary choice of n we have

$$\mathbb{E}\left[\chi_{B_N(T)}\sup_{0\leqslant t\leqslant T}|R_1^\nu(t)|\right]\to 0\,, \quad \nu\to 0\,.$$

Similarly we have

$$\mathbb{E}\left[\chi_{B_N(T)}\sup_{0\leqslant t\leqslant T}|R_2^\nu(t)|\right]\to 0\,, \quad \nu\to 0\,.$$

Now from (5.50), by Gronwall's lemma we have

$$\begin{aligned}&\mathbb{E}\Bigg[\chi_{B_N(T)}\sup_{0\leqslant t\leqslant T}\|(\Psi^{h^\nu,\nu}(t),\Phi^{h^\nu,\nu}(t)\|_{H\times H}\\&\quad+\chi_{B_N(T)}\int_0^T\|(\Psi^{h^\nu,\nu}(s),\Phi^{h^\nu,\nu}(s)\|_{H_0^1\times H_0^1}\,\mathrm{d}s\Bigg]\\&\to 0\,, \quad \nu\to 0\,.\end{aligned}$$

Noticing arbitrary choice of N, by the estimates in Lemma 5.1 and Markov inequality we complete the proof.

To finish the verification of Condition 5.1 we give the following result.

Lemma 5.5 *For every $M>0$, let*

$$K_M=\{(\psi^h,\phi^h)\in E: h\in\mathcal{S}_M\}$$

with (ψ^h,ϕ^h) solving (5.44)–(5.45). Then K_M is compact in E.

Proof Let (ϕ^n,ϕ^n) be a sequence of K_M which are solutions of (5.44)–(5.45) with controls $h^n=(h_1^n,h_2^n)\in\mathcal{S}_M$,

$$\begin{aligned}&\mathrm{i}\psi_t^n+\Delta\psi^n+\phi^n\psi^n=q_1h_1^n\,, \quad \text{on} \quad D\\&\phi_t^n-\Delta\phi^n+\phi^n-|\psi^n|^2=q_2h_2^n\,, \quad \text{on} \quad D\,,\end{aligned}$$

with zero Dirichlet boundary condition and $\psi^n(0)=\psi_0$, $\phi^n(0)=\phi_0$. As $\mathcal{S}_M$ is bounded, there is a subsequence, which we still denote by h^n, weakly converges to some $h=(h_1,h_2)\in\mathcal{S}_M$. If we show that (ψ^n,ϕ^n) converges to (ψ,ϕ) which solves

$$\begin{aligned}&\mathrm{i}\psi_t+\Delta\psi+\phi\psi=q_1h_1\,, \quad \text{on} \quad D\\&\phi_t-\Delta\phi+\phi-|\psi|^2=q_2h_2\,, \quad \text{on} \quad D\,,\end{aligned}$$

with zero Dirichlet boundary condition and $\psi(0)=\psi_0$, $\phi(0)=\phi_0$, the proof will be complete. Let $\Psi^n=\psi^n-\psi$ and $\Phi^n=\phi^n-\phi$, same discussion in Lemma 5.4 yields

$$\begin{aligned}
&\|\Phi^n(t)\|_0^2+2\int_0^t\left[\|\Phi^n(s)\|_1^2+\|\Phi^n(s)\|_0^2\right]\mathrm{d}s\\
=&2\int_0^t\langle|\psi^n(s)|^2-|\psi(s)|^2,\Phi^n(s)\rangle\,\mathrm{d}s+2\int_0^t\langle q_2(h_2^n(s)-h_2(s)),\Phi^n(s)\rangle\,\mathrm{d}s\,,\\
&\|\Psi^n(t)\|_0^2\\
=&2\int_0^t\mathrm{Re}\int_D\mathrm{i}[\psi^n(s)\phi^n(s)-\psi(s)\phi(s)]\Psi^n(s)\,\mathrm{d}x\,\mathrm{d}s\\
&-2\int_0^t\mathrm{Re}\int_D\mathrm{i}q_1(h_1^n(s)-h_1(s))\Psi^n(s)\,\mathrm{d}x\,\mathrm{d}s
\end{aligned}$$

and

$$\begin{aligned}
&\int_0^t\|\Psi^n(s)\|_1^2\,\mathrm{d}s\\
=&\int_0^t\mathrm{Im}\int_D\mathrm{i}[\psi^n(s)\phi^n(s)-\psi(s)\phi(s)]\Psi^n(s)\,\mathrm{d}x\,\mathrm{d}s\\
&-\int_0^t\mathrm{Im}\int_D\mathrm{i}q_1(h_1^n(s)-h_1(s))\Psi^n(s)\,\mathrm{d}x\,\mathrm{d}s\,.
\end{aligned}$$

Then same discussion for the weak convergence result, Lemma 5.4, completes the proof.

Having the above results and by Theorem 5.1 we have the following large deviations result.

Theorem 5.2 *Assume $q_1\in H_0^1$ and $q_2\in H$, (ψ^ν,ϕ^ν) be the solution of stochastic Klein-Gordon-Schrödinger equations (1.5)–(1.8) with $\alpha=0$. Then for any $T>0$, (ψ^ν,ϕ^ν) satisfies the large deviation principle in space $C(0,T;H\times H)\cap L^2(0,T;H_0^1\times H_0^1)$ with good rate function*

$$I((\psi,\phi))=\inf_{\{h\in L^2(0,T;U):(\psi,\phi)=g^0(\int_0^\cdot h(s)\,ds)\}}\frac{1}{2}\int_0^T|h(s)|^2\,\mathrm{d}s$$

and speed $1/\sqrt{\nu}$.

Acknowledgements This work was partially done while Wang Wei was visiting the IAPCM, Beijing. Lv Yan is supported by NSFC (10901083).

References

[1] Baillon J B, Chadam J M. The Cauchy problem for the coupled Schrödinger–Klein–Gordon equations [A]. Medeiros L A J, de la Penha G M ed. Contemporary Develop-

ments in Continuum Mechanics and Partial Differential Equation [C], North-Holland, Amsterdam/New York, 1978, 37–44.

[2] Bao W Z, Dong X C, Wang S. Singular limits of Klein-Gordon-Schrödinger equations to Schrödinger–Yukawa equations [J]. Mult. Model. Simul., 2010, 8(5)1742–1769.

[3] Biler P. Attractors for the system of Schrödinger and Klein–Gordon equations with Yukawa coupling [J]. SIAM J. Math. Anal., 1990, 21: 1190–1212.

[4] Budhiraja A, Dupuis P, Maroulas V. Large deviations for infinite dimensional stochastic dynamical systems [J]. Ann. Prob., 2008, 36(4): 1390–1420.

[5] Chow P L. Large deviation problem for some parabolic Itô equations [J]. Comm. Pure Appl. Math., 1992, 45: 97–120.

[6] Duan J Q, Millet A. Large deviations for the boussinesq equations under random influences [J]. Stoch. Proc. and Appl., 2009, 119(6): 2052–2081.

[7] Dupuis P, Ellis R S. A Weak Convergence Approach to the Theory of Large Deviations [M], J. Wiley & Sons, New York, 1997.

[8] E W, Li X, Vanden-Eijnden E. Some recent progress in multiscale modeling [A]. Attinger S, Koumoutsakos P ed. Multiscale modelling and simulation, Lect. Notes Comput. Sci. Eng., 39, Springer, Berlin, 2004, 3–21.

[9] Freidlin M I, Wentzell A D. Random Perturbations of Dynamical Systems (2 edition) [M]. Springer, New York, 1998.

[10] Fukuda I, Tsutsumi M. On coupled Klein-Gordon-Schrödinger equations II [J]. J. Math. Anal. Appl.,1978, 68: 358–378.

[11] Guo B L, Miao C X. Asymptotic behavior of coupled Klein-Gordon-Schrödinger equations [J]. Sci. China Ser. A, 1995, 25(7): 705–714.

[12] Guo B L, Li Y. Attractors for Klein-Gordon-Schrödinger equations in $\mathbf{R}^3$ [J]. J. Diff. Equa., 1997, 136: 356–377.

[13] Hayashi N, von Wahl W. On the global strong solution of coupled Klein-Gordon-Schrödinger equations [J]. J. Math. Soc. Japan, 1987, 39: 489–497.

[14] Imkeller P, Monahan A ed. Stochastic Climate Dynamics, a Special Issue in the journal *Stoch. and Dyna.*, 2(3), 2002.

[15] Kallianpur G, Xiong J. Large deviations for a class of stochastic partial differential equations [J]. Ann. Prob., 1996, 24: 320–345.

[16] Klainerman S. Uniform decay estimates and the Lorentz invariance of the classical wave equation [J]. Comm. Pure Appl. Math., 1985, 38: 321–332.

[17] Li Y, Guo B L. Asymptotic smoothing effect of solutions to weakly dissipative Klein-Gordon-Schrödinger equations [J]. J. Math. Anal. Appl., 2003, 282: 256–265.

[18] Lu K N, Wang B X. Attractor for dissipative Klein-Gordon-Schrödinger equations in $\mathbf{R}^3$ [J]. J. Diff. Equa., 2001, 170: 281–316.

[19] Lv Y, Guo B L, Yang X P. Dynamics of stochastic Klein-Gordon-Schrödinger equations in unbounded domains [J]. Diff. and Inte. Equa., 2011, 24(3–4): 231–260.

[20] Lv Y, Roberts A J. Large deviation principle for singularly perturbed stochastic damped wave equations [J]. Stoc. Anal. Appl., 2014, 32(1): 50–60.

[21] Ondreját, M. Uniqueness for stochastic evolution equations in Banach spaces [J]. Dissertationes Math., 2004, 426: 1–63.

[22] Ozawa T, Tsutaya K, Tsutsumi Y. Well-posedness in energy space for the Cauchy problem of the Klein-Gordon-Zakharov equations with different propagation speeds in three space dimensions [J]. Math. Ann., 1999,313(1): 127–140.

[23] Peszat S. Large deviation estimates for stochastic evolution equations [J]. Prob. Th. Rela. Fields, 1994, 98: 113–136.

[24] Da Prato G, Zabczyk J. Stochastic Equations in Infinite Dimensions [M], Cambridge University Press, 1992.

[25] Schochet S H, Weinstein M I. The nonlinear Schrödinger limit of the Zakharov equations governing Langmuir turbulence [J]. Comm. Math. Phys., 1986, 106: 173–176.

[26] Sowers R B. Large deviations for a reaction-diffusion equation with non-Gaussian perturbation [J]. Ann. Prob., 1992,20: 504–537.

[27] Sritharan S S, Sundar P. Large deviations for two dimensional Navier–Stokes equations with multiplicative noise [J]. Stoch. Proc. Appl., 2006, 116: 1636–1659.

[28] Tao T. Nonlinear Dispersive Equations: Local and Global Analysis [M], CBMS Regional Conf. Ser. Math. 106, AMS, Providence, RI, 2006.

[29] Wang B X, Lange H. Attractors for Klein-Gordon-Schrödinger equation [J]. J. Math. Phys., 1999, 40: 2445.

[30] Wang W, Duan J Q, Roberts A J. Large deviations for slow-fast stochastic reaction-diffusion equations [J]. J. Diff. Equa., 2012, 253(12): 3501–3522.

[31] Wang W, Duan J Q. Reductions and deviations for stochastic partial differential equations under fast dynamical boundary conditions [J]. Stoch. Anal. Appl., 2009, 27: 431–459.

Global Well-posedness for the Fractional Schrödinger-Boussinesq System*

Han Lijia (韩励佳), Zhang Jingjun (张景军) and Guo Boling (郭柏灵)

Abstract In this paper, we study the fractional Schrödinger-Boussinesq System raised in the laser and plasma physics, where the nonlinear term $f(n)$ can be chosen as a large kind of functions. We obtain that it is global well-posed in $H^s(\mathbb{R})$ in one dimensional space, when $s \geqslant 1$.

Keywords Fractional Schrödinger-Boussinesq System; Cauchy problem; Global well-posedness

1 Introduction

In this paper, we consider the Cauchy problem for the fractional Schrödinger-Boussinesq System (SBS) in one dimensional space:

$$\mathrm{i}E_t + E_{xx} - (nE) = 0, \tag{1.1}$$

$$n_{tt} + |D|^\beta n + \gamma |D|^{\beta+2} n + |D|^\beta f(n) + |D|^\beta (|E|^2) = 0, \tag{1.2}$$

$$E(0,x) = E_0(x), \quad n(0,x) = n_0(x), \quad n_t(0,x) = n_1(x), \tag{1.3}$$

where $E(t,x)$ is a complex valued function of $(t,x) \in \mathbb{R} \times \mathbb{R}$, $\bar{E}$ is the complex conjugation of E. $n(t,x)$ is a real valued function of $(t,x) \in \mathbb{R} \times \mathbb{R}$, $1 < \beta < 2, \gamma > 0$. $f(\cdot)$ is a real valued function. And $|D|^\beta$ means :

$$|D|^\beta u := \mathscr{F}^{-1} |\xi|^\beta (\mathscr{F} u)(\xi), \quad \beta \in \mathbb{R},$$

$\mathscr{F}$ is the Fourier transformation.

Boussinesq equation is derived in the studies of the propagation of long waves on the surface of shallow water[1] , the nonlinear string and the shape-memory alloy[2] and so on. The fractional Schrödinger-Boussinesq system (1.1)–(1.3) was first raised

* Commun Nonlinear Sci Numer Simulat, 2014, 9:2644–2652. DOI: 10.1016/j.cnsns.2013.12.032.

in the laser and plasma physics. It was considered as a model of interactions between short and intermediate long waves, see [10] and the references therein.

When $\beta = 2$, then system (1.1)–(1.3) is derived as

$$\mathrm{i}E_t + E_{xx} - (nE) = 0, \tag{1.4}$$

$$n_{tt} - n_{xx} + \gamma n_{xxxx} - f(n)_{xx} = (|E|^2)_{xx}. \tag{1.5}$$

The existence and uniqueness of the global solutions for the Cauchy problem of system (1.4), (1.5) were researched by many authors. In [4], Guo established the existence and uniqueness of global solution for the system (1.4), (1.5) in $H^s(\mathbb{R})$ $(s \geqslant 4)$, with the condition $\int_0^n f(z)dz \geqslant 0$, $f(z) \in C^4$. In [7], when $f(n) = n^{m+1}$ (m is an even number), Y.Q. Han obtained the local well-posedness, the conservation of energy and the global well-posedness in $H^s(\mathbb{R})$ $(s \geqslant 1)$①. For damped and dissipative Schrödinger-Boussinesq equations with initial boundary value, the existence of global attractors, the finiteness of the Hausdorff and the fractal dimensions of the attractor were established (see [5,10]).

In this paper, we consider the local and global well-posedness for fractional Schrödinger-Boussinesq system (1.1)–(1.3), when the nonlinear term $f(n)$ is undetermined. By using some harmonic analysis techniques, we obtain that (1.1)–(1.3) is global well-posed in $H^s(\mathbb{R}), s \geqslant 1$. System (1.4), (1.5) and the space periodic case can also be solved in this way. Moreover, under the conditions we assume, the undetermined nonlinear term $f(n)$ belongs to a large kind of functions, including $f(n) = n^{m+1}$ (m is an even number) and so on, (see Remark 1.3). This result improves the global well-posedness results in [4] and [7] in one dimension case.

Our method in this paper is also available in solving some other fractional dispersive equations with undetermined nonlinear term $f(n)$, using energy estimates.

Main results

Theorem 1.1 *Assume function $f(z)$ satisfy $f'(z) \in C^1(\mathbb{R})$, $\int_0^n f(z)dz \geqslant 0$, and $\left|\frac{f(z)}{z^2}\right|, \left|\frac{f'(z)}{z^2}\right| \lesssim 1$, when $|z| \leqslant 1$. Then for any $T > 0$, when the initial data (E_0, n_0, n_1) belongs to $(H^1(\mathbb{R}), H^1(\mathbb{R}), H^{-\beta/2}(\mathbb{R}))$, the fractional Schrödinger-*

① when $f(n) = n^{m+1}$, system (1.4)–(1.5) in two- and three- dimensional forms was also considered in [7].

Boussinesq System (1.1)–(1.3) *exists a unique solution*

$$(E, n, n_t) \in (C([0,T], H^1(\mathbb{R}), C([0,T], H^1(\mathbb{R}), C([0,T], H^{-\beta/2}(\mathbb{R})), \tag{1.6}$$

and

$$\begin{aligned}&\|E(t)\|_{L_T^\infty H^1(\mathbb{R})} + \|n(t)\|_{L_T^\infty H^1(\mathbb{R})} + \|n_t(t)\|_{L_T^\infty H^{-\beta/2}(\mathbb{R})}\\ \leqslant\ & C(\|E_0\|_{H^1(\mathbb{R})}, \|n_0\|_{H^1(\mathbb{R})}, \|n_1\|_{H^{-\beta/2}(\mathbb{R})}).\end{aligned}$$

Remark 1.1 *In Theorem 1.1, for any $T > 0$, when the initial data (E_0, n_0, n_1) belongs to $(H^s(\mathbb{R}), H^s(\mathbb{R}), H^{s-1-\beta/2}(\mathbb{R}))$, $s > 1$, the fractional Schrödinger-Boussinesq System* (1.1)–(1.3) *also has a unique solution*

$$(E, n, n_t) \in (C([0,T], H^s(\mathbb{R}), C([0,T], H^s(\mathbb{R}), C([0,T], H^{s-1-\beta/2}(\mathbb{R})).$$

Remark 1.2 *The conditions for $f(z)$ that $\int_0^n f(z)dz \geqslant 0$ and $\left|\frac{f(z)}{z^2}\right|, \left|\frac{f'(z)}{z^2}\right| \lesssim 1$, when $|z| \leqslant 1$ are accordant and a large kind of functions satisfy this condition. This hypothesis can include many nonlinear terms with physical background, for example:*

- *polynomial case: $f(n) = n^{m+1}$, where $m \geqslant 2$ is an even number.*
- *exponential case: $f(n) = (p+1)n^p e^n + n^{p+1}e^n$, where $p \geqslant 3$ is an odd number.*
- *triangle and exponential case: $f(n) = (p+1)\cos n(\sin n)^p e^n + (\sin n)^{p+1}e^n$, where $p \geqslant 3$ is an odd number.*

Remark 1.3 *The conclusion in Theorem 1.1 is also correct when we consider the space periodic problem, where $E(x+2\pi, t) = E(x,t)$, $n(x+2\pi, t) = n(x,t)$, $\forall x, t$.*

In the sequel C will denote a universal positive constant which can be different at each appearance. $x \lesssim y$ (for x, $y > 0$) means that $x \leqslant Cy$, and $x \sim y$ stands for $x \lesssim y$ and $y \lesssim x$.

The Fourier transformation $\mathscr{F}$ is defined as:

$$\mathscr{F}f(\xi) = \hat{f}(\xi) = \int_{\mathbb{R}^d} f(x)e^{-2\pi i x\cdot\xi}dx, \tag{1.7}$$

and the inverse Fourier transformation is defined as:

$$\mathscr{F}^{-1}f(x) = \check{f}(x) = \int_{\mathbb{R}^d} \hat{f}(\xi)e^{2\pi i x\cdot\xi}d\xi. \tag{1.8}$$

We will use the Lebesgue spaces $L^p := L^p(\mathbb{R}^d)$ with the norm $\|\cdot\|_p := \|\cdot\|_{L^p(\mathbb{R}^d)}$. For any $s \in \mathbb{R}$, the norm of the space $H_p^s(\mathbb{R}^d)$ is defined as $\|\cdot\|_{H_p^s(\mathbb{R}^d)} := \|(I-\Delta)^{s/2}\cdot\|_{L^p(\mathbb{R}^d)}$; when $p=2$, we write $H_2^s(\mathbb{R}^d)$ as $H^s(\mathbb{R}^d)$ for short. We also use the function spaces $L_T^q H_p^s$ for which the norms are defined by

$$\|f\|_{L_T^q H_p^s} = \left(\int_0^T \|f\|_{H_p^s}^q \mathrm{d}t\right)^{1/q}.$$

2 Energy estimates

In this section, we deduce some prior energy estimates for the solution of system (1.1)–(1.3) in dimension 1.

Lemma 2.1 *If $E_0 \in L^2$, then the solutions of system (1.1)–(1.3) satisfy :*

$$\|E(\cdot,t)\|_{L_2}^2 = \|E_0(x)\|_{L_2}^2 = \mathcal{E}_1. \tag{2.9}$$

The proof is the same as that in [4].

Lemma 2.2 ([15, 16]) *Let $s \geqslant 0$ and $1 < p < \infty$. If $u,\ v \in \mathcal{S}(\mathbb{R}^d)$, the Schwarz class, then*

$$\|uv\|_{H_p^s} \lesssim \|u\|_{L^{p_1}}\|v\|_{H_{p_2}^s} + \|u\|_{H_{p_3}^s}\|v\|_{L^{p_4}}. \tag{2.10}$$

with $\dfrac{1}{p_1}+\dfrac{1}{p_2}=\dfrac{1}{p_3}+\dfrac{1}{p_4}=\dfrac{1}{p}$ and $p_2, p_3 \in (1,\infty)$. The estimate (2.10) also holds with the inhomogeneous spaces replaced by homogeneous ones.

Proposition 2.1 *Let $f(z) \in C^1(\mathbb{R})$ and $\left|\dfrac{f(z)}{z^2}\right| \lesssim 1$, when $|z| \leqslant 1$, we have for any function $n \in H^s(\mathbb{R})$, $s \geqslant 1$,*

$$\|f(n)\|_{H^s} \leqslant \big(g(\|n\|_{H^1})+1\big)\|n\|_{H^1}\|n\|_{H^s}, \quad s \geqslant 1. \tag{2.11}$$

where $g(x) = \max_{z \leqslant |x|}\{|f(z)|\}$.

Proof Using identity

$$\int_{\mathbb{R}^d} |\xi|^s \hat{f}(\xi)\mathrm{d}\xi = \frac{\pi^{-\frac{s}{2}}\Gamma\left(\dfrac{d+s}{2}\right)}{\pi^{\frac{s+d}{2}}\Gamma\left(-\dfrac{s}{2}\right)}\int_{\mathbb{R}^d}|x|^{-s-d}f(x)\mathrm{d}x, \tag{2.12}$$

see [3] for detailed proof. Then we have

$$|D|^s f(x) = \pi^{-s}\pi^{-\frac{d}{2}}\frac{\Gamma\left(\dfrac{d+s}{2}\right)}{\Gamma\left(-\dfrac{s}{2}\right)}\int_{\mathbb{R}^d} f(x+y)|y|^{-d-s}\mathrm{d}y.$$

Therefor we can estimate $\left\||D|^s f(n)\right\|_{L^2(\mathbb{R})}$ as

$$\begin{aligned}
\left\||D|^s f(n)\right\|_{L_2(\mathbb{R})} &= \pi^{-s}\pi^{-\frac{1}{2}}\frac{\Gamma\left(\dfrac{1+s}{2}\right)}{\Gamma\left(-\dfrac{s}{2}\right)}\left\|\int_{\mathbb{R}} f\Big(n(x+y)\Big)|y|^{-1-s}\mathrm{d}y\right\|_{L_2(\mathbb{R})}\\
&\leqslant C\left\|\int_{\mathbb{R}}\frac{f\Big(n(x+y)\Big)}{n^2(x+y)}n^2(x+y)|y|^{-1-s}\mathrm{d}y\right\|_{L_2(\mathbb{R})}\\
&\leqslant C\left\|\frac{f(n(x))}{n^2}\right\|_{L_x^\infty}\left\|\int_{\mathbb{R}} n^2(x+y)|y|^{-1-s}\mathrm{d}y\right\|_{L_2(\mathbb{R})}\\
&\leqslant C\left\|\frac{f(n(x))}{n^2}\right\|_{L_x^\infty}\||D|^s n^2\|_{L_2(\mathbb{R})}.
\end{aligned} \tag{2.13}$$

And from Lemma 2.2, Sobolev embedding theorem, we have

$$\|n^2\|_{H^s(\mathbb{R})} \lesssim \|n\|_{H^s(\mathbb{R})}\|n\|_{L^\infty(\mathbb{R})} \leqslant \|n\|_{H^1(\mathbb{R})}\|n\|_{H^s(\mathbb{R})}. \tag{2.14}$$

From $\|n\|_{L_x^\infty(\mathbb{R})} \lesssim \|n\|_{H^1(\mathbb{R})}$ and $f \in C^1$, so

$$\|f(n(x))\|_{L_x^\infty} \leqslant g(\|n\|_{L_x^\infty(\mathbb{R})}) \leqslant g(\|n\|_{H^1(\mathbb{R})}).$$

From the assumption $\left|\dfrac{f(z)}{z^2}\right| \leqslant 1$ when $|z| \leqslant 1$, we have

$$\left\|\frac{f(n(x))}{n^2}\right\|_{L_x^\infty} \lesssim 1 + \left\|f\Big(n(x)\Big)\right\|_{L_x^\infty} \leqslant g(\|n\|_{H^1(\mathbb{R})}) + 1. \tag{2.15}$$

Collecting (2.13)–(2.15), we obtain (2.11) as desired. □

Lemma 2.3 *Let* $f(z) \in C^1(\mathbb{R})$, $n(t) \in C^1(\mathbb{R})$, *we have*

$$\int_a^b f(n)n_t\mathrm{d}x = \frac{\mathrm{d}}{\mathrm{d}t}\int_a^b\int_0^{n(t)} f(z)\mathrm{d}z\mathrm{d}x,. \tag{2.16}$$

Proof (2.16) holds from basic calculation.

Proposition 2.2 *If* $E_0, n_0 \in H^1(\mathbb{R})$, $n_1 \in H^{-\frac{\beta}{2}}(\mathbb{R})$, *and*

$$\int_0^n f(z)\mathrm{d}z \geqslant 0, \quad \gamma > 0.$$

Then the solutions of system (1.1)–(1.3) *satisfy the following energy inequality:*

$$\|E_x(\cdot,t)\|_{L_2}^2+\|n(\cdot,t)\|_{L_2}^2+\||D|^{-\frac{\beta}{2}}n_t(\cdot,t)\|_{L_2}^2+\frac{\gamma}{2}\|n_x(\cdot,t)\|_{L_2}^2\leqslant\mathcal{E}_2,\quad\forall t>0,\tag{2.17}$$

where $\mathcal{E}_2$ *is a constant depending only on* $\|E_0\|_{H^1(\mathbb{R})}+\|n_0\|_{H^1(\mathbb{R})}+\|n_1\|_{H^{-\beta/2}(\mathbb{R})}$.

Remark 2.1 *From Proposition 2.2 and embedding theorem, we have*

$$\|E(\cdot,t)\|_{L^\infty(\mathbb{R})},\|n(\cdot,t)\|_{L^\infty(\mathbb{R})}\lesssim\mathcal{E}_2.\tag{2.18}$$

Proof of Proposition 2.2 First we bring in φ. Let $n_t=|D|^\beta\varphi$, then $n_{tt}=|D|^\beta\varphi_t$ and $|D|^{-\beta/2}n_t=|D|^{\beta/2}\varphi$, $\varphi=|D|^{-\beta}n_t$. In this way, system (1.2) was deduced to

$$|D|^\beta\varphi_t+|D|^\beta n+|D|^\beta f(n)+\gamma|D|^{\beta+2}n+|D|^\beta(|E|^2)=0,$$

so we obtain

$$\varphi_t+n+f(n)-\gamma n_{xx}+|E|^2=0,\tag{2.19}$$

Next, we time $\bar{E}_t$ to equation (1.1), and take the real part

$$\mathrm{Re}[(\mathrm{i}E_t,E_t)+(E_{xx},E_t)-(nE,E_t)]=0.\tag{2.20}$$

Since

$$\mathrm{Re}[(E_{xx},E_t)]=-\frac{1}{2}\frac{\mathrm{d}}{\mathrm{d}t}\|E_x\|_{L^2}^2,$$

so from (2.20), we have

$$\frac{1}{2}\frac{\mathrm{d}}{\mathrm{d}t}\|E_x\|_{L^2}^2+\int_a^b n|E|_t^2\mathrm{d}x=0,\tag{2.21}$$

where we choose $a=\infty$ and $b=-\infty$; while in periodic case, we choose $a=0$ and $b=2\pi$. Next

$$\int_a^b n|E|_t^2\mathrm{d}x=-\int_a^b n_t|E|^2\mathrm{d}x+\frac{\mathrm{d}}{\mathrm{d}t}\int_a^b n|E|^2\mathrm{d}x.\tag{2.22}$$

On the other hand, from (2.19) and (2.16)

$$\begin{aligned}\frac{\mathrm{d}}{\mathrm{d}t}\int_a^b\frac{1}{2}n^2\mathrm{d}x&=\int_a^b(-\varphi_t-f(n)+\gamma n_{xx}-|E|^2)n_t\mathrm{d}x\\&=-\frac{1}{2}\frac{\mathrm{d}}{\mathrm{d}t}\||D|^{\beta/2}\varphi\|_{L^2}^2-\frac{\mathrm{d}}{\mathrm{d}t}\int_a^b\int_0^{n(t)}f(z)\mathrm{d}z\mathrm{d}x\\&\quad-\frac{\gamma}{2}\frac{\mathrm{d}}{\mathrm{d}t}\int_a^b n_x^2\mathrm{d}x-\int_a^b n_t|E|^2\mathrm{d}x.\end{aligned}\tag{2.23}$$

From (2.21)–(2.23) we have

$$\frac{1}{2}\frac{\mathrm{d}}{\mathrm{d}t}\|E_x\|_{L^2}^2+\frac{\mathrm{d}}{\mathrm{d}t}\int_a^b\frac{1}{2}n^2\mathrm{d}x+\frac{\gamma}{2}\frac{\mathrm{d}}{\mathrm{d}t}\int_a^b n_x^2\mathrm{d}x+\frac{1}{2}\frac{\mathrm{d}}{\mathrm{d}t}\||D|^{\beta/2}\varphi\|_{L^2}^2$$
$$+\frac{\mathrm{d}}{\mathrm{d}t}\int_a^b n|E|^2+\frac{\mathrm{d}}{\mathrm{d}t}\int_a^b\int_0^{n(t)}f(z)\mathrm{d}z\mathrm{d}x=0. \tag{2.24}$$

Moreover, we have

$$\begin{aligned}\int_a^b n|E|^2\mathrm{d}x&\leqslant\frac{1}{4}\|n\|_{L^2}^2+\|E\|_{L^4}^4\\&\leqslant\frac{1}{4}\|n\|_{L^2}^2+\|E\|_{L^\infty}^2\|E\|_{L^2}^2\\&\leqslant\frac{1}{4}\|n\|_{L^2}^2+\|E\|_{L^2}^2(\epsilon^2\|E_x\|_{L^2}^2+C(\epsilon)^2\|E\|_{L^2}^2)\\&\leqslant\frac{1}{4}\|n\|_{L^2}^2+2\|E_0\|_{L^2}^2\left(\epsilon^2\|E_x\|_{L^2}^2+C(\epsilon)^2\|E_0\|_{L^2}^2\right)\end{aligned} \tag{2.25}$$

and

$$\int_a^b\int_0^n f(z)\mathrm{d}z\mathrm{d}x\geqslant 0. \tag{2.26}$$

Integrate (2.24) from 0 to t, then (2.24)–(2.26) imply that

$$(1-2\epsilon^2\|E_0\|_{L^2}^2)\|E_x\|_{L^2}^2+\frac{1}{4}\|n\|_{L^2}^2+\frac{\gamma}{2}\|n_x\|_{L^2}^2+\frac{1}{2}\||D|^{\beta/2}\varphi\|_{L^2}^2\leqslant\mathcal{E}_2. \tag{2.27}$$

where $\mathcal{E}_2$ is a constant depending only on $\|E_0\|_{H^1(\mathbb{R})}$, $\|n_0\|_{H^1(\mathbb{R})}$ and $\|n_1\|_{H^{-\beta/2}(\mathbb{R})}$. Choose ϵ small enough, such as $1-2\epsilon^2\|E_0\|_{L^2}^2>\frac{1}{4}$, we obtain (2.17) as desired. □

In order to obtain prior estimates for $\|n_t\|_{H^{-\beta/2}}$, We need the estimates for $\|n_t\|_{L_2}$ as well.

Proposition 2.3 *If $E_0,n_0\in H^1(\mathbb{R})$, $n_1\in H^{-\beta/2}(\mathbb{R})$, and $f(n)$ satisfies the condition in Proposition 2.1, then the local solutions of system (1.1)–(1.3) on $[0,T]$ satisfy the following energy inequality:*

$$\|n_t(\cdot,t)\|_{L_2}^2+\||D|^{\beta/2+1}n(\cdot,t)\|_{L_2}^2+\|E_t(\cdot,t)\|_{L_2}^2\leqslant\mathcal{E}_3(T),\quad\forall t\in[0,T], \tag{2.28}$$

where $\mathcal{E}_3(T)$ is a constant depending only on T, $\|E_0\|_{H^1(\mathbb{R})}$, $\|n_0\|_{H^1(\mathbb{R})}$ and $\|n_1\|_{H^{-\beta/2}(\mathbb{R})}$.

Remark 2.2 *From Proposition 2.3, we have $\|E_{xx}(\cdot,t)\|_{L_2},\|n_x(\cdot,t)\|_{L^\infty}\leqslant\mathcal{E}_3(T)$.*

Proof of Proposition 2.3 Let $n_t = A$, $E_t = B$. Notice that n is a real valued function, we time n_t to equation (1.2) and integrate it,

$$(n_t, n_{tt} + |D|^\beta n + |D|^\beta f(n) + \gamma |D|^{\beta+2} n + |D|^\beta(|E|^2)) = 0 \tag{2.29}$$

$$(n_t, n_{tt}) = \frac{1}{2}\frac{\mathrm{d}}{\mathrm{d}t}\|n_t\|_{L^2}^2 = \frac{1}{2}\frac{d}{dt}\|A\|_{L^2}^2,$$

$$(n_t, |D|^\beta n) = \frac{1}{2}\frac{\mathrm{d}}{\mathrm{d}t}\||D|^{\beta/2} n\|_{L^2}^2 \tag{2.30}$$

$$(n_t, \gamma|D|^{\beta+2} n) = \gamma(|D|^{\beta/2+1} n_t, |D|^{\beta/2+1} n) = \frac{\gamma}{2}\frac{\mathrm{d}}{\mathrm{d}t}\||D|^{\beta/2+1} n\|_{L^2}^2. \tag{2.31}$$

Collecting (2.29)–(2.31), we obtain

$$\begin{aligned}&\frac{1}{2}\frac{\mathrm{d}}{\mathrm{d}t}\left(\|D^{\beta/2} n\|_{L^2}^2 + \|A\|_{L^2}^2 + \gamma\||D|^{\beta/2+1} n\|_{L^2}^2\right)\\ &\leqslant \left|(n_t, |D|^\beta(|E|^2)) + (n_t, |D|^\beta f(n))\right|.\end{aligned} \tag{2.32}$$

From Cauchy-Swartz' inequality and Lemma 2.2,

$$\begin{aligned}(n_t, |D|^\beta(|E|^2)) &\lesssim \|n_t\|_{L^2}^2 + \||D|^\beta(|E|^2)\|_{L_2}^2\\ &\lesssim \|B\|_{L^2}^2 + \|E\|_{L^\infty}^2 \||D|^\beta E\|_{L_2}^2\\ &\leqslant C(\|B\|_{L^2}^2 + \||D|^\beta E\|_{L_2}^2),\end{aligned} \tag{2.33}$$

where we used $\|E\|_{L^\infty(\mathbb{R})} \lesssim \|E\|_{H^1(\mathbb{R})} \leqslant C$.

From Proposition 2.1,

$$\begin{aligned}(n_t, |D|^\beta f(n)) \leqslant \|n_t\|_{L^2}^2 + \||D|^\beta f(n)\|_{L_2}^2 &\lesssim \|B\|_{L^2}^2 + \|n\|_{H^1}^2 \|n\|_{H^\beta}^2\\ &\leqslant \|B\|_{L^2}^2 + \||D|^{\beta/2+1} n\|_{L^2}^2,\end{aligned} \tag{2.34}$$

where we used $\|n\|_{H^1} \leqslant C$.

Next, take one order derivative for t on system (1.1), we have

$$(\mathrm{i}B_t + B_{xx} - Bn - EA, B) = 0. \tag{2.35}$$

Notice that $(B_{xx} - Bn, B) = -\|B_x\|_{L_2}^2 - \int_{\mathbb{R}} n|B|^2 \mathrm{d}x$, then take Im part of (2.35),

$$0 = \mathrm{Im}(\mathrm{i}B_t - EA, B) = \frac{1}{2}\frac{\mathrm{d}}{\mathrm{d}t}\|B\|_{L_2}^2 - \mathrm{Im}(EA, B),$$

From Swartz's inequality,

$$\frac{1}{2}\frac{\mathrm{d}}{\mathrm{d}t}\|B\|_{L_2}^2=\mathrm{Im}(EA,B)\leqslant\|E\|_{L^\infty}(\|A\|_{L_2}^2+\|B\|_{L_2}^2)\leqslant C(\|A\|_{L_2}^2+\|B\|_{L_2}^2). \tag{2.36}$$

Collecting (2.36) and (2.32)–(2.34), we have

$$\begin{aligned}&\frac{1}{2}\frac{\mathrm{d}}{\mathrm{d}t}\left(\||D|^{\beta/2}n\|_{L^2}^2+\|A\|_{L_2}^2+\|B\|_{L_2}^2+\gamma\||D|^{\beta/2+1}n\|_{L^2}^2\right)\\&\leqslant K(\|A\|_{L_2}^2+\|B\|_{L_2}^2+\||D|^{\beta}E\|_{L^2}^2+\|D^{\beta/2+1}n\|_{L^2}^2+1).\end{aligned} \tag{2.37}$$

and from equation (1.1), we can estimate $\||D|^{\beta}E\|_{L^2}$

$$\||D|^{\beta}E\|_{L^2}\leqslant\|E_{xx}\|_{L^2}^2+\|E\|_{L^2}^2\leqslant\|E_t\|_{L^2}+\|nE\|_{L^2}^2+C\leqslant\|B\|_{L^2}+K. \tag{2.38}$$

Then from Gronwall's inequality, (2.37) and (2.38) yield

$$\|D^{\beta/2}n\|_{L^2}^2+\|A\|_{L^2}^2+\|B\|_{L_2}^2+\gamma\||D|^{\beta/2+1}n\|_{L^2}^2\leqslant\mathcal{E}_3(T).$$

3 Well-posedness results

In this section, we will apply contraction method to obtain local well-posedness for system (1.1)–(1.2), then from the prior energy estimates we expand the local solutions step by step.

First, using Frourier transformation, we write system (1.1)–(1.2) as integral equations

$$E=S(t)E_0+\int_0^t S(t-\tau)(nE)\mathrm{d}\tau, \tag{3.39}$$

$$n=W_t^{\beta}(t)n_0+W^{\beta}(t)n_1+\int_0^t W^{\beta}(t-\tau)(|D|^{\beta}f(n)+|D|^{\beta}(|E|^2))\mathrm{d}\tau, \tag{3.40}$$

where

$$\begin{aligned}&S(t)u_0=\mathscr{F}^{-1}e^{-\mathrm{i}t|\xi|^2}\mathscr{F}u_0,\\&W^{\beta}(t)u_0=\frac{\sin(|\xi|^{\beta/2}\sqrt{1+|\xi|^2}\ t)}{|\xi|^{\beta/2}\sqrt{1+|\xi|^2}}\mathscr{F}u_0,\quad W_t^{\beta}(t)u_0=\cos(|\xi|^{\beta/2}\sqrt{1+|\xi|^2}\ t)\mathscr{F}u_0.\end{aligned}$$

For the linear operator $W^{\beta}(t)$, we have the following estimates

Lemma 3.1 *Let $s>0$, then*

$$\|W^{\beta}(t)u_0\|_{H^s(\mathbb{R})}\leqslant C(1+|t|)\|u_0\|_{H^{s-1-\beta/2}(\mathbb{R})}, \tag{3.41}$$

$$\|W_t^{\beta}(t)u_0\|_{H^s(\mathbb{R})}\leqslant C\|u_0\|_{H^s(\mathbb{R})}, \tag{3.42}$$

$$\left\|\int_0^t W^{\beta}(t-\tau)f(\tau,x)d\tau\right\|_{L_T^\infty H^s(\mathbb{R})}\leqslant CT(1+T)\|f\|_{L_T^\infty H^{s-1-\beta/2}(\mathbb{R})}. \tag{3.43}$$

Proof Notice that $\lim\limits_{n\to 0}\dfrac{\sin n}{n}=1$, we have

$$\left|\frac{\sin(|\xi|^{\beta/2}\sqrt{1+|\xi|^2}\,t)}{|\xi|^{\beta/2}\sqrt{1+|\xi|^2}}\right|\leqslant\left|\frac{\sin(|\xi|^{\beta/2}\sqrt{1+|\xi|^2}\,t)}{|\xi|^{\beta/2}\sqrt{1+|\xi|^2}\,t}t\right|\lesssim|t|,\quad |\xi|\leqslant 1.$$

$$\left|\frac{\sin(|\xi|^{\beta/2}\sqrt{1+|\xi|^2}\,t)}{|\xi|^{\beta/2}\sqrt{1+|\xi|^2}}\right|\leqslant\frac{1}{|\xi|^{\beta/2}\sqrt{1+|\xi|^2}}\leqslant|\xi|^{-1-\beta/2},\quad |\xi|\geqslant 1.$$

Then from Plancherel' s equality,

$$\|W^\beta(t)u_0\|_{H^s}\leqslant\left\|\frac{\sin(|\xi|^{\beta/2}\sqrt{1+|\xi|^2}\,t)}{|\xi|^{\beta/2}\sqrt{1+|\xi|^2}}\langle\xi\rangle^s\hat{u}_0\right\|_{L^2}\leqslant C(1+|t|)\|u_0\|_{H^{s-1-\beta/2}},\quad(3.44)$$

which is (3.41). (3.42) is trivial.

Using Minkowski' s inequality, we can obtain

$$\begin{aligned}\left\|\int_0^t W^\beta(t-\tau)f(\tau,x)\mathrm{d}\tau\right\|_{L_T^\infty H^s}&\leqslant\int_0^T\|W^\beta(t-\tau)f(\tau,x)\mathrm{d}\tau\|_{L_T^\infty H^s}\\&\leqslant T(1+T)\|f\|_{L_T^\infty H^{s-1-\beta/2}}\end{aligned}$$

as desired. □

For the linear operator $S(t)$, we use the well-known Strichartz estimates

Lemma 3.2 (Strichartz estimates)

$$\begin{aligned}&\|S(t)u_0\|_{L^{\gamma(p)}H_p^s(\mathbb{R}^d)}\leqslant C\|u_0\|_{H^s(\mathbb{R}^d)},\\&\left\|\int_0^t S(t-\tau)f(\tau,x)\mathrm{d}\tau\right\|_{L^{\gamma(p)}H_p^s(\mathbb{R}^d)}\leqslant C\|f\|_{L^{\gamma(r)'}H_r^s(\mathbb{R}^d)},\end{aligned}$$

where $2\leqslant r,p<6$, $\dfrac{1}{\gamma(p)}=\dfrac{d}{2}\left(\dfrac{1}{2}-\dfrac{1}{p}\right)$, $\dfrac{1}{p}+\dfrac{1}{p'}=1$.

3.1 Local well-posedness

Define the following mapping:

$$\begin{aligned}&\mathscr{T}_1:E\to S(t)E_0+\int_0^t S(t-\tau)(nE)\mathrm{d}\tau,\\&\mathscr{T}_2:n\to W_t^\beta(t)n_0+W^\beta(t)n_1+\int_0^t W^\beta(t-\tau)(|D|^\beta f(n)+|D|^\beta(|E|^2))\mathrm{d}\tau.\end{aligned}$$

To prove the local existence and uniqueness part of Theorem 1.1, we use the fixed point argument. We define our resolution spaces X_1^s, X_2^s as:

$$X_1^s = \{u \in \mathscr{S}' : \|u\|_{X_1^s} \leqslant M\}, \qquad \|u\|_{X_1^s} := \|u\|_{L_{T_0}^\infty H^s(\mathbb{R})} + \|u\|_{L_{T_0}^8 H_4^s(\mathbb{R})} \tag{3.45}$$

$$X_2^s = \{u \in \mathscr{S}' : \|u\|_{X_2^s} \leqslant M\}, \quad \|u\|_{X_2^s} := \|u\|_{L_{T_0}^\infty H^s(\mathbb{R})}, \tag{3.46}$$

where $M := 10(\|E_0\|_{H^s(\mathbb{R})} + \|N_0\|_{H^s})$, $\|N_0\|_{H^s} = \|n_0\|_{H^s(\mathbb{R})} + \|n_1\|_{H^{s-1-\beta/2}(\mathbb{R})}$ and T_0 was chosen as:

$$\begin{aligned} T_0 \leqslant \min\Big\{ \Big(\frac{1}{2(\|N_0\|_{H^1} + \|E_0\|_{H^1})}\Big)^{\frac{3}{4}}, \\ \frac{1}{2\|E_0\|_{H^1} + 2\|N_0\|_{H^1}\big(\max\{g(\|N_0\|_{H^1}), \tilde{g}(\|N_0\|_{H^1})\} + 1\big)}, 1\Big\}, \end{aligned} \tag{3.47}$$

where $g(x) = \max_{z \leqslant |x|}\{|f(z)|\}$, $\tilde{g}(x) = \max_{z \leqslant |x|}\{|f'(z)|\}$.

From Lemma 3.2, we have

$$\|S(t)E_0\|_{X_1^s} \leqslant C\|E_0\|_{H^s(\mathbb{R})}. \tag{3.48}$$

Let $s > \dfrac{1}{4}$,

$$\begin{aligned} \Big\| \int_0^t S(t-\tau)(nE)\mathrm{d}\tau \Big\|_{X_1^s} &\leqslant \|nE\|_{L_{T_0}^{\frac{8}{7}} H_{4/3}^s} \\ &\lesssim \|n\|_{L_{T_0}^8 L_4} \|E\|_{L_{T_0}^{\frac{4}{3}} H^s} + \|n\|_{L_{T_0}^8 H_4^s} \|E\|_{L_{T_0}^{\frac{4}{3}} L_2} \\ &\lesssim T_0^{\frac{4}{3}} \Big(\|E\|_{L_{T_0}^8 H_4^s} \|n\|_{L_{T_0}^\infty H^s} + \|E\|_{L_{T_0}^8 H_4^s} \|n\|_{L_{T_0}^\infty L_2} \Big) \\ &\leqslant T_0^{\frac{4}{3}} (\|E\|_{X_1^1} \|n\|_{X_2^s} + \|E\|_{X_1^s} \|n\|_{X_2^1}). \end{aligned} \tag{3.49}$$

So we obtain

$$\|\mathscr{T}_1(E)\|_{X_1^s} \leqslant \|E_0\|_{H^s(\mathbb{R}^n)} + T_0^{\frac{4}{3}} (\|E\|_{X_1^1} \|n\|_{X_2^s} + \|n\|_{X_2^1} \|E\|_{X_1^s}). \tag{3.50}$$

Next, we estimate n, from Lemma 3.1, we have

$$\|W_t^\beta(t) n_0\|_{X_2^s} \leqslant C(1+T_0)\|n_0\|_{H^s(\mathbb{R})}. \tag{3.51}$$

$$\|W^\beta(t) n_1\|_{X_2^s} \leqslant C(1+T_0)\|n_1\|_{H^{s-1-\beta/2}(\mathbb{R})}. \tag{3.52}$$

And notice that $1 < \beta < 2$, $s \geqslant 1$, then

$$\begin{aligned} \Big\| \int_0^t W^\beta(t-\tau)\left(|D|^\beta(|E|^2)\right)\mathrm{d}\tau \Big\|_{X_2^s} &\leqslant CT_0(1+T_0)\left\| |D|^\beta(|E|^2) \right\|_{L_{T_0}^\infty H^{s-1-\beta/2}} \\ &\leqslant CT_0(1+T_0)\left\| E\bar{E} \right\|_{L_{T_0}^\infty H^s} \\ &\leqslant CT_0(1+T_0)\|E\|_{L_{T_0}^\infty H^s} \|E\|_{L_{T_0}^\infty H^1}. \end{aligned} \tag{3.53}$$

From Proposition 2.1, we also have

$$\begin{aligned}
&\left\| \int_0^t W^\beta(t-\tau)\left(|D|^\beta f(n)\right) \mathrm{d}\tau \right\|_{X_2^s} \\
&\leqslant CT_0(1+T_0)\left\| |D|^\beta f(n) \right\|_{L^\infty_{T_0} H^{s-1-\beta/2}} \\
&\leqslant CT_0(1+T_0)\left\| f(n) \right\|_{L^\infty_{T_0} H^s} \\
&\leqslant CT_0(1+T_0)(g(\|n\|_{L^\infty_{T_0}H^1})+1)\|n\|_{L^\infty_{T_0}H^1}\|n\|_{L^\infty_{T_0}H^s} \qquad (3.54)
\end{aligned}$$

From (3.51)–(3.54), we obtain

$$\begin{aligned}
\|\mathscr{T}_2(n)\|_{X_2^s} \leqslant{}& (1+T_0)(\|n_0\|_{H^s(\mathbb{R})}+\|n_1\|_{H^{s-1-\beta/2}(\mathbb{R})}) \\
&+ CT_0(1+T_0)\left(\|E\|_{X_1^s}+\|n\|_{X_2^s}\right)\left(\|E\|_{X_1^1}+\|n\|_{X_2^1}(g\|n\|_{X_2^1}+1)\right).
\end{aligned} \tag{3.55}$$

Next, let

$$\mathscr{T}_1(F) = S(t)E_0 + \int_0^t S(t-\tau)(mF)\mathrm{d}\tau, \tag{3.56}$$

$$\mathscr{T}_2(m) = W_t^\beta(t)n_0 + W^\beta(t)n_1 + \int_0^t W^\beta(t-\tau)(|D|^\beta f(m) + |D|^\beta(|F|^2))\mathrm{d}\tau. \tag{3.57}$$

we can similarly obtain:

$$\begin{aligned}
\|\mathscr{T}_1(E)-\mathscr{T}_1(F)\|_{X_1^s} &\leqslant \|n(E-F)+(n-m)F\|_{L_T^{\frac{8}{8-d}} H^s_{4/3}} \\
&\lesssim T_0^{\frac{4}{3}}\left(\|n\|_{X_2^s}\|E-F\|_{X_1^s}+\|n-m\|_{X_2^s}\|F\|_{X_1^s}\right).
\end{aligned} \tag{3.58}$$

Notice that $\beta<2$

$$\begin{aligned}
&\left\||D|^\beta(f(n)-f(m))\right\|_{L^\infty_{T_0}H^{s-1-\beta/2}} \\
&\leqslant \|f(n)-f(m)\|_{L^\infty_{T_0}H^s} \\
&= \left\| \int_0^1 \frac{\mathrm{d}}{\mathrm{d}\theta} f\left(n\theta+m(1-\theta)\right)\mathrm{d}\theta \right\|_{L^\infty_{T_0}H^s} \\
&\leqslant \sup_{\theta\in[0,1]} \left\|(n-m)f'(n\theta+m(1-\theta))\right\|_{L^\infty_{T_0}H^s} \\
&\leqslant \|(n-m)\|_{L^\infty_{T_0}H^s} \sup_{\theta\in[0,1]} \|f'(n\theta+m(1-\theta))\|_{L^\infty_x} \\
&\quad + \|(n-m)\|_{L^\infty_{T_0}H^s} \sup_{\theta\in[0,1]} \|f'(n\theta+m(1-\theta))\|_{H^s}.
\end{aligned}$$

Similar to (3.53) and (3.54), we have

$$\begin{aligned}&\|\mathscr{T}_2(n)-\mathscr{T}_2(m)\|_{X_2}\\&\leqslant CT_0(1+T_0)\Big(\left(\|E\|_{X_1^s}+\|F\|_{X_1^s}\right)\|E-F\|_{X_1^s}+\\&+\left(\tilde{g}(\|n\|_{H^1}+\|m\|_{H^1})+1\right)(\|n\|_{H^1}+\|m\|_{H^1})\|n-m\|_{X_2^s}\Big).\end{aligned}\tag{3.59}$$

When $s=1$, collecting (3.47), (3.48), (3.50)–(3.52), (3.55) and (3.58), (3.59), from classical iterative method, and fixed point theorem, we can prove that when T_0 satisfied (3.47), then Schrödinger-Boussinesq system (1.1)–(1.3) has unique local solution in $[0,T_0]$ and $(E,n,n_t)\in(C([0,T_0],H^1(\mathbb{R}),C([0,T_0],H^1(\mathbb{R}),C([0,T_0],H^{-\beta/2}(\mathbb{R}))$, where T_0 only depends on $\|E_0\|_{H^1(\mathbb{R})}+\|n_0\|_{H^1(\mathbb{R})}+\|n_1\|_{H^{-\beta/2}(\mathbb{R})}$.

3.2 Global well-posedness

When $s=1$, from the prior energy estimates: Lemma 2.1, Proposition 2.2 and Proposition 2.3, we have for any $t\in[0,T_0]$, the local solutions (E,n,n_t) satisfy

$$\begin{aligned}&\|E(t)\|_{H^1(\mathbb{R})}+\|n(t)\|_{H^1(\mathbb{R})}+\|n_t(t)\|_{H^{-\beta/2}(\mathbb{R})}\\\leqslant&C(\|E_0\|_{H^1(\mathbb{R})},\|n_0\|_{H^1(\mathbb{R})},\|n_1\|_{H^{-\beta/2}(\mathbb{R})}).\end{aligned}\tag{3.60}$$

So, for any $T>0$, we can expand our solution step by step to $[0,T]$. Then we obtain global well-posedness for the solutions of (1.1)–(1.3) in Sobolev space $(H^1(\mathbb{R}^n)$, $H^1(\mathbb{R}^n)$, $H^{-\beta/2}(\mathbb{R}^n))$. Moreover

$$\begin{aligned}&\|E(t)\|_{L_T^\infty H^1(\mathbb{R})}+\|n(t)\|_{L_T^\infty H^1(\mathbb{R})}+\|n_t(t)\|_{L_T^\infty H^{-\beta/2}(\mathbb{R})}\\\leqslant&C(\|E_0\|_{H^1(\mathbb{R})},\|n_0\|_{H^1(\mathbb{R})},\|n_1\|_{H^{-\beta/2}(\mathbb{R})}),\end{aligned}$$

for any $t\in[0,T]$.

For $s>1$, we prove $(E,n,n_t)\in(C([0,T_0],H^s(\mathbb{R}),C([0,T_0],H^s(\mathbb{R}),C([0,T_0]$, $H^{s-1-\beta/2}(\mathbb{R}))$. From (3.50) and (3.55)

$$\begin{aligned}&\|S(t)E_0+\int_0^t S(t-\tau)(nE)\mathrm{d}\tau\|_{L_{T_0}^\infty H^s}\\&\leqslant\|E_0\|_{H^s(\mathbb{R}^n)}+T_0^{\frac{4}{3}}(\|E\|_{X_1^1}\|n\|_{X_2^s}+\|n\|_{X_2^1}\|E\|_{X_1^s}).\end{aligned}$$

$$\begin{aligned}&\|W_t^\beta(t)n_0+W^\beta(t)n_1+\int_0^t W^\beta(t-\tau)(|D|^\beta f(n)+|D|^\beta(|E|^2))\mathrm{d}\tau\|_{L_{T_0}^\infty H^s}\\&\leqslant C(1+T_0)\big(\|n_0\|_{H^s(\mathbb{R})}+\|n_1\|_{H^{s-1-\beta/2}(\mathbb{R})}\\&\quad+T_0\left(\|E\|_{X_1^1}+\|n\|_{X_2^1}(g(\|n\|_{X_2^1}+1))\right)\left(\|E\|_{X_1^s}+\|n\|_{X_2^s}\right)\big).\end{aligned}$$

Since the choice of T_0 and $\|E\|_{X_1^1}$, $\|n\|_{X_2^1}$ are bounded in Section 3.1, from iterative method, we have

$$\begin{aligned}&\|E(t)\|_{L^\infty_{T_0}H^s(\mathbb{R})}, \|n(t)\|_{L^\infty_{T_0}H^s(\mathbb{R})}, \|n_t(t)\|_{L^\infty_{T_0}H^{s-1-\beta/2}(\mathbb{R})}\\ \leqslant\ & C(\|E_0\|_{H^s(\mathbb{R})}, \|n_0\|_{H^s(\mathbb{R})}, \|n_1\|_{H^{s-1-\beta/2}(\mathbb{R})}).\end{aligned}$$

Notice (3.60) and T_0 only depending on $\|E_0\|_{H^1(\mathbb{R})} + \|n_0\|_{H^1(\mathbb{R})} + \|n_1\|_{H^{-\beta/2}(\mathbb{R})}$, so we can also expand our solution step by step to $[0, T]$, $\forall T > 0$, Theorem 1.1 is completed as desired.

Acknowledgements L. Han was supported in part by the National Science Foundation of China (grant 11001022, 11271126, 11271023) and the Fundamental Research Funds for the Central Universities. J. Zhang was supported in part by the National Science Foundation of China (grant 11201185).

References

[1] Boussinesq J. Theorie des ondes et de remous qui se propagent le long d'un canal rectangulaire horizontal, en communiquant au liquide contene dans ce canal des vitesses sensiblement pareilles de la surface au fond, J. Math. Pure Appl., 17(1872), 55–108.

[2] Falk F, Laedkl E W and Spatschek K H. Stability of solitary-wave pulses in shape-memory alloys, Phys. Rev. B., 36(1987), 3031–3041.

[3] Grafakos L. Classical and Modern Fourier analysis. Pearson Education North Asia Limited and China Machine Press., 2004.

[4] Guo B L. The global solution of the system of equations for complex Schrödinger coupled with Boussinesq type self-consistent field, Acta Math. Sin., 26(1983), 297–306.

[5] Guo B L and Chen F X. Finite dimensional behavior of global attractors for weakly damped nonlinear Schrödinger-Boussinesq equations, Phys. D., 93 (1996), 101–118.

[6] Guo B L and Shen L J. The global solution of initial value problem for nonlinear Schrödinger-Boussinesq equation in 3-dimensions, Acta. Math. Appl. Sinica., 6(1990), 11–21.

[7] Han Y Q. The cauchy problem of nonlinear schrödinger equations in $H_s(\mathbb{R}^d)$, J. Partial Diff. Eqs., 18(2005), 1–20.

[8] Kato T. Liapunov Functions and Monotonicity in the Euler and Navier-Stokes Equations, Lecture Notes in Math., vol. 1450, Springer-Verlag, Berlin, 1990.

[9] Kenig C, Ponce G, Vega L. Well-posedness of the initial value problem for the Korteweg-de Vries equation, J. Amer. Math. Soc., 4(1991), 323–347.

[10] Li Y S and Chen Q Y. Finite dimensional global attractor for dissipative Schrödinger-Boussinesq equations, J. Math. Anal. Appl., 205(1997), 107–132.

[11] Makhankov V G. On stationary solutions of Schrödinger equation with a self-consistent potential satisfying Boussinesq's equations, Phys. Lett., 50(1974), 42–44.

[12] Sulem C and Sulem P. The Nonlinear Schrödinger Equation. Self-Focusing and Wave Collapse, Applied Mathematical Sciences, Springer-Verlag, New York, 1999.

[13] Yajima N and Satsuma J. Soliton solutions in a diatomic lattice system, Prog. Theor. Phys., 56(1979), 370–378.

[14] Zakharov V E. On stochastization of one-dimensional chain of nonlinear oscillators, Sov. Phys. JETP., 38(1974), 108–110.

[15] Zakharov V E. Collapse of Langmuir waves, Sov. Phys. JETP., 35(1972), 908–914.

The Stability of Stochastic Generalized Porous Media Equations with General Lévy Noise*

Guo Boling (郭柏灵) and Zhou Guoli (周国立)

Abstract Stochastic generalized porous media equation with jump is considered. The aim is to prove the moment exponential stability and the almost sure exponential stability of the stochastic equation.

Keywords stochastic generalized porous media equations; jump processes; stability

1 Introduction

The stability for stochastic partial differential equations with jump have been studied by many peoples in the recent years. There exists a great amount of literature on the subject, see, for example the monograph [18] and references therein. For the stability of stochastic partial differential equations, there are also lots of literature concerning it. For it's theory and various applications (like applications in finance, automatic control etc), we can see [9], [12], [14] and reference therein. The stability of a linear equation with jump coefficient is studied in [14]. In [3], the stability of a semilinear stochastic differential equation with Wiener process is investigated. The exponential stability of general nonlinear stochastic differential equations with Wiener processes is studied in [15], the asymptotic and exponential stability of the nonlinear stochastic delay differential equations with Wiener processes are considered in [16] and [17] respectively. In [13], stability of infinite dimensional stochastic evolution equations with memory and Lévy jumps is studied.

In this article, the aim is to consider the following stochastic partial differential

* Appl.Math. Mech., 2012,35(8): 1067–1078. DOI: 10.1007/s10483-014-1845-7.

equation

$$
\begin{aligned}
\mathrm{d}X_t = & L\Psi(X_t) + \Phi(X_t)\mathrm{d}t + Q(t,X_t)\mathrm{d}W_t + \int_Z F(t,X_{t-},z)\widetilde{N}(\mathrm{d}t,\mathrm{d}z) \\
& + G(t,X_{t-},z)N(\mathrm{d}t,\mathrm{d}z),
\end{aligned} \tag{1.1}
$$

where L is a partial (or pseudo) differential operator of order (less that or equal to) two, so *e.g.* $L = \Delta$ (or $L = -(-\Delta)^\alpha$, $\alpha \in (0,1]$) on $\mathbb{R}$ or an open subset thereof. The maps $\Psi, \Phi : \mathbb{R} \to \mathbb{R}$ are to fulfill certain monotonicity conditions (cf. (2.3) – (2.5) below). Q is a Hilbert-Schmidt operator valued Lipschitz map, and F denotes a Lipschitz map taking values in a Hilbert space. W stands for the cylindrical wiener process and is defined on a complete probability space $(\Omega, \mathcal{F}, P)$, with normal filtration $\mathcal{F}_t = \sigma\{W(s) : s \leqslant t\}, t \in [0,T]$. $N(\mathrm{d}t, \mathrm{d}z)$ represents Poisson random measure associated with compensated $(\mathcal{F}_t)$-martingale measure defined as $\widetilde{N}(\mathrm{d}t,\mathrm{d}z) := N(\mathrm{d}t,\mathrm{d}z) - \lambda(\mathrm{d}z)\mathrm{d}t$ where $\lambda(\mathrm{d}z)$ defined on separable Banach space Z is the intensity measure of $N(\mathrm{d}t,\mathrm{d}z)$. In particular, if $\Phi = 0, Q = 0, F = 0, G = 0$ and $\Psi(s) := |s|^{r-1}s$ for some $r > 1$, then (1.1) reduces back to the classical porous medium equation(see, e.g., [1])

$$
\frac{\partial u}{\partial t} = \Delta(|u|^{r-1}u).
$$

Furthermore, if L is the Dirichlet Laplacian on an open bounded domain in $\mathbb{R}^d$, it is well known that the solution decays algebraically fast in t. We refer to [1], [2] and [9] and the references therein, also for historical remarks. In this article, under some monotonicity assumptions, we obtain the moment exponential stability and the almost sure exponential stability of the stochastic generalized porous media equations with jump.

In recent years, the stochastic version of the porous medium equation with Brownian motion has been studied intensively, see [4], [6], [7], [8], [10], [19], [11], [20], [21], [22], and the references therein. But as far as we know, there are few papers dealing with porous media equation with jump noise. As jump noise is different from wiener noise, this will bring various new difficulties from calculation and probability when stochastic equations with jump noise is considered. The method using in this article is from [5] and [23]. In [23], the existence and uniqueness of stochastic generalized porous media equations with Lèvy jump were obtained, and in [24], the ergodicity of the same equation were got. In this article, the moment exponential stability is obtained, and then it is used to establish the almost sure exponential stability of equations.

2 Some preliminaries

Let $(\mathbb{E}, \mathcal{M}, v)$ be a separable probability space and $(L, \mathcal{D}(L))$ a negative definite self-adjoint linear operator on $L^2(v)$ having discrete spectrum with eigenvalues

$$0 > -\lambda_1 \geqslant -\lambda_2 \geqslant \cdots \to -\infty$$

and $L^2(v)$-normalized eigenfunctions$\{e_i\}$ such that $e_i \in L^{r+1}(v)$ for any $i \geqslant 1$, where $r \geqslant 1$ is a fixed number throughout this article. A classical example of L is the Laplacian operator on a smooth bounded domain in a complete Riemannian manifold with Dirichlet boundary conditions.

Before stating our equation, we first introduce the state space of the solutions. Let

$$H^1 := \{f \in L^2(v) : \sum_{i=1}^{\infty} \lambda_i \int_{\mathbb{E}} (fe_i)^2 \mathrm{d}v < \infty\}.$$

Define H to be its dual space with inner product $\langle \cdot, \cdot \rangle_H$. Identify $L^2(v)$ with its dual. We get the continuous and dense embedding

$$H^1 \subset L^2(v) \subset H. \tag{2.2}$$

We denote the duality between H and H^1 by $\langle \cdot, \cdot \rangle$. Obviously, when restricted to $L^2(v) \times H^1$ this coincides with the inner product in $L^2(v)$, which we therefore also denote by $\langle \cdot, \cdot \rangle$. It is clear that

$$\langle x, y \rangle_H := \sum_{i=1}^{\infty} \frac{1}{\lambda_i} \langle x, e_i \rangle \langle y, e_i \rangle, \quad x, y \in H.$$

Let $\mathcal{L}_{HS}$ be the space of all Hilbert-Schmidt operators from a separable Hilbert space U to $L^2(v)$. Let W_t denote the cylindrical Brownian motion on U w.r.t. a complete filtered probability space $(\Omega, \mathcal{F}, P)$; Let Ψ and Φ are nonlinear deterministic continuous functions on $\mathbb{R}$ satisfying:

$$|\Psi(s)| + |\Phi(s)| \leqslant c(1 + |s|^r), \tag{2.3}$$

$$\begin{aligned} &\langle \Psi(x) - \Psi(y), y - x \rangle - \langle \Phi(x) - \Phi(y), L^{-1}(x - y) \rangle \\ &\leqslant -\eta \|x - y\|_{r+1}^{r+1} - \theta \|x - y\|_H^2, \end{aligned} \tag{2.4}$$

where $c > 0, s \in \mathbb{R}, x, y \in L^{r+1}(v), \eta \in \mathbb{R}^+$ and $\theta \in \mathbb{R}^+$. Denote $\|\cdot\|_k^k := v(|\cdot|^k), k \in \mathbb{N}$. A very simple example satisfying (2.3)–(2.4) is that $\Psi(t) := |t|^{r-1}t$ and $\Phi(t) := \theta t$, for $t \in \mathbb{R}$. Let

$$Q : [0, T] \times \Omega \to \mathcal{L}_{HS}$$

be progressively measurable such that

$$\|Q(t,x)\|_{\mathcal{L}_{HS}}^2 \leqslant h(t)e^{-\mu t}(1 + \|x\|_H^2), \tag{2.5}$$

$$\|Q(t,x) - Q(t,y)\|_{\mathcal{L}_{HS}}^2 \leqslant h(t)e^{-\mu t}\|x - y\|_H^2, \tag{2.6}$$

where $\mu > 0$ and, for arbitrary $\delta > 0$, the nonnegative continuous function $h(t), t \in \mathbb{R}^+$, satisfies $h(t) = o(e^{\delta t})$, as $t \to +\infty$, i.e., $\lim\limits_{t\to+\infty} \dfrac{h(t)}{e^{\delta t}} = 0$, and $x, y \in H$. From (2.2) Kuratowski's theorem, we know $\mathcal{B}(L^2(v)) \subset \mathcal{B}(H)$. Let

$$F : \mathbb{R}_+ \times H \times Z \to L^2(v)$$

be $\mathcal{B}(\mathbb{R}^+) \times \mathcal{B}(H) \times \mathcal{B}(Z)/\mathcal{B}(L^2(v))$-measurable and satisfy

$$\int_Z \|F(t,y,z)\|_H^2 \lambda(\mathrm{d}z) \leqslant \eta_1(t)e^{-\mu t}(1 + \|y\|_H^2), \tag{2.7}$$

$$\int_Z \|F(t,y_1,z) - F(t,y_2,z)\|_H^2 \lambda(\mathrm{d}z) \leqslant \eta_1(t)e^{-\mu t}\|y_1 - y_2\|_H^2, \tag{2.8}$$

where $\eta_1(t), t \in \mathbb{R}^+$ is nonnegative continuous function and satisfy that for arbitrary $\delta > 0$, $\eta_1(t) = o(e^{\delta t})$, as $t \to +\infty$, i.e., $\lim\limits_{t\to+\infty} \dfrac{\eta_1(t)}{e^{\delta t}} = 0$, and $y, y_1, y_2 \in H$. Let

$$G : \mathbb{R}_+ \times H \times Z \to L^2(v)$$

be $\mathcal{B}(\mathbb{R}^+) \times \mathcal{B}(H) \times \mathcal{B}(Z)/\mathcal{B}(L^2(v))$-measurable and satisfy

$$\int_Z \|G(t,y,z)\|_2 \lambda(\mathrm{d}z) \leqslant \xi_1(t)e^{-\mu t}(1 + \|y\|_H), \tag{2.9}$$

$$\int_Z \|G(t,y,z)\|_2^2 \lambda(\mathrm{d}z) \leqslant \xi_1(t)e^{-\mu t}(1 + \|y\|_H^2), \tag{2.10}$$

$$\int_Z \|G(t,y_1,z) - H(t,y_2,z)\|_2^2 \lambda(\mathrm{d}z) \leqslant \xi_1(t)e^{-\mu t}\|y_1 - y_2\|_H^2, \tag{2.11}$$

where $\xi_1(t), t \in \mathbb{R}^+$ is nonnegative continuous function and satisfy that for arbitrary $\delta > 0$, $\xi_1(t) = o(e^{\delta t})$, as $t \to +\infty$, i.e., $\lim\limits_{t\to+\infty} \dfrac{\xi_1(t)}{e^{\delta t}} = 0$, and $y, y_1, y_2 \in H$.

Remark 2.1 *Since* $h(t) = o(e^{\mu t}), \eta_1(t) = o(e^{\mu t})$, *and* $\xi_1(t) = o(e^{\mu t})$ *there exists a positive constant* γ *such that* $h(t)e^{-\mu t} \leqslant \gamma, \eta_1(t)e^{-\mu t} \leqslant \gamma, \xi_1(t)e^{-\mu t} \leqslant \gamma$ *for all* $t \in \mathbb{R}^+$.

We say mapping $\varphi : [0, \infty[\to H$ is RCLL if it is right continuous and its left limits exists. Then we introduce the definition of the solution to (1.1).

Definition 2.1 *An* H*-valued RCLL* $(\mathcal{F}_t)$*-adapted process* $X(t)$ *is called a solution to* (1.1), *if* $X \in L^{r+1}([0,T] \times \Omega \times \mathbb{E}; ds \times P \times \upsilon)$ *such that for any* $i \in \mathbb{N}$

$$\begin{aligned}\langle X_t, e_i\rangle_H = &\langle X_0, e_i\rangle_H - \int_0^t \int_{\mathbb{E}} (\Psi(X_s)e_i + \Phi(X_s)L^{-1}e_i)\mathrm{d}\upsilon\mathrm{d}s \\ &+ \left\langle \int_0^t Q(s, X_s)\mathrm{d}W_s, e_i \right\rangle_H + \int_0^t \int_Z \langle F(s, X_{s-}, z), e_i\rangle_H \widetilde{N}(\mathrm{d}s, \mathrm{d}z) \\ &+ \int_0^t \int_Z \langle G(s, X_{s-}, z), e_i\rangle_H N(\mathrm{d}s, \mathrm{d}z) \end{aligned} \tag{2.12}$$

We define $V := L^{r+1}(\upsilon)$. As H is Hilbert space, we identify H with H^* via the Riesz isomorphism, then we have

$$V \subset H \equiv H^* \subset V^*$$

continuously and densely. In order to rewrite (2.12) in way of vector form, we need two lemmas from [8]. For the reader's convenience, we cite them in the following .

Lemma 2.1 ([8]) The linear operator

$$Lf := -\sum_{i=1}^{\infty} \lambda_i \left[\int_{\mathbb{E}} (fe_i)\mathrm{d}\upsilon\right] e_i, \quad f \in L^2(\upsilon),$$

defines an isometry from $L^{(r+1)/r}(\upsilon)$ to V^* with dense domain. Its unique continuous extension $\bar{L}$ to all of $L^{(r+1)/r}(\upsilon)$ is an isometric isomorphism from $L^{(r+1)/r}(\upsilon)$ onto V^* such that

$${}_{V^*}\langle -\bar{L}f, g\rangle_V = \int_{\mathbb{E}} (fg)\mathrm{d}\upsilon \quad \textit{for all } f \in L^{(r+1)/r}(\upsilon), g \in L^{(r+1)}(\upsilon).$$

Lemma 2.2([8]) Let $(L^{-1})' : L^{(r+1)/r}(\upsilon) \to L^{(r+1)/r}(\upsilon)$ be the dual operator of $L^{-1} : L^{r+1}(\upsilon) \to L^{r+1}(\upsilon)$. Then the operator

$$J : \bar{L} \circ (L^{-1})' : L^{(r+1)/r}(\upsilon) \to V^*$$

extends the natural inclusion $L^2(v) \subset H \subset V^*$ and for all $f \in L^{(r+1)/r}(v)$

$$_{V^*}\langle Jf, g\rangle_V = -\int_{\mathbb{E}} (fL^{-1}g)dv \quad for\ all\ g \in L^{r+1}(v).$$

By the above lemmas we have

$$\begin{aligned}
{}_{V^*}\langle X_t, e_i\rangle_V = {}& {}_{V^*}\langle X_0, e_i\rangle_V + \int_0^t {}_{V^*}\langle \bar{L}\Psi(X_s) + J\Phi(X_s), e_i\rangle_V \mathrm{d}s \\
& + {}_{V^*}\left\langle \int_0^t Q(s, X_s)\mathrm{d}W_s, e_i\right\rangle_V \\
& + {}_{V^*}\left\langle \int_0^t \int_Z F(X_{s-}, z)\widetilde{N}(\mathrm{d}s, \mathrm{d}z), e_i\right\rangle_V \\
& + {}_{V^*}\left\langle \int_0^t \int_Z G(X_{s-}, z)N(\mathrm{d}s, \mathrm{d}z), e_i\right\rangle_V .
\end{aligned}$$

By (2.3), Lemma 1 and Lemma 2, the integrals

$$\int_0^t (\bar{L}\Psi(X_s) + J\Phi(X_s))\mathrm{d}s, \quad t \geqslant 0,$$

is well defined in V^*. So, (2.12) can be rewritten in way of vector form as

$$\begin{aligned}
X_t = {}& X_0 + \int_0^t (\bar{L}\Psi(X_s) + J\Phi(X_s))\mathrm{d}s \\
& + \int_0^t Q(s, X_s)\mathrm{d}W_s + \int_0^t \int_Z F(s, X_{s-}, z)\widetilde{N}(\mathrm{d}s, \mathrm{d}z) \\
& + \int_0^t \int_Z G(s, X_{s-}, z)N(\mathrm{d}s, \mathrm{d}z).
\end{aligned} \tag{2.13}$$

Before giving our main result, we need a theorem from [23], for reader's convenience, we cite it here.

Theorem 2.1 [23] Assume conditions from (2.3) to (2.11) hold, and let $X_0 \in L^2(\Omega, \mathcal{F}_0, P; H)$. Then there exists a unique solution to (1.1) in the sense of Definition 2.1. Moreover

$$E(\sup_{t\in[0,T]} \|X(t)\|_H^2) < \infty.$$

3 Main results

Theorem 3.1 *In addition to conditions* (2.3)–(2.11), *we assume* $\Psi(0) = 0, \Phi(0) = 0$. *Then, if* X_t *is a global strong solution to equation* (1.1), *there exist constants* $\varepsilon \in (0, \mu)$ *and* $C > 0$ *such that*

$$E\|X_t\|_H^2 \leqslant Ce^{-(\mu-\varepsilon)t}, \quad \forall t \geqslant 0. \tag{3.14}$$

proof Let ε be a very small positive constant such that $\mu-\varepsilon>0$. Then, Itö's formula implies

$$\begin{aligned}
&e^{(\mu-\varepsilon)t}\|X_t\|_H^2-\|X_0\|_H^2\\
=&(\mu-\varepsilon)\int_0^t e^{(\mu-\varepsilon)s}\|X_s\|_H^2\mathrm{d}s\\
&-2\int_0^t e^{(\mu-\varepsilon)s}\int_E \Psi(X_s)X_s\mathrm{d}v\mathrm{d}s\\
&-2\int_0^t e^{(\mu-\varepsilon)s}\int_E \Phi(X_s)L^{-1}X_s\mathrm{d}v\mathrm{d}s\\
&+\int_0^t e^{(\mu-\varepsilon)s}\|Q(s,X_s)\|_{\mathcal{L}_{HS}}^2\mathrm{d}s\\
&+\int_0^t\int_Z e^{(\mu-\varepsilon)s}\|F(s,X_{s-},z)\|_H^2\lambda(\mathrm{d}z)\mathrm{d}s\\
&+\int_0^t\int_Z e^{(\mu-\varepsilon)s}[\|G(s,X_{s-},z)\|_H^2\\
&+2\langle X_{s-},G(s,X_{s-},z)\rangle_H]N(\mathrm{d}s,\mathrm{d}z)\\
&+2\int_0^t e^{(\mu-\varepsilon)s}\langle X_s,Q(s,X_s)\mathrm{d}W_s\rangle_H\\
&+\int_0^t\int_Z e^{(\mu-\varepsilon)s}[\|F(s,X_{s-},z)\|_H^2\\
&+2\langle X_{s-},F(s,X_{s-},z)\rangle_H]\widetilde{N}(\mathrm{d}s,\mathrm{d}z)
\end{aligned}\tag{3.15}$$

By (2.4) and $\Psi(0)=0$ as well as $\Phi(0)=0$, we get

$$\begin{aligned}
&e^{(\mu-\varepsilon)t}\|X_t\|_H^2\\
\leqslant&\|X_0\|_H^2+(\mu-\varepsilon-2\theta)\int_0^t e^{(\mu-\varepsilon)s}\|X_s\|_H^2\mathrm{d}s\\
&-2\eta\int_0^t e^{(\mu-\varepsilon)s}\|X_s\|_{r+1}^{r+1}\mathrm{d}s\\
&+\int_0^t e^{(\mu-\varepsilon)s}(h(s)e^{-\mu s}+\eta_1(s)e^{-\mu s})\mathrm{d}s\\
&+\int_0^t e^{(\mu-\varepsilon)s}(h(s)e^{-\mu s}+\eta_1(s)e^{-\mu s})\|X_s\|_H^2\mathrm{d}s
\end{aligned}$$

$$
\begin{aligned}
&+\int_0^t\int_Z e^{(\mu-\varepsilon)s}[\|G(s,X_{s-},z)\|_H^2\\
&+2\langle X_{s-},G(s,X_{s-},z)\rangle_H]N(\mathrm{d}s,\mathrm{d}z)\\
&+2\int_0^t e^{(\mu-\varepsilon)s}\langle X_s,Q(s,X_s)\mathrm{d}W_s\rangle_H\\
&+\int_0^t\int_Z e^{(\mu-\varepsilon)s}[\|F(s,X_{s-},z)\|_H^2\\
&+2\langle X_{s-},F(s,X_{s-},z)\rangle_H]\widetilde{N}(\mathrm{d}s,\mathrm{d}z) \qquad (3.16)
\end{aligned}
$$

where $C_{L^{-1}}$ denote the normal of L^{-1}. Since

$$
\begin{aligned}
&\int_0^t e^{(\mu-\varepsilon)s}(h(s)e^{-\mu s}+\eta_1(s)e^{-\mu s}+\xi_1(s)e^{-\mu s})\|X_s\|_H^2\mathrm{d}s\\
\leqslant& C\int_0^t e^{(\mu-\varepsilon)s}e^{-(\mu-\frac{\varepsilon}{2})s}\|X_s\|_H^2\mathrm{d}s\\
\leqslant& \eta\int_0^t e^{(\mu-\varepsilon)s}\|X_s\|_{r+1}^{r+1}ds+C\int_0^t e^{-\frac{\varepsilon}{2}s}\mathrm{d}s,
\end{aligned}
$$

where the last inequality follows by Young inequality. Since the last two terms on the right hand side of (3.16) are martingales, we have that

$$
\begin{aligned}
&E(e^{(\mu-\varepsilon)t}\|X_t\|_H^2)\\
\leqslant& E(\|X_0\|_H^2)+(\mu-\varepsilon-2\theta)E\int_0^t e^{(\mu-\varepsilon)s}\|X_s\|_H^2\mathrm{d}s\\
&-\eta E\int_0^t e^{(\mu-\varepsilon)s}\|X_s\|_{r+1}^{r+1}\mathrm{d}s\\
&+\int_0^t e^{-\varepsilon s}(h(s)+\eta_1(s)+\xi_1(s)+1)\mathrm{d}s. \qquad (3.17)
\end{aligned}
$$

Since

$$
C_\varepsilon:=\int_0^{+\infty}e^{-\varepsilon s}(h(s)+\eta_1(s)+\xi_1(s)+1)\mathrm{d}s<\infty,
$$

by (3.17) and the Gronwall inequality, we have

$$
E(e^{(\mu-\varepsilon)t}\|X_t\|_H^2)\leqslant(E(\|X_0\|_H^2)+C_\varepsilon)e^{(\mu-\varepsilon-2\theta)t}.
$$

Finally, we get

$$
E(\|X_t\|_H^2)\leqslant\Big(E(\|X_0\|_H^2)+C_\varepsilon\Big)e^{-2\theta t}.
$$

In the following, we intend to investigate the almost sure stability, which is in most situations the kind of stability one usually wants to have in practical applications, of the trivial solution of (1.1).

Theorem 3.2 *Under the conditions in Theorem* 3.1, *there exist positive constants* M, ϵ *and a subset* $\Omega_0 \subset \Omega$ *with* $P(\Omega_0) = 1$ *such that, for each* $w \in \Omega_0$, *there exists a positive random number* $T(w)$ *such that*

$$\|X_t\|_H^2 \leqslant Me^{-\epsilon t}, \forall t \geqslant T(w).$$

proof For $0 \leqslant s \leqslant t$, by (2.5) and (3.14), we have

$$\begin{aligned}&\int_s^t E\|Q(u, X_u)\|_{\mathcal{L}_{HS}}^2 \mathrm{d}u \\ \leqslant &\int_s^t h(u)e^{-\mu u}(1 + E\|X_u\|_H^2)\mathrm{d}u \\ \leqslant &e^{-(\mu-\varepsilon)s}\int_s^t h(u)e^{-\varepsilon u}(1 + Ce^{-2\theta u})\mathrm{d}u \leqslant C_{h,\varepsilon}e^{-(\mu-\varepsilon)s},\end{aligned} \tag{3.18}$$

where $C_{h,\varepsilon} := \int_0^{+\infty} h(u)e^{-\varepsilon u}(1 + Ce^{-(\mu-\varepsilon)u})du$. Similarly, by conditions (2.7) and (3.14), we have

$$\begin{aligned}&E\int_s^t \int_Z \|F(u, X_u, z)\|_2^2 \lambda(\mathrm{d}z)\mathrm{d}u \\ \leqslant &\int_s^t \eta_1(u)e^{-\mu u}(1 + E\|X_u\|_H^2)\mathrm{d}u \leqslant C_{\eta_1,\varepsilon}e^{-(\mu-\varepsilon)s},\end{aligned} \tag{3.19}$$

where $C_{\eta_1,\varepsilon} := \int_0^{+\infty} \eta_1(u)e^{-\varepsilon u}(1 + Ce^{-2\theta u})\mathrm{d}u$. And we have

$$\begin{aligned}&E\int_s^t \int_Z \|G(u, X_u, z)\|_2^2 \\ &+2|\langle G(u, X_u, z), X(u)\rangle|\lambda(\mathrm{d}z)\mathrm{d}u \\ \leqslant &\int_s^t \xi_1(u)e^{-\mu u}(1 + E\|X_u\|_H^2)\mathrm{d}u \leqslant C_{\xi_1,\varepsilon}e^{-(\mu-\varepsilon)s},\end{aligned} \tag{3.20}$$

where $C_{\xi_1,\varepsilon} := \int_0^{+\infty} \xi_1(u)e^{-\varepsilon u}(1 + Ce^{-2\theta u})\mathrm{d}u$. In the following, we will prove there exists a positive constant $M > 0$ such that

$$E\Big(\sup_{0 \leqslant t < +\infty} \|X_t\|_H^2\Big) \leqslant M.$$

By the Itö formula, we have

$$\begin{aligned}
&\|X_t\|_H^2-\|X_0\|_H^2\\
=&-2\int_0^t\int_E \Psi(X_s)X_s\mathrm{d}\nu\mathrm{d}s-2\int_0^t\int_E \Phi(X_s)L^{-1}X_s\mathrm{d}\nu\mathrm{d}s\\
&+\int_0^t\|Q(s,X_s)\|_{\mathcal{L}_{HS}}^2\mathrm{d}s+\int_0^t\int_Z\|F(s,X_{s-},z)\|_H^2\lambda(\mathrm{d}z)\mathrm{d}s\\
&+\int_0^t\int_Z[\|G(X_{s-},z)\|_H^2+\langle X_{s-},G(X_{s-},z)\rangle_H]N(\mathrm{d}s,\mathrm{d}z)\\
&+2\int_0^t\langle X_s,Q(s,X_s)\mathrm{d}W_s\rangle_H\\
&+\int_0^t\int_Z[\|F(X_{s-},z)\|_H^2+\langle X_{s-},F(X_{s-},z)\rangle_H]\widetilde{N}(\mathrm{d}s,\mathrm{d}z).
\end{aligned}\tag{3.21}$$

By Burkholder-Davis-Gundy's inequality and (3.18), we get for any $T\in\mathbb{R}^+$

$$\begin{aligned}
&2E\Big(\sup_{t\in[0,T]}|\int_0^t\langle X_s,Q(s,X_s)\mathrm{d}W_s\rangle_H|\Big)\\
\leqslant& M_1E\Big[\Big(\int_0^T\|X_s\|_H^2\cdot\|Q(s,X_s)\|_{\mathcal{L}_{HS}}^2\mathrm{d}s\Big)^{1/2}\Big]\\
\leqslant&\frac14E\Big(\sup_{t\in[0,T]}\|X_t\|_H^2\Big)+M_2\int_0^T E\|Q(s,X_s)\|_{\mathcal{L}_{HS}}^2\mathrm{d}s,\\
\leqslant&\frac14E\Big(\sup_{t\in[0,T]}\|X_t\|_H^2\Big)+M_2C_{h,\varepsilon},
\end{aligned}\tag{3.22}$$

where M_1,M_2 are two positive constants and are independent of T. By Burkholder-Davis-Gundy inequality and (3.19), we have

$$\begin{aligned}
&E\Big(\sup_{t\in[0,T]}\int_0^t\int_Z[\|F(X_{s-},z)\|_H^2\\
&+2\langle X_{s-},F(X_{s-},z)\rangle_H]\widetilde{N}(\mathrm{d}s,\mathrm{d}z)\Big)\\
\leqslant& E\Big(\sup_{t\in[0,T]}\int_0^t\int_Z\|F(X_{s-},z)\|_H^2\widetilde{N}(\mathrm{d}s,\mathrm{d}z)\\
&+E\Big(\sup_{t\in[0,T]}\int_0^t\int_Z\langle X_{s-},F(X_{s-},z)\rangle_H\widetilde{N}(\mathrm{d}s,\mathrm{d}z)\Big)\\
\leqslant& M_3E\Big\{\int_0^T\int_Z\|F(X_{s-},z)\|_H^4N(\mathrm{d}s,\mathrm{d}z)\Big\}^{1/2}
\end{aligned}\tag{3.23}$$

$$
\begin{aligned}
&+M_3E\Big(\int_0^T\int_Z\langle X_{s-},F(X_{s-},z)\rangle_H^2N(\mathrm{d}s,\mathrm{d}z)\Big)^{1/2}\\
\leqslant\,& M_3E\int_0^T\int_Z\|F(X_{s-},z)\|_H^2N(\mathrm{d}s,\mathrm{d}z)\\
&+M_3E\Big(\int_0^T\int_Z\langle X_{s-},F(X_{s-},z)\rangle_H^2N(\mathrm{d}s,\mathrm{d}z)\Big)^{1/2}\\
\leqslant\,&(M_3+M_4)C_{\eta_1,\varepsilon}+(1/8)E\Big(\sup_{t\in[0,T]}\|X_t\|_H^2\Big),
\end{aligned}
$$

where M_3 and M_4 are independent of T. Similarly, by (3.23)

$$
\begin{aligned}
&E\Big(\sup_{t\in[0,T]}\int_0^t\int_Z[\|G(X_{s-},z)\|_H^2\\
&+2\langle X_{s-},G(X_{s-},z)\rangle_H]N(\mathrm{d}s,\mathrm{d}z)\Big)\\
\leqslant\,& E\int_0^T\int_Z\|G(X_{s-},z)\|_H^2N(\mathrm{d}s,\mathrm{d}z)\\
&+2E\int_0^T\int_Z|\langle X_{s-},G(X_{s-},z)\rangle_H|N(\mathrm{d}s,\mathrm{d}z)\\
\leqslant\,& M_5C_{\xi_1,\varepsilon}.
\end{aligned}\tag{3.24}
$$

By Itô formula and (2.4), after elementary calculation, we get

$$
\begin{aligned}
\|X_t\|_H^2\leqslant\,&\|X_0\|_H^2+\gamma\int_0^t\|X_s\|_H^2+\int_0^t(h(s)\\
&+\eta_1(s))e^{-\mu s}\mathrm{d}s\\
&+2\int_0^t\langle X_s,Q(s,X_s)\mathrm{d}W_s\rangle_H\\
&+\int_0^t\int_Z[\|F(s,X_{s-},z)\|_H^2\\
&+2\langle X_{s-},F(s,X_{s-},z)\rangle_H]\widetilde{N}(\mathrm{d}s,\mathrm{d}z).
\end{aligned}
$$

By (3.14), (3.21) and (3.22), we have

$$
\begin{aligned}
&E\Big(\sup_{t\in[0,T]}\|X_t\|_H^2\Big)\\
\leqslant\,& E\Big(\|X_0\|_H^2\Big)+C\int_0^T[(h(t)+\eta_1(t))e^{-\mu t}+e^{-2\theta t}]\mathrm{d}t\\
&+(1/4)E\Big(\sup_{t\in[0,T]}\|X_t\|_H^2\Big)+M_2C_{h,\varepsilon}\\
&+(M_3+M_4)C_{\eta_1,\varepsilon}+M_5C_{\xi_1,\varepsilon}+(1/4)E\Big(\sup_{t\in[0,T]}\|X_t\|_H^2\Big)
\end{aligned}
$$

$$\leqslant E\left(\|X_0\|_H^2\right)+C\int_0^T[(h(t)+\eta_1(t))e^{-\mu t}+e^{-2\theta t}]\mathrm{d}t$$
$$+(1/2)E\Big(\sup_{t\in[0,T]}\|X_t\|_H^2\Big)+C_{h,\eta_1,\xi_1,\varepsilon},$$

where $C_{h,\eta_1,\xi_1,\varepsilon}=M_2C_{h,\varepsilon}+(M_3+M_4)C_{\eta_1,\varepsilon}+M_5C_{\xi_1,\varepsilon}$. So, we have

$$\begin{aligned}&E\Big(\sup_{t\in[0,T]}\|X_t\|_H^2\Big)\\ \leqslant&2\Big[E\left(\|X_0\|_H^2\right)+\int_0^{+\infty}[(h(t)+\eta_1(t))e^{-\mu t}+e^{-2\theta t}]\mathrm{d}t+C_{h,\eta_1,\xi_1,\varepsilon}\Big]\\ :=&M<+\infty.\end{aligned}$$

Since, for arbitrary $\delta>0$, $h(t)e^{\delta t},\eta_1(t)e^{\delta t}$ and $\xi_1e^{\delta t}$ are continuous and subexponential growth. By Fatou Lemma, we get

$$E\Big(\sup_{t\in[0,+\infty)}\|X_t\|_H^2\Big)\leqslant\lim_{T\to+\infty}E\Big(\sup_{t\in[0,T]}\|X_t\|_H^2\Big)\leqslant M.$$

Finally, we are going to finish the proof. By Itö formula, after elementary calculation, we get

$$\begin{aligned}\|X_t\|_H^2\leqslant&\|X_N\|_H^2+C\int_N^t\|X_s\|_H^2\mathrm{d}s+\int_N^t(h(s)+\eta_1(s))e^{-\mu s}\mathrm{d}s\\&+2\int_N^t\langle X_s,Q(s,X_s)\mathrm{d}W_s\rangle_H\\&+\int_N^t\int_Z[\|F(s,X_{s-},z)\|_H^2\\&+2\langle X_{s-},F(s,X_{s-},z)\rangle_H]\widetilde{N}(\mathrm{d}s,\mathrm{d}z)\\&+\int_N^t\int_Z[\|G(s,X_{s-},z)\|_H^2\\&+2\langle X_{s-},G(s,X_{s-},z)\rangle_H]N(\mathrm{d}s,\mathrm{d}z).\end{aligned}\tag{3.25}$$

In the following, the positive constant C will change from line to line. By By Burkholder-Davis-Gundy's inequality, and (3.18), we get

$$\begin{aligned}&2E\Big(\sup_{\{N\leqslant t\leqslant N+1\}}\Big|\int_N^t\langle X_s,Q(s,X_s)\mathrm{d}W_s\rangle_H\Big|\Big)\\ \leqslant&\frac{1}{8}E\Big(\sup_{\{N\leqslant t\leqslant N+1\}}\|X_t\|_H^2\Big)+M_2C_{h,\varepsilon}e^{-(\mu-\varepsilon)N}\\ =:&\frac{1}{8}E\Big(\sup_{\{N\leqslant t\leqslant N+1\}}\|X_t\|_H^2\Big)+Ce^{-(\mu-\varepsilon)N}.\end{aligned}\tag{3.26}$$

Analogously to derive (3.23), we get

$$\begin{aligned}&E\Big(\sup_{\{N\leqslant t\leqslant N+1\}}\int_N^t\int_Z[\|F(s,X_{s-},z)\|_H^2\\&+2\langle X_{s-},F(s,X_{s-},z)\rangle_H]\widetilde{N}(\mathrm{d}s,\mathrm{d}z)\Big)\\&\leqslant(M_3+M_4)E\int_N^{N+1}\int_Z\|F(s,X_{s-},z)\|_H^2\lambda(\mathrm{d}z)\mathrm{d}s\\&+\frac{1}{4}E\Big(\sup_{\{N\leqslant t\leqslant N+1\}}\|X_t\|_H^2\Big)\\&\leqslant(M_3+M_4)C_{\eta_1,\varepsilon}e^{-(\mu-\varepsilon)N}+\frac{1}{8}E\Big(\sup_{\{N\leqslant t\leqslant N+1\}}\|X_t\|_H^2\Big)\\&=:\frac{1}{8}E\Big(\sup_{\{N\leqslant t\leqslant N+1\}}\|X_t\|_H^2\Big)+Ce^{-(\mu-\varepsilon)N},\end{aligned}\tag{3.27}$$

where the second inequality follows by (3.19).

Analogously to derive (3.24), we get

$$\begin{aligned}&E\Big(\sup_{\{N\leqslant t\leqslant N+1\}}\int_N^t\int_Z[\|G(s,X_{s-},z)\|_H^2\\&+2\langle X_{s-},G(s,X_{s-},z)\rangle_H]N(\mathrm{d}s,\mathrm{d}z)\Big)\\&\leqslant M_7E\int_N^{N+1}\int_Z\|G(s,X_{s-},z)\|_H^2\lambda(\mathrm{d}z)\mathrm{d}s\\&+\frac{1}{8}E\Big(\sup_{\{N\leqslant t\leqslant N+1\}}\|X_t\|_H^2\Big)\\&\leqslant M_7C_{\xi_1,\varepsilon}e^{-(\mu-\varepsilon)N}\\&+\frac{1}{8}E\Big(\sup_{\{N\leqslant t\leqslant N+1\}}\|X_t\|_H^2\Big)\\&=:\frac{1}{8}E\Big(\sup_{\{N\leqslant t\leqslant N+1\}}\|X_t\|_H^2\Big)+Ce^{-(\mu-\varepsilon)N},\end{aligned}\tag{3.28}$$

where the second inequality follows by (2.9) and the young's inequality. By (3.25)–(3.28), we get

$$\begin{aligned}&E\Big(\sup_{\{N\leqslant t\leqslant N+1\}}\|X(t)\|_H^2\Big)\\&\leqslant E\Big(\|X_N\|_H^2\Big)+C\int_N^{N+1}E\|X_s\|_H^2\mathrm{d}s\\&+\int_N^{N+1}e^{-\mu s}(h(s)+\eta_1(s))\mathrm{d}s\end{aligned}$$

$$
\begin{aligned}
&+2E\Big(\sup_{\{N\leqslant t\leqslant N+1\}}\Big|\int_N^t \langle X_s, Q(s,X_s)\mathrm{d}W(s)\rangle_H\Big|\Big)\\
&+E\Big(\sup_{\{N\leqslant t\leqslant N+1\}}\int_N^t\int_Z[\|F(s,X_{s-},z)\|_H^2\\
&+\langle X_{s-},F(s,X_{s-},z)\rangle_H]\widetilde{N}(\mathrm{d}s,\mathrm{d}z)\Big)\\
&+E\Big(\sup_{\{N\leqslant t\leqslant N+1\}}\int_N^t\int_Z[\|G(s,X_{s-},z)\|_H^2\\
&+\langle X_{s-},G(s,X_{s-},z)\rangle_H]N(\mathrm{d}s,\mathrm{d}z)\Big)\\
\leqslant{}& Ce^{-2\theta N}+e^{-(\mu-\varepsilon)N}\int_N^{N+1}e^{-\varepsilon s}(h(s)+\eta_1(s))\mathrm{d}s\\
&+\frac{1}{4}E\Big(\sup_{\{N\leqslant t\leqslant N+1\}}\|X_t\|_H^2\Big)+Ce^{-(\mu-\varepsilon)N}.
\end{aligned}
$$

So, it follows that

$$
E\Big(\sup_{\{N\leqslant t\leqslant N+1\}}\|X(t)\|_H^2\Big)\leqslant Ce^{-(\mu-\varepsilon)N}+Ce^{-2\theta N}\leqslant Ce^{-\alpha N}, \tag{3.29}
$$

where $\alpha=\min\{2\theta,\mu-\varepsilon\}$. By (3.25), we have

$$
\begin{aligned}
\sup_{\{N\leqslant t\leqslant N+1\}}\|X_t\|_H^2\leqslant{}&\|X_N\|_H^2+C\int_N^{N+1}\|X_s\|_H^2\mathrm{d}s\\
&+\int_N^{N+1}e^{-\mu s}(h(s)+\eta_1(s))\mathrm{d}s\\
&+2\sup_{\{N\leqslant t\leqslant N+1\}}\Big|\int_N^t\langle X_s,Q(s,X_s)\mathrm{d}W_s\rangle_H\Big|\\
&+\sup_{\{N\leqslant t\leqslant N+1\}}\int_N^t\int_Z[\|F(s,X_{s-},z)\|_H^2\\
&+2\langle X_{s-},F(s,X_{s-},z)\rangle_H]\widetilde{N}(\mathrm{d}s,\mathrm{d}z)\\
&+\sup_{\{N\leqslant t\leqslant N+1\}}\int_N^t\int_Z[\|G(s,X_{s-},z)\|_H^2\\
&+2\langle X_{s-},G(s,X_{s-},z)\rangle_H]N(\mathrm{d}s,\mathrm{d}z).
\end{aligned}
$$

Let $\frac{\epsilon_N^2}{5}=e^{-\frac{\alpha N}{8}}\int_0^\infty e^{-\varepsilon s}(h(s)+\eta_1(s)+\xi_1(s))\mathrm{d}s$. By (3.27)–(3.29), we get

$$
\begin{aligned}
&P\Big\{\sup_{\{N\leqslant t\leqslant N+1\}}\|X_t\|_H^2\geqslant\epsilon_N^2\Big\}\\
\leqslant{}&P\Big\{\|X_N\|_H^2\geqslant\frac{\epsilon_N^2}{5}\Big\}+P\Big\{C\int_N^{N+1}E\|X_t\|_H^2\geqslant\frac{\epsilon_N^2}{5}\Big\}
\end{aligned}
$$

$$
\begin{aligned}
&+P\Big\{\int_N^{N+1} e^{-\mu s}(h(s)+\eta_1(s))\mathrm{d}s \geqslant \frac{\epsilon_N^2}{5}\Big\}\\
&+P\Big\{2\sup_{\{N\leqslant t\leqslant N+1\}}\Big|\int_N^t\langle X_s, Q(s,X_s)\mathrm{d}W_s\rangle_H\Big| \geqslant \frac{\epsilon_N^2}{5}\Big\}\\
&+P\Big\{\sup_{\{N\leqslant t\leqslant N+1\}}\int_N^t\int_Z[\|F(s,X_{s-},z)\|_H^2\\
&+2\langle X_{s-}, F(s,X_{s-},z)\rangle_H]\widetilde{N}(\mathrm{d}s,\mathrm{d}z) \geqslant \frac{\epsilon_N^2}{5}\Big\}\\
&+P\Big\{\sup_{\{N\leqslant t\leqslant N+1\}}\int_N^t\int_Z[\|G(s,X_{s-},z)\|_H^2\\
&+2\langle X_{s-}, G(s,X_{s-},z)\rangle_H]N(\mathrm{d}s,\mathrm{d}z) \geqslant \frac{\epsilon_N^2}{5}\Big\}\\
\leqslant{}& \frac{5}{\epsilon_N^2}E(\|X_N\|_H^2) + P\Big\{e^{-(\mu-\varepsilon)N}\int_N^{N+1} e^{-\varepsilon s}(h(s)\\
&+\eta_1(s)+\xi_1(s))\mathrm{d}s \geqslant \frac{\epsilon_N^2}{5}\Big\}\\
&+\frac{10}{\epsilon_N^2}E\Big(\sup_{\{N\leqslant t\leqslant N+1\}}\Big|\int_N^t\langle X_s, Q(s,X_s)\mathrm{d}W_s\rangle_H\Big|\Big)\\
&+\frac{5}{\epsilon_N^2}E\Big(\sup_{\{N\leqslant t\leqslant N+1\}}\int_N^t\int_Z[\|F(s,X_{s-},z)\|_H^2\\
&+\langle X_{s-}, F(s,X_{s-},z)\rangle_H]\widetilde{N}(\mathrm{d}s,\mathrm{d}z)\Big)\\
&+\frac{5}{\epsilon_N^2}E\Big(\sup_{\{N\leqslant t\leqslant N+1\}}\int_N^t\int_Z[\|G(s,X_{s-},z)\|_H^2\\
&+\langle X_{s-}, G(s,X_{s-},z)\rangle_H]N(\mathrm{d}s,\mathrm{d}z)\Big)\\
\leqslant{}& \frac{5}{\epsilon_N^2}Ce^{-2\theta N} + \frac{10}{\epsilon_N^2}\Big(Ce^{-(\mu-\varepsilon)N} + \frac{1}{4}E\Big(\sup_{\{N\leqslant t\leqslant N+1\}}\|X_t\|_H^2\Big)\Big)\\
&+\frac{5}{\epsilon_N^2}\Big(Ce^{-(\mu-\varepsilon)N} + \frac{1}{4}E\Big(\sup_{\{N\leqslant t\leqslant N+1\}}\|X_t\|_H^2\Big)\Big)\\
\leqslant{}& Ce^{-\frac{7}{8}\alpha N}.
\end{aligned}
$$

So, by Borel-Cantelli's Lemma, we complete the proof.

References

[1] Aronson D G. The porous medium equation, Lecture Notes Math., Vol. 1224, Springer, Berlin, 1986, 1–46.

[2] Aronson D G and Peletier L A. Large time behaviour of solutions of the porous medium equation in bounded domains. J. Diff. Equ., 1981, 39: 378–412.

[3] Basak G K, Bisi A and Ghosh M K. Stability of a random diffusion with linear drift. J.Math.Anal.Appl., 1996, 202: 604–622.

[4] Beyn W J, Gess B, Lescot P and Röckner M. The global random attractors for a class of stochastic porous media equations. Communications in Partial Differential Equations, 2011, 36: 446–469.

[5] Caraballo Tomas, Liu K. On the exponential stability criteria of stochastic partial differential equations. Stochastic Processes and their Applications, 1999, 83: 289–301.

[6] Da Prato G and Röckner M. Weak solutions to stochastic Porus media equations. J. Evolution Equ., 2004, 4: 249–271.

[7] Da Prato G and Röckner M. Invariant measures for a stochastic Porus media equation. In: H.Kunita, Y. Takanashi, S.Watanabe,ets. Stochastic Analysis and Related Topics, Advanced in Pure Math, Vol. 41, Japan: Math.Soc, 2004, 13–29.

[8] Da Prato G, Röckner M, Rozovskii B L and Wang F Y. Strong solutions of stochastic generalized Porus media equations: Existence, Uniqueness and Ergodicity, Commnications in Partial Differential Equations, 2006, 31: 277–291.

[9] Da Prato G and Zabczyk J. Stochastic equations in infinite dimensions, Cambridge University Press, 1992.

[10] Ebmeye C. Regularrity in Sobolev spaces for the fast diffusion and the porous medium equation. J. Math.Anal. Appl., 2005, 307: 134–152.

[11] Kim J U. On the stochastic porous medium equation. J. Diff. Equat., 2006, 206: 163–194.

[12] Liu K. Stability of Infinite Dimensional Stochastic Differential Equations with Applications, in: Pitman Monographs series in Pure and Applied Mathematics, vol. 135, Chapman Hall/CRC, 2006.

[13] Luo Jiaowan and Liu Kai. Stability of infinite dimensional stochastic evolution equations with memory and Markovian jumps. Stochastic Processes and their Applications, 2008, 118: 864–895.

[14] Mariton M. Jump Linear systems in Automatic Control, Marcel Dekker, 1990.

[15] Mao X. Stability of stochastic differential equations with Markovian switching. Stochastic Processes and their Applications, 1999, 79: 45–67.

[16] Mao X. Asymptotic stability for stochastic differential delay equations with markovian switching. Funct. Differ. Equa., 2002, 9: 201–220.

[17] Mao X, Matasov A and Piunovskiy A. Stochastic differential delay equations with

markovian switching. Bernoulli, 2002, 6(1): 73–90.

[18] Peszat S and Zabczyk J. Stochastic partial differential equation with Lévy noise: an evolution equation approach, Cambridge University Press, 2007.

[19] Ren J, Röckner M and Wang F Y. Stochastic generalized porous meadia and fast diffusion equation. J. Diff. Equat., 2007, 238: 118–152.

[20] Röckner M and Wang F Y. Non-monotone Stochastic generalized porous media equations. J.Diff. Equat., 2008, 245: 3898–3935.

[21] Röckner M, Wang F Y and Wu Liming. Large deviations for stochastic generalized porous media equations. Stochastic processes and their applications, 2006, 116: 1677–1689.

[22] Wang F Y. Harnack inequality and applications for Stochastic generalized porous media equations. The Annals of Probability, 2006, 35(4): 1333–1350.

[23] Zhou G L and Hou Z T. Stochastic Generalized Porous Media Equations with Lèvy Jump. Acta Mathematica Sinica, 2011, 27(9): 1671–1696.

[24] Zhou G L and Hou Z T. The Ergodicity of Stochastic Generalized Porous Media Equations with Lévy Jump. Acta Mathematica Scientia, 2011, 31(B): 925–933.

The Exponential Decay of Solutions and Traveling Wave Solutions for a Modified Camassa-Holm Equation with Cubic Nonlinearity*

Wu Xinglong (吴兴龙) and Guo Boling (郭柏灵)

Abstract The present paper is devoted to the study of persistence properties, infinite propagation and the traveling wave solutions for a modified Camassa-Holm equation with cubic nonlinearity. We first show that persistence properties of the solution to the equation provided the initial datum is exponential decay and the initial potential satisfies a certain sign condition. Next, we get the infinite propagation if the initial datum satisfies certain compact conditions, while the solution to Eq.(1.1) instantly loses compactly supported, the solution has exponential decay as $|x|$ goes to infinity. Finally, we prove Eq.(1.1) has a family traveling wave solutions.

Keywords A modified Camassa-Holm equation; persistence properties; compact support; exponential decay; infinite propagation speed; traveling wave solutions

1 Introduction

In this paper, we study the Cauchy problem for the modified Camassa-Holm equation with cubic nonlinearity

$$\begin{cases} u_t - u_{txx} + 3u^2u_x - u_x^3 & = (4u - 2u_{xx})u_xu_{xx} + (u^2 - u_x^2)u_{xxx}, \\ & \quad (t,x) \in \mathbb{R}^+ \times \mathbb{R}, \\ u(0,x) = u_0(x), & \quad x \in \mathbb{R}, \end{cases} \tag{1.1}$$

where $u = u(t,x)$ is the fluid velocity and subscripts denote the partial derivatives, u_0 is the initial data.

* JOURNAL OF MATHEMATICAL PHYSICS, 2014, 55, 081504. DOI: 10.1063/1.4891989.

Eq.(1.1) was proposed as a new integrable system by Fuchssteiner [15] and Olver and Rosenau [23] by applying the general method of tri-Hamiltonian duality to the bi-Hamiltonian representation of the modified Korteweg-deVries equation. Recently, Qiao [24] also introduced the modified Camassa-Holm equation from the two dimensional Euler equation by using the approximation procedure

$$u_t + u\cdot\nabla u + \nabla p = 0, \quad divu = 0,$$

where u denotes the fluid velocity, and p is the pressure. He proved the completely integrability of Eq.(1.1) by constructing a Lax pair

$$\begin{cases} \begin{pmatrix} \psi_1 \\ \psi_2 \end{pmatrix}_t = V(u,y,\lambda)\begin{pmatrix} \psi_1 \\ \psi_2 \end{pmatrix}, \\ \begin{pmatrix} \psi_1 \\ \psi_2 \end{pmatrix}_x = U(u,y,\lambda)\begin{pmatrix} \psi_1 \\ \psi_2 \end{pmatrix}, \end{cases}$$

where $y = u - u_{xx}$, λ is a spectral parameter, and

$$U(u,y,\lambda) = \begin{pmatrix} -\frac{1}{2} & \frac{1}{2}y\lambda \\ -\frac{1}{2}y\lambda & \frac{1}{2} \end{pmatrix},$$

$$V(u,y,\lambda) = \begin{pmatrix} \frac{1}{\lambda^2} + \frac{1}{2}(u^2 - u_x^2) & -\frac{1}{\lambda}(u - u_x) - \frac{1}{2}y\lambda(u^2 - u_x^2) \\ \frac{1}{\lambda}(u - u_x) + \frac{1}{2}y\lambda(u^2 - u_x^2) & -\frac{1}{\lambda^2} - \frac{1}{2}(u^2 - u_x^2) \end{pmatrix}.$$

He also showed that it has bi-Hamiltonian structure and an infinite sequence of conserved quantities

$$\begin{aligned} y_t = -[(u^2 - u_x^2)y]_x &= J\frac{\delta H_0}{\delta y} \\ &= K\frac{\delta H_1}{\delta y}, \end{aligned}$$

where

$$J = -\partial y \partial^{-1} y \partial, \quad K = \partial^3 - \partial,$$

and

$$H_0 = 2\int_{\mathbb{R}} uy dx, \qquad H_1 = \frac{1}{4}\int_{\mathbb{R}}\left(u^4 + 2u^2u_x^2 - \frac{1}{3}u_x^4\right)\mathrm{d}x.$$

Moreover Eq.(1.1) has the new peaked solitons [24, 25]

$$u(\pi) = \pm\left\{2 - 3\cosh^2\pi + \left(\cosh\pi + \frac{1}{3}\right)\sqrt{3(3\cosh\pi + 1)(\cosh\pi - 1)}\right\},$$

where the variable $\pi = \frac{x}{2} - \frac{11}{6}t$, and the W/M-shape-peaks solitons

$$u(\pi) = \pm\left\{2 - 3\cosh^2\pi + \left(\cosh\pi + \frac{1}{3}\right)\sqrt{3(3\cosh\pi + 1)(\cosh\pi - 1)}\right\},$$

where the variable $\pi = \frac{\left|x - \frac{11}{3}t\right|}{2} - \ln 2.$

By defining a new dependent variable, the potential y, Eq.(1.1) can be written as

$$y_t + (u^2 - u_x^2)y_x + 2u_x y^2 = 0, \qquad y = u - u_{xx}. \tag{1.2}$$

Eq.(1.2) is similar to the Camassa-Holm equation in form

$$y_t + uy_x + 2u_x y = 0, \qquad y = u - u_{xx}. \tag{1.3}$$

The Camassa-Holm equation models the unidirectional propagation of shallow water waves over a flat bottom [4], and also is a model for the propagation of axially symmetric waves in hyperelastic rods [13]. Eq.(1.3) from an asymptotic approximation to the Hamiltonian for the Green-Naghdi equations in shallow water theory. To begin with, the Camassa-Holm equation approximates unidirectional fluid flow in Euler's equations [17] at the next order beyond the KdV equation [21, 22], it has a bi-Hamiltonian structure [16] and is completely integrable [6], and with a Lax pair based on a linear spectral problem of second order. Also, there are smooth soliton solutions of Eq.(1.3) on a non-zero constant background [5], and there exists blow-up phenomena of the strong solution and global existence of strong solution to Eq.(1.3) [6, 9, 10, 11]. Moreover, the Camassa-Holm equation has peaked solitons at the form $u(t,x) = ce^{-|x-ct|}$, $c \in \mathbb{R}$ [5], which are orbital stable [12], and n-peakon solutions [1]

$$u(t,x) = \sum_{j=1}^{n} p_j(t)\exp(-|x - q_j(t)|),$$

where the positions q_j and amplitudes p_j satisfy the system of ODEs

$$\begin{cases} \dot{q_j} = \sum_{k=1}^{n} p_k \exp(-|q_j - q_k|), \\ \dot{p_j} = p_j \sum_{k=1}^{n} p_k \mathrm{sgn}(q_j - q_k) \exp(-|q_j - q_k|), \end{cases}$$

where $j = 1, \cdots, n$. The Camassa-Holm equation has attracted a lot of interest in the past twenty years for various reasons [2,3,7,9,12].

Recently, the Cauchy problem of the modified Camassa-Holm equation (1.1) has been studied in [14] on the line. The authors established the local well-posedness, derived precise blow-up scenarios and proved the local analytic solution. Moreover, in [18], the authors show the existence of strong solutions which blow-up in finite time and derive the blow-up rate. In this paper, analogous to [26], we mainly study the persistence properties, infinite propagation of the strong solutions to Eq.(1.1). Moreover, we get a family traveling wave solutions of Eq.(1.1). Whether the strong solution exists globally that is an open problem.

The remainder of the paper is organized as follows. In Section 2, we first establish persistence properties of the strong solution to Eq.(1.1) provided the initial datum $u_0(x)$ decays at infinity. Next, if the initial potential $y_0(x) = u_0 - u_{0,xx}$ does not change sign on $\mathbb{R}$ and the initial datum $u_0(x)$ decays at infinity, then the solution $u \equiv 0$ if there exists some t_1 such that $u(t_1, x) \sim o(e^{-x})$, as $x \uparrow \infty$. In Section 3, we derive infinite propagation speed of the solution to Eq.(1.1) provided the initial potential $y_0 = u_0 - u_{0,xx}$ does not change sign on $\mathbb{R}$, and has compact support. In Section 4, we show Eq.(1.1) has a family traveling wave solution.

Notation: For simplicity, we identify all spaces of functions with function spaces over $\mathbb{R}$, we drop $\mathbb{R}$ from our notation. For $1 \leqslant p \leqslant \infty$, the norm in the Banach space $L^p(\mathbb{R})$ will be written by $\|\cdot\|_{L^p}$, while $\|\cdot\|_{H^s}, s \in \mathbb{R}$ will stand for the norm in the classical Sobolev spaces $H^s(\mathbb{R})$. We shall say for some $K > 0$ that

$$f(x) \sim \mathcal{O}(e^{\theta x}) \quad \text{as } x \uparrow \infty, \quad \text{if} \quad \lim_{x\to\infty} \frac{|f(x)|}{e^{\theta x}} \leqslant K,$$

and

$$f(x) \sim o(e^{\theta x}) \quad \text{as } x \uparrow \infty, \quad \text{if} \quad \lim_{x\to\infty} \frac{|f(x)|}{e^{\theta x}} = 0.$$

2 Persistence Properties and Exponential Decay of Solution

In this section, as [8,20], we shall establish the persistence properties of the strong solution to Eq.(2.2), provided the initial datum $u_0(x)$ decays at infinity.

For convenience, we let the potential $y = u - u_{xx}$, then Eq.(1.1) takes the form of a quasi-linear evolution equation of hyperbolic type

$$\begin{cases} u_t + \frac{1}{3}(3u^2 - u_x^2)u_x + \frac{1}{3}\partial_x(1-\partial_x^2)^{-1}(2u^3 + 3uu_x^2) = -\frac{1}{3}(1-\partial_x^2)^{-1}u_x^3, & (t,x) \in \mathbb{R}^+ \times \mathbb{R}, \\ u(0,x) = u_0(x), & x \in \mathbb{R}. \end{cases} \tag{2.1}$$

Note that if $p(x) = \frac{1}{2}e^{-|x|}, x \in \mathbb{R}$, we have $(1-\partial_x^2)^{-1}f = p * f$ for all the $f \in L^2(\mathbb{R})$, and $p * y = u$, here we denote by $*$ the convolution. Then we can rewrite Eq.(2.1) as follows

$$\begin{cases} u_t + \frac{1}{3}(3u^2 - u_x^2)u_x + \frac{1}{3}\partial_x p * (2u^3 + 3uu_x^2) + \frac{1}{3}p * u_x^3 = 0, \\ \qquad (t,x) \in \mathbb{R}^+ \times \mathbb{R}, \\ u(0,x) = u_0(x), \qquad x \in \mathbb{R}. \end{cases} \tag{2.2}$$

Let $p, r = 2$ in paper [14], we have the following local well-posedness result.

Lemma 2.1 *Assume the initial datum $u_0 \in H^s(\mathbb{R}), s > \frac{5}{2}$. Then there exists a unique strong solution $u(t,x)$ to Eq(2.2)(or Eq.(1.1)) and a time $T = T(u_0) > 0$, such that*

$$u(t,x) = u(\cdot, u_0) \in \mathcal{C}([0,T[; H^s) \cap \mathcal{C}^1([0,T[; H^s).$$

Moreover, the solution $u(t,x)$ depends continuously on the initial datum u_0, i.e. the mapping $u_0 \to u(\cdot, u_0) : H^s \to \mathcal{C}([0,T[; H^s) \cap \mathcal{C}^1([0,T[; H^s)$ is continuous.

Consider the ordinary equation of the flow generated by $u^2 - u_x^2$

$$\begin{cases} \frac{\mathrm{d}}{\mathrm{d}t}q(t,x) = (u^2 - u_x^2)(t, q(t,x)), & (t,x) \in \mathbb{R}^+ \times \mathbb{R}, \\ q(0,x) = x, & x \in \mathbb{R}. \end{cases} \tag{2.3}$$

Applying classical results in the theory of ordinary differential equations, from [18], one can obtain the following result on $u(t,x)$ which is crucial in the proof of Theorem 2.2.

Lemma 2.2 *Assume that $u_0 \in H^s(\mathbb{R}), s > \frac{5}{2}$. Given $T = T(u_0)$ be the maximal existence time of the corresponding solution $u(t,x)$ to Eq.(2.2) with the initial datum u_0. Then there exists a unique solution $q \in \mathcal{C}([0,T[\times\mathbb{R}; \mathbb{R})$ to Eq.(2.3) such that the function $q(t,\cdot)$ is an increasing diffeomorphism of $\mathbb{R}$ with*

$$q_x(t,x) = \exp\left(2\int_0^t (yu_x)(s, q(s,x))\mathrm{d}s\right) > 0, \quad \forall (t,x) \in [0,T[\times\mathbb{R},$$

and $y(t,q)q_x = y_0$. *Moreover, if* $y_0 = u_0 - u_{0,xx} \geqslant 0$ (*or* $\leqslant 0$), *i.e.* y_0 *doesn't change sign on* $\mathbb{R}$, *then* $y(t,x) \geqslant 0$ (*or* $\leqslant 0$). *Furthermore, the solution* $u(t,x)$ *satisfies*

$$|u_x(t,x)| \leqslant |u(t,x)| \leqslant \frac{\sqrt{2}}{2}\|u_0\|_{H^1}.$$

Theorem 2.1 *Assume that the initial datum* $u_0 \in H^s(\mathbb{R}), s > \frac{5}{2}$ *and* $T > 0$. *Suppose* $u(t,x) \in C([0,T]; H^s(\mathbb{R}))$ *is the corresponding solution to Eq.(2.2) with the initial datum* u_0. *If there exists* $\alpha \in (0,1)$ *such that*

$$\begin{cases} |u_0(x)| \sim \mathcal{O}(e^{-\alpha x}) & as \quad x \uparrow \infty, \\ |u_{0,x}(x)| \sim \mathcal{O}(e^{-\alpha x}) & as \quad x \uparrow \infty, \end{cases}$$

then, it follows that

$$\begin{cases} |u(t,x)| \sim \mathcal{O}(e^{-\alpha x}) & as \quad x \uparrow \infty, \\ |u_x(t,x)| \sim \mathcal{O}(e^{-\alpha x}) & as \quad x \uparrow \infty, \end{cases}$$

uniformly in the interval $[0,T]$.

Proof For simplicity, let $M = \sup_{t\in[0,T]}\{\|u(t)\|_{H^s}\}$, and

$$G(u,u_x) = \frac{1}{3}\partial_x p * (2u^3 + 3uu_x^2) + \frac{1}{3}p * u_x^3.$$

By the Sobolev embedding theorem, we have $\|u(t)\|_{L^\infty},\ \|u_x(t)\|_{L^\infty},\ \|u_{xx}(t)\|_{L^\infty} \leqslant M$.

Define the weighted function

$$\varphi_N(x) = \begin{cases} 1, & x \leqslant 0, \\ e^{\alpha x}, & 0 < x < N, \\ e^{\alpha N}, & x \geqslant N, \end{cases} \tag{2.4}$$

where $N \in \mathbb{Z}^+$. One can easily check that for all N

$$0 \leqslant \varphi_N'(x) \leqslant \varphi_N(x) \quad a.e. \quad x \in \mathbb{R}. \tag{2.5}$$

Multiplying Eq.(2.2) by φ_N, we obtain that

$$(u\varphi_N)_t + \frac{1}{3}[(3u^2 - u_x^2)\varphi_N]u_x + \varphi_N G(u,u_x) = 0. \tag{2.6}$$

Differentiating Eq.(2.2) with respect to x-variable, after multiply it by φ_N to yield

$$(u_x\varphi_N)_t + 2uu_x^2\varphi_N + (u^2 - u_x^2)u_{xx}\varphi_N + \varphi_N G_x(u,u_x) = 0. \tag{2.7}$$

Multiplying Eq.(2.6) by $(u\varphi_N)^{2n-1}$ with $n \in \mathbb{Z}^+$ and integrating the result on $\mathbb{R}$ with respect to x-variable, we have

$$\int_{\mathbb{R}} (u\varphi_N)_t (u\varphi_N)^{2n-1} \mathrm{d}x = -\int_{\mathbb{R}} \left(\frac{1}{3}[(3u^2 - u_x^2)\varphi_N]u_x + G\varphi_N\right)(u\varphi_N)^{2n-1}\mathrm{d}x. \quad (2.8)$$

Note that

$$\int_{\mathbb{R}} (u\varphi_N)_t (u\varphi_N)^{2n-1} \mathrm{d}x = \|u\varphi_N\|_{L^{2n}}^{2n-1} \frac{\mathrm{d}}{\mathrm{d}t}\|u\varphi_N\|_{L^{2n}},$$

$$\begin{aligned}
&\frac{1}{3}\int_{\mathbb{R}} [(3u^2 - u_x^2)\varphi_N]u_x (u\varphi_N)^{2n-1}\mathrm{d}x \\
&\leqslant \|uu_x\|_{L^\infty}\|u\varphi_N\|_{L^{2n}}^{2n} + \frac{1}{3}\|u_x^3\varphi_N\|_{L^{2n}}\|u\varphi_N\|_{L^{2n}}^{2n-1} \\
&\leqslant M^2\|u\varphi_N\|_{L^{2n}}^{2n} + \frac{M^2}{3}\|u_x\varphi_N\|_{L^{2n}}\|u\varphi_N\|_{L^{2n}}^{2n-1}.
\end{aligned}$$

$$\left|\int_{\mathbb{R}} \varphi_N(p_x * F(u,\eta))(u\varphi_N)^{2n-1}\mathrm{d}x\right| \leqslant \|u\varphi_N\|_{L^{2n}}^{2n-1}\|\varphi_N(p_x * F(u,\eta))\|_{L^{2n}}.$$

In view of (2.8) and the above relations, we have

$$\frac{\mathrm{d}}{\mathrm{d}t}\|u\varphi_N\|_{L^{2n}} \leqslant M^2\|u\varphi_N\|_{L^{2n}} + \frac{M^2}{3}\|u_x\varphi_N\|_{L^{2n}} + \|\varphi_N G(u,u_x)\|_{L^{2n}}. \quad (2.9)$$

Similarly, multiplying Eq.(2.7) by $(u_x\varphi_N)^{2n-1}$, integration by parts, we obtain

$$\begin{aligned}
&\|u_x\varphi_N\|_{L^{2n}}^{2n-1}\frac{\mathrm{d}}{\mathrm{d}t}\|u_x\varphi_N\|_{L^{2n}} \\
=& -\int_{\mathbb{R}} 2uu_x(u_x\varphi_N)^{2n}\mathrm{d}x \\
&- \int_{\mathbb{R}} [(u^2 - u_x^2)u_{xx}\varphi_N + \varphi_N G_x(u,u_x)](u_x\varphi_N)^{2n-1}\mathrm{d}x.
\end{aligned} \quad (2.10)$$

Applying Holder's inequality to (2.10), we end up with

$$\frac{\mathrm{d}}{\mathrm{d}t}\|u_x\varphi_N\|_{L^{2n}} \leqslant M^2\|u\varphi_N\|_{L^{2n}} + 3M^2\|u_x\varphi_N\|_{L^{2n}} + \|G_x\varphi_N\|_{L^{2n}}. \quad (2.11)$$

Add up (2.9) with (2.11). Then by virtue of Gronwall's inequality, we can get

$$\begin{aligned}
&\|u\varphi_N\|_{L^{2n}} + \|u_x\varphi_N\|_{L^{2n}} \leqslant e^{4M^2t}(\|u_0\varphi_N\|_{L^{2n}} + \|u_{0,x}\varphi_N\|_{L^{2n}}) \\
&+ e^{4M^2t}\int_0^t (\|\varphi_N G(u,u_x)(\tau)\|_{L^{2n}} + \|\varphi_N G_x(u,u_x)(\tau)\|_{L^{2n}})\mathrm{d}\tau.
\end{aligned} \quad (2.12)$$

Since $f \in L^1(\mathbb{R}) \cap L^\infty(\mathbb{R})$ implies

$$\lim_{n\uparrow\infty} \|f\|_{L^n} = \|f\|_{L^\infty}.$$

Let $n \uparrow \infty$ in (2.12), it follows that

$$\begin{aligned}&\|u\varphi_N\|_{L^\infty}+\|u_x\varphi_N\|_{L^\infty}\leqslant e^{4M^2t}(\|u_0\varphi_N\|_{L^\infty}+\|u_{0,x}\varphi_N\|_{L^\infty})\\&+e^{4M^2t}\int_0^t(\|\varphi_N G(u,u_x)(\tau)\|_{L^\infty}+\|\varphi_N G_x(u,u_x)(\tau)\|_{L^\infty})\mathrm{d}\tau.\end{aligned}\tag{2.13}$$

A simple calculation shows that there exists c_0 such that $\forall N\in\mathbb{Z}^+$,

$$\varphi_N(x)\int_{\mathbb{R}}\frac{e^{-|x-y|}}{\varphi_N(y)}\mathrm{d}y\leqslant c_0=\frac{4}{1-\alpha}.\tag{2.14}$$

Therefore, in view of (2.14) we obtain

$$\begin{aligned}\frac{2}{3}|\varphi_N(p_x*u^3)|&=\left|\frac{1}{3}\varphi_N(x)\int_{\mathbb{R}}\frac{e^{-|x-y|}}{\varphi_N(y)}\varphi_N(y)u^3(y)\mathrm{d}y\right|\\&\leqslant\frac{c_0}{3}\|u\|_{L^\infty}^2\|u\varphi_N\|_{L^\infty}\leqslant\frac{c_0M^2}{3}\|u\varphi_N\|_{L^\infty}.\end{aligned}\tag{2.15}$$

Similarly, we have

$$\|\varphi_N(p_x*(uu_x^2))\|_{L^\infty}\leqslant\frac{c_0M^2}{2}\|u_x\varphi_N\|_{L^\infty}$$

and

$$\frac{1}{3}\|\varphi_N(p*(u_x^3))\|_{L^\infty}\leqslant\frac{c_0M^2}{6}\|u_x\varphi_N\|_{L^\infty}.$$

Therefore, one can easily get that

$$\|\varphi_N G(u,u_x)(\tau)\|_{L^\infty}\leqslant\frac{c_0M^2}{3}\|\varphi_N u(\tau)\|_{L^\infty}+\frac{2c_0M^2}{3}\|\varphi_N u_x(\tau)\|_{L^\infty}.\tag{2.16}$$

Note that $p_{xx}*f=p*f-f$, similar to (2.16), we obtain

$$\|\varphi_N G_x(u,u_x)(\tau)\|_{L^\infty}\leqslant c_1(\|\varphi_N u(\tau)\|_{L^\infty}+\|\varphi_N u_x(\tau)\|_{L^\infty}),\tag{2.17}$$

where c_1 is constant depending only on c_0, M.

Let $Z(t)=(\|\varphi_N u(t)\|_{L^\infty}+\|\varphi_N u_x(t)\|_{L^\infty})$. Combining (2.13), (2.16) with (2.17), it follows that

$$Z(t)\leqslant e^{4M^2t}Z(0)+c_1e^{4M^2t}\int_0^t Z(\tau)\mathrm{d}\tau.\tag{2.18}$$

Applying Gronwall's inequality to (2.18), for all $t\in[0,T]$, there exists a constant $\tilde{c}=\tilde{c}(c_0,M,T)$ such that

$$Z(t)\leqslant\tilde{c}Z(0)\leqslant\tilde{c}(\|u_0\max(1,e^{\alpha x})\|_{L^\infty}+\|u_{0,x}\max(1,e^{\alpha x})\|_{L^\infty}).\tag{2.19}$$

Letting $N \uparrow \infty$, from (2.19), for all $t \in [0,T]$, we can find that

$$
\begin{aligned}
&(\|u(t,x)e^{\alpha x}\|_{L^\infty} + \|u_x(t,x)\varphi_N\|_{L^\infty}) \\
&\leqslant \tilde{c}(\|u_0 \max(1, e^{\alpha x})\|_{L^\infty} + \|u_{0,x} \max(1, e^{\alpha x})\|_{L^\infty}).
\end{aligned}
$$

This completes the proof of Theorem 2.1. ■

Remark 2.1 *In paper* [26], *as* ρ_0 *does not change sign on* $\mathbb{R}$, *from Lemma* 2.1, *we imply that* $|\eta_x(t,x)| \leqslant |\eta(t,x)|$. *In view of this, we get the similar result to the modified* 2*-component Camassa-Holm equation. However, as* ρ_0 *changes sign on* $\mathbb{R}$, *we can deal with Eq.*(2.2) *as follows: Thanks to* (2.13) *and* (2.16) *in* [26], *it follows that*

$$
\frac{\mathrm{d}}{\mathrm{d}t}\|u\varphi_N\|_{L^\infty} \leqslant M\|u\varphi_N\|_{L^\infty} + \|\varphi_N(p_x * F(u,\eta))\|_{L^\infty}
$$

and

$$
\frac{\mathrm{d}}{\mathrm{d}t}\|u_x\varphi_N\|_{L^\infty} \leqslant 2M\|u_x\varphi_N\|_{L^\infty} + \|\varphi_N(p_{xx} * F(u,\eta))\|_{L^\infty}.
$$

Substituting (2.21) and (2.22) of [26] into the above two inequalities and adding them up to yield

$$
\begin{aligned}
&\frac{\mathrm{d}}{\mathrm{d}t}(\|u\varphi_N\|_{L^\infty} + \|u_x\varphi_N\|_{L^\infty}) \\
&\leqslant C(\|u\varphi_N\|_{L^\infty} + \|u_x\varphi_N\|_{L^\infty} + \|\varphi_N\eta\|_{L^\infty} + \|\varphi_N\eta_x\|_{L^\infty}).
\end{aligned} \tag{2.20}
$$

By virtue of (2.24) and (2.25) in [26], we have

$$
\frac{\mathrm{d}}{\mathrm{d}t}\|\eta\varphi_N\|_{L^\infty} \leqslant M\|u\varphi_N\|_{L^\infty} + \|\varphi_N(p * G(u,\eta))\|_{L^\infty} \tag{2.21}
$$

and

$$
\frac{\mathrm{d}}{\mathrm{d}t}\|\eta_x\varphi_N\|_{L^\infty} \leqslant M\|\eta_x\varphi_N\|_{L^\infty} + \|\varphi_N(p_x * G(u,\eta))\|_{L^\infty}. \tag{2.22}
$$

Define

$$
Z(t) = (\|\varphi_N u(t)\|_{L^\infty} + \|\varphi_N u_x(t)\|_{L^\infty} + \|\varphi_N \eta(t)\|_{L^\infty} + \|\varphi_N \eta_x(t)\|_{L^\infty}).
$$

Add up (2.20), (2.21) *with* (2.22), *in view of* (2.29) *and* (2.30) *in* [26], *we have*

$$
\frac{\mathrm{d}}{\mathrm{d}t}Z(t) \leqslant C_1 Z(t).
$$

Applying Gronwall's inequality to obtain

$$
Z(t) \leqslant Z(0)e^{C_1 t}.
$$

Then, similar to the rest proof of Theorem 2.1, we can get the result of Theorem 2.1.

Remark 2.2 *Under the assumption of Theorem 2.1, if there exists* $\alpha \in (0,1)$ *such that*

$$|u_0(x)|,\ |u_{0,x}(x)| \sim \mathcal{O}(e^{\alpha x}), \qquad as \quad x \downarrow -\infty,$$

by choosing the weight function

$$\varphi_N(x) = \begin{cases} e^{-\alpha N}, & x \leqslant -N, \\ e^{-\alpha x}, & -N < x < 0, \\ 1, & x \geqslant 0, \end{cases}$$

where $N \in \mathbb{Z}^+$. *Analogous to the proof of Theorem 2.1, we have*

$$|u(t,x)|,\ |u_x(t,x)| \sim \mathcal{O}(e^{\alpha x}), \qquad as \quad x \downarrow -\infty,$$

uniformly in the interval $[0,T]$.

Theorem 2.2 *Let the initial datum* $u_0 \in H^s(\mathbb{R}), s > \frac{5}{2}$. *Given* $T = T(u_0) > 0$ *is the maximal existence time of the solution* $u(t,x)$ *to Eq.(2.2) with the initial* u_0. *Assume* $y_0 = u_0 - u_{0,xx} \leqslant 0$ *on* $\mathbb{R}$. *If there exists* $\theta \in \left(\frac{1}{3}, 1\right)$ *such that*

$$\begin{cases} |u_0(x)| \sim o(e^{-|x|}), & as \quad |x| \uparrow \infty, \\ |u_{0,x}(x)| \sim \mathcal{O}(e^{-\theta|x|}), & as \quad |x| \uparrow \infty, \end{cases} \tag{2.23}$$

and there exists $t_1 \in (0, T[$ *such that*

$$|u(t_1,x)| \sim o(e^{-|x|}), \ as \quad |x| \uparrow \infty,$$

then we have the solution $u(t,x) \equiv 0$.

Proof Integrating Eq.(2.2) with respect to t-variable over the interval $[0, t_1]$, we obtain

$$\begin{aligned} &u(t_1,x) - u_0(x) + \frac{1}{3}\int_0^{t_1} (3u^2 - u_x^2)(s,x)\mathrm{d}s \\ &+ \frac{1}{3}\int_0^{t_1} [p_x * (2u^3 + 3uu_x^2) + p * u_x^3](s,x)\mathrm{d}s = 0. \end{aligned} \tag{2.24}$$

Without loss of generality, we only consider $x \uparrow \infty$. By virtue of the assumption of Theorem 2.2 to give by

$$u(t_1,x) - u_0(x) \sim o(e^{-x}), \quad \text{as} \quad x \uparrow \infty. \tag{2.25}$$

In view of Theorem 2.1 and $\theta \in \left(\frac{1}{3}, 1\right)$, we deduce

$$\int_0^{t_1} (3u^2 - u_x^2)(s,x)ds \sim \mathcal{O}(e^{-3\theta x}) \sim o(e^{-x}), \quad \text{as} \quad x \uparrow \infty. \tag{2.26}$$

Since $y_0 = u_0 - u_{0,xx} < 0$, using Lemma 2.2, we have

$$0 \leqslant |u_x| \leqslant -u. \tag{2.27}$$

Therefore,

$$0 \leqslant \frac{1}{2}e^{-x} \int_0^{t_1} \int_{-\infty}^{x} e^z(-2u^3 - 3uu_x^2 + u_x^3)(s,z)\mathrm{d}z\mathrm{d}s. \tag{2.28}$$

On the other hand

$$\begin{aligned}
[p_x * (2u^3 + 3uu_x^2) + p * u_x^3] &= \frac{1}{2}\int_{\mathbb{R}} e^{-|x-z|}u_x^3\mathrm{d}z \\
&\quad - \frac{1}{2}\int_{\mathbb{R}} \mathrm{sgn}(x-z)e^{-|x-z|}(2u^3 + 3uu_x^2)(t,z)\mathrm{d}z \\
&= \frac{1}{2}e^{-x}\int_{-\infty}^{x} e^z(-2u^3 - 2uu_x^2 + u_x^3)\mathrm{d}z \\
&\quad + \frac{1}{2}e^{x}\int_{x}^{\infty} e^{-z}(2u^3 + 3uu_x^2 + u_x^3)\mathrm{d}z.
\end{aligned} \tag{2.29}$$

Thanks to Theorem 2.1 and $\theta \in \left(\frac{1}{3}, 1\right)$, we imply that

$$\begin{aligned}
&\frac{1}{2}e^{x}\int_{x}^{\infty} e^{-z}(2u^3 + 3uu_x^2 + u_x^3)\mathrm{d}z = e^{x}\int_{x}^{\infty} e^{-z}\mathcal{O}(e^{-3\theta z})\mathrm{d}z \\
&\sim o(1)e^{x}\int_{x}^{\infty} e^{-2z}\mathrm{d}z \sim o(e^{-x}), \text{ as } \quad x \uparrow \infty.
\end{aligned} \tag{2.30}$$

Assume $u \not\equiv 0$. By virtue of (2.27). Then $(-2u^3 - 2uu_x^2 + u_x^3) > 0$. Thus, there exists $c > 0$, we obtain for large enough x that

$$\frac{1}{2}e^{-x}\int_{-\infty}^{x} e^z(-2u^3 - 2uu_x^2 + u_x^3)(s,z)\mathrm{d}z \geqslant \frac{c}{2}e^{-x}. \tag{2.31}$$

Combining (2.29), (2.30) with (2.31), for large enough x, it follows that

$$[p_x * (2u^3 + 3uu_x^2) + p * u_x^3] > \frac{c}{3}e^{-x} \sim \mathcal{O}(e^{-x}), \quad \text{as} \quad x \uparrow \infty. \tag{2.32}$$

Plugging (2.25), (2.26) and (2.32) into (2.24) yields a contradiction, therefore, $(-2u^3 - 2uu_x^2 + u_x^3) = 0$, i.e. $u \equiv 0$. This completes the proof of Theorem 2.2. ■

Similarly, we have the following corollary.

Corollary 2.1 *Assume* $y_0 = u_0 - u_{0,xx} \geqslant 0$ *on* $\mathbb{R}$. *Let the initial datum* $u_0 \in H^s(\mathbb{R}), s > \frac{5}{2}$. *Given* $T = T(u_0) > 0$ *is the maximal existence time of the solution* $u(t,x)$ *to Eq.(2.2) with the initial* u_0. *If there exists* $\theta \in \left(\frac{1}{3}, 1\right)$ *such that*

$$\begin{cases} |u_0(x)| \sim o(e^{-|x|}), & as \quad |x| \uparrow \infty, \\ |u_{0,x}(x)| \sim \mathcal{O}(e^{-\theta|x|}), & as \quad |x| \uparrow \infty, \end{cases}$$

and there exists $t_1 \in (0,T)$ *such that*

$$|u(t_1,x)| \sim o(e^{-|x|}), \ as \quad x \uparrow \infty,$$

then we have the solution $u(t,x) \equiv 0$.

Theorem 2.3 *Under the assumption of Theorem 2.2 (or Corollary 2.1). If there exists a* $\theta \in \left(\frac{1}{3}, 1\right)$ *such that*

$$\begin{cases} |u_0(x)| \sim \mathcal{O}(e^{-|x|}), & as \quad |x| \uparrow \infty, \\ |u_{0,x}(x)| \sim \mathcal{O}(e^{-\theta|x|}), & as \quad |x| \uparrow \infty, \end{cases}$$

then

$$|u(t,x)| \sim \mathcal{O}(e^{-|x|}), \ as \quad |x| \uparrow \infty,$$

uniformly in the interval $[0,T[$.

Proof Integrating Eq.(2.2) with respect to t-variable over the interval $[0,t_1]$, we obtain

$$\begin{aligned} &u(t_1,x) - u_0(x) + \frac{1}{3}\int_0^{t_1} (3u^2 - u_x^2)(s,x)\mathrm{d}s \\ &+ \frac{1}{3}\int_0^t [p_x * (2u^3 + 3uu_x^2) + p * u_x^3](s,x)\mathrm{d}s = 0. \end{aligned} \tag{2.33}$$

Due to the assumption of Theorem 2.3, by virtue of Theorem 2.1 and $\theta \in \left(\frac{1}{3}, 1\right)$, similar to the proof of Theorem 2.2, we deduce

$$\begin{cases} (3u^2 - u_x^2) \sim \mathcal{O}(e^{-3\theta x}) \sim o(e^{-|x|}), & \text{as} \quad |x| \uparrow \infty, \\ (p_x * (2u^3 + 3uu_x^2) + p * u_x^3) \sim \mathcal{O}(e^{-|x|}), & \text{as} \quad |x| \uparrow \infty. \end{cases} \tag{2.34}$$

Substituting (2.34) into (2.33) to yield the result. ■

3 Infinite Propagation Speed

In this section, we will prove that the classical solutions of Eq.(2.2) have infinite propagation speed. First, for the convenience of the readers, we recall an useful lemma.

Lemma 3.1 [19] *Let $u(t,x) \in C^2(\mathbb{R}) \cap H^2(\mathbb{R})$ be such that $y(t,x) = u - u_{xx}$ has compact support. Then $u(t,x)$ has compact support if and only if*

$$\int_{\mathbb{R}} e^x y(t,x)dx = \int_{\mathbb{R}} e^{-x} y(t,x)dx = 0. \tag{3.1}$$

Proposition 3.1 *Assume the initial datum $u_0 \in H^s(\mathbb{R}), s \geqslant 3$, such that the initial potential $y_0 = u_0 - u_{0,xx}$ is supported in the interval $[\alpha_{y_0}, \beta_{y_0}]$. If $T = T(u_0)$ is the maximal existence time of the solution $u(t,x)$ to Eq.(2.2), then for any time $t \in [0,T[$, the solution $y(t,\cdot)$ is supported in the interval $[q(t,\alpha_{y_0}), q(t,\beta_{y_0})]$, and such that*

$$y(t,x) = y_0(q^{-1}(t,x)) \exp\left(-2\int_0^t (u_x y)(s,x)ds\right), \qquad \forall\, (t,x) \in [0,T[\times\mathbb{R}.$$

Proof. Since $q(t,x)$ satisfies the ordinary equation (2.3), then $q(t,x)$ is a diffeomorphism, and we have

$$q_x(t,x) = \exp\left(2\int_0^t (u_x y)(s,q(s,x))ds\right) > 0, \qquad \forall\, (t,x) \in [0,T[\times\mathbb{R}.$$

In view of Eq.(1.2), one can easily check that

$$y(t,q)q_x(t,x) = y_0(x).$$

Since y_0 is compactly supported and $q_x(t,x) > 0$, then for any $t \in [0,T[$, the solution $y(t,\cdot)$ is supported in the interval $[q(t,\alpha_{y_0}), q(t,\beta_{y_0})]$, and such that

$$y(t,x) = y_0(q^{-1}(t,x)) \exp\left(-2\int_0^t (u_x y)(s,x)ds\right), \qquad \forall\, (t,x) \in [0,T[\times\mathbb{R}.$$

This completes the proof of Proposition 3.1. ■

Theorem 3.1 *Assume $u_0 \in H^s(\mathbb{R}), s \geqslant 3$, such that the initial potential $y_0 = u_0 - u_{0,xx}$ is compactly supported and doesn't change sign on $\mathbb{R}$. Let $T = T(u_0)$ is the maximal existence time of the solution $u(t,x)$ to Eq.(2.2). If for any time $t \in [0,T[$, the solution $u(t,\cdot)$ is compactly supported, then $u(t,x) \equiv 0$.*

Proof Suppose $u(t,x)$ be compactly supported, using Lemma 3.1 to yield

$$\int_{\mathbb{R}} e^{x} y(t,x)\mathrm{d}x = 0. \tag{3.2}$$

Differentiating (3.2) with respect to t-variable, in view of Eq.(1.1) and integration by parts. Note that $u(t,\cdot)$ is compactly supported, it follows that

$$\begin{aligned} \frac{\mathrm{d}}{\mathrm{d}t}\int_{\mathbb{R}} e^{x} y(t,x)\mathrm{d}x &= -\int_{\mathbb{R}} [(u^2-u_x^2)y]_x e^{x}\mathrm{d}x \\ &= \int_{\mathbb{R}} (u^3 - u^2u_{xx} - uu_x^2 + u_x^2u_{xx})e^{x}\mathrm{d}x \\ &= \int_{\mathbb{R}} \left(\frac{2}{3}u^3 + uu_x^2 - \frac{1}{3}u_x^3\right)e^{x}\mathrm{d}x. \end{aligned} \tag{3.3}$$

Since y_0 doesn't change sign on $\mathbb{R}$. Similar to the proof on page 723 of [27], we imply that

$$\begin{cases} 0 \leqslant |u_x| \leqslant u, & y_0 \geqslant 0, \\ 0 \leqslant |u_x| \leqslant -u, & y_0 \leqslant 0. \end{cases} \tag{3.4}$$

By virtue of (3.4), one can easily check that

$$\begin{cases} \frac{2}{3}u^3 + uu_x^2 - \frac{1}{3}u_x^3 \geqslant \frac{1}{3}u^3 + uu_x^2 \geqslant 0, & y_0 \geqslant 0, \\ \frac{2}{3}u^3 + uu_x^2 - \frac{1}{3}u_x^3 \leqslant \frac{1}{3}u^3 + uu_x^2 \leqslant 0, & y_0 \leqslant 0. \end{cases} \tag{3.5}$$

Inserting (3.5) into (3.3), lead to

$$\begin{cases} \frac{\mathrm{d}}{\mathrm{d}t}\int_{\mathbb{R}} e^{x} y(t,x)\mathrm{d}x \geqslant 0, & y_0 \geqslant 0, \\ \frac{\mathrm{d}}{\mathrm{d}t}\int_{\mathbb{R}} e^{x} y(t,x)\mathrm{d}x \leqslant 0, & y_0 \leqslant 0. \end{cases} \tag{3.6}$$

(3.2) in conjunction with (3.6), we deduce that

$$u(t,x) \equiv 0.$$

This completes the proof of Theorem 3.1. ■

Remark 3.1 *If the initial datum $u_0 \not\equiv 0$ has compact support, then the corresponding solution $u(t,\cdot)$ of Eq.(1.1) must instantly lose the compactness of its support by the same argument as above for any small time interval $[0,\varepsilon]$.*

Although the solution $u(t,\cdot)$ instantly loses compactly supported, the following result tell us some information about the nature of the solution $u(t,\cdot)$ as it evolves over time, i.e. for all $t \in [0,T[$, the solution $u(t,\cdot)$ to Eq.(1.1) has exponential decay as $|x|$ goes to infinity. More precisely, we have the following result.

Theorem 3.2 *Suppose the initial datum* $u_0 \in H^s(\mathbb{R}), s \geqslant 3$. *Given* $T = T(u_0)$ *be the maximal existence time of the solution* $u(t,x)$ *to Eq.(2.2) with the initial data* u_0. *Assume the initial data* u_0 *be compactly supported in the interval* $[\alpha_{u_0}, \beta_{u_0}]$. *Then we have for all* $t \in [0,T[$ *that*

$$2u(t,x) = \begin{cases} e^x E_-(t), & \text{if} \quad x \leqslant q(t,\alpha), \\ e^{-x} E^+(t), & \text{if} \quad x > q(t,\beta), \end{cases} \tag{3.7}$$

where the compact support $[q(t,\alpha), q(t,\beta)]$ *of* $y(t,\cdot)$ *is contained in the interval* $[q(t,\alpha_{u_0}), q(t,\beta_{u_0})]$. *Moreover, if the initial potential* $y_0 = u_0 - u_{0,xx}$ *doesn't change sign on* $\mathbb{R}$ *and for any* $t \in [0,T[$ *the solution* $u \not\equiv 0$, *then* $E_-(t)$ *is nonvanishing and strictly decreasing function and* $E^+(t)$ *is nonvanishing and strictly increasing function, if* $y_0 \geqslant 0$. *Otherwise,* $E_-(t)$ *is nonvanishing and strictly increasing function and* $E^+(t)$ *is nonvanishing and strictly decreasing function, if* $y_0 \leqslant 0$, *with* $E_-(0) = E^+(0) = 0$.

Proof At first, if $u_0(\cdot)$ is compactly supported in the interval $[\alpha_{u_0}, \beta_{u_0}]$, then by the proof Proposition 3.1, we deduce that $y(t,\cdot) = u(t,\cdot) - u_{xx}(t,\cdot)$ has compact support with its support contained in the interval $[q(t,\alpha), q(t,\beta)] \subseteq [q(t,\alpha_{u_0}), q(t,\beta_{u_0})]$. Define the functions

$$E_-(t) = \int_{q(t,\alpha)}^{q(t,\beta)} e^{-z} y(t,z)\mathrm{d}z, \quad \text{and} \quad E^+(t) = \int_{q(t,\alpha)}^{q(t,\beta)} e^{z} y(t,z)\mathrm{d}z. \tag{3.8}$$

Since $u = p * y$, $p = \frac{1}{2}e^{-|x|}$, we have

$$u(t,x) = \frac{1}{2}e^{-x}\int_{-\infty}^{x} e^{z} y(t,z)\mathrm{d}z + \frac{1}{2}e^{x}\int_{x}^{\infty} e^{-z} y(t,z)\mathrm{d}z. \tag{3.9}$$

Combining (3.8) with (3.9). Since $q(t,\cdot)$ is an increasing diffeomorphism, one can easily get that

$$2u(t,x) = \begin{cases} e^x E_-(t) = e^x \displaystyle\int_{q(t,\alpha)}^{q(t,\beta)} e^{-z} y(t,z)\mathrm{d}z, & \text{if} \quad x \leqslant q(t,\alpha), \\ e^{-x} E^+(t) = e^{-x} \displaystyle\int_{q(t,\alpha)}^{q(t,\beta)} e^{z} y(t,z)\mathrm{d}z, & \text{if} \quad x > q(t,\beta). \end{cases} \tag{3.10}$$

Thus, by virtue of (3.8) and (3.10) to give by

$$\begin{cases} u(t,x) = u_x(t,x) = u_{xx}(t,x) = \dfrac{1}{2}e^x E_-(t), & \text{if} \quad x \leqslant q(t,\alpha), \\ u(t,x) = -u_x(t,x) = u_{xx}(t,x) = \dfrac{1}{2}e^{-x} E^+(t), & \text{if} \quad x > q(t,\beta). \end{cases} \tag{3.11}$$

Note that $u_0(x)$ is compactly supported in the interval $[\alpha_{u_0}, \beta_{u_0}]$, integration by parts to (3.8), we can get $E_-(0) = E^+(0) = 0$. In fact

$$\begin{aligned}E^+(0) &= \int_\alpha^\beta e^z y_0(z)\mathrm{d}z = \int_{\mathbb{R}} e^z y_0(z)\mathrm{d}z \\ &= \int_{\mathbb{R}} e^z u_0(z)\mathrm{d}z + \int_{\mathbb{R}} e^z u_{0,x}(z)\mathrm{d}z \\ &= 0.\end{aligned} \tag{3.12}$$

Since $y(t,\cdot)$ is compactly support in the interval $[q(t,\alpha), q(t,\beta)]$ and $u = -u_x$ as $x \geqslant q(t,\beta)$. integrating by parts to (3.8), in view of Eq.(1.2) and (3.11) to imply

$$\begin{aligned}\frac{\mathrm{d}}{\mathrm{d}t}E^+(t) &= \int_{q(t,\alpha)}^{q(t,\beta)} e^z y_t(t,z)\mathrm{d}z \\ &= \int_{\mathbb{R}} e^z y_t(t,z)\mathrm{d}z = -\int_{\mathbb{R}}[(u^2-u_x^2)y]_x e^x\mathrm{d}x \\ &= -(u^2-u_x^2)ye^x|_{-\infty}^{+\infty} + \int_{\mathbb{R}}(u^2-u_x^2)ye^x\mathrm{d}x \\ &= \int_{\mathbb{R}}\left(\frac{2}{3}u^3+uu_x^2-\frac{1}{3}u_x^3\right)e^x\mathrm{d}x + \left(\frac{1}{3}u^3+\frac{1}{3}u_x^3-u^2u_x\right)e^x|_{-\infty}^{+\infty} \\ &\begin{cases}\geqslant \int_{\mathbb{R}}\left(\frac{1}{3}u^3+uu_x^2\right)e^x\mathrm{d}x > 0, & y_0 \geqslant 0, \\ \leqslant \int_{\mathbb{R}}\left(\frac{1}{3}u^3+uu_x^2\right)e^x\mathrm{d}x < 0, & y_0 \leqslant 0,\end{cases}\end{aligned} \tag{3.13}$$

where the strict positively (or negatively) of the relation above follows from our assumption that $u(t,x)$ is nontrivial. Similarly, for all $t \in [0,T[$, we also have that

$$\frac{\mathrm{d}}{\mathrm{d}t}E_-(t) = \begin{cases} -\int_{\mathbb{R}}\left(\frac{2}{3}u^3+uu_x^2+\frac{1}{3}u_x^3\right)e^{-x}\mathrm{d}x < 0, & y_0 \geqslant 0, \\ -\int_{\mathbb{R}}\left(\frac{2}{3}u^3+uu_x^2+\frac{1}{3}u_x^3\right)e^{-x}\mathrm{d}x > 0, & y_0 \leqslant 0.\end{cases}$$

This completes the proof of Theorem 3.2. ■

Remark 3.2 *As long as the solution $u(t,x)$ exists, the result of Theorem* 3.2 *tells us that the solution $u(t,\cdot)$ is positive at $+\infty$ and negative at $-\infty$ if $y_0 \geqslant 0$, regardless of the profile of a fast-decaying data $u \not\equiv 0$.*

Corollary 3.1 *Assume $u_0 \in H^s(\mathbb{R}), s \geqslant 3$. Given $T = T(u_0)$ be the maximal existence time of the solution $u(t,x)$ to Eq.*(2.2) *with the initial datum u_0. If for some $\lambda > 0$,*

$$\partial_x^j u_0 \sim \mathcal{O}(e^{-(1+\lambda)|x|}) \quad as \quad |x| \uparrow \infty \quad j = 0,\ 1,\ 2, \tag{3.14}$$

then for any $t \in (0,T[$, *we obtain*

$$y(t,x) \sim \mathcal{O}(e^{-(1+\lambda)|x|}) \quad as \quad |x| \uparrow \infty$$

and

$$\lim_{x\to+\infty} e^{x}u(t,x) = \frac{1}{2}E^{+}(t), \ \lim_{x\to-\infty} e^{-x}u(t,x) = \frac{1}{2}E_{-}(t),$$

where $E_{-}(t)$ *is defined as (3.8) with* $E_{-}(0) = E^{+}(0) = 0$.

Proof. Multiplying Eq.(1.2) by $e^{(1+\lambda)|x|}$, we deduce

$$(ye^{(1+\lambda)|x|})_t + [(u^2 - u_x^2)y_x]e^{(1+\lambda)|x|} + (u^2 - u_x^2)_x ye^{(1+\lambda)|x|} = 0. \tag{3.15}$$

As the process of the estimation to (2.11), we end up with

$$\|ye^{(1+\lambda)|x|}\|_{L^\infty} \leqslant e^{2M^2(1+\lambda)t}(\|y_0e^{(1+\lambda)|x|}\|_{L^\infty} + 4M^2t). \tag{3.16}$$

By virtue of the assumption (3.14), we can get

$$y_0(x) \sim \mathcal{O}(e^{-(1+\lambda)|x|}) \quad \text{as} \quad |x| \uparrow \infty. \tag{3.17}$$

Combining (3.16) with (3.17), let $|x|$ large enough, we have

$$y(t,x) \sim \mathcal{O}(e^{-(1+\lambda)|x|}) \quad \text{as} \quad |x| \uparrow \infty.$$

On the other hand, by virtue of (3.14) and Theorem 2.1, we deduce for any $\theta \in (0,1)$ that

$$\partial_x^j u \sim \mathcal{O}(e^{-\theta|x|}) \quad \text{as} \quad |x| \uparrow \infty \quad j = 0,\ 1,\ 2. \tag{3.18}$$

In view of (3.7) of Theorem 3.2, it follows for all $t \in [0,T[$ that

$$\lim_{x\to+\infty} e^{x}u(t,x) = \frac{1}{2}E^{+}(t), \ \lim_{x\to-\infty} e^{-x}u(t,x) = \frac{1}{2}E_{-}(t).$$

This completes the proof of Corollary 3.1. ∎

Remark 3.3 *Under the assumption* (3.14) *of Corollary* 3.1, *we imply that* (3.18). *This guarantees that* (3.13) *can be carried out in the same fashion. Therefore,* $E^{+}(t)$ *is strictly increasing (or decreasing)function,* $E_{-}(t)$ *is strictly decreasing function if the initial potential* $y_0 \geqslant 0$, *or* $E^{+}(t)$ *is strictly decreasing function,* $E_{-}(t)$ *is strictly increasing function if the initial potential* $y_0 \leqslant 0$, *and* $E_{-}(0) = E^{+}(0) = 0$. *This is* *different from the Camassa-Holm equation. In fact, in paper [20], for the Camassa-Holm equation, we have $E^+(t)$ is increasing function, and*

$$
\begin{aligned}
\frac{\mathrm{d}}{\mathrm{d}t}E_-(t) &= \int_{\eta(b,t)}^{\eta(a,t)} e^{-y}h_t(y,t)\mathrm{d}y \\
&= \int_{\mathbb{R}} e^{-y}h_t(y,t)\mathrm{d}y = -\int_{\mathbb{R}}[(uh)_x + u_x h]e^{-y}\mathrm{d}y \\
&= \int_{\mathbb{R}}(-u-u_x)(u-u_{xx})e^{-y}\mathrm{d}y \\
&= \int_{\mathbb{R}}(u^2+\frac{1}{2}u_x^2)e^{-y}\mathrm{d}y + (uu_x + \frac{1}{2}u_x^2)e^{-y}|_{-\infty}^{+\infty} \\
&\leqslant 0,
\end{aligned}
\tag{3.19}
$$

where we have used $\lim_{x\to\infty} u(t,x) = 0$, *and* $u(x,t) = u_x(x,t)$, *as* $x \leqslant \eta(a,t)$.

4 Traveling Wave Solutions

In this section, we will prove that Eq.(1.1) have a family traveling wave solutions. First, we present two important definitions and an useful lemma.

Definition 4.1 *The solution $u(t,x)$ to Eq.(1.1) is x-symmetric if there exists a function $b(t) \in C^1(\mathbb{R}^+)$ such that*

$$u(t,x) = u(t, 2b(t) - x), \qquad \forall t \in [0,\infty[,$$

for almost every $x \in \mathbb{R}$. We say that $b(t)$ is the symmetric axis of $u(t,x)$.

Definition 4.2 *Define $\mathscr{X}(\mathbb{R}) = \{u : u \in C(\mathbb{R}^+, H^1(\mathbb{R})),\ u_x \in L^3_{loc}(\mathbb{R})\}$. Assume that $u(t,x) \in \mathscr{X}(\mathbb{R})$ and satisfy*

$$\iint_{\mathbb{R}^+\times\mathbb{R}} \left[u(1-\partial_x^2)\varphi_t + (u_x^3 + uu_x^2)\varphi_x - \frac{1}{3}u_x^3\varphi_{xx} - \frac{1}{3}u_x^3\varphi_{xxx}\right]\mathrm{d}t\mathrm{d}x = 0, \tag{4.1}$$

for all $\varphi \in C_0^\infty(\mathbb{R}^+\times\mathbb{R})$. Then $u(t,x)$ is a weak solution to Eq.(1.1).

Remark 4.1 *Since $C_0^\infty(\mathbb{R}^+\times\mathbb{R})$ is dense in $C_0^1(\mathbb{R}^+, C_0^3(\mathbb{R}))$. By the density argument, we can consider the test functions belong to $C_0^1(\mathbb{R}^+, C_0^3(\mathbb{R}))$. Using the $\langle,\rangle$ notation for distributions, we can rewrite (4.1) as follows*

$$\langle u, (1-\partial_x^2)\varphi_t\rangle + \langle (u_x^3 + uu_x^2), \varphi_x\rangle - \left\langle \frac{1}{3}u_x^3, \varphi_{xx}\right\rangle - \left\langle \frac{1}{3}u_x^3, \varphi_{xxx}\right\rangle = 0. \tag{4.2}$$

Lemma 4.1 *Assume that $U(x) \in \mathscr{X}(\mathbb{R})$ and satisfies*

$$\int_{\mathbb{R}} \left[-cU(1-\partial_x^2)\phi_x + (U^3 + UU_x^2)\phi_x - \frac{1}{3}U_x^3\phi_{xx} - \frac{U^3}{3}\phi_{xxx} \right] \mathrm{d}x = 0, \tag{4.3}$$

for all $\phi \in C_0^\infty(\mathbb{R})$. Then the function u given by

$$u(t,x) = U(x - c(t - t_0)) \tag{4.4}$$

is a weak solution of Eq. (1.1), for any fixed $t_0 \in \mathbb{R}$.

Proof Without loss of generality, we can assume $t_0 = 0$. For $\forall \eta \in C_0^\infty(\mathbb{R}^+ \times \mathbb{R})$, let $\eta_c = \eta(t, x + ct)$, it follows that

$$\begin{cases} \partial_x(\eta_c) = (\eta_x)_c, \\ \partial_t(\eta_c) = (\eta_t)_c + c(\eta_x)_c. \end{cases} \tag{4.5}$$

Assume $u(t,x) = U(x - c(t - t_0))$. One can easily check that

$$\begin{cases} \langle u, \eta \rangle = \langle U, \eta_c \rangle, \ \langle u^3, \eta \rangle = \langle U^3, \eta_c \rangle, \\ \langle uu_x^2, \eta \rangle = \langle UU_x^2, \eta_c \rangle, \ \langle u_x^3, \eta \rangle = \langle U_x^3, \eta_c \rangle, \end{cases} \tag{4.6}$$

where $U = U(x)$. In view of (4.5) and (4.6), we deduce that

$$\begin{aligned} &\langle u, (1-\partial_x^2)\eta_t \rangle + \langle (u^3 + uu_x^2), \eta_x \rangle = \langle U, ((1-\partial_x^2)\eta_t)_c \rangle + \langle (U^3 + UU_x^2), (\eta_x)_c \rangle \\ &= \langle U, (1-\partial_x^2)(\partial_t\eta_c - c\partial_x\eta_c) \rangle + \langle (U^3 + UU_x^2), \partial_x\eta_c \rangle, \end{aligned} \tag{4.7}$$

and

$$\begin{aligned} -\left\langle \frac{u_x^3}{3}, \eta_{xx} \right\rangle - \left\langle \frac{u^3}{3}, \eta_{xxx} \right\rangle &= \left\langle \frac{1}{3}U_x^3, (\eta_{xx})_c \right\rangle - \left\langle \frac{1}{3}U^3, (\eta_{xxx})_c \right\rangle \\ &= \left\langle \frac{1}{3}U_x^3, \partial_x^2\eta_c \right\rangle - \left\langle \frac{1}{3}U^3, \partial_x^3\eta_c \right\rangle. \end{aligned} \tag{4.8}$$

Note that U is independent of time, for T large enough such that it does not belong to the support of η_c, we deduce that

$$\begin{aligned} \langle U, (1-\partial_x^2)\partial_t\eta_c \rangle &= \int_{\mathbb{R}} U(x) \int_{\mathbb{R}_+} \partial_t(1-\partial_x^2)\eta_c \mathrm{d}t\mathrm{d}x \\ &= \int_{\mathbb{R}} U(x)[(1-\partial_x^2)\eta_c(T,x) - (1-\partial_x^2)\eta_c(0,x)]\mathrm{d}x \\ &= 0. \end{aligned} \tag{4.9}$$

Combining (4.7), (4.8) with (4.9), it follows that

$$
\begin{aligned}
&\langle u,(1-\partial_x^2)\eta_t\rangle+\langle (u_x^3+uu_x^2),\eta_x\rangle-\left\langle \frac{1}{3}u_x^3,\eta_{xx}\right\rangle-\left\langle \frac{1}{3}u_x^3,\eta_{xxx}\right\rangle\\
&=\langle U,-c(1-\partial_x^2)\partial_x\eta_c)\rangle+\langle (U^3+UU_x^2),\partial_x\eta_c\rangle-\left\langle \frac{1}{3}U_x^3,\partial_x^2\eta_c\right\rangle-\left\langle \frac{U^3}{3},\partial_x^3\eta_c\right\rangle\\
&=\int_{\mathbb{R}_+}\int_{\mathbb{R}}\left[-cU(1-\partial_x^2)\partial_x\eta_c+(U^3+UU_x^2)\partial_x\eta_c-\frac{1}{3}U_x^3\partial_x^2\eta_c-\frac{U^3}{3}\partial_x^3\eta_c\right]dxdt\\
&=0,
\end{aligned}
$$

where we used (4.3) with $\phi(x)=\eta_c(t,x)$, which belongs to $C_0^\infty(\mathbb{R})$, for every given $t\geqslant 0$. This completes the proof of Lemma 4.1. ■

Finally, for the x-symmetric solutions of Eq.(1.1), the following theorem holds.

Theorem 4.1 *Let $u(t,x)$ be x-symmetric. If $u(t,x)$ be a unique weak solution of Eq.(1.1), then $u(t,x)$ is a traveling wave.*

Proof In view of Remark 4.1, we only to assume $\varphi\in C_0^1(\mathbb{R}^+,C_0^3(\mathbb{R}))$. Let

$$
\varphi_b(t,x)=\varphi(t,2b(t)-x),\qquad b(t)\in C^1(\mathbb{R}).
$$

Then we obtain that $(\varphi_b)_b=\varphi$ and

$$
\begin{cases}
\partial_x u_b=-(\partial_x u)_b,\ \partial_x\varphi_b=-(\partial_x\varphi)_b,\\
\partial_t\varphi_b=(\partial_t\varphi)_b+2\dot{b}(\partial_x\varphi)_b.
\end{cases}
\tag{4.10}
$$

Moreover,

$$
\begin{cases}
\langle u_b,\varphi\rangle=\langle u,\varphi_b\rangle,\ \langle u_b^3,\varphi\rangle=\langle u^3,\varphi_b\rangle,\\
\langle u_b(\partial_x u_b)^2,\varphi\rangle=\langle u(\partial_x u)^2,\varphi_b\rangle,\ \langle(\partial_x u_b)^3,\varphi\rangle=\langle-(\partial_x u)^3,\varphi_b\rangle,
\end{cases}
\tag{4.11}
$$

where $\dot{b}$ denotes the time derivative of b.

Since u is x-symmetric, by virtue of (4.10) and (4.11), we obtain that

$$
\begin{aligned}
&\langle u,(1-\partial_x^2)\partial_t\varphi\rangle+\langle (u^3+u(\partial_x u)^2),\partial_x\varphi\rangle\\
&=\langle u,((1-\partial_x^2)\partial_t\varphi)_b\rangle+\langle (u^3+u(\partial_x u)^2),(\partial_x\varphi)_b\rangle\\
&=\langle u,(1-\partial_x^2)(\partial_t\varphi_b+2\dot{b}\partial_x\varphi_b)\rangle-\langle (u^3+u(\partial_x u)^2),\partial_x\varphi_b\rangle
\end{aligned}
$$

and

$$
-\left\langle \frac{1}{3}u_x^3,\partial_x^2\varphi\right\rangle-\left\langle \frac{u^3}{3},\partial_x^3\varphi\right\rangle=\left\langle \frac{1}{3}u_x^3,\partial_x^2\varphi_b\right\rangle+\left\langle \frac{u^3}{3},\partial_x^3\varphi_b\right\rangle.
$$

In view of (4.2) and above relations to obtain

$$\begin{aligned}&\langle u,(1-\partial_x^2)\varphi_t\rangle+\langle(u^3+uu_x^2),\partial_x\varphi\rangle-\left\langle\frac{1}{3}u_x^3,\varphi_{xx}\right\rangle-\left\langle\frac{1}{3}u^3,\varphi_{xxx}\right\rangle\\&=\langle u,(1-\partial_x^2)(\partial_t\varphi_b+2\dot{b}\partial_x\varphi_b)\rangle-\langle(u^3+uu_x^2),\partial_x\varphi_b\rangle\\&\quad+\left\langle\frac{1}{3}u_x^3,\partial_x^2\varphi_b\right\rangle+\left\langle\frac{u^3}{3},\partial_x^3\varphi_b\right\rangle=0.\end{aligned}\tag{4.12}$$

Therefore, letting φ_b in place of φ in (4.12), as $(\varphi_b)_b=\varphi$, we can get

$$\begin{aligned}&\langle u,(1-\partial_x^2)(\partial_t\varphi+2\dot{b}\partial_x\varphi)\rangle-\langle(u^3+uu_x^2),\partial_x\varphi\rangle\\&+\left\langle\frac{1}{3}u_x^3,\partial_x^2\varphi\right\rangle+\left\langle\frac{u^3}{3},\partial_x^3\varphi\right\rangle=0.\end{aligned}\tag{4.13}$$

Subtracting (4.13) from (4.2), we obtain

$$\langle u,2\dot{b}(1-\partial_x^2)\partial_x\varphi\rangle-2\left\langle(u^3+uu_x^2),\partial_x\varphi\right\rangle+\left\langle\frac{2}{3}u_x^3,\partial_x^2\varphi\right\rangle+\left\langle\frac{2u^3}{3},\partial_x^3\varphi\right\rangle=0.\tag{4.14}$$

For any $\phi\in C_0^\infty(\mathbb{R})$. Let $\varphi_\varepsilon(t,x)=\phi(x)\rho_\varepsilon(t)$, where $\rho_\varepsilon\in C_0^\infty(\mathbb{R}^+)$ is a mollifier with the property that $\rho_\varepsilon\to\delta(t-t_0)$, the Dirac mass at t_0, as $\varepsilon\to0$. From (4.14), by the test function $\varphi_\varepsilon(t,x)$, we have

$$\begin{aligned}&\int_{\mathbb{R}}\left(2(1-\partial_x^2)\partial_x\phi\int_{\mathbb{R}_+}\dot{b}u\rho_\varepsilon(t)\mathrm{d}t\right)\mathrm{d}x-2\int_{\mathbb{R}}\left(\partial_x\phi\int_{\mathbb{R}_+}(u^3+uu_x^2)\rho_\varepsilon(t)\mathrm{d}t\right)\mathrm{d}x\\&+\frac{2}{3}\int_{\mathbb{R}}\left(\partial_x^2\phi\int_{\mathbb{R}_+}u_x^3\rho_\varepsilon(t)\mathrm{d}t\right)\mathrm{d}x+\frac{2}{3}\int_{\mathbb{R}}\left(\partial_x^3\phi\int_{\mathbb{R}_+}u^3\rho_\varepsilon(t)\mathrm{d}t\right)\mathrm{d}x=0.\end{aligned}\tag{4.15}$$

Note that

$$\lim_{\varepsilon\to0}\int_{\mathbb{R}_+}\dot{b}u\rho_\varepsilon(t)\mathrm{d}t=\dot{b}(t_0)u(t_0,x)\quad\text{in } L^2(\mathbb{R}),$$

$$\lim_{\varepsilon\to0}\int_{\mathbb{R}_+}(u^3+uu_x^2)\rho_\varepsilon(t)\mathrm{d}t=(u^3+uu_x^2)(t_0,x)$$

and

$$\lim_{\varepsilon\to0}\int_{\mathbb{R}_+}(u^3+u_x^3)\rho_\varepsilon(t)\mathrm{d}t=(u^3+u_x^3)(t_0,x)$$

in $L^1(\mathbb{R})$. Therefore, letting $\varepsilon\to0$, (4.15) implies that

$$\begin{aligned}&\int_{\mathbb{R}}\left(-\dot{b}(t_0)u(t_0,x)(1-\partial_x^2)\partial_x\phi+(u^3+uu_x^2)(t_0,x)\partial_x\phi\right)\mathrm{d}x\\&-\frac{1}{3}\int_{\mathbb{R}}\left(u_x^3(t_0,x)\partial_x^2\phi+u^3(t_0,x)\partial_x^3\phi\right)\mathrm{d}x=0.\end{aligned}\tag{4.16}$$

Thus, we deduce that $u(t_0, x)$ satisfies (4.3) for $c = \dot{b}(t_0)$. Applying Lemma 4.1, we can get $\tilde{u}(t,x) = u(t_0, x - \dot{b}(t_0)(t - t_0))$ is a traveling wave solution of Eq.(1.1). Since $\tilde{u}(t_0, x) = u(t_0, x)$, by the uniqueness of the solution of Eq.(1.1), we obtain $\tilde{u}(t,x) = u(t,x)$, for all time t. This completes the proof of Theorem 4.1. ■

Acknowledgements This work was partially supported by CPSF (Grant No.: 2012M520007) and CPSF (Grant No.: 2013T60086). The authors thank the references for their valuable comments and constructive suggestions.

References

[1] Beals R, Sattinger D, Szmigielski J. Multipeakons and a theorem of Stieltjes. Inverse Problems, 1999, 15: 1–4.

[2] Bressan A, Constantin A. Global conservative solutions of the Camassa–Holm equation. Arch. Rat. Mech. Anal., 2007, 183: 215–239.

[3] Bressan A, Constantin A. Global dissipative solutions of the Camassa–Holm equation. Anal. Appl., 2007, 5: 1–27.

[4] Camassa R, Holm D. An integrable shallow water equation with peaked solitons. Phys. Rev. Letters, 1993, 71: 1661–1664.

[5] Camassa R, Holm D, Hyman J. An integrable shallow water equation. Adv. Appl. Mech., 1994, 31: 1–33.

[6] Constantin A. Global existence and breaking waves for a shallow water equation: a geometric approch. Ann. Inst. Fourier (Grenoble), 2000, 50: 321–362.

[7] Constantin A. On the scattering problem for the Camassa–Holm equation. Proc. R. Soc. London A, 2001, 457: 953–970.

[8] Constantin A. Finite propagation speed for the Camassa–Holm equation, J. Math. Phys., 2005, 46: 023506.

[9] Constantin A, Escher J. Wave breaking for nonlinear nonlocal shallow water equations. Acta Math., 1998, 181: 229–243.

[10] Constantin A, Escher J. Global existence and blow-up for a shallow water equation. Ann. Scuola Norm. Sup. Pisa, 1998, 26: 303–328.

[11] Constantin A, Escher J. On the blow-up rate and the blow-up set of breaking waves for a shallow water equation. Math. Z., 2000, 233: 75–91.

[12] Constantin A, Strauss W A. Stability of peakons. Comm. Pure Appl. Math., 2000, 53: 603–610.

[13] Dai H H. Model equations for nonlinear dispersive waves in a compressible Mooney–Rivlin rod, Acta Mechanica, 1998, 127: 193–207.

[14] Fu Y, Gui G L, Liu Y, Qu C Z. On the Cauchy problem for the nitegrable Camassa-Holm type equation with cubic nonlinearity. e-print arXiv:1108.5368v3.

[15] B. Fuchssteiner, Some tricks from the symmetry-toolbox for nonlinear equations: generalizations of the Camassa–Holm equation. Physica D, 1996, 4: 229–6243.

[16] Fokas A, Fuchssteiner B. Symplectic structures, their Bäcklund transformation and hereditary symmetries. Physica D, 1981, 4: 47–66.

[17] Green A E, Naghdi P M. A derivation of equations for wave propagation in water of variable depth. J. Fluid Mech., 1976, 78: 237–246.

[18] Gui G L, Liu Y, Olver P L, Qu C Z. Wave-breaking and peakons for a modified Camassa-Holm equation. Comm. Math. Physics, 2012, 526: 199–220.

[19] Henry D. Compactly supported solutions of the Camassa–Holm equation. J. Nonlinear Math. Phys., 2005, 12: 342–347.

[20] Himonas A, Misiolek G, Ponce G, Zhou Y. Persistence properties and unique continuation of solutions of the Camassa–Holm equation. Comm. Math. Phys., 2007, 271: 511–522.

[21] Kato T. On the Korteweg-de Vries equation. Manuscripta Math., 1979, 28: 89–99.

[22] Kato T. On the Cauchy problem for the generalized Korteweg-de Vries equation. in; Studies in Applied Mathematics, in: Adv. Math. Suppl. Stu., Academic Press, New York, 1983, 8: 93–128.

[23] Olver P J, Rosenau P. Tri-Hamiltonian duality between solitons and solitary-wave solutions having compact support. Phys. Rev. E, 1996, 53: 1990–1906.

[24] Qiao Z J. A new integrable equation with cuspons and W/M-shape-peaks solitons. J. Math. Phys., 2006, 47: 112701.

[25] Qiao Z J. New integrable hierarchy, its parametric solutions, cuspons, one-peak solitons, and cuspons and W/M-shape-peaks solitons. J. Math. Phys., 2007, 48: 082701.

[26] Wu X L, Guo B L. Persistence properties and infinite propagation for the modified 2-component Camassa–Holm equation. Discrete Contin. Dyn. Syst. A, 2013, 33: 3211–3223.

[27] Wu X L, Yin Z Y. Well-posedness and global existence for the Novikov equation. Annali Sc. Norm. Sup. Pisa, 2012, XI: 707–727.

Global Smooth Solutions for the Zakharov System with Quantum Effects in Two Space Dimensions *

Gan Zaihui (甘在会), Zhong Ting (钟婷) and Guo Boling (郭柏灵)

Abstract We consider the Zakharov system with quantum effects in the framework of the scalar mode in two space dimensions

$$\begin{aligned} &iE_t + \Delta E = nE + \Gamma\Delta^2 E, \\ &n_t = -\nabla\cdot\mathbf{V}, \\ &\mathbf{V}_t = -\nabla n - \nabla|E|^2 + \Gamma\nabla\Delta n, \end{aligned} \qquad (ZSQ)$$

which is related to quantum corrections to the Zakharov system for Langmuir waves in plasma. We prove the existence and uniqueness of global smooth solutions to the Cauchy problem for (ZSQ) in the Sobolev space through making a priori integral esimates and utilizing the so-called continuity method.

Keywords Zakharov system; Quantum effects; Global solution; Existence; Uniqueness

1 Introduction

We deal with in this paper the Zakharov system with quantum effects in the framework of the scalar model [6]:

$$iE_t + \Delta E = nE + \Gamma\Delta^2 E, \tag{1.1}$$

$$n_t = -\nabla\cdot\mathbf{V}, \tag{1.2}$$

$$\mathbf{V}_t = -\nabla n - \nabla|E|^2 + \Gamma\nabla\Delta n, \tag{1.3}$$

where $(E, n, \mathbf{V}): \quad (t,x)\in[0,\infty)\times\mathbb{R}^2 \to \mathbb{C}\times\mathbb{R}\times\mathbb{R}^2$ and the initial data are taken to be:

$$E(0,x) = E_0(x), \quad n(0,x) = n_0(x), \quad \mathbf{V}(0,x) = \mathbf{V}_0(x). \tag{1.4}$$

* Adv. Nonlinear Stud., 2014, 14(3): 687 – 707. DOI: https://doi.org/10.1515/ans-2014-0310.

In the recent years, special interest has been devoted to quantum corrections to the Zakharov equations for Langmuir waves in a plasma [20]. The system (1.1)–(1.3) is related to the equations that govern the amplitude E of the electric field oscillations and the number density n

$$i\mathbf{E}_t - \alpha\nabla \times (\nabla \times \mathbf{E}) + \nabla(\nabla \cdot \mathbf{E}) = n\mathbf{E} + \Gamma\nabla\Delta(\nabla \cdot \mathbf{E}), \qquad (VZS-1)$$

$$n_{tt} - \Delta n = \Delta|\mathbf{E}|^2 - \Gamma\Delta^2 n. \qquad (VZS-2)$$

The parameter α is defined as the square ratio of the speed of light and the electron Fermi velocity and Γ measures the influence of quantum effects, and (VZS-2) originates from the hydrodynamic system

$$\begin{cases} n_t + \nabla \cdot \mathbf{V} = 0, \\ \mathbf{V}_t = -\nabla n - \nabla|\mathbf{E}|^2 + \Gamma\nabla\Delta n, \end{cases}$$

governing the ion sound waves. As mentioned by [6], it may be suitable to abandon the vector character of the equations (VZS-1)–(VZS-2) for the sake of preserving a non-zero electric field at the center of the cavity, while keeping the rotational symmetry necessary for implementing an asymptotic analysis in the spirit of [7, 8]. We consider in this paper the influence of quantum effects in the framework of the scalar model (1.1)–(1.3).

The system (1.1)–(1.3) describes the quantum corrections to the Zakharov equations for Langmuir waves in a plasma [20] and has been studied by many physicists and mathematicians ([12], [15]). For $\Gamma = 0$, the system (1.1)–(1.3) was derived by Zakharov in [20] to model Langmuir wave in a plasma. From the mathematical side, there has been considerable work on local and global well-posedness of solutions with rough data through the works of Kenig, Ponce and Vega [16], Bourgain and Colliander [4], Ginibre, Tsutsumi and Velo [13], and Bejenaru, Herr, Holmer and Tataru [2]. It was shown in [1,13,14,18] that for initial data small enough, the solution to the Cauchy problem (1.1)–(1.4) remains smooth for all time. In two dimensions, the smallest condition reads $\|E_0\|^2_{L^2(\mathbb{R}^2)} \leqslant \|Q\|^2_{L^2(\mathbb{R}^2)}$ and is optimal, where Q is the ground state of the equation $\Delta u - u + u^3 = 0$. In three dimensions, it requires that the plasma number $\mathcal{N}$ and the Hamiltonian $\mathcal{H}$ satisfy $\mathcal{N}|\mathcal{H}| < \varepsilon$ ($\varepsilon \leqslant 2.6 \times 10^{-4}$) together with $\|\nabla E_0\|^2_{L^2(\mathbb{R}^2)} < |\mathcal{H}|$, where $\mathcal{N}$ and $\mathcal{H}$ are defined as follows:

$$\mathcal{N} = \int_{\mathbb{R}^2} |E|^2 \mathrm{d}x, \qquad (1.5)$$

$$\mathcal{H} = \int_{\mathbb{R}^2} \left\{ |\nabla E|^2 + \frac{1}{2}n^2 + \frac{1}{2}|\mathbf{V}|^2 + n|E|^2 + \Gamma|\Delta E|^2 + \frac{\Gamma}{2}|\nabla n|^2 \right\} \mathrm{d}x. \tag{1.6}$$

Recently, along with the further deep study of physical problems, certain generalized and useful systems of Zakharov equations were proposed. We studied in [9,10,11] some generalized Zakharov system and obtained various conclusions including blow-up, existence and orbital instability of standing waves as well as the sharp threshold for global existence.

In this paper we shall be concerned with the existence and uniqueness of global smooth solutions to the Cauchy problem (1.1)–(1.4). Throughout this paper, we assume that the solution $(E(t,x), n(t,x), \mathbf{V}(t,x))$ of the Cauchy problem (1.1)–(1.4) and its derivatives tend to zero as $|x| \to +\infty$. In addition, for simplicity, we shall denote various positive constants by C and $(f,g) = \int_{\mathbb{R}^2} f(x)\bar{g}(x)\mathrm{d}x$, where $\bar{g}(x)$ denote the complex conjugate function of $g(x)$.

2 A priori estimates

In this section, we mainly make some a priori estimates. Throughout this section, we always assume that (E,n) is a sufficiently smooth solution on $[0,T]$ of the Cauchy problem (1.1)–(1.4) to make the calculations in the following lemmas.

Lemma 2.1 *Let $E_0(x) \in H^1(\mathbb{R}^2)$. Then the solution $E(t,x)$ of the Cauchy problem* (1.1)–(1.4) *satisfies*

$$\|E(t,x)\|_{L^2(\mathbb{R}^2)} = \|E_0(x)\|_{L^2(\mathbb{R}^2)}. \tag{2.1}$$

Lemma 2.1 can be proven by taking the imaginary part of the inner product of (1.1) with E, namely,

$$Im(iE_t + \Delta E - nE - \Gamma\Delta^2 E, E) = 0. \tag{2.2}$$

The proof is omitted here.

Lemma 2.2 [5] *Let $\|u\|_{H^1(\mathbb{R}^2)} \leqslant K$. Then*

$$\|u\|_{L^\infty(\mathbb{R}^2)} \leqslant C\left(1 + \sqrt{\log(1 + \|u\|_{H^2(\mathbb{R}^2)})}\right), \quad u \in H^2(\mathbb{R}^2), \tag{2.3}$$

where the constant C depends only on K.

Lemma 2.3 [3,17,19] *Let $u \in H^1(\mathbb{R}^2)$. Then*

$$\|u\|_{L^4(\mathbb{R}^2)}^4 \leqslant C_0\|\nabla u\|_{L^2(\mathbb{R}^2)}^2\|u\|_{L^2(\mathbb{R}^2)}^2, \tag{2.4}$$

where $C_0 = \dfrac{2}{\|Q\|^2_{L^2(\mathbb{R}^2)}}$ *and* $Q(x)$ *is the ground state solution of the equation*

$$\Delta\psi - \psi + \psi^3 = 0. \tag{2.5}$$

Lemma 2.4 *Assume that*

(i) $E_0(x) \in H^2(\mathbb{R}^2)$, $n_0(x) \in H^1(\mathbb{R}^2)$, $\mathbf{V}_0(x) \in L^2(\mathbb{R}^2)$;

(ii) $\|E_0\|^2_{L^2(\mathbb{R}^2)} < \dfrac{2}{3}\|Q(x)\|^2_{L^2(\mathbb{R}^2)}$,

where $Q(x)$ *is the ground state solution of* (2.5). *Then the solution of the Cauchy problem* (1.1)–(1.4) *satisfies*

$$\|\Delta E\|^2_{L^\infty_t L^2_x} + \|\nabla E\|^2_{L^\infty_t L^2_x} + \|n\|^2_{L^\infty_t L^2_x} + \|\nabla n\|^2_{L^\infty_t L^2_x} + \|\mathbf{V}\|^2_{L^\infty_t L^2_x} \leqslant E_1, \tag{2.6}$$

where the constant E_1 *depends only on* $\|E_0\|^2_{H^2(\mathbb{R}^2)}$, $\|n_0\|^2_{H^1(\mathbb{R}^2)}$ *and* $\|\mathbf{V}_0\|^2_{L^2(\mathbb{R}^2)}$.

Proof Taking the inner product of (1.1) with E_t, one has

$$(iE_t + \Delta E - nE - \Gamma\Delta^2 E, E_t) = 0. \tag{2.7}$$

By a direct calculation, we obtain

$$\begin{aligned}
&Re(iE_t, E_t) = 0,\\
&Re(\Delta E, E_t) = -\frac{1}{2}\frac{\mathrm{d}}{\mathrm{d}t}\|\nabla E\|^2_{L^2(\mathbb{R}^2)},\\
&Re(nE, E_t) = \frac{1}{2}\int_{\mathbb{R}^2} n\frac{\mathrm{d}}{\mathrm{d}t}|E|^2\mathrm{d}x,\\
&Re(\Gamma\Delta^2 E, E_t) = \frac{\Gamma}{2}\frac{\mathrm{d}}{\mathrm{d}t}\|\Delta E\|^2_{L^2(\mathbb{R}^2)},
\end{aligned}$$

which together with (2.7) yield that

$$\frac{\mathrm{d}}{\mathrm{d}t}\left(\|\nabla E\|^2_{L^2(\mathbb{R}^2)} + \int_{\mathbb{R}^2} n|E|^2\mathrm{d}x + \Gamma\|\Delta E\|^2_{L^2(\mathbb{R}^2)}\right) = \int_{\mathbb{R}^2} n_t|E|^2\mathrm{d}x. \tag{2.8}$$

By (1.2) and (1.3)

$$\begin{aligned}
\int_{\mathbb{R}^2} n_t|E|^2\mathrm{d}x &= \int_{\mathbb{R}^2}(-\nabla\cdot\mathbf{V})|E|^2\mathrm{d}x\\
&= \int_{\mathbb{R}^2}\mathbf{V}\nabla|E|^2\mathrm{d}x\\
&= \int_{\mathbb{R}^2}\mathbf{V}(-\nabla n + \Gamma\nabla\Delta n - \mathbf{V}_t)\mathrm{d}x\\
&= -\frac{1}{2}\frac{\mathrm{d}}{\mathrm{d}t}\int_{\mathbb{R}^2}(n^2 + |\mathbf{V}|^2 + \Gamma|\nabla n|^2)\,\mathrm{d}x.
\end{aligned}$$

Combining the above identity with (2.8), one concludes

$$\begin{aligned}\mathcal{H}(t)=&\|\nabla E\|_{L^2(\mathbb{R}^2)}^2+\int_{\mathbb{R}^2} n|E|^2dx+\Gamma\|\Delta E\|_{L^2(\mathbb{R}^2)}^2\\&+\frac{1}{2}\|n\|_{L^2(\mathbb{R}^2)}^2+\frac{1}{2}\|\mathbf{V}\|_{L^2(\mathbb{R}^2)}^2+\frac{1}{2}\Gamma\|\nabla n\|_{L^2(\mathbb{R}^2)}^2\\=&\mathcal{H}(0).\end{aligned}\tag{2.9}$$

By Hölder's inequality, Young's inequality, Lemma 2.3 and (2.1), one observes that

$$\begin{aligned}\int_{\mathbb{R}^2} n|E|^2\mathrm{d}x&\leqslant\frac{1}{3}\int_{\mathbb{R}^2} n^2\mathrm{d}x+\frac{3}{4}\int_{\mathbb{R}^2}|E|^4\mathrm{d}x\\&\leqslant\frac{1}{3}\int_{\mathbb{R}^2} n^2\mathrm{d}x+\frac{3}{4}\frac{2}{\|Q\|_{L^2(\mathbb{R}^2)}^2}\|E_0\|_{L^2(\mathbb{R}^2)}^2\|\nabla E\|_{L^2(\mathbb{R}^2)}^2\\&\leqslant\frac{1}{3}\int_{\mathbb{R}^2} n^2\mathrm{d}x+\|\nabla E\|_{L^2(\mathbb{R}^2)}^2.\end{aligned}\tag{2.10}$$

This together with (2.9) yields the estimate (2.6). □

Lemma 2.5 *Under the conditions of Lemma* 2.4, *let*

$$E_0(x)\in H^3(\mathbb{R}^2),\quad n_0(x)\in H^2(\mathbb{R}^2),\quad \mathbf{V}_0(x)\in H^1(\mathbb{R}^2).$$

Then the solution of the Cauchy problem (1.1)–(1.4) *satisfies*

$$\|\nabla^3E\|_{L_t^\infty L_x^2}^2+\|\Delta n\|_{L_t^\infty L_x^2}^2+\|E_t\|_{L_t^\infty L_x^2}^2+\|n_t\|_{L_t^\infty L_x^2}^2\leqslant E_2,\tag{2.11}$$

where the constant E_2 depends on T, $\|E_0\|_{H^3(\mathbb{R}^2)}^2$, $\|n_0\|_{H^2(\mathbb{R}^2)}^2$ and $\|\mathbf{V}_0\|_{H^1(\mathbb{R}^2)}^2$.

Proof Differentiating (1.1) with respect to t, and taking the inner product of the resulting equation with E_t, we have

$$(iE_{tt}+\Delta E_t-(nE)_t-\Gamma\Delta^2E_t,E_t)=0.\tag{2.12}$$

Noting that

$$\begin{aligned}Im(iE_{tt},E_t)&=Re(E_{tt},E_t)=\frac{1}{2}\frac{\mathrm{d}}{\mathrm{d}t}\|E_t\|_{L^2(\mathbb{R}^2)}^2,\\Im(\Delta E_t,E_t)&=-Im\|\nabla E_t\|_{L^2(\mathbb{R}^2)}^2=0,\\Im((nE)_t,E_t)&=Im(n_tE,E_t)+Im(nE_t,E_t)\\&=Im\int_{\mathbb{R}^2} n_tE\overline{E}_t\mathrm{d}x,\\Im(-\Gamma\Delta^2E_t,E_t)&=Im\Gamma\|\Delta E_t\|_{L^2(\mathbb{R}^2)}^2=0,\end{aligned}$$

we thereby obtain

$$\frac{\mathrm{d}}{\mathrm{d}t}\|E_t\|_{L^2(\mathbb{R}^2)}^2\leqslant2\|E\|_{L^\infty(\mathbb{R}^2)}\|n_t\|_{L^2(\mathbb{R}^2)}\|E_t\|_{L^2(\mathbb{R}^2)}.\tag{2.13}$$

Integrating (2.13) with respect to t yields

$$\begin{aligned}\|E_t(t,.)\|^2_{L^2(\mathbb{R}^2)} \leqslant & \|E_t(0,.)\|^2_{L^2(\mathbb{R}^2)} \\ & +2\int_0^t \|E(\tau,.)\|_{L^\infty(\mathbb{R}^2)}\|n_t(\tau,.)\|_{L^2(\mathbb{R}^2)}\|E_t(\tau,.)\|_{L^2(\mathbb{R}^2)}\mathrm{d}\tau.\end{aligned} \tag{2.14}$$

On the other hand, we take the inner product of (1.1) with ΔE and then obtain

$$(iE_t + \Delta E - nE - \Gamma\Delta^2 E, \Delta E) = 0, \tag{2.15}$$

where

$$\begin{aligned}|(iE_t,\Delta E)| &\leqslant \|E_t\|_{L^2(\mathbb{R}^2)}\|\Delta E\|_{L^2(\mathbb{R}^2)},\\ |(\Delta E,\Delta E)| &= \|\Delta E\|^2_{L^2(\mathbb{R}^2)},\\ |(nE,\Delta E)| &\leqslant \|nE\|_{L^2(\mathbb{R}^2)}\|\Delta E\|_{L^2(\mathbb{R}^2)}\\ &\leqslant \|n\|_{L^4(\mathbb{R}^2)}\|E\|_{L^4(\mathbb{R}^2)}\|\Delta E\|_{L^2(\mathbb{R}^2)}\\ &\leqslant C\|n\|^{\frac{1}{2}}_{L^2(\mathbb{R}^2)}\|\nabla n\|^{\frac{1}{2}}_{L^2(\mathbb{R}^2)}\|\Delta E\|_{L^2(\mathbb{R}^2)}\\ \left|(\Gamma\Delta^2 E,\Delta E)\right| &= \Gamma\|\nabla^3 E\|^2_{L^2(\mathbb{R}^2)}.\end{aligned}$$

By (2.15), Lemma 2.3 and Lemma 2.4, it follows that

$$\|\nabla^3 E\|^2_{L^2(\mathbb{R}^2)} \leqslant C\left(\|E_t\|^2_{L^2(\mathbb{R}^2)} + C_1\right). \tag{2.16}$$

In addition, we take the inner product of (1.1) with E, and achieve

$$(iE_t + \Delta E - nE - \Gamma\Delta^2 E, E) = 0, \tag{2.17}$$

which yields

$$\|E_t\|_{L^2(\mathbb{R}^2)} \leqslant C(\|\Delta E\|_{L^2(\mathbb{R}^2)}\|E\|_{L^2(\mathbb{R}^2)} + \|nE\|_{L^2(\mathbb{R}^2)}\|E\|_{L^2(\mathbb{R}^2)} + \|\Delta E\|^2_{L^2(\mathbb{R}^2)}) \leqslant C. \tag{2.18}$$

Combining (2.14), (2.16) with (2.18), we have by Lemma 2.2 and Lemma 2.4

$$\begin{aligned}\|\nabla^3 E(t,.)\|^2_{L^2(\mathbb{R}^2)} &\leqslant C + C_1\int_0^t \|E(\tau,.)\|_{L^\infty(\mathbb{R}^2)}\|n_t(\tau,.)\|^2_{L^2(\mathbb{R}^2)}\mathrm{d}\tau\\ &\leqslant C + C_2\int_0^t \|n_t(\tau,.)\|^2_{L^2(\mathbb{R}^2)}\mathrm{d}\tau.\end{aligned} \tag{2.19}$$

On the other hand, differentiating (1.2) with respect to t and noting (1.3), we have

$$n_{tt} - \Delta n - \Delta|E|^2 + \Gamma\Delta^2 n = 0. \tag{2.20}$$

By taking the inner product of (2.20) with n_t, one gets

$$(n_{tt} - \Delta n - \Delta|E|^2 + \Gamma\Delta^2 n, n_t) = 0, \tag{2.21}$$

Since

$$(n_{tt} - \Delta n + \Gamma\Delta^2 n, n_t) = \frac{1}{2}\frac{\mathrm{d}}{\mathrm{d}t}\left(\|\nabla n\|^2_{L^2(\mathbb{R}^2)} + \|n_t\|^2_{L^2(\mathbb{R}^2)} + \Gamma\|\Delta n\|^2_{L^2(\mathbb{R}^2)}\right),$$

$$\begin{aligned}
\left|(\Delta|E|^2, n_t)\right| &\leqslant 2\left(\left|(\Delta E\overline{E}, n_t)\right| + \left|(|\nabla E|^2, n_t)\right|\right) \\
&\leqslant 2\|E\|_{L^\infty(\mathbb{R}^2)}\|n_t\|_{L^2(\mathbb{R}^2)}\|\Delta E\|_{L^2(\mathbb{R}^2)} + \|\nabla E\|^4_{L^4(\mathbb{R}^2)} + \|n_t\|^2_{L^2(\mathbb{R}^2)} \\
&\leqslant C\|E\|_{L^\infty(\mathbb{R}^2)}(\|n_t\|^2_{L^2(\mathbb{R}^2)} + \|\Delta E\|^2_{L^2(\mathbb{R}^2)}) \\
&\quad + C_1\|\nabla E\|^2_{L^2(\mathbb{R}^2)}\|\Delta E\|^2_{L^2(\mathbb{R}^2)} + \|n_t\|^2_{L^2(\mathbb{R}^2)} \\
&\leqslant C\|n_t\|^2_{L^2(\mathbb{R}^2)} + C_1,
\end{aligned}$$

noting (2.21) and Lemma 2.4, we have

$$\|\Delta n\|^2_{L^2(\mathbb{R}^2)} + \|n_t\|^2_{L^2(\mathbb{R}^2)} \leqslant C + C_1\int_0^t \|n_t(.,\tau)\|^2_{L^2(\mathbb{R}^2)}\mathrm{d}\tau. \tag{2.22}$$

Let

$$J(t) = \|\nabla^3 E\|^2_{L^2(\mathbb{R}^2)} + \|\Delta n\|^2_{L^2(\mathbb{R}^2)} + \|n_t\|^2_{L^2(\mathbb{R}^2)} + 1. \tag{2.23}$$

Collecting (2.19) with (2.22) yields that

$$J(t) \leqslant C + C\int_0^t J(\tau)\mathrm{d}\tau,$$

which together with Gronwall's inequality implies

$$J(t) \leqslant C(T),\ t \in [0, T]. \tag{2.24}$$

This together with (2.18) yields the estimate (2.11). □

It is easy to obtain the following conclusion by Lemma 2.2 and Lemma 2.5.

Corollary 2.1 *There holds*

$$\|E\|_{L^\infty(\mathbb{R}^2)} + \|\nabla E\|_{L^\infty(\mathbb{R}^2)} + \|n\|_{L^\infty(\mathbb{R}^2)} \leqslant E_3, \tag{2.25}$$

where the constant E_3 *depends on* T, $\|E_0\|^2_{H^3(\mathbb{R}^2)}$, $\|n_0\|^2_{H^2(\mathbb{R}^2)}$, *and* $\|\mathbf{V}_0\|^2_{H^1(\mathbb{R}^2)}$.

Lemma 2.6 *Under the conditions of Lemma 2.4, let*

$$E_0(x) \in H^4(\mathbb{R}^2), \qquad n_0(x) \in H^3(\mathbb{R}^2), \qquad \mathbf{V}_0(x) \in H^2(\mathbb{R}^2).$$

Then the solution of the Cauchy problem (1.1)–(1.4) satisfies

$$\|\Delta^2 E\|^2_{L^\infty_t L^2_x} + \|\nabla^3 n\|^2_{L^\infty_t L^2_x} + \|\nabla E_t\|^2_{L^\infty_t L^2_x} + \|\nabla n_t\|^2_{L^\infty_t L^2_x} \leqslant E_4, \tag{2.26}$$

where the constant E_4 depends on T, $\|E_0\|^2_{H^4(\mathbb{R}^2)}$, $\|n_0\|^2_{H^3(\mathbb{R}^2)}$, and $\|\mathbf{V}_0\|^2_{H^2(\mathbb{R}^2)}$.

Proof Differentiating (1.1) with respect to t, and then taking the inner product of the resulting equation with ΔE_t, we have

$$(iE_{tt} + \Delta E_t - (nE)_t - \Gamma\Delta^2 E_t, \Delta E_t) = 0. \tag{2.27}$$

By a direct calculation one sees

$$\begin{aligned}
&Im(iE_{tt}, \Delta E_t) = -\frac{1}{2}\frac{\mathrm{d}}{\mathrm{d}t}\|\nabla E_t\|^2_{L^2(\mathbb{R}^2)},\\
&Im(\Delta E_t, \Delta E_t) = Im\|\Delta E_t\|^2_{L^2(\mathbb{R}^2)} = 0,\\
&Im(-(nE)_t, \Delta E_t) = Im(\nabla(n_t E + nE_t), \nabla E_t)\\
&\qquad\qquad = Im\left(\nabla n_t E + n_t\nabla E + \nabla n E_t + n\nabla E_t, \nabla E_t\right),\\
&Im(-\Gamma\Delta^2 E_t, \Delta E_t) = Im\Gamma\|\nabla^3 E_t\|^2_{L^2(\mathbb{R}^2)} = 0.
\end{aligned}$$

Therefore, (2.27), Corollary 2.1, Lemma 2.5 and Young's inequality give

$$\begin{aligned}
\frac{1}{2}\frac{\mathrm{d}}{\mathrm{d}t}\|\nabla E_t\|^2_{L^2(\mathbb{R}^2)} \leqslant\ & |(\nabla n_t E, \nabla E_t)| + |(n_t\nabla E, \nabla E_t)|\\
& + |(\nabla n E_t, \nabla E_t)| + |(n\nabla E_t, \nabla E_t)|\\
\leqslant\ & \|E\|_{L^\infty(\mathbb{R}^2)}\|\nabla n_t\|_{L^2(\mathbb{R}^2)}\|\nabla E_t\|_{L^2(\mathbb{R}^2)}\\
& + \|\nabla E\|_{L^\infty(\mathbb{R}^2)}\|n_t\|_{L^2(\mathbb{R}^2)}\|\nabla E_t\|_{L^2(\mathbb{R}^2)}\\
& + \|\nabla n\|^{\frac{1}{2}}_{L^2(\mathbb{R}^2)}\|\Delta n\|^{\frac{1}{2}}_{L^2(\mathbb{R}^2)}\|E_t\|^{\frac{1}{2}}_{L^2(\mathbb{R}^2)}\|\nabla E_t\|^{\frac{1}{2}}_{L^2(\mathbb{R}^2)}\|\nabla E_t\|_{L^2(\mathbb{R}^2)}\\
& + \|n\|_{L^\infty(\mathbb{R}^2)}\|\nabla E_t\|^2_{L^2(\mathbb{R}^2)}\\
\leqslant\ & C\left(\|\nabla n_t\|_{L^2(\mathbb{R}^2)}\|\nabla E_t\|_{L^2(\mathbb{R}^2)} + \|\nabla E_t\|_{L^2(\mathbb{R}^2)} + \|\nabla E_t\|^{\frac{3}{2}}_{L^2(\mathbb{R}^2)}\right)\\
\leqslant\ & C\left(\|\nabla n_t\|^2_{L^2(\mathbb{R}^2)} + \|\nabla E_t\|^2_{L^2(\mathbb{R}^2)} + 1\right).
\end{aligned} \tag{2.28}$$

On the other hand, taking inner product of (2.20) with Δn_t, we obtain

$$(n_{tt} - \Delta n - \Delta|E|^2 + \Gamma\Delta^2 n, \Delta n_t) = 0. \tag{2.29}$$

In view of Corollary 2.1, Lemma 2.4 and Lemma 2.5, the following estimates hold:

$$(n_{tt} - \Delta n + \Gamma\Delta^2 n, \Delta n_t) = -\frac{1}{2}\frac{\mathrm{d}}{\mathrm{d}t}\left(\|\nabla n_t\|^2_{L^2(\mathbb{R}^2)} + \|\Delta n\|^2_{L^2(\mathbb{R}^2)} + \Gamma\|\nabla^3 n\|^2_{L^2(\mathbb{R}^2)}\right),$$

$$
\begin{aligned}
(-\Delta|E|^2, \Delta n_t) = & \left(\nabla^3 E\overline{E} + \Delta E\nabla\overline{E} + 2\nabla^2 E\nabla\overline{E}\right. \\
& \left. + 2\nabla E\nabla^2\overline{E} + \nabla E\Delta\overline{E} + E\nabla^3\overline{E}, \nabla n_t\right) \\
\leqslant & C\left(\|E\|_{L^\infty(\mathbb{R}^2)}\|\nabla^3 E\|_{L^2(\mathbb{R}^2)} + \|\Delta E\|_{L^4(\mathbb{R}^2)}\|\nabla E\|_{L^4(\mathbb{R}^2)}\right. \\
& \left. + \|\nabla^2 E\|_{L^4(\mathbb{R}^2)}\|\nabla E\|_{L^4(\mathbb{R}^2)})\|\nabla n_t\|_{L^2(\mathbb{R}^2)}\right) \\
\leqslant & C\left(1 + \|\nabla n_t\|^2_{L^2(\mathbb{R}^2)}\right).
\end{aligned}
$$

That is,

$$
\frac{\mathrm{d}}{\mathrm{d}t}\left(\|\nabla n_t\|^2_{L^2(\mathbb{R}^2)} + \|\Delta n\|^2_{L^2(\mathbb{R}^2)} + \Gamma\|\nabla^3 n\|^2_{L^2(\mathbb{R}^2)}\right) \leqslant C\left(1 + \|\nabla n_t\|^2_{L^2(\mathbb{R}^2)}\right). \tag{2.30}
$$

Let

$$
G(t) = \|\nabla n_t\|^2_{L^2(\mathbb{R}^2)} + \|\nabla E_t\|^2_{L^2(\mathbb{R}^2)} + \|\Delta n\|^2_{L^2(\mathbb{R}^2)} + \Gamma\|\nabla^3 n\|^2_{L^2(\mathbb{R}^2)} + 1. \tag{2.31}
$$

On account of (1.1), one gets

$$
\begin{aligned}
\|\Delta^2 E\|_{L^2(\mathbb{R}^2)} & \leqslant C\left(\|E_t\|_{L^2(\mathbb{R}^2)} + \|\Delta E\|_{L^2(\mathbb{R}^2)} + \|nE\|_{L^2(\mathbb{R}^2)}\right) \\
& \leqslant C + \|n\|^{\frac{1}{2}}_{L^4(\mathbb{R}^2)}\|E\|^{\frac{1}{2}}_{L^4(\mathbb{R}^2)} \leqslant C.
\end{aligned} \tag{2.32}
$$

Hence, by (2.28), (2.30) and (2.31), we eventually obtain

$$
G(t) \leqslant C + C\int_0^t G(\tau)\mathrm{d}\tau,
$$

which together with Gronwall's inequality yields

$$
\|\nabla E_t\|^2_{L^2(\mathbb{R}^2)} + \|\nabla n_t\|^2_{L^2(\mathbb{R}^2)} + \|\nabla^3 n\|^2_{L^2(\mathbb{R}^2)} \leqslant C(T). \tag{2.33}
$$

Collecting (2.33) with (2.32) yields the estimate (2.26). □

Corollary 2.2 *The following estimate holds:*

$$
\|\Delta E\|_{L^\infty(\mathbb{R}^2)} + \|\nabla n\|_{L^\infty(\mathbb{R}^2)} + \|\mathbf{V}\|_{L^\infty(\mathbb{R}^2)} \leqslant E_5, \tag{2.34}
$$

where the constant E_5 depends on T, $\|E_0\|^2_{H^4(\mathbb{R}^2)}$, $\|n_0\|^2_{H^3(\mathbb{R}^2)}$ and $\|\mathbf{V}_0\|^2_{H^2(\mathbb{R}^2)}$.

Lemma 2.7 *Under the conditions of Lemma 2.4, let*

$$
E_0(x) \in H^5(\mathbb{R}^2), \quad n_0(x) \in H^4(\mathbb{R}^2), \quad \mathbf{V}_0(x) \in H^3(\mathbb{R}^2).
$$

Then the solution of the Cauchy problem (1.1)–(1.4) satisfies

$$
\|\nabla^5 E\|^2_{L^\infty_t L^2_x} + \|\Delta^2 n\|^2_{L^\infty_t L^2_x} + \|\Delta E_t\|^2_{L^\infty_t L^2_x} + \|\Delta n_t\|^2_{L^\infty_t L^2_x} \leqslant E_6, \tag{2.35}
$$

where the constant E_6 *depends on* T, $\|E_0\|^2_{H^5(\mathbb{R}^2)}$, $\|n_0\|^2_{H^4(\mathbb{R}^2)}$, *and* $\|\mathbf{V}_0\|^2_{H^3(\mathbb{R}^2)}$.

Proof Differentiating (1.1) with respect to t, and then taking the inner product of the resulting equation with $\Delta^2 E_t$, we have

$$(iE_{tt} + \Delta E_t - (nE)_t - \Gamma\Delta^2 E_t, \Delta^2 E_t) = 0. \tag{2.36}$$

It is easy to check

$$\begin{aligned}
Im(iE_{tt}, \Delta^2 E_t) &= \frac{1}{2}\frac{\mathrm{d}}{\mathrm{d}t}\|\Delta E_t\|^2_{L^2(\mathbb{R}^2)},\\
Im(\Delta E_t, \Delta^2 E_t) &= -\|\nabla^3 E_t\|^2_{L^2(\mathbb{R}^2)} = 0,\\
(-\Gamma\Delta^2 E_t, \Delta^2 E_t) &= -\Gamma\|\Delta^2 E_t\|^2_{L^2(\mathbb{R}^2)},\\
(-(nE)_t, \Delta^2 E_t) &= -(n_t E + nE_t, \Delta^2 E_t)\\
&= -(\Delta(n_t E + nE_t), \Delta^2 E_t)\\
&= -(\Delta n_t E + 2\nabla n_t \nabla E + n_t \Delta E + \Delta n E_t\\
&\quad +2\nabla n \nabla E_t + n\Delta E_t, \Delta E_t).
\end{aligned}$$

Thus, taking the imaginary part of (2.36), we get from Corollary 2.1, Corollary 2.2, Lemma 2.5 and Lemma 2.6 that

$$\begin{aligned}
\frac{1}{2}\frac{\mathrm{d}}{\mathrm{d}t}\|\Delta E_t\|^2_{L^2(\mathbb{R}^2)} \leqslant{}& |(\Delta n_t E, \Delta E_t)| + 2\,|(\nabla n_t \nabla E, \Delta E_t)|\\
&+ |(n_t\Delta E, \Delta E_t)| + |(\Delta n E_t, \Delta E_t)|\\
&+2\,|(\nabla n \nabla E_t, \Delta E_t)| + |(n\Delta E_t, \Delta E_t)|\\
\leqslant{}& \|E\|_{L^\infty(\mathbb{R}^2)}\|\Delta n_t\|_{L^2(\mathbb{R}^2)}\|\Delta E_t\|_{L^2(\mathbb{R}^2)}\\
&+2\|\nabla E\|_{L^\infty(\mathbb{R}^2)}\|\nabla n_t\|_{L^2(\mathbb{R}^2)}\|\Delta E_t\|_{L^2(\mathbb{R}^2)}\\
&+\|\Delta E\|_{L^\infty(\mathbb{R}^2)}\|n_t\|_{L^2(\mathbb{R}^2)}\|\Delta E_t\|_{L^2(\mathbb{R}^2)}\\
&+\|\Delta n\|_{L^4(\mathbb{R}^2)}\|E_t\|_{L^4(\mathbb{R}^2)}\|\Delta E_t\|_{L^2(\mathbb{R}^2)}\\
&+2\|\nabla n\|_{L^\infty(\mathbb{R}^2)}\|\nabla E_t\|_{L^2(\mathbb{R}^2)}\|\Delta E_t\|_{L^2(\mathbb{R}^2)}\\
&+\|n\|_{L^\infty(\mathbb{R}^2)}\|\Delta E_t\|_{L^2(\mathbb{R}^2)}\\
\leqslant{}& \|E\|_{L^\infty(\mathbb{R}^2)}\|\Delta n_t\|_{L^2(\mathbb{R}^2)}\|\Delta E_t\|_{L^2(\mathbb{R}^2)}\\
&+2\|\nabla E\|_{L^\infty(\mathbb{R}^2)}\|\nabla n_t\|_{L^2(\mathbb{R}^2)}\|\Delta E_t\|_{L^2(\mathbb{R}^2)}\\
&+\|\Delta E\|_{L^\infty(\mathbb{R}^2)}\|n_t\|_{L^2(\mathbb{R}^2)}\|\Delta E_t\|_{L^2(\mathbb{R}^2)}\\
&+\|\Delta n\|^{\frac{1}{2}}_{L^2(\mathbb{R}^2)}\|\nabla^3 n\|^{\frac{1}{2}}_{L^2(\mathbb{R}^2)}\|E_t\|^{\frac{1}{2}}_{L^2(\mathbb{R}^2)}\|\nabla E_t\|^{\frac{1}{2}}_{L^2(\mathbb{R}^2)}\|\Delta E_t\|_{L^2(\mathbb{R}^2)}\\
&+2\|\nabla n\|_{L^\infty(\mathbb{R}^2)}\|\nabla E_t\|_{L^2(\mathbb{R}^2)}\|\Delta E_t\|_{L^2(\mathbb{R}^2)}
\end{aligned}$$

$$
\begin{aligned}
&+\|n\|_{L^\infty(\mathbb{R}^2)}\|\Delta E_t\|_{L^2(\mathbb{R}^2)} \\
&\leqslant C\left(\|\Delta n_t\|_{L^2(\mathbb{R}^2)}\|\Delta E_t\|_{L^2(\mathbb{R}^2)} + \|\Delta E_t\|_{L^2(\mathbb{R}^2)} + \|\Delta E_t\|^2_{L^2(\mathbb{R}^2)}\right) \\
&\leqslant C\left(\|\Delta n_t\|^2_{L^2(\mathbb{R}^2)} + \|\Delta E_t\|^2_{L^2(\mathbb{R}^2)} + 1\right).
\end{aligned} \tag{2.37}
$$

We further obtain through taking inner product of (2.20) with $\Delta^2 n_t$ that

$$
(n_{tt} - \Delta n - \Delta|E|^2 + \Gamma\Delta^2 n, \Delta^2 n_t) = 0. \tag{2.38}
$$

According to Corollary 2.1, Corollary 2.2, Lemma 2.4, Lemma 2.5 and Lemma 2.6, the following estimates hold:

$$
(n_{tt} - \Delta n + \Gamma\Delta^2 n, \Delta^2 n_t) = \frac{1}{2}\frac{\mathrm{d}}{\mathrm{d}t}\left(\|\Delta n_t\|^2_{L^2(\mathbb{R}^2)} + \|\nabla^3 n\|^2_{L^2(\mathbb{R}^2)} + \Gamma\|\Delta^2 n\|^2_{L^2(\mathbb{R}^2)}\right),
$$

$$
\begin{aligned}
(\Delta|E|^2, \Delta^2 n_t) &= (\Delta^2|E|^2, \Delta n_t) \\
&= (\Delta^2 E\overline{E} + E\Delta^2\overline{E} + 4\nabla^3 E\nabla\overline{E} + 4\nabla E\nabla^3\overline{E} \\
&\quad +2\Delta E\Delta\overline{E} + 4\nabla^2 E\nabla^2\overline{E}, \Delta n_t) \\
&\leqslant (\|E\|_{L^\infty(\mathbb{R}^2)}\|\Delta^2 E\|_{L^2(\mathbb{R}^2)} + \|\nabla E\|_{L^\infty(\mathbb{R}^2)}\|\nabla^3 E\|_{L^2(\mathbb{R}^2)} \\
&\quad + \|\Delta E\|_{L^\infty(\mathbb{R}^2)}\|\Delta E\|_{L^2(\mathbb{R}^2)} + \|\nabla^2 E\|^2_{L^4(\mathbb{R}^2)}\Big)\|\Delta n_t\|_{L^2(\mathbb{R}^2)} \\
&\leqslant (C + \|\nabla^2 E\|_{L^2(\mathbb{R}^2)}\|\nabla^3 E\|_{L^2(\mathbb{R}^2)})\,\|\Delta n_t\|_{L^2(\mathbb{R}^2)} \\
&\leqslant C(1 + \|\Delta n_t\|^2_{L^2(\mathbb{R}^2)}).
\end{aligned}
$$

We thus obtain

$$
\frac{\mathrm{d}}{\mathrm{d}t}\left(\|\Delta n_t\|^2_{L^2(\mathbb{R}^2)} + \|\nabla^3 n\|^2_{L^2(\mathbb{R}^2)} + \Gamma\|\Delta^2 n\|^2_{L^2(\mathbb{R}^2)}\right) \leqslant C(1 + \|\Delta n_t\|^2_{L^2(\mathbb{R}^2)}). \tag{2.39}
$$

Let

$$
J(t) = \|\Delta E_t\|^2_{L^2(\mathbb{R}^2)} + \|\Delta n_t\|^2_{L^2(\mathbb{R}^2)} + \|\nabla^3 n\|^2_{L^2(\mathbb{R}^2)} + \Gamma\|\Delta^2 n\|^2_{L^2(\mathbb{R}^2)}. \tag{2.40}
$$

Combining (2.37) and (2.39) with (2.40), one has

$$
J(t) \leqslant C + C\int_0^t J(\tau)\mathrm{d}\tau, \ t \in [0, T],
$$

which together with Gronwall's inequality and Lemma 2.6 yields

$$
\|\Delta E_t\|^2_{L^2(\mathbb{R}^2)} + \|\Delta n_t\|^2_{L^2(\mathbb{R}^2)} + \|\nabla^3 n\|^2_{L^2(\mathbb{R}^2)} + \|\Delta^2 n\|^2_{L^2(\mathbb{R}^2)} \leqslant C(T). \tag{2.41}
$$

On the other hand, from (1.1) it follows that

$$
i\nabla E_t + \nabla^3 E = \nabla(nE) + \Gamma\nabla^5 E.
$$

Furthermore, combining Lemma 2.5 with Lemma 2.6, one easily verifies that

$$\begin{aligned}\|\nabla^5 E\|_{L^2(\mathbb{R}^2)} \leqslant \|\nabla E_t\|_{L^2(\mathbb{R}^2)} + \|\nabla^3 E\|_{L^2(\mathbb{R}^2)} \\ +\|\nabla n E\|_{L^2(\mathbb{R}^2)} + \|n\nabla E\|_{L^2(\mathbb{R}^2)} \leqslant C.\end{aligned} \tag{2.42}$$

Hence (2.35) is valid by (2.41) and (2.42). □

By Lemma 2.2, Lemma 2.7 implies the following estimates:

Corollary 2.3

$$\begin{aligned}\|\nabla^3 E\|_{L^\infty(\mathbb{R}^2)} + \|\Delta n\|_{L^\infty(\mathbb{R}^2)} + \|E_t\|_{L^\infty(\mathbb{R}^2)} + \|n_t\|_{L^\infty(\mathbb{R}^2)} \leqslant E_7, \\ \|n_{tt}\|_{L^2(\mathbb{R}^2)} \leqslant E_8,\end{aligned} \tag{2.43}$$

where E_7 and E_8 depend on T, $\|E_0\|^2_{H^5(\mathbb{R}^2)}$, $\|n_0\|^2_{H^4(\mathbb{R}^2)}$ and $\|\mathbf{V}_0\|^2_{H^3(\mathbb{R}^2)}$.

Lemma 2.8 *Under the conditions of Lemma 2.4, let*

$$E_0(x) \in H^{m+3}(\mathbb{R}^2), \qquad n_0(x) \in H^{m+2}(\mathbb{R}^2), \qquad \mathbf{V}_0(x) \in H^{m+1}(\mathbb{R}^2).$$

Then the solution of the Cauchy problem (1.1)–(1.4) satisfies

$$\begin{aligned}\|\nabla^m E_t\|^2_{L_t^\infty L_x^2} + \|\nabla^m n_t\|^2_{L_t^\infty L_x^2} + \|\nabla^{m+3} E\|^2_{L_t^\infty L_x^2} \\ + \|\nabla^{m+2} n\|^2_{L_t^\infty L_x^2} + \|\nabla^{m+1}\mathbf{V}\|^2_{L_t^\infty L_x^2} \leqslant E_8,\end{aligned} \tag{2.44}$$

where the constant E_9 depends on T, $\|E_0\|^2_{H^{m+3}(\mathbb{R}^2)}$, $\|n_0\|^2_{H^{m+2}(\mathbb{R}^2)}$ and $\|\mathbf{V}_0\|^2_{H^{m+1}(\mathbb{R}^2)}$.

Proof We shall show this lemma by induction on m.

For $m = 0, 1$, (2.44) follows from Lemma 2.5 and Lemma 2.6.

For $m = 2$, (2.44) is also true by Lemma 2.7.

Now, we suppose that the estimate (2.44) is true for $m = k \geqslant 2$, that is,

$$\begin{aligned}\|\nabla^k E_t\|^2_{L_t^\infty L_x^2} + \|\nabla^k n_t\|^2_{L_t^\infty L_x^2} + \|\nabla^{k+3} E\|^2_{L_t^\infty L_x^2} \\ +\|\nabla^{k+2} n\|^2_{L_t^\infty L_x^2} + \|\nabla^{k+1}\mathbf{V}\|^2_{L_t^\infty L_x^2} \leqslant E_9.\end{aligned} \tag{2.45}$$

In the following, we show the estimate (2.44) is also true for $m = k+1$.

Differentiating (1.1) with respect to t, and then taking the inner product of the resulting equation with $\Delta^{k+1}E_t$, we have

$$(iE_{tt} + \Delta E_t - (nE)_t - \Gamma\Delta^2 E_t, \Delta^{k+1}E_t) = 0. \tag{2.46}$$

By a direct calculation, one has

$$\begin{aligned}&Im(iE_{tt}, \Delta^{k+1}E_t) = \frac{1}{2}(-1)^{k+1}\frac{\mathrm{d}}{\mathrm{d}t}\|\nabla^{k+1}E_t\|^2_{L^2(\mathbb{R}^2)}, \\ &Im(\Delta E_t, \Delta^{k+1}E_t) = Im(-1)^k\|\nabla^{k+2}E_t\|^2_{L^2(\mathbb{R}^2)} = 0, \\ &Im(-\Gamma\Delta^2 E_t, \Delta^{k+1}E_t) = Im(-1)^{k-1}\|\nabla^{k+3}E_t\|^2_{L^2(\mathbb{R}^2)} = 0.\end{aligned}$$

(2.46) immediately thereby yields that

$$
\begin{aligned}
&\frac{1}{2}\frac{\mathrm{d}}{\mathrm{d}t}\|\nabla^{k+1}E_t\|^2_{L^2(\mathbb{R}^2)}\\
&\leqslant \left|((nE)_t, \Delta^{k+1}E_t)\right|\\
&= \left|(n_tE+nE_t, \Delta^{k+1}E_t)\right|\\
&\leqslant \left|(\nabla^{k+1}(n_tE+nE_t), \nabla^{k+1}E_t)\right|\\
&\leqslant C\left(\|\nabla^{k+1}E_t\|^2_{L^2(\mathbb{R}^2)}+\|\nabla^{k+1}(n_tE)\|^2_{L^2(\mathbb{R}^2)}+\|\nabla^{k+1}(nE_t)\|^2_{L^2(\mathbb{R}^2)}\right)\\
&\leqslant C\|\nabla^{k+1}E_t\|^2_{L^2(\mathbb{R}^2)}+C\|E\|_{L^\infty(\mathbb{R}^2)}\|\nabla^{k+1}n_t\|^2_{L^2(\mathbb{R}^2)}\\
&\quad +C\|\nabla^{k+1}E\|^2_{L^4(\mathbb{R}^2)}\|n_t\|^2_{L^4(\mathbb{R}^2)}\\
&\quad +C\|n\|_{L^\infty(\mathbb{R}^2)}\|\nabla^{k+1}E_t\|^2_{L^2(\mathbb{R}^2)}+C\|\nabla^{k+1}n\|^2_{L^4(\mathbb{R}^2)}\|E_t\|^2_{L^4(\mathbb{R}^2)}\\
&\quad +C\sum_{i+j=k+1}\|\nabla^jE\|_{L^\infty(\mathbb{R}^2)}\|\nabla^in_t\|^2_{L^4(\mathbb{R}^2)}\\
&\quad +C\sum_{i+j=k+1}\|\nabla^in\|_{L^\infty(\mathbb{R}^2)}\|\nabla^jE_t\|^2_{L^4(\mathbb{R}^2)},
\end{aligned}\tag{2.47}
$$

in which $i, j\in\mathbb{Z}^+$.

Whereas, (2.45) and Lemma 2.2 conclude that for $i\leqslant k+1$, $j\leqslant k$,

$$\|\nabla^iE\|_{L^\infty(\mathbb{R}^2)}+\|\nabla^jn\|_{L^\infty(\mathbb{R}^2)}\leqslant E_{10}.\tag{2.48}$$

Combining (2.45) with (2.47) and (2.48), noting that Lemma 2.4-Lemma 2.7, Corollary 2.1-Corollary 2.3, one gets

$$\frac{\mathrm{d}}{\mathrm{d}t}\|\nabla^{k+1}E_t\|^2_{L^2(\mathbb{R}^2)}\leqslant C(1+\|\nabla^{k+1}E_t\|^2_{L^2(\mathbb{R}^2)}),\tag{2.49}$$

which together with Gronwall's inequality yields

$$\|\nabla^{k+1}E_t\|^2_{L^2(\mathbb{R}^2)}\leqslant C.\tag{2.50}$$

On the other hand, we take inner product of (2.21) with $\Delta^{k+1}n_t$, and then obtain

$$(n_{tt}-\Delta n-\Delta|E|^2+\Gamma\Delta^2n, \Delta^{k+1}n_t)=0,\tag{2.51}$$

which manifests

$$
\begin{aligned}
&\frac{\mathrm{d}}{\mathrm{d}t}\left[\|\nabla^{k+1}n_t\|^2_{L^2(\mathbb{R}^2)}+\|\nabla^{k+2}n\|^2_{L^2(\mathbb{R}^2)}+\|\nabla^{k+3}n\|^2_{L^2(\mathbb{R}^2)}\right]\\
&=(\nabla^{k+3}|E|^2, \nabla^{k+1}n_t)
\end{aligned}
$$

$$
\begin{aligned}
&\leqslant \|\nabla^{k+3}|E|^2\|_{L^2(\mathbb{R}^2)}\|\nabla^{k+1}n_t\|_{L^2(\mathbb{R}^2)}\\
&\leqslant C\left(\|\nabla^{k+3}|E|^2\|^2_{L^2(\mathbb{R}^2)}+\|\nabla^{k+1}n_t\|^2_{L^2(\mathbb{R}^2)}\right)\\
&\leqslant C\left(\|\nabla^{k+3}E\|^2_{L^4(\mathbb{R}^2)}\|E\|^2_{L^4(\mathbb{R}^2)}\right.\\
&\quad\left.+\sum_{i,j\in\mathbb{Z}^+i+j=k+3}\|\nabla^i E\|^2_{L^4(\mathbb{R}^2)}\|\nabla^j E\|^2_{L^4(\mathbb{R}^2)}+\|\nabla^{k+1}n_t\|^2_{L^2(\mathbb{R}^2)}\right)\\
&\leqslant C\left(1+\|\nabla^{k+4}E\|^2_{L^2(\mathbb{R}^2)}+\|\nabla^{k+1}n_t\|^2_{L^2(\mathbb{R}^2)}\right).
\end{aligned}\tag{2.52}
$$

Similarly, on account of (1.1), we can show

$$
\begin{aligned}
\|\nabla^{k+1}E\|^2_{L^2(\mathbb{R}^2)}\leqslant C&\left(\|\nabla^k E_t\|^2_{L^2(\mathbb{R}^2)}+\|\nabla^{k+2}E\|^2_{L^2(\mathbb{R}^2)}\right.\\
&+\|E\|^2_{L^\infty(\mathbb{R}^2)}\|\nabla^k n\|^2_{L^2(\mathbb{R}^2)}+\|n\|_{L^\infty(\mathbb{R}^2)}\|\nabla^k E\|^2_{L^2(\mathbb{R}^2)}\\
&\left.+\sum_{i,j\in\mathbb{Z}^+,i+j=k}\|\nabla n\|^2_{L^2(\mathbb{R}^2)}\|\nabla^j E\|_{L^4(\mathbb{R}^2)}\right),
\end{aligned}\tag{2.53}
$$

which together with (2.45) concludes

$$
\|\nabla^{k+4}E\|^2_{L^2(\mathbb{R}^2)}\leqslant C.\tag{2.54}
$$

Combining (2.52) with (2.54), one easily verifies that

$$
\begin{aligned}
&\frac{\mathrm{d}}{dt}\left[\|\nabla^{k+1}n_t\|^2_{L^2(\mathbb{R}^2)}+\|\nabla^{k+2}n\|^2_{L^2(\mathbb{R}^2)}+\|\nabla^{k+3}n\|^2_{L^2(\mathbb{R}^2)}\right]\\
&\quad\leqslant C\left(1+\|\nabla^{k+1}n_t\|^2_{L^2(\mathbb{R}^2)}\right).
\end{aligned}\tag{2.55}
$$

Let

$$
H(t)=1+\|\nabla^{k+1}n_t\|^2_{L^2(\mathbb{R}^2)}+\|\nabla^{k+2}n\|^2_{L^2(\mathbb{R}^2)}+\|\nabla^{k+3}n\|^2_{L^2(\mathbb{R}^2)}.\tag{2.56}
$$

Gronwall's inequality manifests that

$$
H(t)\leqslant C(T),\ t\in[0,T].\tag{2.57}
$$

Combining (2.50), (2.54), (2.56) with (2.57), we thus obtain the following estimate:

$$
\begin{aligned}
&\|\nabla^{k+1}E_t\|^2_{L^2(\mathbb{R}^2)}+\|\nabla^{k+1}n_t\|^2_{L^2(\mathbb{R}^2)}+\|\nabla^{k+4}E\|^2_{L^2(\mathbb{R}^2)}\\
&\quad+\|\nabla^{k+3}n\|^2_{L^2(\mathbb{R}^2)}+\|\nabla^{k+1}\mathbf{V}\|^2_{L^2(\mathbb{R}^2)}\leqslant C.
\end{aligned}\tag{2.58}
$$

Hence (2.44) is true for $m=k+1$. This completes the proof of Lemma 2.8. □

3 Existence and uniqueness of global smooth solutions

In this section, we consider the existence and uniqueness of global solution to the Cauchy problem (1.1)–(1.4) by using those estimates obtained in Section 2. Theorem 3.1 below states the global existence of solution for the Cauchy problem (1.1)–(1.4). This existence result is proven by using the classical Galerkin method(see e.g., [21], [22]).

Theorem 3.1 *Under the conditions of Lemma* 2.4, *let*

$$E_0(x) \in H^{m+3}(\mathbb{R}^2), \qquad n_0(x) \in H^{m+2}(\mathbb{R}^2), \qquad \mathbf{V}_0(x) \in H^{m+1}(\mathbb{R}^2), \quad m \geqslant 1.$$

The Cauchy problem (1.1)–(1.4) *then admits a global smooth solution such that for* $T > 0$

$$\begin{aligned}
&E(t,x) \in L^\infty\left(0,T;H^{m+3}(\mathbb{R}^2)\right), && E_t(t,x) \in L^\infty\left(0,T;H^{m}(\mathbb{R}^2)\right),\\
&n(t,x) \in L^\infty\left(0,T;H^{m+2}(\mathbb{R}^2)\right), && n_t(t,x) \in L^\infty\left(0,T;H^{m}(\mathbb{R}^2)\right),\\
&\mathbf{V}(t,x) \in L^\infty\left(0,T;H^{m+1}(\mathbb{R}^2)\right), && \mathbf{V}_t(t,x) \in L^\infty\left(0,T;H^{m-1}(\mathbb{R}^2)\right).
\end{aligned} \tag{3.1}$$

Proof. We divide the proof into four steps.

Step 1 We first study an approximate problem of (1.1)–(1.4) in a bounded domain Ω, where Ω is a two-dimensional cube with length $2R$ in each direction, that is,

$$\overline{\Omega} = \{x = (x_1, x_2) : \ |x_i| \leqslant 2R, \ i = 1, 2\}.$$

Let $\{w_s(x)\}$ denote the eigenfunctions of the following problem

$$-\Delta w_s(x) = \lambda_s w_s(x), \qquad w_s(x) \in H^{m+3}(\Omega), \quad s = 1, 2, 3, \cdots, l,$$

where λ_s is the eigenvalue corresponding to the eigenfunction $w_s(x)$.

For any positive integer l, the approximate solution of problem (1.1)–(1.4) can be written as

$$\begin{aligned}
E_l(t,x) &= \sum_{s=1}^{l} \alpha_{sl}(t) w_s(x),\\
n_l(t,x) &= \sum_{s=1}^{l} \beta_{sl}(t) w_s(x),\\
\mathbf{V}_l(t,x) &= \sum_{s=1}^{l} \gamma_{sl}(t) w_s(x),
\end{aligned} \tag{3.2}$$

where

$$\begin{aligned}&\mathbf{V}_l(t,x)=(V_{l1}(t,x),V_{l2}(t,x)),\\&\gamma_{sl}(t,x)=(\gamma_{sl1}(t,x),\gamma_{sl2}(t,x)),\\&V_{lk}(t,x)=\sum_{s=1}^{l}\gamma_{slk}(t)w_s(x),\ k=1,2.\end{aligned}\tag{3.3}$$

The above coefficients α_{sl}, β_{sl} and γ_{sl} need to satisfy the following initial value problem for $s=1,\cdots,l$:

$$i(E_{lt},w_s)+(\Delta E_l,w_s)-(n_lE_l,w_s)-\Gamma(\Delta^2E_l,w_s)=0,\tag{3.4}$$

$$(n_{lt}+\nabla\cdot\mathbf{V}_l,w_s)=0,\tag{3.5}$$

$$(\mathbf{V}_{lt},w_s)+(\nabla n_l,w_s)+(\nabla|E_l|^2,w_s)-\Gamma(\nabla\Delta n_l,w_s)=0,\tag{3.6}$$

$$E_l(0,x)=E_l^0(x),\quad n_l(0,x)=n_l^0(x),\quad \mathbf{V}_l(0,x)=\mathbf{V}_l^0(x),\tag{3.7}$$

where $\mathbf{V}_l(0,x)=(V_{l1}(0,x),V_{l2}(0,x))=(V_{l1}^0(x),V_{l2}^0(x))=\mathbf{V}_l^0(x)$. Suppose that as $l\to\infty$,

$$\begin{aligned}&(E_l^0(x),n_l^0(x),\mathbf{V}_l^0(x))\to(E_0(x),n_0(x),\mathbf{V}_0(x))\\&\text{in } H^{m+3}(\Omega)\times H^{m+2}(\Omega)\times H^{m+1}(\Omega).\end{aligned}\tag{3.8}$$

Indeed, (3.4)–(3.6) is a nonlinear ordinary differential system for the unknown coefficients α_{sl}, β_{sl} and γ_{sl}, and by the classical theory of ODE, we know the initial value problem (3.4)–(3.7) admits a unique local solution $(E_l,n_l,\mathbf{V}_l)$ for $t\in[0,\delta]$. Since the functions $w_s(x)$, $s=1,2,\cdots$ are smooth, the solution $(E_l,n_l,\mathbf{V}_l)$ is also a smooth solution.

Step 2 In order to continue the approximate argument, we have to prove the local existence time of the approximate solution $(E_l,n_l,\mathbf{V}_l)$ is uniform with respect to l. That is, one has to show that δ is independent of l. To this aim, we should estimate the H^{m+3} norm for E_l, the H^{m+2} norm for n_l and the H^{m+1} norm for $\mathbf{V}_l$ from (3.4)–(3.6). Indeed, we can prove that there exist $T_0>0$ and two constants $E_{11},E_{12}>0$ such that

$$\sup_{0\leqslant t\leqslant T_0}\left[\|E_l(t,x)\|_{H^{m+3}(\Omega)}+\|n_l(t,x)\|_{H^{m+2}(\Omega)}+\|\mathbf{V}_l(t,x)\|_{H^{m+1}(\Omega)}\right]\leqslant E_{11},\tag{3.9}$$

$$\sup_{0\leqslant t\leqslant T_0}\left[\|E_{lt}(t,x)\|_{H^{m-1}(\Omega)}+\|n_{lt}(t,x)\|_{H^{m}(\Omega)}+\|\mathbf{V}_{lt}(t,x)\|_{H^{m-1}(\Omega)}\right]\leqslant E_{12}.\tag{3.10}$$

We remark that in (3.9) and (3.10), T_0 depends only on $\|E_0(x)\|_{H^{m+3}(\Omega)}$, $\|n_0(x)\|_{H^{m+2}(\Omega)}$ and $\|\mathbf{V}_0(x)\|_{H^{m+1}(\Omega)}$, E_{11} and E_{12} depend on T_0, $\|E_0(x)\|_{H^{m+3}(\Omega)}$, $\|n_0(x)\|_{H^{m+2}(\Omega)}$ and $\|\mathbf{V}_0(x)\|_{H^{m+1}(\Omega)}$. In particular, T_0, E_{11} and E_{12} are independent of l and R.

Now, we prove (3.9) and (3.10). First, from (3.4), it is easy to see

$$\|E_l\|_{L^2(\Omega)} = \|E_l^0\|_{L^2(\Omega)},$$

and

$$\begin{aligned}\frac{\mathrm{d}}{\mathrm{d}t}\|E_{lt}\|^2_{L^2(\Omega)} &= 2Im\int_\Omega n_{lt}E_l\overline{E_{lt}}\mathrm{d}x\\ &\leqslant C\|n_{lt}\|_{L^2(\Omega)}\|E_l\|_{L^\infty(\Omega)}\|E_{lt}\|_{L^2(\Omega)}\\ &\leqslant C\|n_{lt}\|_{H^m(\Omega)}\|E_l\|_{H^{m+3}(\Omega)}\|E_{lt}\|_{H^{m-1}(\Omega)}.\end{aligned}$$

From (3.5)–(3.6), we have

$$(n_{ltt} - \Delta n_l + \Gamma\Delta^2 n_l, w_s) = (\Delta|E_l|^2, w_s),\ \ s = 1,\cdots,l,$$

which gives

$$\begin{aligned}\frac{\mathrm{d}}{\mathrm{d}t}(\|n_{lt}\|^2_{L^2(\Omega)} + \|\nabla n_l\|^2_{L^2(\Omega)} + \Gamma\|\Delta n_l\|^2_{L^2(\Omega)}) &= 2\int_\Omega \Delta|E_l|^2\cdot n_{lt}\mathrm{d}x\\ &\leqslant C\|E_l\|^2_{H^{m+3}(\Omega)}\|n_{lt}\|_{L^2(\Omega)}.\end{aligned}$$

Next, we give the high order norm estimate for the approximate solution. (3.5) and (3.6) imply that

$$(n_{ltt} - \Delta n_l + \Gamma\Delta^2 n_l, w_s) = (\Delta|E_l|^2, w_s),\ \ s = 1,\cdots,l,$$

which, by multiplying with $\beta'_{sl}(t)\lambda_s^m$ on both sides of the above equality and summing from 1 to l, gives

$$\left(n_{ltt} - \Delta n_l + \Gamma\Delta^2 n_l, \sum_{s=1}^{l}\lambda_s^m\beta'_{sl}w_s\right) = \left(\Delta|E_l|^2, \sum_{s=1}^{l}\lambda_s^m\beta'_{sl}w_s\right),$$

namely,

$$(n_{ltt} - \Delta n_l + \Gamma\Delta^2 n_l, (-\Delta)^m n_{lt}) = (\Delta|E_l|^2, (-\Delta)^m n_{lt}).$$

Hence, we can obtain

$$\frac{1}{2}\frac{\mathrm{d}}{\mathrm{d}t}(\|\nabla^m n_{lt}\|^2_{L^2(\Omega)} + \|\nabla^{m+1}n_l\|^2_{L^2(\Omega)} + \Gamma\|\nabla^{m+2}n_l\|^2_{L^2(\Omega)}) = \int_\Omega \nabla^m\Delta|E_l|^2\nabla^m n_{lt}\mathrm{d}x,$$

and by Cauchy-Schwarz inequality and Sobolev inequality, we have

$$\frac{\mathrm{d}}{\mathrm{d}t}(\|\nabla^m n_{lt}\|^2_{L^2(\Omega)} + \|\nabla^{m+1} n_l\|^2_{L^2(\Omega)} + \Gamma\|\nabla^{m+2} n_l\|^2_{L^2(\Omega)}) \leqslant C\|E_l\|^2_{H^{m+3}(\Omega)}\|n_{lt}\|_{H^m(\Omega)}.$$

Now, we turn to the equation for E_l. Differentiating (3.4) with respect to t gives

$$i(E_{ltt}, w_s) + (\Delta E_{lt}, w_s) - \Gamma(\Delta^2 E_{lt}, w_s) - (n_{lt}E_l + n_l E_{lt}, w_s) = 0.$$

We multiply this equation by $\alpha'_{sl}(t)\lambda_s^{m-1}$ and take the sum for s from 1 to l, then we obtain

$$\begin{aligned}&i(E_{ltt}, (-\Delta)^{m-1}E_{lt}) + (\Delta E_{lt} - \Gamma\Delta^2 E_{lt}, (-\Delta)^{m-1}E_{lt})\\ &-(n_{lt}E_l + n_l E_{lt}, (-\Delta)^{m-1}E_{lt}) = 0.\end{aligned}$$

By taking the imaginary part of the above equality, we have

$$\frac{1}{2}\frac{\mathrm{d}}{\mathrm{d}t}\|\nabla^{m-1}E_{lt}\|^2_{L^2(\Omega)} = Im\int_\Omega \nabla^{m-1}(n_{lt}E_l + n_l E_{lt})\cdot\nabla^{m-1}\overline{E_{lt}}\mathrm{d}x,$$

which gives

$$\begin{aligned}\frac{\mathrm{d}}{\mathrm{d}t}\|\nabla^{m-1}E_{lt}\|^2_{L^2(\Omega)} \leqslant& C(\|\nabla^{m-1}n_{lt}\|_{L^2(\Omega)}\|E_l\|_{L^\infty(\Omega)}\\ &+\|n_{lt}\|_{L^4(\Omega)}\|\nabla^{m-1}E_l\|_{L^4(\Omega)})\|\nabla^{m-1}E_{lt}\|_{L^2(\Omega)}\\ &+C(\|\nabla^{m-1}n_l\|_{L^4(\Omega)}\|E_{lt}\|_{L^4(\Omega)}\\ &+\|n_l\|_{L^\infty(\Omega)}\|\nabla^{m-1}E_{lt}\|_{L^2(\Omega)})\|\nabla^{m-1}E_{lt}\|_{L^2(\Omega)}\\ \leqslant& C(\|n_{lt}\|_{H^m(\Omega)}\|E_l\|_{H^{m+3}(\Omega)}\|E_{lt}\|_{H^{m-1}(\Omega)}\\ &+\|n_l\|_{H^{m+2}(\Omega)}\|E_{lt}\|^2_{H^{m-1}(\Omega)}).\end{aligned}$$

Combining the above estimates, we obtain

$$\begin{aligned}\frac{\mathrm{d}}{\mathrm{d}t}\psi(t) \leqslant& C\|E_l\|^2_{H^{m+3}(\Omega)}\|n_{lt}\|_{L^2(\Omega)}\\ &+C\|n_{lt}\|_{H^m(\Omega)}\|E_l\|_{H^{m+3}(\Omega)}\|E_{lt}\|_{H^{m-1}(\Omega)}\\ &+C\|n_l\|_{H^{m+2}(\Omega)}\|E_{lt}\|^2_{H^{m-1}(\Omega)},\end{aligned}$$

where

$$\begin{aligned}\psi(t) :=& \|E_{lt}\|^2_{L^2(\Omega)} + \|n_{lt}\|^2_{L^2(\Omega)} + \|\nabla n_l\|^2_{L^2(\Omega)} + \Gamma\|\Delta n_l\|^2_{L^2(\Omega)}\\ &+\|\nabla^m n_{lt}\|^2_{L^2(\Omega)} + \|\nabla^{m+1}n_l\|^2_{L^2(\Omega)}\\ &+\Gamma\|\nabla^{m+2}n_l\|^2_{L^2(\Omega)} + \|\nabla^{m-1}E_{lt}\|^2_{L^2(\Omega)}.\end{aligned}$$

Using the inequality $2|ab| \leqslant a^2 + b^2$, we have

$$\|E_l\|^2_{H^{m+3}(\Omega)}\|n_{lt}\|_{H^m(\Omega)} \leqslant \|E_l\|^2_{H^{m+3}(\Omega)}(\|n_{lt}\|^2_{H^m(\Omega)} + 1),$$

$$\|n_l\|_{H^{m+2}(\Omega)}\|E_{lt}\|^2_{H^{m-1}(\Omega)} \leqslant (\|n_l\|^2_{H^{m+2}(\Omega)}+1)\|E_{lt}\|^2_{H^{m-1}(\Omega)},$$

and

$$\begin{aligned}&\|n_{lt}\|_{H^m(\Omega)}\|E_l\|_{H^{m+3}(\Omega)}\|E_{lt}\|_{H^{m-1}(\Omega)}\\ &\leqslant (\|n_{lt}\|^2_{H^m(\Omega)}+\|E_l\|^2_{H^{m+3}(\Omega)})(\|E_{lt}\|^2_{H^{m-1}(\Omega)}+1).\end{aligned}$$

Thus, we obtain

$$\frac{\mathrm{d}}{\mathrm{d}t}\psi(t) \leqslant C(\|E_l\|^2_{H^{m+3}(\Omega)}+\|n_l\|^2_{H^{m+2}(\Omega)}+\|E_{lt}\|^2_{H^{m-1}(\Omega)}+\|n_{lt}\|^2_{H^m(\Omega)}+1)^2.$$

Integrating this inequality with respect to t yields

$$\psi(t) \leqslant C+C\int_0^t (\|E_l\|^2_{H^{m+3}(\Omega)}+\|n_l\|^2_{H^{m+2}(\Omega)}+\|E_{lt}\|^2_{H^{m-1}(\Omega)}+\|n_{lt}\|^2_{H^m(\Omega)}+1)^2(s)\mathrm{d}s.$$

Note that

$$\psi(t) \sim \|E_{lt}\|^2_{H^{m-1}(\Omega)}+\|n_l\|^2_{H^{m+2}(\Omega)}+\|n_{lt}\|^2_{H^m(\Omega)},$$

moreover, from (3.4), we know

$$\|E_l\|_{H^{m+3}(\Omega)} \leqslant C(\|E_{lt}\|_{H^{m-1}(\Omega)}+\|n_l\|_{H^{m+2}(\Omega)}\|E_l\|_{H^{m+3}(\Omega)}).$$

Using these two facts, we see that

$$\widetilde{\psi}(t) \leqslant C+C\int_0^t [\widetilde{\psi}(s)]^3\mathrm{d}s,$$

with

$$\widetilde{\psi}(t) := \|E_{lt}\|^2_{H^{m-1}(\Omega)}+\|n_l\|^2_{H^{m+2}(\Omega)}+\|n_{lt}\|^2_{H^m(\Omega)}+1,$$

where the constant C is independent of l and R. Hence, we have

$$\widetilde{\psi}(t) \leqslant \left(\frac{\widetilde{\psi}^2(0)}{1-2Ct\widetilde{\psi}^2(0)}\right)^{\frac{1}{2}}.$$

Since (3.8) holds, there exists $T_0>0$, depending only on the norm of the initial data, such that $\widetilde{\psi}(t) \leqslant C(T_0)$ for all $t \in [0, T_0]$. From this estimate, the desired estimates (3.9) and (3.10) follow easily.

Step 3 In this step, we show that the Cauchy problem of system (1.1)–(1.4) admits a local smooth solution by approximate argument. In fact, based on the uniform estimates (3.9) and (3.10), by utilizing the Galerkin method, and making the similar procedures to that in [22], we can see that the periodic boundary value problem of the system (1.1)–(1.4) has a local smooth solution

$$
\begin{aligned}
&E(t,x)\in L^\infty\left(0,T_0;H^{m+3}(\Omega)\right), \quad E_t(t,x)\in L^\infty\left(0,T_0;H^{m}(\Omega)\right),\\
&n(t,x)\in L^\infty\left(0,T_0;H^{m+2}(\Omega)\right), \quad n_t(t,x)\in L^\infty\left(0,T_0;H^{m}(\Omega)\right),\\
&\mathbf{V}(t,x)\in L^\infty\left(0,T_0;H^{m+1}(\Omega)\right), \quad \mathbf{V}_t(t,x)\in L^\infty\left(0,T_0;H^{m-1}(\Omega)\right).
\end{aligned}
$$

Therefore, by the uniform boundedness of the initial data in the relative Hilbert spaces, and the independence of the estimates for the approximate solution with respect to R, letting $R\to\infty$, we get the existence of local smooth solution for the Cauchy problem (1.1)–(1.4).

Moreover, for this local smooth solution, we can make the similar argument to that in Step 2, and know that if T^* is the maximal existence time of the solution and $T^*<+\infty$, then there must hold

$$
\lim_{t\to T^*}\left(\|E(t,x)\|_{H^{m+3}(\Omega)}+\|n(t,x)\|_{H^{m+2}(\Omega)}+\|\mathbf{V}(t,x)\|_{H^{m+1}(\Omega)}\right)=+\infty.
$$

Step 4 Finally, we extend the above local solution to be a global one. Indeed, applying the above blow-up criterion, and using the priori estimates obtained in Lemma 2.1-Lemma 2.8 as well as continuity argument, then we can obtain the global existence of smooth solution to the Cauchy problem (1.1)–(1.4).This completes the proof of Theorem 3.1. □

Next, we prove the uniqueness of the smooth solution for (1.1)–(1.4).

Theorem 3.2 *The global smooth solution of the Cauchy problem* (1.1)–(1.4) *is unique.*

Proof. Suppose that $\{E_i(t,x),n_i(t,x),\mathbf{V}_i(t,x)\}$ $(i=1,2)$ is the global solution to the Cauchy problem (1.1)–(1.4). Set

$$
\begin{aligned}
\psi(t,x)&=E_1(t,x)-E_2(t,x),\\
\varphi(t,x)&=n_1(t,x)-n_2(t,x),\\
\mathbf{\Phi}(t,x)&=\mathbf{V}_1(t,x)-\mathbf{V}_2(t,x).
\end{aligned}
\tag{3.11}
$$

$(\psi,\varphi,\mathbf{\Phi})$ then solves the following Cauchy problem according to (1.1)–(1.4):

$$
i\psi_t+\Delta\psi=n_1E_1-n_2E_2+\Gamma\Delta^2\psi, \tag{3.12}
$$

$$
\varphi_t=-\nabla\cdot\mathbf{\Phi}, \tag{3.13}
$$

$$
\mathbf{\Phi}_t=-\nabla\varphi-(\nabla|E_1|^2-\nabla|E_2|^2)+\Gamma\nabla\Delta\varphi, \tag{3.14}
$$

$$\psi(0,x)=0,\quad \varphi(0,x)=0,\quad \mathbf{\Phi}(0,x)=0. \tag{3.15}$$

Differentiating (3.12) with respect to t, and then taking the inner product of the resulting equation with ψ_t, we have

$$(i\psi_{tt}+\Delta\psi_t-(n_1E_1-n_2E_2)_t-\Gamma\Delta^2\psi_t,\psi_t)=0. \tag{3.16}$$

Taking the imaginary part for (3.16), one gets

$$\begin{aligned}\frac{\mathrm{d}}{\mathrm{d}t}\|\psi_t\|^2_{L^2(\mathbb{R}^2)}&=Im\,((n_1E_1-n_2E_2)_t,\psi_t)\\&=Im\,((n_1\psi+\varphi E_2)_t,\psi_t)\\&=Im\,((n_{1t}\psi+n_1\psi_t+\varphi_tE_2+\varphi E_{2t},\psi_t).\end{aligned} \tag{3.17}$$

In addition, the following estimates hold immediately through a direct calculation:

$$\begin{aligned}|Im\,(n_1\psi_t,\psi_t)|&\leqslant\left|Im\int_{\mathbb{R}^2}n_1|\psi_t|^2\mathrm{d}x\right|=0,\\|Im\,(n_{1t}\psi,\psi_t)|&=\left|Im\int_{\mathbb{R}^2}n_{1t}\psi\bar{\psi}_t\mathrm{d}x\right|\\&\leqslant\|n_{1t}\|_{L^\infty(\mathbb{R}^2)}\|\psi\|_{L^2(\mathbb{R}^2)}\|\psi_t\|_{L^2(\mathbb{R}^2)}\\&\leqslant C\|\psi\|_{L^2(\mathbb{R}^2)}\|\psi_t\|_{L^2(\mathbb{R}^2)},\\|Im\,(\varphi_tE_2,\psi_t)|&\leqslant\|E_2\|_{L^\infty(\mathbb{R}^2)}\|\varphi_t\|_{L^2(\mathbb{R}^2)}\|\psi_t\|_{L^2(\mathbb{R}^2)}\\&\leqslant C\|\varphi_t\|_{L^2(\mathbb{R}^2)}\|\psi_t\|_{L^2(\mathbb{R}^2)},\\|Im\,(\varphi E_{2t},\psi_t)|&\leqslant C\|E_{2t}\|_{L^\infty(\mathbb{R}^2)}\|\varphi\|_{L^2(\mathbb{R}^2)}\|\psi_t\|_{L^2(\mathbb{R}^2)}\\&\leqslant C\|\varphi\|_{L^2(\mathbb{R}^2)}\|\psi_t\|_{L^2(\mathbb{R}^2)}.\end{aligned} \tag{3.18}$$

(3.17) and (3.18) thereby conclude

$$\frac{\mathrm{d}}{\mathrm{d}t}\|\psi_t\|^2_{L^2(\mathbb{R}^2)}\leqslant C\left(\|\psi\|^2_{L^2(\mathbb{R}^2)}+\|\psi_t\|^2_{L^2(\mathbb{R}^2)}+\|\varphi_t\|^2_{L^2(\mathbb{R}^2)}\right). \tag{3.19}$$

Similarly, (3.12) manifests

$$(i\psi_t+\Delta\psi_t-(n_1E_1-n_2E_2)-\Gamma\Delta^2\psi,\psi)=0. \tag{3.20}$$

Taking the imaginary part of (3.20), it is easy to obtain

$$\frac{\mathrm{d}}{\mathrm{d}t}\|\psi\|^2_{L^2(\mathbb{R}^2)}\leqslant C\left(\|\psi\|^2_{L^2(\mathbb{R}^2)}+\|\varphi\|^2_{L^2(\mathbb{R}^2)}\right). \tag{3.21}$$

On the other hand, (3.13) and (3.14) yield

$$(\varphi_{tt} - \Delta\varphi - (\Delta|E_1|^2 - \Delta|E_2|^2) + \Gamma\Delta^2\varphi, \varphi_t) = 0 \tag{3.22}$$

which implies

$$\begin{aligned}&\frac{\mathrm{d}}{\mathrm{d}t}\left(\|\varphi_t\|^2_{L^2(\mathbb{R}^2)} + \|\nabla\varphi\|^2_{L^2(\mathbb{R}^2)} + \Gamma\|\Delta\varphi\|^2_{L^2(\mathbb{R}^2)}\right)\\ &\leqslant C\left(\|\varphi\|^2_{L^2(\mathbb{R}^2)} + \|\psi\|^2_{L^2(\mathbb{R}^2)} + \|\varphi_t\|^2_{L^2(\mathbb{R}^2)}\right).\end{aligned} \tag{3.23}$$

It is easy to check

$$\begin{aligned}\frac{\mathrm{d}}{\mathrm{d}t}\|\varphi\|^2_{L^2(\mathbb{R}^2)} &\leqslant 2\|\varphi\|_{L^2(\mathbb{R}^2)}\|\varphi_t\|_{L^2(\mathbb{R}^2)}\\ &\leqslant \|\varphi\|^2_{L^2(\mathbb{R}^2)} + \|\varphi_t\|^2_{L^2(\mathbb{R}^2)}.\end{aligned} \tag{3.24}$$

Thus, (3.19), (3.21), (3.23) and (3.24) conclude

$$\begin{aligned}&\frac{\mathrm{d}}{\mathrm{d}t}\left(\|\psi_t\|^2_{L^2(\mathbb{R}^2)} + \|\psi\|^2_{L^2(\mathbb{R}^2)} + \|\varphi_t\|^2_{L^2(\mathbb{R}^2)}\right.\\ &\left.+\|\nabla\varphi\|^2_{L^2(\mathbb{R}^2)} + \|\Delta\varphi\|^2_{L^2(\mathbb{R}^2)} + \|\varphi\|^2_{L^2(\mathbb{R}^2)}\right)\\ &\leqslant C\left(\|\psi_t\|^2_{L^2(\mathbb{R}^2)} + \|\psi\|^2_{L^2(\mathbb{R}^2)} + \|\varphi_t\|^2_{L^2(\mathbb{R}^2)}\right.\\ &\left.+\|\nabla\varphi\|^2_{L^2(\mathbb{R}^2)} + \|\Delta\varphi\|^2_{L^2(\mathbb{R}^2)} + \|\varphi\|^2_{L^2(\mathbb{R}^2)}\right).\end{aligned} \tag{3.25}$$

Gronwall's inequality together with the zero initial data conditions (3.15) implies that

$$\psi \equiv \varphi \equiv \varphi_t \equiv \psi_t \equiv \nabla\varphi \equiv 0. \tag{3.26}$$

It is also easy to check that $\mathbf{\Phi} \equiv 0$ by $\nabla \cdot \mathbf{\Phi} = 0$, $\mathbf{\Phi}_t = 0$ and $\mathbf{\Phi}(0, x) = 0$. Thus, we obtain the uniqueness of global smooth solution to the Cauchy problem (1.1)–(1.4).□

Acknowledgements The research is supported by National Natural Science Foundation of P.R.China(11171241) and Program for New Century Excellent Talents in University(NCET-12-1058). The authors would like to thank the anonymous referee for many valuable inputs and suggestions.

References

[1] Added H, Added S. *Existence globale de solutions fortes pour les equations de la turbulence de Langmuir en dimension 2*, C. R. Acad. Sci. Paris Ser. I Math. 1984, **200**: 551–554.

[2] Bejenaru I, Herr S, Holmer J and Tataru D. *On the 2D Zakharov system with L^2-Schrödinger data*, Nonlinearity, 2009, **22** (5): 1063–1089.

[3] Berestycki H, Lions P L. *Existence of stationary states in nonlinear scalar field equations*, Arch. Rat. Mech. Anal. 1983, **82**: 313–375.

[4] Bourgain J, Colliander J. *On wellposedness of the Zakharov system*, Internat. Math. Res. Notices, 1996, **11**: 515–546.

[5] Brezis H, Gallonet T. *Nonlinear Schrödinger evolution equation*, Nonlinear Analysis, TMA, 1981, **4(4)**: 677–681.

[6] Degtyarev L M, Kopa-Ovdienko A L. *Scalar model of Langmuir collapse*, Sov. J. Plasma Phys. 1984, **10**: 3–11.

[7] Fibich G, Papanicolaou G C. *Self-focusing in the perturbed and unperturbed nonlinear Schrödinger equation in critical dimension*, SIAM J. Appl. Math. 1999, **60**: 183–240.

[8] Malkin V. *On the analytical theory for stationary self-focusing of radiation*, Physica D, 1993, **64**: 251–266.

[9] Gan Z, Guo B, Han L and Zhang J. *Virial type blow-up solutions for the Zakharov system with magnetic field in a cold plasma*, J. Funct. Anal. 2011, **261**: 2508–2518.

[10] Gan Z, Zhang J. *Blow-up, global existence and standing waves for the magnetic nonlinear Schrodinger equations*, Discrete and Continuous Dynamical Systems-Series A, 2012, **32(3)**: 827–846.

[11] Gan Z, Zhang J. *Nonlocal nonlinear Schrödinger equations in $\mathbb{R}^3$*, Arch. Rational Mech. Anal. 2013 **209**: 1-39.

[12] Garcia L G, Haas F, de Oliveira L P L and Goedert J. *Modified Zakharov equations for plasmas with a quantum correction*, Phys. Plasma, 2005, **12**: 012302.

[13] Ginibre J, Tsutsumi Y, Velo G. *On the Cauchy problem for the Zakharov system*, J. Funct. Anal. 1997, **151(2)**: 384–436.

[14] Glangetas L, Merle F. *Concentration properties of blow-up solutions and instability results for Zakharov equation in dimension two: Part II*, Comm. Math. Phys. 1994, **160**: 349–389.

[15] Haas F, Shukla P K. *Quantum and classical dynamics of Langmuir wave packets*, Phys. Rev, E, 2009, **79**: 066402.

[16] Kenig C E, Ponce G, Vega L. *On the Zakharov and Zakharov-Schulman systems*, J. Funct. Anal. 1995, **127**: 204–234.

[17] Strauss W A. *Existence of solitary waves in higher dimensions*, Commun. Math. Phys. 1977, **55**: 149–162.

[18] Sulem C, Sulem P L. *Quelques résultats de régularité pour les équations de la turbulence de Langmuir*, C.R. Acad. Sci. Paris Ser. 1979, **289** (A-B): 173–176.

[19] Weinstein M I. *Nonlinear Schrödinger equations and sharp interpolation estimates*, Commun. Math. Phys., 1983, **87**: 567–576.

[20] Zakharov V E. *The collapse of Langmuir waves*, Soviet Phys., JETP, 1972, **35**: 908–

914.

[21] Lions J L. Quelques méthodes de résolutions des problèmes aux limites nonlinéaires, Dunod, Paris, 1969.

[22] Zhou Y L, Guo B L. *Periodic boundary problem and initial value problem for the generalized Korteweg-de Vries systems of higher order*, Acta Math. Sinica, 1984, **27**: 154–176.

Ergodicity of the Stochastic Fractional Reaction-Diffusion Equation *

Guo Boling (郭柏灵) and Zhou Guoli (周国立)

Abstract A stochastic reaction-diffusion equation with fractional dissipation is considered. The smaller the exponent of the equation is, the weaker the dissipation of the equation is. The equation is discussed in detail when the exponent changes. The aim is to prove the well-posedness, existence and uniqueness of invariant measure as well as strong law of large numbers and convergence to equilibrium. Without the analytic property of the semigroup, some methods are used to overcome the difficulties to get the energy estimates. The results in this paper can be applied to classic reaction-diffusion equation with Wiener noise.

Keywords Stochastic fractional reaction-diffusion equation; Ergodicity

1 Introduction

As well known, nowadays, it has been universally acknowledged in the physical, chemical and biological communities that the reaction-diffusion equation plays an important role in dissipative dynamical systems. Typical examples are provided by the fact that there are many phenomena in biology where a key element or precursor of a developmental process seems to be the appearance of a traveling wave of chemical concentration (or mechanical deformation). When reaction kinetics and diffusion are coupled, traveling waves of chemical concentration can effect a biochemical change much faster than straight diffusional processes. This usually gives rise to reaction-diffusion equations as follows

$$\frac{\partial u}{\partial t} + \nu \triangle u = f(u), \tag{1.1}$$

where u denotes the chemical concentration, ν is the diffusion coefficient and the function $f(u)$ represents the kinetics. If $f(u)$ is linear, i.e., $f(u) = k_1 u + k_2$, where

* Nonl.Anal.TMA, 2014, 109(1): 1–22. DOI: 10.1016/j.na.2014.06.008.

k_1 and k_2 are real constants, then equation (1.1) can be solved by the separation of variables methods. If $f(u)$ is nonlinear, then the problem is much more intractable. The classic and simplest case of the nonlinear reaction-diffusion equation is the so-called Fisher equation, as the function $f(u) = (k_4u^2 - k_3u)$, which was suggested by Fisher as a deterministic version of a stochastic model for the spatial spread of a favored gene in a population [19]. In the 20th century, the Fisher equation has become the basis for a variety of models for spatial spread. The typical examples are that Aoki discussed gene-culture waves of advance [3] and Ammerman and Cavali-Sforza, in an interesting direct application of the model, applied it to the spread of early farming in Europe [1, 2]. Meanwhile, the solution of the Fisher equation has been widely investigated in [11, 14, 21, 29, 41]. If the nonlinear term $f(u) = k_7u^3 + k_6u^2 + k_5u$, where k_5, k_6 and k_7 are real constants, the equation (1.1) can be regarded as a generalization of the Fisher equation, which is used as a density-dependent diffusion model, in the one-dimensional situation, for studying insect and animal dispersal with growth dynamics [32], and as a genetic model arising from the classical theory of population genetics and combustion [4,5]. More results of equation (1.1) can be found in [15,16, 26, 27, 30] and their numerous references.

Stochastic partial differential equations play an important role in the mathematical modeling of many physical phenomena. These equations not only generalize the models of the deterministic cases, but they lead to new phenomena which is important in physics. For example, Crauel and Flandoli [12] showed that the deterministic pitchfork bifurcation disappears as soon as an additive white noise of arbitrarily small intensity is incorporated the model. Hairer and Mattingly [25] characterized the class of noises for which the 2 dimensional stochastic Navier-Stokes equation is ergodic. In recent series of papers and lectures, Flandoli et. proved that for several examples of deterministic partial differential equations which are illposedness a suitable random noise can restore the illposedness see e.g. [9, 22, 35]. C. Marinelli and M. Rökner [33] studied the ergodicity for stochastic reaction-diffusion equations with multiplicative Poisson noise. Peter W. Bates, Kening Lu and Bixiang Wang [10] proved the existence of the random attractors for stochastic reaction-diffusion equations on unbounded domains. And there are lots of other work, see for short list e.g. [17, 18, 40] and references therein. But the stochastic reaction-diffusion equations do not recover the stochastic fractional reaction-diffusion equation .

More precisely, we consider stochastic fractional reaction-diffusion equation in a bounded domain with Wiener noise as the body forces like this

$$\begin{cases} du + \left((-\Delta)^{\frac{\alpha}{2}} u + u^3 - u\right) \mathrm{d}t = \mathrm{d}W, & t > 0,\ x \in (0,1), \\ u(t,0) = u(t,1) = 0, & t > 0, \\ u(0,x) = u_0(x), & x \in [0,1], \end{cases} \tag{1.2}$$

where $\alpha \in (1,2)$, W stands for an $L^2(0,1)$-cylindrical Wiener process defined on a complete probability space $(\Omega, \mathcal{F}, P)$, with expectation E and normal filtration $\mathcal{F}_t = \sigma\{W(s) : s \leqslant t\}, t \in [0,T]$. And it has the following representation

$$W(t) = \sum_{k=1}^{\infty} e_k \beta_k(t), \quad t \in [0,T],$$

where $e_k, k \in \mathbb{N}$, is an orthonormal basis of $L^2(0,1)$ and $\beta_k, k \in \mathbb{N}$ is a family of independent real-valued Brownian motions.

The fractions of the Laplacian are the infinitesimal generators of Lévy stable diffusion processes and appear in anomalous diffusions in plasma, flames propagation and chemical reactions in liquids, population dynamics, geophysical fluid dynamics, and American options in finance. The equations with fractional diffusion are becoming popular in many areas of applications [24, 28]. In these applications, it is often important to consider boundary value problems. Hence it is useful to study solutions for space fractional diffusion equations on bounded domains with Dirichlet boundary conditions.

Recently, the authors notice that papers [7], [8] are relative to our work. In the two papers, Z. Brzezniak, L.Debbi and Ben Goldys consider the case of multiply noise for the fractional Burgers equation. They use Itö formula, Burkholder-Davis-Gundy inequality, Girsanov theorem and so on techniques in stochastic analysis to prove the global well posedness and ergodicity in $L^2(0,1)$ when the exponent $\alpha \in \left(\frac{3}{2}, 2\right)$. In the present work, we study the case of additive noise for fractional reaction-diffusion equations. We use contraction principle to consider the local existence of solution $u(t)$ in $L^2(0,1)$, but it produces $\|u(t)\|_{L^3}$ which can not be dominated by $\|u(t)\|_{L^2}$. This is different from [7]. So we study the equation in H^σ which satisfies $H^\sigma \subset L^3(0,1)$ with $\sigma \in \left[\frac{1}{6}, \frac{\alpha}{2}\right]$. This will bring more difficulties. Here we relax the assumption of $\alpha \in \left(\frac{3}{2}, 2\right)$ in [7] to $\alpha \in (1,2)$. By changing the stochastic equations into random partial differential equations (PDE), we use techniques from PDE to obtain the global well-posedness in $C([0,T];H^\sigma)$ for $\sigma\in\left[\frac{1}{6},\frac{\alpha}{2}\right]$ as well as ergodicity in $H^{\frac{\alpha}{2}}$. Furthermore, if $\alpha\in\left(\frac{3}{2},2\right)$, we prove the ergodicity in H^σ with arbitrary $\sigma\in\left[\frac{1}{6},\frac{\alpha}{2}\right]$.

The remainder of this paper is organized as follows. In Section 2, we introduce some notations, give the definition of mild solution to (1.2), and then show the local existence to (1.2) in H^σ with $\sigma\in\left[\frac{1}{6},\alpha-\frac{1}{2}\right)$. In Section 3, by the priori estimates, we get the global well-posedness of the solution to (1.2) in H^σ with $\sigma\in\left[\frac{1}{6},\frac{\alpha}{2}\right]$. In Section 4, using Krylov-Bogoliubov method we obtain the existence of the invariant measures to the stochastic fractional reaction-diffusion equation in H^σ with $\sigma\in\left[\frac{1}{6},\frac{\alpha}{2}\right]$. In Section 5, by checking irreducibility property and strong Feller property of the Markov semigroup corresponding to the solution of (1.2) we establish the ergodicity of invariant measures in $H^{\frac{\alpha}{2}}$. Finally, we give a remark that if $\alpha>\frac{3}{2}$, we can impose some appropriate assumptions on $W(t)$ to prove the ergodicity for problem (1.2) in H^σ for $\sigma\in\left[\frac{1}{6},\frac{\alpha}{2}\right]$. As usual, constants C may change from one line to the next, unless, we give a special declaration ; we denote by $C(a)$ a constant which depends on some parameter a.

2 Local existence of the solution

If $p\in[1,\infty)$, we denote by $\|\cdot\|_{L^p}$ the norm in $L^p(0,1)$, if $p=\infty$, we denote $\|\cdot\|_{L^\infty_x}$ the norm in $L^\infty(0,1)$. When $p=2$, we let $\langle\cdot,\cdot\rangle$ be the inner product in $L^2(0,1)$.

For $\sigma>0$, the powers $(-\Delta)^{\frac{\sigma}{2}}$ of the positive operator $-\Delta$, in a bounded domain (0,1) with zero Dirichlet boundary data, are defined through the spectral decomposition using the powers of the eigenvalues of the original operator. Let

$$e_n(x)=\sqrt{\frac{2}{\pi}}\sin n\pi x,\quad n=1,2,\cdots$$

and

$$\lambda_n=\pi^2n^2,\quad n=1,2,\cdots.$$

Then (e_n,λ_n) are the eigenvectors and eigenvalues of $-\Delta$ in $(0,1)$ with zero Dirichlet boundary data. Therefore, $(e_n,\lambda_n^{\frac{\sigma}{2}})$ are the eigenvectors and eigenvalues of $(-\Delta)^{\frac{\sigma}{2}}$, also with Dirichlet boundary conditions. Denote $A=-\Delta$. Then the operator $A^{\frac{\sigma}{2}}$ is well defined in the space of functions

$$H^{\sigma}=\left\{u=\sum u_n e_n \in L^2(0,1); \|u\|_{H^{\sigma}}:=\left(\sum u_n^2\lambda_n^{\sigma}\right)^{\frac{1}{2}}<\infty\right\},$$

and as a consequence

$$A^{\frac{\sigma}{2}}u=\sum u_n\lambda_n^{\frac{\sigma}{2}}e_n.$$

Note that $\|u\|_{H^{\sigma}}=\|A^{\frac{\sigma}{2}}u\|_{L^2}$. The dual space $H^{-\sigma}$ is defined in the standard way, as well as the inverse operator $A^{-\frac{\sigma}{2}}$.

It can be derived from [37] that the solution to the linear part of (1.2)

$$\begin{aligned}
&\mathrm{d}u+(A^{\frac{\alpha}{2}}u-u)\mathrm{d}t=\mathrm{d}W, \quad t>0, x\in(0,1),\\
&u(t,0)=u(t,1)=0, \quad t>0,\\
&u(0,x)=u_0(x), \quad x\in[0,1],
\end{aligned}$$

is unique, and when $u_0=0$, it has the form of

$$z_{\alpha}(t)=\int_0^t e^{(-A^{\frac{\alpha}{2}}+1)(t-s)}\mathrm{d}W(s).$$

It is known (see [37]) that z_{α} is Gaussian process taking values in $L^2(0,1)$. Moreover z_{α} has a version which is, a.s. for $\omega\in\Omega$, Hölder continuous with respect to (t,x). For reader's convenience, we summarize the conclusions as Lemma 2.1 and sketch the proof. Let

$$v(t)=u(t)-z_{\alpha}(t),\ t\geqslant 0.$$

And $\mathcal{A}^{\frac{\alpha}{2}}:=A^{\frac{\alpha}{2}}-I, I$ is the identical operator. Then u is a mild solution, which is defined below, to (1.2) if and only if v solves the following evolution equation:

$$\begin{cases}
\dfrac{\partial v}{\partial t}+\mathcal{A}^{\frac{\alpha}{2}}v+|v+z_{\alpha}|^2(v+z_{\alpha})=0, & t>0, x\in(0,1),\\
v(t,0)=v(t,1)=0, & t>0,\\
v(0)=u_0, & x\in[0,1].
\end{cases}\tag{2.3}$$

Definition 2.1 *We say a $(\mathcal{F}(t))_{t\geqslant 0}$ adapted process $(v(t))_{t\in[0,T]}$ is a mild solution to (2.3), if $(v(t))_{t\in[0,T]}\in C([0,T];H^{\sigma})$ P-a.e. and it satisfies*

$$v(t)=e^{-\mathcal{A}^{\frac{\alpha}{2}}t}u_0+\int_0^t e^{-\mathcal{A}^{\frac{\alpha}{2}}(t-s)}(v+z_{\alpha})^3\mathrm{d}s,\quad t\in[0,T].$$

Equivalently, $(u(t))_{t\in[0,T]}$ is a mild solution to (1.2), if it is a $(\mathcal{F}(t))_{t\geqslant 0}$ adapted process which belongs to $C([0,T];H^\sigma)$ P-a.s. and satisfies

$$u(t)=e^{-\mathcal{A}^{\frac{\alpha}{2}}t}u_0+\int_0^t e^{-\mathcal{A}^{\frac{\alpha}{2}}(t-s)}u^3\mathrm{d}s+\int_0^t e^{-\mathcal{A}^{\frac{\alpha}{2}}(t-s)}\mathrm{d}W(s),\quad t\in[0,T].$$

From now on, we will study the equation of the form (2.3) to get the existence and uniqueness of the solution a.s. $\omega\in\Omega$.

Lemma 2.1 *For $\alpha\in(1,2)$, z_α has a version $z_\alpha(t,\xi),t\geqslant 0,\xi\in[0,1]$, Hölder continuous with respect to $t\geqslant 0,\xi\in[0,1]$.*

Proof For $s,t\in[0,T]$, we have

$$\begin{aligned}&E|z_\alpha(t,x)-z_\alpha(s,x)|^2\\ =&\sum_{k=1}^{\infty}\int_s^t|e^{-(t-\tau)(A^{\frac{\alpha}{2}}-1)}e_k(x)|^2\mathrm{d}\tau\\ &+\sum_{k=1}^{\infty}\int_0^s|(e^{-(t-\tau)(A^{\frac{\alpha}{2}}-1)}-e^{-(s-\tau)(A^{\frac{\alpha}{2}}-1)})e_k(x)|^2\mathrm{d}\tau\\ :=&I_1(t,s,x)+I_2(t,s,x).\end{aligned}$$

Choose $\theta\in(0,1)$, such that

$$\alpha(1-\theta)>1,\ \alpha>1+\theta.$$

Then

$$\begin{aligned}I_1(t,s,x)&\leqslant C\sum_{k=1}^{\infty}\int_s^t|e^{-(t-\tau)(A^{\frac{\alpha}{2}}-1)}e_k(x)|^2\mathrm{d}\tau\\ &\leqslant C\sum_{k=1}^{\infty}\int_s^t e^{-2(t-\tau)(\lambda_k^{\frac{\alpha}{2}}-1)}\mathrm{d}\tau\\ &=C\sum_{k=1}^{\infty}\frac{1-e^{-2(t-s)(\lambda_k^{\frac{\alpha}{2}}-1)}}{2(\lambda_k^{\frac{\alpha}{2}}-1)}\\ &\leqslant C\sum_{k=1}^{\infty}(\pi k)^{-\alpha(1-\theta)}|t-s|^\theta<\infty.\end{aligned}$$

The last inequality follows by

$$\lambda_k^{\frac{\alpha}{2}}\geqslant\lambda_k^{\frac{\alpha}{2}}-1\geqslant\frac{1}{2}\lambda_k^{\frac{\alpha}{2}}$$

for $k=1,2,\cdots$. And

$$\begin{aligned}
&I_2(t,s,x)\\
\leqslant& C\sum_{k=1}^{\infty}\int_0^s \|[e^{-(t-\tau)(\lambda_k^{\frac{\alpha}{2}}-1)}-e^{-(s-\tau)(\lambda_k^{\frac{\alpha}{2}}-1)}]\|^2 ds\\
=&C\sum_{k=1}^{\infty}\int_0^s (e^{-2(t-\tau)(\lambda_k^{\frac{\alpha}{2}}-1)}+e^{-2(s-\tau)(\lambda_k^{\frac{\alpha}{2}}-1)}\\
&-2e^{-(t+s-2\tau)(\lambda_k^{\frac{\alpha}{2}}-1)})d\tau\\
=&C\sum_{k=1}^{\infty}\frac{(1-e^{-(t-s)(\lambda_k^{\frac{\alpha}{2}}-1)})^2-(e^{-t(\lambda_k^{\frac{\alpha}{2}}-1)}-e^{-s(\lambda_k^{\frac{\alpha}{2}}-1)})^2}{\lambda_k^{\frac{\alpha}{2}}-1}\\
\leqslant&\sum_{k=1}^{\infty}(\pi k)^{-\alpha(1-\theta)}|t-s|^{\theta}<\infty.
\end{aligned}$$

Let $t\in[0,T]$ and $x,y\in[0,1]$, it follows that

$$\begin{aligned}
&E|(z_\alpha(t,x)-z_\alpha(t,y))|^2\\
=&\sum_{k=1}^{\infty}\int_0^t e^{-2(t-s)(\lambda_k^{\frac{\alpha}{2}}-1)}|e_k(x)-e_k(y)|^2 ds\\
=&\sum_{k=1}^{\infty}|e_k(x)-e_k(y)|^2\frac{1-e^{-2t(\lambda_k^{\frac{\alpha}{2}}-1)}}{2(\lambda_k^{\frac{\alpha}{2}}-1)}\\
\leqslant&\sum_{k=1}^{\infty}\frac{\lambda_k^{\frac{\theta}{2}}|x-y|^{\theta}}{\lambda_k^{\frac{\alpha}{2}}-1}\\
\leqslant& C\sum_{k=1}^{\infty}(\pi k)^{\theta-\alpha}|x-y|^{\theta}<\infty.
\end{aligned}$$

By Kolmogorov's test theorem [37], we get the conclusion.

Now, we establish the local well-posedness result of (1.2). Let R be a large positive random variable such that $R(\omega)\geqslant\sup\limits_{t\in[0,T]}\|z_\alpha\|_{L_x^\infty}=:\|z_\alpha\|_{L_x^\infty L_t^\infty}$, P-a.e. $\omega\in\Omega$. We will use the classical fixed point theorem for contractions to prove that there exists a positive random variable $T^*(\omega)>0$, P-a.e. $\omega\in\Omega$, such that the local solution of (1.2), in sense of Definition 2.1, is in the space $\mathcal{B}_R^{T^*}$ defined by

$$\mathcal{B}_R^{T^*}=\{u:u\in C([0,T^*];H^\sigma),\|u(t)\|_{H^\sigma}\leqslant R,\quad\forall t\in[0,T^*]\}.$$

Theorem 2.1 *Assume $u_0\in\mathcal{F}_0,\|u_0\|_{H^\sigma}<\dfrac{R}{2}$, and the conditions in Lemma 2.1* *hold. For* $\alpha \in (1,2)$ *and* $\frac{1}{6} \leqslant \sigma < \alpha - \frac{1}{2}$, *there exists a random variable* $T^* > 0$ *such that* (2.3) *has a unique mild solution in* $\mathcal{B}_R^{T^*}$.

Proof Take any v in $\mathcal{B}_R^{T^*}$, and define

$$\mathcal{L}(v) := e^{-t\mathcal{A}^{\frac{\alpha}{2}}} u_0 + \int_0^t e^{-(t-s)\mathcal{A}^{\frac{\alpha}{2}}} |v(s) + z_\alpha(s)|^2 (v(s) + z_\alpha(s)) \mathrm{d}s. \tag{2.4}$$

Then

$$\begin{aligned}\|\mathcal{L}(v)\|_{H^\sigma} \leqslant & \|e^{-t\mathcal{A}^{\frac{\alpha}{2}}} u_0\|_{H^\sigma} \\ & + \left\| \int_0^t e^{-(t-s)\mathcal{A}^{\frac{\alpha}{2}}} |v(s) + z_\alpha(s)|^2 (v(s) + z_\alpha(s)) \mathrm{d}s \right\|_{H^\sigma} \\ := & I_1 + I_2.\end{aligned}$$

Since $v(t) = u(t) - z_\alpha(t)$, we have

$$I_2 \leqslant \int_0^t \|e^{-(t-s)\mathcal{A}^{\frac{\alpha}{2}}} u^3\|_{H^\sigma} \mathrm{d}s = \int_0^t \|A^{\frac{\sigma}{2}} e^{-(t-s)\mathcal{A}^{\frac{\alpha}{2}}} u^3\|_{L^2} \mathrm{d}s.$$

By Sobolev embedding theorem and the fact that $\|e_k\|_{L^\infty} \leqslant 1$, for $k \in \mathbb{N}$, we have for the integrand above

$$\begin{aligned}& \|A^{\frac{\sigma}{2}} e^{-(t-s)\mathcal{A}^{\frac{\alpha}{2}}} u^3\|_{L^2}^2 \\ = & \sum_{k=1}^{\infty} \langle A^{\frac{\sigma}{2}} e^{-(t-s)\mathcal{A}^{\frac{\alpha}{2}}} u^3, e_k \rangle^2 = \sum_{k=1}^{\infty} \langle u^3, e^{-(t-s)\mathcal{A}^{\frac{\alpha}{2}}} A^{\frac{\sigma}{2}} e_k \rangle^2 \\ = & \sum_{k=1}^{\infty} \langle u^3, (\pi k)^\sigma e^{-(t-s)[(\pi k)^\alpha - 1]} e_k \rangle^2 \\ \leqslant & C \sum_{k=1}^{\infty} (\pi k)^{2\sigma - \alpha\theta} (t-s)^{-\theta} (\|v\|_{H^\sigma}^6 + \|z_\alpha\|_{L_x^\infty}^6),\end{aligned}$$

where in order to insure $I_2 < \infty$, we choose θ such that

$$\frac{2\sigma + 1}{\alpha} < \theta < 2.$$

Without loss of generality, we assume $T^*(\omega) \leqslant T, P$- a.e. $\omega \in \Omega$. Therefore,

$$\begin{aligned}I_2 \leqslant & C \sup_{t \in [0,T^*]} (\|v\|_{H^\sigma}^3 + \|z_\alpha\|_{L_x^\infty}^3) t^{1-\frac{\theta}{2}} \\ \leqslant & CR^3 (T^*)^{1-\frac{\theta}{2}} < \infty.\end{aligned}$$

On the other hand, we deduce that

$$
\begin{aligned}
I_1^2 &= \|A^{\frac{\sigma}{2}} e^{-t\mathcal{A}^{\frac{\alpha}{2}}} u_0\|_{L^2}^2 = \sum_{k=1}^{\infty} \langle A^{\frac{\sigma}{2}} e^{-t\mathcal{A}^{\frac{\alpha}{2}}} u_0, e_k\rangle^2 \\
&= \sum_{k=1}^{\infty} \langle u_0, e^{-t\mathcal{A}^{\frac{\alpha}{2}}} A^{\frac{\sigma}{2}} e_k\rangle^2 = \sum_{k=1}^{\infty} \langle u_0, \lambda_k^{\frac{\sigma}{2}} e^{-t(\lambda_k^{\frac{\alpha}{2}}-1)} e_k\rangle^2 \\
&= \sum_{k=1}^{\infty} \langle A^{\frac{\sigma}{2}} u_0, e^{-t(\lambda_k^{\frac{\alpha}{2}}-1)} e_k\rangle^2 \leqslant \|u_0\|_{H^\sigma}^2.
\end{aligned}
$$

By the estimates of I_1 and I_2, we infer that

$$
\|\mathcal{L}(v)\|_{H^\sigma} \leqslant \frac{R}{2} + CR^3 (T^*)^{1-\frac{\theta}{2}}. \tag{2.5}
$$

Now consider $v_1, v_2 \in \mathcal{B}_R^{T^*}$ and set $u_i = v_i + z_\alpha, i = 1, 2$. Then

$$
\mathcal{L}(v_1) - \mathcal{L}(v_2) = \int_0^t e^{-(t-s)\mathcal{A}^{\frac{\alpha}{2}}} (u_1^3 - u_2^3) \mathrm{d}s,
$$

and we derive as above

$$
\|\mathcal{L}(v_1) - \mathcal{L}(v_2)\|_{H^\sigma} \leqslant \int_0^t \|e^{-(t-s)\mathcal{A}^{\frac{\alpha}{2}}} (u_1^3 - u_2^3)\|_{H^\sigma} \mathrm{d}s.
$$

For the integrand, due to

$$
\begin{aligned}
&\|e^{-(t-s)\mathcal{A}^{\frac{\alpha}{2}}} (u_1^3 - u_2^3)\|_{H^\sigma}^2 \\
=& \sum_{k=1}^{\infty} \langle A^{\frac{\sigma}{2}} e^{-(t-s)\mathcal{A}^{\frac{\alpha}{2}}} (u_1^3 - u_2^3), e_k\rangle^2 \\
=& \sum_{k=1}^{\infty} \langle (u_1^3 - u_2^3), e^{-(t-s)\mathcal{A}^{\frac{\alpha}{2}}} A^{\frac{\sigma}{2}} e_k\rangle^2 \\
\leqslant& \sum_{k=1}^{\infty} \lambda_k^{\sigma} e^{-2(t-s)(\lambda_k^{\frac{\alpha}{2}}-1)} \left(\int_0^1 |u_1^3 - u_2^3| dx\right)^2 \\
\leqslant& C \sum_{k=1}^{\infty} (\pi k)^{2\sigma - \alpha\theta} (t-s)^{-\theta} \left(\int_0^1 |u_1^3 - u_2^3| \mathrm{d}x\right)^2 \\
\leqslant& C(t-s)^{-\theta} \left(\int_0^1 |u_1 - u_2| \cdot |u_1^2 + u_2^2 + u_1 u_2| \mathrm{d}x\right)^2 \\
\leqslant& C(t-s)^{-\theta} \|v_1 - v_2\|_{L^3}^2 (\|u_1\|_{L^3}^4 + \|u_2\|_{L^3}^4) \\
\leqslant& C(t-s)^{-\theta} \|v_1 - v_2\|_{H^\sigma}^2 (\|v_1\|_{H^\sigma}^4 + \|v_2\|_{H^\sigma}^4 + \|z_\alpha\|_{L_x^\infty}^4),
\end{aligned}
$$

where we use the Sobolev imbedding theorem in the last inequality. Therefore

$$
\|\mathcal{L}(v_1) - \mathcal{L}(v_2)\|_{H^\sigma} \leqslant C(T^*)^{1-\frac{\theta}{2}} R^2 \|v_1 - v_2\|_{H^\sigma}. \tag{2.6}
$$

Let T^* be small enough such that $2C(T^*)^{1-\frac{\theta}{2}}R^2 < 1$. Then by (2.5) and (2.6), it is easy to check that $\mathcal{L}$ maps $\mathcal{B}_R^{T^*}$ into itself and it is a strict contraction in $\mathcal{B}_R^{T^*}$. Hence, $\mathcal{L}$ has a unique fixed point in $C([0,T^*];H^\sigma)$, which is the unique solution of (2.3) on $[0,T^*]$.

3 Global well-posedness

Theorem 3.1 *Let $1<\alpha<2$ and $\frac{1}{6}\leqslant\sigma\leqslant\frac{\alpha}{2}$. Then for any $T>0$ there exists an $(\mathcal{F}(t))_{t\geqslant 0}$ adapted process $(u(t))_{t\geqslant 0}$ which is the unique global solution of problem (1.2) with initial condition $u_0\in H^\sigma$. Furthermore,*

$$\sup_{t\in[0,T]}\|u(t)\|_{H^\sigma}^2+\int_0^T\|u(t)\|_{H^{\frac{\alpha}{2}+\sigma}}^2\mathrm{d}t\leqslant C(\|z_\alpha\|_{L_x^\infty L_t^\infty},\|v_0\|_{H^\sigma},T).$$

Proof Let P_n be the projection operator in $L^2(0,1)$ onto the space spanned by $e_1,\dots,e_n$. For $u_0\in H^\sigma$, we consider the ordinary differential equation

$$\frac{\partial v^n}{\partial t}+\mathcal{A}^{\frac{\alpha}{2}}v^n+P_n(v^n+z_\alpha)^3=0, \tag{3.7}$$

with initial condition $v_0^n=P_nu_0$. By Theorem 2.1, there exists a positive random variable T_n^* such that for $t\in[0,T_n^*]$, the equation has a unique solution

$$v^n(t)=e^{-\mathcal{A}^{\frac{\alpha}{2}}t}u_0^n+\int_0^t e^{-(t-s)\mathcal{A}^{\frac{\alpha}{2}}}(v^n(s)+z_\alpha^n(s))^3\mathrm{d}s,$$

which is also classical solution to (3.7). So multiplying (3.7) by v^n and integrating over $[0,1]$, we get

$$\left\langle\frac{\partial v^n}{\partial t},v^n\right\rangle+\langle A^{\frac{\alpha}{2}}v^n,v^n\rangle-\langle v^n,v^n\rangle+\langle(v^n+z_\alpha)^3,v^n\rangle=0.$$

Therefore,

$$\begin{aligned}
&\frac{\partial}{\partial t}\|v^n\|_{L^2}^2+\|v^n\|_{H^{\frac{\alpha}{2}}}^2+\|v^n\|_{L^4}^4\\
=&\|v^n\|_{L^2}^2-\langle(z_\alpha)^3,v^n\rangle-3\langle(v^n)^2z_\alpha,v^n\rangle-3\langle v^n(z_\alpha)^2,v^n\rangle\\
\leqslant&C(\|z_\alpha\|_{L_x^\infty})\|v^n\|_{L^1}+C(\|z_\alpha\|_{L_x^\infty})\|v^n\|_{L^3}^3\\
&+(C(\|z_\alpha\|_{L_x^\infty})+1)\|v^n\|_{L^2}^2\\
\leqslant&C(\|z_\alpha\|_{L_x^\infty})+\varepsilon\|v^n\|_{L^4}^4,
\end{aligned}$$

where the last inequality follows by Hölder inequality. Integrating with respect to time t to yield

$$\|v^n(t)\|_{L^2}^2\leqslant C(T,\|u_0^n\|_{L^2},\|z_\alpha\|_{L_x^\infty L_t^\infty})\leqslant C(T,\|u_0\|_{L^2},\|z_\alpha\|_{L_x^\infty L_t^\infty}). \tag{3.8}$$

We now estimate $\|v^n\|_{H^\sigma}$. Multiplying (3.7) by $A^\sigma v^n$, integrating with respect to space variable, to yield

$$\left\langle \frac{\partial v^n}{\partial t}, A^\sigma v^n \right\rangle + \langle \mathcal{A}^{\frac{\alpha}{2}} v^n, A^\sigma v^n\rangle + \langle (v^n + z_\alpha)^3, A^\sigma v^n\rangle = 0,$$

which is equivalent to

$$\frac{1}{2}\frac{\partial}{\partial t}\|v^n\|_{H^\sigma}^2 + \|v^n\|^2_{H^{\frac{\alpha}{2}+\sigma}} - \|v^n\|^2_{H^\sigma} + \langle (v^n + z_\alpha)^3, A^\sigma v^n\rangle = 0. \tag{3.9}$$

If $\sigma = \frac{1}{6}$, by Schwarz inequality and interpolation inequality, dealing with the last term of the left hand side in the above equality, we have

$$\begin{aligned}
&\langle (v^n + z_\alpha)^3, A^\sigma v^n\rangle \\
\leqslant & C\|v^n\|_{H^{\frac{1}{3},4}}\|v^n\|_{L^4} + \|z_\alpha\|^3_{L^\infty_x}\|v^n\|_{H^{\frac{1}{3}}} \\
\leqslant & C\|v^n\|^\theta_{H^{\frac{1}{6}+\frac{\alpha}{2}}}\|v^n\|^{1-\theta}_{L^2}\|v^n\|^{3\theta'}_{H^{\frac{1}{6}+\frac{\alpha}{2}}}\|v^n\|^{3(1-\theta')}_{L^2} + \|z_\alpha\|^3_{L^\infty_x}\|v^n\|_{H^{\frac{1}{3}}} \\
= & C\|v^n\|^{\theta+3\theta'}_{H^{\frac{1}{6}+\frac{\alpha}{2}}}\|v^n\|^{4-\theta-3\theta'}_{L^2} + \|z_\alpha\|^3_{L^\infty_x}\|v^n\|_{H^{\frac{1}{3}}},
\end{aligned}$$

where $\theta = \frac{7}{2+6\alpha} < \frac{7}{8}$ and $\theta' = \frac{3}{6\alpha+2} < \frac{3}{8}$. Since $\theta + 3\theta' < 2$, by (3.8) and Schwarz inequality, we deduce

$$\langle (v^n + z_\alpha)^3, A^\sigma v^n\rangle \leqslant \varepsilon\|v^n\|^2_{H^{\frac{1}{6}+\frac{\alpha}{2}}} + C(T, \|u_0\|_{L^2}, \|z_\alpha\|_{L^\infty_x L^\infty_t}).$$

Consequently, if $\sigma = \frac{1}{6}$, from (3.9) and Gronwall's inequality, we get

$$\sup_{t\in[0,T_n^*]}\|v^n(t)\|^2_{H^{\frac{1}{6}}} + \int_0^T \|v^n(t)\|^2_{H^{\frac{1}{6}+\frac{\alpha}{2}}}dt \leqslant C(T, \|u_0\|_{H^{\frac{1}{6}}}, \|z_\alpha\|_{L^\infty_x L^\infty_t}). \tag{3.10}$$

If $\sigma = \frac{1}{4}$, using Schwarz inequality and Sobolev embedding theorem along with interpolation inequality, we have

$$\begin{aligned}
&\langle (v^n + z_\alpha)^3, A^\sigma v^n\rangle \\
\leqslant & C\|v^n\|_{H^{\frac{1}{2},4}}\|v^n\|^3_{L^4} + \|z_\alpha\|^3_{L^\infty_x}\|v^n\|_{H^{\frac{1}{2}}} \\
\leqslant & C\|v^n\|_{H^{\frac{1}{4}+\frac{\alpha}{2}}}\|v^n\|^{3\theta_1}_{H^{\frac{1}{4}+\frac{\alpha}{2}}}\|v^n\|^{3(1-\theta_1)}_{L^3} + \|z_\alpha\|^3_{L^\infty_x}\|v^n\|_{H^{\frac{1}{2}}} \\
\leqslant & C\|v^n\|_{H^{\frac{1}{4}+\frac{\alpha}{2}}}\|v^n\|^{3\theta_1}_{H^{\frac{1}{4}+\frac{\alpha}{2}}}\|v^n\|^{3(1-\theta_1)}_{H^{\frac{1}{6}}} + \|z_\alpha\|^3_{L^\infty_x}\|v^n\|_{H^{\frac{1}{2}}},
\end{aligned}$$

where $\theta_1 = \dfrac{1}{6\alpha+1} < \dfrac{1}{7}$. Since $1+3\theta_1 < 2$, by Schwarz inequality and (3.10), we infer that

$$\begin{aligned}
&\langle (v^n+z_\alpha)^3, A^\sigma v^n\rangle \\
\leqslant& \varepsilon\|v^n\|^2_{H^{\frac14+\frac{\alpha}{2}}} + C\|v^n\|^{\frac{6(1-\theta_1)}{1-3\theta_1}}_{H^{\frac16}} + C\|z_\alpha\|^6_{L^\infty_x} \\
\leqslant& \varepsilon\|v^n\|^2_{H^{\frac14+\frac{\alpha}{2}}} + C(T, \|u_0\|_{H^{\frac16}}, \|z_\alpha\|_{L^\infty_x L^\infty_t}).
\end{aligned}$$

Consequently, if $\sigma = \dfrac{1}{4}$, from (3.9) and Gronwall's inequality, we get

$$\begin{aligned}
&\sup_{t\in[0,T_n^*]} \|v^n(t)\|^2_{H^{\frac14}} + \int_0^T \|v^n(t)\|^2_{H^{\frac14+\frac{\alpha}{2}}}\mathrm{d}t \\
\leqslant& C(T, \|u_0\|_{H^{\frac14}}, \|z_\alpha\|_{L^\infty_x L^\infty_t}).
\end{aligned} \tag{3.11}$$

If $\sigma = \dfrac{1}{3}$, by applying Schwarz inequality and interpolation inequality to the last term on the left side of (3.9), we have

$$\begin{aligned}
&\langle (v^n+z_\alpha)^3, A^\sigma v^n\rangle \\
\leqslant& \|A^{\frac13}v^n\|_{L^2}\|(v^n+z_\alpha)^3\|_{L^2} \\
\leqslant& C\|v^n\|_{H^{\frac{\alpha}{2}+\frac13}}\|v^n+z_\alpha\|^3_{L^6} \\
\leqslant& C\|v^n\|_{H^{\frac{\alpha}{2}+\frac13}}(\|v^n\|^{3\theta_2}_{H^{\frac{\alpha}{2}+\frac13}}\|v^n\|^{3(1-\theta_2)}_{L^4} + \|z_\alpha\|^3_{L^\infty_x}),
\end{aligned}$$

where $\theta_2 = \dfrac{1}{6\alpha+1} < \dfrac{1}{7}$. Since $1+3\theta_2 < 2$, by Schwarz inequality and (3.11), we infer that

$$\begin{aligned}
\langle (v^n+z_\alpha)^3, A^\sigma v^n\rangle \leqslant& \varepsilon\|v^n\|^2_{H^{\frac{\alpha}{2}+\frac13}} + C\|v^n\|^{\frac{6(1-\theta_2)}{1-3\theta_2}}_{L^4} + \|z_\alpha\|^3_{L^\infty_x} \\
\leqslant& \varepsilon\|v^n\|^2_{H^{\frac{\alpha}{2}+\frac13}} + C\|v^n\|^{\frac{6(1-\theta_2)}{1-3\theta_2}}_{H^{\frac14}} + \|z_\alpha\|^3_{L^\infty_x} \\
\leqslant& \varepsilon\|v^n\|^2_{H^{\frac{\alpha}{2}+\frac13}} + C(T, \|u_0\|_{H^{\frac14}}, \|z_\alpha\|_{L^\infty_x L^\infty_t}).
\end{aligned}$$

Consequently, if $\sigma = \dfrac{1}{3}$, from (3.9) and Gronwall's inequality, we get

$$\sup_{t\in[0,T_n^*]} \|v^n(t)\|^2_{H^{\frac13}} + \int_0^T \|v^n(t)\|^2_{H^{\frac13+\frac{\alpha}{2}}}dt \leqslant C(T, \|u_0\|_{H^{\frac13}}, \|z_\alpha\|_{L^\infty_x L^\infty_t}).$$

In order to get the estimate for (3.9) with arbitrary $\sigma \in \left[\dfrac{1}{3}, \dfrac{\alpha}{2}\right]$, using Schwarz inequality and Sobolev inequality along with the above estimate for $\|v^n(t)\|_{H^{\frac{1}{3}}}$, we have that

$$\begin{aligned}\langle (v^n+z_\alpha)^3, A^\sigma v^n\rangle &\leqslant C\|v^n\|_{H^{\sigma+\frac{\alpha}{2}}}\|v^n+z_\alpha\|_{L^6}^3\\ &\leqslant \varepsilon\|v^n\|_{H^{\sigma+\frac{\alpha}{2}}}^2 + C\|v^n\|_{L^6}^6 + C\|z_\alpha\|_{L^6}^6\\ &\leqslant \varepsilon\|v^n\|_{H^{\sigma+\frac{\alpha}{2}}}^2 + C\|v^n\|_{H^{\frac{1}{3}}}^6 + C\|z_\alpha\|_{L_x^\infty}^6\\ &\leqslant \varepsilon\|v^n\|_{H^{\sigma+\frac{\alpha}{2}}}^2 + C(T, \|u_0\|_{H^{\frac{1}{3}}}, \|z_\alpha\|_{L_x^\infty L_t^\infty})\\ &\leqslant \varepsilon\|v^n\|_{H^{\sigma+\frac{\alpha}{2}}}^2 + C(T, \|u_0\|_{H^\sigma}, \|z_\alpha\|_{L_x^\infty L_t^\infty}).\end{aligned}$$

Thus, plugging the above inequality into (3.10) and regrouping terms, we obtain

$$\begin{aligned}&\frac{1}{2}\frac{\partial}{\partial t}\|v^n\|_{H^\sigma}^2 + (1-\varepsilon)\|v^n\|_{H^{\frac{\alpha}{2}+\sigma}}^2\\ &\leqslant C\|v^n\|_{H^\sigma}^2 + C(T, \|u_0\|_{H^\sigma}, \|z_\alpha\|_{L_x^\infty L_t^\infty}).\end{aligned}$$

Integration with respect to time t, for $t\in[0,T_n^*]$, by Gronwall's inequality, we deduce

$$\sup_{t\in[0,T_n^*]}\|v^n(t)\|_{H^\sigma}^2 + \int_0^{T_n^*}\|v^n(t)\|_{H^{\frac{\alpha}{2}+\sigma}}^2 dt \leqslant C(T, \|u_0\|_{H^\sigma}, \|z_\alpha\|_{L_x^\infty L_t^\infty}).$$

So far, we proved that the above estimate holds for all $\sigma\in\left[\frac{1}{6},\frac{\alpha}{2}\right]$. Repeating the arguments in Theorem 2.1 on $[0,T_n^*], [T_n^*, 2T_n^*], \cdots$, we get the existence of the global solution to (3.7) on $[0,T]$, which also satisfies

$$\sup_{t\in[0,T]}\|v^n(t)\|_{H^\sigma}^2 + \int_0^{T}\|v^n(t)\|_{H^{\frac{\alpha}{2}+\sigma}}^2 dt \leqslant C(T, \|u_0\|_{H^\sigma}, \|z_\alpha\|_{L_t^\infty L_x^\infty}). \tag{3.12}$$

Since the space

$$C([0,T];H^\sigma)\cap L^2([0,T];H^{\frac{\alpha}{2}+\sigma})$$

is complete and $(v_n)_{n\in\mathbb{N}}$ is bounded in this space, it is weekly star convergent in the space to a function $\tilde{v}$ which satisfies

$$\sup_{t\in[0,T]}\|\tilde{v}(t)\|_{H^\sigma} + \int_0^T\|\tilde{v}(t)\|_{H^{\frac{\alpha}{2}+\sigma}} \leqslant C(T, \|u_0\|_{H^\sigma}, \|z_\alpha\|_{L_t^\infty L_x^\infty}).$$

Let us define the mapping $\mathcal{L}_n$ in the same way as $\mathcal{L}$; it is easy to check that $\mathcal{L}_n$ is a strict contraction uniformly in n on $\mathcal{B}_{r(\omega)}^{t(\omega)}$ where

$$r(\omega) = 3(\|z_\alpha\|_{L_t^\infty L_x^\infty} + \sup_{t\in[0,T]}\|\tilde{v}(t)\|_{H^\sigma})$$

and

$$2C(r(\omega))^2(t(\omega))^{1-\frac{\theta}{2}} \leqslant 1,$$

with the constant C is as in (2.6). Then by a standard arguments, we can prove that

$$v_n \to v$$

in $\mathcal{B}^{t(\omega)}_{r(\omega)}$, implying

$$v = \tilde{v} \quad \text{on } [0, t(\omega)]$$

and

$$\|v(t(\omega))\|_{H^\sigma} \leqslant \sup_{s\in[0,T]} \|\tilde{v}(s)\|_{H^\sigma}.$$

Thus we can construct a solution on $[t(\omega), 2t(\omega)]$ starting from $v(t(\omega))$. We get the unique global solution on $[0, T]$ by reiterating this argument.

4 Existence of the invariant measures

In this part, similar to [34], we will prove the existence of an invariant measure for the equation

$$du = (-A^{\frac{\alpha}{2}}u - u^3 + u)dt + GdW(t), \tag{4.13}$$

where $W(t) = \sum_{k=1}^{\infty} e_k\beta_k(t),\ t \in \mathbb{R}$, is cylindrical $L^2(0,1)-$valued two sided Wiener process and $G : L^2(0,1) \to L^2(0,1)$ is a bounded linear operator, injective, with range $R(G)$ dense in $L^2(0,1)$ and satisfying

$$R(G) \subset D(A^{\frac{\alpha+1}{4}+\varepsilon}) \tag{4.14}$$

for some small $\varepsilon > 0$. For each $\gamma \geqslant 0$, we consider the following equation

$$du_\gamma = (-A^{\frac{\alpha}{2}}u_\gamma - u_\gamma^3 + u_\gamma)dt + GdW(t) \tag{4.15}$$

with initial condition

$$u_\gamma(-\gamma) = 0.$$

By Theorem 3.1, there exists a unique solution to (4.15). In order to obtain the invariant measure, we could show the family of laws $\{\mathcal{L}(u_\gamma(0))\}_{\gamma\geqslant 0}$ is tight. Since $H^{\sigma+\varepsilon} \subset H^\sigma$ is compact, for any $\varepsilon > 0$, we only need to prove $\{(u_\gamma(0))\}_{\gamma\geqslant 0}$ is bounded in probability in $H^{\frac{\alpha}{2}+\varepsilon}$. We remark that the estimates obtained in Theorem 3.1 is useless here, since it gives a bound on solutions that grows indefinitely when $t \to \infty$. Similarly as in [13], we introduce a modified stochastic convolution. For any $\beta > 1$ and $t \in \mathbb{R}$, we define

$$z_{\alpha,\beta}(t) := \int_{-\infty}^{t} e^{-(t-s)(A^{\frac{\alpha}{2}}+\beta-1)} G \mathrm{d}W(s). \tag{4.16}$$

$z_{\alpha,\beta}$ is the mild solution of the linear equation

$$\begin{aligned} \mathrm{d}z &= (-A^{\frac{\alpha}{2}} z - (\beta-1)z)\mathrm{d}t + G\mathrm{d}W(t), \\ z(0) &= z_0, \end{aligned} \tag{4.17}$$

where

$$z_0 = \int_{-\infty}^{0} e^{s(A^{\frac{\alpha}{2}}+\beta-1)} G \mathrm{d}W(s).$$

It is well known that $z_{\alpha,\beta}$ is a stationary process. Define

$$v_\gamma(t) := u_\gamma(t) - z_{\alpha,\beta}(t), \quad t \geqslant -\gamma.$$

Then v_γ is the mild solution to the following equation

$$\begin{aligned} \frac{\mathrm{d}v_\gamma(t)}{\mathrm{d}t} &= -\left(A^{\frac{\alpha}{2}} v_\gamma(t) - v_\gamma(t) + (v_\gamma(t) + z_{\alpha,\beta}(t))^3\right) + \beta z_{\alpha,\beta}(t), \\ v_\gamma(-\gamma) &= -z_{\alpha,\beta}(-\gamma). \end{aligned} \tag{4.18}$$

Theorem 4.1 *In addition to conditions in Theorem* 3.1, *assume* (4.14) *holds. Then there exists an invariant measure for problem* (4.13) *on* H^σ *for* $\sigma \in \left[\frac{1}{6}, \frac{\alpha}{2}\right]$.

Proof Multiplying (4.18) by v_γ and integration with respect to x-variable on $[0, 1]$, we obtain

$$\begin{aligned} &\frac{1}{2}\frac{\mathrm{d}}{\mathrm{d}t}\|v_\gamma\|_{L^2}^2 + \|v_\gamma\|_{H^{\frac{\alpha}{2}}}^2 \\ &= -\langle (v_\gamma + z_{\alpha,\beta})^3, v_\gamma \rangle + \|v_\gamma\|_{L^2}^2 + \beta \langle z_{\alpha,\beta}(t), v_\gamma(t) \rangle. \end{aligned} \tag{4.19}$$

The computation, as well as those below, is not rigorous, but the results may be justified by an argument similar to the one in the proof of Theorem 3.1. Since

$$\begin{aligned} \langle (v_\gamma + z_{\alpha,\beta})^3, v_\gamma \rangle = \|v_\gamma\|_{L^4}^4 &+ 3\langle (v_\gamma)^3, z_{\alpha,\beta} \rangle \\ &+ 3\langle (v_\gamma)^2, (z_{\alpha,\beta})^2 \rangle + \langle v_\gamma, (z_{\alpha,\beta})^3 \rangle, \end{aligned}$$

by virtue of (4.19), using Sobolev embedding theorem and Young's inequality, we get

$$\begin{aligned}&\frac{1}{2}\frac{\mathrm{d}}{\mathrm{d}t}\|v_\gamma\|_{L^2}^2+\|v_\gamma\|_{H^{\frac{\alpha}{2}}}^2+\|v_\gamma\|_{L^4}^4\\&\leqslant(1+\vartheta)\|v_\gamma\|_{L^2}^2+\varepsilon\|v_\gamma\|_{L^4}^4+C\|z_{\alpha,\beta}\|_{H^{\frac{1}{4}}}^4+C\|z_{\alpha,\beta}\|_{L^2}^2,\end{aligned}\tag{4.20}$$

where ϑ is small enough such that $1+\vartheta<\lambda_1^{\frac{\alpha}{2}}$. Thus we have

$$\frac{\mathrm{d}}{\mathrm{d}t}\|v_\gamma\|_{L^2}^2+2(\lambda_1^{\frac{\alpha}{2}}-1-\vartheta)\|v_\gamma\|_{L^2}^2+\|v_\gamma\|_{L^4}^4\leqslant C\|z_{\alpha,\beta}\|_{H^{\frac{1}{4}}}^4+C\|z_{\alpha,\beta}\|_{L^2}^2.$$

Then for any $\lambda\in(0,2(\lambda_1^{\frac{\alpha}{2}}-1-\vartheta)]$,

$$\frac{\mathrm{d}}{\mathrm{d}t}\|v_\gamma\|_{L^2}^2+\lambda\|v_\gamma\|_{L^2}^2+\|v_\gamma\|_{L^4}^4\leqslant C\|z_{\alpha,\beta}\|_{H^{\frac{1}{4}}}^4+C\|z_{\alpha,\beta}\|_{L^2}^2.$$

Subsequently, for $t\leqslant 0$, it follows that

$$\begin{aligned}&\|v_\gamma(t)\|_{L^2}^2e^{\lambda t}\\&\leqslant\|z_{\alpha,\beta}(-\gamma)\|_{L^2}^2e^{-\lambda\gamma}+C\int_{-\gamma}^{t}e^{\lambda s}(\|z_{\alpha,\beta}(s)\|_{H^{\frac{1}{4}}}^4+\|z_{\alpha,\beta}\|_{L^2}^2)\mathrm{d}s\\&\leqslant\|z_{\alpha,\beta}(-\gamma)\|_{L^2}^2e^{-\varepsilon\gamma}e^{(\lambda-\varepsilon)t}\\&\quad+Ce^{(\lambda-\varepsilon)t}\int_{-\gamma}^{t}e^{\varepsilon s}(\|z_{\alpha,\beta}(s)\|_{H^{\frac{1}{4}}}^4+\|z_{\alpha,\beta}\|_{L^2}^2)\mathrm{d}s.\end{aligned}\tag{4.21}$$

For $t<0$, we will prove below

$$\|z_{\alpha,\beta}(t)\|_{H^{\frac{\alpha}{2}+\varepsilon}}\leqslant C(w)(|t|+1),\ P-a.e.\ w\in\Omega,\tag{4.22}$$

where $C(\omega)$ is a positive random variable . So, by (4.21) and (4.22), we have

$$\|v_\gamma(t)\|_{L^2}^2\leqslant R(w)e^{-\varepsilon t}.\tag{4.23}$$

It follows that

$$e^{(\lambda_1^{\frac{\alpha}{2}}-1-\vartheta)t}\|v_\gamma(t)\|_{L^2}^{\frac{2(3\alpha-1)}{2\alpha-1}}\leqslant R(w)e^{(\lambda_1^{\frac{\alpha}{2}}-1-\vartheta-\varepsilon\frac{(3\alpha-1)}{2\alpha-1})t}.\tag{4.24}$$

Let ε be so small that

$$\lambda_1^{\frac{\alpha}{2}}-1-\vartheta-\varepsilon\frac{3\alpha-1}{\alpha-1}>0$$

and

$$\frac{\lambda_1^{\frac{\alpha}{2}}-1-\vartheta}{2}-\varepsilon>0.$$

In view of (4.24) to yield

$$\int_{-\infty}^{0} e^{(\lambda_1^{\frac{\alpha}{2}}-1-\vartheta)t}\|v_\gamma(t)\|_{L^2}^{\frac{2(3\alpha-1)}{2\alpha-1}}\mathrm{d}t < R_1(w) < \infty, \tag{4.25}$$

where $R_1(w)$ is a positive random variable. Next, we will prove (4.22). Following the arguments in [37], we have

$$z_{\alpha,\beta}(t) = \int_{-\infty}^{t} (A^{\frac{\alpha}{2}}+\beta-1)e^{-(t-s)(A^{\frac{\alpha}{2}}+\beta-1)}(GW(t)-GW(s))\mathrm{d}s.$$

By condition (4.14), we know $GW(t)$ is a $D(A^{\frac{\alpha}{4}+\frac{\varepsilon}{2}})$ valued Brownian motion.

$$\begin{aligned}
&\|z_{\alpha,\beta}(t)\|_{H^{\frac{\alpha}{2}+\varepsilon}} = \|A^{\frac{\alpha}{4}+\frac{\varepsilon}{2}}z_{\alpha,\beta}(t)\|_{L^2}\\
\leqslant&\int_{-\infty}^{t}\|A^{\frac{\alpha}{4}+\frac{\varepsilon}{2}}(A^{\frac{\alpha}{2}}+\beta-1)e^{-(t-s)(A^{\frac{\alpha}{2}}+\beta-1)}\\
&\times(GW(t)-GW(s))\|_{L^2}\mathrm{d}s\\
=&\int_{-\infty}^{t}\Big(\sum_{k=1}^{\infty}\langle A^{\frac{\alpha}{4}+\frac{\varepsilon}{2}}(A^{\frac{\alpha}{2}}+\beta-1)e^{-(t-s)(A^{\frac{\alpha}{2}}+\beta-1)}\\
&\times(GW(t)-GW(s)),e_k\rangle^2\Big)^{\frac{1}{2}}\mathrm{d}s\\
\leqslant& C\int_{-\infty}^{t}\|GW(t)-GW(s)\|_{H^{\frac{\alpha}{2}+\varepsilon}}\\
&\times e^{-(t-s)(\beta-1)}\Big(\sum_{k=1}^{\infty}((t-s)^{-\theta_1}\lambda_k^{-\frac{\alpha\theta_1}{2}})^2\lambda_k^{\frac{\alpha}{2}}\Big)^{\frac{1}{2}}\mathrm{d}s\\
\leqslant& C\int_{-\infty}^{t}\|GW(t)-GW(s)\|_{H^{\frac{\alpha}{2}+\varepsilon}}(t-s)^{-\theta_1}\\
&\times e^{-(t-s)(\beta-1)}\Big(\sum_{k=1}^{\infty}\lambda_k^{\alpha(1-\theta_1)}\Big)^{\frac{1}{2}}\mathrm{d}s,
\end{aligned}$$

where $\theta_1 = 1+\frac{1}{2}-\frac{\alpha-1}{4\alpha}$ such that $\alpha(1-\theta_1) < -\frac{1}{2}$ implies

$$\sum_{k=1}^{\infty}\lambda_k^{\alpha(1-\theta_1)} < \infty.$$

Therefore,

$$\begin{aligned}
\|z_{\alpha,\beta}(t)\|_{H^{\frac{\alpha}{2}+\varepsilon}}\leqslant& C\int_{-\infty}^{t}\|GW(t)-GW(s)\|_{H^{\frac{\alpha}{2}+\varepsilon}}\\
&\times(t-s)^{-\theta_1}e^{-(t-s)(\beta-1)}\mathrm{d}s.
\end{aligned}\tag{4.26}$$

Using the method in [20], we can estimate $\|GW(t)-GW(s)\|_{H^{\frac{\alpha}{2}+\varepsilon}}$. Let

$$\xi_n=\sup_{n\leqslant s\leqslant t\leqslant n+1}\frac{\|GW(t)-GW(s)\|_{H^{\frac{\alpha}{2}+\varepsilon}}}{|t-s|^{\frac{1}{2}-\frac{\alpha-1}{8\alpha}}},\quad n\in\mathbb{Z}.$$

ξ_n is a sequence of independent random variables. By Kolmogoroff regularity theorem we have $E\xi_0<\infty$. By the law of large numbers, there exists an integer-valued random variable $n_0(\omega)>0$ such that

$$\frac{\xi_{-n}}{n}\leqslant\frac{\xi_{-n}+\cdots+\xi_{-1}}{n}\leqslant E\xi_0+1$$

for all $n>n_0(\omega)$. This implies

$$\xi_{-n}\leqslant C_1(\omega)n$$

for all $n>0$ and for some positive random variable $C_1(\omega)$. Therefore,

$$\|GW(t)-GW(s)\|_{H^{\frac{\alpha}{2}+\varepsilon}}\leqslant C_1(\omega)|[s]||t-s|^{\frac{1}{2}-\frac{\alpha-1}{8\alpha}}$$

for all s and t such that $s\leqslant t\leqslant[s]+1$. By the law of iterated logarithm, we have

$$\|GW(t)\|_{H^{\frac{\alpha}{2}+\varepsilon}}\leqslant C_2(\omega)|t|$$

for $t<-1$ and for some positive random variable $C_2(\omega)$. From (4.26), we summarize that

$$\begin{aligned}
&\|z_{\alpha,\beta}(t)\|_{H^{\frac{\alpha}{2}+\varepsilon}}\\
\leqslant&C\int_{[t]-1}^{t}\|GW(t)-GW(s)\|_{H^{\frac{\alpha}{2}+\varepsilon}}\\
&\times(t-s)^{-\theta_1}e^{-(t-s)(\beta-1)}\mathrm{d}s\\
&+C\int_{-\infty}^{[t]-1}\|GW(t)-GW(s)\|_{H^{\frac{\alpha}{2}+\varepsilon}}\\
&\times(t-s)^{-\theta_1}e^{-(t-s)(\beta-1)}\mathrm{d}s\\
\leqslant&C\int_{[t]-1}^{t}\Big(\|GW(t)-GW([t])\|_{H^{\frac{\alpha}{2}+\varepsilon}}\\
&+\|GW([t])-GW(s)\|_{H^{\frac{\alpha}{2}+\varepsilon}}\Big)\\
&\times(t-s)^{-\theta_1}e^{-(t-s)(\beta-1)}\mathrm{d}s\\
&+C_2(\omega)\int_{-\infty}^{[t]-1}(|t|+|s|)(t-s)^{-\theta_1}e^{-(t-s)(\beta-1)}\mathrm{d}s\\
\leqslant&C_3(\omega)(|t|+1),
\end{aligned}$$

where the fourth inequality follows by the fact that for $a, b, r \in [0,1]$, we have

$$a^r + b^r < 2(a+b)^r,$$

and $C_3(\omega)$ is some positive random variable. Multiplying $e^{\frac{(\lambda_1^{\frac{\alpha}{2}}-1-\vartheta)t}{2}}$ on both sides of (4.20),

$$\begin{aligned}&\left(\frac{1}{2}\frac{\mathrm{d}}{\mathrm{d}t}\|v_\gamma\|_{L^2}^2\right)e^{\frac{(\lambda_1^{\frac{\alpha}{2}}-1-\vartheta)t}{2}} + \|v_\gamma\|_{H^{\frac{\alpha}{2}}}^2 e^{\frac{(\lambda_1^{\frac{\alpha}{2}}-1-\vartheta)t}{2}}\\ \leqslant\, &(1+\vartheta)\|v_\gamma\|_{L^2}^2 e^{\frac{(\lambda_1^{\frac{\alpha}{2}}-1-\vartheta)t}{2}} + C\|z_{\alpha,\beta}\|_{H^{\frac{1}{4}}}^4 e^{\frac{(\lambda_1^{\frac{\alpha}{2}}-1-\vartheta)t}{2}}\\ &+C\|z_{\alpha,\beta}\|_{L^2}^2 e^{\frac{(\lambda_1^{\frac{\alpha}{2}}-1-\vartheta)t}{2}}.\end{aligned}$$

Note that there exists a positive random variable $R_2(w)$ such that

$$\begin{aligned}&\int_{-\gamma}^t \|v_\gamma(s)\|_{H^{\frac{\alpha}{2}}}^2 e^{\frac{(\lambda_1^{\frac{\alpha}{2}}-1-\vartheta)s}{2}}\mathrm{d}s\\ \leqslant\, &\|z_{\alpha,\beta}(-\gamma)\|_{L^2}^2 e^{-\frac{(\lambda_1^{\frac{\alpha}{2}}-1-\vartheta)\gamma}{2}}\\ &+C\int_{-\gamma}^t (\|z_{\alpha,\beta}(s)\|_{H^{\frac{1}{4}}}^4 + \|z_{\alpha,\beta}(s)\|_{L^2}^2)e^{\frac{(\lambda_1^{\frac{\alpha}{2}}-1-\vartheta)s}{2}}\mathrm{d}s\\ &-\frac{3}{4}(\lambda_1^{\frac{\alpha}{2}}-1-\vartheta)\int_{-\gamma}^t \|v_\gamma(s)\|_{L^2}^2 e^{\frac{(\lambda_1^{\frac{\alpha}{2}}-1-\vartheta)s}{2}}\mathrm{d}s \leqslant R_2(\omega) < \infty,\end{aligned} \tag{4.27}$$

the last inequality follows by (4.22). Let $(v_\gamma(t))_{t\in\mathbb{R}}$ be the solution to (4.18). Then

$$\begin{aligned}v_\gamma(0) =& -e^{-(A^{\frac{\alpha}{2}}-1)\gamma} z_{\alpha,\beta}(-\gamma)\\ &-\int_{-\gamma}^0 e^{(A^{\frac{\alpha}{2}}-1)s}(v_\gamma(s)+z_{\alpha,\beta}(s))^3\mathrm{d}s\\ &+\beta\int_{-\gamma}^0 e^{(A^{\frac{\alpha}{2}}-1)s} z_{\alpha,\beta}(s)\mathrm{d}s.\end{aligned} \tag{4.28}$$

In the following, we try to get the priori estimates for $v_\gamma(0)$ in $H^{\frac{\alpha}{2}+\varepsilon}$.

$$\begin{aligned}\|A^{\frac{\alpha}{4}+\frac{\varepsilon}{2}}v_\gamma(0)\|_{L^2} \leqslant\, &\|A^{\frac{\alpha}{4}+\frac{\varepsilon}{2}}e^{-(A^{\frac{\alpha}{2}}-1)\gamma}z_{\alpha,\beta}(-\gamma)\|_{L^2}\\ &+\int_{-\gamma}^0 \|A^{\frac{\alpha}{4}+\frac{\varepsilon}{2}}e^{(A^{\frac{\alpha}{2}}-1)s}(v_\gamma(s)+z_{\alpha,\beta}(s))^3\|_{L^2}\mathrm{d}s\\ &+\beta\int_{-\gamma}^0 \|A^{\frac{\alpha}{4}+\frac{\varepsilon}{2}}e^{(A^{\frac{\alpha}{2}}-1)s}z_{\alpha,\beta}(s)\|_{L^2}\mathrm{d}s\\ :=\, &I_1+I_2+I_3.\end{aligned}$$

Dealing with I_1 as follows

$$\begin{aligned} I_1 &= \left(\sum_{k=1}^{\infty}\langle A^{\frac{\alpha}{4}+\frac{\varepsilon}{2}} e^{-(A^{\frac{\alpha}{2}}-1)\gamma} z_{\alpha,\beta}(-\gamma),\ e_k\rangle^2\right)^{\frac{1}{2}} \\ &= \left(\sum_{k=1}^{\infty}\langle A^{\frac{\alpha}{4}+\frac{\varepsilon}{2}} z_{\alpha,\beta}(-\gamma), e^{-(\lambda_k^{\frac{\alpha}{2}}-1)\gamma} e_k\rangle^2\right)^{\frac{1}{2}} \\ &\leqslant \|z_{\alpha,\beta}(-\gamma)\|_{H^{\frac{\alpha}{2}+\varepsilon}} e^{-(\lambda_1^{\frac{\alpha}{2}}-1)\gamma} < R_3(w) < \infty, \end{aligned}$$

the second inequality follows since $\|z_{\alpha,\beta}(-\gamma)\|_{H^{\frac{\alpha}{2}+\varepsilon}} e^{-(\lambda_1^{\frac{\alpha}{2}}-1)\gamma}$ is bounded a.s.. Note that

$$\begin{aligned} &\|A^{\frac{\alpha}{4}+\frac{\varepsilon}{2}} e^{(A^{\frac{\alpha}{2}}-1)s}(v_\gamma + z_{\alpha,\beta})^3\|_{L^2}^2 \\ =& \sum_{k=1}^{\infty}\langle A^{\frac{\alpha}{4}+\frac{\varepsilon}{2}} e^{(A^{\frac{\alpha}{2}}-1)s}(v_\gamma + z_{\alpha,\beta})^3, e_k\rangle^2 \\ =& \sum_{k=1}^{\infty}\langle (v_\gamma + z_{\alpha,\beta})^3, \lambda_k^{\frac{\alpha}{4}+\frac{\varepsilon}{2}} e^{(\lambda_k^{\frac{\alpha}{2}}-1)s} e_k\rangle^2 \\ \leqslant& \sum_{k=1}^{\infty}\lambda_k^{\frac{\alpha}{2}+\varepsilon} e^{2(\lambda_k^{\frac{\alpha}{2}}-1)s}\|v_\gamma + z_{\alpha,\beta}\|_{L^3}^6 \\ \leqslant& C\sum_{k=1}^{\infty}\lambda_k^{\frac{\alpha}{2}+\varepsilon} e^{2(\lambda_k^{\frac{\alpha}{2}}-1)s}\left(\|v_\gamma\|_{H^{\frac{\alpha}{2}}}^{\frac{2}{\alpha}} \|v_\gamma\|_{L^2}^{\frac{2(3\alpha-1)}{\alpha}} + \|z_{\alpha,\beta}\|_{L^3}^6\right) \\ \leqslant& C\sum_{k=1}^{\infty}\lambda_k^{\frac{\alpha}{2}+\varepsilon}\lambda_k^{-\frac{\alpha\theta_2}{2}}|s|^{-\theta_2}[e^{(\lambda_1^{\frac{\alpha}{2}}-1)s} \\ &\times(\|v_\gamma\|_{H^{\frac{\alpha}{2}}}^4 + \|v_\gamma\|_{L^2}^{\frac{4(3\alpha-1)}{2\alpha-1}} + \|z_{\alpha,\beta}\|_{L^3}^6)], \end{aligned}$$

where we use the Gagliardo-Nirenberg inequality in the third inequality and $\theta_2 \in (0,2)$ in the last inequality. Let $\varepsilon \in \left(0, \dfrac{\alpha-1}{2}\right)$. Then we can choose $\theta_2 \in \left(1+\dfrac{1+2\varepsilon}{\alpha}, 2\right)$ such that

$$\left(\frac{\alpha}{2}+\varepsilon\right) - \frac{\alpha\theta_2}{2} < -\frac{1}{2}$$

implies

$$\sum_{k=1}^{\infty}\lambda_k^{\frac{\alpha}{2}+\varepsilon}\lambda_k^{-\frac{\alpha\theta_2}{2}} < \infty.$$

Therefore,

$$\begin{aligned}&\|A^{\frac{\alpha}{4}+\frac{\varepsilon}{2}}e^{(A^{\frac{\alpha}{2}}-1)s}(v_\gamma+z_{\alpha,\beta})^3\|_{L^2}\\ \leqslant& Cs^{-\frac{\theta_2}{2}}e^{\frac{(\lambda_1^{\frac{\alpha}{2}}-1)s}{2}}(\|v_\gamma\|^2_{H^{\frac{\alpha}{2}}}+\|v_\gamma\|_{L^2}^{\frac{2(3\alpha-1)}{2\alpha-1}}+\|z_{\alpha,\beta}\|^3_{L^3}).\end{aligned}$$

After the preparation, we arrive at

$$\begin{aligned}I_2=&\int_{-\gamma}^0\|A^{\frac{\alpha}{4}+\frac{\varepsilon}{2}}e^{(A^{\frac{\alpha}{2}}-1)s}(v_\gamma(s)+z_{\alpha,\beta}(s))^3\|_{L^2}\mathrm{d}s\\ \leqslant& C\int_{-\gamma}^0 s^{-\frac{\theta_2}{2}}e^{\frac{(\lambda_1^{\frac{\alpha}{2}}-1)s}{2}}\|v_\gamma(s)\|^2_{H^{\frac{\alpha}{2}}}\mathrm{d}s\\ &+C\int_{-\gamma}^0 s^{-\frac{\theta_2}{2}}e^{\frac{(\lambda_1^{\frac{\alpha}{2}}-1)s}{2}}\|v_\gamma(s)\|_{L^2}^{\frac{2(3\alpha-1)}{2\alpha-1}}\mathrm{d}s\\ &+C\int_{-\gamma}^0 s^{-\frac{\theta_2}{2}}e^{\frac{(\lambda_1^{\frac{\alpha}{2}}-1)s}{2}}\|z_{\alpha,\beta}(s)\|^3_{L^3}\mathrm{d}s.\end{aligned}$$

For the first term on the right hand side of the above inequality, we obtain

$$\begin{aligned}&\int_{-\gamma}^0 s^{-\frac{\theta_2}{2}}e^{\frac{(\lambda_1^{\frac{\alpha}{2}}-1)s}{2}}\|v_\gamma(s)\|^2_{H^{\frac{\alpha}{2}}}\mathrm{d}s\\ \leqslant& C\int_{-1}^0|s|^{-\frac{\theta_2}{2}}\mathrm{d}s+C\int_{-\infty}^{-1}e^{\frac{(\lambda_1^{\frac{\alpha}{2}}-1)s}{2}}\|v_\gamma(s)\|^2_{H^{\frac{\alpha}{2}}}\mathrm{d}s\leqslant R_3(\omega),\end{aligned}$$

where the first inequality follows by the fact that by Theorem 3.1 $v_\gamma\in C([-1,0];H^{\frac{\alpha}{2}})$ and the last inequality follows by (4.27). By (4.23), for $s\leqslant 0$, we have

$$\begin{aligned}e^{\frac{(\lambda_1^{\frac{\alpha}{2}}-1)s}{2}}\|v_\gamma(s)\|_{L^2}^{\frac{2(3\alpha-1)}{2\alpha-1}}\leqslant& e^{\frac{(\lambda_1^{\frac{\alpha}{2}}-1)s}{2}}(R(\omega))^{\frac{3\alpha-1}{2\alpha-1}}e^{-\varepsilon\frac{3\alpha-1}{2\alpha-1}s}\\ \leqslant&(R(\omega))^{\frac{3\alpha-1}{2\alpha-1}}e^{(\frac{(\lambda_1^{\frac{\alpha}{2}}-1)}{2}-\varepsilon\frac{3\alpha-1}{2\alpha-1})s}.\end{aligned}$$

Let ε be small enough that

$$\frac{(\lambda_1^{\frac{\alpha}{2}}-1)}{2}-\varepsilon\frac{3\alpha-1}{2\alpha-1}>0.$$

Then, by (4.22), we summarize that,

$$I_2\ \leqslant\ R_3(\omega)+R_4(\omega)<\infty,\ P-a.e.\ \omega\in\Omega,$$

where $R_4(\omega)$ is a positive random variable satisfying

$$\begin{aligned}&C\int_{-\gamma}^0 s^{-\frac{\theta_2}{2}}e^{\frac{(\lambda_1^{\frac{\alpha}{2}}-1)s}{2}}\|v_\gamma(s)\|_{L^2}^{\frac{2(3\alpha-1)}{2\alpha-1}}\mathrm{d}s\\ &+C\int_{-\gamma}^0 s^{-\frac{\theta_2}{2}}e^{\frac{(\lambda_1^{\frac{\alpha}{2}}-1)s}{2}}\|z_{\alpha,\beta}(s)\|^3_{L^3}\mathrm{d}s\leqslant R_4(\omega),\ P-a.e.\ \omega\in\Omega.\end{aligned}$$

By the estimate of I_1 and I_2, we deduce there exists a positive random variable $R_5(\omega)$ such that

$$\|A^{\frac{\alpha}{4}+\frac{\varepsilon}{2}}v_\gamma(0)\|_H \leqslant R_5(\omega) < \infty, \quad P-a.e.\ \omega \in \Omega.$$

Since $u_\gamma(0) = v_\gamma(0) + z_{\alpha,\beta}(0)$ and $\|z_{\alpha,\beta}(0)\|_{H^{\frac{\alpha}{2}+\varepsilon}} < \infty$ by (4.22), it is easy to see $\{u_\gamma(0)\}_{\gamma\geqslant 0}$ is almost surely bounded in $H^{\frac{\alpha}{2}+\varepsilon}$. Since $H^\sigma \subset H^{\frac{\alpha}{2}+\varepsilon}$ is compact, for $\sigma \in \left[\frac{1}{6}, \frac{\alpha}{2}\right]$, the family of laws $\{\mathcal{L}(u_\gamma(0))\}_{\gamma\geqslant 0}$ is tight in H^σ. For $t \geqslant 0$ and $x \in H^\sigma$, set

$$(P_t f)(x) = Ef(u(t,.;0,x))$$

where $f \in C_b(H^\sigma)$. Following the arguments in [37] or [39], for all $t_0 < s < t$, and all $u_{t_0} \in H^\sigma$, by proving

$$E(f(u(t;t_0,u_{t_0}))|\mathcal{F}_s) = P_{t-s}(u(s;t_0,u_{t_0}))$$

we can show u is a Markov process. Here $\mathcal{F}_s$ is the $\sigma-$algebra generated by $W(r)$ for $r \leqslant s$. So $(P_t)_{t\geqslant 0}$ is the Markov semigroup. Define a dual semigroup P_t^* in the space $P(H^\sigma)$ of probability measures on H^σ:

$$\int_{H^\sigma} f\mathrm{d}(P_t^*\mu) = \int_{H^\sigma} P_t f\mathrm{d}\mu.$$

Let ν_τ be the law of $u_\tau(0)$, which is the solution of (4.15) with initial condition $u(-\tau) = 0$. Then we have

$$\begin{aligned}\nu_\tau(f) &= Ef(u_{-\tau}(0)) = Ef(u(\tau,.;0,0))\\ &= (P_\tau f)(0) = \int_{H^\sigma} P_\tau f\mathrm{d}\delta_0 = \int_{H^\sigma} f\mathrm{d}(P_\tau^*\delta_0),\end{aligned}$$

the second equality follows by the fact that $u(\tau,.;0,0)$ and $u_\tau(0)$ have the same law. So obviously

$$P_{\tau_1}^*\nu_\tau = \nu_{\tau+\tau_1}.$$

Let

$$\mu_T = \frac{1}{T}\int_0^T \nu_\tau \mathrm{d}\tau,\ T > 0.$$

The family $\{\mu_T : T \geqslant 0\}$ is tight. By Prohorov theorem, there exists a sequence μ_{t_n} with $t_n \to \infty$, converging in law to some $\mu \in P(H^\sigma)$. We show that μ is invariant.

Indeed, for $\tau, s \geqslant 0$,

$$\begin{aligned}P_s^*\mu_{t_n} &= \frac{1}{t_n}\int_0^{t_n} P_s^*\nu_\tau \mathrm{d}\tau = \frac{1}{t_n}\int_0^{t_n} \nu_{\tau+s}\mathrm{d}\tau \\ &= \mu_{t_n} + \frac{1}{t_n}\int_{t_n-s}^{t_n} \nu_{\tau+s}\mathrm{d}\tau - \frac{1}{t_n}\int_{-s}^{0} \nu_{\tau+s}\mathrm{d}\tau.\end{aligned}$$

Since P_s^* is continuous in the weak star topology of $P(H^\sigma)$, we can pass to the limit and obtain $P_s^*\mu = \mu$.

5 Uniqueness of the invariant measures

The main result of this part is

Theorem 5.1 *In addition to the assumptions in Theorem* 4.1, *we suppose*

$$D(A^{\frac{\alpha}{2}}) \subset R(G) \subset D(A^{\frac{\alpha+1}{4}+\varepsilon}).$$

Then:

(i) *problem* (1.2) *has a unique invariant measure* μ *on* $H^{\frac{\alpha}{2}}$;

(ii) *for all* $u_0 \in H^{\frac{\alpha}{2}}$ *and all Borel measurable functions* $\varphi : H^{\frac{\alpha}{2}} \to \mathbb{R}$, *such that* $\int_{H^{\frac{\alpha}{2}}} |\varphi| d\mu < \infty$,

$$\lim_{T\to\infty}\frac{1}{T}\int_0^T \varphi(u(t;u_0))\mathrm{d}t = \int_{H^{\frac{\alpha}{2}}} \varphi \mathrm{d}\mu \quad a.s.;$$

(iii) *for every Borel measure* μ^* *on* $H^{\frac{\alpha}{2}}$ *we have that*

$$\|P_t^*\mu^* - \mu\|_{TV} \to 0 \quad as\ t \to \infty,$$

where $\|\cdot\|_{TV}$ *deontes the total variation of a measure. In particular, we have that*

$$P_t^*\mu^*(B) \to \mu(B) \quad as\ t \to \infty, \tag{5.29}$$

for every Borel set $B \in \mathcal{B}(H^{\frac{\alpha}{2}})$*(the Borel* $\sigma-$*algebra of* $H^{\frac{\alpha}{2}}$*).*

Define the transition probability measure $P(t,x,B) = P_t^*\delta_x(B) = P(u(t;x) \in B)$ for $t > 0, x \in H^{\frac{\alpha}{2}}$ and $B \in \mathcal{B}(H^{\frac{\alpha}{2}})$. To prove Theorem 5.1, we only need the following result, by Theorem 4.2.1 in [38].

Theorem 5.2 *Assume that the probability measures* $P(t,x,.), t > 0, x \in H^{\frac{\alpha}{2}}$, *are all equivalent, in the sense that they are mutually absolutely continuous. Then Theorem* 5.1 *holds true.*

Next, we will prove the irreducibility and the strong Feller property in $H^{\frac{\alpha}{2}}$ to get the equivalence of the measure $P(t,x,.)$. First, we outline the two properties as follows. Assume $y \in H^{\frac{\alpha}{2}}, \varepsilon > 0$, let

$$B(y,\varepsilon) = \{x \in H^{\frac{\alpha}{2}}; \|x-y\|_{H^{\frac{\alpha}{2}}} < \varepsilon\}.$$

(I) (irreducibility property) Assume $x, y \in H^{\frac{\alpha}{2}}, \varepsilon, t > 0$. Then

$$P(t,x,B(y,\varepsilon)) > 0.$$

(S) (strong Feller property) For all $O \in \mathcal{B}(H^{\frac{\alpha}{2}})$, every $t > 0$, and all $x_n, x \in H^{\frac{\alpha}{2}}$ such that $x_n \to x$ in $H^{\frac{\alpha}{2}}$, it holds

$$P(t,x_n,O) \to P(t,x,O).$$

For $x \in H^{\frac{\alpha}{2}}$ and $\phi : C([0,T]\times[0,1];\mathbb{R})$, set

$$u(t,x,\phi) = v(t,x,\phi) + \phi(t),$$

where $v(t,x,\phi) \in C([0,T];H^{\frac{\alpha}{2}})$ is a solution of the equation

$$\frac{\mathrm{d}v}{\mathrm{d}t} + \mathcal{A}^{\frac{\alpha}{2}}v + (v+\phi)^3 = 0 \tag{5.30}$$

with initial condition $v(0) = x$. In order to check the property (I), we need the following lemma.

Lemma 5.1 *Define* $\Psi(\phi) = u(.,x,\phi)$. *Then*

(a) *the mapping*

$$\Psi : C([0,T]\times[0,1];\mathbb{R}) \to C([0,T];H^{\frac{\alpha}{2}})$$

is continuous;

(b) *for every* $x, y \in H^{\frac{\alpha}{2}}$ *and* $T > 0$, *there exists* $\bar{z} \in C([0,T]\times[0,1];\mathbb{R})$ *such that* $u(T,x,\bar{z}) = y$.

Proof We first consider (a). By Theorem 3.1, there exists solutions v_1 and v_2 of (5.30) with the initial condition $x \in H^{\frac{\alpha}{2}}$ but with different functions ϕ_1 and ϕ_2. Then

$$v_i(t) = \mathrm{e}^{-(A^{\frac{\alpha}{2}}-1)t}x + \int_0^t \mathrm{e}^{-(A^{\frac{\alpha}{2}}-1)(t-s)}(v_i+\phi_i)^3\mathrm{d}s,\ i=1,2.$$

Set $\eta = v_1 - v_2$ and $u_i = v_i + \phi_i, i = 1, 2$. Then

$$\eta(t) = \int_0^t e^{-(A^{\frac{\alpha}{2}}-1)(t-s)}[(v_1+\phi_1)^3 - (v_2+\phi_2)^3]\mathrm{d}s.$$

Therefore,

$$\begin{aligned}
\|\eta(t)\|_{H^{\frac{\alpha}{2}}} \leqslant& \int_0^t \|e^{-(A^{\frac{\alpha}{2}}-1)(t-s)}[(v_1+\phi_1)^3 - (v_2+\phi_2)^3]\|_{H^{\frac{\alpha}{2}}}\mathrm{d}s \\
=& \int_0^t \Big(\sum_{k=1}^{\infty}\langle(u_1^3 - u_2^3), e^{-(\lambda_k^{\frac{\alpha}{2}}-1)(t-s)}\lambda_k^{\frac{\alpha}{4}} e_k\rangle^2\Big)^{\frac{1}{2}}\mathrm{d}s \\
\leqslant& C\int_0^t \|u_1^3 - u_2^3\|_{L^1}\Big(\sum_{k=1}^{\infty}\lambda_k^{\frac{\alpha}{2}-\frac{\alpha\theta}{2}}\Big)^{\frac{1}{2}}(t-s)^{-\frac{\theta}{2}}\mathrm{d}s \\
\leqslant& C\int_0^t \|u_1 - u_2\|_{L^3}(\|u_1\|_{L^3}^2 + \|u_2\|_{L^3}^2)(t-s)^{-\frac{\theta}{2}}\mathrm{d}s \\
\leqslant& C\|\phi_1 - \phi_2\|_{L_t^\infty L_x^\infty}(\|\phi_1\|_{L_t^\infty L_x^\infty}^2 + \|\phi_2\|_{L_t^\infty L_x^\infty}^2 \\
&+ \sup_{t\in[0,T]}\|v_1\|_{H^{\frac{\alpha}{2}}}^2 + \sup_{t\in[0,T]}\|v_2\|_{H^{\frac{\alpha}{2}}}^2)T^{1-\frac{\theta}{2}} \\
&+ C(\|\phi_1\|_{L_t^\infty L_x^\infty}^2 + \|\phi_2\|_{L_t^\infty L_x^\infty}^2 \\
&+ \sup_{t\in[0,T]}\|v_1\|_{H^{\frac{\alpha}{2}}}^2 + \sup_{t\in[0,T]}\|v_2\|_{H^{\frac{\alpha}{2}}}^2) \\
&\times \int_0^t \sup_{\tau\in[0,s]}\|\eta(\tau)\|_{H^{\frac{\alpha}{2}}}(t-s)^{-\frac{\theta}{2}}\mathrm{d}s,
\end{aligned}$$

where $1 + \dfrac{1}{\alpha} < \theta < 2$. From the Gronwall inequality we have

$$\sup_{t\in[0,T]}\|v_1 - v_2\|_{H^{\frac{\alpha}{2}}} \leqslant \|\phi_1 - \phi_2\|_{L_t^\infty L_x^\infty} C(v_1, v_2, \phi_1, \phi_2, T).$$

Next we will prove (b). Let $x, y, \in H^{\frac{\alpha}{2}}$ and $T > 0$, choose any $0 < t_0 < t_1 < T$. Define $\bar{u}$ as:

$$\begin{aligned}
\bar{u}(t) =& e^{-tA^{\frac{\alpha}{2}}}x, \quad t \in [0, t_0], \\
\bar{u}(t) =& e^{-(T-t)A^{\frac{\alpha}{2}}}y, \quad t \in [t_1, T], \\
\bar{u}(t) =& \bar{u}(t_0) + \frac{t - t_0}{t_1 - t_0}(\bar{u}(t_1) - \bar{u}(t_0)), \quad t \in (t_0, t_1).
\end{aligned}$$

Obviously $\bar{u}(t) \in C([0, T]; H^{\frac{\alpha}{2}})$. Define $\bar{v}$ as the solution of the equation

$$\frac{\mathrm{d}\bar{v}}{\mathrm{d}t} + \mathcal{A}^{\frac{\alpha}{2}}\bar{v} + \bar{u}^3 = 0$$

with initial condition $\bar{v}(0) = x$, then $\bar{v} \in C([0,T]; H^{\frac{\alpha}{2}})$. Set $\bar{z} = \bar{u} - \bar{v}$, then $\bar{z} \in C([0,T]; H^{\frac{\alpha}{2}})$. Since $H^{\frac{\alpha}{2}} \subset C([0,1];\mathbb{R})$, $\bar{z}$ satisfies all the requirements of the lemma, part (b).

Proposition 5.1 *With conditions in Theorem* 5.1, *the irreducibility property* (I) *is satisfied.*

Proof Let $x \in H^{\frac{\alpha}{2}}, z \in C([0,T]\times[0,1];\mathbb{R})$ and $\bar{z}$ be given by part (b) of Lemma 5.1. By the above lemma, for $\varepsilon > 0$, there exists $\delta > 0$ such that

$$\|z - \bar{z}\|_{L_t^\infty L_x^\infty} < \delta$$

implies

$$\|u(.,x,z) - u(.,x,\bar{z})\|_{C([0,T];H^{\frac{\alpha}{2}})} < \varepsilon.$$

Let $\delta_1 \in (0,\delta)$ and

$$U_{\delta_1} =: \{z \in C([0,T]\times[0,1];\mathbb{R}); \|z-\bar{z}\|_{L_t^\infty L_x^\infty} < \delta_1\}.$$

Then for $z \in U_{\delta_1}$, we have that

$$\|u(T,x,z) - y\|_{H^{\frac{\alpha}{2}}} < \varepsilon.$$

Recall now that the solution u of (4.13) is equal to $\Psi(z), z$ being the Ornstein-Uhlenbeck process. Then it remains to show that

$$P\{z(.,w) \in U_{\delta_1}\} > 0,$$

which follows from Proposition 2.11 in [31]. So far we have proved that for $x, y \in H^{\frac{\alpha}{2}}$ and $t, \varepsilon > 0$, we have

$$P(t,x,B(y,\varepsilon)) > 0.$$

In this part, we prove that the property (S) holds true, which will complete the proof of Theorem 5.1. At first we prove the strong Feller property on $H^{\frac{\alpha}{2}}$ for modified fractional reaction-diffusion equation (5.31) below. Then let $R \to \infty$ to check the property (S).

Fix $R > 0$, let $K_R : [0,\infty[\to [0,\infty[$ satisfy $K_R \in C^1(\mathbb{R}_+)$ such that $|K_R| \leqslant 1, |K_R'| \leqslant 2$ and

$$K_R(x) = 1, \quad \text{if } x < R; \quad K_R(x) = 0, \quad \text{if } x \geqslant R+1.$$

Consider the equation

$$\begin{cases} \mathrm{d}u_R(t) + A^{\frac{\alpha}{2}} u_R(t)\mathrm{d}t \\ \quad + (K_R(\|u_R(t)\|^2_{H^{\frac{\alpha}{2}}}))^3 (u_R(t))^3 \mathrm{d}t = G\mathrm{d}W(t), & t > 0,\ x \in (0,1), \\ u_R(t,0) = u_R(t,1) = 0, & t > 0, \\ u_R(0,x) = u_R(x) \in H^{\frac{\alpha}{2}}, & x \in [0,1]. \end{cases} \tag{5.31}$$

Proposition 5.2 *There exists a unique mild solution* $u_R(.,\omega) \in C([0,T]; H^{\frac{\alpha}{2}})$ *for* (5.31) *which is Markov process with the Feller property in* $H^{\frac{\alpha}{2}}$, *i.e. for every* $R > 0, t > 0$, *there exists a constant* $L = L(t,R) > 0$ *such that*

$$|P_t^{(R)}\phi(x) - P_t^{(R)}\phi(y)| \leqslant L\|x-y\|_{H^{\frac{\alpha}{2}}}$$

holds for all $x, y \in H^{\frac{\alpha}{2}}$, *and all* $\phi \in C_b(H^{\frac{\alpha}{2}}) \leqslant 1$, *where* $P_t^{(R)}\phi(x) := \int_{H^{\frac{\alpha}{2}}} \phi(y) P_R(t, x, dy)$, $P_R(t,x,\cdot)$ *is the transition probabilities corresponding to* (5.31).

Proof Analogously to the arguments in section 3 and section 4, we can prove the existence, uniqueness and Markov property of the solution to (5.31).

To prove the Fell property, we first consider the following Galerkin approximations of (5.31). Let P_n be the orthogonal projection in $L^2(0,1)$ defined as $P_n x = \sum_{j=1}^n \langle x, e_j\rangle e_j, x \in L^2(0,1)$. Denote $H_n := P_n L^2(0,1)$ for every $n \in \mathbb{N}$. Consider the equation in H_n :

$$\frac{\partial u_n^{(R)}(t)}{\partial t} + A^{\frac{\alpha}{2}} u_n^{(R)}(t) + (K_R(\|u_n^{(R)}(t)\|^2_{H^{\frac{\alpha}{2}}}))^3$$

$$P_n(u_n^{(R)}(t))^3 - P_n u_n^{(R)}(t) = P_n G\mathrm{d}W(t) \tag{5.32}$$

with initial condition $u_n^{(R)}(0) = P_n u_0$. This is a finite dimensional equation with globally Lipschitz nonlinear functions. Thus it has a unique progressively measurable solution with P-a.e. trajectory $u_n^{(R)}(.,\omega) \in C([0,T]; H_n)$, which is also a Markov process in H_n with associated semigroup $P_{n,t}^{(R)}$ defined as

$$P_{n,t}^{(R)}\phi(x) = E\phi(u_n^{(R)}(t;x))$$

for all $x \in H_n$ and $\phi \in C_b(H_n)$. For every $R > 0, t > 0$, we can prove that there exists a constant $L = L(t,R) > 0$ such that

$$|P_{n,t}^{(R)}\phi(x) - P_{n,t}^{(R)}\phi(y)| \leqslant L\|x-y\|_{H^{\frac{\alpha}{2}}} \tag{5.33}$$

holds for all $n \in \mathbb{N}, x, y \in H_n$, and all $\phi \in C_b(H_n)$ with $\|\phi\|_{H_n} \leqslant 1$. Indeed, the following remarkable formula holds true for the differential in x of $P_{n,t}^{(R)}\phi$ ([37]):

$$\begin{aligned} D_x P_{n,t}^{(R)}\phi(x)\cdot h = &\frac{1}{t}E\Big(\phi(u_n^{(R)}(t;x)) \\ &\times \int_0^t \langle (P_nGG^*P_n)^{-\frac{1}{2}} D_x u_n^{(R)}(s;x)\cdot h, d\beta_n(s)\rangle\Big) \end{aligned}$$

for all $h \in H_n$, where β_n is a n-dimensional standard Wiener process. We have used the fact that $P_nGW(t)$ is a n-dimensional Wiener process with incremental covariance $P_nGG^*P_n$. Therefore,

$$|D_x P_{n,t}^{(R)}\phi(x)\cdot h| \leqslant \frac{1}{t}E\Big(\int_0^t \|(P_nGG^*P_n)^{-\frac{1}{2}} D_x u_n^{(R)}(s;x)\cdot h\|_{L^2}^2 \mathrm{d}s\Big)^{\frac{1}{2}}.$$

Since $D(A^{\frac{\alpha}{2}}) \subset R(G)$, we can follow the arguments in [23] to arrive at

$$\begin{aligned} \|(P_nGG^*P_n)^{-\frac{1}{2}}y\|_{L^2}^2 = &\langle (P_nGG^*P_n)^{-1}y, y\rangle \\ = &\langle (A^{\frac{\alpha}{2}}P_nGG^*P_nA^{\frac{\alpha}{2}})^{-1}A^{\frac{\alpha}{2}}y, A^{\frac{\alpha}{2}}y\rangle \leqslant C\|y\|_{H^\alpha}^2. \end{aligned}$$

Next, we will prove

$$\begin{aligned} |D_x P_{n,t}^{(R)}\phi(x)\cdot h| \leqslant &\frac{1}{t}CE\Big(\int_0^t \|D_x u_n^{(R)}(s;x)\cdot h\|_{H^\alpha}^2 \mathrm{d}s\Big)^{\frac{1}{2}} \\ \leqslant &\frac{1}{t}C(R,T)\|h\|_{H^{\frac{\alpha}{2}}}, \end{aligned}$$

where $C(R,T)$ is independent of $x \in H_n$ and $n \in \mathbb{N}$. Indeed, let ξ_n be the differential of the mapping $x \to u_n$ in the direction h at point x, for given $x, h \in H^{\frac{\alpha}{2}}$:

$$\xi_n(t) = D_x u_n^{(R)}(t;x)\cdot h.$$

Thus ξ_n satisfies

$$\begin{aligned} &\frac{\mathrm{d}}{\mathrm{d}t}\xi_n + A^{\frac{\alpha}{2}}\xi_n \\ = &-6(K_R(\|u_n^{(R)}\|_{H^{\frac{\alpha}{2}}}^2))^2 K_R'(\|u_n^{(R)}\|_{H^{\frac{\alpha}{2}}}^2) \\ &\times \langle A^{\frac{\alpha}{4}}u_n^{(R)}, A^{\frac{\alpha}{4}}\xi_n\rangle (u_n^{(R)})^3 \\ &-3(K_R(\|u_n^{(R)}\|_{H^{\frac{\alpha}{2}}}^2))^3 |u_n^{(R)}|^2\xi_n + \xi_n. \end{aligned}$$

We have

$$\begin{aligned}&\frac{\mathrm{d}}{\mathrm{d}t}\|\xi_n\|^2_{H^{\frac{\alpha}{2}}}+\|\xi_n\|^2_{H^\alpha}\\=&-6(K_R(\|u_n^{(R)}\|^2_{H^{\frac{\alpha}{2}}}))^2K_R'(\|u_n^{(R)}\|^2_{H^{\frac{\alpha}{2}}})\\&\times\langle A^{\frac{\alpha}{4}}u_n^{(R)},A^{\frac{\alpha}{4}}\xi_n\rangle\langle(u_n^{(R)})^3,A^{\frac{\alpha}{2}}\xi_n\rangle\\&-3(K_R(\|u_n^{(R)}\|^2_{H^{\frac{\alpha}{2}}}))^3\langle|u_n^{(R)}|^2\xi_n,A^{\frac{\alpha}{2}}\xi_n\rangle+\|\xi_n\|^2_{H^{\frac{\alpha}{2}}}.\end{aligned}$$

Therefore, recalling that K_R has compact support in $[-R-1,R+1]$,

$$\begin{aligned}&\frac{\mathrm{d}}{\mathrm{d}t}\|\xi_n\|^2_{H^{\frac{\alpha}{2}}}+\|\xi_n\|^2_{H^\alpha}\\\leqslant&\,6K_R'(\|u_n^{(R)}\|^2_{H^{\frac{\alpha}{2}}})\|u_n^{(R)}\|_{H^{\frac{\alpha}{2}}}\|\xi_n\|_{H^{\frac{\alpha}{2}}}\|u_n^{(R)}\|^3_{L^6}\|\xi_n\|_{H^\alpha}\\&+3K_R(\|u_n^{(R)}\|^2_{H^{\frac{\alpha}{2}}})\|u_n^{(R)}\|^2_{L^\infty_x}\|\xi_n\|_{L^2}\|\xi_n\|_{H^\alpha}+\|\xi_n\|^2_{H^{\frac{\alpha}{2}}}\\\leqslant&\,C(R)\|\xi_n\|^2_{H^{\frac{\alpha}{2}}}+\varepsilon\|\xi_n\|^2_{H^\alpha},\end{aligned}$$

where the seconde inequality follows by Sobolev embedding theorem and the Young inequality. From the Gronwall inequality, we have now

$$\|\xi_n(t)\|^2_{H^{\frac{\alpha}{2}}}\leqslant C(R,T)\|h\|^2_{H^{\frac{\alpha}{2}}},\ \forall t\in[0,T],$$

and therefore, using again the previous inequality,

$$\int_0^T\|\xi_n(t)\|^2_{H^\alpha}\mathrm{d}t\leqslant C(R,T)\|h\|^2_{H^{\frac{\alpha}{2}}}.$$

Therefore,

$$\begin{aligned}&|P_{n,t}^{(R)}\phi(x)-P_{n,t}^{(R)}\phi(y)|\\\leqslant&\sup_{\|h\|_{H^{\frac{\alpha}{2}}}\leqslant1,k\in H_n}|D_xP_{n,t}^{(R)}\phi(k)\cdot h|\|x-y\|_{H^{\frac{\alpha}{2}}}\\\leqslant&\frac{1}{t}C(R,T)\|x-y\|_{H^{\frac{\alpha}{2}}}.\end{aligned}$$

In the following step, we will let $n\to\infty$ to get the Fell property for equation (5.31). Let $x\in H^{\frac{\alpha}{2}}$ and $\phi\in C_b(H^{\frac{\alpha}{2}})$ be given. Following the arguments in Section 3, we can also get the same estimates as (3.12) for $u_n^{(R)}$. So there exists a subsequence, still denoted by $(u_n^{(R)})_{n\in\mathbb{N}}$, which converges to $u^{(R)}(,;x)$ strongly in $L^2(0,T;H^{\frac{\alpha}{2}}),P$-a.s.. Fix $\omega\in\Omega$ such that this property holds. For this ω, by a simple argument on subsequences that converge a.s. in t, we see that $\phi(u_n^{(R)}(,;x))$ converges to $\phi(u^{(R)}(,;x))$ in $L^1(0,T)$. By the boundedness and continuous of ϕ as well as Lebesgue dominated convergence theorem, we have

$$E\int_0^T |\phi(u_n^{(R)}(,;x)) - \phi(u^{(R)}(,;x))|\mathrm{d}t \to 0,$$

which implies that for some subsequence n_k

$$E\phi(u_{n_k}^{(R)}(,;x)) \to E\phi(u^{(R)}(,;x))$$

for *a.e.* $t \in [0,T]$. Take $x, y \in H^{\frac{\alpha}{2}}$, by the previous argument, we can find a subsequence n_k such that the previous almost sure convergence in $t \in [0,T]$ holds true for both x and y. Thus, from (5.33), we have

$$|P_t^{(R)}\phi(x) - P_t^{(R)}\phi(y)| \leqslant L\|x-y\|_{H^{\frac{\alpha}{2}}}$$

for *a.e.*$t \in [0,T]$. As $u^{(R)}(t;x)$ has continuous trajectories with values in $H^{\frac{\alpha}{2}}$, the above inequality holds for all $t \in [0,T]$.

Proposition 5.3 *Under conditions of Theorem* 5.1, (S) *holds true.*

Proof Take $t>0, x_n, x \in H^{\frac{\alpha}{2}}$ satisfying $x_n \to x$ in $H^{\frac{\alpha}{2}}$. For every $R>0$ we have that

$$\begin{aligned}
&\|P_R(t,x_n,.) - P_R(t,x,.)\|_{TV} \\
=&\sup_{\|\phi\|_{C_b(H^{\frac{\alpha}{2}})}\leqslant 1} |P_t^{(R)}\phi(x_n) - P_t^{(R)}\phi(x)| \\
\leqslant& L\|x_n - x\|_{H^{\frac{\alpha}{2}}} \to 0,
\end{aligned}$$

as $n\to\infty$ by proposition 5.2. Then

$$\begin{aligned}
&\|P_R(t,x_n,.) - P(t,x_n,.)\|_{TV} + \|P_R(t,x,.) - P(t,x,.)\|_{TV} \\
=&\sup_{\|\phi\|_{C_b(H^{\frac{\alpha}{2}})}\leqslant 1} |P_t^{(R)}\phi(x_n) - P_t\phi(x_n)| \\
&+\sup_{\|\phi\|_{C_b(H^{\frac{\alpha}{2}})}\leqslant 1} |P_t^{(R)}\phi(x) - P_t\phi(x)| \\
=&\sup_{\|\phi\|_{C_b(H^{\frac{\alpha}{2}})}\leqslant 1} |E\phi(u_R(t;x_n)) - E\phi(u(t;x_n))|
\end{aligned}$$

$$
\begin{aligned}
&+\sup_{\|\phi\|_{C_b(H^{\frac{\alpha}{2}})}\leqslant 1}|E\phi(u_R(t;x))-E\phi(u(t;x))|\\
&\leqslant 2\int_\Omega I_{\{\sup_{n\in\mathbb{N}}\|u(t;x_n)\|_{H^{\frac{\alpha}{2}}}>R\}}P(\mathrm{d}w)\\
&+2\int_\Omega I_{\{\|u(t;x)\|_{H^{\frac{\alpha}{2}}}>R\}}P(\mathrm{d}w)\to 0,\quad as\ R\to\infty,
\end{aligned}
$$

where the inequality follows by the consistency of $u(t;x)$ and $u^{(R)}(t;x)$, when $\|u(t;x)\|^2_{H^{\frac{\alpha}{2}}}\leqslant R$, and the limit follows by Theorem 3.1. Therefore,

$$
\begin{aligned}
&\|P(t,x_n,.)-P(t,x,.)\|_{TV}\\
\leqslant&\|P(t,x_n,.)-P_R(t,x_n,.)\|_{TV}+\|P_R(t,x_n,.)-P_R(t,x,.)\|_{TV}\\
&+\|P_R(t,x,.)-P(t,x,.)\|_{TV}\to 0
\end{aligned}
$$

as $n\to\infty$.

6 Some final remarks

In Theorem 5.1, we establish the ergodicity for problem (1.2) in $H^{\frac{\alpha}{2}}$. In fact, if $\alpha\in\left(\frac{3}{2},2\right)$, we can also obtain the same result in H^σ with $\sigma\in\left[\frac{1}{6},\frac{\alpha}{2}\right]$.

Similarly to the arguments in Lemma 2.1, we can conclude that if

$$
R(G)\subset D(A^{\frac{\sigma}{2}}) \tag{6.34}
$$

with $\sigma\in\left[\frac{1}{6},\frac{\alpha}{2}\right]$, then

$$
z_\alpha(.,\omega)\in C([0,T];H^\sigma)
$$

for P-a.e. $\omega\in\Omega$. By Theorem 3.1, we have that $(u(t))_{t\geqslant 0}$the solution of (1.2) belongs to $C([0,T];H^\sigma)$. In the procedure of proving (4.22), we find that if $\alpha\in\left(\frac{3}{2},2\right)$, we can choose $\theta_1\in\left(1,\frac{3}{2}\right)$ such that

$$
\alpha(1-\theta_1)<-\frac{3}{4}.
$$

Therefore, under condition (6.34), following the steps in the proof of (4.22) we can deduce

$$
\|z_{\alpha,\beta}(t)\|_{H^{\sigma+\varepsilon}}\leqslant C_1(w)(|t|+1),\ P-a.e.\ \omega\in\Omega,
$$

where $\varepsilon \leqslant \frac{1}{8}$ and $C_1(\omega)$ is a positive random variable. Hence, in addition to conditions of Theorem 4.1, we assume $\alpha \in \left(\frac{3}{2}, 2\right)$, then proceeding as in section 4 we can also obtain the existence of the invariant measure in H^σ where $\sigma \in \left[\frac{1}{6}, \frac{\alpha}{2}\right]$. Next we will make some modifications for Lemma 5.1. All the notations emerging in the following have the same meaning as in Section 5.

Lemma 6.1 *Define* $\Psi(\phi) = u(., x, \phi)$. *Then*

(a) *the mapping*

$$\Psi : C([0,T]; H^\sigma) \to C([0,T]; H^\sigma)$$

is continuous;

(b) *for every* $x, y \in H^\sigma$ *and* $T > 0$, *there exists* $\bar{z} \in C([0,T]; H^\sigma)$ *such that* $u(T, x, \bar{z}) = y$.

The proof is similar to that in Lemma 5.1. We will not give the proof here. Recalling that for $x, y, \in H^\sigma$, $T > 0$, and $0 < t_0 < t_1 < T$. $\bar{u}$ is defined as:

$$\begin{aligned}
\bar{u}(t) &= e^{-tA^{\frac{\alpha}{2}}} x, \quad t \in [0, t_0], \\
\bar{u}(t) &= e^{-(T-t)A^{\frac{\alpha}{2}}} y, \quad t \in [t_1, T], \\
\bar{u}(t) &= \bar{u}(t_0) + \frac{t - t_0}{t_1 - t_0}(\bar{u}(t_1) - \bar{u}(t_0)), \quad t \in (t_0, t_1).
\end{aligned}$$

Obviously $\bar{u}(t) \in C([0,T]; H^\sigma)$. In this part, we only should point out that by energy estimates the solution $\bar{v}$ of the equation

$$\frac{d\bar{v}}{dt} + A^{\frac{\alpha}{2}}\bar{v} + \bar{u}^3 = 0$$

with initial condition $\bar{v}(0) = x$, satisfies $\bar{v} \in C([0,T]; H^\sigma)$ provided $\bar{u}^3 \in L^2([0,T]; D(A^{-\frac{\alpha}{4}+\frac{\sigma}{2}}))$. In the following we will check the fact

$$\bar{u}^3 \in L^2([0,T]; D(A^{-\frac{\alpha}{4}+\frac{\sigma}{2}})). \tag{6.35}$$

(i) If $\sigma \in \left[\frac{1}{3}, \frac{\alpha}{2}\right]$, then $-\frac{\alpha}{4} + \frac{\sigma}{2} \leqslant 0$ and

$$\|\bar{u}^3\|^2_{D(A^{-\frac{\alpha}{4}+\frac{\sigma}{2}})} \leqslant \|\bar{u}^3\|^2_{L^2} \leqslant \|\bar{u}\|^6_{L^6} \leqslant \|\bar{u}\|^6_{H^{\frac{1}{3}}}.$$

Then (6.35) follows by $\bar{u}(t) \in C([0,T]; H^\sigma)$.

(ii) If $\sigma \in \left[\frac{1}{6}, \frac{1}{4}\right]$, then

$$\|\bar{u}^3\|^2_{D(A^{-\frac{\alpha}{4}+\frac{\sigma}{2}})} = \sum_{k=1}^{\infty} \langle \bar{u}^3, \lambda_k^{-\frac{\alpha}{4}+\frac{\sigma}{2}} e_k \rangle^2 \leqslant \sum_{k=1}^{\infty} \lambda_k^{-\frac{\alpha}{2}+\sigma} \|\bar{u}\|^6_{H^{\frac{1}{6}}}. \tag{6.36}$$

Since $\alpha > \frac{3}{2}$ and $\sigma \leqslant \frac{1}{4}$, it follows $\frac{\alpha}{2} - \sigma > \frac{1}{2}$. Therefore,

$$\sum_{k=1}^{\infty} \lambda_k^{-\frac{\alpha}{2}+\sigma} < \infty.$$

By (6.36),

$$\int_0^T \|\bar{u}^3\|^2_{D(A^{-\frac{\alpha}{4}+\frac{\sigma}{2}})} \mathrm{d}s \leqslant C \int_0^T \|\bar{u}\|^6_{H^{\frac{1}{6}}} ds \leqslant C \sup_{t\in[0,T]} \|\bar{u}\|^6_{H^{\frac{1}{6}}}.$$

Thus, (6.35) is proved in this case.

(iii) We consider $\sigma \in \left(\frac{1}{4}, \frac{1}{3}\right)$.

$$\begin{aligned}
&\int_0^T \|A^{\frac{1}{2}}\bar{u}\|^2_{L^2} \mathrm{d}s \\
&= \int_0^{t_0} \|A^{\frac{1}{2}}\bar{u}\|^2_{L^2} \mathrm{d}s + \int_{t_0}^{t_1} \|A^{\frac{1}{2}}\bar{u}\|^2_{L^2} \mathrm{d}s + \int_{t_1}^{T} \|A^{\frac{1}{2}}\bar{u}\|^2_{L^2} \mathrm{d}s \\
&=: J_1 + J_2 + J_3.
\end{aligned}$$

For J_1, we have

$$\begin{aligned}
J_1 &= \int_0^{t_0} \|A^{\frac{1}{4}} e^{-sA^{\frac{\alpha}{2}}} x\|^2_{L^2} \mathrm{d}s \\
&= \int_0^{t_0} \sum_{k=1}^{\infty} \lambda_k^{\frac{1}{2}-\sigma} e^{-2s\lambda_k^{\frac{\alpha}{2}}} \langle A^{\frac{\sigma}{2}} x, e_k\rangle^2 \mathrm{d}s \leqslant \|x\|^2_{H^\sigma} \sum_{k=1}^{\infty} \lambda_k^{\frac{1}{2}-\sigma-\frac{\alpha}{2}}.
\end{aligned}$$

Since $\alpha > \frac{3}{2}$ and $\sigma > \frac{1}{4}$, then

$$\frac{1}{2} - \sigma - \frac{\alpha}{2} < -\frac{1}{2}$$

implies

$$\sum_{k=1}^{\infty} \lambda_k^{\frac{1}{2}-\sigma-\frac{\alpha}{2}} < \infty.$$

Hence, we deduce

$$J_1 \leqslant C\|x\|^2_{H^\sigma}.$$

Analogously to J_1, we can get the same estimates for J_2 and J_3. Hence, we proved that $\bar{u} \in L^2([0,T]; H^{\frac{1}{2}})$. By Interpolation inequality and Sobolev embedding theorem,

$$\|\bar{u}\|_{L^6} \leqslant C\|\bar{u}\|_{H^{\frac{1}{2}}}^{\frac{1}{3}} \|\bar{u}\|_{L^4}^{\frac{2}{3}} \leqslant C\|\bar{u}\|_{H^{\frac{1}{2}}}^{\frac{1}{3}} \|\bar{u}\|_{H^{\frac{1}{4}}}^{\frac{2}{3}}.$$

Therefore,

$$\begin{aligned}\int_0^T \|\bar{u}^3\|^2_{D(A^{-\frac{\alpha}{4}+\frac{\sigma}{2}})} ds &\leqslant \int_0^T \|\bar{u}\|^6_{L^6} ds \leqslant C\int_0^T \|\bar{u}\|^2_{H^{\frac{1}{2}}} \|\bar{u}\|^4_{H^{\frac{1}{4}}} ds \\ &\leqslant C \sup_{t\in[0,T]} \|\bar{u}\|^4_{H^{\frac{1}{4}}} \int_0^T \|\bar{u}\|^2_{H^{\frac{1}{2}}} ds < \infty.\end{aligned}$$

So far, we can see that (6.35) is indeed true.

Having these preparation we can follow the steps in this article without major modifications to obtain the results below.

(a) If $G = A^{-\frac{3}{4}}$, we can obtain the ergodicity for (1.2) in H^σ with $\sigma = \frac{3}{4}$.

(b) If

$$D(A^{\frac{\alpha}{4}+\frac{\sigma}{2}}) \subset R(G) \subset D(A^{\frac{1}{2}})$$

for $\sigma \in \left[\frac{1}{4}, \frac{3}{4}\right)$ and $\alpha \in \left(\frac{3}{2}, 2\right)$, we can establish the ergodicity for (1.2) in H^σ where $\sigma \in \left[\frac{1}{4}, \frac{3}{4}\right)$.

(c) If

$$D(A^{\frac{\alpha}{4}}) \subset R(G) \subset D(A^{\frac{1}{4}})$$

for $\alpha \in \left(\frac{3}{2}, 2\right)$, we can also get the ergodicity for (1.2) in H^σ for $\sigma \in \left[\frac{1}{6}, \frac{1}{4}\right)$. If $\alpha = 2$ in (1.2), the ergodicity property for the stochastic dynamics is classical(see e.g. [33, 36, 37] and references therein). So far, in our article we prove the ergodicity for problem (1.2) in H^σ for $\sigma \in \left[\frac{1}{6}, \frac{\alpha}{2}\right]$.

References

[1] Ammerman A J and Cavalli-Sforza L L. Measuring the rate of spread of early farming, Man, 1971, 6: 674–688.

[2] Ammerman A J and Cavalli-Sforza L L. The Neolithic Transition and the Genetics of Populations in Europe, Princeton University Press, Princeton, 1983.

[3] Aoki K. Gene-culture waves of advance, J. Math. Biol., 1987, 25: 453–464.

[4] Aronson D G and Weinberger H F. Multidimensional nonlinear diffusion arising in population genetics, Adv. Math., 1978, 30: 33–76.

[5] Aronson D G and Weinberger H F. Nonlinear diffusion in population genetics, combustion, and nerve pulse propagation in "Partial Differential Equations and Related Topics (J.A. Goldstein, ed.)," Lecture Notes in Math., Springer, Berlin, 1975, 446: 5–49.

[6] Ablowitz M J and Zeppetella A. Explicit solution of Fisher's equation for a special wave speed, Bull. Math. Biol., 1979, 41: 835–840.

[7] Brzezniak Z and Debbi L. On stochastic Burgers equation driven by a fractional Laplacian and space-time white noise, Stochastic differential equations: Theorem and applications, pp.135-167, Interdiscip Math.Sci. 2, World Sci. Publ.,Hackensack, NJ. 2007.

[8] Brzezniak Z, Debbi L and Ben Goldys. Ergodic properties of fractional stochas- tic Burgers equation, Preprint, (2011), arXiv:1106.1918v1.

[9] Barbato D, Flandoli F and Morandin F. Uniqueness for a stochastic inviscid dyadic model, Proc. Amer. Math. Soc., 2010, 7.

[10] Peter W. Bates, Kening Lu and Bixiang Wang, Random attractors for stochastic reaction–diffusion equations on unbounded domains, J.Differential Equations, 2009, 246: 845–869.

[11] Britton N F. Reaction–Diffusion Equations and Their Applications to Biology, Academic Press, New York, 1986.

[12] Crauel H and Flandoli F. Hausdorff dimension of invariant sets for random dynamical systems, J. Dynam. Differential Equations, 1998, 3: 449–474.

[13] Crauel H and Flandoli F. Attractors for random dynamical systems, Probab. Theory Relat. Fields, 1994, 100: 365–393.

[14] Chen Z X and Guo B Y. Analytic solutions of carrier flow equations via the Painlev'e analysis approach, J. Phys. A (Math. Gen.), 1989, 22: 5187–5194.

[15] Clarkson P A and Mansfield E L. Symmetry reductions and exact solutions of a class of nonlinear heat equations, Phys. D, 1994, 70: 250–288.

[16] Cohen H. Nonlinear diffusion problems, In "Studies in Applied Mathematics (A.H. Taub Ed.)," The Mathematical Association of America, 1971, 27–64.

[17] Cerrai S and Rökner M. Large deviations for stochastic reaction-diffusion systems with multiplicative noise and non-lipschitz reaction term, The Annals of Probability, 2004, 32: No. 1B, 1100–1139.

[18] Freidlin M I. Random perturbations of reaction-diffusion equations: The quasi deterministic approximation, Trans. Amer. Math. Soc., 1988, 305: 665–697.

[19] Fisher R A. The wave of advance of advantageous genes, Ann. Eugenics, 1937, 7: 353–369.

[20] Flandoli F. Dissipativity and invariant measures for stochastic Navier-Stokes equations, NoDEA, 1994, 1: 403-423.

[21] Fife P C. Mathematical aspects of reacting and diffusing systems, Lecture Notes in Biomathematics, Springer–Verlag, Berlin, 1979.

[22] Flandoli F, Gubinelli M and Priola E. Well-posedness of the transport equation by stochastic perturbation, Invent. Math., 2010, 1: 1–53.

[23] Flandoli F and Maslowski B. Ergodicity of the 2-D Navier-Stokes equation under random perturbations, Commun.Math.Phys., 1995, 171: 119–141.

[24] Gorenflo R and Mainardi F. Fractional diffusion processes: probability distribution and continuous time random walk, Lecture Notes in Phys., 2003, 621: 148–166.

[25] Hairer M and Mattingly J C. Ergodicity of the 2D Navier-Stokes equations with degenerate stochastic forcing, Ann. Math., 2006, 3: 993–1032.

[26] Hereman W. Application of a Macsyma program for the Painlev'e test to the FitzHugh–Nagumo equation, In "Partially Integrable Evolution Equations in Physics (R. Conte and N. Boccara)," Kluwer, Dordrecht, 1990, 585–586.

[27] Herrera J, Minzoni A and Ondarza R. Reaction-diffusion equations in one dimension: particular solutions and relaxation, Phys. D, 1992, 57: 249–266.

[28] Metzler R, Klafter J. The restaurant at the end of the random walk: recent developments in the description of anomalous transport by fractional dynamics, J. Phys. A, 2004, 37: R161–R208.

[29] Kolmogorov A. I. Petrovsky and N. Piskunov, Etude de l'equation de la diffusion avec croissance de la quantite de matiere et son application a un probleme biologique, Moscow Bull. Univ. Math., 1937, 1: 1–25.

[30] Krishnan E V. On some diffusion equations, J. Phys. Soc. Jpn., 1994, 63: 460–465.

[31] Maslowski B. On probability distributions of solutions of semilinear stochastic evolution equations, Stochastics, 1993, 45: 17–44.

[32] Murray J D. Mathematical Biology, Springer–Verlag, New York, 1993.

[33] Marinelli C and Rökner M. Well-posedness and asymptotic behavior for stochastic reaction-diffusion equations with multiplicative Poisson noise, Electronic Journal of Probability, 2010, 15: 1528–1555.

[34] Prato G D, Debussche A and Temam R. Stochastic Burgers' equation, NoDEA, 1994, 1: 389–402.

[35] Prato G D and Flandoli F. Pathwise uniqueness for a class of SDE in Hilbert spaces and applications, J. Funct. Anal., 2010, 1: 243–267.

[36] Prato G D and Rökner M. Singular dissipative stochastic equations in Hilbert spaces, Probab. Theory Related Fields, 2002, 124(2): 261–303.

[37] Prato G D and Zabczyk J. Stochastic equations in infinite dimensions, Cambridge University Press, Cambridge, 1992.

[38] Prato G D and Zabczyk J. Ergodicity for Infinite Dimensional Systems, London Math-

ematical Society Lecture Note Series. 229, Cambridge University Press, Cambridge, 1996.

[39] Röckner M. Intruduction to stochastic partial differential equation, Purdue University, Fall 2005 and Spring 2006, Version: April 26, 2007.

[40] Sowers R. Large deviations for a reaction-diffusion equation with non-Gaussian perturbation, Ann. Probab., 1992, 20: 504–537.

[41] Zhao S and Wei G W. Comparison of the discrete singular convolution and three other numerical schemes for solving Fisher's equation, SIAM J. Sci. Comput., 2003, 25 : 127–147.

Existence of Generalized Heteroclinic Solutions of the Coupled Schrödinger System under a Small Perturbation*

Deng Shengfu (邓圣福), Guo Boling (郭柏灵) and Wang Tingchun (王廷春)

Abstract The coupled Schrödinger system with a small perturbation

$$
\begin{aligned}
&u_{xx} + u - u^3 + \beta uv^2 + \epsilon f(\epsilon, u, u_x, v, v_x) = 0 \quad \text{in } \mathbf{R}, \\
&v_{xx} - v + v^3 + \beta u^2 v + \epsilon g(\epsilon, u, u_x, v, v_x) = 0 \quad \text{in } \mathbf{R}
\end{aligned}
$$

is considered where β and ϵ are small parameters. The whole system has a periodic solution with the aid of a Fourier series expansion technique, and its dominant system has a heteroclinic solution. Then adjusting some appropriate constants and applying the fixed point theorem and the perturbation method yield that this heteroclinic solution deforms to a heteroclinic solution exponentially approaching the obtained periodic solution (called generalized heteroclinic solution, thereafter).

Keywords coupled Schrödinger system; heteroclinic solutions; reversibility

1 Introduction

The coupled nonlinear Schrödinger system was first derived in [4] for two interacting nonlinear wave packets in a dispersive and conservative system, which can be written as

$$
\begin{aligned}
&\mathrm{i}\partial_t \phi + \Delta\phi + \mu_1 |\phi|^2 \phi + \beta |\psi|^2 \phi = 0, \\
&\mathrm{i}\partial_t \psi + \Delta\psi + \mu_2 |\psi|^2 \psi + \beta |\phi|^2 \psi = 0,
\end{aligned} \tag{1.1}
$$

where μ_j ($j = 1, 2$) is a constant and β is a coupling constant. In general, the sign of the parameter μ_j discriminates between the focusing and defocusing behavior of

* Chin. Ann. Math. Ser. B, 2014, 35B(6): 857–872. DOI: 10.1007/s11401-014-0867-3.

a single component, and the sign of β determines the type of interplay between the two states. The system (1.1) has applications in many physical problems such as semiconductor electronics [6], optics in nonlinear media [19], photonics [17], plasmas [13], fundamentation of quantum mechanics [30], dynamics of accelerators [14] or mean-field theory of Bose-Einstein condensates [11]. In some of these fields and many others, the system (1.1) appears as an asymptotic limit for a slowly varying dispersive wave envelope propagating in a nonlinear medium [31]. In recent years the system (1.1) has been broadly investigated in many aspects like concentration and multi-bump phenomena for semiclassical states [1,2,16,25], bounded solutions [23,24], blow up [10,15] and positive periodic solutions with variable coefficients and more general nonlinear terms [3].

It is very important to stress that, in the particular case of standing wave solutions of (1.1), namely special solutions of (1.1) of the form

$$\phi(x,t) = \mathrm{e}^{\mathrm{i}\lambda_1 t}u(x), \qquad \psi(x,t) = \mathrm{e}^{\mathrm{i}\lambda_2 t}v(x), \tag{1.2}$$

where u and v are real functions on $\mathbf{R}$, there is also an enormous literature regarding the corresponding system

$$\begin{aligned} &u_{xx} - \lambda_1 u + \mu_1 u^3 + \beta uv^2 = 0, \\ &v_{xx} - \lambda_2 v + \mu_2 v^3 + \beta u^2 v = 0. \end{aligned} \tag{1.3}$$

For instance, Yang [35] discussed the classification of the solitary waves. Pelinovsky and Yang [28] analytically and numerically studied internal modes of vector solitons. The stability of solitary waves can be found in [21,27]. The existence of generalized homoclinic solutions (homoclinic solutions exponentially approaching the periodic solutions) under a small perturbations was proved by Deng and Guo [12] when $\lambda_1 = -\lambda_2 = \mu_1 = -\mu_2 = 1$.

In this paper, we take $\lambda_2 = \mu_2 = -\mu_1 = -\lambda_1 = 1$ and investigate the following system

$$u_{xx} + u - u^3 + \beta uv^2 + \epsilon f(\epsilon, u, u_x, v, v_x) = 0, \tag{1.4}$$

$$v_{xx} - v + v^3 + \beta u^2 v + \epsilon g(\epsilon, u, u_x, v, v_x) = 0, \tag{1.5}$$

where β and ϵ are small parameters and the general nonlinear terms f and g satisfy the conditions given in (2.2) so that this system is reversible. For $\beta = \epsilon = 0$, this system has three saddle-center equilibriums $(u, u_x, v, v_x) = (0,0,0,0), (1,0,1,0)$ and $(-1,0,-1,0)$ (a positive eigenvalue, a negative eigenvalue and a pair of purely imaginary eigenvalues). It is easy to check that (1.4) has two heteroclinic solutions exponentially approaching $(1,0,)$ and $(-1,0)$ while (1.5) has a family of periodic solutions around $(1,0)$ and $(-1,0)$, respectively. This implies that (1.4) and (1.5) may have a heteroclinic solution exponentially approaching a periodic solution at infinity (i.e., generalized heteroclinic solution). In this paper, we will rigorously prove this. Our result is new.

There are a lot of results about the saddle-center problems if the system is conservative and Hamiltonian in particular. We mention the work: homoclinic solutions [7-9, 20, 22, 26, 29, 33, 35], generalized homoclinic solutions [5, 12, 32, 36-39] and heteroclinic orbits to invariant tori [36]).

Our system might not be conservative. We will use a dynamic approach given in [12], which is more general and can be applied to a number of systems like the Schrödinger-KdV system since it does not require that the system has a Hamiltonian structure.

This paper is organized as follows. In Section 2, we derive the properties of heteroclinic solutions of (1.4) and (1.5) for $\beta = \epsilon = 0$. In Section 3, we use the a Fourier series expansion technique to prove that the system of (1.4) and (1.5) has a periodic solution. In Section 4, we apply the fixed point theorem and the perturbation method to demonstrate that this heteroclinic solution deforms to a heteroclinic solution exponentially approaching the periodic solution obtained in Section 3 when small perturbation terms are added. This gives the existence of a generalized heteroclinic solution of (1.4) and (1.5). Section 5 is an appendix which solves an equation left in Section 4.

Throughout this paper, M denotes a positive constant and $B = O(C)$ means that $|B| \leqslant M|C|$.

2 Preliminary

Let $u_1 = u_x$ and $v_1 = v_x$ which change (1.4) and (1.5) into

$$u_x = u_1,$$
$$u_{1x} = -u + u^3 - \beta uv^2 - \epsilon f(\epsilon, u, u_1, v, v_1),$$

$$\begin{aligned} &v_x = v_1, \\ &v_{1x} = v - v^3 - \beta u^2 v - \epsilon g(\epsilon, u, u_1, v, v_1). \end{aligned} \tag{2.1}$$

In this paper, we assume that f and g satisfy

$$\begin{aligned} &f(\epsilon, u, -u_1, v, -v_1) = f(\epsilon, u, u_1, v, v_1), \quad g(\epsilon, u, -u_1, v, -v_1) = g(\epsilon, u, u_1, v, v_1), \\ &f(\epsilon, -u, u_1, v, -v_1) = -f(\epsilon, u, u_1, v, v_1), \quad g(\epsilon, -u, u_1, v, -v_1) = g(\epsilon, u, u_1, v, v_1), \end{aligned} \tag{2.2}$$

and define two operators S_1 and S_2 by

$$S_1(u, u_1, v, v_1) = (u, -u_1, v, -v_1), \quad S_2(u, u_1, v, v_1) = (-u, u_1, v, -v_1). \tag{2.3}$$

From (2.2), the system (2.1) is reversible with the reverser S_k, that is, $S_kU(-x)$ is also a solution whenever $U(x) = (u(x), u_1(x), v(x), v_1(x))^{\mathrm{T}}$ is a solution for $k = 1, 2$. A solution $U(x)$ is reversible if $S_kU(-x) = U(x)$ for $k = 1, 2$. We will use the first reversibility to look for periodic solutions and the second one to construct the generalized heteroclinic solutions of the system (2.1), respectively.

When $\beta = \epsilon = 0$, the first two equations of the system (2.1) have three equilibriums $(-1, 0)$, $(0, 0)$ and $(1, 0)$. It is easy to check that $(-1, 0)$ and $(1, 0)$ are saddle points and $(0, 0)$ is a center. There are two heteroclinic solutions

$$H_1(x) = \left(\tanh\left(\frac{x}{\sqrt{2}}\right), \frac{1}{\sqrt{2}}\mathrm{sech}^2\left(\frac{x}{\sqrt{2}}\right)\right)^{\mathrm{T}} \tag{2.4}$$

and

$$H_2(x) = \left(-\tanh\left(\frac{x}{\sqrt{2}}\right), -\frac{1}{\sqrt{2}}\mathrm{sech}^2\left(\frac{x}{\sqrt{2}}\right)\right)^{\mathrm{T}} \tag{2.5}$$

connecting two saddle points $(-1, 0)$ and $(1, 0)$. The last two equations of the system (2.1) also have three equilibriums $(-1, 0)$, $(0, 0)$ and $(1, 0)$. Clearly, $(-1, 0)$ and $(1, 0)$ are centers and $(0, 0)$ is a saddle point. In the following we will prove that the heteroclinic solution (2.4) will deform to a generalized heteroclinic solution. By the same method, the deformation of the other heteroclinic solution (2.5) can be obtained.

Let

$$\begin{aligned} &u = \tilde{u} + 1, \quad u_1 = \tilde{u}_1, \quad v = \tilde{v} + 1, \quad v_1 = \tilde{v}_1, \\ &\tilde{f}(\epsilon, \tilde{u}, \tilde{u}_1, \tilde{v}, \tilde{v}_1) = f(\epsilon, \tilde{u} + 1, \tilde{u}_1, \tilde{v} + 1, \tilde{v}_1), \\ &\tilde{g}(\epsilon, \tilde{u}, \tilde{u}_1, \tilde{v}, \tilde{v}_1) = g(\epsilon, \tilde{u} + 1, \tilde{u}_1, \tilde{v} + 1, \tilde{v}_1), \end{aligned} \tag{2.6}$$

and we have from (2.2)

$$\tilde{f}(\epsilon,\tilde{u},-\tilde{u}_1,\tilde{v},-\tilde{v}_1)=\tilde{f}(\epsilon,\tilde{u},\tilde{u}_1,\tilde{v},\tilde{v}_1),\quad \tilde{g}(\tilde{\epsilon},\tilde{u},-\tilde{u}_1,\tilde{v},-\tilde{v}_1)=\tilde{g}(\epsilon,\tilde{u},\tilde{u}_1,\tilde{v},\tilde{v}_1). \tag{2.7}$$

Note that (2.6) changes the system (2.1) into

$$\begin{aligned}
&\tilde{u}_x=\tilde{u}_1,\\
&\tilde{u}_{1x}=2\tilde{u}+3\tilde{u}^2+\tilde{u}^3-\beta(\tilde{u}+1)(\tilde{v}+1)^2-\epsilon\tilde{f}(\epsilon,\tilde{u},\tilde{u}_1,\tilde{v},\tilde{v}_1),\\
&\tilde{v}_x=\tilde{v}_1,\\
&\tilde{v}_{1x}=-2\tilde{v}-3\tilde{v}^2-\tilde{v}^3-\beta(\tilde{u}+1)^2(\tilde{v}+1)-\epsilon\tilde{g}(\epsilon,\tilde{u},\tilde{u}_1,\tilde{v},\tilde{v}_1).
\end{aligned} \tag{2.8}$$

Symbolically, it can be written as

$$\frac{\mathrm{d}\tilde{U}}{\mathrm{d}x}=L\tilde{U}+N(\tilde{U})+\tilde{N}(\beta,\tilde{U})+\epsilon R(\epsilon,\tilde{U}) \tag{2.9}$$

where $\tilde{U}=(\tilde{u},\tilde{u}_1,\tilde{v},\tilde{v}_1)^{\mathrm{T}}$,

$$L=\begin{pmatrix}0&1&0&0\\2&0&0&0\\0&0&0&1\\0&0&-2&0\end{pmatrix},\qquad N(\tilde{U})=\begin{pmatrix}0\\3\tilde{u}^2+\tilde{u}^3\\0\\-3\tilde{v}^2-\tilde{v}^3\end{pmatrix},$$

$$\tilde{N}(\beta,\tilde{U})=\begin{pmatrix}0\\-\beta(\tilde{u}+1)(\tilde{v}+1)^2\\0\\-\beta(\tilde{u}+1)^2(\tilde{v}+1)\end{pmatrix},\qquad R(\epsilon,\tilde{U})=\begin{pmatrix}0\\-\tilde{f}(\epsilon,\tilde{u},\tilde{u}_1,\tilde{v},\tilde{v}_1)\\0\\-\tilde{g}(\epsilon,\tilde{u},\tilde{u}_1,\tilde{v},\tilde{v}_1)\end{pmatrix}. \tag{2.10}$$

Note that from (2.7) the system (2.9) is still reversible with the reverser S_1 if

$$S_1(\tilde{u},\tilde{u}_1,\tilde{v},\tilde{v}_1)=(\tilde{u},-\tilde{u}_1,\tilde{v},-\tilde{v}_1), \tag{2.11}$$

where we avoid the introduction of a new notation. We write the dominant system of (2.9) as

$$\frac{\mathrm{d}\tilde{U}}{\mathrm{d}x}=L\tilde{U}+N(\tilde{U}) \tag{2.12}$$

which has a heteroclinic solution $H(x)$ given by

$$H(x)=\left(\tanh\left(\frac{x}{\sqrt{2}}\right)-1,\frac{1}{\sqrt{2}}\mathrm{sech}^2\left(\frac{x}{\sqrt{2}}\right),0,0\right)^{\mathrm{T}} \tag{2.13}$$

approaching $(0,0,0,0)^{\mathrm{T}}$ as $x\to\infty$ and $(-2,0,0,0)^{\mathrm{T}}$ as $x\to-\infty$. Moreover,

$$H(0)=\left(-1,\frac{1}{\sqrt{2}},0,0\right)^{\mathrm{T}} \tag{2.14}$$

and $H(x)$ satisfies the following inequality

$$\left|H(x)\right| \leqslant M\mathrm{e}^{-\sqrt{2}x} \quad \text{for } x \in [0, +\infty). \tag{2.15}$$

In Section 4, we will prove the deformation of this heteroclinic solution $H(x)$ for the whole system (2.9). This demonstrates that the original system (2.1) has a generalized heteroclinic solution.

3 Periodic solutions

Using the a Fourier series expansion technique, we will shows that (2.9) has periodic solutions which determine the forms of the generalized heteroclinic solutions at infinity. The general theory for reversible systems can be found in [18].

Let

$$C = \sqrt{2}\tilde{v} - \mathrm{i}\tilde{v}_1, \quad \tau = \sqrt{2}(1 + r_1)x, \tag{3.1}$$

where r_1 is a small real constant to be determined later. Using the fact

$$\tilde{v} = \frac{C + \bar{C}}{2\sqrt{2}}, \qquad \tilde{v}_1 = \mathrm{i}\frac{C - \bar{C}}{2}, \tag{3.2}$$

we can write (2.9) as

$$\begin{aligned}
\tilde{u}_\tau &= \frac{1}{\sqrt{2}(1 + r_1)}\tilde{u}_1, \\
\tilde{u}_{1\tau} &= \frac{\sqrt{2}}{1 + r_1}\tilde{u} + h_1(\beta, \epsilon, \tilde{u}, \tilde{u}_1, C, \bar{C}), \\
C_\tau &= \frac{\mathrm{i}}{1 + r_1}C + h_2(\beta, \epsilon, \tilde{u}, \tilde{u}_1, C, \bar{C}), \\
\bar{C}_\tau &= \frac{-\mathrm{i}}{1 + r_1}\bar{C} - h_2(\beta, \epsilon, \tilde{u}, \tilde{u}_1, C, \bar{C}),
\end{aligned} \tag{3.3}$$

where h_1 is a real function, h_2 is a purely imaginary function and

$$\begin{aligned}
h_1(\beta, \epsilon, \tilde{u}, \tilde{u}_1, C, \bar{C}) = {} & \frac{1}{\sqrt{2}(1 + r_1)}\left(3\tilde{u}^2 + \tilde{u}^3 - \beta(\tilde{u} + 1)\left(\frac{C + \bar{C}}{2\sqrt{2}} + 1\right)^2\right. \\
& \left. -\epsilon\tilde{f}\left(\epsilon, \tilde{u}, \tilde{u}_1, \frac{C + \bar{C}}{2\sqrt{2}}, \mathrm{i}\frac{C - \bar{C}}{2}\right)\right), \\
h_2(\beta, \epsilon, \tilde{u}, \tilde{u}_1, C, \bar{C}) = {} & \frac{-i}{\sqrt{2}(1 + r_1)}\left(-3\left(\frac{C + \bar{C}}{2\sqrt{2}}\right)^2 - \left(\frac{C + \bar{C}}{2\sqrt{2}}\right)^3\right. \\
& \left. - \beta(\tilde{u} + 1)^2\left(\frac{C + \bar{C}}{2\sqrt{2}} + 1\right) - \epsilon\tilde{g}\left(\epsilon, \tilde{u}, \tilde{u}_1, \frac{C + \bar{C}}{2\sqrt{2}}, \mathrm{i}\frac{C - \bar{C}}{2}\right)\right).
\end{aligned} \tag{3.4}$$

From (2.7) and (2.11), we may define

$$S_1(\tilde{u}, \tilde{u}_1, C, \bar{C}) = (\tilde{u}, -\tilde{u}_1, \bar{C}, C) \tag{3.5}$$

such that the system (3.3) is reversible where we avoid again the introduction of a new notation. Assume

$$\left(\tilde{u}(\tau), \tilde{u}_1(\tau), C(\tau), \bar{C}(\tau)\right) = \left(\sum_n \tilde{u}_n \mathrm{e}^{\mathrm{i}n\tau}, \sum_n \tilde{u}_{1,n} \mathrm{e}^{\mathrm{i}n\tau}, \sum_n C_n \mathrm{e}^{\mathrm{i}n\tau}, \sum_n \bar{C}_n \mathrm{e}^{-\mathrm{i}n\tau}\right). \tag{3.6}$$

Plugging (3.6) into (3.3) and making the coefficient of each term in the Fourier series equal yield

$$\begin{aligned}
\tilde{u}_n &= \frac{-(1+r_1)}{\sqrt{2}(1+r_1)^2 n^2 + \sqrt{2}} \left[h_1(\beta, \epsilon, \tilde{u}, \tilde{u}_1, C, \bar{C})\right]_n, \\
\tilde{u}_{1,n} &= \frac{-\mathrm{i}(1+r_1)^2 n}{(1+r_1)^2 n^2 + 1} \left[h_1(\beta, \epsilon, \tilde{u}, \tilde{u}_1, C, \bar{C})\right]_n, \\
C_n &= \frac{-\mathrm{i}(1+r_1)}{n(1+r_1) - 1} \left[h_2(\beta, \epsilon, \tilde{u}, \tilde{u}_1, C, \bar{C})\right]_n \qquad \text{for } n \neq 1, \\
\bar{C}_n &= \frac{-\mathrm{i}(1+r_1)}{n(1+r_1) - 1} \left[h_2(\beta, \epsilon, \tilde{u}, \tilde{u}_1, C, \bar{C})\right]_{-n} \qquad \text{for } n \neq 1,
\end{aligned} \tag{3.7}$$

and for $n = 1$

$$r_1 C_1 = -\mathrm{i}(1+r_1)\left[h_2(\beta, \epsilon, \tilde{u}, \tilde{u}_1, C, \bar{C})\right]_1, \tag{3.8}$$

$$r_1 \bar{C}_1 = -\mathrm{i}(1+r_1)\left[h_2(\beta, \epsilon, \tilde{u}, \tilde{u}_1, C, \bar{C})\right]_{-1}, \tag{3.9}$$

where $[f]_k$ denotes the k-th Fourier coefficient of f.

Now we activate C_1, i.e., consider C_1 as a free constant to be chosen later. We first solve (3.7) for $\tilde{u}_n$, $\tilde{u}_{1,n}$, C_n and $\bar{C}_n$ $(n \neq 1)$, and then solve (3.8) for r_1.

Let $H^m(0, 2\pi)$ be a space of periodic functions of τ with a period 2π such that their derivatives up to order m are in $L^2(0, 2\pi)$, which norm is denoted by $\|\cdot\|_m$. Fix C_1 and define two spaces

$$H_1^1(0, 2\pi) = \left\{ f(\tau) = \sum_n f_n \mathrm{e}^{\mathrm{i}n\tau} \in H^1(0, 2\pi) \mid f_1 = 0 \right\},$$

$$H_{-1}^1(0, 2\pi) = \left\{ f(\tau) = \sum_n f_n \mathrm{e}^{\mathrm{i}n\tau} \in H^1(0, 2\pi) \mid f_{-1} = 0 \right\}.$$

For $A, B \in H^1(0, 2\pi) \times H^1(0, 2\pi)$ and $D \in H_1^1(0, 2\pi)$, use (3.7) and we define a mapping $\Theta(A, B, D, \bar{D}; \varpi)$ from $H^1(0, 2\pi) \times H^1(0, 2\pi) \times H_1^1(0, 2\pi) \times H_{-1}^1(0, 2\pi)$ to itself by

$$\Theta(A,B,D,\bar{D};\varpi) = \begin{pmatrix} \displaystyle\sum_n \frac{-(1+r_1)}{\sqrt{2}(1+r_1)^2n^2+\sqrt{2}}\left[h_1(\beta,\epsilon,\tilde{u},\tilde{u}_1,C,\bar{C})\right]_n \mathrm{e}^{\mathrm{i}n\tau} \\ \displaystyle\sum_n \frac{-\mathrm{i}(1+r_1)^2 n}{(1+r_1)^2n^2+1}\left[h_1(\beta,\epsilon,\tilde{u},\tilde{u}_1,C,\bar{C})\right]_n \mathrm{e}^{\mathrm{i}n\tau} \\ \displaystyle\sum_{n\neq 1} \frac{-\mathrm{i}(1+r_1)}{n(1+r_1)-1}\left[h_2(\beta,\epsilon,\tilde{u},\tilde{u}_1,C,\bar{C})\right]_n \mathrm{e}^{\mathrm{i}n\tau} \\ \displaystyle\sum_{n\neq 1} \frac{-\mathrm{i}(1+r_1)}{n(1+r_1)-1}\left[h_2(\beta,\epsilon,\tilde{u},\tilde{u}_1,C,\bar{C})\right]_{-n} \mathrm{e}^{-\mathrm{i}n\tau} \end{pmatrix}, \tag{3.10}$$

where $\varpi = (\beta, \epsilon, r_1, C_1, \bar{C}_1)$. Assume that $B_r(0)$ is a ball with a radius r in the space $H^1(0, 2\pi) \times H^1(0, 2\pi) \times H_1^1(0, 2\pi) \times H_{-1}^1(0, 2\pi)$. It is easy to check the following lemma.

Lemma 3.1 *For $(A, B, D, \bar{D}), (A_1, B_1, D_1, \bar{D}_1), (A_2, B_2, D_2, \bar{D}_2) \in \bar{B}_r(0)$ and any small bounded ϖ and r, Θ is smooth in its arguments and satisfies*

$$\|\Theta(A,B,D,\bar{D};\varpi)\|_1 \leqslant M\left(|\beta|+|\epsilon|+\|A\|_1^2+\|B\|_1^2+\|D\|_1^2+|C_1|^2\right),$$

$$\begin{aligned} &\|\Theta(A_1,B_1,D_1,\bar{D}_1;\varpi)-\Theta(A_2,B_2,D_2,\bar{D}_2;\varpi)\|_1 \\ &\leqslant M\big(|\beta|+|\epsilon|+|C_1|+\|A_1\|_1 \\ &\quad +\|A_2\|_1+\|B_1\|_1+\|B_2\|_1+\|D_1\|_1+\|D_2\|_1\big)\big(\|A_1-A_2\|_1 \\ &\quad +\|B_1-B_2\|_1+\|D_1-D_2\|_1\big). \end{aligned}$$

Take $r = |C_1|$ and

$$\beta = \beta_1|C_1|^{\alpha_1}, \quad \epsilon = \epsilon_1|C_1|^{\alpha_2}, \quad \alpha_1 > 1, \quad \alpha_2 > 1, \tag{3.11}$$

where $\beta_1, \epsilon_1, \alpha_1$ and α_2 are fixed constants. Lemma 3.1 yields that Θ is a contraction mapping on $\bar{B}_r(0)$ for small C_1. Thus, Θ has a unique fixed point which is a smooth function of ϖ. Write this fixed point as

$$(u_p^0, u_{1p}^0, C_p^0, \bar{C}_p^0)(\beta, \epsilon, r_1, C_1, \bar{C}_1)(\tau) \tag{3.12}$$

which satisfies

$$\|u_p^0\|_1 + \|u_{1p}^0\|_1 + \|C_p^0\|_1 + \|\bar{C}_p^0\|_1 \leqslant M\big(|\beta| + |\epsilon| + |C_1|^2\big). \tag{3.13}$$

Using the same argument we can show that (3.12) is in $H^m(0,2\pi)$ and satisfies (3.13) with $H^m(0,2\pi)$-norm for any integer $m>0$. We use $(\tilde{u}_p,\tilde{u}_{1p},C_p,\bar{C}_p)(\tau)$ to denote

$$\left(u_p^0(\tau),u_{1p}^0(\tau),C_p^0(\tau)+C_1\mathrm{e}^{\mathrm{i}\tau},\bar{C}_p^0(\tau)+\bar{C}_1\mathrm{e}^{-i\tau}\right).$$

Now we solve (3.8) for r_1. Substitute (3.12) into (3.8) and obtain

$$-r_1C_1+g_1(\beta,\epsilon,r_1,C_1,\bar{C}_1)=0, \tag{3.14}$$

where

$$g_1(\beta,\epsilon,r_1,C_1,\bar{C}_1)=-\mathrm{i}(1+r_1)\left[h_1(\beta,\epsilon,\tilde{u},\tilde{u}_1,C,\bar{C})\right]_1$$

is smooth when $\beta,\epsilon,r_1,C_1,\bar{C}_1$ are near 0.

If $\left(\tilde{u},\tilde{u}_1,C,\bar{C}\right)(\tau)$ is a solution of (3.3), then

$$S_1\left(\tilde{u},\tilde{u}_1,C,\bar{C}\right)(-\tau),\quad \left(\tilde{u},\tilde{u}_1,C,\bar{C}\right)(\tau+\theta)$$

are also solutions of (3.3) for any real number θ since (3.3) has the reversibility property by (3.5) and the translation invariance. Using these, we may take

$$C_1=I>0 \tag{3.15}$$

so that (3.11) becomes

$$\beta=\beta_1I^{\alpha_1},\quad \epsilon=\epsilon_1I^{\alpha_2},\quad \alpha_1>1,\quad \alpha_2>1, \tag{3.16}$$

and (3.14) is equivalent to the following equation

$$r_1=\tilde{g}_1(\beta,\epsilon,r_1,I)$$

where $\tilde{g}_1$ is real and smooth in its arguments and is a contraction mapping satisfying $|\tilde{g}_1|\leqslant M(|\beta|+|\epsilon|+I)$ under the condition (3.16) (More details can be found in Deng and Guo [12]). By the fixed point theorem, $\tilde{g}_1$ has a unique fixed point

$$r_1=r_1(\beta,\epsilon,I) \tag{3.17}$$

as a smooth real function for small (β,ϵ,I) satisfying

$$|r_1|\leqslant M(|\beta|+|\epsilon|+I). \tag{3.18}$$

Therefore, (3.3) has a periodic solution

$$\big(\tilde{u}_p(\beta,\epsilon,I)(\tau),\tilde{u}_{1p}(\beta,\epsilon,I)(\tau),C_p(\beta,\epsilon,I)(\tau),\bar{C}_p(\beta,\epsilon,I)(\tau)\big)$$

in $H^m(0,2\pi)$ if $I\in(0,I_1]$ and (3.16) holds where I_1 is a fixed small positive constant.

By the relation $\tau=\sqrt{2}(1+r_1)x$, we write the periodic solution $\big(\tilde{u}_p,\tilde{u}_{1p},C_p,\bar{C}_p\big)(\tau)$ as $\big(\tilde{u}_p(\beta,\epsilon,I)(x),\tilde{u}_{1p}(\beta,\epsilon,I)(x),C_p(\beta,\epsilon,I)(x),\bar{C}_p(\beta,\epsilon,I)\big)(x)$ with the frequency

$$\omega_1(\beta,\epsilon,I)=\sqrt{2}\big(1+r_1(\beta,\epsilon,I)\big) \tag{3.19}$$

for $I\in(0,I_1]$. Moreover, this solution is reversible since $C_1=I$ is real, i.e.,

$$\begin{aligned}&S_1\big(\tilde{u}_p(\beta,\epsilon,I),\tilde{u}_{1p}(\beta,\epsilon,I),C_p(\beta,\epsilon,I),\bar{C}_p(\beta,\epsilon,I)\big)(-x)\\&=\big(\tilde{u}_p(\beta,\epsilon,I),-\tilde{u}_{1p}(\beta,\epsilon,I),\bar{C}_p(\beta,\epsilon,I),C_p(\beta,\epsilon,I)\big)(-x)\\&=\big(\tilde{u}_p(\beta,\epsilon,I),\tilde{u}_{1p}(\beta,\epsilon,I),C_p(\beta,\epsilon,I),\bar{C}_p(\beta,\epsilon,I)\big)(x).\end{aligned}$$

Letting $C_p=\sqrt{2}\tilde{v}_p-\mathrm{i}\tilde{v}_{1p}$, we have

$$\tilde{v}_p(-x)=\tilde{v}_p(x),\quad \tilde{v}_{1p}(-x)=-\tilde{v}_{1p}(x). \tag{3.20}$$

Define

$$X_{\beta,\epsilon,I}(x)=(\tilde{u}_p,\tilde{u}_{1p},\tilde{v}_p,\tilde{v}_{1p})^{\mathrm{T}}(x)=\left(\tilde{u}_p,\tilde{u}_{1p},\frac{C_p+\bar{C}_p}{2\sqrt{2}},\mathrm{i}\frac{C_p-\bar{C}_p}{2}\right)^{\mathrm{T}}(x) \tag{3.21}$$

which is smooth for x and small (β,ϵ,I) with the condition (3.16). Then, $X_{\beta,\epsilon,I}(x)$ is a reversible periodic solution of (2.9) under the reversor S_1 with frequency $\omega_1(\beta,\epsilon,I)$, which from (3.13) satisfies that for any integer $m>0$

$$\big\|X_{\beta,\epsilon,I}(x)\big\|_m\leqslant M\big(|\beta|+|\epsilon|+I\big). \tag{3.22}$$

The Sobolev embedding theorem gives that (3.22) holds also in $C_B^m(\mathbf{R})$-norm, which is a space of continuously differentiable functions up to order m with a supreme norm.

4 Generalized heteroclinic solutions

In this section we demonstrate that (2.9) has a generalized heteroclinic solution exponentially approaching the periodic solution $X_{\beta,\epsilon,I}$ obtained in Section 3.

Theorem 4.1 *Suppose that the assumption* (2.2) *holds. There exist constants* $I_0>0$, β_1 *and* ϵ_1 *such that for* $I\in(0,I_0]$, *if the small parameters* $\beta=\beta_1 I^{3/2}$ *and* $\epsilon=\epsilon_1 I^{3/2}$, *then* (2.1) *has a generalized heteroclinic solution, i.e.* (2.1) *has a solution which is reversible and exponentially approaches the periodic solution* $(1,0,1,0)^{\mathrm{T}}+X_{\beta,\epsilon,I}(x+\theta)$ *as* $x\to\infty$ *and the periodic solution* $(-1,0,1,0)^{\mathrm{T}}+S_2\big(X_{\beta,\epsilon,I}(-x+\theta)\big)$ *as* $x\to-\infty$ *where the phase shift* θ *is a continuous function in* I, *and the operator* S_2 *is defined in* (2.3).

We divide the proof into two steps. Using the relationship between (2.1) and (2.9), we will first prove that (2.9) has a solution for $x\in[0,\infty)$, which exponentially approaches the periodic solution $X_{\beta,\epsilon,I}(x+\theta)$ for some phase shift θ as $x\to\infty$. Then we solve for θ as a function of β,ϵ and I. This yields that this solution can be extended to $x\in(-\infty,0]$ by using the reversibility.

Step 1 Solution of (2.9) for $x\in[0,\infty)$.

Assume that the solution $\mathcal{U}(x)$ of (2.9) has the following form

$$\mathcal{U}(x)=H(x)+Z(x)+\zeta(x)X_{\beta,\epsilon,I}(x+\theta),\tag{4.1}$$

where $H(x)$ and $X_{\beta,\epsilon,I}(x)$ are defined in (2.13) and (3.21) respectively, the phase shift $\theta\in S^1=[0,2\pi]$ is a constant, the cut-off function $\zeta(x)$ is in $C^{\infty}(\mathbf{R},\mathbf{R})$ satisfying $0\leqslant\zeta(x)\leqslant 1$ and

$$\zeta(x)=\begin{cases}1, & |x|\geqslant 2,\\ 0, & |x|\leqslant 1,\end{cases}\tag{4.2}$$

and $Z(x)$ is a perturbation term to be determined, which exponentially tends to 0 as $x\to\infty$ so that $\mathcal{U}(x)$ is a solution of (2.9) that approaches the periodic solution $X_{\beta,\epsilon,I}(x+\theta)$ as $x\to\infty$.

Since $H(x)$ is a solution of (2.12) and $X_{\beta,\epsilon,I}(x)$ is a solution of (2.9), plugging (4.1) into (2.9) yields

$$\frac{\mathrm{d}Z}{\mathrm{d}x}=\mathcal{L}(x)Z+\mathcal{N}(x,Z)+\epsilon\mathcal{R}(x,\epsilon,Z),\tag{4.3}$$

where

$$\mathcal{L}(x)=L+\mathrm{d}N[H(x)]=\begin{pmatrix}0 & 1 & 0 & 0\\ 3\tanh^2\left(\dfrac{x}{\sqrt{2}}\right)-1 & 0 & 0 & 0\\ 0 & 0 & 0 & 1\\ 0 & 0 & -2 & 0\end{pmatrix},$$

$$
\begin{aligned}
\mathcal{N}(x,Z) = &N(H(x)+Z(x)+\zeta(x)X_{\beta,\epsilon,I}(x+\theta))-N(H(x))\\
&-\zeta(x)N(X_{\beta,\epsilon,I}(x+\theta))-\mathrm{d}N[H(x)]Z(x)\\
&+\tilde{N}(\beta,H(x)+Z(x)+\zeta(x)X_{\beta,\epsilon,I}(x+\theta))-\zeta(x)\tilde{N}(\beta,X_{\beta,\epsilon,I}(x+\theta)),\\
\mathcal{R}(x,\epsilon,Z) = &R(\epsilon,H(x)+Z(x)+\zeta(x)X_{\beta,\epsilon,I}(x+\theta))-\zeta(x)R(\epsilon,X_{\beta,\epsilon,I}(x+\theta))\\
&-\frac{1}{\epsilon}\zeta'(x)X_{\beta,\epsilon,I}(x+\theta),
\end{aligned}
\tag{4.4}
$$

and d means taking the Fréchet derivative.

Now we first consider $x\in[0,\infty)$ and have the following lemma by using (2.15) and (3.22).

Lemma 4.1 *If β, ϵ and I are small and $|Z|+|Z_1|+|Z_2|\leqslant M_0$ for some positive constant M_0, then $\mathcal{N}$ and $\mathcal{R}$ satisfy for $x\geqslant 0$*

$$
\begin{aligned}
&\big|\mathcal{N}(x,Z)\big|\leqslant M\big[(\mathrm{e}^{-\sqrt{2}x}+|Z|)(|\beta|+|\epsilon|+I)+|Z|^2\big],\\
&\big|\mathcal{N}(x,Z_1)-\mathcal{N}(x,Z_2)\big|\leqslant M(|\beta|+|\epsilon|+I+|Z_1|+|Z_2|)|Z_1-Z_2|,\\
&\big|\mathcal{R}(x,\epsilon,Z)\big|\leqslant M\Big(\mathrm{e}^{-\sqrt{2}x}+|Z|+\frac{|\beta|+|\epsilon|+I}{|\epsilon|}\mathrm{e}^{-\sqrt{2}x}\Big),\\
&\big|\mathcal{R}(x,\epsilon,Z_1)-\mathcal{R}(x,\epsilon,Z_2)\big|\leqslant M|Z_1-Z_2|.
\end{aligned}
\tag{4.5}
$$

Note that

$$
\frac{\mathrm{d}Z(x)}{\mathrm{d}x}=\mathcal{L}(x)Z(x) \tag{4.6}
$$

has four linearly independent solutions

$$
\begin{aligned}
s_1(x) &= \frac{1}{\sqrt{2}}\Big(\mathrm{sech}^2\Big(\frac{x}{\sqrt{2}}\Big),-\sqrt{2}\mathrm{sech}^2\Big(\frac{x}{\sqrt{2}}\Big)\tanh\Big(\frac{x}{\sqrt{2}}\Big),0,0\Big)^{\mathrm{T}},\\
s_2(x) &= \frac{1}{16}\Big(\mathrm{sech}^2\Big(\frac{x}{\sqrt{2}}\Big)(6\sqrt{2}x+8\sinh(\sqrt{2}x)+\sinh(2\sqrt{2}x)),\\
&\qquad 4(\sqrt{2}\cosh(\sqrt{2}x)+3\mathrm{sech}^2\Big(\frac{x}{\sqrt{2}}\Big)(\sqrt{2}-x\tanh\Big(\frac{x}{\sqrt{2}}\Big))),0,0\Big)^{\mathrm{T}},\\
s_3(x) &= \Big(0,0,\cos(\sqrt{2}x),-\sqrt{2}\sin(\sqrt{2}x)\Big)^{\mathrm{T}},\\
s_4(x) &= \Big(0,0,\sin(\sqrt{2}x),\sqrt{2}\cos(\sqrt{2}x)\Big)^{\mathrm{T}},
\end{aligned}
\tag{4.7}
$$

which satisfy

$$
|s_1(x)|\leqslant M\mathrm{e}^{-\sqrt{2}x},\quad |s_2(x)|\leqslant M\mathrm{e}^{\sqrt{2}x},\quad |s_3(x)|+|s_4(x)|\leqslant M \tag{4.8}
$$

for $x \in [0,\infty)$. Moreover,

$$\begin{aligned} s_1(0) &= \left(\frac{1}{\sqrt{2}},0,0,0\right)^{\mathrm{T}}, & s_2(0) &= \left(0,\sqrt{2},0,0\right)^{\mathrm{T}},\\ s_3(0) &= \left(0,0,1,0\right)^{\mathrm{T}}, & s_4(0) &= \left(0,0,0,\sqrt{2}\right)^{\mathrm{T}}. \end{aligned} \tag{4.9}$$

The adjoint equation of (4.6) has four linearly independent solutions given by

$$\begin{aligned} s_1^*(x) &= \frac{1}{16}\Bigg(4\left(\sqrt{2}\cosh(\sqrt{2}x)+3\mathrm{sech}^2\left(\frac{x}{\sqrt{2}}\right)\left(\sqrt{2}-x\tanh\left(\frac{x}{\sqrt{2}}\right)\right)\right),\\ &\quad -\mathrm{sech}^2\left(\frac{x}{\sqrt{2}}\right)\left(6\sqrt{2}x+8\sinh(\sqrt{2}x)+\sinh(2\sqrt{2}x)\right),0,0\Bigg)^{\mathrm{T}},\\ s_2^*(x) &= \frac{1}{\sqrt{2}}\left(\sqrt{2}\mathrm{sech}^2\left(\frac{x}{\sqrt{2}}\right)\tanh\left(\frac{x}{\sqrt{2}}\right),\mathrm{sech}^2\left(\frac{x}{\sqrt{2}}\right),0,0\right)^{\mathrm{T}},\\ s_3^*(x) &= \frac{1}{\sqrt{2}}\left(0,0,\sqrt{2}\cos(\sqrt{2}x),-\sin(\sqrt{2}x)\right)^{\mathrm{T}},\\ s_4^*(x) &= \frac{1}{\sqrt{2}}\left(0,0,\sqrt{2}\sin(\sqrt{2}x),\cos(\sqrt{2}x)\right)^{\mathrm{T}}, \end{aligned} \tag{4.10}$$

which satisfy

$$|s_1^*(x)| \leqslant M\mathrm{e}^{\sqrt{2}x}, \quad |s_2^*(x)| \leqslant M\mathrm{e}^{-\sqrt{2}x}, \quad |s_3^*(x)|+|s_4^*(x)| \leqslant M \tag{4.11}$$

for $x \in [0,\infty)$ and

$$\langle s_k(x), s_j^*(x)\rangle = 0 \ \text{ for } \ k \neq j, \qquad \langle s_k(x), s_k^*(x)\rangle = 1, \quad k,j=1,2,3,4 \tag{4.12}$$

for each $x \in [0,\infty)$ where $\langle\cdot,\cdot\rangle$ denotes the Euclidean inner product on $\mathbf{R}^4$.

The solution of (4.3) that decays to zero at infinity can be found as

$$\begin{aligned} Z = \mathcal{F}(Z) &\triangleq \int_0^x \langle \mathcal{N}(t,Z)+\epsilon\mathcal{R}(t,\epsilon,Z), s_1^*(t)\rangle\,\mathrm{d}t\, s_1(x)\\ &\quad -\sum_{j=2}^{4}\int_x^{\infty}\langle \mathcal{N}(t,Z)+\epsilon\mathcal{R}(t,\epsilon,Z), s_j^*(t)\rangle\,\mathrm{d}t\, s_j(x). \end{aligned} \tag{4.13}$$

Fix $\nu \in (0,\sqrt{2})$ and consider (4.13) as a fixed point problem in a Banach space

$$E_\nu = \left\{Z \in C\big([0,\infty)\times S^1\big) \mid \sup_{x\in[0,\infty)} \{|Z(x,\theta)|\mathrm{e}^{\nu x}\} < \infty\right\}$$

with the norm

$$\|Z\|_\nu = \sup\left\{|Z(x,\theta)|\ \mathrm{e}^{\nu x} \mid x\in[0,\infty), \theta\in S^1\right\}.$$

It is easy to check the following lemma by using (2.15), (3.22), (4.8), (4.11) and Lemma 4.1.

Lemma 4.2 *The function $\mathcal{F}$ satisfies*

$$\|\mathcal{F}(Z)\|_\nu \leqslant M\big[(1+\|Z\|_\nu)(|\beta|+|\epsilon|+I)+\|Z\|_\nu^2\big],$$
$$\|\mathcal{F}(Z_1)-\mathcal{F}(Z_2)\|_\nu \leqslant M(|\beta|+|\epsilon|+I+\|Z_1\|_\nu+\|Z_2\|_\nu)\|Z_1-Z_2\|_\nu$$

for $Z, Z_1, Z_2 \in E_v$.

For any fixed constant

$$\gamma \in (0,1), \tag{4.14}$$

we let $r = MI^\gamma$ and

$$\beta=\beta_1 I^{\alpha_1}, \qquad \epsilon=\epsilon_1 I^{\alpha_2}, \qquad \alpha_k=\gamma+\tilde{\alpha}_k>1, \tag{4.15}$$

where $\tilde{\alpha}_k$ are positive constants for $k=1,2$. Thus, (3.16) is satisfied. We can show from Lemma 4.2 that $\mathcal{F}$ is a contraction on $\bar{B}_r(0)\subset E_\nu$ for small I. Therefore, (4.13) has a unique solution $Z(x;\theta,\beta,\epsilon,I)$ satisfying

$$|Z(x;\theta,\beta,\epsilon,I)|\leqslant MI^\gamma, \quad x\in[0,\infty). \tag{4.16}$$

Using the same argument as that for (4.16) and an extension of a contraction mapping principle [34], we can show that Z is smooth in its arguments. Obviously, the solution Z of (4.3) exists if x is in a finite interval and an initial condition is given. Thus, we have showed that $\mathcal{U}(x;\theta,\beta,\epsilon,I)$ defined in (4.1) exists for $x\geqslant\tilde{x}_0$ with any fixed $\tilde{x}_0\in(-\infty,\infty)$.

Step 2 Solution of (2.9) for $x\in(-\infty,\infty)$.

Using (2.3), (4.1) and the relationship between u and $\tilde{u}$ in (2.6), we may define

$$\tilde{\mathcal{U}}(x)=\begin{cases}\mathcal{U}(x;\theta,\beta,\epsilon,I)+U_1 & \text{for } x\geqslant 0,\\ S_2U_1+S_2\left(\mathcal{U}(-x;\theta,\beta,\epsilon,I)\right) & \text{for } x\leqslant 0,\end{cases} \tag{4.17}$$

where $U_1=(1,0,1,0)^{\mathrm{T}}$. If the following equation

$$(I-S_2)\big(\mathcal{U}(0;\theta,\beta,\epsilon,I)+U_1\big)=0 \tag{4.18}$$

is true (The basic idea is to solve this equation for θ, which is given in Section 5), the uniqueness of the solution for an initial value problem implies that $\tilde{\mathcal{U}}$ is a generalized heteroclinic solution of (2.1) and $S_2\tilde{\mathcal{U}}(-x)=\tilde{\mathcal{U}}(x)$, which exponentially approaches the periodic solution $U_1+X_{\beta,\epsilon,I}(x+\theta)=(1,0,1,0)^{\mathrm{T}}+X_{\beta,\epsilon,I}(x+\theta)$ as $x\to\infty$ and the periodic solution $S_2\big(U_1+X_{\beta,\epsilon,I}(-x+\theta)\big)=(-1,0,1,0)^{\mathrm{T}}+S_2\big(X_{\beta,\epsilon,I}(-x+\theta)\big)$ as $x\to-\infty$. This completes the proof of Theorem 4.1.

5 Appendix

In this section, we will solve (4.18) for θ. By (2.14), the definition of $\zeta(x)$ in (4.2) and $Z=(\tilde{u},\tilde{u}_1,\tilde{v},\tilde{v}_1)^{\mathrm{T}}$, it is easy to check that (4.18) is equivalent to

$$\tilde{u}(0)=0, \tag{5.1}$$

$$\tilde{v}_1(0)=0. \tag{5.2}$$

Using (4.9) and (4.13), we know that (5.1) holds automatically. Thus, we only have to study (5.2) which can be transformed to

$$\int_0^{\infty}\langle\mathcal{N}(t,Z)+\epsilon\mathcal{R}(t,\epsilon,Z),s_4^*(t)\rangle\,\mathrm{d}t=0. \tag{5.3}$$

Lemma 5.1 *Under the assumption in Theorem* 4.1, *the equation* (5.3) *can be transformed to*

$$\theta=I^{1/2}\Theta(\theta,\beta,\epsilon,I) \tag{5.4}$$

where Θ is differentiable with respect to its arguments, and Θ and its derivative with respect to θ are uniformly bounded for small bounded β, ϵ and I.

Using the fixed point theorem, we can solve (5.4) for θ as a smooth function of β, ϵ and I so the equation (5.3) is true.

Proof Let

$$C_p=\sqrt{2}\tilde{v}_p-\mathrm{i}\tilde{v}_{1p},\quad \tau=\sqrt{2}(1+r_1)x \tag{5.5}$$

where r_1, $\tilde{v}_p$ and $\tilde{v}_{1p}$ are given in (3.17) and (3.21) respectively, which yields

$$\tilde{v}_p=\frac{C_p+\bar{C}_p}{2\sqrt{2}},\qquad \tilde{v}_{1p}=\mathrm{i}\frac{C_p-\bar{C}_p}{2}. \tag{5.6}$$

Thus, $(\tilde{u}_p,\tilde{u}_{1p},C_p,\bar{C}_p)^{\mathrm{T}}(\tau)$ is a 2π–periodic solution of the following system

$$\begin{aligned}
\tilde{u}_{p\tau}&=\frac{1}{\sqrt{2}(1+r_1)}\tilde{u}_{1p},\\
\tilde{u}_{1p\tau}&=\frac{\sqrt{2}}{1+r_1}\tilde{u}_p+h_1(\beta,\epsilon,\tilde{u}_p,\tilde{u}_{1p},C_p,\bar{C}_p),\\
C_{p\tau}&=\frac{\mathrm{i}}{1+r_1}C_p+h_2(\beta,\epsilon,\tilde{u}_p,\tilde{u}_{1p},C_p,\bar{C}_p),\\
\bar{C}_{p\tau}&=\frac{-\mathrm{i}}{1+r_1}\bar{C}_p-h_2(\beta,\epsilon,\tilde{u}_p,\tilde{u}_{1p},C_p,\bar{C}_p),
\end{aligned} \tag{5.7}$$

where h_1 and h_2 are given in (3.4). We can express $C_p(\tau)$ as

$$C_p(\tau) = \mathrm{e}^{\mathrm{i}\frac{\tau}{1+r_1}} C_p(0) + w(\tau), \tag{5.8}$$

where

$$w(\tau) = \int_0^{\tau} \mathrm{e}^{\mathrm{i}\frac{\tau-s}{1+r_1}} h_2(\beta, \epsilon, \tilde{u}_p, \tilde{u}_{1p}, C_p, \bar{C}_p)\,\mathrm{d}s. \tag{5.9}$$

Note that the coefficient of $\mathrm{e}^{\mathrm{i}\tau}$ in $C_p(\tau)$ is $C_1 = I$ (see (3.15)). Thus,

$$\begin{aligned} I &= \frac{1}{2\pi}\int_0^{2\pi} C_p(s)\mathrm{e}^{-\mathrm{i}s}\,\mathrm{d}s \\ &= \frac{1}{2\pi}\int_0^{2\pi} \mathrm{e}^{-\mathrm{i}s}\big(\mathrm{e}^{\mathrm{i}\frac{s}{1+r_1}} C_p(0) + w(s)\big)\,\mathrm{d}s \\ &= \big(1+\kappa(r_1)\big)C_p(0) + \frac{1}{2\pi}\int_0^{2\pi} \mathrm{e}^{-\mathrm{i}s} w(s)\,\mathrm{d}s, \end{aligned} \tag{5.10}$$

where $\kappa(r_1) = \dfrac{1+r_1}{\mathrm{i}2\pi r_1}(1-\mathrm{e}^{-\mathrm{i}\frac{2\pi r_1}{1+r_1}}) - 1 = O(r_1)$ and $\kappa(0) = 0$, which yields

$$C_p(0) = \frac{1}{1+\kappa(r_1)}\Big(I - \frac{1}{2\pi}\int_0^{2\pi} \mathrm{e}^{-\mathrm{i}s} w(s)\,\mathrm{d}s\Big). \tag{5.11}$$

Thus,

$$C_p(\tau) = \frac{\mathrm{e}^{\mathrm{i}\frac{\tau}{1+r_1}}}{1+\kappa(r_1)}\Big(I - \frac{1}{2\pi}\int_0^{2\pi} \mathrm{e}^{-\mathrm{i}s} w(s)\,\mathrm{d}s\Big) + w(\tau), \tag{5.12}$$

or

$$C_p(x) = \frac{\mathrm{e}^{\mathrm{i}\sqrt{2}x}}{1+\kappa(r_1)}\Big(I - \frac{1}{2\pi}\int_0^{2\pi} \mathrm{e}^{-\mathrm{i}s} w(s)\,\mathrm{d}s\Big) + w(\sqrt{2}(1+r_1)x). \tag{5.13}$$

(3.18), (3.22), (4.15) and the expression of h_2 in (3.4) show

$$w(x) = O(|\beta| + |\epsilon| + I^2)$$

so that $C_p(x) = O(I)$. Therefore, we obtain by (5.6)

$$\tilde{v}_p(x) = \frac{1}{\sqrt{2}}\operatorname{Re} C_p(x) = \frac{1}{\sqrt{2}}\cos(\sqrt{2}x)I + V_1(x, \beta, \epsilon, I), \tag{5.14}$$

$$\tilde{v}_{1p}(x) = -\operatorname{Im} C_p(x) = -\sin(\sqrt{2}x)I + V_2(x, \beta, \epsilon, I), \tag{5.15}$$

where $V_1(x) = O(|\beta| + |\epsilon| + I^2)$ and $V_2(x) = O(|\beta| + |\epsilon| + I^2)$.

From (2.13), (3.21), (4.4) and $Z=(\tilde{u},\tilde{u}_1,\tilde{v},\tilde{v}_1)^{\mathrm{T}}$, we know that the equation (5.3) becomes

$$\begin{aligned}0=&\int_0^{\infty}\frac{1}{\sqrt{2}}\Big(-3\big(\tilde{v}(s)+\zeta(s)\tilde{v}_p(s+\theta)\big)^2-\big(\tilde{v}(s)+\zeta(s)\tilde{v}_p(s+\theta)\big)^3\\&+\zeta(s)\big(3\tilde{v}_p^2(s+\theta)+\tilde{v}_p^3(s+\theta)\big)\\&-\beta\left(\tanh\left(\frac{s}{\sqrt{2}}\right)+\tilde{u}(s)+\zeta(s)\tilde{u}_p(s+\theta)\right)^2\big(\tilde{v}(s)+\zeta(s)\tilde{v}_p(s+\theta)+1\big)\\&+\beta\zeta(s)\big(\tilde{u}_p(s+\theta)+1\big)^2\big(\tilde{v}_p(s+\theta)+1\big)\Big)\cos(\sqrt{2}x)\\&-\zeta'(s)\tilde{v}_p(s+\theta)\sin(\sqrt{2}s)-\frac{1}{\sqrt{2}}\zeta'(s)\tilde{v}_{1p}(s+\theta)\cos(\sqrt{2}s)\\&-\frac{1}{\sqrt{2}}\epsilon\Big(\tilde{g}\Big(\epsilon,\tanh\left(\frac{s}{\sqrt{2}}\right)-1+\tilde{u}(s)+\zeta(s)\tilde{u}_p(s+\theta),\\&\frac{1}{2}\mathrm{sech}^2\left(\frac{s}{\sqrt{2}}\right)+\tilde{u}_1(s)+\zeta(s)\tilde{u}_{1p}(s+\theta),\\&\tilde{v}(s)+\zeta(s)\tilde{v}_p(s+\theta),\tilde{v}_1(s)+\zeta(s)\tilde{v}_{1p}(s+\theta)\Big)\\&-\zeta(s)\tilde{g}\big(\epsilon,\tilde{u}_p(s+\theta),\tilde{u}_{1p}(s+\theta),\tilde{v}_p(s+\theta),\tilde{v}_{1p}(s+\theta)\big)\Big)\cos(\sqrt{2}s)\,\mathrm{d}s.\end{aligned}\tag{5.16}$$

For computational simplicity, we take $\gamma=\tilde{\alpha}_k=\dfrac{3}{4}$ for $k=1,2$ such that (4.14) and (4.15) are satisfied. Thus, (5.16) is changed into

$$\begin{aligned}0=&I\int_0^{\infty}-\frac{1}{\sqrt{2}}\zeta'(s)\Big(\cos\big(\sqrt{2}(s+\theta)\big)\sin(\sqrt{2}s)-\sin(\sqrt{2}(s+\theta))\cos(\sqrt{2}s)\Big)\,\mathrm{d}s\\&+\mathcal{P}(\theta,\beta,\epsilon,I)\\=&\frac{1}{\sqrt{2}}\sin(\sqrt{2}\theta)I+\mathcal{P}(\theta,\beta,\epsilon,I),\end{aligned}\tag{5.17}$$

where $\mathcal{P}(\theta,\beta,\epsilon,I)=O(I^{3/2})$, which is equivalent to

$$\theta=I^{1/2}\Psi(\theta,\beta,\epsilon,I),\tag{5.18}$$

where $\Psi(\theta,\beta,\epsilon,I)=-\dfrac{1}{\sqrt{2I}}\arcsin\big(\sqrt{2}\mathcal{P}(\theta,\beta,\epsilon,I)/I\big)$ is uniformly bounded for small β, ϵ and I. We can also check that $\Psi(\theta,\beta,\epsilon,I)$ is differentiable with respect to its arguments, and Ψ and its derivative with respect to θ are uniformly bounded for small bounded β, ϵ and I. This completes the proof of Lemma 5.1.

Acknowledgements This work is supported by the National Natural Science Foundation of China under grant number 11126292, 11201239, 11371314, the Guangdong

Natural Science Foundation No. S2013010015957 and the Project of Department of Education of Guangdong Province No. 2012KJCX0074.

References

[1] Ambrosetti A, Malchiodi A. Ni W M. Singularly perturbed elliptic equations with symmetry: existence of solutions concentrating on spheres, Part I [J]. Comm. Math. Phys., 2003, 235: 427–466.

[2] Ambrosetti A, Malchiodi A, Secchi S. Multiplicity results for some nonlinear Schrödinger equations with potentials [J]. Arch. Rat. Mech. Anal., 2001, 159: 253–271.

[3] Belmonte-Beitia J. A note on radial nonlinear Schrödinger systems with nonlinearity spatially modulated [J]. Electron. J. Diff. Equ., 2008, 148: 1–6.

[4] Benney D J, Newell A C. The propagation of nonlinear wave envelopes [J]. J. Math. and Phys., 1967, 46: 133–139.

[5] Bernard P. Homoclinic orbit to a center manifold [J]. Calc. Var. Partial Differential Equations, 2003, 17: 121–157.

[6] Brezzi F, Markowich P A. The three-dimensional Wigner-Poisson problem: existence, uniqueness and approximation [J]. Math. Mod. Meth. Appl. Sci., 1991, 14: 35–61.

[7] Champneys A R. Homoclinic orbits in reversible systems and their applications in mechanics, fluids and optics [J]. Phys. D, 1998, 112: 158–186.

[8] Champneys A R. Homoclinic orbits in reversible systems II: Multi-bumps and saddle-centres [J]. CWI. Quart., 1999, 12: 185–212.

[9] Champneys A R, Malomed B A, Yang J, Kaup D J. Embedded solitons: solitary waves in resonance with the linear spectrum [J]. Phys. D, 2001, 152-153: 340–354.

[10] Chen X, Lin T, Wei J. Blowup and solitary wave solutions with ring profiles of two-component nonlinear Schrödinger systems [J]. Phys. D, 2010, 239: 613–616.

[11] Dalfovo F, Giorgini S, Pitaevskii L P, Stringari S. Theory of Bose-Einstein condensation in trapped gases [J]. Rev. Mod. Phys., 1997, 71: 463–512.

[12] Deng S, Guo B. Generalized homoclinic solutions of a coupled Schrödinger system under a small perturbation [J]. J. Dyn. Diff. Equat., 2012, 24: 761–776.

[13] Dodd R K, Eilbeck J C, Gibbon J D, Morris H C. Solitons and Nonlinear Wave Equations [M]. New York: Academic Press, 1982.

[14] Fedele R, Miele G, Palumbo L, Vaccaro V G. Thermal wave model for nonlinear longitudinal dynamics in particle accelerators [J]. Phys. Lett. A, 1993, 173: 407–413.

[15] Fibich G, Gavish N, Wang X P. Singular ring solutions of critical and supercritical nonlinear Schrödinger equations [J]. Phys. D, 2007, 231: 55–86.

[16] Floer A, Weinstein A. Nonspreading wave packets for the cubic Schrödinger equation with a bounded potential [J]. J. Funct. Anal., 1986, 69: 397–408.

[17] Hasegawa A. Optical Solitons in Fibers [M]. Berlin: Springer-Verlag, 1989.

[18] Kielhöfer H. Bifurcation Theory: An Introduction with Applications to PDEs [M]. New York: Springer-Verlag, 2003.

[19] Kivshar Y, Agrawal G P. Optical Solitons: From Fibers to Photonic Crystals [M]. San Diego: Academic Press, 2003.

[20] Lerman L M. Hamiltonian systems with a separatrix loop of a saddle-center [J]. Selecta. Math. Sov., 1991, 10: 297–306.

[21] Mesentsev V K, Turitsyn S K. Stability of vector solitons in optical fibers [J]. Opt. Lett., 1992, I7: 1497–1500.

[22] Mielke A, Holmes P, O'Reilly O. Cascades of homoclinic orbits to, and chaos near, a Hamiltonian saddle-center [J]. J. Dyn. Diff. Equat., 1992, 4: 95–126.

[23] Noris B, Ramos M. Existence and bounds of positive solutions for a nonlinear Schrödinger system [J]. Proc. Amer. Math. Soc., 2010, 138: 1681–1692.

[24] Noris B, Trvares H, Terracini S, Verzini G. Uniform Hölder bounds for nonlinear Schrödinger systems with strong competition [J]. Comm. Pure. Appl. Math., 2010, 6: 267–302.

[25] Oh Y-G. On positive multi-lump bound states of nonlinear Schrödinger equations under multiple well potentials [J]. Commun. Math. Phys., 1990, 131: 223–253.

[26] Peletier L A, Rodrígues J A. Homoclinic orbits to a saddle-center in a fourth-order differential equation [J]. J. Differential Equations, 2004, 203: 185–215.

[27] Pelinovsky D E, Yang J. Instabilities of multihump vector solitons in coupled nonlinear Schrödinger equations [J]. Stud. Appl. Math., 2005, 115: 109–137.

[28] Pelinovsky D E, Yang J. Internal oscilations and radiation damping of vector solitions [J]. Stud. Appl. Math., 2000, 105: 245–276.

[29] Ragazzo C G. Irregular dynamics and homoclinic orbits to Hamiltonian saddle-centers [J]. Comm. Pure. App. Math., 1997, 50: 105–147.

[30] Rosales J L, Sánchez-Gómez J L. Nonlinear Schödinger equation coming from the action of the particles gravitational field on the quantum potential [J]. Phys. Lett. A, 1992, 66: 111–115.

[31] Scott A. Nonlinear Science: Emergence and Dynamics of Coherent Structures [M]. Oxford: Oxford University Press, 1999.

[32] Shatah J, Zeng C. Orbits homoclinic to center manifolds of conservative PDEs [J]. Nonlinearity, 2003, 16: 591–614.

[33] Wagenknecht T, Champneys A R. When gap solitons become embedded solitons: a generic unfolding [J]. Phys. D, 2003, 177: 50–70.

[34] Walter W. Gewöhnliche Differentialgleichungen [M]. Berlin: Springer-Verlag, 1992.

[35] Yang J. Classification of the solitary waves in coupled nonlinear Schröinger equations [J]. Phys. D, 1997, 108: 92-112.

[36] Yagasaki K. Homoclinic and heteroclinic orbits to invariant tori in multi-degree-of-freedom Hamiltonian systems with saddle-centres [J]. Nonlinearity, 2005, 18: 1331–

1350.

[37] Coti Zelati V, Macrì M. Homoclinic solutions to invariant tori in a center manifold [J]. Atti Accad. Naz. Lincei Cl. Sci. Fis. Mat. NaturṘend. Lincei (9) Mat. Appl., 2008, 19: 103–134.

[38] Coti Zelati V, Macrì M. Multibump solutions homoclinic to periodic orbits of large energy in a centre manifold [J]. Nonlinearity, 2005, 18: 2409–2445.

[39] Coti Zelati V, Macrì M. Existence of homoclinic solutions to periodic orbits in a center manifold [J]. J. Differential Equations, 2004, 202: 158–182.

On the Equations of Thermally Radiative Magnetohydrodynamics*

Li Xiaoli (李晓莉) and Guo Boling (郭柏灵)

Abstract An initial-boundary value problem is considered for the viscous compressible thermally radiative magnetohydrodynamic (MHD) flows coupled to self-gravitation describing the dynamics of gaseous stars in a bounded domain of $\mathbb{R}^3$. The conservative boundary conditions are prescribed. Compared to Ducomet-Feireisl [13] (also see, for instance, Feireisl [18], Feireisl-Novotný [20]), a rather more general constitutive relationship is given in this paper. The analysis allows for the initial density with vacuum. Every transport coefficient admits a certain temperature scaling. The global existence of a variational (weak) solution with any finite energy and finite entropy data is established through a three-level approximation and methods of weak convergence.

Keywords magnetohydrodynamic (MHD) flows; compressible; thermally radiative; global existence; variational (weak) solution

1 Introduction

Magnetohydrodynamics (MHD) concerns the motion of conducting fluids (cf. gases) in an electromagnetic field with a very broad range of applications in physical areas from liquid metals to cosmic plamas. In moving conducting magnetic fluids, magnetic fields can induce electric fields, and electric currents are developed, which create forces on the fluids and considerably effect changes in the magnetic fields. The dynamic motion of the fluids and the magnetic field interacts strongly on each other and both the hydrodynamic and electrodynamic effects have to be taken into account. Except for this, considerable attention has been put to study the effects of thermal radiation recently, because the radiation field significantly affects the dynamics of fluids,

* J. Differential Equations, 2014, 257: 3334–3381. DOI: 10.1016/j.jde. 2014.06.015.

for example, certain re-entry of space vehicles, astrophysical phenomena and nuclear fusion, and hydrodynamics with explicit account of radiation energy and momentum contribution constitutes the character of radiation hydrodynamics. In this paper, we consider the viscous compressible thermally radiative conducting fluids driven by the self-gravitation in the full magnetohydrodynamic setting. The equations to the three-dimensional full magnetohydrodynamic flows have the following form ([3, 13, 29, 30]):

$$\begin{cases} \rho_t + \nabla \cdot (\rho \mathbf{u}) = 0, \qquad \mathbf{x} \in \Omega \subset R^3, \ t > 0, \\ (\rho \mathbf{u})_t + \nabla \cdot (\rho \mathbf{u} \otimes \mathbf{u}) + \nabla p = \nabla \cdot \mathbb{S} + \rho \nabla \Psi + (\nabla \times \mathbf{H}) \times \mathbf{H}, \\ \mathcal{E}_t + \nabla \cdot \left(\left(\rho e + \frac{1}{2} \rho |\mathbf{u}|^2 + p \right) \mathbf{u} \right) + \nabla \cdot \mathbf{q} \\ \quad = \nabla \cdot ((\mathbf{u} \times \mathbf{H}) \times \mathbf{H} + \nu \mathbf{H} \times (\nabla \times \mathbf{H}) + \mathbb{S} \mathbf{u}) + \rho \nabla \Psi \cdot \mathbf{u}, \\ \mathbf{H}_t - \nabla \times (\mathbf{u} \times \mathbf{H}) = -\nabla \times (\nu \nabla \times \mathbf{H}), \qquad \nabla \cdot \mathbf{H} = 0, \end{cases} \tag{1.1}$$

where $\rho \in \mathbb{R}$ denotes the density, $\mathbf{u} \in \mathbb{R}^3$ the fluid velocity and $\mathbf{H} \in \mathbb{R}^3$ the magnetic field, $p \in \mathbb{R}$ the pressure.

$$\mathcal{E} = \rho e + \frac{1}{2}(\rho |\mathbf{u}|^2 + |\mathbf{H}|^2)$$

is the total energy with e being the specific internal energy. $\mathbb{S}$ stands for the viscous stress tensor, given by Newton's law of viscosity:

$$\mathbb{S} = \mu(\nabla \mathbf{u} + \nabla^\top \mathbf{u}) + \lambda(\nabla \cdot \mathbf{u})\, \mathbb{I}_3 \tag{1.2}$$

with μ the shear viscosity coefficient and $\eta = \lambda + \frac{2}{3}\mu$ the bulk viscosity coefficient of the flow (while μ should be positive for any "genuinely" viscous fluid, η may vanish, e.g. for a monoatomic gas), $\mathbb{I}_3$ the 3×3 identity matrix and $\nabla^\top \mathbf{u}$ the transpose of the matrix $\nabla \mathbf{u}$. Note that

$$\nabla \cdot \mathbb{S} = \left(\eta + \frac{1}{3}\mu \right) \nabla(\nabla \cdot \mathbf{u}) + \mu \Delta \mathbf{u},$$

$$\mathbb{S} : \nabla \mathbf{u} = \mu |\nabla \mathbf{u}|^2 + \mu \nabla \mathbf{u} : \nabla^\top \mathbf{u} + \left(\eta - \frac{2}{3}\mu \right) (\nabla \cdot \mathbf{u})^2.$$

$\mathbf{q}$ is the heat flux obeying the classical Fourier's law:

$$\mathbf{q} = -\kappa \nabla \vartheta, \quad \kappa \geqslant 0, \tag{1.3}$$

where ϑ means the absolute temperature, κ is the heat conductivity coefficient. The term $\rho \nabla \Psi$ is the gravitational force where the potential Ψ obeys Poisson's equation on the whole physical space $\mathbb{R}^3$ which is

$$-\Delta\Psi = G\rho \quad \text{with a constant } G > 0,$$

where ρ was extended to be zero outside Ω. The coefficient $\nu > 0$ is termed the magnetic diffusivity of the fluid. Usually, we refer to equation $(1.1)_1$ as the continuity equation (mass conservation equation), $(1.1)_2$ and $(1.1)_3$ as the momentum and the total energy conservation equation, respectively. It is well-known that the electromagnetic fields are governed by the Maxwell's equations. In magnetohydrodynamics, the displacement currents can be neglected in the time dependent Maxwell equations (see [22, 29, 30]), which transforms the hyperbolic Maxwell's system into a parabolic equation from a mathematical viewpoint. Accordingly, equation $(1.1)_4$ is called the induction equation, and, the electric field $\mathbf{E}$ is related to the magnetic induction vector $\mathbf{H}$ and the fluid velocity $\mathbf{u}$ via Ampère's law:

$$\nu\nabla \times \mathbf{H} = \mathbf{E} + \mathbf{u} \times \mathbf{H}.$$

As for the constraint $\nabla \cdot \mathbf{H} = 0$, it can be seen just as a restriction on the initial value $\mathbf{H}_0$, since $(\nabla \cdot \mathbf{H})_t \equiv 0$. The equations in (1.1) describe the macroscopic behavior of the magnetohydrodynamic flow with dissipative mechanisms. Magnetic reconnection is thought to be the mechanism responsible for the conversion of magnetic energy into heat and fluid motion(c.f. [3, 8]).

Next, we turn to the pressure-density-temperature (pdt) state equation. The well-known case is the ideal gas flow provided by Boyle's law:

$$p_G(\rho, \vartheta) = R\rho\vartheta,$$

where R is a constant inversely proportional to the mean molecular weight of the gas(cf. [18]). However, Boyle's law is definitely not satisfactory in the high temperature and density regime physically relevant to general viscous fluids in the full thermodynamical setting. For example, it is known the pressure of highly condensed cold matter is proportional to $\rho^{\frac{5}{3}}$ (see Chapters 3, 11 of [43]), also the isentropic state equation for a perfect monoatomic gas. In this paper, we will consider a much more general constitutive relationship than that introduced in [18], the so-called constitutive law for pressure, i.e., $p_G(\rho, \vartheta)$ will be determined via

$$p_G(\rho, \vartheta) = p_e(\rho) + \vartheta p_\vartheta(\rho) + \vartheta^2 p_{\vartheta^2}(\rho) \tag{1.4}$$

with the elastic pressure p_e and the thermal pressure components p_ϑ, p_{ϑ^2} being C^1 functions of the density. In particular, for the so-called electronic pressure, one has $p_G(\rho,\vartheta)=p_e(\rho)+R\rho\vartheta+\sqrt{\rho}\vartheta^2$ (cf. [43]). From the mathematical point of view, (1.4) can be understood as the first three terms in the Taylor expansion:

$$p_G(\rho,\vartheta)=p_G(\rho,\Theta)+(\vartheta-\Theta)\frac{\partial p_G}{\partial\vartheta}(\rho,\Theta)+\frac{(\vartheta-\Theta)^2}{2}\frac{\partial^2 p_G}{\partial^2\vartheta}(\rho,\Theta)+\text{higher order terms}$$

for a given $\Theta>0$.

In addition, it is worth-noting that the regularizing effect of radiation has been already observed in [9]. The radiation pressure is attributed to photons of very high energy, for example, the radiation energy associated with Planck distribution varies as the fourth power of the temperature, and the importance of the thermal radiation increases as the temperature is raised. Especially, at high temperatures, a completely different mechanism of heat energy transfer appears due to radiation, the energy and momentum densities of radiation field may become comparable to or even dominate the corresponding fluid quantities, for example, the heat conductivity coefficient κ becomes a rather sensitive function of temperature. As a consequence, the total pressure in fluid is augmented through the effect of high temperature radiation, by a radiation component $p_R(\vartheta)$ related to the absolute temperature through

$$p_R(\vartheta)=\frac{a}{3}\vartheta^4$$

with the Stefan-Boltzmann constant $a>0$ (see [2,11,37], also see Chapter 15 in [15]).

To conclude, we have the equation of state

$$p=p(\rho,\vartheta)=p_G(\rho,\vartheta)+p_R(\vartheta)=p_e(\rho)+\vartheta p_\vartheta(\rho)+\vartheta^2 p_{\vartheta^2}(\rho)+\frac{a}{3}\vartheta^4 \tag{1.5}$$

in this paper, which relates the pressure with the density and the absolute temperature of the flow.

Given the (pdt) state equation discussed above, note that the basic principle of the second law of thermodynamics implies that the internal energy and pressure are interrelated through Maxwell's relationship, we define the specific entropy s, up to an additive constant, through the thermodynamics equation:

$$\vartheta Ds(\rho,\vartheta)=De(\rho,\vartheta)+p(\rho,\vartheta)D\left(\frac{1}{\rho}\right). \tag{1.6}$$

The quantity $\frac{1}{\theta}\left(De+pD\left(\frac{1}{\rho}\right)\right)$ must be a perfect gradient, which is the well-known

Gibbs' relation on p, e and s, implying that e and p are interrelated through

$$\frac{\partial e}{\partial \rho}=\frac{1}{\rho^2}\left(p-\vartheta\frac{\partial p}{\partial \vartheta}\right)=\frac{1}{\rho^2}\left(p_e(\rho)-\vartheta^2 p_{\vartheta^2}(\rho)-a\vartheta^4\right) \tag{1.7}$$

(see e.g. Chapter 3 in [1]). In fact, the equality (1.7) comes from $\dfrac{\partial^2 s}{\partial\rho\partial\vartheta}=\dfrac{\partial^2 s}{\partial\vartheta\partial\rho}$.

Accordingly, e can be written in the form:

$$e=P_e(\rho)-\vartheta^2 P_{\vartheta^2}(\rho)+\frac{a\vartheta^4}{\rho}+Q(\vartheta),$$

where

$$P_e(\rho)=\int_1^\rho \frac{p_e(z)}{z^2}\,\mathrm{d}z \ \text{ is the elastic potential,}$$

$$P_{\vartheta^2}(\rho)=\int_1^\rho \frac{p_{\vartheta^2}(z)}{z^2}\,\mathrm{d}z,$$

and the thermal energy contribution Q is a non-decreasing function of ϑ. Here

$$Q(\vartheta)=\int_0^\vartheta c_v(\xi)\,\mathrm{d}\xi,$$

where $c_v(\vartheta)$ denotes the specific heat at constant volume such that

$$c_v\in C^1([0,\infty)),\quad \inf_{\vartheta\in[0,\infty)} c_v(\vartheta)>0.$$

The subsequent analysis leans essentially on thermodynamic stability of the fluid system expressed through

$$\frac{\partial p}{\partial\rho}>0,\quad \frac{\partial e}{\partial\vartheta}>0 \ \text{ for all } \ \rho,\vartheta>0.$$

Taking the high temperature and density regime physically relevant to our model equation into account, we can suppose

$$\begin{cases} p_e(0)=p_\vartheta(0)=p_{\vartheta^2}(0)=0,\\ p_e'(\rho)\geqslant a_1\rho^{\gamma-1}-b_1,\ p_\vartheta'(\rho)\geqslant 0,\ p_{\vartheta^2}'(\rho)\geqslant 0,\\ p_e(\rho)\leqslant a_2\rho^\gamma+b_2,\ p_\vartheta(\rho)\leqslant a_3\rho^\varsigma+b_3,\ p_{\vartheta^2}(\rho)\leqslant a_4\rho^\varsigma+b_4, \end{cases} \tag{1.8}$$

with $a_1>0$, $\gamma\geqslant 2$, $\gamma>\frac{4}{3}\varsigma$, $\gamma>2\zeta$. We remark here that p_e need not be a non-decreasing function of ρ.

Many theoretical studies have been devoted to the global-in-time existence of solutions with large data for the multidimensional continuum isothermal or isentropic fluid mechanics and electrodynamics (see [17,21,30,32,35,36,42]), especially for the magnetohydrodynamics because of its physical importance, complexity, rich phenomena and mathematical challenges; see [3,6,13,14,16,22,24,25,29,39,40] and the references cited therein. Note that the existence problem for a general full system including the energy equation is far from being solved. It is not know whether there is a classical (smooth) solution of system (1.1) with large initial data on an arbitrary time interval $(0,T)$ or not, even for the one-dimensional full perfect MHD equations with large data when all the viscosity, heat conductivity and magnetic diffusivity coefficients are constants, or for the three-dimensional Navier-Stokes equations describing the motion of compressible (incompressible) fluids. The simplest and most interesting case of the ideal gas flow with the viscosity coefficients and the heat conductivity coefficient being constants is completely open. P.-L. Lions [32] gives a formal proof of weak stability under the additional hypothesis of boundedness of $\rho, \mathbf{u}$ and ϑ in $L^\infty(\Omega \times (0,T))$. The corresponding problem for the one-dimensional Navier-Stokes was solved in [27] in the seventies last century. For the gases in one-dimension with small smooth initial data, the existence of global solutions was proved in [26], and the large-time behavior was studied in [33]. For large initial data, additional difficulties appear because of the presence of the magnetic field and its interaction with the hydrodynamic motion of the flow of large oscillation. Chen and Wang [5] investigated a free boundary problem for plane magnetohydrodynamic flows with general large initial data in 1-D and established the existence, uniqueness, and regularity of global solutions in H^1. Taking the effect of self-gravitation and the influence of high temperature radiation into account, global existence and uniqueness of a classical solution with large initial data was proved in [44] under a general assumption on the heat conductivity while all the viscosity, and magnetic diffusivity coefficients are constants. Based on the concept of variational (weak) solutions in the spirit of Leray's pioneering work (see [31]) in the context of incompressible, linearly viscous fluids, the existence theory was extended to the full Navier-Stokes system, including the thermal energy equation, under certain mostly technical hypothesis imposed on the quantities appearing in the constitutive equations (see Theorem 7.1 in [18]) by Ducomet and Feireisl. They first developed the global existence of variational solution in [18] under the assumptions that the viscosity coefficients μ and λ must be constant, while the heat conductivity coefficient κ depends on the temperature ϑ, and later they extended the result in [19] when μ and λ depend on ϑ. More complex, the effects of self-gravitation as well as the influence of radiations on the dynamics at high temperature regimes were included in [12]. Using the similar technique as in [12, 18, 19], Ducomet and Feireisl [13] studied the full compressible MHD equations while considering the effects of self-gravitation and the influence of radiations on the dynamics at high temperature regimes. Under the assumption that the viscosity coefficients depend on the temperature and the magnetic field, the pressure behaves like the power law $\rho^{\frac{5}{3}}$ for large density (reminiscent of the isentropic state equation for a perfect monoatomic gas), and all the transport coefficients satisfy certain $(1+\vartheta^{\alpha})$-growth conditions for any $\alpha \in \left[1, \frac{65}{27}\right)$, they introduced the total entropy balance as one of main field equations and proved the global existence of variational solution to any finite energy data on a bounded spatial domain in $\mathbb{R}^3$, supplemented with conservative boundary conditions. The reader is also referred to the monograph [20] for more details about the system and the methods. Hu and Wang [24] considered a 3-D model problem for full compressible MHD flows with more general pressure, by using the thermal equation as in [19] instead of the entropy equation employed in [13,20], they proved the existence of a global variational weak solution to the MHD equations with large data.

We shall study the global existence of the variational (weak) solutions to the real magnetohydrodynamic flows, with general pressure and internal energy while permitting the generation of heat by the magnetic field as well as its interaction with the fluid motion, in a bounded domain Ω in $\mathbb{R}^3$. Supplementing system (1.1) with the following initial and boundary conditions:

$$\begin{cases} (\rho, \rho\mathbf{u}, \vartheta, \mathbf{H}) \mid_{t=0} = (\rho_0, \mathbf{m}_0, \vartheta_0, \mathbf{H}_0), \quad \text{for } \mathbf{x} \in \Omega, \\ \rho_0 \geqslant 0, \ \operatorname{ess\,inf}_{\Omega} \vartheta_0 > 0, \\ \rho_0 \in L^{\gamma}(\Omega), \ \mathbf{m}_0 \in L^1(\Omega), \ \dfrac{1}{\rho_0}|\mathbf{m}_0|^2 \in L^1(\Omega), \ (\rho Q(\vartheta))_0 = \rho_0 Q(\vartheta_0) \in L^1(\Omega), \\ \vartheta_0 \in L^{\infty}(\Omega), \ \mathbf{H}_0 \in L^2(\Omega), \ \nabla \cdot \mathbf{H}_0 = 0 \text{ in } \mathcal{D}'(\Omega), \end{cases} \tag{1.9}$$

$$\mathbf{u} \mid_{\partial\Omega} = \mathbf{0}, \quad \mathbf{q} \cdot \mathbf{n} \mid_{\partial\Omega} = 0, \quad \text{and,} \quad \mathbf{H} \cdot \mathbf{n} \mid_{\partial\Omega} = 0, \quad (\nabla \times \mathbf{H}) \times \mathbf{n} \mid_{\partial\Omega} = \mathbf{0}, \tag{1.10}$$

where $\mathbf{n}$ denotes the unit outward normal on $\partial\Omega$. The boundary condition prescribed on the velocity is the so-called non-slip boundary condition, on the temperature is the conservation boundary condition, which means the system is thermally insulated (isolated), and on the magnetic field is known as the perfectly conducting wall condition which describes the case where the wall of container is made of perfectly conductive materials. Such boundary conditions are classical in the theory of magnetohydrodynamics and conform to that the physical system (1.1) is energetically isolated. Note here

(i) $\mathcal{D}$ denotes C_0^∞ , and $\mathcal{D}'$ for the sense of distributions;

(ii) $\frac{1}{\rho_0}|\mathbf{m}_0|^2 \in L^1(\Omega)$ indicates $\mathbf{m}_0 = \mathbf{0}$ a.a. $\mathbf{x} \in \{\rho_0 = 0\}$.

The problem considered in our paper seems more rational and physically valid in many astrophysical models, since it is well known that the dynamics of gaseous stars in astrophysics is dominated by intense magnetic fields, self-gravitation, and high temperature radiation(c.f. [7,41]).

Given the rather poor *a priori* estimates (ensuring equi-integrability, or weak L^1 compactness of the quantities appearing in the corresponding balance laws) available for the MHD equations, approximate (or even exact) solutions are bounded only in the Lebesgue spaces of integrable functions, and, consequently, any existence theory must be build up on the methods of weak convergence. The idea of approximation was used in [12, 20], where detailed existence proofs for simpler systems were given. In addition, the constitutive relations concerning the pressure in this paper is more general, we need to deal with the new terms in the (pdt) state equation, also for the general form of the thermal energy contribution $Q(\vartheta)$; and overcome the difficulty arising from the presence of the magnetic field and its coupling and interaction with the fluid variables. The heat conductivity is more complicated, not depending solely on the temperature. Except for the total energy conservation, we will formally obtain an entropy-type energy estimate involving the dissipative effects of viscosity, magnetic diffusion, and heat diffusion, which is essential to deduce the required available *a priori* estimates on the velocity, the magnetic induction vector and the temperature from boundedness of the initial total energy and the initial total entropy of the system by our careful analysis.

We introduce a suitable variational formulation of the problem and state the main existence result following a series of *a priori* estimates on the formal solution in Section 2, employ a three-level approximation scheme (see, for instance, [12,13,18,19,20,24]) to construct a sequence of approximation solutions in Section 3, and show the existence of global variational (weak) solution with large initial data in the last four sections. Our main result will be proved successively through the Galerkin method, a vanishing viscosity and vanishing artificial pressure limit passage using the methods of weak convergence.

2 Notations and results

Notations:

(1) $\Omega_T = \Omega \times (0, T)$ for some fixed time $T > 0$.

(2) For $k \geqslant 1$ and $p \geqslant 1$, denote $W^{k,p} = W^{k,p}(\Omega)$ for the Sobolev space, whose norm is denoted as $\|\cdot\|_{W^{k,p}}$, and $H^k = W^{k,2}(\Omega)$. For $T > 0$ and a function space X, denote by $L^p(0, T; X)$ the set of Bochner measurable X-valued time dependent functions f such that $t \to \|f\|_X$ belongs to $L^p(0, T)$, and the corresponding Lebesgue norm is denoted by $\|\cdot\|_{L^p_T(X)}$.

Let us consider first a classical solution $(\rho, \mathbf{u}, \vartheta, \mathbf{H})$ of the problem (1.1),(1.9),(1.10) in Ω_T. Observe from the continuity equation that the total mass is a constant of motion, i.e., we obtain the conservation of mass in the integral form:

$$\int_\Omega \rho(t)\, \mathrm{d}\mathbf{x} = \int_\Omega \rho_0\, \mathrm{d}\mathbf{x} \quad \text{for all} \quad t \in [0, T]. \tag{2.11}$$

Note that if we multiply the continuity equation by $b'(\rho)$, where $b \in C^1((0, \infty))$ and usually its derivative vanishes for large arguments (see, for instance, [10]), the renormalized continuity equation is obtained:

$$(b(\rho))_t + \nabla \cdot (b(\rho)\mathbf{u}) + (b'(\rho)\rho - b(\rho))\nabla \cdot \mathbf{u} = 0. \tag{2.12}$$

Multiplying the momentum equation by $\mathbf{u}$, the induction equation by $\mathbf{H}$, and inserting the results to the total energy equation, we get the following internal energy balance

$$(\rho e)_t + \nabla \cdot (\rho e \mathbf{u}) + p\nabla \cdot \mathbf{u} = \mathbb{S} : \nabla \mathbf{u} - \nabla \cdot \mathbf{q} + \nu |\nabla \times \mathbf{H}|^2, \tag{2.13}$$

where $A : B$ denotes the scalar product of the two matrices A and B.

Recalling $\mathbf{q} = -\kappa \nabla \vartheta$ and the state equation

$$p(\rho, \vartheta) = p_e(\rho) + \vartheta p_\vartheta(\rho) + \vartheta^2 p_{\vartheta^2}(\rho) + \frac{a}{3}\vartheta^4,$$

$$e(\rho, \vartheta) = P_e(\rho) - \vartheta^2 P_{\vartheta^2}(\rho) + \frac{a}{\rho}\vartheta^4 + Q(\vartheta),$$

we get the thermal energy equation

$$\begin{aligned} &\left(a\vartheta^4 + \rho Q(\vartheta) - \rho\vartheta^2 P_{\vartheta^2}(\rho)\right)_t + \nabla \cdot \left((a\vartheta^4 + \rho Q(\vartheta) - \rho\vartheta^2 P_{\vartheta^2}(\rho))\mathbf{u}\right) - \nabla \cdot (\kappa \nabla \vartheta) \\ &= \mathbb{S} : \nabla \mathbf{u} + \nu |\nabla \times \mathbf{H}|^2 - \left(\vartheta p_\vartheta(\rho) + \vartheta^2 p_{\vartheta^2}(\rho) + \frac{a}{3}\vartheta^4\right) \nabla \cdot \mathbf{u}, \end{aligned} \tag{2.14}$$

where $\mathbb{S} : \nabla \mathbf{u}$ is termed the dissipation function responsible for the irreversible transfer of the mechanical energy into heat. Here we have used the fact that

$$(\rho P_e(\rho))_t + \nabla \cdot (\rho P_e(\rho)\mathbf{u}) + p_e(\rho)\nabla \cdot \mathbf{u} = 0.$$

Moreover, if the temperature is strictly positive, multiplying (2.14) by $\frac{1}{\vartheta}$ and using the continuity equation, we obtain the entropy equation

$$(\rho s)_t + \nabla \cdot (\rho s \mathbf{u}) - \nabla \cdot \left(\frac{\kappa \nabla \vartheta}{\vartheta}\right) = \frac{\mathbb{S} : \nabla \mathbf{u} + \nu |\nabla \times \mathbf{H}|^2}{\vartheta} + \frac{\kappa |\nabla \vartheta|^2}{\vartheta^2}, \tag{2.15}$$

where the entropy

$$s = s(\rho, \vartheta) = \frac{4}{3}\frac{a\vartheta^3}{\rho} + \int_1^{\vartheta} \frac{c_v(\xi)}{\xi}\, d\xi - P_\vartheta(\rho) - 2\vartheta P_{\vartheta^2}(\rho)$$

with

$$P_\vartheta(\rho) = \int_1^{\rho} \frac{p_\vartheta(z)}{z^2}\, dz.$$

According to the Clausius-Duhem inequality (the second law of thermodynamics), the right hand of (2.15) must be non-negative for any possible motion, thus in particular, the viscosity coefficients μ, η for the Newtonian fluid and the magnetic diffusivity coefficient ν must be non-negative. Experiments show that the viscosity of fluids is quite sensitive to changes in temperature, for example, viscosity of gases increases with temperature, also of liquids decreases. The total heat-conductivity

$$\kappa := \kappa(\rho, \vartheta, \mathbf{H}) = \kappa_G(\rho, \vartheta, \mathbf{H}) + \kappa_R \vartheta^3$$

where $\kappa_R > 0$ is a constant (see [2]), and $\kappa_G > 0$ satisfies certain growth conditions. For the sake of simplicity, but not without certain physical background, also in agreement with numerous practical experiments, we shall assume all the transport coefficients admit some temperature scalings. Specifically, we assume

$$0 < c_1(1 + \vartheta^\alpha) \leqslant \mu(\vartheta, \mathbf{H}) \leqslant c_2(1 + \vartheta)^\alpha,$$
$$0 < c_3\vartheta^\alpha \leqslant \eta(\vartheta, \mathbf{H}) \leqslant c_4(1 + \vartheta)^\alpha$$

for some constant $\alpha \geqslant \dfrac{1}{2}$, and, set

$$0 < c_5(1 + \vartheta^\beta) \leqslant \nu(\rho, \vartheta, \mathbf{H}),\ \kappa_G(\rho, \vartheta, \mathbf{H}) \leqslant c_6(1 + \vartheta^\beta), \text{ and } c_v(\vartheta) \leqslant c_7(1 + \vartheta^{\frac{\beta}{2}-1}) \tag{2.16}$$

with $\beta \geqslant 1$ to be specified below. Note that we only consider the case when the viscosity coefficients are independent of the density (though being physically relevant) to avoid unsurmountable technical details in mathematics. The condition on $\kappa_G(\rho,\vartheta,\mathbf{H})$ is physically reasonable as experiments predict the value of $\beta \approx 4.5-5.5$ while Q should behave like $\vartheta^{1.5}$ for large arguments, which is in good agreement with the hypothesis on $c_v(\vartheta)$(see c.f [43]). We remark here that, if the magnetic field is absent, it has been shown by methods of statistical thermodynamics that $\mu = c\vartheta^{\frac{1}{2}}$ for a gas under normal conditions, and meanwhile, the coefficients of viscosity in gases show only little dependence on the density (see, for instance, Chapter 10 in [4]). The idea to impose several kinds of temperature scalings on the transport coefficients was inspired by [12,13,18,23]. The effect of the magnetic field is indeed very complicated because the viscous stress becomes unisotropic depending effectively on the direction of $\mathbf{H}$ (see Section 19.44 in [4]).

Since the gravitational potential Ψ can be determined by the boundary value problem:

$$\begin{cases} -\Delta\Psi = G\rho, & (t,\mathbf{x}) \in \Omega_T, \\ \Psi\mid_{\partial\Omega} = 0. \end{cases} \tag{2.17}$$

By using the maximum principle, $\Psi \geqslant 0$ in Ω_T, and

$$\Psi = G(-\Delta)^{-1}[\rho] \quad \text{with} \quad (-\Delta)^{-1}[\rho](\mathbf{x}) = \mathcal{F}_{\xi\to\mathbf{x}}\left[|\xi|^2\mathcal{F}_{\mathbf{x}\to\xi}[\rho]\right], \tag{2.18}$$

where $\mathcal{F}$ stands for the Fourier transform.

Moreover, taking advantage of the continuity equation, we have

$$\begin{aligned} \int_\Omega \rho\nabla\Psi\cdot\mathbf{u}\,\mathrm{d}\mathbf{x} &= -\int_\Omega \Psi\nabla\cdot(\rho\mathbf{u})\,\mathrm{d}\mathbf{x} = \int_\Omega \Psi\rho_t\,\mathrm{d}\mathbf{x} = \frac{1}{2G}\frac{\mathrm{d}}{\mathrm{d}t}\int_\Omega |\nabla\Psi|^2\,\mathrm{d}\mathbf{x} \\ &= \frac{1}{2}\frac{\mathrm{d}}{\mathrm{d}t}\int_\Omega \rho\Psi\,\mathrm{d}\mathbf{x} = -\frac{G}{2}\frac{\mathrm{d}}{\mathrm{d}t}\int_\Omega \Delta^{-1}[\rho]\rho\,\mathrm{d}\mathbf{x}. \end{aligned}$$

From the total energy equation and the boundary conditions (1.10), we deduce that the total energy of the system is a constant of motion, i.e., the total energy is conserved,

$$\frac{\mathrm{d}}{\mathrm{d}t}\int_\Omega \left(\rho P_e(\rho) - \rho\vartheta^2 P_{\vartheta^2}(\rho) + a\vartheta^4 + \rho Q(\vartheta) + \frac{1}{2}\rho|\mathbf{u}|^2 + \frac{1}{2}|\mathbf{H}|^2 + \frac{G}{2}\Delta^{-1}[\rho]\rho\right)\mathrm{d}\mathbf{x} = 0,$$

$$\begin{aligned} E(t) &= \int_\Omega \left(\rho P_e(\rho) - \rho\vartheta^2 P_{\vartheta^2}(\rho) + a\vartheta^4 + \rho Q(\vartheta) + \frac{1}{2}\rho|\mathbf{u}|^2 + \frac{1}{2}|\mathbf{H}|^2 + \frac{G}{2}\Delta^{-1}[\rho]\rho\right)\mathrm{d}\mathbf{x} \\ &= E_0 \end{aligned} \tag{2.19}$$

for a.a. $t \in (0,T)$, where

$$E_0=\int_\Omega \rho_0P_e(\rho_0)-\rho_0\vartheta_0^2P_{\vartheta^2}(\rho_0)+a\vartheta_0^4+\rho_0Q(\vartheta_0)+\frac{1}{2\rho_0}|\mathbf{m}_0|^2+\frac{1}{2}|\mathbf{H}_0|^2+\frac{G}{2}\Delta^{-1}[\rho_0]\rho_0 \,\mathrm{d}\mathbf{x}.$$

Note that p_e, p_{ϑ^2} are continuous functions vanishing at zero, thus

$$\rho\mapsto\rho P_e(\rho)\in C[0,\infty),\quad \lim_{\rho\to0+}\rho P_e(\rho)=0,$$

$$\rho\mapsto\rho P_{\vartheta^2}(\rho)\in C[0,\infty),\quad \lim_{\rho\to0+}\rho P_{\vartheta^2}(\rho)=0.$$

As for the energy contribution related to the term $\frac{G}{2}\int_\Omega\Delta^{-1}[\rho]\rho \,d\mathbf{x}$ is, in fact, negative. Using the fact that the total mass is a constant of motion, i.e., (2.11), the Hölder inequality and the classical elliptic estimate, we obtain

$$\begin{aligned}\frac{G}{2}\int_\Omega|\Delta^{-1}[\rho]\rho| \,\mathrm{d}\mathbf{x}&\leqslant\frac{G}{2}\|\rho\|_{L^\gamma}\|(-\Delta)^{-1}[\rho]\|_{L^{\frac{\gamma}{\gamma-1}}}\\&\leqslant C\|\rho\|_{L^\gamma}\|\rho\|_{L^1}\leqslant C\|\rho\|_{L^\gamma},\quad\gamma\geqslant2.\end{aligned}$$

Next we shall obtain sufficient *a priori* estimates on the solution by virtue of the total energy conservation (2.19). Firstly, the assumption (1.8) implies that

$$p_e(\rho)\geqslant\frac{a_1}{\gamma}\rho^\gamma-b_1\rho.$$

Furthermore, there are two positive constants $\tilde{c_1}$ and $\tilde{c_2}$ such that

$$\rho P_e(\rho)\geqslant\tilde{c_1}\rho^\gamma-\tilde{c_2}\quad\text{for any }\ \rho\geqslant0,$$

in particular,

$$\rho P_e(\rho)\geqslant\tilde{c_3}|p_e(\rho)|-\tilde{c_4}\quad\text{for }\rho\geqslant0. \tag{2.20}$$

By using the Cauchy-Schwarz inequality and the Hölder inequality, there are three positive constants $\tilde{c_5}(\leqslant\tilde{c_1})$, $\tilde{c_6}(\leqslant a)$ and $\tilde{c_7}$ such that

$$|\rho\vartheta^2P_{\vartheta^2}(\rho)|\leqslant\tilde{c_5}\rho^\gamma+\tilde{c_6}\vartheta^4+\tilde{c_7}.$$

From (2.11),(2.19), we have

$$\rho^\gamma,\ \vartheta^4,\ \rho P_e(\rho),\ \rho\vartheta^2P_{\vartheta^2}(\rho),\ \rho Q(\vartheta),\ \frac{1}{2}\rho|\mathbf{u}|^2,\ \frac{1}{2}|\mathbf{H}|^2\in L^\infty(0,T;L^1(\Omega)). \tag{2.21}$$

Obviously, the elastic pressure component $p_e(\rho)$ is integrable as a result of (2.20). Moreover, by virtue of the Hölder inequality, $\rho\mathbf{u}\in L^\infty(0,T;L^{\frac{2\gamma}{\gamma+1}}(\Omega))$.

Secondly, in order to get estimates on the temperature, we integrate (2.15) over Ω_T,

$$\int_{\Omega_T}\left(\frac{\mathbb{S}:\nabla\mathbf{u}+\nu|\nabla\times\mathbf{H}|^2}{\vartheta}+\frac{\kappa|\nabla\vartheta|^2}{\vartheta^2}\right)\mathrm{d}\mathbf{x}\mathrm{d}t=S(T)-S_0, \tag{2.22}$$

where $\frac{\nabla\vartheta}{\vartheta}$ will be interpreted as $\nabla\ln\vartheta$ in the spirit of Lemma 5.3 in [12](see also Lemma 2.1 in [13]), $S(t)=\int_\Omega \rho s\,\mathrm{d}\mathbf{x}$, and,

$$\begin{aligned}
S_0=&\int_\Omega \rho_0 s(\rho_0,\vartheta_0)\,\mathrm{d}\mathbf{x}\\
=&\int_\Omega\left(\frac{4}{3}a{\vartheta_0}^3+\rho_0\int_1^{\vartheta_0}\frac{c_v(\xi)}{\xi}\mathrm{d}\xi-\rho_0P_\vartheta(\rho_0)-2\rho_0\vartheta_0P_{\vartheta^2}(\rho_0)\right)\mathrm{d}\mathbf{x}.
\end{aligned}$$

Moreover, the presence of ϑ in the denominator indicates that this quantity must be positive on a set of full measure for the above arguments to make sense.

It follows from (1.8) that for some certain $C>0$,

$$|\rho P_\vartheta(\rho)|\leqslant C(1+\rho P_e(\rho)),\quad \rho^2P_{\vartheta^2}^2(\rho)\leqslant C(1+\rho P_e(\rho)),$$

then

$$\begin{aligned}
\rho s&\leqslant\frac{4a}{3}\vartheta^3+\rho Q(\vartheta)+|\rho P_\vartheta(\rho)|+2|\rho\vartheta P_{\vartheta^2}(\rho)|\\
&\leqslant\frac{4a}{3}\vartheta^3+\rho Q(\vartheta)+|\rho P_\vartheta(\rho)|+\rho^2P_{\vartheta^2}^2(\rho)+\vartheta^2\\
&\leqslant C\left(\rho Q(\vartheta)+\rho P_e(\rho)+\vartheta^4+1\right).
\end{aligned}$$

Here we have also used

$$\int_1^\vartheta\frac{c_v(\xi)}{\xi}\,\mathrm{d}\xi\leqslant 0,\quad 0<\vartheta\leqslant 1,$$

$$\int_1^\vartheta\frac{c_v(\xi)}{\xi}\,\mathrm{d}\xi\leqslant\int_1^\vartheta c_v(\xi)\,\mathrm{d}\xi=Q(\vartheta)-Q(1)\leqslant Q(\vartheta),\quad \vartheta>1,$$

and hence

$$\int_1^\vartheta\frac{c_v(\xi)}{\xi}\,\mathrm{d}\xi\leqslant Q(\vartheta),\quad \text{for any } \vartheta>0.$$

Then from (2.19),(2.22), we have

$$\begin{aligned}
&\int_{\Omega_T}\left(\frac{\mathbb{S}:\nabla\mathbf{u}+\nu|\nabla\times\mathbf{H}|^2}{\vartheta}+\frac{\kappa|\nabla\vartheta|^2}{\vartheta^2}\right)\mathrm{d}\mathbf{x}\mathrm{d}t-\operatorname*{ess\ inf}_{t\in[0,T]}\int_\Omega\rho(t)\ln\vartheta(t)\,\mathrm{d}\mathbf{x}\\
\leqslant& C(E_0,T)-S_0.
\end{aligned}$$

Recalling the assumption (2.16) on $\kappa_G(\rho,\vartheta,\mathbf{H})$, we have

$$\begin{aligned}\int_{\Omega_T}\frac{\kappa|\nabla\vartheta|^2}{\vartheta^2}\,\mathrm{d}\mathbf{x}\mathrm{d}t&=\int_{\Omega_T}\frac{(\kappa_R\vartheta^3+\kappa_G(\rho,\vartheta,\mathbf{H}))|\nabla\vartheta|^2}{\vartheta^2}\,\mathrm{d}\mathbf{x}\mathrm{d}t\\&\geqslant c\int_{\Omega_T}\frac{1+\vartheta^\beta+\vartheta^3}{\vartheta^2}|\nabla\vartheta|^2\,\mathrm{d}\mathbf{x}\mathrm{d}t.\end{aligned}$$

Consequently, on the one hand,

$$\begin{aligned}&\operatorname*{ess\ sup}_{t\in[0,T]}\int_\Omega\rho(t)|\ln\vartheta(t)|\,\mathrm{d}\mathbf{x}+\int_{\Omega_T}\left(|\nabla\vartheta^{\frac{\beta}{2}}|^2+|\nabla\vartheta^{\frac{3}{2}}|^2+|\frac{\nabla\vartheta}{\vartheta}|^2\right)\mathrm{d}\mathbf{x}\mathrm{d}t\\&\leqslant C(E_0,S_0,T),\end{aligned}\tag{2.23}$$

which yields

$$\rho|\ln\vartheta|\in L^\infty(0,T;L^1(\Omega)),\quad\frac{\nabla\vartheta}{\vartheta},\ \nabla\vartheta^{\frac{\beta}{2}},\ \nabla\vartheta^{\frac{3}{2}}\in L^2(\Omega_T).$$

First of all, $\nabla\vartheta^{\frac{3}{2}}\in L^2(\Omega_T)$ together with (2.21) give rise to

$$\vartheta^{\frac{3}{2}}\in L^2(0,T;H^1(\Omega)).$$

Next, since $\vartheta>0$, then

$$\begin{aligned}\int_\Omega|\nabla\vartheta|^2\,\mathrm{d}\mathbf{x}&=\int_\Omega\frac{\vartheta\nabla\vartheta}{\sqrt{\kappa}}\frac{\sqrt{\kappa}\nabla\vartheta}{\vartheta}\,\mathrm{d}\mathbf{x}\\&\leqslant\left(\int_\Omega\frac{\vartheta^2|\nabla\vartheta|^2}{\kappa}\,\mathrm{d}\mathbf{x}\right)^{\frac{1}{2}}\left(\int_\Omega\frac{\kappa|\nabla\vartheta|^2}{\vartheta^2}\,\mathrm{d}\mathbf{x}\right)^{\frac{1}{2}}\\&\leqslant C\left(\int_\Omega|\nabla\vartheta|^2\,\mathrm{d}\mathbf{x}\right)^{\frac{1}{2}}\left(\int_\Omega\frac{\kappa|\nabla\vartheta|^2}{\vartheta^2}\,\mathrm{d}\mathbf{x}\right)^{\frac{1}{2}}\\&\leqslant\epsilon\int_\Omega|\nabla\vartheta|^2\,\mathrm{d}\mathbf{x}+C_\epsilon\int_\Omega\frac{\kappa|\nabla\vartheta|^2}{\vartheta^2}\,\mathrm{d}\mathbf{x},\end{aligned}$$

where $\epsilon>0$ small enough. Thus,

$$\nabla\vartheta\in L^2(\Omega_T).$$

Recalling $\vartheta\in L^\infty(0,T;L^4(\Omega))$ again, we have

$$\vartheta\in L^2(0,T;H^1(\Omega)).$$

Now, taking advantage of Lemma 5.3 in [12](see also Lemma 2.1 in [13]) and the estimates in (2.23), we know

$$\ln\vartheta\in H^1(\Omega),\quad\nabla\ln\vartheta=\frac{\nabla\vartheta}{\vartheta}\quad\text{a.e. on }\Omega,$$

$$\|\ln\vartheta\|_{L^2}^2 \leqslant C(\|\rho\ln\vartheta\|_{L^1}^2 + \|\frac{\nabla\vartheta}{\vartheta}\|_{L^2}^2),$$

and furthermore,

$$\ln\vartheta \text{ is bounded in } L^2(\Omega_T)$$

by a constant depending only on the data and T. This estimate can be seen as "weak positivity" of the temperature ϑ. Finally, we conclude

$$\ln\vartheta \in L^2(0,T;H^1(\Omega)).$$

On the other hand, since

$$\begin{aligned}\frac{\mathbb{S}:\nabla\mathbf{u}}{\vartheta} &= \frac{\mu(\vartheta,\mathbf{H})}{\vartheta}\left(|\nabla\mathbf{u}|^2 + \nabla\mathbf{u}:\nabla^\top\mathbf{u} - \frac{2}{3}(\nabla\cdot\mathbf{u})^2\right) + \frac{\eta(\vartheta,\mathbf{H})}{\vartheta}(\nabla\cdot\mathbf{u})^2\\ &= \frac{\mu(\vartheta,\mathbf{H})}{2\vartheta}|\nabla\mathbf{u} + \nabla^\top\mathbf{u} - \frac{2}{3}\nabla\cdot\mathbf{u}\,\mathbb{I}_3|^2 + \frac{\eta(\vartheta,\mathbf{H})}{\vartheta}(\nabla\cdot\mathbf{u})^2\\ &\geqslant c\vartheta^{\alpha-1}|\nabla\mathbf{u} + \nabla^\top\mathbf{u}|^2,\end{aligned}$$

and, by virtue of Young's inequality,

$$|\nabla\mathbf{u} + \nabla^\top\mathbf{u}|^r \leqslant C(\vartheta^{\alpha-1}|\nabla\mathbf{u} + \nabla^\top\mathbf{u}|^2 + \vartheta^4) \text{ with } r = \frac{8}{5-\alpha}.$$

Here we need $\alpha \leqslant 1$. Hence $r \leqslant 2$, and

$$\mathbf{u} \in L^r(0,T;W_0^{1,r}(\Omega)).$$

In view of the entropy equation (2.22) again, combining with the assumption on the magnetic diffusivity coefficient, we have

$$(1+\vartheta)^{\frac{\beta-1}{2}}\nabla\times\mathbf{H} \in L^2(\Omega_T).$$

Bearing in mind the fact that $\|\nabla\times\mathbf{H}\|_{L^2} = \|\nabla\mathbf{H}\|_{L^2}$ when $\nabla\cdot\mathbf{H} = 0$, together with (2.21), we conclude

$$\mathbf{H} \in L^2(0,T;H^1(\Omega)).$$

Therefore, for the initial-boundary value problem, based on our assumptions on p_e, p_ϑ, p_{ϑ^2}, i.e.,(1.8), on all the transport coefficients μ,η,ν,κ_G and on the initial data, we have *a priori* estimates resulting from boundedness of the initial total energy and the initial total entropy as follows:

$$\rho^\gamma \in L^\infty(0,T;L^1(\Omega)),\ \rho\mathbf{u} \in L^\infty(0,T;L^{\frac{2\gamma}{\gamma+1}}(\Omega)),$$

$$\rho P_e(\rho),\ \rho\vartheta^2 P_{\vartheta^2}(\rho),\ \rho Q(\vartheta) \in L^\infty(0,T;L^1(\Omega)),$$

$$\vartheta \in L^\infty(0,T;L^4(\Omega))\cap L^2(0,T;H^1(\Omega)),\ \ \nabla\vartheta^{\frac{\beta}{2}} \in L^2(\Omega_T),\ \vartheta^{\frac{3}{2}},\ \ln\vartheta \in L^2(0,T;H^1(\Omega)),$$

$$\mathbf{u} \in L^r(0,T;W_0^{1,r}(\Omega)),\ \ r=\frac{8}{5-\alpha},$$

$$\mathbf{H} \in L^\infty(0,T;L^2(\Omega))\cap L^2(0,T;H^1(\Omega)).$$

We remark that (i) the velocity gradient $\nabla\mathbf{u}$ is not known to be square integrable.

(ii) a variational (weak) formulation of the momentum equation may not yield the full amount of mechanical energy dissipated by a (non-smooth) motion, then it may only satisfy the inequality

$$(\rho e)_t + \nabla\cdot(\rho e\mathbf{u}) + p\nabla\cdot\mathbf{u} \geqslant \mathbb{S}:\nabla\mathbf{u} - \nabla\cdot\mathbf{q} + \nu|\nabla\times\mathbf{H}|^2$$

instead of the internal energy balance (2.13). And consequently, equation (2.14) becomes

$$\begin{aligned}&\left(a\vartheta^4+\rho Q(\vartheta)-\rho\vartheta^2 P_{\vartheta^2}(\rho)\right)_t+\nabla\cdot\left(\left(a\vartheta^4+\rho Q(\vartheta)-\rho\vartheta^2 P_{\vartheta^2}(\rho)\right)\mathbf{u}\right)-\nabla\cdot(\kappa\nabla\vartheta)\\&\geqslant \mathbb{S}:\nabla\mathbf{u}+\nu|\nabla\times\mathbf{H}|^2-\left(\vartheta p_\vartheta(\rho)+\vartheta^2 p_{\vartheta^2}(\rho)+\frac{a}{3}\vartheta^4\right)\nabla\cdot\mathbf{u}.\end{aligned} \tag{2.24}$$

Using the same argument as the production of the the entropy equation (2.15), the thermal energy inequality (2.24) can be "equivalently" expressed through the variational principle of entropy

$$\begin{aligned}&\int_{\Omega_T}\left(\rho s\varphi_t+\rho s\mathbf{u}\cdot\nabla\varphi-\frac{\kappa\nabla\vartheta}{\vartheta}\cdot\nabla\varphi\right)\mathrm{d}\mathbf{x}\mathrm{d}t\\&\leqslant -\int_{\Omega_T}\left(\frac{\mathbb{S}:\nabla\mathbf{u}+\nu|\nabla\times\mathbf{H}|^2}{\vartheta}+\frac{\kappa|\nabla\vartheta|^2}{\vartheta^2}\right)\varphi\,\mathrm{d}\mathbf{x}\mathrm{d}t,\end{aligned}$$

for any $0\leqslant\varphi\in\mathcal{D}(\Omega_T;\mathbb{R})$.

The above arguments suggest us what we mean by a variational (weak) solution of the system (1.1),(1.9),(1.10) based on the second law of thermodynamics and the integral representation of balance laws.

Definition 2.1 *Given the initial distribution of the state variables*

$$\rho\mid_{t=0}=\rho_0,\ \rho\mathbf{u}\mid_{t=0}=\mathbf{m}_0,\ \vartheta\mid_{t=0}=\vartheta_0,\ \mathbf{H}\mid_{t=0}=\mathbf{H}_0,\ \ \rho_0\geqslant 0,\ \vartheta_0>0.$$

Let $T>0$ be given, $(\rho,\mathbf{u},\vartheta,\mathbf{H})$ is called a variational solution of (1.1),(1.9)–(1.10), *if*

- *$\rho \geqslant 0$, $\mathbf{u} \in L^r(0,T;W_0^{1,r}(\Omega))$ with $r > 1$ and $\mathbf{H} \in C([0,T];L^2_{weak}(\Omega)) \cap L^2(0,T;H^1(\Omega))$ satisfy the continuity equation in $\mathcal{D}'(\mathbb{R}^3 \times [0,T))$, the momentum conservation equation and the induction equation in $\mathcal{D}'(\Omega \times [0,T))$, which are*

$$\int_{\Omega_T} (\rho\psi'\phi + \psi\rho\mathbf{u} \cdot \nabla\phi)\, \mathrm{d}\mathbf{x}\mathrm{d}t + \psi(0)\int_\Omega \rho_0\phi\, \mathrm{d}\mathbf{x} = 0,$$

for any $\psi \in C^\infty([0,T])$ with $\psi(T) = 0$ and $\phi \in \mathcal{D}(\mathbb{R}^3;\mathbb{R})$, $ess\lim_{t\to 0+}\int_\Omega \rho\phi\, \mathrm{d}\mathbf{x}$ $= \int_\Omega \rho_0\phi\, \mathrm{d}\mathbf{x}$;

$$\int_{\Omega_T} (\psi'\rho\mathbf{u} \cdot \phi + \psi(\rho\mathbf{u} \otimes \mathbf{u}) : \nabla\phi + \psi p\nabla \cdot \phi)\, \mathrm{d}\mathbf{x}\mathrm{d}t$$
$$= \int_{\Omega_T} (\psi\mathbb{S} : \nabla\phi - \rho\psi\nabla\Psi \cdot \phi - \psi((\nabla \times \mathbf{H}) \times \mathbf{H}) \cdot \phi)\, \mathrm{d}\mathbf{x}\mathrm{d}t - \psi(0)\int_\Omega \mathbf{m}_0 \cdot \phi\, \mathrm{d}\mathbf{x},$$

where $\Psi = G(-\Delta)^{-1}[1_\Omega\rho]$, and

$$\int_{\Omega_T} (\psi'\mathbf{H} \cdot \phi + \psi(\mathbf{u} \times \mathbf{H}) \cdot (\nabla \times \phi) - \psi\nu(\nabla \times \mathbf{H}) \cdot (\nabla \times \phi))\, \mathrm{d}\mathbf{x}\mathrm{d}t$$
$$= -\psi(0)\int_\Omega \mathbf{H}_0 \cdot \phi\, \mathrm{d}\mathbf{x},$$

for any $\psi \in C^\infty([0,T])$ with $\psi(T) = 0$, $\phi \in \mathcal{D}(\Omega;\mathbb{R}^3)$, and

$$ess\lim_{t\to 0+}\int_\Omega \rho\mathbf{u} \cdot \phi\, \mathrm{d}\mathbf{x} = \int_\Omega \mathbf{m}_0 \cdot \phi\, dx,\ ess\lim_{t\to 0+}\int_\Omega \mathbf{H} \cdot \phi\, \mathrm{d}\mathbf{x}$$
$$= \int_\Omega \mathbf{H}_0 \cdot \phi\, \mathrm{d}\mathbf{x}.$$

- *The propagation of density oscillations is described by (2.12), i.e., the continuity equation is satisfied in the sense of renormalized solutions introduced in [10], which is, (2.12) holds in $\mathcal{D}'(\mathbb{R}^3 \times [0,T))$ with any $b \in C^1(\mathbb{R}^+)$ satisfying*

$$b'(z) = 0 \text{ for all } z \in \mathbb{R}^+ \text{ large enough, e.g., } z \geqslant z_b, \tag{2.25}$$

where the constant z_b depends on the choice of function b, that means,

$$\begin{aligned}&\int_{\Omega_T} (b(\rho)\psi'\phi + \psi b(\rho)\mathbf{u} \cdot \nabla\phi + \psi(b(\rho) - b'(\rho)\rho)\nabla \cdot \mathbf{u}\phi)\, \mathrm{d}\mathbf{x}\mathrm{d}t\\ &= -\psi(0)\int_\Omega b(\rho_0)\phi\, \mathrm{d}\mathbf{x},\end{aligned} \tag{2.26}$$

for any $\psi \in C^\infty([0,T])$ with $\psi(T) = 0$ and $\phi \in \mathcal{D}(\Omega;\mathbb{R})$, $ess\lim_{t\to 0+}\int_\Omega b(\rho)\phi\, \mathrm{d}\mathbf{x}$ $= \int_\Omega b(\rho_0)\phi\, \mathrm{d}\mathbf{x}$.

- $\vartheta > 0$ *satisfies the variational principle of entropy production*

$$\int_{\Omega_T}\left(\rho s\psi'\phi + \psi\rho s\mathbf{u}\cdot\nabla\phi - \psi\frac{\kappa\nabla\vartheta}{\vartheta}\cdot\nabla\phi\right)\,\mathrm{d}\mathbf{x}\mathrm{d}t$$
$$\leqslant -\int_{\Omega_T}\left(\frac{\mathbb{S}:\nabla\mathbf{u}+\nu|\nabla\times\mathbf{H}|^2}{\vartheta}+\frac{\kappa|\nabla\vartheta|^2}{\vartheta^2}\right)\psi\phi\,\mathrm{d}\mathbf{x}\mathrm{d}t - \psi(0)\int_\Omega \rho_0 s(\rho_0,\vartheta_0)\phi\,\mathrm{d}\mathbf{x},$$

for any $0\leqslant\psi\in C^\infty([0,T])$ *with* $\psi(T)=0$ *and* $0\leqslant\phi\in\mathcal{D}(\mathbb{R}^3;\mathbb{R})$, *ess* $\lim\limits_{t\to 0+}$

$$\int_\Omega \rho s\phi \mathrm{d}\mathbf{x} \geqslant \int_\Omega \rho_0 s(\rho_0,\vartheta_0)\phi\,\mathrm{d}\mathbf{x}.$$

- *The total energy* $E(t)$ *defined in* (2.19) *is a constant of motion:*

$$\int_0^T E\psi'\,\mathrm{d}t = -E_0\psi(0),$$

for any $\psi\in C^\infty([0,T])$ *with* $\psi(T)=0$.

- $(\rho,\mathbf{u},\vartheta,\mathbf{H})$ *satisfies* (1.10) *in the sense of trace a.a. in* $(0,T)$.

Note that all the choice of the test functions agree with the boundary condition (1.10). We remark here, if the magnetic field $\mathbf{H}$ is absent, the system (1.1) with the constitutive relations (1.2)(1.3) is called the full Navier-Stokes-Fourier system, and, a variational formulation of such a system with conservative boundary conditions was introduced in [12]. Now, we are ready to state our main theorem of this paper, which is the existence of global variational (weak) solutions for (1.1),(1.9),(1.10). More precisely, we prove

Theorem 2.1 *Let* $\Omega\subset\mathbb{R}^3$ *be a bounded domain of class* $C^{2+\iota}, \iota\in(0,1]$. *Assume that the pressure* p *determined by* (1.5), *the internal energy* e *and the specific entropy* s *are interrelated by* (1.6). *Furthermore, suppose that the temperature scaling on* μ,η *satisfies* $\frac{1}{2}\leqslant\alpha\leqslant 1$, *on* ν, κ_G *and* c_v *satisfies* $1\leqslant\beta\leqslant 4$. *Then the system* (1.1),(1.9),(1.10) *has at least one global variational (weak) solution for all* $T\in(0,\infty)$ *such that*

$$\begin{cases}\rho\in C([0,T];L^1(\Omega))\cap L^\infty(0,T;L^\gamma(\Omega)), \quad \rho\mathbf{u}\in C([0,T];L^{\frac{2\gamma}{\gamma+1}}_{weak}(\Omega)),\\ \mathbf{u}\in L^r(0,T;W_0^{1,r}(\Omega)) \ \ with\ r>1,\\ \vartheta\in L^\infty(0,T;L^4(\Omega))\cap L^2(0,T;H^1(\Omega)), \quad \rho Q(\vartheta)\in L^\infty(0,T;L^1(\Omega)),\\ \vartheta^{\frac{3}{2}},\ \vartheta^{\frac{\beta}{2}},\ \ln\vartheta\in L^2(0,T;H^1(\Omega)),\\ \mathbf{H}\in C([0,T];L^2_{weak}(\Omega))\cap L^2(0,T;H^1(\Omega)).\end{cases}$$

Remark 2.1 *1. $\rho \in L^\infty(0,T;L^\gamma(\Omega))$ can be strengthened to $\rho \in C([0,T];L^\gamma_{weak}(\Omega))$, in particular, $\rho(t) \rightharpoonup \rho_0$ in $L^\gamma(\Omega)$ as $t \to 0$, and, $\int_\Omega \rho(t)\,\mathrm{d}\mathbf{x} = \int_\Omega \rho_0\,\mathrm{d}\mathbf{x}$ is a constant of motion (independent of $t \in [0,T]$).*

2. It will be shown that ϑ satisfies the initial condition

$$\mathrm{ess}\lim_{t\to 0+}\int_\Omega \vartheta\phi\,\mathrm{d}\mathbf{x} = \int_\Omega \vartheta_0\phi\,\mathrm{d}\mathbf{x},\ \textit{for any } \phi \in \mathcal{D}(\Omega;\mathbb{R})$$

if there exists a sequence of times $t_n \to 0$ such that $\{\vartheta_{t_n}\}$ is precompact in $L^1(\Omega)$.

3 Approximation scheme associated to (1.1)

We consider the following regularized problem

$$\begin{cases}\rho_t + \nabla\cdot(\rho\mathbf{u}) = \varepsilon\Delta\rho,\\ (\rho\mathbf{u})_t + \nabla\cdot(\rho\mathbf{u}\otimes\mathbf{u}) + \nabla(p_m(\rho)+p_b(\rho)+\vartheta p_\vartheta(\rho)+\vartheta^2 p_{\vartheta^2}(\rho)+\frac{a}{3}\vartheta^4)\\ \qquad +\delta\nabla\rho^\Gamma + \varepsilon\nabla\mathbf{u}\cdot\nabla\rho = \nabla\cdot\mathbb{S} + \rho\nabla\Psi + (\nabla\times\mathbf{H})\times\mathbf{H},\\ (a\vartheta^4 + \rho Q(\vartheta) - \rho\vartheta^2 P_{\vartheta^2}(\rho))_t + \nabla\cdot((a\vartheta^4+\rho Q(\vartheta)-\rho\vartheta^2 P_{\vartheta^2}(\rho))\mathbf{u})\\ \qquad -\nabla\cdot((\kappa_G(\rho,\vartheta,\mathbf{H})+\kappa_R\vartheta^3)\nabla\vartheta) = \mathbb{S}:\nabla\mathbf{u} + \nu(\rho,\vartheta,\mathbf{H})|\nabla\times\mathbf{H}|^2\\ \qquad +\varepsilon|\nabla\rho|^2\left(\delta\Gamma\rho^{\Gamma-2}+\dfrac{p_m'(\rho)}{\rho}\right) - \left(\vartheta p_\vartheta(\rho)+\vartheta^2 p_{\vartheta^2}(\rho)+\dfrac{a}{3}\vartheta^4\right)\nabla\cdot\mathbf{u},\\ \mathbf{H}_t - \nabla\times(\mathbf{u}\times\mathbf{H}) = -\nabla\times(\nu(\rho,\vartheta,\mathbf{H})\nabla\times\mathbf{H}),\quad \nabla\cdot\mathbf{H}=0,\end{cases} \tag{3.27}$$

with the initial-boundary conditions

$$\begin{cases}\nabla\rho\cdot\mathbf{n}\,|_{\partial\Omega} = 0,\ \rho\,|_{t=0} = \rho_{0,\delta},\\ \mathbf{u}\,|_{\partial\Omega} = 0,\ (\rho\mathbf{u})\,|_{t=0} = \mathbf{m}_{0,\delta} = \begin{cases}\mathbf{m}_0, & \text{if } \rho_{0,\delta} \geqslant \rho_0,\\ 0, & \text{if } \rho_{0,\delta} < \rho_0,\end{cases}\\ \nabla\vartheta\cdot\mathbf{n}\,|_{\partial\Omega} = 0\ \text{(no-flux)},\ \vartheta\,|_{t=0} = \vartheta_{0,\delta},\\ \mathbf{H}\cdot\mathbf{n}\,|_{\partial\Omega} = (\nabla\times\mathbf{H})\times\mathbf{n}\,|_{\partial\Omega} = 0,\ \mathbf{H}\,|_{t=0} = \mathbf{H}_{0,\delta},\end{cases} \tag{3.28}$$

where "the elastic pressure component" p_e has been decomposed as

$$p_e(\rho) = p_m(\rho) + p_b(\rho)$$

with $p_m, p_b \in C^1[0,\infty)$, $p_m'(\rho) \geqslant 0$, $|p_b| \leqslant M$. The parameters $\varepsilon, \delta > 0$ and $\delta\nabla\rho^\Gamma$ is the artificial pressure with $\Gamma > 0$ (a constant to be determined when facilitate the limit passage $\varepsilon \to 0$).

Here the approximate initial density, temperature and magnetic induction vector distributions are chosen in such a way that

$$
\begin{cases}
\rho_{0,\delta} \in C^{2+\iota}(\overline{\Omega}), \ \nabla \rho_{0,\delta} \cdot \mathbf{n} \,|_{\partial\Omega} = 0, \ \inf_{\mathbf{x}\in\Omega} \rho_{0,\delta} > 0, \\
\rho_{0,\delta} \to \rho_0 \text{ in } L^\gamma(\Omega), \ |\{\mathbf{x} \in \Omega \mid \rho_{0,\delta} < \rho_0\}| \to 0 \text{ as } \delta \to 0; \\
\vartheta_{0,\delta} \in C^{2+\iota}(\overline{\Omega}), \ \nabla \vartheta_{0,\delta} \cdot \mathbf{n} \,|_{\partial\Omega} = 0, \ \inf_{\mathbf{x}\in\Omega} \vartheta_{0,\delta} > 0, \\
\vartheta_{0,\delta} \to \vartheta_0 \text{ in } L^1(\Omega) \text{ as } \delta \to 0; \\
\mathbf{H}_{0,\delta} \in \mathcal{D}(\Omega;\mathbb{R}^3), \ \nabla \cdot \mathbf{H}_{0,\delta} = 0, \ \mathbf{H}_{0,\delta} \cdot \mathbf{n} \,|_{\partial\Omega} = (\nabla \times \mathbf{H}_{0,\delta}) \times \mathbf{n} \,|_{\partial\Omega} = 0, \\
\mathbf{H}_{0,\delta} \to \mathbf{H}_0 \text{ in } L^2(\Omega;\mathbb{R}^3) \ \text{ as } \delta \to 0.
\end{cases}
\tag{3.29}
$$

Taking $\varepsilon \to 0$ and $\delta \to 0$ in (3.27) will give the solution of system (1.1),(1.9),(1.10) in Theorem 2.1. Note that the most important principle we want to conform to is that the total energy is a constant of motion at every step of approximation. In particular, denoting

$$
P_m(\rho) = \int_1^\rho \frac{p_m(z)}{z^2}\,\mathrm{d}z,
$$

the initial value of the regularized total energy

$$
\begin{aligned}
E_{0,\delta} = \int_\Omega \Big(& \rho_{0,\delta} P_m(\rho_{0,\delta}) - \rho_{0,\delta}\vartheta_{0,\delta}^2 P_{\vartheta^2}(\rho_{0,\delta}) + \rho_{0,\delta} Q(\vartheta_{0,\delta}) + a\vartheta_{0,\delta}^4 + \frac{1}{2\rho_{0,\delta}}|\mathbf{m}_{0,\delta}|^2 \\
& + \frac{1}{2}|\mathbf{H}_{0,\delta}|^2 + \frac{\delta}{\Gamma - 1}\rho_{0,\delta}^\Gamma \Big)\,\mathrm{d}\mathbf{x}
\end{aligned}
$$

is bounded by a constant independent of $\delta > 0$.

Moreover, it is easy to check that the corresponding approximate solutions satisfy the energy balance:

$$
\begin{aligned}
& \frac{\mathrm{d}}{\mathrm{d}t}\int_\Omega \Big(\rho P_m(\rho) - \rho\vartheta^2 P_{\vartheta^2}(\rho) + \rho Q(\vartheta) + \frac{1}{2}\rho|\mathbf{u}|^2 + a\vartheta^4 + \frac{1}{2}|\mathbf{H}|^2 + \frac{\delta}{\Gamma-1}\rho^\Gamma \Big)\,\mathrm{d}\mathbf{x} \\
= & \int_\Omega \rho\nabla\Psi \cdot \mathbf{u}\,\mathrm{d}\mathbf{x} + \int_\Omega \nabla p_b \cdot \mathbf{u}\,\mathrm{d}\mathbf{x} \quad \text{in } \mathcal{D}'(0,T).
\end{aligned}
$$

Here we have used

$$
(\rho P_m(\rho))_t + \nabla\cdot(\rho P_m(\rho)\mathbf{u}) + p_m(\rho)\nabla\cdot\mathbf{u} = \varepsilon P_m(\rho)\Delta\rho + \varepsilon\frac{p_m(\rho)}{\rho}\Delta\rho
$$

which leads to

$$
\frac{\mathrm{d}}{\mathrm{d}t}\int_\Omega \rho P_m(\rho)\,\mathrm{d}\mathbf{x} = \int_\Omega \nabla p_m(\rho)\cdot\mathbf{u}\,\mathrm{d}\mathbf{x} - \varepsilon\int_\Omega \frac{p_m'(\rho)}{\rho}|\nabla\rho|^2\,\mathrm{d}\mathbf{x},
$$

and

$$\begin{aligned}
&\delta\int_\Omega \nabla\rho^\Gamma\cdot\mathbf{u}\,\mathrm{d}\mathbf{x}-\varepsilon\delta\Gamma\int_\Omega|\nabla\rho|^2\rho^{\Gamma-2}\,\mathrm{d}\mathbf{x}\\
&=\delta\Gamma\int_\Omega\rho^{\Gamma-1}\nabla\rho\cdot\mathbf{u}\,\mathrm{d}\mathbf{x}-\frac{\varepsilon\delta\Gamma}{\Gamma-1}\int_\Omega\nabla\rho\cdot\nabla(\rho^{\Gamma-1})\,\mathrm{d}\mathbf{x}\\
&=\delta\Gamma\int_\Omega\rho^{\Gamma-1}\nabla\rho\cdot\mathbf{u}\,\mathrm{d}\mathbf{x}+\frac{\varepsilon\delta\Gamma}{\Gamma-1}\int_\Omega\rho^{\Gamma-1}\Delta\rho\,\mathrm{d}\mathbf{x}\\
&=\delta\Gamma\int_\Omega\rho^{\Gamma-1}\nabla\rho\cdot\mathbf{u}\,\mathrm{d}\mathbf{x}+\frac{\delta\Gamma}{\Gamma-1}\int_\Omega\rho^{\Gamma-1}(\rho_t+\nabla\cdot(\rho\mathbf{u}))\,\mathrm{d}\mathbf{x}\\
&=\delta\Gamma\int_\Omega\rho^{\Gamma-1}\nabla\rho\cdot\mathbf{u}\,\mathrm{d}\mathbf{x}+\frac{\delta}{\Gamma-1}\frac{d}{dt}\int_\Omega\rho^\Gamma\,\mathrm{d}\mathbf{x}+\frac{\delta\Gamma}{\Gamma-1}\int_\Omega\rho^{\Gamma-1}\nabla\cdot(\rho\mathbf{u})\,\mathrm{d}\mathbf{x}\\
&=\delta\Gamma\int_\Omega\rho^{\Gamma-1}\nabla\rho\cdot\mathbf{u}\,\mathrm{d}\mathbf{x}+\frac{\delta}{\Gamma-1}\frac{d}{dt}\int_\Omega\rho^\Gamma\,\mathrm{d}\mathbf{x}-\delta\Gamma\int_\Omega\rho^{\Gamma-1}\nabla\rho\cdot\mathbf{u}\,\mathrm{d}\mathbf{x}\\
&=\frac{\delta}{\Gamma-1}\frac{\mathrm{d}}{\mathrm{d}t}\int_\Omega\rho^\Gamma\,\mathrm{d}\mathbf{x}.
\end{aligned}$$

Moreover,

$$\begin{aligned}
&\lim_{t\to0}\int_\Omega\left(\rho P_m(\rho)-\rho\vartheta^2P_{\vartheta^2}(\rho)+\rho Q(\vartheta)+a\vartheta^4+\frac12\rho|\mathbf{u}|^2+\frac12|\mathbf{H}|^2+\frac{\delta}{\Gamma-1}\rho^\Gamma\right)\mathrm{d}\mathbf{x}\\
=&\int_\Omega\Big(\rho_{0,\delta}P_m(\rho_{0,\delta})-\rho_{0,\delta}\vartheta_{0,\delta}^2P_{\vartheta^2}(\rho_{0,\delta})+\rho_{0,\delta}Q(\vartheta_{0,\delta})+a\vartheta_{0,\delta}^4+\frac{1}{2\rho_{0,\delta}}|\mathbf{m}_{0,\delta}|^2\\
&+\frac12|\mathbf{H}_{0,\delta}|^2+\frac{\delta}{\Gamma-1}\rho_{0,\delta}^\Gamma\Big)\,\mathrm{d}\mathbf{x}.
\end{aligned}$$

After the above modification, the proof of Theorem 2.1 consists of the following steps:

Step 1: Solving problem for fixed parameters $\varepsilon>0$, $\delta>0$ and $\Gamma>0$ by the Galerkin method, deriving estimates independent of the dimension k of the Galerkin approximation and carrying out the limit as $k\to\infty$ provided Γ has been chosen large enough.

Step 2: Passing to the limit $\varepsilon\to0$.

Step 3: Letting $\delta\to0$.

4 Proof of Theorem 2.1

In this section we introduce the chain of approximations which we use to solve the original problem (1.1),(1.9),(1.10). At any level of approximations we formulate the statements about the existence of variational (weak) solutions and their properties which are needed to carry about the proof of existence for the original system.

To begin with, the goal proposed in Step 1 can be achieved via a Schauder-Tychonoff-type fixed point argument. More precisely, we first establish that ρ,Ψ,ϑ, and $\mathbf{H}$ can be computed successively from the first equation of (3.27), (2.18), and the last two equations of (3.27) as functions of $\mathbf{u}$, then the approximation problem for fixed parameters ε and δ can be easily solved by means of a modified Faedo-Galerkin method in the same way as in Chapter 7 in [18].

4.1 Solvability of continuity equation with dissipation

Given velocity field $\mathbf{u} \in C([0,T]; C_0^2(\overline{\Omega}, \mathbb{R}^3))$, the density $\rho := \rho[\mathbf{u}]$ is determined uniquely as the solution of the Neumann (suggested by the fact that conservation of mass in the form $\left(\int_\Omega \rho \,\mathrm{d}\mathbf{x}\right)_t = 0$ should hold) initial-boundary value problem:

$$\begin{cases} \rho_t + \nabla \cdot (\rho \mathbf{u}) = \varepsilon \Delta \rho, \ \varepsilon > 0, \\ \rho\mid_{t=0} = \rho_{0,\delta}, \\ \nabla \rho \cdot \mathbf{n}\mid_{\partial\Omega} = 0, \end{cases} \tag{4.30}$$

with $\rho_{0,\delta}$ satisfying (3.29). More precisely, since this is a linear parabolic Neumann problem in ρ, the existence and uniqueness of a classical solution

$$\rho \in C([0,T]; C^{2+\iota}(\overline{\Omega})), \quad \rho_t \in C([0,T]; C^{\iota}(\overline{\Omega})),$$

$$\rho(t,\mathbf{x}) \geqslant \inf_{\mathbf{x}\in\Omega} \rho_{0,\delta}(\mathbf{x}) \exp(-\|\nabla \cdot \mathbf{u}\|_{L_t^1(L^\infty)}) > 0 \ \text{ on } \overline{\Omega}_T$$

can be obtained by the Galerkin method(Theorem 5.1.2 in [34], also see Section 7.6 in [38] for details), the solution mapping $\mathbf{u} \mapsto \rho[\mathbf{u}]$ is bounded and

$$\mathbf{u} \in C([0,T]; C_0^2(\overline{\Omega}, \mathbb{R}^3)) \mapsto \rho[\mathbf{u}] \in C^1(\overline{\Omega}_T)$$

is continuous (Proposition 7.1 in [18]) . The gravitational potential Ψ will be solved by (2.17) and (2.18) by extending ρ to be zero outside of Ω.

4.2 Solvability of both the magnetic field and the temperature

In this section we show that the following system can be uniquely solved in terms of $\mathbf{u}$.

$$\begin{cases} \mathbf{H}_t - \nabla \times (\mathbf{u} \times \mathbf{H}) = -\nabla \times (\nu(\rho,\vartheta,\mathbf{H}) \nabla \times \mathbf{H}), \quad \nabla \cdot \mathbf{H} = 0, \\ (a\vartheta^4 + \rho Q(\vartheta) - \rho\vartheta^2 P_{\vartheta^2}(\rho))_t + \nabla \cdot ((a\vartheta^4 + \rho Q(\vartheta) - \rho\vartheta^2 P_{\vartheta^2}(\rho))\mathbf{u}) \\ = \mathbb{S} : \nabla \mathbf{u} + \nu(\rho,\vartheta,\mathbf{H})|\nabla \times \mathbf{H}|^2 + \varepsilon|\nabla\rho|^2 \left(\delta\Gamma\rho^{\Gamma-2} + \dfrac{p_m'(\rho)}{\rho}\right) \\ \quad + \nabla \cdot ((\kappa_G(\rho,\vartheta,\mathbf{H}) + \kappa_R\vartheta^3)\nabla\vartheta) - \left(\vartheta p_\vartheta(\rho) + \vartheta^2 p_{\vartheta^2}(\rho) + \dfrac{a}{3}\vartheta^4\right) \nabla \cdot \mathbf{u}, \\ \mathbf{H}\mid_{t=0} = \mathbf{H}_{0,\delta}, \ \vartheta\mid_{t=0} = \vartheta_{0,\delta}, \\ \mathbf{H} \cdot \mathbf{n}\mid_{\partial\Omega} = (\nabla \times \mathbf{H}) \times \mathbf{n}\mid_{\partial\Omega} = 0, \ \nabla\vartheta \cdot \mathbf{n}\mid_{\partial\Omega} = 0, \end{cases} \tag{4.31}$$

with $\mathbf{H}_{0,\delta}$ and $\vartheta_{0,\delta}$ satisfying (3.29).

For given $\mathbf{u} \in C([0,T]; C_0^2(\overline{\Omega}; \mathbb{R}^3))$, ρ has already been given by (4.30), equation $(4.31)_1$ is a quasilinear parabolic-type structure in $\mathbf{H}$ and equation $(4.31)_2$ is indeed a non-degenerate parabolic-type system in terms of ϑ^4, since

$$-\nabla \times (\nu(\rho, \vartheta, \mathbf{H}) \nabla \times \mathbf{H}) = -\nabla \nu(\rho, \vartheta, \mathbf{H}) \times (\nabla \times \mathbf{H}) + \nu(\rho, \vartheta, \mathbf{H}) \Delta \mathbf{H},$$

$$\nabla \cdot \big((\kappa_G(\rho, \vartheta, \mathbf{H}) + \kappa_R \vartheta^3) \nabla \vartheta \big) = \frac{\kappa_R}{4} \Delta \vartheta^4 + \kappa_G(\rho, \vartheta, \mathbf{H}) \Delta \vartheta + \nabla \kappa_G(\rho, \vartheta, \mathbf{H}) \cdot \nabla \vartheta.$$

Thus, $\mathbf{H}, \vartheta$ can be solved by means of the standard Faedo-Galerkin methods. More explicitly, the boundary value problem of (4.31) has a unique solution ($\vartheta := \vartheta[\mathbf{u}]$, $\mathbf{H} := \mathbf{H}[\mathbf{u}]$) defined on the whole time interval $(0,T)$ satisfying the following properties:

- ϑ is a strong solution to (4.31) and strictly positive on $\overline{\Omega}_T$. In fact, the existence of a weak solution $\vartheta \in L^2(0,T; H^1(\Omega))$ can be obtained by the standard iterational process as Chapter 1.2 in [28]. And, the regularity of weak solutions(i.e. the Hölder continuity of weak solutions in a strictly interior subdomain) can be established as Chapter 5.2 in [28]. As the first three terms on the right-hand side of $(4.31)_2$ are always non-negative, and the function $\vartheta = 0$ is a subsolution, by using the comparison theorem, $\vartheta(t, \mathbf{x}) \geqslant 0$ for all $t \in [0,T]$, $\mathbf{x} \in \Omega$. In agreement with the physical background and as required in the variational formulation introduced in Section 2, the absolute temperature must be positive a.a. on Ω_T.
- $\mathbf{H} \in C([0,T]; L^2_{\text{weak}}(\Omega)) \cap L^2(0,T; H^1(\Omega))$.

5 The Faedo-Galerkin approximation scheme

In this section, we establish the existence of solutions to (3.27). Although $(3.27)_2$ and $(3.27)_3$ are of parabolic type, the unknowns $\mathbf{u}$ and ϑ appear to be multiplied by ρ in the leading terms, we have to use a more complicated approach based on the Faedo-Galerkin approximation technique to obtain the first level approximate solutions. In order to do this, assume the vector functions $\mathbf{w}_j = \mathbf{w}_j(\mathbf{x})$ $(j = 1, 2, \cdots)$ are smooth, $\{\mathbf{w}_j\}_{j=1}^\infty$ is an orthogonal basis of $H_0^1(\Omega)$, and $\{\mathbf{w}_j\}_{j=1}^\infty$ is an orthonormal basis of $L^2(\Omega)$. Define k-D Euclidean space $Y_k = \text{span}\{\mathbf{w}_j\}_{j=1}^k$ with scalar product $\langle \mathbf{v}, \mathbf{w} \rangle = \int_\Omega \mathbf{v} \cdot \mathbf{w} \, d\mathbf{x}$, $\mathbf{v}, \mathbf{w} \in Y_k$ and let $P_k : (L^2(\Omega))^3 \to Y_k$ be the orthonormal projection. The approximate velocity field $\mathbf{u}_k \in C([0,T];Y_k)$, we may write $\mathbf{u}_k(t,\mathbf{x}) = \sum_{j=1}^k g_k^j(t)\mathbf{w}_j(\mathbf{x})$, satisfies

$$\begin{aligned}&\langle(\rho_k\mathbf{u}_k)_t,\mathbf{w}_j\rangle + \langle\nabla\cdot(\rho_k\mathbf{u}_k\otimes\mathbf{u}_k)+\nabla p_k+\delta\nabla\rho_k^\Gamma+\varepsilon\nabla\mathbf{u}_k\cdot\nabla\rho_k,\mathbf{w}_j\rangle\\ &=\langle\nabla\cdot\mathbb{S}_k+\rho_k\nabla\Psi_k+(\nabla\times\mathbf{H}_k)\times\mathbf{H}_k,\mathbf{w}_j\rangle\end{aligned}\tag{5.32}$$

with the initial conditions

$$\langle(\rho_k\mathbf{u}_k)(0),\mathbf{w}_j\rangle=\langle\mathbf{m}_{0,\delta},\mathbf{w}_j\rangle,$$

for all $t\in[0,T]$, $j=1,\cdots,k$, where $\{(\rho_k,\Psi_k,\mathbf{H}_k,\vartheta_k)\}_{k=1}^\infty$ are determined as the unique solution of (4.30), (2.18), (4.31) in terms of $\{\mathbf{u}_k\}_{k=1}^\infty$ on $[0,T]$. Here $\rho_k=\rho_{\delta,\varepsilon}[\mathbf{u}_k]$, etc., $p_k=p(\rho_k,\vartheta_k)$, $\Psi_k=G(-\Delta)^{-1}[\rho_k]$, and $\varepsilon,\delta,\Gamma$ are fixed positive parameters.

Given

$$f\in C([0,T];L^1(\Omega)),\ f_t\in L^1(\Omega_T),\ \operatorname{ess\,inf}_{(t,\mathbf{x})\in\Omega_T}f(t,\mathbf{x})\geqslant a>0,$$

define an operator

$$\mathcal{O}_{f(t)}:Y_k\to Y_k,\ \langle\mathcal{O}_{f(t)}\mathbf{u},\mathbf{v}\rangle\equiv\int_\Omega f(t)\mathbf{u}\cdot\mathbf{v}\,\mathrm{d}\mathbf{x},\ \text{for }\mathbf{u},\mathbf{v}\in Y_k,\ t\in[0,T].$$

It is easy to derive that $\mathcal{O}_{f(t)}^{-1}$ exists for all $t\in[0,T]$ and $\|\mathcal{O}_{f(t)}^{-1}\|_{\mathcal{L}(Y_k,Y_k)}\leqslant\dfrac{1}{a}$.

Taking advantage of the operator $\mathcal{O}_{f(t)}$, and since

$$\rho_k(t)\geqslant\inf_{\mathbf{x}\in\Omega}\rho_{0,\delta}(\mathbf{x})\exp(-\|\nabla\cdot\mathbf{u}_k\|_{L_t^1(L^\infty)})>0,$$

then (5.32) can be rephrased as

$$\begin{aligned}\mathbf{u}_k(t)=&\mathcal{O}_{\rho_k(t)}^{-1}\Big(P_k\mathbf{m}_{0,\delta}+\int_0^t P_k\big(-\nabla\cdot(\rho_k\mathbf{u}_k\otimes\mathbf{u}_k)-\nabla p_k-\delta\nabla\rho_k^\Gamma-\varepsilon\nabla\mathbf{u}_k\cdot\nabla\rho_k\\ &+\nabla\cdot\mathbb{S}_k+\rho_k\nabla\Psi_k+(\nabla\times\mathbf{H}_k)\times\mathbf{H}_k\big)\,\mathrm{d}\tau\Big).\end{aligned}$$

The local existence of the velocity $\mathbf{u}_k$ can be obtained by fixed point argument and the uniform estimates obtained from (5.33)(5.34) furnish the possibility of repeating the fixed point argument to extend the solution to the whole time interval $[0,T]$(see [38], Chapter 7.7 for details). Thus, for any fixed $k=1,2,\cdots$, we solve first the regularized system for positive values of the parameters ε and δ, and, the solution $(\rho_k,\mathbf{u}_k,\vartheta_k,\mathbf{H}_k)$ defined on the whole time interval.

Our plan is hereafter to send $k\to\infty$, and so we will need to obtain uniform estimates that are independent of the dimension k of Y_k. We start with the energy estimates which can be derived as follows: multiplying (5.32) by $g_k^j(t)$, summing $j = 1, \cdots, k$, and repeating the procedure for *a priori* estimates in Section 2. It yields the approximate kinetic energy and total energy balance:

$$\begin{aligned}
&\frac{\mathrm{d}}{\mathrm{d}t}\int_\Omega \rho_k\left(\frac{1}{2}|\mathbf{u}_k|^2+\frac{\delta}{\Gamma-1}\rho_k^{\Gamma-1}+P_m(\rho_k)\right)\mathrm{d}\mathbf{x}\\
&+\int_\Omega (\mathbb{S}_k:\nabla\mathbf{u}_k+\varepsilon(\delta\Gamma\rho_k^{\Gamma-2}+\frac{p_m'(\rho_k)}{\rho_k})|\nabla\rho_k|^2)\mathrm{d}\mathbf{x}\\
=&\int_\Omega(\rho_k\nabla\Psi_k+(\nabla\times\mathbf{H}_k)\times\mathbf{H}_k)\cdot\mathbf{u}_k\,\mathrm{d}\mathbf{x}\\
&+\int_\Omega (p_b(\rho_k)+\vartheta_k p_\vartheta(\rho_k)+\vartheta_k^2 p_{\vartheta^2}(\rho_k)+\frac{a}{3}\vartheta_k^4)\nabla\cdot\mathbf{u}_k\mathrm{d}\mathbf{x},
\end{aligned}\tag{5.33}$$

and

$$\begin{aligned}
&\frac{\mathrm{d}}{\mathrm{d}t}\int_\Omega\left(\rho_k\left(\frac{1}{2}|\mathbf{u}_k|^2+P_m(\rho_k)+Q(\vartheta_k)-\vartheta_k^2P_{\vartheta^2}(\rho_k)+\frac{\delta}{\Gamma-1}\rho_k^{\Gamma-1}\right)\right.\\
&\left.+a\vartheta_k^4+\frac{1}{2}|\mathbf{H}_k|^2\right)\mathrm{d}\mathbf{x}\\
=&\int_\Omega\rho_k\nabla\Psi_k\cdot\mathbf{u}_k\,\mathrm{d}\mathbf{x}+\int_\Omega p_b(\rho_k)\nabla\cdot\mathbf{u}_k\,\mathrm{d}\mathbf{x},
\end{aligned}\tag{5.34}$$

where

$$\begin{aligned}
&\int_\Omega \mathbb{S}_k:\nabla\mathbf{u}_k\,\mathrm{d}\mathbf{x}\\
=&\int_\Omega\left(\left(\eta(\vartheta_k,\mathbf{H}_k)+\frac{1}{3}\mu(\vartheta_k,\mathbf{H}_k)\right)(\nabla\cdot\mathbf{u}_k)^2+\mu(\vartheta_k,\mathbf{H}_k)|\nabla\mathbf{u}_k|^2\right)\mathrm{d}\mathbf{x}.
\end{aligned}$$

Since ϑ_k is strictly positive on $\overline{\Omega}_T$, then multiplying the regularized thermal energy equation by $\dfrac{1}{\vartheta_k}$ and using the continuity equation with dissipation, we have

$$\begin{aligned}
&\left(\frac{4a}{3}\vartheta_k^3\right)_t+\nabla\cdot\left(\frac{4a}{3}\vartheta_k^3\mathbf{u}_k\right)+\varepsilon\Delta\rho_k\left(\frac{Q(\vartheta_k)}{\vartheta_k}+\vartheta_k\left(P_{\vartheta^2}(\rho_k)+\rho_kP_{\vartheta^2}'(\rho_k)\right)\right)\\
&+p_\vartheta(\rho_k)\nabla\cdot\mathbf{u}_k+\rho_k\frac{c_v(\vartheta_k)}{\vartheta_k}\vartheta_{kt}+\rho_k\frac{c_v(\vartheta_k)\nabla\vartheta_k\cdot\mathbf{u}_k}{\vartheta_k}-(2\rho_kP_{\vartheta^2}(\rho_k)\vartheta_k)_t\\
&-\nabla\cdot(2\rho_kP_{\vartheta^2}(\rho_k)\vartheta_k\mathbf{u}_k)\\
=&\frac{1}{\vartheta_k}\Big(\mathbb{S}_k:\nabla\mathbf{u}_k+\nabla\cdot\left((\kappa_G(\rho_k,\vartheta_k,\mathbf{H}_k)+\kappa_R\vartheta_k^3)\nabla\vartheta_k\right)+\nu(\rho_k,\vartheta_k,\mathbf{H}_k)|\nabla\times\mathbf{H}_k|^2\\
&+\varepsilon|\nabla\rho_k|^2\left(\delta\Gamma\rho_k^{\Gamma-2}+\frac{p_m'(\rho_k)}{\rho_k}\right)\Big),
\end{aligned}$$

$$\int_\Omega\rho_k(t)\,\mathrm{d}\mathbf{x}=\int_\Omega\rho_{0,\delta}\,\mathrm{d}\mathbf{x}\quad\text{for any }\ t\geqslant 0,\tag{5.35}$$

$$\frac{1}{2}\frac{\mathrm{d}}{\mathrm{d}t}\int_\Omega \rho_k^2\,\mathrm{d}\mathbf{x}+\varepsilon\int_\Omega |\nabla\rho_k|^2\,\mathrm{d}\mathbf{x}=-\frac{1}{2}\int_\Omega \rho_k^2\nabla\cdot\mathbf{u}_k\,\mathrm{d}\mathbf{x}, \tag{5.36}$$

and hence, on the one hand, the regularized thermal energy equation can be rewritten as an "entropy inequality":

$$\begin{aligned}
&\left(\frac{4a}{3}\vartheta_k^3+\rho_k\int_1^{\vartheta_k}\frac{c_v(\xi)}{\xi}\,d\xi-2\rho_k P_{\vartheta^2}(\rho_k)\vartheta_k\right)_t\\
&+\nabla\cdot\left(\left(\frac{4a}{3}\vartheta_k^3+\rho_k\int_1^{\vartheta_k}\frac{c_v(\xi)}{\xi}\,d\xi-2\rho_k P_{\vartheta^2}(\rho_k)\vartheta_k\right)\mathbf{u}_k\right)\\
&-\nabla\cdot\left(\frac{\kappa_G(\rho_k,\vartheta_k,\mathbf{H}_k)+\kappa_R\vartheta_k^3}{\vartheta_k}\nabla\vartheta_k\right)\\
\geqslant{}&\varepsilon\left(\int_1^{\vartheta_k}\frac{c_v(\xi)}{\xi}\,\mathrm{d}\xi-\frac{Q(\vartheta_k)}{\vartheta_k}-\vartheta_k\left(P_{\vartheta^2}(\rho_k)+\rho_k P'_{\vartheta^2}(\rho_k)\right)\right)\Delta\rho_k-p_\vartheta(\rho_k)\nabla\cdot\mathbf{u}_k\\
&+\frac{\kappa_G(\rho_k,\vartheta_k,\mathbf{H}_k)+\kappa_R\vartheta_k^3}{\vartheta_k^2}|\nabla\vartheta_k|^2+\frac{\mathbb{S}_k:\nabla\mathbf{u}_k+\nu(\rho_k,\vartheta_k,\mathbf{H}_k)|\nabla\times\mathbf{H}_k|^2}{\vartheta_k}.
\end{aligned} \tag{5.37}$$

On the other hand, combining (5.34), (5.36), (5.37), it yields

$$\begin{aligned}
&\frac{\mathrm{d}}{\mathrm{d}t}\int_\Omega\Bigg(\rho_k\left(\frac{1}{2}|\mathbf{u}_k|^2+P_m(\rho_k)+Q(\vartheta_k)-\vartheta_k^2P_{\vartheta^2}(\rho_k)+\frac{\delta}{\Gamma-1}\rho_k^{\Gamma-1}\right)\\
&+a\vartheta_k^4+\frac{1}{2}|\mathbf{H}_k|^2\Bigg)\mathrm{d}\mathbf{x}+\frac{\mathrm{d}}{\mathrm{d}t}\int_\Omega\left(\frac{1}{2}\rho_k^2-\frac{4a}{3}\vartheta_k^3-\rho_k\int_1^{\vartheta_k}\frac{c_v(\xi)}{\xi}\,\mathrm{d}\xi+2\rho_kP_{\vartheta^2}(\rho_k)\vartheta_k\right)\mathrm{d}\mathbf{x}\\
&+\int_\Omega\Bigg(\frac{\kappa_G(\rho_k,\vartheta_k,\mathbf{H}_k)+\kappa_R\vartheta_k^3}{\vartheta_k^2}|\nabla\vartheta_k|^2+\varepsilon|\nabla\rho_k|^2\\
&+\frac{\mathbb{S}_k:\nabla\mathbf{u}_k+\nu(\rho_k,\vartheta_k,\mathbf{H}_k)|\nabla\times\mathbf{H}_k|^2}{\vartheta_k}\Bigg)\,\mathrm{d}\mathbf{x}\\
\leqslant{}&\varepsilon\int_\Omega\left(\left(\frac{Q(\vartheta_k)}{\vartheta_k^2}+P_{\vartheta^2}(\rho_k)+\rho_kP'_{\vartheta^2}(\rho_k)\right)\nabla\vartheta_k\cdot\nabla\rho_k+\frac{p'_{\vartheta^2}(\rho_k)}{\rho_k}\vartheta_k|\nabla\rho_k|^2\right)\mathrm{d}\mathbf{x}\\
&+\int_\Omega\rho_k\nabla\Psi_k\cdot\mathbf{u}_k\,\mathrm{d}\mathbf{x}+\int_\Omega\left(p_\vartheta(\rho_k)+p_b(\rho_k)-\frac{1}{2}\rho_k^2\right)\nabla\cdot\mathbf{u}_k\,\mathrm{d}\mathbf{x}.
\end{aligned} \tag{5.38}$$

The classical elliptic estimate yields

$$\|\nabla\Psi_k\|_{L^\infty}\leqslant C\|(-\Delta)^{-1}[\rho_k]\|_{W^{2,\Gamma}}\leqslant C\|\rho_k\|_{L^\Gamma}$$

provided $\Gamma>3$. Taking advantage of (5.35), one has

$$\int_\Omega\rho_k\nabla\Psi_k\cdot\mathbf{u}_k\,\mathrm{d}\mathbf{x}\leqslant C\|\rho_k\|_{L^\Gamma}\left(\int_\Omega\rho_{0,\delta}\,\mathrm{d}\mathbf{x}\right)^{\frac{1}{2}}\left(\int_\Omega\rho_k|\mathbf{u}_k|^2\,\mathrm{d}\mathbf{x}\right)^{\frac{1}{2}}.$$

By virtue of definition of $\mathbf{m}_{0,\delta}$ in (3.28), we have

$$\int_\Omega \mathbf{m}_{0,\delta}\cdot\mathbf{u}_{0,\delta,k}\,\mathrm{d}\mathbf{x} \leqslant \frac{1}{2}\int_\Omega\left(\frac{|\mathbf{m}_{0,\delta}|^2}{\rho_{0,\delta}}+\rho_{0,\delta}|\mathbf{u}_{0,\delta,k}|^2\right)\,\mathrm{d}\mathbf{x}$$
$$=\frac{1}{2}\int_\Omega\left(\frac{|\mathbf{m}_0|^2}{\rho_{0,\delta}}+\mathbf{m}_{0,\delta}\mathbf{u}_{0,\delta,k}\right)\,\mathrm{d}\mathbf{x},$$

and

$$\int_\Omega \mathbf{m}_{0,\delta}\cdot\mathbf{u}_{0,\delta,k}\,\mathrm{d}\mathbf{x}\leqslant\int_\Omega\frac{|\mathbf{m}_0|^2}{\rho_{0,\delta}}\,\mathrm{d}\mathbf{x},$$

where the value of $\mathbf{u}_{0,\delta,k}\in Y_k$ is uniquely determined by

$$\int_\Omega \rho_{0,\delta}\mathbf{u}_{0,\delta,k}\cdot\mathbf{w}_j\,\mathrm{d}\mathbf{x}=\int_\Omega\mathbf{m}_{0,\delta}\cdot\mathbf{w}_j\,\mathrm{d}\mathbf{x},\quad j=1,\cdots,k.$$

Recalling the fact that

$$\frac{\mathbb{S}_k:\nabla\mathbf{u}_k}{\vartheta_k}\geqslant c\vartheta_k^{\alpha-1}|\nabla\mathbf{u}_k+\nabla^\top\mathbf{u}_k|^2,$$

and, by virtue of Young's inequality,

$$|\nabla\mathbf{u}_k+\nabla^\top\mathbf{u}_k|^r\leqslant C(\vartheta_k^{\alpha-1}|\nabla\mathbf{u}_k+\nabla^\top\mathbf{u}_k|^2+\vartheta_k^4)\text{ with }r=\frac{8}{5-\alpha},$$

combining with

$$\int_\Omega\rho_{0,\delta}P_m(\rho_{0,\delta})\,\mathrm{d}\mathbf{x}\leqslant\int_\Omega(\rho_{0,\delta}P_e(\rho_{0,\delta})+C(1+\rho_{0,\delta}))\,\mathrm{d}\mathbf{x},$$

the hypothesis (2.16) on c_v and κ_G, we are ready to apply Gronwall's lemma to (5.38) provided $\Gamma=\Gamma(r)$ is large enough to handle the last integral on the right-hand of (5.38). Here the Cauchy-Schwarz inequality has been used repeatedly. Moreover,

$$\int_\Omega\left(|\nabla\ln\vartheta_k|^2+|\nabla\vartheta_k^{\frac{3}{2}}|^2\right)\,\mathrm{d}\mathbf{x}\leqslant C\int_\Omega\frac{\kappa_G(\rho_k,\vartheta_k,\mathbf{H}_k)+\kappa_R\vartheta_k^3}{\vartheta_k^2}|\nabla\vartheta_k|^2\,\mathrm{d}\mathbf{x}.$$

Thus, the following estimates holds:

$$\sup_{t\in[0,T]}\left(\|\rho_k\|_{L^\Gamma}+\|\rho_k|\mathbf{u}_k|^2\|_{L^1}\right)\leqslant C(\delta),\tag{5.39}$$

$$\sup_{t\in[0,T]}(\|\rho_kQ(\vartheta_k)\|_{L^1}+\|\vartheta_k\|_{L^4}+\|\mathbf{H}_k\|_{L^2})+\text{ess}\sup_{t\in[0,T]}\int_\Omega\rho_k|\ln\vartheta_k|\,\mathrm{d}\mathbf{x}\leqslant C(\delta),\tag{5.40}$$

$$\int_{\Omega_T}\frac{\mathbb{S}_k:\nabla\mathbf{u}_k+\nu|\nabla\times\mathbf{H}_k|^2}{\vartheta_k}+|\nabla\ln\vartheta_k|^2+|\nabla\vartheta_k^{\frac{3}{2}}|^2+|\nabla\vartheta_k^{\frac{\beta}{2}}|^2+\varepsilon|\nabla\rho_k|^2\,\mathrm{d}\mathbf{x}\mathrm{d}t\leqslant C(\delta),\tag{5.41}$$

and

$$\|\mathbf{u}_k\|_{L^r(0,T;W_0^{1,r})} \leqslant C(\delta) \text{ with } r=\frac{8}{5-\alpha}, \quad \|(1+\vartheta_k)^{\frac{\beta-1}{2}}\nabla\times\mathbf{H}_k\|_{L^2(0,T;L^2)} \leqslant C(\delta). \tag{5.42}$$

Note that all constants in (5.39)–(5.42) are independent of k and ε.

In order to identify a limit for $k\to\infty$ of the approximate solutions $\{(\rho_k,\mathbf{u}_k,\vartheta_k,\mathbf{H}_k)\}_{k=1}^{\infty}$ obtained above as a solution of problem (3.27) (3.28), additional estimates are needed. Firstly, we have the uniform estimates of the artificial pressure which is proportional to ρ^Γ and the density gradient estimates the same as Section 5.2 and Section 5.3 in [9], we state without proof the following result:

Lemma 5.1 *Under the hypothesis of Theorem 2.1, let $\Gamma=\Gamma(r)$ be large enough, then the density sequence $\{\rho_k\}_{k=1}^{\infty}$ satisfies the following properties:*

$$\|\rho_k\|_{L^{\frac{\Gamma}{r'}}(0,T;L^{\frac{3\Gamma}{r'}})} \leqslant C(\varepsilon,\delta), \quad r':=\frac{r}{r-1}=\frac{8}{3+\alpha},$$

$$\varepsilon\|\nabla\rho_k\|_{L^q(\Omega_T)} \leqslant C(\delta) \text{ for a certain } q>r',$$

$$\|\rho_{kt}\|_{L^\pi(\Omega_T)},\ \|\Delta\rho_k\|_{L^\pi(\Omega_T)} \leqslant C(\varepsilon,\delta) \text{ for a certain } \pi>1.$$

Now, using the same arguments as in Section 2.4 in [21], for the sequences $\{\rho_k\}_{k=1}^{\infty}$ and $\{\mathbf{u}_k\}_{k=1}^{\infty}$, we have (at least for some chosen subsequences)

$$\rho_k\to\rho \text{ in } C([0,T];L^\Gamma_{\text{weak}})\cap L^1(\Omega_T), \quad \rho\geqslant 0, \tag{5.43}$$

$$\rho_t,\ \Delta\rho\in L^\pi(\Omega_T) \text{ for a certain } \pi>1,$$

$$\mathbf{u}_k\rightharpoonup\mathbf{u} \text{ in } L^r(0,T;W_0^{1,r}(\Omega)),\ r=\frac{8}{5-\alpha}, \tag{5.44}$$

where the limit velocity $\mathbf{u}$ satisfies the non-slip boundary condition in the sense of traces. And, ρ is the unique strong solution to (4.30), i.e., the functions $\rho,\mathbf{u}$ satisfying the continuity equation with dissipation a.e. in Ω_T, the initial condition a.e. in Ω and the homogeneous Neumann boundary condition in the sense of traces a.e. in $(0,T)$. In particular,

$$\rho_t+\nabla\cdot(\rho\mathbf{u})=\varepsilon\nabla\cdot(1_\Omega\nabla\rho) \text{ in } \mathcal{D}'(\mathbb{R}^3\times(0,T))$$

provided $\rho,\mathbf{u}$ were extended to be zero outside Ω. Moreover, in accordance with (5.41) and Lemma 5.1, one has

$$\int_0^T \varepsilon\|\nabla\rho\|_{L^2}^2+\varepsilon^q\|\nabla\rho\|_{L^q}^q\,dt \leqslant C(\delta) \text{ for a certain } q>r'. \tag{5.45}$$

By using interpolation, the estimate (5.39) and Lemma 5.1 lead to

$$\rho_k \to \rho \ \text{ in } L^{\pi}(\Omega_T) \ \text{ for some } \ \pi > \Gamma. \tag{5.46}$$

Combining the estimate (5.39) and the strong convergence (5.46), we also have

$$\rho_k \mathbf{u}_k \overset{*}{\rightharpoonup} \rho \mathbf{u} \ \text{ in } \ L^{\infty}(0,T; L^{\frac{2\Gamma}{\Gamma+1}}(\Omega)).$$

The estimates obtained in Lemma 5.1 can be used to deduce from $(4.30)_1$ that the integral mean functions

$$t \mapsto \int_{\Omega} \rho_k \mathbf{u}_k \cdot \mathbf{w}_j \, \mathrm{d}\mathbf{x} \quad \text{form a precompact system in } C([0,T])$$

for any fixed j. This implies that

$$\rho_k \mathbf{u}_k \to \rho \mathbf{u} \ \text{ in } \ C([0,T]; L^{\frac{2\Gamma}{\Gamma+1}}_{\text{weak}}(\Omega)),$$

with the limit function satisfying $\rho\mathbf{u}(0) = \mathbf{m}_{0,\delta}$. As the space $L^{\frac{2\Gamma}{\Gamma+1}}(\Omega)$ is compactly imbedded into the dual space $W^{-1,r'}(\Omega)$ for suitable (large) Γ, and, consequently,

$$\rho_k \mathbf{u}_k \otimes \mathbf{u}_k \rightharpoonup \rho \mathbf{u} \otimes \mathbf{u} \ \text{ in } \ L^r(0,T; L^{\pi}(\Omega)), \quad 1 < \pi \leqslant \frac{6\Gamma r}{3r + \Gamma r + 6\Gamma}. \tag{5.47}$$

In accordance with Lemma 5.1, it yields strong convergence

$$\nabla \rho_k \to \nabla \rho \ \text{ in } L^{\pi}(\Omega_T) \ \text{ for a certain } \ r' \leqslant \pi < q,$$

in particular,

$$\nabla \mathbf{u}_k \cdot \nabla \rho_k \to \nabla \mathbf{u} \cdot \nabla \rho \ \text{ in } \ \mathcal{D}'(\Omega_T).$$

Secondly, we need to show pointwise convergence of the sequence $\{\vartheta_k\}_{k=1}^{\infty}$. From estimates (5.40) and (5.41), repeating the procedure for *a priori* estimates in Section 2, we know

$$\{\vartheta_k\}_{k=1}^{\infty} \ \text{ is bounded in } L^{\infty}(0,T; L^4(\Omega)) \cap L^2(0,T; H^1(\Omega)), \tag{5.48}$$

$$\{\ln \vartheta_k\}_{k=1}^{\infty} \ \text{ is bounded in } \ L^2(0,T; H^1(\Omega)), \tag{5.49}$$

$$\{\vartheta_k^{\frac{3}{2}}\}_{k=1}^{\infty} \ \text{ is bounded in } \ L^2(0,T; H^1(\Omega)), \tag{5.50}$$

$$\{\nabla \vartheta_k^{\frac{\beta}{2}}\}_{k=1}^{\infty} \ \text{ is bounded in } \ L^2(\Omega_T). \tag{5.51}$$

This implies that, by selecting a subsequence if necessary, there exists a function ϑ such that

$$\vartheta_k \overset{*}{\rightharpoonup} \vartheta \text{ in } L^\infty(0,T;L^4(\Omega)), \text{ and } \vartheta_k \rightharpoonup \vartheta \text{ in } L^2(0,T;H^1(\Omega)),$$

$$\ln\vartheta_k \rightharpoonup \overline{\ln\vartheta} \text{ in } L^2(0,T;H^1(\Omega)),$$

and

$$f(\vartheta_k) \rightharpoonup \overline{f(\vartheta)} \text{ in } L^2(0,T;H^1(\Omega))$$

for any

$$f\in C^1(0,\infty),\ |f'(\xi)|\leqslant C\left(\frac{1}{\xi}+\xi^{\frac{1}{2}}\right),\ \xi>0.$$

Here and in what follows, the symbol $\overline{F(v)}$ stands for a weak limit of a composition $\{F(v_k)\}_{k=1}^\infty$ in $L^1(\Omega_T)$.

Next, we claim that $\{\vartheta_{kt}\}_{k=1}^\infty$ satisfies the "entropy inequality" (5.37). Note that, according to (5.39) and (5.49), by virtue of the Sobolev imbedding theorem $H^1(\Omega)\hookrightarrow L^6(\Omega)$, one has,

$$\{\rho_k\ln\vartheta_k\}_{k=1}^\infty \quad \text{bounded in } L^2(0,T;L^{\frac{6\Gamma}{6+\Gamma}}(\Omega)).$$

Together with (5.40), we have

$$\rho_k\ln\vartheta_k\in L^\infty(0,T;L^1(\Omega))\cap L^2(0,T;L^{\frac{6\Gamma}{6+\Gamma}}(\Omega)).$$

Similarly, (5.39) leads to

$$\sup_{t\in[0,T]}\|\rho_k\mathbf{u}_k\|_{L^{\frac{2\Gamma}{\Gamma+1}}}\leqslant C(\delta), \tag{5.52}$$

together with (5.49), it yields

$$\rho_k\mathbf{u}_k\ln\vartheta_k\in L^2(0,T;L^{\frac{6\Gamma}{3+4\Gamma}}(\Omega)).$$

Recalling the hypothesis (2.16) on c_v, we know, from (5.39) and (5.48),

$$\rho_k\vartheta_k^{\frac{\beta}{2}-1}\in L^\infty(0,T;L^1(\Omega))\cap L^2(0,T;L^{\frac{6\Gamma}{6+\Gamma}}(\Omega)) \text{ for } \beta\leqslant 4, \tag{5.53}$$

provided Γ large enough, say, $\Gamma>3$, then

$$\rho_k\int_1^{\vartheta_k}\frac{c_v(\xi)}{\xi}\,\mathrm{d}\xi\in L^\infty(0,T;L^1(\Omega))\cap L^2(0,T;L^{\frac{6\Gamma}{6+\Gamma}}(\Omega)),$$

and

$$\rho_k \mathbf{u}_k \int_1^{\vartheta_k} \frac{c_v(\xi)}{\xi}\, \mathrm{d}\xi \in L^2(0,T; L^{\frac{6\Gamma}{3+4\Gamma}}(\Omega)).$$

In accordance with hypothesis (1.8) and estimate (5.39),

$$\rho_k P_{\vartheta^2}(\rho_k) \ \text{ is bounded in } L^\infty(0,T; L^{\frac{\Gamma}{\varsigma}}(\Omega)),$$

and furthermore, together with (5.48),

$$\rho_k P_{\vartheta^2}(\rho_k)\vartheta_k \in L^\infty(0,T; L^1(\Omega)) \cap L^2(0,T; L^{\frac{6\Gamma}{6\varsigma+\Gamma}}(\Omega)).$$

According to Lemma 6.3 of Chapter 6 in [18], we have

$$\begin{aligned} &\frac{4a}{3}\vartheta_k^3 + \rho_k \int_1^{\vartheta_k} \frac{c_v(\xi)}{\xi}\, \mathrm{d}\xi - 2\rho_k P_{\vartheta^2}(\rho_k)\vartheta_k \\ &\to \frac{4a}{3}\overline{\vartheta^3} + \overline{\rho\int_1^{\vartheta} \frac{c_v(\xi)}{\xi}\, \mathrm{d}\xi} - 2\rho P_{\vartheta^2}(\rho)\vartheta \ \text{ in } \ L^2(0,T; H^{-1}(\Omega)). \end{aligned}$$

In addition, we employ (5.48) to get

$$\begin{aligned} &\int_{\Omega_T} \left(\frac{4a}{3}\vartheta_k^3 + \rho_k \int_1^{\vartheta_k} \frac{c_v(\xi)}{\xi}\, \mathrm{d}\xi - 2\rho_k P_{\vartheta^2}(\rho_k)\vartheta_k \right) \vartheta_k \, \mathrm{d}\mathbf{x}\mathrm{d}t \\ &\to \int_{\Omega_T} \left(\frac{4a}{3}\overline{\vartheta^3} + \overline{\rho\int_1^{\vartheta} \frac{c_v(\xi)}{\xi}\, \mathrm{d}\xi} - 2\rho P_{\vartheta^2}(\rho)\vartheta \right) \vartheta \, \mathrm{d}\mathbf{x}\mathrm{d}t. \end{aligned} \tag{5.54}$$

Since

$$\lim_{k\to\infty} \int_{\Omega_T} \rho_k P_{\vartheta^2}(\rho_k)\vartheta_k^2 \, \mathrm{d}\mathbf{x}\mathrm{d}t = \int_{\Omega_T} \rho P_{\vartheta^2}(\rho)\vartheta^2 \, \mathrm{d}\mathbf{x}\mathrm{d}t,$$

then (5.54) reduces to

$$\begin{aligned} &\int_{\Omega_T} \left(\frac{4a}{3}\vartheta_k^3 + \rho_k \int_1^{\vartheta_k} \frac{c_v(\xi)}{\xi}\, \mathrm{d}\xi \right) \vartheta_k \, \mathrm{d}\mathbf{x}\mathrm{d}t \\ &\to \int_{\Omega_T} \left(\frac{4a}{3}\overline{\vartheta^3} + \overline{\rho\int_1^{\vartheta} \frac{c_v(\xi)}{\xi}\, \mathrm{d}\xi} \right) \vartheta \, \mathrm{d}\mathbf{x}\mathrm{d}t. \end{aligned} \tag{5.55}$$

As the function $\xi \mapsto \frac{4a}{3}\xi^3 + \rho \int_1^{\xi} \frac{c_v(y)}{y}\, \mathrm{d}y$ is non-decreasing, we have, from (5.55), the following strong (pointwise) convergence

$$\vartheta_k \to \vartheta \ \text{ in } L^1(\Omega_T).$$

Employing (5.48) again, by virtue of a simple interpolation argument, we obtain

$$\vartheta_k \to \vartheta \text{ in } L^{\pi}(\Omega_T) \text{ for some } \pi > 4.$$

Finally, using $\nabla \cdot \mathbf{H}_k = 0$, the estimates (5.40), (5.42) on the magnetic field $\{\mathbf{H}_k\}_{k=1}^{\infty}$ imply that

$$\mathbf{H}_k \overset{*}{\rightharpoonup} \mathbf{H} \text{ in } L^{\infty}(0,T;L^2(\Omega)),\ \mathbf{H}_k \rightharpoonup \mathbf{H} \text{ in } L^2(0,T;H^1(\Omega)) \text{ and } \mathbf{H}_k \to \mathbf{H} \text{ in } L^2(\Omega_T),$$

and consequently, via interpolation,

$$(\nabla \times \mathbf{H}_k) \times \mathbf{H}_k \rightharpoonup (\nabla \times \mathbf{H}) \times \mathbf{H} \text{ in } L^{\pi}(\Omega_T) \text{ for a certain } \pi > 1.$$

Moreover,

$$\mathbb{S}_k \rightharpoonup \mathbb{S} \text{ in } L^q(\Omega_T) \text{ for some } q > 1,$$

where $\mathbb{S} = \mu(\vartheta, \mathbf{H})(\nabla \mathbf{u} + \nabla^{\top}\mathbf{u}) + \eta(\vartheta, \mathbf{H})(\nabla \cdot \mathbf{u})\, \mathbb{I}_3$. we also have

$$\mathbf{u}_k \times \mathbf{H}_k \rightharpoonup \mathbf{u} \times \mathbf{H} \text{ in } L^{\pi}(\Omega_T) \text{ for a certain } \pi > 1,$$

in accordance with (5.44).

Now, we can pass to the limit for $k \to \infty$ in (5.32) to get

$$\begin{aligned}&(\rho\mathbf{u})_t + \nabla \cdot (\rho\mathbf{u} \otimes \mathbf{u}) + \nabla \left(p_e(\rho) + \vartheta p_{\vartheta}(\rho) + \vartheta^2 p_{\vartheta^2}(\rho) + \frac{a}{3}\vartheta^4 + \delta\rho^{\Gamma}\right) + \varepsilon \nabla \mathbf{u} \cdot \nabla \rho \\ &= \nabla \cdot \mathbb{S} + \rho \nabla \Psi + (\nabla \times \mathbf{H}) \times \mathbf{H} \text{ in } \mathcal{D}'(\Omega_T),\end{aligned} \tag{5.56}$$

where the potential Ψ satisfies (2.17).

Moreover, due to the estimates (5.41) and (5.48), as β satisfies the hypothesis in (5.53), we know

$$\nu(\rho_k, \vartheta_k, \mathbf{H}_k)\nabla \times \mathbf{H}_k = \sqrt{\vartheta_k \nu(\rho_k, \vartheta_k, \mathbf{H}_k)}\sqrt{\frac{\nu(\rho_k, \vartheta_k, \mathbf{H}_k)}{\vartheta_k}}\ \nabla \times \mathbf{H}_k$$

are bounded in $L^{\pi}(\Omega_T)$ for a certain $\pi > 1$.

Thus the limit quantities satisfy

$$\begin{aligned}&\int_{\Omega_T} (\psi'\mathbf{H} \cdot \phi + \psi(\mathbf{u} \times \mathbf{H}) \cdot (\nabla \times \phi) - \psi\nu(\nabla \times \mathbf{H}) \cdot (\nabla \times \phi))\, \mathrm{d}\mathbf{x}\mathrm{d}t \\ &= -\psi(0) \int_{\Omega} \mathbf{H}_0 \cdot \phi\, \mathrm{d}\mathbf{x},\end{aligned}$$

for any $\psi \in C^\infty([0,T])$ with $\psi(T)=0$ and $\phi \in \mathcal{D}(\Omega;\mathbb{R}^3)$.

In the same way, we let $k\to\infty$ in the energy equality (5.34) to get

$$
\begin{aligned}
&-\int_{\Omega_T}\psi'\left(\rho\left(\frac{1}{2}|\mathbf{u}|^2+P_m(\rho)+Q(\vartheta)-\vartheta^2P_{\vartheta^2}(\rho)+\frac{\delta}{\Gamma-1}\rho^{\Gamma-1}\right)+a\vartheta^4+\frac{1}{2}|\mathbf{H}|^2\right)\mathrm{d}\mathbf{x}\mathrm{d}t\\
=&\int_\Omega\left(\frac{1}{2}\frac{|\mathbf{m}_{0,\delta}|^2}{\rho_{0,\delta}}+\rho_{0,\delta}P_m(\rho_{0,\delta})+\rho_{0,\delta}Q(\vartheta_{0,\delta})-\rho_{0,\delta}\vartheta_{0,\delta}^2P_{\vartheta^2}(\rho_{0,\delta})\right.\\
&\left.+\frac{\delta}{\Gamma-1}\rho_{0,\delta}^\Gamma+a\vartheta_{0,\delta}^4+\frac{1}{2}|\mathbf{H}_{0,\delta}|^2\right)\mathrm{d}\mathbf{x}+\int_{\Omega_T}\psi\left(\rho\nabla\Psi\cdot\mathbf{u}+p_b(\rho)\nabla\cdot\mathbf{u}\right)\mathrm{d}\mathbf{x}\mathrm{d}t.
\end{aligned}
\tag{5.57}
$$

for any $\psi\in C^\infty([0,T])$ with $\psi(0)=1,\ \psi(T)=0$. Here we have used

$$
\int_{\Omega_\tau}(\rho\nabla\Psi\cdot\mathbf{u}+p_b(\rho)\nabla\cdot\mathbf{u})\,\mathrm{d}\mathbf{x}\mathrm{d}t=\lim_{k\to\infty}\int_{\Omega_\tau}(\rho_k\nabla\Psi_k\cdot\mathbf{u}_k+p_b(\rho_k)\nabla\cdot\mathbf{u}_k)\,\mathrm{d}\mathbf{x}\mathrm{d}t
$$

for any $\tau\geqslant 0$ in accordance with (5.44),(5.46). By virtue of the hypothesis (2.16) on $c_v(\vartheta)$, (5.53) on β and the estimates (5.48),(5.51), we have

$$
Q(\vartheta_k)\rightharpoonup\overline{Q(\vartheta)}\ \text{ in }\ L^2(0,T;H^1(\Omega)),
$$

combining with (5.39),

$$
\rho_kQ(\vartheta_k)\rightharpoonup\rho\overline{Q(\vartheta)}\ \text{ in }\ L^2(0,T;L^q(\Omega)),\quad 1<q<\frac{6\Gamma}{\Gamma+6}.
$$

Now since $Q(\vartheta_k)$ tends to $Q(\vartheta)$ a.e. on Ω, we infer $Q(\vartheta)=\overline{Q(\vartheta)}$, and,

$$
\int_\Omega\rho Q(\vartheta)(\tau+)\,\mathrm{d}\mathbf{x}=\lim_{h\to0+}\frac{1}{h}\left(\lim_{k\to\infty}\int_\tau^{\tau+h}\int_\Omega\rho_kQ(\vartheta_k)\,\mathrm{d}\mathbf{x}\mathrm{d}t\right)
$$

for a.e. $\tau\in[0,T]$. We also have used (5.47) to deduce

$$
\begin{aligned}
\int_\Omega\rho|\mathbf{u}|^2(\tau+)\phi\,\mathrm{d}\mathbf{x}&=\lim_{h\to0+}\frac{1}{h}\int_\tau^{\tau+h}\int_\Omega\rho|\mathbf{u}|^2\phi\,\mathrm{d}\mathbf{x}\mathrm{d}t\\
&=\lim_{h\to0+}\frac{1}{h}\left(\lim_{k\to\infty}\int_\tau^{\tau+h}\int_\Omega\rho_k|\mathbf{u}_k|^2\phi\,\mathrm{d}\mathbf{x}\mathrm{d}t\right)\quad\text{for any }\phi\in\mathcal{D}(\Omega;\mathbb{R}),
\end{aligned}
$$

and, similarly,

$$
\int_\Omega\rho P_m(\rho)(\tau+)\phi\,\mathrm{d}\mathbf{x}=\lim_{h\to0+}\frac{1}{h}\left(\lim_{k\to\infty}\int_\tau^{\tau+h}\int_\Omega\rho_kP_m(\rho_k)\phi\,\mathrm{d}\mathbf{x}\mathrm{d}t\right),
$$

$$\int_{\Omega} \rho\vartheta^2 P_{\vartheta^2}(\rho)(\tau+)\phi \,\mathrm{d}\mathbf{x} = \lim_{h\to 0+}\frac{1}{h}\left(\lim_{k\to\infty}\int_{\tau}^{\tau+h}\int_{\Omega}\rho_k\vartheta_k^2 P_{\vartheta^2}(\rho_k)\phi \,\mathrm{d}\mathbf{x}\mathrm{d}t\right),$$

$$\int_{\Omega} \rho^{\Gamma}(\tau+)\phi \,\mathrm{d}\mathbf{x} = \lim_{h\to 0+}\frac{1}{h}\left(\lim_{k\to\infty}\int_{\tau}^{\tau+h}\int_{\Omega}\rho_k^{\Gamma}\phi \,\mathrm{d}\mathbf{x}\mathrm{d}t\right),$$

for a.e. $\tau \in [0,T]$.

The final aim in this section is to pass to the limit $k\to\infty$ in the "entropy inequality ". Note it is enough to show that one can pass to the limit in all nonlinear terms contained in (5.37). To this end, we start with the observation that

$$\begin{aligned}\frac{\mathbb{S}_k:\nabla\mathbf{u}_k}{\vartheta_k} &= \frac{\mu(\vartheta_k,\mathbf{H}_k)}{\vartheta_k}\left(|\nabla\mathbf{u}_k|^2+\nabla\mathbf{u}_k:\nabla^{\top}\mathbf{u}_k-\frac{2}{3}(\nabla\cdot\mathbf{u}_k)^2\right)+\frac{\eta(\vartheta_k,\mathbf{H}_k)}{\vartheta_k}(\nabla\cdot\mathbf{u}_k)^2\\ &= \left|\sqrt{\frac{\mu(\vartheta_k,\mathbf{H}_k)}{2\vartheta_k}}(\nabla\mathbf{u}_k+\nabla^{\top}\mathbf{u}_k-\frac{2}{3}\nabla\cdot\mathbf{u}_k\,\mathbb{I}_3)\right|^2+\left(\sqrt{\frac{\eta(\vartheta_k,\mathbf{H}_k)}{\vartheta_k}}\nabla\cdot\mathbf{u}_k\right)^2,\end{aligned}$$

then, by the estimate (5.41) and the weak lower semi-continuity,

$$\int_{\Omega_\tau}\frac{\mathbb{S}:\nabla\mathbf{u}}{\vartheta}\,\mathrm{d}\mathbf{x}\mathrm{d}t \leqslant \liminf_{k\to\infty}\int_{\Omega_\tau}\frac{\mathbb{S}_k:\nabla\mathbf{u}_k}{\vartheta_k}\,\mathrm{d}\mathbf{x}\mathrm{d}t$$

for any $\tau \geqslant 0$.

Since one can estimate the entropy flux

$$\left|\frac{\kappa_G(\rho_k,\vartheta_k,\mathbf{H}_k)+\kappa_R\vartheta_k^3}{\vartheta_k}\nabla\vartheta_k\right| \leqslant C\left(|\nabla\ln\vartheta_k|+(\vartheta_k^2+\vartheta_k^{\beta-1})\,|\nabla\vartheta_k|\right),$$

where

$$(\vartheta_k^2+\vartheta_k^{\beta-1})\,|\nabla\vartheta_k| = \frac{2}{3}\vartheta_k^{\frac{3}{2}}|\nabla\vartheta_k^{\frac{3}{2}}|+\frac{2}{\beta}\vartheta_k^{\frac{\beta}{2}}|\nabla\vartheta_k^{\frac{\beta}{2}}|.$$

By virtue of the hypothesis (5.53) on β and the estimates (5.48)–(5.51), one can show that

$$\left\{\frac{\kappa_G(\rho_k,\vartheta_k,\mathbf{H}_k)+\kappa_R\vartheta_k^3}{\vartheta_k}\nabla\vartheta_k\right\}_{k=1}^{\infty} \quad \text{is bounded in } L^{\pi}(\Omega_T) \text{ for a certain } \pi>1.$$

Furthermore,

$$\left\{\frac{\kappa_G(\rho_k,\vartheta_k,\mathbf{H}_k)+\kappa_R\vartheta_k^3}{\vartheta_k^2}|\nabla\vartheta_k|^2\right\}_{k=1}^{\infty} \quad \text{is bounded in } L^{\pi}(\Omega_T) \text{ for some } \pi>1.$$

And, consequently,

$$\frac{\kappa_G(\rho_k,\vartheta_k,\mathbf{H}_k)+\kappa_R\vartheta_k^3}{\vartheta_k}\nabla\vartheta_k \rightharpoonup \frac{\kappa_G(\rho,\vartheta,\mathbf{H})+\kappa_R\vartheta^3}{\vartheta}\nabla\vartheta \text{ in } L^{\pi}(\Omega_T),$$

for a certain $\pi > 1$.

$$\frac{\kappa_G(\rho_k, \vartheta_k, \mathbf{H}_k) + \kappa_R\vartheta_k^3}{\vartheta_k^2}|\nabla\vartheta_k|^2 \rightharpoonup \frac{\kappa_G(\rho, \vartheta, \mathbf{H}) + \kappa_R\vartheta^3}{\vartheta^2}|\nabla\vartheta|^2 \ \text{ in } L^\pi(\Omega_T)$$

for a certain $\pi > 1$.

Making use of Lemmas 5.3, 5.4 of Chapter 5 in [12], the estimates (5.39)–(5.42), the above relations together with (5.37), it yields the desired variational form of the entropy inequality:

$$\begin{aligned}
&\int_{\Omega_T} \psi' \left(\frac{4a}{3}\vartheta^3 + \rho \int_1^\vartheta \frac{c_v(\xi)}{\xi}\,\mathrm{d}\xi - 2\rho P_{\vartheta^2}(\rho)\vartheta \right) \phi \,\mathrm{d}\mathbf{x}\mathrm{d}t \\
&+ \int_{\Omega_T} \psi \left(\frac{4a}{3}\vartheta^3 + \rho \int_1^\vartheta \frac{c_v(\xi)}{\xi}\,\mathrm{d}\xi - 2\rho P_{\vartheta^2}(\rho)\vartheta \right) \mathbf{u}\cdot\nabla\phi \,\mathrm{d}\mathbf{x}\mathrm{d}t \\
&- \int_{\Omega_T} \psi \left(\frac{\kappa_G(\rho, \vartheta, \mathbf{H}) + \kappa_R\vartheta^3}{\vartheta}\nabla\vartheta \right) \cdot \nabla\phi \,\mathrm{d}\mathbf{x}\mathrm{d}t \\
&\leqslant \int_{\Omega_T} \varepsilon\nabla \left(\psi\phi \left(\int_1^\vartheta \frac{c_v(\xi)}{\xi}\,\mathrm{d}\xi - \frac{Q(\vartheta)}{\vartheta} - \vartheta(P_{\vartheta^2}(\rho) + \rho P'_{\vartheta^2}(\rho)) \right) \right) \cdot \nabla\rho \\
&+ \psi\phi p_\vartheta(\rho)\nabla\cdot\mathbf{u} \,\mathrm{d}\mathbf{x}\mathrm{d}t \\
&- \int_{\Omega_T} \psi\phi \Big(\frac{\kappa_G(\rho, \vartheta, \mathbf{H}) + \kappa_R\vartheta^3}{\vartheta^2}|\nabla\vartheta|^2 + \frac{\mathbb{S} : \nabla\mathbf{u} + \nu(\rho, \vartheta, \mathbf{H})|\nabla\times\mathbf{H}|^2}{\vartheta} \Big) \mathrm{d}\mathbf{x}\mathrm{d}t \\
&- \psi(0) \int_\Omega \phi \left(\frac{4a}{3}\vartheta_{0,\delta}^3 + \rho_{0,\delta} \int_1^{\vartheta_{0,\delta}} \frac{c_v(\xi)}{\xi}\,\mathrm{d}\xi \right) \mathrm{d}\mathbf{x}
\end{aligned} \tag{5.58}$$

for any $0 \leqslant \psi \in C^\infty([0, T])$ with $\psi(T) = 0$ and $0 \leqslant \phi \in \mathcal{D}(\Omega; \mathbb{R})$.

6 Passing to the limit $\varepsilon \to 0$

Our goal in this section is to take the vanishing limit of the artificial viscosity $\varepsilon \to 0$ for the family of approximate solutions $\{(\rho_{\delta,\varepsilon}, \Psi_{\delta,\varepsilon}, \mathbf{H}_{\delta,\varepsilon}, \vartheta_{\delta,\varepsilon})\}$ constructed in Section 5, i.e., to get rid of the artificial viscosity in $(3.27)_1$. In other words, we are to show the weak sequential stability (compactness) for the approximate solutions. Denote $\rho_\varepsilon = \rho_{\delta,\varepsilon}$, etc. in this section. Due to the bounds of the density estimates in Lemma 5.1 depend on ε, we definitely loose boundedness of $\nabla\rho_\varepsilon$ and, consequently, strong compactness of the sequence of $\{\rho_\varepsilon\}_{\varepsilon>0}$ in $L^1(\Omega_T)$ becomes a central issue now, so more refined estimates are needed to make sure the limit passage. At this stage, we first point that it is easy to check that the sequences $\{\rho_\varepsilon\mathbf{u}_\varepsilon\}_{\varepsilon>0}, \{\rho_\varepsilon\mathbf{u}_\varepsilon \otimes \mathbf{u}_\varepsilon\}_{\varepsilon>0}, \{\rho_\varepsilon\nabla\Psi_\varepsilon\}_{\varepsilon>0}, \{(\nabla\times\mathbf{H}_\varepsilon)\times\mathbf{H}_\varepsilon\}_{\varepsilon>0}$ are bounded in $L^\pi(\Omega_T)$ for a certain $\pi > 1$

because of the estimates (5.39)(5.42). Moreover, since

$$\mathbb{S}_\varepsilon = \sqrt{\vartheta_\varepsilon \mu(\vartheta_\varepsilon, \mathbf{H}_\varepsilon)} \sqrt{\frac{\mu(\vartheta_\varepsilon, \mathbf{H}_\varepsilon)}{\vartheta_\varepsilon}} \left(\nabla \mathbf{u}_\varepsilon + \nabla^\top \mathbf{u}_\varepsilon - \frac{2}{3}(\nabla \cdot \mathbf{u}_\varepsilon)\, \mathbb{I}_3 \right)$$
$$+ \sqrt{\vartheta_\varepsilon \eta(\vartheta_\varepsilon, \mathbf{H}_\varepsilon)} \sqrt{\frac{\eta(\vartheta_\varepsilon, \mathbf{H}_\varepsilon)}{\vartheta_\varepsilon}} (\nabla \cdot \mathbf{u}_\varepsilon)\, \mathbb{I}_3.$$

By virtue of (5.40)(5.41), we know that $\{\mathbb{S}_\varepsilon\}_{\varepsilon>0}$ is bounded in $L^\pi(\Omega_T)$for a certain$\pi > 1$. As already pointed out in [18], both classical stumbling blocks of this approach- the phenomena of oscillations and concentrations- are likely to appear. In order to deal with the non-linear constitutive equations for the pressure and other quantities, the density oscillations as well as concentrations in the temperature must be excluded. Therefore, we have to find a bound (independent of ε) for the the pressure term in a reflexive space $L^\pi(\Omega_T)$ with $\pi > 1$. To this end, similar as Section 6.1 in [13], let us introduce an operator $\mathcal{B} = (\mathcal{B}_1, \mathcal{B}_2, \mathcal{B}_3)$ with the following properties:

- $$\mathcal{B} : \left\{ f \in L^\pi(\Omega) \mid \int_\Omega f \, \mathrm{d}\mathbf{x} = 0 \right\} \mapsto W^{1,\pi}(\Omega)^3$$
  is a bounded linear operator, i.e.,
  $$\|\mathcal{B}[f]\|_{W^{1,\pi}(\Omega)^3} \leqslant C(\pi)\|f\|_{L^\pi} \quad \text{for any} \quad 1 < \pi < \infty; \tag{6.59}$$
- the function $\mathbf{v} = \mathcal{B}[f]$ solves the BVP:
  $$\nabla \cdot \mathbf{v} = f \ \text{ in } \ \Omega, \qquad \mathbf{v}\,|_{\partial\Omega} = \mathbf{0}; \tag{6.60}$$
  Usually, the symbol $\mathcal{B} \approx (\nabla \cdot)^{-1}$ is called the Bogovskii operator.
- if $f \in L^\pi(\Omega)$ can be written as $f = \nabla \cdot \mathbf{g}$ with $\mathbf{g} \in L^q(\Omega)^3$, $\mathbf{g} \cdot \mathbf{n}\,|_{\partial\Omega} = \mathbf{0}$, then
  $$\|\mathcal{B}[f]\|_{L^\pi} \leqslant C(\pi)\|\mathbf{g}\|_{L^q} \quad \text{for any} \quad 1 < \pi < \infty.$$

Considering the regularity of the approximate density functions, we can use the quantities

$$\left\{ \psi(t)\ \mathcal{B}[\rho_\varepsilon - \frac{1}{|\Omega|} \int_\Omega \rho_\varepsilon \, \mathrm{d}\mathbf{x}] \right\}_{\varepsilon>0}, \quad \psi \in \mathcal{D}(0,T), \quad 0 \leqslant \psi \leqslant 1$$

as test functions in the momentum equation (5.56). Bearing in mind property (6.60) of the linear operator $\mathcal{B}$, we have the following integral identity:

$$\int_0^T \psi \left(\int_\Omega (p_e(\rho_\varepsilon) + \vartheta_\varepsilon p_\vartheta(\rho_\varepsilon) + \vartheta_\varepsilon^2 p_{\vartheta^2}(\rho_\varepsilon) + \frac{a}{3}\vartheta_\varepsilon^4 + \delta \nabla \rho_\varepsilon^\Gamma)\rho_\varepsilon \, \mathrm{d}\mathbf{x} \right) \mathrm{d}t \tag{6.61}$$
$$= I + II + \cdots + IX,$$

where

$$I = \frac{\displaystyle\int_\Omega \rho_\varepsilon \,\mathrm{d}\mathbf{x}}{|\Omega|} \int_0^T \psi \left(\int_\Omega p_e(\rho_\varepsilon) + \vartheta_\varepsilon p_\vartheta(\rho_\varepsilon) + \vartheta_\varepsilon^2 p_{\vartheta^2}(\rho_\varepsilon) + \frac{a}{3}\vartheta_\varepsilon^4 + \delta \nabla \rho_\varepsilon^\Gamma \,\mathrm{d}\mathbf{x} \right) \mathrm{d}t,$$

$$II = \int_0^T \psi \left(\int_\Omega \mathbb{S}_\varepsilon : \nabla \mathcal{B}[\rho_\varepsilon - \frac{1}{|\Omega|}\int_\Omega \rho_\varepsilon \,\mathrm{d}\mathbf{x}] \,\mathrm{d}\mathbf{x} \right) \mathrm{d}t,$$

$$III = -\int_0^T \psi \left(\int_\Omega (\rho_\varepsilon \mathbf{u}_\varepsilon \otimes \mathbf{u}_\varepsilon) : \nabla \mathcal{B}[\rho_\varepsilon - \frac{1}{|\Omega|}\int_\Omega \rho_\varepsilon \,\mathrm{d}\mathbf{x}] \,\mathrm{d}\mathbf{x} \right) \mathrm{d}t,$$

$$IV = \varepsilon \int_0^T \psi \left(\int_\Omega (\nabla \mathbf{u}_\varepsilon \cdot \nabla \rho_\varepsilon \cdot \mathcal{B}[\rho_\varepsilon - \frac{1}{|\Omega|}\int_\Omega \rho_\varepsilon \,\mathrm{d}\mathbf{x}] \,\mathrm{d}\mathbf{x} \right) \mathrm{d}t,$$

$$V = -\int_0^T \psi \left(\int_\Omega (\rho_\varepsilon \nabla \Psi_\varepsilon \cdot \mathcal{B}[\rho_\varepsilon - \frac{1}{|\Omega|}\int_\Omega \rho_\varepsilon \,\mathrm{d}\mathbf{x}] \,\mathrm{d}\mathbf{x} \right) \mathrm{d}t,$$

$$VI = -\int_0^T \psi' \left(\int_\Omega (\rho_\varepsilon \mathbf{u}_\varepsilon \cdot \mathcal{B}[\rho_\varepsilon - \frac{1}{|\Omega|}\int_\Omega \rho_\varepsilon \,\mathrm{d}\mathbf{x}] \,\mathrm{d}\mathbf{x} \right) \mathrm{d}t,$$

$$VII = -\varepsilon \int_0^T \psi \left(\int_\Omega (\rho_\varepsilon \mathbf{u}_\varepsilon \cdot \mathcal{B}[\nabla \cdot (\nabla \rho_\varepsilon)] \,\mathrm{d}\mathbf{x} \right) \mathrm{d}t,$$

$$VIII = \int_0^T \psi \left(\int_\Omega (\rho_\varepsilon \mathbf{u}_\varepsilon \cdot \mathcal{B}[\nabla \cdot (\rho_\varepsilon \mathbf{u}_\varepsilon)] \,\mathrm{d}\mathbf{x} \right) \mathrm{d}t,$$

$$IX = -\int_0^T \psi \left(\int_\Omega (\nabla \times \mathbf{H}) \times \mathbf{H} \cdot \mathcal{B}[\rho_\varepsilon - \frac{1}{|\Omega|}\int_\Omega \rho_\varepsilon \,\mathrm{d}\mathbf{x}] \,\mathrm{d}\mathbf{x} \right) \mathrm{d}t.$$

By virtue of the hypothesis (1.8), the fact that

$$\int_\Omega \rho_\varepsilon \,\mathrm{d}\mathbf{x} = \int_\Omega \rho_{0,\delta} \,\mathrm{d}\mathbf{x} \quad \text{independent of} \quad t,$$

and the property (6.59), one has the integrals I, II, III, V and IX bounded uniformly with respect to $\varepsilon > 0$.

Using the same argument as Section 6.1 in [13], we know the left terms IV, VI, VII and VIII are bounded uniformly for any small ε, and furthermore, from (6.61), the resulting estimate reads

$$\int_{\Omega_T} \rho_\varepsilon^{\Gamma+1} \,\mathrm{d}\mathbf{x} \,\mathrm{d}t \leqslant C(\delta), \quad \text{with } C(\delta) \text{ independent of } \varepsilon.$$

Now, we have, at least for some subsequences

$$\rho_\varepsilon \to \rho \quad \text{in} \quad C([0,T]; L^\Gamma_{\text{weak}}(\Omega)), \tag{6.62}$$

$$\mathbf{u}_\varepsilon \rightharpoonup \mathbf{u} \ \text{ in } \ L^r(0,T;W_0^{1,r}(\Omega)),\ r=\frac{8}{5-\alpha}, \tag{6.63}$$

and

$$\rho_\varepsilon \mathbf{u}_\varepsilon \to \rho\mathbf{u} \ \text{ in } \ C([0,T];L_{\text{weak}}^{\frac{2\Gamma}{\Gamma+1}}(\Omega)), \tag{6.64}$$

as $L_{\text{weak}}^{\Gamma}(\Omega)$ is continuously imbedded into the dual space $W^{-1,r'}(\Omega)$ for suitable (large) $\Gamma\left(>\dfrac{24}{17+3\alpha}\right)$;

$$\vartheta_\varepsilon \overset{*}{\rightharpoonup} \vartheta \ \text{ in } \ L^\infty(0,T;L^4(\Omega)), \ \text{ and } \vartheta_k \rightharpoonup \vartheta \ \text{ in } \ L^2(0,T;H^1(\Omega)),$$

$$\begin{aligned} \ln\vartheta_\varepsilon &\rightharpoonup \overline{\ln\vartheta} \ \text{ in } \ L^2(0,T;H^1(\Omega)),\\ \vartheta_\varepsilon^{\frac{\beta}{2}} &\rightharpoonup \overline{\vartheta^{\frac{\beta}{2}}} \ \text{ in } \ L^2(0,T;H^1(\Omega)), \end{aligned} \tag{6.65}$$

and

$$f(\vartheta_\varepsilon) \rightharpoonup \overline{f(\vartheta)} \ \text{ in } \ L^2(0,T;H^1(\Omega))$$

for any

$$f\in C^1(0,\infty),\ \ |f'(\xi)|\leqslant C\left(\frac{1}{\xi}+\xi^{\frac{1}{2}}\right),\ \xi>0,$$

$$\mathbf{H}_\varepsilon \overset{*}{\rightharpoonup} \mathbf{H} \ \text{ in } \ L^\infty(0,T;L^2(\Omega)),$$

$$\mathbf{H}_\varepsilon \rightharpoonup \mathbf{H} \ \text{ in } \ L^2(0,T;H^1(\Omega)),\ \mathbf{H}_\varepsilon \to \mathbf{H} \ \text{ in } \ L^2(0,T;L^2(\Omega)),$$

$$(\nabla\times\mathbf{H}_\varepsilon)\times\mathbf{H}_\varepsilon \rightharpoonup (\nabla\times\mathbf{H})\times\mathbf{H} \ \text{ in } \ L^\pi(\Omega_T) \ \text{ for a certain } \ \pi>1,$$

$$\mathbf{u}_\varepsilon\times\mathbf{H}_\varepsilon \rightharpoonup \mathbf{u}\times\mathbf{H} \ \text{ in } \ L^\pi(\Omega_T) \ \text{ for a certain } \ \pi>1,$$

and

$$\nu(\rho_\varepsilon,\vartheta_\varepsilon,\mathbf{H}_\varepsilon)\nabla\times\mathbf{H}_\varepsilon \rightharpoonup \nu(\rho,\vartheta,\mathbf{H})\nabla\times\mathbf{H} \ \text{ in } \ L^\pi(\Omega_T) \ \text{ for a certain } \ \pi>1.$$

Furthermore, based on (6.62) and (6.63), if $\Gamma>r'$, we can show the density $\rho\geqslant 0$ and the velocity $\mathbf{u}$ solve the original continuity equation $(1.1)_1$ on the whole space $\mathbb{R}^3\times[0,T)$ via extending both of them to be zero outside Ω, where, in fact,

$$\rho_\varepsilon\mathbf{u}_\varepsilon \to \rho\mathbf{u} \ \text{ in } C([0,T];L_{\text{weak}}^{\frac{2\Gamma}{\Gamma+1}}(\mathbb{R}^3))$$

provided $\rho_\varepsilon,\mathbf{u}_\varepsilon$ were extended to be zero outside Ω. Thus $\rho,\mathbf{u}$ satisfy the integral identity:

$$\int_{\Omega_T}(\rho\psi'\phi+\psi\rho\mathbf{u}\cdot\nabla\phi)\,\mathrm{d}\mathbf{x}\mathrm{d}t+\psi(0)\int_\Omega\rho_0\phi\,\mathrm{d}\mathbf{x}=0,$$

for any $\psi \in C^\infty([0,T])$ with $\psi(T) = 0$ and $\phi \in \mathcal{D}(\mathbb{R}^3;\mathbb{R})$. Meanwhile, by using the celebrated regularization technique of DiPerna and P.-L. Lions [10], $\rho, \mathbf{u}$ (being extended to be zero outside Ω) solve the renormalized continuity equation (2.12) in $\mathcal{D}'(\mathbb{R}^3 \times [0,T))$ for any continuously differentiable function b satisfying (2.25).

By the Hölder inequality, together with (6.63) and (6.64), we have

$$\rho_\varepsilon \mathbf{u}_\varepsilon \otimes \mathbf{u}_\varepsilon \rightharpoonup \rho\mathbf{u}\otimes\mathbf{u} \ \text{ in } \ L^r(0,T;L^\pi(\Omega)), \quad 1 < \pi \leqslant \frac{6\Gamma r}{3r + \Gamma r + 6\Gamma}.$$

Here we have used the continuous imbedding

$$L^{\frac{2\Gamma}{\Gamma+1}}_{\text{weak}}(\Omega) \subset W^{-1,r'}(\Omega) \ \text{ provieded } \Gamma \geqslant \frac{12}{5+3\alpha}.$$

Furthermore, by the same token, in accordance with (6.64) and (6.65), we have

$$\rho_\varepsilon \mathbf{u}_\varepsilon \ln\vartheta_\varepsilon \rightharpoonup \rho\mathbf{u}\,\overline{\ln\vartheta} \ \text{ in } L^\pi(\Omega_T) \ \text{ for a certain } \ \pi > 1.$$

Using the same argument as in Section 5, we know

$$\rho_\varepsilon \mathbf{u}_\varepsilon \int_1^{\vartheta_\varepsilon} \frac{c_v(\xi)}{\xi}\,\mathrm{d}\xi \rightharpoonup \rho\mathbf{u}\overline{\int_1^{\vartheta_\varepsilon} \frac{c_v(\xi)}{\xi}\,\mathrm{d}\xi} \ \text{ in } L^\pi(\Omega_T) \ \text{ for a certain } \ \pi > 1.$$

By virtue of the Hölder inequality and (5.45), we have

$$\begin{aligned} &\varepsilon\int_\Omega \nabla\left(\varpi\left(\int_1^{\vartheta_\varepsilon} \frac{c_v(\xi)}{\xi}\,\mathrm{d}\xi - \frac{Q(\vartheta_\varepsilon)}{\vartheta_\varepsilon} - \vartheta_\varepsilon(P_{\vartheta^2}(\rho_\varepsilon) + \rho_\varepsilon P'_{\vartheta^2}(\rho_\varepsilon))\right)\right)\cdot\nabla\rho_\varepsilon\,\mathrm{d}\mathbf{x} \\ &\to 0 \ \text{ in } L^\pi(0,T) \end{aligned} \tag{6.66}$$

for any $\pi \geqslant 1$ and any $\varpi \in C^1(\overline{\Omega})$.

Hence, for the quantities $\frac{4a}{3}\vartheta_\varepsilon^3 + \rho_\varepsilon\int_1^{\vartheta_\varepsilon}\frac{c_v(\xi)}{\xi}\,\mathrm{d}\xi - 2\rho_\varepsilon P_{\vartheta^2}(\rho_\varepsilon)\vartheta_\varepsilon$ in (5.58), by the same argument as in Section 5,

$$\begin{aligned} &\int_{\Omega_T}\left(\frac{4a}{3}\vartheta_\varepsilon^3 + \rho_\varepsilon\int_1^{\vartheta_\varepsilon}\frac{c_v(\xi)}{\xi}\,\mathrm{d}\xi - 2\rho_\varepsilon P_{\vartheta^2}(\rho_\varepsilon)\vartheta_\varepsilon\right)\vartheta_\varepsilon\,\mathrm{d}\mathbf{x}\mathrm{d}t \\ \to &\int_{\Omega_T}\left(\frac{4a}{3}\overline{\vartheta^3} + \rho\overline{\int_1^{\vartheta}\frac{c_v(\xi)}{\xi}\,\mathrm{d}\xi} - 2\overline{\rho P_{\vartheta^2}(\rho)\vartheta}\right)\vartheta\,\mathrm{d}\mathbf{x}\mathrm{d}t. \end{aligned}$$

Note that, in contrast to Section 5, we do not know yet if the densities $\{\rho_\varepsilon\}_{\varepsilon>0}$ converge strongly in $L^1(\Omega_T)$. Fortunately, it holds that

$$\lim_{\varepsilon\to 0}\int_{\Omega_T}\rho_\varepsilon P_{\vartheta^2}(\rho_\varepsilon)\,\vartheta_\varepsilon^2\,\mathrm{d}\mathbf{x}\mathrm{d}t = \int_{\Omega_T}\overline{\rho P_{\vartheta^2}(\rho)}\,\vartheta^2\,\mathrm{d}\mathbf{x}\mathrm{d}t.$$

Thus, exactly as in Section 5,

$$\int_{\Omega_T}\left(\frac{4a}{3}\vartheta_\varepsilon^3+\rho_\varepsilon\int_1^{\vartheta_\varepsilon}\frac{c_v(\xi)}{\xi}\,\mathrm{d}\xi\right)\vartheta_\varepsilon\,\mathrm{d}\mathbf{x}\mathrm{d}t$$
$$\to\int_{\Omega_T}\left(\frac{4a}{3}\overline{\vartheta^3}+\overline{\rho\int_1^{\vartheta}\frac{c_v(\xi)}{\xi}\,\mathrm{d}\xi}\right)\vartheta\,\mathrm{d}\mathbf{x}\mathrm{d}t.$$

which implies strong convergence of $\{\vartheta_\varepsilon\}_{\varepsilon>0}$, and especially, $\vartheta_\varepsilon\to\vartheta$ in $L^2(\Omega_T)$. Then it comes from Lemma 5.4 in in [12] that ϑ is strictly positive a.e. on Ω_T, and $\ln\vartheta=\overline{\ln\vartheta}$.

Letting $\varepsilon\to 0$ in (5.56), due to the estimate (5.45), the extra terms

$$\varepsilon\nabla\mathbf{u}_\varepsilon\cdot\nabla\rho_\varepsilon\to 0 \text{ in } L^\pi(\Omega_T)\quad\text{for a certain }\ \pi>1,$$

indeed note that

$$\varepsilon\int_\Omega\nabla\mathbf{u}_\varepsilon\cdot\nabla\rho_\varepsilon\cdot\phi\,d\mathbf{x}\to 0\ \text{ for any } \phi\in C(\overline{\Omega};\mathbb{R}^3)\ \text{ as } \varepsilon\to 0$$

uniformly in $t\in[0,T]$, we know the limit quantities satisfy an "averaged" momentum equation

$$\begin{aligned}&(\rho\mathbf{u})_t+\nabla\cdot(\rho\mathbf{u}\otimes\mathbf{u})+\nabla(\overline{p_e(\rho)}+\vartheta\overline{p_\vartheta(\rho)}+\vartheta^2\overline{p_{\vartheta^2}(\rho)}+\frac{a}{3}\vartheta^4+\delta\overline{\rho^\Gamma})\\&=\nabla\cdot\mathbb{S}+\rho\nabla\Psi+(\nabla\times\mathbf{H})\times\mathbf{H}\ \text{ in } \mathcal{D}'(\Omega_T),\end{aligned}\tag{6.67}$$

where the potential Ψ satisfies (2.17).

In order to commute the limits with the composition operators in the "averaged" momentum equation (6.67) , we need to show strong (pointwise) convergence of the sequence $\{\rho_\varepsilon\}_{\varepsilon>0}$, i.e.,

$$\rho_\varepsilon\to\rho\ \text{ in }\ L^1(\Omega_T).$$

This is a lengthy but nowadays formal procedure, by making good use of the special function $b(\rho)=\rho\ln\rho$ in the renormalized continuity equation and, the weak continuity of the effective viscous pressure "$p-(\lambda+2\mu)\nabla\cdot\mathbf{u}$". We omit it here (readers can refer to Section 6.3 in [9] for the details). Consequently, the limit functions $\rho,\mathbf{u}$ satisfy the momentum equation $(1.1)_2$, where p is replaced by $p_e(\rho)+\vartheta p_\vartheta(\rho)+\vartheta^2p_{\vartheta^2}(\rho)+\frac{a}{3}\vartheta^4+\delta\rho^\Gamma$, in $\mathcal{D}'(\Omega_T)$.

The magnetic induction vector $\mathbf{H}$ satisfies

$$\begin{aligned}&\int_{\Omega_T}(\psi'\mathbf{H}\cdot\phi+\psi(\mathbf{u}\times\mathbf{H})\cdot(\nabla\times\phi)-\psi\nu(\nabla\times\mathbf{H})\cdot(\nabla\times\phi))\,\mathrm{d}\mathbf{x}\mathrm{d}t\\&=-\psi(0)\int_\Omega\mathbf{H}_0\cdot\phi\,\mathrm{d}\mathbf{x},\end{aligned}\tag{6.68}$$

for any $\psi \in C^\infty([0,T])$ with $\psi(T) = 0$ and $\phi \in \mathcal{D}(\Omega; \mathbb{R}^3)$.

To conclude, we have to let $\varepsilon \to 0$ in the energy equality (5.57) and the entropy inequality (5.58). Since $\rho, \mathbf{u}$ satisfy the renormalized continuity equation in $\mathcal{D}'(\mathbb{R}^3 \times [0,T))$, then

$$\int_{\Omega_T} \psi p_b(\rho) \nabla \cdot \mathbf{u} \, \mathrm{d}\mathbf{x}\mathrm{d}t = \int_{\Omega_T} \psi' \rho P_b(\rho) \, \mathrm{d}\mathbf{x}\mathrm{d}t + \int_\Omega \rho_{0,\delta} P_b(\rho_{0,\delta}) \, \mathrm{d}\mathbf{x}$$

for any $\psi \in C^\infty([0,T])$ with $\psi(0) = 1$, $\psi(T) = 0$. Here $P_b(\rho) = \displaystyle\int_1^\rho \frac{p_b(z)}{z^2} \, \mathrm{d}z$.

Thus passing the limit for $\varepsilon \to 0$ in (5.57), one has the total energy balance:

$$\begin{aligned}
&-\int_{\Omega_T} \psi' \Big(\rho \left(\frac{1}{2} |\mathbf{u}|^2 + P_e(\rho) + Q(\vartheta) - \vartheta^2 P_{\vartheta^2}(\rho) + \frac{G}{2} \Delta^{-1}[\rho] + \frac{\delta}{\Gamma - 1} \rho^{\Gamma - 1} \right) \\
&+ a\vartheta^4 + \frac{1}{2} |\mathbf{H}|^2 \Big) \mathrm{d}\mathbf{x}\mathrm{d}t \\
=& \int_\Omega \Big(\rho_{0,\delta} \Big(\frac{1}{2} \frac{|\mathbf{m}_{0,\delta}|^2}{\rho_{0,\delta}^2} + P_e(\rho_{0,\delta}) + Q(\vartheta_{0,\delta}) - \vartheta_{0,\delta}^2 P_{\vartheta^2}(\rho_{0,\delta}) + \frac{G}{2} \Delta^{-1}[\rho_{0,\delta}] \\
&+ \frac{\delta}{\Gamma - 1} \rho_{0,\delta}^{\Gamma - 1} \Big) + a\vartheta_{0,\delta}^4 + \frac{1}{2} |\mathbf{H}_{0,\delta}|^2 \Big) \, \mathrm{d}\mathbf{x} \triangleq E_{0,\delta}
\end{aligned} \tag{6.69}$$

for any $\psi \in C^\infty([0,T])$ with $\psi(0) = 1$, $\psi(T) = 0$. Note that,

$$\Delta^{-1}[\rho_\varepsilon] \to \Delta^{-1}[\rho] \text{ in } C(\overline{\Omega}_T)$$

by combining (6.62) with the standard elliptic theory. Here we have used the fact that

$$\begin{aligned}
\int_\Omega \rho_\varepsilon \nabla \Psi_\varepsilon \cdot \mathbf{u}_\varepsilon \, \mathrm{d}\mathbf{x} &= -\int_\Omega \Psi_\varepsilon \nabla \cdot (\rho_\varepsilon \mathbf{u}_\varepsilon) \, \mathrm{d}\mathbf{x} = \int_\Omega \Psi_\varepsilon (\rho_{\varepsilon t} - \varepsilon \Delta \rho_\varepsilon) \, \mathrm{d}\mathbf{x} \\
&= \frac{1}{2} \frac{\mathrm{d}}{\mathrm{d}t} \int_\Omega \rho_\varepsilon \Psi_\varepsilon \, \mathrm{d}\mathbf{x} + \varepsilon G \int_\Omega \rho_\varepsilon^2 \, \mathrm{d}\mathbf{x} \\
&= \frac{G}{2} \frac{\mathrm{d}}{\mathrm{d}t} \int_\Omega \rho_\varepsilon (-\Delta)^{-1}[\rho_\varepsilon] \, \mathrm{d}\mathbf{x} + \varepsilon G \int_\Omega \rho_\varepsilon^2 \, \mathrm{d}\mathbf{x}.
\end{aligned}$$

The limit entropy inequality reads

$$\begin{aligned}
&\int_{\Omega_T} \psi' \left(\frac{4a}{3} \vartheta^3 + \rho \int_1^\vartheta \frac{c_v(\xi)}{\xi} \, \mathrm{d}\xi - \rho P_\vartheta(\rho) - 2\rho\vartheta P_{\vartheta^2}(\rho) \right) \phi \, \mathrm{d}\mathbf{x}\mathrm{d}t \\
&+ \int_{\Omega_T} \psi \left(\frac{4a}{3} \vartheta^3 + \rho \int_1^\vartheta \frac{c_v(\xi)}{\xi} \, \mathrm{d}\xi - \rho P_\vartheta(\rho) - 2\rho\vartheta P_{\vartheta^2}(\rho) \right) \mathbf{u} \cdot \nabla\phi \, \mathrm{d}\mathbf{x}\mathrm{d}t \\
&- \int_{\Omega_T} \psi \Big(\frac{\kappa_G(\rho, \vartheta, \mathbf{H}) + \kappa_R \vartheta^3}{\vartheta} \nabla\vartheta \Big) \cdot \nabla\phi \, \mathrm{d}\mathbf{x}\mathrm{d}t
\end{aligned}$$

$$
\begin{aligned}
&\leqslant -\int_{\Omega_T} \psi\phi \left(\frac{\kappa_G(\rho,\vartheta,\mathbf{H}) + \kappa_R \vartheta^3}{\vartheta^2} |\nabla\vartheta|^2 + \frac{\mathbb{S} : \nabla \mathbf{u} + \nu(\rho,\vartheta,\mathbf{H}) |\nabla \times \mathbf{H}|^2}{\vartheta} \right) \mathrm{d}\mathbf{x}\mathrm{d}t \\
&\quad - \psi(0) \int_\Omega \phi \left(\frac{4a}{3} \vartheta_{0,\delta}^3 + \rho_{0,\delta} \int_1^{\vartheta_{0,\delta}} \frac{c_v(\xi)}{\xi} \,\mathrm{d}\xi - \rho_{0,\delta} P_\vartheta(\rho_{0,\delta}) - 2\rho_{0,\delta}\vartheta_{0,\delta} P_{\vartheta^2}(\rho_{0,\delta}) \right) \mathrm{d}\mathbf{x}
\end{aligned} \tag{6.70}
$$

for any $0 \leqslant \psi \in C^\infty([0,T])$ with $\psi(T) = 0$ and $0 \leqslant \phi \in \mathcal{D}(\Omega;\mathbb{R})$. Here we have used (6.66), the fact that

$$
\begin{aligned}
&\int_{\Omega_T} \psi\phi p_\vartheta(\rho) \nabla \cdot \mathbf{u} \,\mathrm{d}\mathbf{x}\mathrm{d}t \\
&= \int_{\Omega_T} \psi' \phi\rho P_\vartheta(\rho) \,\mathrm{d}\mathbf{x}\mathrm{d}t + \psi(0) \int_\Omega \phi\rho_{0,\delta} P_\vartheta(\rho_{0,\delta}) \,\mathrm{d}\mathbf{x} + \int_{\Omega_T} \psi\rho P_\vartheta(\rho) \mathbf{u} \cdot \nabla\phi \,\mathrm{d}\mathbf{x}\mathrm{d}t.
\end{aligned}
$$

7 Passing to the limit $\delta \to 0$

Our final task is to carry out the lime process when the parameter $\delta \to 0$ to recover the original system (1.1) by evoking the full strength of the pressure and temperature estimates, in other words, we will establish the weak sequential stability property for the approximate solutions set $\{\rho_\delta, \mathbf{u}_\delta, \vartheta_\delta, \mathbf{H}_\delta\}_{\delta>0}$ constructed in Section 6. To begin with, note that, from hypothesis (3.29) on the approximate initial data, we have

$$
E_{0,\delta} \to E_0 \quad \text{as } \delta \to 0,
$$

and

$$
\begin{aligned}
&\int_\Omega \phi \left(\frac{4a}{3} \vartheta_{0,\delta}^3 + \rho_{0,\delta} \int_1^{\vartheta_{0,\delta}} \frac{c_v(\xi)}{\xi} \,\mathrm{d}\xi - \rho_{0,\delta} P_\vartheta(\rho_{0,\delta}) - 2\rho_{0,\delta}\vartheta_{0,\delta} P_{\vartheta^2}(\rho_{0,\delta}) \right) \mathrm{d}\mathbf{x} \\
&\to \int_\Omega \phi \left(\frac{4a}{3} \vartheta_0^3 + \rho_0 \int_1^{\vartheta_0} \frac{c_v(\xi)}{\xi} \,\mathrm{d}\xi - \rho_0 P_\vartheta(\rho_0) - 2\rho_0\vartheta_0 P_{\vartheta^2}(\rho_0) \right) \mathrm{d}\mathbf{x} \quad \text{as } \delta \to 0
\end{aligned}
$$

for any $\phi \in C^\infty(\overline{\Omega})$, $\phi \geqslant 0$.

In light with the total energy balance (6.69), the following uniform bounds by a constant depending only on E_0 hold:

$$
\rho_\delta \quad \text{bounded in } L^\infty(0,T;L^\gamma(\Omega)), \tag{7.71}
$$

$$
\rho_\delta |\mathbf{u}_\delta|^2,\ \rho_\delta Q(\vartheta_\delta) \quad \text{bounded in } L^\infty(0,T;L^1(\Omega)),
$$

and, consequently, by Hölder's inequality,

$$
\rho_\delta \mathbf{u}_\delta \quad \text{bounded in } L^\infty(0,T;L^{\frac{2\gamma}{\gamma+1}}(\Omega)); \tag{7.72}
$$

$$\vartheta_\delta \ \text{ bounded in } L^\infty(0,T;L^4(\Omega)),$$

$$\mathbf{H}_\delta \ \text{ bounded in } L^\infty(0,T;L^2(\Omega)), \tag{7.73}$$

and

$$\delta \int_{\Omega_T} \rho_\delta^\Gamma \, \mathrm{d}\mathbf{x}\mathrm{d}t \leqslant C \quad \text{uniformly with respect to } \delta > 0.$$

As for the term $\int_\Omega \Delta^{-1}[\rho]\rho \, \mathrm{d}\mathbf{x}$ in the energy equality (6.69) related to the gravitational potential, from the elliptic estimates and the fact that

$$\int_\Omega \rho_\delta(t) \, \mathrm{d}\mathbf{x} = \int_\Omega \rho_{0,\delta} \, \mathrm{d}\mathbf{x} = M_0 \quad \text{for a.a } t \in (0,T),$$

we know, by virtue of Hölder's inequality,

$$\int_\Omega |\Delta^{-1}[\rho]\rho| \, \mathrm{d}\mathbf{x} \leqslant \|\rho\|_{L^2} \|\Delta^{-1}[\rho]\|_{L^2} \leqslant M_0 \|\rho\|_{L^2}.$$

Similarly, the entropy production inequality (6.70) gives rise to

$$\operatorname*{ess\ sup}_{t\in[0,T]} \int_\Omega \rho_\delta |\ln \vartheta_\delta| \, \mathrm{d}\mathbf{x} \leqslant C(S_0),$$

$$\int_{\Omega_T} \frac{\mathbb{S}_\delta : \nabla \mathbf{u}_\delta + \nu |\nabla \times \mathbf{H}_\delta|^2}{\vartheta_\delta} + |\nabla \ln \vartheta_\delta|^2 + |\nabla \vartheta_\delta^{\frac{3}{2}}|^2 + |\nabla \vartheta_\delta^{\frac{\beta}{2}}|^2 \, \mathrm{d}\mathbf{x}\mathrm{d}t \leqslant C(S_0), \tag{7.74}$$

together with

$$\int_{\Omega_T} \frac{\kappa_G(\rho_\delta, \vartheta_\delta, \mathbf{H}_\delta) + \kappa_R \vartheta_\delta^3}{\vartheta_\delta^2} |\nabla \vartheta_\delta|^2 \, \mathrm{d}\mathbf{x}\mathrm{d}t \leqslant C(S_0),$$

and, consequently, by using the same arguments in Section 2 or in Section 5,

$$\|(1+\vartheta_\delta)^{\frac{\beta-1}{2}} \nabla \times \mathbf{H}_\delta\|_{L^2(L^2)} \leqslant C(S_0),$$

$$\|\vartheta_\delta\|_{L^2(H^1)} + \|\ln \vartheta_\delta\|_{L^2(H^1)} + \|\vartheta_\delta^{\frac{3}{2}}\|_{L^2(H^1)} + \|\vartheta_\delta^{\frac{\beta}{2}}\|_{L^2(H^1)} \leqslant C(E_0, S_0),$$

$$\|\mathbf{u}_\delta\|_{L^r(W_0^{1,r})} \leqslant C(E_0, S_0), \quad r = \frac{8}{5-\alpha},$$

furthermore, utilizing estimates (7.71),(7.72),(7.73) and the hypothesis (1.8),

$$\vartheta_\delta p_\vartheta(\rho_\delta),\ \vartheta_\delta^2 p_{\vartheta^2}(\rho_\delta),\ (\nabla\times\mathbf{H}_\delta)\times\mathbf{H}_\delta,\ \mathbf{u}_\delta\times\mathbf{H}_\delta,\ \nu(\rho_\delta,\vartheta_\delta,\mathbf{H}_\delta)\nabla\times\mathbf{H}_\delta \ \text{ bounded in } L^\pi(\Omega_T)$$

for some certain $\pi > 1$,

$$\rho_\delta \mathbf{u}_\delta \otimes \mathbf{u}_\delta \ \text{ bounded in } \ L^r(0,T;L^{\frac{6\gamma r}{3r+\gamma r+6\gamma}}(\Omega)).$$

And, the bounds are independent of $\delta > 0$. Here we have used $\vartheta_\delta^{\frac{3}{2}} \in L^2(0,T;H^1(\Omega))$, and a simple interpolation argument

$$L^\infty(0,T;L^4(\Omega)) \cap L^3(0,T;L^9(\Omega)) \subset L^\pi(\Omega_T), \quad \text{with } \pi = \frac{17}{3}.$$

Note that $\ln \vartheta_\delta$ is bounded in $L^2(\Omega_T)$ by a constant independent of $\delta > 0$, which implies the strict positivity of the temperature.

Since

$$\begin{aligned}\mathbb{S}_\delta =& \sqrt{\vartheta_\delta \mu(\vartheta_\delta, \mathbf{H}_\delta)}\sqrt{\frac{\mu(\vartheta_\delta, \mathbf{H}_\delta)}{\vartheta_\delta}}\left(\nabla \mathbf{u}_\delta + \nabla^\top \mathbf{u}_\delta - \frac{2}{3}(\nabla \cdot \mathbf{u}_\delta)\, \mathbb{I}_3\right) \\ &+ \sqrt{\vartheta_\delta \eta(\vartheta_\delta, \mathbf{H}_\delta)}\sqrt{\frac{\eta(\vartheta_\delta, \mathbf{H}_\delta)}{\vartheta_\delta}}(\nabla \cdot \mathbf{u}_\delta)\, \mathbb{I}_3,\end{aligned}$$

the estimate (7.74) yields

$$\mathbb{S}_\delta \;\text{ bounded in }\; L^2(0,T;L^{\frac{4}{3}}(\Omega))$$

uniformly with respect to δ.

Moreover, we can repeat step by step the proof of the refined pressure estimates in [12], the resulting estimate reads

$$\int_{\Omega_T} \left(p_e(\rho_\delta) + \vartheta_\delta p_\vartheta(\rho_\delta) + \vartheta_\delta^2 p_{\vartheta^2}(\rho_\delta) + \frac{a}{3}\vartheta_\delta^4 + \delta \rho_\delta^\Gamma\right) \rho_\delta^\omega \; \mathrm{d}\mathbf{x}\mathrm{d}t \leqslant C(E_0, S_0),$$

in particular,

$$\{\rho_\delta^{\gamma+\omega}\}_{\delta>0},\ \{\delta \rho_\delta^{\Gamma+\omega}\}_{\delta>0} \;\text{ are bounded in }\; L^1(\Omega_T). \tag{7.75}$$

In view of the above estimates, we may assume that, up to a subsequence,

$$\rho_\delta \to \rho \;\text{ in }\; C([0,T];L^\gamma_{\text{weak}}(\Omega)),$$

$$\mathbf{u}_\delta \rightharpoonup \mathbf{u} \;\text{ in }\; L^r(0,T;W_0^{1,r}(\Omega)),\ r = \frac{8}{5-\alpha},$$

and

$$\rho_\delta \mathbf{u}_\delta \to \rho \mathbf{u} \;\text{ in }\; C([0,T];L^{\frac{2\gamma}{\gamma+1}}_{\text{weak}}(\Omega)),$$

as $L^\gamma_{\text{weak}}(\Omega)$ is continuously imbedded into the dual space $W^{-1,r'}(\Omega)$ since $\gamma \geqslant 2$. Moreover, due to the choice of initial data $\rho_{0,\delta}$ and $\mathbf{m}_{0,\delta}$,

$$\rho(0, \mathbf{x}) = \rho_0(\mathbf{x}) \;\text{ a.e. on } \Omega,$$

$$\frac{\delta}{\Gamma-1}\int_{\Omega}\rho_{0,\delta}^{\Gamma}\,\mathrm{d}\mathbf{x}\to 0 \text{ as } \delta\to 0$$

and

$$\rho\mathbf{u}(0,\mathbf{x})=\mathbf{m}_0(\mathbf{x}) \text{ a.e. on } \Omega.$$

Consequently, in accordance with the hypothesis $\gamma\geqslant 2$, the space $L^{\frac{2\gamma}{\gamma+1}}(\Omega)$ is compactly imbedded into $W^{-1,r'}(\Omega)$, which yields compactness of the convective terms:

$$\rho_\delta\mathbf{u}_\delta\otimes\mathbf{u}_\delta\rightharpoonup\rho\mathbf{u}\otimes\mathbf{u} \text{ in } L^r(0,T;L^{\frac{6\gamma r}{3r+\gamma r+6\gamma}}(\Omega)).$$

Here we have used the fact that $\rho_\delta,\mathbf{u}_\delta$ satisfy $(1.1)_1$ and $(\rho_\delta\mathbf{u}_\delta)_t$ can be expressed by (6.67). Since $\gamma\geqslant 2$, we can use the regularization technique developed in [10] to show $\rho,\mathbf{u}$ satisfy the (2.26), and furthermore, $\rho\in C([0,T];L^1(\Omega))$.

$$\vartheta_\delta\overset{*}{\rightharpoonup}\vartheta \text{ in } L^\infty(0,T;L^4(\Omega)), \text{ and } \vartheta_\delta\rightharpoonup\vartheta \text{ in } L^2(0,T;H^1(\Omega)), \tag{7.76}$$

$$\ln\vartheta_\delta\rightharpoonup\overline{\ln\vartheta} \text{ in } L^2(0,T;H^1(\Omega)),$$

furthermore,

$$\rho_\delta\ln\vartheta_\delta\rightharpoonup\rho\overline{\ln\vartheta} \text{ in } L^2(0,T;L^{\frac{6\gamma}{6+\gamma}}(\Omega)),$$

$$\rho_\delta\ln\vartheta_\delta\mathbf{u}_\delta\rightharpoonup\rho\overline{\ln\vartheta}\,\mathbf{u} \text{ in } L^2(0,T;L^{\frac{6\gamma}{3+4\gamma}}(\Omega)),$$

$$\vartheta_\delta^{\frac{\beta}{2}}\rightharpoonup\overline{\vartheta^{\frac{\beta}{2}}} \text{ in } L^2(0,T;H^1(\Omega)),$$

$$Q(\vartheta_\delta)\rightharpoonup\overline{Q(\vartheta)} \text{ in } L^2(0,T;H^1(\Omega)),$$

$$\rho_\delta Q(\vartheta_\delta)\rightharpoonup\rho\overline{Q(\vartheta)} \text{ in } L^2(0,T;L^{\frac{6\gamma}{6+\gamma}}(\Omega)),$$

$$\rho_\delta\mathbf{u}_\delta Q(\vartheta_\delta)\rightharpoonup\rho\mathbf{u}\overline{Q(\vartheta)} \text{ in } L^2(0,T;L^{\frac{6\gamma}{3+4\gamma}}(\Omega)),$$

and

$$f(\vartheta_\delta)\rightharpoonup\overline{f(\vartheta)} \text{ in } L^2(0,T;H^1(\Omega))$$

for any

$$f\in C^1(0,\infty),\ |f'(\xi)|\leqslant C\left(\frac{1}{\xi}+\xi^{\frac{1}{2}}\right),\ \xi>0,$$

$$\mathbf{H}_\delta\overset{*}{\rightharpoonup}\mathbf{H} \text{ in } L^\infty(0,T;L^2(\Omega)),$$

which can be improved to

$$\mathbf{H}_\delta\to\mathbf{H} \text{ in } C([0,T];L^2_{\text{weak}}(\Omega)),$$

since $\mathbf{H}_{\delta t}$ can be expressed through equation (6.68);

$$\mathbf{H}_\delta \rightharpoonup \mathbf{H} \ \text{ in } \ L^2(0,T;H^1(\Omega)), \ \mathbf{H}_\delta \to \mathbf{H} \ \text{ in } \ L^2(\Omega_T),$$

$$(\nabla\times\mathbf{H}_\delta)\times\mathbf{H}_\delta \rightharpoonup (\nabla\times\mathbf{H})\times\mathbf{H} \ \text{ in } \ L^\pi(\Omega_T) \ \text{ for a certain } \ \pi>1,$$

$$\mathbf{u}_\delta\times\mathbf{H}_\delta \rightharpoonup \mathbf{u}\times\mathbf{H} \ \text{ in } \ L^\pi(\Omega_T) \ \text{ for a certain } \ \pi>1,$$

$$\nu(\rho_\delta,\vartheta_\delta,\mathbf{H}_\delta)\nabla\times\mathbf{H}_\delta \rightharpoonup \nu(\rho,\vartheta,\mathbf{H})\nabla\times\mathbf{H} \ \text{ in } \ L^\pi(\Omega_T) \ \text{ for a certain } \ \pi>1,$$

and,

$$\mathbf{H}(0,\mathbf{x}) = \mathbf{H}_0(\mathbf{x}) \ \text{ a.e. on } \Omega.$$

As usual as in Section 5 and Section 6, our next duty is to show strong (pointwise) convergence of the temperature. By virtue of hypothesis (1.8), we have

$$\rho_\delta P_\vartheta(\rho_\delta) \ \text{ is bounded in } \ L^\infty(0,T;L^{\frac{\gamma}{\varsigma}}(\Omega)), \ \text{ with } \ \frac{\gamma}{\varsigma} > \frac{4}{3},$$

$$\rho_\delta P_{\vartheta^2}(\rho_\delta) \ \text{ is bounded in } \ L^\infty(0,T;L^{\frac{\gamma}{\zeta}}(\Omega)), \ \text{ with } \ \frac{\gamma}{\zeta} > 2,$$

and

$$\rho_\delta\vartheta_\delta P_{\vartheta^2}(\rho_\delta) \ \text{ is bounded in } \ L^\infty(0,T;L^{\frac{4\gamma}{\gamma+4\zeta}}(\Omega)), \ \text{ with } \ \frac{4\gamma}{\gamma+4\zeta} > \frac{4}{3},$$

in accordance with (7.76).

Using the entropy inequality (6.70) together with Lemma 6.3 of Chapter 6 in [18], we have

$$\begin{aligned}
&\frac{4a}{3}\vartheta_\delta^3 + \rho_\delta\int_1^{\vartheta_\delta}\frac{c_v(\xi)}{\xi}\,\mathrm{d}\xi - \rho_\delta P_\vartheta(\rho_\delta) - 2\rho_\delta\vartheta_\delta P_{\vartheta^2}(\rho_\delta) \\
&\to \frac{4a}{3}\overline{\vartheta^3} + \overline{\rho\int_1^{\vartheta}\frac{c_v(\xi)}{\xi}\,\mathrm{d}\xi} - \overline{\rho P_\vartheta(\rho)} - 2\vartheta\overline{\rho P_{\vartheta^2}(\rho)} \ \text{ in } \ L^2(0,T;H^{-1}(\Omega)),
\end{aligned}$$

and, in particular,

$$\begin{aligned}
&\int_{\Omega_T}\left(\frac{4a}{3}\vartheta_\delta^3 + \rho_\delta\int_1^{\vartheta_\delta}\frac{c_v(\xi)}{\xi}\,\mathrm{d}\xi - \rho_\delta P_\vartheta(\rho_\delta) - 2\rho_\delta\vartheta_\delta P_{\vartheta^2}(\rho_\delta)\right)\vartheta_\delta\,\mathrm{d}\mathbf{x}\mathrm{d}t \\
&\to \int_{\Omega_T}\left(\frac{4a}{3}\overline{\vartheta^3} + \overline{\rho\int_1^{\vartheta}\frac{c_v(\xi)}{\xi}\,\mathrm{d}\xi} - \overline{\rho P_\vartheta(\rho)} - 2\vartheta\overline{\rho P_{\vartheta^2}(\rho)}\right)\vartheta\,\mathrm{d}\mathbf{x}\mathrm{d}t.
\end{aligned} \tag{7.77}$$

Since ρ_δ satisfies the renormalized equation (2.26), we have

$$b(\rho_\delta) \to \overline{b(\rho)} \ \text{ in } \ C([0,T];L^\gamma_{\text{weak}}(\Omega))$$

provided b is a bounded and continuously differential function. Thus, a simple approximation argument yields

$$\rho_\delta P_\vartheta(\rho_\delta) \to \overline{\rho P_\vartheta(\rho)} \text{ in } C([0,T]; L^{\frac{\gamma}{\varsigma}}_{\text{weak}}(\Omega)),$$

$$\rho_\delta P_{\vartheta^2}(\rho_\delta) \to \overline{\rho P_{\vartheta^2}(\rho)} \text{ in } C([0,T]; L^{\frac{\gamma}{\varsigma}}_{\text{weak}}(\Omega)),$$

whence, using (7.76) again,

$$\lim_{\delta\to 0}\int_{\Omega_T} \rho_\delta P_\vartheta(\rho_\delta)\vartheta_\delta \,\mathrm{d}\mathbf{x}\mathrm{d}t = \lim_{\delta\to 0}\int_{\Omega_T} \overline{\rho P_\vartheta(\rho)}\vartheta \,\mathrm{d}\mathbf{x}\mathrm{d}t,$$

$$\lim_{\delta\to 0}\int_{\Omega_T} \rho_\delta P_{\vartheta^2}(\rho_\delta)\vartheta_\delta^2 \,\mathrm{d}\mathbf{x}\mathrm{d}t = \lim_{\delta\to 0}\int_{\Omega_T} \overline{\rho P_\vartheta(\rho)}\vartheta^2 \,\mathrm{d}\mathbf{x}\mathrm{d}t.$$

Consequently, (7.77) reduces to

$$\int_{\Omega_T}\left(\frac{4a}{3}\vartheta_\delta^3 + \rho_\delta\int_1^{\vartheta_\delta}\frac{c_v(\xi)}{\xi}\,\mathrm{d}\xi\right)\vartheta_\delta \,\mathrm{d}\mathbf{x}\mathrm{d}t \to \int_{\Omega_T}\left(\frac{4a}{3}\overline{\vartheta^3} + \overline{\rho\int_1^{\vartheta}\frac{c_v(\xi)}{\xi}\,\mathrm{d}\xi}\right)\vartheta \,\mathrm{d}\mathbf{x}\mathrm{d}t.$$

Exactly as in Section 6, we obtain

$$\vartheta_\delta \to \vartheta \text{ in } L^2(\Omega_T).$$

Thanks to Lemma 5.4 in [12], ϑ is positive a.a. on Ω_T and $\ln\vartheta = \overline{\ln\vartheta}$.

In order to establish strong convergence of the sequence $\{\rho_\delta\}_{\delta>0}$, we pursue the approach similarly as in Sections 7.5,7.6 in [12]. The results on propagation of oscillations stated in Section 7.6 in [12] yield

$$\rho_\delta \to \rho \text{ in } L^1(\Omega_T),$$

which can be strengthened to

$$\rho_\delta \to \rho \text{ in } C([0,T]; L^1(\Omega))$$

(see [10] or Section 6.7 in [18]).

Consequently, the limit function $\rho, \mathbf{u}$ satisfy the continuity equation

$$\rho_t + \nabla\cdot(\rho\mathbf{u}) = 0 \text{ in } \mathcal{D}'(\mathbb{R}^3\times[0,T))$$

as well as its renormalized version (2.26).

Similarly, the momentum equation

$$(\rho\mathbf{u})_t + \nabla\cdot(\rho\mathbf{u}\otimes\mathbf{u}) + \nabla\left(p_e(\rho) + \vartheta p_\vartheta(\rho) + \vartheta^2 p_{\vartheta^2}(\rho) + \frac{a}{3}\vartheta^4\right)$$
$$= \nabla\cdot\mathbb{S} + \rho\nabla\Psi + (\nabla\times\mathbf{H})\times\mathbf{H}$$

is satisfied in $\mathcal{D}'(\Omega_T)$. One can handle the induction equation as well, i.e.,

$$\mathbf{H}_t - \nabla \times (\mathbf{u} \times \mathbf{H}) = -\nabla \times (\nu \nabla \times \mathbf{H})$$

is satisfied in $\mathcal{D}'(\Omega_T)$.

By the same token, we can pass to the limit in the energy equality (6.69) in order to obtain (6.69). Note that

$$\delta \rho_\delta^\Gamma \to 0 \ \text{ in } \ L^1(\Omega_T)$$

as a consequence of (7.75). Hence, the regularizing $\delta-$ dependent terms on the left-hand disappear. And, it is a routine matter to deal with the entropy inequality (6.70) based on the above estimates. We remark here that it is standard to pass to the limit in the production rate keeping the correct sense of the inequality as all terms are convex with respect to the spatial gradients of $\mathbf{u}, \vartheta$ and $\mathbf{H}$.

Last but not least, we need to show that the temperature ϑ tends to its prescribed initial distribution ϑ_0 for $t \to 0$. Since the total energy of the system is conserved, we have

$$\begin{aligned}
&E(t)\\
&= \int_\Omega \left(\rho P_e(\rho) - \rho\vartheta^2 P_{\vartheta^2}(\rho) + a\vartheta^4 + \rho Q(\vartheta) + \frac{1}{2}\rho|\mathbf{u}|^2 + \frac{1}{2}|\mathbf{H}|^2 + \frac{G}{2}\Delta^{-1}[1_\Omega \rho]\rho\right)\mathrm{d}\mathbf{x}\\
&= \int_\Omega \Big(\rho_0 P_e(\rho_0) - \rho_0\vartheta_0^2 P_{\vartheta^2}(\rho_0) + a\vartheta_0^4 + \rho_0 Q(\vartheta_0)\\
&\quad + \frac{1}{2\rho_0}|\mathbf{m}_0|^2 + \frac{1}{2}|\mathbf{H}_0|^2 - \frac{1}{2}\rho_0\Psi_0\Big)\mathrm{d}\mathbf{x}.
\end{aligned}$$

with $\Psi_0 = G(-\Delta)^{-1}[1_\Omega \rho_0]$.

Bearing in mind,

$$\begin{aligned}
\rho(t) &\rightharpoonup \rho_0 \ \text{ in } L^\gamma(\Omega) \ \text{ for } t \to 0,\\
\rho\mathbf{u}(t) &\rightharpoonup \mathbf{m}_0 \ \text{ in } L^{\frac{2\gamma}{\gamma+1}}(\Omega) \ \text{ for } t \to 0,\\
\mathbf{H}(t) &\rightharpoonup \mathbf{H}_0 \ \text{ in } L^2(\Omega) \ \text{ for } t \to 0,
\end{aligned}$$

using the same argument as Section 7.7 in [12] to derive

$$\begin{aligned}
&\operatorname*{ess\,lim}_{t\to 0+} \int_\Omega \left(\frac{4a}{3}\vartheta^3 + \rho\int_1^\vartheta \frac{c_v(\xi)}{\xi}\,\mathrm{d}\xi - \rho P_\vartheta(\rho) - 2\rho\vartheta P_{\vartheta^2}(\rho)\right)\phi\,\mathrm{d}\mathbf{x}\\
\geqslant &\int_\Omega \left(\frac{4a}{3}\vartheta_0^3 + \rho_0\int_1^{\vartheta_0} \frac{c_v(\xi)}{\xi}\,\mathrm{d}\xi - \rho_0 P_\vartheta(\rho_0) - 2\rho_0\vartheta_0 P_{\vartheta^2}(\rho_0)\right)\phi\,\mathrm{d}\mathbf{x}
\end{aligned}$$

for any $0 \leqslant \phi \in \mathcal{D}(\Omega;\mathbb{R})$. And,

$$\vartheta(t) \to \vartheta_0 \quad \text{in } L^4(\Omega) \quad \text{as } t \to 0+ .$$

Theorem 2.1 has been proved.

Acknowledgements X. Li's research was supported in part by the China Postdoctoral Science Foundation funded project under Grant No.:2013T60085, the National Natural Science Foundation of China (11271052).

References

[1] Batchelor G K. An introduction to fluid dynamics. Cambridge University Press, Cambridge, 1967.

[2] Buet C, Després B. Asymptotic analysis of fluid models for the coupling of radiation and hydrodynamics. Journal of Quantitative Spectroscopy and Radiative Transfert, 2004, 85: 385–418.

[3] Cabannes H. Theoretical Magnetofluiddynamics. Academic Press, New York, 1970.

[4] Chapman S, Cowling T. G. Mathematical theory of non-uniform gases. Cambridge University Press, Cambridge, 1990.

[5] Chen G -Q, Wang D. Global solutions of nonlinear magnetohydrodynamics with large initial data. J. Differential Equations, 2002, 182: 344–376.

[6] Cowling T G. Magnetohydrodynamics. Interscience Tracts on Physics and Astronomy, New York, 1957.

[7] Cox J P, Giuli R T. Principles of stellar structure, I, II. Gordon and Breach, New York, 1968.

[8] Davidson P A. Turbulence: An introduction for scientists and engineers. Oxford Univ. Prss, Oxford, 2004.

[9] Desjardins B. Regularity of weak solutions of the compressible isentropic Navier-Stokes equations. Comm. Partial Differential Equations, 1997, 22: 977–1008.

[10] DiPerna R J, Lions P -L. Ordinary differential equations, transport theory and Sobolev spaces. Invent. Math., 1989, 98: 511–547.

[11] Ducomet B. Simplified models of quantum fluids of nuclear physics. Mathematica Bohemica, 2001, 126: 323–336.

[12] Ducomet B, Feireisl E. On the dynamics of gaseous stars. Arch. Rational Mech. Anal., 2004, 174: 221–266.

[13] Ducomet B, Feireisl E. The equations of magnetohydrodynamics: on the interaction between matter and radiation in the evolution of gaseous. Commun. Math. Phys., 2006, 266: 595–629.

[14] Duvaut G, Lions J. L. Inéquations en thermoélasticité et magnétohydrodynamique. Arch. Rational Mech. Anal., 1972, 46: 241–279.

[15] Eliezer S, Hora A, Hora H. An introduction to equations of states, theory and applications. Cambridge: Cambridge University Press, 1986.

[16] Fan J, Yu W. Global variational solutions to the compressible magnetohydrodynamic equations. Nonlinear Analysis, 2008, 69: 3637–3660.

[17] Feireisl E. On compactness of solutions to the compressible isentropic Navier-Stokes equations when the density is not square integrable. Comment. Math. Univ. Carolin, 2001, 42: 83–98.

[18] Feireisl E. Dynamics of viscous compressible fluids. Oxford Lecture Series in Mathematics and its Applications, 26. Oxford University Press, Oxford, 2004.

[19] Feireisl E. On the motion of a viscous compressible, and heat conduting fluid. Indiana Univ. Math. J., 2004, 53: 1707–1740.

[20] Feireisl E, Novotný A. Singular limits in thermodynamics of viscous fluids. Advances in Mathematical Fluid Mechanics, Birkhäuser-Verlag, Basel, 2009.

[21] Feireisl E, Novotný A, Petzeltová H. On the existence of globally defined weak solutions to the Navier-Stokes equations of compressible isentropic fluids. J. Math. Fluid Dynamics, 2001, 3: 358–392.

[22] Gerbeau J F, Le Bris C. Existence of solution for a density-dependent magnetohydrodynamic equation. Differential Equations, 1997, 2: 427–452.

[23] Giovangigli V. Multicomponent flow modeling. Basel: Birkhäuser, 1999.

[24] Hu X, Wang D. Global solutions to the three-dimensional full compressible magnetohydrodynamic flows. Commun. Math. Phys., 2008, 283: 255–284.

[25] Hu X, Wang D. Global existence and large-time behavior of solutions to the three-dimensional equations of compressible magnetohydrodynamic flows. Arch. Rational Mech. Anal., 2010, 197: 203–238.

[26] Kawashima S, Okada M. Smooth global solutions for the one-dimensional equaitons in magnetohydrodynamics. Proc. Japan Acad. Ser. A, Math. Sci., 1982, 58: 384–387.

[27] Kazhikhov A V, Shelukhin V V. Unique global solution with respect to time of initial-boundary-value problems for one-dimensional equations of a viscous gas. J. Appl. Math., 1977, 41: 273–282.

[28] Koshelev A. Regularity problem for quasilinear elliptic and parabolic systems. Lecture notes im mathematics, 1614. Springer, 1995.

[29] Kulikovskiy A G, Lyubimov G A. Magnetohydrodynamics. Addison-Wesley, Reading, Massachusetts, 1965.

[30] Landau L D, Lifchitz E M. Electrodynamics of continuous media. 2ND ED. Pergamon, New York, 1984.

[31] Leray J. Sur le mouvement d'un liquide visqueux emplissant l'espace. Acta Math., 1934, 63: 193–248.

[32] Lions P L. Mathematical topics in fluid mechanics. Vol. 2. Compressible models. Oxford Lecture Series in Mathematics and its Applications, 10. Oxford Science Publications. The Clarendon Press, Oxford University Press, New York, 1998.

[33] Liu T -P, Zeng Y. Large time behavior of solutions for general quasilinear hyperbolic-parabolic systems of conservation laws. Memoirs Amer. Math. Soc. 599, 1997.

[34] Lunardi A. Analytic semigroups and optimal regularity in parabolic problems. Birkhäuser, Berlin, 1995.

[35] Matsumura A, Nishida T. The initial value problem for the equations of motion of viscous and heat-conductive gases. J. Math. Kyoto Univ., 1980, 20: 67–104.

[36] Matsumura A, Nishida T. Initial-boundary value problems for the equations of motion of compressible viscous and heat-conductive fluids. Comm. Math. Phys., 1983, 89: 445–464.

[37] Mihalas D, Weibel-Mihalas B. Foundations of radiation hydrodynamics. Oxford University Press, New York, 1984.

[38] Novotný A, Straśkraba I. Introduction to the mathematical theory of compressible flow. Oxford Lecture Series in Mathematics and its Applications, 27. Oxford University Press, Oxford, 2004.

[39] Schmidt P. On a magnetohydrodynamic problem of Euler type. J. Differential Equation, 1988, 74: 318–335.

[40] Schonbek M E, Schonbek T P, Süli E. Large-time behaviour of solutions to the magneto-hydrodynamics equations. Springer-Verlag, Math. Ann., 1996, 304: 717–759.

[41] Shore S N. An introduction to astrophusical hydrodynamics. Academic Press, New York, 1992.

[42] Vaigant V A, Kazhikhov A V. On the existence of global solutions to two-dimensional Navier-Stokes equations of a compressible viscous fluid (in Russian). Sibirskij Mat. Z., 1995, 36: 1283–1316.

[43] Zel'dovich Y B, Raizer Y P. Physics of shock waves and high-temperature hydrodynamic phenomena. Academic Press, New York, 1966.

[44] Zhang J, Xie F. Global solution for a one-dimensional model problem in thermally radiative magnetohydrodynamics. J. Differential Equation, 2008, 245: 1853–1882.

大气、海洋动力学中一些非线性偏微分方程的研究*

郭柏灵　黄代文　黄春研

摘　要　本文简要综述大气、海洋动力学中一些非线性偏微分及其无穷维动力系统的研究进展，其中包括我们近年来取得的一些研究成果. 首先，我们介绍（无耗散的和耗散的）表面准地砖方程近二十来年的数学理论的研究成果. 接着，我们综述了描述大尺度大气、海洋运动的大气、海洋原始方程组的一些定性理论的研究成果，含适定性、解的长时间行为和整体吸引子的存在性等. 最后，我们介绍大气、海洋随机动力学的一些数学模型的研究结果，主要是带随机力的海洋原始方程组的适定性及其对应的无穷维随机动力系统的全局吸引子的存在性的结果.

关键词　表面准地砖方程 (the surface quasi-geostrophic equations); 原始方程组 (the primitive equations); 随机偏微分方程; 整体适定性; 全局吸引子

1　引言

由于大气是特殊的可压流体, 海洋是特殊的不可压流体, 人们可以应用动力学方法研究支配大气、海洋运动的非线性偏微分方程组 (见文献 [1-5]), 从而理解天气预报和气候变化机制. 气象学的先驱之一 V. Bjerknes 曾经说:“天气预报可以被看作数学物理中一个初边值问题”, 参见文献 [6]. 也就是说, 人们可以建立以数学物理方法为基础的数值天气预报.

支配大气、海洋运动的数学方程组主要由下面方程组成：质量、动量、能量守恒方程和状态方程 (还有其他方程, 对于大气是水汽守恒方程, 而对于海洋是盐度守恒方程). 大气的质量和动量守恒方程从本质上看是可压的 (对于海洋, 取 Boussinesq 近似, 为不可压的)Navier- Stokes 方程. 因为大气、海洋运动的基本方程组含有太多复杂的信息 (例如, 空间尺度从 10 km 的小尺度, 100 km 的中尺度一直到 1000 km 的大尺度都有), 所以人们在可预见的未来可能无法从数值和理论上彻底地解决它们. 为此, 人

* 中国科学：物理学、力学、天文学, 非线性专辑，2014, 44(12), 1275-1285.

们必须略去一些中小尺度的因素, 合理地简化大气、海洋运动的基本方程组, 从而才能实现数值天气预报. 由于全球大气 (海洋) 的垂直方向上的尺度比水平方向上的尺度小得多, 最自然的简化方法就是取静力近似, 即把竖直方向上的动量守恒方程用流体静力平衡方程 (它满足大尺度大气、海洋的观测数据) 来代替. 1922 年, Richardson 为了实现数值天气预报首次引进了大气原始方程组和海洋原始方程组. 它们包括带科氏力的流体力学方程组、热动力学方程和状态方程等, 参见文献 [4]. 当时, 因为这些方程组还是比较复杂, 又因为计算能力还较为落后, 所以人们仍然无法从理论和数值上解决它, 也就无法真正实现数值天气预报的梦想.

为了克服这一困难, 人们后来引进了一些简化数值模式, 比如 Charney, Fjortaft 和 Neumann 提出的正压模式 (见文献 [7]), 和 Charney 和 Philips 在文献 [8] 中引进的准地转模式. 1950 年, Charney, Fjortaft 和 Neumann 利用他们的二维正压模式 (即球面上带科里奥利力的 Euler 方程) 在普林斯顿的高等研究所的 ENIAC 计算机上成功地实现数值天气预报, 这被称为天气预报的一场重大革命. 随后, 作为大气科学发展的一个重要成果, 数值天气预报已在世界各国得到了广泛的应用.

20 世纪 50 年代后, 气象科学有了蓬勃发展. 同时, 一些近代科学技术的成就大大地促进了数值天气预报的快速发展: 利用空间科学技术 (如雷达) 和气象卫星等现代化工具, 人们能够及时地获得全球大气从底到高的气象资料; 大型计算机提供了加工处理资料和进行数值计算的强有力工具. 因此, 研究人员又开始转向利用大气原始方程组和海洋原始方程组来预报天气和气候. 大气原始方程组和海洋原始方程组分别是美国的大气研究国家中心 (NCAR) 使用的全球大气环流模式 (AGCM) 和全球海洋环流模式 (OGCM) 的核心组成部分. 许多数值天气预报的模式都是基于大气原始方程组和海洋原始方程组, 见文献 [9-16] 和其中的引文. 数值天气预报的物质条件方面已经比较成熟了, 理论研究的重要性就越来越突出, 这方面的文章见 [5,17,18]. 人们有必要深入分析各种各样的大气、海洋动力学模式, 用合理的数学方法来研究这些模式, 建立它们的定性理论 (定性理论在数值预报的可靠性分析中就发挥着至关重要的作用). 从而, 人们可以选择较为准确的数值预报模式.

起初, 人们主要研究简化二维和三维大气、海洋的数学模型, 包括二维和三维准地砖方程. 这些模式着重于突出研究大气海洋的某些特征. 由于突出重点, 数学理论比较成熟, 问题可以研究得更为透彻, 比如说带黏性的二维和三维准地砖方程的适定性已经基本解决, 见文献 [19-24]. 20 世纪 80 年代起, 二维和三维准地砖方程已经成为数学研究的一个主题 (人们可以研究对应问题的适定性、波的非线性稳定性, 也可以研究各种各样的简化模型的数学合理性), 相关研究结果参见文献 [25-30]. 穆穆院士在准地砖方程研究方面做了许多重要的、很有理论意义和实际应用价值的工作. 为了描述一致位涡场 ($q = 0$) 的下表面的位温 (或浮力) 的演化, 人们得到了表面准地砖方程, 参

见文献 [31]. Constantin 等在文献 [32] 中研究了表面准地砖方程, 揭示了该方程与三维不可压缩的 Euler 方程有一些重要的相似之处, 而且通过计算猜测表面准地砖方程可能出现奇性解. 后来表面准地砖方程的适定性问题已经成为非线性偏微分方程研究的一个主题, 相关的研究工作可以参见文献 [33-42] 和其中的引文. 20 世纪 90 年代初, 许多数学家 (例如 J. L. Lions, R. Temam 和 S. Wang) 开始从数学上研究大气、海洋和耦合的大气海洋原始方程组 (见文献 [17,18,43-46,49] 和其中的引文). 近十年来, 人们也开始重视随机大气、海洋动力学中一些数学模型的理论和数值研究.

这里, 我们主要介绍大气、海洋动力学中表面准地砖方程、大气海洋原始方程组和一些随机非线性偏微分方程的研究的进展. 本文的安排如下：首先, 介绍表面准地砖方程的一些研究进展；其次, 给出大气、海洋原始方程组的定性理论的一些研究成果；最后, 介绍大气、海洋动力学的一些随机数学模型的研究结果.

2 表面准地砖方程

2.1 无耗散的表面准地砖方程

下面, 我们首先给出表面准地砖方程的推导. 把在静力近似和 Boussinesq 近似下的绝热无摩擦的大气或海洋基本方程组关于小 Rossby 数展开, 可以得到一阶近似

$$\partial_t \xi + J(\psi, \xi) = f \partial_z w,$$

$$\partial_t \theta + J(\psi, \theta) = -N^2 w,$$

其中 $\xi = \psi_{xx} + \psi_{yy}$, ψ 为流函数, $J(\psi, \xi) = \psi_x \xi_y - \psi_y \xi_x$, w 为垂直方向上的速度, $\theta = f\psi_z$, 对于大气而言, θ 是位温 (它是将空气微团通过绝热过程移动到压强为 1 个标准大气压处的温度), 对于海洋而言, θ 是浮力, f 为科氏参数, N 为浮力频率, 为了简单起见, 假设 f 和 N 均为常数. 消去前面两个方程中的 w, 关于 z 做一个变换, 得到

$$\partial_t q + J(\psi, q) = 0, \tag{2.1}$$

其中 $q = (\partial_{xx} + \partial_{yy} + \partial_{zz})\psi$, q 表示位涡. 如果边界上没有热交换, 也不考虑边界上的摩擦, 底面为平面, 方程 (2.1) 的下边界 ($z = 0$) 条件可以写为 $\dfrac{\mathrm{d}}{\mathrm{d}t}(\dfrac{\partial \psi}{\partial z}) = 0$, 即

$$\partial_t \theta + J(\psi, \theta) = 0, \tag{2.2}$$

这里 $\theta(x, y) = \partial_z \psi|_{z=0}$. (2.1) 和 (2.2) 的一个特殊例子是如下的系统:

$$(\partial_{xx} + \partial_{yy} + \partial_{zz})\psi = 0, \quad z > 0, \tag{2.3}$$

$$\partial_t\theta + J(\psi,\theta) = 0, \quad z = 0, \tag{2.4}$$

$$\psi \to 0, \quad z \to \infty. \tag{2.5}$$

系统 (2.3)–(2.5) 描述一致位涡场 ($q = 0$) 的下表面的位温 (或浮力) 的演化, 它可以用于模拟冷暖气团交界处的锋生 (frontogenesis). 关于系统 (2.3)–(2.5) 的应用背景的详细描述可参见文献 [31].

如果 $\theta(x, y)$ 是具体给定的, 那么通过解椭圆方程的边值问题

$$(\partial_{xx} + \partial_{yy} + \partial_{zz})\psi = 0, \quad z > 0,$$

$$\partial_z\psi|_{z=0} = \theta(x, y),$$

$$\psi \to 0, \quad z \to \infty,$$

就可以得到 ψ 在 $z > 0$ 中的分布. 因此, 为了研究系统 (2.3)–(2.5), 可以考虑下面方程的 Cauchy 问题: $\partial_t\theta + J(\psi,\theta) = 0$, 即

$$\partial_t\theta + u\cdot\nabla\theta = 0, \quad 在 \ \mathbb{R}^2 \ 中, \tag{2.6}$$

其中

$$-\theta = (-\Delta)^{\frac{1}{2}}\psi, \tag{2.7}$$

$$u = (u_1, u_2) = (-\psi_y, \psi_x), \tag{2.8}$$

$\mathbb{R}^2$ 可以换为二维的换面 $\mathbb{T}^2$, 非局部算子 $(-\Delta)^{\frac{1}{2}}$ 是通过如下的 Fourier 变换定义的,

$$\widehat{(-\Delta)^{\frac{1}{2}}\psi}(\xi) = |\xi|\hat{\psi}(\xi), \tag{2.9}$$

这里 $\hat{\psi}(\xi) = \dfrac{1}{(2\pi)^2}\displaystyle\int_{\mathbb{R}^2} e^{-i\eta\cdot\xi}\psi(\eta)\mathrm{d}\eta$. (2.8) 可以重写为

$$u = (\partial_y(-\Delta)^{-\frac{1}{2}}\theta, -\partial_x(-\Delta)^{-\frac{1}{2}}\theta) = (-R_2\theta, R_1\theta), \tag{2.10}$$

其中 R_1, R_2 表示 Riesz 变换, $\widehat{R_j f}(\xi) = -\dfrac{i\xi_j}{|\xi|}\hat{f}(\xi)$. 方程 (2.6) 称为**表面准地砖方程**(Surface quasi-geostrophic equation).

人们研究方程 (2.6) 的另一个重要原因是: (2.6) 与三维不可压 Euler 方程有一些重要的相似之处. 事实上, 方程 (2.6) 关于 y 和 x 求导, 得到

$$\frac{\partial\nabla^{\perp}\theta}{\partial t} + (u\cdot\nabla)\nabla^{\perp}\theta = (\nabla u)\nabla^{\perp}\theta. \tag{2.11}$$

其中 $\nabla^{\perp}\theta=\begin{pmatrix}-\theta_y\\ \theta_x\end{pmatrix}$, $(\nabla u)=\begin{pmatrix}u_{1x} & u_{1y}\\ u_{2x} & u_{2y}\end{pmatrix}$, 而三维不可压缩的 Euler 方程的涡度形式为

$$\frac{\partial \omega}{\partial t}+(v\cdot\nabla)\omega=(\nabla v)\omega, \tag{2.12}$$

其中 $\omega=\mathrm{curl}v$, $v=(v_1,v_2,v_3)$, $(\nabla v)=(\partial_j v_i),\partial_j=\partial_x,\partial_y,\partial_z,i=1,2,3$. (2.11) 和 (2.12) 具有一种相似的结构: 旋涡的拉伸 (vortex stretching).

在文献 [32] 中, Constantin 等通过数学理论和数值实验的结合研究了无耗散的表面准地砖方程的 Cauchy 问题, 即 (2.6) 带初始条件:

$$\theta(t,x,y)|_{t=0}=\theta_0(x,y). \tag{2.13}$$

首先, 他们得到 (2.6) 和 (2.13) 的光滑解的局部存在性. 同时, 他们也得到了无耗散的表面准地砖方程的 Cauchy 问题的爆破 (Blow up) 准则.

虽然无耗散的表面准地砖方程的 Cauchy 问题和周期问题都存在局部光滑解, 但是他们是否存在有限时间爆破解仍然是一个公开问题. 为了解决这一问题, 人们试图通过数值模拟寻找有限时间爆破解. 在文献 [32] 中, Constantin 等选取

$$\theta_0=\sin x\sin y+\cos y, \tag{2.14}$$

通过数值模拟得到了无耗散的表面准地砖方程的周期问题一个形成强锋面的解, 由此猜测该解可能产生有限时间的爆破. 而且, 他们受到数值模拟结果的启发, 研究了产生爆破的位温张量的水平集的拓扑特征. 文献 [50, 51] 和文献 [34] 分别给出了表面准地砖方程以 (2.14) 的 θ_0 为初始数据的解不会在有限时间爆破的数值和理论证据.

2.2 耗散的表面准地砖方程

带耗散的表面准地砖方程为

$$\partial_t\theta+u\cdot\nabla\theta+\kappa(-\Delta)^{-\alpha}\theta=0,\ \text{在}\ \Omega\ \text{中}, \tag{2.15}$$

其中 $-\theta=(-\Delta)^{\frac12}\psi$, $u=(-\psi_y,\psi_x)=(-R_2\theta,R_1\theta)$, $\kappa>0$, 非局部算子 $(-\Delta)^{\alpha}(1\geqslant\alpha\geqslant0)$ 是通过如下的 Fourier 变换定义的: $\widehat{(-\Delta)^{\alpha}}\theta(\xi)=|\xi|^{2\alpha}\hat{\theta}(\xi)$, R_j 是 Riesz 变换, Ω 可以取 $\mathbb{R}^2$ 或二维环面 $\mathbb{T}^2$.

当 $\alpha=\dfrac12$ 时, 方程 (2.15) 由下面的系统导出,

$$(\partial_{xx}+\partial_{yy}+\partial_{zz})\psi=0,z>0,$$

$$\frac{\mathrm{d}\frac{\partial\psi}{\partial z}}{\mathrm{d}t}=\frac{\mathrm{d}\theta}{\mathrm{d}t}=\frac{\partial\theta}{\partial t}+J(\psi,\theta)=-\frac{E_v^{\frac12}}{2\varepsilon}(\partial_{xx}+\partial_{yy})\psi, \tag{2.16}$$

$$z = 0, \psi \to 0, z \to \infty,$$

其中 (2.16) 中第二个式子是文献 [52] 中的方程 (6.6.10)(当 $\eta_B = 0, H = 0$ 时, 即下边界是平面, 且边界上没有热变换). E_v 是 Ekman 数, ε 是 Rossby 数, $-\frac{E_v^{\frac{1}{2}}}{2\varepsilon}(\partial_{xx} + \partial_{yy})\psi$ 表示摩擦.

下面, 我们介绍关于带耗散的表面准地砖方程的适定性的研究成果. Resnick 在文献 [53] 中得到了带外力和耗散的表面准地砖方程的弱解的整体存在性. Constantin 和 Wu 在文献 [35] 中得到当 $\frac{1}{2} < \alpha \leqslant 1$ 时, 带耗散的表面准地砖方程的光滑解的整体存在性.

注记 2.1 当 $1 \geqslant \alpha > \frac{1}{2}$ 时, 带耗散的表面准地砖方程 (2.15) 的光滑解整体存在. 人们把当 $\alpha > \frac{1}{2}$ 时的 (2.15) 称为**带次临界耗散的表面准地砖方程**; 把当 $\alpha = \frac{1}{2}$ 时的 (2.15) 称为**带临界耗散的表面准地砖方程**; 把当 $0 \leqslant \alpha < \frac{1}{2}$ 时的 (2.15) 称为**带超临界耗散的表面准地砖方程**.

带临界耗散的表面准地砖方程的整体适定性问题曾经是一个公开问题. 这里我们介绍一下关于这个问题的几个结果:

1. Constantin 等在文献 [36] 中得到了在 $\|\theta_0\|_{L^\infty}$ 充分小的条件下的光滑解的整体存在性, 他们主要的定理如下:

定理 2.1 如果 $\theta_0 \in H^2(\Omega)$, $\Omega = [0, 2\pi]^2$, θ_0 满足周期的边界条件, 而且存在正的常数 C_∞, 使得 $\|\theta_0\|_{L^\infty} \leqslant C_\infty k$, 则带临界耗散的表面准地砖方程的初值问题

$$\begin{cases} \theta_t + u \cdot \nabla\theta + \kappa(-\Delta)^{\frac{1}{2}}\theta = 0, & \text{in } \Omega, \\ u = (u_1, u_2) = (-R_2\theta, R_1\theta), & \text{in } \Omega, \\ \theta(x, 0) = \theta_0(x) \end{cases} \tag{2.17}$$

的解整体存在, 而且是唯一的, 对于 $\forall t \geqslant 0$, $\|\theta(\cdot, t)\|_{H^2} \leqslant \|\theta_0\|_{H^2}$.

2. 最近, 有两个关于带临界耗散的表面准地砖方程的整体适定性的重要成果:

(1) Kiselev 等在文献 [40] 中利用非局部最大值原理构造适当的连续模, 从而证明带临界耗散的表面准地砖方程的周期问题的整体适定性, 即

定理 2.2 带临界耗散的表面准地砖方程的周期问题: $\theta_t + u \cdot \nabla\theta + \kappa(-\Delta)^{\frac{1}{2}}\theta = 0$, $\theta(x, y, 0) = \theta_0$, θ_0 是光滑且周期的, 则该问题整体存在唯一且光滑的解, $\|\nabla\theta\|_\infty \leqslant C\|\nabla\theta_0\|_\infty \exp\exp c\|\theta_0\|_\infty$.

(2) 在文献 [42] 中, Caffarelli 和 Vasseur 应用 De Giorgi Nash Moser 方法和调和延拓证明了弱解的正则性, 从而得到了带临界耗散的表面准地砖方程的 Cauchy 问题的整体适定性.

带超临界耗散的表面准地砖方程的整体适定性问题至今仍然是一个公开问题. 人们得到了该方程的 Cauchy 问题或周期问题在 Sobolev 空间和 Besov 空间的小初值解的整体存在性或大初值解的局部存在性, 这些结果可以参见文献 [54-59].

3 大气、海洋原始方程组

在这一节中, 我们介绍大气、海洋原始方程组的定性理论方面的研究进展, 主要包括整体适定性以及大气、海洋无穷维动力系统吸引子的存在性的结果.

人们利用大气、海洋原始方程组进行数值天气预报, 首先关心的是这些方程组在数学上是否具有内在的逻辑统一性, 即适定性. 所以, 本节的主要任务之一就是介绍这些方程组的适定性方面的一些结果. 如果大气、海洋原始方程组的初边值问题解存在而且唯一, 那么就存在与这些方程组对应的无穷维动力系统. 为了理解大气、海洋无穷维动力系统的长时间行为 (这是大气、海洋无穷维动力系统的主要研究内容), 人们需要研究这些无穷维动力系统的吸引子的存在性和维数. 如果大气、海洋原始方程组对应的无穷维动力系统的全局吸引子存在, 而且是有限维的, 也就是说这些大气、海洋无穷维系统的解集收缩到有限维的流形上, 那么人们可以采用低阶谱截断方法将复杂的大气、海洋偏微分方程组化为常微分方程组, 这样问题变得简化多了. 研究大气、海洋吸引子具有重要的实际意义, 例如, 它有助于阐明气候系统的适应和演变过程, 有助于人们提出长期预报模式设计的准则等, 详细内容见参考文献 [17,18,60-65].

早在 1979 年, 曾庆存院士就在文献 [5] 中应用 Galerkin 方法讨论了不带黏性的大气原始方程组的适定性问题, 得到了弱解的存在性. 20 世纪 90 年代初, 许多数学家开始从数学上研究大气、海洋原始方程组和耦合的大气海洋原始方程组. 在文献 [43] 中, 通过引入黏性和一些技术处理, Lions, Temam 和 Wang 得到了干大气原始方程组的新表示. 在气压坐标系下, 这一原始方程组的新表述与不可压流体的 Navier-Stokes 方程类似 (当然还是有些不同, 主要是: Navier-Stokes 方程的非线性项是 $(u\cdot\nabla)u$, 而大气原始方程组的新表示的非线性项含有 $\left(\int_\xi^1 \mathrm{div} v \mathrm{d}\xi'\right)\dfrac{\partial v}{\partial \xi}$, u 是 Navier-Stokes 方程的三维速度场, v 是大气中水平方向上的速度场). 利用解决 Navier-Stokes 方程的方法 (即通常称的 Leray-Hopf 方法), 他们得到了干大气原始方程组新表述的初边值问题整体弱解的存在性. 而且, 在带垂直黏性的大气原始方程组的初边值问题存在整体强解的假设下, 他们做出了大气全局吸引子的 Hausdorff 和分形维数的估计. 在文献 [44,46] 中, 他们分别建立了海洋原始方程组和文献 [45] 引入的耦合大气–海洋模型的一些数学理论 (他们主要解决了整体弱解的存在性, 在强解存在的假设下研究了全局吸引子的 Hausdorff 和分形维数的估计). 在文献 [47,48] 中, 丑纪范院士和李建平教授研究了干和湿大气基本方程组解的渐近行为. 2008 年, 李建平教授和汪守宏教授对大气和海洋非线性无穷维动力系统的研究做了一个比较系统的回顾, 参见文献 [66].

近十年来, 一些数学家开始考虑大气、海洋三维黏性原始方程组的强解的整体存在性、唯一性和关于初始值的连续依赖性 (见文献 [49,67-72] 和其中的引文). 在文献 [70] 中, Guillén-González 等巧妙地利用各向异性估计来处理非线性项 $\left(\int_\xi^1 \mathrm{div} v \mathrm{d}\xi'\right)\frac{\partial v}{\partial \xi}$, 从而在初始数据充分小的假设下得到了海洋原始方程组的强解的整体存在性, 而且还证明了对所有初始数据 (没有充分小的假设) 强解的局部存在性. 在文献 [49] 中, Temam 和 Ziane 研究了大气、海洋和耦合的大气–海洋原始方程组的强解的局部存在性. 文献 [67-69] 主要用于考虑无量刚 Boussinesq 方程组及其修正模式 (这些模式可见文献 [3, 72]). 在文献 [67] 中, Cao 和 Titi 考虑了三维全球地转模式的整体适定性和有限维全局吸引子的存在性. 文献 [69] 的作者考虑了如下柱形区域中海洋三维黏性原始方程组强解的整体适定性,

$$\begin{aligned}
&\frac{\partial v}{\partial t} + (v\cdot\nabla)v + w\frac{\partial v}{\partial z} + fk\times v + \nabla p - \frac{1}{Re_1}\Delta v - \frac{1}{Re_2}\frac{\partial^2 v}{\partial z^2} = 0,\\
&\frac{\partial p}{\partial z} + T = 0,\\
&\mathrm{div} v + \frac{\partial w}{\partial z} = 0,\\
&\frac{\partial T}{\partial t} + v\cdot\nabla T + w\frac{\partial T}{\partial z} - \frac{1}{Rt_1}\Delta T - \frac{1}{Rt_2}\frac{\partial^2 T}{\partial z^2} = Q,\\
&\frac{\partial v}{\partial z} = 0,\ w = 0,\ \frac{\partial T}{\partial z} = -\alpha_s T \qquad \text{在 } M\times\{0\} = \Gamma_u \text{ 上},\\
&\frac{\partial v}{\partial z} = 0,\ w = 0,\ \frac{\partial T}{\partial z} = 0 \qquad \text{在 } M\times\{-1\} = \Gamma_b \text{ 上},\\
&v\cdot\vec{n} = 0,\ \frac{\partial v}{\partial \vec{n}}\times\vec{n} = 0,\ \frac{\partial T}{\partial \vec{n}} = 0 \qquad \text{在 } \partial M\times[-1,0] = \Gamma_l \text{ 上},
\end{aligned}$$

其中未知函数是 v, w, p, T, $v = (v^{(1)}, v^{(2)})$ 是水平方向上的速度, w 是垂直方向上的速度, p 是压力, T 是温度, $f = f_0(\beta + y)$ 是 Coriolis 参数, k 是垂直方向上的单位向量, Re_1, Re_2 是 Reynolds 数, Rt_1, Rt_2 是水平和垂直方向上热扩散系数, Q 是给定的 Ω 上的函数, Ω 是柱形区域, $\Omega = M\times(-1,0)$, M 是 $\mathbb{R}^2$ 中边界光滑的区域. Cao 和 Titi 在文献 [69] 中利用静力近似, 把水平方向上的速度场 v 分解为正压流 $\bar{v}$ $\left(\bar{v} = \int_0^1 v\mathrm{d}\xi\right)$ 和斜压流 $\tilde{v}$ $(\tilde{v} = v - \bar{v})$, 再经过一些复杂的计算得到了斜压流 $\tilde{v}$ 的 L^6-范数关于时间局部一致的有界性. 利用斜压流 $\tilde{v}$ 的 L^6-范数关于时间一致的有界性,

Cao 和 Titi 完整地证明了海洋三维黏性原始方程组的强解的整体存在性、唯一性和关于初始值的连续依赖性. 从而, 人们从某种意义上可以说海洋三维黏性原始方程组比不可压 Navier-Stokes 方程来得简单, 这是与物理的观点一致的, 因为海洋的原始方程组是经过取静力近似而得到的.

2006 年, Guo 和 Huang 在文献 [73] 中研究了如下气压坐标系下的湿大气原始方程组

$$\frac{\partial v}{\partial t}+\nabla_v v+W(v)\frac{\partial v}{\partial \xi}+\frac{f}{R_0}k\times v+\int_\xi^1 \frac{bP}{p}\mathrm{grad}[(1+cq)T]\,\mathrm{d}\xi' +\mathrm{grad}\Phi_s-\Delta v-\frac{\partial^2 v}{\partial \xi^2}=0,$$

$$\frac{\partial T}{\partial t}+\nabla_v T+W(v)\frac{\partial T}{\partial \xi}-\frac{bP}{p}(1+cq)W(v)-\Delta T-\frac{\partial^2 T}{\partial \xi^2}=Q_1,$$

$$\frac{\partial q}{\partial t}+\nabla_v q+W(v)\frac{\partial q}{\partial \xi}-\Delta q-\frac{\partial^2 q}{\partial \xi^2}=Q_2,$$

$$\int_0^1 \mathrm{div}v\ \mathrm{d}\xi=0,$$

其中未知函数是 v, ω, Φ, T, q, $v=(v_\theta,v_\varphi)$ 是水平方向上的速度, $\omega=\frac{\mathrm{d}p}{\mathrm{d}t}=W(v)$ 是气压 $p-$ 坐标系下垂直方向上的速度, $\Phi=gz$ 是地势, T 是温度, $q=\frac{\rho_1}{\rho}$ 是空气中水汽的混合比, ρ_1 是空气中水汽的密度, $f=2\cos\theta$ 是科氏参数, k 是垂直方向上的单位向量, $\overline{T}$ 是参考的空气温度, μ_i,ν_i,c 是正的常数 ($i=1,2,3,c\approx 0.618$), Q_1 是热源, Q_2 是水汽的源, $W(v)(t;\theta,\varphi,\xi)=\int_\xi^1 \mathrm{div}v(t;\theta,\varphi,\xi')\ d\xi'$. Guo 和 Huang 证明了湿大气原始方程组弱解的整体存在性, 并利用上述方程组关于时间平移不变的特性, 在弱解的基础上得到了关于时间平移半群的大气吸引子 (弱意义下的大气吸引子). 2008 年, 利用流体静力近似, Guo 和 Huang 在文献 [74] 得到了干大气原始方程组强解的整体存在性、唯一性, 同时也得到了其对应无穷维动力系统关于解半群的全局吸引子存在性; 2009 年, 通过精细的能量估计, 我们在文献 [75] 证明了干大气原始方程组光滑解的整体存在性, 从而比较完整地解决了三维干大气原始方程组的定性理论. 在文献 [76], 我们解决了湿大气原始方程组的整体适定性问题, 也研究了强解的长时间行为, 从而得到了湿大气的存在性.

近几年来, 人们研究了带其他边界条件的大气、海洋原始方程组的整体适定性, 例如文献 [77, 78], 同时也研究了带部分耗散的大气、海洋原始方程组的定性理论, 参见文献 [79].

4 大气、海洋随机动力学中的一些数学模型

在长期天气预报和气候预测中, 随机因素的作用是至关重要的, 把大气过程看作随机过程比较合理, 此时, 人们只能预报大气相关物理量的期望、方差等. 类似地, 在长期海洋预报中, 可以把海洋过程看作随机过程. 为了更加客观地预测气候变化, 1975 年后, 人们提出了一些随机气候模式, 建立了描述气候随机变化的 Langevin 方程及相应的 Fokker-Planck 方程, 这方面的工作可以参见文献 [80-82]. 1980 后, 人们建立了简化随机气候模式, 揭示了随机力对气候系统变化的影响, 可以参见文献 [83-86].

近十年来, 人们又开始重视随机气候模式的研究. Majda 及其合作者从数学上对随机气候模式做了大量理论和数值计算方面的工作, 取得了许多重要的成果, 参见文献 [87-92]. 周秀骥院士在文献 [93] 中指出：起源于分子热运动的宏观微尺度随机力是大气本身固有的属性；太阳辐射作为决定大气运动与变化的主要因子, 它的变化具有随机性, 是大气随机强迫因子, 它对气候变化具有决定性影响；地–气相互作用是一个时变的非线性相互反馈的耦合过程, 形成了下边界对大气复杂的随机强迫作用. 所以, 人们在长期天气预报和气候预测中有必要考虑随机力的因素、太阳辐射和地表的随机强迫.

下面, 主要介绍大气、海洋随机动力学中的一些数学模型 (主要包括随机准地砖方程、随机大气海洋原始方程组) 的定性理论的研究进展. 2001 年起, 段金桥教授等就在文献 [94-96] 中考虑了大气、海洋动力学中一些随机偏微分方程 (如随机的二维和三维准地砖方程) 的定性理论. 2008 年, Guo、Huang 和 Han 在文献 [97] 中考虑带随机外力、摩擦和耗散的二维准地砖方程,

$$\left(\frac{\partial}{\partial t}+\frac{\partial\psi}{\partial x}\frac{\partial}{\partial y}-\frac{\partial\psi}{\partial y}\frac{\partial}{\partial x}\right)(\Delta\psi-F\psi+\beta_0 y)$$

$$=\frac{1}{R_e}\Delta^2\psi-\frac{r}{2}\Delta\psi+f(x,y,t),$$

其中 F 是 Froude 数, β_0 是 Rossby 参数, R_e 为 Reynolds 数, r 是 Ekman 耗散常数. $f(x,y,t)=-\dfrac{\mathrm{d}W}{\mathrm{d}t}$ 是随机力, 随机过程 W 关于时间是双边的 Wiener 过程, 它的形式为

$$W(t)=\sum_{i=1}^{+\infty}\mu_i\omega_i(t)e_i,$$

其中 $\omega_1,\omega_2,\cdots$ 是完备概率空间 $(\Omega,\mathcal{F},P)$ (它的期望记为 E) 中独立布朗运动系列, μ_i 满足

$$\sum_{j=1}^{+\infty}\frac{\mu_i^2}{\lambda_i^{\frac{1}{2}-2\beta_1}}<+\infty,\ 对某一\ \beta_1>0.$$

我们证明了上述随机准地砖方程对应的动力系统的随机吸引子的存在性. 后来, 文献LLXH研究了无界带形区域中随机准地砖方程, 得到了同样的结果. 关于大气、海洋科学中一些二维的随机数学模型的研究, 可以参见文献 [99,100] 及其中的引文.

我们研究了带随机外力的海洋原始方程组的适定性及其对应的无穷维随机动力系统的全局吸引子的存在性, 我们的结论适用于带随机外力且受到太阳辐射的随机强迫的大气原始方程组; 在 4.3 节, 我们研究了带随机边界的海洋原始方程组的定性理论, 随机边界表示大气对海洋的随机强迫作用, 这是符合把大气运动看作随机过程的.

2009 年, Guo 和 Huang 在文献 [101] 中研究了带随机力的海洋三维原始方程组:

$$
\begin{aligned}
&\frac{\partial v}{\partial t}+(v\cdot\nabla)v+\Phi(v)\frac{\partial v}{\partial z}+fk\times v+\nabla p_b-\int_{-1}^{z}\nabla T\mathrm{d}z' \quad -\Delta v-\frac{\partial^2 v}{\partial z^2}=\Psi,\\
&\frac{\partial T}{\partial t}+(v\cdot\nabla)T+\Phi(v)\frac{\partial T}{\partial z}-\Delta T-\frac{\partial^2 T}{\partial z^2}=Q,\\
&\int_{-1}^{0}\nabla\cdot v\,\mathrm{d}z=0.\\
&\frac{\partial v}{\partial z}=\tau,\ \frac{\partial T}{\partial z}=-\alpha_u T \qquad \text{在 } \Gamma_u \text{ 上},\\
&\frac{\partial v}{\partial z}=0,\ \ \frac{\partial T}{\partial z}=0 \qquad \text{在 } \Gamma_b \text{ 上},\\
&v\cdot\boldsymbol{n}=0,\ \frac{\partial v}{\partial \vec{n}}\times\boldsymbol{n}=0,\ \frac{\partial T}{\partial \boldsymbol{n}}=0 \qquad \text{在 } \Gamma_l \text{上},
\end{aligned}
$$

其中 $\Psi=\dfrac{\mathrm{d}W}{\mathrm{d}t}$, W 是关于时间双边的 Wiener 过程, 我们得到了该随机偏微分方程组的整体适定性, 及其对应的无穷维随机动力系统的随机吸引子存在性, 该结果也适用于带加性白噪声型随机边界的三维海洋原始方程组, 参见文献 [102]. 上述两个结果适用于该节的结论适用于带随机外力和受到太阳辐射的随机强迫的大气原始方程组. 后来, Gao 和 Sun 在文献 [103, 104] 中研究了带加性白噪声的三维大气、海洋原始方程组的不变测度的存在性和随机吸引子的维数估计, 改进了文献 [101] 的结果. Debussche 等也研究了一些带乘性白噪声的三维大气、海洋原始方程组的定性理论, 参见文献 [105].

5 小结

本文主要介绍了大气、海洋动力学中一些非线性偏微分方程及其 (随机) 无穷维动力系统的研究进展, 其中包括我们取得的一些研究成果. 这里我们要特别感谢曾庆存院士、穆穆院士、刘式适教授、李建平教授等许多专家学者的大力支持、指导和帮助. 我们希望本文有助于人们了解大气、海洋动力学的一些数学模型的研究成果和尚未解决的关键问题, 希望更多大气、海洋动力学中数学模型的定性理论被建立. 这些理论将在数值预报的可靠性分析中发挥着重要的作用, 也对设计更好的大气、海洋数值预报模式提供理论参考.

致谢 我们非常感谢审稿人的宝贵意见和建议.

参考文献

[1] Gill A E. Atmosphere-ocean Dynamics. New York: Academic Press, 1982.

[2] Holton J R. An Introduction to Dynamic Meteorology. 3rd Edition. New York: Academic Press, 1992.

[3] Pedlosky J. Geophysical Fluid Dynamics. 2nd Edition, Berlin, New York: Springer-Verlag, 1987.

[4] Richardson L F. Weather Prediction by Numerical Press. Cambridge: Cambridge University Press, 1922.

[5] Zeng Q. Mathematical and Physical Foundations of Numerical Weather Prediction. Science Press, Beijing, 1979, in Chinese.

[6] Bjerknes V. Das Problem von der Wettervorhersage, betrachtet vom Standpunkt der Mechanik under der Physik, Meter. Z., 1904, 21: 1–7.

[7] Charney J G, Fjortaft R, Neumann J Von. Numerical integration of the barotropic vorticity equation. Tellus, 1950, 2: 237–254.

[8] Charney J G, Philips N A. Numerical integration of the quasi-geostrophic equations for barotropic simple baroclinic flows. J Meteor, 1953, 10: 71–99.

[9] Cox M D. A Primitive Equation, Three-Dimensional Model(GFDL Ocean Group, 1984). Tech. Rept. No. 1.

[10] Itoo H, Kurihara Y, Asai T, et al. Numerical test of finite-difference form of primitive equations for barotrophic case. J Meteo Soc Japn, 40(1962), No. 2.

[11] Kasahara A, Washington W M. NCAR global general circulation model of the atmosphere. Mon Wea Rec,**95**(1967), 389–402, 958–968.

[12] Lorenz E N. Energy and numerical weather prediction. Tellus, 1960, 12: 364–373.

[13] Reiser H. Barochinic forecasting with the primitive equation. Proc Int Symp on Num Weather Pred. in Tokyo, Tokyo, 1962.

[14] Smagorinsky J. General circulation experiments with the primitive equations, I. The basic experiment. Mon Wea Rev, 1963, 91: 98–164.

[15] Shuman F G, Hovermale J B. A Six-level primitive equation Model. J Appl Meteor, 1968, 7: 525–531.

[16] Washington W M, Parkinson C L. An introduction to three-dimensional climate modeling. Oxford, New York: Oxford Univ. Press, 1986.

[17] Li J Chou J. The qualitative theory on the dynamical equations of atmospheric motion and its applications. Chin J Atmos Sci, 1998, 22(4): 348–360.

[18] Li J Chou J. The global analysis theory of climate system and its applications, Chin Sci Bull, 2003, 48(10): 1034–1039.

[19] Bennett A F, Kloeden P E. The simplified quasi-geostrophic equations : Existence and uniqueness of strong solutions. Mathematika, 1980, 27: 287–311.

[20] Bennett A F, Kloeden P E. The dissipative quasi-geostrophic equations, Mathematika, 1981, 28: 265–285.

[21] Mu M. Global classical solutions of initial-boundary value problems for nonlinear vorticity equation and its applications. Acta Math Sci, 1986, 6: 201–218.

[22] Mu M. Global classical solutions of initial-boundary value problems for generalized vorticity equations. Sci Sin (Ser. A), 1987, 30: 359–371.

[23] Wang S. On the 2-D model of large-scale atmospheric motion: well-posedness and attractors. Nonl Anal TMA, 1992, 18(1): 17–60.

[24] Wang S. Attractors for the 3-D baroclinic quasi-geostrophic equations of large-scale atmosphere. J Math Anal Appl, 1992, 165(1): 266–283.

[25] Bourgeois A J, Beale J T. Validity of the quasigeostrophic model for large-scale flow in the atmosphere and ocean. SIAM J Math Anal, 1994, 25: 1023–1068.

[26] Majda A, Wang X. Validity of the one and one-half layer quasi-geostrophic model and effective topography. Comm PDE, 2005, 30: 1305–1314.

[27] Liu Y, Mu M, Shepherd T. Nonlinear stability of continuously stratified quasi-geostrophic flow. J Fluid Mech, 1996, 325: 419–439.

[28] Mu M. Nonlinear stability of two-dimensional quasigeostrophic motions. Geophys, Astrophys, Fluid Dyn, 1992, 65: 57–76.

[29] Wolansky G. Existence, uniqueness and stability of stationary barotropic flow with forcing and dissipation. Comm Pure Appl Math, 1988, 41: 19–46.

[30] Wolansky G. The barotropic vorticity equation under forcing and dissipation: Bifurcations of nonsymmetric responses and multiplicity of solutions. SIAMJ Appl Math, 1989, 49(6): 1585–1607.

[31] Held I M, Pierrehumbert R T, Gerner S, et al. Surface quasi-geostrophic dynamics. J Fluid Mech, 1995, 282: 1–20.

[32] Constantin P, Majda A, Tabak E. Formation of strong fronts in the 2-D quasi-geostrophic thermal active scalar. Nonlinearity, 1994, 7: 1495–1533.

[33] Constantin P, Majda A, Tabak E. Singular front formation in a model for quasi-geostrophic flow. Phys Fluids, 1994, 6: 9–11.

[34] Cordoba D. Nonexistence of simple hyperbolic blow-up for the quasi-geostrophic equation. Ann Math, 1998, 148: 1135–1152.

[35] Constantin P, Wu J. Behavior of solutions of 2D quasi-geostrophic equations. SIAM J Math Anal, 1999, 30: 937–948.

[36] Constantin P, Cordoba D, Wu J. On the critical dissipative quasi-geostrophic equation. Indiana Univ Math J, 2001, 50(Special Issue): 97–107.

[37] Wu J. Global solutions of the 2D dissipative quasi-geostrophic equations in Besov spaces, SIAM J Math Anal, 2004, 36(3): 1014–1030.

[38] Wu J. The two-dimensional quasi-geostrophic equation with critical or supercritical dissipation. Nonlinearity, 2005, 18: 139–154.

[39] Chae D., On the regularity conditions for the dissipative quasi-geostrophic equations, SIAM J Math Anal, 2006, 37(5): 1649–1656.

[40] Kiselev A, Nazarov F, Volberg A. Global well-posedness for the critical 2D dissipative quasi-geostrophic equation. Invent Math, 2007, 167(3): 445–453.

[41] Abidi H Hmidi T. On the global well -posedness of the critical quasi-geostrophic equation,. SIAM J Math Anal, 2008, 40(1): 167–185.

[42] Caffarelli L, Vasseur A. Drift diffusion equations with fractional diffusion and the quasi-geostroophic equation, Ann Math, 171(3): 1903–1930.

[43] Lions J L, Temam R, Wang S. New formulations of the primitive equations of atmosphere and applications. Nonlinearity, 1992, 5, 237–288.

[44] Lions J L, Temam R, Wang S. On the equations of the large-scale ocean. Nonlinearity, 1992, 5: 1007–1053.

[45] Lions J L, Temam R, Wang S. Models of the coupled atmosphere and ocean(CAO I). Comput Mech Adv, 1993, 1, 1–54.

[46] Lions J L, Temam R, Wang S. Mathematical theory for the coupled atmosphere-ocean models(CAO III). J Math Pures Appl, 1995, 74, 105–163.

[47] Li J, Chou J. Existence of atmosphere attractors. Science in China, Series D, 1997, 40(2): 215–224.

[48] Li J, Chou J. Asymptotic behavior of solutions of the moist atmospheric equations. Acta Meteor Sin, 1998, 56(2): 61–72, in Chinese.

[49] Temam R, Ziane M. Some mathematical problems in geophysical fluid dynamics. In: Handbook of Mathematical Fluid Dynamics, 2003.

[50] Contantin P, Nie Q, Schorghofer N. Nonsingular surface quasi-geostrophic flows. Phys Lett A, 1998, 241(3): 168–172.

[51] Contantin P, Nie Q, Schorghofer N. Front formation in atctive Scalar. Phys Rev E, 1999, 60(3): 2858–2863.

[52] Pedlosky J. Geophysical Fluid Dynamics. 1st Edition. Berlin/New York: Springer-Verlag, 1979.

[53] Resnick S. Dynamical problems in nolinear edvective partial differeatial equation. Ph.D. Thesis, University of Chicago, 1995.

[54] Chae D, Lee J. Global well-posedness in the super-critical dissipative quasi-geostrophic equations. Comm Math Phys, 2003, 233(2): 297–311.

[55] Cordoba A, Cordoba D. A maximum principle applied to quasi-geostrophic equations.

Comm Math Phys, 2004, 249: 511–528.

[56] Hmidi T, Keraani S. Global solutions of the super-critical 2D quasi-geostrophic equations in Besov spaces. Adv Math, 2007, 214: 618–638.

[57] Ju N., Existence and Uniqueness of the solution to the dissipative 2D quasi-geostrophic equations in the Sobolev Space, Comm Math Phys, 2004, 251: 365–376.

[58] Ju N. Global solutions to the-two-dimensional quasi-geostrophic equation with critical or super-critical dissipation. Math Ann, 2006, 334: 627–642.

[59] Chen Q, Miao C, Zhang Z. A new Bernstein inequality and the 2D dissipative quasi-geostrophic Equation. Comm Math Phys, 2007, 271: 821–838.

[60] Nicolis C, Nicolis G. Is there a climatic attractor? Nature, 1984, 311: 529–532.

[61] Nicolis C, Nicolis G. Evidence for climatic attractors. Nature, 1987, 326: 523.

[62] Essex C, Lookman T, Nerenberg M A H. The climate attractor over short timescales. Nature, 1987, 326: 64–66.

[63] Tsonis A A, Elsner J B. The weather attractor over very short Timescales. Nature, 1988, 333: 545–547.

[64] Lorenz E N. Dimension of weather and climate attractors. Nature, 1991, 353: 241–244.

[65] Lorenz E N. An attractor embedded in the atmosphere. Tellus, 2006, 58A: 425–429.

[66] Li J, Wang S. Some mathematical and numerical issues in geophysical fluid dynamics and climate dynamics. Comm Comp Phys, 2008, 3(4): 759–793.

[67] Cao C, Titi E S. Global well-posedness and finite dimensional global attractor for a 3-D planetary geostrophic viscous model. Comm Pure Appl Math, 2003, 56: 198–233.

[68] Cao C, Titi E S, Ziane M. A "horizontal" hyper-diffusion 3-D thermocline planetary geostrophic model: Well-posedness and long-time behavior. Nonlinearity, 2004, 17: 1749–1776.

[69] Cao C, Titi E S. Global well-posedness of the three-dimensional viscous primitive equations of large-scale ocean and atmosphere dynamics. Ann Math, 2007, 166: 245–267.

[70] Guillén-González F., Masmoudi N, Rodríguez-Bellido M. A., Anisotropic estimates and strong solutions for the primitive equations. Diff Int Equ, 2001, 14: 1381–1408.

[71] Hu C, Temam R, Ziane M. The primimitive equations of the large scale ocean under the small depth hypothesis. Disc Cont Dyn Sys, 2003, 9(1): 97–131.

[72] Samelson R, Temam R, Wang S. Some mathematical properties of the planetary geostrophic equations for large-scale ocean circulation. Appl Anal, 1998, 70(1-2): 147–173.

[73] Guo B, Huang D. Existence of weak solutions and trajectory attractors for the moist atmospheric equations in geophysics. J Math Phys, 2006, 47: 083508.

[74] Huang D, Guo B. On the existence of atmospheric attractors. Sci. in China Ser. D: Earth Sci, 2008, 51(3): 469–480.

[75] Guo B, Huang D. On the 3D viscous primitive equations of the large-scale atmosphere.

Acta Math Sci, 2009, 29(4): 846-866.

[76] Guo B, Huang D. Existence of the universal attractor for the 3-D viscous primitive equations of large-scale moist atmosphere. J Diff Equ, 2011, 251(3): 457–491.

[77] Kukavica I, Ziane M. On the regularity of the primitive equations of the ocean. Nonlinearity, 2007, 20(12): 2739–2753.

[78] Evans L C, Gastler R. Some results for the primitive equations with physical boundary conditions. Zeitschrift f ü r angewandte Mathematik und Physik, 2013, 64(6): 1729–1744.

[79] Cao C, Titi E S. Global well-posedness of the 3D Primitive equations with partial vertical turbulence mixing heat diffusion. Comm Math Phys, 2012, 310(2): 537–568.

[80] Leith C E. Climate response and fluctuation dissipation. J Atmos Sci, 1975, 32: 2022–2025.

[81] Hasselmann K. Stochastic climate models. Part I: Theory. Tellus, 1976, 28: 473–485.

[82] Frankignoul C, Hasselmann K. Stochastic climate models. Part II: Application to sea-surface temperature anomalies and thermocline Variability. Tellus, 1977, 29: 289–305.

[83] 李麦村, 黄嘉佑. 大气关于海温准三年及准半年周期震荡的随机气候模式. 气象学报, 1984, 42(2): 168–176.

[84] Penland C, Matrosova L. A balance condition for stochastic numerical-models with applications to El Niño-Southern oscillation. J Climate, 1994, 7(9): 1352–1372.

[85] Griffies S, Tziperman E. A linear thermohaline oscillator driven by stochastic atmospheric forcing J Climate, 1995, 8(10): 2440–2453.

[86] Kleeman R, Moore A. A theory for the limitation of ENSO predictability due to stochastic atmospheric transients. J Atmos Sci, 1997, 54(6): 753–767.

[87] Majda A, Timofeyev I, Vanden-Eijnden E. Models for stochastic climate prediction. Proc Natl Acad Sci USA, 1999, 96: 14687–14691.

[88] Majda A, Timofeyev I, Vanden-Eijnden E. A mathematical framework for stochastic climate models. Comm Pure Appl Math, 2001, 54: 891–974.

[89] Majda A, Wang X. The emergence of large-scale coherent structure under small-scale random bombardments. Comm Pure Appl Math, 2001, 59: 467–500.

[90] Majda A, Timofeyev I, Vanden-Eijnden E. A priori tests of a stochastic mode reduction strategy. Phys D., 2002, 170: 206–252.

[91] Majda A, Timofeyev I, Vanden-Eijnden E. Systematic strategies for stochastic mode reduction in climate. J Atmos Sci, 2003, 60: 1705–1722.

[92] Franzke C, Majda A, Vanden-Eijnden E. Low-order stochastic mode reduction for a realistic barotropic model climate. J Atmos Sci, 2005, 62: 1722–1745.

[93] 周秀骥. 大气随机动力学与可预报性. 气象学报, 2005, 63(5): 806–811.

[94] Duan J, Gao H, Schmalfuss B. Stochastic dynamics of a coupled atmosphere-ocean model. Stoch Dynam, 2002, 2(3): 357–380.

[95] Duan J, Kloeden P E, Schmalfuss B. Exponential stability of the quasi-geostrophic

equation under random perturbations. Prog Probability, 2001, 49: 241–256.

[96] Duan J, Schmalfuss B. The 3D quasi-geostrophic fluid dynamics under random forcing on boundary. Comm Math Sci, 2003, 1: 133–151.

[97] Huang D, Guo B, Han Y. Random attractors for a quasi-geostrophic dynamical system under stochastic forcing. Int J Dyn Syst Differ Equ, 2008, 1(3): 147–154.

[98] Lu H, Lv S, Xin J, et al. A random attractor for the stochastic quasi-geostrophic dynamical system on unbounded domains. Nonl Anal TMA, 2013, 90: 96–112.

[99] Glatt-Holtz N, Temam R. Pathwise Solutions of the 2-D Stochastic Primitive Equations. Appl Math Opti, 2011, 63(3): 401-433.

[100] Gao H, Sun C. Large Deviations for the Stochastic Primitive Equations in Two Space Dimensions. Comm Math Sci, 2012, 10: 575–593.

[101] Guo B, Huang D. 3D stochastic primitive equations of the large-scale ocean: Global well-posedness and attractors. Comm Math Phys, 2009, 286: 697–723.

[102] Guo B, Huang D. On the primitive equations of large-scale ocean with stochastic boundary. Diff Int Equ, 2010, 23: 373–398.

[103] Gao H, Sun C. Random attractor for the 3D viscous stochastic primitive equations with additive noise. Stoch Dyn, 2009, 9: 293–313.

[104] Gao H, Sun C. Hausdorff dimension of random attractor for stochastic Navier-Stokes-Voight equations and primitive equations. Dyn PDE, 2010, 7(4): 307–326.

[105] Debussche A, Glatt-Holtz N, Temam R, et al. Global existence and regularity for the 3D stochastic primitive equations of the ocean and atmosphere with multiplicative white noise. Nonlinearity, 2012, 25(7): 2093–2118.

连串反应中控制火焰的耦合 GKS-CGL 方程组解的极限行为*

郭昌洪　房少梅　郭柏灵

摘　要　本文考虑连串反应中控制火焰的耦合广义 Kuramoto Sivashinsky-Ginzburg Landau (GKS-CGL) 方程组的周期初值问题, 主要研究其解在系数 $g \to 0$ 以及 $\delta \to 0$ 时的极限行为. 首先, 采用 Galerkin 方法, 通过构造一系列精细的先验估计, 得到 GKS-CGL 方程组周期初值问题整体光滑解的存在唯一性. 其次, 利用一致有界估计证得 GKS-CGL 方程组极限解收敛, 并给出了解的收敛率估计.

关键词　极限行为; Kuramoto Sivashinsky-Ginzburg Landau (KS-CGL) 方程组; 连串反应; 光滑解; 最优收敛率

1　引言

本文主要研究连串反应中控制火焰的耦合广义 Kuramoto Sivashinsky-Ginzburg Landau (GKS-CGL) 方程组的周期初值问题:

$$P_t = \xi P + (1+i\mu)P_{xx} - (1+i\nu)|P|^2P - P_xQ_x - r_1PQ_{xx} - gr_2PQ_{xxxx}, \quad (x,t)\in\Omega\times\mathbb{R}_+, \tag{1.1}$$

$$Q_t = mQ_{xx} - gQ_{xxxx} + \delta Q_{xxxxxx} - \frac{1}{2}|Q_x|^2 - \eta|P|^2, \quad (x,t)\in\Omega\times\mathbb{R}_+, \tag{1.2}$$

$$P(x+2L,t) = P(x,t), \quad Q(x+2L,t) = Q(x,t), \quad (x,t)\in\Omega\times\mathbb{R}_+, \tag{1.3}$$

$$P(x,0) = P_0(x), \quad Q(x,0) = Q_0(x), \quad x\in\Omega, \tag{1.4}$$

其中复函数 $P(x,t)$ 是火焰振荡的振幅, 实函数 $Q(x,t)$ 是第一个火焰锋面的热变形, 定义域 $\Omega = [-L, L] (L > 0)$, $2L$ 是周期, $t \geqslant 0$. $\xi > 0$, μ, ν 是 Landau 系数, $r_1 = r_{1r} + ir_{1i}, r_2 = r_{2r} + ir_{2i}$ 为复数, $m > 0$, $g > 0$ 是与振荡模式临界值成比例的系数, δ 是耗散系数, $\eta > 0$ 为耦合系数. 除系数 r_1, r_2 为复数外, 其他系数均为实数.

* 中国科学: 数学, 2014, 44(4): 329–348. DOI: 10.1360/012014-22

在方程 (1.2) 中, 若 $\delta=0$, 则耦合的 GKS-CGL 方程组 (1.1)–(1.2) 将化简为经典的 KS-CGL 方程组 [1], KS-CGL 方程组描述了两步连串反应中, 两个火焰锋面之间的非线性相互作用, 即具有固定长度距离的, 且是均匀传播的两个火焰之间的单调不稳定性和振荡不稳定性. 如果在方程组 (1.1)–(1.4) 取 $g=\delta=0$, 那么 GKS-CGL 方程组 (1.1)–(1.4) 将简化为耦合的 Burgers-Ginzburg Landau(Burgers-CGL) 方程组 [1]:

$$P_t=\xi P+(1+i\mu)P_{xx}-(1+i\nu)|P|^2P-P_xQ_x-r_1PQ_{xx}, \tag{1.5}$$

$$Q_t=mQ_{xx}-\frac{1}{2}|Q_x|^2-\eta|P|^2, \tag{1.6}$$

$$P(x+2L,t)=P(x,t),\quad Q(x+2L,t)=Q(x,t), \tag{1.7}$$

$$P(x,0)=P_0(x),\quad Q(x,0)=Q_0(x). \tag{1.8}$$

自然就有这样的问题出现: 一是问题 (1.1)–(1.4) 与问题 (1.5)–(1.8) 同样是周期初值问题, 那它们之间存在怎样的关系? 二是 GKS-CGL 方程组 (1.1)–(1.4) 中系数 $g\to 0$ 和 $\delta\to 0$ 时, 其解是否收敛于 Burgers-CGL 方程组 (1.5)–(1.8) 的解? 如果收敛, 那么其收敛率是多少? 本文, 我们将研究这两个问题.

耦合 GKS-CGL 方程组 (1.1)–(1.4) 以及 Burgers-CGL 方程组 (1.5)–(1.8) 都用来描述均匀传播火焰单调不稳定性和振荡不稳定性之间的相互作用的. 具体而言, GKS-CGL 方程组 (1.1)–(1.4) 描述活跃单调不稳定性与活跃振荡不稳定性, 以及活跃单调不稳定性与受阻振荡不稳定性之间的相互作用. 而 Burgers-CGL 方程组 (1.5)–(1.8) 描述受阻单调不稳定性与活跃振荡不稳定性之间的相互作用. 如果方程 (1.1) 和 (1.5) 中不含有刻画单调性的项 $Q(x,t)$, 那么方程 (1.1) 和 (1.5)) 就变为经典的复 Ginzburg-Landau 方程 [2,3]. 如果方程 (1.2) 中的系数 $\delta=0$ 以及 $\eta=0$, 那么方程 (1.2) 就为经典的 Kuramoto-Sivashinsky 方程 [4]. 同样地, 如果方程 (1.6) 中的耦合系数 $\eta=0$, 那么方程 (1.6) 也简化为 Burgers 方程 [5]. 到目前为止, 关于这些方程的研究, 已有许多工作, 例如 CGL 方程 [6–15], KS 方程 [16–19] 以及 Burgers 方程 [20–23].

对于耦合的 GKS-CGL 方程组 (1.1)–(1.4) 以及 Burgers-CGL 方程组 (1.5)–(1.8), 我们已经做了一些工作, 如得到广义 GKS-CGL 方程组在低维情形时整体解的存在唯一性 [24], 以及耦合 Burgers-CGL 方程组解的长时间行为 [25]. 在此基础上, 本文将讨论 GKS-CGL 方程组 (1.1)–(1.4) 周期初值问题解的存在唯一性, 更主要地是研究其解的极限行为. 具体而言, 就是当方程组 (1.1)–(1.2) 中的系数 $g\to 0,\delta\to 0$ 时其解的收敛性问题, 以及收敛率估计问题.

本文主要内容安排如下: 在第二部分, 给出一些基础知识准备和主要结果; 第三部分, 对 GKS-CGL 方程组 (1.1)–(1.4) 周期初值问题解做先验估计, 得到其解的存在唯一性和收敛性 (定理 1, 定理 2 和定理 3 的证明). 在第四部分中, 利用一致有界估计得到最优收敛率估计 (定理 4, 定理 5 和定理 6 的证明); 在最后的第五部分, 对所得结果进行总结.

2 基础知识及主要结果

在本文中, 我们对实值函数和复值函数空间都采用相同的记号. 一般地, $\|\cdot\|_X$ 表示 Banach 空间 X 上的范数. 定义常见的 Sobolev 空间为 $W^{k,p}(\Omega)$, 其相应的范数为

$$\|u\|_{W^{k,p}(\Omega)}=\left(\sum_{|\alpha|\leqslant k}\|D^\alpha u\|^p\right)^{\frac{1}{p}}(1\leqslant p<\infty) \text{ 以及 } \|u\|_{W^{k,\infty}(\Omega)}=\sum_{|\alpha|\leqslant k}\operatorname*{ess\,sup}_{\Omega}|D^\alpha u|.$$

当 $p=2$ 时, 我们记 $\|\cdot\|_{H^k}=\|\cdot\|_{W^{k,2}(\Omega)}$. 当 $k=0$ 时, $W^{k,p}(\Omega)$ 就简化为我们常见的 L^p 空间, 即 $L^p(\Omega)=W^{0,p}(\Omega)$, 为方便起见, 常记 $\|\cdot\|=\|\cdot\|_{L^2(\Omega)}$. 同样字母 C 每处所代表的值可能有所不同, 但都表示依赖于已知常量的正常数, 其他字母 $C_i, E_i, M_i, (i=1,2,\cdots)$ 等均表示依赖于已知量的常数. 在本文中, 还假定系数满足

$$0<g\leqslant 1,\ 0<\delta\leqslant\delta_0,\quad \lim_{g\to 0,\delta\to 0}\frac{g^2}{\delta}=0. \tag{2.9}$$

本文第一个主要结果, 是考虑周期初值问题 (1.1)–(1.4) 解的存在唯一性. 即在假设 (2.9) 下, 有

定理 1 (整体解存在唯一性) 假设 $2-\dfrac{(1+2\eta-2r_{1r})^2}{4m}-\dfrac{g^2r_{2r}^2}{\delta}>0$, $P_0(x)\in H^5(\Omega)$ 以及 $Q_0(x)\in H^6(\Omega)$, 那么周期初值问题 (1.1)–(1.4) 存在唯一整体解 $P(x,t)$, $Q(x,t)$, 且满足

$$P(x,t)\in L^\infty(0,T;H^5(\Omega)),\quad P_t(x,t)\in L^\infty(0,T;H^1(\Omega)),$$

$$Q(x,t)\in L^\infty(0,T;H^6(\Omega)),\quad Q_t(x,t)\in L^\infty(0,T;H^2(\Omega)).$$

定理 2(整体光滑解存在唯一性) 假设 $2-\dfrac{(1+2\eta-2r_{1r})^2}{4m}-\dfrac{g^2r_{2r}^2}{\delta}>0$, $P_0(x)\in H^k(\Omega)$ 以及 $Q_0(x)\in H^{k+1}(\Omega)(k\geqslant 5)$, 那么周期初值问题 (1.1)–(1.4) 存在唯一整体光滑解 $P(x,t), Q(x,t)$, 且满足

$$P(x,t)\in L^\infty(0,T;H^k(\Omega)),\quad P_t(x,t)\in L^\infty(0,T;H^{k-4}(\Omega)),$$

$$Q(x,t)\in L^\infty(0,T;H^{k+1}(\Omega)),\quad Q_t(x,t)\in L^\infty(0,T;H^{k-3}(\Omega)).$$

本文第二个主要结果, 是考虑 GKS-CGL 方程组 (1.1)–(1.4) 在初值 $P_{\varepsilon0}, Q_{\varepsilon0}$ 下其解 $P_\varepsilon(x,t), Q_\varepsilon(x,t)$ 的收敛性问题, 当 $\varepsilon=\max(g,\delta)$ 时, 其解收敛到 Burgers-CGL

方程组 (1.5)–(1.8) 在初值 (P_0, Q_0) 条件下的解 $P(x,t), Q(x,t)$, 其中 $\lim_{\varepsilon\to 0} P_\varepsilon(x,t) = P_0, \lim_{\varepsilon\to 0} Q_\varepsilon(x,t) = Q_0$. 即有

定理 3 在定理 1 条件下, 并设 $P_\varepsilon(x,t), Q_\varepsilon(x,t)$ 是 GKS-CGL 方程组 (1.1)–(1.4) 的解, 那么存在一个收敛子列（仍记为）$(P_\varepsilon(x,t), Q_\varepsilon(x,t))$, 使得当 $\varepsilon \to 0$ 时, 收敛到 $(P(x,t), Q(x,t))$, 且是 Burgers-CGL 方程组 (1.5)–(1.8) 的解.

关于收敛率结果, 假定 $P_{\varepsilon 0}, P_0 \in H^5(\Omega)$ 以及 $Q_{\varepsilon 0}, Q_0 \in H^6(\Omega)$, 于是有

定理 4 在定理 3 条件下, 对于任意的 $0 \leqslant t \leqslant T(T < \infty)$, 有

$$\begin{aligned}\|P_\varepsilon - P\|^2 + \|Q_{\varepsilon x} - Q_x\|^2 \quad &\leqslant 3(\|P_{\varepsilon 0} - P_0\|^2 + \|Q_{\varepsilon x0} - Q_{x0}\|^2) \\ &\quad +3g^2 E_3(|r_2|M_0 + 1)^2 e^{2C_7 t} + 3\delta^2 E_5 e^{2C_7 t},\end{aligned} \tag{2.10}$$

其中正常数 E_3, M_0, C_7 和 E_5 均与 g 和 δ 无关.

特别地, 若 $\|P_{\varepsilon 0} - P_0\|^2$ 和 $\|Q_{\varepsilon x0} - Q_{x0}\|^2$ 均与 g^2, δ^2 同阶, 则

$$\|P_\varepsilon - P\| + \|Q_{\varepsilon x} - Q_x\| = O(g) + O(\delta), \quad \varepsilon \to 0. \tag{2.11}$$

定理 5 在定理 3 条件下. 对于任意的 $0 \leqslant t \leqslant T(T < \infty)$, 有

$$\begin{aligned}&\|P_{\varepsilon x} - P_x\|^2 + \|Q_{\varepsilon xx} - Q_{xx}\|^2 \\ \leqslant &(\|P_{\varepsilon x0} - P_{x0}\|^2 + \|Q_{\varepsilon xx0} - Q_{xx0}\|^2)e^{C_8 t} \\ &+3C_9(\|P_{\varepsilon 0} - P_0\|^2 + \|Q_{\varepsilon x0} - Q_{x0}\|^2)te^{C_8 t} \\ &+g^2 C_{10} te^{C_8 t} + \delta^2 C_{11} te^{C_8 t},\end{aligned} \tag{2.12}$$

其中正常数 C_8, C_9, C_{10} 和 C_{11} 均与 g 和 δ 无关.

特别地, 如果 $\|P_{\varepsilon 0} - P_0\|_{H^1}^2$ 和 $\|Q_{\varepsilon 0} - Q_0\|_{H^2}^2$ 均与 g^2, δ^2 同阶, 则

$$\|P_{\varepsilon x} - P_x\| + \|Q_{\varepsilon xx} - Q_{xx}\| = O(g) + O(\delta), \quad \varepsilon \to 0. \tag{2.13}$$

推论 1 在定理 4 和定理 5 条件下, 关于 GKS-CGL 方程组 (1.1)–(1.4) 解 $P_\varepsilon(x,t)$ 和 $Q_\varepsilon(x,t)$ 有以下收敛率估计 (1.1)–(1.4)

$$\|Q_\varepsilon - Q\| = O(g) + O(\delta), \quad \varepsilon \to 0, \tag{2.14}$$

以及

$$\|P_\varepsilon - P\|_{L^\infty} + \|Q_\varepsilon - Q\|_{L^\infty} = O(g) + O(\delta), \quad \varepsilon \to 0. \tag{2.15}$$

更一般地, 对于 GKS-CGL 方程组 (1.1)–(1.4) 的解 $P_\varepsilon(x,t), Q_\varepsilon(x,t)$, 有

定理 6 假设 $2-\dfrac{(1+2\eta-2r_{1r})^2}{4m}-\dfrac{g^2r_{2r}^2}{\delta}>0$, $P_{\varepsilon 0},P_0\in H^k(\Omega)$, $Q_{\varepsilon 0},Q_0\in H^{k+1}(\Omega)$, 并且 $\|P_{\varepsilon 0}-P_0\|_{H^{k-4}}^2$ 和 $\|Q_{\varepsilon 0}-Q_0\|_{H^{k-3}}^2$ are of order $O(g^2)+O(\delta^2)(k\geqslant 5)$ 均与 g^2,δ^2 同阶. 同时假设 $P_\varepsilon(x,t),Q_\varepsilon(x,t)$ 是 GKS-CGL 方程组 (1.1)–(1.4) 在初始值为 $P_{\varepsilon 0},Q_{\varepsilon 0}$ 的解, 而 $P(x,t),Q(x,t)$ 是 Burgers-CGL 方程组 (1.5)–(1.8) 在初始值为 (P_0,Q_0) 的解. 则有

$$\|P_\varepsilon-P\|_{H^{k-4}}+\|Q_\varepsilon-Q\|_{H^{k-3}}=O(g)+O(\delta),\ \text{当}\ \varepsilon\to 0, \tag{2.16}$$

其中 $\varepsilon=\max(g,\delta)$.

在下面部分, 我们将用到如下的不等式.

引理 1(广义 Hölder 不等式)[26] 设 $1\leqslant p_1,p_2,\cdots,p_m\leqslant\infty$, 满足 $\dfrac{1}{p_1}+\dfrac{1}{p_2}+\cdots+\dfrac{1}{p_m}=1$, 并且令所有的 $k=1,2,\cdots,m$ 有 $u_k\in L^{p_k}(\Omega)$, 那么则有

$$\int_\Omega |u_1\cdots u_m|dx\leqslant\prod_{k=1}^m\|u_k\|_{L^{p_k}(\Omega)}.$$

引理 2 (Gagliardo-Nirenberg 不等式)[27] 假设 Ω 是一有界区域, 且 $\partial\Omega\in C^m$, 设函数 $u\in W^{m,r}(\Omega)\cap L^q(\Omega)$, $1\leqslant q,r\leqslant\infty$. 对于任何整数 j, $0\leqslant j<m$, 以及任何数 a, $j/m\leqslant a\leqslant 1$, 满足

$$\frac{1}{p}=\frac{j}{n}+a\left(\frac{1}{r}-\frac{m}{n}\right)+(1-a)\frac{1}{q}.$$

如果 $m-j-n/r$ 不是一个非负正整数, 那么则有

$$\|D^ju\|_{L^p}\leqslant C\|u\|_{W^{m,r}}^a\|u\|_{L^q}^{1-a}. \tag{2.17}$$

如果 $m-j-n/r$ 是一个非负正整数, 那么 (2.17) 对于 $a=j/m$ 成立. 其中常数 C 仅依赖于 Ω,r,q,j,a.

引理 3 (Agmon 不等式)[28] 假设 $\Omega\subset\mathbb{R}^n$ 且属于 C^n, 那么存在一个仅依赖于 Ω 的常数 C 使得下面两个式子成立:

$$\|u\|_{L^\infty(\Omega)}\leqslant C\|u\|_{H^{(n/2)-1}(\Omega)}^{1/2}\|u\|_{H^{(n/2)+1}(\Omega)}^{1/2},\quad \forall u\in H^{(n/2)+1}(\Omega)\quad \text{如果}\ n\ \text{是偶数},$$

以及

$$\|u\|_{L^\infty(\Omega)}\leqslant C\|u\|_{H^{(n-1)/2}(\Omega)}^{1/2}\|u\|_{H^{(n+1)/2}(\Omega)}^{1/2},\quad \forall u\in H^{(n+1)/2}(\Omega)\quad \text{如果}\ n\ \text{是奇数}.$$

特别地, 当 $n=1$ 时, 我们有:

$$\|u\|_{L^\infty}\leqslant C\|u\|^{1/2}\|u\|_{H^1}^{1/2},\quad \forall u\in H^1(\Omega). \tag{2.18}$$

3 存在性和收敛性

本节主要对 GKS-CGL 方程组 (1.1)–(1.4) 的解做先验估计, 利用 Galerkin 方法和 Arzela Ascoli 定理得到其解的存在唯一性和收敛性.

3.1 先验估计

引理 4 假设 $P_0(x)\in L^2(\Omega), Q_0(x)\in H^1(\Omega)$, 以及 $2-\dfrac{(1+2\eta-2r_{1r})^2}{4m}-\dfrac{g^2r_{2r}^2}{\delta}>0$, 那么对于所有的 $0\leqslant t\leqslant T$, 周期初值问题 (1.1)–(1.4) 的解有以下估计

$$\|P\|^2+\|Q_x\|^2\leqslant E_0(T), \tag{3.19}$$

其中 $E_0(T)=e^{K_0T}(\|P(0)\|^2+\|Q_x(0)\|^2)$ 是与 g 和 δ 都无关的常数.

证明 首先, 方程 (1.2) 关于 x 求导一次, 并令

$$W=Q_x. \tag{3.20}$$

那么方程组 (1.1)(1.2) 可化为

$$P_t=\xi P+(1+i\mu)P_{xx}-(1+i\nu)|P|^2P-P_xW-r_1PW_x-gr_2PW_{xxx}, \tag{3.21}$$

$$W_t=mW_{xx}-gW_{xxxx}+\delta W_{xxxxxx}-WW_x-\eta(|P|^2)_x. \tag{3.22}$$

方程 (3.21) 两边同时乘以 $\overline{P}$, 并在 Ω 上关于 x 积分, 取其实部, 则有

$$\begin{aligned}\frac{1}{2}\frac{\mathrm{d}}{\mathrm{d}t}\|P\|^2=&\xi\|P\|^2-\|P_x\|^2-\int_\Omega|P|^4\mathrm{d}x-\mathrm{Re}\int_\Omega WP_x\overline{P}\mathrm{d}x\\&-r_{1r}\int_\Omega W_x|P|^2\mathrm{d}x-gr_{2r}\int_\Omega W_{xxx}|P|^2\mathrm{d}x,\end{aligned} \tag{3.23}$$

其中

$$-\mathrm{Re}\int_\Omega WP_x\overline{P}\mathrm{d}x=-\mathrm{Re}\left(\frac{1}{2}W|P|^2|_{-L}^{L}-\frac{1}{2}\int_\Omega W_x|P|^2\mathrm{d}x\right)=\frac{1}{2}\int_\Omega W_x|P|^2\mathrm{d}x. \tag{3.24}$$

同时, 方程 (3.22) 两边同时乘以 W, 并在 Ω 上积分有

$$\begin{aligned}\frac{1}{2}\frac{\mathrm{d}}{\mathrm{d}t}\|W\|^2=&-m\|W_x\|^2-g\|W_{xx}\|^2-\delta\|W_{xxx}\|^2\\&-\int_\Omega W^2W_x\mathrm{d}x-\eta\int_\Omega W(|P|^2)_x\mathrm{d}x,\end{aligned} \tag{3.25}$$

其中利用函数周期性有

$$\int_\Omega W^2W_x\mathrm{d}x=\frac{1}{3}W^3|_{-L}^{L}=0, \tag{3.26}$$

$$-\eta\int_\Omega W\left(|P|^2\right)_x\mathrm{d}x=-\eta W|P|^2|_{-L}^{L}+\eta\int_\Omega W_x|P|^2\mathrm{d}x=\eta\int_\Omega W_x|P|^2\mathrm{d}x. \tag{3.27}$$

联合 (3.23)–(3.27) 可得

$$\begin{aligned}\frac{\mathrm{d}}{\mathrm{d}t}\left(\|P\|^2+\|W\|^2\right)=&2\xi\|P\|^2-2\|P_x\|^2-2\int_\Omega|P|^4\mathrm{d}x-2m\|W_x\|^2-2g\|W_{xx}\|^2\\&-2\delta\|W_{xxx}\|^2+(1+2\eta-2r_{1r})\int_\Omega W_x|P|^2\mathrm{d}x\\&-2gr_{2r}\int_\Omega W_{xxx}|P|^2\mathrm{d}x.\end{aligned} \tag{3.28}$$

首先, 根据 Hölder 不等式以及带 ε 的 Young 不等式, 有

$$\begin{aligned}\left|(1+2\eta-2r_{1r})\int_\Omega W_x|P|^2\mathrm{d}x\right|&\leqslant|1+2\eta-2r_{1r}|\left(\int_\Omega|P|^4\mathrm{d}x\right)^{\frac{1}{2}}\|W_x\|\\&\leqslant\varepsilon_1\|W_x\|^2+\frac{(1+2\eta-2r_{1r})^2}{4\varepsilon_1}\int_\Omega|P|^4\mathrm{d}x,\end{aligned} \tag{3.29}$$

以及

$$\begin{aligned}\left|-2gr_{2r}\int_\Omega W_{xxx}|P|^2\mathrm{d}x\right|&\leqslant2g|r_{2r}|\|W_{xxx}\|\left(\int_\Omega|P|^4\mathrm{d}x\right)^{\frac{1}{2}}\\&\leqslant\varepsilon_2\|W_{xxx}\|^2+\frac{g^2r_{2r}^2}{\varepsilon_2}\int_\Omega|P|^4\mathrm{d}x.\end{aligned} \tag{3.30}$$

由 (3.28)–(3.30), 则有

$$\begin{aligned}\frac{\mathrm{d}}{\mathrm{d}t}\left(\|P\|^2+\|W\|^2\right)\leqslant&2\xi\|P\|^2-2\|P_x\|^2-(2m-\varepsilon_1)\|W_x\|^2\\&-2g\|W_{xx}\|^2-(2\delta-\varepsilon_2)\|W_{xxx}\|^2\\&-\left(2-\frac{(1+2\eta-2r_{1r})^2}{4\varepsilon_1}-\frac{g^2r_{2r}^2}{\varepsilon_2}\right)\int_\Omega|P|^4\mathrm{d}x.\end{aligned} \tag{3.31}$$

在 $2-\dfrac{(1+2\eta-2r_{1r})^2}{4m}-\dfrac{g^2r_{2r}^2}{\delta}>0$ 条件下, 可选取 ε_1, ε_2 满足 $m<\varepsilon_1<2m$ 以及 $\delta<\varepsilon_2<2\delta$, 从而使得 $2m-\varepsilon_1>0$, $2\delta-\varepsilon_2>0$ 以及 $2-\dfrac{(1+2\eta-2r_{1r})^2}{4\varepsilon_1}-\dfrac{g^2r_{2r}^2}{\varepsilon_2}>0$ 都成立. 于是可得

$$\frac{\mathrm{d}}{\mathrm{d}t}\left(\|P\|^2+\|W\|^2\right)\leqslant2\xi\|P\|^2+\|W\|^2\leqslant K_1(\|P\|^2+\|W\|^2), \tag{3.32}$$

其中 $K_0=\max(2\xi,1)$.

利用 Gronwall 不等式, 以及变换 (3.20), 可有

$$\|P\|^2+\|Q_x\|^2\leqslant e^{K_0T}(\|P(0)\|^2+\|Q_x(0)\|^2)=E_0, \tag{3.33}$$

其中 E_0 是与 g 和 δ 无关的常数. 于是引理 4 得证.

引理 5 在引理 4 条件下, 并设 $P_0(x)\in H^1(\Omega), Q_0(x)\in H^2(\Omega)$, 那么对任意的 $0\leqslant t\leqslant T$, 都有

$$\|P_x\|^2+\|Q_{xx}\|^2\leqslant E_1(T), \tag{3.34}$$

其中 $E_1(T)=e^{K_1T}(\|P_x(0)\|^2+\|Q_{xx}(0)\|^2+CT)$ 是与 g 和 δ 无关的常数.

证明 类似于引理 4 证明的第一步, 利用变换得到方程组. 首先在方程 (3.21) 两端乘以 $(-\overline{P_{xx}})$, 在 Ω 积分, 并取实部, 则有

$$\begin{aligned}\frac{1}{2}\frac{\mathrm{d}}{\mathrm{d}t}\|P_x\|^2=&\xi\|P_x\|^2-\|P_{xx}\|^2+\mathrm{Re}\int_\Omega 1+i\nu)|P|^2P\overline{P_{xx}}\mathrm{d}x+\mathrm{Re}\int_\Omega WP_x\overline{P_{xx}}\mathrm{d}x\\&+\mathrm{Re}\int_\Omega r_1W_xP\overline{P_{xx}}\mathrm{d}x+\mathrm{Re}\int_\Omega gr_2W_{xxx}P\overline{P_{xx}}\mathrm{d}x.\end{aligned} \tag{3.35}$$

其次, 方程 (3.22) 与 $(-W_{xx})$ 在 Ω 上做内积, 即得

$$\begin{aligned}\frac{1}{2}\frac{\mathrm{d}}{\mathrm{d}t}\|W_x\|^2=&-m\|W_{xx}\|^2-g\|W_{xxx}\|^2-\delta\|W_{xxxx}\|^2+\int_\Omega WW_xW_{xx}\mathrm{d}x\\&+\eta\int_\Omega(|P|^2)_xW_{xx}\mathrm{d}x.\end{aligned} \tag{3.36}$$

将 (3.35) 和 (3.36) 相加可得

$$\begin{aligned}\frac{\mathrm{d}}{\mathrm{d}t}(\|P_x\|^2+\|W_x\|^2)=&2\xi\|P_x\|^2-2\|P_{xx}\|^2-2m\|W_{xx}\|^2-2g\|W_{xxx}\|^2\\&-2\delta\|W_{xxxx}\|^2+2\mathrm{Re}\int_\Omega(1+i\nu)|P|^2P\overline{P_{xx}}\mathrm{d}x\\&+2\mathrm{Re}\int_\Omega WP_x\overline{P_{xx}}\mathrm{d}x+2\mathrm{Re}\int_\Omega r_1W_xP\overline{P_{xx}}\mathrm{d}x\\&+2\mathrm{Re}\int_\Omega gr_2W_{xxx}P\overline{P_{xx}}\mathrm{d}x+2\int_\Omega WW_xW_{xx}\mathrm{d}x\\&+2\eta\int_\Omega(|P|^2)_xW_{xx}\mathrm{d}x.\end{aligned} \tag{3.37}$$

现在需要估计 (3.37) 右端各项. 首先, 利用 Gagliardo-Nirenberg 不等式以及引理 4, 有

$$\|W_{xxx}\|\quad\leqslant C+C(\|W_x\|^{\frac{3}{4}}+\|W_{xx}\|^{\frac{3}{4}}+\|W_{xxxx}\|^{\frac{3}{4}}). \tag{3.38}$$

同样地, 利用 Gagliardo-Nirenberg 不等式, 可得

$$\begin{aligned}\left|2\mathrm{Re}\int_\Omega(1+i\nu)|P|^2P\overline{P_{xx}}dx\right| &\leqslant\leqslant 2|1+i\nu|\|P\|_{L^8}^2\|P\|_{L^4}\|P_{xx}\| \\ &\leqslant C\|P\|_{H^2}^{\frac{3}{8}}\|P\|^{\frac{13}{8}}\|P\|_{H^2}^{\frac{1}{8}}\|P\|^{\frac{7}{8}}\|P_{xx}\| \\ &\leqslant\frac{2}{5}\|P_{xx}\|^2+\xi\|P_x\|^2+C,\end{aligned}\tag{3.39}$$

$$\begin{aligned}\left|2\mathrm{Re}\int_\Omega WP_x\overline{P_{xx}}dx\right| &\leqslant 2\|P_x\|_{L^\infty}\|W\|\|P_{xx}\| \\ &\leqslant C((\|P\|^2+\|P_x\|^2+\|P_{xx}\|^2)^{\frac{1}{2}})^{\frac{3}{4}}\|P\|^{\frac{1}{4}}\|P_{xx}\| \\ &\leqslant\frac{2}{5}\|P_{xx}\|^2+\xi\|P_x\|^2+C.\end{aligned}\tag{3.40}$$

同样地, 易有

$$\|W_x\|_{L^\infty}\leqslant C\|W\|_{H^2}^{\frac{3}{4}}\|W\|^{\frac{1}{4}}\leqslant C(C+C\|W_x\|^{\frac{3}{4}}+C\|W_{xx}\|^{\frac{3}{4}}).\tag{3.41}$$

于是由 (3.41), 可得

$$\begin{aligned}\left|2\mathrm{Re}\int_\Omega r_1W_xP\overline{P_{xx}}\mathrm{d}x\right| &\leqslant 2|r_1|\|W_x\|_{L^\infty}\|P\|\|P_{xx}\| \\ &\leqslant C(C+C\|W_x\|^{\frac{3}{4}}+C\|W_{xx}\|^{\frac{3}{4}})\|P\|\|P_{xx}\| \\ &\leqslant\frac{m}{2}\|W_{xx}\|^2+\|W_x\|^2+\frac{2}{5}\|P_{xx}\|^2+C.\end{aligned}\tag{3.42}$$

对于接下来的一项, 注意到 $0<g\leqslant 1$, 于是结合 (3.38), 有

$$\begin{aligned}&\left|2\mathrm{Re}\int_\Omega gr_2W_{xxx}P\overline{P_{xx}}\mathrm{d}x\right| \\ &\leqslant 2g|r_2|\|P\|_{L^\infty}\|W_{xxx}\|\|P_{xx}\| \\ &\leqslant 2g|r_2|C(\|P\|^{\frac{1}{2}}+\|P_x\|^{\frac{1}{2}}+\|P_{xx}\|^{\frac{1}{2}}) \\ &\quad\cdot(C+\|W_x\|^{\frac{3}{4}}+\|W_{xx}\|^{\frac{3}{4}}+\|W_{xxxx}\|^{\frac{3}{4}})\|P_{xx}\| \\ &\leqslant 2\delta\|W_{xxxx}\|^2+\frac{m}{2}\|W_{xx}\|^2+\|W_x\|^2+\frac{2}{5}\|P_{xx}\|^2+\xi\|P_x\|^2+C.\end{aligned}\tag{3.43}$$

对于 (3.37) 的最后两项, 我们有如下估计

$$\begin{aligned}\left|2\int_\Omega WW_xW_{xx}\mathrm{d}x\right| &\leqslant 2\|W_x\|_{L^\infty}\|W\|\|W_{xx}\| \\ &\leqslant C(C+C\|W_x\|^{\frac{3}{4}}+C\|W_{xx}\|^{\frac{3}{4}})\|W\|\|W_{xx}\| \\ &\leqslant\frac{m}{2}\|W_{xx}\|^2+\|W_x\|^2+C,\end{aligned}\tag{3.44}$$

以及

$$\begin{aligned}\left|2\eta\int_\Omega(|P|^2)_xW_{xx}\mathrm{d}x\right| &= \left|2\eta\int_\Omega P_x\overline{P}W_{xx}dx+2\eta\int_\Omega P\overline{P_x}W_{xx}\mathrm{d}x\right|\\ &\leqslant 4\omega\|P_x\|_{L^\infty}\|P\|\|W_{xx}\|\\ &\leqslant C((\|P\|^2+\|P_x\|^2+\|P_{xx}\|^2)^{\frac{1}{2}})^{\frac{1}{2}}\|P\|^{\frac{3}{2}}\|W_{xx}\|\\ &\leqslant \frac{m}{2}\|W_{xx}\|^2+\frac{2}{5}\|P_{xx}\|^2+\xi\|P_x\|^2+C.\end{aligned} \tag{3.45}$$

将 (3.39)–(3.45) 代入 (3.37) 中, 可得

$$\frac{\mathrm{d}}{\mathrm{d}t}(\|P_x\|^2+\|W_x\|^2)\leqslant 6\xi\|P_x\|^2+3\|W_x\|^2+C\leqslant K_1(\|P_x\|^2+\|W_x\|^2)+C, \tag{3.46}$$

其中 $K_1=\max(6\xi,3)$. 利用 (3.20) 以及 Gronwall 不等式, 可有

$$\|P_x\|^2+\|Q_{xx}\|^2\leqslant e^{K_1T}(\|P_x(0)\|^2+\|Q_{xx}(0)\|^2+CT)=E_1, \tag{3.47}$$

其中 E_1 是依赖于 $\|P_x(0)\|$ 和 $\|Q_{xx}(0)\|$ 的常数, 并且与 g 和 δ 无关. 从而引理 5 证毕.

由引理 4, 引理 5 以及 Agmon 不等式 (2.18), 易得以下推论

推论 2 在引理 4 和引理 5 条件下, 周期初值问题 (1.1)–(1.4) 的解满足

$$\|P\|_{L^\infty}\leqslant M_0(T),\quad \|Q_x\|_{L^\infty}\leqslant M_0(T), \tag{3.48}$$

其中常数仅依赖于 $M_0(t)$ $E_0(T)$ 和 $E_1(T)$.

引理 6 在引理 4 和引理 5 条件下, 对于问题 (1.1)–(1.4) 的解, 有以下估计

$$\|Q\|^2\leqslant C(T),\quad \|Q\|_{L^\infty}\leqslant C(T), \tag{3.49}$$

其中正常数 $C(T)$ 与 g 和 δ 都无关.

证明 考虑方程 (1.2), 在其两端分别乘以 Q, 并在 Ω 上积分

$$\frac{1}{2}\frac{\mathrm{d}}{\mathrm{d}t}\|Q\|^2 = -m\|Q_x\|^2-g\|Q_{xx}\|^2-\delta\|Q_{xxx}\|^2-\frac{1}{2}\int_\Omega|Q_x|^2Q\mathrm{d}x-\eta\int_\Omega|P|^2Q\mathrm{d}x. \tag{3.50}$$

利用前面引理结果以及推论 (3.48), 即有

$$\left|-\frac{1}{2}\int_\Omega|Q_x|^2Q\mathrm{d}x\right|\leqslant\frac{1}{2}\|Q_x\|_{L^\infty}\|Q_x\|\|Q\|\leqslant m\|Q_x\|^2+C_0\|Q\|^2, \tag{3.51}$$

以及

$$\left|-\eta\int_\Omega|P|^2Q\mathrm{d}x\right|\leqslant\eta\|P\|_{L^\infty}\|P\|\|Q\|\leqslant C_1\|Q\|^2+C. \tag{3.52}$$

结合 (3.50)–(3.52), 可得

$$\frac{\mathrm{d}}{\mathrm{d}t}\|Q\|^2 \leqslant 2(C_0+C_1)\|Q\|^2+C. \tag{3.53}$$

利用 Gronwall 不等式, 则有 $\|Q\|^2 \leqslant C(T)$. 进一步地, 利用 Agmon 不等式 (2.18), 有 $\|Q\|_{L^\infty} \leqslant C$, 于是引理 6 得证.

引理 7 在引理 5 条件下, 并假设 $P_0(x) \in H^2(\Omega), Q_0(x) \in H^3(\Omega)$, 则对于任意的 $0 \leqslant t \leqslant T$, 问题 (1.1)–(1.4) 的解有以下估计

$$\|P_{xx}\|^2+\|Q_{xxx}\|^2 \leqslant E_2(T), \tag{3.54}$$

其中 $E_2(T)=e^{K_2T}(\|P_{xx}(0)\|^2+\|Q_{xxx}(0)\|^2+CT)$ 是与 g 和 δ 无关的常数.

证明 利用函数变换 (3.20), 将方程 (3.21) 与 $\overline{P_{xxxx}}$ 在 Ω 上做内积, 并取其实部, 则有

$$\begin{aligned}\frac{1}{2}\frac{\mathrm{d}}{\mathrm{d}t}\|P_{xx}\|^2=&\xi\|P_{xx}\|^2-\|P_{xxx}\|^2-\mathrm{Re}\int_\Omega(1+i\nu)|P|^2P\overline{P_{xxxx}}\mathrm{d}x\\&-\mathrm{Re}\int_\Omega WP_x\overline{P_{xxxx}}\mathrm{d}x-\mathrm{Re}\int_\Omega r_1W_xP\overline{P_{xxxx}}\mathrm{d}x\\&-\mathrm{Re}WP_x\int_\Omega gr_2W_{xxx}P\overline{P_{xxxx}}\mathrm{d}x.\end{aligned} \tag{3.55}$$

另一方面, 方程 (3.22) 两边乘以 W_{xxxx}, 并在 Ω 上积分, 则有

$$\begin{aligned}\frac{1}{2}\frac{\mathrm{d}}{\mathrm{d}t}\|W_{xx}\|^2=&-m\|W_{xxx}\|^2-g\|W_{xxxx}\|^2-\delta\|W_{xxxxx}\|^2\\&-\int_\Omega WW_xW_{xxxx}\mathrm{d}x-\eta\int_\Omega(|P|^2)_xW_{xxxx}\mathrm{d}x.\end{aligned} \tag{3.56}$$

将 (3.55) 和 (3.56) 相加可得

$$\begin{aligned}&\frac{\mathrm{d}}{\mathrm{d}t}\left(\|P_{xx}\|^2+\|W_{xx}\|^2\right)\\=&2\xi\|P_{xx}\|^2-2\|P_{xxx}\|^2-2m\|W_{xxx}\|^2-2g\|W_{xxxx}\|^2-2\delta\|W_{xxxxx}\|^2\\&-2\mathrm{Re}\int_\Omega(1+i\nu)|P|^2P\overline{P_{xxxx}}\mathrm{d}x-2\mathrm{Re}\int_\Omega WP_x\overline{P_{xxxx}}\mathrm{d}x\\&-2\mathrm{Re}\int_\Omega r_1W_xP\overline{P_{xxxx}}\mathrm{d}x-2\mathrm{Re}\int_\Omega gr_2W_{xxx}P\overline{P_{xxxx}}\mathrm{d}x\\&-2\int_\Omega WW_xW_{xxxx}\mathrm{d}x-2\eta\int_\Omega(|P|^2)_xW_{xxxx}\mathrm{d}x.\end{aligned} \tag{3.57}$$

首先由 Gagliardo-Nirenberg 不等式以及前面引理结果, 有

$$\|W_{xxxx}\| \leqslant C + C\|W_{xx}\|^{\frac{4}{5}} + C\|W_{xxx}\|^{\frac{4}{5}} + C\|W_{xxxxx}\|^{\frac{4}{5}}. \tag{3.58}$$

同样有 (3.38), 有 $\|W_{xxx}\| \leqslant C + C(\|W_x\|^{\frac{3}{4}} + \|W_{xx}\|^{\frac{3}{4}} + \|W_{xxxx}\|^{\frac{3}{4}})$, 再由 (3.58) 可得

$$\|W_{xxx}\| \leqslant C + C\|W_{xx}\|^{\frac{4}{5}} + C\|W_{xxxxx}\|^{\frac{3}{5}}, \tag{3.59}$$

$$\|W_{xxxx}\| \leqslant C + C\|W_{xx}\|^{\frac{4}{5}} + C\|W_{xxxxx}\|^{\frac{4}{5}}. \tag{3.60}$$

同时, 利用引理结果和推论 2, 则有

$$\begin{aligned}\left|-2\mathrm{Re}\int_\Omega (1+i\nu)|P|^2 P\overline{P_{xxxx}}\mathrm{d}x\right| &= \left|2\mathrm{Re}\int_\Omega (1+i\nu)(2PP_x\overline{P} + P^2\overline{P_x})\overline{P_{xxx}}\mathrm{d}x\right| \\ &\leqslant \frac{1}{2}\|P_{xxx}\|^2 + C(6|1+i\nu|)^2\|P\|_{L^\infty}^4\|P_x\|^2 \\ &\leqslant \frac{1}{2}\|P_{xxx}\|^2 + C, \end{aligned} \tag{3.61}$$

$$\begin{aligned}&\left|-2\mathrm{Re}\int_\Omega WP_x\overline{P_{xxxx}}\mathrm{d}x\right| \\ =&\left|2\mathrm{Re}\int_\Omega W_xP_x\overline{P_{xxx}}dx + 2\mathrm{Re}\int_\Omega WP_{xx}\overline{P_{xxx}}\mathrm{d}x\right| \\ \leqslant& 2\|P_x\|_{L^\infty}\|W_x\|\|P_{xxx}\| + 2\|W\|_{L^\infty}\|P_{xx}\|\|P_{xxx}\| \\ \leqslant& C((\|P\|^2 + \|P_x\|^2 + \|P_{xx}\|^2)^{\frac{1}{2}})^{\frac{3}{4}}\|P\|^{\frac{1}{4}}\|P_{xxx}\| + C\|P_{xx}\|\|P_{xxx}\| \\ \leqslant& \frac{1}{2}\|P_{xxx}\|^2 + C_2\|P_{xx}\|^2 + C, \end{aligned} \tag{3.62}$$

以及

$$\begin{aligned}&\left|-2\mathrm{Re}\int_\Omega r_1W_xP\overline{P_{xxxx}}\mathrm{d}x\right| \\ =&\left|2\mathrm{Re}\int_\Omega r_1W_{xx}P\overline{P_{xxx}}dx + 2\mathrm{Re}\int_\Omega r_1W_xP_x\overline{P_{xxx}}\mathrm{d}x\right| \\ \leqslant& 2|r_1|\|P\|_{L^\infty}\|W_{xx}\|\|P_{xxx}\| + 2|r_1|\|P_x\|_{L^\infty}\|W_x\|\|P_{xxx}\| \\ \leqslant& C\|W_{xx}\|\|P_{xxx}\| + C(\|P\|^{\frac{3}{4}} + \|P_x\|^{\frac{3}{4}} + \|P_{xx}\|^{\frac{3}{4}})\|P_{xxx}\| \\ \leqslant& \frac{1}{2}\|P_{xxx}\|^2 + \xi\|P_{xx}\|^2 + C_3\|W_{xx}\|^2 + C. \end{aligned} \tag{3.63}$$

另一方面, 由 (3.59)(3.60) 以及 $0 < g \leqslant 1$, 可有

$$
\begin{aligned}
&\left|-2\mathrm{Re}\int_\Omega gr_2W_{xxx}P\overline{P_{xxxx}}\mathrm{d}x\right|\\
=&\left|2\mathrm{Re}\int_\Omega gr_2W_{xxxx}P\overline{P_{xxx}}\mathrm{d}x+2\mathrm{Re}\int_\Omega gr_2W_{xxx}P_x\overline{P_{xxx}}\mathrm{d}x\right|\\
\leqslant& 2g|r_2|\|P\|_{L^\infty}\|W_{xxxx}\|\|P_{xxx}\|+2g|r_2|\|P_x\|_{L^\infty}\|W_{xxx}\|\|P_{xxx}\|\\
\leqslant& 2g|r_2|(C+C\|W_{xx}\|^{\frac{4}{5}}+C\|W_{xxxxx}\|^{\frac{4}{5}})\|P_{xxx}\|\\
&+C(C+C\|W_{xx}\|^{\frac{4}{5}}+C\|W_{xxxxx}\|^{\frac{3}{5}})\|P_{xxx}\|\\
\leqslant& \delta\|W_{xxxxx}\|^2+2m\|W_{xxx}\|^2+\|W_{xx}\|^2+\frac{1}{2}\|P_{xxx}\|^2+\xi\|P_{xx}\|^2+C. \qquad (3.64)
\end{aligned}
$$

对于最后两项的估计, 我们有

$$
\begin{aligned}
&\left|-2\int_\Omega WW_xW_{xxxx}\mathrm{d}x-2\eta\int_\Omega(|P|^2)_xW_{xxxx}\mathrm{d}x\right|\\
=&\left|\int_\Omega W^2W_{xxxxx}\mathrm{d}x+2\eta\int_\Omega|P|^2W_{xxxxx}\mathrm{d}x\right|\\
\leqslant&\|W\|_{L^\infty}\|W\|\|W_{xxxxx}\|+2\eta\|P\|_{L^\infty}\|P\|\|W_{xxxxx}\|\\
\leqslant&\delta\|W_{xxxxx}\|^2+C. \qquad (3.65)
\end{aligned}
$$

结合 (3.57)–(3.65), 则有

$$
\begin{aligned}
\frac{\mathrm{d}}{\mathrm{d}t}(\|P_{xx}\|^2+\|W_{xx}\|^2)&\leqslant(4\xi+C_2)\|P_{xx}\|^2+(C_3+1)\|W_{xx}\|^2+C\\
&\leqslant K_2(\|P_{xx}\|^2+\|W_{xx}\|^2)+C, \qquad (3.66)
\end{aligned}
$$

其中 $K_2=\max(4\xi+C_2,C_3+1)$. 利用变换 (3.20) 以及 Gronwall 不等式, 得

$$
\|P_{xx}\|^2+\|Q_{xxx}\|^2\leqslant e^{K_2T}(\|P_{xx}(0)\|^2+\|Q_{xxx}(0)\|^2+CT)=E_2, \qquad (3.67)
$$

其中 E_2 是不依赖于 g 和 δ 的正常数. 综上, 引理 7 证毕.

推论 3 在引理 5 和引理 7 条件下, 有如下估计

$$
\|P_x\|_{L^\infty}\leqslant M_1(T), \quad \|Q_{xx}\|_{L^\infty}\leqslant M_1(T), \qquad (3.68)
$$

其中 $M_1(t)$ 是依赖于 $E_0(T),E_1(T)$ 和 $E_2(T)$ 的正常数, 且与 g 和 δ 无关.

引理 8 在引理 7 条件下, 令 $P_0(x)\in H^3(\Omega)$, $Q_0(x)\in H^4(\Omega)$, 那么对于任意的 $0\leqslant t\leqslant T$, 问题 (1.1)–(1.4) 的解具有以下估计

$$
\|P_{xxx}\|^2+\|Q_{xxxx}\|^2\leqslant E_3(T), \qquad (3.69)
$$

其中 $E_3(T)=e^{K_3T}(\|P_{xxx}(0)\|^2+\|Q_{xxxx}(0)\|^2+CT)$ 是与 g 和 δ 无关的正常数.

证明 首先, 方程 (3.21) 与 $(-\overline{P_{xxxxxx}})$ 在 Ω 做内积, 并取其实部则有

$$\begin{aligned}\frac{1}{2}\frac{\mathrm{d}}{\mathrm{d}t}\|P_{xxx}\|^2=&\xi\|P_{xxx}\|^2-\|P_{xxxx}\|^2+\mathrm{Re}\int_\Omega(1+i\nu)|P|^2P\overline{P_{xxxxxx}}\mathrm{d}x\\&+\mathrm{Re}\int_\Omega WP_x\overline{P_{xxxxxx}}\mathrm{d}x+\mathrm{Re}\int_\Omega r_1W_xP\overline{P_{xxxxxx}}\mathrm{d}x\\&+\mathrm{Re}\int_\Omega gr_2W_{xxx}P\overline{P_{xxxxxx}}\mathrm{d}x.\end{aligned}\tag{3.70}$$

同时, 方程 (3.22) 两端乘以 $(-W_{xxxxxx})$, 并在 Ω 上积分, 则有

$$\begin{aligned}\frac{1}{2}\frac{\mathrm{d}}{\mathrm{d}t}\|W_{xxx}\|^2=&-m\|W_{xxxx}\|^2-g\|W_{xxxxx}\|^2-\delta\|W_{xxxxxx}\|^2\\&+\int_\Omega WW_xW_{xxxxxx}\mathrm{d}x+\eta\int_\Omega(|P|^2)_xW_{xxxxxx}\mathrm{d}x.\end{aligned}\tag{3.71}$$

将 (3.70) 和 (3.71) 相加起来, 则有

$$\begin{aligned}&\frac{\mathrm{d}}{\mathrm{d}t}(\|P_{xxx}\|^2+\|W_{xxx}\|^2)\\=&2\xi\|P_{xxx}\|^2-2\|P_{xxxx}\|^2-2m\|W_{xxxx}\|^2-2g\|W_{xxxxx}\|^2-2\delta\|W_{xxxxxx}\|^2\\&+2\mathrm{Re}\int_\Omega(1+i\nu)|P|^2P\overline{P_{xxxxxx}}\mathrm{d}x+2\mathrm{Re}\int_\Omega WP_x\overline{P_{xxxxxx}}\mathrm{d}x\\&+2\mathrm{Re}\int_\Omega r_1W_xP\overline{P_{xxxxxx}}\mathrm{d}x+2\mathrm{Re}\int_\Omega gr_2W_{xxx}P\overline{P_{xxxxxx}}\mathrm{d}x\\&+2\int_\Omega WW_xW_{xxxxxx}\mathrm{d}x+2\eta\int_\Omega(|P|^2)_xW_{xxxxxx}\mathrm{d}x.\end{aligned}\tag{3.72}$$

利用 Gagliardo-Nirenberg 不等式以及 (3.60), 有

$$\begin{aligned}\|W_{xxxxx}\|&\leqslant C+C\|W_{xxx}\|^{\frac{3}{4}}+C\|W_{xxxx}\|^{\frac{3}{4}}+C\|W_{xxxxxx}\|^{\frac{3}{4}}\\&\leqslant C+C\|W_{xxx}\|^{\frac{3}{4}}+C\|W_{xxxxxx}\|^{\frac{3}{4}},\end{aligned}\tag{3.73}$$

$$\|W_{xxxx}\|\leqslant C+C\|W_{xxx}\|^{\frac{4}{5}}+C\|W_{xxxxxx}\|^{\frac{3}{5}}.\tag{3.74}$$

同时, 利用前面引理结果有

$$\begin{aligned}&\left|2\mathrm{Re}\int_\Omega(1+i\nu)|P|^2P\overline{P_{xxxxxx}}\mathrm{d}x\right|\\=&\left|2\mathrm{Re}\int_\Omega(1+i\nu)(2P_x^2\overline{P}+2|P|^2P_{xx}+4P|P_x|^2+P^2\overline{P_{xx}})\overline{P_{xxxx}}\mathrm{d}x\right|\\\leqslant&2|1+i\nu|(6\|P_x\|_{L^\infty}^2\|P\|+3\|P\|_{L^\infty}^2\|P_{xx}\|)\|P_{xxxx}\|\\\leqslant&\frac{1}{2}\|P_{xxxx}\|^2+C,\end{aligned}\tag{3.75}$$

以及

$$
\begin{aligned}
&\left|2\mathrm{Re}\int_{\Omega} WP_x\overline{P_{xxxxxx}}\mathrm{d}x\right| \\
=&\left|2\mathrm{Re}\int_{\Omega}(W_{xx}P_x+2W_xP_{xx}+WP_{xxx})\overline{P_{xxxx}}\mathrm{d}x\right| \\
\leqslant&2(\|P_x\|_{L^\infty}\|W_{xx}\|+2\|W_x\|_{L^\infty}\|P_{xx}\|+\|W\|_{L^\infty}\|P_{xxx}\|)\|P_{xxxx}\| \\
\leqslant&\frac{1}{2}\|P_{xxxx}\|^2+C_4\|P_{xxx}\|^2+C.
\end{aligned}
\tag{3.76}
$$

类似于估计 (3.76), 同样有

$$
\left|2\mathrm{Re}\int_{\Omega} r_1W_xP\overline{P_{xxxxxx}}\mathrm{d}x\right|\leqslant\frac{1}{2}\|P_{xxxx}\|^2+C_5\|W_{xxx}\|^2+C. \tag{3.77}
$$

另一方面, 由 Gagliardo-Nirenberg 不等式以及 (3.73)(3.74), 同样可得

$$
\begin{aligned}
&\left|2\mathrm{Re}\int_{\Omega} gr_2W_{xxx}P\overline{P_{xxxxxx}}\mathrm{d}x\right| \\
=&\left|2\mathrm{Re}\int_{\Omega} gr_2(W_{xxxxx}P+2W_{xxxx}P_x+W_{xxx}P_{xx})\overline{P_{xxxx}}\mathrm{d}x\right| \\
\leqslant&2g|r_2|(\|P\|_{L^\infty}\|W_{xxxxx}\|+2\|P_x\|_{L^\infty}\|W_{xxxx}\|)\|P_{xxxx}\| \\
&+2g|r_2|\|P_{xx}\|_{L^\infty}\|W_{xxx}\|\|P_{xxxx}\| \\
\leqslant&2g|r_2|(C+C\|W_{xxx}\|^{\frac{3}{4}}+C\|W_{xxxxxx}\|^{\frac{3}{4}})\|P_{xxxx}\| \\
&+C(C+C\|W_{xxx}\|^{\frac{4}{5}}+C\|W_{xxxxxx}\|^{\frac{3}{5}})\|P_{xxxx}\|+C\|W_{xxx}\|\|P_{xxxx}\| \\
\leqslant&\delta\|W_{xxxxxx}\|^2+C_6\|W_{xxx}\|^2+\frac{1}{2}\|P_{xxxx}\|^2+C.
\end{aligned}
\tag{3.78}
$$

对于最后两项, 有以下估计

$$
\begin{aligned}
&\left|2\int_{\Omega} WW_xW_{xxxxxx}\mathrm{d}x+2\eta\int_{\Omega}(|P|^2)_xW_{xxxxxx}\mathrm{d}x\right| \\
\leqslant&2\|W\|_{L^\infty}\|W_x\|\|W_{xxxxxx}\|+4\|P\|_{L^\infty}\|P_x\|\|W_{xxxxxx}\| \\
\leqslant&\delta\|W_{xxxxxx}\|^2+C.
\end{aligned}
\tag{3.79}
$$

综合 (3.72)–(3.79), 则有

$$
\begin{aligned}
\frac{\mathrm{d}}{\mathrm{d}t}(\|P_{xxx}\|^2+\|W_{xxx}\|^2)&\leqslant(2\xi+C_4)\|P_{xxx}\|^2+(C_5+C_6)\|W_{xxx}\|^2+C \\
&\leqslant K_3(\|P_{xxx}\|^2+\|W_{xxx}\|^2)+C,
\end{aligned}
\tag{3.80}
$$

其中 $K_3=\max(2\xi+C_4,C_5+C_6+1)$ 是一个正常数. 由变换 (3.20) 以及 Gronwall 不等式, 可以得到

$$
\|P_{xxx}\|^2+\|Q_{xxxx}\|^2\leqslant e^{K_3T}(\|P_{xxx}(0)\|^2+\|Q_{xxxx}(0)\|^2+CT)=E_3, \tag{3.81}
$$

其中 E_3 是不依赖于 g 和 δ 的一个正常数. 因此, 引理 8 得证.

推论 4 在引理 8 条件下, 对于问题 (1.1)–(1.4) 的解, 有如下估计

$$\|P_{xx}\|_{L^\infty} \leqslant M_2(T), \quad \|Q_{xxx}\|_{L^\infty} \leqslant M_2(T), \tag{3.82}$$

其中常数 $M_2(T)$ 仅依赖于 $E_0(T), E_1(T), E_2(T)$ 以及 $E_3(T)$.

更一般地, 对于耦合的 GKS-CGL 方程组 (1.1)–(1.4) 周期初值问题的解, 有以下估计.

引理 9 假设 $P_0(x) \in H^k(\Omega), Q_0(x) \in H^{k+1}(\Omega)$ $(k \geqslant 3)$, 并设 $2 - \dfrac{(1+2\eta-2r_{1r})^2}{4m} - \dfrac{g^2 r_0^2}{\delta} > 0$, 则对于任意的 $0 \leqslant t \leqslant T$, 周期初值问题 (1.1)–(1.4) 的解有以下估计

$$\left\|\frac{\mathrm{d}^k P}{\mathrm{d}x^k}\right\|^2 + \left\|\frac{\mathrm{d}^{k+1} Q}{\mathrm{d}x^{k+1}}\right\|^2 \leqslant E_k(T), \tag{3.83}$$

其中正常数 $E_k(T)$ 不依赖于 g 以及 δ.

证明 在此, 将利用数学归纳法证明. 为简单起见, 分别用 $P^{(k)}$ 和 $Q^{(k)}$ 表示 $\dfrac{\mathrm{d}^k P}{\mathrm{d}x^k}$ 以及 $\dfrac{\mathrm{d}^k Q}{\mathrm{d}x^k}$. 当 $k=3$ 时, 由引理 8 可得. 当 $k>3$, 首先假设当 $k=n-1(n\geqslant 4)$ 时, 估计 (3.83) 成立. 即有

$$\|P^{(n-1)}\|^2 \leqslant E_{n-1}, \quad \|Q^{(n)}\|^2 \leqslant E_{n-1}, \tag{3.84}$$

以及

$$\|P^{(n-2)}\|_{L^\infty} \leqslant C, \quad \|Q^{(n-1)}\|_{L^\infty} \leqslant C, \tag{3.85}$$

其中 C 均为正常数. 注意到 (3.85) 也等同于

$$\|P^{(n-2)}\|_{L^\infty} \leqslant C, \quad \|W^{(n-2)}\|_{L^\infty} \leqslant C. \tag{3.86}$$

接下来, 将要证明估计式 (3.83) 对于 $k=n$ 也是成立的. 首先, 利用变换方程组 (3.21)(3.22), 将方程 (3.21) 与 $(-1)^n\overline{P^{(2n)}}$ 在 Ω 上做内积, 并取实部则有

$$\begin{aligned}\frac{1}{2}\frac{\mathrm{d}}{\mathrm{d}t}\|P^{(n)}\|^2 =& \xi\|P^{(n)}\|^2 - \|P^{(n+1)}\|^2 - (-1)^n \mathrm{Re}\int_\Omega (1+i\nu)|P|^2 P\overline{P^{(2n)}}\mathrm{d}x\\ &-(-1)^n\mathrm{Re}\int_\Omega WP_x\overline{P^{(2n)}}\mathrm{d}x - (-1)^n\mathrm{Re}\int_\Omega r_1 W_x P\overline{P^{(2n)}}\mathrm{d}x\\ &-(-1)^n\mathrm{Re}\int_\Omega g r_2 W_{xxx} P\overline{P^{(2n)}}\mathrm{d}x.\end{aligned} \tag{3.87}$$

其次, 方程 (3.22) 两边同时乘以 $(-1)^n W^{(2n)}$, 并在 Ω 上关于 x 积分有

$$\begin{aligned}\frac{1}{2}\frac{\mathrm{d}}{\mathrm{d}t}\|W^{(n)}\|^2 =& -m\|W^{(n+1)}\|^2 - g\|W^{(n+2)}\|^2 - \delta\|W^{(n+3)}\|^2\\ &-(-1)^n\int_\Omega WW_x W^{(2n)}\mathrm{d}x - (-1)^n\eta\int_\Omega (|P|^2)_x W^{(2n)}\mathrm{d}x.\end{aligned} \tag{3.88}$$

将 (3.87) 和 (3.88) 相加起来则有

$$
\begin{aligned}
&\frac{\mathrm{d}}{\mathrm{d}t}(\|P^{(n)}\|^2+\|W^{(n)}\|^2)\\
=&2\xi\|P^{(n)}\|^2-2\|P^{(n+1)}\|^2-2m\|W^{(n+1)}\|^2-2g\|W^{(n+2)}\|^2\\
&-2\delta\|W^{(n+3)}\|^2-2(-1)^n\mathrm{Re}\int_\Omega(1+i\nu)|P|^2P\overline{P^{(2n)}}\mathrm{d}x\\
&-2(-1)^n\mathrm{Re}\int_\Omega WP_x\overline{P^{(2n)}}\mathrm{d}x-2(-1)^n\mathrm{Re}\int_\Omega r_1W_xP\overline{P^{(2n)}}\mathrm{d}x\\
&-2(-1)^n\mathrm{Re}\int_\Omega gr_2W_{xxx}P\overline{P^{(2n)}}\mathrm{d}x-2(-1)^n\int_\Omega WW_xW^{(2n)}\mathrm{d}x\\
&-2(-1)^n\eta\int_\Omega(|P|^2)_xW^{(2n)}\mathrm{d}x.
\end{aligned}\tag{3.89}
$$

首先利用 (3.84), (3.86) 以及前面引理结果, 即有

$$
\begin{aligned}
\left|-2(-1)^n\mathrm{Re}\int_\Omega(1+i\nu)|P|^2P\overline{P^{(2n)}}\mathrm{d}x\right|&=\left|2\mathrm{Re}\int_\Omega(1+i\nu)(|P|^2P)^{(n-1)}\overline{P^{(n+1)}}\mathrm{d}x\right|\\
&\leqslant\frac{1}{2}\|P^{(n+1)}\|^2+C,
\end{aligned}\tag{3.90}
$$

以及

$$
\begin{aligned}
\left|-2(-1)^n\mathrm{Re}\int_\Omega WP_x\overline{P^{(2n)}}\mathrm{d}x\right|&=\left|2\mathrm{Re}\int_\Omega(WP_x)^{(n-1)}\overline{P^{(n+1)}}\mathrm{d}x\right|\\
&\leqslant\frac{1}{2}\|P^{(n+1)}\|^2+C_{n1}\|P^{(n)}\|^2+C.
\end{aligned}\tag{3.91}
$$

类似于估计 (3.91), 可得到

$$
\left|-2(-1)^n\mathrm{Re}\int_\Omega r_1W_xP\overline{P^{(2n)}}\mathrm{d}x\right|\leqslant\frac{1}{2}\|P^{(n+1)}\|^2+C_{n2}\|W^{(n)}\|^2+C.\tag{3.92}
$$

另一方面, 利用 Gagliardo-Nirenberg 不等式, 可有

$$
\begin{aligned}
&\left|-2(-1)^n\mathrm{Re}\int_\Omega gr_2W_{xxx}P\overline{P^{(2n)}}\mathrm{d}x\right|\\
=&\left|\mathrm{Re}\int_\Omega gr_2(W_{xxx}P)^{(n-1)}\overline{P^{(n+1)}}\mathrm{d}x\right|\\
\leqslant&\, g|r_2|(C\|W_{x^{n+2}}\|+C\|W^{(n+1)}\|+C\|W^{(n)}\|+C)\|P^{(n+1)}\|\\
\leqslant&\, g|r_2|\Big(C\|W^{(n+3)}\|^{\frac{n+2}{n+3}}\|W\|^{\frac{1}{n+3}}+C\|W^{(n+3)}\|^{\frac{n+1}{n+3}}\|W\|^{\frac{2}{n+3}}
\end{aligned}
$$

$$
\begin{aligned}
&+C\|W^{(n)}\|+C\Big)\|P^{(n+1)}\| \\
&\leqslant \frac{1}{2}\|P^{(n+1)}\|^2+\delta\|W^{(n+3)}\|^2+C_{n3}\|W^{(n)}\|^2+C.
\end{aligned}
\tag{3.93}
$$

对于最后的两项, 同样有估计

$$
\begin{aligned}
&\left|-2(-1)^n\int_\Omega WW_xW^{(2n)}\mathrm{d}x-2(-1)^n\eta\int_\Omega(|P|^2)_xW^{(2n)}\mathrm{d}x\right| \\
\leqslant&\left|2\int_\Omega(WW_x)^{(n-3)}W^{(n+3)}\right|+\left|2\eta\int_\Omega(|P|^2)^{(n-2)}W^{(n+3)}\mathrm{d}x\right| \\
\leqslant&\frac{\delta}{2}\|W^{(n+3)}\|^2+\frac{\delta}{2}\|W^{(n+3)}\|^2+C \\
\leqslant&\delta\|W^{(n+3)}\|^2+C.
\end{aligned}
\tag{3.94}
$$

结合 (3.89)–(3.94), 则有

$$
\begin{aligned}
\frac{\mathrm{d}}{\mathrm{d}t}(\|P^{(n)}\|^2+\|W^{(n)}\|^2)&\leqslant(2\xi+C_{n1})\|P^{(n)}\|^2+(C_{n2}+C_{n3})\|W^{(n)}\|^2+C \\
&\leqslant K_n(\|P^{(n)}\|^2+\|W^{(n)}\|^2)+C,
\end{aligned}
\tag{3.95}
$$

其中 $K_n=\max(2\xi+C_{n1},C_{n2}+C_{n3}+1)$ 是一个正常数.

利用变换 (3.20) 以及 Gronwall 不等式, 可证得估计 (3.83) 对于 $k=n$ 也是成立的. 因此, 利用数学归纳法可证得引理 9.

推论 5 在引理引理 9 条件下, 有如下估计

$$
\|P^{(k)}\|_{L^\infty}\leqslant M_k(T),\quad \|Q^{(k+1)}\|_{L^\infty}\leqslant M_k(T),\quad (k\geqslant 2)
\tag{3.96}
$$

其中正常数 $M_k(T)$ 仅依赖于 $E_i(T)(i=1,2,\cdots,k+1)$.

作为引理 9 的一个结果, 即当 $k=5$ 时, 关于周期初值问题 (1.1)–(1.4) 的解, 有估计

引理 10 假定 $P_0(x)\in H^5(\Omega),Q_0(x)\in H^6(\Omega)$, 以及假设 $2-\dfrac{(1+2\eta-2r_{1r})^2}{4m}-\dfrac{g^2r_0^2}{\delta}>0$, 则对于任意的 $0\leqslant t\leqslant T$, 问题 (1.1)–(1.4) 的解有如下估计

$$
\|P_{xxxxx}\|^2+\|Q_{xxxxxx}\|^2\leqslant E_5(T),
\tag{3.97}
$$

以及

$$
\|P_{xxxx}\|_{L^\infty}\leqslant M_4(T),\quad \|Q_{xxxxx}\|_{L^\infty}\leqslant M_4(T),
\tag{3.98}
$$

其中 $E_5(T)$ 和 $M_4(T)$ 都是与 g 和 δ 无关的常数.

3.2 存在性和收敛性定理证明

首先, 利用 Galerkin 方法构建问题 (1.1)–(1.4) 的近似解, 因此选取周期基函数 $\{\omega_j\} \subset H^6(\Omega), (j = 1, 2, \cdots, N)$, 且满足 $-\Delta\omega_j = \lambda_j\omega_j$. 因此其近似解可表示为

$$P_N(x,t) = \sum_{j=1}^{N} \alpha_{jN}(t)\omega_j(x), \quad Q_N(x,t) = \sum_{j=1}^{N} \beta_{jN}(t)\omega_j(x). \tag{3.99}$$

根据 Galerkin 方法的思想, 待定系数 $\alpha_{jN}(t)$ 和 $\beta_{jN}(t)$ 满足如下的常微分方程组

$$\begin{aligned}(P_{Nt}, \omega_j) =& \xi(P_N, \omega_j) + (1 + i\mu)(P_{Nxx}, \omega_j) - (1 + i\nu)(|P_N|^2 P_N, \omega_j) \\ & -(P_{Nx}Q_{Nx}, \omega_j) - r_1(P_N Q_{Nxx}, \omega_j) - gr_2((Q_{Nxxxx})P_N, \omega_j),\end{aligned} \tag{3.100}$$

$$\begin{aligned}(Q_{Nt}, \omega_j) =& m(Q_{Nxx}, \omega_j) - g(Q_{Nxxxx}, \omega_j) + \delta(Q_{Nxxxxxx}, \omega_j) \\ & -\frac{1}{2}(|Q_{Nx}|^2, \omega_j) - \eta(|P_N|^2, \omega_j),\end{aligned} \tag{3.101}$$

其初始条件为

$$P_N(x,0) = P_{0N}(x), \quad Q_N(x,0) = Q_{0N}(x), \tag{3.102}$$

其中 $0 \leqslant t \leqslant T$ 以及 $j = 1, 2, \cdots, N$.

在此, 假定

$$P_{0N}(x) \stackrel{H^5(\Omega)}{\longrightarrow} P_0(x), \quad Q_{0N}(x) \stackrel{H^6(\Omega)}{\longrightarrow} Q_0(x), \quad N \to \infty. \tag{3.103}$$

从而类似于引理 4–引理 10 的证明, 可以得到问题 (1.1)–(1.4) 关于 N 的一致有界估计. 利用致密性定理以及连续延拓方法, 可得到定理 1 和定理 2 的证明.

其次, 同样利用前面引理的证明可证得 GKS-CGL 方程组 (1.1)–(1.4) 光滑解的序列 $\{P_{\varepsilon t}\}, \{P_\varepsilon\}$, $\{P_{\varepsilon x}\}, \{P_{\varepsilon xx}\}$, $\{Q_{\varepsilon t}\}$, $\{Q_{\varepsilon xx}\}$, $\{Q_{\varepsilon xxxx}\}$, $\{Q_{\varepsilon xxxxxxx}\}$ 关于 $\varepsilon = \max(g, \delta)$ 是一致有界和等度连续的, 于是利用 Arzela Ascoli 定理, 可证得当 $\varepsilon \to 0$ 时, 存在一个收敛子列收敛到相应 Burgers-CGL 方程组 (1.5)–(1.8) 的解. 因此, 定理 3 证毕.

4 收敛定理的证明

4.1 定理 4 的证明

设 $P_\varepsilon(x,t), Q_\varepsilon(x,t)$ 是 GKS-CGL 方程组 (1.1)–(1.4) 在初始值 $P_{\varepsilon 0}, Q_{\varepsilon 0}$ 的解, 而 $P(x,t), Q(x,t)$ 是 Burgers-CGL 方程组 (1.5)–(1.8) 在初始值 P_0, Q_0 的解, 那么两者解

之间的差 $X(x,t)=P_\varepsilon(x,t)-P(x,t)$ 和 $Y(x,t)=Q_{\varepsilon x}(x,t)-Q_x(x,t)$ 满足以下周期初值问题:

$$X_t=\xi X+(1+i\mu)X_{xx}-(1+i\nu)(|P_\varepsilon|^2P_\varepsilon-|P|^2P)-(P_{\varepsilon x}Q_{\varepsilon x}-P_xQ_x)$$
$$-r_1(P_\varepsilon Q_{\varepsilon xx}-PQ_{xx})-gr_2P_\varepsilon Q_{\varepsilon xxxx}, \tag{4.104}$$

$$Y_t=mY_{xx}-gQ_{\varepsilon xxxx}+\delta Q_{\varepsilon xxxxxx}-(Q_{\varepsilon x}Q_{\varepsilon xx}-Q_xQ_{xx})-\eta((|P_\varepsilon|^2)_x-(|P|^2)_x). \tag{4.105}$$

$$X(x+2L,t)=X(x,t),\quad Y(x+2L,t)=Y(x,t), \tag{4.106}$$

$$X(x,0)=X_0(x)=P_{\varepsilon 0}-P_0,\quad Y(x,0)=Y_0(x)=Q_{\varepsilon x0}-Q_{x0}. \tag{4.107}$$

方程 (4.104) 两边乘以 $\overline{X}$, 方程 (4.105) 两边乘以 Y, 并在 Ω 积分, 取其实部, 最后两个方程相加起来得

$$\begin{aligned}
&\frac{\mathrm{d}}{\mathrm{d}t}\left(\|X\|^2+\|Y\|^2\right)\\
=&2\xi\|X\|^2-2\|X_x\|^2-2m\|Y_x\|^2-2\mathrm{Re}\int_\Omega(1+i\nu)(|P_\varepsilon|^2P_\varepsilon-|P|^2P)\overline{X}\mathrm{d}x\\
&-2\mathrm{Re}\int_\Omega(P_{\varepsilon x}Q_{\varepsilon x}-P_xQ_x)\overline{X}\mathrm{d}x-2\mathrm{Re}\int_\Omega r_1(P_\varepsilon Q_{\varepsilon xx}-PQ_{xx})\overline{X}\mathrm{d}x\\
&-2\mathrm{Re}\int_\Omega gr_2P_\varepsilon Q_{\varepsilon xxxx}\overline{X}\mathrm{d}x-2g\int_\Omega Q_{\varepsilon xxxx}Y dx+2\delta\int_\Omega Q_{\varepsilon xxxxxx}Y\mathrm{d}x\\
&-2\int_\Omega(Q_{\varepsilon x}Q_{\varepsilon xx}-Q_xQ_{xx})Y\mathrm{d}x-2\eta\mathrm{Re}\int_\Omega((|P_\varepsilon|^2)_x-(|P|^2)_x)Y\mathrm{d}x.
\end{aligned} \tag{4.108}$$

现在需要对 (4.108) 右端各项进行估计. 首先利用推论 2 和推论 3, 有

$$\begin{aligned}
&\left|-2\mathrm{Re}\int_\Omega(1+i\nu)(|P_\varepsilon|^2P_\varepsilon-|P|^2P)\overline{X}\mathrm{d}x\right|\\
=&\left|-2\mathrm{Re}\int_\Omega(1+i\nu)((P_\varepsilon+P)\overline{P_\varepsilon}X-P^2\overline{X})\overline{X}\mathrm{d}x\right|\\
\leqslant&2|1+i\nu|(\|P_\varepsilon\|_{L^\infty}^2+\|P\|_{L^\infty}^2+\|P_\varepsilon\|_{L^\infty}\|P\|_{L^\infty})\|X\|^2\\
\leqslant&6|1+i\nu|M_0^2\|X\|^2,
\end{aligned} \tag{4.109}$$

$$\begin{aligned}
\left|-2\mathrm{Re}\int_\Omega(P_{\varepsilon x}Q_{\varepsilon x}-P_xQ_x)\overline{X}\mathrm{d}x\right|=&\left|-2\mathrm{Re}\int_\Omega(Q_{\varepsilon x}X_x+P_xY)\overline{X}\mathrm{d}x\right|\\
\leqslant&2\|Q_{\varepsilon x}\|_{L^\infty}\|X_x\|\|X\|+2\|P_x\|_{L^\infty}\|Y\|\|X\|\\
\leqslant&\|X_x\|^2+(M_0^2+M_1)\|X\|^2+M_1\|Y\|^2,
\end{aligned} \tag{4.110}$$

以及

$$
\begin{aligned}
&\left|-2\mathrm{Re}\int_\Omega r_1(P_\varepsilon Q_{\varepsilon xx}-PQ_{xx})\overline{X}\mathrm{d}x\right| \\
=&\left|-2\mathrm{Re}\int_\Omega r_1(Q_{\varepsilon xx}X+PY_x)\overline{X}\mathrm{d}x\right| \\
\leqslant& 2|r_1|\|Q_{\varepsilon xx}\|_{L^\infty}\|X\|^2+2|r_1|\|P\|_{L^\infty}\|Y_x\|\|X\| \\
\leqslant& m\|Y_x\|^2+\left(2|r_1|M_1+\frac{|r_1|^2}{m}M_0^2\right)\|X\|^2. \qquad (4.111)
\end{aligned}
$$

对于接下来的三项, 利用引理 8 和引理 10, 可有

$$
\begin{aligned}
\left|-2\mathrm{Re}\int_\Omega gr_2P_\varepsilon Q_{\varepsilon xxxx}\overline{X}\mathrm{d}x\right| &\leqslant 2g|r_2|\|P_\varepsilon\|_{L^\infty}\|Q_{\varepsilon xxxx}\|\|X\| \\
&\leqslant 2g|r_2|M_0E_3^{\frac{1}{2}}\|X\|, \qquad (4.112)
\end{aligned}
$$

$$
\left|-2g\int_\Omega Q_{\varepsilon xxxx}Y\mathrm{d}x\right|\leqslant 2g\|Q_{\varepsilon xxxx}\|\|Y\|\leqslant 2gE_3^{\frac{1}{2}}\|Y\|, \qquad (4.113)
$$

以及

$$
\left|2\delta\int_\Omega Q_{\varepsilon xxxxxx}Y\mathrm{d}x\right|\leqslant 2\delta\|Q_{\varepsilon xxxxxx}\|\|Y\|\leqslant 2\delta E_5^{\frac{1}{2}}\|Y\|. \qquad (4.114)
$$

最后, 结合推论 2 以及推论 3, 则有

$$
\begin{aligned}
\left|-2\int_\Omega(Q_{\varepsilon x}Q_{\varepsilon xx}-Q_xQ_{xx})Y\mathrm{d}x\right| &= \left|-2\int_\Omega(Q_{\varepsilon xx}Y+Q_xY_x)Y\mathrm{d}x\right| \\
&\leqslant 2\|Q_{\varepsilon xx}\|_{L^\infty}\|Y\|^2+2\|Q_x\|_{L^\infty}\|Y_x\|\|Y\| \\
&\leqslant m\|Y_x\|^2+\left(\frac{M_0^2}{m}+2M_1\right)\|Y\|^2, \qquad (4.115)
\end{aligned}
$$

以及

$$
\begin{aligned}
&\left|-2\eta\mathrm{Re}\int_\Omega((|P_\varepsilon|^2)_x-(|P|^2)_x)Y dx\right| \\
=&\left|-2\eta\mathrm{Re}\int_\Omega(\overline{P_\varepsilon}X_x+P_x\overline{X}+P_\varepsilon\overline{X_x}+\overline{P_x}X)Y\mathrm{d}x\right| \\
\leqslant& 4\eta\|P_\varepsilon\|_{L^\infty}\|X_x\|\|Y\|+4\eta\|P_x\|_{L^\infty}\|X\|\|Y\| \\
\leqslant& \|X_x\|^2+2\eta M_1\|X\|^2+(4\eta^2M_0^2+2\eta M_1)\|Y\|^2. \qquad (4.116)
\end{aligned}
$$

综上, 联合 (4.108)–(4.116) 可得

$$
\begin{aligned}
&\frac{\mathrm{d}}{\mathrm{d}t}\left(\|X\|^2+\|Y\|^2\right)\\
&\leqslant\left(2\xi+\left(6|1+i\nu|+1+\frac{|r_1|^2}{m}\right)M_0^2+(2|r_1|+2\eta+1)M_1\right)\|X\|^2\\
&\quad+\left(\left(\frac{1}{m}+4\eta^2\right)M_0^2+(2\eta+3)M_1\right)\|Y\|^2\\
&\quad+2g|r_2|M_0E_3^{\frac{1}{2}}\|X\|+2gE_3^{\frac{1}{2}}\|Y\|+2\delta E_5^{\frac{1}{2}}\|Y\|\\
&\leqslant 2C_7(\|X\|^2+\|Y\|^2)+2gE_3^{\frac{1}{2}}(|r_2|M_0+1)(\|X\|^2+\|Y\|^2)^{\frac{1}{2}}\\
&\quad+2\delta E_5^{\frac{1}{2}}(\|X\|^2+\|Y\|^2)^{\frac{1}{2}},
\end{aligned}
\tag{4.117}
$$

其中 $C_7=\max(\xi+\left(3|1+i\nu|+1/2+|r_1|^2/2m\right)M_0^2+(|r_1|+\eta+1/2)M_1,\ (1/2m+2\eta^2)M_0^2+(\eta+3/2)M_1)$ 是一个正常数.

如果记 $Z^2=\|X\|^2+\|Y\|^2$, 那么 (4.117) 可写成

$$
\frac{\mathrm{d}}{\mathrm{d}t}Z^2\leqslant 2C_7Z^2+2gE_3^{\frac{1}{2}}(|r_2|M_0+1)Z+2\delta E_5^{\frac{1}{2}}Z,
\tag{4.118}
$$

因此,

$$
\frac{\mathrm{d}}{\mathrm{d}t}Z\leqslant C_7Z+gE_3^{\frac{1}{2}}(|r_2|M_0+1)+\delta E_5^{\frac{1}{2}}.
\tag{4.119}
$$

利用 Gronwall 引理, 可以得到

$$
Z\leqslant Z(0)+gE_3^{\frac{1}{2}}(|r_2|M_0+1)e^{C_7t}+\delta E_5^{\frac{1}{2}}e^{C_7t},
\tag{4.120}
$$

其中 $0\leqslant t\leqslant T$.

根据 Z,X,Y 的定义, 有

$$
\|X\|^2+\|Y\|^2\leqslant 3(\|X(0)\|^2+\|Y(0)\|^2)+3g^2E_3(|r_2|M_0+1)^2e^{2C_7t}+3\delta^2E_5e^{2C_7t},
\tag{4.121}
$$

以及

$$
\begin{aligned}
\|P_\varepsilon-P\|^2+\|Q_{\varepsilon x}-Q_x\|^2\leqslant{}&3(\|P_{\varepsilon 0}-P_0\|^2+\|Q_{\varepsilon x0}-Q_{x0}\|^2)\\
&+3g^2E_3(|r_2|M_0+1)^2e^{2C_7t}+3\delta^2E_5e^{2C_7t},
\end{aligned}
\tag{4.122}
$$

其中 $0\leqslant t\leqslant T$. 定理 4 证毕.

4.2 定理 5 的证明

同样地, 方程 (4.104) 两边乘以 $-\overline{X_{xx}}$, 在 Ω 积分, 并取实部. 方程 (4.105) 与 $-Y_{xx}$ 在 Ω 做内积, 最后将两个等式相加起来有

$$
\begin{aligned}
\frac{\mathrm{d}}{\mathrm{d}t}\left(\|X_x\|^2+\|Y_x\|^2\right)=&2\xi\|X_x\|^2-2\|X_{xx}\|^2-2m\|Y_{xx}\|^2\\
&+2\mathrm{Re}\int_\Omega(1+i\nu)(|P_\varepsilon|^2P_\varepsilon-|P|^2P)\overline{X_{xx}}\mathrm{d}x\\
&+2\mathrm{Re}\int_\Omega(P_{\varepsilon x}Q_{\varepsilon x}-P_xQ_x)\overline{X_{xx}}\mathrm{d}x\\
&+2\mathrm{Re}\int_\Omega r_1(P_\varepsilon Q_{\varepsilon xx}-PQ_{xx})\overline{X_{xx}}\mathrm{d}x\\
&+2\mathrm{Re}\int_\Omega gr_2P_\varepsilon Q_{\varepsilon xxxx}\overline{X_{xx}}\mathrm{d}x+2g\int_\Omega Q_{\varepsilon xxxx}Y_{xx}\mathrm{d}x\\
&-2\delta\int_\Omega Q_{\varepsilon xxxxxx}Y_{xx}\mathrm{d}x+2\int_\Omega(Q_{\varepsilon x}Q_{\varepsilon xx}-Q_xQ_{xx})Y_{xx}\mathrm{d}x\\
&+2\eta\mathrm{Re}\int_\Omega((|P_\varepsilon|^2)_x-(|P|^2)_x)Y_{xx}\mathrm{d}x. \qquad (4.123)
\end{aligned}
$$

对于 (4.123) 右端各项的估计, 首先利用推论 2 以及推论 3, 则有

$$
\begin{aligned}
&\left|2\mathrm{Re}\int_\Omega(1+i\nu)(|P_\varepsilon|^2P_\varepsilon-|P|^2P)\overline{X_{xx}}\mathrm{d}x\right|\\
=&\left|2\mathrm{Re}\int_\Omega(1+i\nu)((P_\varepsilon+P)\overline{P_\varepsilon}X-P^2\overline{X})\overline{X_{xx}}\mathrm{d}x\right|\\
\leqslant&2|1+i\nu|(\|P_\varepsilon\|_{L^\infty}^2+\|P\|_{L^\infty}^2+\|P_\varepsilon\|_{L^\infty}\|P\|_{L^\infty})\|X\|\|X_{xx}\|\\
\leqslant&\frac{1}{2}\|X_{xx}\|^2+18|1+i\nu|^2M_0^4\|X\|^2, \qquad (4.124)
\end{aligned}
$$

$$
\begin{aligned}
\left|2\mathrm{Re}\int_\Omega(P_{\varepsilon x}Q_{\varepsilon x}-P_xQ_x)\overline{X_{xx}}\mathrm{d}x\right|=&\left|2\mathrm{Re}\int_\Omega(Q_{\varepsilon x}X_x+P_xY)\overline{X_{xx}}\mathrm{d}x\right|\\
\leqslant&2\|Q_{\varepsilon x}\|_{L^\infty}\|X_x\|\|X_{xx}\|+2\|P_x\|_{L^\infty}\|Y\|\|X_{xx}\|\\
\leqslant&\frac{1}{2}\|X_{xx}\|^2+4M_0^2\|X_x\|^2+4M_1^2\|Y\|^2, \qquad (4.125)
\end{aligned}
$$

以及

$$
\begin{aligned}
&\left|2\mathrm{Re}\int_\Omega r_1(P_\varepsilon Q_{\varepsilon xx}-PQ_{xx})\overline{X_{xx}}\mathrm{d}x\right|\\
=&\left|2\mathrm{Re}\int_\Omega r_1(Q_{\varepsilon xx}X+PY_x)\overline{X_{xx}}\mathrm{d}x\right|\\
\leqslant&\, 2|r_1|\|Q_{\varepsilon xx}\|_{L^\infty}\|X\|\|X_{xx}\|+2|r_1|\|P\|_{L^\infty}\|Y_x\|\|X_{xx}\|\\
\leqslant&\,\frac{1}{2}\|X_{xx}\|^2+4|r_1|^2M_1^2\|X\|^2+4|r_1|^2M_0^2\|Y_x\|^2.
\end{aligned}\tag{4.126}
$$

对于接下来的三项, 由推论 2, 引理 8 和引理 10, 可以得到

$$
\begin{aligned}
\left|2\mathrm{Re}\int_\Omega gr_2P_\varepsilon Q_{\varepsilon xxxx}\overline{X_{xx}}\mathrm{d}x\right|&\leqslant 2g|r_2|\|P_\varepsilon\|_{L^\infty}\|Q_{\varepsilon xxxx}\|\|X_{xx}\|\\
&\leqslant\frac{1}{2}\|X_{xx}\|^2+2g^2|r_2|^2M_0^2E_3^2,
\end{aligned}\tag{4.127}
$$

$$
\left|2g\int_\Omega Q_{\varepsilon xxxx}Y_{xx}\mathrm{d}x\right|\leqslant 2g\|Q_{\varepsilon xxxx}\|\|Y_{xx}\|\leqslant\frac{m}{2}\|Y_{xx}\|^2+2g^2E_3^2,\tag{4.128}
$$

以及

$$
\left|-2\delta\int_\Omega Q_{\varepsilon xxxxxx}Y_{xx}\mathrm{d}x\right|\leqslant 2\delta\|Q_{\varepsilon xxxxxx}\|\|Y_{xx}\|\leqslant\frac{m}{2}\|Y_{xx}\|^2+2\delta^2E_5^2.\tag{4.129}
$$

最后, 同样可以有

$$
\begin{aligned}
\left|2\int_\Omega(Q_{\varepsilon x}Q_{\varepsilon xx}-Q_xQ_{xx})Y_{xx}\mathrm{d}x\right|&=\left|2\int_\Omega(Q_{\varepsilon xx}Y+Q_xY_x)Y_{xx}\mathrm{d}x\right|\\
&\leqslant 2\|Q_{\varepsilon xx}\|_{L^\infty}\|Y\|\|Y_{xx}\|+2\|Q_x\|_{L^\infty}\|Y_x\|\|Y_{xx}\|\\
&\leqslant\frac{m}{2}\|Y_{xx}\|^2+\frac{4M_0^2}{m}\|Y_x\|^2+\frac{4M_1^2}{m}\|Y\|^2,
\end{aligned}\tag{4.130}
$$

以及

$$
\begin{aligned}
&\left|2\eta\mathrm{Re}\int_\Omega((|P_\varepsilon|^2)_x-(|P|^2)_x)Y_{xx}\mathrm{d}x\right|\\
=&\left|2\eta\mathrm{Re}\int_\Omega(\overline{P_\varepsilon}X_x+P_x\overline{X}+P_\varepsilon\overline{X_x}+\overline{P_x}X)Y_{xx}\mathrm{d}x\right|\\
\leqslant&\,4\eta\|P_\varepsilon\|_{L^\infty}\|X_x\|\|Y_{xx}\|+4\eta\|P_x\|_{L^\infty}\|X\|\|Y_{xx}\|\\
\leqslant&\,\frac{m}{2}\|Y_{xx}\|^2+\frac{16\eta^2M_0^2}{m}\|X_x\|^2+\frac{16\eta^2M_1^2}{m}\|X\|^2.
\end{aligned}\tag{4.131}
$$

将 (4.124)–(4.131) 代入 (4.123), 可得

$$
\begin{aligned}
&\frac{\mathrm{d}}{\mathrm{d}t}\left(\|X_x\|^2+\|Y_x\|^2\right)\\
\leqslant&\left(2\xi+4M_0^2+\frac{16\eta^2M_0^2}{m}\right)\|X_x\|^2+\left(4|r_1|^2M_0^2+\frac{4M_0^2}{m}\right)\|Y_x\|^2\\
&+\left(18|1+i\nu|^2M_0^4+4|r_1|^2M_1^2+\frac{16\eta^2M_1^2}{m}\right)\|X\|^2\\
&+\left(4M_1^2+\frac{4M_1^2}{m}\right)\|Y\|^2+2g^2E_3^2(|r_2|^2M_0^2+1)+2\delta^2E_5^2\\
\leqslant&C_8(\|X_x\|^2+\|Y_x\|^2)+C_9(\|X\|^2+\|Y\|^2)+2g^2E_3^2(|r_2|^2M_0^2+1)+2\delta^2E_5^2.
\end{aligned}
\tag{4.132}
$$

其中 $C_8=\max\left(2\xi+4M_0^2+16\eta^2M_0^2/m,\ 4|r_1|^2M_0^2+4M_0^2/m\right)$ 和 $C_9=\max(18|1+i\nu|^2M_0^4+4|r_1|^2M_1^2+16\eta^2M_1^2/m,\ 4M_1^2+4M_1^2/m)$ 是两个与 g 和 δ 都无关的正常数.

利用 (4.121) 的结果, 可以得到

$$
\begin{aligned}
\frac{\mathrm{d}}{\mathrm{d}t}\left(\|X_x\|^2+\|Y_x\|^2\right)\leqslant&C_8(\|X_x\|^2+\|Y_x\|^2)+3C_9(\|X(0)\|^2+\|Y(0)\|^2)\\
&+g^2\left(9C_9E_3(|r_2|M_0+1)^2e^{2C_7T}+2E_3^2(|r_2|^2M_0^2+1)\right)\\
&+\delta^2\left(9C_9E_5e^{2C_7T}+2E_5^2\right)\\
\leqslant&C_8(\|X_x\|^2+\|Y_x\|^2)+3C_9(\|X(0)\|^2+\|Y(0)\|^2)\\
&+g^2C_{10}+\delta^2C_{11},
\end{aligned}
\tag{4.133}
$$

其中正常数 $C_{10}=9C_9E_3(|r_2|M_0+1)^2e^{2C_7T}+2E_3^2(|r_2|^2M_0^2+1)$ 和 $C_{11}=9C_9E_5e^{2C_7T}+2E_5^2$ 均不依赖于 g 和 δ.

应用 Gronwall 不等式, 并注意到 X,Y 的定义, 即有

$$
\begin{aligned}
\|P_{\varepsilon x}-P_x\|^2+\|Q_{\varepsilon xx}-Q_{xx}\|^2\leqslant&(\|P_{\varepsilon x0}-P_{x0}\|^2+\|Q_{\varepsilon xx0}-Q_{xx0}\|^2)e^{C_8t}\\
&+3C_9(\|P_{\varepsilon 0}-P_0\|^2+\|Q_{\varepsilon x0}-Q_{x0}\|^2)te^{C_8t}\\
&+g^2C_{10}te^{C_8t}+\delta^2C_{11}te^{C_8t},
\end{aligned}
\tag{4.134}
$$

其中 $0\leqslant t\leqslant T$. 因此, 定理 5 证毕.

4.3 定理 6 的证明

利用定理 2 的结果, 类似于定理 4 和定理 5 的证明, 同样可得到定理 6 的证明.

5 结论

耦合 GKS-CGL 方程组描述了两步连串反应中, 两个火焰锋面之间的非线性相互作用, 即描述具有固定长度距离的, 且是均匀传播的两个火焰之间的单调不稳定性和振荡不稳定性. 因此, 对于 GKS-CGL 方程组的研究, 引起了数学学者和其他研究者的兴趣. 本文主要从数学上研究了耦合 GKS-CGL 方程组的周期初值问题. 首先, 利用巧妙的函数变换, 精细的先验估计和 Galerkin 方法得到其整体解的存在唯一性. 其次重点研究了 GKS-CGL 方程组解的极限行为, 当系数 $g \to 0$ 和 $\delta \to 0$ 时, 其解收敛到 Burgers-CGL 方程组的解, 并且其收敛估计为

$$\|P_\varepsilon - P\|_{H^{k-4}} + \|Q_\varepsilon - Q\|_{H^{k-3}} = O(g) + O(\delta), \quad g \to 0, \delta \to 0, \quad (k \geqslant 5). \quad (4.135)$$

致谢 衷心感谢审稿人对初稿提出的宝贵修改意见和建议.

参考文献

[1] Golovin A A, Matkowsky B J, Bayliss A, Nepomnyashchy A A. Coupled KS–CGL and coupled Burgers-CGL equations for flames governed by a sequential reaction [J]. Phys. D, 1999, 129: 253–298.

[2] Kirrmann P, Schneider G, Mielke A. The validity of modulation equations for extended systems with cubic nonlinearities[J]. Proc. Roy. Soc. Edinb. Sect. A, 1992, 122: 85–91.

[3] Nepomnyashchy A A. Order parameter equations for long wavelength instabilities [J]. Phys. D, 1995, 86: 90–95.

[4] Sivashinsky G I. Nonlinear analysis of hydrodynamic instability in laminar flames I. Derivation of basic equations [J]. Acta Astro., 1977, 4(11-12): 1177–1206.

[5] Burgers J M. A mathematical model illustrating the theory of turbulence[J]. Adv. Appl. Mech., 1948, 1: 171–199.

[6] Ghidaglia J M, Héron B. Dimension of the attractor associated to the Ginzburg-Landau equation [J]. Phys. D, 1987, 28(3): 282–304.

[7] Doering C R, Gibbon J D, Holm D D, Nicolaenko B. Low-dimensional behavior in the complex Ginzburg-Landau equation [J]. Nonlinearity, 1988, 1(2): 279–309.

[8] Promislow K. Induced trajectories and approximate inertial manifolds for the Ginzburg-Landau partial differential equation [J]. Phys. D, 1990, 41(2): 232–252.

[9] Li D, Guo B, Liu X. Regularity of the attractor for 3-D complex Ginzburg-Landau equation [J]. Acta Math. Appl. Sin. Engl. Ser., 2011, 27(2): 289–302.

[10] Li D, Guo B, Liu X. Existence of global solution for complex Ginzburg Landau equation in three dimensions [J]. Appl. Math. J. Chinese Univ. Ser. A, 2004, 19(4): 409–416.

[11] Montgomery K A, Silber M. Feedback control of travelling wave solutions of the complex Ginzburg-Landau equation [J]. Nonlinearity, 2004, 17(6): 2225–2248.

[12] Sherratt J A, Smith M J, Rademacher J D M. Patterns of sources and sinks in the complex Ginzburg-Landau equation with zero linear dispersion [J]. SIAM J. Appl. Dyn. Syst., 2010, 9(3): 883–918.

[13] Wu J. The inviscid limit of the complex Ginzburg-Landau equation [J]. J. Differential Equations, 1998, 142: 413–433.

[14] Bechouche P, Jüngel A. Inviscid limits of the complex Ginzburg-Landau equation [J]. Commun. Math. Phys., 2000, 214: 201–226.

[15] Wang B. The limit behavior of solutions for the Cauchy problem of the complex Ginzburg-Landau equation [J]. Commu. Pure. Appl. Math., 2002, 55: 0481–0508.

[16] Guo B. The existence and nonexistence of a global smooth solution for the initial value problem of generalized Kuramoto-Sivashinsky type equations [J]. J. Math. Res. Exposition, 1991, 11(1): 57–70.

[17] Guo B. The nonlinear Galerkin methods for the generalized Kuramoto-Sivashinsky type equations [J]. Adv. Math., 1993, 22(2): 182–184.

[18] Doronin1 G G, Larkin N A. Kuramoto-Sivashinsky model for a dusty medium [J]. Math. Methods Appl. Sci., 2003, 26(3): 179–192.

[19] MacKenzie T, Roberts A J. Accurately model the Kuramoto-Sivashinsky dynamics with holistic discretization[J]. SIAM J. Appl. Dynam. Sys., 2006, 5(3): 365–402.

[20] Hopf E. The partial differential equation $u_t + uu_x = \mu u_{xx}$ [J]. Comm. Pure Appl. Math., 1950, 3: 201–230.

[21] Wood W L. An exact solution for Burgers equation[J]. Commun. Numer. Meth. Eng., 2006, 22: 797–798.

[22] Zhu C, Wang R. Numerical solution of Burgers' equation by cubic B-spline quasi-interpolation [J]. Appl. Math. Comput., 2009, 208: 260–272.

[23] Biagioni H A, Oberguggenberger M. Generalized solutions to Burgers equation [J]. J. Differential Equations, 1992, 97: 263–287.

[24] Guo C, Fang S, Guo, B. Global smooth solutions of the generalized GKS-CGL equations for flames governed by a sequential reaction [J]. Commun. Math. Sci., 2014, 12(8): 1457–1474.

[25] Guo C, Fang S, Guo, B. Long time behavior of the solutions for the coupled Burgers-CGL equations [J]. Appl. Math. Mech. -Engl. Ed., 2014, 35(4): 515–534.

[26] Evans L C. Partial differential equations, Graduate studies in mathematics [M]. Providence, Rhode Island, American Mathematical Society, 1988.

[27] Friedman A. Partial differential equations [M]. New York: Dover Publications, Inc, 1969.

[28] Temam R. Infinite dimensional dynamical systems in mechanics and physics [M]. New York: Springer-Verlag, 1997.

A Class of Stable and Conservative Finite Difference Schemes for the Cahn-Hilliard Equation*

Wang Tingchun (王廷春), Zhao Limei (赵丽梅) and Guo Boling (郭柏灵)

Abstract In this work, we propose a class of stable finite difference schemes for the initial-boundary value problem of the Cahn-Hilliard equation. These schemes are proved to inherit the total energy dissipation property and mass conservation property from the associated continuous problem. The dissipation of the total energy implies boundedness of the numerical solutions in the discrete H^1 norm. This in turn implies, by discretized Sobolev's lemma, the boundedness ofthe numerical solutions in the max norm, and hence the stability of the difference schemes. Unique existence of the numerical solutions is proved by fixed-point theorem. Convergence rate of the class of finite difference schemes is studied by using discrete energy method. An efficient iterative algorithm for solving these nonlinear schemes is proposed and discussed in detail.

Keywords Cahn-Hilliard equation; Finite difference scheme; Conservation of mass; Dissipation of energy; Convergence; Iterative algorithm

1 Introduction

The Cahn-Hilliard equation

$$\frac{\partial u}{\partial t} - q\frac{\partial^4 u}{\partial x^4} = \frac{\partial^2}{\partial x^2}\phi(u), \quad (x,t) \in (0,L) \times (0,T], \tag{1.1}$$

$$\phi(u) = pu + ru^3, \quad (x,t) \in (0,L) \times (0,T], \tag{1.2}$$

arises in the study of phase separation in cooling binary solutions such as alloys, glasses and polymer mixtures(see [1,2,3] and the references cited therein). Here $p < 0, q < 0$

* Acta Math. Appl. Sin., Engl. Ser., 2015, 3(4): 863 – 878. DOI: 10.1007/s10255-015-0536-7.

and $r>0$ are constants, $u(x,t)$ is a perturbation of the concentration of one of the phases. Initial and boundary conditions are given as follows

$$u(x,0)=u_0(x), \quad x\in(0,L), \tag{1.3}$$

$$\frac{\partial}{\partial x}u(x,t)=\frac{\partial^3}{\partial x^3}u(x,t)=0, \quad (x,t)\in\{0,L\}\times(0,T]. \tag{1.4}$$

(1.4) leads to

$$\frac{\partial}{\partial x}\phi(u(x,t))=0, \quad (x,t)\in\{0,L\}\times(0,T].$$

By simple calculation we can see that $\phi(u)+q\dfrac{\partial^2 u}{\partial x^2}$ is the variational derivative of

$$G(u(x,t))=\widetilde{\phi}(u)-\frac{1}{2}q\left(\frac{\partial u}{\partial x}\right)^2,$$

with respect to $u(x,t)$, i.e., $\phi(u)+q\dfrac{\partial^2 u}{\partial x^2}=\dfrac{\delta G}{\delta u}$, where $\widetilde{\phi}(u)=\dfrac{1}{2}pu^2+\dfrac{1}{4}ru^4$, and the functional G means a local free energy called a Ginzburg-Landau free energy.

The important features of the Cahn-Hilliard equation are that the total mass $\displaystyle\int_0^L u(x,t)\mathrm{d}x$ is conserved and the total free energy $\displaystyle\int_0^L G(u(x,t))\mathrm{d}x$ decreases with time. Namely,

$$\int_0^L u(x,t)\mathrm{d}x=\int_0^L u_0(x)\mathrm{d}x=\mathcal{M}, \quad t>0, \tag{1.5}$$

$$\frac{\mathrm{d}}{\mathrm{d}t}\mathcal{F}(u)\leqslant 0, \quad \mathcal{F}(u)=\int_0^L G(u(x,t))\mathrm{d}x. \tag{1.6}$$

The conservation of mass (1.5) and the dissipation of the total energy (1.6) can be shown easily as follows,

$$\frac{\mathrm{d}}{\mathrm{d}t}\int_0^L u(x,t)\mathrm{d}x=\int_0^L\frac{\partial u(x,t)}{\partial t}\mathrm{d}x=\int_0^L\frac{\partial^2}{\partial x^2}\frac{\delta G}{\delta u}\mathrm{d}x=\left[\frac{\partial}{\partial x}\frac{\delta G}{\delta u}\right]_0^L=0. \tag{1.7}$$

$$\frac{\mathrm{d}}{\mathrm{d}t}\int_0^L G(u(x,t))\mathrm{d}x=\int_0^L\frac{\delta G}{\delta u}\frac{\partial u}{\partial t}\mathrm{d}x=-\int_0^L\left[\frac{\partial}{\partial x}\frac{\delta G}{\delta u}\right]^2\mathrm{d}x\leqslant 0. \tag{1.8}$$

Remark 1.1 *We see from* (1.8) *that the total energy can be employed as a Lyapunov functional of the system* (see [4]).

Since the pioneering work of Cahn and Hilliard[1], the Cahn-Hilliard equation has been extensively studied by Wang and Shi[5], Jabbari and Peppas[6], Puri and

Binder[7] for the study of interfaces. Global existence and uniquiness of the solution have been shown by Elliott and Zheng[8]. Yin[9] has shown the existence of the continuous solution for the problem with degenerate mobility. Finite element Galerkin solutions have been obtained by Elliott and French[10,11] and French and Jensen[12]. Elliott et al.[13] have obtained optimal order bounds using a second order splitting method. Elliott and Larsson[14] discuss the error estimates with smooth and nonsmooth data for a finite element method for the Cahn-Hilliard equation. Mixed finite element method is applied by Dean et al.[15]. A finite difference scheme has been studied by Furihata et al.[17] who have examined the boundedness of the solution. Sun[17] proposed an interesting linearized conservative finite difference scheme which is uniquely solvable and convergent with the convergence rate of order two in discrete L_2 norm. Choo et al.[18,19] have proposed a nonlinear difference scheme based on the Crank-Nicolson scheme for the Cahn-Hilliard equation. Their schemes are proved to be unconditional stable and conserve the total mass. Dehghan and Mirzae[20] describe a numerical method based on the boundary integral equation and dual reciprocity methods for solving the one-dimensional Cahn-Hilliard equation. A time-stepping method and a predictor-corrector scheme are employed to deal with the time derivative and the nonlinearity respectively. In [21,22], a combined spectral and large time-stepping methods were proposed and studied for the nonlinear diffusion equations for thin film epitaxy. In [23], the convergence of the spatial discretization of the Cahn-Hilliard is considered. In resent study[24,25,26], the unconditionally stable algorithms were developed for Cahn-Hillard equation. These algorithms allow for an increasing time step in Cahn-Hillard systems as time proceeds. He and Liu[27] has proposed a class of fully discrete dissipative Fourier spectral schemes for solving the two-dimensional Cahn-Hilliard equation, and presented semi-implicit prediction-correction schemes. Ye[28,29,30] has studied numerically the Cahn-Hilliard equation by using the Fourier collocation method, Fourier spectral method and Legendre spectral method, respectively. Both the semi-discrete and the fully discrete schemes derived in [28,29,30] are uniquely solvable and inherit the energy dissipation property and the mass conservation property. The optimal error bounds of numerical solutions has also been obtained. Ceniceros and Roma[31] present a nonstiff, fully adaptive mesh refinement-based method for the Cahn-Hilliard equation. Yinhua Xia et al.[32] develop local discontinuous Galerkin methods for the fourth order nonlinear Cahn-Hilliard equation and system. In [33], Furihata has designed a difference scheme which inherits the conservation of the total mass and the decrease of the total energy, and proved that the designed scheme preserving characteristic properties of the original equation are numerically stable. Adopting the idea of Furihata[33] Choo, Chung and Lee[34] propose a nonlinear difference scheme for the viscous Cahn-Hilliard equations with nonconstant gradient energy coefficient q and showed the scheme preserves the energy dissipation property and mass conservation as for the classical solution.

In this paper, we mainly do three things. Firstly, we propose a class of finite difference schemes which are stable and preserve both of the two properties (1.5) and (1.6). Secondly, we prove the unique existence and convergence of the numerical solutions. Lastly, we construct and discuss in detail an iterative algorithm for solving the proposed nonlinear schemes.

The remainder of this paper is arranged as follows. In Section 2, we propose a class of finite difference schemes which are proved to inherit the properties (1.5) and (1.6), and consequently the stability of them is obtained. In Section 3, the unique existence of the numerical solutions is discussed by *Brouwer* fixed-point theorem. In Section 4, the convergence of the class of finite difference schemes is proved. In Second 5, an iterative algorithm for solving the proposed nonlinear schemes is constructed and discussed in detail, and then a prediction-correction scheme is proposed based on the iterative algorithm. In Second 6, some numerical results demonstrate the effectiveness of the proposed schemes.

2 Finite difference schemes

Let $h=\dfrac{L}{J}$ be the uniform step in spatial direction for a positive integer J and $\Omega_h=\{x_j=jh|j=-2,-1,0,1,\cdots,J,J+1,J+2\}$. Let $\tau=\dfrac{T}{N}$ be the uniform step in the temporal direction for any positive integer N. Denote $V_j^n=V(x_j,t_n)$ for $x_j=jh, j=-2,-1,0,\cdots,J+1,J+2$ and $t_n=n\tau, n=0,1,2,\cdots,N$. For mesh functions $U^n=(U_{-2}^n,U_{-1}^n,U_0^n,\cdots,U_{J+1}^n,U_{J+2}^n)$ and $V^n=(V_{-2}^n,V_{-1}^n,V_0^n,\cdots,V_{J+1}^n,V_{J+2}^n)$ defined on Ω_h, define the difference operators as for $j=0,1,\cdots,J$,

$$\delta_x^+V_j^n=\frac{V_{j+1}^n-V_j^n}{h},\quad \delta_x^-V_j^n=\frac{V_j^n-V_{j-1}^n}{h},\quad \delta_x^{\langle 1\rangle}V_j^n=\frac{V_{j+1}^n-V_{j-1}^n}{2h},$$

$$\delta_x^{\langle 2\rangle}V_j^n=\frac{V_{j+1}^n-2V_j^n+V_{j-1}^n}{h^2},\quad \delta_x^{\langle 3\rangle}V_j^n=\frac{V_{j+2}^n-2V_{j+1}^n+2V_{j-1}^n-V_{j+2}^n}{2h^3},$$

$$\delta_x^{\langle 4\rangle} V_j^n = \frac{V_{j+2}^n - 4V_{j+1}^n + 6V_j^n - 4V_{j-1}^n + V_{j+2}^n}{h^4}, \quad \delta_t^+ V_j^n = \frac{V_j^{n+1} - V_j^n}{\tau},$$

$$\delta_t^- V_j^n = \frac{V_j^n - V_j^{n-1}}{\tau}, \quad (U^n, V^n)_h = h\left[\frac{1}{2}U_0^n V_0^n + \sum_{j=1}^{J-1} U_j^n V_j^n + \frac{1}{2}U_J^n V_J^n\right],$$

$$||V^n||^2 = (V^n, V^n)_h, \quad ||V^n||_\infty = \max_{0\leqslant j\leqslant J} |V_j^n|, \quad ||\delta_x^+ V^n||^2 = h\sum_{j=0}^{J-1} \left(\delta_x^+ V_j^n\right)^2.$$

In this paper we denote $C_k, \widetilde{C}_k, k = 0, 1, 2, \cdots$, as general positive constants which may have different values in different occurrences.

Denote

$$\varphi(u, v, \alpha) = \left(\frac{u+v}{2}\right)(r(\alpha u^2 + (1-\alpha)v^2) + p),$$

which can be seen in [27]. Clearly, $\varphi_0(u, u, \alpha) = \phi(u)$. In particular,

$$\varphi\left(U_j^{n+1}, U_j^n, \frac{1}{2}\right) = \left(\frac{U_j^{n+1} + U_j^n}{2}\right)\left(r\frac{(U_j^{n+1})^2 + (U_j^n)^2}{2} + p\right)$$

.

We propose the following finite difference scheme

$$\begin{aligned}
&\delta_t^+ U_j^n + \beta h^2 \delta_x^{\langle 2\rangle} \delta_t^+ U_j^n - \frac{q}{2}\delta_x^{\langle 4\rangle}(U_j^{n+1} + U_j^n) = \delta_x^{\langle 2\rangle} \varphi(U_j^{n+1}, U_j^n, \alpha), \\
&\qquad 0 \leqslant j \leqslant J, \quad n = 0, 1, 2, \cdots, N-1, &(2.1)\\
&U_j^0 = u_0(x_j), \quad j = -2, -1, 0, \cdots, J, J+1, J+2, &(2.2)\\
&\delta_x^{\langle 1\rangle} U_0^n = \delta_x^{\langle 1\rangle} U_J^n = \delta_x^{\langle 3\rangle} U_0^n = \delta_x^{\langle 3\rangle} U_J^n = 0, \quad n = 0, 1, 2, \cdots, N. &(2.3)
\end{aligned}$$

Obviously, the scheme (2.1)–(2.3) is a nonlinear implicit one. In order to obtain the solution U_j^{n+1} in the level $n+1$, an outer nonlinear iteration for U_j^{n+1} needs to be done and the iterative values of U_j^{n+1} are solved by an inner linear system. Therefore, an efficient iterative algorithm is required to solve the scheme (2.1)–(2.3), and we will construct one in Section 5.

The boundary condition (2.3) leads to

$$\left[\delta_x^{\langle 1\rangle} \varphi(U_j^{n+1}, U_j^n, \alpha)\right]_0^J = 0.$$

We now turn to establish the discrete analogues of (1.5)–(1.6). Let $\alpha \geqslant \frac{1}{2}$ and $\beta \leqslant \frac{1}{4}$. If U_j^n is the numerical solution of the scheme (2.1)–(2.3), we can obtain the following Lemma.

Lemma 2.1 *The solution of the scheme* (2.1)–(2.3) *satifies*

$$h\sum_{j=0}^{J}{}''U_j^n = h\sum_{j=0}^{J}{}''U_j^0 = h\sum_{j=0}^{J}{}''u_0(x_j), \qquad n=0,1,2,\cdots,N, \tag{2.4}$$

$$\frac{1}{\tau}\left[h\sum_{j=0}^{J}{}''G_d(U_j^{n+1}) - h\sum_{j=0}^{J}{}''G_d(U_j^n)\right] \leqslant 0, \qquad n=0,1,2,\cdots,N-1, \tag{2.5}$$

where

$$G_d(U_j^n) = \widetilde{\phi}(U_j^n) - \frac{q}{2}\frac{(\delta_j^+U_j^n)^2+(\delta_j^-U_j^n)^2}{2}.$$

Proof

$$\begin{aligned}
&\frac{1}{\tau}\left\{h\sum_{j=0}^{J}{}''U_j^{n+1} - h\sum_{j=0}^{J}{}''U_j^n\right\} = h\sum_{j=0}^{J}{}''\delta_t^+U_j^n \\
=&h\sum_{j=0}^{J}{}''\left[-\beta h^2\delta_x^{\langle 2\rangle}\delta_t^+U_j^n + \frac{q}{2}\delta_x^{\langle 4\rangle}(U_j^{n+1}+U_j^n) + \delta_x^{\langle 2\rangle}\varphi(U_j^{n+1},U_j^n,\alpha)\right] \\
=&h\sum_{j=0}^{J}{}''\delta_x^{\langle 2\rangle}\left[-\beta h^2\delta_t^+U_j^n + \frac{q}{2}\delta_x^{\langle 2\rangle}(U_j^{n+1}+U_j^n) + \varphi(U_j^{n+1},U_j^n,\alpha)\right] \\
=&\left[\delta_x^{\langle 1\rangle}\left[-\beta h^2\delta_t^+U_j^n + \frac{q}{2}\delta_x^{\langle 2\rangle}(U_j^{n+1}+U_j^n) + \varphi(U_j^{n+1},U_j^n,\alpha)\right]\right]_0^J = 0.
\end{aligned} \tag{2.6}$$

Then (2.4) follows from (2.6).

$$\begin{aligned}
&\frac{1}{\tau}\left[h\sum_{j=0}^{J}{}''G_d(U_j^{n+1}) - h\sum_{j=0}^{J}{}''G_d(U_j^n)\right] \\
=&h\sum_{j=0}^{J}{}''\left[\widetilde{\phi}(U_j^{n+1}) - \frac{q}{2}\frac{(\delta_j^+U_j^{n+1})^2+(\delta_j^-U_j^{n+1})^2}{2} - \widetilde{\phi}(U_j^n) + \frac{q}{2}\frac{(\delta_j^+U_j^n)^2+(\delta_j^-U_j^n)^2}{2}\right] \\
=&h\sum_{j=0}^{J}{}''\left[\varphi(U_j^{n+1},U_j^n,\frac{1}{2}) + \frac{q}{2}\delta_x^{\langle 2\rangle}(U_j^{n+1}+U_j^n)\right]\delta_t^+U_j^n \\
=&h\sum_{j=0}^{J}{}''\left[\varphi(U_j^{n+1},U_j^n,\alpha) + \frac{q}{2}\delta_x^{\langle 2\rangle}(U_j^{n+1}+U_j^n)\right]\delta_t^+U_j^n \\
&-\left(\alpha-\frac{1}{2}\right)\frac{1}{2\tau}h\sum_{j=0}^{J}{}''\left[(U_j^{n+1})^2-(U_j^n)^2\right]^2
\end{aligned}$$

$$
\begin{aligned}
&= -\left(\alpha - \frac{1}{2}\right)\frac{1}{2\tau} h \sum_{j=0}^{J}{}'' \left[(U_j^{n+1})^2 - (U_j^n)^2\right]^2 \\
&\quad + h \sum_{j=0}^{J}{}'' \left[\varphi(U_j^{n+1}, U_j^n, \alpha) + \frac{q}{2}\delta_x^{\langle 2\rangle}(U_j^{n+1} + U_j^n)\right] \\
&\quad \times \delta_x^{\langle 2\rangle}\left[-\beta h^2 \delta_t^+ U_j^n + \frac{q}{2}\delta_x^{\langle 2\rangle}(U_j^{n+1} + U_j^n) + \varphi(U_j^{n+1}, U_j^n, \alpha)\right]
\end{aligned} \tag{2.7}
$$

When $0 \leqslant \beta \leqslant \dfrac{1}{4}$, it follows from (2.7) that

$$
\begin{aligned}
&\frac{1}{\tau}\left[h \sum_{j=0}^{J}{}'' G_d(U_j^{n+1}) - h \sum_{j=0}^{J}{}'' G_d(U_j^n)\right] \\
&= -\left(\alpha - \frac{1}{2}\right)\frac{1}{2\tau} h \sum_{j=0}^{J}{}'' \left[(U_j^{n+1})^2 - (U_j^n)^2\right]^2 \\
&\quad - h \sum_{j=0}^{J}{}'' \left(\delta_x^+ \left[\varphi(U_j^{n+1}, U_j^n, \alpha) + \frac{q}{2}\delta_x^{\langle 2\rangle}(U_j^{n+1} + U_j^n)\right]\right)^2 \\
&\quad - h \sum_{j=0}^{J}{}'' \left[\varphi(U_j^{n+1}, U_j^n, \alpha) + \frac{q}{2}\delta_x^{\langle 2\rangle}(U_j^{n+1} + U_j^n)\right] \times \beta \delta_x^{\langle 2\rangle} \delta_t^+ U_j^n \\
&= -\left(\alpha - \frac{1}{2}\right)\frac{1}{2\tau} h \sum_{j=0}^{J}{}'' \left[(U_j^{n+1})^2 - (U_j^n)^2\right]^2 \\
&\quad - h \sum_{j=0}^{J}{}'' \left(\delta_x^+ \left[\varphi(U_j^{n+1}, U_j^n, \alpha) + \frac{q}{2}\delta_x^{\langle 2\rangle}(U_j^{n+1} + U_j^n)\right]\right)^2 \\
&\quad - h \sum_{j=0}^{J}{}'' \delta_x^{\langle 2\rangle} \left[\varphi(U_j^{n+1}, U_j^n, \alpha) + \frac{q}{2}\delta_x^{\langle 2\rangle}(U_j^{n+1} + U_j^n)\right] \times \beta h^2 \delta_t^+ U_j^n \\
&= -\left(\alpha - \frac{1}{2}\right)\frac{1}{2\tau} h \sum_{j=0}^{J}{}'' \left[(U_j^{n+1})^2 - (U_j^n)^2\right]^2 \\
&\quad - h \sum_{j=0}^{J}{}'' \left(\delta_x^+ \left[\varphi(U_j^{n+1}, U_j^n, \alpha) + \frac{q}{2}\delta_x^{\langle 2\rangle}(U_j^{n+1} + U_j^n)\right]\right)^2 \\
&\quad - h \sum_{j=0}^{J}{}'' \left[\delta_t^+ U_j^n + \beta h^2 \delta_x^{\langle 2\rangle} \delta_t^+ U_j^n\right] \times \beta h^2 \delta_t^+ U_j^n
\end{aligned}
$$

$$
\begin{aligned}
&= -\left(\alpha-\frac{1}{2}\right)\frac{1}{2\tau}h\sum_{j=0}^{J}{}''\left[(U_j^{n+1})^2-(U_j^n)^2\right]^2\\
&\quad -h\sum_{j=0}^{J}{}''\left(\delta_x^+\left[\varphi(U_j^{n+1},U_j^n,\alpha)+\frac{q}{2}\delta_x^{\langle 2\rangle}(U_j^{n+1}+U_j^n)\right]\right)^2\\
&\quad -h\sum_{j=0}^{J}{}''\left[\beta h^2(\delta_t^+U_j^n)^2-\beta^2h^4(\delta_x^+\delta_t^+U_j^n)^2\right]\\
&\leqslant -\left(\alpha-\frac{1}{2}\right)\frac{1}{2\tau}h\sum_{j=0}^{J}{}''\left[(U_j^{n+1})^2-(U_j^n)^2\right]^2\\
&\quad -h\sum_{j=0}^{J}{}''\left(\delta_x^+\left[\varphi(U_j^{n+1},U_j^n,\alpha)+\frac{q}{2}\delta_x^{\langle 2\rangle}(U_j^{n+1}+U_j^n)\right]\right)^2\\
&\quad -h\sum_{j=0}^{J}{}''\left[\beta h^2(\delta_t^+U_j^n)^2-\beta^2h^4\frac{4}{h^2}(\delta_t^+U_j^n)^2\right]\\
&\leqslant -\left(\alpha-\frac{1}{2}\right)\frac{1}{2\tau}h\sum_{j=0}^{J}{}''\left[(U_j^{n+1})^2-(U_j^n)^2\right]^2\\
&\quad -h\sum_{j=0}^{J}{}''\left(\delta_x^+\left[\varphi(U_j^{n+1},U_j^n,\alpha)+\frac{q}{2}\delta_x^{\langle 2\rangle}(U_j^{n+1}+U_j^n)\right]\right)^2\\
&\quad -h\sum_{j=0}^{J}{}''\beta h^2(1-4\beta)(\delta_t^+U_j^n)^2\leqslant 0.
\end{aligned}\tag{2.8}
$$

When $\beta\leqslant 0$, denote

$$
\varphi_0(U_j^{n+1},U_j^n,\alpha)=-\beta h^2\delta_t^+U_j^n+\varphi(U_j^{n+1},U_j^n,\alpha),
$$

then it follows from (2.7) that

$$
\begin{aligned}
&\frac{1}{\tau}\left[h\sum_{j=0}^{J}{}''G_d(U_j^{n+1})-h\sum_{j=0}^{J}{}''G_d(U_j^n)\right]\\
&= -\left(\alpha-\frac{1}{2}\right)\frac{1}{2\tau}h\sum_{j=0}^{J}{}''\left[(U_j^{n+1})^2-(U_j^n)^2\right]^2\\
&\quad +h\sum_{j=0}^{J}{}''\left[\varphi_0(U_j^{n+1},U_j^n,\alpha)+\frac{q}{2}\delta_x^{\langle 2\rangle}(U_j^{n+1}+U_j^n)\right]\\
&\quad \times\delta_x^{\langle 2\rangle}\left[\frac{q}{2}\delta_x^{\langle 2\rangle}(U_j^{n+1}+U_j^n)+\varphi_0(U_j^{n+1},U_j^n,\alpha)\right]-h\sum_{j=0}^{J}{}''\beta h^2\delta_x^{\langle 2\rangle}\delta_t^+U_j^n\delta_t^+U_j^n
\end{aligned}
$$

$$
\begin{aligned}
&=-\left(\alpha-\frac{1}{2}\right)\frac{1}{2\tau}h\sum_{j=0}^{J}{}''\left[(U_j^{n+1})^2-(U_j^n)^2\right]^2+h\sum_{j=0}^{J}{}''\beta h^2\left(\delta_x^+\delta_t^+U_j^n\right)^2\\
&\quad-h\sum_{j=0}^{J}{}''\left(\delta_x^+\left[\frac{q}{2}\delta_x^{\langle 2\rangle}(U_j^{n+1}+U_j^n)+\varphi_0(U_j^{n+1},U_j^n,\alpha)\right]\right)^2\leqslant 0.
\end{aligned}\tag{2.9}
$$

Then (2.5) is obtained from (2.8) and (2.9).

Based on Lemma 2.1 we turn to estimate the numerical solution of the scheme (2.1)–(2.3) by using the similar method in [16].

Lemma 2.2 *For the solution of the fintie diference scheme* (2.1)–(2.3), *we have the following estimate*

$$
||U^n||^2_{d(1,2)}\leqslant\frac{1}{min\left(\lambda,-\frac{q}{2}\right)}\left\{\frac{(p-2\lambda)^2}{4r}L+h\sum_{j=0}^{J}{}''G_d\left(U_j^0\right)\right\},\tag{2.10}
$$

where λ *is any a positive number and* $||U^n||_{d(1,2)}$ *is a discrete first-order Sobolev-Hilbert norm which is defined as*

$$
||f||^2_{d(1,2)}\triangleq h\sum_{j=0}^{J}{}''(f_j)^2+h\sum_{j=0}^{J-1}{}''(\delta_j^+f_j)^2,\quad f=(f_j)_{j=-l}^{J+l}\in\mathcal{R}^{J+1+2l},\quad l\geqslant 0.
$$

Proof It follows from the dissipation of the total energy (2.5) that

$$
\begin{aligned}
&h\sum_{j=0}^{J}{}''G_d(U_j^0)\geqslant h\sum_{j=0}^{J}{}''G_d(U_j^n)\\
&\geqslant h\sum_{j=0}^{J}{}''\left\{\lambda(U_j^n)^2-\frac{(p-2\lambda)^2}{4r}-\frac{q}{2}\frac{(\delta_j^+U_j^n)^2+(\delta_j^-U_j^n)^2}{2}\right\}\\
&\quad\cdot\left(since\ \ \frac{1}{2}pX^2+\frac{1}{4}rX^4\geqslant\lambda X^2-\frac{(p-2\lambda)^2}{4r}\right)\\
&\geqslant\min\left(\lambda,-\frac{q}{2}\right)\sum_{j=0}^{J}{}''\left\{(U_j^n)^2+\frac{(\delta_j^+U_j^n)^2+(\delta_j^-U_j^n)^2}{2}\right\}-\frac{(p-2\lambda)^2}{4r}L\\
&=\min\left(\lambda,-\frac{q}{2}\right)||U^n||^2_{d(1,2)}-\frac{(p-2\lambda)^2}{4r}L,
\end{aligned}\tag{2.11}
$$

where the boundary condition (2.3) was used. Then (2.10) is obtained from (2.11).

Lemma 2.3 [35]

$$
||f||_\infty\leqslant C_5\sqrt{||f||}\sqrt{||f||+||\delta_x^+f||}.
$$

Lemma 2.4 [33]

$$\|f\|_\infty \leqslant 2\max\left(\frac{1}{\sqrt{L}},\sqrt{L}\right)\|f\|_{d(1,2)}.$$

Applying Lemma 2.4 to (2.10), we obtain the following inequality.

Theorem 2.1 *The solution of the fintie diference scheme* (2.1)–(2.3) *is bounded in the discrete* L^∞ *norm, i.e.,*

$$\|U^n\|_\infty \leqslant 2\sqrt{\frac{\max\left(\frac{1}{L},L\right)}{\min\left(\lambda,-\frac{q}{2}\right)}\left\{\frac{(p-2\lambda)^2}{4r}L+h\sum_{j=0}^{J}{}''G_d\left(U_j^0\right)\right\}}. \tag{2.12}$$

Theorem 2.1 implies that the proposed difference scheme (2.1)–(2.3) is stable.

3 Unique existence of the numerical solutions

To show the existence of the numerical solutions $U^1,U^2,\cdots,U^N$ for the scheme (2.1)–(2.3), we shall use the following Brouwer-type theorem (see [36,37]).

Lemma 3.1 *Let* $(H,(\cdot,\cdot))$ *be a finite-dimensional inner product space,* $\|\cdot\|$ *the associated norm, and* $g:H\to H$ *be continuous. Assume, moreover, that*

$$\exists\kappa>0,\forall z\in H,\|z\|=\kappa, Re(g(z),z)\geqslant 0.$$

Then, there exists a $z^*\in H$ *such that* $g(z^*)=0$ *and* $\|z^*\|\leqslant\kappa$.

For fixed j, we rewrite (2.1) in the form

$$\begin{aligned}&\frac{U_j^{n+1}+U_j^n}{2}-U_j^n+\beta h^2\delta_x^{\langle 2\rangle}\left(\frac{U_j^{n+1}+U_j^n}{2}-U_j^n\right)-\frac{\tau}{2}q\delta_x^{\langle 4\rangle}\frac{U_j^{n+1}+U_j^n}{2}\\&-\frac{\tau}{2}\delta_x^{\langle 2\rangle}\left[p\frac{U_j^{n+1}+U_j^n}{2}+4r\alpha(1-\alpha)\left(\frac{U_j^{n+1}+U_j^n}{2}\right)^3\right]\\&-\frac{\tau}{2}\delta_x^{\langle 2\rangle}\left[r\frac{U_j^{n+1}+U_j^n}{2}\left(2\alpha\frac{U_j^{n+1}+U_j^n}{2}-U_j^n\right)^2\right]=0.\end{aligned}$$

The mapping $F:\mathbb{R}\to\mathbb{R}$

$$\begin{aligned}(F(V))_j=&V_j-U_j^n+\beta h^2\delta_x^{\langle 2\rangle}V_j-\beta h^2\delta_x^{\langle 2\rangle}U_j^n-\frac{\tau}{2}q\delta_x^{\langle 4\rangle}V_j\\&-\frac{\tau}{2}\delta_x^{\langle 2\rangle}\left[pV_j+4r\alpha(1-\alpha)(V_j)^3+rV_j\left(2\alpha V_j-U_j^n\right)^2\right],\quad 0\leqslant j\leqslant J,\end{aligned} \tag{3.1}$$

is obviously continuous. In (3.1), $V=\{V_j\}_{j=0}^{J}$ and operators are defined under the boundary condition (2.3), i.e.,

$$V_{-1}=V_1,\ \ V_{-2}=V_2,\ \ V_{J+1}=V_{J-1},\ \ V_{J+2}=V_{J-2}. \tag{3.2}$$

If the mapping F has a zero-point V^*, then $2V^*-U^n$ is the solution U^{n+1} of the proposed scheme (2.1)–(2.3). From Theorem 2.1 we know that the $||V^*||_\infty$ is bounded if there exists a numerical solution for the scheme (2.1)–(2.3). In the mapping (3.1), we assume $||V||_\infty \leqslant C(U^n)$ where $C(U^n)>\dfrac{10}{\sqrt{L}}||U^n||$.

Theorem 3.1 *If τ is sufficiently small, then the scheme* (2.1)–(2.3) *has an unique solution.*

Proof Computing the inner product of (3.2) with V, since

$$\left(\delta_x^{\langle 2\rangle}V,U\right)_h=-h\sum_{j=0}^{J-1}\delta_x^{+}U_j\delta_x^{+}V_j=\left(V,\delta_x^{\langle 2\rangle}U\right)_h, \tag{3.3}$$

$$\begin{aligned}&-\frac{\tau}{2}\left(\delta_x^{\langle 2\rangle}\left[4r\alpha(1-\alpha)\left(V_j\right)^3\right],V\right)_h\\ =&2\tau r\alpha(1-\alpha)h\sum_{j=0}^{J-1}\left(\delta_j^{+}V_j\right)^2\left[\left(V_{j+1}\right)^2+V_{j+1}V_j^n+\left(V_j^n\right)^2\right]\end{aligned} \tag{3.4}$$

$$\begin{aligned}&-\frac{\tau}{2}\left(\delta_x^{\langle 2\rangle}\left[rV_j\left(2\alpha V_j-U_j^n\right)^2\right],V\right)_h\\ \geqslant&-\frac{\tau|q|}{4}\left|\left|\delta_x^{\langle 2\rangle}V\right|\right|^2-\frac{\tau r}{4|q|}\left|\left|V\left(2\alpha V-U^n\right)^2\right|\right|^2\\ \geqslant&-\frac{\tau|q|}{4}\left|\left|\delta_x^{\langle 2\rangle}V\right|\right|^2-\frac{\tau r}{4|q|}\left|\left|\left(2\alpha V-U^n\right)^2\right|\right|_\infty^2||V||^2\\ \geqslant&-\frac{\tau|q|}{4}\left|\left|\delta_x^{\langle 2\rangle}V\right|\right|^2-C_0\frac{r}{4|q|}\tau\,||V||^2\end{aligned} \tag{3.5}$$

where $(\delta_x^{+}V)_j=\delta_j^{+}V_j$, $\left(\delta_x^{\langle 2\rangle}V\right)_j=\delta_j^{\langle 2\rangle}V_j$ and $C_0=(2\alpha+1)^4[C\left(U^n\right)]^4$, we have

$$\begin{aligned}(F(V),V)_h=&(V,V)_h-(U^n,V)_h+\beta h^2\left(\delta_x^{\langle 2\rangle}V,V\right)_h\\ &-\beta h^2\left(\delta_x^{\langle 2\rangle}U^n,V\right)_h-\frac{\tau}{2}q\left(\delta_x^{\langle 4\rangle}V,V\right)_h\\ &-\frac{\tau}{2}\left(\delta_x^{\langle 2\rangle}\left[pV_j+4r\alpha(1-\alpha)\left(V_j\right)^3+rV_j\left(2\alpha V_j-U_j^n\right)^2\right],V\right)_h\end{aligned}$$

$$
\begin{aligned}
&= ||V||^2 - (U^n, V)_h - \beta h^2 ||\delta_x^+ V||^2 \\
&\quad + \frac{\beta}{2} h^2 \left(\delta_x^+ U^n, \delta_x^+ V\right)_h + \frac{\beta}{2} h^2 \left(\delta_x^- U^n, \delta_x^- V\right)_h - \frac{\tau}{2} q \left|\left|\delta_x^{\langle 2\rangle} V\right|\right|^2 \\
&\quad + 2\tau r\alpha(1-\alpha) h \sum_{j=0}^{J-1} \left(\delta_x^+ V_j\right)^2 \left[(V_{j+1})^2 + V_{j+1} V_j^n + (V_j^n)^2\right] \\
&\quad - \frac{\tau}{2} \left(pV_j + rV_j \left(2\alpha V_j - U_j^n\right)^2, \delta_x^{\langle 2\rangle} V\right)_h \\
&\geqslant ||V||^2 - \frac{1}{2} ||V||^2 - \frac{1}{2} ||U^n||^2 - \frac{5|\beta|}{4} h^2 ||\delta_x^+ V||^2 \\
&\quad - |\beta| h^2 ||\delta_x^+ U^n||^2 - \frac{\tau}{2} q \left|\left|\delta_x^{\langle 2\rangle} V\right|\right|^2 - \frac{\tau|q|}{4} \left|\left|\delta_x^{\langle 2\rangle} V\right|\right|^2 \\
&\quad - C_0 \frac{r}{4|q|} \tau ||V||^2 - \frac{\tau|q|}{4} \left|\left|\delta_x^{\langle 2\rangle} V\right|\right|^2 - \frac{p^2}{4|q|} \tau ||V||^2 \\
&\geqslant \frac{1}{2} ||V||^2 - \frac{1}{2} ||U^n||^2 - 5|\beta| h^2 ||V||^2 - 4|\beta| h^2 ||U^2||^2 - C_1 \tau ||V||^2 \\
&\geqslant \left(\frac{1}{2} - 5|\beta| - C_1 \tau\right) ||V||^2 - \left(\frac{1}{2} + 4|\beta|\right) ||U^n||^2,
\end{aligned} \tag{3.6}
$$

where $C_1 = C_0 \frac{r}{4|q|} + \frac{p^2}{4|q|}$. Taking $\beta \leqslant \frac{1}{12}, \tau \leqslant \frac{1}{24C_1}$ and $C(U^n) > \frac{10}{\sqrt{L}} ||U^n||$, we obtain $(F(V), V)_h \geqslant 0$ for $||V|| \geqslant 5 ||U^n||$. The existence of $U^{n+\frac{1}{2}}$ satisfying $\left|\left|U^{n+\frac{1}{2}}\right|\right|_\infty \leqslant C(U^n)$ follows from Lemma 3.1 and consequently the existence of U^{n+1} is obtained.

Using the similar proof of Theorem 4.1 in the next section, we can obtain the uniqueness of the numerical solution.

4 Convergence of the numerical solutions

The purpose of this section is to discuss the convergence of the numerical solutions. We denote $u_j^n = u(x_j, t_n)$ and define the error as

$$
e_j^n \triangleq u_j^n - U_j^n, \quad j = -1, 0, 1, \cdots, J, J+1, \tag{4.1}
$$

where $u(x_j, t_n)$ is the solution to the Cahn-Hilliard equation at the point (x_j, t_n). We define an extension of u by

$$
u(x,t) = \begin{cases} u(x - 2mL, t) & \text{for} \quad 2mL \leqslant x \leqslant (2m+1)L, \\ u(2mL - x, t) & \text{for} \quad (2m-1)L \leqslant x \leqslant 2mL, \end{cases}
$$

where $m \in \mathcal{Z}$. Define the truncation error of the scheme (2.1)–(2.3) as follows

$$r_j^{n+1/2} = \delta_t^+ u_j^n + \beta h^2 \delta_x^{\langle 2\rangle} \delta_t^+ u_j^n - \delta_x^{\langle 2\rangle} v_j^{n+1/2}, \tag{4.2}$$

$$\eta_j^{n+1/2} = v_j^n - \varphi(u_j^{n+1}, u_j^n, \alpha) - \frac{q}{2}\delta_x^{\langle 2\rangle}(u_j^{n+1} + u_j^n), \tag{4.3}$$

where

$$v_j^{n+\frac{1}{2}} = \left\{pu + ru^3 + q\frac{\partial^2 u}{\partial x^2}\right\}\Big|_{(x,t)=(x_j,t_{n+1/2})} \quad \text{for} \quad j = 0, 1, \cdots, J.$$

Using Taylor expansion, we can obtain

$$r_j^{n+1/2} = O(h^2 + \tau^2), \quad \eta_j^{n+1/2} = O(h^2 + \tau). \tag{4.4}$$

Especially for $\beta = \dfrac{1}{12}$ and $\alpha = \dfrac{1}{2}$, there is

$$r_j^{n+1/2} = O(h^4 + \tau^2), \quad \eta_j^{n+1/2} = O(h^2 + \tau^2). \tag{4.5}$$

Lemma 4.1

$$\begin{aligned}
&\frac{1}{\tau}\left\{\left(\left\|e^{n+1}\right\|^2 - \beta\left\|\delta_x^+ e^{n+1}\right\|^2\right) - \left(\left\|e^n\right\|^2 - \beta\left\|\delta_x^+ e^n\right\|^2\right)\right\} \\
\leqslant& \frac{1}{2}\left\{\left\|e^{n+1}\right\|^2 + \left\|e^n\right\|^2\right\} - \frac{1}{q}\left\|\varphi(u^{n+1}, u^n, \alpha) - \varphi(U^{n+1}, U^n, \alpha)\right\|^2 \\
&+ \left\|r^n\right\|^2 - \frac{1}{q}\left\|\eta^n\right\|^2,
\end{aligned} \tag{4.6}$$

where

$$\phi_j^{n+\frac{1}{2}} = \{pu + ru^3\}\Big|_{(x,t)=(x_j,t_{n+1/2})} \quad \text{for} \quad j = 0, 1, \cdots, J.$$

Proof Denote

$$V_j^{n+\frac{1}{2}} = \varphi(U_j^{n+1}, U_j^n, \alpha) + \frac{q}{2}\delta_x^{\langle 2\rangle}(U_j^{n+1} + U_j^n), \tag{4.7}$$

then (2.1) can be written as

$$\delta_t^+ U_j^n + \beta h^2 \delta_x^{\langle 2\rangle} \delta_t^+ U_j^n = \delta_x^{\langle 2\rangle} V_j^{n+\frac{1}{2}}. \tag{4.8}$$

Denote

$$\xi_j^n \triangleq v_j^n - V_j^n, \quad j = -1, 0, 1, \cdots, J, J+1. \tag{4.9}$$

then it follows from (4.7 and (4.1)–(4.3) that

$$\delta_t^+ e_j^n + \beta h^2 \delta_x^{\langle 2\rangle} \delta_t^+ e_j^n = \delta_x^{\langle 2\rangle} \xi_j^{n+1/2} + r_j^{n+1/2}, \tag{4.10}$$

$$\xi_j^{n+\frac{1}{2}} = \widetilde{\varphi}^{n+1/2} + \frac{q}{2}\delta_x^{\langle 2\rangle}(e_j^{n+1} + e_j^n) + \eta_j^{n+1/2}. \tag{4.11}$$

where $\widetilde{\varphi}^{n+1/2} = \varphi(u_j^{n+1}, u_j^n, \alpha) - \varphi(U_j^{n+1}, U_j^n, \alpha)$.

Taking the inner product of (4.10) and (4.11) with $\frac{1}{2}\left(e^{n+1}+e^n\right)$ and $\frac{1}{q}\xi^{n+\frac{1}{2}}$ respectively, then adding the results together, we obtain

$$\begin{aligned}
&\frac{1}{2\tau}\left\{\left(\left\|e^{n+1}\right\|^2 - \beta h^2\left\|\delta_x^+ e^{n+1}\right\|^2\right) - \left(\left\|e^n\right\|^2 - \beta h^2\left\|\delta_x^+ e^n\right\|^2\right)\right\} - \frac{1}{q}\left\|\xi^{n+\frac{1}{2}}\right\|^2\\
=&\frac{1}{2}\left(\delta_x^{\langle 2\rangle}\xi^{n+1/2}, e^{n+1}+e^n\right)_h + \frac{1}{2}\left(r^{n+1/2}, e^{n+1}+e^n\right)_h - \frac{1}{q}\left(\widetilde{\varphi}^{n+1/2}, \xi^{n+\frac{1}{2}}\right)_h\\
&-\frac{1}{2}\left(\delta_x^{\langle 2\rangle}e^{n+1}+e^n, \xi^{n+1/2}\right)_h - \frac{1}{q}\left(\eta^{n+1/2}, \xi^{n+1/2}\right)_h\\
=&-\frac{1}{q}\left(\widetilde{\varphi}^{n+1/2}, \xi^{n+\frac{1}{2}}\right)_h + \frac{1}{2}\left(r^{n+1/2}, e^{n+1}+e^n\right)_h - \frac{1}{q}\left(\eta^{n+1/2}, \xi^{n+1/2}\right)_h\\
\leqslant&-\frac{1}{2q}\left\|\widetilde{\varphi}^{n+1/2}\right\|^2 - \frac{1}{2q}\left\|\xi^{n+\frac{1}{2}}\right\|^2 + \frac{1}{2}\left\|r^{n+1/2}\right\|^2\\
&+\frac{1}{4}\left(\left\|e^{n+1}\right\|^2 + \left\|e^n\right\|^2\right) - \frac{1}{2q}\left\|\eta^{n+1/2}\right\|^2 - \frac{1}{2q}\left\|\xi^{n+1/2}\right\|^2.
\end{aligned} \tag{4.12}$$

Hence the inequality (4.6) is obtain from (4.12).

Lemma 4.2

$$-\frac{1}{q}\left\|\varphi(u^{n+1}, u^n, \alpha) - \varphi(U^{n+1}, U^n, \alpha)\right\|^2 \leqslant C_3\left(\left\|e^{n+1}\right\|^2 + \left\|e^n\right\|^2\right), \tag{4.13}$$

where

$$C_3 = -\frac{1}{2q}\left(-p + 2r(C_2)^2\right)^2,\ \ C_2 = \max_{0\leqslant n\leqslant N}\{\|U^n\|_\infty, \|u^n\|_{L^\infty},\}.$$

Proof

$$\begin{aligned}
&\varphi(u_j^{n+1}, u_j^n, \alpha) - \varphi(U_j^{n+1}, U_j^n, \alpha)\\
=&p\frac{u_j^{n+1}+u_j^n}{2} + r\frac{u_j^{n+1}+u_j^n}{2}\left[\alpha\left(u_j^{n+1}\right)^2 + (1-\alpha)\left(u_j^n\right)^2\right]
\end{aligned}$$

$$
\begin{aligned}
&-p\frac{U_j^{n+1}+U_j^n}{2}-r\frac{U_j^{n+1}+U_j^n}{2}\left[\alpha\left(U_j^{n+1}\right)^2+(1-\alpha)\left(U_j^n\right)^2\right]\\
=&p\frac{e_j^{n+1}+e_j^n}{2}+r\frac{e_j^{n+1}+e_j^n}{2}\left[\alpha\left(u_j^{n+1}\right)^2+(1-\alpha)\left(u_j^n\right)^2\right]\\
&+r\frac{U_j^{n+1}+U_j^n}{2}\left[\alpha\left(u_j^{n+1}+U_j^{n+1}\right)e_j^{n+1}+(1-\alpha)\left(u_j^n+U_j^n\right)e_j^n\right]. \qquad (4.14)
\end{aligned}
$$

It follows from (4.13) that

$$
\left|\varphi(u_j^{n+1},u_j^n,\alpha)-\varphi(U_j^{n+1},U_j^n,\alpha)\right|\leqslant\left(-p+2r(C_2)^2\right)\left|\frac{e_j^{n+1}+e_j^n}{2}\right|. \qquad (4.15)
$$

Hence the inequality (4.13) is obtain from (4.15).

Theorem 4.1 *If (1.1) has a solution such that $u(x,t)\in C^{6,3}([0,L]\times[0,T])$, $\beta\leqslant\frac{1}{12}$ and τ is sufficiently small, then the solution of the difference scheme (2.1) converges to the solution of (1.1) in the sense of discrete L_2-norm, and the convergence rate is $O\left(h^2+\tau^2\right)$ for $\alpha=1/2$ and $O\left(h^2+\tau\right)$ for $\alpha\neq 1/2$.*

Proof It follows from Lemma 4.1 and lemma 4.2 that

$$
\begin{aligned}
&\frac{1}{\tau}\left\{\left(\left\|e^{n+1}\right\|^2-\beta h^2\left\|\delta_x^+e^{n+1}\right\|^2\right)-\left(\left\|e^n\right\|^2-\beta h^2\left\|\delta_x^+e^n\right\|^2\right)\right\}\\
\leqslant&\left(\frac{1}{2}+C_3\right)\left(\left\|e^{n+1}\right\|^2+\left\|e^n\right\|^2\right)+\left\|r^n\right\|^2-\frac{1}{q}\left\|\eta^n\right\|^2. \qquad (4.16)
\end{aligned}
$$

Denote

$$
B^{n+1}=\left\|e^{n+1}\right\|^2-\beta h^2\left\|\delta_x^+e^{n+1}\right\|^2. \qquad (4.17)
$$

Since

$$
\left\|\delta_x^+e^{n+1}\right\|^2\leqslant\frac{4}{h^2}\left\|e^{n+1}\right\|^2,
$$

we obtain that $B^{n+1}\geqslant\frac{2}{3}\left\|e^{n+1}\right\|^2$ if $\beta\leqslant\frac{1}{12}$. Hence, it follows from (4.16) that

$$
B^{n+1}-B^n\leqslant\left(\frac{3}{4}+\frac{3}{2}C_3\right)\tau\left(B^{n+1}+B^n\right)+\tau\left(\left\|r^n\right\|^2-\frac{1}{q}\left\|\eta^n\right\|^2\right). \qquad (4.18)
$$

Using Gronwall's inequality, we obtain

$$
\begin{aligned}
\max_{0\leqslant n\leqslant N}B^n&\leqslant\left(B^0+\tau\sum_{k=1}^{N}\left(\left\|r^n\right\|^2-\frac{1}{q}\left\|\eta^n\right\|^2\right)\right)e^{(3+6C_3)T}\\
&=e^{(3+6C_3)T}\sum_{k=1}^{N}\left(\left\|r^n\right\|^2-\frac{1}{q}\left\|\eta^n\right\|^2\right)\tau,\ \text{ for }\ \tau\leqslant\frac{N+1}{(3+6C_3)N}. \qquad (4.19)
\end{aligned}
$$

It follows from (4.2)–(4.3),(4.17) and (4.19) that Theorem 4.1 holds.

5 Iterative algorithm

To compute the numerical solutions $U^1, U^2, \cdots, U^n$ satisfying (2.1)–(2.3), we require to solve a nonlinear system of algebraic equations with $J+1$ unknowns at each time step, so it is necessary to construct an effective iterative algorithm to solve it in implementation. In this section, we will construct two iterative algorithms based on which two predication-correction schemes will be proposed. The first iterative algorithm is as follows:

$$\begin{aligned}&\delta_t^- U_j^{n+1(s+1)} + \beta h^2 \delta_x^{\langle 2\rangle} \delta_t^- U_j^{n+1(s+1)} - \frac{q}{2}\delta_x^{\langle 4\rangle}(U_j^{n+1(s+1)} + U_j^n)\\ =&\delta_x^{\langle 2\rangle}\varphi(U_j^{n+1(s)}, U_j^n, \alpha), \quad 0 \leqslant j \leqslant J, \quad n = 0,1,2,\cdots,N-1,\end{aligned} \tag{5.1}$$

$$U_j^0 = u_0(x_j), \quad j = -2,-1,0,\cdots,J,J+1,J+2, \tag{5.2}$$

$$\delta_x^{\langle 1\rangle} U_0^n = \delta_x^{\langle 1\rangle} U_J^n = \delta_x^{\langle 3\rangle} U_0^n = \delta_x^{\langle 3\rangle} U_J^n = 0, \quad n = 0,1,2,\cdots,N. \tag{5.3}$$

with

$$U_j^{n+1(0)} = \begin{cases} U_j^n & \text{for} \quad n = 0, \\ 2U_j^n - U_j^{n-1} & \text{for} \quad n \geqslant 1, \end{cases}$$

where

$$\delta_t^- U_j^{n+1(s+1)} = \frac{U_j^{n+1(s+1)} - U_j^n}{\tau}.$$

Theorem 5.1 *Assume* $u \in C^{6,3}([0,L]\times[0,T])$, $\beta \leqslant \dfrac{1}{12}$ *and* τ, h *are sufficiently small, then the solution of the iterative method* (5.1)–(5.3) *converges to the numerical solution of the scheme* (2.1)–(2.3).

Proof Denote

$$\theta_j^{n+1(s)} = U_j^{n+1(s)} - U_j^{n+1}, \quad n = 0,1,2,\cdots,N-1;\ s = 0,1,2,\cdots.$$

Then subtracting (2.1) from (5.1), we obtain

$$\begin{aligned}&\frac{1}{\tau}\theta_j^{n+1(s+1)} + \frac{1}{\tau}\beta h^2 \delta_x^{\langle 2\rangle}\theta_j^{n+1(s+1)} - \frac{q}{2}\delta_x^{\langle 4\rangle}\theta_j^{n+1(s+1)}\\ =&\delta_x^{\langle 2\rangle}\varphi(U_j^{n+1(s)}, U_j^n, \alpha) - \delta_x^{\langle 2\rangle}\varphi(U_j^{n+1}, U_j^n, \alpha)\\ =&\frac{p}{2}\delta_x^{\langle 2\rangle}\theta_j^{n+1(s)} + \frac{r}{2}\delta_x^{\langle 2\rangle}\left\{\theta_j^{n+1(s)}\left[\alpha\left(U_j^{n+1}\right)^2 + (1-\alpha)\left(U_j^n\right)^2\right]\right.\\ &\left.+\frac{r}{2}\alpha\left(U_j^{n+1(s)} + U_j^n\right)\left(U_j^{n+1(s)} + U_j^{n+1}\right)\theta_j^{n+1(s)}\right\}.\end{aligned} \tag{5.4}$$

Noting, when $n=0$,

$$\begin{aligned}\theta_j^{1(0)} =& U_j^{1(0)} - U_j^1 = U_j^0 - U_j^1 \\ =& \left[(U_j^0 - u_j^0) + (u_j^0 - u_j^1) + (u_j^1 - U_j^1)\right] \\ =& 0 + O(\tau) + O(h^2+\tau) = O(h^2+\tau),\end{aligned} \tag{5.5}$$

and when $n \geqslant 1$,

$$\begin{aligned}\theta_j^{n+1(0)} =& U_j^{n+1(0)} - U_j^{n+1} = 2U_j^n - U_j^{n-1} - U_j^{n+1} \\ =& \left[2\left(U_j^n - u_j^n\right) + \left(u_j^{n-1} - U_j^{n-1}\right) + \left(u_j^{n+1} - U_j^{n+1}\right) + \left(2u_j^n - u_j^{n-1} - u_j^{n+1}\right)\right] \\ =& O(h^2+\tau) + O(h^2+\tau) + O(h^2+\tau) = O(h^2+\tau).\end{aligned} \tag{5.6}$$

If taking $\alpha = 1/2$, we have

$$\theta_j^{n+1(0)} = O(h^2+\tau^2), \quad \text{for } n \geqslant 1. \tag{5.7}$$

It follows from (5.5)–(5.7) that

$$\left\|\theta^{n+1(0)}\right\|_\infty \leqslant \widetilde{C}_0(h^2+\tau), \quad \text{for } 1/2 < \alpha \leqslant 1, \tag{5.8}$$

and

$$\left\|\theta^{n+1(0)}\right\|_\infty \leqslant \begin{cases} \widetilde{C}_0(h^2+\tau) & \text{for } \ n=0, \\ \widetilde{C}_0(h^2+\tau^2) & \text{for } \ n \geqslant 1, \end{cases} \tag{5.9}$$

if we take $\alpha = 1/2$. In the next study, we just only discuss the case $\alpha = 1/2$, the case $1/2 < \alpha \leqslant 1$ can be discussed by the similar method.

Now, suppose

$$\left\|\theta^{n+1(s)}\right\| \leqslant \widetilde{C}_s(h^2+\tau^2), \quad n=1,2,\cdots,N-1; \ s=0,1,\cdots. \tag{5.10}$$

It follows from *Sobolev* estimate, we obtain

$$\begin{aligned}\left\|\theta^{n+1(s)}\right\|_\infty \leqslant& C_5\sqrt{\left\|\theta^{n+1(s)}\right\|}\sqrt{\left\|\delta_x^+\theta^{n+1(s)}\right\| + \left\|\theta^{n+1(s)}\right\|} \\ \leqslant& C_5\sqrt{\left\|\theta^{n+1(s)}\right\|}\sqrt{\frac{2}{h}\left\|\theta^{n+1(s)}\right\| + \left\|\theta^{n+1(s)}\right\|} \\ \leqslant& C_5\sqrt{1+\frac{2}{h}}\left\|\theta^{n+1(s)}\right\| \leqslant C_5\widetilde{C}_s\left(1+2h^{-1/2}\right)(h^2+\tau^2) \\ & n=1,2,\cdots,N-1; \ \ s=0,1,2,\cdots.\end{aligned} \tag{5.11}$$

Thus

$$\begin{aligned}\left\|U^{n+1(s)}\right\|_\infty &\leqslant \left\|U^{n+1}\right\|_\infty + \left\|\theta^{n+1(s)}\right\|_\infty \\ &\leqslant C_4 + C_5\widetilde{C}_s\left(1+2h^{-1/2}\right)(h^2+\tau^2), \\ &n=1,2,\cdots,N-1;\quad s=0,1,2,\cdots,\end{aligned} \tag{5.12}$$

where

$$C_4 = 2\sqrt{\frac{\max\left(\frac{1}{L},L\right)}{\min(\lambda,-\frac{q}{2})}\left\{\frac{(p-2\lambda)^2}{4r}L + h\sum_{j=0}^{J}{}''G_d\left(U_j^0\right)\right\}}.$$

Computing the inner product of (5.4) with $\theta^{n+1(s+1)}$

$$\begin{aligned}&\left\|\theta^{n+1(s+1)}\right\|^2 - \beta h^2\left\|\delta_x^+\theta^{n+1(s+1)}\right\|^2 - \frac{q}{2}\tau\left\|\delta_x^{\langle 2\rangle}\theta^{n+1(s+1)}\right\|^2 \\ =&\frac{p}{2}\tau\left(\theta^{n+1(s)},\delta_x^{\langle 2\rangle}\theta^{n+1(s+1)}\right)_h \\ &+\frac{r}{2}\tau\left(\left[\alpha\left(U^{n+1}\right)^2+(1-\alpha)\left(U^n\right)^2\right]\theta^{n+1(s)},\delta_x^{\langle 2\rangle}\theta^{n+1(s+1)}\right)_h \\ &+\frac{r}{2}\tau\left(\alpha\left(U^{n+1(s)}+U^n\right)\left(U^{n+1(s)}+U^{n+1}\right)\theta^{n+1(s)},+\delta_x^{\langle 2\rangle}\theta^{n+1(s+1)}\right)_h \\ \leqslant&-\frac{q}{4}\tau\left\|\delta_x^{\langle 2\rangle}\theta^{n+1(s+1)}\right\|^2 - \frac{p^2}{4q}\tau\left\|\theta^{n+1(s)}\right\|^2 - \frac{q}{8}\tau\left\|\delta_x^{\langle 2\rangle}\theta^{n+1(s+1)}\right\|^2 \\ &-\frac{r^2}{2q}(C_4)^4\tau\left\|\theta^{n+1(s)}\right\|^2 - \frac{q}{8}\tau\left\|\delta_x^{\langle 2\rangle}\theta^{n+1(s+1)}\right\|^2 \\ &-\frac{r^2}{2q}\left[2C_4+C_5\widetilde{C}_s\left(1+2h^{-1/2}\right)(h^2+\tau^2)\right]^4\tau\left\|\theta^{n+1(s)}\right\|^2.\end{aligned} \tag{5.13}$$

When $\beta \leqslant \frac{1}{12}$, we obtain from (5.13) that

$$\left\|\theta^{n+1(s+1)}\right\|^2 \leqslant C_6\tau\left\|\theta^{n+1(s)}\right\|^2, \tag{5.14}$$

where

$$C_6 = -\frac{3p^2}{8q} - \frac{3r^2}{4q}\left[(C_4)^4 + \left(2C_4 + C_5\widetilde{C}_s\left(1+2h^{-1/2}\right)(h^2+\tau^2)\right)^4\right].$$

Thus, for sufficiently small τ and h such that $C_6\tau < 1$, then the solution of the iterative algorithm (5.1)–(5.3) converges to the solution of the nonlinear scheme (2.1)–(2.3). The convergence in the case of $n = 0$ can be proved by the similar method.

In the case of $s = 1$, we obtain from the iterative algorithm (5.1)–(5.3) the linearized implicit prediction-correction scheme

$$\delta_t^- \widetilde{U}_j^{n+1} + \beta h^2 \delta_x^{\langle 2\rangle} \delta_t^- \widetilde{U}_j^{n+1} - \frac{q}{2}\delta_x^{\langle 4\rangle}(\widetilde{U}_j^{n+1} + U_j^n) = \delta_x^{\langle 2\rangle}\varphi(U_j^n, U_j^n, \alpha),$$
$$0 \leqslant j \leqslant J, \ \ n = 0, \tag{5.15}$$

$$\delta_t^- \widetilde{U}_j^{n+1} + \beta h^2 \delta_x^{\langle 2\rangle} \delta_t^- \widetilde{U}_j^{n+1} - \frac{q}{2}\delta_x^{\langle 4\rangle}(\widetilde{U}_j^{n+1} + U_j^n) = \delta_x^{\langle 2\rangle}\varphi(2U_j^n - U_j^{n-1}, U_j^n, \alpha),$$
$$0 \leqslant j \leqslant J, \ \ n = 1, 2, \cdots, N, \tag{5.16}$$

$$\delta_t^- U_j^{n+1} + \beta h^2 \delta_x^{\langle 2\rangle} \delta_t^- U_j^{n+1} - \frac{q}{2}\delta_x^{\langle 4\rangle}(U_j^{n+1} + U_j^n) = \delta_x^{\langle 2\rangle}\varphi(\widetilde{U}_j^n, U_j^n, \alpha),$$
$$0 \leqslant j \leqslant J, \ \ n = 0, 1, 2, \cdots, N, \tag{5.17}$$

$$U_j^0 = u_0(x_j), \qquad j = -2, -1, 0, \cdots, J, J+1, J+2, \tag{5.18}$$

$$\delta_x^{\langle 1\rangle} U_0^n = \delta_x^{\langle 1\rangle} U_J^n = \delta_x^{\langle 3\rangle} U_0^n = \delta_x^{\langle 3\rangle} U_J^n = 0, \qquad n = 0, 1, 2, \cdots, N, \tag{5.19}$$

where

$$\delta_t^- \widetilde{U}_j^{n+1} = \frac{\widetilde{U}_j^{n+1} - U_j^n}{\tau}.$$

Now, we give the second iterative algorithm as follows:

$$\begin{aligned}&\delta_t^- U_j^{n+1(s+1)} + \beta h^2 \delta_x^{\langle 2\rangle} \delta_t^- U_j^{n+1(s+1)} - \frac{q}{2}\delta_x^{\langle 4\rangle}(U_j^{n+1(s+1)} + U_j^n) \\ =&\delta_x^{\langle 2\rangle}\left(\frac{p}{2}\left(U_j^{n+1(s+1)} + U_j^n\right) + \frac{r}{2}\left(U_j^{n+1(s+1)} + U_j^n\right)\right. \\ &\times \left.\left(\alpha(U_j^{n+1(s)})^2 + (1-\alpha)(U_j^n)^2\right)\right), \quad 0 \leqslant j \leqslant J, \quad 0 \leqslant n < N,\end{aligned} \tag{5.20}$$

$$U_j^0 = u_0(x_j), \ \ j = -2, -1, 0, \cdots, J, J+1, J+2, \tag{5.21}$$

$$\delta_x^{\langle 1\rangle} U_0^n = \delta_x^{\langle 1\rangle} U_J^n = \delta_x^{\langle 3\rangle} U_0^n = \delta_x^{\langle 3\rangle} U_J^n = 0, \quad n = 0, 1, 2, \cdots, N, \tag{5.22}$$

with

$$U_j^{n+1(0)} = \begin{cases} U_j^n & \text{for} \ \ n = 0, \\ 2U_j^n - U_j^{n-1} & \text{for} \ \ n \geqslant 1. \end{cases}$$

By the similar proof, we can obtain the following theorem:

Theorem 5.2 *Suppose that $u \in C^{6,3}([0, L] \times [0, T])$ is the solution of (1.1)–(1.4), $\beta \leqslant \dfrac{1}{12}$ and τ, h are sufficiently small, then the solution of the iterative method (5.20)–(5.22) converges to the numerical solution of the scheme (2.1)–(2.3).*

Similarly, in the case of $s=1$, we obtain from the iterative algorithm (5.20)–(5.22) the linearized implicit prediction-correction scheme:

$$\begin{aligned}
&\delta_t^- \widetilde{U}_j^{n+1} + \beta h^2 \delta_x^{\langle 2\rangle} \delta_t^- \widetilde{U}_j^{n+1} - \frac{q}{2}\delta_x^{\langle 4\rangle}(\widetilde{U}_j^{n+1} + U_j^n) \\
=&\delta_x^{\langle 2\rangle}\left(\frac{p}{2}\left(\widetilde{U}_j^{n+1} + U_j^n\right) + \frac{r}{2}\left(\widetilde{U}_j^{n+1} + U_j^n\right)\left(U_j^n\right)^2\right), \quad 0 \leqslant j \leqslant J, \ \ n=0,
\end{aligned} \tag{5.23}$$

$$\begin{aligned}
&\delta_t^- \widetilde{U}_j^{n+1} + \beta h^2 \delta_x^{\langle 2\rangle} \delta_t^- \widetilde{U}_j^{n+1} - \frac{q}{2}\delta_x^{\langle 4\rangle}(\widetilde{U}_j^{n+1} + U_j^n) \\
=&\delta_x^{\langle 2\rangle}\left(\frac{p}{2}\left(\widetilde{U}_j^{n+1} + U_j^n\right) + \frac{r}{2}\left(\widetilde{U}_j^{n+1} + U_j^n\right)\left(\alpha(2U_j^n - U_j^{n-1})^2 + (1-\alpha)(U_j^n)^2\right)\right), \\
&\qquad 0 \leqslant j \leqslant J, \quad n = 1, 2, \cdots, N-1,
\end{aligned} \tag{5.24}$$

$$\begin{aligned}
&\delta_t^- U_j^{n+1} + \beta h^2 \delta_x^{\langle 2\rangle} \delta_t^- U_j^{n+1} - \frac{q}{2}\delta_x^{\langle 4\rangle}(U_j^{n+1} + U_j^n) \\
=&\delta_x^{\langle 2\rangle}\left(\frac{p}{2}\left(U_j^{n+1} + U_j^n\right) + \frac{r}{2}\left(U_j^{n+1} + U_j^n\right)\left(\alpha(\widetilde{U}_j^{n+1})^2 + (1-\alpha)(U_j^n)^2\right)\right), \\
&\qquad 0 \leqslant j \leqslant J, \quad n = 0, 1, 2, \cdots, N-1,
\end{aligned} \tag{5.25}$$

$$U_j^0 = u_0(x_j), \qquad j = -2, -1, 0, \cdots, J, J+1, J+2, \tag{5.26}$$

$$\delta_x^{\langle 1\rangle} U_0^n = \delta_x^{\langle 1\rangle} U_J^n = \delta_x^{\langle 3\rangle} U_0^n = \delta_x^{\langle 3\rangle} U_J^n = 0, \qquad n = 0, 1, 2, \cdots, N. \tag{5.27}$$

Obviously, the above implicit scheme is linearized in the practical computation, i.e. at each time step, we just only apply Thomas algorithm to solve two five-diagonal linear systems. Thus, the prediction-correction scheme can be expected to be more efficient in the practical computation.

Acknowledgements This work is supported by the National Natural Science Foundation of China under grant numbers 11201239, 11571181.

References

[1] Cahn J W, Hilliard J E. Free energy of non-uniform system. I. Interfacial free erengy [J]. J. Chem. Phys,1958, 28: 258–267.

[2] Novick-Cohen A, Segel L A. Nonlinear aspects of the Cahn-Hilliard equation [J]. Physica D,1984, 10: 277–298.

[3] Tanaka H, Nishi T. Direct Detemination of the Probability Distribution Function of Concentration in Polymer Mixtures Undergoing Phase Separation [J]. Phys. Rev. Lett,1987, 59: 692–695.

[4] Du Q, Nicolaides R A. Numerical analysis of a continuum Model of phase transition [J]. SIAM J. Numer. Anal., 1991, 28: 1310–1322.

[5] Wang S, Shi Q. Interdiffusion in binary polymer mixtures [J]. Macromolecules, 1993, 26: 1091–1096.

[6] Jabbari E, Peppas N A. A model for interdiffusion at interfaces of polymers with dissimilar physical properties [J]. Polymer, 1995, 36: 575–586.

[7] Puri S, Binder K. Phenomenological theory for the formation of interfaces via the interdiffusion of layers [J]. Phys. Rev. B.,1991, 44: 9735–9738.

[8] Elliott C M, Zheng S. On the Cahn-Hilliard equation [J]. Arch. Rat. Mech. Anal.,1986, 96: 339–357.

[9] Yin J. On the existence of nonnegative continuous solutions of the Cahn.Hilliard equation [J]. J. Differ. Equations ,1992, 97: 310–327.

[10] Elliott C M, French D A. Numerical studies of the Cahn-Hilliard equation for phase separation [J]. IMA J. Appl. Math.,1987, 38: 97–128.

[11] Elliott C M, French D A. A nonconforming finite-element method for the two-dimensional Cahn-Hilliard equation [J]. SIAM J. Numer. Anal.,1989, 26: 884–903.

[12] French D A, Jensen S. Long-time behaviour of arbitrary order continuous time Galerkin schemes for some onedimensional phase transition problems [J]. IMA J. Numer. Anal., 1994, 14: 421–442.

[13] Elliott C M, French D A, Milner F A. A second-order splliting method for the Cahn-Hilliard equation [J]. Numer. Math.,1989, 54: 575–590.

[14] Elliott C M, Larsson S. Error estimates with smooth and nonsmooth data for a finite element method for the Cahn-Hilliard equation [J]. Math. Comp.,1992, 58: 603–630.

[15] Dean E J, Glowinski R, Trevas D A. An approximate factorization/least squares solution method for a mixed finite element approximation of the Cahn-Hilliard equation [J]. Japan J. Indust. Appl. Math.,1996, 13: 495–517.

[16] Furihata D, Onda T, Mori M. A finite difference scheme for the Cahn-Hilliard equation based on a Lyapunov functional [J]. GAKUTO Internat. Ser. Math. Sci. Appl.,1993, 2: 347–358.

[17] Sun Z. A second-order accurate linearized difference scheme for the two-dimensional Cahn-Hilliard equation [J]. Math. Comp., 1995, 64: 1463–1471.

[18] Choo S M, Chung S K. Conservative nonlinear difference scheme for the Cahn-Hilliard equation [J]. Comput. Math. Appl., 1998, 36: 31–39.

[19] Choo S M, Chung S K, Kim K I. Conservative nonlinear difference scheme for the Cahn-Hilliard equation: II. Comput [J]. Math. Appl., 2000, 39: 229–243.

[20] Dehghan M, Mirzae D. A numerical method based on the boundary integral equation and dual reciprocity methods for one-dimensional Cahn-Hilliard equation [J]. Eng. Anal. Bound. Elem., 2009, 33: 522–528.

[21] He Y, Liu Y, Tang T. On large time-stepping methods for the Cahn-Hilliard equation [J]. Appl. Numer. Math., 2007, 57: 616–628.

[22] Xu C, Tang T. Stability analysis of large time-stepping methods for epitaxial growth models [J]. SIAM J. Numer. Anal., 2006, 44: 1759–1779.

[23] He Y, Liu Y. Stability and convergence of the spectral Galerkin method for the Cahn-Hilliard equation [J]. Numer. Meth. PDEs., 2008, 24: 1485–1500.

[24] Cheng M, Warren J A. An efficient algorithm for solving the phase field crystal model [J]. J. Comput. Phys., 2008, 227: 6241–6248.

[25] Vollmayr-Lee B P, Rutenberg A D. Fast and accurate coarsening simulation with an unconditionally stable time step [J]. Phys. Rev. E., 2003, 68: 13.

[26] Eyre D J. < *http : //www.math.utah.edu/eyre/research/methods/stable.ps* >.

[27] He L, Liu Y. A class of stable spectral methods for the Cahn-Hilliard equation [J]. J. Comput. Phys., 2009, 228: 5101–5110.

[28] Ye X. The Fourier collocation method for the Cahn-Hilliard equation [J]. Comput. Math. Appl., 2002, 44: 213–229.

[29] Ye X, Cheng X L. Legendre spectral approximation for the Cahn-Hilliard equation [J]. Math. Numer. Sinica., 2003, 25: 157–170.

[30] Ye X. The Fourier spectral method for the Cahn-Hilliard equation [J]. Appl. Math. Comput., 2005, 171: 345–357.

[31] Ceniceros H D, Roma A M. A nonstiff, adaptive mesh refinement-based method for the Cahn-Hilliard equation [J]. J. Comput. Phys., 2007, 225: 1849–1862.

[32] Xia Y, Xu Y, Shu C. Local discontinuous Galerkin methods for the Cahn C Hilliard type equations [J]. J. Comput. Phys., 2007, 227: 472–491.

[33] Furihata D. A stable and conservative finite difference scheme for the Cahn-Hilliard equation [J]. Numer. Math., 2001, 87: 675–699.

[34] Choo S M, Chung S K, Lee Y J. A conservative difference scheme for the viscous Cahn C Hilliard equation with a nonconstant gradient energy coefficient [J]. Appl. Numer. Math., 2004, 51: 207–219.

[35] Zhou Y. Application of Discrete Functional Analysis to the Finite Difference Methods [M]. International Academic Publishers, Beijing, 1990.

[36] Akrives G D. Finite difference discretization of the cubic Schrödinger equation [J]. IMA J. Numer. Anal., 1993, 13: 115–124.

[37] Akrives G D, Dogalis V A, karakashina O A, On fully discrete Galerkin methods of second-order temporal accuracy for the nonlinear Schrödinger equation [J]. Numer. Math., 1991, 59: 31–53.

Diffusion Limit of 3D Primitive Equations of the Large-scale Ocean under Fast Oscillating Random Force*

Guo Boling (郭柏灵), Huang Daiwen (黄代文) and Wang Wei (王伟)

Abstract The three-dimensional (3D) viscous primitive equations describing the large-scale oceanic motions under fast oscillating random perturbation are studied. Under some assumptions on the random force, the solution to the initial boundary value problem (IBVP) of the 3D random primitive equations converges in distribution to that of IBVP of the limiting equations, which are the 3D stochastic primitive equations describing the large-scale oceanic motions under a white in time noise forcing. This also implies the convergence of the stationary solution of the 3D random primitive equations.

Keywords Random primitive equations; stationary solution; martingale; statistical solution

1 Introduction

The important 3D viscous primitive equations of the large-scale ocean in a Cartesian coordinate system, are written as the following system on a cylindrical domain

$$\frac{\partial v}{\partial t}+(v\cdot\nabla)v+\Phi(v)\frac{\partial v}{\partial z}+fk\times v+\nabla p_b-\int_{-1}^{z}\nabla T\mathrm{d}z'-\Delta v-\frac{\partial^2 v}{\partial z^2}=\Psi_1\,, \tag{1.1}$$

$$\frac{\partial T}{\partial t}+(v\cdot\nabla)T+\Phi(v)\frac{\partial T}{\partial z}-\Delta T-\frac{\partial^2 T}{\partial z^2}=\Psi_2\,, \tag{1.2}$$

$$\int_{-1}^{0}\nabla\cdot v\,\mathrm{d}z=0. \tag{1.3}$$

* J. Diff. Equ., 2015, 259: 2388-2407.

with boundary value conditions

$$\frac{\partial v}{\partial z}=0,\quad \frac{\partial T}{\partial z}=-\alpha_u T \qquad \text{on} \quad M\times\{0\}=\Gamma_u\,, \tag{1.4}$$

$$\frac{\partial v}{\partial z}=0\,,\quad \frac{\partial T}{\partial z}=0 \qquad \text{on} \quad M\times\{-1\}=\Gamma_b\,, \tag{1.5}$$

$$v\cdot \boldsymbol{n}=0,\quad \frac{\partial v}{\partial \boldsymbol{n}}\times \boldsymbol{n}=0,\quad \frac{\partial T}{\partial \boldsymbol{n}}=0 \quad \text{on } \partial M\times[-1,0]=\Gamma_l\,, \tag{1.6}$$

and the initial value conditions

$$u|_{t=t_0}=(v|_{t=t_0}\,,T|_{t=t_0})=u_{t_0}=(v_{t_0}\,,T_{t_0})\,, \tag{1.7}$$

where the unknown functions are v, p_b, T, $v=(v^{(1)},v^{(2)})$ the horizontal velocity, p_b the pressure, T temperature, $\Phi(v)(t,x,y,z)=-\int_{-1}^{z}\nabla\cdot v(t,x,y,z')\,\mathrm{d}z'$ vertical velocity, $f=f_0(\beta+y)$ the Coriolis parameter, k vertical unit vector, Ψ_1 a given forcing field, Ψ_2 a given heat source, $\nabla=(\partial_x,\ \partial_y)$, $\Delta=\partial_x^2+\partial_y^2$, α_u a positive constant, $\boldsymbol{n}$ the norm vector to Γ_l and M a smooth bounded domain in $\mathbb{R}^2$. For more details for (1.1)–(1.7), see [3, 23] and references therein.

In the past two decades, there are several research works about the well-posedness of the above 3D deterministic primitive equations of the large-scale ocean. In [18], Lions, Temam and Wang obtained the global existence of weak solutions for the primitive equations. In [15], Guillén-González etc. obtained the global existence of strong solutions to the primitive equations with small initial data. Moreover, they proved the local existence of strong solutions to the equations. In [3], Cao and Titi developed a beautiful approach to proving that L^6-norm of the horizontal velocity is uniformly in t bounded, and obtained the global well-posedness for the 3D viscous primitive equations.

In study of the primitive equations of the large-scale ocean or atmosphere, taking the stochastic external factors into account is reasonable and necessary. There are many works about mathematical study of some stochastic climate models, see, e.g., [7,8,9,21,22]. [9] is one of the first works on a 3D stochastic quasi-geostrophic model. Guo and Huang in [14] considered the global well-posedness and long-time dynamics for the 3D stochastic primitive equations of the large-scale ocean under a white in time noise forcing.

In realistic model, random fluctuation always exits. We consider the following 3D primitive equations with fast oscillating random force

$$\frac{\partial v^\epsilon}{\partial t}+(v^\epsilon\cdot\nabla)v^\epsilon+\Phi(v^\epsilon)\frac{\partial v^\epsilon}{\partial z}+fk\times v^\epsilon+\nabla p_b^\epsilon-\int_{-1}^{z}\nabla T^\epsilon dz'-\Delta v^\epsilon-\frac{\partial^2 v^\epsilon}{\partial z^2}=\frac{1}{\sqrt{\epsilon}}\eta_1^\epsilon, \tag{1.8}$$

$$\frac{\partial T^\epsilon}{\partial t}+(v^\epsilon\cdot\nabla)T^\epsilon+\Phi(v^\epsilon)\frac{\partial T^\epsilon}{\partial z}-\Delta T^\epsilon-\frac{\partial^2 T^\epsilon}{\partial z^2}=\Psi+\frac{1}{\sqrt{\epsilon}}\eta_2^\epsilon, \tag{1.9}$$

$$\int_{-1}^{0}\nabla\cdot v^\epsilon\,\mathrm{d}z=0. \tag{1.10}$$

with boundary value conditions

$$\frac{\partial v^\epsilon}{\partial z}=0,\quad \frac{\partial T^\epsilon}{\partial z}=-\alpha_u T^\epsilon,\quad \text{on}\quad \Gamma_u, \tag{1.11}$$

$$\frac{\partial v^\epsilon}{\partial z}=0,\quad \frac{\partial T^\epsilon}{\partial z}=0,\quad \text{on}\quad \Gamma_b, \tag{1.12}$$

$$v^\epsilon\cdot\boldsymbol{n}=0,\quad \frac{\partial v^\epsilon}{\partial \boldsymbol{n}}\times\boldsymbol{n}=0,\quad \frac{\partial T}{\partial \boldsymbol{n}}=0,\quad \text{on}\quad \Gamma_l, \tag{1.13}$$

and initial value conditions

$$u^\epsilon|_{t=t_0}=(v^\epsilon|_{t=t_0},T^\epsilon|_{t=t_0})=u_{t_0}^\epsilon=(v_{t_0}^\epsilon,T_{t_0}^\epsilon), \tag{1.14}$$

where $(v_0^\epsilon,T_0^\epsilon)\in V$ is $\mathcal{F}_0^0$-measurable random variable. Here η_1^ϵ and η_2^ϵ are random forces which are stationary processes satisfying some assumptions given in subsection, and Ψ is a given heat source defined on $\Omega=M\times(-1,0)$.

One reason of considering such fast oscillating random force in the primitive equation is that white noise is an idealistic model. On the other hand the random model (1.8)–(1.10) converges in some sense to the white noise driven primitive equations as $\epsilon\to 0$ which implies that the random model (1.8)–(1.10) is more appropriate to describe some physical phenomena if stochastic primitive equations do.

There are lots of models of such random forces η_1 and η_2. Here, for simplicity, we assume that these random forces are some Gaussian processes solving linear stochastic differential equation (see (2.4)). Classical result [1, Chapter 10, e.g.] shows that

$$\frac{1}{\sqrt{\epsilon}}\int_0^t\eta(s/\epsilon)\,\mathrm{d}s\quad\text{converges in distribution to}\quad W\quad\text{as}\quad\epsilon\to 0$$

for some scalar Wiener process W provided that process η has some mixing property. Then for small $\epsilon>0$, the limit of the above 3D random primitive equations is expected to be 3D primitive equations driven by white noise. In fact by a weak convergence method, we show that the limit of 3D random primitive equations is the limiting model, which is the 3D stochastic primitive equations driven by white in time noise (see Theorem 4.1) as $\epsilon \to 0$. To show the limit of stationary solution, we work on the statistical solution on $[0,\infty)$. The limit of the statistical solution of the 3D random primitive equations is shown to be that of the 3D primitive equations driven by a white noise, which implies the limit of stationary solution of the 3D random primitive equations is that of the 3D stochastic primitive equations. One difficulty here is the singularity caused by the randomly fast oscillation. For this we define a martingale to replace this fast oscillation term, which eliminates the ϵ^{-1} term. This method is applied in both energy estimates for solutions and passing limit of $\epsilon \to 0$, see the proofs of Lemma 4.1 and Theorem 4.1.

The above limit approach is also called a diffusion limit which has been applied to study the asymptotic behavior of stochastic Burgers type equations with stochastic advection [26]. There are also some works on diffusion approximation for some random PDEs with different approachs [5, 16, e.g.].

The paper is organized as follows. In section 2, the 3D random primitive equations are introduced. Our working spaces and a new formulation of the initial boundary value problem for the primitive equations with fast oscillating random force are given in this section. We obtain the global well-posedness to 3D primitive equations of the large-scale ocean under fast oscillating random force in section 3. We prove main results of our paper in section 4.

2 New formulation for the 3D random primitive equations

2.1 New formulation for the 3D random primitive equations

Before formulating a new formulation for the 3D random primitive equations, notations for some function spaces, functionals and operators are given.

Let $L^p(\Omega)$ be the usual Lebesgue space with the norm $|\cdot|_p$, $1 \leqslant p \leqslant \infty$. $H^m(\Omega)$ is the usual Sobolev space(m is a positive integer) with the norm

$$\|h\|_m = \left[\int_\Omega \Big(\sum_{1\leqslant k\leqslant m} \sum_{i_j=1,2,3; j=1,\cdots,k} |\nabla_{i_1}\cdots\nabla_{i_k} h|^2 + |h|^2\Big)\right]^{\frac{1}{2}},$$

where $\nabla_1 = \dfrac{\partial}{\partial x}, \nabla_2 = \dfrac{\partial}{\partial y}$ and $\nabla_3 = \dfrac{\partial}{\partial z}$. $\int_\Omega \cdot \mathrm{d}\Omega$ and $\int_M \cdot \mathrm{d}M$ are denoted by $\int_\Omega \cdot$

and $\int_M \cdot$ respectively.

Define our working spaces for (1.8)–(1.14) as

$$\mathcal{V}_1 := \Big\{ v \in (C^\infty(\Omega))^2; \frac{\partial v}{\partial z}|_{\Gamma_u,\Gamma_b} = 0, v\cdot \boldsymbol{n}|_{\Gamma_l} = 0, \frac{\partial v}{\partial \boldsymbol{n}} \times \boldsymbol{n}|_{\Gamma_l} = 0,$$
$$\int_{-1}^{0} \nabla \cdot v dz = 0 \Big\},$$
$$\mathcal{V}_2 := \Big\{ T \in C^\infty(\Omega); \frac{\partial T}{\partial z}|_{\Gamma_u} = -\alpha_u T, \frac{\partial T}{\partial z}|_{\Gamma_b} = 0, \frac{\partial T}{\partial \boldsymbol{n}}|_{\Gamma_l} = 0 \Big\},$$
$V_1 =$ the closure of $\mathcal{V}_1$ with respect to the norm $\|\cdot\|_1$,

$V_2 =$ the closure of $\mathcal{V}_2$ with respect to the norm $\|\cdot\|_1$,

$H_1 =$ the closure of $\mathcal{V}_1$ with respect to the norm $|\cdot|_2$,

$$V = V_1 \times V_2, \qquad H = H_1 \times L^2(\Omega).$$

The inner products and norms on $V,\ H$ are given by

$$\langle u, u_1\rangle_V = \langle v, v_1\rangle_{V_1} + \langle T, T_1\rangle_{V_2},$$
$$\langle u, u_1\rangle = \langle v^{(1)}, (v_1)^{(1)}\rangle + \langle v^{(2)}, (v_1)^{(2)}\rangle + \langle T, T_1\rangle,$$
$$\|u\| = \langle u, u\rangle_V^{\frac{1}{2}} = \langle v, v\rangle_{V_1}^{\frac{1}{2}} + \langle T, T\rangle_{V_2}^{\frac{1}{2}} = \|v\| + \|T\|, \qquad |u|_2 = \langle u, u\rangle^{\frac{1}{2}},$$

where $u = (v, T),\ u_1 = (v_1, T_1) \in V$, and $\langle\cdot,\cdot\rangle$ denotes the inner product in $L^2(\Omega)$.

Then, we define the functionals $a : V\times V \to \mathbb{R},\ a_1 : V_1\times V_1 \to \mathbb{R},\ a_2 : V_2\times V_2 \to \mathbb{R}$, and their corresponding linear operators $A : V \to V',\ A_1 : V_1 \to V_1',\ A_2 : V_2 \to V_2'$ by

$$a(u, u_1) = \langle Au, u_1\rangle = a_1(v, v_1) + a_2(T, T_1),$$

where

$$a_1(v, v_1) = \langle A_1 v, v_1\rangle = \int_\Omega \left(\nabla v \cdot \nabla v_1 + \frac{\partial v}{\partial z}\cdot\frac{\partial v_1}{\partial z}\right),$$
$$a_2(T, T_1) = \langle A_2 T, T_1\rangle = \int_\Omega \left(\nabla T \cdot \nabla T_1 + \frac{\partial T}{\partial z}\frac{\partial T_1}{\partial z}\right) + \alpha_u \int_{\Gamma_u} TT_1.$$

According to Lemma 3.1 in [14], we know that a and A have the following properties. a is coercive and continuous, and $A : V \to V'$ is isomorphism. Moreover,

$$a(u, u_1) \leqslant c\|v\|\|v_1\| + c\|T\|\|T_1\| \leqslant c\|u\|\|u_1\|,$$

$$a(u, u) \geqslant c\|v\|^2 + c\|T\|^2 \geqslant c\|u\|^2.$$

The isomorphism $A : V \to V'$ can be extended to a self-adjoint unbounded linear operator on H with a compact inverse $A^{-1} : H \to H$ and with the domain of definition of the operator $D(A) = V \cap [(H^2(\Omega))^2 \times H^2(\Omega)]$. Denote by $0 < \lambda_1 \leqslant \lambda_2 \leqslant \cdots$ the eigenvalues of A and by $e_1, e_2 \cdots$ the corresponding complete orthonormal system of eigenvectors.

We define a nonlinear operator $N = (N_1, N_2) : V \times V \to V'$ by

$$\langle N(u_1, u_1), u_2\rangle_H = \langle N_1(v_1, v_1), v_2\rangle + \langle N_2(v_1, T_1), T_2\rangle_{H_2},$$

where

$$\langle N_1(v, v_1), v_2\rangle = \int_\Omega \left((v \cdot \nabla)v_1 + \Phi(v)\frac{\partial v_1}{\partial z}\right) \cdot v_2,$$

$$\langle N_2(v, T_1), T_2\rangle = \int_\Omega \left((v_1 \cdot \nabla)T_1 + \Phi(v_1)\frac{\partial T_1}{\partial z}\right) T_2,$$

Related to the linear terms, we define a bilinear functional $l : V \to \mathbb{R}$ and its corresponding operator $L : V \to V$ by

$$l(u, u_1) = \langle Lu, u_1\rangle_H = \int_\Omega f(k \times v) \cdot v_1 - \int_\Omega \left(\int_{-1}^z \nabla T dz'\right) T_1.$$

Now, we rewrite (1.8)–(1.14) as the following abstract stochastic evolution equations

$$\frac{\partial v^\epsilon}{\partial t} + N_1(v^\epsilon, v^\epsilon) + Lu^\epsilon + A_1 v^\epsilon = \frac{1}{\sqrt{\epsilon}}\eta_1^\epsilon, \tag{2.1}$$

$$\frac{\partial T^\epsilon}{\partial t} + N_2(v^\epsilon, T^\epsilon) + A_2 T^\epsilon = \Psi + \frac{1}{\sqrt{\epsilon}}\eta_2^\epsilon, \tag{2.2}$$

$$u^\epsilon(0) = (v^\epsilon(0), T^\epsilon(0)) = (v_0^\epsilon, T_0^\epsilon). \tag{2.3}$$

For the above random system we give the following definition.

Definition 2.1 *For any $\mathcal{T} > t_0$, a process $u^\epsilon(t, \omega) = (v^\epsilon, T^\epsilon)$ is called a strong solution to* (2.1)–(2.3) *in* $[t_0, \mathcal{T}]$, *if, for $\mathbb{P}$-a.e. $\omega \in \mathbf{\Omega}$, u^ϵ satisfies*

$$\langle v^\epsilon(t), \varphi_1\rangle - \int_{t_0}^t [\langle N_1(v^\epsilon, \varphi_1), v\rangle - \langle Lu^\epsilon, \varphi\rangle_H] + \int_{t_0}^t \langle v^\epsilon, A_1\varphi_1\rangle$$

$$= \langle v_{t_0}^\epsilon, \varphi_1\rangle + \frac{1}{\sqrt{\epsilon}}\int_{t_0}^t \langle \eta_1^\epsilon(s, \omega), \varphi_1\rangle,$$

$$\langle T^\epsilon(t), \varphi_2\rangle - \int_{t_0}^t [\langle N_2(v^\epsilon, \varphi_2), T^\epsilon\rangle - \langle T^\epsilon, A_2\varphi_2\rangle]$$

$$= \langle T_{t_0}^\epsilon, \varphi_2\rangle + \int_{t_0}^t \langle \Psi, \varphi_2\rangle + \frac{1}{\sqrt{\epsilon}}\int_{t_0}^t \langle \eta_2^\epsilon(s, \omega), \varphi_2\rangle,$$

for all $t \in [t_0, \mathcal{T}]$ and $\varphi = (\varphi_1, \varphi_2) \in D(A_1) \times D(A_2)$, moreover $u^\epsilon \in L^\infty(t_0, \mathcal{T}; V) \cap L^2(t_0, \mathcal{T}; (H^2(\Omega))^2)$ and is progressively measurable in these topologies.

2.2 A model for $(\eta_1^\epsilon, \eta_2^\epsilon)$

To detail the random model (1.8)–(1.14), we assume that the stationary process $\eta^\epsilon(t) = (\eta_1^\epsilon(t), \eta_2^\epsilon(t))$ solves the following linear stochastic system

$$\epsilon d\frac{\eta^\epsilon(t)}{\sqrt{\epsilon}} = -\frac{\eta^\epsilon(t)}{\sqrt{\epsilon}}\mathrm{d}t + \mathrm{d}W(t), \tag{2.4}$$

where $W(t) = (W_1(t), W_2(t))$ is H-valued Wiener process with covariance operator $\mathbb{Q} = (Q_1, Q_2)$. Here we assume that $\dot{W}_i$, $i = 1, 2$, have the following form

$$\frac{\mathrm{d}W_i(t)}{\mathrm{d}t} = \sqrt{Q_i}\frac{\mathrm{d}\tilde{W}_i(t)}{\mathrm{d}t},$$

where $\tilde{W}_i(t)$ is a cylindrical Wiener process in H_i defined on a complete probability space $(\boldsymbol{\Omega}, \mathcal{F}, \mathbb{P})$ with expectation denoted by $\mathbb{E}$, and $\sqrt{Q_i}$ is a linear operator.

Assumption $(\mathbf{H}_1)$ Q_i, $i = 1, 2$, satisfy

$$\mathrm{Tr} A^3 Q_i < +\infty .$$

Remark 2.1 *An example for W is a two-sided in time finite-dimensional Brownian motion with the form*

$$W = \sum_{i=1}^{m} \delta_i \beta_i(t, \omega) e_i.$$

In the above formula, $\beta_1, \cdots, \beta_m$ are independent standard one-dimensional Brownian motions on a complete probability space $(\boldsymbol{\Omega}, \mathcal{F}, P)$, and δ_i are real coefficients.

Remark 2.2 *An another example for W is a two-sided in time infinite dimensional Brownian motion with the form*

$$W(t) = \sum_{i=1}^{+\infty} \mu_i \beta_i(t, \omega) e_i.$$

Here $\beta_1, \beta_2, \cdots$ is a sequence of independent standard one-dimensional Brownian motions on a complete probability space $(\boldsymbol{\Omega}, \mathcal{F}, P)$ and the coefficients μ_i satisfy $\sum_{i=1}^{+\infty} \lambda_i^3 \mu_i^2 < +\infty$.

By the stationarity and the assumption $(\mathbf{H}_1)$, a simple application of Itô formula yields

$$\mathbb{E}\|\eta^\epsilon(t)\|_3^2 = \mathrm{Tr}(A^3\mathbb{Q}) < \infty. \tag{2.5}$$

Moreover, the system (2.4) is strong mixing. To describe, this we define

$$\mathcal{F}_{s/\epsilon}^{t/\epsilon} = \sigma\{\eta^\epsilon(\tau) : s \leqslant \tau \leqslant t\}$$

and

$$\phi(t/\epsilon) = \sup_{s\geqslant 0} \sup_{A\in\mathcal{F}_0^{s/\epsilon}, B\in\mathcal{F}_{s/\epsilon+t/\epsilon}^{\infty}} |\mathbb{P}(AB) - \mathbb{P}(A)\mathbb{P}(B)|.$$

Then

$$\int_0^\infty \phi^k(t)\,\mathrm{d}t < \infty$$

for any $k > 0$. In fact by the exponentially stable of the linear system (2.4), we have for $t > 0$

$$\phi(t/\epsilon) < Ce^{-t/\epsilon}$$

for some constant $C > 0$. Moreover for $s \geqslant t$

$$\mathbb{E}[\eta^\epsilon(s)|\mathcal{F}_0^{t/\epsilon}] = \eta^\epsilon(t)e^{-(s-t)/\epsilon}, \tag{2.6}$$

$$\mathbb{E}\eta_i^\epsilon(x,t)\eta_i^\epsilon(y,s) = \frac{1}{2}q_i(x,y)\exp\left(-\frac{|t-s|}{\epsilon}\right), \quad i = 1, 2, \tag{2.7}$$

where $q_i(x,y)$ satisfy $q_i(x,y) = q_i(y,x)$ and

$$Q_i f(x) = \int_\Omega q_i(x,y)f(y)\,\mathrm{d}y, \quad i = 1, 2.$$

Notice that (v^ϵ, T^ϵ) is not Markovian. So in the following we consider the Markov process $(v^\epsilon, T^\epsilon, \eta_1^\epsilon, \eta_2^\epsilon)$.

3 The global well-posedness and existence of stationary measure

Let $U^\epsilon = (u^\epsilon, \eta^\epsilon)$, with $u^\epsilon = (v^\epsilon, T^\epsilon)$. To show the global well-posedness to (2.1)–(2.3), we introduce process $\mathcal{Z}^\epsilon$ solving

$$\dot{\mathcal{Z}}^\epsilon = -A\mathcal{Z}^\epsilon + \frac{1}{\sqrt{\epsilon}}\eta^\epsilon, \quad \mathcal{Z}^\epsilon(0) = 0.$$

One can see that

$$\mathcal{Z}^\epsilon(t) = \frac{1}{\sqrt{\epsilon}}\int_0^t e^{-A(t-s)}\eta^\epsilon(s)\,\mathrm{d}s.$$

Then we have the following estimates.

Lemma 3.1 *For any integer $k \geqslant 0$, there is a constant $c > 0$, such that for any $t > 0$,*

$$\mathbb{E}\|\mathcal{Z}^\epsilon(t)\|_k^2 \leqslant c\mathrm{Tr}A^k\mathbb{Q}.$$

We need some estimates on $Z_\alpha^\epsilon(t)$. First

$$\mathbb{E}\|\mathcal{Z}^\epsilon(t)\|_0^2 = \frac{1}{\epsilon}\mathbb{E}\int_0^t\int_0^t \left\langle e^{-A(t-s)}\eta^\epsilon(s), e^{-A(t-r)}\eta^\epsilon(r)\right\rangle \mathrm{d}r\,\mathrm{d}s,$$

then by (2.6)–(2.7) we have for any $t > 0$

$$\mathbb{E}\|\mathcal{Z}^\epsilon(t)\|_0^2 \leqslant c\left[\|q_1\|_{L^2}^2 + \|q_2\|_{L^2}^2\right]$$

for some constant $c > 0$. Similarly by the regularity of η^ϵ, for $k > 0$,

$$\mathbb{E}\|\mathcal{Z}^\epsilon(t)\|_k^2 \leqslant c\left[\|q_1\|_{H^k}^2 + \|q_2\|_{H^k}^2\right].$$

Now define

$$w^\epsilon = v^\epsilon - \mathcal{Z}^{\epsilon,1}, \quad \theta^\epsilon = T^\epsilon - \mathcal{Z}^{\epsilon,2}, \quad \text{and} \quad \tilde{u}^\epsilon = u^\epsilon - \mathcal{Z}^\epsilon,$$

where $\mathcal{Z}^\epsilon = (\mathcal{Z}^{\epsilon,1}, \mathcal{Z}^{\epsilon,2})$ Then, to obtain the global existence of strong solutions to the system (2.1)–(2.3), we just consider that of the following system

$$\frac{\partial w^\epsilon}{\partial t} + N_1(w^\epsilon + \mathcal{Z}^{\epsilon,1}, w^\epsilon + \mathcal{Z}^{\epsilon,1}) + L(\tilde{u}^\epsilon + \mathcal{Z}^\epsilon) + A_1 w^\epsilon = 0, \tag{3.1}$$

$$\frac{\partial \theta^\epsilon}{\partial t} + N_2(w^\epsilon + \mathcal{Z}^{\epsilon,1}, \theta^\epsilon + \mathcal{Z}^{\epsilon,2}) + A_2\theta^\epsilon = \Psi, \tag{3.2}$$

$$\tilde{u}^\epsilon(0) = (w^\epsilon(0), \theta^\epsilon(0)) = u_0^\epsilon = (v_0^\epsilon, T_0^\epsilon). \tag{3.3}$$

Theorem 3.1 (Global well-posedness of (2.1)–(2.3)) *Assume that $\Psi \in V_2$ and $(\mathbf{H}_1)$ hold, then*

(1) For any initial data $u_0^\epsilon \in V$, there exists globally a unique strong solution u^ϵ to (2.1)–(2.3), i.e., for any $\mathcal{T} > 0$,

$$u^\epsilon \in C(0, \mathcal{T}; V) \cap L^2(0, \mathcal{T}; (H^2(\Omega))^2)$$

for $\mathbb{P}$ *a.e.* $\omega \in \mathbf{\Omega}$.

(2) The process U^ϵ is a Markov process in $\mathbb{V} := V \times H^1 \times H^1$.

Remark 3.1 *Result (1) is proved by Lemma 3.1 and the method for 3D stochastic primitive equations [14]. Result (2) can be justified by a standard argument [24, Theorem 9.8].*

Proposition 3.1 *Assume $\Psi \in V_2$ and $B_\rho = \{u; \|u\| \leqslant \rho, u \in V\}$. Then there exist $r_0(\omega, \|\Psi\|_1)$ and $t(\omega,\rho) \leqslant -1$ such that for any $t_0 \leqslant t(\omega,\rho)$, $u^\epsilon_{t_0} \in B_\rho$,*

$$\|u^\epsilon(0,\omega,u^\epsilon_{t_0})\| \leqslant r_0(\omega),$$

where $u^\epsilon(t,\omega,u^\epsilon_{t_0})$ is the strong solution of (2.1)–(2.3) *with initial data $u^\epsilon(t_0) = u^\epsilon_{t_0}$. Moreover, the family of random variables $\{u^\epsilon(0,\omega,u^\epsilon_{t_0}) : -\infty < t_0 < t(\omega,\rho)\}$* is tight in V.

Remark 3.2 *The estimate in the above proposition can be followed by a similar discussion for* 3D *stochastic primitive equations* [14] *and the tight result can be derived by the similar method* [13].

Denote by P^ϵ_t, $t \geqslant 0$, the associated Markov semigroup to U^ϵ. Let $P^{\epsilon *}_t$ be the dual semigroup which is defined, in space $Pr(\mathbb{V})$ consisting of probability measures on $\mathbb{V}$, as

$$\int_{\mathbb{V}} \varphi d P^{\epsilon *}_t \mu = \int_{\mathbb{V}} P^\epsilon_t \varphi d\mu$$

for all $\varphi \in C_b(\mathbb{V})$ and $\mu \in Pr(\mathbb{V})$. Then a probability measure $\mu \in Pr(\mathbb{V})$ is called a stationary one if

$$P^{\epsilon *}_t \mu = \mu.$$

Then we have the following result.

Theorem 3.2 (The existence of stationary measure) *Under the assumption* ($\mathbf{H}_1$), *the system of* 3D *random primitive equations* (2.1)–(2.3) *coupled with* (2.4) *has at least a stationary measure.*

Proof By the Proposition 3.1, the distribution of U^ϵ in space $\mathbb{V}$ is tight, then by the Kryloff–Bogoliubov procedure [25] one can construct one stationary measure. □

Let $\mathfrak{y}^{*\epsilon}$ be a stationary measure of 3D random primitive equations (2.1)–(2.3) coupled with (2.4). Then $U^{*\epsilon} = (v^{*\epsilon}, T^{*\epsilon}, \eta^\epsilon)$, the solution with initial value distributes as $\mathfrak{y}^{*\epsilon}$, is a stationary solution to 3D random primitive equations (2.1)–(2.3) coupled with (2.4). Then we call $(v^{*\epsilon}, T^{*\epsilon})$ a stationary solution to 3D random primitive equations (2.1)–(2.3).

4 The diffusion limit for the 3D random primitive equations

Notice that the distributions of η_1^ϵ and η_2^ϵ are independent of ϵ. We just consider the limit of (v^ϵ, T^ϵ). We have showed the tightness of the distributions of $\{(v^\epsilon, T^\epsilon)\}_{0<\epsilon\leqslant 1}$ in space $C([0,\infty);V)$. Further the following result determine the limit of (v^ϵ, T^ϵ) in the sense of distribution. For this we first give the following limiting equation, 3D stochastic primitive equations

$$\frac{\partial v}{\partial t} + (v\cdot\nabla)v + \Phi(v)\frac{\partial v}{\partial z} + fk\times v + \nabla p_b - \int_{-1}^{z}\nabla T\mathrm{d}z' - \Delta v - \frac{\partial^2 v}{\partial z^2} = \dot{W}_1, \quad (4.1)$$

$$\frac{\partial T}{\partial t} + (v\cdot\nabla)T + \Phi(v)\frac{\partial T}{\partial z} - \Delta T - \frac{\partial^2 T}{\partial z^2} = \Psi + \dot{W}_2, \quad (4.2)$$

$$\int_{-1}^{0}\nabla\cdot v\,\mathrm{d}z = 0. \quad (4.3)$$

with boundary value conditions (1.4)–(1.6) and initial value

$$u(0) = (v_0, T_0). \quad (4.4)$$

Remark 4.1 *The above equations are called as* 3D *stochastic primitive equations of the large-scale ocean which are considered in the article* [14]. *There are two main reasons for considering the stochastic model. Firstly, it is impossible to accurately predict long-term oceanic motions in deterministic frameworks since the space scale and time scale of the forecast target are bounded in the deterministic oceanic forecasting. To predict long-term oceanic motions more objectively, it is suitable to apply some stochastic modes. Secondly, it is suitable and useful to consider the sochastic primitive equations of the large-scale ocean with a white in time noise since the primitive oceanic equations are usually used to understand the mechanism of long-term oceanic motions.*

We need some estimates on the solution.

Lemma 4.1 *Under the assumption, for any $T>0$ the following estimate holds*

$$\sup_{0\leqslant t\leqslant T}\mathbb{E}\|u^\epsilon(t)\| + \mathbb{E}\int_0^T\|\nabla u^\epsilon(s)\|^2\,\mathrm{d}s \leqslant C_T$$

for some constant $C_T>0$.

Proof The difficulty is the existence of singular terms. To overcome this, we introduce the following processes

$$
\begin{aligned}
M_t^{1,\epsilon} = & \epsilon[\langle \eta_1^\epsilon(t), v^\epsilon(t)\rangle - \langle \eta_1^\epsilon(0), v^\epsilon(0)\rangle] + \frac{1}{\sqrt{\epsilon}}\int_0^t \langle \eta_1^\epsilon(s), v^\epsilon(s)\rangle \mathrm{d}s \\
& + \epsilon\int_0^t \langle \eta_1^\epsilon(s), -A_1 v^\epsilon(s) - Lu^\epsilon(s) - N_1(v^\epsilon(s), v^\epsilon(s))\rangle\, \mathrm{d}s \\
& - \sqrt{\epsilon}\int_0^t \langle \eta_1^\epsilon(s), \eta_1^\epsilon(s)\rangle\, \mathrm{d}s
\end{aligned}
$$

and

$$
\begin{aligned}
M_t^{2,\epsilon} = & \epsilon[\langle \eta_2^\epsilon(t), T^\epsilon(t)\rangle - \langle \eta_2^\epsilon(0), T^\epsilon(0)\rangle] + \frac{1}{\sqrt{\epsilon}}\int_0^t \langle \eta_2^\epsilon(s), T^\epsilon(s)\rangle \mathrm{d}s \\
= & +\epsilon\int_0^t \langle \eta_2^\epsilon(s), -A_2 T^\epsilon(s) - N_2(v^\epsilon(s), T^\epsilon(s)) + \Psi\rangle\, \mathrm{d}s \\
= & -\sqrt{\epsilon}\int_0^t \langle \eta_2^\epsilon(s), \eta_2^\epsilon(s)\rangle\, \mathrm{d}s\,.
\end{aligned}
$$

For fixed $\epsilon > 0$, by direct calculation, we have

$$
\sup_{0\leqslant t\leqslant T} \mathbb{E}\|u^\epsilon(t)\| + \int_0^T \|\nabla u^\epsilon(t)\|^2\, \mathrm{d}t \leqslant C_{\epsilon,T},
$$

which implies that, by a direct verification [6, Lemma 2], $M_t^{1,\epsilon}$ and $M_t^{2,\epsilon}$ are square integrable martingale with respect to $\mathcal{F}_0^{t/\epsilon}$. Multiplying equation (2.1) with v^ϵ on both sides in H_1 yields

$$
\begin{aligned}
& \mathrm{d}\left[\frac{1}{2}\|v^\epsilon(t)\|^2 + \epsilon\langle \eta_1^\epsilon(t), v^\epsilon(t)\rangle\right] \\
= & \langle -A_1 v^\epsilon(t) - Lu^\epsilon(t) - N_1(v^\epsilon, v^\epsilon), v^\epsilon(t)\rangle\, \mathrm{d}t \\
& - \epsilon\langle -A_1 v^\epsilon(t) - Lu^\epsilon(t) - N_1(v^\epsilon, v^\epsilon), \eta_1^\epsilon\rangle\, \mathrm{d}t \\
& - \sqrt{\epsilon}\|\eta_1^\epsilon\|^2\, \mathrm{d}t + \mathrm{d}M_t^{1,\epsilon}
\end{aligned}
$$

and

$$
\begin{aligned}
& \mathrm{d}\left[\frac{1}{2}\|T^\epsilon(t)\|^2 + \epsilon\langle \eta_2^\epsilon(t), T^\epsilon(t)\rangle\right] \\
= & \langle -A_2 T^\epsilon(t) - N_2(v^\epsilon, T^\epsilon) + \Psi, T^\epsilon(t)\rangle\, \mathrm{d}t \\
& - \epsilon\langle -A_2 T^\epsilon(t) - N_2(v^\epsilon, T^\epsilon), \eta_2^\epsilon\rangle\, \mathrm{d}t \\
& - \sqrt{\epsilon}\|\eta_2^\epsilon\|^2\, \mathrm{d}t + \mathrm{d}M_t^{2,\epsilon}\,.
\end{aligned}
$$

Then for $\epsilon > 0$ small enough, by the martingale property of $M_t^{1,\epsilon}$, $M_t^{2,\epsilon}$ and Gronwall lemma, we have the estimate of this lemma. □

Theorem 4.1 *Assume* $(v_0^\epsilon, T_0^\epsilon) \in V$ *converges in distribution to* (v_0, T_0) *as* $\epsilon \to 0$. *The solution* (v^ϵ, T^ϵ) *of* 3D *random primitive equations* (1.8)–(1.14) *converges in distribution, as* $\epsilon \to 0$, *to the solution of* 3D *stochastic primitive equations* (4.1)–(4.4) *in space* $C([0,\infty); V)$.

Remark 4.2 *We apply a weak convergence method developed by Kushner* [17] *to prove this result. Such method is also applied to study the stochastic self-similarity in stochastic Burgers equation* [26].

Proof Denote by (v, T) one limit point in the sense of distribution of (v^ϵ, T^ϵ) as $\epsilon \to 0$ in space $C([0,\infty); V)$. For simplicity we assume (v^ϵ, T^ϵ) converges in distribution to (v, T) as $\epsilon \to 0$ in $C([0,\infty); V)$. Notice that convergence in distribution is not enough to pass limit $\epsilon \to 0$, by Skorohod theorem we can construct a new probability space and new random variables without changing distributions in $C([0,\infty); V)$, of which for simplicity we do not change the notations, such that $(v_0^\epsilon, T_0^\epsilon)$ converges almost surely to (v_0, T_0) in V and (v^ϵ, T^ϵ) converges almost surely to (v, T) in $C([0,\infty); V)$.

In the following, we shall prove that (v, T) is a solution to (4.1)–(4.4). Now for any $\varphi = (\varphi_1, \varphi_2) \in C_0^\infty$ and C^3-differentiable compactly supported real valued function F, we consider processes

$$\{F(\langle (v^\epsilon, T^\epsilon), \varphi \rangle)\}_{0<\epsilon\leqslant 1}. \tag{4.5}$$

We derive from equations (1.8)–(1.14)

$$\begin{aligned}
&F(\langle (v^\epsilon(t), T^\epsilon(t)), \varphi \rangle) - F(\langle (v_0^\epsilon, T_0^\epsilon), \varphi \rangle) \\
=&\int_0^t F'(\langle (v^\epsilon(s), T^\epsilon(s)), \varphi \rangle)\big[\langle G_1^\epsilon(s), \varphi_1 \rangle + \langle G_2^\epsilon(s), \varphi_2 \rangle\big]\,\mathrm{d}s \\
&+\frac{1}{\sqrt{\epsilon}}\int_0^t F'(\langle (v^\epsilon(s), T^\epsilon(s)), \varphi \rangle)\langle \eta_1^\epsilon(s), \varphi_1 \rangle\,\mathrm{d}s \\
&+\frac{1}{\sqrt{\epsilon}}\int_0^t F'(\langle (v^\epsilon(s), T^\epsilon(s)), \varphi \rangle)\langle \eta_2^\epsilon(s), \varphi_2 \rangle\,\mathrm{d}s,
\end{aligned} \tag{4.6}$$

where

$$G_1^\epsilon = \Delta v^\epsilon - (v^\epsilon \cdot \nabla)v^\epsilon - \Phi(v^\epsilon)\frac{\partial v^\epsilon}{\partial z} - fk \times v^\epsilon - \nabla p_b^\epsilon + \int_{-1}^z \nabla T^\epsilon \mathrm{d}z'$$

and

$$G_2^\epsilon = \Delta T^\epsilon - (v^\epsilon \cdot \nabla)T^\epsilon - \Phi(v^\epsilon)\frac{\partial T^\epsilon}{\partial z} + \Psi.$$

To treat the singular terms in (4.6) we introduce

$$F_1^\epsilon(t) := \frac{1}{\sqrt{\epsilon}}\mathbb{E}\left[\int_t^\infty F'(\langle (v^\epsilon(t), T^\epsilon(t)), \varphi \rangle)\langle \eta_1^\epsilon(s), \varphi_1 \rangle\,\mathrm{d}s\Big|\mathcal{F}_0^{t/\epsilon}\right]$$

and

$$F_2^\epsilon(t) := \frac{1}{\sqrt{\epsilon}}\mathbb{E}\left[\int_t^\infty F'(\langle(v^\epsilon(t), T^\epsilon(t)), \varphi\rangle)\langle\eta_2^\epsilon(s), \varphi_2\rangle \,\mathrm{d}s \Big| \mathcal{F}_0^{t/\epsilon}\right].$$

Then by the property (2.6) of η^ϵ we have

$$F_1^\epsilon(t) = \sqrt{\epsilon}F'(\langle(v^\epsilon(t), T^\epsilon(t)), \varphi\rangle)\langle\eta_1^\epsilon(t), \varphi_1\rangle$$

and

$$F_2^\epsilon(t) = \sqrt{\epsilon}F'(\langle(v^\epsilon(t), T^\epsilon(t)), \varphi\rangle)\langle\eta_2^\epsilon(t), \varphi_2\rangle.$$

Moreover

$$\mathbb{E}|F_i^\epsilon(t)| \leqslant C\sqrt{\epsilon}, \quad i = 1, 2$$

for some constant $C > 0$.

Next we construct a martingale depends on ϵ and pass the limit $\epsilon \to 0$ in this martingale. For this we introduce the pseudo-differential operator A^ϵ defined by

$$A^\epsilon X(t) = \mathbb{P} - \lim_{\delta\to 0}\frac{1}{\delta}\mathbb{E}\left[X(t+\delta) - X(t) \mid \mathcal{F}_0^{t/\epsilon}\right] \tag{4.7}$$

for any $\mathcal{F}_0^{t/\epsilon}$ measurable function X with $\sup_t \mathbb{E}|X(t)| < \infty$. Then Ethier and Kurtz's proposition [10, Proposition 2.7.6] yields that

$$X(t) - \int_0^t A^\epsilon X(s)\,\mathrm{d}s$$

is a martingale with respect to $\mathcal{F}_0^{t/\epsilon}$. Now define (Y^ϵ, Z^ϵ) as

$$Y^\epsilon(t) = F(\langle(v^\epsilon(t), T^\epsilon(t)), \varphi\rangle) + F_1^\epsilon(t) + F_2^\epsilon(t), \quad Z^\epsilon(\tau) = A^\epsilon Y^\epsilon(t).$$

Then

$$\begin{aligned}
&Z^\epsilon(t)\\
=&F'(\langle(v^\epsilon(t), T^\epsilon(t)), \varphi\rangle)\left[\langle G_1^\epsilon(t), \varphi_1\rangle + \langle G_2^\epsilon(t), \varphi_2\rangle\right]\\
&+\sqrt{\epsilon}F''(\langle(v^\epsilon(t), T^\epsilon(t)), \varphi\rangle)\langle\eta_1^\epsilon(t), \varphi_1\rangle\left[\langle G_1^\epsilon(t), \varphi_1\rangle + \langle G_2^\epsilon(t), \varphi_2\rangle\right]\\
&+\sqrt{\epsilon}F''(\langle(v^\epsilon(t), T^\epsilon(t)), \varphi\rangle)\langle\eta_2^\epsilon(t), \varphi_2\rangle\left[\langle G_1^\epsilon(t), \varphi_1\rangle + \langle G_2^\epsilon(t), \varphi_2\rangle\right]\\
&+F''(\langle(v^\epsilon(t), T^\epsilon(t)), \varphi\rangle)\langle\eta_1^\epsilon(t), \varphi_1\rangle^2 + F''(\langle(v^\epsilon(t), T^\epsilon(t)), \varphi\rangle)\langle\eta_2^\epsilon(t), \varphi_2\rangle^2\\
&+2F''(\langle(v^\epsilon(t), T^\epsilon(t)), \varphi\rangle)\langle\eta_1^\epsilon(t), \varphi_1\rangle\langle\eta_2^\epsilon(t), \varphi_2\rangle.
\end{aligned} \tag{4.8}$$

In fact first by equaton (4.6) and the definition of A^ϵ, we have

$$\begin{aligned}&A^\epsilon F(\langle(v^\epsilon(t),T^\epsilon(t)),\varphi\rangle)\\&=F'(\langle(v^\epsilon(t),T^\epsilon(t)),\varphi\rangle)\big[\langle G_1^\epsilon(t),\varphi_1\rangle+\langle G_2^\epsilon(t),\varphi_2\rangle\big]\\&+\frac{1}{\sqrt{\epsilon}}F'(\langle(v^\epsilon(t),T^\epsilon(t)),\varphi\rangle)\langle\eta_1^\epsilon(t),\varphi_1\rangle\\&+\frac{1}{\sqrt{\epsilon}}F'(\langle(v^\epsilon(t),T^\epsilon(t)),\varphi\rangle)\langle\eta_2^\epsilon(t),\varphi_2\rangle\,.\end{aligned}$$

Now by the construction of η^ϵ and F_1^ϵ,

$$\begin{aligned}&\mathbb{E}\big[F_1^\epsilon(t+\delta)|\mathcal{F}_0^{t/\epsilon}\big]\\&=\sqrt{\epsilon}\mathbb{E}\big\{\big[F'(\langle(v^\epsilon(t+\delta),T^\epsilon(t+\delta)),\varphi\rangle)\\&\quad-F'(\langle(v^\epsilon(t),T^\epsilon(t)),\varphi\rangle)\big]\langle\eta_1^\epsilon(t+\delta),\varphi_1\rangle\big|\mathcal{F}_0^{t/\epsilon}\big\}\\&\quad+\sqrt{\epsilon}F'(\langle(v^\epsilon(t),T^\epsilon(t)),\varphi\rangle)\langle\eta_1^\epsilon(t),\varphi_1\rangle e^{-\delta/\epsilon}\,.\end{aligned}$$

Then

$$\begin{aligned}A^\epsilon F_1^\epsilon(t)=&\sqrt{\epsilon}F''(\langle(v^\epsilon(t),T^\epsilon(t)),\varphi\rangle)\langle\eta_1^\epsilon(t),\varphi_1\rangle\big[\langle G_1^\epsilon(t),\varphi_1\rangle+\langle G_2^\epsilon(t),\varphi_2\rangle\big]\\&+F''(\langle(v^\epsilon(t),T^\epsilon(t)),\varphi\rangle)\langle\eta_1^\epsilon(t),\varphi_1\rangle^2\\&+F''(\langle(v^\epsilon(t),T^\epsilon(t)),\varphi\rangle)\langle\eta_1^\epsilon(t),\varphi_1\rangle\langle\eta_2^\epsilon(t),\varphi_2\rangle\\&-\frac{1}{\sqrt{\epsilon}}F'(\langle(v^\epsilon(t),T^\epsilon(t)),\varphi\rangle)\langle\eta_1^\epsilon(t),\varphi_1\rangle\,.\end{aligned}$$

Similarly

$$\begin{aligned}A^\epsilon F_2^\epsilon(t)=&\sqrt{\epsilon}F''(\langle(v^\epsilon(t),T^\epsilon(t)),\varphi\rangle)\langle\eta_2^\epsilon(t),\varphi_1\rangle\big[\langle G_1^\epsilon(t),\varphi_1\rangle+\langle G_2^\epsilon(t),\varphi_2\rangle\big]\\&+F''(\langle(v^\epsilon(t),T^\epsilon(t)),\varphi\rangle)\langle\eta_2^\epsilon(t),\varphi_2\rangle^2\\&+F''(\langle(v^\epsilon(t),T^\epsilon(t)),\varphi\rangle)\langle\eta_1^\epsilon(t),\varphi_1\rangle\langle\eta_2^\epsilon(t),\varphi_2\rangle\\&-\frac{1}{\sqrt{\epsilon}}F'(\langle(v^\epsilon(t),T^\epsilon(t)),\varphi\rangle)\langle\eta_2^\epsilon(t),\varphi_2\rangle\end{aligned}$$

which shows (4.8).

Now denote by $Z_i^\epsilon(t)$, $i=1,\cdots,5$, the five terms on the righthand side of $Z^\epsilon(t)$, then

$$\mathbb{E}\big[|Z_2^\epsilon(t)|+|Z_3^\epsilon(t)|\big]=\mathcal{O}(\sqrt{\epsilon}),\quad\epsilon\to 0\,.$$

To pass the limit $\epsilon \to 0$, we further need more processes. Define

$$F_3^\epsilon(t) = F''(\langle(v^\epsilon(t), T^\epsilon(t)), \varphi\rangle) \int_t^\infty \mathbb{E}\left[\langle \eta_1^\epsilon(s), \varphi_1\rangle^2 - \frac{1}{2}\langle Q_1\varphi_1, \varphi_1\rangle \middle| \mathcal{F}_0^{t/\epsilon}\right] \mathrm{d}s\,,$$

$$F_4^\epsilon(t) = F''(\langle(v^\epsilon(t), T^\epsilon(t)), \varphi\rangle) \int_t^\infty \mathbb{E}\left[\langle \eta_2^\epsilon(s), \varphi_2\rangle^2 - \frac{1}{2}\langle Q_2\varphi_2, \varphi_2\rangle \middle| \mathcal{F}_0^{t/\epsilon}\right] \mathrm{d}s$$

and

$$F_5^\epsilon(t) = F''(\langle(v^\epsilon(t), T^\epsilon(t)), \varphi\rangle) \int_t^\infty \mathbb{E}\left[\langle \eta_1^\epsilon(s), \varphi_1\rangle\langle \eta_2^\epsilon(s), \varphi_2\rangle \middle| \mathcal{F}_0^{t/\epsilon}\right] \mathrm{d}s\,.$$

Then by the property of η^ϵ we have

$$F_3^\epsilon(t) = \frac{\epsilon}{2} F''(\langle(v^\epsilon(t), T^\epsilon(t)), \varphi\rangle)\left[\langle \eta_1^\epsilon(t), \varphi_1\rangle^2 - \frac{1}{2}\langle Q_1\varphi_1, \varphi_1\rangle\right],$$

$$F_4^\epsilon(t) = \frac{\epsilon}{2} F''(\langle(v^\epsilon(t), T^\epsilon(t)), \varphi\rangle)\left[\langle \eta_2^\epsilon(t), \varphi_2\rangle^2 - \frac{1}{2}\langle Q_2\varphi_2, \varphi_2\rangle\right]$$

and

$$F_5^\epsilon(t) = \frac{\epsilon}{2} F''(\langle(v^\epsilon(t), T^\epsilon(t)), \varphi\rangle)\left[\langle \eta_1^\epsilon(t), \varphi_1\rangle\langle \eta_2^\epsilon(t), \varphi_2\rangle\right].$$

Further by direct calculation and estimates on (v^ϵ, T^ϵ) we have

$$\sup_{t\geqslant 0} \mathbb{E} F_i^\epsilon(t) = \mathcal{O}(\epsilon)\,, \quad i = 3, 4, 5\,, \quad \text{as} \quad \epsilon \to 0\,.$$

Moreover

$$A^\epsilon F_3^\epsilon(t) = F''(\langle(v^\epsilon(t), T^\epsilon(t)), \varphi\rangle)\left[\frac{1}{2}\langle Q_1\varphi_1, \varphi_1\rangle - \langle \eta_1^\epsilon(t), \varphi_1\rangle^2\right] + R_3^\epsilon(t)\,,$$

and

$$A^\epsilon F_4^\epsilon(t) = F''(\langle(v^\epsilon(t), T^\epsilon(t)), \varphi\rangle)\left[\frac{1}{2}\langle Q_2\varphi_2, \varphi_2\rangle - \langle \eta_2^\epsilon(t), \varphi_2\rangle^2\right] + R_4^\epsilon(t)$$

with

$$\sup_{t\geqslant 0} \mathbb{E}|R_3^\epsilon(t)| = \mathcal{O}(\epsilon) \quad \text{and} \quad \sup_{t\geqslant 0} \mathbb{E}|R_4^\epsilon(t)| = \mathcal{O}(\epsilon)$$

and

$$\sup_{t\geqslant 0} \mathbb{E}|A^\epsilon F_5^\epsilon(t)| = \mathcal{O}(\epsilon) \quad \text{as} \quad \epsilon \to 0\,.$$

Now we have the following $\mathcal{F}_0^{t/\epsilon}$ martingale

$$
\begin{aligned}
&\mathcal{M}^\epsilon(t)\\
=&F(\langle(v^\epsilon(t),T^\epsilon(t)),\varphi\rangle)-F(\langle(v_0^\epsilon,T_0^\epsilon),\varphi\rangle)+F_1^\epsilon(t)+F_2^\epsilon(t)+F_3^\epsilon(t)+F_4^\epsilon(t)\\
&+F_5^\epsilon(t)-\int_0^t F'(\langle(v^\epsilon(s),T^\epsilon(s)),\varphi\rangle)\big[\langle G_1^\epsilon(s),\varphi_1\rangle+\langle G_2^\epsilon(s),\varphi_2\rangle\big]\,\mathrm{d}s\\
&-\frac{1}{2}\int_0^t F''(\langle(v^\epsilon(s),T^\epsilon(s)),\varphi\rangle)\langle Q_1\phi_1,\phi_1\rangle\,\mathrm{d}s\\
&-\frac{1}{2}\int_0^t F''(\langle(v^\epsilon(s),T^\epsilon(s)),\varphi\rangle)\langle Q_2\phi_2,\phi_2\rangle\,\mathrm{d}s+R^\epsilon(t)
\end{aligned}
$$

where

$$
R^\epsilon(t)=\int_0^t[Z_2^\epsilon(s)+Z_3^\epsilon(s)+R_3^\epsilon(s)+R_4^\epsilon(s)+A^\epsilon F_5^\epsilon(s)]\,\mathrm{d}s
$$

with $\mathbb{E}|R^\epsilon(t)|=\mathcal{O}(\epsilon)$ as $\epsilon\to 0$. Now passing to the limit $\epsilon\to 0$ in $\mathcal{M}^\epsilon(t)$ shows the distribution of the limit (v,T) solves the following martingale problem

$$
\begin{aligned}
\mathcal{M}(t)=&F(\langle(v(t),T(t)),\varphi\rangle)-F(\langle(v_0,T_0),\varphi\rangle)\\
&-\int_0^t F'(\langle(v(s),T(s)),\varphi\rangle)\big[\langle G_1(s),\varphi_1\rangle+\langle G_2(s),\varphi_2\rangle\big]\,\mathrm{d}s\\
&-\tfrac{1}{2}\int_0^t F''(\langle(v(s),T(s)),\varphi\rangle)\langle Q_1\phi_1,\phi_1\rangle\,\mathrm{d}s\\
&-\tfrac{1}{2}\int_0^t F''(\langle(v(s),T(s)),\varphi\rangle)\langle Q_2\phi_2,\phi_2\rangle\,\mathrm{d}s
\end{aligned}
$$

which is equivalent to the martingale solution to the 3D stochastic primitive equations (4.1)–(4.4)([20]). By the global well-posedness of equations (4.1)–(4.4), we complete the proof. □

By Theorem 4.1 we have the following result on convergence of stationary solution of the 3D random primitive equations. Denote by $\mathfrak{P}^{*\epsilon}=\mathcal{D}(\hat{v}^{*\epsilon},\hat{T}^{*\epsilon},\hat{\eta}_1^\epsilon,\hat{\eta}_2^\epsilon)$, a stationary statistical solution (see Appendix A) to the system of random primitive equations (1.8)–(1.14) coupled with (2.4). Let $\mathbb{P}^{*\epsilon}=\mathcal{D}(\hat{v}^{*\epsilon},\hat{T}^{*\epsilon})$, then we have

Corollary 4.1 *For $\epsilon\to 0$, there is sequence $\epsilon_n\to 0$, as $n\to\infty$, such that*

$$
\mathbb{P}^{*\epsilon_n}\to\mathbb{P}^*\quad\text{weakly as}\quad n\to\infty
$$

where $\mathbb{P}^*$ is a probability measure on $C([0,\infty);V)$, which is a stationary statistical solution to 3D stochastic primitive equations (4.1)–(4.4).

Proof By the tightness of $\{\mathbb{P}^{*\epsilon}\}_\epsilon = \{\mathcal{D}(\hat{v}^{*\epsilon}, \hat{T}^{*\epsilon})\}_\epsilon$ in $C((0,\infty];V)$, there is a sequence $\epsilon_n \to 0$, such that $\mathcal{D}(\hat{v}^{*\epsilon_n}(0), \hat{T}^{*\epsilon_n}(0))$ converges weakly to $\mathcal{D}(\hat{v}^*(0), \hat{T}^*(0))$ for some random variable $(\hat{v}^*(0), \hat{T}^*(0))$. Denote by $(\hat{v}^*, \hat{T}^*)$ the solution to 3D stochastic primitive equations (4.1)–(4.4) with initial data distributes as $(\hat{v}^*(0), \hat{T}^*(0))$. Then by Theorem 4.1,

$$\mathbb{P}^{*\epsilon_n} \to \mathbb{P} \quad \text{weakly as} \quad n \to \infty .$$

Here $\mathbb{P} = \mathcal{D}(\hat{v}^*, \hat{T}^*)$, by the stationary property of $\mathbb{P}^{*\epsilon_n}$, is a statistical stationary solution to 3D stochastic primitive equations (4.1)–(4.4). □

The above result shows that for $\epsilon \to 0$ and any stationary solution $(v^{*\epsilon}, T^{*\epsilon})$ to 3D random primitive equations, there is a subsequence $\epsilon_n \to 0$, as $n \to \infty$, such that $(v^{*\epsilon_n}, T^{*\epsilon_n})$ converges in distribution to a stationary solution (v^*, T^*) to 3D stochastic primitive equations.

A Statistical solution

We give an introduction of statistical solution of 3D random primitive equations (1.8)–(1.14) coupled with (2.4).

Statistical solution was introduced to study universal properties of turbulent flows [11, 12, 27, e.g.]. We say the system of 3D random primitive equations (1.8)–(1.14) coupled with (2.4) has a statistical solution in space $C([0,\infty);\mathbb{V})$ if there is a probability measure $\mathfrak{P}^\epsilon$ supported on $C([0,\infty);\mathbb{V})$, and there are processes $(\hat{v}^\epsilon, \hat{T}^\epsilon, \hat{\eta}_1^\epsilon, \hat{\eta}_2^\epsilon) \in C([0,\infty);\mathbb{V})$, $\hat{W} = (\hat{W}_1, \hat{W}_2)$ defined on a new probability space, such that

(1) $\mathcal{D}(\hat{v}^\epsilon, \hat{T}^\epsilon, \hat{\eta}_1^\epsilon, \hat{\eta}_2^\epsilon) = \mathfrak{P}^\epsilon$;

(2) $\hat{W}_1$ and $\hat{W}_2$ are Wiener processes distribute same as W_1 and W_2 respectively;

(3) $\mathcal{D}(\hat{v}^\epsilon(0), \hat{T}^\epsilon(0)) = \mathcal{D}(v_0^\epsilon, T_0^\epsilon)$, $\mathcal{D}(\hat{\eta}_1^\epsilon, \hat{\eta}_2^\epsilon) = \mathcal{D}(\eta_1^\epsilon, \eta_2^\epsilon)$ and $(\hat{v}^\epsilon(0), \hat{T}^\epsilon(0))$ are independent from $\hat{W}_1$ and $\hat{W}_2$;

(4) The process $(\hat{v}^\epsilon, \hat{T}^\epsilon)$ is a weak solution of 3D random primitive equations (1.8)–(1.14) with η_1^ϵ, η_2^ϵ replaced by $\hat{\eta}_1^\epsilon$, $\hat{\eta}_2^\epsilon$ respectively. Here $\hat{\eta}^\epsilon = (\hat{\eta}_1^\epsilon, \hat{\eta}_\epsilon^2)$ is stationary process solving (2.4) with W replaced by $\hat{W}$.

The above definition of statistical solutions are also used in [4] to study stochastic 3D Navier–Stokes equations.

A stationary statistical solution is a statistical solution, a Borel measure $\mathfrak{P}^\epsilon$, which is invariant under the following translation on $C([0,\infty);\mathbb{V})$

$$(v(\cdot),T(\cdot),\eta_1(\cdot),\eta_2(\cdot)) \mapsto (v(\cdot+t),T(\cdot+t),\eta_1(\cdot+t),\eta_2(\cdot+t)), \quad t\geqslant 0$$

for $(v,T,\eta_1,\eta_2)\in C([0,\infty);\mathbb{V})$. For a statistical solution of the system of random 3D primitive equations (1.8)–(1.14) coupled with (2.4), we denote by $\mathfrak{P}_t^\epsilon=\mathcal{D}(\hat{v}^\epsilon(\cdot+t),$ $\hat{T}^\epsilon(\cdot+t),\hat{\eta}_1^\epsilon(\cdot+t),\hat{\eta}_2^\epsilon(\cdot+t))$, which is also a statistical solution of the random 3D primitive equations (1.8)–(1.14) coupled with (2.4) . For a stationary statistical solution $\mathfrak{P}^*$ we have

$$\mathfrak{P}_t^*=\mathfrak{P}^*, \quad t\geqslant 0.$$

The following result shows that the relation between the stationary measure and stationary statistical solution.

Lemma A.1 *The* 3D *random primitive equations* (1.8)–(1.14) *coupled with* (2.4) *has a stationary measure supported on* $\mathbb{V}$, *then there is a stationary statistical solution in* $C([0,\infty);\mathbb{V})$.

Proof The proof is direct by the following observation [4]: Let $\mathfrak{P}^{*\epsilon}=\mathcal{D}(\hat{v}^{*\epsilon},$ $\hat{T}^{*\epsilon},\hat{\eta}_1^{*\epsilon},\hat{\eta}_2^{*\epsilon})$ be a stationary statistical solution to the 3D random primitive equations coupled with (2.4), then

$$\mathfrak{y}^{*\epsilon}=\mathcal{D}(\hat{v}^{*\epsilon}(0),\hat{T}^{*\epsilon}(0),\hat{\eta}_1^{*\epsilon}(0),\hat{\eta}_2^{*\epsilon}(0))$$

is a stationary measure for the Markov process defined by the 3D random primitive equations coupled with (2.4); Conversely, assume $\mathfrak{y}^{*\epsilon}$ is a stationary measure of the random 3D primitive equations coupled with (2.4), let $(v^{*\epsilon},T^{*\epsilon},\eta_1^\epsilon,\eta_2^\epsilon)$ be a solution of the 3D random primitive equations coupled with (2.4) with $\mathcal{D}(v^{*\epsilon}(0),T^{*\epsilon}(0),\eta_1^\epsilon(0),$ $\eta_2^\epsilon(0))=\mathfrak{y}^{*\epsilon}$, then $\mathfrak{P}^{*\epsilon}=\mathcal{D}(v^{*\epsilon},T^{*\epsilon},\eta_1^\epsilon,\eta_2^\epsilon)$ is a stationary statistical solution of the 3D random primitive equations coupled with (2.4). □

For stochastic 3D primitive equation (4.1)–(4.4), a statistical solution in space $C([0,\infty;V)$ is a probability measure $\mathbb{P}$ supported on $C([0,\infty;V)$ and there are processes $(\hat{v},\hat{T})\in C([0,\infty;V)$, $\hat{W}=(\hat{W}_1,\hat{W}_2)$ defined on a new probability space such that

(i) $\mathcal{D}(\hat{v},\hat{T})=\mathbb{P}$;

(ii) $\hat{W}_1$ and $\hat{W}_2$ are Wiener processes distribute same as W_1 and W_2 respectively;

(iii) $\mathcal{D}(\hat{v}(0),\hat{T}(0))=\mathcal{D}(v_0,T_0)$ and $(\hat{v}(0),\hat{T}(0))$ are independent from $\hat{W}_1$ and $\hat{W}_2$;

(iv) The process $(\hat{v}, \hat{T})$ is a weak solution of stochastic 3D primitive equations (4.1)–(4.4) with W replaced by $\hat{W}$.

Notice the above definition of statistical solution is in fact a solution to a martingale problem [20, Chapter V].

Similarly we also have stationary statistical solution and relation in Lemma holds.

Acknowledgements The authors would like to express my heartful thanks to the referee for the valuable comments and suggestions. This work was supported by 973 Program (grant No. 2013CB834100), and National Natural Science Foundation of China (91130005,11271052).

References

[1] Arnold L. *Stochastic Differential Equations: Theory and Applications*, Wiley, New York–London–Sydney, 1974.

[2] Billingsley P. *Convergence of Probability Measures* 2nd, John Wiley and Sons, New York, 1999.

[3] Cao C and Titi E S. Global well-posedness of the three-dimensional viscous primitive equations of large-scale ocean and atmosphere dynamics, *Ann. of Math.*, **166** (2007), 245–267.

[4] Chueshov I and Kuksin S. Stochastic 3D Navier–Stokes equations in a thin domain and its α-approximation, *Physica D*, **10–12** (2008), 1352–1367.

[5] Debussche A and Vovelle J. Diffusion limit for a stochastic kinetic problem, preprinted, 2011.

[6] Diop M A, Iftimie B, Pardoux E, Piatnitski A L. Singular homogenization with stationary in time and periodic in space coefficients, *J. Funct. Anal.*, **231** (2006), 1–46.

[7] Duan J, Gao H and Schmalfuss B. Stochastic dynamics of a coupled atmosphere-ocean model, Stoch. and Dynam., **2**(2002), no. 3, 357–380.

[8] Duan J, Kloeden P E and Schmalfuss B. Exponential stability of the quasi-geostrophic equation under random perturbations, Prog. in Probability, **49**(2001), 241–256.

[9] Duan J and Schmalfuss B. The 3D quasi-geostrophic fluid dynamics under random forcing on boundary, *Comm. Math. Sci.*, **1** (2003), 133–151.

[10] Ethier S N and Kurtz T G. *Markov Processes: Characterization and Convergence*, John Wiley and Sons, 1986.

[11] Foias C. Statistical study of Navier–Stokes equations I., *Rend. Sem. Mat. Univ. Padova*, **48** (1972), 219–348.

[12] Foias C. Statistical study of Navier–Stokes equations II., *Rend. Sem. Mat. Univ. Padova*, **49** (1973), 9–123.

[13] Gao H and Sun C. Random attractor for the 3D viscous stochastic primitive equations with additive noise, *Stoch. Dyn.*, **9** (2009), 293–313.

[14] Guo B and Huang D. 3D stochastic primitive equations of the large-scale ocean: global well-posedness and attractors, *Comm. Math. Phys.*, **286** (2009), 697–723.

[15] Guillén-González F, Masmoudi N and Rodríguez-Bellido M A. Anisotropic estimates and strong solutions for the primitive equations, *Diff. Int. Equ.*, **14** (2001), 1381–1408.

[16] Kifer Y. L^2-diffusion approximation for slow motion in averaging, *Stoch. and Dynam.*, **3** (2003), 213–246.

[17] Kushner H. *Approximation and Weak Convergence Methods for Random Processes with Applications to Stochastic Systems Theory*, MIT Press, 1984.

[18] Lions J L, Temam R and Wang S. On the equations of the large scale ocean, *Nonlinearity*, **5** (1992), 1007–1053.

[19] Maslowski B. On probability distributions of solutions of semilinear stochastic evolution equations, *Stoc. Stoc. Rep.*, **45** (1993), 265–289.

[20] Metivier M. *Stochastic Partial Differential Equations in Infinite Dimensional Spaces*, Scuola Normale Superiore, Pisa, 1988.

[21] Majda A and Eijnden E V. A mathematical framework for stochastic climate models, *Comm. Pure Appl. Math.*, **54** (2001), 891–974.

[22] Müller P. Stochastic forcing of quasi-geostrophic eddies, Stochastic Modelling in Physical Oceanography, edited by R. J. Adler, P. Müller and B. Rozovskii, Birkhäuser, Basel, 1996.

[23] Pedlosky J. Geophysical Fluid Dynamics, 2nd Edition, Springer- Verlag, Berlin/New York, 1987.

[24] Da Prato G and Zabczyk J. *Stochastic Equations in Infinite Dimensions*, Cambridge University Press, 1992.

[25] Da Prato G and Zabczyk J. *Ergodicitiy for Infinite Dimensions*, Cambridge University Press, 1996.

[26] Wang W and Roberts A J. Diffusion approximation for self-similarity of stochastic advection in Burgers' equation, *Comm. Math. Phys.*, to appear, 2014. http://dx.doi.org/10.1007/s00220-014-2117-7.

[27] Vishik M I, Fursikov A V. *Mathematical problems of statistical hydromechanics*, Kluwer Academic Publishers, Dordrecht, 1988.

[28] Watanabe H. Averaging and fluctuations for parabolic equations with rapidly oscillating random coefficients, *Probab. The. Rel. Fields*, **77** (1988), 359–378.

Existence of Topological Vortices in an Abelian Chern-Simons Model*

Guo Boling (郭柏灵) and Li Fangfang (李方方)

Abstract In this paper, we prove the existence of topological vortices by variational method applied on an Abelian Chern-Simons model with a generic renormalizable potential. We also establish some properties of the solutions.

Keywords Chern-Simons model; energy-minimizing solutions; variational method

1 Introduction

Gauge field theories in two spatial dimensions have long been recognized as important for understanding several physical phenomena, like high temperature superconductors and the fractional quantum Hall effect. In particular, Chern-Simons gauge theories have a special place in understanding these phenomena. Chern-Simons gauge theories display a number of interesting properties. For example, it provides a mass term for the gauge field, while keeping renormalizability, without evoking spontaneous symmetry breaking. It can have the effect of statistical transmutation, attaching magnetic fluxes to a fermion or boson coupled to the gauge field and making them anyons[1].

Vortices arise as static solutions to gauge field equations in two-space dimensions and have important applications in many fundamental areas of physics. For example, in particle physics, vortices allow one to generate dually (electrically and magnetically) charged particle-like solitons[2,3] known as dyons[4,5]; in cosmology, vortices generate topological defects known as cosmic strings[6] which give rise to useful mechanisms for matter formation in the early universe. Besides, charged vortices arise in a wide range of areas including high-temperature superconductivity[7], optics[8] and the quantum Hall effect[9] and superfluids.

* Journal of Mathematical Physics, 2015, 56: 101505. DOI: 10.1063/1.4933222.

Mathematically, the Chern-Simons models are hard to approach even in the radially symmetric static cases. However, since the discovery of the self-dual structure in the Abelian Chern-Simons model[10−12] in 1990, there have been a wide range of fruitful works on Chern-Simons vortex equations, non-relativistic and relativistic[13], Abelian and non-Abelian[14], and a rich spectrum of mathematical existence results for the Bogomol'nyi-type Chern-Simons vortex equations have been obtained[15−19].

The Lagrangian density for the Chern-Simons model[20] we consider here is defined over the Minkowski spacetime $\mathbf{R}^{2,1}$ in terms of a real-valued vector field $A_\mu(\mu = 0, 1, 2)$ and a complex-valued Higgs field ϕ as follows:

$$\mathcal{L} = \frac{1}{4}\kappa\epsilon^{\mu\nu\rho}A_\mu F_{\nu\rho} + |D_\mu\phi|^2 - V(|\phi|), \tag{1.1}$$

with a generic renormalizable potential

$$V(|\phi|) = \frac{\alpha e^4}{\kappa^2}(|\phi|^2 - v^2)^2[|\phi|^2 - \beta(|\phi|^2 - v^2)], \tag{1.2}$$

covariant derivative $D_\mu = \partial_\mu - ieA_\mu$, the Maxwell field strength $F_{\mu\nu} = \partial_\mu A_\nu - \partial_\nu A_\mu$, spacetime metric $\eta_{\mu\nu} = \mathrm{diag}(1, -1, -1)$, the constant $\kappa > 0$ being a coupling coefficient of the Chern-Simons term, $\varepsilon^{\mu\nu\rho}$ the Kronecker skew-symmetric tensor with $\varepsilon^{012} = 1$, and e, v some other positive coupling constants, and α, β couplings parameters that can be varied.

This non self-dual model contains Chern-Simons terms expressed in terms of an Abelian gauge field. We shall see that for particular values of the potential parameters, interesting new properties appear. As a consequence the study of vortex solutions when there is a lack of self-dual structure has not been carried out for the full range of models due to the difficulties involved. In [21], the existence of vortex solutions in the Abelian Chern-Simons model with no Maxwell term in non-self-dual regimes has been established. In [22], the authors have established the existence of charged vortices in the full Chern-Simons-Higgs theory with the Maxwell term in both Abelian and non-Abelian cases.

When $\alpha > 0$ and $\beta = 0$ the model corresponds to the following Bogomol'nyi-Prasad-Sommerfield (BPS) potential

$$V_{\mathrm{BPS}}(|\phi|) = \frac{\alpha e^4}{\kappa^2}(|\phi|^2 - v^2)^2|\phi|^2. \tag{1.3}$$

In this case the potential still has two degenerate vacua, we are only changing the height of the potential. We note the model with the potential (1.3) have a self-dual structure. As well known, a self-dual structure is always advantageous in gauge theory, because it permits the identification of a special class of static solutions (e.g., monopoles, vortices, domain wall, etc.) by solving appropriate first-order equations which are also referred to as the Bogomol'nyi-Prasad-Sommerfield (BPS) equations[12,23]. In [24 – 26] and [28], the authors have obtained many results on the model which have different parameters from potential (1.3), so we will not discuss this situation in this paper.

The contents of the rest of the paper are outlined as follows. In Section 2, we describe the Abelian Chern-Simons theory which we aim to solve and then state our main existence theorem for vortices solutions. In Sections 3, we prove the existence of a finite-energy critical point of the action functional by the variational method. In Sections 4, we establish the properties of the energy-minimizing solutions.

2 Abelian Chern-Simons-Higgs Model

The extremals of the Lagrangian density (1.1) formally satisfy the Abelian Chern-Simons equations[20]

$$D_\mu D^\mu \phi = \frac{\alpha e^4}{\kappa^2}(v^2 - |\phi|^2)(3|\phi|^2 - 3\beta(|\phi|^2 - v^2) - v^2)\phi, \tag{2.1}$$

$$\frac{1}{2}\kappa\varepsilon^{\mu\nu\rho}F_{\nu\rho} = -J^\mu. \tag{2.2}$$

The current density J^μ is given by

$$J^\mu = \mathrm{i}e(\bar{\phi}D^\mu\phi - \phi\overline{D^\mu\phi}), \quad \mu = 0, 1, 2. \tag{2.3}$$

We will consider static configurations only. Thus

$$\kappa F_{12} = \rho = J^0 = \mathrm{i}e(\bar{\phi}D^0\phi - \phi\overline{D^0\phi}) = -2e^2 A_0|\phi|^2, \tag{2.4}$$

where ρ represents electric charge density and which implies that the total magnetic charge Φ and electric charge Q are related by

$$\kappa\Phi = \kappa\int_{\mathbf{R}^2} F_{12}\,\mathrm{d}x = \int_{\mathbf{R}^2}\rho\,\mathrm{d}x = Q. \tag{2.5}$$

So the component A_0 of the gauge field A_μ is essential for the presence of electric charge. Besides, the electric field $E = E^j$ and magnetic field B induced from the gauge field A_μ are

$$E^j = \partial_j A_0, \quad j = 1, 2; \quad B = F_{12},$$

respectively.

The Hamiltonian H or the energy density of the theory is given by

$$H(A,\phi)=|D_i\phi|^2+\frac{\kappa^2B^2}{4e^2|\phi|^2}+V(|\phi|).$$

Thus, the finite-energy condition

$$\begin{aligned}E(A,\phi)&=\int_{\mathbf{R}^2}H(A,\phi)(x)\mathrm{d}x<\infty,\\&=\int_{\mathbf{R}^2}\left\{|D_i\phi|^2+\frac{\kappa^2B^2}{4e^2|\phi|^2}+V(|\phi|)\right\}\mathrm{d}x<\infty\end{aligned}\tag{2.6}$$

leads us to impose the following natural asymptotic behavior for the fields $A_0,\ A_j,\ \phi$:

$$|\phi|\to v,\ |D_i\phi|\to 0,\ A_0\to 0,\tag{2.7}$$

as $|x|\to\infty$.

The static version of the Maxwell-Chern-Simons equations (2.1)-(2.2) takes the explicit form

$$D_j^2\phi=\frac{\alpha e^4}{\kappa^2}(|\phi|^2-v^2)(3|\phi|^2-3\beta(|\phi|^2-v^2)-v^2)\phi-e^2A_0^2\phi,\tag{2.8}$$

$$\kappa\varepsilon_{jk}\partial_kA_0=\mathrm{i}e(\bar{\phi}D^\mu\phi-\phi\overline{D^\mu\phi}),\tag{2.9}$$

$$\kappa F_{12}+2e^2|\phi|^2A_0=0,\tag{2.10}$$

where $j,k=1,2$. Removing the electic field potential A_0, we obtain a system of coupled elliptic equations

$$D_j^2\phi=\frac{\alpha e^4}{\kappa^2}(|\phi|^2-v^2)(3|\phi|^2-3\beta(|\phi|^2-v^2)-v^2)\phi-e^2A_0^2\phi,\tag{2.11}$$

$$\kappa\varepsilon_{jk}\partial_k\left(\frac{-\kappa F_{12}}{2e^2|\phi|^2}\right)=ie(\bar{\phi}D^\mu\phi-\phi\overline{D^\mu\phi}).\tag{2.12}$$

Consider the topological vortex in a cylindrical symmetric ansatz

$$\phi=v\mathrm{e}^{\mathrm{i}n\theta}h(r),\quad \mathrm{e}A_j=na(r)\varepsilon_{kj}\frac{x_k}{r^2},\quad j,k=1,2,\tag{2.13}$$

with

$$h,a\to 1,\ \text{as}\ \ r\to\infty.\tag{2.14}$$

After insertion of the ansatz into the (2.11)–(2.12), we get

$$\partial_r^2a-\frac{1}{r}\partial_ra-\frac{2(\partial_ra)(\partial_rh)}{h}+m^2(1-a)h^4=0,\tag{2.15}$$

$$\frac{1}{r}\partial_r(r\partial_rh)-\frac{n^2}{r^2}(1-a)^2h+\frac{n^2(\partial_ra)^2}{m^2r^2h^3}-\frac{1}{2v^2}\frac{\partial V}{\partial h}=0.\tag{2.16}$$

For a generic potential we obtain

$$\frac{1}{2v^2}\frac{\partial V}{\partial h}=\frac{\alpha}{4}m^2(h^2-1)[3h^2-3\beta(h^2-1)-1]h, \tag{2.17}$$

where $m=2e^2v^2/\kappa$.

We note that equations (2.15)–(2.16) are the Euler-Lagrange equations of the action functional

$$T(h,a)=2\pi v^2\int \mathrm{d}r\,r\left\{\frac{n^2}{m^2}\frac{(\partial_r a)^2}{r^2h^2}+(\partial_r h)^2+\frac{n^2}{r^2}(1-a)^2h^2\right.$$
$$\left.+\frac{\alpha}{4}m^2(h^2-1)^2[h^2-\beta(h^2-1)]\right\}. \tag{2.18}$$

We note that the first term is difficult to deal with. We will take use of the study of planar vortex minimizers of the Ginzburg-Landau energy[27] to solve our problems.

With the above preparation, we can state our main result as follows.

Theorem 2.1 *For any given integer n, $\alpha>0, 0<\beta<1$, the Chern-Simons equations* (2.11)-(2.12) *over $\mathbf{R}^2\backslash\{0\}$ have a finite-energy solution (A,ϕ) satisfying the asymptotic properties* (2.7) *as $|x|\to\infty$. Moreover, the integer n is the winding number of ϕ near infinity.*

The proof of Theorem 2.1 is established via a variational method applied on action functionals. In doing so, we extend the methods of [21, 27, 28] for the charged vortex problem. Furthermore, we also established some properties of the solution.

Theorem 2.2 *For any given integer n, the radial solutions (h,a) obtained in Theorem* 2.1 *are in $C^\infty(0,\infty)$, and satisfying the following properties:*

$$0<h<1, 0<a\leqslant 1,\ \ \forall\ \ r\in(0,\infty),$$

and $a(r)$ is a non-decreasing function, $h(r)$ is an increasing function.

In the section 3 and 4, we will give proofs of these theorems respectively.

3 Minimization of $T(h,a)$

We find that the model have some similarities to the rigorous study of planar vortex minimizers of the Ginzburg-Landau energy[27]

$$I(\phi,A)=\frac{1}{2}\int_{\mathbf{R}^2}|D_i\phi|^2+|\mathrm{curl}A|^2+\lambda^2(1-|\phi|^2)^2, \tag{3.1}$$

which originated with the work of Berger-Chen[27] and Robin-Chen [21]. And the radial Ginzburg-Landau energy is given by

$$I_r(h,a)=\frac{1}{2}\int_{\mathbf{R}^2}|h'|^2+\frac{n^2}{r^2}(1-a)^2h^2+\frac{n^2(a')^2}{r^2}+\lambda^2(1-h^2)^2\,\mathrm{d}x. \tag{3.2}$$

According to the method of [27], we now define the function spaces for the radially symmetric functions h and a. Let

$S_h=$ the set of real-valued radially symmetric functions $h(|x|)$ defined on $\mathbf{R}^2$, such that h is nonnegative almost everywhere and $1-h\in W^{1,2}(\mathbf{R}^2)$;

$S_a=$ the set of real-valued radially symmetric functions $a(|x|)$ defined on $\mathbf{R}^2$, such that $(1/r)a(r)\in L^2_{loc}(\mathbf{R}^2)$ with $(1/r)\big(a'(r)\big)\in L^2(\mathbf{R}^2)$, where the derivative a' is in the distributional sense.

We now give some properties of the S_h and S_a.

Lemma 3.1 [27] (i) *For $h\in S_h$ and $a\in S_a$, we have $h(r)\in C(0,\infty)$ and $a(r)\in C[0,\infty), a(0)=0, a(r)=\int_0^r a'(s)\,\mathrm{d}s$, $\|a\|_{S_a}=\|(1/r)a'(r)\|_{L^2(\mathbf{R}^2)}$, and*

$$\sup_{r\in(0,\infty)}\left|\frac{a}{r}\right|\leqslant\|a\|_{S_a}.$$

(ii) *Assume that $h\in S_h$, $a\in S_a$, and $I_r(h,a)<\infty$. Then $h\in C[0,\infty)$ and $h(0)=0$.*

Recall the following results for Ginzburg-Landau equations.

Theorem 3.1 [27] *For any integer n and λ, there is a radial solution (h,a) to the Ginzburg-Landau equations. In particular, $(h,a)\in S_h\oplus S_a$ minimizes the radial Ginzburg-Landau energy $I_r(h,a)$.*

The proof of Theorem 2.1 relies on Lemma 3.1 and Theorem 3.1 for the Ginzburg-Landau energy; we will choose to minimize the functional (2.18) over a constraint set determined by Ginzburg-Landau vortices. The most difficult term comes from $(a')^2/(r^2h^2)$ in the functional (2.18).

First, we note that $T(1,1)=0$, therefore, $h=a=1$ give a trivial solution to the Chern-Simons equations. We note that $T(h,a)\geqslant 0$ as $\alpha>0, 0<\beta<1$, if we take (h,a) of functional (2.18) with the condition $h,a\to 1$ at ∞, and if

$$\eta=\inf_{S_h\oplus S_a}T(h,a)$$

is attained at (h_0, a_0), then $\eta > 0$. Otherwise

$$\int_{\mathbf{R}^2} (1-a_0)^2 h_0^2 \frac{1}{r^2}\,\mathrm{d}x = \int_{\mathbf{R}^2} (h_0^2-1)^2[h_0^2 - \beta(h_0^2-1)]\,\mathrm{d}x = 0.$$

From Lemma 3.1, we obtain $h_0 \equiv 1$ and $a_0 \equiv 1$, which contradicts that $a_0/r \in L^2_{loc}(\mathbf{R}^2)$. Similarly, we know that

$$\inf_{S_h \oplus S_a} I_r(h,a) > 0.$$

Let

$$\eta_0 = \inf_{S_h \oplus S_a} I_r(h,a) > 0.$$

From Theorem 3.1, we know that η_0 is attained in $S_h \oplus S_a$. For any $C > 0$, we choose the admissible space $\mathcal{A}$,

$$\mathcal{A} = \{(h,a) \in S_h \oplus S_a, I_r(h,a) < \eta_0 + C\}. \tag{3.3}$$

Next, we will consider the minimization problem,

$$T_m = \inf\{T(h,a) \mid (h,a) \ \in \mathcal{A}\}. \tag{3.4}$$

Let $\{(h_k, a_k)\}$ be a minimizing sequence of (3.4). Without loss of generality, we may assume that

$$T(h_k, a_k) \leqslant M, \ k = 1, 2, \cdots \tag{3.5}$$

and $I_r(h_k, a_k) < \eta_0 + C$.

So we obtain

$$\begin{aligned} M &> \int_{\mathbf{R}^2} |h_k'|^2 + \lambda^2(1-h_k^2)^2\,\mathrm{d}x = \int_{\mathbf{R}^2} |h_k'|^2 + \lambda^2(1-h_k)^2(1+h_k)^2\,\mathrm{d}x \\ &\geqslant \min\{1,\lambda^2\} \int_{\mathbf{R}^2} |h_k'|^2 + \lambda^2(1-h_k)^2\,\mathrm{d}x \\ &= \min\{1,\lambda^2\}\|1-h_k\|^2_{W^{1,2}(\mathbf{R}^2)}, \\ M &> \int_{\mathbf{R}^2} \frac{n^2(a')^2}{r^2}\,\mathrm{d}x = \|a_k\|^2_{S_a}. \end{aligned}$$

Therefore, $\|1-h_k\|_{W^{1,2}(\mathbf{R}^2)}$ and $\|a_k\|^2_{S_a}$ are uniformly bounded. So we find that $1-h_k$(in fact, a subsequence in it) is weakly convergent in $W^{1,2}(\mathbf{R}^2)$ and a_k(in fact, a subsequence in it) is weakly convergent in S_a. So

$$1-h_k \rightharpoonup 1-h \text{ in } W^{1,2}(\mathbf{R}^2), \ a_k \rightharpoonup a \text{ in } S_a,$$

with $(h,a)\in S_h\oplus S_a$. According to the Rellich-Kondrachov embedding theorem, it implies strong convergence in $L^p_{loc}(\mathbf{R}^2)$. And from [29], we have

$$h_k\to h,\ \ a_k\to a \qquad a.e. \text{ on } (0,\infty). \tag{3.6}$$

It is easy to see from the Ginzburg-Landau energy $I_r(h,a)$ that

$$I_r(h,a)\leqslant \liminf_{k\to\infty} I_r(h_k,a_k)\leqslant \eta_0+C.$$

Hence $(h,a)\in\overline{\mathcal{A}}$.

In view of $T(h_k,a_k)\leqslant T_m+1$, we know that

$$\frac{a_k'}{rh_k}\in L^2(\mathbf{R}^2).$$

Let $f_k'=a_k'/h_k$. So $f_k'/r\in L^2(\mathbf{R}^2)$, and

$$\begin{aligned}\int_0^r |f_k'|\,\mathrm{d}\rho &\leqslant \left(\int_0^r \rho\,\mathrm{d}\rho\right)^{\frac12}\left(\int_0^r \frac{1}{\rho}|f_k'|^2\,\mathrm{d}\rho\right)^{\frac12}\\ &\leqslant \frac{r}{\sqrt2}\left(\int_{\mathbf{R}^2}\left(\frac{1}{\rho}f_k'\right)^2\mathrm{d}x\right)^{\frac12}<\infty,\end{aligned}$$

we have $f_k'\in L^1_{loc}(0,\infty)$ and

$$f_k(r)-f_k(0)=\int_0^r f_k'(\rho)\,\mathrm{d}\rho.$$

So

$$\begin{aligned}\frac1r|f_k(r)-f_k(0)| &\leqslant \frac1r\int_0^r |f_k'(\rho)|\,\mathrm{d}\rho\\ &\leqslant \frac{1}{\sqrt2}\left(\int_{\mathbf{R}^2}\left(\frac{1}{\rho}f_k'\right)^2\mathrm{d}x\right)^{\frac12},\quad \text{for all } r\in[0,\infty),\end{aligned}$$

we get $(1/r)(f_k(r)-f_k(0))\in L^2_{loc}(\mathbf{R}^2)$. Let $g_k=f_k(r)-f_k(0)$, we obtain $g_k\in S_a$, $\|g_k\|_{S_a}<M$. So there is a subsequence, still denoted g_k, so that

$$g_k\rightharpoonup g, \text{ in } S_a,$$

and

$$g_k\to g \qquad \text{a.e. on } (0,\infty). \tag{3.7}$$

Because of $a_k\in S_a$, we know that $a_k'\in L^1_{loc}(0,\infty)$. So for any $0<l<p$, we have

$$a_k(p)-a_k(l)=\int_l^p a_k'(\rho)\,\mathrm{d}\rho.$$

In view of $a_k' = g_k' h_k$ By integrating by parts, we have

$$a_k(p) - a_k(l) = h_k g_k \Big|_l^p - \int_l^p h_k'(\rho) g_k(\rho)\,\mathrm{d}\rho.$$

Letting $k \to \infty$ and taking advantage of (3.6) and (3.7), we obtain

$$a_k(p) - a_k(l) \to a(p) - a(l), \quad h_m g_m \Big|_l^p \to hg\Big|_l^p$$

and

$$\begin{aligned}
&\left| \int_l^p h_k'(\rho) g_k(\rho)\,\mathrm{d}\rho - \int_l^p h'(\rho) g(\rho)\,\mathrm{d}\rho \right| \\
=&\left| \int_l^p h_k'(g_k - g)\,\mathrm{d}\rho + \int_l^p (h_k' - h) g\,\mathrm{d}\rho \right| \\
\leqslant&\, \|g_k - g\|_{L^\infty[l,p]} \int_l^p |h_k'(\rho)|\,\mathrm{d}\rho + \int_l^p (h_k' - h) g\,\mathrm{d}\rho \\
\leqslant&\, \|g_k - g\|_{L^\infty[l,p]} \left(\int_l^p \frac{1}{\rho}\,\mathrm{d}\rho \right)^{\frac{1}{2}} \left(\int_{\mathbf{R}^2} |h_k'|^2\,\mathrm{d}x \right)^{\frac{1}{2}} + \int_l^p (h_k' - h) g\,\mathrm{d}\rho \\
\to&\, 0 \text{ as } k \to \infty.
\end{aligned}$$

So we obtain

$$a(p) - a(l) = \int_l^p a'(\rho)\,\mathrm{d}\rho,$$

which gives that $a' = g'h$. Therefore,

$$\frac{a_k'}{rh_k} \rightharpoonup \frac{a'}{rh}, \text{ in } L^2(\mathbf{R}^2). \tag{3.8}$$

With

$$\mathcal{H} = \frac{n^2}{m^2} \frac{(\partial_r a)^2}{r^2 h^2} + (\partial_r h)^2 + \frac{n^2}{r^2}(1-a)^2 h^2 + \frac{\alpha}{4} m^2 (h^2 - 1)^2 [h^2 - \beta(h^2 - 1)]$$

denoting the energy density, and by using the weak lower semi-continuity of the norm and Fatou's lemma, we obtain

$$\begin{aligned}
\int_{\mathbf{R}^2} \mathcal{H}(h, a)\,\mathrm{d}x &= \int_{\mathbf{R}^2} \liminf_{k\to\infty} \mathcal{H}(h_k, a_k)\,\mathrm{d}x \\
&\leqslant \liminf_{k\to\infty} \int_{\mathbf{R}^2} \mathcal{H}(h_k, a_k)\,\mathrm{d}x \\
&\leqslant \liminf_{k\to\infty} T(h_k, a_k).
\end{aligned}$$

Thus, we have $T(h,a) = \inf\limits_{\mathcal{A}} T(h,a)$. To show that (h,a) is a critical point of $T(h,a)$, it suffices to see that it is an interior point. Set

$$\mathcal{A}_{\frac{1}{k}}=\left\{(h,a)\in S_h\oplus S_a, I_r(h,a)<\eta_0+C+\frac{1}{k}\right\}.$$

Let (h^k,a^k) be minimizers of $T(h,a)$ over $\overline{\mathcal{A}_{\frac{1}{k}}}$ for $k=1,2,\cdots$. Suppose they are not interior minimizers, we have

$$T(h^k,a^k)=\inf_{\mathcal{A}_{\frac{1}{k}}}T(h,a)\leqslant T(h^{k-1},a^{k-1}),\ I_r(h^k,a^k)=\eta_0+C+\frac{1}{k},\tag{3.9}$$

particularly, $I_r(h^k,a^k)\leqslant\eta_0+C+1$. So $I_r(h^k,a^k)$ is uniformly bounded and we obtain

$$1-h^k\rightharpoonup 1-h \text{ in } W^{1,2}(\mathbf{R}^2),\ a^k\rightharpoonup a \text{ in } S_a.$$

According to the Rellich-Kondrachov embedding theorem and by passing to a diagonal subsequence, we have

$$h^k\to h,\ a^k\to a\quad a.e.\text{ on }(0,\infty),\ \frac{(a')^k}{rh^k}\rightharpoonup\frac{a'}{rh},\ \text{in } L^2(\mathbf{R}^2)$$

In view of using the weak lower semi-continuity of the norm, Fatous lemma and (3.9), we obtain

$$\begin{aligned}T(h,a)&\leqslant\liminf_{k\to\infty}T(h^k,a^k)=T(h^1,a^1),\\ I_r(h,a)&\leqslant\liminf_{k\to\infty}I_r(h^k,a^k)=\eta_0+C.\end{aligned}$$

So (h,a) is an interior minimizers of $\mathcal{A}_{\frac{1}{k}}$ for any k. Thus, we find (h,a) is found to be a solution of (3.4).

Consequently, we have the following.

Theorem 3.2 *For $\alpha>0, 0<\beta<1$, there is a $C>0$ such that the minimization problem*

$$\inf\{T(h,a)\mid(h,a)\ \in\mathcal{A}\}\tag{3.10}$$

has a solution.

And it is easy to check that minimizing solution (h,a) solves the equations (2.15)–(2.16) on $\mathbf{R}^2\backslash\{0\}$. So the Theorem 2.1 is proved.

4 Some Properties of the Minimizing Solution

Next we will establish some properties of the minimizing solutions.

Lemma 4.1 *Suppose* (h, a) *is the minimizing solution of* (3.4), *Then* $h, a \in C^{\infty}(0, \infty)$.

Proof From Lemma 3.1 and Theorem 3.2, we know that $h, a \in C[0, \infty), h(0) = a(0) = 0$. And since $h, a \to 1$ as $r \to \infty$, we have $h, a \in L^{\infty}$. We write (2.15) as follows:

$$\left(\frac{a'}{rh^2}\right)' = -\frac{m^2}{r}(1-a)h^2. \tag{4.1}$$

For any $l > 0$, we have

$$\begin{aligned}\int_{1/l}^{l}\left(\frac{a'}{rh^2}\right)' \mathrm{d}r &= m^2\int_{1/l}^{l}\frac{|1-a|h^2}{r}\,\mathrm{d}r \\ &= m^2\|f\|_{L^\infty}\left(\int_{1/l}^{l}\frac{1}{r}\,\mathrm{d}r\right)^{\frac{1}{2}}\left(\frac{1}{2\pi v^2}T(h,a)\right)^{\frac{1}{2}} < \infty.\end{aligned}$$

So $\left(a'/(rh^2)\right)' \in L^1[1/l, l]$ for any $l > 0$, we obtain $a'/(rh^2) \in C(0, \infty)$. Therefore, $a \in C^1(0, \infty)$. By using standard elliptic theory applied in (2.16), we have $h \in C^2(0, \infty)$. We also have $h, a \in C^{\infty}(0, \infty)$ by a bootstrap argument.

Lemma 4.2 *Suppose* (h, a) *is the minimizing solution of* (3.4), *we have*

$$0 < a(r) \leqslant 1, \qquad for \quad r \in (0, \infty),$$

and $a(r)$ *is a non-decreasing function.*

Proof First, we show $0 \leqslant a(r) \leqslant 1$ for $0 < r < \infty$. Otherwise, since $a(r) \in C^2(0, \infty)$, define

$$\widetilde{a}(x) = \begin{cases} 1, & a(r) > 1, \\ a(r), & \text{otherwise}. \end{cases}$$

Then $(h(r), \widetilde{a}(r)) \in \mathcal{A}$ and $T(h, \widetilde{a}) < T(h, a)$, which contradicts the definition of $T(h, a)$, So $a(r) \leqslant 1$. Similarly, we also have $a(r) \geqslant 0$.

Next we make use of the equation (2.15), which can be written as follows:

$$a''\left(\frac{1}{rh^2}\right) + a'\left(\frac{1}{rh^2}\right)' = \frac{m^2(a-1)h^2}{r} \leqslant 0. \tag{4.2}$$

Therefore, we obtain that either $a(r) > 0$ on $(0, \infty)$ or $a(r) \equiv 0$ by applying the maximum principle on (2.15).

If $a(r) \equiv 0$, from the energy we know

$$\int_{\mathbf{R}^2}\frac{n^2}{r^2}h^2\,\mathrm{d}x < \infty.$$

We know $(h,a)\in\mathcal{A}$, so we have

$$\int_{r\geqslant 1}\frac{n^2(1-h^2)}{r^2}\,\mathrm{d}x\leqslant\left(\int_{r\geqslant 1}\frac{n^4}{r^4}\,\mathrm{d}x\right)^{\frac{1}{2}}\left(\int_{r\geqslant 1}(1-h^2)^2\,\mathrm{d}x\right)^{\frac{1}{2}}<\infty,$$

which implies that

$$\int_{r\geqslant 1}\frac{n^2}{r^2}\,\mathrm{d}x<\infty,$$

a contradiction. So we have $a(r)>0$ on $(0,\infty)$.

Next we will prove $a(r)$ is a nondecreasing function. Suppose that there is a point $r_1>0$ such that $a'(r_1)<0$. Since $0<a(r)<1$ and $a(\infty)=1$, we may deduce the existence of a pair of points around $r_1, 0<r_2<r_3<\infty$ such that

$$a(r)\leqslant a(r_3),\ r\in[r_2,r_3],\ a(r_2)=a(r_3),\ a(r)<a(r_3)\ \text{ for some } r\in(r_2,r_3).$$

Now define

$$\widetilde{a}(x)=\begin{cases} a(r_3), & r\in[r_2,r_3],\\ a(r), & r\notin[r_2,r_3].\end{cases}\tag{4.3}$$

Then $(h,\widetilde{a})\in\mathcal{A}$, So $T(h,\widetilde{a})<T(h,a)$, which is a contradiction. So $a'(r)\geqslant 0$.

Lemma 4.3 *Suppose (h,a) is the minimizing solution of (3.4), we have*

$$0<h(r)<1,\qquad for\quad r\in(0,\infty),$$

and $h(r)$ is a increasing function.

Proof From section 3, we know that (h,a) is the minimizer of $T(h,a)$ over $\mathcal{A}$, so we have

$$T'(h,a)=0,\ \ T''(h,a)\geqslant 0,$$

where

$$\begin{aligned}&T'(h,a)\\&=\begin{pmatrix} -\frac{1}{r}(rh')'-\frac{n^2(a')^2}{m^2r^2h^3}+\frac{n^2}{r^2}(1-a)^2h+\frac{\alpha}{4}m^2(h^2-1)h[3h^2-3\beta(h^2-1)-1]\\ -\frac{n^2}{m^2r}(\frac{a'}{rh^2})'-\frac{n^2}{r^2}(1-a)h^2\end{pmatrix}\\&=\begin{pmatrix}0\\0\end{pmatrix},\end{aligned}\tag{4.4}$$

$$T''(h,a)=\begin{pmatrix} T_{hh} & T_{ha}\\ T_{ah} & T_{aa}\end{pmatrix}\geqslant 0,\ T_{hh}\geqslant 0,\tag{4.5}$$

where

$$
\begin{aligned}
T_{hh} &= -\Delta_r + \frac{3n^2(a')^2}{m^2r^2h^4} + \frac{n^2}{r^2}(1-a)^2 + \frac{\alpha}{4}m^2\big(15(1-\beta)h^4 + 6(3\beta-2)h^2 - 3\beta + 1\big),\\
T_{ha} &= \frac{2n^2}{rm^2}\left(\frac{a'}{rh^3}\right)' - \frac{2n^2}{r^2}(1-a)h,\\
T_{aa} &= -\frac{n^2}{m^2r^2}\left(\frac{1}{h^2}\partial^2 + \left(\frac{1}{rh^2} + \frac{2}{h^3}\right)\partial\right) + \frac{2n^2}{r^2}h^2.
\end{aligned}
$$

We also have

$$
\frac{\mathrm{d}}{\mathrm{d}r}T'(h,a) = \begin{pmatrix} 0 \\ 0 \end{pmatrix}. \tag{4.6}
$$

So the first row of (4.6) is as follows:

$$
\left(T_{hh} + \frac{1}{r^2}\right)h' + \left(T_{ha} - \frac{4n^2}{rm^2}\left(\frac{a'}{rh^3}\right)'\right)a' = \frac{2n^2}{r^3}(1-a)^2h + \frac{6n^2(a')^2}{m^2r^2h^4}, \tag{4.7}
$$

where

$$
\begin{aligned}
\left(T_{ha} - \frac{4n^2}{rm^2}\left(\frac{a'}{rh^3}\right)'\right)a' &= \left(-\frac{2n^2}{rm^2}\left(\frac{a'}{rh^3}\right)' - \frac{2n^2}{r^2}(1-a)h\right)a'\\
&= \left(-\frac{2n^2}{rm^2}\left(\frac{a'}{rh^3}\right)' + \frac{2n^2}{m^2rh}\left(\frac{a'}{rh^2}\right)'\right)a'\\
&= \frac{2n^2a'h'}{m^2r^2h^4}.
\end{aligned}
$$

So (4.7) becomes

$$
\left(T_{hh} + \frac{1}{r^2} + \frac{2n^2a'}{m^2r^2h^4}\right)h' = \frac{2n^2}{r^3}(1-a)^2h + \frac{6n^2(a')^2}{m^2r^2h^4}. \tag{4.8}
$$

In view of Lemma 4.2, (4.5) and $h \geqslant 0, h \neq 0$, the right side of (4.8) is non-negative. By using the maximum principle, we obtain $h' > 0$. So $h(r)$ is a increasing function. Combining the result and Lemma 4.1, we obtain $0 < h < 1$. The proof of Lemma 4.3 is completed. From Lemma 4.1–4.3, we establish Theorem 2.2.

Remark 4.1 *When $0 < a(r) < 1$, from Lemma 4.2, we know that the right hand side of (4.2) is negative. Therefore, by a maximum principle, we get that $a'(r) > 0$.*

Acknowledgements This work was supported in part by the National Natural Science Foundation of China (11471099).

References

[1] Rao S. An Anyon Primer[J]. Physics, 2001 arXiv:hep-th/9209066.

[2] Paul S K, Khare A. Charged vortices in an abelian Higgs model with Chern-Simons term[J]. Phys. Lett. B, 1986, 174(4): 420–422. Errata: Phys. Lett. B, 1986, 177: 453.

[3] de Vega H J, Schaposnik F A. Electrically Charged Vortices in Non-Abelian Gauge Theories with Chern-Simons Term[J]. Phys. Rev. Lett., 1986, 56(24):2564–2566.

[4] Julia B, Zee A. Poles with both magnetic and electric charges in non-Abelian gauge theory[J]. Phys. Rev. D., 1975, 11(11): 2227–2232.

[5] Schwinger J. A Magnetic Model of Matter[J]. Science, 1969, 165: 757–761.

[6] Orlov S. Foundation of Vortex Gravitation, Cosmology and Cosmogony[J]. Kathmandu University J. Sci., Eng. Technol. 2011, 3(7): 226–245.

[7] Matsuda Y, Nozakib K, Kumagaib K. Charged vortices in high temperature superconductors probed by nuclear magnetic resonance[J]. J. Phys. Chem. Solids, 2002, 63: 1061–1063.

[8] Bezryadina A, Eugenieva E, Chen Z. Self-trapping and flipping of double-charged vortices in optically induced photonic lattices[J]. Opt. Lett. B, 2006, 31(16): 2456–2458.

[9] Sokoloff J B. Charged vortex excitations in quantum Hall systems[J]. Phys. Rev. B, 1985, 31(4): 1924–1928.

[10] Hong J, Kim Y, Pac P Y. Multivortex solutions of the Abelian Chern–Simons–Higgs theory[J]. Phys. Rev. Lett., 1990, 64: 2230–2233.

[11] Jackiw R, Weinberg E J. Self-dual Chern–Simons vortices[J]. Phys. Rev. Lett., 1990, 64: 2234–2237.

[12] Bogomol'nyi E B. The stability of classical solutions[J]. Sov. J. Nucl. Phys., 1976, 24: 449–454.

[13] Dunne G. Aspects of Chern–Simons theory[J], Les Houches-Ecole d'Ete de Theorique, 1999, 69: 179–263.

[14] Dunne G. Self-dual Chern–Simons theories[M]. New York: Springer, 1995.

[15] Lin C, Ponce A C, Yang Y. A syatem of elliptic equations arising in Chern–Simons field theory[J]. J. Funct. Anal., 2007, 247: 289–350.

[16] Spruck J, Yang Y. The existencee of non-topological solitons in the self-dual Chern-Simons theory[J]. Commun. Math. Phys., 1992, 149: 361–376.

[17] Chae D, Kim N. Topological multivortex solutions of the self-dual Maxwell Chern-Simons-Higgs system[J]. J. Diff. Equ.,1997, 134: 154–182.

[18] Chan H, Fu C C, Lin C. Non-topological multivortex solutions to the self-dual Chern-Simons-Higgs equation[J]. Commun. Math. Phys., 2002, 231: 189–221.

[19] Yang Y. The relativistic non-Abelian Chern-Simons equations[J]. Commun. Math. Phys., 1997, 186: 199–218.

[20] Bolognesi S, Gudnason S B. A note on Chern–Simons solitons -a type III vortex from the wall vortex[J] Nucl. Phys. B, 2008, 805: 104–123.

[21] Chen R M, Spirn D. Symmetric Chern-Simons-Higgs vortices[J]. Commun. Math. Phys. D, 2009, 285: 1005–1031.

[22] Chen R M, Guo Y, Spirn D, Yang Y. Electrically and magnetically charged vortices in the Chern-Simons-Higgs theory[J]. Proc. R. Soc. A, 2009, 465: 3089–3516.

[23] Prasad M K, Sommerfeld C M. Exact classical solutions for the 't Hooft monopole and the Julia–Zee dyon[J]. Phys. Rev. Lett., 1975, 35: 760–762.

[24] Jaffe A, Taubes C. Vortices and Monopoles[M]. Boston: Birkhauser, 1990.

[25] Wang R. The existence of Chern-Simons vortices[J]. Comm. Math. Phys., 1991, 137: 587–597.

[26] Spruck J, Yang Y. Topological solitons in the self-dual Chern-Simons theory: Existence and approximation[J]. Ann. Herri Poincare, 1991, 1: 75–97.

[27] Berger M S, Chen Y Y. Symmetric vortices for the Ginzburg-Landau equations of superconductivity and the nonlinear desingularization phenomenon[J]. J. Funct. Anal., 1989, 82: 259–295.

[28] Yang Y. Solitons in field theory and nonlinear analysis[M]. Springer Monographs on Mathematics. New York: Springer-Verlag, 2001.

[29] Strauss W A. Existence of solitary waves in higher dimensions[J]. Comm. Math. Phys., 1997, 55: 149–162.

Radiative Transfer in Two Adjoining Media: The Continuity of Thermal Flux and the Existence and Uniqueness of Solutions*

Guo Boling (郭柏灵) and Han Yongqian (韩永前)

Abstract The spatial transport of radiation in two adjoining media and the continuity of flux at the interface are considered. As usual, we assume that the material is in local thermodynamic equilibrium. First in the absence of hydrodynamic motion and thermal diffusion, the existence and uniqueness of global solution of the classical radiative transfer equations with absorbing boundary condition is derived. But the continuity of thermal flux at the interface of two materials can not be preserve in this classical model. To preserve the continuity of thermal flux at the interface, we consider a radiative transfer equations modified by thermal diffusion. The existence and uniqueness of global solution of this modified radiative transfer equations is also derived.

Keywords radiative transfer equations; transport equation; two adjoining media; multimedia; multigroup equations; continuity of flux; absorbing boundary condition

1 Introduction

As well known, the spatial transport of radiation in material medium is described (see [6, 27, 32, 38]) by the radiative transfer equations which consist of the following time-dependent transfer equation (TrEq)

$$\frac{1}{c}\partial_t u + \omega\partial_x u + \sigma(v)u = \sigma_s(v)\bar{u} + \sigma_a(v)B(v,\nu), \tag{1.1}$$

and the associated energy balance equation

* J. Comp. Theor. Transport, 2015, 44(1): 24–67. Doi: 10.1080/23324309.2014.991973.

$$C_h\partial_t v + 4\pi\int_0^\infty \sigma_a(v)(B(v,\nu)-\bar{u})\mathrm{d}\nu = 0. \tag{1.2}$$

Here the material is in local thermodynamic equilibrium, material motion and thermal diffusion are ignored. $u(t,x,\nu,\omega)$ denotes the specific intensity of radiation at location $x\in[-1,1]$, $v(t,x)$ denotes the material temperature at location $x\in[-1,1]$, $t\geqslant 0$ and ν are the temporal and frequency variables respectively, $\omega=\cos\theta$ and θ is the angle between the direction of travel of the radiation and the positive x coordinate axis, c is the vacuum speed of light, $\sigma=\sigma_s+\sigma_a$ is the transport coefficient, $\sigma_s(v)=\sigma_s(v,x,\nu)>0$ is the scattering cross section, $\sigma_a(v)=\sigma_a(v,x,\nu)>0$ is the absorption cross section, constant $C_h>0$ is the material pseudo-heat capacity, $B(v,\nu)$ is the Planck function given by

$$B(v,\nu)=\frac{2h\nu^3}{c^2}(e^{h\nu/kv}-1)^{-1},\ \ B(v,0)=0,\ \ B(0,\nu)=0,\qquad \forall v,\nu\geqslant 0, \tag{1.3}$$

$$\int_0^\infty B(v,\nu)d\nu=\beta v^4, \tag{1.4}$$

$$\partial_v B(v,\nu)\geqslant 0,\ \int_0^\infty |\partial_v B(v,\nu)|\mathrm{d}\nu=\left|\partial_v\int_0^\infty B(v,\nu)\mathrm{d}\nu\right|=4\beta|v^3|, \tag{1.5}$$

$h>0$ is Planck constant, $k>0$ is the usual Boltzmann constant, $\pi\beta>0$ is called the Stefan-Boltzmann constant, the scalar density defined by

$$\bar{u}(t,x,\nu)=\frac{1}{2}\int_{-1}^1 u(t,x,\nu,\omega)\mathrm{d}\omega. \tag{1.6}$$

Now let us consider the spatial transport of radiation in two adjoining media, i.e., there are two materials in domains $[-1,0)$ and $(0,1]$ respectively, and $x=0$ is the interface between two adjoining media. The radiative transfer equations in this medium can be rewritten as the following transfer equations (TrEq)

$$\partial_t u_1 + c\omega\partial_x u_1 + \sigma_1(v_1)u_1 = \sigma_{1s}(v_1)\bar{u}_1 + \sigma_{1a}(v_1)B(v_1,\nu), \tag{1.7}$$

$$\partial_t u_2 + c\omega\partial_x u_2 + \sigma_2(v_2)u_2 = \sigma_{2s}(v_2)\bar{u}_2 + \sigma_{2a}(v_2)B(v_2,\nu), \tag{1.8}$$

and the associated energy balance equations

$$\partial_t v_1 + \frac{1}{C_{1h}}\int_0^\infty \sigma_{1a}(v_1)(B(v_1,\nu)-\bar{u}_1)\mathrm{d}\nu = 0, \tag{1.9}$$

$$\partial_t v_2 + \frac{1}{C_{2h}}\int_0^\infty \sigma_{2a}(v_2)(B(v_2,\nu)-\bar{u}_2)\mathrm{d}\nu = 0. \tag{1.10}$$

Here $u_1(t,x,\nu,\omega)$ and $u_2(t,x,\nu,\omega)$ denote the specific intensities of radiation at location $x \in [-1,0)$ and $x \in (0,1]$ respectively, $v_1(t,x)$ and $v_2(t,x)$ denote the material temperatures at location $x \in [-1,0)$ and $x \in (0,1]$ respectively, C_{1h} and C_{2h} are two materials pseudo-heat capacities respectively, $\sigma_j = \sigma_{ja} + \sigma_{js}$ $(j = 1,2)$.

Employing equations (1.7)–(1.10), we can see clear the interaction between u_1, u_2, v_1 and v_2 across the interface $x = 0$. Equations (1.7) (1.8) (1.9) and (1.10) must be supplemented with initial condition

$$\begin{aligned} &u_1(0,x,\nu,\omega) = u_1^0(x,\nu,\omega), \quad x \in [-1,0], \\ &u_2(0,x,\nu,\omega) = u_2^0(x,\nu,\omega), \quad x \in [0,1], \end{aligned} \tag{1.11}$$

$$\begin{aligned} &v_1(0,x) = v_1^0(x), \quad x \in [-1,0], \\ &v_2(0,x) = v_2^0(x), \quad x \in [0,1], \end{aligned} \tag{1.12}$$

and boundary condition

$$\begin{aligned} &u_1(t,x,\nu,\omega)|_{x=-1} = u_L(t,\nu,\omega), \quad \text{for } \omega > 0, \\ &u_2(t,x,\nu,\omega)|_{x=1} = u_R(t,\nu,\omega), \quad \text{for } \omega < 0. \end{aligned} \tag{1.13}$$

The boundary functions $u_L(t,\nu,\omega)$ and $u_R(t,\nu,\omega)$ specify the photon entering the domain at the left and right boundaries, respectively. u_1^0, u_2^0, u_L and u_R should be compatible, i.e.

$$\begin{aligned} &u_1^0(x,\nu,\omega)|_{x=0} = u_2^0(x,\nu,\omega)|_{x=0}, \\ &u_1^0(x,\nu,\omega)|_{x=-1} = u_L(t,\nu,\omega)|_{t=0}, \quad \text{for } \omega > 0, \\ &u_2^0(x,\nu,\omega)|_{x=1} = u_R(t,\nu,\omega)|_{t=0}, \quad \text{for } \omega < 0. \end{aligned} \tag{1.14}$$

At the interface between two adjoining media, boundary conditions are imposed to preserve continuity of flux:

$$u_1(t,x,\nu,\omega)|_{x=0} \leftleftarrows u_2(t,x,\nu,\omega)|_{x=0}, \quad \text{for } \omega < 0, \tag{1.15}$$

$$u_2(t,x,\nu,\omega)|_{x=0} \leftleftarrows u_1(t,x,\nu,\omega)|_{x=0}, \quad \text{for } \omega > 0, \tag{1.16}$$

Equation (1.15) means that $u_1(t,0,\nu,\omega)$ is given by $u_2(t,0,\nu,\omega)$ and $u_2(t,0,\nu,\omega)$ is given by equation (1.8) with initial data $u_2^0(x,\nu,\omega)$ for $\omega < 0$. Equation (1.16) means that $u_2(t,0,\nu,\omega)$ is given by $u_1(t,0,\nu,\omega)$ and $u_1(t,0,\nu,\omega)$ is given by equation (1.7) with initial data $u_1^0(x,\nu,\omega)$ for $\omega > 0$.

The flux continuity (1.15) (1.16) was first introduced by Bensoussan, Lions and Papanicolaou [5] in considering neutron transport in two adjoining media.

Since $x = 0$ is the interface between two adjoining media, it is reasonable to assume that

$$\begin{aligned} &\sigma_1(v_1, x, \nu)|_{x=0} \neq \sigma_2(v_2, x, \nu)|_{x=0}, \\ &\sigma_{1a}(v_1, x, \nu)|_{x=0} \neq \sigma_{1a}(v_2, x, \nu)|_{x=0} \\ &\sigma_{1s}(v_1, x, \nu)|_{x=0} \neq \sigma_{1s}(v_2, x, \nu)|_{x=0}. \end{aligned} \tag{1.17}$$

Therefore for the solution (u_1, u_2, v_1, v_2) of problem (1.7)– (1.13) (1.15) (1.16), the thermal flux may be discontinuous at the interface $x = 0$, i.e. $v_1|_{x=0} \neq v_2|_{x=0}$. In some situation, for example at longer time scales, if we want to preserve continuity of flux:

$$\begin{aligned} v_1(t, x)|_{x=0} &= v_2(t, x)|_{x=0}, \\ d_1\partial_x v_1(t, x)|_{x=0} &= d_2\partial_x v_2(t, x)|_{x=0}, \end{aligned} \tag{1.18}$$

the thermal diffusion in this two adjoining media should be considered, and the associated energy balance equations (1.9) (1.10) are modified by the following equations.

$$\partial_t v_1 - d_1\partial_x^2 v_1 + \frac{1}{C_{1h}} \int_0^\infty \sigma_{1a}(v_1)(B(v_1, \nu) - \bar{u}_1)\mathrm{d}\nu = 0, \tag{1.19}$$

$$\partial_t v_2 - d_2\partial_x^2 v_2 + \frac{1}{C_{2h}} \int_0^\infty \sigma_{2a}(v_2)(B(v_2, \nu) - \bar{u}_2)\mathrm{d}\nu = 0. \tag{1.20}$$

Here positive constants d_1 and d_2 are the thermal diffusion coefficients of two materials respectively.

The energy balance equation including thermal diffusion term has been employed to treat the time-evolution of coupled radiation, electron and ion energies [10], and to describe the time-evolution of coupled radiation diffusion and material conduction [35].

Equations (1.19) (1.20) must be supplemented with initial condition (1.12) and boundary condition

$$v_1(t, x)|_{x=-1} = v_L(t), \qquad v_2(t, x)|_{x=1} = v_R(t). \tag{1.21}$$

Here v_1^0, v_2^0, v_L and v_R should be compatible, i.e.

$$\begin{aligned} &v_1^0|_{x=0} = v_2^0|_{x=0}, \quad d_1\partial_x v_1^0|_{x=0} = d_2\partial_x v_2^0|_{x=0} \\ &v_1^0|_{x=-1} = v_L|_{t=0}, \quad v_2^0|_{x=1} = v_R|_{t=0}. \end{aligned} \tag{1.22}$$

From the equations (1.19) (1.20), we can see clear the loss of regularity of v_1 and v_2 across the interface $x = 0$. Therefore the normal method of establishing the maximum-minimum principle to parabolic equation is not working. To establish existence and uniqueness of the global solution, this is the big difficult, which has to overcome.

Equations (1.1) and (1.2) are extremely difficult to solve, either analytically or numerically. For this reason, many asymptotic analysis [15, 16, 27], approximate models and computational methods have been investigated in many recent papers [1, 2, 6, 20, 21, 24, 28, 33, 34, 38, 39] and therein references. In the absence of scattering, the time-discreted equations of (1.1) (1.2) is studied by Larsen, et al [24], the existence and uniqueness of solution for the time-discreted equations are established. The asymptotic analysis in the paper [27], which emphasized physical intuition and physical arguments, is mathematically formal. In [15, 16], some mathematically rigorous results of asymptotic analysis are established for one group equation in single medium, which transport coefficients are constants.

Well-posedness of solutions of equations is fundamental to get mathematically rigorous results of asymptotic analysis. So this paper is first devoted to investigate the well-posedness of solutions for the transfer equations and the associated energy balance equations, which describe the spatial transport of radiation in the two adjoining media.

In order to establish the existence and uniqueness of the global solution of the radiative transfer equations, we need to estimate the lower bound and upper bound of the solution, which is fulfilled by using comparison principle. To establish comparison principle of the radiative transfer equations, we need comparison principle of the corresponding linear equations, especially the comparison principle of linear energy balance equations. The comparison principle of linear equation is equivalent to that the solution is nonnegative if the data is nonnegative. The comparison principle of linear energy balance equations is derived by using corresponding ordinary differential-difference equations. This is the main ingredient.

The relative results of linear transport equations and linear energy balance equations in the two adjoining media, which the continuity of flux is preserved, are important. They are used many times in this article. The relative topic of the linear neutron transport equation is extensively studied in [4, 5, 7, 11, 12, 13, 14, 22, 23, 25, 26, 31, 36, 38] and therein references. The linear equations of parabolic type, which the continuity of flux is not concerned, is extensively studied in [3, 17, 19] and therein references.

Throughout this paper, space $C(I; B) \cap L^\infty(I; B)$ denotes by $C_b(I; B)$, where I is any domain, B is any Banach space.

The outline of this paper is as follows. In Section 2, we consider the existence and uniqueness of the global solution of linear transport equations and linear energy balance equations in the two adjoining media. The estimates of solutions and comparison principle will be used many times. In Section 3, we study the existence and uniqueness of the global solution of the radiative transfer equations ignoring thermal diffusion. In Section 4, we investigate the existence and uniqueness of the global solution of the radiative transfer equations with thermal diffusion. A few concluding remarks are given in Section 5.

2 Existence and uniqueness of solution of linear equations

2.1 Linear transport equations

To solve equations (1.7) (1.8), we shall consider the existence and uniqueness of solution of the following linear transport equations

$$\partial_t W_1 + c\omega\partial_x W_1 = f_1(t,x,\nu,\omega), \quad x \in (-1,0), \tag{2.1}$$

$$\partial_t W_2 + c\omega\partial_x W_2 = f_2(t,x,\nu,\omega), \quad x \in (0,1), \tag{2.2}$$

with initial-boundary condition

$$\begin{aligned} W_1(0,x,\nu,\omega) &= u_1^0(x,\nu,\omega), \quad x \in [-1,0], \\ W_2(0,x,\nu,\omega) &= u_2^0(x,\nu,\omega), \quad x \in [0,1], \end{aligned} \tag{2.3}$$

$$\begin{aligned} W_1(t,x,\nu,\omega)|_{x=-1} &= u_L(t,\nu,\omega), \quad \text{for } \omega > 0, \\ W_2(t,x,\nu,\omega)|_{x=1} &= u_R(t,\nu,\omega), \quad \text{for } \omega < 0, \end{aligned} \tag{2.4}$$

and continuous condition at interface

$$W_1(t,x,\nu,\omega)|_{x=0} \leftleftharpoons W_2(t,x,\nu,\omega)|_{x=0}, \quad \text{for } \omega < 0, \tag{2.5}$$

$$W_2(t,x,\nu,\omega)|_{x=0} \leftleftharpoons W_1(t,x,\nu,\omega)|_{x=0}, \quad \text{for } \omega > 0. \tag{2.6}$$

Here u_1^0, u_2^0, u_L and u_R satisfy the compatible condition (1.14).

The existence, uniqueness and principle of maximum of linear transport equation, which dose not concern the continuity of flux, are established in [7,8,11,5], etc.

As in [5], let

$$t_{1d} = t_{1d}(x,\omega) = \begin{cases} \dfrac{x+1}{c\omega}, & \omega > 0 \\ \dfrac{x}{c\omega}, & \omega < 0 \end{cases} \tag{2.7}$$

denote the time for a photon to reach the boundary of domain $[-1,0]$ starting from $x\in[-1,0]$ and moving with velocity $-c$ in a direction $\theta=\cos^{-1}\omega$, and

$$t_{2d}=t_{2d}(x,\omega)=\begin{cases}\dfrac{x}{c\omega}, & \omega>0\\ \dfrac{x-1}{c\omega}, & \omega<0\end{cases}\tag{2.8}$$

denote the time for a photon to reach the boundary of domain $[0,1]$ starting from $x\in[0,1]$ and moving with velocity $-c$ in a direction $\theta=\cos^{-1}\omega$.

Then we rewrite equations (2.1) and (2.2) as integral equations by the usual method of characteristics

$$\begin{aligned}W_1(t,x,\nu,\omega)=&\chi(t<t_{1d})u_1^0(x_{\omega t},\nu,\omega)+\int_{t-t\wedge t_{1d}}^{t}f_1(s,x_{\omega s},\nu,\omega)\mathrm{d}s\\&+\chi(t\geqslant t_{1d})\{\chi(\omega>0)u_L(t-t_{1d},\nu,\omega)\\&+\chi(\omega<0)W_2(t-t_{1d},0,\nu,\omega)\},\qquad -1<x<0,\end{aligned}\tag{2.9}$$

$$\begin{aligned}W_2(t,x,\nu,\omega)=&\chi(t<t_{2d})u_2^0(x_{\omega t},\nu,\omega)+\int_{t-t\wedge t_{2d}}^{t}f_2(s,x_{\omega s},\nu,\omega)\mathrm{d}s\\&+\chi(t\geqslant t_{2d})\{\chi(\omega<0)u_R(t-t_{2d},\nu,\omega)\\&+\chi(\omega>0)W_1(t-t_{2d},0,\nu,\omega)\},\qquad 0<x<1.\end{aligned}\tag{2.10}$$

Here $x_{\omega t}=x-c\omega t$, $t\wedge t_{jd}=\min(t,t_{jd})$, $\chi(t<t_{jd})$ equals one if $t<t_{jd}$ and zero otherwise, $j=1,2$.

Moreover, at the interface $x=0$, we have

$$\begin{aligned}W_1(t,0,\nu,\omega)=&\chi(t<t_{1d}^0)u_1^0(x_{\omega t}^0,\nu,\omega)+\chi(t\geqslant t_{1d}^0)u_L(t-t_{1d}^0,\nu,\omega)\\&+\int_{t-t\wedge t_{1d}^0}^{t}f_1(s,x_{\omega s}^0,\nu,\omega)\mathrm{d}s,\qquad \forall\omega>0,\end{aligned}\tag{2.11}$$

$$\begin{aligned}W_2(t,0,\nu,\omega)=&\chi(t<t_{2d}^0)u_2^0(x_{\omega t}^0,\nu,\omega)+\chi(t\geqslant t_{2d}^0)u_R(t-t_{2d}^0,\nu,\omega)\\&+\int_{t-t\wedge t_{2d}^0}^{t}f_2(s,x_{\omega s}^0,\nu,\omega)\mathrm{d}s,\qquad \forall\omega<0,\end{aligned}\tag{2.12}$$

where $x_{\omega t}^0=-c\omega t$, $t_{1d}^0=\dfrac{1}{c\omega}$ and $t_{2d}^0=\dfrac{-1}{c\omega}$.

Putting (2.9)–(2.12) together, it is proved that there exists a unique solution (W_1,W_2) of problem (2.1) –(2.6). Moreover, we have

$$\|W_j\|_{L_T^\infty L_x^\infty L_\nu^1 L_\omega^\infty}\leqslant\sum_{k=1}^{2}\left\{\|u_k^0\|_{L_x^\infty L_\nu^1 L_\omega^\infty}+\int_0^T\|f_k(t,\cdot)\|_{L_x^\infty L_\nu^1 L_\omega^\infty}\mathrm{d}t\right\}$$

$$+\|u_L\|_{L_T^\infty L_\nu^1 L_\omega^\infty}+\|u_R\|_{L_T^\infty L_\nu^1 L_\omega^\infty},\qquad j=1,2. \tag{2.13}$$

2.2 Linear energy balance equations with thermal diffusion

To solve equations (1.19) (1.20), we shall consider the existence and uniqueness of solution of the following linear thermal diffusion equations

$$\partial_t W_1-d_1\partial_x^2 W_1+a(t,x)W_1=F(t,x),\quad x\in(-1,0), \tag{2.14}$$

$$\partial_t W_2-d_2\partial_x^2 W_2+b(t,x)W_2=G(t,x),\quad x\in(0,1), \tag{2.15}$$

with initial-boundary condition

$$\begin{aligned} W_1(0,x)&=v_1^0(x),\quad x\in[-1,0],\\ W_2(0,x)&=v_2^0(x),\quad x\in[0,1], \end{aligned} \tag{2.16}$$

$$W_1(t,x)|_{x=-1}=v_L(t),\ \ W_2(t,x)|_{x=1}=v_R(t) \tag{2.17}$$

and continuous condition at the interface

$$\begin{aligned} W_1(t,x)|_{x=0}&=W_2(t,x)|_{x=0},\\ d_1\partial_x W_1(t,x)|_{x=0}&=d_2\partial_x W_2(t,x)|_{x=0}. \end{aligned} \tag{2.18}$$

Here v_1^0, v_2^0, v_L and v_R satisfy the compatible condition (1.22). We assume that $a(t,x)\in L^\infty([0,T];L^\infty(-1,0))$ and $b(t,x)\in L^\infty([0,T];L^\infty(0,1))$ for any $T>0$. Moreover there exists a constant $A>0$ such that

$$\begin{aligned} |a(t,x)|&\leqslant A,\quad \forall t\in[0,T],\ \ x\in[-1,0],\\ |b(t,x)|&\leqslant A,\quad \forall t\in[0,T],\ \ x\in[0,1]. \end{aligned} \tag{2.19}$$

The existence and uniqueness of linear equations of parabolic type, which dose not concern the continuity of flux, are extensively studied in [17, 3, 19] and therein references.

We employ the finite difference method to establish the existence of solution of problem (2.14)–(2.18).

Let us divide the intervals $[-1,0]$ and $[0,1]$ into small segment grids by the points $x_j=-1+jh$ and $y_j=jh$ $(j=0,1,\cdots,J)$ respectively, where $Jh=1$, J is an integer and h is the step size. Corresponding to the problem (2.14)–(2.18), let $U_j=W_1(t,x_j)$, $V_j=W_2(t,y_j)$, $a_j=a(t,x_j)$, $b_j=b(t,y_j)$, $F_j=F(t,x_j)$, $G_j=G(t,y_j)$ and construct the following ordinary differential-difference system

$$\frac{\mathrm{d}U_j}{\mathrm{d}t}-d_1\frac{D_+D_-U_j}{h^2}+a_jU_j=F_j,\quad j=1,\cdots,J-1, \tag{2.20}$$

$$\frac{\mathrm{d}V_j}{\mathrm{d}t} - d_2\frac{D_+D_-V_j}{h^2} + b_jV_j = G_j, \quad j = 1, \cdots, J-1, \tag{2.21}$$

$$U_0 = v_L, \qquad V_J = v_R \tag{2.22}$$

with initial condition

$$U_j|_{t=0} = U_j^0, \; V_j|_{t=0} = V_j^0, \quad j = 0, 1, \cdots, J \tag{2.23}$$

and continuous condition at the interface

$$U_J = V_0, \qquad d_1(U_J - U_{J-1}) = d_2(V_1 - V_0), \tag{2.24}$$

where D_+ and D_- denote the forward and backward difference operators respectively, $U_j^0 = v_1^0(x_j)$, $V_j^0 = v_2^0(y_j)$.

It is well-known that there exists a unique solution of problem (2.20)–(2.24) $U_h = \{U_j | j = 0, 1, \cdots, J\}$ and $V_h = \{V_j | j = 0, 1, \cdots, J\}$. In order to verify the existence of solution (W_1, W_2) of the problem (2.14)– (2.18), it suffices to get a priori uniform estimates of (U_h, V_h) with respect to h. Then the solution (W_1, W_2) is the limit of sequence $\{(U_h, V_h)\}$, when h tends to zero.

Let us define $\delta u_h = \left\{\frac{D_+u_j}{h}\middle| j = 0, 1, \cdots, J-1\right\} = \left\{\frac{D_-u_j}{h}\middle| j = 1, \cdots, J\right\}$. Similarly, the discrete functions $\delta^k u_h$ $(k \geqslant 2)$ can be defined. The norms of discrete functions $\delta^k u_h$ $(k \geqslant 0)$ are defined as follows:

$$\|\delta^k u_h\|_2 = \left(\sum_{j=0}^{J-k}\left|\frac{D_+^k u_j}{h^k}\right|^2 h\right)^{1/2} = \left(\sum_{j=l}^{J-k+l}\left|\frac{D_+^{k-l}D_-^l u_j}{h^k}\right|^2 h\right)^{1/2}.$$

Now we get a priori uniform estimates of (U_h, V_h) with respect to h.

Lemma 2.1 (Discrete Version of Energy Estimate) *Assume that v_1^0, v_2^0, v_L and v_R satisfy (1.22), a and b satisfy (2.19), $v_1^0 \in H^1(-1, 0)$, $v_2^0 \in H^1(0, 1)$. For any given $T \geqslant 0$, $v_L, v_R \in W^{1,\infty}(0, T)$, $a \in L^\infty([0, T]; L^\infty(-1, 0))$, $b \in L^\infty([0, T]; L^\infty(0, 1))$, $F \in C([0, T];\, L^2(-1, 0))$ and $G \in C([0, T]; L^2(-1, 0))$, then there exists a unique solution $U_j, V_j \in C([0, T])$ $(j = 0, 1, \cdots, J)$ of the problem (2.20)–(2.24) such that discrete function (U_h, V_h) holds the following estimate*

$$\begin{aligned}&\|U_h(t)\|_2 + \|\delta U_h(t)\|_2 + \|V_h(t)\|_2 + \|\delta V_h(t)\|_2 \\ &\leqslant Ce^{(1+2A)t}\Big\{\|U_h(0)\|_2 + \|\delta U_h(0)\|_2 + \|V_h(0)\|_2 + \|\delta V_h(0)\|_2 + (1+T) \\ &(\|v_L\|_{W_T^{1,\infty}} + \|v_R\|_{W_T^{1,\infty}}) + \int_0^t \{\|F_h(s)\|_2 + \|G_h(s)\|_2\}\mathrm{d}s\Big\}, \quad \forall t \in [0, T],\end{aligned} \tag{2.25}$$

where constant C is independent of t and h. Moreover, if v_1^0, v_2^0, v_L and v_R are nonnegative, $0 \leqslant F \in C([0,\infty);L^\infty(-1,0))$ and $0 \leqslant G \in C([0,\infty);L^\infty(-1,0))$, then we have

$$U_j(t) \geqslant 0, \quad V_j(t) \geqslant 0, \quad j=0,1,\cdots,J, \ \forall t \geqslant 0. \tag{2.26}$$

Proof By using (2.24), we have

$$U_J = V_0 = \frac{d_1 U_{J-1} + d_2 V_1}{d_1 + d_2}. \tag{2.27}$$

Equations (2.20) (2.21) (2.23) is equivalent to

$$\begin{aligned} U_j(t) =& U_j^0 \exp\left\{ -\int_0^t \left(2d_1/h^2 + a_j(\tau)\right)\mathrm{d}\tau \right\} \\ &+ \int_0^t \exp\left\{ -\int_0^{t-\tau} \left(2d_1/h^2 + a_j(s)\right)\mathrm{d}s \right\}\{F_j + d_1(U_{j-1}+U_{j+1})/h^2\}(\tau)\mathrm{d}\tau, \end{aligned} \tag{2.28}$$

$$\begin{aligned} V_j(t) =& V_j^0 \exp\left\{ -\int_0^t \left(2d_2/h^2 + b_j(\tau)\right)\mathrm{d}\tau \right\} \\ &+ \int_0^t \exp\left\{ -\int_0^{t-\tau} \left(2d_2/h^2 + b_j(s)\right)\mathrm{d}s \right\}\{G_j + d_2(V_{j-1}+V_{j+1})/h^2\}(\tau)\mathrm{d}\tau, \end{aligned} \tag{2.29}$$

where $j=1,\cdots,J-1$. U_0 and V_J have been derived by (2.22). U_j $(j=1,\cdots,J)$ and V_j $(j=0,1,\cdots,J-1)$ can be derived by solving (2.27)–(2.29). Now we construct iteration

$$U_J^{(0)} = V_0^{(0)} = \frac{d_1 U_{J-1}^{(0)} + d_2 V_1^{(0)}}{d_1 + d_2}, \tag{2.30}$$

$$\begin{aligned} U_1^{(0)}(t) =& U_1^0 \exp\left\{ -\int_0^t \left(2d_1/h^2 + a_j(\tau)\right)\mathrm{d}\tau \right\} \\ &+ \int_0^t \exp\left\{ -\int_0^{t-\tau} \left(2d_1/h^2 + a_j(s)\right)\mathrm{d}s \right\}\{F_1 + d_1 U_0/h^2\}(\tau)\mathrm{d}\tau, \end{aligned} \tag{2.31}$$

$$\begin{aligned} U_j^{(0)}(t) =& U_j^0 \exp\left\{ -\int_0^t \left(2d_1/h^2 + a_j(\tau)\right)\mathrm{d}\tau \right\} \\ &+ \int_0^t \exp\left\{ -\int_0^{t-\tau} \left(2d_1/h^2 + a_j(s)\right)\mathrm{d}s \right\} F_j(\tau)\mathrm{d}\tau, \end{aligned} \tag{2.32}$$

$$V_j^{(0)}(t) = V_j^0 \exp\left\{ -\int_0^t \left(2d_2/h^2 + b_j(\tau)\right)\mathrm{d}\tau \right\}$$

$$+\int_0^t \exp\left\{-\int_0^{t-\tau}(2d_2/h^2+b_j(s))\mathrm{d}s\right\}G_j(\tau)\mathrm{d}\tau, \tag{2.33}$$

$$\begin{aligned}V_{J-1}^{(0)}(t)=&V_{J-1}^0\exp\left\{-\int_0^t(2d_2/h^2+b_j(\tau))\mathrm{d}\tau\right\}\\&+\int_0^t\exp\left\{-\int_0^{t-\tau}(2d_2/h^2+b_j(s))\mathrm{d}s\right\}\{G_{J-1}+d_2V_J/h^2\}(\tau)\mathrm{d}\tau,\end{aligned} \tag{2.34}$$

$$U_J^{(n)}=V_0^{(n)}=\frac{d_1U_{J-1}^{(n)}+d_2V_1^{(n)}}{d_1+d_2}, \tag{2.35}$$

$$U_1^{(n)}(t)=\int_0^t\exp\left\{-\int_0^{t-\tau}(2d_1/h^2+a_j(s))\mathrm{d}s\right\}\{d_1U_2^{(n-1)}/h^2\}(\tau)\mathrm{d}\tau, \tag{2.36}$$

$$U_j^{(n)}(t)=\int_0^t\exp\left\{-\int_0^{t-\tau}(2d_1/h^2+a_j(s))\mathrm{d}s\right\}\{d_1(U_{j-1}^{(n-1)}+U_{j+1}^{(n-1)})/h^2\}(\tau)\mathrm{d}\tau, \tag{2.37}$$

$$V_j^{(n)}(t)=\int_0^t\exp\left\{-\int_0^{t-\tau}(2d_2/h^2+b_j(s))\mathrm{d}s\right\}\{d_2(V_{j-1}^{(n-1)}+V_{j+1}^{(n-1)})/h^2\}(\tau)\mathrm{d}\tau, \tag{2.38}$$

$$V_{J-1}^{(n)}(t)=\int_0^t\exp\left\{-\int_0^{t-\tau}(2d_2/h^2+b_j(s))\mathrm{d}s\right\}\{d_2V_{J-2}^{(n-1)}/h^2\}(\tau)\mathrm{d}\tau, \tag{2.39}$$

where $n=1,2,\cdots$.

$v_1^0\geqslant 0,\ v_2^0\geqslant 0,\ v_L\geqslant 0,\ v_R\geqslant 0,\ F(t)\geqslant 0$ and $G(t)\geqslant 0$ imply that

$$U_j^{(0)},V_j^{(0)}\geqslant 0,\ \ j=0,1,\cdots,J. \tag{2.40}$$

By induction we have

$$U_j^{(n)},V_j^{(n)}\geqslant 0,\ \ j=0,1,\cdots,J,\ \ n=0,1,\cdots. \tag{2.41}$$

For any $T>0$ and $t\in[0,T]$, (2.31)–(2.34) imply that

$$|U_1^{(0)}(t)|\leqslant\exp(AT)\{|U_1^0|+T\|F_1\|_{L^\infty}+d_1T\|v_L\|_{L_T^\infty}/h^2\}, \tag{2.42}$$

$$|U_j^{(0)}(t)|=\exp(AT)\{|U_j^0|+T\|F_j\|_{L^\infty}\},\ \ j=2,\cdots,J-1, \tag{2.43}$$

$$|V_j^{(0)}(t)|\leqslant\exp(AT)\{|V_j^0|+T\|G_j\|_{L^\infty}\},\ \ j=1,\cdots,J-2, \tag{2.44}$$

$$|V_{J-1}^{(0)}(t)|=\exp(AT)\{|V_{J-1}^0|+T\|G_{J-1}\|_{L^\infty}+d_2T\|v_R\|_{L_T^\infty}/h^2\}. \tag{2.45}$$

By induction we derive ($\forall t \in [0,T]$)

$$|U_j^{(n)}(t)| \leqslant \frac{\{2(d_1+d_2)\exp(AT)/h^2\}^n t^n}{n!}\exp(AT)f_0, \quad j=1,\cdots,J, \tag{2.46}$$

$$|V_j^{(n)}(t)| \leqslant \frac{\{2(d_1+d_2)\exp(AT)/h^2\}^n t^n}{n!}\exp(AT)f_0, \ \ j=0,1,\cdots,J-1, \tag{2.47}$$

where $n=0,1,2,\cdots$ and

$$\begin{aligned} f_0 = &\max_{j=0,1,\cdots,J}\{|U_j^0|,|V_j^0|\} + \frac{T}{h^2}(d_1\|v_L\|_{L_T^\infty} + d_2\|v_R\|_{L_T^\infty}) \\ &+T\max_{j=0,1,\cdots,J}\{\|F_j\|_{L_T^\infty}, \|G_j\|_{L_T^\infty}\}. \end{aligned}$$

Since the solution U_j ($j=1,\cdots,J$) and V_j ($j=0,1,\cdots,J-1$) of problem (2.27)–(2.29) are formally

$$U_j(t) = \sum_{n=0}^\infty U_j^{(n)}(t), \ \ V_j(t) = \sum_{n=0}^\infty V_j^{(n)}(t), \quad \forall t \in [0,T],$$

our estimates above show that these series converge, $U_j(t)$ ($j=1,\cdots,J$) and $V_j(t)$ ($j=0,1,\cdots,J-1$) exist for all $t\in[0,T]$, and

$$\begin{aligned} &|U_j(t)| \leqslant \exp\{2(d_1+d_2)T\exp(AT)/h^2\}\exp(AT)f_0, \ \ j=1,\cdots,J, \\ &|V_j(t)| \leqslant \exp\{2(d_1+d_2)T\exp(AT)/h^2\}\exp(AT)f_0, \ \ j=0,1,\cdots,J-1. \end{aligned} \tag{2.48}$$

Estimate (2.48) implies that the solution (U_j,V_j) is unique.

Moreover, estimate (2.41) implies that $U_j(t)\geqslant 0$ and $V_j(t)\geqslant 0$ for all $t\geqslant 0$, $j=0,1,\cdots,J$, if $v_1^0\geqslant 0$, $v_2^0\geqslant 0$, $v_L\geqslant 0$, $v_R\geqslant 0$, $F(t)\geqslant 0$ and $G(t)\geqslant 0$.

To prove estimate (2.25), we let

$$g^*(t,x) = \begin{cases} -x^3 v_L(t), & x\in[-1,0] \\ x^3 v_R(t), & x\in[0,1] \end{cases}. \tag{2.49}$$

Employing the following variation

$$\begin{aligned} U_j^*(t) =& U_j(t) - g^*(t,x_j), \ \ j=0,1,\cdots,J, \\ V_j^*(t) =& V_j(t) - g^*(t,y_j), \ \ j=0,1,\cdots,J, \end{aligned} \tag{2.50}$$

we have

$$\frac{\mathrm{d}U_j^*}{\mathrm{d}t} - d_1\frac{D_+D_-U_j^*}{h^2} + a_jU_j^* = F_j^*, \quad j=1,\cdots,J-1, \tag{2.51}$$

$$\frac{\mathrm{d}V_j^*}{\mathrm{d}t} - d_2\frac{D_+D_-V_j^*}{h^2} + b_jV_j^* = G_j^*, \quad j=1,\cdots,J-1, \tag{2.52}$$

$$U_0^* = 0, \qquad V_J^* = 0, \tag{2.53}$$

$$U_j^*|_{t=0} = U_j^{*0}, \quad V_j^*|_{t=0} = V_j^{*0}, \quad j = 0, 1, \cdots, J, \tag{2.54}$$

$$U_J^* = V_0^*, \qquad d_1(U_J^* - U_{J-1}^*) = d_2(V_1^* - V_0^*), \tag{2.55}$$

where $U_j^{*0} = U_j^0 - g^*(0, x_j)$, $V_j^{*0} = V_j^0 - g^*(0, y_j)$,

$$\begin{aligned} F_j^* =& F_j - \frac{dg^*(t, x_j)}{dt} + d_1 \frac{D_+D_-g^*(t, x_j)}{h^2} - a_j g^*(t, x_j), \\ G_j^* =& G_j - \frac{dg^*(t, y_j)}{dt} + d_2 \frac{D_+D_-g^*(t, y_j)}{h^2} - b_j g^*(t, y_j). \end{aligned} \tag{2.56}$$

Taking the scalar product of (2.51) and U_j^*, and the scalar product of (2.52) and V_j^*, summing the result over j from 1 to $J-1$, we have

$$\begin{aligned} &\frac{1}{2}\frac{\mathrm{d}}{\mathrm{d}t}\sum_{j=1}^{J-1}\{(U_j^*)^2 + (V_j^*)^2\}h + \sum_{j=0}^{J-1}\left\{d_1\left(\frac{D_+U_j^*}{h}\right)^2 + d_2\left(\frac{D_+V_j^*}{h}\right)^2\right\}h \\ =& -\sum_{j=1}^{J-1}\{a_j(U_j^*)^2 + b_j(V_j^*)^2\}h + \sum_{j=1}^{J-1}\{F_j^*U_j^* + G_j^*V_j^*\}h \\ \leqslant& A\sum_{j=1}^{J-1}\{(U_j^*)^2 + (V_j^*)^2\}h + \frac{1}{2}\sum_{j=1}^{J-1}\{(F_j^*)^2 + (U_j^*)^2 + (G_j^*)^2 + (V_j^*)^2\}h. \end{aligned} \tag{2.57}$$

By Gronwall inequality, one gets

$$\begin{aligned} &\sum_{j=1}^{J-1}\{(U_j^*)^2(t) + (V_j^*)^2(t)\}h \\ \leqslant& e^{(1+2A)t}\left\{\sum_{j=1}^{J-1}\{(U_j^{*0})^2 + (V_j^{*0})^2\}h + \int_0^t\sum_{j=1}^{J-1}\{(F_j^*)^2(s) + (G_j^*)^2(s)\}h\,\mathrm{d}s\right\}. \end{aligned} \tag{2.58}$$

Taking the scalar product of (2.51) and U_{jt}^*, and the scalar product of (2.52) and V_{jt}^*, summing the result over j from 1 to $J-1$, we have

$$\begin{aligned} &\sum_{j=1}^{J-1}\{(U_{jt}^*)^2 + (V_{jt}^*)^2\}h + \frac{1}{2}\frac{\mathrm{d}}{\mathrm{d}t}\sum_{j=0}^{J-1}\left\{d_1\left(\frac{D_+U_j^*}{h}\right)^2 + d_2\left(\frac{D_+V_j^*}{h}\right)^2\right\}h \\ =& \sum_{j=1}^{J-1}\{-a_jU_j^*U_{jt}^* - b_jV_j^*V_{jt}^* + F_j^*U_{jt}^* + G_j^*V_{jt}^*\}h \\ \leqslant& \sum_{j=1}^{J-1}\left\{A^2(U_j^*)^2 + A^2(V_j^*)^2 + (F_j^*)^2 + (G_j^*)^2 + \frac{1}{2}(U_{jt}^*)^2 + \frac{1}{2}(V_{jt}^*)^2\right\}h. \end{aligned} \tag{2.59}$$

By Gronwall inequality, one gets

$$\begin{aligned}&\sum_{j=0}^{J-1}\left\{d_1\left(\frac{D_+U_j^*(t)}{h}\right)^2+d_2\left(\frac{D_+V_j^*(t)}{h}\right)^2\right\}h\\&\leqslant\sum_{j=0}^{J-1}\left\{d_1\left(\frac{D_+U_j^{*0}}{h}\right)^2+d_2\left(\frac{D_+V_j^{*0}}{h}\right)^2\right\}h\\&\quad+2\int_0^t\sum_{j=1}^{J-1}\{A^2(U_j^*)^2(s)+A^2(V_j^*)^2(s)+(F_j^*)^2(s)+(G_j^*)^2(s)\}h\,\mathrm{d}s,\quad\forall t\geqslant 0.\end{aligned}\tag{2.60}$$

Putting together (2.58) (2.60) (2.50) (2.56) (2.27), we can prove (2.25).

Applying the standard method [41], we construct a set of piecewise constant functions $U_h^{(k)}(t,x)=\dfrac{\Delta_+^kU_j}{h^k}$ and $V_h^{(k)}(t,x)=\dfrac{\Delta_+^kV_j}{h^k}$ for all $(t,x)\in\mathbf{R}_+\times(jh,(j+1)h]$, $j=0,1,\cdots,J$ and $k=0,1$. The estimate (2.25) also implies that $U_h^{(k)}$ and $V_h^{(k)}$ weakly$*$ converge to $U^{(k)}$ and $V^{(k)}$ in $L^\infty(\mathbf{R}_+;L^2(-1,0))$ and $L^\infty(\mathbf{R}_+;L^2(0,1))$ as $h\to 0$, respectively. Moreover $U^{(1)}(t,x)=\partial_xU^{(0)}(t,x)$, $V^{(1)}(t,x)=\partial_xV^{(0)}(t,x)$, and $(W_1,W_2)=(U^{(0)},V^{(0)})$ is the global solution of problem (2.14)–(2.18).

The estimate (2.26) implies that $U_h^{(0)}(t,x),V_h^{(0)}(t,x)\geqslant 0$ for any t and x. Therefore $W_1(t,x),W_2(t,x)\geqslant 0$ for any t and x.

Theorem 2.1 (Existence and Uniqueness of Solution) *Assume that v_1^0, v_2^0, v_L and v_R satisfy (1.22), a and b satisfy (2.19), $v_1^0\in H^1(-1,0)$, $v_2^0\in H^1(0,1)$. For any given $T\geqslant 0$, $v_L,v_R\in W^{1,\infty}(0,T)$, $a\in L^\infty([0,T];L^\infty(-1,0))$, $b\in L^\infty([0,T];L^\infty(0,1))$, $F\in L^\infty([0,T];\,L^2[-1,0])$ and $G\in L^\infty([0,T];L^2[0,1])$. Then there exits a unique solution $W_1\in C([0,T];H^1(-1,0))\cap L^2(0,T;H^2(-1,0))$ and $W_2\in C([0,T];\,H^1(0,1))\cap L^2(0,T;H^2(0,1))$ of problem (2.14)– (2.18). Moreover, if $0\leqslant F\in C([0,T];\,L^\infty(-1,0))$, $0\leqslant G\in C([0,T];\,L^\infty(-1,0))$, and $v_1^0\geqslant 0$, $v_2^0\geqslant 0$, $v_L\geqslant 0$, $v_R\geqslant 0$, then we have*

$$W_k(t,x)\geqslant 0,\quad\forall t\in[0,T],\ x\in I_k,\ k=1,2,\tag{2.61}$$

where $I_1=[-1,0]$ and $I_2=[0,1]$.

The existence of solution (W_1,W_2) has been proved, the uniqueness can be proved by the energy estimate in the following lemma.

Lemma 2.2 (Energy Estimate) *Assume that v_1^0, v_2^0, v_L and v_R satisfy (1.22), a and b satisfy (2.19), $v_1^0\in H^1(-1,0)$, $v_2^0\in H^1(0,1)$. For any given $T\geqslant 0$, $v_L,v_R\in$*

$W^{1,\infty}(0,T)$, $a \in L^\infty([0,T];\ L^\infty(-1,0))$, $b \in L^\infty([0,T];L^\infty(0,1))$, $F \in L^\infty([0,T];$ $L^2[-1,0])$ *and* $G \in L^\infty([0,T];\ L^2[0,1])$. *Then for the solution* (W_1, W_2) *of the problem* (2.14)–(2.18), *the following estimate is valid:*

$$\begin{aligned}
&\|W_1(t,\cdot)\|^2_{H^1_x} + \|W_2(t,\cdot)\|^2_{H^1_x} + \int_0^t \{\|\partial_x^2 W_1(\tau,\cdot)\|^2_{L^2_x} + \|\partial_x^2 W_2(\tau,\cdot)\|^2_{L^2_x}\}\mathrm{d}\tau \\
\leqslant & m_0(A,t)\left\{2\|v_1^0\|^2_{H^1_x} + 2\|v_1^0\|^2_{H^1_x} + m_b(A,t)\left(\|v_L\|^2_{W_T^{1,\infty}} + \|v_R\|^2_{W_T^{1,\infty}}\right)\right.\\
&\left. + 2\int_0^t \{\|F(s,\cdot)\|^2_{L^2_x} + \|G(s,\cdot)\|^2_{L^2_x}\}\mathrm{d}s\right\}, \qquad \forall t \in [0,T],
\end{aligned} \tag{2.62}$$

where $m_0(A,t) = 15e^{(1+2A)t}(A^2t + d_M)/d_m$, $m_b(A,t) = (A^2 + 73d_M^2)t + 4$, $d_M =$ $\max\{1, d_1, d_2\}$, $d_m = \min\{1, d_1, d_2\}$.

Proof Employing the following variation

$$\begin{aligned}
U(t,x) =& W_1(t,x) - g^*(t,x), \quad x \in [-1,0], \\
V(t,x) =& W_2(t,x) - g^*(t,x), \quad x \in [0,1],
\end{aligned} \tag{2.63}$$

we have

$$\partial_t U - d_1\partial_x^2 U + a(t,x)U = F^*(t,x), \quad x \in (-1,0) \tag{2.64}$$

$$\partial_t V - d_2\partial_x^2 V + b(t,x)V = G^*(t,x), \quad x \in (0,1) \tag{2.65}$$

$$\begin{aligned}
U(0,x) = U^0 = v_1^0 - g^*(0,x), \quad x \in [-1,0], \\
V(0,x) = V^0 = v_2^0 - g^*(0,x), \quad x \in [0,1],
\end{aligned} \tag{2.66}$$

$$U(t,x)|_{x=-1} = 0, \quad V(t,x)|_{x=1} = 0, \tag{2.67}$$

$$\begin{aligned}
U(t,x)|_{x=0} = V(t,x)|_{x=0}, \\
d_1\partial_x U(t,x)|_{x=0} = d_2\partial_x V(t,x)|_{x=0},
\end{aligned} \tag{2.68}$$

where g^* is defined in (2.49),

$$\begin{aligned}
F^* =& F - \partial_t g^* + d_1\partial_x^2 g^* - a(t,x)g^*, \\
G^* =& G - \partial_t g^* + d_2\partial_x^2 g^* - b(t,x)g^*.
\end{aligned} \tag{2.69}$$

Taking the scalar product of (U, V) and the equations (2.64) (2.65), and then integrating the result over $[-1,0]$ and $[0,1]$, respectively, we have

$$\begin{aligned}
&\frac{1}{2}\frac{\mathrm{d}}{\mathrm{d}t}\{\|U(t,\cdot)\|^2_{L^2_x} + \|V(t,\cdot)\|^2_{L^2_x}\} + d_1\|\partial_x U(t,\cdot)\|^2_{L^2_x} + d_2\|\partial_x V(t,\cdot)\|^2_{L^2_x} \\
=& -\int_{-1}^0 aU^2\mathrm{d}x - \int_0^1 bV^2\mathrm{d}x + \int_{-1}^0 F^*U\mathrm{d}x + \int_0^1 G^*V\mathrm{d}x
\end{aligned}$$

$$\leqslant \frac{(1+2A)}{2}\{\|U(t,\cdot)\|_{L_x^2}^2+\|V(t,\cdot)\|_{L_x^2}^2\}+\frac{1}{2}\{\|F^*(t,\cdot)\|_{L_x^2}^2+\|G^*(t,\cdot)\|_{L_x^2}^2\}. \tag{2.70}$$

By Gronwall inequality, one has

$$\begin{aligned}\|U(t,\cdot)\|_{L_x^2}^2+\|V(t,\cdot)\|_{L_x^2}^2 \leqslant & e^{(1+2A)t}\{\|U^0\|_{L_x^2}^2+\|V^0\|_{L_x^2}^2\\ &+\int_0^t\{\|F^*(s,\cdot)\|_{L_x^2}^2+\|G^*(s,\cdot)\|_{L_x^2}^2\}\mathrm{d}s\},\quad \forall t\in[0,T].\end{aligned} \tag{2.71}$$

Taking the scalar product of (U_t, V_t) and the equations (2.64) (2.65), and then integrating the result over $[-1,0]$ and $[0,1]$, respectively, we have

$$\begin{aligned}&\frac{1}{2}\frac{\mathrm{d}}{\mathrm{d}t}\{d_1\|\partial_x U(t,\cdot)\|_{L_x^2}^2+d_2\|\partial_x V(t,\cdot)\|_{L_x^2}^2\}+\|U_t(t,\cdot)\|_{L_x^2}^2+\|V_t(t,\cdot)\|_{L_x^2}^2\\ =&\int_{-1}^0\{-aU+F^*\}U_t\mathrm{d}x+\int_0^1\{-bV+G^*\}V_t\mathrm{d}x\\ \leqslant&\frac{1}{2}\{\|U_t(t,\cdot)\|_{L_x^2}^2+\|V_t(t,\cdot)\|_{L_x^2}^2\}+A^2\{\|U(t,\cdot)\|_{L_x^2}^2\\ &+\|V(t,\cdot)\|_{L_x^2}^2\}+\|F^*(t,\cdot)\|_{L_x^2}^2+\|G^*(t,\cdot)\|_{L_x^2}^2.\end{aligned} \tag{2.72}$$

By Gronwall inequality, one has

$$\begin{aligned}&d_1\|\partial_x U(t,\cdot)\|_{L_x^2}^2+d_2\|\partial_x V(t,\cdot)\|_{L_x^2}^2+\int_0^t\{\|U_t(s,\cdot)\|_{L_x^2}^2+\|V_t(s,\cdot)\|_{L_x^2}^2\}\mathrm{d}s\\ \leqslant&d_1\|\partial_x U^0\|_{L_x^2}^2+d_2\|\partial_x V^0\|_{L_x^2}^2\\ &+2\int_0^t\{A^2\|U(s,\cdot)\|_{L_x^2}^2+A^2\|V(s,\cdot)\|_{L_x^2}^2+\|F^*(s,\cdot)\|_{L_x^2}^2+\|G^*(s,\cdot)\|_{L_x^2}^2\}\mathrm{d}s\\ \leqslant&2e^{(1+2A)t}(A^2t+d_M)\{\|U^0\|_{H_x^1}^2+\|V^0\|_{H_x^1}^2\\ &+\int_0^t\{\|F^*(s,\cdot)\|_{L_x^2}^2+\|G^*(s,\cdot)\|_{L_x^2}^2\}ds\},\quad \forall t\geqslant 0.\end{aligned} \tag{2.73}$$

Equations (2.64) (2.65) and estimates (2.71) (2.73) imply that

$$\begin{aligned}&\int_0^t\{d_1\|\partial_x^2 U(t,\cdot)\|_{L_x^2}^2+d_2\|\partial_x^2 V(t,\cdot)\|_{L_x^2}^2\}\mathrm{d}s\\ \leqslant&3\int_0^t\{A^2\|U(s,\cdot)\|_{L_x^2}^2+A^2\|V(s,\cdot)\|_{L_x^2}^2+\|U_t(s,\cdot)\|_{L_x^2}^2\\ &+\|V_t(s,\cdot)\|_{L_x^2}^2+\|F^*(s,\cdot)\|_{L_x^2}^2+\|G^*(s,\cdot)\|_{L_x^2}^2\}\mathrm{d}s\\ \leqslant&12e^{(1+2A)t}(A^2t+d_M)\{\|U^0\|_{H_x^1}^2+\|V^0\|_{H_x^1}^2\\ &+\int_0^t\{\|F^*(s,\cdot)\|_{L_x^2}^2+\|G^*(s,\cdot)\|_{L_x^2}^2\}\mathrm{d}s\},\quad \forall t\geqslant 0.\end{aligned} \tag{2.74}$$

Employing estimates (2.71) (2.73) (2.74) and variations (2.63) (2.69), we can prove estimate (2.62).

3 Radiative transfer equations ignoring thermal diffusion

3.1 Local solution ignoring thermal diffusion

We shall consider the existence and uniqueness of solution of radiative transfer equations (1.7)–(1.10) with initial-boundary conditions (1.11)–(1.13) and continuous conditions (1.15) (1.16) at the interface. Here we assume that there exist constants M_{ja}, M_{js}, N_{ja} and N_{js} $(j = 1, 2)$ such that

$$\begin{aligned} 0 \leqslant \sigma_{ja}(v,x,\nu) \leqslant M_{ja},\quad 0 \leqslant \sigma_{js}(v,x,\nu) \leqslant M_{js}, \\ |\sigma_{ja}(v,x,\nu) - \sigma_{ja}(v',x,\nu)| \leqslant N_{ja}|v-v'|, \\ |\sigma_{js}(v,x,\nu) - \sigma_{js}(v',x,\nu)| \leqslant N_{js}|v-v'|,\ j=1,2. \end{aligned} \tag{3.1}$$

Let $M_j = M_{ja} + M_{js}$ and $N_j = N_{ja} + N_{js}$. It is obvious that

$$\begin{aligned} 0 \leqslant \sigma_j(v,x,\nu) = \sigma_{ja}(v,x,\nu) + \sigma_{js}(v,x,\nu) \leqslant M_j, \\ |\sigma_j(v,x,\nu) - \sigma_j(v',x,\nu)| \leqslant N_j|v-v'|,\ j=1,2. \end{aligned} \tag{3.2}$$

First we establish the existence and uniqueness of local solution of radiative transfer equations ignoring thermal diffusion.

Theorem 3.1 (Existence and Uniqueness of Local Solution) *Assume that condition* (3.1) *and the following conditions are satisfied.*

(1) u_1^0, u_2^0, u_L *and* u_R *satisfy* (1.14).

(2) $u_1^0 \in L^\infty([-1,0]; L^1(\mathbf{R}_+; L^\infty[-1,1]))$, $u_2^0 \in L^\infty([0,1]; L^1(\mathbf{R}_+; L^\infty[-1,1]))$, $u_L \in C_b(\mathbf{R}_+; L^1(\mathbf{R}_+; L^\infty(0,1]))$, $u_R \in C_b(\mathbf{R}_+; L^1(\mathbf{R}_+; L^\infty[-1,0)))$.

(3) $v_1^0 \in L^\infty(-1,0)$, $v_2^0 \in L^\infty(0,1)$.

Then there exists $T_{max} > 0$ *such that the solution* (u_1,u_2,v_1,v_2) *of equations* (1.7)–(1.10) *with initial-boundary conditions* (1.11)–(1.13) *and continuous conditions* (1.15) (1.16) *is existent and unique. Moreover we have*

$$\begin{aligned} u_1 &\in C([0,T_{max}); L^\infty([-1,0]; L^1(\mathbf{R}_+; L^\infty[-1,1]))), \\ u_2 &\in C([0,T_{max}); L^\infty([0,1]; L^1(\mathbf{R}_+; L^\infty[-1,1]))), \\ v_1 &\in C([0,T_{max}); L^\infty(-1,0)), \qquad v_2 \in C([0,T_{max}); L^\infty(0,1)). \end{aligned}$$

If $T_{max} < \infty$, *then*

$$\limsup_{t\uparrow T_{max}} \sum_{j=1}^{2} \left\{ \|u_j(t,\cdot)\|_{L_x^\infty L_\nu^1 L_\omega^\infty} + + \|v_j(t,\cdot)\|_{L_x^\infty} \right\} = \infty. \tag{3.3}$$

Proof We use the the principle of contraction mapping to prove the existence and uniqueness of solution. Let $T > 0$ and set

$$B = \Big\{ (u_1,u_2,v_1,v_2) \Big| u_1 \in C([0,T]; L^\infty([-1,0]; L^1(\mathbf{R}_+; L^\infty[-1,1]))),$$

$$u_2 \in C([0,T]; L^\infty([0,1]; L^1(\mathbf{R}_+; L^\infty[-1,1]))),$$
$$v_1 \in C([0,T]; L^\infty(-1,0)), \quad v_2 \in C([0,T]; L^\infty(0,1)),$$
$$\sum_{j=1}^{2} \left(\|u_j\|_{L_T^\infty L_x^\infty L_\nu^1 L_\omega^\infty} + \|v_j\|_{L_T^\infty L_x^\infty}\right) \leqslant 2r_0\Big\}, \tag{3.4}$$

$$r_0 = 2\left\{\sum_{j=1}^{2} \|u_j^0\|_{L_x^\infty L_\nu^1 L_\omega^\infty} + \|v_j^0\|_{L_x^\infty}\right\} + 2\|u_L\|_{L_T^\infty L_\nu^1 L_\omega^\infty} + 2\|u_R\|_{L_T^\infty L_\nu^1 L_\omega^\infty}. \tag{3.5}$$

For all $(u_1, u_2, v_1, v_2) \in B$, we define a map

$$\Phi : (u_1, u_2, v_1, v_2) \to (U_1, U_2, V_1, V_2),$$

where $U_1 = \Phi_1(u_1, u_2, v_1, v_2)$ and $U_2 = \Phi_2(u_1, u_2, v_1, v_2)$ are the solution of problem (2.1)–(2.6) with

$$f_j = f_j(u_j, v_j) = \sigma_{js}(v_j)\bar{u}_j + \sigma_{ja}(v_1)B(v_j, \nu) - \sigma_j(v_j)u_j, \qquad j = 1, 2, \tag{3.6}$$

$V_1 = \Phi_3(u_1, u_2, v_1, v_2)$ and $V_2 = \Phi_4(u_1, u_2, v_1, v_2)$ are defined as follows:

$$V_j = v_j^0 + \int_0^t g_j(s, x)\mathrm{d}s, \qquad j = 1, 2. \tag{3.7}$$

$$g_j = g_j(u_j, v_j) = -\frac{1}{C_{jh}} \int_0^\infty \sigma_{ja}(v_j)(B(v_j, \nu) - \bar{u}_j)\mathrm{d}\nu, \qquad j = 1, 2. \tag{3.8}$$

We wish to find T such that the map $\Phi : B \to B$ is a strict contraction.

For $0 \leqslant t \leqslant T$, using (1.4) (2.13) (3.1) (3.2) and (3.6), we have

$$\begin{aligned}
&\sum_{j=1}^{2} \|U_j(t, \cdot)\|_{L_x^\infty L_\nu^1 L_\omega^\infty} \\
\leqslant & 2\sum_{j=1}^{2} \left\{\|u_j^0\|_{L_x^\infty L_\nu^1 L_\omega^\infty} + \int_0^T \|f_j(t, \cdot)\|_{L_x^\infty L_\nu^1 L_\omega^\infty}\mathrm{d}t\right\} \\
& + 2\|u_L\|_{L_T^\infty L_\nu^1 L_\omega^\infty} + 2\|u_R\|_{L_T^\infty L_\nu^1 L_\omega^\infty} \\
\leqslant & 2\sum_{j=1}^{2} \left\{\|u_j^0\|_{L_x^\infty L_\nu^1 L_\omega^\infty} + 2TM_j\|u_j\|_{L_T^\infty L_x^\infty L_\nu^1 L_\omega^\infty} + TM_j\beta\|v_j\|_{L_T^\infty L_x^\infty}^4\right\} \\
& + 2\|u_L\|_{L_T^\infty L_\nu^1 L_\omega^\infty} + 2\|u_R\|_{L_T^\infty L_\nu^1 L_\omega^\infty}.
\end{aligned} \tag{3.9}$$

Employing (1.4) (3.1) (3.2) (3.7) and (3.8), we obtain

$$\sum_{j=1}^{2} \|V_j(t, \cdot)\|_{L^\infty}$$

$$\leqslant \sum_{j=1}^{2} \|v_j^0\|_{L^\infty} + T \sum_{j=1}^{2} \frac{M_j}{C_{jh}} \{\|u_j\|_{L_T^\infty L_x^\infty L_\nu^1 L_\omega^\infty} + \beta \|v_j\|_{L_T^\infty L_x^\infty}^4\}, \qquad \forall t \in [0, T]. \quad (3.10)$$

Estimates (3.9) (3.10) imply

$$\sum_{j=1}^{2} \Big(\|U_j(t,\cdot)\|_{L_x^\infty L_\nu^1 L_\omega^\infty} + \|V_j(t,\cdot)\|_{L_x^\infty} \Big)$$
$$\leqslant r_0 + T5M_m \Big\{ \sum_{j=1}^{2} \|u_j\|_{L_T^\infty L_x^\infty L_\nu^1 L_\omega^\infty} + \beta \Big(\sum_{j=1}^{2} \|v_j\|_{L_T^\infty L_x^\infty} \Big)^4 \Big\}, \quad (3.11)$$

where $M_m = \max\{M_1, M_2, M_1/C_{1h}, M_1/C_{1h}\}$. Let us take $T \leqslant \dfrac{1}{10M_m(1+8\beta r_0^3)}$, then $\Phi: B \to B$.

For any (u_1, u_2, v_1, v_2), $(u_1', u_2', v_1', v_2') \in B$, we have $(0 \leqslant t \leqslant T)$

$$\sum_{j=1}^{2} \|\{\Phi_j(u_1,u_2,v_1,v_2) - \Phi_j(u_1',u_2',v_1',v_2')\}(t,\cdot)\|_{L_x^\infty L_\nu^1 L_\omega^\infty}$$
$$+ \sum_{j=3}^{4} \|\{\Phi_j(u_1,u_2,v_1,v_2) - \Phi_j(u_1',u_2',v_1',v_2')\}(t,\cdot)\|_{L_x^\infty}$$
$$\leqslant \sum_{j=1}^{2} \int_0^T \{2\|(f_j(u_j,v_j) - f_j(u_j',v_j'))(t,\cdot)\|_{L_x^\infty L_\nu^1 L_\omega^\infty}$$
$$+ \|(g_j(u_j,v_j) - g_j(u_j',v_j'))(t,\cdot)\|_{L_x^\infty}\}\mathrm{d}t. \quad (3.12)$$

Applying (1.3) (3.1) (3.2) and (3.6), we get

$$\big|f_j(u_j,v_j) - f_j(u_j',v_j')\big| \leqslant 2M_j|u_j - u_j'| + \Big\{ 2N_j|u_j| + N_jB(v_j,\nu)$$
$$+ M_j\Big|\partial_v \int_0^1 B\big(v_j + \theta(v_j' - v_j), \nu\big)\mathrm{d}\theta\Big| \Big\} |v_j - v_j'|. \quad (3.13)$$

Similarly, using (1.3) (1.4) (1.5) (3.1) (3.2) and (3.8), we obtain

$$\big|g_j(u_j,v_j) - g_j(u_j',v_j')\big|$$
$$\leqslant \frac{M_j}{C_{jh}} \|(u_j - u_j')(t,x,\cdot)\|_{L_\nu^1 L_\omega^\infty} + \frac{1}{C_{jh}} \Big\{ N_j\|u_j(t,x,\cdot)\|_{L_\nu^1 L_\omega^\infty} + N_j\beta v_j^4$$
$$+ M_j\beta\big(|v_j|^3 + |v_j' - v_j||v_j|^2 + |v_j' - v_j|^2|v_j| + |v_j' - v_j|^3\big) \Big\} |v_j - v_j'|. \quad (3.14)$$

Inserting (3.13) (3.14) into (3.12), we have

$$\sum_{j=1}^{2} \|\{\Phi_j(u_1,u_2,v_1,v_2) - \Phi_j(u_1',u_2',v_1',v_2')\}(t,\cdot)\|_{L_x^\infty L_\nu^1 L_\omega^\infty}$$

$$+\sum_{j=3}^{4}\|\{\Phi_j(u_1,u_2,v_1,v_2)-\Phi_j(u_1',u_2',v_1',v_2')\}(t,\cdot)\|_{L_x^\infty}$$

$$\leqslant T\sum_{j=1}^{2}\{5M_m\|u_j-u_j'\|_{L_T^\infty L_x^\infty L_\nu^1 L_\omega^\infty}+N_0\|v_j-v_j'\|_{L_T^\infty L_x^\infty}\}, \tag{3.15}$$

where $N_0=10N_mr_0+48N_m\beta r_0^4+360M_m\beta r_0^3$ and $N_m=\max\{N_1,N_2,N_1/C_{1h},N_2/C_{2h}\}$.

Let

$$T_0=\min\left\{\frac{1}{10M_m(1+8\beta r_0^3)},\frac{1}{2N_0}\right\},$$

and $T\leqslant T_0$. Then $\Phi: B\to B$ is a strict contraction. By the principle of contraction mapping, there exists a unique solution (u_1,u_2,v_1,v_2) of equations (1.7)–(1.10) with initial-boundary conditions (1.11)–(1.13) and continuous conditions (1.15) (1.16). Moreover we have

$$\begin{aligned}
u_1 &\in C\big([0,T_0];L^\infty([-1,0];L^1(\mathbf{R}_+;L^\infty[-1,1]))\big),\\
u_2 &\in C\big([0,T_0];L^\infty([0,1];L^1(\mathbf{R}_+;L^\infty[-1,1]))\big),\\
v_1 &\in C\big([0,T_0];L^\infty(-1,0)\big),\qquad v_2\in C\big([0,T_0];L^\infty(0,1)\big),
\end{aligned}$$

and

$$\max_{0\leqslant t\leqslant T_0}\sum_{j}^{2}\left(\|u_j(t,\cdot)\|_{L_x^\infty L_\nu^1 L_\omega^\infty}+\|v_j(t,\cdot)\|_{L_x^\infty}\right)\leqslant 2r_0. \tag{3.16}$$

Now we can solve the problem (1.7)–(1.10) (1.13) and (1.15) (1.16) at the initial time moment $t=T_0$ instead of $t=0$ with initial data $(u_1(T_0,x,\omega),u_2(T_0,x,\omega),v_1(T_0,x),v_2(T_0,x))$ instead of $(u_1^0,u_2^0,v_1^0,v_2^0)$ again. Then the time domain $[0,T_0]$ can be extended, which denotes by $[0,T_1]$, and the unique solution (u_1,u_2,v_1,v_2) of equations (1.7)–(1.10) is well defined for any $t\in[0,T_1]$. Here $T_1>T_0$. Repeating the above procedure again and again, we can obtain a series $\{T_n\}$ and the existence of $T_{max}=\sup T_n$.

If $T_{max}<\infty$ and

$$\overline{\lim}_{t\uparrow T_{\max}}\sum_{j}^{2}\left(\|u_j(t,\cdot)\|_{L_x^\infty L_\nu^1 L_\omega^\infty}+\|v_j(t,\cdot)\|_{L_x^\infty}\right)<\infty,$$

then

$$\begin{aligned}
K_0=&2\sup_{t\in[0,T_{\max})}\left\{\sum_{j=1}^{2}\|u_j(t,\cdot)\|_{L_x^\infty L_\nu^1 L_\omega^\infty}+\|u_L(t,\cdot)\|_{L_\nu^1 L_\omega^\infty}+\|u_R(t,\cdot)\|_{L_\nu^1 L_\omega^\infty}\right\}\\
&+2\sup_{t\in[0,T_{\max})}\sum_{j=1}^{2}\|v_j(t,\cdot)\|_{L_x^\infty}
\end{aligned}$$

$<\infty,$

Let

$$T_\delta = \min\left\{\frac{1}{10M_m(1+8K_0^3)}, \frac{1}{2(10N_mK_0+48N_m\beta K_0^4+360N_m\beta K_0^3)}\right\}.$$

Now we solve the problem (1.7)–(1.10) (1.13) and (1.15) (1.16) at the initial time moment $t=T_{\max}-\frac{1}{2}T_\delta$ instead of $t=0$ with initial data $\left(u_1\left(T_{\max}-\frac{1}{2}T_\delta,x,\omega\right),\right.$ $u_2(T_{\max}-\frac{1}{2}T_\delta,x,\omega), v_1\left(T_{\max}-\frac{1}{2}T_\delta,x\right), v_2\left(T_{\max}-\frac{1}{2}T_\delta,x\right)\Big)$ instead of $(u_1^0,u_2^0,v_1^0,v_2^0)$. Then the unique solution (u_1,u_2,v_1,v_2) of equations (1.7)– (1.10) is well defined for any $t\in\left[0,T_{\max}+\frac{1}{2}T_\delta\right]$. This is contradictory.

Therefore (3.3) is proved.

3.2 Global solution ignoring thermal diffusion

For some special initial and boundary conditions, we should prove that the solution of (1.7)–(1.10), which is constructed in Theorem 3.1, is global.

First we prove that the solution of the following problem

$$\partial_t w_1+\frac{M_{1a}}{C_{1h}}\int_0^\infty B(w_1,\nu)\mathrm{d}\nu=h_1(t,x),\quad -1<x<0, \tag{3.17}$$

$$\partial_t w_2+\frac{M_{2a}}{C_{2h}}\int_0^\infty B(w_2,\nu)\mathrm{d}\nu=h_2(t,x),\quad 0<x<1, \tag{3.18}$$

$$\begin{aligned} w_1(0,x)&=v_1^0(x),\quad x\in[-1,0],\\ w_2(0,x)&=v_2^0(x),\quad x\in[0,1],\end{aligned} \tag{3.19}$$

is global, where v_1^0, v_2^0, $h_1\geqslant 0$ and $h_2\geqslant 0$.

By the same argument as in the proof of Theorem 3.1, we can establish the following result.

Theorem 3.2 (Existence and Uniqueness of Global Solution) *Assume that v_1^0, v_2^0, h_1 and h_2 are non-negative, $v_1^0\in L^\infty(-1,0)$, $v_2^0\in L^\infty(0,1)$. For any $T>0$, if $h_1\in L^\infty\big(0,T;L^\infty(-1,0)\big)$, $h_2\in L^\infty\big(0,T;L^\infty(0,1)\big)$, then there exists a unique solution (w_1,w_2) of problem* (3.17)– (3.19). *Moreover we have $w_j\geqslant 0$ $(j=1,2)$ and*

$$w_1\in C\big([0,T];L^\infty(-1,0)\big),\qquad w_2\in C\big([0,T];L^\infty(0,1)\big). \tag{3.20}$$

Proof By equations (3.17)–(3.19), we derive that

$$\begin{aligned}w_j(t,x)=&v_j^0(x)\exp\Big\{-\frac{M_{ja}\beta}{C_{jh}}\int_0^t w_j^3(s,x)\mathrm{d}s\Big\}\\&+\int_0^t\exp\Big\{-\frac{M_{ja}\beta}{C_{jh}}\int_s^t w_j^3(\tau,x)\mathrm{d}\tau\Big\}h_j(s,x)\mathrm{d}s,\quad j=1,2,\end{aligned}\tag{3.21}$$

where we have used the fact (1.4). v_j^0, $h_j\geqslant 0$ and equation (3.21) imply that

$$0\leqslant w_j(t,x)\leqslant\|v_j^0\|_{L_x^\infty}+T\|h_j\|_{L_T^\infty L_x^\infty},\ j=1,2,\ \ \forall t\in[0,T].\tag{3.22}$$

By the standard method, we can prove that there exists a unique solution (w_1,w_2) of problem (3.17)– (3.19). Moreover w_1 and w_2 satisfy (3.20) and (3.22).

Now we prove the solution of (1.7)–(1.10) is non-negative for the non-negative initial and boundary conditions.

Lemma 3.1 (Non-negative) *Assume that u_L, u_R, u_j^0, $v_j^0\geqslant 0$ $(j=1,2)$, and the assumptions in Theorem 3.1 are satisfied. Then the solution (u_1,u_2,v_1,v_2), which is constructed in Theorem 3.1, satisfies the following estimate:*

$$\begin{aligned}&u_j(t,x,\nu,\omega)\geqslant 0,\quad v_j(t,x)\geqslant 0,\\&\quad\forall\, t\in[0,T_{\max}),\ \ x\in I_j,\ \ \nu\in[0,\infty),\ \ \omega\in[-1,1],\ \ j=1,2,\end{aligned}\tag{3.23}$$

Where $I_1=[-1,0]$ and $I_2=[0,1]$

Proof Let (u_1,u_2,v_1,v_2) be the solution of equations (1.7)– (1.10) with initial-boundary conditions (1.11)– (1.13) and continuous conditions (1.15) (1.16) solved in Theorem 3.1.

Considering $v_j(j=1,2)$ as known function, we re-solve the equations (1.7) (1.8). That is to solve the following integral equations

$$\begin{aligned}&w_1(t,x,\nu,\omega)\\=&\chi(t<t_{1d})u_1^0(x_{\omega t},\nu,\omega)E_1(t,0)+\chi(t\geqslant t_{1d})\{\chi(\omega>0)u_L(t-t_{1d},\nu,\omega)\\&+\chi(\omega<0)w_2(t-t_{1d},0,\nu,\omega)\}E_1(t_{1d},0)\\&+\int_{t-t\wedge t_{1d}}^t E_1(t,\tau)\{\sigma_{1s}(v_1)\bar{w}_1+\sigma_{1a}(v_1)B(v_1,\nu)\}(\tau,x_{\omega\tau},\nu)\mathrm{d}\tau,\end{aligned}\tag{3.24}$$

$$\begin{aligned}&w_2(t,x,\nu,\omega)\\=&\chi(t<t_{2d})u_2^0(x_{\omega t},\nu,\omega)E_2(t,0)+\chi(t\geqslant t_{2d})\{\chi(\omega<0)u_R(t-t_{2d},\nu,\omega)\end{aligned}$$

$$+\chi(\omega>0)w_1(t-t_{2d},0,\nu,\omega)\}E_2(t_{2d},0)$$
$$+\int_{t-t\wedge t_{2d}}^{t}E_2(t,\tau)\{\sigma_{2s}(v_2)\bar{w}_2+\sigma_{2a}(v_2)B(v_2,\nu)\}(\tau,x_{\omega\tau},\nu)\mathrm{d}\tau. \tag{3.25}$$

Here w_j is unknown function and v_j is known function, t_{1d} and t_{2d} are defined in (2.7) and (2.8) with $c_j=c$, $x_{\omega t}=x-c\omega t$, $t\wedge t_{jd}=\min(t,t_{jd})$, $\chi(t<t_{jd})$ equals one if $t<t_{jd}$ and zero otherwise, $j=1,2$,

$$E_j(t,\tau)=\exp\left\{-\int_{\tau}^{t}\sigma_j\big(v_j(\eta,x),x,\nu\big)\mathrm{d}\eta\right\},\qquad j=1,2. \tag{3.26}$$

Moreover, at the interface $x=0$, we have

$$\begin{aligned}&w_1(t,0,\nu,\omega)\\=&\chi(t<t_{1d}^0)u_1^0(x_{\omega,t}^0,\nu,\omega)E_1(t,0)+\chi(t\geqslant t_{1d}^0)u_L(t-t_{1d}^0,\nu,\omega)E_1(t_{1d}^0,0)\\&+\int_{t-t\wedge t_{1d}^0}^{t}E_1(t,\tau)\{\sigma_{1s}(v_1)\bar{w}_1+\sigma_{1a}(v_1)B(v_1,\nu)\}(\tau,x_{\omega,\tau}^0,\nu)\mathrm{d}\tau,\ \ \forall\omega>0,\end{aligned} \tag{3.27}$$

$$\begin{aligned}&w_2(t,0,\nu,\omega)\\=&\chi(t<t_{2d}^0)u_2^0(x_{\omega,t}^0,\nu,\omega)E_2(t,0)+\chi(t\geqslant t_{2d}^0)u_R(t-t_{2d}^0,\nu,\omega)E_2(t_{2d}^0,0)\\&+\int_{t-t\wedge t_{2d}^0}^{t}E_2(t,\tau)\{\sigma_{2s}(v_2)\bar{w}_2+\sigma_{2a}(v_2)B(v_2,\nu)\}(\tau,x_{\omega,\tau}^0,\nu)\mathrm{d}\tau,\ \ \forall\omega<0,\end{aligned} \tag{3.28}$$

where $x_{\omega,t}^0=-c\omega t$, $t_{1d}^0=\dfrac{1}{c\omega}$ and $t_{2d}^0=\dfrac{-1}{c\omega}$.

It is obvious that $(w_1,w_2)=(u_1,u_2)$ is the unique solution of equation (3.24)–(3.28), which the uniqueness follows as in the proof of Theorem 3.1 by the principle of contraction mapping.

On the other hand, we can re-solve equation (3.24)– (3.28) by iteration. We let

$$\begin{aligned}&w_1^{(1)}(t,x,\nu,\omega)\\=&\chi(t<t_{1d})u_1^0(x_{\omega t},\nu,\omega)E_1(t,0)+\chi(t\geqslant t_{1d})\chi(\omega>0)E_1(t_{1d},0)u_L(t-t_{1d},\nu,\omega)\\&+\chi(t\geqslant t_{1d})\chi(\omega<0)\{\chi(t-t_{1d}<t_{2d}^0)u_2^0(x_{\omega,t-t_{1d}}^0,\nu,\omega)E_2(t-t_{1d},0)\\&+\chi(t-t_{1d}\geqslant t_{2d}^0)u_R(t-t_{1d}-t_{2d}^0,\nu,\omega)E_2(t_{2d}^0,0)\\&+\int_{t-t_{1d}-(t-t_{1d})\wedge t_{2d}^0}^{t-t_{1d}}E_2(t-t_{1d},\tau)\{\sigma_{2a}(v_2)B(v_2,\nu)\}(\tau,x_{\omega,\tau}^0,\nu)d\tau\}E_1(t_{1d},0)\\&+\int_{t-t\wedge t_{1d}}^{t}E_1(t,\tau)\{\sigma_{1a}(v_1)B(v_1,\nu)\}(\tau,x_{\omega\tau},\nu)\mathrm{d}\tau,\end{aligned} \tag{3.29}$$

$$
\begin{aligned}
& w_2^{(1)}(t,x,\nu,\omega) \\
=& \chi(t<t_{2d})u_2^0(x_{\omega t},\nu,\omega)E_2(t,0)+\chi(t\geqslant t_{2d})\chi(\omega<0)E_2(t_{2d},0)u_R(t-t_{2d},\nu,\omega) \\
& +\chi(t\geqslant t_{2d})\chi(\omega>0)\{\chi(t-t_{2d}<t_{1d}^0)u_1^0(x_{\omega,t-t_{2d}}^0,\nu,\omega)E_1(t-t_{2d},0) \\
& +\chi(t-t_{2d}\geqslant t_{1d}^0)u_L(t-t_{2d}-t_{1d}^0,\nu,\omega)E_1(t_{1d}^0,0) \\
& +\int_{t-t_{2d}-(t-t_{2d})\wedge t_{1d}^0}^{t-t_{2d}} E_1(t-t_{2d},\tau)\{\sigma_{1a}(v_1)B(v_1,\nu)\}(\tau,x_{\omega,\tau}^0,\nu)d\tau\}E_2(t_{2d},0) \\
& +\int_{t-t\wedge t_{2d}}^{t} E_2(t,\tau)\{\sigma_{2a}(v_2)B(v_2,\nu)\}(\tau,x_{\omega\tau},\nu)\mathrm{d}\tau,
\end{aligned} \tag{3.30}
$$

$$
\begin{aligned}
& w_1^{(n+1)}(t,x,\nu,\omega) \\
=& \int_{t-t\wedge t_{1d}}^{t} E_1(t,\tau)\{\sigma_{1s}(v_1)\bar{w}_1^{(n)}\}(\tau,x_{\omega\tau},\nu)d\tau+\chi(t\geqslant t_{1d})\chi(\omega<0)\cdot \\
& E_1(t_{1d},0)\int_{t-t_{1d}-(t-t_{1d})\wedge t_{2d}^0}^{t-t_{1d}} E_2(t-t_{1d},\tau)\{\sigma_{2s}(v_2)\bar{w}_2^{(n)}\}(\tau,x_{\omega,\tau}^0,\nu)\mathrm{d}\tau,
\end{aligned} \tag{3.31}
$$

$$
\begin{aligned}
& w_2^{(n+1)}(t,x,\nu,\omega) \\
=& \int_{t-t\wedge t_{2d}}^{t} E_2(t,\tau)\{\sigma_{2s}(v_2)\bar{w}_2^{(n)}\}(\tau,x_{\omega\tau},\nu)d\tau+\chi(t\geqslant t_{2d})\chi(\omega>0)\cdot \\
& E_2(t_{2d},0)\int_{t-t_{2d}-(t-t_{2d})\wedge t_{1d}^0}^{t-t_{2d}} E_1(t-t_{2d},\tau)\{\sigma_{1s}(v_1)\bar{w}_1^{(n)}\}(\tau,x_{\omega,\tau}^0,\nu)\mathrm{d}\tau,
\end{aligned} \tag{3.32}
$$

where $n=1,\ 2,\ \cdots$. u_L, u_R, $u_1^0\geqslant 0$ and $u_2^0\geqslant 0$ imply that

$$
w_j^{(n)}\geqslant 0,\qquad j=1,\ 2,\ \ n=1,\ 2,\ \cdots. \tag{3.33}
$$

For any $T\in(0,T_{\max})$, we find that

$$
\|w_j^{(1)}(t,\cdot)\|_{L_x^\infty L_\nu^1 L_\omega^\infty}\leqslant C(T),\quad \forall t\in[0,T],\ \ j=1,2 \tag{3.34}
$$

and inductively

$$
\|w_j^{(n)}(t,\cdot)\|_{L_x^\infty L_\nu^1 L_\omega^\infty}\leqslant \frac{(C'(T)t)^n}{n!}C(T),\quad \forall t\in[0,T],\ \ j=1,2, \tag{3.35}
$$

where $C(T)$ and $C'(T)$ are constants. Since the solutions w_1^* and w_2^* of equations (3.24)– (3.28) are formally

$$
w_j^*(t,x,\nu,\omega)=\sum_{n=1}^{\infty} w_j^{(n)}(t,x,\nu,\omega),\quad \forall t\in[0,T],\ \ j=1,2,
$$

our estimates above show that these series converge and $w_j^*(t,x,\nu,\omega) \geqslant 0$ exists for all $t \in [0,T]$, $x \in I_j$, $\nu \in \mathbf{R}_+$ and $\omega \in [-1,1]$, where $I_1 = [-1,0]$, $I_2 = [0,1]$, $j = 1,2$.

By the uniqueness of solution of equation (3.24)– (3.28), we have $u_j(t,x,\nu,\omega) = w_j^*(t,x,\nu,\omega) \geqslant 0$ for all $t \in [0,T]$, $j = 1,2$. Since $T \in (0,T_{\max})$ is arbitrary, we can obtain that $u_j(t,x,\nu,\omega) \geqslant 0$ for all $t \in [0,T_{\max})$, $j = 1,2$.

Let (w_1,w_2) be the solution of problem (3.17)– (3.19) with

$$h_j(t,x) = \frac{1}{C_{jh}} \int_0^\infty \sigma_{ja}(v_j)\bar{u}_j(t,x,\nu)\mathrm{d}\nu, \qquad j = 1,2.$$

By Theorem 3.2, we derive that

$$w_j(t,x) \geqslant 0, \ \ j = 1,2, \ \ \forall t \in [0,T_{\max}).$$

Let $V_j = v_j - w_j$, $j = 1,2$. Then V_j satisfies the following equations

$$\begin{aligned} &\partial_t V_1 + \frac{1}{C_{1h}} \int_0^\infty \sigma_{1a}(v_1) \int_0^1 \partial_v B\big(w_1 + \theta(v_1 - w_1),\nu\big)\mathrm{d}\theta\mathrm{d}\nu V_1 \\ =&\frac{1}{C_{1h}} \int_0^\infty \big(M_{1a} - \sigma_{1a}(v_1)\big)B(w_1,\nu)\mathrm{d}\nu, \ \ -1 < x < 0, \end{aligned} \tag{3.36}$$

$$\begin{aligned} &\partial_t V_2 + \frac{1}{C_{2h}} \int_0^\infty \sigma_{2a}(v_2) \int_0^1 \partial_v B\big(w_2 + \theta(v_2 - w_2),\nu\big)\mathrm{d}\theta\mathrm{d}\nu V_2 \\ =&\frac{1}{C_{2h}} \int_0^\infty \big(M_{2a} - \sigma_{2a}(v_2)\big)B(w_2,\nu)\mathrm{d}\nu, \ \ 0 < x < 1, \end{aligned} \tag{3.37}$$

$$\begin{aligned} V_1(0,x) = 0, \quad x \in [-1,0], \\ V_2(0,x) = 0, \quad x \in [0,1]. \end{aligned} \tag{3.38}$$

Moreover we derive

$$\begin{aligned} V_j(t,x) = &\int_0^t \exp\left\{ - \int_\eta^t \frac{1}{C_{jh}} \left(\int_0^\infty \sigma_{ja}(v_j) \int_0^1 \partial_v B(w_{j\theta},\nu)\mathrm{d}\theta\mathrm{d}\nu \right)(\tau,x)\mathrm{d}\tau \right\} \\ &\cdot \frac{1}{C_{jh}} \left(\int_0^\infty \big(M_{ja} - \sigma_{ja}(v_j)\big)B(w_j,\nu)\mathrm{d}\nu \right)(\eta,x)\mathrm{d}\eta \geqslant 0, \qquad j = 1,2, \end{aligned} \tag{3.39}$$

where $w_{j\theta} = w_j + \theta(v_j - w_j)$, $j = 1,2$. Equation (3.39) implies that

$$v_j(t,x) \geqslant w_j(t,x) \geqslant 0, \ \ \forall t \in [0,T_{\max}).$$

(3.23) is proved.

To prove the solution of (1.7)–(1.10) is global, the estimate from above is needed. Therefore we assume that there exists a constant $v_0 > 0$ such that

$$\max\{\|u_L\|_{L_t^\infty L_\omega^\infty},\ \|u_R\|_{L_t^\infty L_\omega^\infty},\ \|u_1^0\|_{L_x^\infty L_\omega^\infty},\ \|u_2^0\|_{L_x^\infty L_\omega^\infty}\} \leqslant B(v_0, \nu). \tag{3.40}$$

Remark 3.1 *The Planckian initial and boundary conditions, which is discussed in* [27] *by Larsen et al., satisfies this condition* (3.40).

Let

$$u_s = B(v_s, \nu),\quad v_s = \max\{\|v_1^0\|_{L_x^\infty L_\omega^\infty},\ \|v_2^0\|_{L_x^\infty L_\omega^\infty},\ v_0\}. \tag{3.41}$$

For the solution (u_1, u_2, v_1, v_2) of (1.7)–(1.10), we will prove that u_j is bounded by u_s from above and v_j is bounded by v_s from above.

Lemma 3.2 (Upper Bound) *Assume that u_L, u_R, $u_j^0 \geqslant 0$ and $v_j^0 \geqslant 0$ $(j = 1, 2)$ satisfy* (3.40) *and the assumptions in Theorem 3.1. Then the solution (u_1, u_2, v_1, v_2), which is constructed in Theorem 3.1, satisfies the following estimate:*

$$\begin{aligned} &0 \leqslant u_j(t, x, \nu, \omega) \leqslant u_s,\quad 0 \leqslant v_j(t, x) \leqslant v_s,\\ &\forall t \in [0, T_{\max}),\ x \in I_j,\ \nu \in \mathbf{R}_+,\ \omega \in [-1, 1],\ j = 1, 2, \end{aligned} \tag{3.42}$$

where $I_1 = [-1, 0]$ and $I_2 = [0, 1]$.

Proof Let (u_1, u_2, v_1, v_2) be the solution of equations (1.7)– (1.10) with initial-boundary conditions (1.11)– (1.13) and continuous conditions (1.15) (1.16) solved in Theorem 3.1.

We define

$$\begin{aligned} U_j(t, x, \omega) = u_s - u_j(t, x, \nu, \omega),\ \ &V_j(t, x) = v_s - v_j(t, x),\\ &j = 1, 2,\ \ \forall t \in [0, T_{\max}). \end{aligned} \tag{3.43}$$

Then (U_1, U_2, V_1, V_2) is the unique solution of the following equations

$$\partial_t U_1 + c\omega\partial_x U_1 + \sigma_1(v_1)U_1 = \sigma_{1s}(v_1)\bar{U}_1 + \sigma_{1a}(v_1)\int_0^1 \partial_v B(v_{1\theta}, \nu)\mathrm{d}\theta V_1, \tag{3.44}$$

$$\partial_t U_2 + c\omega\partial_x U_2 + \sigma_2(v_2)U_2 = \sigma_{2s}(v_2)\bar{U}_2 + \sigma_{2a}(v_2)\int_0^1 \partial_v B(v_{2\theta}, \nu)\mathrm{d}\theta V_2, \tag{3.45}$$

$$\partial_t V_1 + \frac{1}{C_{1h}}\int_0^\infty \sigma_{1a}(v_1)\left\{\int_0^1 \partial_v B(v_{1\theta}, \nu)\mathrm{d}\theta V_1 - \bar{U}_1\right\}\mathrm{d}\nu = 0, \tag{3.46}$$

$$\partial_t V_2 + \frac{1}{C_{2h}}\int_0^\infty \sigma_{2a}(v_2)\left\{\int_0^1 \partial_v B(v_{2\theta}, \nu)\mathrm{d}\theta V_2 - \bar{U}_2\right\}\mathrm{d}\nu = 0, \tag{3.47}$$

$$\begin{aligned} U_1(0,x,\nu,\omega) &= U_1^0 = u_s - u_1^0(x,\nu,\omega), \quad x\in[-1,0],\\ U_2(0,x,\nu,\omega) &= U_2^0 = u_s - u_2^0(x,\nu,\omega), \quad x\in[0,1], \end{aligned} \tag{3.48}$$

$$\begin{aligned} V_1(0,x) &= V_1^0 = v_s - v_1^0(x), \quad x\in[-1,0],\\ V_2(0,x) &= V_2^0 = v_s - v_2^0(x), \quad x\in[0,1], \end{aligned} \tag{3.49}$$

$$\begin{aligned} U_1(t,x,\nu,\omega)|_{x=-1} &= U_L = u_s - u_L(t,\nu,\omega), \quad \text{for } \omega>0,\\ U_2(t,x,\nu,\omega)|_{x=1} &= U_R = u_s - u_R(t,\nu,\omega), \quad \text{for } \omega<0, \end{aligned} \tag{3.50}$$

$$U_1(t,x,\nu,\omega)|_{x=0} \leftleftarrows U_2(t,x,\nu,\omega)|_{x=0}, \quad \text{for } \omega<0, \tag{3.51}$$

$$U_2(t,x,\nu,\omega)|_{x=0} \leftleftarrows U_1(t,x,\nu,\omega)|_{x=0}, \quad \text{for } \omega>0, \tag{3.52}$$

where $v_{j\theta} = v_j + \theta(v_s - v_j)$, $j=1,2$. By the principle of contraction mapping, the uniqueness of (U_1,U_2,V_1,V_2) follows as in the proof of Theorem 3.1.

It implies (3.42) that U_j and V_j $(j=1,2)$ are non- negative. The iteration argument below, which is similar to the iteration in the proof of Lemma 3.1, leads directly to the desired result as we now show. Let $U_j^{(n)} = U_j^{(n)}(t,x,\nu,\omega)$, $V_j^{(n)} = V_j^{(n)}(t,x)$ and

$$\begin{aligned} &U_1^{(0)}(t,x,\nu,\omega)\\ =&\chi(t<t_{1d})U_1^0(x_{\omega t},\cdot)E_1(t,0) + \chi(t\geqslant t_{1d})\chi(\omega>0)E_1(t_{1d},0)U_L(t-t_{1d},\cdot)\\ &+\chi(t\geqslant t_{1d})\chi(\omega<0)\{\chi(t-t_{1d}<t_{2d}^0)U_2^0(x_{\omega,t-t_{1d}}^0,\cdot)E_2(t-t_{1d},0)\\ &+\chi(t-t_{1d}\geqslant t_{2d}^0)U_R(t-t_{1d}-t_{2d}^0,\cdot)E_2(t_{2d}^0,0)\}E_1(t_{1d},0), \end{aligned} \tag{3.53}$$

$$\begin{aligned} &U_2^{(0)}(t,x,\nu,\omega)\\ =&\chi(t<t_{2d})U_2^0(x_{\omega t},\cdot)E_2(t,0) + \chi(t\geqslant t_{2d})\chi(\omega<0)E_2(t_{2d},0)U_R(t-t_{2d},\cdot)\\ &+\chi(t\geqslant t_{2d})\chi(\omega>0)\{\chi(t-t_{2d}<t_{1d}^0)U_1^0(x_{\omega,t-t_{2d},\cdot}^0)E_1(t-t_{2d},0)\\ &+\chi(t-t_{2d}\geqslant t_{1d}^0)U_L(t-t_{2d}-t_{1d}^0,\cdot)E_1(t_{1d}^0,0)\}E_2(t_{2d},0), \end{aligned} \tag{3.54}$$

$$\begin{cases} \partial_t V_1^{(0)} + \dfrac{1}{C_{1h}}\displaystyle\int_0^\infty \sigma_{1a}(v_1)\int_0^1 \partial_v B(v_{1\theta},\nu)\mathrm{d}\theta V_1^{(0)}\mathrm{d}\nu = 0,\\ V_1^{(0)}(t,x)|_{t=0} = V_1^0, \end{cases} \tag{3.55}$$

$$\begin{cases} \partial_t V_2^{(0)} + \dfrac{1}{C_{2h}}\displaystyle\int_0^\infty \sigma_{2a}(v_2)\int_0^1 \partial_v B(v_{2\theta},\nu)\mathrm{d}\theta V_2^{(0)}\mathrm{d}\nu = 0,\\ V_2^{(0)}(t,x)|_{t=0} = V_2^0, \end{cases} \tag{3.56}$$

$$U_1^{(n+1)}(t,x,\nu,\omega)$$

$$
\begin{aligned}
&= \int_{t-t\wedge t_{1d}}^{t} E_1(t,\tau)\left(\sigma_{1s}(v_1)\bar{U}_1^{(n)}+\sigma_{1a}(v_1)\int_0^1 \partial_v B(v_{1\theta},\nu)\mathrm{d}\theta V_1^{(n)}\right)(\tau,x_{\omega\tau},\nu)\mathrm{d}\tau\\
&\quad+\chi(t\geqslant t_{1d})\chi(\omega<0)E_1(t_{1d},0)\int_{t-t_{1d}-(t-t_{1d})\wedge t_{2d}^0}^{t-t_{1d}} E_2(t-t_{1d},\tau)\\
&\quad\cdot\left(\sigma_{2s}(v_2)\bar{U}_2^{(n)}+\sigma_{2a}(v_2)\int_0^1 \partial_v B(v_{2\theta},\nu)d\theta V_2^{(n)}\right)(\tau,x_{\omega,\tau}^0,\nu)\mathrm{d}\tau,
\end{aligned}
\tag{3.57}
$$

$$
\begin{aligned}
&U_2^{(n+1)}(t,x,\nu,\omega)\\
&= \int_{t-t\wedge t_{2d}}^{t} E_2(t,\tau)\left(\sigma_{2s}(v_2)\bar{U}_2^{(n)}+\sigma_{2a}(v_2)\int_0^1 \partial_v B(v_{2\theta},\nu)\mathrm{d}\theta V_2^{(n)}\right)(\tau,x_{\omega\tau},\nu)\mathrm{d}\tau\\
&\quad+\chi(t\geqslant t_{2d})\chi(\omega>0)E_2(t_{2d},0)\int_{t-t_{2d}-(t-t_{2d})\wedge t_{1d}^0}^{t-t_{2d}} E_1(t-t_{2d},\tau)\\
&\quad\cdot\left(\sigma_{1s}(v_1)\bar{U}_1^{(n)}+\sigma_{1a}(v_1)\int_0^1 \partial_v B(v_{1\theta},\nu)\mathrm{d}\theta V_1^{(n)}\right)(\tau,x_{\omega,\tau}^0,\nu)\mathrm{d}\tau,
\end{aligned}
\tag{3.58}
$$

$$
\begin{cases}
\partial_t V_1^{(n+1)} = \dfrac{1}{C_{1h}}\displaystyle\int_0^\infty \sigma_{1a}(v_1)\left\{\overline{U_1^{(n)}}-\int_0^1 \partial_v B(v_{1\theta},\nu)\mathrm{d}\theta V_1^{(n+1)}\right\}\mathrm{d}\nu,\\
V_1^{(n+1)}(t,x)|_{t=0}=0,
\end{cases}
\tag{3.59}
$$

$$
\begin{cases}
\partial_t V_2^{(n+1)} = \dfrac{1}{C_{2h}}\displaystyle\int_0^\infty \sigma_{2a}(v_2)\left\{\overline{U_2^{(n)}}-\int_0^1 \partial_v B(v_{2\theta},\nu)\mathrm{d}\theta V_2^{(n+1)}\right\}\mathrm{d}\nu,\\
V_2^{(n+1)}(t,x)|_{t=0}=0,
\end{cases}
\tag{3.60}
$$

where $E_j(t,\tau)$ is defined in (3.26), $n=0,1,2,\cdots$. U_L, U_R, $U_j^0\geqslant 0$ and $V_j^0\geqslant 0$ imply that

$$
U_j^{(0)}\geqslant 0,\qquad j=1,\ 2,
\tag{3.61}
$$

$$
V_j^{(0)} = V_j^0 e_j(t,0)\geqslant 0,\qquad j=1,2.
\tag{3.62}
$$

Let

$$
e_j(t,\tau)=\exp\left\{-\frac{1}{C_{jh}}\int_\tau^t\left(\int_0^\infty \sigma_{ja}(v_j)\int_0^1\partial_v B(v_{j\theta},\nu)\mathrm{d}\theta\mathrm{d}\nu\right)(\eta,x)\mathrm{d}\eta\right\},\quad j=1,2.
$$

By induction, we have

$$
\begin{aligned}
&U_j^{(n)}(t,x,,\nu,\omega)\geqslant 0,\\
&V_j^{(n)}(t,x)=\int_0^t\left(e_j(t,\tau)\int_0^\infty\frac{\sigma_{ja}(v_j)}{C_{jh}}\overline{U_j^{(n-1)}}d\nu\right)(\tau,x)\mathrm{d}\tau\geqslant 0,\\
&\forall t\in[0,t_{\max}),\ x\in I_j,\ \nu\in\mathbf{R}_+,\ \omega\in[-1,1],\ j=1,2,\ n=1,2,\cdots.
\end{aligned}
\tag{3.63}
$$

For any $T \in (0, T_{\max})$ and $t \in [0, T]$, we find that

$$\|U_j^{(0)}(t,\cdot)\|_{L_x^\infty L_\nu^1 L_\omega^\infty} \leqslant 4\|u_s\|_{L_\nu^1}, \quad j=1,2, \tag{3.64}$$

$$\|V_j^{(0)}(t,\cdot)\|_{L_x^\infty} \leqslant v_s, \quad j=1,2, \tag{3.65}$$

$$\|U_j^{(1)}(t,\cdot)\|_{L_{x,\omega}^\infty} \leqslant (M_{1s}+M_{2s})4\|u_s\|_{L_\nu^1}t + (M_{1B}+M_{2B})v_s t, \quad j=1,2, \tag{3.66}$$

$$\|V_j^{(1)}(t,\cdot)\|_{L_x^\infty} \leqslant \frac{M_{ja}}{C_{jh}}4\|u_s\|_{L_\nu^1}t, \quad j=1,2, \tag{3.67}$$

and inductively

$$\|U^{(n)}(t,\cdot)\|_{L_{x,\omega}^\infty} \leqslant \frac{C_1^n t^n}{n!}(4\|u_s\|_{L_\nu^1}+v_s), \quad j=1,2,\ n\geqslant 2, \tag{3.68}$$

$$\|V_j^{(n+1)}(t,\cdot)\|_{L_x^\infty} \leqslant \frac{M_{ja}}{C_{jh}}\int_0^t \|U_j^{(n)}(s,\cdot)\|_{L_x^\infty L_\nu^1 L_\omega^\infty}\mathrm{d}s, \quad j=1,2,\ n\geqslant 1, \tag{3.69}$$

where constants $M_{jB} = M_{ja}\|\int_0^1 \partial_v B(v_{j\theta},\nu)\mathrm{d}\theta\|_{L_x^\infty L_\nu^1}$, $j=1,2$,

$$C_1 = \sum_{j=1}^2\left(M_{js}+M_{jB}+\frac{M_{ja}}{C_{jh}}\right).$$

Since the solution $(U_1^*, U_2^*, V_1^*, V_2^*)$ of the problem (3.44)–(3.52) is formally

$$U_j^*(t,x,\nu,\omega) = \sum_{n=0}^\infty U_j^{(n)}(t,x,\nu,\omega), \quad j=1,2,\ \forall t\in[0,T],$$

$$V_j^*(t,x) = \sum_{n=0}^\infty V_j^{(n)}(t,x), \quad j=1,2,\ \forall t\in[0,T],$$

our estimates above show that these four series converge and $(U_1^*, U_2^*, V_1^*, V_2^*)$ exists for all $t\in[0,T]$. Moreover, $U_j^*(t,x,\nu,\omega)\geqslant 0$ and $V_j^*(t,x)\geqslant 0$ for all $t\in[0,T]$, $x\in I_j$, $\nu\in\mathbf{R}_+$, $\omega\in[-1,1]$, $j=1,2$.

By the uniqueness of solution of problem (3.44)– (3.52), we have $U_j(t,x,\nu,\omega) = U_j^*(t,x,\nu,\omega)\geqslant 0$ and $V_j(t,x)=V_j^*(t,x)\geqslant 0$ for all $t\in[0,T]$, $x\in I_j$, $\nu\in\mathbf{R}_+$, $\omega\in[-1,1]$, $j=1,2$. Since $T\in(0,T_{\max})$ is arbitrary, we can derive that $U_j(t,x,\nu,\omega)\geqslant 0$ and $V_j(t,x)\geqslant 0$ for all $t\in[0,T_{\max})$, $x\in I_j$, $\nu\in\mathbf{R}_+$, $\omega\in[-1,1]$, $j=1,2$. It implies (3.42) that U_j and V_j $(j=1,2)$ are non-negative.

As a consequence of (3.42), the solution of equations (1.7)–(1.10) with initial-boundary conditions (1.11)–(1.13) and continuous conditions (1.15) (1.16) is global.

Theorem 3.3 (Global Solution)*quad Assume that* u_L, u_R, u_j^0 *and* v_j^0 $(j=1,2)$ *satisfy the assumptions in Lemma 3.2. Then the solution* (u_1,u_2,v_1,v_2), *which is constructed in Theorem 3.1, is global. Moreover, if*

$$u_j^0(x,\nu,\omega)\in C_B(I_j;L^1([0,\infty);L^\infty[-1,1])),\quad v_j^0(x)\in C_B(I_j),\quad j=1,2,$$

where $I_1=[-1,0]$ *and* $I_2=[0,1]$, *then*

$$\begin{aligned}&u_j(t,x,\nu,\omega)\in C([0,\infty);C_B(I_j;L^1([0,\infty);L^\infty[-1,1]))),\\&v_j(t,x)\in C([0,\infty);C_B(I_j)),\qquad j=1,2.\end{aligned}$$

4 Radiative transfer equations with thermal diffusion

4.1 Local solution with thermal diffusion

We shall consider the existence and uniqueness of solution of radiative transfer equations (1.7) (1.8) (1.19) (1.20) with initial-boundary conditions (1.11)–(1.13) (1.21) and continuous conditions (1.15) (1.16) (1.18) at the interface.

First we establish the existence and uniqueness of local solution of radiative transfer equations with thermal diffusion.

Theorem 4.1 (Existence and Uniqueness of Local Solution) *Assume that condition* (3.1) *and the following conditions are satisfied:*

(1) u_1^0, u_2^0, u_L *and* u_R *satisfy* (1.14).

(2) v_1^0, v_2^0, v_L *and* v_R *satisfy* (1.22).

(3) $u_1^0\in L^\infty([-1,0];L^1(\mathbf{R}_+;L^\infty[-1,1]))$, $u_2^0\in L^\infty([0,1];L^1(\mathbf{R}_+;L^\infty[-1,1]))$, $u_L\in C_b(\mathbf{R}_+;L^1(\mathbf{R}_+;L^\infty(0,1]))$, $u_R\in C_b(\mathbf{R}_+;L^1(\mathbf{R}_+;L^\infty[-1,0)))$.

(4) $v_1^0\in H^1(-1,0)$, $v_2^0\in H^1(0,1)$, $v_L\in W^{1,\infty}(\mathbf{R}_+)$, $v_R\in W^{1,\infty}(\mathbf{R}_+)$.

Then there exists $T_{max}>0$ *such that the solution* (u_1,u_2,v_1,v_2) *of equations* (1.7) (1.8) (1.19) (1.20) *with initial-boundary conditions* (1.11)– (1.13) (1.21) *and continuous conditions* (1.15) (1.16) (1.18) *is existent and unique. Moreover we have*

$$\begin{aligned}u_1\in&C\big([0,T_{\max});L^\infty([-1,0];L^1(\mathbf{R}_+;L^\infty[-1,1]))\big),\\u_2\in&C\big([0,T_{\max});L^\infty([0,1];L^1(\mathbf{R}_+;L^\infty[-1,1]))\big),\\v_1\in&C\big([0,T_{\max});H^1(-1,0)\big)\cap L^2(0,T;H^2(-1,0)),\quad\forall T\in[0,T_{\max}),\\v_2\in&C\big([0,T_{\max});H^1(0,1)\big)\cap L^2(0,T;H^2(0,1)),\quad\forall T\in[0,T_{\max}).\end{aligned}$$

If $T_{\max}<\infty$, *then*

$$\limsup_{t\uparrow T_{\max}}\sum_{j=1}^{2}\{\|u_j(t,\cdot)\|_{L_x^\infty L_\nu^1 L_\omega^\infty}+\|v_j(t,\cdot)\|_{H_x^1}\}=\infty.\tag{4.1}$$

Proof We use the the principle of contraction mapping to prove the existence and uniqueness of solution. Let $T > 0$ and set

$$B = \Big\{(u_1, u_2, v_1, v_2)\Big| u_j \in C([0,T]; L^\infty(I_j; L^1(\mathbf{R}_+; L^\infty[-1,1]))),\\ v_j \in C([0,T]; H^1(I_j)) \cap L^2(0,T; H^2(I_j)),\\ \sum_{j=1}^{2}\left(\|u_j\|_{L_T^\infty L_x^\infty L_\nu^1 L_\omega^\infty} + \|v_j\|_{L_T^\infty H_x^1} + \|\partial_x^2 v_j\|_{L_T^2 L_x^2}\right) \leqslant 2r_0\Big\}, \tag{4.2}$$

$$\begin{aligned} r_0 =& 2\Big\{\sum_{j=1}^{2}\|u_j^0\|_{L_x^\infty L_\nu^1 L_\omega^\infty} + \|u_L\|_{L_T^\infty L_\nu^1 L_\omega^\infty} + \|u_R\|_{L_T^\infty L_\nu^1 L_\omega^\infty}\Big\} \\ &+ 60\frac{d_M^2}{d_m}\Big\{\sum_{j=1}^{2}\|v_j^0\|_{H^1} + \|v_L\|_{W^{1,\infty}(0,T)} + \|v_R\|_{W^{1,\infty}(0,T)}\Big\}, \end{aligned} \tag{4.3}$$

where $I_1 = [-1, 0]$ and $I_2 = [0, 1]$. For all $(u_1, u_2, v_1, v_2) \in B$, we define a map

$$\Phi: (u_1, u_2, v_1, v_2) \to (U_1, U_2, V_1, V_2).$$

Here $U_1 = \Phi_1(u_1, u_2, v_1, v_2)$ and $U_2 = \Phi_2(u_1, u_2, v_1, v_2)$ are the solution of problem (2.1)–(2.6), $V_1 = \Phi_3(u_1, u_2, v_1, v_2)$ and $V_2 = \Phi_4(u_1, u_2, v_1, v_2)$ are the solution of problem (2.14)–(2.18) with $a(t,x) = b(t,x) = 0$, where f_1 and f_2 are defined in (3.6), $F = g_1$, $G = g_2$, g_1 and g_2 are defined in (3.8). We want to find T such that the map $\Phi: B \to B$ is a strict contraction.

For $0 \leqslant t \leqslant T$, using (2.13) (3.6) (3.1) (3.2) and (1.4), we have

$$\begin{aligned} &\sum_{j=1}^{2}\|U_j(t,\cdot)\|_{L_x^\infty L_\nu^1 L_\omega^\infty} \\ \leqslant& 2\sum_{k=1}^{2}\Big\{\|u_k^0\|_{L_x^\infty L_\nu^1 L_\omega^\infty} + \int_0^T \|f_k(t,\cdot)\|_{L_x^\infty L_\nu^1 L_\omega^\infty}\mathrm{d}t\Big\} \\ &+ 2\|u_L\|_{L_T^\infty L_\nu^1 L_\omega^\infty} + 2\|u_R\|_{L_T^\infty L_\nu^1 L_\omega^\infty} \\ \leqslant& 2\sum_{k=1}^{2}\{\|u_k^0\|_{L_x^\infty L_\nu^1 L_\omega^\infty} + 2TM_k\|u_k\|_{L_T^\infty L_x^\infty L_\nu^1 L_\omega^\infty} + TM_k\beta\|v_k\|_{L_T^\infty L_x^\infty}^4\} \\ &+ 2\|u_L\|_{L_T^\infty L_\nu^1 L_\omega^\infty} + 2\|u_R\|_{L_T^\infty L_\nu^1 L_\omega^\infty}. \end{aligned} \tag{4.4}$$

Employing (2.62) (3.8) (3.1) (3.2) and (1.4), we obtain ($\forall t \in [0,T]$)

$$\sum_{j=1}^{2}\{\|V_j(t,\cdot)\|_{H_x^1}^2 + \|\partial_x^2 V_j\|_{L_t^2 L_x^2}^2\}$$

$$
\begin{aligned}
&\leqslant \frac{15d_M}{d_m} e^T \left\{ 2\sum_{j=1}^{2} \|v_j^0\|_{H_x^1}^2 + (73d_M^2 T + 4)\left(\|v_L\|_{W_T^{1,\infty}}^2 + \|v_R\|_{W_T^{1,\infty}}^2\right) \right\} \\
&\quad + \frac{30d_M}{d_m} e^T \int_0^T \{\|g_1(s,\cdot)\|_{L_x^2}^2 + \|g_2(s,\cdot)\|_{L_x^2}^2\} ds \\
&\leqslant \frac{15d_M}{d_m} e^T \left\{ 2\sum_{j=1}^{2} \|v_j^0\|_{H_x^1}^2 + (73d_M^2 T + 4)\left(\|v_L\|_{W_T^{1,\infty}}^2 + \|v_R\|_{W_T^{1,\infty}}^2\right) \right\} \\
&\quad + \frac{60d_M}{d_m} e^T T \sum_{j=1}^{2} \frac{M_j^2}{C_{jh}^2} \{\|u_j\|_{L_T^\infty L_x^\infty L_\nu^1 L_\omega^\infty} + \beta^2 \|v_j\|_{L_T^\infty L_x^\infty}^8\}, \qquad \forall t \in [0,T]. \qquad (4.5)
\end{aligned}
$$

Note that $H^1 \subset L^\infty$ and $\exists S_\infty \geqslant 1$ such that

$$
\|v\|_{L^\infty} \leqslant S_\infty \|v\|_{H^1}. \tag{4.6}
$$

Let $T \leqslant 1$, then $e^T \leqslant e$, $T \leqslant T^{1/2}$ and estimates (4.4) (4.5) (4.6) imply

$$
\begin{aligned}
&\sum_{j=1}^{2} \left(\|U_j(t,\cdot)\|_{L_x^\infty L_\nu^1 L_\omega^\infty} + \|V_j(t,\cdot)\|_{H_x^1} + \|\partial_x^2 V_j\|_{L_t^2 L_x^2} \right) \\
&\leqslant r_0 + 18M_m T^{1/2} \left\{ \sum_{j=1}^{2} \|u_j\|_{L_T^\infty L_x^\infty L_\nu^1 L_\omega^\infty} + \beta S_\infty^4 \left(\sum_{j=1}^{2} \|v_j\|_{L_T^\infty H_x^1} \right)^4 \right\}, \quad \forall t \in [0,T].
\end{aligned} \tag{4.7}
$$

where $M_m = \dfrac{d_M}{d_m} \max\{M_1, M_2, M_1/C_{1h}, M_1/C_{1h}\}$. Let us take

$$
T \leqslant \min\left\{1, \frac{1}{\{36M_m(1 + 8\beta S_\infty^4 r_0^3)\}^2}\right\},
$$

then $\Phi:\ B \to B$.

For any (u_1, u_2, v_1, v_2), $(u_1', u_2', v_1', v_2') \in B$, we have $(0 \leqslant t \leqslant T \leqslant 1)$

$$
\begin{aligned}
&\sum_{j=1}^{2} \left(\|\{U_j - U_j'\}(t,\cdot)\|_{L_x^\infty L_\nu^1 L_\omega^\infty} + \|\{V_j - V_j'\}(t,\cdot)\|_{H_x^1} + \|\partial_x^2 \{V_j - V_j'\}\|_{L_t^2 L_x^2} \right) \\
&\leqslant \sum_{j=1}^{2} \left\{ \int_0^T 2\|(f_j(u_j, v_j) - f_j(u_j', v_j'))(t,\cdot)\|_{L_x^\infty L_\nu^1 L_\omega^\infty} dt \right. \\
&\quad \left. + \left(\int_0^T 90\frac{d_M}{d_m} \|(g_j(u_j, v_j) - g_j(u_j', v_j'))(t,\cdot)\|_{L_x^2}^2 dt \right)^{1/2} \right\},
\end{aligned} \tag{4.8}
$$

where

$$
U_j = \Phi_j(u_1, u_2, v_1, v_2), \ \ U_j' = \Phi_j(u_1', u_2', v_1', v_2'), \ \ j = 1, 2,
$$

$$V_j = \Phi_{j+2}(u_1, u_2, v_1, v_2), \quad V_j' = \Phi_{j+2}(u_1', u_2', v_1', v_2'), \quad j = 1, 2.$$

Inserting (3.13) (3.14) into (4.8), we have

$$\sum_{j=1}^{2} \left(\|\{U_j - U_j'\}(t,\cdot)\|_{L_x^\infty L_\nu^1 L_\omega^\infty} + \|\{V_j - V_j'\}(t,\cdot)\|_{H_x^1} + \|\partial_x^2\{V_j - V_j'\}\|_{L_t^2 L_x^2}\right)$$

$$\leqslant T^{1/2} \sum_{j=1}^{2} \{18M_m\|u_j - u_j'\|_{L_T^\infty L_x^\infty L_\nu^1 L_\omega^\infty} + N_r\|v_j - v_j'\|_{L_T^\infty H_x^1}\}, \quad \forall t \leqslant T \leqslant 1, \tag{4.9}$$

where

$$N_r = 36N_m r_0 + 256N_m\beta S_\infty^4 r_0^4 + 1920M_m\beta S_\infty^4 r_0^3,$$

$$N_m = \frac{d_M}{d_m}\max\{N_1, N_2, N_1/C_{1h}, N_2/C_{2h}\}.$$

Let

$$T_0 = \min\left\{1, \frac{1}{\{36M_m(1 + 8\beta S_\infty^4 r_0^3)\}^2}, \frac{1}{4N_r^2}\right\}$$

and $T \leqslant T_0$. Then $\Phi: B \to B$ is a strict contraction. By the principle of contraction mapping, there exists a unique solution (u_1, u_2, v_1, v_2) of equations (1.7) (1.8) (1.19) (1.20) with initial-boundary conditions (1.11)–(1.13) (1.21) and continuous conditions (1.15) (1.16) (1.18). Moreover we have

$$\begin{aligned}
u_1 &\in C([0, T_0]; L^\infty([-1, 0]; L^1(\mathbf{R}_+; L^\infty[-1, 1]))),\\
u_2 &\in C([0, T_0]; L^\infty([0, 1]; L^1(\mathbf{R}_+; L^\infty[-1, 1]))),\\
v_1 &\in C([0, T_0]; H^1(-1, 0)) \cap L^2(0, T_0; H^2(-1, 0)),\\
v_2 &\in C([0, T_0]; H^1(0, 1)) \cap L^2(0, T_0; H^2(0, 1)),
\end{aligned}$$

and

$$\max_{0\leqslant t\leqslant T_0} \sum_{j}^{2} \left(\|u_j(t,\cdot)\|_{L_x^\infty L_\nu^1 L_\omega^\infty} + \|v_j(t,\cdot)\|_{H_x^1} + \|\partial_x^2 v_j\|_{L_t^2 L_x^2}\right) \leqslant 2r_0. \tag{4.10}$$

Now we can solve the problem (1.7) (1.8) (1.11)–(1.13) (1.15) (1.16) (1.18)–(1.21) at the initial time moment $t = T_0$ instead of $t = 0$ with the initial data $(u_1(T_0, x, \omega), u_2(T_0, x, \omega), v_1(T_0, x), v_2(T_0, x))$ instead of $(u_1^0, u_2^0, v_1^0, v_2^0)$ again. Then the time domain $[0, T_0]$ can be extended, which denotes by $[0, T_1]$, and the unique solution (u_1, u_2, v_1, v_2) of equations (1.7) (1.8) (1.19) (1.20) is well defined for any $t \in [0, T_1]$. Here $T_1 > T_0$. Repeating the above procedure again and again, we can obtain a series $\{T_n\}$ and the existence of $T_{max} = \sup T_n$.

If $T_{max} < \infty$ and

$$\overline{\lim}_{t\uparrow T_{max}} \sum_{j}^{2} \left(\|u_j(t,\cdot)\|_{L_x^\infty L_\nu^1 L_\omega^\infty} + \|v_j(t,\cdot)\|_{H_x^1}\right) < \infty,$$

then

$$\begin{aligned} K_0 =& 2 \sup_{t\in[0,T_{\max})} \left\{\sum_{j=1}^{2} \|u_j(t,\cdot)\|_{L_x^\infty L_\nu^1 L_\omega^\infty} + \|u_L(t,\cdot)\|_{L_\nu^1 L_\omega^\infty} + \|u_R(t,\cdot)\|_{L_\nu^1 L_\omega^\infty}\right\} \\ &+ 60\frac{d_M^2}{d_m} \sup_{t\in[0,T_{\max})} \left\{\sum_{j=1}^{2} \|v_j(t,\cdot)\|_{H_x^1} + \|v_L\|_{W_t^{1,\infty}} + \|v_R\|_{W_t^{1,\infty}}\right\} \\ &<\infty. \end{aligned}$$

Let

$$\begin{aligned} N_k &= 36N_m K_0 + 256 N_m \beta S_\infty^4 K_0^4 + 1920 M_m \beta S_\infty^4 K_0^3, \\ T_\delta &= \min\left\{1, \frac{1}{\{36M_m(1+8\beta S_\infty^4 K_0^3)\}^2}, \frac{1}{4N_k^2}\right\}. \end{aligned}$$

We solve the problem (1.7) (1.8) (1.11)– (1.13) (1.15) (1.16) (1.18)–(1.21) at the initial time moment $t = T_{\max} - \frac{1}{2}T_\delta$ instead of $t = 0$ with initial data $\left(u_1\left(T_{\max} - \frac{1}{2}T_\delta, x, \omega\right), u_2\left(T_{\max} - \frac{1}{2}T_\delta, x, \omega\right), v_1\left(T_{\max} - \frac{1}{2}T_\delta, x\right), v_2\left(T_{\max} - \frac{1}{2}T_\delta, x\right)\right)$ instead of $(u_1^0, u_2^0, v_1^0, v_2^0)$. Then the unique solution (u_1, u_2, v_1, v_2) of equations (1.7) (1.8) (1.19) (1.20) is well defined for any $t \in \left[0, T_{\max} + \frac{1}{2}T_\delta\right]$. This is contradictory.

Therefore (4.1) is proved.

4.2 Global solution with thermal diffusion

If u_L, u_R, v_L, v_R, $u_j^0 \geqslant 0$ and $v_j^0 \geqslant 0$ $(j = 1, 2)$, we will prove that the solution of equations (1.7) (1.8) (1.19) (1.20), which is constructed in Theorem 4.1, is global.

First we prove that the solution is nonnegative if the data is non-negative.

Lemma 4.1 (Non-negative) *Assume that* u_L, u_R, v_L, v_R, u_j^0, $v_j^0 \geqslant 0$ $(j = 1, 2)$, *and the assumption in Theorem* 4.1 *is satisfied. Then the solution* (u_1, u_2, v_1, v_2), *which is constructed in Theorem* 4.1, *satisfies the following estimate:*

$$\begin{aligned} &u_j(t,x,\nu,\omega) \geqslant 0, \quad v_j(t,x) \geqslant 0, \\ &\forall\, t \in [0, T_{\max}), \;\; x \in I_j, \;\; \nu \in [0,\infty), \;\; \omega \in [-1,1], \;\; j = 1,2, \end{aligned} \tag{4.11}$$

where $I_1 = [-1, 0]$ *and* $I_2 = [0, 1]$.

Proof Let (u_1, u_2, v_1, v_2) be the solution of equations (1.7) (1.8) (1.19) (1.20) with initial- boundary conditions (1.11)–(1.13) (1.21) and continuous conditions (1.15) (1.16) (1.18) solved in Theorem 4.1. By the same argument as in the proof of Lemma 3.1, we derive $u_j \geqslant 0$ $(j = 1, 2)$.

Let $W_j = v_j,\ \ j = 1, 2$. Then W_j satisfies problem (2.14)–(2.18) with

$$a(t, x) = \frac{1}{C_{1h}} \int_0^\infty \sigma_{1a}(v_1) \int_0^1 \partial_v B(\theta v_1), \nu) \mathrm{d}\theta \mathrm{d}\nu,$$

$$b(t, x) = \frac{1}{C_{2h}} \int_0^\infty \sigma_{2a}(v_2) \int_0^1 \partial_v B(\theta v_2, \nu) \mathrm{d}\theta \mathrm{d}\nu,$$

$$F(t, x) = \frac{1}{C_{1h}} \int_0^\infty \sigma_{1a}(v_1) \bar{u}_1(t, x, \nu) \mathrm{d}\nu \geqslant 0,$$

$$G(t, x) = \frac{1}{C_{2h}} \int_0^\infty \sigma_{ja}(v_2) \bar{u}_2(t, x, \nu) \mathrm{d}\nu \geqslant 0.$$

For any $T \in [0, T_{\max})$, by Theorem 2.1, we know that there exists a unique solution (W_1, W_2) of problem (2.14) –(2.18) such that

$$W_j(t, x) \geqslant 0, \quad \forall t \in [0, T],\ \ x \in I_j,\ \ j = 1, 2.$$

Since $T \in [0, T_{\max})$ is arbitrary, we have

$$v_j(t, x) \geqslant 0,\ \ \forall t \in [0, T_{\max}).$$

(4.11) is proved.

For the solution (u_1, u_2, v_1, v_2) of (1.7) (1.8) (1.19) (1.20), we will prove that u_j is bounded by u_s from above and v_j is bounded by v_s from above, where (u_s, v_s) is defined in (3.41).

Lemma 4.2 (Upper bound) *Assume that* u_L, u_R, $u_j^0 \geqslant 0$ *and* $v_j^0 \geqslant 0$ $(j = 1, 2)$ *satisfy* (3.40), *and the assumption in Theorem* 4.1 *is satisfied. Then the solution* (u_1, u_2, v_1, v_2), *which is constructed in Theorem* 4.1, *satisfies the following estimate:*

$$\begin{gathered} 0 \leqslant u_j(t, x, \nu, \omega) \leqslant u_s,\ \ 0 \leqslant v_j(t, x) \leqslant v_s, \\ \forall t \in [0, T_{\max}),\ x \in I_j,\ \nu \in \mathbf{R}_+,\ \omega \in [-1, 1],\ j = 1, 2, \end{gathered} \tag{4.12}$$

where $I_1 = [-1, 0]$ *and* $I_2 = [0, 1]$.

Proof Let (u_1, u_2, v_1, v_2) be the solution of equations (1.7) (1.8) (1.19) (1.20) with initial-boundary conditions (1.11)–(1.13) (1.21) and continuous conditions (1.15) (1.16) (1.18) solved in Theorem 4.1.

We define

$$U_j(t,x,\omega) = u_s - u_j(t,x,\nu,\omega), \quad V_j(t,x) = v_s - v_j(t,x), \quad j=1,2, \quad \forall t \in [0, T_{\max}). \tag{4.13}$$

Then (U_1, U_2, V_1, V_2) is the unique solution of the following equations

$$\partial_t U_1 + c\omega\partial_x U_1 + \sigma_1(v_1)U_1 = \sigma_{1s}(v_1)\bar{U}_1 + \sigma_{1a}(v_1)\int_0^1 \partial_v B(v_{1\theta},\nu)\mathrm{d}\theta V_1, \tag{4.14}$$

$$\partial_t U_2 + c\omega\partial_x U_2 + \sigma_2(v_2)U_2 = \sigma_{2s}(v_2)\bar{U}_2 + \sigma_{2a}(v_2)\int_0^1 \partial_v B(v_{2\theta},\nu)\mathrm{d}\theta V_2, \tag{4.15}$$

$$\partial_t V_1 - d_1\partial_x^2 V_1 + \frac{1}{C_{1h}}\int_0^\infty \sigma_{1a}(v_1)\left\{\int_0^1 \partial_v B(v_{1\theta},\nu)\mathrm{d}\theta V_1 - \bar{U}_1\right\}\mathrm{d}\nu = 0, \tag{4.16}$$

$$\partial_t V_2 - d_2\partial_x^2 V_2 + \frac{1}{C_{2h}}\int_0^\infty \sigma_{2a}(v_2)\left\{\int_0^1 \partial_v B(v_{2\theta},\nu)\mathrm{d}\theta V_2 - \bar{U}_2\right\}\mathrm{d}\nu = 0, \tag{4.17}$$

$$\begin{aligned} U_1(0,x,\nu,\omega) &= U_1^0 = u_s - u_1^0(x,\nu,\omega), \quad x \in [-1,0], \\ U_2(0,x,\nu,\omega) &= U_2^0 = u_s - u_2^0(x,\nu,\omega), \quad x \in [0,1], \end{aligned} \tag{4.18}$$

$$\begin{aligned} V_1(0,x) &= V_1^0 = v_s - v_1^0(x), \quad x \in [-1,0], \\ V_2(0,x) &= V_2^0 = v_s - v_2^0(x), \quad x \in [0,1], \end{aligned} \tag{4.19}$$

$$\begin{aligned} U_1(t,x,\nu,\omega)|_{x=-1} &= U_L = u_s - u_L(t,\nu,\omega), \quad \text{for } \omega > 0, \\ U_2(t,x,\nu,\omega)|_{x=1} &= U_R = u_s - u_R(t,\nu,\omega), \quad \text{for } \omega < 0, \end{aligned} \tag{4.20}$$

$$V_1(t,x)|_{x=-1} = V_L = v_s - v_L(t), \quad V_2(t,x)|_{x=1} = V_R = v_s - v_R(t), \tag{4.21}$$

$$U_1(t,x,\nu,\omega)|_{x=0} \Leftarrow U_2(t,x,\nu,\omega)|_{x=0}, \quad \text{for } \omega < 0, \tag{4.22}$$

$$U_2(t,x,\nu,\omega)|_{x=0} \Leftarrow U_1(t,x,\nu,\omega)|_{x=0}, \quad \text{for } \omega > 0, \tag{4.23}$$

$$\begin{aligned} V_1(t,x)|_{x=0} &= V_2(t,x)|_{x=0}, \\ d_1\partial_x V_1(t,x)|_{x=0} &= d_2\partial_x V_2(t,x)|_{x=0}, \end{aligned} \tag{4.24}$$

where $v_{j\theta} = v_j + \theta(v_s - v_j)$, $j = 1,2$. By the principle of contraction mapping, the uniqueness of (U_1, U_2, V_1, V_2) follows as in the proof of Theorem 4.1.

It implies (4.12) that U_j and V_j $(j = 1,2)$ are non-negative. The iteration argument below, which is similar to the iteration in the proof of Lemma 3.2, leads directly to the desired result as we now show. Let $U_j^{(n)} = U_j^{(n)}(t,x,\nu,\omega)$, $V_j^{(n)} = V_j^{(n)}(t,x)$ and

$$\begin{aligned}
&U_1^{(0)}(t,x,\nu,\omega)\\
=&\chi(t<t_{1d})U_1^0(x_{\omega t},\cdot)E_1(t,0)+\chi(t\geqslant t_{1d})\chi(\omega>0)E_1(t_{1d},0)U_L(t-t_{1d},\cdot)\\
&+\chi(t\geqslant t_{1d})\chi(\omega<0)\{\chi(t-t_{1d}<t_{2d}^0)U_2^0(x_{\omega,t-t_{1d}}^0,\cdot)E_2(t-t_{1d},0)\\
&+\chi(t-t_{1d}\geqslant t_{2d}^0)U_R(t-t_{1d}-t_{2d}^0,\cdot)E_2(t_{2d}^0,0)\}E_1(t_{1d},0),
\end{aligned}\tag{4.25}$$

$$\begin{aligned}
&U_2^{(0)}(t,x,\nu,\omega)\\
=&\chi(t<t_{2d})U_2^0(x_{\omega t},\cdot)E_2(t,0)+\chi(t\geqslant t_{2d})\chi(\omega<0)E_2(t_{2d},0)U_R(t-t_{2d},\cdot)\\
&+\chi(t\geqslant t_{2d})\chi(\omega>0)\{\chi(t-t_{2d}<t_{1d}^0)U_1^0(x_{\omega,t-t_{2d},}^0\cdot)E_1(t-t_{2d},0)\\
&+\chi(t-t_{2d}\geqslant t_{1d}^0)U_L(t-t_{2d}-t_{1d}^0,\cdot)E_1(t_{1d}^0,0)\}E_2(t_{2d},0),
\end{aligned}\tag{4.26}$$

$$\partial_t V_1^{(0)}-d_1\partial_x^2 V_1^{(0)}+\frac{1}{C_{1h}}\int_0^\infty\sigma_{1a}(v_1)\int_0^1\partial_v B(v_{1\theta},\nu)\mathrm{d}\theta\mathrm{d}\nu V_1^{(0)}=0,\tag{4.27}$$

$$\partial_t V_2^{(0)}-d_2\partial_x^2 V_2^{(0)}+\frac{1}{C_{2h}}\int_0^\infty\sigma_{2a}(v_2)\int_0^1\partial_v B(v_{2\theta},\nu)\mathrm{d}\theta\mathrm{d}\nu V_2^{(0)}=0,\tag{4.28}$$

$$V_1^{(0)}(t,x)|_{t=0}=V_1^0,\quad V_2^{(0)}(t,x)|_{t=0}=V_2^0,\tag{4.29}$$

$$V_1^{(0)}(t,x)|_{x=-1}=V_L,\quad V_2^{(0)}(t,x)|_{x=1}=V_R,\tag{4.30}$$

$$\begin{aligned}
V_1^{(0)}(t,x)|_{x=0}=&V_2^{(0)}(t,x)|_{x=0},\\
d_1\partial_x V_1^{(0)}(t,x)|_{x=0}=&d_2\partial_x V_2^{(0)}(t,x)|_{x=0},
\end{aligned}\tag{4.31}$$

$$\begin{aligned}
&U_1^{(n+1)}(t,x,\nu,\omega)\\
=&\int_{t-t\wedge t_{1d}}^t E_1(t,\tau)\left(\sigma_{1s}(v_1)\bar U_1^{(n)}+\sigma_{1a}(v_1)\int_0^1\partial_v B(v_{1\theta},\nu)\mathrm{d}\theta V_1^{(n)}\right)(\tau,x_{\omega\tau},\nu)\mathrm{d}\tau\\
&+\chi(t\geqslant t_{1d})\chi(\omega<0)E_1(t_{1d},0)\int_{t-t_{1d}-(t-t_{1d})\wedge t_{2d}^0}^{t-t_{1d}}E_2(t-t_{1d},\tau)\cdot\\
&\left(\sigma_{2s}(v_2)\bar U_2^{(n)}+\sigma_{2a}(v_2)\int_0^1\partial_v B(v_{2\theta},\nu)\mathrm{d}\theta V_2^{(n)}\right)(\tau,x_{\omega,\tau}^0,\nu)\mathrm{d}\tau,
\end{aligned}\tag{4.32}$$

$$\begin{aligned}
&U_2^{(n+1)}(t,x,\nu,\omega)\\
=&\int_{t-t\wedge t_{2d}}^t E_2(t,\tau)\left(\sigma_{2s}(v_2)\bar U_2^{(n)}+\sigma_{2a}(v_2)\int_0^1\partial_v B(v_{2\theta},\nu)\mathrm{d}\theta V_2^{(n)}\right)(\tau,x_{\omega\tau},\nu)\mathrm{d}\tau
\end{aligned}$$

$$+\chi(t\geqslant t_{2d})\chi(\omega>0)E_2(t_{2d},0)\int_{t-t_{2d}-(t-t_{2d})\wedge t_{1d}^0}^{t-t_{2d}} E_1(t-t_{2d},\tau)$$

$$\cdot\left(\sigma_{1s}(v_1)\bar{U}_1^{(n)}+\sigma_{1a}(v_1)\int_0^1 \partial_v B(v_{1\theta},\nu)\mathrm{d}\theta V_1^{(n)}\right)(\tau,x_{\omega,\tau}^0,\nu)\mathrm{d}\tau, \tag{4.33}$$

$$\partial_t V_1^{(n+1)}-d_1\partial_x^2 V_1^{(n+1)}+\frac{1}{C_{1h}}\int_0^\infty \sigma_{1a}(v_1)\int_0^1 \partial_v B(v_{1\theta},\nu)\mathrm{d}\theta\mathrm{d}\nu V_1^{(n+1)}$$
$$=\frac{1}{C_{1h}}\int_0^\infty \sigma_{1a}(v_1)\overline{U_1^{(n)}}\mathrm{d}\nu, \tag{4.34}$$

$$\partial_t V_2^{(n+1)}-d_2\partial_x^2 V_2^{(n+1)}+\frac{1}{C_{2h}}\int_0^\infty \sigma_{2a}(v_2)\int_0^1 \partial_v B(v_{2\theta},\nu)\mathrm{d}\theta\mathrm{d}\nu V_2^{(n+1)}$$
$$=\frac{1}{C_{2h}}\int_0^\infty \sigma_{2a}(v_2)\overline{U_2^{(n)}}\mathrm{d}\nu, \tag{4.35}$$

$$V_1^{(n+1)}(t,x)|_{t=0}=0,\quad V_2^{(n+1)}(t,x)|_{t=0}=0, \tag{4.36}$$

$$V_1^{(n+1)}(t,x)|_{x=-1}=0,\quad V_2^{(n+1)}(t,x)|_{x=1}=0, \tag{4.37}$$

$$\begin{aligned} V_1^{(n+1)}(t,x)|_{x=0}=&V_2^{(n+1)}(t,x)|_{x=0},\\ d_1\partial_x V_1^{(n+1)}(t,x)|_{x=0}=&d_2\partial_x V_2^{(n+1)}(t,x)|_{x=0}, \end{aligned} \tag{4.38}$$

where $E_j(t,\tau)$ is defined in (3.26), $n=0,\ 1,\ 2,\ \cdots$.

U_L, $U_R\geqslant 0$ and $U_j^0\geqslant 0$ imply that

$$U_j^{(0)}\geqslant 0,\qquad j=1,\ 2. \tag{4.39}$$

Problem (4.27)–(4.31) can be transformed into problem (2.14)–(2.18) by transformation

$$W_j=V_j^{(0)},\ v_1^0=V_1^0,\ v_2^0=V_2^0,\ v_L=V_L,\ v_R=V_R,\ F=G=0,$$

$$a=\frac{1}{C_{1h}}\int_0^\infty \sigma_{1a}(v_1)\int_0^1 \partial_v B(v_{1\theta},\nu)\mathrm{d}\theta\mathrm{d}\nu,$$

$$b=\frac{1}{C_{2h}}\int_0^\infty \sigma_{2a}(v_2)\int_0^1 \partial_v B(v_{2\theta},\nu)\mathrm{d}\theta\mathrm{d}\nu.$$

By Theorem 2.1, V_L, $V_R\geqslant 0$ and $V_j^0\geqslant 0$ imply that there exists unique solution $V_j^{(0)}$ of problem (4.27)– (4.31) such that

$$V_j^{(0)}\geqslant 0,\qquad j=1,\ 2. \tag{4.40}$$

By induction, we have

$$U_j^{(n)}(t,x,\nu,\omega) \geqslant 0, \quad j=1,2,\ n=1,2,\cdots. \tag{4.41}$$

Problem (4.34)–(4.38) can be transformed into problem (2.14)–(2.18) by transformation

$$W_j = V_j^{(n+1)},\ v_1^0 = v_2^0 = v_L = v_R = 0,$$

$$F = \frac{1}{C_{1h}}\int_0^\infty \sigma_{1a}(v_1)\overline{U_1^{(n)}}\mathrm{d}\nu,\ G = \frac{1}{C_{2h}}\int_0^\infty \sigma_{2a}(v_2)\overline{U_2^{(n)}}\mathrm{d}\nu,$$

$$a = \frac{1}{C_{1h}}\int_0^\infty \sigma_{1a}(v_1)\int_0^1 \partial_v B(v_{1\theta},\nu)\mathrm{d}\theta\mathrm{d}\nu,$$

$$b = \frac{1}{C_{2h}}\int_0^\infty \sigma_{2a}(v_2)\int_0^1 \partial_v B(v_{2\theta},\nu)\mathrm{d}\theta\mathrm{d}\nu.$$

By induction and Theorem 2.1, we derive that there exists unique solution $V_j^{(n+1)}$ of problem (4.34)– (4.38) such that

$$V_j^{(n+1)}(t,x) \geqslant 0, \quad j=1,2,\ n=0,1,2,\cdots. \tag{4.42}$$

For any $T\in(0,T_{\max})$ and $t\in[0,T]$, we find that

$$\|U_j^{(0)}(t,\cdot)\|_{L_x^\infty L_\nu^1 L_\omega^\infty} \leqslant 4\|u_s\|_{L_\nu^1}, \qquad j=1,2. \tag{4.43}$$

By Lemma 2.2, we get

$$\sum_{j=1}^{2}\{\|V_j^{(0)}(t,\cdot)\|_{H_x^1} + \|\partial_x^2 V_j^{(0)}\|_{L_t^2L_x^2}\} \leqslant C_0(A,T)C_1(A,T)v_s, \quad \forall t\in[0,T], \tag{4.44}$$

where $C_0(A,T) = \left\{30e^{(1+2A)T}(A^2T+d_M)/d_m\right\}^{1/2}$, $C_1(A,T) = \{(A^2+73d_M^2)T+6\}^{1/2}$, $d_M=\max\{1,d_1,d_2\}$, $d_m=\min\{1,d_1,d_2\}$, $A=\max\{A_1,A_2\}$,

$$A_1 = \max_{0\leqslant t\leqslant T}\left\|\frac{M_{1a}}{C_{1h}}\int_0^\infty\int_0^1 \partial_v B(v_{1\theta},\nu)\mathrm{d}\theta\mathrm{d}\nu\right\|_{L_x^\infty},$$

$$A_2 = \max_{0\leqslant t\leqslant T}\left\|\frac{M_{2a}}{C_{2h}}\int_0^\infty\int_0^1 \partial_v B(v_{2\theta},\nu)\mathrm{d}\theta\mathrm{d}\nu\right\|_{L_x^\infty}.$$

By Sobolev imbedding theorem, we have

$$\sum_{j=1}^{2}\|V_j^{(0)}(t,\cdot)\|_{L_x^\infty} \leqslant S_\infty C_0(A,T)C_1(A,T)v_s, \qquad \forall t\in[0,T], \tag{4.45}$$

By induction and Lemma 2.2, we derive that

$$
\begin{aligned}
&\|U_j^{(n)}(t,\cdot)\|_{L_x^\infty L_\nu^1 L_\omega^\infty}\\
\leqslant&\sum_{j=1}^{2}\int_0^t \{M_{js}\|U_j^{(n-1)}(\tau,\cdot)\|_{L_x^\infty L_\nu^1 L_\omega^\infty}+A\|V_j^{(n-1)}(\tau,\cdot)\|_{L_x^\infty}\}\mathrm{d}\tau\\
\leqslant&\frac{C_2^n t^n}{n!}(4\|u_s\|_{L_\nu^1}+C_0(A,T)C_1(A,T)v_s),\quad j=1,2,\ n\geqslant 1,
\end{aligned}\tag{4.46}
$$

$$
\begin{aligned}
&\sum_{j=1}^{2}\{\|V_j^{(n)}(t,\cdot)\|_{H_x^1}+\|\partial_x^2 V_j^{(n)}\|_{L_t^2L_x^2}\}\\
\leqslant&C_0(A,T)\sum_{j=1}^{2}\frac{M_{ja}}{C_{jh}}\int_0^t\|U_j^{(n-1)}(\tau,\cdot)\|_{L_x^\infty L_\nu^1 L_\omega^\infty}\mathrm{d}\tau\\
\leqslant&\frac{C_2^n t^n}{n!}(4\|u_s\|_{L_\nu^1}+C_0(A,T)C_1(A,T)v_s),\qquad n\geqslant 1,
\end{aligned}\tag{4.47}
$$

$$
\begin{aligned}
&\sum_{j=1}^{2}\|V_j^{(n)}(t,\cdot)\|_{L_x^\infty}\\
\leqslant&S_\infty\frac{C_2^n t^n}{n!}(4\|u_s\|_{L_\nu^1}+C_0(A,T)C_1(A,T)v_s),\qquad n\geqslant 1,\ \forall t\in[0,T],
\end{aligned}\tag{4.48}
$$

where constant $C_2=\sum_{j=1}^{2}\left(M_{js}+2C_0(A,T)\dfrac{M_{ja}}{C_{jh}}\right)+S_\infty A$.

Since the solution $(U_1^*,U_2^*,V_1^*,V_2^*)$ of the problem (4.14)–(4.24) is formally

$$
U_j^*(t,x,\nu,\omega)=\sum_{n=0}^{\infty}U_j^{(n)}(t,x,\nu,\omega),\quad j=1,2,\ \ \forall t\in[0,T],
$$

$$
V_j^*(t,x)=\sum_{n=0}^{\infty}V_j^{(n)}(t,x),\quad j=1,2,\ \ \forall t\in[0,T],
$$

our estimates above show that these four series converge and $(U_1^*,U_2^*,V_1^*,V_2^*)$ exists for all $t\in[0,T]$. Moreover, $U_j^*(t,x,\nu,\omega)\geqslant 0$ and $V_j^*(t,x)\geqslant 0$ for all $t\in[0,T]$, $x\in I_j$, $\nu\in\mathbf{R}_+$, $\omega\in[-1,1]$, $j=1,2$.

By the uniqueness of solution of problem (4.14)– (4.24), we have $U_j(t,x,\nu,\omega)=U_j^*(t,x,\nu,\omega)\geqslant 0$ and $V_j(t,x)=V_j^*(t,x)\geqslant 0$ for all $t\in[0,T]$, $x\in I_j$, $\nu\in\mathbf{R}_+$, $\omega\in[-1,1]$, $j=1,2$. Since $T\in(0,T_{\max})$ is arbitrary, we can derive that $U_j(t,x,\nu,\omega)\geqslant 0$ and $V_j(t,x)\geqslant 0$ for all $t\in[0,T_{\max})$, $x\in I_j$, $\nu\in\mathbf{R}_+$, $\omega\in[-1,1]$, $j=1,2$. It implies (4.12) that U_j and V_j $(j=1,2)$ are non-negative.

To prove existence of global solution, we need the following energy estimate.

Lemma 4.3 (Energy Estimate) *Assume that the assumption in Lemma 4.2 is satisfied. Then for the solution (u_1, u_2, v_1, v_2), which is constructed in Theorem 4.1, the following estimate is valid:*

$$\begin{aligned}
&\|v_1(t,\cdot)\|_{H_x^1}^2 + \|v_2(t,\cdot)\|_{H_x^1}^2 + \int_0^t \{\|\partial_x^2 v_1(\tau,\cdot)\|_{L_x^2}^2 + \|\partial_x^2 v_2(\tau,\cdot)\|_{L_x^2}^2\} \mathrm{d}\tau \\
\leqslant & m_0^e(t)\Big\{2\|v_1^0\|_{H_x^1}^2 + 2\|v_1^0\|_{H_x^1}^2 + m_b^e(t)\Big(\|v_L\|_{W_T^{1,\infty}}^2 + \|v_R\|_{W_T^{1,\infty}}^2\Big) \\
& + 2\Big(\frac{M_{1a}^2}{C_{1h}^2} + \frac{M_{2a}^2}{C_{2h}^2}\Big)(\|u_s\|_{L_\nu^1} + \beta v_s^4)^2 t\Big\}, \qquad \forall t \in [0, T_{\max}),
\end{aligned} \tag{4.49}$$

where $m_0^e(t) = 15e^t d_M/d_m$, $m_b^e(t) = 73d_M^2 t + 4$, $d_M = \max\{1, d_1, d_2\}$, *and* $d_m = \min\{1, d_1, d_2\}$.

Proof Let (u_1, u_2, v_1, v_2) be the solution established in Theorem 4.1. For any $T \in (0, T_{\max})$ and $t \in [0, T]$, $(W_1, W_2) = (v_1, v_2)$ satisfies the problem (2.14)– (2.18) with $a(t,x) = b(t,x) = 0$ and

$$F(t,x) = \frac{1}{C_{1h}} \int_0^\infty \sigma_{1a}(v_1)(\bar{u}_1 - B(v_1,\nu))\mathrm{d}\nu,$$

$$G(t,x) = \frac{1}{C_{2h}} \int_0^\infty \sigma_{2a}(v_2)(\bar{u}_2 - B(v_2,\nu))\mathrm{d}\nu.$$

Then the energy estimate (2.62) hold. Applying (4.12), we have

$$\begin{aligned}
\|F(t,\cdot)\|_{L_x^\infty} &\leqslant \frac{M_{1a}}{C_{1h}}(\|u_s\|_{L_\nu^1} + \beta v_s^4), \quad \forall t \in [0, T_{\max}), \\
\|G(t,\cdot)\|_{L_x^\infty} &\leqslant \frac{M_{2a}}{C_{2h}}(\|u_s\|_{L_\nu^1} + \beta v_s^4), \quad \forall t \in [0, T_{\max}).
\end{aligned} \tag{4.50}$$

Employing estimates (2.62) and (4.50), we can prove estimate (4.49).

As a consequence of (4.12) and (4.49), the solution of equations (1.7) (1.8) (1.19) (1.20) with initial-boundary conditions (1.11)– (1.13) (1.21) and continuous conditions (1.15) (1.16) (1.18) is global.

Theorem 4.2 (Global Solution) *Assume that the assumption in Lemma 4.2 is satisfied. Then the solution (u_1, u_2, v_1, v_2), which is constructed in Theorem 4.1, is global. Moreover, if*

$$u_j^0(x,\nu,\omega) \in C_B(I_j; L^1([0,\infty); L^\infty[-1,1])), \quad j = 1, 2,$$

where $I_1 = [-1, 0]$ and $I_2 = [0, 1]$, then

$$u_j(t,x,\nu,\omega) \in C([0,\infty); C_B(I_j; L^1([0,\infty); L^\infty[-1,1]))), \qquad j = 1, 2.$$

5 Conclusion

In this article the spatial transport of radiation in two adjoining media is considered. As usual, we assume that the material is in local thermodynamic equilibrium. In the absence of hydrodynamic motion and heat conduction, we establish the existence and uniqueness of global solution of the classical radiative transfer problem (1.7)- (1.10) (1.11)–(1.13) (1.15) (1.16), see Theorem 3.3. On the other hand, we introduce new radiative transfer equations (1.7) (1.8) (1.19) (1.20) modified by thermal diffusion. The existence and uniqueness of global solution of this modified radiative transfer problem (1.7) (1.8) (1.19) (1.20) (1.11)–(1.13) (1.21) (1.15) (1.16) (1.18) is also derived, see Theorem 4.2. We think that our new model is more reasonable than the classical radiative transfer equations due to this new model can preserve continuity of flux at the interface of two materials.

Here the methods and results still hold for the spatial transport of radiation in multimedia (more than two materials). The existence and uniqueness of global solution of multigroup equations is the corollary of here results.

Acknowledgements This work is supported by National Natural Science Foundation of China–NSAF, Grant No.10976003 and NSF No.11271052, and also supported by Beijing Center for Mathematics and InformationInterdisciplinary Sciences.

The authors would like to thank referees for their helpful suggestions.

References

[1] Adams M L, Larsen E W. Fast iterative methods for discrete-ordinates particle transport calculations [J]. Prog. Nucl. Energy, 2002, 40: 3–159.

[2] Anistratov D Y, Larsen E W. Nonlinear and linear α-weighted methods for particle transport problems [J]. J. Comput. Phys., 2001, 173(2): 664–684.

[3] Aronson D, Besala P. Parabolic equations with unbounded coefficients [J]. J. Diff. Equations, 1967, 3: 1–14.

[4] Bardos C, Santos R, Sentis R. Diffusion approximation and computation of the critical size [J]. Trans. Amer. Math. Soc., 1984, 284(2): 617–649.

[5] Bensoussan A, Lions J L, Papanicolaou G C. Boundary layers and homogenization of transport processes [J]. J. Publ. RIMS Kyoto Univ., 1979, 15: 53–157.

[6] Bowers R L, Wilson J R. Numerical Modeling in Applied Physics and Astrophysics [M]. Boston: Jones and Bartlett Publishers, 1991.

[7] Case K M, Zweifel P F. Existence and uniqueness theorems for neutron transport equation [J]. J. Math. Phys., 1963, 4: 1376–1385.

[8] Case K M, Zweifel P F. Linear Transport Theory [M]. Addison-Wesley, Reading, Mass., 1967.

[9] Chandrasekhar S. Radiative Transfer [M]. New York: Dover, 1960.

[10] Evans T M, Densmore J D. Methods for coupling radiation, ion, and electron energies in grey Implicit Monte Carlo [J]. J. Comput. Phys., 2007, 225: 1695–1720.

[11] Germogenova T. The principle of maximum for the transport equation [J]. USSR Comput. Math. Math. Phys., 1962, 2: 169–174.

[12] Golse F, Shi Jin, Levermore C D. The convergence of numerical transfer schemes in diffusive regimes I: discrete-ordinate method [J]. SIAM J. Numer. Anal., 1999, 36(5): 1333–1369.

[13] Golse F, Shi Jin, Levermore C D. A domain decomposition analysis for a two-scale linear transport problem [J]. Math. Model Numer. Anal., 2003, 37: 869–892.

[14] Habetler G J, Matkowsky B J. Uniform asymptotic expansions in transport theory with small mean free paths and the diffusion approximation [J]. J. Math. Phys., 1975, 16(4): 846–854.

[15] Boling Guo, Yongqian Han. Diffusion limit of small mean free path of transfer equation in $\mathbf{R}^3$ [J]. Transport Theory and Stat. Phys., 2011, 40(5): 243–281.

[16] Boling Guo, Yongqian Han. Diffusion limit of a small mean free path of radiative transfer equations with absorbing boundary condition [J]. Transport Theory and Statistical Physics, 2012, 41(7): 552–582.

[17] Il'in A M, Kalashnikov A S, Oleinik O A. Linear equations of the second order of parabolic type [J]. Russian Math. Surveys (Uspahi Mat. Nauk), 1962, 17(3): 1–143.

[18] Ladysenskaya O A. The Boundary Value Problems of Mathematical Physics [M]. Applied Mathematical Sciences 49. Berlin, Heidelberg, New York: Springer, 1985.

[19] Ladysenskaya O A, Solonnikov V A, Ural'ceva N U. Linear and Quasi-linear Equations of Parabolic type [M]. AMS Trans. 23, Providence, Rhode Island, 1968.

[20] Larsen E W. Unconditionally stable diffusion synthetic acceleration methods for the slab geometry discrete ordinates equation [J]. Nuclear Sci. Engineering, 1982, 82: 47–68.

[21] Larsen E W. A grey transport acceleration method for time-dependent radiative transfer problems [J]. J. Comput. Phys., 1988, 78: 459–480.

[22] Larsen E W. The asymptotic diffusion limit of discretized transport problems [J]. Nuclear Sci. Engrg., 1992, 112: 336–346.

[23] Larsen E W, Keller J B. Asymptotic solution of neutron transport problems for small mean free paths [J]. J. Math. Phys., 1974, 15(1): 75–81.

[24] Larsen E W, Kumar A, Morel J E. Properties of the implicitly time-differenced equations of thermal radiation transport [J]. J. Comput. Phys., 2013, 238: 82–96.

[25] Larsen E W, Miller W F Jr. Convergence rates of spatial difference equations for the discrete-ordinate neutron transport equations in slab geometry [J]. Nuclear Sci.

Engrg., 1980, 73: 76–83.

[26] Larsen E W, Morel J E, Miller W F Jr. Asymptotic solutions of numerical transport problems in optically thick, diffusive regimes [J]. J. Comput. Phys., 1987, 69: 283–324.

[27] Larsen E W, Pomraning G C, Badham V C. Asymptotic analysis of radiative transfer problems [J]. J. Quant. Spectrosc. Radiat. Transfer, 1983, 29(4): 285–310.

[28] Shuanggui Li, Tinggui Feng. Diffusion-synthetic acceleration method for diamond-differenced discrete-ordinates radiative transfer equations [J]. Chinese J. Comput. Phys., 2008, 25(1): 1–6.

[29] Liebman G M. Second Order Parabolic Partial Differential Equations [M]. World Scientific, Singapore, 1996.

[30] Lunardi A. Analytic Semigroups and Optimal Regularity in Parabolic Problem [M]. Progress in nonlinear differential equations and their applications 16. Birkhauser Verlag, 1995.

[31] Malvagi F, Pomraning G C. Initial and boundary conditions for diffusive linear transport problems [J]. J. Math. Phys., 1991, 32(3): 805–820.

[32] Mihalas D, Mihalas B. Foundation of Radiation Hydrodynamics [M]. Oxford University Press, 1984.

[33] Morel J E, Larsen E W, Matzen M K. A synthetic acceleration scheme for radiative diffusion calculations [J]. J. Quant. Spectrosc. Radiat. Transfer, 1985, 34: 243–261.

[34] Morel J E. Discrete-ordinates methods for radiative transfer in the non-relativistic stellar regime. Computational Methods in Transport [M]. Lecture in Computational Science and Engineering 48, 69–81. Berlin Heidelberg: Springer-Verlag, 2006.

[35] Olson G L. Efficient solution of multi-dimensional flux-limited nonequilibrium radiation diffusion coupled to material conduction with second-order time discretization [J]. J. Comput. Phys., 2007, 226: 1181–1195.

[36] Papanicolaou G C. Asymptotic analysis of transport processes [J]. Bull. Amer. Math. Soc., 1975, 81(2): 330–392.

[37] Pazy A. Semigroups of Linear Operators and Applications to Partial Differential Equations [M]. Applied mathematical sciences 44. NewYork: Springer-Verlag, 1983.

[38] Pomraning G C. The Equations of Radiation Hydrodynamics [M]. New York: Pergamon Press, 1973.

[39] Roberts L, Anistratov D Y. Nonlinear weighted flux methods for particle transport problems [J]. Transp. Theory Stat. Phys., 2007, 36(7): 589–608 .

[40] Sulem P L, Sulem C, Bardos C. On the continuous limit for a system of classical spins [J]. Comm. Math. Phys., 1986, 107: 431–454.

[41] Zhou Yulin. Applications of Discrete Functional Analysis to the Finite Difference Method [M]. International Academic Publishers, 1991.

Darboux Transformation and Multi-dark Soliton for N-component Nonlinear Schrödinger Equations*

Ling Liming (凌黎明), Zhao Lichen (赵立臣) and Guo Boling (郭柏灵)

Abstract In this paper, we obtain a uniform Darboux transformation for multi-component coupled NLS equations, which can be reduced to all previous presented Darboux transformation. As a direct application, we derive the single dark soliton and multi-dark soliton solutions for multi-component NLS equations with defocusing case and mixed focusing and defocusing case. Some exact single and two-dark solitons of three-component NLS equations are investigated explicitly. The results are meaningful for vector dark soliton studies in many physical systems, such as Bose-Einstein condensate, nonlinear optics, etc.

Keywords Darboux transformation; Dark soliton; N-component NLS equtions

1 Introduction

It is well known that nonlinear Schrödinger (NLS)-type equations play a prominent role in nonlinear physical systems, such as nonlinear optics [1], Bose-Einstein condensates [2]. In these physical systems, the nonlinear coefficient can be positive or negative, depending on the physical situations [3]. For example, the nonlinearity can be positive or negative when the interaction between atoms is repulsive or attractive in Bose-Einstein condensates [2]. For nonlinear optics, it corresponds to the focusing or defocusing case. For one-component system, enormous studies have been done [4], which demonstrate that the corresponding scalar NLS equation admits bright solitons [4], breather [5,6] and rogue wave [7,8] in the focusing case, and dark solitons [9] in the defocusing case.

* Nonlinearity, 2015, 28, 3243-3261. Doi: 10.1088/0951-7715/28/9/3243.

Since a variety of complex systems, such as Bose-Einstein condensates, nonlinear optical fibers, etc., usually involve more than one component, the studies should be extended to multi-component NLS equations cases [10, 11, 12]. For two-component coupled system with both focusing case, the coupled NLS equations admit bright-bright solitons, bright-dark, breather, rogue wave, bright-dark-breather and bright-dark-rogue wave solution [11,12,13,14,15,16]. With both defocusing case, the coupled NLS equations admit bright-dark, dark-dark solitons and breather solution [15, 16, 17, 18]. With defocusing and focusing coexisting case, the coupled NLS equations admit bright-bright solitons, bright-dark solitons, dark-dark soliton and breather solution [16, 19, 20, 21]. For the three-component NLS equations with all focusing, the "four-petaled flower" structure rogue wave was presented recently by Darboux transformation (DT) [22].

The DT method is a very effective and convenient way to derive kinds of localized waves, such as bright soliton, breather, rogue wave [11, 12, 13, 14, 16]. However, the dark soliton can not be obtained by classical DT method directly. The first time to obtain the dark soliton of single NLS equation through DT method was appeared in 1996 [23]. The single dark soliton formula of N-component NLS type equation appeared in 2006 and 2009 [24, 25]. The dark soliton of inverse scattering method in the coupled NLS system is an open problem until 2006 [26]. Even the soliton solutions obtained in reference [26], can be degenerated into scalar NLS equation. Therefore, it is still desirable to study how to obtain multi-dark soliton through the DT method, which would be very meaningful for vector dark soliton studies in related physical systems.

In this paper, we would like to derive a simple multi-soliton formula for dark soliton of integrable N-component NLS system (2.6) through generalizing DT method. In order to present our work clearly and readably, we revisit the method in 1996 and 2006 [23, 24]. In 1996 [23], Mañas proposed a method to derive the dark soliton for defocusing NLS equation. The Darboux matrix can be represented as

$$T = I + \frac{\mu_1 - \lambda_1}{\lambda - \mu_1} \frac{\Phi_1 \Psi_1}{\Psi_1 \Phi_1},$$

where μ_1 and λ_1 are real numbers, $\Phi_1 = (\phi_1, \bar{\phi}_1)^T$ is a solution with spectral parameter $\lambda = \lambda_1$, $\Psi_1 = (\psi_1, -\bar{\psi}_1)$ is a solution of conjugation system with spectral parameter $\lambda = \mu_1$. One fold DT could yields two dark soliton for defocusing NLS equation. The symmetry relation is given as

$$T^{\dagger}(\bar{\lambda})\sigma_3 T(\lambda) = f(\lambda)\sigma_3,$$

where σ_3 is the third standard Pauli matrix, $f(\lambda)$ is a function of λ. However, try our best, Manãns' method can not be applied to multi-component case besides the degenerate case. Thus we give up generalizing the Manãns's method.

In [24], Degasperis and Lambardo presented a one fold DT method for dark soliton. The Darboux matrix is

$$D(x,t;\lambda) = I + \mathrm{i}\frac{\hat{z}\hat{z}^{\dagger}\Sigma}{(\lambda-\lambda_1)P_1}, \quad \hat{z}^{\dagger}\Sigma\hat{z} = 0, \quad P_1 = \int \hat{z}^{\dagger}\Sigma\hat{z}\mathrm{d}x, \tag{1.1}$$

where $\hat{z}$ is a solution with spectral parameter $\lambda = \lambda_1$, Σ is diagonal matrix with elements ± 1. Evidently, it is not convenient to calculate the integration. Besides, the integration restrict us to do iterate the above DT (1.1) step by step. In order to look for a simpler method, we continue to search some valuable information.

In 2009 [27], Cieśliński revisited different types of DT methods. He pointed out the classical binary DT can be represented as a nilpotent gauge matrix. Indeed, we can see that the DT (1.1) is nothing but nilpotent Darboux matrix. Thus we can check whether or not the DT (1.1) can be converted into the classical binary DT. The answer is affirmative. Firstly, we know that the most important property for DT is the kernel for $D(x,t;\lambda)$:

$$\lim_{\lambda\to\lambda_1} D(x,t;\lambda_1)\hat{z}(\lambda) = 0.$$

On the other hand, we have the following equality

$$\lim_{\lambda\to\lambda_1}\frac{\mathrm{i}\hat{z}_1^{\dagger}\Sigma\hat{z}(\lambda)}{\lambda-\lambda_1} = -\int \hat{z}_1^{\dagger}\Sigma\hat{z}_1\mathrm{d}x.$$

Thus the DT (1.1) can be represented as the classical DT [29]

$$\hat{z}(\lambda) \to D(x,t;\lambda)\hat{z}(\lambda) = \hat{z}(\lambda) - \frac{\hat{z}_1\Omega(\hat{z}_1,\hat{z})}{\Omega(\hat{z}_1,\hat{z}_1)}, \quad \mathrm{d}\Omega(f,g) = f^{\dagger}\Sigma g\mathrm{d}x.$$

Generally speaking, DT is considered as a special gauge transformation [28, 29]. That is the reason that we underestimate the classical binary DT [29, 30, 31]. The binary DT was first proposed by Babich, Matveev and Salle in [30]. The details for binary DT can refer to the reference [29] (and references therein). Indeed, the binary DT is the consistent transformation for AKNS system. With the binary DT method, we can reduce it to Zakharov-Shabat dressing operator [32] or the loop group representation [33]. Particularly, we can obtain the DT which can be used to derive dark soliton very effectively and conveniently. It's worth mentioning that there are some other methods to derive the dark soliton for multi-component NLS equations recently, such as the algebraic-geometry reduction method, KP equation reduction method and dressing-Hirota method [34, 19, 35].

Here, we need to remark that the single step DT for focusing and defocusing multi-component NLS equations was given by Degasperis and Lombardo in 2006. However, they did not given a uniform representation. Very recently before we submitted our paper, Tsuchida studied on the subject by Darboux-Bäcklund transformation [36], in which both bright-soliton solutions and dark-soliton solutions can be obtained, depending on the signs of the nonlinear terms. Indeed, their results are the generalization of Park and Shin's results [37] and Degasperis and Lombardo's results [24, 25] based on a special calculation technique.

In general, if we have one fold DT, it follows that the n-fold DT can be obtained by the iteration. However, it is not fitted for the dark soliton's DT. For the Degasperis and Lombardo's method, the difficulties to iterate the DT is how to calculate integration. As the time of iteration increases, we can not solve the integration even though by computer soft, since the integrands would become very complicated. Indeed, the same thing to be meet on the Tsuchida's method, the first step DT $T(\lambda)$ with special function $\Phi_1(\lambda_1)$ for linear spectral problem, we can deal with the integration by the limit technique. However, if we iterate the DT directly, we can not use the calculation trick again, since there is no the similar equation (3.28). And if we need to iterate the DT, we need to use a new special function $\Phi_2(\lambda_2)$ calculate the limit

$$\lim_{\lambda\to\lambda_2}\frac{[T(\lambda_2)\Phi_2(\lambda_2)]^\dagger\Lambda T(\lambda)\Phi_2(\lambda)}{\lambda-\lambda_2}$$

to replace the integration of Degasperis and Lombardo's integration. However, it is almost impossible to calculate limitation directly even more difficult than the Degasperis and Lombardo's integration.

Based on above mentioned problems, we propose a systematical method to construct the multi-fold uniform DT based on the Degasperis and Lombardo's method [24, 25] and Cieśliński's method [27]. The aim of our work is two fold. Firstly, we reduce the binary DT of AKNS system to obtain a uniform transformation for AKNS system, which can be reduced to all previous presented DT. Secondly, we use the binary DT of AKNS system to derive dark soliton and multi-dark soliton for N-component NLS equations (2.6). The multi-dark soliton for N-component NLS equation (2.6) through DT method is obtained for the first time. In section , we introduce some basic knowledge about AKNS system. Then we give the binary DT for AKNS system. Based on the binary DT, we reduce it as the different transformations of AKNS system. In section , by the uniform transformation and limit technique [8, 38, 39], we derive the single dark soliton and multi-dark soliton for N-component NLS equations (2.6). In order to give us a clear understanding of our formula, we plot the explicit dark soliton and two dark solitons picture of three-component NLS equations. Final section involves some conclusions and discussions.

2 The AKNS system and the Binary Darboux transformation

In this section, we firstly recall some results about the AKNS system and its reduction for multi-component NLS equations. Secondly, we introduce the binary DT for AKNS system. Finally, we reduce the binary DT into different transformation by conjugation equation and limit technique. The integration for high-order solution is automatic through the limit technique.

2.1 The AKNS system

We recall the classical results about the AKNS system [40]. Let $a = \mathrm{diag}(a_1, a_2, \cdots, a_n)$ be a fixed nonzero diagonal matrix in $\mathrm{gl}(n, \mathbb{C})$, and denote

$$\begin{aligned}\mathrm{sl}(n)_a =& \{y \in \mathrm{sl}(n, \mathbb{C}) | [a, y] = 0\},\\ \mathrm{sl}(n)_a^{\perp} =& \{y \in \mathrm{sl}(n, \mathbb{C}) | \mathrm{tr}(zy) = 0 \quad \text{for any } z \in \mathrm{sl}(n)_a\}.\end{aligned}$$

Let $L^{\infty}(\mathbb{R}, \mathrm{sl}(n)_a^{\perp})$ denotes the space of maps in the $L^{\infty}(\mathbb{R})$ class. For the spectral problem

$$\Phi_x = (a\lambda + u(x))\Phi, \tag{2.2}$$

when $\lambda \to \infty$, we have the formal asymptotical behavior

$$\Phi \to \exp[a\lambda x].$$

Thus we can suppose $\Phi = m(x; \lambda)e^{a\lambda x}$, where $m(x; \lambda)$ is an analytical function and possesses the following formal expansion

$$m(x; \lambda) = I + m_1(x)\lambda^{-1} + m_2(x)\lambda^{-2} + \cdots .$$

Substituting it into (2.2), we can obtain

$$\Phi_x\Phi^{-1}=mam^{-1}\lambda+m_xm^{-1}=a\lambda+[m_1,a]+O(\lambda^{-1})$$

and $\Phi_x\Phi^{-1}$ being holomorphic in $\lambda\in\mathbb{C}$ implies that $\Phi_x\Phi^{-1}=a\lambda+[m_1,a]$.

Let $b\in \mathrm{gl}(n,\mathbb{C})$ such that $[b,a]=0$, we have the allowing formal expansion of mbm^{-1} at $\lambda=\infty$,

$$mbm^{-1}\sim Q_{b,0}+Q_{b,1}\lambda^{-1}+Q_{b,2}\lambda^{-2}+\cdots.$$

Since $\Phi_x\Phi^{-1}=\lambda a+u$ and $\Phi b\Phi^{-1}=mbm^{-1}$, we can obtain that

$$\left[\partial_x+a\lambda+u,\Phi b\Phi^{-1}\right]=0.$$

It follows that

$$(Q_{b,j}(u))_x+[u,Q_{b,j}(u)]=[Q_{b,j+1}(u),a]. \tag{2.3}$$

Write

$$Q_{b,j}=T_{b,j}+P_{b,j}\in sl(n)_a+\mathrm{sl}(n)_a^{\perp}.$$

Then equation (2.3) gives

$$\begin{aligned}P_{b,j}&=-\mathrm{ad}(a)^{-1}((P_{b,j-1})_x+\pi_1([u,Q_{b,j-1}])),\\(T_{b,j})_x&=-\pi_0([u,P_{b,j}]),\end{aligned}$$

where π_0 and π_1 denote the projection of $sl(n,\mathbb{C})$ onto $sl(n)_a^{\perp}$ and $sl(n)_a^{\perp}$ with respect to $sl(n,\mathbb{C})=sl(n)_a+sl(n)_a^{\perp}$ respectively. In reference [33], they proved that if b is a polynomial of a, then $Q_{b,j}$ is an order-(j-1) polynomial differential operator in u.

Then we have the following proposition:

Proposition 2.1 (Terng and Uhlenbeck, [33]) *Suppose $u(\cdot,t)\in L^\infty(\mathbb{R},\mathrm{sl}(n)_a^{\perp})$ for all t,*

$$\left[\partial_x+a\lambda+u,\partial_t+b\lambda^j+v_1\lambda^{j-1}+\cdots+v_j\right]=0$$

for some $v_1,\cdots,v_j$, and $\lim\limits_{x\to-\infty}v_k(x,t)=\lim\limits_{x\to-\infty}Q_{b,k}(u(x,t))$ for all $1\leqslant k\leqslant j$. Then we have $v_k=Q_{b,k}(u)$, and

$$u_t=(Q_{b,j}(u))_x+[u,Q_{b,j}(u)]=[Q_{b,j+1}(u),a]. \tag{2.4}$$

In what following, we consider the reality conditions. The details for reality conditions are given in reference [33]. A Lax pair $[\partial_x+A(x,t;\lambda),\partial_t+B(x,t;\lambda)]=0$ is said to satisfy the reality condition if $\sigma(A(x,t;\bar{\lambda}))=A(x,t;\lambda)$ and $\sigma(B(x,t;\bar{\lambda}))=B(x,t;\lambda)$, where the overbar denotes the complex conjugation and σ is complex conjugate linear Lie algebra involution in $\mathrm{sl}(n,\mathbb{C})$.

In this paper, we merely consider the $u(N+1-k,k)$ hierarchy on $L^\infty(\mathbb{R},u_a^\perp)$ [33]. Here

$$u(N+1-k,k)=\{y\in gl(N,\mathbb{C})|y^\dagger\Lambda+\Lambda y=0\},$$

where the symbol † denotes the hermite conjugation, $\Lambda=\mathrm{diag}(1,\Lambda_1)$ and

$$\Lambda_1=\mathrm{diag}(\epsilon_1,\epsilon_2,\cdots,\epsilon_N), \tag{2.5}$$

$\epsilon_i=-1$, for $1\leqslant i\leqslant k$; $\epsilon_i=1$, for $k+1\leqslant i\leqslant N$. Let $a=\mathrm{diag}(\mathrm{i},-\mathrm{i}I_N)$. Then

$$u_a^\perp=\{y\in u(k,N+1-k)|[a,y]=0\}=\left\{\begin{pmatrix}0 & \mathrm{i}q^\dagger\Lambda_1\\ \mathrm{i}q & 0\end{pmatrix}|q\in\mathbb{C}^N\right\}.$$

The second flow in $u(N+1-k,k)$ hierarchy on $L^\infty(\mathbb{R},u_a^\perp)$ with $b=a$ is the following N-component NLS equations:

$$\mathrm{i}\mathbf{q}_t+\frac{1}{2}\mathbf{q}_{xx}+\mathbf{q}\mathbf{q}^\dagger\Lambda_1\mathbf{q}=0, \tag{2.6}$$

where

$$\mathbf{q}=(q_1,\,q_2,\cdots,q_N)^{\mathrm{T}},$$

which admits the following Lax pair

$$\begin{aligned}\Phi_x&=(\mathrm{i}\lambda\sigma_3+\mathrm{i}Q)\,\Phi,\\ \Phi_t&=\left(\mathrm{i}\lambda^2\sigma_3+\mathrm{i}\lambda Q-\frac{1}{2}(\mathrm{i}\sigma_3Q^2-\sigma_3Q_x)\right)\Phi,\end{aligned} \tag{2.7}$$

where

$$Q=\begin{bmatrix}0 & \mathbf{q}^\dagger\Lambda_1\\ \mathbf{q} & \mathbf{0}_{N\times N}\end{bmatrix},\quad \sigma_3=\begin{bmatrix}1 & \mathbf{0}_{1\times N}\\ \mathbf{0}_{N\times 1} & -\mathbf{I}_{N\times N}\end{bmatrix}.$$

If all $\epsilon_i=1$, which corresponds to the focusing case, if all $\epsilon_i=-1$, which corresponds to the defocusing case, otherwise the mixed case. In this paper, we would like to focus on the DT and multi-dark soliton solution for the N-component NLS system (2.6).

2.2 The Binary DT for AKNS system

We consider the binary DT for AKNS system with symmetry reduction. Firstly, we give some lemmas.

Lemma 2.1 *Suppose Φ_1 and Φ are the special vector solutions for system (2.7) at $\lambda=\lambda_1$ and λ respectively, then we can have the following total differential*

$$\mathrm{d}\Omega(\Phi_1,\Phi)=\Phi_1^\dagger\Lambda\sigma_3\Phi\mathrm{d}x+\left[(\lambda+\bar{\lambda}_1)\Phi_1^\dagger\Lambda\sigma_3\Phi+\Phi_1^\dagger\Lambda Q\Phi\right]\mathrm{d}t. \tag{2.8}$$

In addition, we have

$$\Omega(\Phi_1,\Phi)=\frac{\Phi_1^\dagger\Lambda\Phi}{\mathrm{i}(\lambda-\bar{\lambda}_1)}+C. \tag{2.9}$$

If $\lambda_1\in\mathbb{R}$, we have

$$\Omega(\Phi_1,\Phi_1)=\lim_{\lambda\to\lambda_1}\frac{\Phi_1^\dagger\Lambda\Phi}{\mathrm{i}(\lambda-\lambda_1)}+C, \tag{2.10}$$

where C is a complex constant.

Proof Taking complex conjugation to (2.7) both sides, we have

$$\begin{aligned}\Phi_{1,x}^\dagger\Lambda&=-\Phi_1^\dagger\Lambda[\mathrm{i}\bar{\lambda}_1\sigma_3+\mathrm{i}Q],\\ \Phi_{1,t}^\dagger\Lambda&=-\Phi_1^\dagger\Lambda[\mathrm{i}\bar{\lambda}_1^2\sigma_3+\mathrm{i}\bar{\lambda}_1Q-\frac{1}{2}\sigma_3(\mathrm{i}Q^2-Q_x)].\end{aligned} \tag{2.11}$$

Left multiplying by $\Phi_1^\dagger\Lambda$ into both sides of (2.7) and right multiplying by Φ into both sides of (2.11), then we can obtain

$$\begin{aligned}\left[\frac{\Phi_1^\dagger\Lambda\Phi}{\mathrm{i}(\lambda-\bar{\lambda}_1)}\right]_x&=\Phi_1^\dagger\Lambda\sigma_3\Phi,\\ \left[\frac{\Phi_1^\dagger\Lambda\Phi}{\mathrm{i}(\lambda-\bar{\lambda}_1)}\right]_t&=(\lambda+\bar{\lambda}_1)\Phi_1^\dagger\Lambda\sigma_3\Phi+\Phi_1^\dagger\Lambda Q\Phi.\end{aligned}$$

It follows that equations (2.8), (2.9) and (2.10) are verified. □

In what following, to keep the uniqueness of the constants C, we choose it as zero. Following the idea in the introduction, we can obtain that the one fold binary DT for N-component NLS equations (2.6) is

$$\begin{aligned}\Phi\to\Phi[1]&=\Phi-\frac{\Phi_1\Omega(\Phi_1,\Phi)}{\Omega(\Phi_1,\Phi_1)},\\ Q\to Q[1]&=Q-\mathrm{i}\left[\sigma_3,\frac{\Phi_1\Phi_1^\dagger\Lambda}{\Omega(\Phi_1,\Phi_1)}\right].\end{aligned} \tag{2.12}$$

In the following, we verify the validity of the above transformation (2.12).

Theorem 2.1 *Suppose Φ satisfies the system (2.7), and Φ_1 is a special solution for Lax pair (2.7) at $\lambda=\lambda_1$, and $\Phi_1^\dagger\Lambda\Phi_1=0$ if $\lambda_1\in\mathbb{R}$, then we have*

$$
\begin{aligned}
&\Phi[1]_x=(\mathrm{i}\lambda\sigma_3+\mathrm{i}Q[1])\,\Phi[1],\\
&\Phi[1]_t=\left(\mathrm{i}\lambda^2\sigma_3+\mathrm{i}\lambda Q[1]-\frac{1}{2}(\mathrm{i}\sigma_3Q[1]^2-\sigma_3Q[1]_x)\right)\Phi[1].
\end{aligned}\tag{2.13}
$$

Proof We firstly verify the first equation of (2.13). By lemma 2.1, we have $\Omega(\Phi_1,\Phi)=\dfrac{\Phi_1^\dagger\Lambda\Phi}{\mathrm{i}(\lambda-\bar\lambda_1)}$. It follows that

$$
\begin{aligned}
\Phi[1]_x=&(\mathrm{i}\lambda\sigma_3+\mathrm{i}Q)\Phi-\frac{(\mathrm{i}\lambda_1\sigma_3+\mathrm{i}Q)\Phi_1\Omega(\Phi_1,\Phi)}{\Omega(\Phi_1,\Phi_1)}-\frac{\Phi_1\Phi_1^\dagger\Lambda\sigma_3\Phi}{\Omega(\Phi_1,\Phi_1)}+\frac{\Phi_1\Phi_1^\dagger\Lambda\sigma_3\Phi_1\Omega(\Phi_1,\Phi)}{\Omega^2(\Phi_1,\Phi_1)}\\
=&\left[\mathrm{i}\lambda\sigma_3+\mathrm{i}\left(Q-\mathrm{i}\frac{\sigma_3\Phi_1\Phi_1^\dagger\Lambda}{\Omega(\Phi_1,\Phi_1)}+\mathrm{i}\frac{\Phi_1\Phi_1^\dagger\Lambda\sigma_3}{\Omega(\Phi_1,\Phi_1)}\right)\right]\left(\Phi-\frac{\Phi_1\Omega(\Phi_1,\Phi)}{\Omega(\Phi_1,\Phi_1)}\right),
\end{aligned}
$$

where the second equality use the relation $\Phi_1^\dagger\Lambda\Phi_1=0$ if $\lambda_1\in\mathbb{R}$. And it is readily to verify the validity of symmetry relation for $Q[1]$. Then the first equation of (2.13) is verified. Besides, we can obtain the following relation. Since $\Phi[1]=T\Phi$,

$$
T=I-\frac{\Phi_1\Phi_1^\dagger\Lambda}{\mathrm{i}(\lambda-\bar\lambda_1)\Omega(\Phi_1,\Phi_1)},
$$

it follows that

$$
T_x+\mathrm{i}T(\lambda\sigma_3+Q)=\mathrm{i}(\lambda\sigma_3+Q[1])T.\tag{2.14}
$$

Expanding the matrix T with $T=I+\dfrac{T_1}{\lambda}+\dfrac{T_2}{\lambda^2}+\cdots$, substituting into equation (2.14), and comparing the coefficient of λ, we can get

$$
Q[1]T_1=T_1Q+[T_2,\sigma_3]-\mathrm{i}T_{1,x}.\tag{2.15}
$$

Finally, we consider the time evolution equation of (2.13). Since matrix T is a special gauge transformation, by directly calculating we have

$$
\left[T_t+T\left(\mathrm{i}\lambda^2\sigma_3+\mathrm{i}\lambda Q-\frac{1}{2}\sigma_3(\mathrm{i}Q^2-Q_x)\right)\right]T^{-1}=\mathrm{i}\lambda^2\sigma_3+\mathrm{i}\lambda Q[1]+V_1[1],\tag{2.16}
$$

where $V_1[1]=-\dfrac{1}{2}(\mathrm{i}\sigma_3Q^2-\sigma_3Q_x)+\mathrm{i}[T_2,\sigma_3]+\mathrm{i}(T_1Q-Q[1]T_1)$. By equation (2.15), we have $V_1[1]^o=\dfrac{1}{2}\sigma_3Q[1]_x$, and then we have $V_1[1]^d=-\dfrac{1}{2}\sigma_3Q[1]^2$ by directly calculating. The superscript ^{o} and ^{d} denote the off-diagonal and diagonal part of the block matrix respectively. This completes the theorem. □

In the following, we consider the n-fold binary DT based on above theorem.

Theorem 2.2 *Suppose we have n different solutions Φ_is for Lax pair (2.7) at $\lambda = \lambda_i$ $(i = 1, 2, \cdots, n)$, and $\Phi_j^\dagger \Lambda \Phi_j = 0$ if $\lambda_j \in \mathbb{R}$, then the n-fold binary DT can be presented as*

$$\Phi \to \Phi[n] = \Phi - \Theta M^{-1}\Omega, \quad \Theta = [\Phi_1, \Phi_2, \cdots, \Phi_n], \tag{2.17}$$

where

$$M = \begin{bmatrix} \Omega(\Phi_1, \Phi_1) & \Omega(\Phi_1, \Phi_2) & \cdots & \Omega(\Phi_1, \Phi_n) \\ \Omega(\Phi_2, \Phi_1) & \Omega(\Phi_2, \Phi_2) & \cdots & \Omega(\Phi_2, \Phi_n) \\ \vdots & \vdots & & \vdots \\ \Omega(\Phi_n, \Phi_1) & \Omega(\Phi_n, \Phi_2) & \cdots & \Omega(\Phi_n, \Phi_n) \end{bmatrix}, \quad \Omega = \begin{bmatrix} \Omega(\Phi_1, \Phi) \\ \Omega(\Phi_2, \Phi) \\ \vdots \\ \Omega(\Phi_n, \Phi) \end{bmatrix}.$$

The transformation between fields is

$$Q \to Q[n] = Q - \mathrm{i}\left[\sigma_3, \Theta M^{-1}\Theta^\dagger \Lambda\right]. \tag{2.18}$$

Proof In the first place, we have

$$\begin{aligned}
&\Phi[n]_x \\
&= \Phi_x - \Theta_x M^{-1}\Omega - \Theta M^{-1}\Omega_x + \Theta M^{-1} M_x M^{-1}\Omega \\
&= (\mathrm{i}\lambda\sigma_3 + \mathrm{i}Q)\Phi - \mathrm{i}(Q\Theta + \sigma_3\Theta D)M^{-1}\Omega - \Theta M^{-1}\Theta^\dagger\Lambda\sigma_3\Phi + \Theta M^{-1}\Theta^\dagger\Lambda\sigma_3\Theta M^{-1}\Omega \\
&= (\mathrm{i}\lambda\sigma_3 + \mathrm{i}Q)\Phi - \mathrm{i}Q\Theta M^{-1}\Omega - \Theta M^{-1}\Theta^\dagger\Lambda\sigma_3\Phi + \Theta M^{-1}\Theta^\dagger\Lambda\sigma_3\Theta M^{-1}\Omega \\
&\quad + \mathrm{i}\sigma_3\Theta M^{-1}(-\lambda + D^\dagger)\Omega + \sigma_3\Theta M^{-1}\Theta^\dagger\Lambda\Phi - \mathrm{i}\sigma_3\Theta D^\dagger M^{-1}\Omega \\
&= (\mathrm{i}\lambda\sigma_3 + \mathrm{i}Q)\Phi - \mathrm{i}Q\Theta M^{-1}\Omega - \Theta M^{-1}\Theta^\dagger\Lambda\sigma_3\Phi + \Theta M^{-1}\Theta^\dagger\Lambda\sigma_3\Theta M^{-1}\Omega \\
&\quad - \mathrm{i}\lambda\sigma_3\Theta M^{-1}\Omega + \sigma_3\Theta M^{-1}\Theta^\dagger\Lambda\Phi - \sigma_3\Theta M^{-1}(\mathrm{i}MD - \mathrm{i}D^\dagger M)M^{-1}\Omega \\
&= (\mathrm{i}\lambda\sigma_3 + \mathrm{i}Q)\Phi - \mathrm{i}Q\Theta M^{-1}\Omega - \Theta M^{-1}\Theta^\dagger\Lambda\sigma_3\Phi + \Theta M^{-1}\Theta^\dagger\Lambda\sigma_3\Theta M^{-1}\Omega \\
&\quad - \mathrm{i}\lambda\sigma_3\Theta M^{-1}\Omega + \sigma_3\Theta M^{-1}\Theta^\dagger\Lambda\Phi - \sigma_3\Theta M^{-1}\Theta^\dagger\Lambda\Theta M^{-1}\Omega \\
&= (\mathrm{i}\lambda\sigma_3 + \mathrm{i}Q[n])\Phi[n],
\end{aligned}$$

where the third equality uses the relation $\mathrm{i}(-\lambda + D^\dagger)\Omega + \Theta^\dagger\Lambda\Phi = 0$, the fifth equality uses the relation $\Theta^\dagger\Lambda\Theta = \mathrm{i}(MD - D^\dagger M)$ and $D = \mathrm{diag}(\lambda_1, \lambda_2, \cdots, \lambda_n)$. Thus the spectral problem is valid.

Based on the above theorem, we merely need to verify the spectral problem. The time evolution part is similar as above theorem. Thus we omit it. □

2.3 The uniform transformation through binary DT

In this subsection, we consider reduction from the binary DT. For convenience, we merely consider the one-fold binary DT, since the n-fold binary DT is nothing but

n-times iteration of one-fold DT. Firstly, we consider how to reduce the binary DT into Zakharov-Shabat dressing operator [4]. If the spectral parameters $\lambda_1 \neq \bar{\lambda}_1$, we use the relation

$$
\begin{aligned}
&\left(\frac{\Phi_1^\dagger \Lambda \Phi_1}{\lambda_1 - \bar{\lambda}_1}\right)_x = \mathrm{i}\Phi_1^\dagger \sigma_3 \Lambda \Phi_1, \\
&\left(\frac{\Phi_1^\dagger \Lambda \Phi_1}{\lambda_1 - \bar{\lambda}_1}\right)_t = \mathrm{i}\left[(\lambda_1 + \bar{\lambda}_1)\Phi_1^\dagger \Lambda \sigma_3 \Phi_1 + \Phi_1^\dagger \sigma_3 \Lambda Q \Phi_1\right].
\end{aligned} \tag{2.19}
$$

It follows that

$$
\Omega(\Phi_1, \Phi_1) = -\mathrm{i}\left(\frac{\Phi_1^\dagger \Lambda \Phi_1}{\lambda_1 - \bar{\lambda}_1}\right).
$$

Then we have

$$
\Phi[1] = \left(I + \frac{\bar{\lambda}_1 - \lambda_1}{\lambda - \bar{\lambda}_1}\frac{\Phi_1 \Phi_1^\dagger \Lambda}{\Phi_1^\dagger \Lambda \Phi_1}\right)\Phi, \quad Q[1] = Q + (\lambda_1 - \bar{\lambda}_1)\left[\sigma_3, \frac{\Phi_1 \Phi_1^\dagger \Lambda}{\Phi_1^\dagger \Lambda \Phi_1}\right].
$$

By this transformation, we can obtain bright soliton, breather and rogue wave solution. The high-order DT of this type was obtained in reference [8] by the limit technique.

If the spectral parameters $\lambda_1 = \bar{\lambda}_1$. In this case, we use the limit technique to deal with this problem. Suppose Ψ_1 and Φ_1 are two mutual dependent solutions for Lax pair at $\lambda = \lambda_1$ such that $\Phi_1^\dagger \Lambda \Psi_1 \equiv C_1 = \text{const} \neq 0$ and $\Phi_1^\dagger \Lambda \Phi_1 = 0$, set $\Theta_1(\nu) = \Phi_1(\nu) + \dfrac{\beta(\nu - \lambda_1)}{C_1}\Psi_1(\lambda_1)$, then we can obtain

$$
\begin{aligned}
&\lim_{\nu \to \lambda_1}\left(\frac{\Phi_1^\dagger \Lambda \Theta_1(\nu)}{\nu - \lambda_1}\right)_x = \mathrm{i}\Phi_1^\dagger \sigma_3 \Lambda \Phi_1, \\
&\lim_{\nu \to \lambda_1}\left(\frac{\Phi_1^\dagger \Lambda \Theta_1(\nu)}{\nu - \lambda_1}\right)_t = \mathrm{i}\left[2\lambda_1 \Phi_1^\dagger \Lambda \sigma_3 \Phi_1 + \Phi_1^\dagger \sigma_3 \Lambda Q \Phi_1\right].
\end{aligned} \tag{2.20}
$$

It follows that

$$
\Omega(\Phi_1, \Phi_1) = -\mathrm{i}\lim_{\nu \to \lambda_1}\left(\frac{\Phi_1^\dagger \Lambda \Theta_1(\lambda)}{\nu - \lambda_1}\right).
$$

Then we have

$$
\begin{aligned}
&\Phi[1] = \lim_{\nu \to \lambda_1}\left(I + \frac{\lambda_1 - \nu}{\lambda - \lambda_1}\frac{\Phi_1 \Phi_1^\dagger \Lambda}{\Phi_1^\dagger \Lambda \Theta_1(\nu)}\right)\Phi, \\
&Q[1] = Q + \lim_{\nu \to \lambda_1}\left[\sigma_3, \frac{(\nu - \lambda_1)\Phi_1 \Phi_1^\dagger \Lambda}{\Phi_1^\dagger \Lambda \Theta_1(\nu)}\right].
\end{aligned} \tag{2.21}
$$

To keep the non-singularity of above transformation, we have

$$\beta + \lim_{\nu \to \lambda_1} \left(\frac{\Phi_1^\dagger \Lambda \Phi_1(\nu)}{\nu - \lambda_1} \right) \neq 0, \text{ for any } (x,t) \in \mathbb{R}^2.$$

In the following section, we would like to use the above transformation to derive the single dark soliton and multi-dark soliton for N-component NLS equations (2.6).

3 Dark soliton and multi-dark soliton

In this section, we consider the application for binary DT. A direct application is using the DT to derive some special solutions. By the DT (2.21), we can obtain the dark soliton and multi-dark soliton for N-component NLS equations (2.6) in a simple way.

3.1 Single dark soliton for N-component NLS equations

To obtain the dark soliton, we use the seed solutions

$$q_i = c_i \, e^{\mathrm{i}\theta_i}, \ \theta_i = a_i \, x - \left(\frac{{a_i}^2}{2} - \sum_{l=1}^{N} \epsilon_l {c_l}^2 \right) t, \ \ i = 1, 2, \cdots, N. \tag{3.22}$$

In the first place, we need to solve the Lax pair equation (2.7) with above seed solutions. In order to solve the Lax pair equation, we use the gauge transformation

$$D = \mathrm{diag}\left(1, \, e^{\mathrm{i}\theta_1}, \cdots, e^{\mathrm{i}\theta_N}\right)$$

converts the variable coefficient differential equation into constant coefficient equation. Then we can obtain

$$\begin{aligned} &\Phi_{0,x} = \mathrm{i} U_0 \Phi_0, \quad \Phi = D\Phi_0, \\ &\Phi_{0,t} = \mathrm{i} \left(\frac{1}{2} U_0^2 + \lambda U_0 - \sum_{l=1}^{N} \epsilon_l {c_l}^2 - \frac{1}{2} \lambda^2 \right) \Phi_0, \end{aligned} \tag{3.23}$$

where

$$U_0 = \begin{bmatrix} \lambda & C\Lambda_1 \\ C^T & -\lambda I_N - A \end{bmatrix}, \quad A = \mathrm{diag}(a_1, \cdots, a_N), \quad C = (c_1, \cdots, c_N). \tag{3.24}$$

In the following, we consider the property of the matrix U_0. Firstly we can obtain the characteristic equation of matrix U_0:

$$\det(\mu - U_0) = 0. \tag{3.25}$$

Then we have the vector solution for (3.23):

$$\Phi_0 = \begin{bmatrix} e^{\mathrm{i}X} \\ (\lambda + a_1 + \mu)^{-1} e^{\mathrm{i}X} \\ \vdots \\ (\lambda + a_N + \mu)^{-1} e^{\mathrm{i}X} \end{bmatrix}, \quad X = \mu x + \left(\frac{1}{2}\mu^2 + \lambda\mu - \frac{1}{2}\lambda^2 - \sum_{l=1}^{N} \epsilon_l {c_l}^2 \right) t.$$

Since the coefficient of algebraic equation (3.25) is real number, the roots of equation (3.25) are either real roots or complex conjugation root pairs. Thus if the number of real roots is less than the order of algebraic equation, then there exists complex conjugation root pair. The real roots of algebraic equation can be found by existence of theorem of zero root. To obtain the dark soliton, we need to choose the pair of conjugate complex roots of the characteristic equation (3.25). If μ and $\bar{\mu}$ are the roots of characteristic equation (3.25), then we have

$$\mu_j - \lambda_j - \sum_{l=1}^{N} \frac{\epsilon_l {c_l}^2}{\lambda_j + \mu_j + a_l} = 0, \tag{3.26}$$

and

$$\bar{\mu}_i - \lambda_i - \sum_{l=1}^{N} \frac{\epsilon_l {c_l}^2}{\lambda_i + \bar{\mu}_i + a_l} = 0, \tag{3.27}$$

where μ_j and $\bar{\mu}_i$ are the roots of characteristic equation (3.25) with $\lambda = \lambda_j$ and $\lambda = \lambda_i$ respectively $(i, j = 1, 2, \cdots, n)$, the overbar denotes the complex conjugation. It follows that

$$\mu_j - \lambda_j - (\bar{\mu}_i - \lambda_i) + \sum_{l=1}^{N} \frac{\epsilon_l {c_l}^2 [(\lambda_j + \mu_j) - (\lambda_i + \bar{\mu}_i)]}{(\lambda_j + \mu_j + a_l)(\lambda_i + \bar{\mu}_i + a_l)} = 0.$$

Then we can obtain that

$$\frac{\mu_j - \lambda_j - (\bar{\mu}_i - \lambda_i)}{[(\lambda_j + \mu_j) - (\lambda_i + \bar{\mu}_i)]} + \sum_{l=1}^{N} \frac{\epsilon_l {c_l}^2}{(\lambda_j + \mu_j + a_l)(\lambda_i + \bar{\mu}_i + a_l)} = 0. \tag{3.28}$$

Thus the formula

$$\frac{\Phi_i^\dagger \Lambda \Phi_j}{\lambda_j - \lambda_i} = \frac{2 e^{\mathrm{i}(X_j - \bar{X}_i)}}{\lambda_j - \lambda_i + \mu_j - \bar{\mu}_i}, \tag{3.29}$$

where

$$\begin{aligned} &\Phi_i = D\Phi_0|_{\lambda=\lambda_i, \mu=\mu_i}, \quad \Phi_j = D\Phi_0|_{\lambda=\lambda_j, \mu=\mu_j} \\ &X_j = \left[\mu_j x + \left(\lambda_j \mu_j + \frac{1}{2}{\mu_j}^2 - \frac{1}{2}{\lambda_j}^2 \right) t \right], \\ &\bar{X}_i = \left[\bar{\mu}_i x + \left(\lambda_i \bar{\mu}_i + \frac{1}{2}\bar{\mu}_i^2 - \frac{1}{2}{\lambda_i}^2 \right) t \right]. \end{aligned}$$

With this formula, we can readily take limit $\lambda_j \to \lambda_i$, it follows that

$$\lim_{\lambda_j \to \lambda_i} \frac{\Phi_i^\dagger \Lambda \Phi_j}{\lambda_j - \lambda_i} = \frac{2\mathrm{e}^{\mathrm{i}(X_i - \bar{X}_i)}}{\mu_i - \bar{\mu}_i}, \tag{3.30}$$

which also implies that $\Phi_i^\dagger \Lambda \Phi_i = 0$. Then we come back to the DT (2.21). Through above explicit expression and set $\beta = \dfrac{2\mathrm{e}^{2\alpha\mu_{1I}}}{\mu_1 - \bar{\mu}_1}$, where $\mu_{1I} = \mathrm{Im}(\mu_1)$ and $\alpha \in \mathbb{R}$, we have

$$\lim_{\nu \to \lambda_1} \frac{\Phi_1^\dagger \Lambda \Theta_1(\nu)}{\nu - \lambda_1} = \frac{2[\mathrm{e}^{\mathrm{i}(X_1 - \bar{X}_1)} + \mathrm{e}^{2\alpha\mu_{1I}}]}{\mu_1 - \bar{\mu}_1}. \tag{3.31}$$

Then the DT (2.21) can be constructed explicitly as

$$T = I - \frac{(\mu_1 - \bar{\mu}_1)\Phi_1 \Phi_1^\dagger \Lambda}{2(\lambda - \lambda_1)(\mathrm{e}^{\mathrm{i}(X_1 - \bar{X}_1)} + \mathrm{e}^{2\alpha\mu_{1I}})}. \tag{3.32}$$

It follows that single dark soliton solutions for N-component NLS equations (2.6) are

$$q_i[1] = c_i \left[1 - \frac{B_i}{2} + \frac{B_i}{2}\tanh(Y_1)\right] e^{\mathrm{i}\theta_i}, \tag{3.33}$$

where

$$B_i = \frac{\mu_1 - \bar{\mu}_1}{\lambda_1 + a_i + \mu_1}, \quad i = 1, 2, \cdots, N,$$
$$Y_1 = -\mu_{1I}\left[x + (\lambda + \mu_{1R})t + \alpha\right], \ \mu_{1I} = \mathrm{Im}(\mu_1), \ \mu_{1R} = \mathrm{Re}(\mu_1).$$

Without losing of generality, we suppose $\mu_{1I} > 0$. When $x \to -\infty$, we have

$$q_i[1] \to c_i\, e^{\mathrm{i}\theta_i}.$$

When $x \to +\infty$, we have

$$q_i[1] \to c_i\, e^{\mathrm{i}(\theta_i + \omega_i)}, \quad \ln\left(\frac{\lambda_1 + a_i + \bar{\mu}_1}{\lambda_1 + a_i + \mu_1}\right) = \mathrm{i}\omega_i.$$

The center of dark soliton $q_i[1]$ is along the line $x + (\lambda + \mu_{1R})t + \alpha = 0$. The velocity of dark soliton $|q_i[1]|^2$ is $v = -(\lambda_1 + \mu_{1R})$. The depth of cavity of $|q_i[1]|^2$ is

$$\frac{c_i^2 \mu_{1I}^2}{(\lambda_1 + a_i + \mu_{1R})^2 + \mu_{1I}^2}.$$

In the following, we consider how to determine whether or not exist the dark bound states [19]. Through the relation (3.28), we have

$$-\sum_{l=1}^{N} \frac{\epsilon_l {c_l}^2}{|\lambda_1 + \mu_1 + a_l|^2} = 1. \tag{3.34}$$

Indeed, through the expression for single dark soliton (3.33), to obtain the velocity of dark soliton, we need to know the parameter $\lambda_1+\mu_1$. And the velocity of soliton is controlled by $-(\lambda_1+\mu_{1R})$. Thus if we need to find the soliton with velocity equals to zero, we merely to solve the following equation.

$$-\sum_{l=1}^{N}\frac{\epsilon_l {c_l}^2}{a_l^2+\mu_{1I}^2}=1. \tag{3.35}$$

If $\epsilon_l=-1$ for all l, which corresponds to the defocusing case, then the function $F(\mu_{1I})=\sum\limits_{l=1}^{N}\frac{{c_l}^2}{a_l^2+\mu_{1I}^2}$ is an increasing function in positive half axis. Then the equation (3.35) merely has a positive solution. Thus in defocusing case, there exist no dark bound state. So the dark bound state merely maybe exists in the mixed case.

In what following, we illustrate some exact examples to the single dark soliton. Since the velocity of soliton possesses the exact physical meaning, we can obtain the soliton by the velocity $v=-(\lambda_1+\mu_{1R})$. Firstly we solve the following equation about μ_{1I}:

$$\sum_{l=1}^{N}\frac{-\epsilon_l {c_l}^2}{(a_l-v)^2+\mu_{1I}^2}=1.$$

Substituting $\mu_1=-(v+\lambda_1)+\mathrm{i}\mu_{1I}^2$ into characteristic equation (3.25), we can obtain an algebraic equation about λ_1. Solving the algebraic equation, we can obtain all of parameters about single dark soliton. For instance, we consider the three-component NLS equtions with the defocusing case (i.e. $N=3$, $\epsilon_1=\epsilon_2=\epsilon_3=-1$). If we need to find the soliton with velocity $v=0$, then we choose the parameters:

$$\begin{aligned}&a_1=1,\ a_2=-1,\ a_3=0,\ c_1=1,\ c_2=2,\ c_3=\frac{3}{2},\ \alpha=0,\\&\lambda_1=-\frac{12}{33+\sqrt{769}},\ \mu_1=\frac{99-3\sqrt{769}}{80}+\frac{\mathrm{i}}{4}\sqrt{50+2\sqrt{769}}.\end{aligned} \tag{3.36}$$

We can plot the picture of single dark soliton by Maple (Fig. 1). Since the soliton are stationary, we merely plot the picture at $t=0$.

3.2 Multi-dark soliton for N-component NLS equations

In order to give the multi-dark soliton solution, we first adapt the binary DT with the limit technique. The n-fold binary DT (2.17) can be written with the following form

$$\Phi[n]=\Phi-\sum_{i=1}^{n}s_i\Omega(\Phi_i,\Phi). \tag{3.37}$$

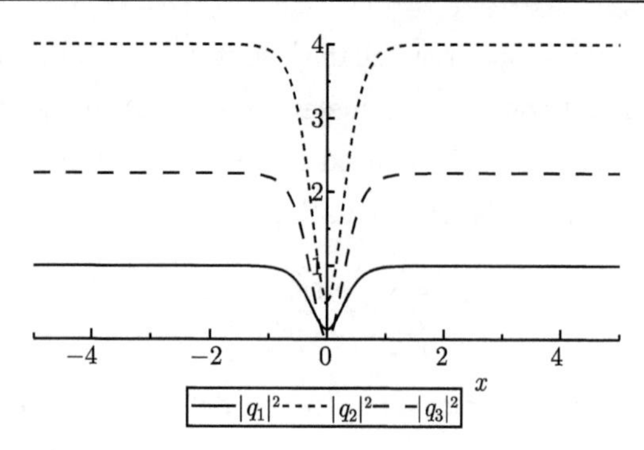

Figure 1 The figure at $t=0$: Solid line $|q_1|^2$, Dot line $|q_2|^2$, Dash line $|q_3|^2$. The parameters are given in equations (3.36)

Thus we can suppose that

$$\Phi[n]=T_n\Phi,\quad T_n=I-\sum_{i=1}^{n}s_i\frac{\Phi_i^{\dagger}\Lambda}{\lambda-\lambda_i}. \tag{3.38}$$

The explicit expression for Darboux matrix T_n can be determined by the following equations

$$\lim_{\lambda\to\lambda_j}T_n\left(\Phi_j+\frac{\beta_j}{C_j}(\lambda-\lambda_j)\Psi_j\right)=0,$$

where Ψ_j is mutual independent solution with Φ_j at $\lambda=\lambda_j$, $\beta_j=\dfrac{2}{\mu_j-\bar{\mu}_j}\mathrm{e}^{2\alpha_j\mu_{jI}}$, $\mu_{jI}=\mathrm{Im}(\mu_j)$, $\alpha_j\in\mathbb{R}$ and $C_j\equiv\Phi_j^{\dagger}\Lambda\Psi_j=\text{const}$. By linear algebra, we have the following expression for T_n:

$$T_n=I-\Theta M^{-1}(\lambda-D)^{-1}\Theta^{\dagger}\Lambda,\quad \Theta=[\Phi_1,\Phi_2,\cdots,\Phi_n], \tag{3.39}$$

where

$$M=\begin{bmatrix}\lim\limits_{\nu\to\lambda_1}\dfrac{\Phi_1^{\dagger}\Lambda\Phi_1(\nu)}{\nu-\lambda_1}+\beta_1 & \dfrac{\Phi_1^{\dagger}\Lambda\Phi_2}{\lambda_2-\lambda_1} & \cdots & \dfrac{\Phi_1^{\dagger}\Lambda\Phi_n}{\lambda_n-\lambda_1}\\ \dfrac{\Phi_2^{\dagger}\Lambda\Phi_1}{\lambda_1-\lambda_2} & \lim\limits_{\nu\to\lambda_2}\dfrac{\Phi_2^{\dagger}\Lambda\Phi_2(\nu)}{\nu-\lambda_2}+\beta_2 & \cdots & \dfrac{\Phi_2^{\dagger}\Lambda\Phi_n}{\lambda_n-\lambda_2}\\ \vdots & \vdots & & \vdots\\ \dfrac{\Phi_n^{\dagger}\Lambda\Phi_1}{\lambda_1-\lambda_n} & \dfrac{\Phi_n^{\dagger}\Lambda\Phi_2}{\lambda_2-\lambda_n} & \cdots & \lim\limits_{\nu\to\lambda_n}\dfrac{\Phi_n^{\dagger}\Lambda\Phi_n(\nu)}{\nu-\lambda_n}+\beta_n\end{bmatrix},$$

$D=\mathrm{diag}(\lambda_1,\lambda_2,\cdots,\lambda_n)$.

By the equality (3.29) and (3.30) in the above subsection, together with transformation (2.18), then the n-dark soliton solutions for equations (2.6) can be represented as following:

$$q_i[n] = 2\frac{\begin{vmatrix} M & X^\dagger \\ \Theta_i & \frac{c_i}{2} \end{vmatrix}}{|M|}\mathrm{e}^{\mathrm{i}\theta_i},\ i = 1, 2, \cdots, N, \tag{3.40}$$

where

$$M = \left(\frac{2\left[\mathrm{e}^{\mathrm{i}(X_j - \bar{X}_m)} + \delta_{mj}\right]}{(\lambda_j + \mu_j) - (\lambda_m + \bar{\mu}_m)}\right)_{1\leqslant m,j\leqslant n},$$

$$X = \left[\mathrm{e}^{\mathrm{i}X_1}, \mathrm{e}^{\mathrm{i}X_2}, \cdots, \mathrm{e}^{\mathrm{i}X_n}\right],$$

$$\Theta_i = \left[\beta_{1,i}\mathrm{e}^{\mathrm{i}X_1}, \beta_{2,i}\mathrm{e}^{\mathrm{i}X_2}, \cdots, \beta_{N,i}\mathrm{e}^{\mathrm{i}X_n}\right],$$

and

$$\delta_{mj} = \begin{cases} 0, & m \neq j, \\ e^{2\alpha_m \mu_{mI}}, & m = j, \end{cases}, \quad \beta_{j,i} = \frac{1}{\lambda_j + a_i + \mu_j},\ j = 1, 2, \cdots, N.$$

Furthermore, the above formula (3.40) can be reduced as the following compact expression

$$q_i[n] = c_i\frac{|M_i|}{|M|}\mathrm{e}^{\mathrm{i}\theta_i}, \tag{3.41}$$

where

$$M_i = \left(\frac{2\left[\mathrm{e}^{\mathrm{i}(X_j - \bar{X}_m)} + \delta_{mj}\right]}{(\lambda_j + \mu_j) - (\lambda_m + \bar{\mu}_m)} - \frac{2\beta_{j,i}}{c_i}\mathrm{e}^{\mathrm{i}(X_j - \bar{X}_m)}\right)_{1\leqslant m,j\leqslant n}.$$

In what following, we consider some dynamics for two dark soliton (3.41) of three-component NLS equations. Firstly, we consider the defocusing case $\epsilon_i = -1$, $i = 1, 2, 3$. By the method in the above subsection (we choose the velocity $v_1 = 1$ and $v_2 = -1$), the parameters are choosing as following:

$$\begin{aligned} & a_1 = 1,\ a_2 = -1,\ a_3 = 0,\ c_1 = 1,\ c_2 = 2,\ c_3 = \frac{3}{2}, \\ & \lambda_1 \approx -1.121588903,\ \lambda_2 \approx 0.7430861497,\ \alpha_1 = \alpha_2 = 0, \\ & \mu_1 \approx 0.1215889040 + 2.265094396\mathrm{i},\ \mu_2 \approx 0.2569138501 + 2.564117194\mathrm{i}. \end{aligned} \tag{3.42}$$

Then we can show the dynamics of two dark soliton in figure 2. It is seen that the two dark solitons in each component collide elastically. The "phase shift" still emerge after the two soliton interaction, which is similar to the bright soliton interactions in scalar case.

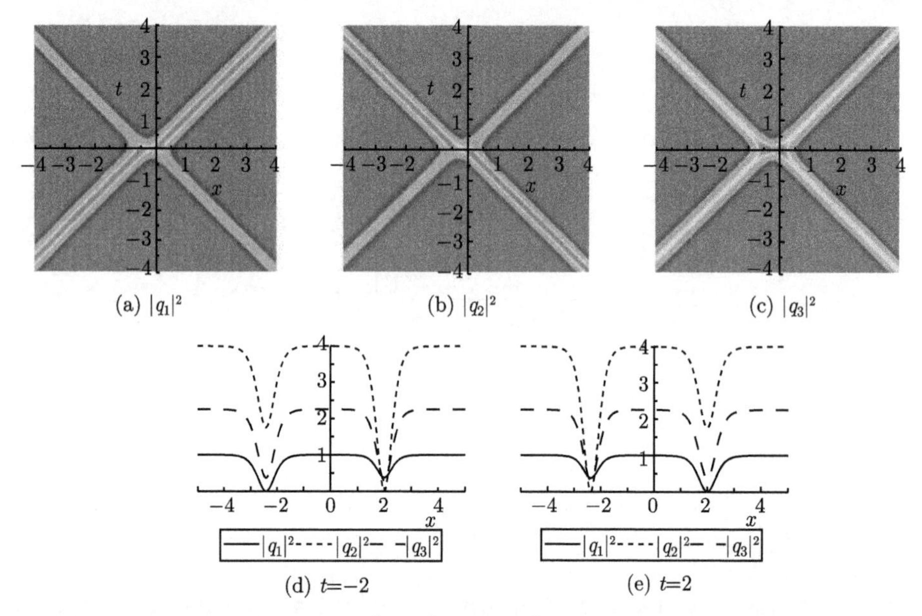

(a) $|q_1|^2$ (b) $|q_2|^2$ (c) $|q_3|^2$

(d) t=−2 (e) t=2

Figure 2 (a)–(e) Density plot of $|q_1|^2, |q_2|^2$ and $|q_3|^2$ pectively. (d)–(e) Solid line $|q_1|^2$, Dot line $|q_2|^2$, Dash line $|q_3|^2$ $t=-2$ in (d) and $t=2$ in (e). The parameters are given in equations (3.43)

Secondly, we consider the mixed focusing and defocusing case $\epsilon_i=-1$, $i=1,2$ and $\epsilon_3=1$. In the first place, we consider the two-pole two-dark soliton. Since the characteristic equation (3.25) for three-component NLS equations is a quartic equation, there maybe exists two pairs of conjugate complex roots. These kinds of soliton can not exist in the scalar or two-component NLS system, since the characteristic equation is not allowed to exist two pairs of conjugation complex roots. For instance, we choose the parameters as following

$$\begin{aligned}&a_1=1,\, a_2=-1,\, a_3=0,\, c_1=c_2=c_3=1,\\&\lambda_1=\lambda_2=0,\, \mu_1=\frac{\sqrt{2}}{2}(1+\mathrm{i}),\, \mu_2=-\frac{\sqrt{2}}{2}(1+\mathrm{i}),\, \alpha_1=\alpha_2=0.\end{aligned}\tag{3.43}$$

The figures are given in Fig 3. However, it is seen that there is not evident different dynamics behavior with ordinary two-dark soliton solution.

Then, we consider the two bound state of three-component NLS equations, namely, the two solitons have the same velocity. The parameters are choosing by the method of above subsection. Firstly, we choose the background parameters a_i, c_i and $v=0$, we can obtain two different values $\mathrm{Im}(\mu_1), \mathrm{Im}(\mu_2)$. And then substituting them into characteristic equation (3.25), we can obtain the two different spectral parameters λ_1, λ_2. Since the parameters α_1, α_2 depend the initial position of soliton, we choose different values to distinguish two soliton. For instance, we choose the parameters as following

$$
\begin{aligned}
&a_1 = 1,\ a_2 = -1,\ a_3 = 0,\ c_1 = 2,\ c_2 = c_3 = 1,\\
&\lambda_1 = \frac{3}{2}\frac{\sqrt{5}-1}{3\sqrt{5}-5},\ \lambda_2 = \frac{3}{2}\frac{\sqrt{5}+1}{3\sqrt{5}-5},\ \alpha_1 = \frac{5}{\sqrt{5}-1},\ \alpha_2 = \frac{-5}{\sqrt{5}+1},\\
&\mu_1 = -\frac{15+3\sqrt{5}}{20} + \frac{\mathrm{i}}{2}(\sqrt{5}-1),\ \mu_2 = -\frac{15-3\sqrt{5}}{20} + \frac{\mathrm{i}}{2}(\sqrt{5}+1),
\end{aligned}
\tag{3.44}
$$

10 t 5 −10 −5 0 5 10 x −5 −10

(a) $|q_1|^2$

10 t 5 −10 −5 0 5 10 x −5 −10

(b) $|q_2|^2$

10 t 5 −10 −5 0 5 10 x −5 −10

(c) $|q_3|^2$

0.9 0.8 0.7 0.6 0.5 0.4 0.3 0.2 −10 −5 0 5 10 x

$|q_1|^2$ - - - - $|q_2|^2$ — — $|q_3|^2$

(d) $t=-5$

1.0 0.9 0.8 0.7 0.6 0.5 0.4 0.3 0.2 −10 −5 0 5 10 x

$|q_1|^2$ - - - - $|q_2|^2$ — — $|q_3|^2$

(e) $t=5$

Figure 3 (a)–(c), Density plot of $|q_1|^2$, $|q_2|^2$ and $|q_3|^2$; (d),(e) solid line $|q_1|^2$, dot line $|q_2|^2$, dash line $|q_3|^2$, (d) $t = -5$, (e) $t = 5$. The parameters are given in equations (3.44)

The dynamical evolution of corresponding bound state dark solitons are shown in Fig. 4.

4 Conclusions and Discussions

In this paper, we obtain the uniform transformation for N-component NLS equations, which can be used to derive multi-dark soliton solutions and many other type localized wave solutions conveniently. To our knowledge, the transformation has two-fold meaning as follows.

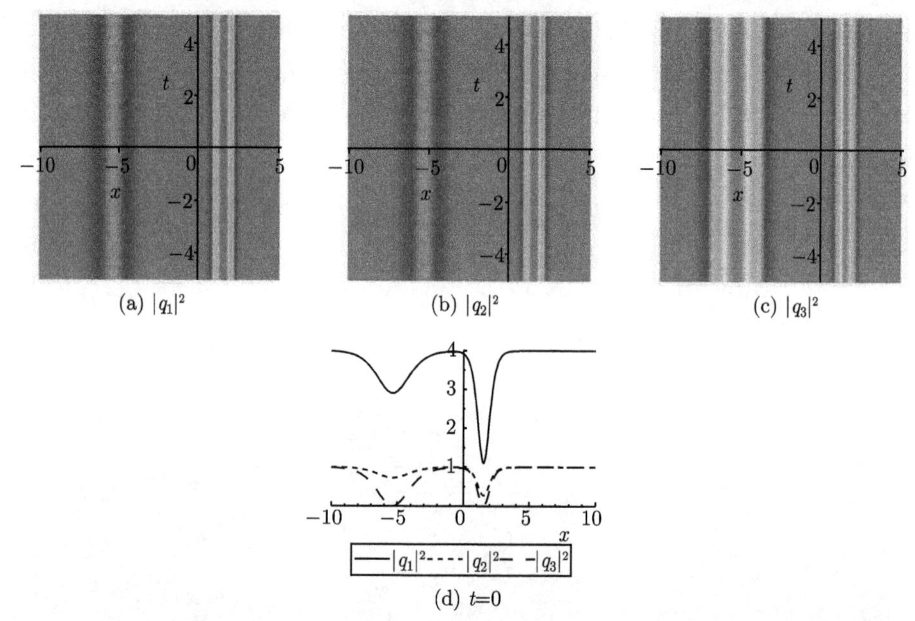

Figure 4 (a)–(c), Density plot of $|q_1|^2$, $|q_2|^2$ and $|q_3|^2$; (d)$t=0$: Solid line $|q_1|^2$, Dot line $|q_2|^2$, Dash line $|q_3|^2$. The parameters are given in equations (3.45)

Firstly, the DT is related with inverse scattering transformation which is a method to solve the initial value problem of integrable PDE. The inverse scattering method of coupled NLS equations is an open problem in soliton theory. In 2006, Abolowitz et.al solved this problem with the special background [26]. The DT method presented here provides a way to solve this problem at least for the discrete spectrum without restricting the background.

There is another open question in the famous book of Faddeev and Takhtajan (P. 145 the end of second paragraph). The authors deem that the solution of Riemann problem with zeros cannot be expressed as a product of Blaschke-Potapov factors and a solution of the regular Riemann problem with same continuous spectrum data. Indeed, by the above binary DT, we can construct the $L^2(\mathbb{R})$ eigenfunction and the potential function for the spectral problem:

$$L\Phi=\lambda_1\Phi,\quad L=-\mathrm{i}\sigma_3\partial_x-\sigma_3 Q,\quad Q=\begin{bmatrix}0 & -\bar{q}\\ q & 0\end{bmatrix},\ q\to \mathrm{e}^{\mathrm{i}(-c^2t\pm\theta_\pm)} \text{ as } x\to\pm\infty,$$

here

$$\Phi=\frac{\Phi_1}{2[e^{2Y_1}+1]},\quad q=c\left(1-\frac{B_1}{2}+\frac{B_1}{2}\tanh(Y_1)\right)\mathrm{e}^{-\mathrm{i}c^2t},$$

where $\theta_\pm$ are asymptotical phase and

$$B_1 = \frac{2i\mu_1}{\lambda_1 + i\mu_1},\ Y_1 = -\mu_1(x + \lambda_1 t + \alpha),\ \mu_1 = \sqrt{c^2 - \lambda_1^2},\ -c < \lambda_1 < c,\ \alpha \in \mathbb{R},$$

$$\Phi_1 = \begin{bmatrix} 1 \\ (\lambda_1 + i\mu_1)^{-1} e^{-ic^2 t} \end{bmatrix} e^{X_1},\ X_1 = -\mu_1(x + \lambda_1 t) + \frac{i}{2}c^2 t. \tag{3.45}$$

Thus this transformation can be used to add the discrete spectrum of the above spectral problem. The detail research for this transformation applied to inverse scattering transformation will be proceeded in the future.

Secondly, the direct and simple application for this transformation is to derive the dark and multi-dark soliton solution, which are meaningful for many different physical systems. Besides, the method in our paper can be generalized to look for some other type nonlinear localized wave solutions. We would like to explore them in the future as well.

Acknowledgements This work is supported by National Natural Science Foundation of China (Contact No. 11271052) and Fundamental Research Funds for the Central Universities (Contact No. 2014ZB0034).

References

[1] Agrawal G P. Nonlinear Fiber Optics [M]. San Diego: Academic Press, 1989.

[2] Dalfovo F, Giorgini S, Pitaevskii L P, Staingari S. Theory of Bose-Einstein condensation in trapped gases [J]. Rev. Mod. Phys., 1999, 71: 463–512.

[3] Ablowitz M J, Segure H. Solitons and the Inverse Scattering Transform [M]. Philadelphia: SIAM, 1981.

[4] Zakharov V E, Shabat A B. Exact theory of two-dimensional self-focusing and one-dimensional self-modulation of waves in nonlinear media [J]. Sov. Phys. JETP, 1972, 34: 62–69.

[5] Ma Y C, Ablowitz M J. The periodic cubic Schrödinger equation [J]. Stud. Appl. Math., 1981, 65: 113–158.

[6] Gelash A A, Zakharov V E. Super-regular solitonic solutions: a novel scenario for nonlinear stage of modulation instability [J]. Nonlinearity, 2014, 27: R1-R39.

[7] Peregrine D H. Water waves, nonlinear Schrödinger equations and their solutions [J]. J. Aust. Math. Soc. Series B, Appl. Math., 1983, 25: 16.

[8] Guo B, Ling L, Liu Q P. Nonlinear Schrödinger equation: Generalized Darboux transformation and rogue wave solutions [J]. Phys. Rev. E, 2012, 85: 026607.

[9] Faddeev L D, Takhtajan L A. Hamiltonian Methods in the Theory of Solitons. Berlin: Springer, 1987.

[10] Manakov S V. On the theory of two-dimensional stationary self-focusing of electromagnetic waves [J]. Sov. Phys. JETP, 1974, 38: 248.

[11] Baronio F, Degasperis A, Conforti M, Wabnitz S. Solutions of the Vector Nonlinear Schrödinger Equations: Evidence for Deterministic Rogue Waves [J]. Phys. Rev. Lett., 2012, 109: 044102.

[12] Zhao L C, Liu J. Localized nonlinear waves in a two-mode nonlinear fiber [J]. Journ. Opt. Soc. Am. B, 2012, 29: 3119–3127.

[13] Wang D S, Zhang D, Yang J. Integrable properties of the general coupled nonlinear Schrödinger equations [J]. J. Math. Phys., 2010, 51: 023510.

[14] Guo B, Ling L.Rogue Wave, Breathers and Bright-Dark-Rogue Solutions for the Coupled Schrodinger Equations [J]. Chin. Phys. Lett., 2011, 28: 110202.

[15] Forest M G, McLaughlin D W, Muraki D J, Wright O. C. Nonfocusing instabilities in coupled, integrable nonlinear Schrdinger pdes [J]. J.Nonlinear Sci., 2000, 10: 291–331.

[16] Wright O C, Forest M G. On the Bäcklund-gauge transformation and homoclinic orbits of a coupled nonlinear Schrödinger system [J]. Phys. D, 2000, 141: 104–116.

[17] Sheppard A P, Kivshar Y S. Polarized dark solitons in isotropic Kerr media [J]. Phys. Rev. E, 1997, 55: 4773–4782.

[18] Radhakrishnan R, Lakshmanan M. Bright and dark soliton solutions to coupled nonlinear Schrödinger equations [J]., J. Phys. A: Math. Gen., 1995, 28: 2683–2692.

[19] Ohta Y, Wang D-S, Yang J. General N-Dark-Dark Solitons in the Coupled Nonlinear Schrödinger Equations [J]. Stud. Appl. Math., 2011, 127: 345–371.

[20] Forest M G, Wright O C. An integrable model for stable:unstable wave coupling phenomena [J]. Phys. D, 2003, 178: 173–189.

[21] Vijayajayanthi M, Kanna T, Lakshmanan M. Bright-dark solitons and their collisions in mixed N-coupled nonlinear Schrödinger equations [J]., Phys. Rev. A, 2008, 77: 013820.

[22] Zhao L, Liu J. Rogue-wave solutions of a three-component coupled nonlinear Schrödinger equation [J]. Phys. Rev. E, 2013, 87: 013201.

[23] Mañas M. Darboux transformations for nonlinear Schrödinger equations [J]. J. Phys. A: Math. Theor., 1996, 29 7721–7737.

[24] Degasperis A, Lombardo S. Multicomponent integrable wave equations: I. Darboux-dressing transformation [J]. J. Phys. A: Math. Theor., 2007, 40: 961–977.

[25] Degasperis A, Lombardo S. Multicomponent integrable wave equations: II. Soliton solutions [J]. J. Phys. A: Math. Theor., 2009, 42: 385206.

[26] Ablowitz M J, Biondini G, Prinari B. Inverse scattering transform for the vector nonlinear Schrödinger equation with nonvanishing boundary conditions [J]. J. Math. Phys., 2006, 47: 063508.

[27] Cieśliński J L. Algebraic construction of the Darboux matrix revisited [J]. J. Phys. A: Math. Theor., 2009, 42: 404003.

[28] Gu C H, Hu H S, Zhou Z X. Darboux Transformation in Soliton Theory, and its Geometric Applications [M]. Shanghai: Shanghai Science and Technology Publishers, 2005.

[29] Matveev V, Salle M. Darboux transformation and solitons [M]. Berlin: Springer-Verlag, 1991.

[30] Babich M V, Matveev V B, Salle M A. Binary transformation for the Toda lattice [Zapiski Nauchnyh seminarov Lomi v. 1985, 145: 44-61 (in Russian), English translation: Springer Journal of Mathematical sciences v. 1986, 35: 2582–2589.

[31] Matveev V B, Salle M A. New families of the explicit solutions of the Kadomtsev-Petviashvily equation and their application to the Johnson equation pp.304-315, in "Some topics on Inverse problems" (Proceedings of the XVIth Workshop on Interdisciplinary Study of Inverse Problems) ed. by P.Sabatier, World Scientific Publishing., pp.1-419, 1988.

[32] Novikov S P, Manakov S V, Pitaevskii L, Zakharov V E. Theory of Solitons: The Inverse Scattering Method [M]. Berlin: Springer, 1984.

[33] Terng C-L, Uhlenbeck K. Bäcklund transformations and loop group actions [J]. Comm. Pure Appl. Math., 2000, 53: 1–75.

[34] Assuncao A de O, Blas H, Silva M J B F da. New derivation of soliton solutions to the $AKNS_2$ system via dressing transformation methods [J]. J. Phys. A: Math. Theor., 2012, 45: 085205.

[35] Kalla C. Breathers and solitons of generalized nonlinear Schrödinger equations as degenerations of algebro-geometric solutions [J]. J. Phys. A: Math. Theor., 2011, 44: 335210.

[36] Tsuchida T. Exact solutions of multicomponent nonlinear Schrödinger equations under general plane-wave boundary conditions [J]. 2013, arXiv:1308.6623.

[37] Park Q-Han, Shin H J. Systematic construction of multicomponent optical solitons [J]. Phys. Rev. E, 2000, 61: 3093–3106.

[38] Guo B, Ling L, Liu Q P. High-Order Solutions and Generalized Darboux Transformations of Derivative Nonlinear Schrödinger Equations [J]. Stud. Appl. Math., 2013, 130: 317–344.

[39] Bian D, Guo B, Ling L. High-Order Soliton Solution of Landau-Lifshitz Equation [J]. Stud. Appl. Math., 2015, 134: 181–214.

[40] Ablowitz M J, Kaup D J, Newell A C, Segur H. The inverse scattering transform Fourier analysis for nonlinear problems [J]. Stud. Appl. Math., 1974 53: 249–315.

The Cauchy Problem of the Modified CH and DP Equations*

Wu Xinglong (吴兴龙) and Guo Boling (郭柏灵)

Abstract This paper is concerned with the Cauchy problem of modified Camassa-Holm (CH) and Degasperis-Procesi (DP) equations. Firstly, we establish the local well-posedness for the equation in Besov space. Secondly, we derive the conservation laws and a precise blow-up scenario. Moreover, we prove the existence of blow-up solutions and obtain its blow-up rate, providing the initial data satisfies certain conditions. Finally, we present the persistence properties of the equation.

Keywords The modified CH and DP equations; local well-posedness; Besov space; conservation laws; blow-up scenario; blow-up phenomena; blow-up rate; persistence properties

1 Introduction

Recently, the b-family of equations

$$y_t + uy_x + bu_x y = 0, \quad y = u - u_{xx},\ b \in \mathbb{R}, \tag{1.1}$$

has been investigated [1, 13, 14]. If b = 2, Eq.(1.1) becomes the CH equation

$$u_t - u_{txx} + 3uu_x = 2u_x u_{xx} + uu_{xxx}, \tag{1.2}$$

which has been investigated by many researchers, cf. [4, 6, 7, 8]. The CH equation is considered as a shallow water equation and is originally derived as an approximation to the incompressible Euler equation. It approximates unidirectional fluid flow in Euler's equations at the next order beyond the KdV equation, has a bi-Hamiltonian structure [17] and is completely integrable [5], with a Lax pair based on a linear spectral problem of second order. Also, there are smooth soliton solutions of Eq.(1.2)

* IMA Journal of Applied Mathematics, 2015, 80, 906–930. DOI: 10.1093/imamat/hxu032.

on a non-zero constant background [1] and peakon solutions. When $b = 3$, Eq.(1.1) reduces to the DP equation [18, 23, 26, 28]

$$u_t - u_{txx} + 4uu_x = 3u_x u_{xx} + uu_{xxx}. \tag{1.3}$$

The DP equation can be regarded as a model for nonlinear shallow water dynamics [20]. Degasperis, Holm and Hone [13] prove the formal integrability of Eq.(1.3) by constructing a Lax pair. They also show [13] that it has a bi-Hamiltonian structure and an infinite sequence of conserved quantities, and admits exact peakon solutions.

Despite the form is similar to the CH equation, it should be emphasized that these two equations are truly different. One of the important features of Eq.(1.3) is that it has not only peakon solitons [13], i.e. solutions at the form $u(t,x) = ce^{-|x-ct|}$ and periodic peakon solutions [37], but also shock peakons [3, 27] which are given by

$$u(t,x) = -\frac{1}{t+k}\operatorname{sgn}(x)e^{-|x|}, \qquad k > 0,$$

and periodic shock peakons [16].

Since the CH and DP equations have rich structures, Wazwaz [31] suggested a modified form of the CH equation

$$u_t - u_{txx} + 3u^2 u_x = 2u_x u_{xx} + uu_{xxx}, \tag{1.4}$$

and a modified form of the DP equation

$$u_t - u_{txx} + 4u^2 u_x = 3u_x u_{xx} + uu_{xxx}. \tag{1.5}$$

They are obtained by changing the nonlinear convection term uu_x in Eqs.(1.2) and (1.3) to $u^2 u_x$, respectively. Wazwaz [31] obtained two bell-shaped travelling wave solutions of wave speed $c = 2$ for Eq.(1.4), and two bell-shaped travelling wave solutions of wave speed $c = 5/2$ for Eq.(1.5). In [32], he found more solutions of wave speeds $c = 1$ and $c = 2$ for the modified CH equation, and more solutions of wave speed $c = 5/2$ for the modified DP equation. Recently, Wang and Tang [33] obtained some traveling wave solutions of wave speed $c = 1/3$ and some peakon solutions of wave speed $c = 3$ for the modified CH equation, and some traveling wave solutions of wave speed $c = 1/4$ and some peakon solutions of wave speed $c = 4$ for the modified DP equation. Liu and Ouyang [25] found the coexistence of bell-shaped solution and peakon solution of the same wave speed $c = 2$ for the modified CH equation, and the coexistence of bell-shaped solution and peakon solution of the same wave speed $c = 5/2$ for the modified DP equation. It is interesting to note that Ma et al. uses a generalized auxiliary equation method to study the modified DP equation differently [29]. Shen and Xue [30] considered the bifurcations of the smooth and non-smooth traveling waves of Eq.(1.4), while Zhang et al. [38] analyzed Eq.(1.5). Recently, Liang and Jeffrey [24] obtained many new traveling wave solutions to Eqs.(1.4) and (1.5), most importantly, these solutions are in general forms.

In this paper, we consider the Cauchy problem for the following generalized modified CH and DP equations [31]

$$\begin{cases} u_t - u_{txx} + (b+1)u^2 u_x & = bu_x u_{xx} + uu_{xxx}, \\ & t > 0,\ x \in \mathbb{R}, \\ u(0,x) = u_0(x), & x \in \mathbb{R}, \end{cases} \tag{1.6}$$

where $b \neq -1$ is any real number. It is obtained by changing the nonlinear convection term uu_x in Eq.(1.1) to u^2u_x. When $b = 2,\ 3$, Eq.(1.6) becomes Eqs.(1.4) and (1.5) respectively. By the sin-cosin method, Wazwaz [31] obtained two bell-shaped traveling wave solutions of wave speed $c = \dfrac{b+2}{2}$ to Eq.(1.6), i.e.

$$u(t,x) = -\left(\frac{3(b+2)}{2(b+1)}\right)\operatorname{sech}^2\frac{1}{2}\left(x - \frac{b+2}{2}t\right), \tag{1.7}$$

and

$$u(t,x) = \left(\frac{3(b+2)}{2(b+1)}\right)\operatorname{csch}^2\frac{1}{2}\left(x - \frac{b+2}{2}t\right). \tag{1.8}$$

One can easily check that (1.7) is a global smooth solution of Eq.(1.6). The aim of this paper is to establish the local well-posedness of Eq.(1.6) in Besov space by the transport theory: First step, by Lemma 2.4, we construct approximate solutions of Eq.(2.2) which are smooth solutions of some linear equation, the approximate solutions are uniformly bounded in $E^s_{p,r}(T)$ and are Cauchy sequences in $C([0,T); B^{s-1}_{p,r})$. Second step, by Lemma 2.6 and Lemma 2.7, we check the limit of solutions is indeed a solution and has the desired regularity and continuity. Final step, by Lemma 2.5, we show the solution is unique with the initial data. Thank to the local well-posedness, we get the blow-up scenario, blow-up phenomena and blow-up rate. Moreover, we get the persistence properties of the solution, these results will perfectly explain the physical phenomena. The readers can find the detail of proof in the following sections.

The remainder of the paper is organized as follows. In Section 2, we establish local well-posedness of Eq.(1.6) . In Section 3, we derive conservation laws and a precise blow-up scenario to Eq.(1.6). In Section 4, we prove the existence of blow-up solutions and obtain its blow-up rate, providing the initial data satisfies certain conditions. Finally, in Section 5, we present the persistence properties of the solution to Eq.(1.6).

2 Local well-posedness

In this section, we apply transport equations theory to establish the local well-posedness for the Cauchy problem of Eq.(1.6). Note that if $p(x)=\frac{1}{2}e^{-|x|}, x\in\mathbb{R}$, we have $(1-\partial_x^2)^{-1}f=p*f$ for all the $f\in L^2(\mathbb{R})$, and $p*y=u$, here we denote by $*$ the convolution. Then we can rewrite Eq.(1.6) as follows

$$\begin{cases} u_t+uu_x=-p*\left((b+1)u^2u_x-uu_x+(3-b)u_xu_{xx}\right), \\ u(0,x)=u_0(x), \quad x\in\mathbb{R}. \end{cases} \tag{2.1}$$

Or in the equivalent form

$$\begin{cases} u_t+uu_x=-(1-\partial_x^2)^{-1}\left((b+1)u^2u_x-uu_x+(3-b)u_xu_{xx}\right), \\ u(0,x)=u_0(x), \quad x\in\mathbb{R}. \end{cases} \tag{2.2}$$

First, we shall establish the local well-posedness for the Cauchy problem of Eq.(2.2) in the Besov spaces. For the convenience of the readers, we recall some facts on the Littlewood-Paley decomposition and some useful lemmas.

Proposition 2.1 [12] *There exists a couple of C^∞ functions (χ,φ) valued in $[0,1]$, such that χ is supported in the ball $\mathcal{B}=\left\{\xi\in\mathbb{R}:|\xi|\leqslant\frac{4}{3}\right\}$, and φ is supported in the annulus $\mathcal{C}=\left\{\xi\in\mathbb{R}:\frac{3}{4}\leqslant|\xi|\leqslant\frac{8}{3}\right\}$. Moreover,*

$$\chi(\xi)+\sum_{q\in\mathbb{N}}\varphi(2^{-q}\xi)=1, \qquad \forall\xi\in\mathbb{R}^n,$$

$$q\geqslant 1\Longrightarrow \operatorname{supp}\chi(\cdot)\cap\operatorname{supp}\varphi(2^{-q}\cdot)=\varnothing,$$

$$\operatorname{supp}\varphi(2^{-q}\cdot)\cap\operatorname{supp}\varphi(2^{-p}\cdot)=\varnothing, \quad \textit{if } |p-q|\geqslant 2$$

and

$$\frac{1}{3}\leqslant\chi^2(\xi)+\sum_{q\in\mathbb{N}}\varphi^2(2^{-q}\xi)\leqslant 1, \quad \forall\xi\in\mathbb{R}^n.$$

Let $\tilde{h} = \mathcal{F}^{-1}\chi$ and $h = \mathcal{F}^{-1}\varphi$. Then the dyadic operators Δ_q and S_q can be defined as follows

$$\Delta_{-1}u = S_0 u \quad \text{and} \quad \Delta_q u = 0, \quad \text{if} \quad q \leqslant -2,$$

$$\Delta_q u = \varphi(2^{-q}D)u = 2^{qn}\int_{\mathbb{R}^n} h(2^q y)u(x-y)\mathrm{d}y, \quad \text{if } q \geqslant 0,$$

$$S_q u = \sum_{p\geqslant -1}^{q-1} \Delta_p u = \chi(2^{-q}D)u = 2^{qn}\int_{\mathbb{R}^n} \tilde{h}(2^q y)u(x-y)\mathrm{d}y.$$

Moreover, if $u, v \in \mathcal{S}'(\mathbb{R}^n)$, then we have

$$\Delta_p\Delta_q = 0 \quad \text{if} \quad |p-q| \geqslant 2,$$

$$\Delta_q(S_{p-1}u\Delta_p v) = 0 \quad \text{if} \quad |p-q| \geqslant 5.$$

Furthermore, for all $u \in \mathcal{S}'(\mathbb{R}^n)$, one can easily check that

$$u = \sum_{q\in\mathbb{Z}} \Delta_q u \quad \text{in} \quad \mathcal{S}'(\mathbb{R}^n).$$

By virtue of the dyadic operators Δ_k, the nonhomogeneous Besov space $B^s_{p,r}$ is defined as follows.

Definition 2.1 *Let $s \in \mathbb{R}$, $p, r \in [1, \infty]$ and $T > 0$, we set*

$$B^s_{p,r}(\mathbb{R}^n) = \{u \in \mathcal{S}'(\mathbb{R}^n); \|u\|_{B^s_{p,r}} = \|2^{ks}\Delta_k u\|_{l^r(L^p)} < \infty\},$$

where Δ_k is the dyadic operator. If $s = \infty$, $B^\infty_{p,r} = \bigcap_{s\in\mathbb{R}} B^s_{p,r}$.

The following lemmas and definition will be used in the proof of Theorem 2.1.

Lemma 2.1 [10, 12] *The following properties hold:*

(i) *Density: if $p, r < \infty$, then $\mathcal{S}(\mathbb{R}^n)$ is dense in $B^s_{p,r}$.*

(ii) *Generalized derivatives: Let $f \in C^\infty(\mathbb{R}^n)$ be a homogeneous function of degree $m \in \mathbb{R}$ away from a neighborhood of the origin. There exists a constant c depending only on f such that $\|f(D)u\|_{B^{s-m}_{p,r}} \leqslant c\|u\|_{B^s_{p,r}}$.*

(iii) *Sobolev embeddings: if $p_1 \leqslant p_2$ and $r_1 \leqslant r_2$, then $B^s_{p_1,r_1} \hookrightarrow B^{s-n(\frac{1}{p_1}-\frac{1}{p_2})}_{p_2,r_2}$. If $s_1 < s_2, 1 \leqslant p \leqslant \infty$ and $1 \leqslant r_1, r_2 \leqslant \infty$, then the embedding $B^{s_2}_{p,r_2} \hookrightarrow B^{s_1}_{p,r_1}$ is locally compact.*

(iv) *Algebraic properties: for $s > 0$, $B^s_{p,r} \cap L^\infty$ is an algebra. Moreover ($B^s_{p,r}$ is an algebra) $\Leftrightarrow$ ($B^s_{p,r} \hookrightarrow L^\infty$) $\Leftrightarrow$ ($s > n/p$ or ($s \geqslant n/p$ and $r = 1$)).*

(v) *Fatou property: if* $\{u_n\}_{n\in\mathrm{N}}$ *is a bounded sequence of* $B^s_{p,r}$ *which tends to* u *in* $\mathcal{S}'$, *then* $u \in B^s_{p,r}$ *and* $\|u\|_{B^s_{p,r}} \leqslant \lim_{n\to\infty}\inf \|u_n\|_{B^s_{p,r}}$.

(vi) *Complex interpolation: if* $u \in B^s_{p,r} \cap B^{s_1}_{p,r}$ *and* $\theta \in [0,1]$, $p,r \in [1,\infty]$, *then* $u \in B^{\theta s+(1-\theta)s_1}_{p,r}$ *and*

$$\|u\|_{B^{\theta s+(1-\theta)s_1}_{p,r}} \leqslant \|u\|^{\theta}_{B^s_{p,r}} \|u\|^{1-\theta}_{B^{s_1}_{p,r}}.$$

(vii) *Let* $m \in \mathbb{R}$ *and* f *be a* S^m*-multiplier, i.e.* $f: \mathbb{R}^n \mapsto \mathbb{R}$ *is smooth function and satisfies that for all multi-index* α, *there exists a constant* C_α *such that* $\forall \xi \in \mathbb{R}^n$, $|\partial^\alpha f(\xi)| \leqslant C_\alpha(1+|\xi|)^{m-|\alpha|}$. *Then for all* $s \in \mathbb{R}$ *and* $p,r \in [1,\infty]$, *the operator* $f(D)$ *is continuous from* $B^s_{p,r}$ *to* $B^{s-m}_{p,r}$.

Lemma 2.2 [10, 12] *Assume that* $p,r \in [1,\infty]$ *and* $s > -n/p$. *Let* v *be a vector field such that* ∇v *belongs to* $L^1([0,T]; B^{s-1}_{p,r})$ *if* $s > n/p+1$ *or to* $L^1([0,T]; B^{n/p}_{p,r} \cap L^\infty)$ *otherwise. Suppose also that* $f_0 \in B^s_{p,r}$, $F \in L^1([0,T]; B^s_{p,r})$ *and that* $f \in L^\infty([0,T]; B^s_{p,r}) \cap C([0,T]; \mathcal{S}')$ *solves the n-dimension linear transport equation*

$$\begin{cases} \partial_t f + v\cdot\nabla f = F, & t>0,\ x\in\mathbb{R}^n, \\ f(0,x) = f_0(x), & x\in\mathbb{R}^n. \end{cases} \tag{2.3}$$

Then there exists a constant c *depending only on* s,p *and* n, *and such that the following statements hold:*

(1) *If* $r=1$ *or* $s \neq n/p+1$,

$$\|f(t)\|_{B^s_{p,r}} \leqslant e^{cV(t)}\left(\|f_0\|_{B^s_{p,r}} + \int_0^t e^{-cV(\tau)}\|F(\tau)\|_{B^s_{p,r}}\mathrm{d}\tau\right) \tag{2.4}$$

with $V(t) = \displaystyle\int_0^t \|\nabla v(\tau)\|_{B^{n/p}_{p,r}\cap L^\infty}\mathrm{d}\tau$ *if* $s < n/p+1$ *and* $V(t) = \displaystyle\int_0^t \|\nabla v(\tau)\|_{B^{s-1}_{p,r}}\mathrm{d}\tau$ *else.*

(2) *If* $s \leqslant n/p+1$ *and, in addition,* $\nabla f_0 \in L^\infty$, $\nabla f \in L^\infty([0,T)\times\mathbb{R}^n)$ *and* $\nabla F \in L^1([0,T]; L^\infty)$, *then*

$$\begin{aligned} &\|f(t)\|_{B^s_{p,r}} + \|\nabla f(t)\|_{L^\infty} \leqslant e^{cV(t)} \\ &\left(\|f_0\|_{B^s_{p,r}} + \|\nabla f_0\|_{L^\infty} + \int_0^t e^{-cV(\tau)}(\|F(\tau)\|_{B^s_{p,r}} + \|\nabla F(\tau)\|_{L^\infty})\mathrm{d}\tau\right) \end{aligned} \tag{2.5}$$

with $V(t) = \displaystyle\int_0^t \|\nabla v(\tau)\|_{B^{n/p}_{p,r}\cap L^\infty}\mathrm{d}\tau$.

(3) *If* $r<\infty$, *then* $f \in C([0,T]; B^s_{p,r})$. *If* $r=\infty$, *then* $f \in C([0,T]; B^{s_1}_{p,r})$ *for all* $s_1 < s$. *If* $\mathrm{div} v = 0$ *and* $v\cdot\nabla f$ *stands for* $\mathrm{div}(vf)$, *then all the above results hold true for* $s > -(n/p+1)$.

Lemma 2.3 [2] *Assume that $p, r \in [1, \infty]$, the following estimates hold:*

(i) *If $s > 0$, $\|uv\|_{B^s_{p,r}} \leqslant c(\|f\|_{B^s_{p,r}}\|g\|_{L^\infty} + \|f\|_{L^\infty}\|g\|_{B^s_{p,r}})$.*

(ii) *If $s_1 \leqslant 1/p, s_2 > 1/p$ ($s_2 \geqslant 1/p$ if $r = 1$) and $s_1 + s_2 > 0$, then*

$$\|uv\|_{B^{s_1}_{p,r}} \leqslant c\|u\|_{B^{s_1}_{p,r}}\|v\|_{B^{s_2}_{p,r}},$$

where c does not depends on u and v.

Definition 2.2 *Let $s \in \mathbb{R}$, $p, r \in [1, \infty]$ and $T > 0$, we set*

$$E^s_{p,r}(T) = C([0,T]; B^s_{p,r}) \bigcap C^1([0,T]; B^{s-1}_{p,r}), if \quad r < \infty,$$

$$E^s_{p,\infty}(T) = L^\infty([0,T]; B^s_{p,\infty}) \bigcap Lip([0,T]; B^{s-1}_{p,\infty}).$$

Now let us present the main result, which is motivated by the proof of local existence about CH equation [10].

Theorem 2.1 *Let $p, r \in [1, \infty]$ and $s > \max\left\{\frac{3}{2}, 1 + \frac{1}{p}\right\}$. Assume that $u_0 \in B^s_{p,r}$. There exists a time $T > 0$ and a unique solution $u \in E^s_{p,r}(T)$ to Eq.(2.2) such that the map $u_0 \mapsto u : B^s_{p,r} \mapsto C([0,T]; B^{s'}_{p,r}) \cap C^1([0,T]; B^{s'-1}_{p,r})$ is continuous for every $s' < s$ when $r = \infty$ and $s' = s$ whereas $r < \infty$.*

We will break the argument into several lemmas. First, using the classical Friedrichs regularization method to construct the approximate solutions to Eq.(2.2).

Lemma 2.4 *Assume that $u^0 = 0$, Let $p, r \in [1, \infty]$, $s > \left\{\frac{3}{2}, 1 + \frac{1}{p}\right\}$ and $u_0 \in B^s_{p,r}$. Then there exists a sequence of smooth functions $\{u^n\}_{n \in \mathbb{N}} \in C(\mathbb{R}; B^\infty_{p,r})$ solving the following linear transport equation by induction*

$$\begin{cases} (\partial_t + u^n\partial_x)u^{n+1} = P(D)\left((b+1)(u^n)^2u^n_x - u^nu^n_x + (3-b)u^n_xu^n_{xx}\right), \\ u^{n+1}(0,x) = u^{n+1}_0(x) = S_{n+1}u_0, \ x \in \mathbb{R}, \end{cases} \tag{2.6}$$

where the operator $P(D) = -(1-\partial_x^2)^{-1}$. Moreover, there is maximal existence time $T > 0$ such that the solutions u^n satisfying the following conditions:

(a) *$\{u^n\}_{n \in \mathbb{N}}$ is uniformly bounded in $E^s_{p,r}(T)$.*

(b) *$\{u^n\}_{n \in \mathbb{N}}$ is a Cauchy sequence in $C([0,T]; B^{s-1}_{p,r})$.*

Proof Due to the initial data $S_{n+1}u_0 \in B^\infty_{p,r}$, Lemma 2.2 enables us to show by induction that for all $n \in \mathbb{N}$, Eq.(2.6) has a global solution which belongs to

$C(\mathbb{R}^+; B^{\infty}_{p,r})$. Note that $P(D)$ is S^{-2}-multiplier, and $s > 1 + \frac{1}{p}$, $B^{s-1}_{p,r} \subset L^\infty$ is an algebra, by virtue of Lemma 2.1, we obtain

$$\begin{aligned}
&\|P(D)\left((b+1)(u^n)^2 u^n_x - u^n u^n_x + (3-b) u^n_x u^n_{xx}\right)\|_{B^s_{p,r}} \\
&\leqslant \left\| \frac{b+1}{3}(u^n)^3 - \frac{1}{2}(u^n)^2 + \frac{3-b}{2}(u^n)^2_x \right\|_{B^{s-1}_{p,r}} \\
&\leqslant c(\|u^n\|^3_{B^s_{p,r}} + \|u^n\|^2_{B^s_{p,r}}).
\end{aligned} \tag{2.7}$$

In view of Lemma 2.2 and (2.7), it follows that

$$\begin{aligned}
&\exp\left\{-c\int_0^t \|\partial_x(u^n)(\tau)\|_{B^{s-1}_{p,r}} d\tau\right\} \|u^{n+1}(t)\|_{B^s_{p,r}} \\
\leqslant &\|u^{n+1}_0\|_{B^s_{p,r}} + c\int_0^t e^{-c\int_0^\tau \|\partial_x u^n(\tau')\|_{B^{s-1}_{p,r}} d\tau'} \left(\|u^n(\tau)\|^3_{B^s_{p,r}} + \|u^n(\tau)\|^2_{B^s_{p,r}}\right) d\tau.
\end{aligned} \tag{2.8}$$

If we choose a $T > 0$ such that

$$T \leqslant \min\left\{ \frac{1}{4c\|u_0\|_{B^s_{p,r}}}, \frac{1 - \left(\dfrac{4\|u_0\|^2_{B^s_{p,r}}}{3\|u_0\|_{B^s_{p,r}} + 4\|u_0\|^2_{B^s_{p,r}}}\right)^{2/3}}{4c\|u_0\|_{B^s_{p,r}}} \right\}$$

and suppose by induction that for all $t \in [0, T)$

$$\|u^n(t)\|_{B^s_{p,r}} \leqslant \frac{2\|u_0\|_{B^s_{p,r}}}{1 - 4c\|u_0\|_{B^s_{p,r}} t}. \tag{2.9}$$

Inserting (2.9) into (2.8) yields that

$$\begin{aligned}
&\|u^{n+1}(t)\|_{B^s_{p,r}} \leqslant \frac{1}{\sqrt{1 - 4c\|u_0\|_{B^s_{p,r}} t}} \\
&\times \left(\|u_0\|_{B^s_{p,r}} + \int_0^t \left(\frac{8\|u_0\|^3_{B^s_{p,r}}}{(1 - 4c\|u_0\|_{B^s_{p,r}}\tau)^{5/2}} + \frac{4\|u_0\|^2_{B^s_{p,r}}}{(1 - 4c\|u_0\|_{B^s_{p,r}}\tau)^{3/2}} \right) d\tau \right) \\
= &\frac{-\|u_0\|_{B^s_{p,r}} - \frac{4}{3}\|u_0\|^2_{B^s_{p,r}}}{\sqrt{1 - 4c\|u_0\|_{B^s_{p,r}} t}} + \frac{4\|u_0\|^2_{B^s_{p,r}}}{3(1 - 4c\|u_0\|_{B^s_{p,r}} t)^2} + \frac{2\|u_0\|_{B^s_{p,r}}}{1 - 4c\|u_0\|_{B^s_{p,r}} t} \\
\leqslant &\frac{2\|u_0\|_{B^s_{p,r}}}{1 - 4c\|u_0\|_{B^s_{p,r}} t}.
\end{aligned} \tag{2.10}$$

Thus $\{u^n\}_{n\in\mathbb{N}}$ is uniformly bounded in $C([0, T); B^s_{p,r})$. In view of Lemma 2.1 and Lemma 2.3, we have

$$\|u^n \partial_x u^{n+1}\|_{B^{s-1}_{p,r}} \leqslant c\|u^n\|_{B^s_{p,r}} \|\partial_x u^{n+1}\|_{B^{s-1}_{p,r}} \leqslant c\|u^n\|_{B^s_{p,r}} \|u^{n+1}\|_{B^s_{p,r}}. \tag{2.11}$$

Combining (2.7), (2.11) with Eq.(2.6), we deduce that $\partial_t u^{n+1} \in C([0,T); B^{s-1}_{p,r})$ is uniformly bounded. Thus we get a).

Next we shall show b). Note that Eq.(2.6), for all $m, n \in \mathbb{N}$, we get

$$(\partial_t + u^{m+n}\partial_x)(u^{m+n+1} - u^{n+1}) = -(u^{m+n} - u^n)u_x^{n+1} - \partial_x(1-\partial_x^2)^{-1}f, \quad (2.12)$$

where $f = \dfrac{b+1}{3}[(u^{m+n})^3 - (u^n)^3] - \dfrac{1}{2}[(u^{m+n})^2 - (u^n)^2] + \dfrac{3-b}{2}[(u^{m+n})_x^2 - (u^n)_x^2]$. Due to $s > 1 + \dfrac{1}{p}$, by virtue of Lemma 2.1, $B^{s-1}_{p,r} \subset L^\infty$ is an algebra. Thus we obtain

$$\|(u^{m+n} - u^n)u_x^{n+1}\|_{B^{s-1}_{p,r}} \leqslant c\|u^{m+n} - u^n\|_{B^{s-1}_{p,r}}\|u^{n+1}\|_{B^s_{p,r}}. \quad (2.13)$$

On the other hand, since $\partial_x(1-\partial_x^2)^{-1}$ is S^{-1}-multiplies, by Lemma 2.1 and Lemma 2.3, we have

$$\begin{aligned}
&\|-\partial_x(1-\partial_x^2)^{-1}f\|_{B^{s-1}_{p,r}} \leqslant \|f\|_{B^{s-2}_{p,r}} \\
\leqslant\ & c\|u^{m+n} - u^n\|_{B^{s-1}_{p,r}}(\|u^{m+n}\|^2_{B^{s-1}_{p,r}} + \|u^n\|^2_{B^{s-1}_{p,r}} \\
&+ \|u^{m+n}\|_{B^{s-1}_{p,r}}\|u^n\|_{B^{s-1}_{p,r}} + \|u^{m+n}\|_{B^{s-1}_{p,r}} + \|u^n\|_{B^{s-1}_{p,r}}) \\
=\ & cM\|u^{m+n} - u^n\|_{B^{s-1}_{p,r}}. \qquad (2.14)
\end{aligned}$$

Note that

$$\begin{aligned}
&\|u_0^{m+n+1} - u_0^{n+1}\|_{B^{s-1}_{p,r}} = \|S_{m+n+1}u_0 - S_{n+1}u_0\|_{B^{s-1}_{p,r}} \\
=\ & \Big\|\sum_{k=n+1}^{m+n} \Delta_k u_0\Big\|_{B^{s-1}_{p,r}} \leqslant c2^{-n}. \qquad (2.15)
\end{aligned}$$

In view of Lemma 2.2 and (2.13)–(2.15), for all $t \in [0,T]$, we obtain

$$\begin{aligned}
&\|(u^{m+n+1} - u^{n+1})(t)\|_{B^{s-1}_{p,r}} \leqslant e^{c\int_0^t \|u^{m+n}(\tau)\|_{B^{s-1}_{p,r}}\mathrm{d}\tau} \\
&\left(\|u_0^{m+n+1} - u_0^{n+1}\|_{B^{s-1}_{p,r}} + \int_0^t e^{-c\int_0^\tau \|u^{m+n}(\tau')\|_{B^{s-1}_{p,r}}\mathrm{d}\tau'}\right. \\
&\left.\times (M + \|u^{n+1}(\tau)\|_{B^s_{p,r}})\|(u^{m+n} - u^n)(\tau)\|_{B^{s-1}_{p,r}}\mathrm{d}\tau\right) \\
\leqslant\ & C\left(2^{-n} + \int_0^t \|(u^{m+n} - u^n)(\tau)\|_{B^{s-1}_{p,r}}\mathrm{d}\tau\right).
\end{aligned}$$

By induction with respect to the index n, one can easily get that

$$\|(u^{m+n+1} - u^{n+1})(t)\|_{B^{s-1}_{p,r}} \leqslant \frac{TC}{(n+1)!}(\|u^m\|_{B^s_{p,r}}) + C\sum_{k=0}^{n} 2^{k-n}\frac{(TC)^k}{k!}.$$

Due to $\|u^m\|_{B^s_{p,r}}$ and C be bounded independently of m,n, then there exists constant C_1 independently of m,n such that

$$\|(u^{m+n+1}-u^{n+1})(t)\|_{B^{s-1}_{p,r}}\leqslant C_1 2^{-n}.$$

Thus $\{u^n\}_{n\in\mathbb{N}}$ is a Cauchy sequence in $C([0,T);B^{s-1}_{p,r})$. □

Now let us start another important lemma, the uniqueness of the solution to Eq.(2.2) is only a corollary.

Lemma 2.5 *Assume that $p,r\in[1,\infty]$ and $s>\max\left\{\frac{3}{2},1+\frac{1}{p}\right\}$. Let $u,\ v\in L^\infty([0,T);\ B^s_{p,r})\cap C([0,T);\mathcal{S}')$ be two given solutions to Eq.(2.2) with the initial data $u_0,\ v_0\in B^s_{p,r}$. Then for every $t\in[0,T]$, we have*

$$\|u(t)-v(t)\|_{B^{s-1}_{p,r}}\leqslant\|u_0-v_0\|_{B^{s-1}_{p,r}}\times\exp\left\{c\int_0^t(\|u(\tau)\|^2_{B^s_{p,r}}\right.$$
$$\left.+\|v(\tau)\|^2_{B^s_{p,r}}+\|u(\tau)\|_{B^s_{p,r}}\|v(\tau)\|_{B^s_{p,r}}+\|u(\tau)\|_{B^s_{p,r}}+\|v(\tau)\|_{B^s_{p,r}})\mathrm{d}\tau\right\}.$$

Proof Let $w=u-v$. It is obvious that w solves the transport equation

$$\begin{cases} w_t+uw_x=-v_xw-\partial_x(1-\partial_x^2)^{-1}g, & t>0,\ x\in\mathbb{R},\\ w(0,x)=w_0(x)=u_0-v_0, & x\in\mathbb{R},\end{cases}\tag{2.16}$$

where $g=\frac{b+1}{3}(u^3-v^3)-\frac{1}{2}(u^2-v^2)+\frac{3-b}{2}(u_x^2-v_x^2)$. In view of Lemma 2.2, it follows that

$$e^{-c\int_0^t\|u(\tau)\|_{B^{s-1}_{p,r}}\mathrm{d}\tau}\|w(t)\|_{B^{s-1}_{p,r}}$$
$$\leqslant\|w_0\|_{B^{s-1}_{p,r}}+c\int_0^t e^{-c\int_0^\tau\|u(\tau')\|_{B^{s-1}_{p,r}}\mathrm{d}\tau'}\times\left(\|v_xw(\tau)\|_{B^{s-1}_{p,r}}+\|g(\tau)\|_{B^{s-1}_{p,r}}\right)\mathrm{d}\tau.\tag{2.17}$$

Similar to (2.13) and (2.14), we obtain

$$\|v_xw\|_{B^{s-1}_{p,r}}\leqslant c\|v\|_{B^s_{p,r}}\|w\|_{B^{s-1}_{p,r}}\ \text{and}$$
$$\|\partial_x(1-\partial_x^2)^{-1}g\|_{B^{s-1}_{p,r}}\leqslant\|g\|_{B^{s-2}_{p,r}}\leqslant c\|w\|_{B^{s-1}_{p,r}}\times$$
$$(\|u\|^2_{B^s_{p,r}}+\|v\|^2_{B^s_{p,r}}+\|u\|_{B^s_{p,r}}\|v\|_{B^s_{p,r}}+\|u\|_{B^s_{p,r}}+\|v\|_{B^s_{p,r}}).$$

From the above two relations, applying Gronwall's inequality to (2.17), we have

$$\exp\left(-c\int_0^t\|u(\tau)\|_{B^{s-1}_{p,r}}\mathrm{d}\tau\right)\|w(t)\|_{B^{s-1}_{p,r}}$$
$$\leqslant\|w_0\|_{B^{s-1}_{p,r}}\times\exp\left(c\int_0^t(\|u(\tau)\|^2_{B^s_{p,r}}+\|v(\tau)\|^2_{B^s_{p,r}}+\|u(\tau)\|_{B^s_{p,r}}\|v(\tau)\|_{B^s_{p,r}}\right.$$
$$\left.+\|u(\tau)\|_{B^s_{p,r}}+\|v(\tau)\|_{B^s_{p,r}})\mathrm{d}\tau\right).$$

This completes the proof of Lemma 2.5. □

Remark 2.1 *While $s-1=1+\frac{1}{p}$, we can not use Lemma 2.2. However, applying interpolation theorem as follows*

$$\begin{aligned}
\|u(t)-v(t)\|_{B^{s-1}_{p,r}} &= \|u(t)-v(t)\|_{B^{1+\frac{1}{p}}_{p,r}} \\
&\leqslant \|u(t)-v(t)\|^{\theta}_{B^{s_1}_{p,r}}\|u(t)-v(t)\|^{1-\theta}_{B^{s_2}_{p,r}} \\
&\leqslant (\|u(t)\|_{B^s_{p,r}}+\|v(t)\|_{B^s_{p,r}})^{1-\theta}\|u(t)-v(t)\|^{\theta}_{B^{s_1}_{p,r}} \\
&\leqslant C(\|u(t)\|_{B^s_{p,r}},\|v(t)\|_{B^s_{p,r}})\|u_0-v_0\|^{\theta}_{B^{s_1}_{p,r}} \\
&\leqslant C(\|u(t)\|_{B^s_{p,r}},\|v(t)\|_{B^s_{p,r}})\|u_0-v_0\|^{\theta}_{B^{s-1}_{p,r}},
\end{aligned}$$

where $\max\left\{\frac{3}{2},1+\frac{1}{p}\right\}-1<s_1<1+\frac{1}{p}$, $1+\frac{1}{p}<s_2<2+\frac{1}{p}$ *and* $\theta s_1+(1-\theta)s_2=1+\frac{1}{p}$. *Thus we also get the result of Lemma 2.5 in the condition of* $s-1=1+\frac{1}{p}$.

Lemma 2.6 *Let $\bar{\mathbb{N}}=\mathbb{N}\cup\{\infty\}$. Assume that $\{v^n\}_{n\in\bar{\mathbb{N}}}$ be a sequence of functions belongs to $C([0,T);B^{s-1}_{p,r})$, $p,r\in[1,\infty]$, $s>\left\{\frac{3}{2},1+\frac{1}{p}\right\}$. If v^n is the solution to*

$$\partial_t v^n + a^n\partial_x v^n = f, \quad v^n(0,x)=v_0,$$

with the initial data $v_0\in B^{s-1}_{p,r}$, $f\in L^1([0,T);B^{s-1}_{p,r})$, and such that

$$\sup_{n\in\mathbb{N}}\|\partial_x a^n(t)\|_{B^{s-1}_{p,r}}\leqslant \alpha(t), \quad \textit{for some } \alpha(t)\in L^1(0,T).$$

In addition a^n tends to a^∞ in $L^1([0,T);B^{s-1}_{p,r})$, then v^n tends to v^∞ in $C([0,T);B^{s-1}_{p,r})$.

Proof Let $w^n=v^n-v^\infty$. It is obvious that w^n solves the transport equation

$$\partial_t w^n + a^n\partial_x w^n = (a^\infty - a^n)\partial_x v^\infty.$$

Assume that $v_0\in B^s_{p,r}$ and $f\in L^1([0,T);B^s_{p,r})$. By Lemma 2.2, we obtain

$$\|v^n\|_{B^s_{p,r}}\leqslant e^{c\int_0^t\alpha(\tau)\mathrm{d}\tau}\|v_0\|_{B^s_{p,r}}+\int_0^t e^{c\int_\tau^t\alpha(\tau')\mathrm{d}\tau'}\|f(\tau)\|_{B^s_{p,r}}\mathrm{d}\tau. \tag{2.18}$$

On the other hand, by Lemma 2.2, we also have

$$\|w^n(t)\|_{B^{s-1}_{p,r}}\leqslant\int_0^t e^{c\int_\tau^t\|a^n(\tau')\|_{B^{s-1}_{p,r}}\mathrm{d}\tau'}\|(a^\infty-a^n)\partial_x v^\infty(\tau)\|_{B^{s-1}_{p,r}}\mathrm{d}\tau. \tag{2.19}$$

Since $B^{s-1}_{p,r}, s>\frac{3}{2}$ is an algebra, in view of (2.18) and (2.19), we deduce

$$\|w^n(t)\|_{B^{s-1}_{p,r}} \leqslant C\left(\|v_0\|_{B^s_{p,r}} + \int_0^t \|f(\tau)\|_{B^s_{p,r}}\mathrm{d}\tau\right)\int_0^t \|(a^\infty - a^n)(\tau)\|_{B^{s-1}_{p,r}}\mathrm{d}\tau, \quad (2.20)$$

which yields the desired result of convergence.

While $v_0 \in B^{s-1}_{p,r}$, one can processed as follows for $n\in\bar{\mathbb{N}}$ and $k\in\mathbb{N}$

$$\|w^n\|_{B^{s-1}_{p,r}} \leqslant \|v^n - v^n_k\|_{B^{s-1}_{p,r}} + \|v^n_k - v^\infty_k\|_{B^{s-1}_{p,r}} + \|v^\infty_k - v^\infty\|_{B^{s-1}_{p,r}}, \quad (2.21)$$

where v^n_k is the solution to equation

$$\partial_t v^n_k + a^n\partial_x v^n_k = S_k f, \quad v^n(0,x) = S_k v_0 = \sum_{q<k}\Delta_q v_0.$$

Due to $S_k v_0 \in B^s_{p,r}$ and $S_k f\in L^1([0,T];B^s_{p,r})$, thanks to (2.20), we have

$$\|v^n_k - v^\infty_k\|_{B^{s-1}_{p,r}} \leqslant C\left(\|S_k v_0\|_{B^s_{p,r}} + \int_0^t \|S_k f(\tau)\|_{B^s_{p,r}}\mathrm{d}\tau\right)\int_0^t \|(a^\infty - a^n)(\tau)\|_{B^{s-1}_{p,r}}\mathrm{d}\tau. \quad (2.22)$$

On the other hand, $(v^m - v^m_k)$ solves the following equation

$$\partial_t u + a^m\partial_x u = f - S_k f, \quad (v^n - v^n_k)(0,x) = v_0 - S_k v_0.$$

By Lemma 2.2 once again, it follows that

$$\|v^m_k - v^m\|_{B^{s-1}_{p,r}} \leqslant e^{c\int_0^t \alpha(\tau)\mathrm{d}\tau}\left(\|v_0 - S_k v_0\|_{B^{s-1}_{p,r}} + \int_0^t \|f(\tau) - S_k f(\tau)\|_{B^{s-1}_{p,r}}\mathrm{d}\tau\right). \quad (2.23)$$

Plugging (2.22) and (2.23) into (2.21), we can get

$$\begin{aligned}\|w^n(t)\|_{B^{s-1}_{p,r}} &\leqslant C_1\left(\|v_0 - S_k v_0\|_{B^{s-1}_{p,r}} + \int_0^t \|f(\tau) - S_k f(\tau)\|_{B^{s-1}_{p,r}}\mathrm{d}\tau\right)\\ &+ C_1\left(\|S_k v_0\|_{B^s_{p,r}} + \int_0^t \|S_k f(\tau)\|_{B^s_{p,r}}\mathrm{d}\tau\right)\int_0^t \|(a^\infty - a^n)(\tau)\|_{B^{s-1}_{p,r}}\mathrm{d}\tau.\end{aligned}$$

For fixed k, let n tend to infinity so that the last term of right tends to zero. By Lebesgue dominated convergence theorem, the front two terms maybe made arbitrary small for k large enough. This completes the proof of Lemma 2.6. □

Thanks to Lemma 2.6, we can deduce the following result, which is necessary to proof the continuity of the solution to Eq.(2.2).

Lemma 2.7 *If $u_0 \in B^s_{p,r}$, $p,r \in [1,\infty]$, $s > \left\{\frac{3}{2}, 1+\frac{1}{p}\right\}$, there exists a $T>0$ and a neighborhood V of u_0 in $B^s_{p,r}$ such that for any solution v to Eq.(2.2) with the initial v_0, the map*

$$\Phi : v_0 \mapsto v : V \subset B^s_{p,r} \mapsto C([0,T); B^s_{p,r}) \qquad \textit{is continuous.}$$

Proof Let $\{u_0^n\}_{n\in\mathbb{N}}$ tend to $u_0^\infty \in B^s_{p,r}$. According to Lemma 2.4, there exists $M>0$, such that $\sup_{n\in\mathbb{N}} \|u^n\|_{B^s_{p,r}} \leqslant M$. Set $v^n = \partial_x u^n$, $v^\infty = \partial_x u^\infty$. One can check that v^n solves the following equation

$$\partial_t v^n + u^n \partial_x v^n = f^n, \quad v^n(0,x) = \partial_x u_0^n,$$

where $f^n = -(u_x^n)^2 + \partial_x^2 P(D)\left(\frac{b+1}{3}(u^n)^3 - \frac{1}{2}(u^n)^2 + \frac{3-b}{3}(u^n)_x^2\right)$. By virtue of Section 10 in [21], we decompose $v^n = z^n + w^n$ such that

$$\partial_t z^n + u^n \partial_x z^n = f^n - f^\infty, \quad z^n(0,x) = \partial_x u_0^n - \partial_x u_0^\infty$$

$$\text{and} \qquad \partial_t w^n + u^n \partial_x w^n = f^\infty, \quad w^n(0,x) = \partial_x u_0^\infty.$$

Note that

$$\begin{aligned} \|f^n - f^\infty\|_{B^{s-1}_{p,r}} \leqslant & c\|u^n - u^\infty\|_{B^s_{p,r}} \\ & + c(\|u^n\|_{B^s_{p,r}} + \|u^\infty\|_{B^s_{p,r}})\|\partial_x u^n - \partial_x u^\infty\|_{B^{s-1}_{p,r}}. \end{aligned}$$

Thus, by Lemma 2.2, we can get

$$\begin{aligned} \|z^n(t)\|_{B^{s-1}_{p,r}} \leqslant & e^{c\int_0^t \|u^n(\tau)\|_{B^{s-1}_{p,r}} \mathrm{d}\tau} \|\partial_x u_0^n - \partial_x u_0^\infty\|_{B^{s-1}_{p,r}} \\ & + c\int_0^t (\|u^n(\tau) - u^\infty(\tau)\|_{B^{s-1}_{p,r}} + \|\partial_x u^n(\tau) - \partial_x u^\infty(\tau)\|_{B^{s-1}_{p,r}}) \mathrm{d}\tau. \end{aligned} \tag{2.24}$$

On the other hand, applying Lemma 2.5, $\{u^n\}_{n\in\mathbb{N}}$ tends to u^∞ in $C([0,T); B^{s-1}_{p,r})$, by Lemma 2.6, we have $w^n \mapsto w^\infty$ in $C([0,T); B^{s-1}_{p,r})$.

Set $\epsilon > 0$. Since $\{u^n\}_{n\in\mathbb{N}}$ is uniformly bounded in $C([0,T); B^s_{p,r})$. From (2.24) for large enough $n \in \mathbb{N}$, it follows that

$$\begin{aligned} & \|\partial_x u^n(t) - \partial_x u^\infty(t)\|_{B^{s-1}_{p,r}} \\ \leqslant & \|z^n(t)\|_{B^{s-1}_{p,r}} + \|w^n(t) - w^\infty(t)\|_{B^{s-1}_{p,r}} \\ \leqslant & \epsilon + C(\|\partial_x u_0^n - \partial_x u_0^\infty\|_{B^{s-1}_{p,r}} + \int_0^t (\|u^n(\tau) - u^\infty(\tau)\|_{B^{s-1}_{p,r}} \\ & + \|\partial_x u^n(\tau) - \partial_x u^\infty(\tau)\|_{B^{s-1}_{p,r}}) \mathrm{d}\tau) \leqslant \epsilon + \|\partial_x u_0^n - \partial_x u_0^\infty\|_{B^{s-1}_{p,r}} \\ & + C\int_0^t \|\partial_x u^n(\tau) - \partial_x u^\infty(\tau)\|_{B^{s-1}_{p,r}} \mathrm{d}\tau. \end{aligned} \tag{2.25}$$

Applying Gronwall's inequality to (2.25), we obtain

$$\|\partial_x u^n(t) - \partial_x u^\infty(t)\|_{L^\infty([0,T);B^{s-1}_{p,r})} \leqslant C_1(\epsilon + \|\partial_x u^n_0 - \partial_x u^\infty_0\|_{B^{s-1}_{p,r}}).$$

where C_1 depends only on M, T. This completes the proof of Lemma 2.7. □

Proof of Theorem 2.1. Thanks to Lemma 2.4, $\{u^n\}_{n\in\mathrm{N}}$ is a Cauchy sequence in $C([0,T);B^{s-1}_{p,r})$, therefore, it converges to a function u in $C([0,T);B^{s-1}_{p,r})$, and $\{u^n\}_{n\in\mathrm{N}}$ is uniformly bounded in $L^\infty([0,T);B^s_{p,r})$, in view of v) of Lemma 2.1, we have $u \in L^\infty([0,T);B^s_{p,r})$.

Due to $\{u^n\}_{n\in\mathrm{N}}$ converges to u in $C([0,T);B^{s-1}_{p,r})$, applying interpolation theorem, one can easily get that the convergence holds in $C([0,T);B^{s_1}_{p,r})$, $\forall s_1 < s$. Passing to limit in Eq.(2.6), we obtain that u is a solution to Eq.(2.2). Note that the right-hand of the equation

$$u_t + u_x u = P(D)\left((b+1)u^2 u_x - uu_x + (3-b)u_x u_{xx}\right)$$

belongs to $L^\infty([0,T);B^s_{p,r})$. In view of Lemma 2.2 again, we have $u \in C([0,T);B^s_{p,r})$, by Eq.(2.2), we have $u_t \in C([0,T);B^{s-1}_{p,r})$. Thus $u \in E^s_{p,r}(T)$. Thanks to Lemma 2.5 and Lemma 2.7, we obtain the uniqueness and continuity with the initial data u_0 in $C([0,T);B^s_{p,r})$. This completes the proof of Theorem 2.1. □

Remark 2.2 *In the proof of Theorem 2.1, while $s = 1/p+2$ and $s = 1/p+1$, we can not use Lemma 2.5. However, applying interpolation theorem as Remark 2.1, we also get the result. Moreover, if $u_0 \in B^{3/2}_{2.1}$, similar to [11], we obtain the well-posedness of Eq.(2.2) in $B^{3/2}_{2.1}$. Due to $H^s = B^s_{2,2}$ and*

$$H^s \hookrightarrow B^{3/2}_{2,1} \hookrightarrow H^{3/2} \hookrightarrow B^{3/2}_{2,\infty} \hookrightarrow H^{s'}, \quad \forall s' < \frac{3}{2} < s.$$

Therefore, the space H^s and $B^s_{2,r}$ are very close and $s = \dfrac{3}{2}$ is critical for H^s.

3 Conservation laws and Blow-up scenario

At first, by Sobolev's embedding theorem and Theorem 2.1, we can get the local well-posedness result in Sobolev space $H^s(\mathbb{R})$.

Theorem 3.1 *Assume that $u_0 \in H^s(\mathbb{R})$, $s > \dfrac{3}{2}$. Then there exists a unique solution u to Eq.(1.6) (or Eq.(2.2)), and a $T = T(\|u_0\|_s)$ such that*

$$u = u(\cdot, u_0) \in C([0,T);H^s(\mathbb{R})) \cap C^1([0,T);H^{s-1}(\mathbb{R})).$$

Moreover, the solution depends continuously on the initial data, i.e. the map $u_0 \to u(\cdot, u_0) : H^s(\mathbb{R}) \to C([0,T]; H^s(\mathbb{R})) \cap C^1([0,T]; H^{s-1}(\mathbb{R}))$ *is continuous.*

Remark 3.1 *Similar to the proof of Theorem* 2.1 *in* [34], *the local well-posedness of the Cauchy problem to Eq.*(1.6) *in* $H^s(\mathbb{R})$, $s > \dfrac{3}{2}$ *also can be established by the Kato's semigroup theory.*

Next, we shall begin deriving conservation laws for the strong solutions to Eq.(1.6).

Theorem 3.2 *Assume the initial data* $u_0 \in H^s(\mathbb{R}), s \geqslant 3$, *and the potential* $y_0 = u_0 - u_{0,xx} \in L^1(\mathbb{R})$. *Then as long as the solution* $u(t,x)$ *to Eq.(1.6) given by Theorem 3.1 exists, we have*

$$\int_{\mathbb{R}} y(t,x)\mathrm{d}x = \int_{\mathbb{R}} u(t,x)\mathrm{d}x = \int_{\mathbb{R}} y_0(x)\mathrm{d}x = \int_{\mathbb{R}} u_0(x)\mathrm{d}x.$$

Moreover, if $b = 2$, *it follows that*

$$\int_{\mathbb{R}} \left(u^2(t,x) + u_x^2(t,x)\right) \mathrm{d}x = \int_{\mathbb{R}} \left(u_0^2 + u_{0,x}^2\right) \mathrm{d}x,$$

$$|u(t,x)| \leqslant \frac{\sqrt{2}}{2}\|u_0\|_1,$$

where $u_0 = u(0,x), u_{0,x} = u_x(0,x)$, *and* $y = u - u_{xx}$.

Proof Since $u_0 \in H^s(\mathbb{R}), s \geqslant 3$. Using Eq.(1.6), we obtain

$$\begin{aligned}\frac{\mathrm{d}}{\mathrm{d}t}\int_{\mathbb{R}} y(t,x)\mathrm{d}x &= \int_{\mathbb{R}} \left(bu_x u_{xx} + uu_{xxx} - (b+1)u^2 u_x\right) \mathrm{d}x \\ &= 0.\end{aligned}$$

Thus

$$\int_{\mathbb{R}} y(t,x)\mathrm{d}x = \int_{\mathbb{R}} y_0(x)\mathrm{d}x.$$

Similarly,

$$\int_{\mathbb{R}} u(t,x)\mathrm{d}x = \int_{\mathbb{R}} y_0(x)\mathrm{d}x = \int_{\mathbb{R}} u_0(x)\mathrm{d}x.$$

If $b = 2$, multiplying Eq.(1.6) by u, integrating by parts on $\mathbb{R}$, we can get

$$\begin{aligned}&\frac{1}{2}\frac{\mathrm{d}}{\mathrm{d}t}\int_{\mathbb{R}} \left(u^2 + u_x^2\right) \mathrm{d}x = \int_{\mathbb{R}} \left(uu_t - uu_{txx}\right) \mathrm{d}x \\ &= \int_{\mathbb{R}} \left(-3u^3 u_x + 2uu_x u_{xx} + u^2 u_{xxx}\right) \mathrm{d}x \\ &= \int_{\mathbb{R}} \left(-\frac{3}{4}(u^4)_x - u^2 u_{xxx} + u^2 u_{xxx}\right) \mathrm{d}x \\ &= 0.\end{aligned} \tag{3.1}$$

This yields

$$\int_{\mathbb{R}} (u^2 + u_x^2)\,\mathrm{d}x = \int_{\mathbb{R}} (u_0^2 + u_{0,x}^2)\,\mathrm{d}x. \tag{3.2}$$

From the above conservation law (3.2), it follows that

$$\begin{aligned}
\sqrt{2}|u(t,x)| &= \left(\int_{-\infty}^{x} 2uu_x\mathrm{d}x - \int_{x}^{\infty} 2uu_x\mathrm{d}x\right)^{\frac{1}{2}} \\
&\leqslant \left(\int_{\mathbb{R}} 2|uu_x|\mathrm{d}x\right)^{\frac{1}{2}} \leqslant \left(\int_{\mathbb{R}} (u^2 + u_x^2)\mathrm{d}x\right)^{\frac{1}{2}} \\
&= \left(\int_{\mathbb{R}} (u_0^2 + u_{0,x}^2)\mathrm{d}x\right)^{\frac{1}{2}} = \|u_0\|_1.
\end{aligned}$$

This completes the proof of Theorem 3.2. □

For the convenience of presentation, we first give the following useful lemmas, which will be used in the proof of Theorem 3.3.

Lemma 3.1 [22] *Assume that $s > 0$. Then we have*

$$\|[\Lambda^s, g]f\|_{L^2} \leqslant c(\|\partial_x g\|_{L^\infty}\|\Lambda^{s-1}f\|_{L^2} + \|\Lambda^s g\|_{L^2}\|f\|_{L^\infty}),$$

where c is constant depending only on s.

Lemma 3.2 [9] *Assume that $F \in C^{m+2}(\mathbb{R})$ with $F(0) = 0$. Then for every $\frac{1}{2} < s \leqslant m$, we have*

$$\|F(u)\|_s \leqslant \tilde{F}(\|u\|_{L^\infty})\|u\|_s, \quad u \in H^s(\mathbb{R}),$$

where $\tilde{F}$ is a monotone increasing function depending only on F and s.

Lemma 3.3 [22] *Assume that $s > 0$. Then $H^s(\mathbb{R}) \cap L^\infty(\mathbb{R})$ is an algebra. Moreover*

$$\|fg\|_s \leqslant c(\|f\|_{L^\infty}\|g\|_s + \|f\|_s\|g\|_{L^\infty}),$$

where c is a constant depending only on s.

Lemma 3.4 [7] *Let $T > 0$ and $u \in C^1([0,T); H^2(\mathbb{R}))$. Then for every $t \in [0,T)$, there exist at least one pair points $\xi(t), \zeta(t) \in \mathbb{R}$, such that*

$$m(t) = \inf_{x\in\mathbb{R}} u(t,x) = u(t,\xi(t)), \quad M(t) = \sup_{x\in\mathbb{R}} u(t,x) = u(t,\zeta(t)),$$

and $m(t), M(t)$ are absolutely continuous in $[0,T)$. Moreover,

$$\frac{\mathrm{d}m(t)}{\mathrm{d}t} = u_t(t,\xi(t)), \quad \frac{\mathrm{d}M(t)}{\mathrm{d}t} = u_t(t,\zeta(t)), \qquad a.e. \ \ on \ \ [0,T).$$

By virtue of the above conservation laws and lemmas, we can derive a blow-up scenario.

Theorem 3.3 *Let $u_0 \in H^s(\mathbb{R}), s \geqslant 3$. If T is the maximal existence time of corresponding solution $u(t,x)$ to Eq.(1.6) with the initial data u_0, then the $H^s(\mathbb{R})$-norm of $u(t,x)$ to Eq.(1.6) (or (2.2)) blows up on $[0,T)$ if and only if*

$$\overline{\lim_{t\uparrow T}}\|u(t,x)\|_{L^\infty} + \|u_x(t,x)\|_{L^\infty} = \infty.$$

Moreover, if $b=2$, the $H^s(\mathbb{R})$-norm of $u(t,x)$ blows up on $[0,T)$ if and only if

$$\liminf_{t\uparrow T}\{\inf_{x\in\mathbb{R}} u_x(t,x)\} = -\infty.$$

Proof Let $u(t,x)$ be the solution to Eq.(1.6) with the initial data $u_0 \in H^s(\mathbb{R}), s \geqslant 3$, which is guaranteed by Theorem 3.1.

If $\overline{\lim}_{t\uparrow T}\|u_x(t,x)\|_{L^\infty} = \infty$, by Sobolev's embedding theorem, we obtain the solution $u(t,x)$ will blow up in finite time.

Next, applying the operator Λ^s to Eq.(2.2), multiplying by $\Lambda^s u$, and integrating by parts on $\mathbb{R}$, we deduce that

$$\frac{\mathrm{d}}{\mathrm{d}t}\langle u,u\rangle_s = -2\langle uu_x,u\rangle_s + 2\langle f(u),u\rangle_s, \tag{3.3}$$

where $f(u) = -(1-\partial_x^2)^{-1}\partial_x\left(\frac{b+1}{3}u^3 - \frac{1}{2}u^2 + \frac{3-b}{2}u_x^2\right)$.
Assume there exists a $M>0$, such that $\overline{\lim}_{t\uparrow T}\|u(t,x)\|_{L^\infty} + \|u_x(t,x)\|_{L^\infty} \leqslant M$. Then we obtain

$$\begin{aligned}
&|\langle uu_x,u\rangle_s| = |\langle\Lambda^s uu_x,\Lambda^s u\rangle_0|\\
=&\ |\langle[\Lambda^s,u]u_x,\Lambda^s u\rangle_0 + \langle u\Lambda^s u_x,\Lambda^s u\rangle_0|\\
\leqslant&\ \|[\Lambda^s,u]u_x\|_{L^2}\|u\|_s + \|u_x\|_{L^\infty}\|u\|_s^2\\
\leqslant&\ c\left(\|u_x\|_{L^\infty}\|\Lambda^{s-1}u_x\|_{L^2} + \|\Lambda^s u\|_{L^2}\|u_x\|_{L^\infty}\right)\|u\|_s + cM\|u\|_s^2\\
\leqslant&\ cM(\|u\|_s + \|u\|_s)\|u\|_s + cM\|u\|_s^2\\
\leqslant&\ 3cM\|u\|_s^2 \leqslant c\|u\|_s^2,
\end{aligned} \tag{3.4}$$

where we applied Lemma 3.1 with $s \geqslant 3$.

On the other hand, we estimate the second term of the right hand side of Eq.(3.3)

$$
\begin{aligned}
|\langle f(u), u\rangle_s| &= \left|\left\langle (1-\partial_x^2)^{-1}\partial_x\left(\frac{b+1}{3}u^3 - \frac{1}{2}u^2 + \frac{3-b}{2}u_x^2\right), u\right\rangle_s\right| \\
&\leqslant c\|u\|_s \left\|\frac{b+1}{3}u^3 - \frac{1}{2}u^2 + \frac{3-b}{2}u_x^2\right\|_{s-1} \\
&\leqslant c\|u\|_s \left(\frac{|b|+1}{3}\|u^3\|_{s-1} + \frac{1}{2}\|u^2\|_{s-1} + \frac{3+|b|}{2}\|u_x^2\|_{s-1}\right) \\
&\leqslant c\|u\|_s (c\|u\|_{s-1} + c\|u\|_{s-1} + c\|u_x\|_{s-1}) \\
&\leqslant 3c\|u\|_s^2, \qquad (3.5)
\end{aligned}
$$

here we applied Lemma 3.2 with $F(u) = u^2, u^3, u_x^2$, and $s = s-1$.

Combining (3.3), (3.4) with (3.5), we obtain

$$
\frac{\mathrm{d}}{\mathrm{d}t}\|u\|_s^2 \leqslant c\|u\|_s^2.
$$

Thus using Gronwall's inequality, we get

$$
\|u(t)\|_s^2 \leqslant \|u_0\|_s^2 \exp(ct).
$$

Finally, if $b=2$, the slope of solution is bounded from below in finite time, then we deduce that the solution will not blow up in finite time. Differentiating Eq.(2.1) with respect to x, in view of the identity $\partial_x^2(p*f) = p*f - f$, it follows that

$$
u_{tx} + uu_{xx} = -\frac{1}{2}u_x^2 + u^3 - \frac{1}{2}u^2 + p*\left(-\frac{1}{2}u_x^2 - u^3 + \frac{1}{2}u^2\right).
$$

Then, we deduce that

$$
u_{tx} + uu_{xx} \leqslant -\frac{1}{2}u_x^2 + u^3 + p*\left(-u^3 + \frac{1}{2}u^2\right).
$$

By Young's inequality and Theorem 3.2, we obtain that

$$
\begin{aligned}
&\left\|u^3 + p*\left(-u^3 + \frac{1}{2}u^2\right)\right\|_{L^\infty} \\
&\leqslant \|u\|_{L^\infty}^3 + \|p\|_{L^1}\left\|-u^3 + \frac{1}{2}u^2\right\|_{L^\infty} \\
&\leqslant \|u\|_1^3 + \|u\|_1^3 + \frac{1}{2}\|u\|_1^2 \\
&\leqslant 2\|u_0\|_1^3 + \frac{1}{2}\|u_0\|_1^2. \qquad (3.6)
\end{aligned}
$$

Define $M(t) = \sup_{x\in\mathbb{R}} u_x(t,x)$. By Lemma 3.4, we can check that the function $M(t)$ is locally Lipschitz, and $M(t) = u_x(t,\zeta(t))$. Since $u_{xx}(t,\zeta(t)) = 0$ for all $t \in$

$[0,T)$, it follows that a.e. on $[0,T)$

$$\frac{\mathrm{d}}{\mathrm{d}t}M(t) \leqslant -\frac{1}{2}M^2(t) + 2\|u_0\|_1^3 + \frac{1}{2}\|u_0\|_1^2.$$

Letting $C = \left(2\|u_0\|_1^3 + \frac{1}{2}\|u_0\|_1^2\right)^{\frac{1}{2}}$, we can get

$$M'(t) \leqslant -\frac{1}{2}M^2(t) + C^2.$$

If $M(t) > \sqrt{2}C$, then $M'(t) < 0$ and $M(t)$ is decreasing. Otherwise, $M(t) \leqslant \sqrt{2}C$. Thus we get

$$M(t) \leqslant \max\{M(0), \sqrt{2}C\}.$$

Therefore, we obtain the slope of solution is bounded, which guarantees the solution will not blow up in finite time. □

Remark 3.2 *Theorem* 3.3 *shows that the blow-up scenario of the modified CH and DP equations is different with that of the CH equation and the DP equation* [7, 36]. *However, if* $b = 2$, *the conservation law is similar to the CH equation. Since the* H^1*–norm of Eq.*(1.4) *is conserved, we see that the slope of the solution to Eq.*(1.4) *becomes unbounded whereas its amplitude remains bounded.*

Remark 3.3 *By Theorem* 3.3, *we find that* (1.7) *is the unique smooth solution of Eq.*(1.6) *with corresponding initial data. Moreover, when* $b = 2$ *and given*

$$u(t,x) = -2\mathrm{sech}^2\frac{1}{2}(x-ct), \quad u_0 = -2\mathrm{sech}^2\frac{1}{2}x.$$

Similarly, (1.8) is the blow-up strong solution to Eq.(1.6). *While* $b = 2,\ 3$, *Lang et. get many general form solutions of Eqs.*(1.4) *and* (1.5) [24]. *These arguments are consisted with the result of Theorem* 3.3.

4 Blow-up phenomena and blow-up rate

In this section, we consider $b = 2$, Eq.(1.6) becomes the modified CH equation, we prove that the solutions of Eq.(1.4) have some properties, provided the initial data satisfies certain conditions.

Consider the following differential equation

$$\begin{cases} q_t = u(t,q), & t > 0,\ x \in \mathbb{R}, \\ q(0,x) = x, & x \in \mathbb{R}. \end{cases} \tag{4.1}$$

Applying classical results in the theory of ordinary differential equations, one can obtain the following useful result on the above initial value problem.

Lemma 4.1 [26] *Let $u_0 \in H^s(\mathbb{R}), s \geqslant 3$, and T be the maximal existence time of the corresponding solution $u(t,x)$ to Eq.(2.1). Then Eq.(4.1) has a unique solution $q(t,x) \in C^1([0,T) \times \mathbb{R}, \mathbb{R})$. Moreover, the map $q(t,\cdot)$ is an increasing diffeomorphism of $\mathbb{R}$ with*

$$q_x = \exp\left(\int_0^t u_x(s, q(s,x))\mathrm{d}s\right) > 0, \qquad \forall (t,x) \in [0,T) \times \mathbb{R}.$$

In view of Theorem 3.2 and Lemma 4.1, we obtain the blow-up phenomena, i.e. the wave-breaking.

Theorem 4.1 *Assume that $u_0 \in H^s(\mathbb{R}), s \geqslant 3$. If there exists a $x_0 \in \mathbb{R}$ such that*

$$u_0^{'}(x_0) < -\sqrt{4\|u_0\|_1^3 + \|u_0\|_1^2}.$$

Then the corresponding solution $u(t,x)$ of Eq.(2.1) with the initial data u_0 blows up in finite time. Moreover, the maximal time of existence is estimated above by

$$\frac{1}{\sqrt{4\|u_0\|_1^3 + \|u_0\|_1^2}} \ln\left(\frac{u_0^{'}(x_0) - (4\|u_0\|_1^3 + \|u_0\|_1^2)^{\frac{1}{2}}}{u_0^{'}(x_0) + (4\|u_0\|_1^3 + \|u_0\|_1^2)^{\frac{1}{2}}}\right).$$

Proof Let T be the maximal time of existence of the solution $u(t,x)$ to Eq.(2.1) with the initial data u_0. Differentiating Eq.(2.1) with respect to x, in view of $\partial_x^2(p * f) = p * f - f$, we obtain from $b = 2$ that

$$u_{tx} + uu_{xx} = -\frac{1}{2}u_x^2 + u^3 - \frac{1}{2}u^2 + p * \left(-\frac{1}{2}u_x^2 - u^3 + \frac{1}{2}u^2\right). \tag{4.2}$$

Note that

$$\begin{aligned}\frac{\mathrm{d}u_x(t,q(t,x))}{\mathrm{d}t} &= u_{tx}(t,q(t,x)) + u_{xx}(t,q(t,x))\frac{\mathrm{d}q(t,x)}{\mathrm{d}t} \\ &= u_{tx}(t,q(t,x)) + u(t,q(t,x))u_{xx}(t,q(t,x)).\end{aligned} \tag{4.3}$$

By (3.6), (4.2) and (4.3), it follows that

$$\begin{aligned}\frac{\mathrm{d}u_x(t,q(t,x))}{\mathrm{d}t} &= -\frac{1}{2}u_x^2(t,q(t,x)) + \left(u^3 - \frac{1}{2}u^2\right)(t,q(t,x)) \\ &\quad + p * \left[\left(-\frac{1}{2}u_x^2 - u^3 + \frac{1}{2}u^2\right)(t,q(t,x))\right] \\ &\leqslant -\frac{1}{2}u_x^2(t,q(t,x)) + u^3(t,q(t,x)) \\ &\quad + p * \left[\left(-u^3 + \frac{1}{2}u^2\right)(t,q(t,x))\right] \\ &\leqslant -\frac{1}{2}u_x^2(t,q(t,x)) + 2\|u_0\|_1^3 + \frac{1}{2}\|u_0\|_1^2.\end{aligned} \tag{4.4}$$

Letting $m(t)=u_x(t,q(t,x_0))$ and $K=\left(2\|u_0\|_1^3+\frac{1}{2}\|u_0\|_1^2\right)^{\frac{1}{2}}$, we have

$$\frac{\mathrm{d}m(t)}{\mathrm{d}t}\leqslant-\frac{1}{2}m^2(t)+K^2. \tag{4.5}$$

By the assumption of the theorem to yield

$$m'(0)\leqslant-\frac{1}{2}m^2(0)+K^2<0.$$

In view of (4.5), the standard argument of continuity shows

$$m'(t)<0, \text{and} \quad m(t)<-\sqrt{2}K, \qquad \forall t\in[0,T).$$

By solving the inequality (4.5), we have

$$\frac{m(0)+\sqrt{2}K}{m(0)-\sqrt{2}K}e^{\sqrt{2}Kt}-1\leqslant\frac{2\sqrt{2}K}{m(t)-\sqrt{2}K}<0.$$

Since $0<\dfrac{m(0)+\sqrt{2}K}{m(0)-\sqrt{2}K}<1$, it follows that

$$0<T\leqslant\frac{1}{\sqrt{2}K}\ln\frac{m(0)-\sqrt{2}K}{m(0)+\sqrt{2}K},$$

such that $\lim_{t\uparrow T}m(t)=-\infty$. This completes the proof of the theorem. □

We are now concerned with the blow-up rate of the blow-up solutions to Eq.(1.4).

Theorem 4.2 *Assume that $T<\infty$ be the blow-up time of the corresponding solution $u(t,x)$ to Eq.(2.1) with initial data $u_0\in H^s(\mathbb{R}), s\geqslant 3$. Then we have*

$$\lim_{t\uparrow T}\left(\inf_{x\in\mathbb{R}}\{u_x(t,x)\}(T-t)\right)=-2,$$

while the solution $u(t,x)$ satisfies

$$\|u\|_{L^\infty}\leqslant\frac{\sqrt{2}}{2}\|u_0\|_1.$$

Proof Define $m(t)=\inf_{x\in\mathbb{R}}\{u_x(t,x)\}$, for $t\in[0,T)$. Obviously, we can check that the function $m(t)$ is locally Lipschitz. By Lemma 3.4, it follows that

$$m(t)=u_x(t,\xi(t)), \qquad t\in[0,T).$$

Moreover, in view of Theorem 3.3, we have

$$\liminf_{t\uparrow T}m(t)=-\infty. \tag{4.6}$$

Differentiating Eq.(2.1) with respect to x, in view of the identity $\partial_x^2(p*f)=p*f-f$, it follows that

$$u_{tx}+uu_{xx}=-\frac{1}{2}u_x^2+u^3-\frac{1}{2}u^2+p*\left(-\frac{1}{2}u_x^2-u^3+\frac{1}{2}u^2\right). \tag{4.7}$$

By Young's inequality and Theorem 3.2, we obtain that

$$\begin{aligned}
&\left\|u^3-\frac{1}{2}u^2+p*\left(-\frac{1}{2}u_x^2-u^3+\frac{1}{2}u^2\right)\right\|_{L^\infty}\\
\leqslant{}&\|u\|_{L^\infty}^3+\frac{1}{2}\|u\|_{L^\infty}^2+\|p*u^3\|_{L^\infty}+\frac{1}{2}\|p*(u^2+u_x^2)\|_{L^\infty}\\
\leqslant{}&\|u\|_1^3+\frac{1}{2}\|u\|_1^2+\|u\|_1^3+\frac{1}{2}\|p\|_{L^\infty}\|u^2+u_x^2)\|_{L^1}\\
\leqslant{}&2\|u_0\|_1^3+\frac{1}{2}\|u_0\|_1^2+\frac{1}{4}\|u_0\|_1^2\\
={}&2\|u_0\|_1^3+\frac{3}{4}\|u_0\|_1^2.
\end{aligned} \tag{4.8}$$

Note that $u_{xx}(t,\xi(t))=0$, for a.e. $t\in[0,T)$. Then, by (4.7) and (4.8) we have

$$\begin{aligned}
\frac{dm(t)}{dt}&\leqslant-\frac{1}{2}m^2(t)+\left\|u^3-\frac{1}{2}u^2+p*\left(-\frac{1}{2}u_x^2-u^3+\frac{1}{2}u^2\right)\right\|_{L^\infty}\\
&\leqslant-\frac{1}{2}m^2(t)+2\|u_0\|_1^3+\frac{3}{4}\|u_0\|_1^2\\
&=-\frac{1}{2}m^2(t)+K^2,
\end{aligned} \tag{4.9}$$

where $K=\left(2\|u_0\|_1^3+\frac{3}{4}\|u_0\|_1^2\right)^{\frac{1}{2}}$.

Choose now $\varepsilon\in(0,1)$. Applying (4.6), we can find $t_0\in[0,T)$ such that

$$m(t_0)<-\left(K^2+\frac{K^2}{\varepsilon}\right)^{\frac{1}{2}}.$$

In view of (4.9), we deduce that $m(t)$ is decreasing on $[t_0,T)$ and

$$m(t)<-\left(K^2+\frac{K^2}{\varepsilon}\right)^{\frac{1}{2}}.$$

Again, applying (4.7) and (4.8), it follows that for a.e. $t\in[t_0,T)$

$$-\frac{1}{2}m^2(t)-K^2\leqslant\frac{dm(t)}{dt}\leqslant-\frac{1}{2}m^2(t)+K^2.$$

Note that $m(t)$ is locally Lipschitz and $m(t)<0$ on (t_0,T). Solving the above inequality, we have

$$\frac{1}{2}-\varepsilon\leqslant\frac{d}{dt}\left(\frac{1}{m(t)}\right)\leqslant\frac{1}{2}+\varepsilon.$$

Integrating the above inequality on (t,T) with $t\in[t_0,T)$ and noticing that $\lim\limits_{t\to T} m(t)=-\infty$, we get

$$\left(\frac{1}{2}-\varepsilon\right)(T-t)\leqslant-\frac{1}{m(t)}\leqslant\left(\frac{1}{2}+\varepsilon\right)(T-t).$$

It follows that

$$\frac{2}{1+2\varepsilon}\leqslant(t-T)m(t)\leqslant\frac{2}{1-2\varepsilon}.$$

Since $\varepsilon\in(0,1)$ is arbitrary, in view of the definition of $m(t)$, we get the result of Theorem 4.2. □

Remark 4.1 *Note that the blow-up rates of breaking waves to the CH equation and the DP equation are* -2, -1, *respectively* [8,15].

5 Persistence properties

In this section, as [19,35], we shall establish the persistence properties of the strong solution to Eq.(2.1), providing the initial data $u_0(x)$ decays at infinity.

Theorem 5.1 *Let* $u_0\in H^s(\mathbb{R})$, $s>\frac{3}{2}$ *and* $T>0$. *Assume that* $u\in C([0,T];H^s(\mathbb{R}))$ *is the corresponding solution to Eq.(2.1) with the initial data* u_0. *If there exists* $\alpha\in(0,1)$ *such that*

$$|u_0(x)|,\ |u_{0,x}(x)|\sim\mathcal{O}(e^{-\alpha x}),\qquad as\ x\uparrow\infty,$$

then, we have

$$|u(t,x)|,\ |u_x(t,x)|\sim\mathcal{O}(e^{-\alpha x})\qquad as\ x\uparrow\infty$$

uniformly in the interval $[0,T]$.

Proof For simplicity, let $M=\sup_{t\in[0,T]}\|u(t)\|_s$, and

$$F(u)=\frac{b+1}{3}u^3-\frac{1}{2}u^2+\frac{3-b}{2}u_x^2.$$

By the Sobolev embedding theorem, we have $\|u(t)\|_{L^\infty},\ \|u_x(t)\|_{L^\infty}\leqslant M$.
Let the weighted function

$$\varphi_N(x)=\begin{cases}1, & x\leqslant 0,\\ e^{\alpha x}, & 0<x<N,\\ e^{\alpha N}, & x\geqslant N,\end{cases}\tag{5.1}$$

where $N\in\mathbb{Z}^+$. One can easily check that for all N

$$0\leqslant\varphi_N'(x)\leqslant\varphi_N(x)\quad a.e.\quad x\in\mathbb{R}.\tag{5.2}$$

By Eq.(2.1), we obtain that

$$(u\varphi_N)_t + (u\varphi_N)u_x + (p_x * F(u))\varphi_N = 0, \quad \text{and} \tag{5.3}$$

$$(u_x\varphi_N)_t + u_x^2\varphi_N + uu_{xx}\varphi_N + (p_{xx} * F(u))\varphi_N = 0. \tag{5.4}$$

Multiplying Eq.(5.3) by $(u\varphi_N)^{2m-1}$ with $m \in \mathbb{Z}^+$ and integrating the result in $\mathbb{R}$ with respect to the x-variable, we have

$$\int_{\mathbb{R}} (u\varphi_N)_t(u\varphi_N)^{2m-1}\mathrm{d}x = -\int_{\mathbb{R}} ((u\varphi_N)^{2m}u_x + \varphi_N(p_x * F(u))(u\varphi_N)^{2m-1})\mathrm{d}x. \tag{5.5}$$

Note that

$$\int_{\mathbb{R}} (u\varphi_N)_t(u\varphi_N)^{2m-1}\mathrm{d}x = \|u\varphi_N\|_{L^{2m}}^{2m-1}\frac{\mathrm{d}}{\mathrm{d}t}\|u\varphi_N\|_{L^{2m}},$$

$$\left|\int_{\mathbb{R}} (u\varphi_N)^{2m}u_x\mathrm{d}x\right| \leqslant M\|u\varphi_N\|_{L^{2m}}^{2m}, \text{ and}$$

$$\left|\int_{\mathbb{R}} \varphi_N(p_x * F(u))(u\varphi_N)^{2m-1}\mathrm{d}x\right| \leqslant \|u\varphi_N\|_{L^{2m}}^{2m-1}\|\varphi_N(p_x * F(u))\|_{L^{2m}}.$$

In view of (5.5) and the above relations, we have

$$\frac{\mathrm{d}}{\mathrm{d}t}\|u\varphi_N\|_{L^{2m}} \leqslant M\|u\varphi_N\|_{L^{2m}} + \|\varphi_N(p_x * F(u))\|_{L^{2m}}. \tag{5.6}$$

Applying Gronwall's inequality to (5.6), we can get

$$\|u\varphi_N\|_{L^{2m}} \leqslant \left(\|u_0\varphi_N\|_{L^{2m}} + \int_0^t \|\varphi_N(p_x * F(u))(\tau)\|_{L^{2m}}\mathrm{d}\tau\right)e^{Mt}. \tag{5.7}$$

Since $f \in L^1(\mathbb{R}) \cap L^\infty(\mathbb{R})$ implies

$$\lim_{n\uparrow\infty} \|f\|_{L^n} = \|f\|_{L^\infty}.$$

Let $m \uparrow \infty$ in (5.7), it follows that

$$\|u\varphi_N\|_{L^\infty} \leqslant \left(\|u_0\varphi_N\|_{L^\infty} + \int_0^t \|\varphi_N(p_x * F(u))(\tau)\|_{L^\infty}\mathrm{d}\tau\right)e^{Mt}. \tag{5.8}$$

Similarly, multiplying Eq.(5.4) by $(u_x\varphi_N)^{2m-1}$, integration by parts, we obtain

$$\begin{aligned}\|u_x\varphi_N\|_{L^{2m}}^{2m-1}\frac{\mathrm{d}}{\mathrm{d}t}\|u_x\varphi_N\|_{L^{2m}} =& -\int_{\mathbb{R}} \varphi_N(u_x^2 + uu_{xx})(u_x\varphi_N)^{2m-1}\mathrm{d}x \\ &- \int_{\mathbb{R}} \varphi_N(p_{xx} * F(u))(u_x\varphi_N)^{2m-1}\mathrm{d}x. \end{aligned} \tag{5.9}$$

In view of (5.2), we only estimate the following term

$$\begin{aligned}
&\left|\int_{\mathbb{R}} uu_{xx}\varphi_N(u_x\varphi_N)^{2m-1}\mathrm{d}x\right| = \left|\int_{\mathbb{R}} u(u_x\varphi_N)^{2m-1}[(u_x\varphi_N)_x - u_x\varphi_N']\mathrm{d}x\right| \\
=&\left|\int_{\mathbb{R}}\left(\frac{u_x}{2m}(u_x\varphi_N)^{2m} + uu_x\varphi_N'(u_x\varphi_N)^{2m-1}\right)\mathrm{d}x\right| \\
\leqslant&\left(\frac{\|u_x\|_{L^\infty}}{2m} + \|u\|_{L^\infty}\right)\int_{\mathbb{R}}|u_x\varphi_N|^{2m}\mathrm{d}x \\
\leqslant&\frac{2m+1}{2m}M\|u_x\varphi_N\|_{L^{2m}}^{2m}.
\end{aligned}$$

By virtue of (5.9), Holder's inequality and Gronwall's inequality, letting $m \uparrow \infty$, we have

$$\|u_x\varphi_N\|_{L^\infty} \leqslant \left(\|u_{0,x}\varphi_N\|_{L^\infty} + \int_0^t \|\varphi_N(p_{xx} * F(u))(\tau)\|_{L^\infty}d\tau\right)e^{2Mt}. \tag{5.10}$$

Combining (5.8) with (5.10), it follows that

$$\begin{aligned}
&\|u\varphi_N\|_{L^\infty} + \|u_x\varphi_N\|_{L^\infty} \\
\leqslant& e^{2Mt}\left(\|u_0\varphi_N\|_{L^\infty} + \|u_{0,x}\varphi_N\|_{L^\infty}\right) \\
&+ e^{2Mt}\left(\int_0^t (\|\varphi_N(p_x * F(u))(\tau)\|_{L^\infty} + \|\varphi_N(p_{xx} * F(u))(\tau)\|_{L^\infty})\mathrm{d}\tau\right).
\end{aligned}$$

A simple calculation shows that there exists c_0 such that $\forall N \in \mathbb{Z}^+$,

$$\varphi_N(x)\int_{\mathbb{R}}\frac{e^{-|x-y|}}{\varphi_N(y)}\mathrm{d}y \leqslant c_0 = \frac{4}{1-\alpha}. \tag{5.11}$$

Therefore, in view of (5.12) we obtain

$$\begin{aligned}
|\varphi_N(p_x * u^3)| &= \left|\frac{1}{2}\varphi_N(x)\int_{\mathbb{R}}\frac{e^{-|x-y|}}{\varphi_N(y)}\varphi_N(y)u^3(y)\mathrm{d}y\right| \\
&\leqslant \frac{c_0}{2}\|\varphi_N u\|_{L^\infty}\|u\|_{L^\infty}^2 \leqslant \frac{c_0M^2}{2}\|\varphi_N u\|_{L^\infty}.
\end{aligned} \tag{5.12}$$

Similarly, we have

$$\|\varphi_N(p_x * u^2)\|_{L^\infty} \leqslant \frac{c_0M}{2}\|\varphi_N u\|_{L^\infty},$$

$$\|\varphi_N(p_x * u_x^2)\|_{L^\infty} \leqslant \frac{c_0M}{2}\|\varphi_N u_x\|_{L^\infty}.$$

Thus, one can easily get that

$$\|\varphi_N(p_x * F(u))(\tau)\|_{L^\infty} \leqslant c_1(\|\varphi_N u(\tau)\|_{L^\infty} + \|\varphi_N u_x(\tau)\|_{L^\infty}). \tag{5.13}$$

Note that $p_{xx} * f = p * f - f$, similar to (5.14), we obtain

$$\begin{aligned}\|\varphi_N(p_{xx} * F(u))(\tau)\|_{L^\infty} &= \|\varphi_N(F(u) - p * F(u))(\tau)\|_{L^\infty} \\ &\leqslant c_1(\|\varphi_N u(\tau)\|_{L^\infty} + \|\varphi_N u_x(\tau)\|_{L^\infty}),\end{aligned} \tag{5.14}$$

where c_1 is constant depending only on c_0, M.

Combining (5.11), (5.14) with (5.15), we can get

$$\begin{aligned}\|u\varphi_N\|_{L^\infty} + \|u_x\varphi_N\|_{L^\infty} \leqslant & e^{2Mt}(\|u_0\varphi_N\|_{L^\infty} + \|u_{0,x}\varphi_N\|_{L^\infty}) \\ &+ 2c_1 e^{2Mt}\left(\int_0^t (\|u(\tau)\varphi_N\|_{L^\infty} + \|u_x(\tau)\varphi_N\|_{L^\infty})\mathrm{d}\tau\right).\end{aligned} \tag{5.15}$$

Applying Gronwall's inequality, for all $t \in [0,T]$, there exists a constant $\tilde{c} = \tilde{c}(c_0, M, T)$ such that

$$\begin{aligned}\|u\varphi_N\|_{L^\infty} + \|u_x\varphi_N\|_{L^\infty} &\leqslant \tilde{c}\,(\|u_0\varphi_N\|_{L^\infty} + \|u_{0,x}\varphi_N\|_{L^\infty}) \\ &\leqslant \tilde{c}\,(\|u_0 \max(1, e^{\alpha x})\|_{L^\infty} + \|u_{0,x} \max(1, e^{\alpha x})\|_{L^\infty}).\end{aligned} \tag{5.16}$$

Letting $N \uparrow \infty$, from (5.17), for all $t \in [0,T]$, we can find that

$$\begin{aligned}&(\|u(t,x)e^{\alpha x}\|_{L^\infty} + \|u_x(t,x)\varphi_N\|_{L^\infty}) \\ &\qquad \leqslant \tilde{c}\,(\|u_0 \max(1, e^{\alpha x})\|_{L^\infty} + \|u_{0,x} \max(1, e^{\alpha x})\|_{L^\infty}).\end{aligned}$$

This completes the proof of Theorem 5.1. □

Acknowledgements This work was partially supported by CPSF (Grant No.: 2012M520007) and CPSF (Grant No.: 2013T60086). The authors thank the references for their valuable comments and constructive suggestions.

References

[1] Camassa R, Holm D. An integrable shallow water equation with peaked solitons. Phys. Rev. Letters, 1993, 71: 1661–1664.

[2] Chemin J Y. Localization in Fourier space and Navier-Stokes system. Phase Space Analysis of Partial Differential Equations. Proceedings, CRM series, Pisa, 2004, 53–136.

[3] Coclite G M, Karlsen K H, Risebro N H. Numberical schemes for computing discontinuous solutions of the Degasperis-Procesi equation. IMA J. Numer. Anal., 2008, 28: 80–105.

[4] Constantin A. Global existence and breaking waves for a shallow water equation: a geometric approch. Ann. Inst. Fourier (Grenoble), 2000, 50: 321–362.

[5] Constantin A. On the scattering problem for the Camassa-Holm equation. Proc. R. Soc. London A, 2001, 457: 953–970.

[6] Constantin A. Finite propagation speed for the Camassa-Holm equation, J. Math. Phys., 2005, 46: 023506.

[7] Constantin A, Escher J. Wave breaking for nonlinear nonlocal shallow water equations. Acta Math., 1998, 181: 229–243.

[8] Constantin A, Escher J. On the blow-up rate and the blow-up set of breaking waves for a shallow water equation. Math. Z., 2000, 233: 75–91.

[9] Constantin A, Molinet L. The initial value problem for a generalized Boussinesq equation. Differential and Integral equations, 2002, 15: 1061–1072.

[10] Danchin R. A few remarks on the Camassa-Holm equation. Differential and Integral equations, 2001, 14: 953–988.

[11] Danchin R. A note on well-posedness for Camassa-Holm equation. J. Differential Equations, 2003, 192: 953–988.

[12] Danchin R. Fourier Analysis Methods for PDEs. Lecture Notes, 14 November, 2005.

[13] Degasperis A, Holm D D, Hone A N W. A new integrable equation with peakon solution. Theoret. and Math. Phys., 2002, 133: 1463–1474.

[14] Dullin H R, Gottwald G A, Holm D D. On asymptotically equivalent shallow water wave equations. Physica D, 2004, 190: 1–14.

[15] Escher J, Liu Y, Yin Z Y. Global weak solutions and blow-up structure for the Degasperis-Procesi equation. J. Funct. Anal., 2006, 241: 457–485.

[16] Escher J, Liu Y, Yin Z Y. Shock waves and blow–up phenomena forthe periodic Degasperis-Procesi equation, Indiana Univ. Math. J., 2007, 56: 87–117.

[17] Fokas A, Fuchssteiner B. Symplectic structures, their Bäcklund transformation and hereditary symmetries. Physica D, 1981, 4: 47–66.

[18] Henry D. Compactly supported solutions of the Camassa-Holm equation. J. Nonlinear Math. Phys., 2005, 12: 342–347.

[19] Himonas A, Misiolek G, Ponce G, Zhou Y. Persistence properties and unique continuation of solutions of the Camassa-Holm equation. Comm. Math. Phys., 2007, 271: 511–522.

[20] Holm D D, Staley M F. Wave structure and nonlinear balances in a family of evolutionary PDEs. SIAM J. Appl. Dyn. Syst., 2003, 3: 323–380.

[21] Kato T. Quasi-linear equation of evolution, with applications to partical differential equations. in: Spectral Theorey and Differential Equation, in: Lecture Notes in Math., Spring-Verlag, Berlin, 1975, 488: 25–70.

[22] Kato T, Ponce G. Commutator estimation and the Euler and Navier–Stokes Equation. Comm. Pure Appl. Math., 1998, 41: 891–907.

[23] Lenells J. Traveling wave solutions of the Degasperis-Procesi equation. J. Math. Anal. Appl., 2005, 306: 72–82.

[24] Liang S, Jeffrey D J. New travelling wave solutions to modified CH and DP equations. Comput. Phys. Comm., 2009, 180: 1429–1433.

[25] Liu Z, Ouyang Z. A note on solitary waves for modified forms of Camassa-Holm and Degasperis-Procesi equations. Phys. Lett. A, 2007, 366: 377–381.

[26] Liu Y, Yin Z Y. Global existence and blow-up phenomena for Degasperis-Procesi equation. Comm. Math. Phys., 2006, 267: 801–820.

[27] Lundmark H. Formation and dynamics of shock waves in the Degasperis-Procesi equation. J. Nonlinear Sci., 2007, 17: 169–198.

[28] Lundmark H, Szmigielski J. Multi-peakon solutions of the Degasperis-Procesi equation. Inverse Problems, 2003, 19: 1241–1245.

[29] Ma H, Yu Y, Ge D. New exact traveling wave solutions for the modified form of Degasperis-Procesi equation. Appl. Math. Comput., 2008, 203: 792–798.

[30] Shen J, Xu W. Bifurcation of smooth and non-smooth traveling wave solutions in the generalized Camassa-Holm equation. Chaos, Solitons and Fractals, 2005, 26: 1149–1162.

[31] Wazwaz A M. Solitary wave solutions for modified forms of Degasperis-Procesi and Camassa-Holm equations. Phys. Lett. A, 2006, 352: 500–504.

[32] Wazwaz A M. New solitary wave solutions to the modified forms of Degasperis-Procesi and Camassa-Holm equations. Appl. Math. Comput., 2007, 186: 130–141.

[33] Wang Q, Tang M. New exact solutions for two nonlinear equations. Phys. Lett. A, 2008, 372: 2995–3000.

[34] Wu X L, Yin Z Y. Well-posedness and blow-up phenomena for the generalized Degasperis-Procesi equation. Nonlinear Anal., 2010, 73: 136–146.

[35] Wu X L, Guo B L. Persistence properties and infinite propagation for the modified 2-component Camassa-Holm equation. Discrete Contin. Dyn. Syst. A, 2013, 33: 3211–3223.

[36] Yin Z Y. On the Cauchy problem for an integrable equation with peakon solutins. Illinois J. Math., 2003, 47: 649–666.

[37] Yin Z Y. Global weak solutions to a new periodic integrable equation with peakon solutions. J. Funct. Anal., 2004, 212: 182–194.

[38] Zhang L, Chen L, Huo X. Bifurcation of smooth and non-smooth traveling wave solutions in the generalized Degasperis-Procesi equation. J. Comp. Appl. Math., 2007, 205: 174–185.

Stochastic Korteweg-de Vries Equation Driven by Fractional Brownian Motion*

Wang Guolian (王国联) and Guo Boling (郭柏灵)

Abstract We consider the Cauchy problem for the Korteweg-de Vries equation driven by a cylindrical fractional Brownian motion (fBm) in this paper. With Hurst parameter $H \geqslant \frac{7}{16}$ of the fBm, we obtain the local existence results with initial value in classical Sobolev spaces H^s with $s \geqslant -\frac{9}{16}$. Furthermore, we give the relation between the Hurst parameter H and the index s to the Sobolev spaces H^s, which finds out the regularity between the driven term fBm and the initial value for the stochastic Korteweg-de Vries equation.

Keywords Korteweg-de Vries equation; fractional Brownian motion; Hurst parameter; stochastic convolution; bilinear estimate

1 Introduction

The Korteweg-de Vries (KdV) equation models long, unidirectional weakly nonlinear waves propagating at the surface of a fluid. When the pressure above is not constant or the bottom is not flat, a forcing term is added which is either the pressure gradient or the gradient of the function whose graph defines the bottom. The case where the forcing term is Brownian motion type noise has been discussed by many documents (see [6, 8, 15, 26, 27]). Since uncertainty may have some correlation with the evolution of the time, we focus on the case where the forcing term is a fractional Brownian motion type noise. Recently, many documents considered stochastic system driven by fBm, such as [5, 10–13, 16, 17, 23, 29, 30].

The stochastic KdV equation studied here is written in Itô form

$$du + (\partial_x^3 u + \partial_x u^2)dt = \Phi dB^H, \tag{1.1}$$

* DISCRETE CONT DYN, 2015, 35(11): 5255 – 5272. DOI: 10.3934/dcds.2015.35.5255.

with initial condition

$$u(0,x) = u_0(x). \tag{1.2}$$

The initial u_0 is a function of some Sobolev space, $u = u(t,x)$ is a random process defined on $(t,x) \in \mathbb{R}^+ \times \mathbb{R}$, Φ is a linear operator given below in (1.6), $B^H = (B_t^H, t \geqslant 0)$ is a cylindrical fBm in $L^2(\mathbb{R})$. We emphasize the fact here Φ does not depend on the unknown u, i.e. the noise is assumed to be additive.

The cylindrical fBm in $L^2(\mathbb{R})$ is given by

$$B^H = \sum_{i=0}^{\infty} \beta_i^H(t) e_i(x),$$

where $(e_i(x))_{i\in\mathbb{N}}$ is an orthonormal basis in $L^2(\mathbb{R})$ and $(\beta_i^H(t))_{i\in\mathbb{N}}$ is a sequence of mutually independent fBm in a fixed probability space $(\Omega, \mathcal{F}^H, \mathbb{P}^H)$ endowed with the natural filtration $(\mathcal{F}_t^H)_{t\in[0,T]}$; for the detail, one can refer to [11, 13, 29].

We will solve equation (1.1) supplemented with the initial condition (1.2) by considering its mild form

$$u(t) = U(t)u_0 - \int_0^t U(t-s)\partial_x(u^2(s))\mathrm{d}s + \int_0^t U(t-s)\Phi \mathrm{d}B^H(s), \tag{1.3}$$

where $U(\cdot)$ is the Airy group associated to linear equation

$$\partial_t u + \partial_x^3 u = 0. \tag{1.4}$$

We denote the stochastic convolution in (1.3) by $Z(t)$

$$Z(t) \equiv \int_0^t U(t-s)\Phi \mathrm{d}B^H(s), \tag{1.5}$$

which is solved the linear problem

$$\begin{cases} \mathrm{d}Z + \partial_x^3 Z \mathrm{d}t = \Phi \mathrm{d}B^H, \\ Z(0,x) = 0. \end{cases}$$

Our aim in this paper is to study the existence and uniqueness of solution to (1.1), (1.2). Before describing our work, we recall some facts about the KdV equation. The Cauchy problem for the deterministic KdV equation, that is, with $\Phi = 0$ in (1.1), was studied in many works (see [4, 18, 19]). Bourgain introduced new space-time function spaces based on linear Airy group in [4], which allowed him to prove a global well posedness result for the KdV equation in $L^2(\mathbb{R})$. Then, following the theory, Kenig, Ponce and Vega in [18, 19] obtained a bilinear estimate in Bourgain-type spaces $X_{s,b}$ (given below) with $b > 1/2$, which allowed them to get existence and uniqueness in Sobolev spaces of negative indices, namely in $H^s(\mathbb{R})$ with $s > -\frac{3}{4}$. The Cauchy problem for stochastic KdV equation driven by Brownian motion i.e. fBm with $H = \frac{1}{2}$ was considered by de Bouardet et al in [8] in Bourgain-type spaces $X_{s,b}$ with $b < \frac{1}{2}$ and close enough to $\frac{1}{2}$. They obtained local well-posedness in Sobolev spaces of negative indices, namely in $H^s(\mathbb{R})$ with $s > -\frac{5}{8}$.

We consider (1.1), (1.2) in Bourgain-type spaces $X_{s,b}$ with $b \geqslant \frac{7}{16}$ in the present paper. It seems that here $b \geqslant \frac{7}{16}$ gives an optimal lower bound. It is difficult to solve the problem due to the irregularity of $X_{s,b}$ as the indices b is less than $\frac{1}{2}$. We recall that de Bouard et al in [8] introduced weighted Bourgain-type spaces as bridge to deal with irregularity of $X_{s,b}$ as $b < \frac{1}{2}$ and close enough to $\frac{1}{2}$ to get the bilinear estimate. However, this method cannot apply to the lower indices b. Inspired by the Fourier restriction method in [14] which works in Bourgain-type spaces $X_{s,b}$ with $b > \frac{1}{2}$, we overcome the difficulty in the case of the lower indices $b < \frac{1}{2}$ by generalizing some inequalities and with the use of multi-linear operator from [28]. The method used in this paper provides another proof for the bilinear estimate in [8]. We can also use the method to analysis other stochastic partial differential equations driven by fBm, such as Schrödinger equation, long-short wave equation, Benjemin-Ono equation.

In the following paper, we work on the Bourgain-type spaces $X_{s,b}$ with $b \geqslant \frac{7}{16}$ and focus on the case that $\frac{7}{16} \leqslant b < \frac{1}{2}$. To explain the reason, let's recall the definition of the fBm, and some properties related to the Hurst parameter.

The fBm was introduced by Mandelbrot & van Ness in [22]. It was first studied by Kolmogorov in [20] and later by Mandelbrot and Van Ness in [21], where a stochastic integral representation in terms of an ordinary Brownian motion was established. The fBm is a continuous-time Gaussian process $\beta^H(t)$ on $[0, T]$, which starts at zero, has expectation zero for all t in $[0, T]$, and has the following covariance function:

$$\mathbb{E}(\beta_t^H \beta_s^H) = \frac{1}{2}\{|t|^{2H} + |s|^{2H} - |t-s|^{2H}\}$$

where H is a real number in $(0, 1)$, called Hurst parameter.

The value of H determines what kinds of processes the $\beta^H(t)$ is. If $H=\frac{1}{2}$ the process is in fact a Brownian motion. If $\frac{1}{2}<H<1$, it has higher regularity than Brownian motion. So some properties related to the Hurst index with $H=\frac{1}{2}$ can easily be extended to the Hurst indices with $\frac{1}{2}<H<1$ in mathematics. However, it is significant in physics, for the increments of the process are positively correlated. If $0<H<\frac{1}{2}$, it has lower regularity than Brwonian motion. So properties adapted to the Hurst index with $H=\frac{1}{2}$ are extended to the case $0<H<\frac{1}{2}$, rigorous proofs are necessary.

Roughly speaking, the indices b in $X_{s,b}$ represents the smoothness in time (see [30]). We know that the stochastic system (1.1), (1.2) has the same time regularity as fBm, which leads us to work in $X_{s,b}$ with $b\leqslant H$. de Bouard et al in [8] dealt with the case $X_{s,b}$ as $b<\frac{1}{2}$ and close enough $\frac{1}{2}$. We will concentrate on the case that the lower b in the following paper.

Before stating our results precisely, we introduce some special Sobolev spaces and notations.

Given two separable Hilbert spaces $\bar{H}$ and $\tilde{H}$, We denote by $L_2^0(\bar{H};\tilde{H})$ the space of Hilbert-Schmidt operators from $\bar{H}$ into $\tilde{H}$. Its norm is given by

$$\|\Phi\|^2_{L_2^0(\bar{H};\tilde{H})}=\sum_{i\in\mathbb{N}}\|\Phi e_i\|^2_{\tilde{H}},\quad \Phi\in L_2^0(\bar{H};\tilde{H}),\tag{1.6}$$

where $(e_i)_{i\in\mathbb{N}}$ is any orthonormal basis of $\bar{H}$. When $\bar{H}=L^2(\mathbb{R}),\tilde{H}=H^s$, $L_2^0(L^2(\mathbb{R});H^s)$ is simply denoted by $L_2^{0,s}$.

Denote $\hat{u}(\tau,\xi)=\mathscr{F}u$ by the Fourier transform in t and x of u and $\mathscr{F}_{(\cdot)}u$ by the Fourier transform in the $(\cdot)$ variable. Denoted by $\langle\cdot,\cdot\rangle$ the L^2 space-time duality product, i.e.

$$\langle f,g\rangle=\iint_{\mathbb{R}^2}f(t,x)\overline{g(t,x)}\mathrm{d}t\mathrm{d}x.$$

As in [4,8,18,19], for $s,b\in\mathbb{R}$, we define the spaces $X_{s,b}$ to be the completion of the Schwartz function space on $\mathbb{R}^2$ with respect to the norm

$$\|u\|_{X_{s,b}}=\|U(-t)u\|_{H^s_xH^b_t}=\|\langle\xi\rangle^s\langle\tau-\xi^3\rangle^b\mathscr{F}u\|_{L^2_\xi L^2_\tau},$$

where $U(t)=\mathscr{F}_x^{-1}\mathrm{e}^{\mathrm{i}t\xi^3}\mathscr{F}_x$ denotes the unitary operator associated to linear equation (1.4) and $\langle\cdot\rangle=(1+|\cdot|)$.

For $T>0$, we consider the spaces $X_{s,b}^T$ as functions in $X_{s,b}$ restriction to $[0,T]$, which are endowed with norm

$$\|u\|_{X_{s,b}^T}=\inf\{\|\tilde{u}\|_{X_{s,b}},\tilde{u}\in X_{s,b}\text{ and }u=\tilde{u}|_{[0,T]}\}.$$

Next, we will give some useful notations for multi-linear expressions from [28]. Let Z be any Abelian additive group with an invariant measure $d\xi$. For any integer $k\geqslant 2$, we denote $\Gamma_k(Z)$ by the "hyperplane"

$$\Gamma_k(Z)=\{(\xi_1,...,\xi_k)\in Z^k:\xi_1+...+\xi_k=0\},$$

and define a $[k;Z]$-multiplier to be any function m: $\Gamma_k(Z)\to\mathbb{C}$.

If m is a [k; Z]-multiplier, we define $\|m\|_{[k;Z]}$ to be the best constant, such that the inequality

$$\Big|\int_{\Gamma_k(Z)}m(\xi)\prod_{j=1}^k f_j(\xi_j)\Big|\leqslant\|m\|_{[k;Z]}\prod_{j=1}^k\|f_j\|_{L_2(Z)}$$

holds for all test functions f_j defined on Z.

It is clear that $\|m\|_{[k;Z]}$ determines a norm on m for test functions at least; We are interested in obtaining the good boundedness on the norm. In this paper, let $Z=\mathbb{R}\times\mathbb{R}$. We will use the following Lemma from [28].

Lemma 1.1 *If m and M are $[k;Z]$ multipliers and satisfy $|m(\xi)|\leqslant|M(\xi)|$ for all $\xi\in\Gamma_k(Z)$, then $\|m\|_{[k;Z]}\leqslant\|M\|_{[k;Z]}$.*

Now, we give the statement of the main results.

Theorem 1.1 *Let $u_0(\omega,x)$ be such that $u_0\in H^s(\mathbb{R})$ for a.s. $\omega\in\Omega$, and u_0 is $\mathcal{F}_0$-measurable. Assume that:*

(i) *for some s with $s\geqslant-\dfrac{9}{16}$ and $a>\dfrac{1}{16}$, $\Phi\in L_2^{0,s+\frac{3}{2}a}$ if $\dfrac{7}{16}\leqslant H<\dfrac{1}{2}$; or*

(ii) *for some s with $s>-\dfrac{5}{8}$, $\Phi\in L_2^{0,s}$ if $\dfrac{1}{2}\leqslant H<1$.*

Then, for a.s. $\omega\in\Omega$, there is a $T_\omega>0$ and a unique solution $u(t)$ of the initial value problem (1.1), (1.2) *on $[0,T_\omega]$ which satisfies*

$$u\in C([0,T_\omega];H^s(\mathbb{R}))\cap X_{s,b}^{T_\omega}.$$

Remark 1.1 *In the case $H=\dfrac{1}{2}$, the result in the Theorem* 1.1 *corresponds to the one in* [8].

Remark 1.2 *In fact, Theorem* 1.1 *holds in the following domain with the indices* s, H *satisfying*

if $\frac{7}{16} \leqslant H < \frac{1}{2}$,

$$-s - H \leqslant \frac{1}{8}, 1 + s - H \leqslant 0;$$

if $\frac{1}{2} \leqslant H < 1$,

$$-s - H < \frac{1}{8}, 1 + s - H < 0.$$

Remark 1.3 *In the case* $H = \frac{1}{2}$ *and* $u_0 \in L^2(\mathbb{R})$, *the global existence result in* [8] *can be extended to the case* $\frac{1}{2} \leqslant H < 1$ *and* $u_0 \in L^2(\mathbb{R})$ *to the Cauchy problem* (1.1), (1.2). *However, in other cases, due to the lack of global a priori estimates, the global existence result can not be obtained.*

It is standard to prove Theorem 1.1 with the use of a fixed point argument. First using the newly developing Wiener integral with respect to fBm to consider the stochastic integral (1.5), then by Bourgain restriction method to consider the second term in (1.3), we get some a priori estimates about (1.5) and the second term in (1.3). We prove the local existence results of Theorem 1.1 by a fixed point argument.

The paper is arranged as follows. In Section 2, we first introduce the stochastic calculus with fBm in brief in Subsection 2.1, which is preliminary knowledge to derive the estimate on (1.5) in Subsection 2.2; then we turn our main attention to derive a bilinear estimate of the term $\partial_x u^2$ concerned with the stochastic problem (1.1)–(1.2) in Section 3. With the preliminary estimates given in Section 2 and Section 3, we end the paper by proving the main result Theorem 1.1 in Section 4.

2 Stochastic convolution

Our aim in this section is to estimate the stochastic convolution (1.5) in Bourgain-type spaces $X_{s,b}$. There are several ways to define a stochastic calculus with respect to fBm which are summarized in [3, 24]. For our purpose we select the newly developing Wiener integrals for fBm in [1, 2, 29]. In the following, we first introduce the Wiener integrals and some related properties in Subsection 2.1. By this definition, we can obtain an expression of the Wiener integral with respect to fBm in terms of an integral with respect to the Brownian motion, which is the key to deduce a priori estimates about the (1.5) in Subsection 2.2.

2.1 The Wiener integral with fBm

Fix an interval $[0,T]$, let $\beta^H(t), t\in[0,T]$, be a fBm of Hurst parameter $H\in(0,1)$ on the probability space $(\Omega,\mathcal{F}^H,\mathbb{P}^H)$ endowed with the natural filtration $(\mathcal{F}_t^H)_{t\in[0,T]}$ and the law $\mathbb{P}^H$ of $\beta^H(t)$. This means by definition that $\beta^H(t)$ is a centered Gaussian process with covariance

$$R(t,s)=E(\beta^H(s)\beta^H(t))=\frac{1}{2}(t^{2H}+s^{2H}-|t-s|^{2H}).$$

Note that $\beta^{\frac{1}{2}}$ is standard Brownian motion. Moreover β^H has the following Wiener integral representation in [1]:

$$\beta^H(t)=\int_0^t K^H(t,s)\mathrm{d}W(s), \tag{2.1}$$

where $W=\{W(t):t\in[0,T]\}$ is a Wiener process, and $K^H(t,s)$ is the kernel given by

$$\begin{aligned} K^H(t,s)=&c_H(t-s)^{H-\frac{1}{2}}\\ &+c_H\left(\frac{1}{2}-H\right)\int_s^t(u-s)^{H-\frac{3}{2}}\left(1-\left(\frac{s}{u}\right)^{\frac{1}{2}-H}\right)\mathrm{d}u, \end{aligned} \tag{2.2}$$

where

$$c_H=\left(\frac{2H\Gamma\left(\frac{3}{2}-H\right)}{\Gamma\left(H+\frac{1}{2}\right)\Gamma(2-2H)}\right)^{\frac{1}{2}}$$

and $H\in\left(0,\frac{1}{2}\right)$.

If $H\in\left(\frac{1}{2},1\right)$ the kernel $K^H(t,s)$ has the simpler expression

$$K^H(t,s)=c_H\left(H-\frac{1}{2}\right)s^{\frac{1}{2}-H}\int_s^t(u-s)^{H-\frac{3}{2}}u^{H-\frac{1}{2}}\mathrm{d}u. \tag{2.3}$$

From (2.2) we obtain

$$\frac{\partial K^H}{\partial t}(t,s)=c_H\left(H-\frac{1}{2}\right)(t-s)^{H-\frac{3}{2}}\left(\frac{s}{t}\right)^{\frac{1}{2}-H}. \tag{2.4}$$

We remark that in the sequel we will usually use the notation K instead of K^H, unless we need to specify the associated Hurst parameter H.

We will denote by $\mathscr{E}$ the linear space of step functions on $[0,T]$ of the form

$$\varphi(t)=\sum_{i=1}^{n-1} a_i 1_{(t_i,t_{i+1}]}(t), \tag{2.5}$$

where $0=t_1<t_2<\cdots<t_n=T, n\in\mathbb{N}, a_i\in\mathbb{R}$ and by $\mathcal{H}$ the closure of $\mathscr{E}$ with respect to the scalar product

$$\langle 1_{[0,t]}, 1_{[0,s]}\rangle_{\mathcal{H}}=R(t,s).$$

For φ of the form (2.5) we define its Wiener integral with respect to the fBm as

$$\int_0^T \varphi_s \mathrm{d}\beta^H(s)=\sum_{i=1}^{n} a_i\left(\beta^H(t_{i+1})-\beta^H(t_i)\right).$$

Obviously, the mapping

$$\varphi=\sum_{i=1}^{n} a_i 1_{(t_i,t_{i+1}]}(t)\to\int_0^T \varphi_s \mathrm{d}\beta^H(s)$$

is an isometry between $\mathscr{E}$ and the the linear space $span\{\beta^H(t), t\in[0,T]\}$ viewed as a subspace of $L^2(\Omega)$ and it can be extended to an isometry between $\mathcal{H}$ and the first Wiener chaos of the fBm $\overline{span}^{L^2(\Omega)}\{\beta^H(t), t\in[0,T]\}$. The image on an element $\Phi\in\mathcal{H}$ by this isometry is called the Wiener integral of Φ with respect to β^H.

Let $K_T^*:\mathscr{E}\to L^2([0,T])$ be the linear map given by

$$(K_T^*\varphi)(s)=\varphi(s)K(T,s)+\int_s^T(\varphi(t)-\varphi(s))\frac{\partial K}{\partial t}(t,s)\mathrm{d}t. \tag{2.6}$$

We refer to [1] for the proof of the fact that K_T^* is a isometry and can be extended to the closure $\mathcal{H}$ for the norm of the inner product.

When $H>\frac{1}{2}$, the operator K_T^* has the simpler expression

$$(K_T^*\varphi)(s)=\int_s^T\varphi(r)\frac{\partial K}{\partial r}(r,s)\mathrm{d}r.$$

For any $t\in[0,T]$ we can define K_t^* similarly. Also, for φ in $\mathscr{E}$ and h in $L^2(0,T)$ the following duality holds

$$\int_s^T(K_T^*\varphi)(t)h(t)\mathrm{d}t=\int_0^T\varphi(t)(Kh)(\mathrm{d}t), \tag{2.7}$$

and extends integration of step function with respect to $Kh(dt)$ to integrands in $\mathcal{H}$.

As a consequence, we have the following relation between the Wiener integral with respect to fBm and the Itô integral with respect to the Wiener process

$$\int_0^T \varphi(s)\mathrm{d}\beta^H(s) = \int_0^T (K_T^*\varphi)(s)\mathrm{d}W(s), \tag{2.8}$$

which holds for every $\varphi \in \mathcal{H}$ if and only if $K_T^*\varphi \in L^2([0,T])$, where $W(s)$ denotes the Wiener process. For any $s,t \in [0,T]$, one can check the relation

$$K_T^*[\varphi 1_{[0,t]}](s) = K_T^*[\varphi](s)1_{[0,t]}(s).$$

Then if one defines the definite stochastic integral $\int_0^t \varphi(s)\mathrm{d}\beta^H(s)$, as it should be, by $\int_0^t \varphi(s)1_{[0,t]}(s)\mathrm{d}\beta^H(s)$, we obtain

$$\int_0^t \varphi(s)\mathrm{d}\beta^H(s) = \int_0^t (K_T^*\varphi)(s)\mathrm{d}W(s) \tag{2.9}$$

for every $t \in [0,T]$ and $\varphi 1_{[0,t]} \in \mathcal{H}$ if and only if $K_T^*\varphi \in L^2(0,T)$.

Note that in the general theory of Skorohod integration with respect to fBm with values in a Hilbert space V, a relation such as (2.9) requires careful justification of the existence of its right-hand side (see [25], Section 5.1). But we will work only with Wiener integrals over Hilbert spaces; in this case we note that, if $u \in L^2([0,T];V)$ is a deterministic function, then relation (2.9) holds, the Wiener integral on the righthand side being well defined in $L^2(\Omega;V)$ if K^*u belongs to $L^2([0,T]\times V)$.

The stochastic integral with respect to a fractional Wiener processes in Hilbert space V when the integrand is a bounded operator Λ from a Hilbert space E to V is defined for t positive as, see for example [29],

$$\int_0^t \Lambda(s)\mathrm{d}\beta^H(s) = \sum_{j\in\mathbb{N}}\int_0^t \Lambda(s)\Phi e_j\mathrm{d}\beta_j^H(s) = \sum_{j\in\mathbb{N}}\int_0^t (K_t^*\Lambda(\cdot)\Phi e_j)(s)\mathrm{d}\beta_j(s), \tag{2.10}$$

where $(\Lambda(t))_{t\in[0,T]}$ is such that

$$\sum_{j\in\mathbb{N}}\int_0^t \|(K_t^*\Lambda(\cdot)\Phi e_j)(s)\|_F^2\mathrm{d}s < \infty,$$

where $(e_j)_{j\in\mathbb{N}}$ is a complete orthonormal system of E, here assumed to be $L^2(E)$. We also check from (2.10) that $\Lambda(t)$ commutes with K_T^*.

2.2 Estimate on the stochastic convolution

The stochastic convolution is

$$Z(t) \equiv \int_0^t U(t-s)\Phi\mathrm{d}B^H(s). \tag{2.11}$$

In the case $H = \dfrac{1}{2}$, de Bouard et al [8] established the estimate of $Z(t)$ in the $X_{s,b}$. We will establish the estimate of $Z(t)$ in Proposition 2.1. Due to the suitable Wiener integral, the proof of this proposition is similar to that of Proposition 2.1 given in [8]. However, a basic inequality given below plays an important role in the proof of Proposition 2.1.

Lemma 2.1 *For arbitrary $x, y \in \mathbb{R}$ and $a \in (0,1)$, we have*

$$|\mathrm{e}^{-\mathrm{i}x} - \mathrm{e}^{-\mathrm{i}y}| \leqslant C(a)|x-y|^a.$$

Let ψ be a function such that $\psi = 0$ for $t < 0$, $\psi(t) = 1$ for $t \in [0,1]$ and $\psi(t) = 0$ for $t \geqslant 2$ with $\psi \in C_0^\infty(\mathbb{R}^+)$. Note that such ψ belongs to $H^b(\mathbb{R})$ for any $b \leqslant H$. We give the following result.

Proposition 2.1 *Let $s, b \in \mathbb{R}$ with $0 < b \leqslant H < \dfrac{1}{2}$, and we assume that $\Phi \in L_2^{0,s+\frac{3}{2}a}$ for some $a \in (0,1)$; then $Z(t)$ defined by (2.11) satisfies*

$$\psi Z \in L^2(\Omega, X_{s+\frac{3}{2}a,b})$$

and

$$\mathbb{E}(\|\psi Z\|^2_{X_{s,b}}) \leqslant M(b,\psi)\|\Phi\|^2_{L_2^{0,s+\frac{3}{2}a}},$$

where $M(a,b,H,\psi)$ is a constant depending only on $a, b, H, \|\psi\|_{H^b}, \||t|^{\frac{1}{2}}\psi\|_{L^2}$, and $\||t|^{\frac{1}{2}}\psi\|_{L^\infty}$.

Proof Define $g(t,\cdot) = \psi(t)\displaystyle\int_0^t U(-s)\Phi \mathrm{d}\beta^H(s)$, so that $\psi(t)Z(t) = U(t)g(t,\cdot)$, then by the definition of $X_{s,b}$, we get

$$\begin{aligned}
&\|\psi Z\|^2_{X_{s,b}} = \|U(t)g(t,\cdot)\|^2_{X_{s,b}} \\
=&\|(1+|\xi|)^s(1+|\tau+\xi^3|)^b(\mathrm{e}^{-\mathrm{i}t\xi^3}g(t,\xi))^{\wedge_t}\|_{L^2_\tau L^2_\xi} \\
=&\|(1+|\xi|)^s(1+|\tau+\xi^3|)^b\hat{g}(\tau+\xi^3,\xi)\|^2_{L^2_\tau L^2_\xi} \\
=&\|(1+|\xi|)^s(1+|\tau|)^b\hat{g}(\tau,\xi)\|^2_{L^2_\tau L^2_\xi}.
\end{aligned}$$

Hence

$$\begin{aligned}
\mathbb{E}\left(\|\psi Z\|^2_{X_{s,b}}\right) &= \mathbb{E}\iint_{\mathbb{R}^2}(1+|\xi|)^{2s}(1+|\tau|)^{2b}|\hat{g}(\tau,\xi)|^2\mathrm{d}\tau\mathrm{d}\xi \\
&= \int_{\mathbb{R}}(1+|\xi|)^{2s}\mathbb{E}\left(|(\mathscr{F}_x g)(\cdot,\xi)|^2_{H_t^b}\right)\mathrm{d}\xi,
\end{aligned} \tag{2.12}$$

where

$$(\mathscr{F}_x g)(t,\xi)=\psi(t)\sum_{k=0}^{\infty}\int_0^t \mathrm{e}^{-\mathrm{i}s\xi^3}\widehat{\Phi e_k}(\xi)\mathrm{d}\beta_k^H(s).$$

Recalling the equivalent norm on $H_t^b(\mathbb{R})$

$$\|h\|_{H_t^b}^2=\int\int_{\mathbb{R}^2}\frac{|h(t)-h(s)|^2}{|t-s|^{1+2b}}\mathrm{d}t\mathrm{d}s+\|h\|_{L^2(\mathbb{R})}^2,\ \forall h\in H_t^b(\mathbb{R}).$$

Then we have

$$\begin{aligned}
&\mathbb{E}\left(\|\mathscr{F}_x g(\cdot,\xi)\|_{H_t^b}^2\right)=\sum_{k=0}^{\infty}|\widehat{\Phi e_k}|^2\mathbb{E}\left(\left\|\psi(t)\int_0^t \mathrm{e}^{-\mathrm{i}s\xi^3}\mathrm{d}\beta_k^H\right\|_{H_t^b}^2\right)\\
=&\sum_{k=0}^{\infty}|\widehat{\Phi e_k}|^2\left(\mathbb{E}\left(\left\|(\psi(t)\int_0^t \mathrm{e}^{-\mathrm{i}s\xi^3}\mathrm{d}\beta_k^H(s)\right\|_{L^2(\mathbb{R})}^2\right)\right.\\
&\left.+\mathbb{E}\left(\iint_{\mathbb{R}^2}\frac{|\psi(t_1)\int_0^{t_1}\mathrm{e}^{-\mathrm{i}s\xi^3}\mathrm{d}\beta_k^H(s)-\psi(t_2)\int_0^{t_2}\mathrm{e}^{-\mathrm{i}s\xi^3}\mathrm{d}\beta_k^H(s)|^2}{|t_1-t_2|^{1+2b}}\mathrm{d}t_1\mathrm{d}t_2\right)\right)\\
=&\sum_{k=0}^{\infty}|\widehat{\Phi e_k}|^2(I_1+I_2).
\end{aligned}\tag{2.13}$$

By the relationship between the fBm and the Wiener process (2.9), (2.10) and Itô isometry formula, we get

$$\begin{aligned}
I_1&=\int_0^2|\psi(t)|^2\mathbb{E}\left(\left|\int_0^t \mathrm{e}^{-\mathrm{i}s\xi^3}\mathrm{d}\beta_k^H\right|^2\right)\mathrm{d}t\\
&=\int_0^2|\psi(t)|^2\mathbb{E}\left(\left|\int_0^t K^*\mathrm{e}^{-\mathrm{i}(\cdot)\xi^3}\mathrm{d}\beta_k(s)\right|^2\right)\mathrm{d}t\\
&=\int_0^2|\psi(t)|^2\int_0^t\left|K^*(\mathrm{e}^{-\mathrm{i}(\cdot)\xi^3})(s)\right|^2\mathrm{d}s\mathrm{d}t.
\end{aligned}\tag{2.14}$$

Noticing the expression of K^* given in (2.6), we have

$$\begin{aligned}
&\int_0^t\left|K^*(\mathrm{e}^{-\mathrm{i}(\cdot)\xi^3})(s)\right|^2\mathrm{d}s\\
=&\int_0^t\left|\mathrm{e}^{-\mathrm{i}s\xi^3}K(t,s)+\int_s^t\left(\mathrm{e}^{-\mathrm{i}\sigma\xi^3}-\mathrm{e}^{-\mathrm{i}s\xi^3}\right)\frac{\partial K}{\partial\sigma}(\sigma,s)\mathrm{d}\sigma\right|^2\mathrm{d}s\\
\leqslant&2\left(\int_0^t\left|\mathrm{e}^{-\mathrm{i}s\xi^3}K(t,s)\right|^2\mathrm{d}s+\int_0^t\left|\int_s^t\left(\mathrm{e}^{-\mathrm{i}\sigma\xi^3}-\mathrm{e}^{-\mathrm{i}s\xi^3}\right)\frac{\partial K}{\partial\sigma}(\sigma,s)\mathrm{d}\sigma\right|^2\mathrm{d}s\right)\\
\leqslant&2(T_1+T_2).
\end{aligned}\tag{2.15}$$

The integral in T_1 is

$$T_1 = \mathbb{E}((\beta^H(t))^2) = t^{2H}.$$

Using Lemma 2.1, we obtain

$$\begin{aligned}T_2 &\leqslant C(a)|\xi|^{3a}\int_0^t\left|\int_s^t(\sigma-s)^{H-\frac{3}{2}+a}\left(\frac{\sigma}{s}\right)^{H-\frac{1}{2}}\mathrm{d}\sigma\right|^2\mathrm{d}s\\&\leqslant C(a)|\xi|^{3a}\int_0^t\left|\int_s^t(\sigma-s)^{H-\frac{3}{2}+a}\mathrm{d}\sigma\right|^2\mathrm{d}s,\end{aligned}$$

which is well-defined if $H-\dfrac{3}{2}+a>-1$ for some $a\in(0,1)$. We finally obtain that

$$T_2 \leqslant C(a)|\xi|^{3a}t^{2H+2a}.$$

Then we have

$$I_1 \leqslant C\left(|t^{2H}\psi(t)|^2_{L^2_t}+|\xi|^{3a}|t^{2H+2a}\psi(t)|^2_{L^2_t}\right).$$

Now we begin to estimate I_2.

$$\begin{aligned}I_2 =&2\int_0^\infty\int_{t_1<t_2}\frac{\mathbb{E}\left(\left|\psi(t_1)\int_0^{t_1}\mathrm{e}^{-\mathrm{i}s\xi^3}\mathrm{d}\beta_k^H(s)-\psi(t_2)\int_0^{t_2}\mathrm{e}^{-\mathrm{i}s\xi^3}\mathrm{d}\beta_k^H(s)\right|^2\right)}{|t_1-t_2|^{1+2b}}\mathrm{d}t_1\mathrm{d}t_2\\\leqslant&2\int_0^\infty\int_{t_1<0}\frac{|\psi(t_2)|^2\mathbb{E}\left(\left|\int_0^{t_2}\mathrm{e}^{-\mathrm{i}s\xi^3}\mathrm{d}\beta_k^H(s)\right|^2\right)}{|t_1-t_2|^{1+2b}}\mathrm{d}t_1\mathrm{d}t_2\\&+4\int_0^\infty\int_{0<t_1<t_2}\frac{|\psi(t_1)-\psi(t_2)|^2\mathbb{E}\left(\left|\int_0^{t_1}\mathrm{e}^{-\mathrm{i}s\xi^3}\mathrm{d}\beta_k^H(s)\right|^2\right)}{|t_1-t_2|^{1+2b}}\mathrm{d}t_1\mathrm{d}t_2\\&+4\int_0^\infty\int_{0<t_1<t_2}\frac{|\psi(t_2)|^2\mathbb{E}\left(\left|\int_{t_1}^{t_2}\mathrm{e}^{-\mathrm{i}s\xi^3}\mathrm{d}\beta_k^H(s)\right|^2\right)}{|t_1-t_2|^{1+2b}}\mathrm{d}t_1\mathrm{d}t_2\\\equiv&2I_{2,1}+4I_{2,2}+4I_{2,3}\end{aligned}$$

Similarly to I_1, we estimate the above three terms individually. For $I_{2,1}$, and for some $a\in(0,1)$ such that $H-\dfrac{3}{2}+a>-1$, we have

$$\begin{aligned}I_{2,1} &\leqslant C(a)\int_0^2\left(t_2^{2H}+t_2^{2H+2a}|\xi|^{3a}\right)|\psi(t_2)|^2\int_{-\infty}^0\frac{\mathrm{d}t_1}{|t_1-t_2|^{1+2b}}\mathrm{d}t_2\\&\leqslant C(a,b)\int_0^2\left(t_2^{2H-2b}+t_2^{2(H+a-b)}|\xi|^{3a}\right)|\psi(t_2)|^2\mathrm{d}t_2\\&\leqslant C(a,b)|\xi|^{3a}\left\||t|^{H-b}\psi\right\|^2_{L^2_t},\end{aligned}$$

where we use (2.6), (2.9), (2.10) and Lemma 2.1.

Noticing the definition of the norm on H_t^b, and $0 < b < \frac{1}{2}$, and for some $a \in (0,1)$ such that $H - \frac{3}{2} + a > -1$, we have

$$\begin{aligned}
I_{2,2} &\leqslant C(a)\int_0^\infty\int_0^{t_2}\frac{\left(t_1^{2H}+t_1^{2H+2a}|\xi|^{3a}\right)|\psi(t_1)-\psi(t_2)|^2}{|t_1-t_2|^{1+2b}}\mathrm{d}t_1\mathrm{d}t_2\\
&\leqslant C(a)\left(\int_0^2\int_0^{t_2}\frac{\left(t_1^{2H}+t_1^{2H+2a}|\xi|^{3a}\right)|\psi(t_1)-\psi(t_2)|^2}{|t_1-t_2|^{1+2b}}\mathrm{d}t_1\mathrm{d}t_2\right.\\
&\quad\left.+\int_2^\infty\int_0^2\frac{\left(t_1^{2H}+t_1^{2H+2a}|\xi|^{3a}\right)|\psi(t_1)|^2}{|t_1-t_2|^{1+2b}}\mathrm{d}t_1\mathrm{d}t_2\right)\\
&\leqslant C(a,H)|\xi|^{3a}\left(\int_0^2\int_0^{t_2}\frac{|\psi(t_1)-\psi(t_2)|^2}{|t_1-t_2|^{1+2b}}\mathrm{d}t_1\mathrm{d}t_2\right.\\
&\quad\left.+\left\||t|^H\psi\right\|_{L^\infty}^2\int_2^\infty\int_0^2\frac{1}{|t_1-t_2|^{1+2b}}\mathrm{d}t_1\mathrm{d}t_2\right)\\
&\leqslant C(a,H)|\xi|^{3a}\left(\|\psi\|_{H_t^b}^2+C_b\||t|^{1/2}\psi\|_{L^\infty}^2\right),
\end{aligned}$$

where we use (2.6), (2.9), (2.10) and Lemma 2.1.

In the same way, and for some $a \in (0,1)$ such that $H - \frac{3}{2} + a > -1$, we obtain

$$\begin{aligned}
I_{2,3} &\leqslant C(a)\int_0^2\int_0^{t_2}|\psi(t_2)|^2\Big(|t_1-t_2|^{2H-1-2b}\\
&\quad+|t_1-t_2|^{2H+2a-1-2b}|\xi|^{3a}\Big)\mathrm{d}t_1\mathrm{d}t_2\\
&\leqslant C(a,b,H)|\xi|^{3a}\left\||t|^{H-b}\psi\right\|_{L_t^2}^2.
\end{aligned}$$

Then, we get

$$I_2 \leqslant M(a,b,H,\psi)|\xi|^{3a}, \tag{2.16}$$

with $M(b,\psi) = C(a,b,H)\left(\||t|^{1/2}\psi\|_{L^2}^2 + |\psi|_{H_t^b}^2 + \||t|^{1/2}\psi\|_{L^\infty}^2\right)$.

Combining (2.13)–(2.16) with (2.12), we complete the proof of Proposition 2.1.

Remark 2.1 *Assume that $H > \frac{5}{12}$ and is sufficiently close to $\frac{5}{12}$; then it is enough to choose $a = \frac{1}{12}$ during the proof of Proposition 2.1. Note that we need $\frac{5}{12} < b \leqslant H$ and b, H close enough to $\frac{5}{12}$ in Section 3. That is the reason why we need the condition that $\Phi \in L_2^{0,s+\frac{1}{8}}$ with Hurst parameter $\frac{5}{12} < H < \frac{1}{2}$ in Theorem 1.1.*

Proposition 2.2 *Let $s, b \in \mathbb{R}$ with $\frac{1}{2} < b \leqslant H < 1$, and we assume that $\Phi \in L_2^{0,s}$; then $Z(t)$ defined by* (2.11) *satisfies*

$$\psi Z \in L^2(\Omega, X_{s,b})$$

and

$$\mathbb{E}(\|\psi Z\|_{X_{s,b}}^2) \leqslant M(b, H, \psi)\|\Phi\|_{L_2^{0,s}}^2,$$

where $M(b,\psi)$ is a constant depending only on $b, H, |\psi|_{H^b}, ||t|^{\frac{1}{2}}\psi|_{L^2}, ||t|^{\frac{1}{2}}\psi|_{L^\infty}$.

The proof of this proposition is similar to that of Proposition 2.1. One can check it easily. Note that when $H > \frac{1}{2}$, the assumption on a is not necessary, indeed the derivative of the kernel is integrable; one can see it from the expression of (2.4).

Remark 2.2 *For $H = \frac{1}{2}$, the fBm is then a standard Brownian motion. In this case, the estimate on the stochastic convolution under $b < \frac{1}{2}$ was proved in* [8].

3 Bilinear estimates

The bilinear estimate will allow us to bound the second term in the right hand side of (1.3). Under the Hurst parameter $\frac{1}{2} < H < 1$ the bilinear estimate was proved in [19]. We will prove the bilinear estimate under the Hurst parameter $H \geqslant \frac{7}{16}$. The method used here is motivated by [14], where they considered deterministic KdV-BO equation by Fourier restriction method in $X_{s,b}$ with $b > \frac{1}{2}$.

Before proving bilinear estimate under the Hurst parameter $H \geqslant \frac{7}{16}$, we first recall some useful preliminary estimates.

Lemma 3.1 *If f, f_1, f_2, belong to Schwartz space on $\mathscr{S}(\mathbb{R}^2)$, then*

$$\int_\star \bar{\hat{f}}(\xi,\tau)\hat{f}_1(\xi_1,\tau_1)\hat{f}_2(\xi_2,\tau_2)\mathrm{d}\delta = \int \bar{f} f_1 f_2(x,t)\mathrm{d}x\mathrm{d}t.$$

For this equality, one can refer to [14].

Lemma 3.2 *Denote*

$$\mathscr{F}F_\rho(\xi,\tau) \equiv \frac{f(\xi,\tau)}{(1+|\tau-\xi^3|)^\rho};$$

then we have

(i) *If* $\rho > \frac{1}{3}$, *then* $\|F_\rho\|_{L_x^4 L_t^4} \leqslant C\|f\|_{L_\xi^2 L_\tau^2}$;

(ii) *If* $\rho > \frac{3}{8}$, *then* $\|D_x^{\frac{1}{8}} F_\rho\|_{L_x^4 L_t^4} \leqslant C\|f\|_{L_\xi^2 L_\tau^2}$;

(iii) *If* $\rho > \frac{5}{12}$, *then* $\|F_\rho\|_{L_x^4 L_t^6} \leqslant C\|f\|_{L_\xi^2 L_\tau^2}$;

(iv) *If* $\rho > \frac{1}{3}$, *then* $\|D_x^{\frac{1}{4}} F_\rho\|_{L_x^4 L_t^3} \leqslant C\|f\|_{L_\xi^2 L_\tau^2}$;

(v) *If* $\rho > \frac{\theta}{2}$ *with* $\theta \in [0,1]$, *then* $\| D_x^\theta F_\rho \|_{L_x^{\frac{2}{1-\theta}} L_t^2} \leqslant C \| f \|_{L_\xi^2 L_\tau^2}$.

Here D_x^s *denote the homogeneous derivative of order* s *with respect to* x.

For the proof of these inequalities, one can refer to Lemma 2.4-2.7 in [19].

Lemma 3.3 *Assume* $\frac{1}{4} < l_1 < \frac{1}{2}, l_2 > \frac{1}{2}$. *For some constant* $C > 0$, *we have*

$$\int_{-\infty}^{+\infty} \frac{\mathrm{d}x}{(1+|x-\alpha|)^{2l_1}(1+|x-\beta|)^{2b}} \leqslant \frac{C}{(1+|\alpha-\beta|)^{4l_1-1}}; \tag{3.1}$$

$$\int_{-\infty}^{+\infty} \frac{\mathrm{d}x}{(1+|x|)^{2l_2}|\sqrt{\alpha-x}|} \leqslant \frac{C}{(1+|\alpha|)^{1/2}}. \tag{3.2}$$

The two basic inequalities have been used in [8] and [19] respectively. One can prove it directly. We omit it.

With the help of the equality and inequalities given by the above Lemmas, we begin to focus on the proof of the bilinear estimate and state it as follows.

Proposition 3.1 *Assume* $\frac{7}{16} \leqslant a, b < \frac{1}{2}$. *For* $s \geqslant -\frac{9}{16}$, $-s-a \leqslant \frac{1}{8}$ *and* $-s-b \leqslant \frac{1}{8}$, *we have*

$$\| \partial_x(u_1 u_2) \|_{X_{s,-a}} \leqslant C \| u_1 \|_{X_{s,b}} \| u_2 \|_{X_{s,b}}, \tag{3.3}$$

for $u_1, u_2 \in \mathscr{S}(\mathbb{R}^2)$ *such that the right hand side is finite.*

Proof We first introduce some variables.

$$\sigma = \tau - \xi^3, \quad \sigma_1 = \tau_1 - \xi_1^3, \quad \sigma_2 = \tau_2 - \xi_2^3.$$

In the following, denote $\int_\star \mathrm{d}\delta$ the convolution integral by

$$\int_{\xi=\xi_1+\xi_2;\tau=\tau_1+\tau_2} \mathrm{d}\tau_1 \mathrm{d}\tau_2 \mathrm{d}\xi_1 \mathrm{d}\xi_2.$$

Let $r=-s$, $\varphi\in X_{r,a}$; $f_0=\langle\xi\rangle^r\langle\sigma\rangle^a\hat{\varphi}$, $f_j=\langle\xi_j\rangle^s\langle\sigma_j\rangle^b\hat{u}_j$, $j=1,2$; $\xi=\xi_1+\xi_2,\tau=\tau_1+\tau_2$. One easily obtains that $\| f_j \|_{L^2}=\| u_j \|_{X_{s,b}}$, $j=1,2$. By duality and Plancheral Theorem, we will estimate

$$
\begin{aligned}
|\langle\varphi,\partial_x(u_1u_2)\rangle| &\leqslant \left|\int_\star |\xi|\overline{\mathscr{F}\varphi(\tau,\xi)}\mathscr{F}u_1(\tau_1,\xi_1)\mathscr{F}u_2(\tau_2,\xi_2)\mathrm{d}\delta\right| \\
&= \left|\int_\star |\xi|\frac{\overline{f_0(\tau,\xi)}}{\langle\xi\rangle^r\langle\sigma\rangle^a}\frac{\langle\xi_1\rangle^r f_1(\tau_1,\xi_1)}{\langle\sigma_1\rangle^b}\frac{\langle\xi_2\rangle^r f_2(\tau_2,\xi_2)}{\langle\sigma_2\rangle^b}\mathrm{d}\delta\right|.
\end{aligned}
\tag{3.4}
$$

Denote

$$
\mathscr{F}F_\rho^j(\xi,\tau)\equiv\frac{f_j(\xi,\tau)}{(1+|\tau-\xi^3|)^\rho},\quad j=0,1,2.
$$

In order to bound (3.4), we split the domain of integration into several pieces.

We consider the most interesting case $s\leqslant 0$. By symmetry it suffices to estimate the integral in the domain

$$
|\xi_1|\leqslant|\xi_2|.
$$

We recall the identity (see [19])

$$
\tau-\xi^3-[(\tau_1-\xi_1^3)+(\tau_2-\xi_2^3)]=3\xi\xi_1\xi_2
$$

which implies that one of the following cases always occurs:

(a) $|\tau-\xi^3|\geqslant\xi\xi_1\xi_2$;

(b) $|\tau_1-\xi_1^3|\geqslant\xi\xi_1\xi_2$;

(c) $|\tau_2-\xi_2^3|\geqslant\xi\xi_1\xi_2$.

To bound the inegral (3.4), we split the domain of integration as follows.

Case I Assume: $|\xi|\leqslant 2$.

Subcase 1 If $|\xi_1|\leqslant 1$, then we have $|\xi_2|=|\xi-\xi_1|\leqslant 3$. Consequently, using Hölder inequality, Lemma 3.2 and Lemma 3.1, (3.4) restricted to this domain is bounded by

$$
\begin{aligned}
&\int_\star \frac{|\xi|\bar{f}_0(\tau,\xi)}{\langle\xi\rangle^r\langle\sigma\rangle^a}\frac{\langle\xi_1\rangle^r f_1(\tau_1,\xi_1)}{\langle\sigma_1\rangle^b}\frac{\langle\xi_2\rangle^r f_2(\tau_2,\xi_2)}{\langle\sigma_2\rangle^b}\mathrm{d}\delta \\
&\leqslant C\int_\star \frac{\bar{f}_0(\tau,\xi)}{\langle\sigma\rangle^a}\frac{f_1(\tau_1,\xi_1)}{\langle\sigma_1\rangle^b}\frac{f_2(\tau_2,\xi_2)}{\langle\sigma_2\rangle^b}\mathrm{d}\delta \\
&\leqslant C\int \overline{F_a^0}\cdot F_b^1\cdot F_b^2(x,t)\mathrm{d}x\mathrm{d}t \\
&\leqslant C\|F_a^0\|_{L_x^2L_t^2}\|F_b^1\|_{L_x^4L_t^4}\|F_b^2\|_{L_x^4L_t^4} \\
&\leqslant C\|f_0\|_{L_\xi^2L_\tau^2}\|f_1\|_{L_\xi^2L_\tau^2}\|f_2\|_{L_\xi^2L_\tau^2}.
\end{aligned}
$$

Subcase 2 If $1 \leqslant |\xi_1| \leqslant |\xi_2|$, then we consider the three cases (a),(b) and (c) separately.

If (a) holds, then we have $r - a \leqslant \frac{1}{8}$. We get

$$\begin{aligned}
&\int_\star \frac{|\xi|\bar{f}_0(\tau,\xi)}{\langle\xi\rangle^r(|\xi||\xi_1||\xi_2|)^a} \frac{\langle\xi_1\rangle^r f_1(\tau_1,\xi_1)}{\langle\sigma_1\rangle^b} \frac{\langle\xi_2\rangle^r f_2(\tau_2,\xi_2)}{\langle\sigma_2\rangle^b} \mathrm{d}\delta \\
\leqslant{}& C\int_\star \bar{f}_0(\tau,\xi) \frac{\langle\xi_1\rangle^{r-a}\chi_{|\xi_1|\geqslant 1} f_1(\tau_1,\xi_1)}{\langle\sigma_1\rangle^b} \frac{\langle\xi_2\rangle^{r-a}\chi_{|\xi_2|\geqslant 1} f_2(\tau_2,\xi_2)}{\langle\sigma_2\rangle^b} \mathrm{d}\delta \\
\leqslant{}& C\int \overline{F_0^0} \cdot D_x^{\frac18} F_b^1 \cdot D_x^{\frac18} F_b^2(x,t)\mathrm{d}x\mathrm{d}t \\
\leqslant{}& C\|F_a^0\|_{L_x^2 L_t^2} \|D_x^{\frac18} F_b^1\|_{L_x^4 L_t^4} \|D_x^{\frac18} F_b^2\|_{L_x^4 L_t^4} \\
\leqslant{}& C\|f_0\|_{L_\xi^2 L_\tau^2} \|f_1\|_{L_\xi^2 L_\tau^2} \|f_2\|_{L_\xi^2 L_\tau^2},
\end{aligned}$$

which follows by using Hölder inequality and Lemma 3.2 and Lemma 3.1.

If (b) holds, then we have $r - b \leqslant \frac{1}{8}$. We get

$$\begin{aligned}
&\int_\star \frac{|\xi|\bar{f}_0(\tau,\xi)}{\langle\xi\rangle^r\langle\sigma\rangle^a} \frac{\langle\xi_1\rangle^r f_1(\tau_1,\xi_1)}{(|\xi||\xi_1||\xi_2|)^b} \frac{\langle\xi_2\rangle^r f_2(\tau_2,\xi_2)}{\langle\sigma_2\rangle^b} \mathrm{d}\delta \\
\leqslant{}& C\int_\star \frac{|\xi|\bar{f}_0(\tau,\xi)}{\langle\xi\rangle^r\langle\sigma\rangle^a} \chi_{|\xi_1|\geqslant 1} f_1(\tau_1,\xi_1) \frac{\langle\xi_2\rangle^{2(r-b)}\chi_{|\xi_2|\geqslant 1} f_2(\tau_2,\xi_2)}{\langle\sigma_2\rangle^b} \mathrm{d}\delta \\
\leqslant{}& C\int \overline{F_a^0} \cdot F_0^1 \cdot D_x^{\frac14} F_b^2(x,t)\mathrm{d}x\mathrm{d}t \\
\leqslant{}& C\|F_a^0\|_{L_x^4 L_t^6} \|F_0^1\|_{L_x^2 L_t^2} \|D_x^{\frac14} F_b^2\|_{L_x^4 L_t^3} \\
\leqslant{}& C\|f_0\|_{L_\xi^2 L_\tau^2} \|f_1\|_{L_\xi^2 L_\tau^2} \|f_2\|_{L_\xi^2 L_\tau^2},
\end{aligned}$$

which follows by using Hölder inequality and Lemma 3.2 and Lemma 3.1.

The argument of (c) is similar to (b).

Case II Assume: $|\xi| \geqslant 2$.

Subcase 1 If $1 \leqslant |\xi_1| \leqslant |\xi_2|$, then we consider the three cases (a), (b) and (c) separately in this domain.

If (a) occurs, for $1 - r - a \leqslant 0$ and $r - a \leqslant \frac{1}{8}$, we obtain

$$\begin{aligned}
&\int_\star \frac{|\xi|^{1-r}\chi_{|\xi|\geqslant 2}\bar{f}_0(\tau,\xi)}{(|\xi_1||\xi_2||\xi|)^a} \frac{\langle\xi_1\rangle^r\chi_{|\xi_1|\geqslant 1} f_1(\tau_1,\xi_1)}{\langle\sigma_1\rangle^b} \frac{\langle\xi_2\rangle^r\chi_{|\xi_2|\geqslant 1} f_2(\tau_2,\xi_2)}{\langle\sigma_2\rangle^b} \mathrm{d}\delta \\
\leqslant{}& C\int_\star |\xi|^{1-r-a}\chi_{|\xi|\geqslant 2}\bar{f}_0(\tau,\xi)
\end{aligned}$$

$$
\begin{aligned}
&\times \frac{|\xi_1|^{r-a}\chi_{|\xi_1|\geqslant 1a}f_1(\tau_1,\xi_1)}{\langle\sigma_1\rangle^b}\frac{|\xi_2|^{r-a}\chi_{|\xi_2|\geqslant 1}f_2(\tau_2,\xi_2)}{\langle\sigma_2\rangle^b}\mathrm{d}\delta \\
&\leqslant C\int \overline{F_0^0}\cdot D_x^{\frac{1}{8}}F_b^1\cdot D_x^{\frac{1}{8}}F_b^2(x,t)\mathrm{d}x\mathrm{d}t \\
&\leqslant C\|F_0^0\|_{L_x^2L_t^2}\|D_x^{\frac{1}{8}}F_a^1\|_{L_x^4L_t^4}\|D_x^{\frac{1}{8}}F_b^2\|_{L_x^4L_t^4} \\
&\leqslant C\|f_0\|_{L_\xi^2L_\tau^2}\|f_1\|_{L_\xi^2L_\tau^2}\|f_2\|_{L_\xi^2L_\tau^2},
\end{aligned}
$$

which follows by using Hölder inequality and Lemma 3.2 and Lemma 3.1.

We know that (3.4) can be written as

$$
\int_\star \frac{|\xi|\langle\xi_1\rangle^r\langle\xi_2\rangle^r}{\langle\sigma\rangle^b\langle\xi\rangle^r\langle\sigma_1\rangle^a\langle\sigma_2\rangle^a}\bar{f}_0(\tau,\xi)f_1(\tau_1,\xi_1)f_2(\tau_2,\xi_2)\mathrm{d}\delta.
$$

Using the multi-linear expression [28], if $1-r-a\leqslant 0$, $r-a\leqslant \frac{1}{8}$, we have

$$
\left\|\frac{|\xi|\langle\xi_1\rangle^r\langle\xi_2\rangle^r}{\langle\sigma\rangle^b\langle\xi\rangle^r\langle\sigma_1\rangle^a\langle\sigma_2\rangle^a}\right\|_{[3;\mathbb{R}\times\mathbb{R}]}\leqslant C.
$$

However, if $r\leqslant\frac{9}{16}$, then by Lemma 1.1 we have

$$
\left\|\frac{|\xi|\langle\xi_1\rangle^r\langle\xi_2\rangle^r}{\langle\sigma\rangle^a\langle\xi\rangle^r\langle\sigma_1\rangle^b\langle\sigma_2\rangle^b}\right\|_{[3;\mathbb{R}\times\mathbb{R}]}\leqslant C.
$$

In fact, if $r_1\leqslant r_2$, by $\xi=\xi_1+\xi_2$, we obtain

$$
\frac{|\xi|\langle\xi_1\rangle^{r_1}\langle\xi_2\rangle^{r_1}}{\langle\sigma\rangle^a\langle\xi\rangle^{r_1}\langle\sigma_1\rangle^b\langle\sigma_2\rangle^b}\leqslant C\frac{|\xi|\langle\xi_1\rangle^{r_2}\langle\xi_2\rangle^{r_2}}{\langle\sigma\rangle^a\langle\xi\rangle^{r_2}\langle\sigma_1\rangle^b\langle\sigma_2\rangle^b}.
$$

Then by Lemma 1.1, the boundedness of (3.4) in this case holds.

If (b) holds, by Lemma 3.2 and Lemma 3.1, for $1-r-b\leqslant 0$, $r-b\leqslant\frac{1}{8}$, we get

$$
\begin{aligned}
&\int_\star \frac{|\xi|^{1-r}\chi_{|\xi|\geqslant 2}\bar{f}_0(\tau,\xi)}{\langle\sigma\rangle^a}\frac{|\xi_1|^r\chi_{|\xi_1|\geqslant 1}f_1(\tau_1,\xi_1)}{(|\xi_1||\xi_2||\xi|)^b}\frac{\langle\xi_2\rangle^r\chi_{|\xi_2|\geqslant 1}f_2(\tau_2,\xi_2)}{\langle\sigma_2\rangle^b}\mathrm{d}\delta \\
&\leqslant C\int_\star \frac{|\xi|^{1-r-b}\chi_{|\xi|\geqslant 2}\bar{f}_0(\tau,\xi)}{\langle\sigma\rangle^a}\chi_{|\xi_1|\geqslant 1}f_1(\tau_1,\xi_1)\frac{|\xi_2|^{2(r-b)}\chi_{|\xi_2|\geqslant 1}f_2(\tau_2,\xi_2)}{\langle\sigma_2\rangle^b}\mathrm{d}\delta \\
&\leqslant C\int \overline{F_a^0}\cdot F_0^1\cdot D_x^{\frac{1}{4}}F_b^2(x,t)\mathrm{d}x\mathrm{d}t \\
&\leqslant C\|F_a^0\|_{L_x^4L_t^6}\|F_0^1\|_{L_x^2L_t^2}\|D_x^{\frac{1}{4}}F_b^2\|_{L_x^4L_t^3} \\
&\leqslant C\|f_0\|_{L_\xi^2L_\tau^2}\|f_1\|_{L_\xi^2L_\tau^2}\|f_2\|_{L_\xi^2L_\tau^2},
\end{aligned}
$$

which follows Hölder inequality and Lemma 3.2 and Lemma 3.1.

If $r \leqslant \frac{9}{16}$, then the boundedness of this case is similar to the case (a).

The argument of (c) is similar to (a), (b).

Subcase 2 If $|\xi_1| \leqslant 1$ such that $|\xi-\xi_1| \geqslant 1$. Due to the singularity of the variable $|\xi_1|$ at zero, the above method doesn't work in this case. Fortunately, we can prove it by using the basic inequalities provided by Lemma 3.3. We will prove the case $s=0$. The proof of $s>-\frac{5}{8}$ follows from this case, therefore it will be omitted. Then

$$
\begin{aligned}
&\|\partial_x(u_1u_2)\|_{X_{0,-a}}\\
=&\left\|\frac{\xi}{(1+|\tau-\xi^3|)^a}\widehat{u_1}*\widehat{u_2}\right\|_{L^2_\tau L^2_\xi}\\
\leqslant&\left\|\frac{|\xi|}{(|1+|\tau-\xi^3|)^a}\right.\\
&\times\left.\left(\int_{-\infty}^{+\infty}\int_{-\infty}^{+\infty}\frac{\mathrm{d}\xi_1\mathrm{d}\tau_1}{(1+|\tau_1-\xi_1^3|)^{2b}(1+|\tau_2+\xi_2^3|)^{2b}}\right)^{1/2}\right\|_{L^\infty_\xi L^\infty_\tau}\\
&\times\|u_1\|_{X_{0,b}}\|u_2\|_{X_{0,b}}\\
\leqslant& C\|u_1\|_{X_{0,b}}\|u_2\|_{X_{0,b}}.
\end{aligned}\tag{3.5}
$$

Thus we turn our attention to bound

$$
\frac{|\xi|}{(|1+|\tau-\xi^3|)^a}\left(\int_{-\infty}^{+\infty}\int_{-\infty}^{+\infty}\frac{\mathrm{d}\xi_1\mathrm{d}\tau_1}{(1+|\tau_1-\xi_1^3|)^{2b}(1+|\tau_2-\xi_2^3|)^{2b}}\right)^{1/2}.
$$

Since $b>\frac{1}{4}$, from Lemma 3.2 it follows that

$$
\begin{aligned}
&\int_{-\infty}^{+\infty}\frac{\mathrm{d}\xi_1\mathrm{d}\tau_1}{(1+|\tau_1-\xi_1^3|)^{2b}(1+|\tau_2-\xi_2^3|)^{2b}}\\
\leqslant&\frac{C}{(1+|\tau-\xi^3+3\xi\xi_1(\xi-\xi_1)|)^{4b-1}}.
\end{aligned}
$$

To integrate with respect to ξ_1 we change variable

$$
\mu=\tau-\xi^3+3\xi\xi_1(\xi-\xi_1);
$$

then

$$
\mathrm{d}\mu=3\xi(\xi-2\xi_1)\mathrm{d}\xi_1\quad\text{and}\quad\xi_1=\frac{1}{2}\left\{\xi\pm\sqrt{\frac{4\tau-\xi^3-4\mu}{3\xi}}\right\}.
$$

Therefore

$$|\xi(\xi-2\xi_1)| = C\sqrt{|\xi|}\sqrt{4\tau-\xi^3-4\mu} \quad \text{and} \quad \mathrm{d}\xi_1 = \frac{C\mathrm{d}\mu}{\sqrt{|\xi|}\sqrt{4\tau-\xi^3-4\mu}}.$$

Now combining these identities with Lemma 3.2 we find

$$\begin{aligned}
&\int_{-\infty}^{+\infty} \frac{\mathrm{d}\xi_1}{(1+|\tau-\xi^3+3\xi\xi_1(\xi-\xi_1)|)^{4b-1}} \\
\leqslant\ & \frac{C}{|\xi|^{\frac{1}{2}}}\int_{-\infty}^{+\infty} \frac{\mathrm{d}\mu}{(1+|\mu|)^{4b-1}\sqrt{4\tau-\xi^3-4\mu}} \\
\leqslant\ & \frac{C}{|\xi|^{\frac{1}{2}}(1+|4\tau-\xi^3|)^{\frac{1}{2}}}.
\end{aligned}$$

Then we get

$$\begin{aligned}
&\frac{|\xi|}{(|1+|\tau-\xi^3|)^a}\left(\int_{-\infty}^{+\infty}\int_{-\infty}^{+\infty} \frac{\mathrm{d}\xi_1\mathrm{d}\tau_1}{(1+|\tau_1-\xi_1^3|)^{2b}(1+|\tau_2-\xi_2^3|)^{2b}}\right)^{1/2} \\
\leqslant\ & \frac{C|\xi|^{\frac{3}{4}}}{(1+|\tau-\xi^3|)^{1-b}(1+|4\tau-\xi^3|)^{\frac{1}{4}}}.
\end{aligned}$$

Thus the boundedness of (3.5) follows.

Then we complete the proof of the Proposition 3.1.

Remark 3.1 *We can read more information from the proof of Proposition* 3.1. *The indices s and b of Bourgain-type spaces $X_{s,b}$ satisfy*

$$-s-b\leqslant\frac{1}{8},1+s-b\leqslant 0,\frac{5}{12}<b<\frac{1}{2}.$$

For the stochastic system (1.1), (1.2) *has the same time regularity as fBm, we can substitute H for b. Then we get the relationship between the regularity of the fBm indicated by H and the spacial regularity of system* (1.1), (1.2) *indicated by s in the convex region defined by the above three inequalities.*

4 Existence result

In this section, we proceed exactly as in [8], and work pathwise on (1.3). With the use of Proposition 2.1 and Proposition 3.1, we prove the local existence result of Theorem 1.1 by using a fixed point argument in the space $X_{s,b}^T$ with Hurst parameter $\frac{7}{16}\leqslant H<\frac{1}{2}$; and with the help of Proposition 2.2 and Proposition 3.1, we can prove Theorem 1.1 when parameter $\frac{1}{2}\leqslant H<1$.

Lemma 4.1 [8, 18] *If* $\frac{7}{16} \leqslant a, b < \frac{1}{2}$, $c > \frac{1}{2}$, $s \in \mathbb{R}$, $u_0 \in H^s(\mathbb{R})$, *and* $f \in X_{s,-a}^T$, *then we have*

$$\|U(t)u_0\|_{X_{s,c}^T} \leqslant C\|u_0\|_{H^s}, \tag{4.1}$$

$$\left\|\int_0^t U(t-t')f(t')\mathrm{d}t'\right\|_{X_{s,b}} \leqslant CT^{1-a-b}\|f\|_{X_{s,-a}}. \tag{4.2}$$

Now we prove Theorem 1.1 Let u_0 be $\mathcal{F}_0$-measurable with $u_0 \in H^s(\mathbb{R})$ for *a.s.* $\omega \in \Omega$. For we work pathwise on equation (1.3), we fix $\omega \in \Omega$ such that (4.5) and $u_0(\omega, \cdot) \in H^s(\mathbb{R})$ hold. Let $Z(t)$ be defined by (2.11), and we set $z(t) = U(t)u_0$. Then we may rewrite equation (1.3) in terms of $v(t) = u(t) - Z(t) - z(t)$ as

$$\begin{aligned} v(t) &= \mathcal{T}v(t) \\ &= -\frac{1}{2}\int_0^t U(t-s)\partial_x\left(v^2 + Z^2 + z^2 + 2vZ + 2vz + 2zZ\right)(s)\mathrm{d}s. \end{aligned} \tag{4.3}$$

We show that $\mathcal{T}$ is a contraction mapping in

$$K_R^T = \left\{v \in X_{s,b}^T, \|v\|_{X_{s,b}^T} \leqslant R\right\}, \tag{4.4}$$

for $R > 0$ sufficiently large, provided that T is chosen sufficiently small.

From Proposition 2.1, we get

$$\|Z(t)\|_{X_{s,b}^T} \leqslant \|\psi Z(t)\|_{X_{s,b}}, \tag{4.5}$$

for $Z(t)$ defined in (2.11), ψ given before and $T \in [0, 1]$ for almost each $\omega \in \Omega$.

Noticing Proposition 2.1, Proposition 3.1 and Lemma 4.1, we easily get

$$\|\mathcal{T}v\|_{X_{s,b}^T} \leqslant C'T^{1-a-b}\left(R^2 + \|Z\|_{X_{s,b}^T}^2 + \|u_0\|_{H^s}^2\right).$$

For $v_1, v_2 \in K_R^T$, we get

$$\|\mathcal{T}v_1 - \mathcal{T}v_2\|_{X_{s,b}^T} \leqslant C'T^{1-a-b}\left(R + \|Z\|_{X_{s,b}^T} + \|u_0\|_{H^s}\right)\|v_1 - v_2\|_{X_{s,b}^T}.$$

First setting

$$R_\omega^t = \|Z\|_{X_{s,b}^T}^2 + \|u_0\|_{H^s}^2,$$

and then defining the stopping time T_ω by

$$T_\omega = \inf\left\{t > 0, 2C't^{1-a-b}R_\omega^t \leqslant \frac{1}{2}\right\}.$$

Then $\mathcal{T}$ maps $K_{R_\omega}^{T_\omega}$ in $X_{s,b}^{T_\omega}$ into itself, and

$$\|\mathcal{T}v_1 - \mathcal{T}v_2\|_{X_{s,b}^{T_\omega}} \leqslant \frac{3}{4}\|v_1 - v_2\|_{X_{s,b}^{T_\omega}}.$$

Hence $\mathcal{T}$ has a unique fixed point, which is the unique solution of (4.3) in$X_{s,b}^{T_\omega}$ on $[0, T_\omega]$.

It is checked that $\mathcal{T}$ maps K_R^T int itself and is a strict contraction in K_R^T for the norm $\|v\|_{X_{s,b}^{T_\omega}}$. So it remains to show that $u = z + v + Z \in X_{s,c}^{T_\omega} + X_{s,b}^{T_\omega}$ in $C([0, T_\omega], H^s(\mathbb{R}))$ $\left(\text{note that here } b < \frac{1}{2}, c > \frac{1}{2}\right)$.

Since $c > \frac{1}{2}$, we have $z \in C([0, T_\omega], H^s(\mathbb{R}))$ by using the Sobolev imbedding Theorem in time.

Since $\Phi \in L_2^{0,s}$ and $U(\cdot)$ is a unitary group in $H^s(\mathbb{R})$, we have $Z(t)$ has a continuous modification with values in H^s similar to Theorem 6.10 in [7].

By Proposition 3.1, we have in this case, $\partial_x(\tilde{u}^2) \in X_{s,-a}$, for $\tilde{u}$ any prolongation of u in $X_{s,c} + X_{s,b}$, it follows that (see [18])

$$\left\|\varphi_T \int_0^t U(t-s)\partial_x(\tilde{u}^2(s))\mathrm{d}s\right\|_{X_{s,1-a}} \leqslant C\left\|\partial_x(\tilde{u}^2(s))\right\|_{X_{s,-a}},$$

since $1 - a > \frac{1}{2}$, then $\tilde{u} \in X_{s,b} \subset C([0, T_\omega], H^s(\mathbb{R}))$, where φ_T is a cut-off function defined by $\varphi \in C_0^\infty(\mathbb{R})$ with $\varphi = 1$ on $[0, 1]$, supp$\varphi \subset [-1, 2]$ and $\varphi = 0$ on $t \leqslant -1, t \geqslant 2$. Denote $\varphi_\delta(\cdot) = \varphi(\delta^{-1}(\cdot))$ for some $\delta \in \mathbb{R}$.

With the help of Proposition 2.2 and Proposition 3.1, the proof of Theorem 1.1 with Hurst parameter $\frac{1}{2} \leqslant H < 1$ is simpler and similar to the case with Hurst parameter $\frac{7}{16} \leqslant H < \frac{1}{2}$. So we omit it.

This completes the proof of Theorem 1.1.

References

[1] Alòs E, Mazet O, Nualart D. Stochastic calculus with respect to Gaussian processes [J]. *Annals of Probab.*, 2001, 29: 766–801.

[2] Alòs E, Nualart D. Stochastic calculus with respect to fractional Brownian motion [J]. *Stoch. Stoch. Rep.*, 2003, 75: 129–152.

[3] Biagini F, Hu Y, Oksendal B, Zhang T. Stochastic Calculus for Fractional Brownian Motion and Applications [M]. Springer, 2008.

[4] Bourgain J. Fourier restriction phenomena for certain lattice subsets and applications to nonlinear evolution equations, part I: Schrödinger equation, part II: the KdV equation [J]. *Geom. Funct. Anal.*, 1993, 2: 107–156, 209–262.

[5] Caithamer P. The stochastic wave equation driven by fractional Brownian noise and temporally correlated smooth noise [J]. *Stoch. Dyn.*, 2005, 5: 45–64.

[6] Chang H Y, Lien C, Sukarto S, Raychaudhury S, Hill J, Tsikis E K, Lonngren K E. Propagation of ion-acoustic solitons in a non-quiescent plasma [J]. *Plasma Phys. Control. Fusion*, 1986, 28: 675–681.

[7] Da Prato G, Zabczyk J. Stochastic Equations in infinite Dimensions [M]. Cambridge University Press, 1992.

[8] de Bouard A, Debussche A, Tsutsumi Y. White noise driven Korteweg-de Veris equation [J]. *J. Funct. Anal.*, 1999, 169: 532–558.

[9] de Bouard A, Debussche A, Tsutsumi Y. Periocic solutions of the Korteweg-de Veris equation driven by white noise [J]. *SIAM J. Math. Anal.*, 2004, 36: 818–855.

[10] Duncan T E, Jakubowski J, Pasik-Duncan B. Stochastic integration for fractional Brownian motion in a Hilbert space [J]. *Stoch. Dyn.*, 2006, 6: 53–75.

[11] Duncan T E, Maslowski B, Pasik-Duncan B. Fractional Brownian motion and stochastic equations in Hilbert spaces [J]. *Stoch. Dyn.*, 2002, 2: 225–250.

[12] Erraoui M, Nualart D, Ouknine Y. Hyperbolic stochastic partial differential equations with additive fractional Brownian sheet [J]. *Stoch. Dyn.*, 2003, 3: 121–139.

[13] Grecksch W, Ahn V V. A parabolic stochastic differential equation with fractional Brownian motion input [J]. *Stat. Probab. Lett.*, 1999, 41: 337–346.

[14] Guo B, Huo Z. The well-posedness of the Korteweg-de Vries-Benjamin-Ono equation, *J. Math. Anal. Appl.*, 2004, 295: 444–458.

[15] Herman R. The stochastic, damped Korteweg-de Vries equation [J]. *J. Phys. A.*, 1990, 23: 1063–1084.

[16] Hu Y. Heat equation with fractional white noise potential [J]. *Appl. Math. Optim.*, 2001, 43: 221–243.

[17] Hu Y, Nualart D. Stochastic heat equation driven by fractional noise and local time [J]. *Probab. Theory Related Fields*, 2009, 143: 285–328.

[18] Kenig C E, Ponce G, Vega L. A bilinear estimate with applications to the Kdv equation [J]. *J. Amer. Math. Soc.*, 1996, 9: 573–603.

[19] Kenig C E, Ponce G, Vega L. The Cauchy problem for the Korteweg-de Vries equation in Sobolev spaces of negative indices [J]. *Duke Math. J.*, 1993, 71: 1–21.

[20] Kolmogorov A N. Wienersche Spiralen und einige andere interessante Kurven im Hilbertschen Raum [J]. *C. R. (Doklady) Acad. URSS (N.S.)*, 1940, 26: 115–118.

[21] Mandelbrot B B. The fractal geometry of nature, Freeman [M]. San Francisco, CA, 1983.

[22] Mandelbrot B B, Van Ness J W. Fractional Brownian motions, fractional noises and

applications [J]. *SIAM Rev.* , 1968, 10: 422–437.

[23] Maslowski B, Nualart D. Evolution equations driven by a fractional Brownian motion [J]. *J. Funct. Anal.*, 2003, 202: 277–305.

[24] Mishura Y. Stochastic Calculus for Fractional Brownian Motion and Related Processes [M]. Berlin: Springer, 2008.

[25] Nualart D. Malliavin Calculus and Related topics [M]. Springer Verlag, 1995.

[26] Printems J. The stochastic Korteweg-de Vries equation in $L^2(\mathbb{R})$ [J]. *J. Differ. Equations*, 1999, 153: 338–373.

[27] Scalerandi M, Romano A, Condat C A. Korteweg-de Vries solitons under additive stochastic perturbations [J]. *Phys. Rev. E*, 1998, 58: 4166–4173.

[28] Tao T. Multilinear weighted convolution of L^2 functions, and applications to nonlinear dispersive equation [J]. *Amer. J. Math.*, 2001, 123: 839–908.

[29] Tindel S, Tudor C A, Viens F. Stochastic evolution equations with fractional Brownian motion [J]. *Probab. Theory Related Fields*, 2003, 127: 186–204.

[30] Wang G, Zeng M, Guo B. Stochastic Burgers' equation driven by fractional Brownian motion [J]. *J. Math. Anal. Appl.* , 2010, 371: 210–222.

Global Strong Spherically Symmetric Solutions to the Full Compressible Navier-Stokes Equations with Stress Free Boundary*

Bian Dongfen (边东芬), Guo Boling (郭柏灵) and Zhang Jingjun (张景军)

Abstract In this paper, we are concerned with the global strong spherically symmetric solutions to the full compressible Navier-Stokes equations with large initial data in the case that the viscosity coefficients μ, λ are both constants and the heat conductivity coefficient $\kappa(\theta) \sim \text{const}(1+\theta^q)$, $q \geqslant 1$. We show that the three dimensional full compressible Navier-Stokes equations away from symmetry center with the free boundary condition has a unique global strong solution.

Keywords compressible Navier-Stokes equations; global strong solution; stress free boundary

1 Introduction

We consider the full compressible Navier-Stokes equations in $\mathbf{R}^N$, $N=2,\ 3$, for $t>0$,

$$\begin{cases} \partial_t\rho + \operatorname{div}(\rho U) = 0, \\ \partial_t(\rho U) + \operatorname{div}(\rho U\otimes U) + \nabla p - \operatorname{div}\mathbb{S} = 0, \\ \partial_t(\rho E) + \operatorname{div}[(\rho E + p)U] = \operatorname{div}(\mathbb{S}U) + \operatorname{div}(\kappa(\theta)\nabla\theta). \end{cases} \tag{1.1}$$

Here ρ, U, θ and

$$p = a\rho^\gamma + R\rho\theta \tag{1.2}$$

denote the density, velocity, absolute temperature and pressure, respectively. $\mathbb{S}$ denotes the viscous stress tensor which can be expressed as

$$\mathbb{S} = \lambda(\operatorname{div}U)\mathrm{I} + \mu\mathbb{D}(U), \tag{1.3}$$

* J. Math. Phys., 2015, 56(023509): 1–23. DOI: 10.1063/1.4908283.

where the strain tensor $\mathbb{D}(U) = \frac{1}{2}(\nabla U + (\nabla U)^T)$. The total energy ρE takes the form

$$\rho E = \frac{1}{2}\rho|U|^2 + \rho e \text{ with } e = \frac{a}{\gamma - 1}\rho^{\gamma-1} + C_\nu\theta. \tag{1.4}$$

The physical constants $a > 0$, $C_\nu > 0$ and $\gamma > 1$ are the entropy constant, the heat capacity at constant volume and the adiabatic index, respectively. $\kappa > 0$ is the coefficient of heat conductivity. $R > 0$ is a constant and satisfies $R = C_\nu(\gamma - 1)$.

The global existence of smooth solutions to initial boundary value problems or the Cauchy problem of the full compressible Navier-Stokes equations (1.1) has been investigated by many authors. In one dimension, it is well known that global smooth solutions exist for smooth (large) initial data (e.g., see [2], [17], [27]–[29] for initial boundary value problems, and [2], [9], [18]–[19] for the Cauchy problem; also cf. [13]–[16], [20] for real gases). In particular, Kawohl in [20] got the global classical solution with $\rho_0 > 0$ and the viscosity coefficient $\mu = \mu(\rho)$ satisfying $0 < \mu_0 \leqslant \mu(\rho) \leqslant \mu_1$ for $\rho \geqslant 0$, where μ_0 and μ_1 are constants. In more than one dimension, the global existence of smooth solutions has been investigated for general domains only in the case of sufficiently small initial data (e.g., see [25],[26], [33] for initial boundary value problems, and [23], [24] for the Cauchy problem; also see the survey article [32]). For large initial data, the global existence of solutions to (1.1) has been studied in the case of a bounded annular domain. Nikolaev[30] in 1983 considered the initial boundary value problem of (1.1) with vanishing velocity and constant temperature on the boundary and proved that for (smooth) spherically symmetric initial data a (smooth) spherically symmetric solution exists globally in time if the initial density and temperature are strictly positive. Yashima and Benabidallah[37] dealt with the case of non-negative initial density and temperature. They showed the global existence of spherically symmetric solutions to (1.1). Jiang in [11] proved the global existence of spherically symmetric smooth solutions in Hölder spaces to the equations of a viscous polytropic ideal gas in the domain exterior to a ball in $\mathbf{R}^N$ ($N = 2$ or $N = 3$) when $\rho_0 > 0$. The global well-posedness of the full compressible Navier-Stokes equations (1.1) in $\mathbf{R}^N$ ($N = 2, 3$) for the general initial data is open. For the large time behavior, in the absence of heat conduction, it was proved by Xin[34] that any non-zero smooth solution to the Cauchy problem of the non-barotropic compressible Navier-Stokes system with initially compact supported density would blow up in finite time. This result was generalized to the cases for the non-barotropic compressible Navier-Stokes system with heat conduction[5] and for non-compact but rapidly decreasing at far field initial densities[31]. Recently, Xin-Yan[35] improved this result and proved that any classical solutions of viscous compressible fluids without heat conduction would blow up in finite time, as long as the initial data has an isolated mass group.

For the isentropic compressible Navier-Stokes equations with large initial data and vacuum, when the viscosity coefficients λ and μ are constants, the free boundary problems for one-dimensional case were investigated in [1], [2], [12], [22] and among others, where global existence of weak solutions was proved. Ding-Wen-Yao-Zhu in [8] have obtained the existence and uniqueness of global spherically symmetric classical solutions with large initial data and vacuum in a bounded domain or exterior domain Ω of $\mathbf{R}^N$, $N \geqslant 2$. Chen and Kratka in [4] obtained global solutions of the multidimensional Navier-Stokes equations for compressible heat- conducting flow with spherically symmetric initial data of large oscillation. When the the viscosity coefficients λ and μ satisfy $\lambda(\rho) = 0$, $\mu(\rho) = \rho$, Guo-Li-Xin in [10] first studied the free boundary value problem and obtained the global existence of spherically symmetric strong solutions away from the symmetry center to the isentropic flow in $\mathbf{R}^N$, $N = 2, 3$, furthermore, the authors of [10] considered the problem of vacuum formation. We also refer to [41] for the asymptotic behavior of the weak solutions to the compressible Navier-Stokes equations when the viscosity depends on the density in the case $N = 1$. For the free boundary problem of the full compressible Navier-Stokes equations, Yashima and Benabidallah[36] obtained both weak and strong solution for ideal gas when the coefficients λ, μ and κ are positive constants. In this paper, we study the non-isentropic compressible multi-dimensional Navier-Stokes equations (1.1) with stress free across the free surface and focus on the existence of global strong solutions, moreover, the heat conductivity coefficient in our discussion depends on the temperature.

The rest of the paper is arranged as follows. In Section 2, the main result about global existence of the strong solution to the full compressible Navier-Stokes equations is stated. In Section 3, we prove our main result through some crucial *a priori* estimates.

Notation: The notations $L^p(1 \leqslant p \leqslant \infty)$ and $W^{k,p}$ (in particular, $W^{k,2}$ is also denoted by H^k) stand for the usual Lebesgue spaces and the usual Sobolev spaces, respectively. For the sake of clarity, we sometimes use $\|u\|_{L^\infty_x}$ for $u = u(x,t)$ to mean that the L^∞-norm is taken with respect to the space variable, so $\|u\|_{L^2_x}$, $\cdots$ can be understood in a similar way.

2 The main results

We present the main results about the global existence of solution to the free boundary value problem (2.3), (2.5)–(2.6) in this section. For the sake of convenience, we mainly study the case $N=3$, since the two-dimensional case can be treated in a similar way. Throughout the paper, we make the following assumptions on the viscous coefficients and the heat conductivity coefficient:

$$\begin{gathered}\mu \text{ is a positive constant, } \lambda=0,\\ C^{-1}(1+\theta^q)\leqslant\kappa=\kappa(\theta)\leqslant C(1+\theta^q),\ q\geqslant 1,\\ |\kappa'(\theta)|\leqslant C(1+\theta^{q-1}).\end{gathered}\tag{2.1}$$

We remark that the assumption $\lambda=0$ in (2.1) is only for the sake of simplicity, and the main result as well as the argument stated in this paper still hold when $\lambda\neq 0$. The growth condition on the heat conductivity coefficient κ is motivated by physical facts. κ may depend on the temperature, especially for important physical regimes. For example, at high temperatures, κ is proportional to a power of the temperature, and for large θ, $\kappa(\theta)\sim\theta^q$ with $q\approx 4.5-5.5$ in the multiply ionized region (see e.g. P. 655 in [38]).

Consider a spherically symmetry solution (ρ,U,θ) to (1.1) in $\mathbf{R}^3$ so that

$$\rho(x,t)=\rho(r,t),\ \rho U(x,t)=\rho u(r,t)\frac{x}{r},\ \theta(x,t)=\theta(r,t),\ r=|x|,\ x\in\mathbf{R}^3,\tag{2.2}$$

and the full compressible Navier-Stokes equations (1.1) are changed into

$$\begin{cases}(r^2\rho)_t+(r^2\rho u)_r=0,\\ (r^2\rho u)_t+(r^2\rho u^2)_r+r^2p_r-r^2\left[\mu u_r+\dfrac{2\mu u}{r}\right]_r=0,\\ (r^2\rho E)_t+[r^2(\rho E+p)u]_r=(r^2\mu u u_r)_r+(r^2\kappa(\theta)\theta_r)_r,\end{cases}\tag{2.3}$$

where $(r,t)\in\Omega_T$ with

$$\Omega_T=\{(r,t)|0<r_0\leqslant r\leqslant a(t),0\leqslant t\leqslant T\},\tag{2.4}$$

and the data $a(0)=:a_0$ is given. The initial data is taken as

$$(\rho,\rho u,\theta)(r,0)=(\rho_0,m_0,\theta_0)(r)=:(\rho_0,\rho_0u_0,\theta_0)(r),\ r\in[r_0,a_0].\tag{2.5}$$

The boundary conditions are given by

$$\begin{aligned} &u(r_0,t)=0,\ \left(p-\mu\left(u_r+\frac{2}{r}u\right)\right)(a(t),t)=0,\ t>0,\\ &\theta_r(a(t),t)=\theta_r(r_0,t)=0,\ t>0, \end{aligned} \tag{2.6}$$

with $a'(t)=u(a(t),t)(t>0)$. Also, it is assumed in this paper that the initial data (2.5) is consistent with the boundary values (2.6) to high order.

The main result for the FBVP (2.3) and (2.5)–(2.6) is stated in Theorem 2.1 below. For a value r_1 in the initial domain $[r_0,a_0]$, we compute $x_1=\displaystyle\int_{r_0}^{r_1} r^2\rho_0(r)\,\mathrm{d}r$, then $r_{x_1}(t)$ denotes the particle path uniquely determined by

$$\frac{\mathrm{d}}{\mathrm{d}t}r_{x_1}(t)=u(r_{x_1}(t),t),\qquad r_{x_1}(t)=r_1. \tag{2.7}$$

Theorem 2.1 *Let $T>0$, $\gamma>1$ and (2.1) holds. Assume that the initial data (ρ_0, u_0,θ_0) satisfies $\inf_{r\in[r_0,a_0]}\rho_0(r)>0$, $\inf_{r\in[r_0,a_0]}\theta_0(r)>0$, $\rho_0\in W^{1,\infty}([r_0,a_0])$ and $(u_0,\theta_0)\in H^2([r_0,a_0])$. Then, there exists a unique global strong solution (ρ,u,θ,a) to the FBVP (2.3) and (2.5)–(2.6) satisfying*

$$\|(u,\theta)(t)\|_{H^2([r_0,a(t)])}+\|(\rho_r,u_t,\theta_t)(t)\|_{L^2([r_0,a(t)])}+\int_0^t\|(u_t,\theta_t)(s)\|^2_{H^1([r_0,a(s)])}\,\mathrm{d}s\leqslant C \tag{2.8}$$

for any $t\in[0,T]$. In addition, for all $t\in[0,T]$, there hold

$$\begin{aligned} &C^{-1}x_1^{\frac{\gamma}{3(\gamma-1)}}\leqslant r_{x_1}(t)\leqslant a(t),\\ &C^{-1}(x_2-x_1)^{\frac{\gamma}{\gamma-1}}\leqslant r^3_{x_2}(t)-r^3_{x_1}(t),\\ &C^{-1}\leqslant a(t)\leqslant C,\\ &C^{-1}\leqslant \rho(r,t)\leqslant C,\ r\in[r_0,a(t)], \end{aligned} \tag{2.9}$$

where $r_{x_i}(t)$ is the particle path defined by (2.7), $i=1,2$. The constants C in (2.8) and (2.9) depend on the initial data, r_0, T and the physical coefficients and parameters in the system (2.3).

For simplicity, we assume that $\displaystyle\int_{r_0}^{a_0}\rho_0 r^2\,\mathrm{d}r=1$, which, by the first equation of (2.3), implies

$$\int_{r_0}^{a(t)}\rho r^2\,\mathrm{d}r=\int_{r_0}^{a_0}\rho_0 r^2\,\mathrm{d}r=1.$$

For $r\in[r_0,a(t)]$ and $t\in[0,T]$, we define the Lagrangian coordinates transformation

$$x(r,t)=\int_{r_0}^{r}\rho y^2\,\mathrm{d}y=1-\int_{r}^{a(t)}\rho y^2\,\mathrm{d}y,\ \tau=t, \tag{2.10}$$

which translates the domain $[0,T]\times[r_0,a(t)]$ into $[0,T]\times[0,1]$ and satisfies

$$\frac{\partial x}{\partial r}=\rho r^2,\ \frac{\partial x}{\partial t}=-\rho u r^2,\ \frac{\partial \tau}{\partial r}=0,\ \frac{\partial \tau}{\partial t}=1, \tag{2.11}$$

and

$$r^3(x,\tau)=r_0^3+3\int_0^x\frac{1}{\rho}(y,\tau)\,\mathrm{d}y=a^3(\tau)-3\int_x^1\frac{1}{\rho}(y,\tau)\,\mathrm{d}y,\ \frac{\partial r}{\partial \tau}=u. \tag{2.12}$$

Then, the equations (2.3) are changed into

$$\begin{cases}\rho_\tau+\rho^2(r^2u)_x=0,\\ u_\tau+r^2p_x-r^2[\rho\mu(r^2u)_x]_x=0,\\ E_\tau+(r^2pu)_x=[\mu r^4\rho u u_x]_x+[\rho r^4\kappa(\theta)\theta_x]_x,\end{cases} \tag{2.13}$$

for $(x,\tau)\in[0,1]\times[0,T]$, with the initial data given by

$$(\rho,u,\theta)(x,0)=(\rho_0,u_0,\theta_0)(x),\ x\in[0,1], \tag{2.14}$$

and the boundary conditions given by

$$\begin{aligned}&u(0,\tau)=0,\ (p-\mu\rho(r^2u)_x)(1,\tau)=0,\ \tau>0,\\ &\theta_x(0,\tau)=\theta_x(1,\tau)=0,\ \tau>0,\end{aligned} \tag{2.15}$$

where $r=r(x,\tau)$ is defined by

$$\frac{\mathrm{d}}{\mathrm{d}\tau}r(x,\tau)=u(x,\tau),\ (x,\tau)\in[0,1]\times[0,T]. \tag{2.16}$$

Note that the fixed boundary $x=1$ corresponds to the free boundary $a(\tau)=r(1,\tau)$ in Eulerian form determined by

$$\frac{\mathrm{d}}{\mathrm{d}\tau}a(\tau)=u(1,\tau),\ a(0)=a_0. \tag{2.17}$$

It is clear that the initial data (2.14) is consistent with the boundary data (2.15) to high order provided that (2.5) and (2.6) satisfy the compatible conditions. Besides, from (1.4) and (2.13), the third equation in (2.13) can be replaced by the following equation:

$$e_\tau+p(r^2u)_x=\mu r^4\rho u_x^2+[\rho r^4\kappa(\theta)\theta_x]_x+\frac{2\mu u^2}{\rho r^2}. \tag{2.18}$$

Now, in Lagrangian form, Theorem 2.1 can be restated as follows.

Theorem 2.2 *Let $T > 0$, $\gamma > 1$ and (2.1) holds. Assume that the initial data (ρ_0, u_0, θ_0) satisfies $\inf_{x\in[0,1]} \rho_0(x) > 0$, $\inf_{x\in[0,1]} \theta_0(x) > 0$, $\rho_0 \in W^{1,\infty}([0,1])$ and $(u_0, \theta_0) \in H^2([0,1])$. Then, there exists a unique global strong solution (ρ, u, θ, a) to the FBVP (2.13)–(2.17) satisfying*

$$\|(u,\theta)(\tau)\|_{H^2([0,1])} + \|(\rho_x, u_\tau, \theta_\tau)(\tau)\|_{L^2([0,1])} + \int_0^\tau \|(u_\tau, \theta_\tau)(s)\|^2_{H^1([0,1])}\,\mathrm{d}s \leqslant C, \tag{2.19}$$

for any $\tau \in [0,T]$. In addition, for all $\tau \in [0,T]$, there hold

$$\begin{aligned} & C^{-1} x_1^{\frac{\gamma}{3(\gamma-1)}} \leqslant r(x_1,\tau) \leqslant a(\tau), \ \forall x_1 \in [0,1], \\ & C^{-1}(x_2 - x_1)^{\frac{\gamma}{\gamma-1}} \leqslant r^3(x_2,\tau) - r^3(x_1,\tau), \ 0 \leqslant x_1 < x_2 \leqslant 1, \\ & C^{-1} \leqslant a(\tau) \leqslant C, \\ & C^{-1} \leqslant \rho(x,\tau) \leqslant C, \ x \in [0,1], \end{aligned} \tag{2.20}$$

where $r = r(x_i,\tau)$ is the particle path defined by (2.16) with $r(x_i,0) = r_i \in [r_0, a_0]$ and $x_i = 1 - \int_{r_i}^{a_0} r^2 \rho_0(r)\, dr$, $i = 1,2$. The constants C in (2.19) and (2.20) depend on the initial data, r_0, T and the physical coefficients and parameters in the system (2.13).

To our best knowledge, there have no results on existence and uniqueness of global solutions to the free boundary value problem of full compressible Navier-Stokes equations (1.1) with general initial data. This paper shows the global strong solution to the 3-dimensional full compressible Navier-Stokes equations with large data in the spherically symmetric case. However, the well-posedness of global strong solution to the high-dimensional full compressible Navier-Stokes equations with large initial data and vacuum is still open.

We prove Theorem 2.2 (thus, Theorem 2.1) through a series of *a priori* estimates for the solution of the FBVP (2.13)–(2.17), where (2.20) is proved by adapting the method used in [10]. More precisely, Lemma 3.1 is the basic energy estimate for the third equation of (2.13). The first two estimates in (2.20) as well as the low bound for $a(\tau)$ are given in Lemma 3.2. Lemmas 3.3–3.4 are prepared to prove the lower and upper bounds for ρ, which is presented in Lemma 3.5. The $L^\infty L^2$-norm for ρ_x and the upper bound of $a(\tau)$ are obtained in Lemma 3.6.

Due to the coupled structure of the pressure and the free boundary condition, the main difficulty is to obtain the $L^\infty L^2$-norm for u_x. To overcome this obstacle, we use some ideas given in the work [3], [6], [7], [17], [40] (see also [39]). We first show that the L^2L^2-norm for u_x and $L^\infty L^2$-norm for θ can be controlled by the $L^\infty_{t,x}$-norm of the temperature, which in turn, by Sobolev's inequality, controlled by $\|\theta_x\|_{L^\infty L^2}$ (see Lemmas 3.7–3.8). Through delicate analysis, in Lemma 3.9, we prove that both $\|\theta_x\|_{L^\infty L^2}$ and $\|\theta_\tau\|_{L^2L^2}$ (or $\|\theta_{xx}\|_{L^2L^2}$) can be bounded by the $L^\infty L^2$-norm of u_{xx}. Then combining these estimates, we can obtain the desired bounds for the first and second derivatives of u (thus, the $L^\infty H^1$-norm of θ) in Lemma 3.10. Using these bounds, we finally get the lower bound of θ in Lemma 3.11 and the $L^\infty L^2$-norm of θ_{xx} in Lemma 3.12.

3 A priori estimates

In this section, we establish the *a priori* estimates (see Lemmas 3.1–3.12 below) for any solution (ρ, u, θ, a) to the FBVP (2.13)–(2.17). Then, with the help of these estimates, we can finish the proofs of Theorems 2.1 and 2.2. Throughout this section, $T > 0$ is fixed, and $C > 0$ is a generic constant, which may depend on the initial data, T, r_0 as well as the physical coefficients and parameters appearing in the full compressible Navier-Stokes equations. Note that the value of C may be different in each appearance.

Lemma 3.1 *Let $\gamma > 1$, $T > 0$, and (ρ, u, θ, a) with $\rho > 0$ and $\theta > 0$ be any regular solution to the FBVP* (2.13)–(2.17) *for $\tau \in [0, T]$ under the assumptions of Theorem 2.2. Then, there holds*

$$\int_0^1 \left(\frac{1}{2}u^2 + C_\nu\theta + \frac{a}{\gamma-1}\rho^{\gamma-1}\right) \mathrm{d}x + 2\mu \int_0^\tau (ru^2)(1,s)\,\mathrm{d}s$$
$$\leqslant \int_0^1 \left(\frac{1}{2}u_0^2 + C_\nu\theta_0 + \frac{a}{\gamma-1}\rho_0^{\gamma-1}\right) \mathrm{d}x \leqslant C. \tag{3.1}$$

Proof Integrating the third equation of (2.13) with respect to x over $[0,1]$ and using (2.15), we get

$$\frac{\mathrm{d}}{\mathrm{d}\tau}\int_0^1 E\,\mathrm{d}x + 2\mu(ru^2)(1,\tau) = 0.$$

Integrating about τ, it is easy to get

$$\int_0^1 E\,\mathrm{d}x + 2\mu\int_0^\tau (ru^2)(1,s)\,\mathrm{d}s = \int_0^1 E_0\,\mathrm{d}x.$$

By the definition of E, the estimate (3.1) follows immediately. This ends the proof of Lemma 3.1.

Lemma 3.2 *Under the same assumptions as Lemma 3.1, then*

$$3^{\frac{1}{3}}\left(\frac{\gamma-1}{a}\right)^{-\frac{1}{3(\gamma-1)}}E_0^{-\frac{1}{3(\gamma-1)}}x^{\frac{\gamma}{3(\gamma-1)}}\leqslant r(x,\tau)\leqslant a(\tau),\ (x,\tau)\in[0,1]\times[0,T], \tag{3.2}$$

$$3\left(\frac{\gamma-1}{a}\right)^{-\frac{1}{\gamma-1}}E_0^{-\frac{1}{\gamma-1}}(x_2-x_1)^{\frac{\gamma}{\gamma-1}}\leqslant r^3(x_2,\tau)-r^3(x_1,\tau),\ 0\leqslant x_1<x_2\leqslant 1, \tag{3.3}$$

where

$$E_0=\int_0^1\left(\frac{1}{2}u_0^2+C_\nu\theta_0+\frac{a}{\gamma-1}\rho_0^{\gamma-1}\right)\mathrm{d}x.$$

In particular, it holds for $x=1$ that

$$3^{\frac{1}{3}}\left(\frac{\gamma-1}{a}\right)^{-\frac{1}{3(\gamma-1)}}E_0^{-\frac{1}{3(\gamma-1)}}\leqslant a(\tau)=r(1,\tau),\ \tau\in[0,T]. \tag{3.4}$$

Proof First, for any $x\in(0,1)$ and $r_0\leqslant r(x,\tau)\leqslant a(\tau)$, it follows from (2.10) and (3.1) that

$$\begin{aligned}x=\int_{r_0}^{r(x,\tau)}\rho y^2\,\mathrm{d}y&\leqslant\left(\int_{r_0}^{r(x,\tau)}\rho^\gamma y^2\,\mathrm{d}y\right)^{\frac{1}{\gamma}}\left(\int_{r_0}^{r(x,\tau)}y^2\,\mathrm{d}y\right)^{\frac{\gamma-1}{\gamma}}\\&\leqslant 3^{\frac{1-\gamma}{\gamma}}r^{\frac{3(\gamma-1)}{\gamma}}(x,\tau)\left(\int_0^1\rho^\gamma r^2(z,\tau)\frac{\partial r}{\partial z}\,\mathrm{d}z\right)^{\frac{1}{\gamma}}.\end{aligned} \tag{3.5}$$

Note that

$$\begin{aligned}\int_0^1\rho^\gamma r^2(z,\tau)\frac{\partial r}{\partial z}\,\mathrm{d}z&=\int_0^1\rho^\gamma r^2(z,\tau)\frac{1}{\rho r^2(z,\tau)}\,\mathrm{d}z=\int_0^1\rho^{\gamma-1}(z,\tau)\,\mathrm{d}z\\&=\frac{\gamma-1}{a}\int_0^1\frac{a}{\gamma-1}\rho^{\gamma-1}\,\mathrm{d}x\leqslant\frac{\gamma-1}{a}E_0.\end{aligned}$$

Inserting this estimate into (3.5), we prove that (3.2) holds. Similarly, for any $0<x_1\leqslant x_2<1$, there holds

$$x_2-x_1=\int_{r(x_1,\tau)}^{r(x_2,\tau)}\rho y^2\,\mathrm{d}y\leqslant 3^{\frac{1-\gamma}{\gamma}}\left(\frac{\gamma-1}{a}E_0\right)^{\frac{1}{\gamma}}\left(r^3(x_2,\tau)-r^3(x_1,\tau)\right)^{\frac{\gamma-1}{\gamma}},$$

which implies (3.3). The proof of Lemma 3.2 is finished.

Lemma 3.3 *Under the same assumptions as Lemma 3.1, there holds that*

$$\int_0^T\int_0^1\frac{\rho r^4u_x^2}{\theta}\,\mathrm{d}x\,\mathrm{d}s+\int_0^T\int_0^1\frac{\kappa(\theta)\rho r^4\theta_x^2}{\theta^2}\,\mathrm{d}x\,\mathrm{d}s+\int_0^T\int_0^1\frac{u^2}{\rho r^2\theta}\,\mathrm{d}x\,\mathrm{d}s\leqslant C. \tag{3.6}$$

Proof By (2.13) and (2.18), we can obtain that θ satisfies

$$C_\nu\theta_\tau + R\rho\theta(r^2u)_x = \mu r^4\rho u_x^2 + (r^4\kappa(\theta)\theta_x\rho)_x + \frac{2\mu u^2}{\rho r^2}. \tag{3.7}$$

Multiplying the above equation by θ^{-1}, it is easy to see that

$$C_\nu(\ln\theta)_\tau + R\rho(r^2u)_x = \frac{\mu\rho r^4u_x^2}{\theta} + \frac{\rho r^4\kappa(\theta)\theta_x^2}{\theta^2} + \left(\frac{\rho r^4\kappa(\theta)\theta_x}{\theta}\right)_x + \frac{2\mu u^2}{\rho r^2\theta}. \tag{3.8}$$

Integrating (3.8) over $[0,1]\times[0,\tau]$ and applying (2.15), one gets

$$\begin{aligned}&\int_0^\tau\int_0^1\left(\frac{\mu\rho r^4u_x^2}{\theta} + \frac{\rho r^4\kappa(\theta)\theta_x^2}{\theta^2} + \frac{2\mu u^2}{\rho r^2\theta}\right)\mathrm{d}x\,\mathrm{d}s\\ &= \int_0^1 C_\nu\ln\theta\,\mathrm{d}x - \int_0^1 C_\nu\ln\theta_0\,\mathrm{d}x + \int_0^\tau\int_0^1 R\rho(r^2u)_x\,\mathrm{d}x\,\mathrm{d}s.\end{aligned} \tag{3.9}$$

Due to Lemma 3.1 and the initial condition on θ_0, we can easily estimate the first and second term in the right hand side of (3.9). So it remains to estimate the last term in (3.9). Fix a cut off function $\chi(z)\in C^\infty(0,\infty)$ satisfying $0\leqslant\chi(z)\leqslant 1$, $\chi(z)=1$ if $0<z\leqslant 1$ and $\chi(z)=0$ if $z\geqslant 2$, then we have

$$\int_0^\tau\int_0^1 \rho(r^2u)_x\,\mathrm{d}x\,\mathrm{d}s = \int_0^\tau\int_0^1 \rho(r^2u)_x\chi(\rho)\,\mathrm{d}x\,\mathrm{d}s + \int_0^\tau\int_0^1 \rho(r^2u)_x(1-\chi(\rho))\,\mathrm{d}x\,\mathrm{d}s.$$

Note that $\chi(\rho)\neq 0$ implies $0<\rho\leqslant 2$, so the first term can be estimated by using Cauchy-Schwarz inequality and (3.1),

$$\begin{aligned}\int_0^\tau\int_0^1 R\rho(r^2u)_x\chi(\rho)\,\mathrm{d}x\,\mathrm{d}s &= \int_0^\tau\int_0^1 R\rho r^2u_x\chi(\rho)\,\mathrm{d}x\,\mathrm{d}s\\ &\quad+\int_0^\tau\int_0^1 \frac{2Ru}{r}\chi(\rho)\,\mathrm{d}x\,\mathrm{d}s\\ &\leqslant \frac{\mu}{4}\int_0^\tau\int_0^1\frac{\rho r^4u_x^2}{\theta}\,\mathrm{d}x\,\mathrm{d}s + \frac{\mu}{2}\int_0^\tau\int_0^1\frac{u^2}{\rho r^2\theta}\,\mathrm{d}x\,\mathrm{d}s\\ &\quad+C\int_0^\tau\int_0^1\rho\chi^2(\rho)\theta\,\mathrm{d}x\,\mathrm{d}s\\ &\leqslant \frac{\mu}{4}\int_0^\tau\int_0^1\frac{\rho r^4u_x^2}{\theta}\,\mathrm{d}x\,\mathrm{d}s + \frac{\mu}{2}\int_0^\tau\int_0^1\frac{u^2}{\rho r^2\theta}\,\mathrm{d}x\,\mathrm{d}s\\ &\quad+C.\end{aligned} \tag{3.10}$$

Using the first equation of (2.13), we see

$$\begin{aligned}\int_0^T\int_0^1 R\rho(r^2u)_x(1-\chi(\rho))\,\mathrm{d}x\,\mathrm{d}s &= -\int_0^T\int_0^1 R\rho\frac{\rho_\tau}{\rho^2}(1-\chi(\rho))\,\mathrm{d}x\,\mathrm{d}s\\ &= -R\int_0^1(1-\chi(\rho))\ln\rho\,\mathrm{d}x\\ &\quad +R\int_0^1(1-\chi(\rho_0))\ln\rho_0\,\mathrm{d}x\\ &\quad -\int_0^T\int_0^1 R\ln\rho\chi'(\rho)\rho_\tau\,\mathrm{d}x\,\mathrm{d}s,\end{aligned}$$

where

$$-\int_0^T\int_0^1 R\ln\rho\chi'(\rho)\rho_\tau\,\mathrm{d}x\,\mathrm{d}s = \int_0^T\int_0^1 R\rho^2\ln\rho(r^2u)_x\chi'(\rho)\,\mathrm{d}x\,\mathrm{d}s.$$

According to the definition of χ, we know $\chi'(\rho)\neq 0$ implies $1\leqslant\rho\leqslant 2$. Applying similar argument as (3.10), we can obtain

$$\int_0^T\int_0^1 R\rho^2\ln\rho(r^2u)_x\chi'(\rho)\,\mathrm{d}x\,\mathrm{d}s \leqslant \frac{\mu}{4}\int_0^T\int_0^1\frac{\rho r^4u_x^2}{\theta}\,\mathrm{d}x\,\mathrm{d}s+\frac{\mu}{2}\int_0^T\int_0^1\frac{u^2}{\rho r^2\theta}\,\mathrm{d}x\,\mathrm{d}s+C.$$

Hence, there holds that

$$\begin{aligned}&\int_0^T\int_0^1 R\rho(r^2u)_x(1-\chi(\rho))\,\mathrm{d}x\,\mathrm{d}s\\ &\leqslant -R\int_0^1(1-\chi(\rho))\ln\rho\,\mathrm{d}x+R\int_0^1(1-\chi(\rho_0))\ln\rho_0\,\mathrm{d}x\\ &\quad +\frac{\mu}{4}\int_0^T\int_0^1\frac{\rho r^4u_x^2}{\theta}\,\mathrm{d}x\,\mathrm{d}s+\frac{\mu}{2}\int_0^T\int_0^1\frac{u^2}{\rho r^2\theta}\,\mathrm{d}x\,\mathrm{d}s+C.\end{aligned}\tag{3.11}$$

Coming (3.9)–(3.11) and using (3.1) as well as the fact $(1-\chi(\rho))\ln\rho\geqslant 0$, we have

$$\begin{aligned}&\int_0^T\int_0^1\frac{\mu\rho r^4u_x^2}{2\theta}\,\mathrm{d}x\,\mathrm{d}s+\int_0^T\int_0^1\frac{\kappa(\theta)\rho r^4\theta_x^2}{\theta^2}\,\mathrm{d}x\,\mathrm{d}s\\ &\quad+\int_0^T\int_0^1\frac{\mu u^2}{\rho r^2\theta}\,\mathrm{d}x\,\mathrm{d}s+R\int_0^1(1-\chi(\rho))\ln\rho\,\mathrm{d}x\\ &\leqslant\int_0^1 C_\nu\ln\theta\,\mathrm{d}x-\int_0^1 C_\nu\ln\theta_0\,\mathrm{d}x+R\int_0^1(1-\chi(\rho_0))\ln\rho_0\,\mathrm{d}x+C\\ &\leqslant C.\end{aligned}$$

Thus the proof of Lemma 3.3 is completed.

Lemma 3.4 *Under the same assumptions as Lemma* 3.1, *there holds that*

$$\int_0^T\|\theta(s)\|_{L^\infty([0,1])}^q\,\mathrm{d}s\leqslant C.$$

Proof Choosing $x^* \in [0,1]$ such that $\theta(x^*,\tau) = \int_0^1 \theta(x,\tau)\,\mathrm{d}x \leqslant C$, by Cauchy-Schwarz inequality, the facts $\kappa(\theta) \leqslant C(1+\theta^q)$, $r_0 \leqslant r(x,\tau)$ and Lemma 3.1, then we have

$$\begin{aligned}
\theta^q(x,\tau) &= \theta^q(x^*,\tau) + q\int_{x^*}^{x} (\theta^{q-1}\theta_x)(z,\tau)\,\mathrm{d}z \\
&\leqslant C + C\Big(\int_0^1 \frac{\theta^{2q}}{\rho r^4\kappa(\theta)}\,\mathrm{d}x\Big)^{\frac{1}{2}}\Big(\int_0^1 \frac{\rho r^4\kappa(\theta)\theta_x^2}{\theta^2}\,\mathrm{d}x\Big)^{\frac{1}{2}} \\
&\leqslant C + C\Big(\|\theta(t)\|_{L_x^\infty}^q\Big)^{\frac{1}{2}}\Big(\int_0^1 \frac{1}{\rho r^4}\,\mathrm{d}x\int_0^1 \frac{\rho r^4\kappa(\theta)\theta_x^2}{\theta^2}\,\mathrm{d}x\Big)^{\frac{1}{2}} \\
&\leqslant C + \frac{1}{2}\|\theta(t)\|_{L_x^\infty}^q + C\int_0^1 \frac{1}{\rho r^4}\,\mathrm{d}x\int_0^1 \frac{\rho r^4\kappa(\theta)\theta_x^2}{\theta^2}\,\mathrm{d}x. \qquad (3.12)
\end{aligned}$$

Note that

$$\int_0^1 \frac{1}{\rho r^4}\,\mathrm{d}x = \int_{r_0}^{a(\tau)} \frac{1}{r^2}\,\mathrm{d}r = \frac{1}{r_0} - \frac{1}{a(\tau)} \leqslant \frac{1}{r_0} \leqslant C.$$

Now, taking the L^∞ norm with respect to x on both sides of (3.12), then integrating the resulted inequality with respect to time and using Lemma 3.3, we thus obtain Lemma 3.4. Now, we can give the lower and upper bounds for the density ρ.

Lemma 3.5 *Under the same assumptions as Lemma* 3.1, *there holds that*

$$C^{-1} \leqslant \rho(x,\tau) \leqslant C, \ \textit{for}\ (x,\tau) \in [0,1]\times[0,T].$$

Proof By the mass equation in (2.13), we get

$$\rho(r^2u)_x = -(\ln\rho)_\tau, \qquad (3.13)$$

which, combining with the momentum equation of (2.13), gives that

$$u_\tau + r^2p_x + \mu r^2(\ln\rho)_{x\tau} = 0. \qquad (3.14)$$

That is,

$$\Big(\frac{u}{r^2}\Big)_\tau + \frac{2u^2}{r^3} + p_x + \mu(\ln\rho)_{x\tau} = 0.$$

Integrating the above equation over $[x,1]\times[0,\tau]$, we have

$$\begin{aligned}
&\int_x^1 \frac{u}{r^2}\,\mathrm{d}x - \int_x^1 \frac{u}{r^2}(x,0)\,\mathrm{d}x + \int_0^\tau\int_x^1 \frac{2u^2}{r^3}\,\mathrm{d}x\,\mathrm{d}s + \int_0^\tau p(1,s)\,\mathrm{d}s - \int_0^\tau p(x,s)\,\mathrm{d}s \\
&= -\mu\int_0^\tau (\ln\rho(1,s))_s\,\mathrm{d}s + \mu\int_0^\tau (\ln\rho)_s\,\mathrm{d}s,
\end{aligned}$$

which implies that

$$\begin{aligned}\mu\ln\rho =&\mu\ln\rho_0(x)+\mu\ln\rho(1,\tau)-\mu\ln\rho_0(1)+\int_x^1\frac{u}{r^2}\,\mathrm{d}x-\int_x^1\frac{u}{r^2}(x,0)\,\mathrm{d}x\\&+\int_0^\tau\int_x^1\frac{2u^2}{r^3}\,\mathrm{d}x\,\mathrm{d}s+\int_0^\tau(a\rho^\gamma+R\rho\theta)(1,s)\,\mathrm{d}s-\int_0^\tau(a\rho^\gamma+R\rho\theta)\,\mathrm{d}s.\end{aligned}$$

Since the first equation of (2.13) and the condition (2.15) imply that

$$(a\rho^\gamma+R\rho\theta)(1,s)+\mu(\ln\rho(1,s))_s=0,$$

we have

$$\mu\ln\rho =\mu\ln\rho_0+\int_x^1\frac{u}{r^2}\,\mathrm{d}x-\int_x^1\frac{u}{r^2}(x,0)\,\mathrm{d}x+\int_0^\tau\int_x^1\frac{2u^2}{r^3}\,\mathrm{d}x\,\mathrm{d}s-\int_0^\tau(a\rho^\gamma+R\rho\theta)\,\mathrm{d}s.$$

Hence, it follows from Lemma 3.1 and the above equality that

$$\mu\ln\rho \leqslant C+\int_x^1\frac{u}{r^2}\,\mathrm{d}x+\int_0^\tau\int_x^1\frac{2u^2}{r^3}\,\mathrm{d}x\,\mathrm{d}s\leqslant C,$$

which yields the upper bound for ρ. On the other hand, it follows from Lemma 3.4 that

$$\begin{aligned}\mu\ln\rho&\geqslant C^{-1}+\int_x^1\frac{u}{r^2}\,\mathrm{d}x-\int_x^1\frac{u}{r^2}(0,x)\,\mathrm{d}x-\int_0^\tau(a\rho^\gamma+R\rho\theta)\,\mathrm{d}s\\&\geqslant C^{-1}-C-a\tau\|\rho\|^\gamma_{L^\infty_{x,\tau}}-R\|\rho\|_{L^\infty_{x,\tau}}\int_0^\tau\|\theta\|_{L^\infty_x}\,\mathrm{d}s\\&\geqslant -C,\end{aligned}$$

which yields the lower bound for ρ. Thus, the proof of Lemma 3.5 is completed.

Lemma 3.6 *Under the same assumptions as Lemma* 3.1, *there holds that*

$$\begin{aligned}&\int_0^1(u+r^2(\mu\ln\rho)_x)^2\,\mathrm{d}x+\int_0^\tau\int_0^1[(\rho^{\frac{\gamma}{2}})_xr^2]^2\,\mathrm{d}x\,\mathrm{d}s\\&+\int_0^\tau[p\rho^\gamma r^3](1,s)\,\mathrm{d}s+[\rho^\gamma r^3](1,\tau)\leqslant C.\end{aligned}$$

In particular, this estimate implies

$$\int_0^1\rho_x^2\,\mathrm{d}x\leqslant C,\quad r(1,\tau)=a(\tau)\leqslant C.$$

Proof Multiplying the equation (3.14) by $u+r^2(\mu\ln\rho)_x$ and integrating the resulted equation over $[0,1]$ gives that

$$
\begin{aligned}
&\frac{1}{2}\frac{\mathrm{d}}{\mathrm{d}\tau}\int_0^1 (u+r^2(\mu\ln\rho)_x)^2\,\mathrm{d}x\\
&=\int_0^1 2r\mu(\ln\rho)_x u^2\,\mathrm{d}x+\int_0^1 2r^3\mu^2(\ln\rho)_x^2 u\,\mathrm{d}x-\int_0^1 ar^2(\rho^\gamma)_x u\,\mathrm{d}x\\
&\quad-\int_0^1 ar^2(\rho^\gamma)_x r^2(\mu\ln\rho)_x\,\mathrm{d}x-\int_0^1 Rr^2(\rho\theta)_x(u+r^2(\mu\ln\rho)_x)\,\mathrm{d}x\\
&=:\sum_{i=1}^5 I_i. \qquad (3.15)
\end{aligned}
$$

So, one has to estimate the five terms on the right-hand side of (3.15). For the term I_1, by using Cauchy-Schwarz inequality, Sobolev inequality and the fact $(a+b)^2\leqslant 2(a^2+b^2)$, we have

$$
\begin{aligned}
|I_1|&\leqslant C\int_0^1 (r^2(\mu\ln\rho)_x+u-u)^2\,\mathrm{d}x+C\int_0^1\frac{u^4}{r^2}\,\mathrm{d}x\\
&\leqslant C\int_0^1 (r^2(\mu\ln\rho)_x+u)^2\,\mathrm{d}x+C\int_0^1 u^2\,\mathrm{d}x+C\|u\|_{L_x^\infty}^2\int_0^1 u^2\,\mathrm{d}x\\
&\leqslant C\int_0^1 (r^2(\mu\ln\rho)_x+u)^2\,\mathrm{d}x+C+C\|u\|_{L_x^\infty}^2\\
&\leqslant C\int_0^1 (r^2(\mu\ln\rho)_x+u)^2\,\mathrm{d}x+C+C\|u_x\|_{L_x^2}^2.
\end{aligned}
$$

For the term I_2, we estimate it as follows:

$$
\begin{aligned}
|I_2|&\leqslant C\left\|\frac{u}{r}\right\|_{L_x^\infty}\int_0^1 (r^2(\mu\ln\rho)_x+u-u)^2\,\mathrm{d}x\\
&\leqslant C\|u\|_{L_x^\infty}\int_0^1 (r^2(\mu\ln\rho)_x+u)^2\,\mathrm{d}x+C\|u\|_{L_x^\infty}\int_0^1 u^2\,\mathrm{d}x\\
&\leqslant C(1+\|u_x\|_{L_x^2})\int_0^1 (r^2(\mu\ln\rho)_x+u)^2\,\mathrm{d}x+C(1+\|u_x\|_{L_x^2}).
\end{aligned}
$$

The term I_3 is treated as follows:

$$
\begin{aligned}
I_3&=-[ar^2\rho^\gamma u](1,\tau)+[ar^2\rho^\gamma u](0,\tau)+\int_0^1 a\rho^\gamma(r^2u)_x\,\mathrm{d}x\\
&=-[ar^2\rho^\gamma u](1,\tau)-\int_0^1 a\rho^{\gamma-2}\rho_\tau\,\mathrm{d}x\\
&=-[ar^2\rho^\gamma u](1,\tau)-\frac{\mathrm{d}}{\mathrm{d}\tau}\int_0^1\frac{a\rho^{\gamma-1}}{\gamma-1}\,\mathrm{d}x.
\end{aligned}
$$

For I_4, it is easy to see that

$$
I_4=-\int_0^1 a\gamma\mu r^4\rho^{\gamma-2}\rho_x^2\,\mathrm{d}x=-\frac{4\mu a}{\gamma}\int_0^1 [(\rho^{\frac{\gamma}{2}})_x r^2]^2\,\mathrm{d}x.
$$

For the last term I_5, one has

$$
\begin{aligned}
|I_5| &= \left|\int_0^1 (Rr^2\rho_x\theta + Rr^2\rho\theta_x)(u + r^2(\mu\ln\rho)_x)\,\mathrm{d}x\right| \\
&\leqslant C\int_0^1 \theta(u + r^2(\mu\ln\rho)_x)^2\,\mathrm{d}x + C\int_0^1 \theta(r^2\mu\rho_x + u - u)^2\,\mathrm{d}x \\
&\quad + C\int_0^1 \frac{\kappa(\theta)\rho r^4\theta_x^2}{\theta^2}\,\mathrm{d}x + C\int_0^1 \frac{\theta^2}{\kappa(\theta)}(u + r^2(\mu\ln\rho)_x)^2\,\mathrm{d}x \\
&\leqslant C(1 + \|\theta\|_{L_x^\infty})\int_0^1 (u + r^2(\mu\ln\rho)_x)^2\,\mathrm{d}x + C\int_0^1 \frac{\kappa(\theta)\rho r^4\theta_x^2}{\theta^2}\,\mathrm{d}x + C\|\theta\|_{L_x^\infty},
\end{aligned}
$$

where, in the last step we have used the fact

$$
\left\|\frac{\theta^2}{\kappa(\theta)}\right\|_{L_x^\infty} \leqslant C\left\|\frac{\theta^2}{1+\theta^q}\right\|_{L_x^\infty} \leqslant C(1 + \|\theta\|_{L_x^\infty}),\ q \geqslant 1.
$$

Collecting the estimates for $I_1, \cdots, I_5$, we obtain

$$
\begin{aligned}
&\frac{\mathrm{d}}{\mathrm{d}\tau}\left(\int_0^1 (u + r^2(\mu\ln\rho)_x)^2\,\mathrm{d}x + \int_0^1 \frac{a\rho^{\gamma-1}}{\gamma-1}\,\mathrm{d}x\right) \\
&\quad + \frac{4\mu a}{\gamma}\int_0^1 [(\rho^{\frac{\gamma}{2}})_x r^2]^2\,\mathrm{d}x + ar^2\rho^\gamma u(1,\tau) \\
&\leqslant C(1 + \|u_x\|_{L_x^2} + \|\theta\|_{L_x^\infty})\int_0^1 (u + r^2(\mu\ln\rho)_x)^2\,\mathrm{d}x + C\int_0^1 u_x^2\,\mathrm{d}x \\
&\quad + C\int_0^1 \frac{\kappa(\theta)\rho r^4\theta_x^2}{\theta^2}\,\mathrm{d}x + C\|\theta\|_{L_x^\infty} + C.
\end{aligned} \tag{3.16}
$$

Note that

$$
\begin{aligned}
&-\int_0^\tau [ar^2\rho^\gamma u](1,s)\,\mathrm{d}s = -\frac{a}{3}\int_0^\tau [(\rho^\gamma r^3)_s - (\rho^\gamma)_s r^3](1,s)\,\mathrm{d}s \\
&= -\frac{a}{3}[\rho^\gamma r^3](1,\tau) + \frac{a}{3}[\rho^\gamma r^3](1,0) + \frac{a\gamma}{3}\int_0^\tau [\rho^{\gamma-1}\rho_s r^3](1,s)\,\mathrm{d}s \\
&= -\frac{a}{3}[\rho^\gamma r^3](1,\tau) + \frac{a}{3}[\rho^\gamma r^3](1,0) - \frac{a\gamma}{3\mu}\int_0^\tau [\rho^\gamma(R\rho\theta + a\rho^\gamma)r^3](1,s)\,\mathrm{d}s.
\end{aligned}
$$

Now, integrating (3.16) with respect to time, then using the above equality and recalling that

$$
\int_0^\tau\int_0^1 u_x^2 dxds \leqslant C,\quad \int_0^\tau \|\theta\|_{L_x^\infty}^q ds \leqslant C,\quad \int_0^\tau\int_0^1 \frac{\kappa(\theta)\rho r^4\theta_x^2}{\theta^2} dxds \leqslant C,
$$

hence, with the help of Gronwall's inequality, the desired estimate of Lemma 3.6 follows easily.

Lemma 3.7 *Under the same assumptions as Lemma* 3.1, *there holds that*

$$\int_0^\tau \int_0^1 \rho(r^2 u)_x^2 \,\mathrm{d}x\,\mathrm{d}s \leqslant C,$$

that is,

$$\int_0^\tau \int_0^1 r^4 u_x^2 \,\mathrm{d}x\,\mathrm{d}s + \int_0^\tau \int_0^1 \frac{u^2}{r^2} \,\mathrm{d}x\,\mathrm{d}s \leqslant C.$$

Moreover, there holds

$$\int_0^\tau \int_0^1 u_x^4 \,\mathrm{d}x\,\mathrm{d}s \leqslant C + C \sup_{(x,s)\in[0,1]\times[0,\tau]} |\theta(x,s)|^{(3-q)_+} + C \sup_{s\in[0,\tau]} |u(1,s)|^2, \quad (3.17)$$

where $(3-q)_+ := \max\{0, 3-q\}$.

Proof Multiplying the momentum equation of (2.13) by u and applying the boundary conditions (2.15), we get

$$\int_0^1 \frac{1}{2} u^2 \,\mathrm{d}x - \int_0^1 \frac{1}{2} u_0^2 \,\mathrm{d}x - \int_0^\tau \int_0^1 p(r^2 u)_x \,\mathrm{d}x\,\mathrm{d}s + \int_0^\tau \int_0^1 \rho\mu(r^2 u)_x^2 \,\mathrm{d}x\,\mathrm{d}s = 0.$$

A direct computation gives that

$$\begin{aligned}
\left|\int_0^\tau \int_0^1 p(r^2 u)_x \,\mathrm{d}x\,\mathrm{d}s\right| &\leqslant \left|\int_0^\tau \int_0^1 a\rho^\gamma (r^2 u)_x \,\mathrm{d}x\,\mathrm{d}s\right| + \left|\int_0^\tau \int_0^1 R\rho\theta (r^2 u)_x \,\mathrm{d}x\,\mathrm{d}s\right| \\
&\leqslant \left|\int_0^1 \frac{a\rho^{\gamma-1}}{\gamma-1} \,\mathrm{d}x - \int_0^1 \frac{a\rho_0^{\gamma-1}}{\gamma-1} \,\mathrm{d}x\right| \\
&\quad + \frac{\mu}{2} \int_0^\tau \int_0^1 \rho(r^2 u)_x^2 \,\mathrm{d}x\,\mathrm{d}s + C \int_0^\tau \int_0^1 \rho\theta^2 \,\mathrm{d}x\,\mathrm{d}s \\
&\leqslant C + \frac{\mu}{2} \int_0^\tau \int_0^1 \rho(r^2 u)_x^2 \,\mathrm{d}x\,\mathrm{d}s + C \int_0^\tau \|\theta\|_{L_x^\infty} \left(\int_0^1 \theta \,\mathrm{d}x\right) \mathrm{d}s.
\end{aligned}$$

By Lemma 3.1, Lemma 3.4 and Lemma 3.5, one can prove that

$$\int_0^\tau \int_0^1 \rho\mu(r^2 u)_x^2 \,\mathrm{d}x\,\mathrm{d}s \leqslant C. \quad (3.18)$$

To prove (3.17), we set $v(x,\tau) = \int_x^1 u(y,\tau)\,\mathrm{d}y$, by integrating the second equation of (2.13) with respect to space variable over $[x,1]$, we can obtain

$$v_\tau - \mu\rho r^4 v_{xx} = f(x,\tau), \quad (3.19)$$

where

$$\begin{aligned}f(x,\tau):=&r^2p-2\mu ru+\int_x^1 2(\rho r)^{-1}[p-\mu\rho(r^2u)_x]\,\mathrm{d}x\\=&r^2p+\int_x^1 2(\rho r)^{-1}p\,\mathrm{d}x-2\mu ru-\int_x^1 2\mu r^{-1}(r^2u)_x\,\mathrm{d}x\\=&r^2p+\int_x^1 2(\rho r)^{-1}p\,\mathrm{d}x-2\mu[ru](1,\tau)-2\mu\int_x^1(\rho r^2)^{-1}u\,\mathrm{d}x.\end{aligned}$$

Note that $\mu\rho r^4\geqslant C>0$, so Eq. (3.19) is a parabolic equation for v satisfying the following initial-boundary conditions:

$$v(x,0)=\int_x^1 u_0(y)\,\mathrm{d}y,\quad v(0,\tau)=\int_0^1 u(y,\tau)\,\mathrm{d}y,\quad v(1,\tau)=0.$$

Using Lemma 3.1, Lemma 3.2, Lemma 3.5 and (3.18), it is easy to see that

$$|f(x,\tau)|\leqslant C+C\theta+2\mu|ru|(1,\tau).$$

Hence, by standard parabolic theory(eg., see [21]), we have

$$\begin{aligned}\int_0^\tau\int_0^1 u_x^4\,\mathrm{d}x\,\mathrm{d}s=&\int_0^\tau\int_0^1 v_{xx}^4\,\mathrm{d}x\,\mathrm{d}s\leqslant\int_0^\tau\int_0^1 f^4(x,\tau)\,\mathrm{d}x\,\mathrm{d}s+C\\\leqslant&C+C\int_0^\tau\int_0^1\theta^4\,\mathrm{d}x\,\mathrm{d}s+C\int_0^\tau|ru|^4(1,s)\,\mathrm{d}s\\\leqslant&C+C\sup_{(x,s)\in[0,1]\times[0,\tau]}|\theta|^{(3-q)_+}\int_0^\tau\|\theta\|_{L_x^\infty}^{\min\{q,3\}}\left(\int_0^1\theta\,\mathrm{d}x\right)\mathrm{d}s\\&+C\sup_{s\in[0,\tau]}|u(1,s)|^2\int_0^\tau[ru^2](1,s)\,\mathrm{d}s\\\leqslant&C+C\sup_{(x,s)\in[0,1]\times[0,\tau]}|\theta(x,s)|^{(3-q)_+}+C\sup_{s\in[0,\tau]}|u(1,s)|^2,\end{aligned}$$

where $(3-q)_+:=\max\{0,3-q\}$, and we have used Lemma 3.1 and Lemma 3.4 in the last step. This ends the proof of the lemma.

In the following arguments , we define $\Theta(\tau):=\sup_{(x,s)\in[0,1]\times[0,\tau]}\theta(x,s)$.

Lemma 3.8 *Under the same assumptions as Lemma* 3.1, *there holds that*

$$\int_0^1\theta^2\,\mathrm{d}x+\int_0^\tau\int_0^1\rho r^4\kappa(\theta)\theta_x^2\,\mathrm{d}x\,\mathrm{d}s\leqslant C+C\Theta(\tau)+C\Theta^{(3-q)_+}(\tau).$$

Proof Multiplying (3.7) by θ, we can show that

$$\begin{aligned}&\frac{\mathrm{d}}{\mathrm{d}\tau}\int_0^1\frac{1}{2}C_\nu\theta^2\,\mathrm{d}x+\int_0^1 R\rho\theta^2(r^2u)_x\,\mathrm{d}x\\=&\int_0^1\mu\rho r^4u_x^2\theta\,\mathrm{d}x-\int_0^1\rho r^4\kappa(\theta)\theta_x^2\,\mathrm{d}x+\int_0^1\frac{2\mu u^2\theta}{\rho r^2}\,\mathrm{d}x.\end{aligned}$$

Integrating the above equation on $[0,\tau]$, one gets

$$\begin{aligned}
&\int_0^1 \frac{1}{2}C_\nu\theta^2 \,\mathrm{d}x + \int_0^\tau\int_0^1 \rho r^4\kappa(\theta)\theta_x^2 \,\mathrm{d}x\,\mathrm{d}s \\
=&-\int_0^\tau\int_0^1 R\rho\theta^2(r^2u)_x \,\mathrm{d}x\,\mathrm{d}s + \int_0^\tau\int_0^1 \mu\rho r^4u_x^2\theta \,\mathrm{d}x\,\mathrm{d}s \\
&+\int_0^\tau\int_0^1 \frac{2\mu u^2\theta}{\rho r^2} \,\mathrm{d}x\,\mathrm{d}s + \int_0^1 \frac{1}{2}C_\nu\theta_0^2 \,\mathrm{d}x. \qquad (3.20)
\end{aligned}$$

By Lemma 3.1, Lemma 3.4, Lemma 3.5, Lemma 3.7 and Cauchy-Schwarz inequality, it is easy to show that

$$\begin{aligned}
\left|\int_0^\tau\int_0^1 R\rho\theta^2(r^2u)_x \,\mathrm{d}x\,\mathrm{d}s\right| &\leqslant C\int_0^\tau\int_0^1 \rho(r^2u)_x^2 \,\mathrm{d}x\,\mathrm{d}s + C\int_0^\tau\int_0^1 \rho\theta^4 \,\mathrm{d}x\,\mathrm{d}s \\
&\leqslant C + C\Theta^{(3-q)+}\int_0^\tau \|\theta\|_{L_x^\infty}^{\min\{q,3\}}\left(\int_0^1 \theta \,\mathrm{d}x\right)\mathrm{d}s \\
&\leqslant C + C\Theta^{(3-q)+},
\end{aligned}$$

$$\left|\int_0^\tau\int_0^1 \mu\rho r^4u_x^2\theta \,\mathrm{d}x\,\mathrm{d}s\right| \leqslant \Theta\int_0^\tau\int_0^1 \mu\rho r^4u_x^2 \,\mathrm{d}x\,\mathrm{d}s \leqslant C\Theta$$

and

$$\left|\int_0^\tau\int_0^1 \frac{2\mu u^2\theta}{\rho r^2} \,\mathrm{d}x\,\mathrm{d}s\right| + \left|\int_0^1 \frac{1}{2}C_\nu\theta_0^2 \,\mathrm{d}x\right| \leqslant C\Theta + C.$$

Inserting the above four estimates into (3.20), we thus obtain the desired estimate.

In order to give the $L_T^\infty L_x^2$-norm estimate for u_x and u_{xx}, as in [3], [7], [17] and [40], we now define

$$\begin{aligned}
X(T) &:= \int_0^T\int_0^1 (1+\theta^q)\theta_\tau^2(x,\tau) \,\mathrm{d}x\,\mathrm{d}\tau, \\
Y(T) &:= \sup_{\tau\in[0,T]} \tilde{Y}(\tau) := \sup_{\tau\in[0,T]}\int_0^1 (1+\theta^{2q})\theta_x^2(x,\tau) \,\mathrm{d}x, \\
Z(T) &:= \sup_{\tau\in[0,T]} \tilde{Z}(\tau) := \sup_{\tau\in[0,T]}\int_0^1 u_{xx}^2 dx.
\end{aligned}$$

Also, we recall that

$$\Theta(\tau) := \sup_{(x,s)\in[0,1]\times[0,\tau]} \theta(x,s)$$

which is already used in Lemma 3.8. First, we claim that

$$\Theta^{2q+3}(T) \leqslant C + CY(T). \qquad (3.21)$$

In fact, by choosing $x_0 \in [0,1]$ such that $\theta(x_0,\tau) = \int_0^1 \theta(x,\tau)\,\mathrm{d}x \leqslant C$, we have

$$\begin{aligned}
\theta^{2q+3}(x,\tau) &= \theta^{2q+3}(x_0,\tau) + (2q+3)\int_{x_0}^{x} \theta^{2q+2}\theta_x\,\mathrm{d}x \\
&\leqslant C + C\int_0^1 \theta^{2q}\theta_x^2\,\mathrm{d}x + \varepsilon\|\theta(\tau)\|_{L^\infty}^{2q+3}\left(\int_0^1 \theta\,\mathrm{d}x\right) \\
&\leqslant C + C\tilde{Y}(\tau) + \varepsilon C\|\theta(\tau)\|_{L^\infty}^{2q+3}.
\end{aligned}$$

Hence, (3.21) follows immediately if we set $\varepsilon = \dfrac{1}{2C}$ and take the supremum with respect to (x,τ). Next, by Sobolev interpolation inequality and (3.1), we note that

$$\int_0^1 u_x^2\mathrm{d}x \leqslant C\int_0^1 u^2\mathrm{d}x + C\left(\int_0^1 u^2 dx\right)^{\frac12}\left(\int_0^1 u_{xx}^2\mathrm{d}x\right)^{\frac12} \leqslant C + CZ^{\frac12}(\tau)$$

and

$$u_x^2 \leqslant C\int_0^1 u_x^2\mathrm{d}x + C\left(\int_0^1 u_x^2\mathrm{d}x\right)^{\frac12}\left(\int_0^1 u_{xx}^2\mathrm{d}x\right)^{\frac12} \leqslant C + CZ^{\frac12}(\tau) + CZ^{\frac34}(\tau),$$

so we have

$$\begin{aligned}
&\sup_{s\in[0,\tau]}\int_0^1 u_x^2(x,s)\mathrm{d}x \leqslant C + CZ^{\frac12}(\tau), \\
&\sup_{(x,s)\in[0,1]\times[0,\tau]} |u_x(x,s)| \leqslant C + CZ^{\frac38}(\tau).
\end{aligned} \tag{3.22}$$

Now, we give the following lemma which states that $X(T)$ and $Y(T)$ (thus, $\Theta(T)$) can be estimated in terms of $Z(T)$.

Lemma 3.9 *Under the same assumptions as Lemma* 3.1, *there holds that*

$$X(T) + Y(T) \leqslant C + CZ^{\alpha}(T),$$

where $\alpha := \max\left\{\dfrac{2q+3}{2q+6}, \dfrac12\right\} < 1.$

Proof From (2.13) and (2.18), we see that θ satisfies

$$C_\nu\theta_\tau - [\rho r^4\kappa(\theta)\theta_x]_x = -R\rho\theta(r^2u)_x + \mu r^4\rho u_x^2 + \frac{2\mu u^2}{\rho r^2}. \tag{3.23}$$

Multiplying this equation by $(\rho r^4)^{-1}\kappa(\theta)\theta_\tau$, then integrating with respect to (x,τ) gives that $(\tau \leqslant T)$

$$\int_0^\tau\int_0^1 C_\nu(\rho r^4)^{-1}\kappa(\theta)\theta_\tau^2\,\mathrm{d}x\,\mathrm{d}s - \int_0^\tau\int_0^1 [\rho r^4\kappa(\theta)\theta_x]_x(\rho r^4)^{-1}\kappa(\theta)\theta_\tau\,\mathrm{d}x\,\mathrm{d}s$$

$$= -\int_0^\tau \int_0^1 R\theta(r^2u)_x r^{-4}\kappa(\theta)\theta_\tau \,\mathrm{d}x\,\mathrm{d}s + \int_0^\tau \int_0^1 \mu u_x^2 \kappa(\theta)\theta_\tau \,\mathrm{d}x\,\mathrm{d}s$$

$$+ \int_0^\tau \int_0^1 2\mu\rho^{-2}r^{-6}u^2\kappa(\theta)\theta_\tau \,\mathrm{d}x\,\mathrm{d}s. \tag{3.24}$$

For simplicity, we rewritten this equality as

$$J_1 + J_2 = J_3 + J_4 + J_5.$$

So, we have to estimate these five terms. For J_1, it is easy to see that

$$J_1 = \int_0^\tau \int_0^1 C_\nu(\rho r^4)^{-1}\kappa(\theta)\theta_\tau^2 \,\mathrm{d}x\,\mathrm{d}s \geqslant C^{-1}\int_0^\tau \int_0^1 \kappa(\theta)\theta_\tau^2 \,\mathrm{d}x\,\mathrm{d}s \geqslant C^{-1}X(\tau),$$

since both ρ and r have positive upper and lower bounds, and $\kappa(\theta) \geqslant C^{-1}(1+\theta^q)$. For J_2, using integration by parts and the fact $\dfrac{\partial r}{\partial x} = (\rho r^2)^{-1}$, we have

$$\begin{aligned}
J_2 &= \int_0^\tau \int_0^1 \rho r^4 \kappa(\theta)\theta_x[(\rho r^4)^{-1}\kappa(\theta)\theta_\tau]_x \,\mathrm{d}x\,\mathrm{d}s \\
&= -\int_0^\tau \int_0^1 \rho^{-1}\kappa^2(\theta)\rho_x\theta_x\theta_\tau \,\mathrm{d}x\,\mathrm{d}s - 4\int_0^\tau \int_0^1 \rho^{-1}r^{-3}\kappa^2(\theta)\theta_x\theta_\tau \,\mathrm{d}x\,\mathrm{d}s \\
&\quad + \int_0^\tau \int_0^1 \kappa(\theta)\kappa'(\theta)\theta_x^2\theta_\tau \,\mathrm{d}x\,\mathrm{d}s + \int_0^\tau \int_0^1 \kappa^2(\theta)\theta_x\theta_{\tau x} \,\mathrm{d}x\,\mathrm{d}s \\
&= -\int_0^\tau \int_0^1 \rho^{-1}\kappa^2(\theta)\rho_x\theta_x\theta_\tau \,\mathrm{d}x\,\mathrm{d}s - 4\int_0^\tau \int_0^1 \rho^{-1}r^{-3}\kappa^2(\theta)\theta_x\theta_\tau \,\mathrm{d}x\,\mathrm{d}s \\
&\quad + \frac{1}{2}\int_0^1 [\kappa^2(\theta)\theta_x^2](x,\tau)\,\mathrm{d}x - \frac{1}{2}\int_0^1 [\kappa^2(\theta)\theta_x^2](x,0)\,\mathrm{d}x.
\end{aligned}$$

Note that

$$\begin{aligned}
\left|\int_0^\tau \int_0^1 \rho^{-1}\kappa^2(\theta)\rho_x\theta_x\theta_\tau \,\mathrm{d}x\,\mathrm{d}s\right| &\leqslant \epsilon X(\tau) + C\int_0^\tau \int_0^1 (1+\theta^q)\rho_x^2|\kappa(\theta)\theta_x|^2 \,\mathrm{d}x\,\mathrm{d}s \\
&\leqslant C\int_0^\tau \left(\max[(1+\theta^q)|\kappa(\theta)\theta_x|^2]\int_0^1 \rho_x^2 \,\mathrm{d}x\right)\mathrm{d}s \\
&\quad + \epsilon X(\tau) \\
&\leqslant \epsilon X(\tau) + C\int_0^\tau \max[(1+\theta^q)|\kappa(\theta)\theta_x|^2]\,\mathrm{d}s
\end{aligned}$$

and

$$\begin{aligned}
\left|4\int_0^\tau \int_0^1 \rho^{-1}r^{-3}\kappa^2(\theta)\theta_x\theta_\tau \,\mathrm{d}x\,\mathrm{d}s\right| &\leqslant \epsilon X(\tau) + C\int_0^\tau \int_0^1 (1+\theta^q)|\kappa(\theta)\theta_x|^2 \,\mathrm{d}x\,\mathrm{d}s \\
&\leqslant \epsilon X(\tau) + C\int_0^\tau \max[(1+\theta^q)|\kappa(\theta)\theta_x|^2]\,\mathrm{d}s.
\end{aligned}$$

Combining the above three estimates, we can obtain

$$J_2 \geqslant \frac{1}{2}\tilde{Y}(\tau) - C - 2\epsilon X(\tau) - C\int_0^\tau \max[(1+\theta^q)|\kappa(\theta)\theta_x|^2]\,\mathrm{d}s.$$

Now, we estimate the term J_3 as follows:

$$\begin{aligned}|J_3| &= \left|\int_0^\tau\int_0^1 R\theta(r^2u)_x r^{-4}\kappa(\theta)\theta_\tau\,\mathrm{d}x\,\mathrm{d}s\right| \\ &= \left|\int_0^\tau\int_0^1 2Rr^{-5}\rho^{-1}\theta u\kappa(\theta)\theta_\tau\,\mathrm{d}x\,\mathrm{d}s + \int_0^\tau\int_0^1 R\theta r^{-2}u_x\kappa(\theta)\theta_\tau\,\mathrm{d}x\,\mathrm{d}s\right| \\ &\leqslant \epsilon X(\tau) + C\int_0^\tau\int_0^1 \theta^2u^2\kappa(\theta)\,\mathrm{d}x\,\mathrm{d}s + C\int_0^\tau\int_0^1 \theta^2u_x^2\kappa(\theta)\,\mathrm{d}x\,\mathrm{d}s \\ &\leqslant \epsilon X(\tau) + C\Theta^{q+2}(\tau) + C.\end{aligned}$$

For the term J_4, we treat it as follows:

$$\begin{aligned}|J_4| &= \left|\int_0^\tau\int_0^1 \mu u_x^2\kappa(\theta)\theta_\tau\,\mathrm{d}x\,\mathrm{d}s\right| \\ &\leqslant \epsilon X(\tau) + C\int_0^\tau\int_0^1 u_x^4\kappa(\theta)\,\mathrm{d}x\,\mathrm{d}s \\ &\leqslant \epsilon X(\tau) + C(1+\Theta^q(\tau))\int_0^\tau\int_0^1 u_x^4\,\mathrm{d}x\,\mathrm{d}s \\ &\leqslant \epsilon X(\tau) + C(1+\Theta^q(\tau))(1+\Theta^{(3-q)+}(\tau) + Z^{\frac{1}{2}}(\tau)),\end{aligned}$$

where we have used (3.17) and

$$\sup_{s\in[0,\tau]}|u(1,s)| \leqslant \sup_{(x,s)\in[0,1]\times[0,\tau]}|u(x,s)| \leqslant C\sup_{s\in[0,\tau]}(\|u\|_{L_x^2}^2 + \|u_x\|_{L_x^2}^2) \leqslant C(1+Z^{\frac{1}{2}}(\tau))$$

in the last step. For the term J_5, it is easy to see that

$$\begin{aligned}|J_5| &= \left|\int_0^\tau\int_0^1 2\mu\rho^{-2}r^{-6}u^2\kappa(\theta)\theta_\tau\,\mathrm{d}x\,\mathrm{d}s\right| \\ &\leqslant \epsilon X(\tau) + C\int_0^\tau\int_0^1 u^4\kappa(\theta)\,\mathrm{d}x\,\mathrm{d}s \\ &\leqslant \epsilon X(\tau) + C\Theta^q(\tau) + C,\end{aligned}$$

due to

$$\begin{aligned}\int_0^\tau\int_0^1 u^4\,\mathrm{d}x\,\mathrm{d}s &= \int_0^\tau \|u\|_{L_x^\infty}^2\left(\int_0^1 u^2\,\mathrm{d}x\right)\mathrm{d}s \leqslant C\int_0^\tau \|u\|_{L_x^\infty}^2\,\mathrm{d}s \\ &\leqslant C\int_0^\tau (\|u\|_{L_x^2}^2 + \|u_x\|_{L_x^2}^2)\,\mathrm{d}s \leqslant C.\end{aligned}$$

Without loss of generality, we assume $\Theta(\tau) \geqslant 1$. Inserting the estimates for J_1, $J_2, \cdots, J_5$ into (3.24), we can obtain (note that $(3-q)_+ \leqslant 2$ since $q \geqslant 1$)

$$\begin{aligned} X(\tau)+\tilde{Y}(\tau) \leqslant\; & 5\epsilon X(\tau)+C\Theta^{q+2}(\tau)+CZ^{\frac{1}{2}}(\tau)+C\Theta^{q}(\tau)Z^{\frac{1}{2}}(\tau)+C \\ & +C\int_0^\tau \max[(1+\theta^q)|\kappa(\theta)\theta_x|^2]\,\mathrm{d}s. \end{aligned} \tag{3.25}$$

Now, we estimate the last term in (3.25). From (3.23), we see

$$\begin{aligned} \left|[\rho r^4\kappa(\theta)\theta_x]_x\right|^2 & \leqslant |C_\nu\theta_\tau|^2+|R\rho\theta(r^2u)_x|^2+|\mu r^4\rho u_x^2|^2+\left|\frac{2\mu u^2}{\rho r^2}\right|^2 \\ & \leqslant C(\theta_\tau^2+\theta^2u^2+\theta^2u_x^2+u_x^4+u^4), \end{aligned}$$

so we have

$$\begin{aligned} & \int_0^\tau\int_0^1 (1+\theta^q)\left|[\rho r^4\kappa(\theta)\theta_x]_x\right|^2\,\mathrm{d}x\,\mathrm{d}s \\ \leqslant\; & C\int_0^\tau\int_0^1 (1+\theta^q)(\theta_\tau^2+\theta^2u^2+\theta^2u_x^2+u_x^4+u^4)\,\mathrm{d}x\,\mathrm{d}s \\ \leqslant\; & CX(\tau)+C+C\Theta^{q+2}(\tau)+CZ^{\frac{1}{2}}(\tau)+C\Theta^q(\tau)Z^{\frac{1}{2}}(\tau). \end{aligned}$$

Multiplying (3.25) by $C+1$, then adding the resulted inequality to the above estimate, one gets

$$\begin{aligned} & X(\tau)+\tilde{Y}(\tau)+\int_0^\tau\int_0^1 (1+\theta^q)\left|[\rho r^4\kappa(\theta)\theta_x]_x\right|^2\,\mathrm{d}x\,\mathrm{d}s \\ \leqslant\; & 5\epsilon X(\tau)+C\Theta^{q+2}(\tau)+CZ^{\frac{1}{2}}(\tau)+C\Theta^q(\tau)Z^{\frac{1}{2}}(\tau)+C \\ & +C\int_0^\tau \max[(1+\theta^q)|\kappa(\theta)\theta_x|^2]\,\mathrm{d}s. \end{aligned} \tag{3.26}$$

Using the fact $\|f\|^2_{L^\infty(0,1)} \leqslant 2\int_0^1 |ff_x|\,\mathrm{d}x$ if $f|_{x=0,1}=0$, we have

$$\begin{aligned} \int_0^\tau \max[(1+\theta^q)|\kappa(\theta)\theta_x|^2]\,\mathrm{d}s & \leqslant C(1+\Theta^q(\tau))\int_0^\tau \max|\rho r^4\kappa(\theta)\theta_x|^2\,\mathrm{d}s \\ & \leqslant C(1+\Theta^q(\tau))\int_0^\tau\int_0^1 |\rho r^4\kappa(\theta)\theta_x|\cdot|[\rho r^4\kappa(\theta)\theta_x]_x|\,\mathrm{d}s \\ & \leqslant \frac{1}{2}\int_0^\tau\int_0^1 (1+\theta^q)\left|[\rho r^4\kappa(\theta)\theta_x]_x\right|^2\,\mathrm{d}x\,\mathrm{d}s \\ & \quad +C(1+\Theta^{2q}(\tau))\int_0^\tau\int_0^1 \kappa(\theta)\theta_x^2\,\mathrm{d}x\,\mathrm{d}s \\ & \leqslant \frac{1}{2}\int_0^\tau\int_0^1 (1+\theta^q)\left|[\rho r^4\kappa(\theta)\theta_x]_x\right|^2\,\mathrm{d}x\,\mathrm{d}s \\ & \quad +C(1+\Theta^{2q+1}(\tau)+\Theta^{2q+(3-q)_+}(\tau)), \end{aligned}$$

where we have used Lemma 3.8 in the last step. With the help of this estimate, (3.26) can be further improved as

$$X(\tau)+\tilde{Y}(\tau)+\frac{1}{2}\int_0^\tau\int_0^1(1+\theta^q)\left|[\rho r^4\kappa(\theta)\theta_x]_x\right|^2\,\mathrm{d}x\,\mathrm{d}s$$
$$\leqslant 5\epsilon X(\tau)+C\Theta^{2q+1}(\tau)+C\Theta^{2q+(3-q)_+}(\tau)+CZ^{\frac{1}{2}}(\tau)+C\Theta^q(\tau)Z^{\frac{1}{2}}(\tau)+C.$$

Taking supremum of the above estimate with respect to $\tau\in[0,T]$ gives

$$\begin{aligned}X(T)+Y(T)\leqslant&5\epsilon X(T)+C\Theta^{2q+1}(T)+C\Theta^{2q+(3-q)_+}(T)\\&+CZ^{\frac{1}{2}}(T)+C\Theta^q(T)Z^{\frac{1}{2}}(T)+C,\end{aligned}\tag{3.27}$$

here, we have used the fact $q+2\leqslant 2q+1$. In virtue of (3.21) and Young's inequality, it is easy to see that

$$\Theta^{2q+1}(T)\leqslant C+CY^{\frac{2q+1}{2q+3}}(T)\leqslant C+\epsilon Y(T),\tag{3.28}$$

$$\Theta^{2q+(3-q)_+}(T)\leqslant C+CY^{\frac{2q+(3-q)_+}{2q+3}}(T)\leqslant C+\epsilon Y(T),\tag{3.29}$$

$$\begin{aligned}\Theta^q(T)Z^{\frac{1}{2}}(T)&\leqslant CZ^{\frac{1}{2}}(T)+CY^{\frac{q}{2q+3}}(T)Z^{\frac{1}{2}}(T)\\&\leqslant CZ^{\frac{1}{2}}(T)+\epsilon Y(T)+CZ^{\frac{2q+3}{2q+6}}(T).\end{aligned}\tag{3.30}$$

By choosing ϵ sufficiently small, it follows from (3.27) and (3.28)–(3.30) that

$$X(T)+Y(T)\leqslant C+CZ^{\alpha}(T),\ \alpha:=\max\left\{\frac{2q+3}{2q+6},\frac{1}{2}\right\}.$$

Thus, we obtain the desired estimate.

Now, we are going to estimate $Z(T)$, namely, the $L^\infty L^2$-norm for u_{xx}.

Lemma 3.10 *Under the same assumptions as Lemma* 3.1, *it holds that*

$$\sup_{\tau\in[0,T]}\int_0^1(u_x^2+u_{xx}^2+u_\tau^2)\,\mathrm{d}x+\int_0^T\int_0^1u_{\tau x}^2\,\mathrm{d}x\,\mathrm{d}\tau\leqslant C.\tag{3.31}$$

In particular, this estimate, (3.21) *and Lemma 3.9 imply that*

$$X(T)+Y(T)+\Theta(T)\leqslant C.\tag{3.32}$$

Proof Differentiating the momentum equation of (2.13) with respect to τ gives

$$u_{\tau\tau}+2rup_x-2ru[\rho\mu(r^2u)_x]_x+r^2p_{x\tau}-r^2[\rho\mu(r^2u)_x]_{x\tau}=0.$$

Taking the inner product of the above equation with u_τ over $[0,1]$, we get

$$
\begin{aligned}
&\frac{1}{2}\frac{\mathrm{d}}{\mathrm{d}\tau}\int_0^1 u_\tau^2\,\mathrm{d}x+\int_0^1\left(r^2p_{x\tau}-r^2[\rho\mu(r^2u)_x]_{x\tau}\right)u_\tau\,\mathrm{d}x\\
&=-\int_0^1 2rup_xu_\tau\,\mathrm{d}x+\int_0^1 2ru[\rho\mu(r^2u)_x]_xu_\tau\,\mathrm{d}x.
\end{aligned}\tag{3.33}
$$

Using Lemma 3.6, it is easy to see that

$$
\begin{aligned}
\left|\int_0^1 2rup_xu_\tau\,\mathrm{d}x\right|&=\left|\int_0^1 2ru(R\rho\theta+a\rho^\gamma)_xu_\tau\,\mathrm{d}x\right|\\
&\leqslant C\int_0^1(|\rho_x|+|\theta_x|+|\rho_x||\theta|)|u||u_\tau|\,\mathrm{d}x\\
&\leqslant C\int_0^1|u_\tau|^2\,\mathrm{d}x+C\|u\|_{L_x^\infty}^2\int_0^1(|\rho_x|^2+|\theta_x|^2)\,\mathrm{d}x\\
&\leqslant C\int_0^1|u_\tau|^2\,\mathrm{d}x+C\|u\|_{L_x^\infty}^2(1+Y(T))+C\|u\|_{L_x^\infty}^2\|\theta\|_{L_x^\infty}^2.
\end{aligned}\tag{3.34}
$$

Note that

$$
[\rho(r^2u)_x]_x=\left[\frac{2u}{r}+\rho r^2u_x\right]_x=\frac{4u_x}{r}-\frac{2u}{\rho r^4}+\rho_xr^2u_x+\rho r^2u_{xx},
$$

so by (3.22) we have

$$
\begin{aligned}
\left|\int_0^1 2ru[\rho\mu(r^2u)_x]_xu_\tau\,\mathrm{d}x\right|&\leqslant C\int_0^1(|uu_xu_\tau|+|u^2u_\tau|+|u\rho_xu_xu_\tau|+|uu_{xx}u_\tau|)\,\mathrm{d}x\\
&\leqslant C\int_0^1|u_\tau|^2\,\mathrm{d}x+C\|u\|_{L_x^\infty}^2\int_0^1(|u|^2+|u_x|^2)\,\mathrm{d}x\\
&\quad+C\|u\|_{L_x^\infty}^2\|u_x\|_{L_x^\infty}^2\int_0^1|\rho_x|^2\,\mathrm{d}x+\|u\|_{L_x^\infty}^2\int_0^1|u_{xx}|^2\,\mathrm{d}x\\
&\leqslant C\int_0^1|u_\tau|^2\,\mathrm{d}x+C\|u\|_{L_x^\infty}^2(1+\tilde{Z}^{\frac{1}{2}}(\tau)+\tilde{Z}^{\frac{3}{4}}(\tau)+\tilde{Z}(\tau))\\
&\leqslant C\int_0^1|u_\tau|^2\,\mathrm{d}x+C\|u\|_{L_x^\infty}^2(1+\tilde{Z}(\tau)).
\end{aligned}\tag{3.35}
$$

Using the boundary condition (2.15), we have

$$
\begin{aligned}
&\int_0^1 r^2\left(p_{x\tau}-[\rho\mu(r^2u)_x]_{x\tau}\right)u_\tau\,\mathrm{d}x\\
&=-\int_0^1 2r(\rho r^2)^{-1}\left(p_\tau-[\rho\mu(r^2u)_x]_\tau\right)u_\tau\,\mathrm{d}x-\int_0^1 r^2\left(p_\tau-[\rho\mu(r^2u)_x]_\tau\right)u_{\tau x}\,\mathrm{d}x\\
&=:K_1+K_2.
\end{aligned}\tag{3.36}
$$

For the term K_1, note that

$$
\begin{aligned}
p_\tau - [\rho\mu(r^2u)_x]_\tau =& R\rho_\tau\theta + R\rho\theta_\tau + a\gamma\rho^{\gamma-1}\rho_\tau \\
& - \frac{2\mu u_\tau}{r} + \frac{2\mu u^2}{r^2} - \mu\rho_\tau r^2 u_x - \mu\rho 2ruu_x - \mu\rho r^2 u_{\tau x},
\end{aligned}
$$

so we can obtain (using $\rho_\tau = -\rho^2(r^2u)_x$)

$$
\begin{aligned}
|K_1| \leqslant& \epsilon \int_0^1 |u_{\tau x}|^2 \,\mathrm{d}x + C\int_0^1 |u_\tau|^2 \,\mathrm{d}x \\
& + C\int_0^1 (|\theta\rho_\tau|^2 + |\theta_\tau|^2 + |\rho_\tau|^2 + u^4 + |\rho_\tau u_x|^2 + |uu_x|^2) \,\mathrm{d}x \\
\leqslant& \epsilon \int_0^1 |u_{\tau x}|^2 \,\mathrm{d}x + C\int_0^1 |u_\tau|^2 \,\mathrm{d}x + C\Theta^2(T)\int_0^1 |(r^2u)_x|^2 \,\mathrm{d}x + C\int_0^1 |\theta_\tau|^2 \,\mathrm{d}x \\
& + C(1+\tilde{Z}(\tau))\int_0^1 (|(r^2u)_x|^2 + u^2) \,\mathrm{d}x
\end{aligned} \tag{3.37}
$$

and

$$
\begin{aligned}
K_2 =& -\int_0^1 r^2[R\rho_\tau\theta + R\rho\theta_\tau + a\gamma\rho^{\gamma-1}\rho_\tau - \frac{2\mu u_\tau}{r} + \frac{2\mu u^2}{r^2} - \mu\rho_\tau r^2 u_x - \mu\rho 2ruu_x]u_{\tau x} \,\mathrm{d}x \\
& + \int_0^1 \rho\mu r^4 u_{\tau x}^2 \,\mathrm{d}x \\
\geqslant& -C\int_0^1 (|\theta\rho_\tau|^2 + |\theta_\tau|^2 + |u_\tau|^2 + |\rho_\tau|^2 + u^4 + |\rho_\tau u_x|^2 + |uu_x|^2) \,\mathrm{d}x \\
& + (C^{-1} - \epsilon)\int_0^1 u_{\tau x}^2 \,\mathrm{d}x \\
\geqslant& -C\int_0^1 |u_\tau|^2 \,\mathrm{d}x - C\Theta^2(T)\int_0^1 |(r^2u)_x|^2 \,\mathrm{d}x - C\int_0^1 |\theta_\tau|^2 \,\mathrm{d}x \\
& + (C^{-1} - \epsilon)\int_0^1 |u_{\tau x}|^2 \,\mathrm{d}x \\
& - C(1+\tilde{Z}(\tau))\int_0^1 (|(r^2u)_x|^2 + u^2) \,\mathrm{d}x.
\end{aligned} \tag{3.38}
$$

Now, combining the estimates (3.33)–(3.38) and choosing ϵ small enough, we get

$$
\frac{1}{2}\frac{\mathrm{d}}{\mathrm{d}\tau}\int_0^1 u_\tau^2 \,\mathrm{d}x + \frac{1}{2C}\int_0^1 |u_{\tau x}|^2 \,\mathrm{d}x \leqslant C\int_0^1 |u_\tau|^2 \,\mathrm{d}x + \varphi(\tau)\tilde{Z}(\tau) + \psi(\tau), \tag{3.39}
$$

where

$$
\varphi(\tau) := C\|u\|_{L_x^\infty}^2 + C\int_0^1 (|(r^2u)_x|^2 + u^2) \,\mathrm{d}x
$$

and

$$\begin{aligned}\psi(\tau) :=& C\|u\|_{L_x^\infty}^2(1+Y(T)) + C\|u\|_{L_x^\infty}^2\|\theta\|_{L_x^\infty}^2\\&+C\int_0^1 |\theta_\tau|^2\,\mathrm{d}x + C(1+\Theta^2(T))\int_0^1 |(r^2u)_x|^2\,\mathrm{d}x + C.\end{aligned}$$

Since u satisfies the second equation of (2.13), namely,

$$\mu\rho r^4 u_{xx} = u_\tau + r^2(a\rho^\gamma + R\rho\theta)_x - 4\mu r u_x + 2\mu(\rho r^2)^{-1}u - \mu r^4\rho_x u_x,$$

one sees that

$$\begin{aligned}\tilde{Z}(\tau) = \int_0^1 u_{xx}^2\,\mathrm{d}x &\leqslant C\int_0^1 u_\tau^2\,\mathrm{d}x + C\int_0^1 (\rho_x^2+\theta_x^2+\theta^2\rho_x^2+u^2+u_x^2+\rho_x^2u_x^2)\,\mathrm{d}x\\&\leqslant C\int_0^1 u_\tau^2\,\mathrm{d}x + C\|\theta_x\|_{L_x^2}^2 + C\|u_x\|_{L_x^2}^2 + C\|u_x\|_{L_x^\infty}^2 + C\Theta^2(T) + C\\&\leqslant C\int_0^1 u_\tau^2\,\mathrm{d}x + C\tilde{Z}^{\frac12}(\tau) + C\tilde{Z}^{\frac34}(\tau) + C\|\theta_x\|_{L_x^2}^2 + C\Theta^2(T) + C\\&\leqslant C\int_0^1 u_\tau^2\,\mathrm{d}x + \frac12\tilde{Z}(\tau) + CY(T) + C\Theta^2(T) + C,\end{aligned}$$

from which we get

$$\tilde{Z}(\tau) \leqslant C\int_0^1 u_\tau^2\,\mathrm{d}x + CY(T) + C\Theta^2(T) + C. \tag{3.40}$$

Hence, (3.39) can be further written as

$$\begin{aligned}\frac{\mathrm{d}}{\mathrm{d}\tau}\int_0^1 u_\tau^2\,\mathrm{d}x + \int_0^1 |u_{\tau x}|^2\,\mathrm{d}x \leqslant& C(1+\varphi(\tau))\int_0^1 |u_\tau|^2\,\mathrm{d}x\\&+C\varphi(\tau)(1+Y(T)+\Theta^2(T)) + \psi(\tau).\end{aligned} \tag{3.41}$$

Note that, by Lemma 3.1, Lemma 3.7 and Lemma 3.9, we have

$$\int_0^T \varphi(\tau)\,\mathrm{d}\tau \leqslant \int_0^T (\|u\|_{L_x^2}^2 + \|u_x\|_{L_x^2}^2)\,\mathrm{d}\tau \leqslant C$$

and

$$\int_0^T \psi(\tau)\,\mathrm{d}\tau \leqslant C + CX(T) + CY(T) + C\Theta^2(T),$$

then (3.41) and Gronwall's inequality give

$$\sup_{\tau\in[0,T]}\int_0^1 u_\tau^2\,\mathrm{d}x + \int_0^T\int_0^1 u_{\tau x}^2\,\mathrm{d}x\mathrm{d}\tau \leqslant C + CX(T) + CY(T) + C\Theta^2(T). \tag{3.42}$$

Taking the supremum with respect to $\tau \in [0,T]$ in (3.40) and using (3.42), one has

$$Z(T) \leqslant C + CX(T) + CY(T) + C\Theta^2(T) \leqslant C + CX(T) + CY(T) \leqslant C + CZ^{\alpha}(T),$$

where we have used (3.21) and Young's inequality in the second inequality, and Lemma 3.9 in the last inequality. Since $\alpha < 1$, Young's inequality gives

$$Z(T) \leqslant C + \frac{1}{2}Z(T),$$

then this estimate and (3.42) lead to the desired estimates. This ends the proof of the lemma.

Lemma 3.11 *Under the same assumptions as Lemma* 3.1, *it holds that*

$$\theta \geqslant C^{-1} > 0.$$

Proof Multiplying (3.7) by $\frac{1}{\theta^2}$, and taking

$$\frac{(r^4\kappa(\theta)\theta_x\rho)_x}{\theta^2} = -\left[\rho r^4\kappa(\theta)\left(\frac{1}{\theta}\right)_x\right]_x + \frac{2\rho r^4\kappa(\theta)\theta_x^2}{\theta^3},$$

into account, we can obtain

$$C_\nu\left(\frac{1}{\theta}\right)_\tau \leqslant \frac{R\rho(r^2u)_x}{\theta} + \left[\rho r^4\kappa(\theta)\left(\frac{1}{\theta}\right)_x\right]_x. \tag{3.43}$$

If we multiply (3.43) by $\chi\left(\frac{1}{\theta}\right)^{\chi-1}$ ($\chi \geqslant 4$ integer), then integrate over $[0,1]\times[0,\tau]$ and employ a partial integration with respect to x, we thus find by using (2.15)

$$C_\nu\|\frac{1}{\theta}\|_\chi^\chi \leqslant C_\nu\|\frac{1}{\theta_0}\|_\chi^\chi + Cr\int_0^\tau \max_{x\in[0,1]}\{(r^2u)_x\}\|\frac{1}{\theta}\|_\chi^\chi \,\mathrm{d}s, \tag{3.44}$$

where C is independent of χ. Note that Lemma 3.7 and Lemma 3.10 imply

$$\int_0^\tau |\max_{x\in[0,1]}\{(r^2u)_x\}|\,\mathrm{d}s \leqslant \int_0^\tau\int_0^1 ([(r^2u)_x]^2 + [(r^2u)_{xx}]^2)\,\mathrm{d}s \leqslant C.$$

Hence, an application of Gronwall's inequality to (3.44) yields

$$\|\frac{1}{\theta(\tau)}\|_\chi^\chi \leqslant C\mathrm{e}^{C(1+T)\chi}\|\frac{1}{\theta_0}\|_\chi^\chi, \ \forall\, \tau \in [0,T],$$

which, by taking the $\left(\frac{1}{\chi}\right)^{th}$ power and then passing to the limit as $\chi \to \infty$, implies

$$\|\frac{1}{\theta(\tau)}\|_{L^\infty} \leqslant C\|\frac{1}{\theta_0}\|_{L^\infty} \leqslant C,$$

for any $\tau \in [0,T]$. Thus, Lemma 3.11 is proved.

The last lemma below gives the $L^\infty L^2$-norm estimate for θ_{xx}.

Lemma 3.12 *Under the same assumptions as Lemma* 3.1, *it holds that*

$$\int_0^1 (\theta_\tau^2+\theta_{xx}^2)\,\mathrm{d}x+\int_0^T\int_0^1 \theta_{x\tau}^2\,\mathrm{d}x\,\mathrm{d}\tau\leqslant C.$$

Proof Differentiating the equation (3.7) with respect to τ gives that

$$\begin{aligned}C_\nu\theta_{\tau\tau}&+R\rho_\tau\theta(r^2u)_x+R\rho\theta_\tau(r^2u)_x+R\rho\theta(r^2u)_{x\tau}=4\mu r^3\rho u u_x^2+\mu r^4\rho_\tau u_x^2\\&+2\mu r^4u_xu_{x\tau}+(r^4\kappa(\theta)\theta_x\rho)_{x\tau}+\frac{4\mu u u_\tau}{\rho r^2}-\frac{2\mu u^2\rho_\tau}{\rho^2r^2}-\frac{4\mu u^3}{\rho r^3}.\end{aligned}$$

Multiplying the above equation by θ_τ, and integrating the resulted equation with respect to space variable x over $[0,1]$, one gets

$$\begin{aligned}\frac{C_\nu}{2}\frac{\mathrm{d}}{\mathrm{d}\tau}\int_0^1\theta_\tau^2\,\mathrm{d}x=&-\int_0^1(R\rho_\tau\theta(r^2u)_x+R\rho\theta_\tau(r^2u)_x)\theta_\tau\,\mathrm{d}x\\&-\int_0^1 R\rho\theta(r^2u)_{x\tau}\theta_\tau\,\mathrm{d}x\\&+\int_0^1(4\mu r^3\rho u u_x^2+\mu r^4\rho_\tau u_x^2+2\mu r^4u_xu_{x\tau})\theta_\tau\,\mathrm{d}x\\&+\int_0^1\left(\frac{4\mu u u_\tau}{\rho r^2}-\frac{2\mu u^2\rho_\tau}{\rho^2r^2}-\frac{4\mu u^3}{\rho r^3}\right)\theta_\tau\,\mathrm{d}x\\&+\int_0^1(r^4\kappa(\theta)\theta_x\rho)_{x\tau}\theta_\tau\,\mathrm{d}x\\=&\sum_{i=1}^5 L_i.\end{aligned}\tag{3.45}$$

So, we have to estimate the five terms $L_1,\cdots,L_5$ in (3.45). Note that the following estimates have already been obtained in the previous lemmas:

$$C^{-1}\leqslant\rho,r,\theta\leqslant C,\ \|\rho_x\|_{L^\infty L^2}\leqslant C,$$

$$\|(u,u_x,u_{xx},u_\tau,\theta,\theta_x)\|_{L^\infty L^2}+\|(u,u_x,\theta)\|_{L^\infty_{x,\tau}}+\|u_{x\tau}\|_{L^2L^2}\leqslant C.$$

Hence, the terms on the right-hand side of (3.45) can be estimated as follows:

$$\begin{aligned}|L_1|&\leqslant C\|\rho_\tau\theta(r^2u)_x\|_{L_x^\infty}\int_0^1|\theta_\tau|\,\mathrm{d}x+C\|\rho(r^2u)_x\|_{L_x^\infty}\int_0^1|\theta_\tau|^2\,\mathrm{d}x\\&\leqslant C\int_0^1|\theta_\tau|^2\,\mathrm{d}x+C,\end{aligned}$$

$$\begin{aligned}|L_2|&=\left|\int_0^1 R\rho\theta\left(\frac{2u_\tau}{\rho r}-\frac{2u\rho_\tau}{\rho^2 r}-\frac{2u^2}{\rho r^2}+2ruu_x+r^2u_{x\tau}\right)\theta_\tau\,\mathrm{d}x\right|\\&\leqslant C+C\int_0^1\theta_\tau^2\,\mathrm{d}x+C\int_0^1 u_{x\tau}^2\,\mathrm{d}x,\end{aligned}$$

$$|L_3|+|L_4| \leqslant C+C\int_0^1 \theta_\tau^2\,\mathrm{d}x+C\int_0^1 u_{x\tau}^2\,\mathrm{d}x,$$

$$\begin{aligned}
L_5 &= -\int_0^1 (r^4\kappa(\theta)\theta_x\rho)_\tau\theta_{x\tau}\,\mathrm{d}x \\
&= -\int_0^1 r^4\kappa(\theta)\rho\theta_{x\tau}^2\,\mathrm{d}x-\int_0^1 (4r^3\kappa(\theta)\theta_x\rho u+r^4\kappa'(\theta)\theta_x\theta_\tau\rho+r^4\kappa(\theta)\theta_x\rho_\tau)\theta_{x\tau}\,\mathrm{d}x \\
&\leqslant -\int_0^1 r^4\kappa(\theta)\rho\theta_{x\tau}^2\,\mathrm{d}x+\epsilon\int_0^1 \theta_{x\tau}^2\,\mathrm{d}x+C\int_0^1 \theta_\tau^2\theta_x^2\,\mathrm{d}x+C \\
&\leqslant -\int_0^1 r^4\kappa(\theta)\rho\theta_{x\tau}^2\,\mathrm{d}x+\epsilon\int_0^1 \theta_{x\tau}^2\,\mathrm{d}x+C\|\theta_\tau\|_{L^\infty}^2+C \\
&\leqslant -\int_0^1 r^4\kappa(\theta)\rho\theta_{x\tau}^2\,\mathrm{d}x+2\epsilon\int_0^1 \theta_{x\tau}^2\,\mathrm{d}x+C\int_0^1 \theta_\tau^2\,\mathrm{d}x+C.
\end{aligned}$$

Inserting these estimates into (3.45) leads to

$$\int_0^1 \theta_\tau^2\,\mathrm{d}x+\int_0^T\int_0^1 \theta_{x\tau}^2\,\mathrm{d}x\,\mathrm{d}\tau \leqslant C+\int_0^T\int_0^1 \theta_\tau^2\,\mathrm{d}x\,\mathrm{d}\tau.$$

Consequently, Gronwall's inequality shows that

$$\int_0^1 \theta_\tau^2\,\mathrm{d}x+\int_0^T\int_0^1 \theta_{x\tau}^2\,\mathrm{d}x\,\mathrm{d}\tau \leqslant C.$$

On the other hand, from (3.7), we have

$$|\rho r^4\kappa(\theta)\theta_{xx}|^2 \leqslant C+C(\theta_\tau^2+\theta_x^4+\rho_x^2\theta_x^2),$$

which, by Sobolev inequalities, gives that

$$\begin{aligned}
\int_0^1 \theta_{xx}^2\,\mathrm{d}x &\leqslant C+\int_0^1 \theta_\tau^2\,\mathrm{d}x+C\int_0^1 \theta_x^4\,\mathrm{d}x+C\int_0^1 \rho_x^2\theta_x^2\,\mathrm{d}x \\
&\leqslant C+\frac{1}{2}\int_0^1 \theta_{xx}^2\,\mathrm{d}x.
\end{aligned}$$

Thus, there holds that

$$\int_0^1 \theta_{xx}^2\,\mathrm{d}x \leqslant C.$$

Hence, we complete the proof of Lemma 3.12.

Now, in virtue of the above lemmas, we can present the proof of Theorems 2.1–2.2.

Proof With the help of Lemmas 3.1–3.12, Theorem 2.2 can be proved quite easily in terms of short time existence, *a priori* estimates, and a continuity argument. Indeed, the short time existence of the unique strong solution (ρ, u, θ, a) to the FBVP

(2.13)-(2.17) under the assumptions of Theorem 2.2 can be shown by the standard argument as in [12]. By the *a priori* estimates established in Lemmas 3.1-3.12 for (ρ, u, θ, a) and a continuity argument, we show that it is indeed a global strong solution to the FBVP (2.13)-(2.17) satisfying (2.20)–(2.19).

The proof of Theorem 2.1 follows from Theorem 2.2 and the coordinates transform (2.11)–(2.12). The proofs thus are completed.

Acknowledgements The authors would like to thank the referee's valuable suggestions which improved the presentation considerably. Also, the authors would like to thank Prof. Z. P. Xin and Prof. S. Jiang for their value discussions. This work was supported by "The Institute of Mathematical Sciences, The Chinese University of Hong Kong" and "Department of Mathematics, City University of Hong Kong". The first author is supported by NSFC under Grant No. 11471323, 11271052. The second author is supported by NSFC under Grant No. 11271052. The third author is supported by NSFC under Grant No. 11201185 and Zhejiang Provincial Natural Science Foundation of China under Grant No. LQ12A01013.

References

[1] Amosov A A, Zlotnik A A. Global generalized solutions of the equations of the one-dimensional motion of a viscous heat-conducting gas [J]. Soviet Math. Dokl., 1989, 38: 1–5.

[2] Antontsev S N, Kazhikhov A V, Monakhov V N. Boundary value probtems in mechanics of nonhomogeneous fluids [M]. Amsterdam, New York: North-Holland, 1990.

[3] Chen G Q, Wang D H. Global solutions of nonlinear magnetohydrodynamics with large initial data [J]. J. Differential Equations, 2002, 182: 344–376.

[4] Chen G Q, Kratka M. Global solutions to the Navier-Stokes equations for compressible heat-conducting flow with symmetry and free boundary [J]. Commun. Partial Differ. Equations, 2002, 27: 907–943.

[5] Cho Y, Jin B J. Blow-up of viscous heat-conducting compressible flows [J]. J. Math. Anal. Appl., 2006, 320: 819–826.

[6] Dafermos C M. Global smooth solutions to the initial-boundary value problem for the equations of one-dimensional nonlinear thermoviscoelasticity [J]. SIAM J. Math. Anal., 1982, 13: 397–408.

[7] Dafermos C M, Hsiao L. Global smooth thermomechanical processes in onedimensional nonlinear thermoviscoelasticity [J]. Nonlinear Anal. TMA, 1982, 6: 435–454.

[8] Ding S J, Wen H Y, Yao L, et al. Global sperically symmetric classical solution to compressible Navier-Stokes equations with large initial data and vacuum [J]. SIAM J.

Math. Anal., 2012, 44: 1257–1278.

[9] Liu H X, Yang T, Zhao H J, et al. One-dimensional compressible Navier-Stokes equations with temperature dependent transport coefficients and large Data [J]. SIAM J. Math. Anal., 2014, 46(3): 2185–2228.

[10] Guo Z H, Li H L, Xin Z P. Lagrange structure and dynamics for solutions to the spherically symmetric compressible Navier-Stokes equations [J]. Commun. Math. Phys., 2012, 309: 371–412.

[11] Jiang S. Global spherically symmetric solutins to the equations of a viscous polytropic ideal gas in an exterior domain [J]. Commun. Math. Phys., 1996, 178: 339–374.

[12] Jiang S, Xin Z P, Zhang P. Global weak solutions to 1D compressible isentropic Navier-Stokes with density-dependent viscosity [J]. Methods Appl. Anal., 2005, 12: 239–252.

[13] Jiang S. On initial boundary value problems for a viscous, heat-conducting, one-dimensional real gas [J]. J. Differential Equations, 1994, 110: 157–181.

[14] Jiang S. On the asymptotic behavior of the motion of a viscous, heat-conducting, onedimensional real gas [J]. Math. Z., 1994, 216: 317–336.

[15] Jiang S. Remarks on the global existence in the dynamics of a viscous, heat-conducting, one-dimensional gas [C]. Proc. of the Workshop on Qualitative Aspects and Appl. of Nonlinear Evol. Eqns., Beirao da Veiga H, Li Ta-Tsien (Eds.), Singapore: World Scientific Publ., 1994, 156–162.

[16] Jiang S. Global smooth solutions to the equations of a viscous, heat-conducting, one-dimensional gas with density-dependent viscosity [J]. Math. Nachr., 1998, 190: 169–183.

[17] Kazhikhov A V,Shelukhin V V. Unique global solution with respect to time of initial-boundary value problems for one-dimensional equations of a viscous gas [J]. J. Appl. Math. Mech., 1977, 41: 273–282.

[18] Kazhikhov A V. To a theory of boundary value problems for equations of one-dimensional non-stationary motion of viscous heat-conduction gases [C]. Boundary Value Problems for Hydrodynamical Equations, Inst. Hydrodynamics, Siberian Branch Akad., USSR, 1981, 50: 37–62 (Russian).

[19] Kawashima S, Nishida T. Global solutions to the initial value problem for the equations of one-dimensional motion of viscous polytropic gases [J]. J. Math. Kyoto Univ., 1981, 21: 825–837.

[20] Kawohl B. Global existence of large solutions to initial boundary value problems for a viscous, heat-conducting, one-dimensional real gas [J]. J. Differential Equations, 1985, 58: 76–103.

[21] Lieberman G M. Second order parabolic partial differential equations [M]. River Edge (NJ), World Scientific, 1996.

[22] Luo T, Xin Z P, Yang T. Interface behavior of compressible Navier-Stokes equations with vacuum [J]. SIAM J. Math. Anal., 2000, 31: 1175–1191.

[23] Matsumura A, Nishida T. The initial value problem for the equations of motion of viscous

and heat-conductive gases [J]. J. Math. Kyoto Univ., 1980, 20: 67–104.

[24] Matsumura A, Nishida T. The initial value problem for the equations of motion of compressible viscous and heat-conductive fluids [J]. Proc. Japan Acad. Ser. A, 1979, 55: 337–342.

[25] Matsumura A, Nishida T. Initial boundary value problems for the equations of motion of general fluids [C]. Computing Meth. in Appl. Sci. and Engin., Glowinski V R, Lions J L (Eds.), Amsterdam: North-Holland, 1982, 389–406.

[26] Matsumura A, Nishida T. Initial boundary value problems for the equations of motion of compressible viscous and heat-conductive fluids [J]. Commun. Math. Phys., 1983, 89: 445–464.

[27] Nagasawa T. On the one-dimensional motion of the polytropic ideal gas non-fixed on the boundary [J]. J. Diffenrential Equations, 1986, 65: 49–67.

[28] Nagasawa T. On the outer pressure problem of the one-dimensional polytropic ideal gas [J]. Japan J. Appl. Math., 1988, 5: 53–85.

[29] Nagasawa T. On the one-dimensional free boundary problem for the heat conductive compressible viscous gas [C]. Lecture Notes in Num. Appl. Anal., Mimura M, Nishida T (Eds.), Vol. 10, Tokyo: Kinoktmiya/North-Holland, 1989, 83–99.

[30] Nikolaev V B. On the solvability of mixed problem for one-dimensional axisymmetrical viscous gas flow [J]. Dinamicheskie zadachi Mekhaniki sploshnoj sredy, 63 Sibirsk. Otd. Acad. Nauk SSSR, Inst. Gidrodinamiki, 1983 (Russian).

[31] Rozanova O. Blow up of smooth solutions to the compressible Navier-Stokes equations with the data highly decreasing at infinity [J]. J. Differential Equations, 2008, 245: 1762–1774.

[32] Valli A. Mathematical results for compressible flows [C]. Mathematical Topics in Fluid Mechanics, Rodrigues J F, Sequeira A (Eds.) Pitman Research Notes in Math. Set. 274, New York, John Wiley, 1992, 193–229.

[33] Valli A, Zajaczkowski W M. Navier-Stokes Equations for compressible fluids: global existence and qualitative properties of the solutions in the general case [J]. Commun. Math. Phys., 1986, 103: 259–296.

[34] Xin Z P. Blow up of smooth solutions to the compressible Navier-Stokes equation with compact density [J]. Comm. Pure Appl. Math., 1998, 51: 229–240.

[35] Xin Z P, Yan W. On blowup of classical solutions to the compressible Navier-Stokes equations [J]. Commun. Math. Phys., 2012, 321: 529–541.

[36] Yashima H F, Benabidallah R. Equation à symétrie sphérique d'un gaz visqueux et calorifère avec la surface libre [J]. Annali Mat. Pura Applicata, 1995, CLXVIII: 75–117.

[37] Yashima H F, Benabidallah R. Unicite' de la solution de l'équation monodimensionnelle ou à symétrie sphérique d'un gaz visqueux et calorifère [J]. Rendi. del Circolo Mat. di Palermo, 1993, XLII: 195–218.

[38] Zel'dovich Y B, Raizer Y P. Physics of shock waves and high-temperature hydrodynamic

phenomena [M]. Academic Press, New York, 1967.
[39] Zhang J J, Guo B L. Global existence of solution for thermally radiative magnetohydrodynamic equations with the displacement current [J]. J. Math. Phys., 2013, 54: 013519-1–013519-18.
[40] Zhang J W, Xie F. Global solution for a one-dimensional model problem in thermally radiative magnetohydrodynamics [J]. J. Differential Equations, 2008, 245: 1853–1882.
[41] Zhu C J. Asymptotic behavior of compressible Navier-Stokes equations with density dependent viscosity and vacuum [J]. Commun. Math. Phys., 2010, 293: 279–299.

Solutions of Ginzburg-Landau Theory for Atomic Fermi Gases near the BCS-BEC Crossover*

Chen Shuhong (陈淑红) and Guo Boling (郭柏灵)

Abstract We are concerned with a time-dependent Ginzburg-Landau equations come from the superfluid atomic Fermi-gases near the Feshbach resonance from the fermion-boson model. We obtain the global existence and uniqueness of solutions to the TDGL equations near the BCS-BEC crossover.

Keywords Global existence; Uniqueness; Time-dependent Ginzburg-Landau theory; BCS-BEC crossover

1 Introduction

In this paper we study the global existence and uniqueness to solutions of the coupled time-dependent Ginzburg-Landau equations for atomic Fermi gases near the BCS-BEC crossover with the form of:

$$
\begin{aligned}
-\mathrm{i}du_t =& \left(-\frac{dg^2+1}{U}+a\right)u + g\left[a+d(2\nu-2\mu)\right]\varphi_B + \frac{c}{4m}u_{xx} \\
&+\frac{g}{4m}(c-d)\varphi_{Bxx} - b|u+g\varphi_B|^2(u+g\varphi_B),
\end{aligned} \tag{1.1}
$$

$$
\mathrm{i}\frac{\partial\varphi_B}{\partial t} = -\frac{g}{U}u + (2\nu-2\mu)\varphi_B - \frac{1}{4m}\varphi_{Bxx}, \tag{1.2}
$$

$$
u(x,0)=u_0(x),\quad \varphi_B(x,0)=\varphi_{B0}(x),\qquad x\in\Omega=[-L,L], \tag{1.3}
$$

$$
u(x,t)=u(x+L,t),\quad \varphi_B(x,t)=\varphi_B(x+L,t),\qquad x\in\Omega=[-L,L], \tag{1.4}
$$

where the fermion-pair field $u(x,t)$ and the condensed boson field $\varphi_B(x,t)$ are L-periodic, $t\geqslant 0$, μ is a real coefficient standing for the chemical potential, 2ν standing

* Acta Math. Appl. Sin. Engl. Ser., 2015, 31(3): 665–676. DOI: 10.1007/s10255-015-0492-2.

for the threshold energy of the Feshbach resonance is a real constant, g being the coupling constant in the Feshbach resonance is a real coefficient, d is generally complex, and in the BCS limit d can be considered to be pure imaginary, in the BEC region, the imaginary of part of d vanished; $U > 0, a, b$ and c are all real numbers.

The Ginzburg-Landau(GL) theory has played an important role in the history of superconductivity research, because it has captured almost every unique feature that the superfluid exhibits macroscopically[3], though its mathematical framework is simple. A two-component time dependent Ginburg-Landau (TDGL)[8] equations from the fermion-boson model (double-channel model)-one of the microscopic models-has so far been extensively studied, since it can describe a large variety of nonequilibrium dynamics observed in physical systems[1]. In the superfluid atomic Fermi gases near the Feshbach resonance the strong attractive interaction is realized between fermion atoms, which can cause the BCS-BEC crossover [7][9][13]. To our knowledge the GL theory has not yet been fully studied in the BCS-BEC crossover regime except for a few pioneering works [5][11] and recent related ones [2][4][12][14] in the single-component fermion systems (single channel model). Though rich research has been done in the TDGL equation derived from the single-channel model. There without any results be got for a two-component time-dependent Ginzburg-Landau(TDGL)[8]. In this paper we are concerned with the global existence and uniqueness theory for a TDGL equations for superfluid atomic Fermi-gases near the Feshbach resonance showing the BCS-BEC crossover[8].

Let L^k_{per} and $H^k_{\text{per}}, k = 1, 2, \cdots$ denote the Hilbert and Sobolev spaces of L-periodic, complex-valued functions endowed with the usual L^2 inner product $(f, g) = \displaystyle\int f\bar{g}dx$ and norms $\|f\|_{L^2} = \sqrt{(f,g)}, \|f\|_{H^k} = \left(\sum_0^k \|\partial^j u/\partial x^j\|\right)^{\frac{1}{2}}$. Here $\bar{g}$ denotes the complex conjugate of g. For brevity we write $\|f\|_{L^2} = \|f\|$ and denote the L^p-norm by $\|f\|_p = \left(\displaystyle\int |f|^p \mathrm{d}x\right)^{1/p}$.

Let $d = d_r + id_i$, and $|d|^2 = d_r^2 + d_i^2$, we can state the main result.

Theorem 1.1 *Assume that* $U, c, d_i, m, b > 0$. *Then for every* $u_0(x) \in H^1_{per}, \varphi_{B0}(x) \in H^1_{per}, t \geqslant 0$, *the initial value problem* (1.1)–(1.4) *has an unique global solution*

$$
\begin{aligned}
& u(x,t) + g\varphi_B(x,t) \in L^\infty\left(0,T;H^1_{per}\right) \cap L^4\left(0,T;L^4_{per}\right), \\
& u(x,t) \in L^\infty\left(0,T;H^1_{per}\right), \quad \varphi_B(x,t) \in L^\infty\left(0,T;H^1_{per}\right) \\
\text{and} \quad & u_t(x,t) \in L^\infty\left(0,T;L^2_{per}\right), \quad \varphi_{Bt}(x,t) \in L^\infty\left(0,T;L^2_{per}\right)
\end{aligned}
$$

2 Approximate Solution and a Priori Estimate

We use Galerkin method to establish a priori estimate in this section. At first, we definite weak solution to (1.1)–(1.4).

Definition 2.1 The vector functions $u(x,t)+g\varphi_B(x,t) \in L^\infty(0,T;H^1_{per}) \cap L^4(0,T;L^4_{per})$, $u(x,t) \in L^\infty(0,T;H^1_{per})$, $\varphi_B(x,t) \in L^\infty(0,T;H^1_{per})$ and $u_t(x,t) \in L^\infty(0,T;L^2_{per})$, $\varphi_{Bt}(x,t) \in L^\infty(0,T;L^2_{per})$ are called weak solutions to (1.1)–(1.4), if for any vector valued test function $\psi \in H^1_{per} \bigcap L^\infty(\Omega)$ and for every $[t_1,t_2] \subset [0,T)$, the following equations hold

$$
\begin{aligned}
-\mathrm{i}d(u_t,\psi) &= \left(-\frac{dg^2+1}{U}+a\right)(u,\psi)+g\left[a+d(2\nu-2\mu)\right](\varphi_B,\psi)\\
&-\frac{c}{4m}(u_x,\psi_x)-\frac{g}{4m}(c-d)(\varphi_{Bx},\psi_x)-b(|u+g\varphi_B|^2(u+g\varphi_B),\psi),
\end{aligned}
\tag{2.1}
$$

$$
\mathrm{i}(\varphi_{Bt},\psi) = -\frac{g}{U}(u,\psi)+(2\nu-2\mu)(\varphi_B,\psi)+\frac{1}{4m}(\varphi_{Bx},\psi_x), \tag{2.2}
$$

$$
(u(x,0),\psi)=(u_0(x),\psi), \quad (\varphi_B(x,0),\psi)=(\varphi_{B0}(x),\psi), \tag{2.3}
$$

$$
(u(x+L,t),\psi)=(u(x,t),\psi), \quad (\varphi_B(x+L,t),\psi)=(\varphi_B(x,t),\psi), \tag{2.4}
$$

where

$$
(u,v)=\int_0^L u(x)\bar{v}(x)\mathrm{d}x.
$$

Now we use Galerkin method to establish the existence of approximation solutions to (1.1)–(1.4).

Let $w_j(x)(j=1,2,\cdots)$ be the unit eigenfunctions satisfying the equations

$$
\Delta w_j+\lambda_j w_j=0, \quad j=1,2,\cdots, \quad w_j \in H^1_0(\Omega)\cap L^4(\Omega), \tag{2.5}
$$

and $\lambda_j(j=1,2,\cdots)$ the corresponding eigenvalues different from each other. $\{w_j(x)\}$ consists of the orthogonal base in L^2.

Denote the approximate solutions to (1.1)–(1.4) by $u_l(x,t)$ and $\varphi_{Bl}(x,t)$ in the form

$$
u_l(x,t)=\sum_{j=1}^{l}\alpha_{jl}(t)w_j(x), \qquad \varphi_{Bl}(x,t)=\sum_{j=1}^{l}\beta_{jl}(t)w_j(x),
$$

where $\alpha_{jl}(t),\beta_{jl}(t)(t\in R^+)(j=1,2,\cdots,l;l=1,2,\cdots)$ are vector-valued functions satisfying the following system of ordinary differential equations of first order

$$-id(u_{lt},w_j)=\left(-\frac{dg^2+1}{U}+a\right)(u_l,w_j)+g\left[a+d(2\nu-2\mu)\right](\varphi_{Bl},w_j)$$
$$-\frac{c}{4m}(u_{lx},w_{jx})-\frac{g}{4m}(c-d)(\varphi_{Blx},w_{jx})$$
$$-b(|u_l+g\varphi_{Bl}|^2(u_l+g\varphi_{Bl}),w_j), \tag{2.6}$$

$$i(\varphi_{Blt},w_j)=-\frac{g}{U}(u_l,w_j)+(2\nu-2\mu)(\varphi_{Bl},w_j)+\frac{1}{4m}(\varphi_{Blx},w_{jx}), \tag{2.7}$$

$$(u_l(x+L,t),\psi)=(u_l(x,t),\psi),\quad (\varphi_{Bl}(x+L,t),\psi)=(\varphi_{Bl}(x,t),\psi),$$

and

$$u_l(x,0)=u_{0l}(x)\in[w_1,w_2,\cdots,w_l],\quad \varphi_{Bl}(x,0)=\varphi_{B0l}(x)\in[w_1,w_2,\cdots,w_l]$$
$$u_{0l}=\sum_{i=1}^{l}\alpha_{il}(t)w_i(x)\to u_0,\varphi_{B0l}=\sum_{i=1}^{l}\beta_{il}(t)w_i(x)\to\varphi_{B0}\quad\text{in}\quad H_0^1(\Omega)\cap L^4(\Omega). \tag{2.8}$$

Thus following from the local existence theory for nonlinear ordinary differential equations, the problem (2.6)–(2.8) has solutions on $[0,t_m]$. And the following priori estimates ensure that they are global on $[0,T]$.

Obviously there holds

$$\int_R u_{lt}w_j\mathrm{d}x=\alpha'_{jl}(t),\quad \int_r \varphi_{Blt}w_j\mathrm{d}x=\beta'_{jl}(t).$$

By following priori estimate, we know that there exists a global solution to (2.6)–(2.8) on $[0,T]$.

Lemma 2.1 *Assume that $(u_{0l},\varphi_{B0l})\in(H^1_{per},H^1_{per})$. Then for $U,c,d_i,m,b>0$, we have the following priori uniformly estimates*

$$\|u_l+g\varphi_{Bl}\|^2\leqslant\left(\|u_{0l}+g\varphi_{B0l}\|^2+\|\varphi_{B0l}\|^2\right)e^{C_1t},$$
$$\|\varphi_{Bl}\|^2\leqslant\left(\|u_{0l}+g\varphi_{B0l}\|^2+\|\varphi_{B0l}\|^2\right)e^{C_1t},$$
$$\|u_l(x,t)\|^2\leqslant 2(1+g^2)\left(\|u_{0l}+g\varphi_{B0l}\|^2+\|\varphi_{B0l}\|^2\right)e^{C_1t},$$

for $C_1=\max\left\{\frac{|g|}{U}+\frac{|g|}{|d|U}+\frac{2ad_i}{|d|^2}-\frac{2d_i}{|d|^2U},\ \frac{|g|}{U}+\frac{|g|}{|d|U}\right\}$ and

$$\|u_l+g\varphi_{Bl}\|_4\leqslant C_2(u_{0l},\varphi_{B0l},a,d,U,g).$$

Proof Combining the equation (2.6) with (2.7), we get

$$\int(u_l+g\varphi_{Bl})_t\overline{w_j}\mathrm{d}x=\frac{ig}{dU}\int\varphi_{Bl}\overline{w_j}\mathrm{d}x+\left(\frac{ia}{d}-\frac{i}{dU}\right)\int(u_l+g\varphi_{Bl})\overline{w_j}\mathrm{d}x$$

$$-\frac{ic}{4md}\int(u_l+g\varphi_{Bl})_x\overline{w_j}_x\mathrm{d}x-\frac{ib}{d}\int|u_l+g\varphi_{Bl}|^2(u_l+g\varphi_{Bl})\overline{w_j}\mathrm{d}x. \quad (2.9)$$

Multiplying the equation(2.9) by $(\alpha_{jl}+g\beta_{jl})$ and equation (2.7) by β_{jl}, respectively, and then summing up the products for $j=1,2,\cdots,l$, and then taking the real part of the result inequality to obtain:

$$\begin{aligned}\frac{1}{2}\frac{\mathrm{d}}{\mathrm{d}t}\|u_l+g\varphi_{Bl}\|^2=&Re\left(\frac{ig}{dU}\int\varphi_{Bl}(\overline{u_l+g\varphi_{Bl}})\mathrm{d}x\right)\\&+\left(\frac{ad_i}{|d|^2}-\frac{d_i}{|d|^2U}\right)\int|u_l+g\varphi_{Bl}|^2\mathrm{d}x\\&-\frac{cd_i}{4m|d|^2}\int|(u_l+g\varphi_{Bl})_x|^2\mathrm{d}x-\frac{bd_i}{|d|^2}\int|u_l+g\varphi_{Bl}|^4\mathrm{d}x,\end{aligned} \quad (2.10)$$

$$\frac{1}{2}\frac{\mathrm{d}}{\mathrm{d}t}\|\varphi_{Bl}\|^2=Re\left(-\frac{ig}{U}\int\varphi_{Bl}(\overline{u_l+g\varphi_{Bl}})\mathrm{d}x\right). \quad (2.11)$$

From (2.10) and (2.11), and noting that $U,c,d_i,m,b>0$, we have

$$\begin{aligned}&\frac{1}{2}\frac{\mathrm{d}}{\mathrm{d}t}\left(\|u_l+g\varphi_{Bl}\|^2+\|\varphi_{Bl}\|^2\right)\\\leqslant&\left(\frac{|g|}{2U}+\frac{|g|}{2|d|U}+\frac{ad_i}{|d|^2}-\frac{d_i}{|d|^2U}\right)\int|u_l+g\varphi_{Bl}|^2\mathrm{d}x+\left(\frac{|g|}{2U}+\frac{|g|}{2|d|U}\right)\int|\varphi_{Bl}|^2\mathrm{d}x\\\leqslant&\frac{1}{2}C_1\left(\|u_l+g\varphi_{Bl}\|^2+\|\varphi_{Bl}\|^2\right),\end{aligned} \quad (2.12)$$

for $C_1=\max\left\{\frac{|g|}{U}+\frac{|g|}{|d|U}+\frac{2ad_i}{|d|^2}-\frac{2d_i}{|d|^2U},\ \frac{|g|}{U}+\frac{|g|}{|d|U}\right\}$.

$$\text{i.e.,}\quad\frac{\mathrm{d}}{\mathrm{d}t}\left(\|u_l+g\varphi_{Bl}\|^2+\|\varphi_{Bl}\|^2\right)\leqslant C_1\left(\|u_l+g\varphi_{Bl}\|^2+\|\varphi_{Bl}\|^2\right). \quad (2.13)$$

Applying the Gronwall lemma, we can see

$$\|u_l+g\varphi_{Bl}\|^2+\|\varphi_{Bl}\|^2\leqslant\left(\|u_{0l}+g\varphi_{B0l}\|^2+\|\varphi_{B0l}\|^2\right)e^{C_1t}. \quad (2.14)$$

Thus,

$$\|u_l+g\varphi_{Bl}\|^2\leqslant\left(\|u_{0l}+g\varphi_{B0l}\|^2+\|\varphi_{B0l}\|^2\right)e^{C_1t}, \quad (2.15)$$

$$\|\varphi_{Bl}\|^2\leqslant\left(\|u_{0l}+g\varphi_{B0l}\|^2+\|\varphi_{B0l}\|^2\right)e^{C_1t}, \quad (2.16)$$

$$\|u_l\|^2\leqslant2\left(\|u_l+g\varphi_{Bl}\|^2+g^2\|\varphi_{Bl}\|^2\right)\leqslant2(1+g^2)\left(\|u_{0l}+g\varphi_{B0l}\|^2+\|\varphi_{B0l}\|^2\right)e^{C_1t}. \quad (2.17)$$

and then (2.10) gives

$$\|u_l+g\varphi_{Bl}\|_4\leqslant C_2.$$

This proves lemma 2.1.

Lemma 2.2 *Under the assumptions of lemma* 2.1, *we also have a priori estimate*

$$\frac{cd_i}{2m|d|^2}\int_s^t\|(u_l+g\varphi_{Bl})_x\|^2dt\leqslant\|u_l(x,s)+g\varphi_{Bl}(x,s)\|^2$$
$$+2C_1\left(\frac{|g|}{|d|U}+\frac{|ad_i|}{|d|^2}\right)\left(\|u_{0l}+g\varphi_{B0l}\|^2+\|\varphi_{B0l}\|^2\right)\left(e^{C_1t}-e^{C_1s}\right),$$

in particular,

$$\frac{cd_i}{2m|d|^2}\int_0^T\|(u_l+g\varphi_{Bl})_x\|^2dt\leqslant\|u_{0l}+g\varphi_{B0l}\|^2$$
$$+2C_1\left(\frac{|g|}{|d|U}+\frac{|ad_i|}{|d|^2}\right)\left(\|u_{0l}+g\varphi_{B0l}\|^2+\|\varphi_{B0l}\|^2\right)\left(e^{C_1T}-1\right),$$

for $0\leqslant s\leqslant t$ *and* $T\geqslant 0$.

Proof Now, from (2.10), we have

$$\begin{aligned}&\frac{1}{2}\frac{\mathrm{d}}{\mathrm{d}t}\|u_l+g\varphi_{Bl}\|^2\\&\leqslant\left(\frac{|g|}{2|d|U}+\frac{ad_i}{|d|^2}-\frac{d_i}{|d|^2U}\right)\int|u_l+g\varphi_{Bl}|^2\mathrm{d}x\\&\quad+\frac{|g|}{2|d|U}\int|\varphi_{Bl}|^2\mathrm{d}x-\frac{cd_i}{4m|d|^2}\int|(u_l+g\varphi_{Bl})_x|^2\mathrm{d}x\\&\leqslant\left(\frac{|g|}{|d|U}+\frac{|ad_i|}{|d|^2}\right)\left(\|u_{0l}+g\varphi_{B0l}\|^2+\|\varphi_{B0l}\|^2\right)e^{C_1t}-\frac{cd_i}{4m|d|^2}\int|(u_l+g\varphi_{Bl})_x|^2\mathrm{d}x.\end{aligned}$$

Which means that,

$$\begin{aligned}&\frac{\mathrm{d}}{\mathrm{d}t}\|u_l+g\varphi_{Bl}\|^2+\frac{cd_i}{2m|d|^2}\int|(u_l+g\varphi_{Bl})_x|^2\mathrm{d}x\\&\leqslant 2\left(\frac{|g|}{|d|U}+\frac{|ad_i|}{|d|^2}\right)\left(\|u_{0l}+g\varphi_{B0l}\|^2+\|\varphi_{B0l}\|^2\right)e^{C_1t}.\end{aligned}\tag{2.18}$$

Integrating the inequality (2.18) on (s,t) (or $(0,T)$), we induce

$$\begin{aligned}&\frac{cd_i}{2m|d|^2}\int_s^t\|(u_l+g\varphi_{Bl})_x\|^2\mathrm{d}t+\|u_l(x,t)+g\varphi_{Bl}(x,t)\|^2-\|u_l(x,s)+g\varphi_{Bl}(x,s)\|^2\\&\leqslant 2C_1\left(\frac{|g|}{|d|U}+\frac{|ad_i|}{|d|^2}\right)\left(\|u_{0l}+g\varphi_{B0l}\|^2+\|\varphi_{B0l}\|^2\right)\left(e^{C_1t}-e^{C_1s}\right),\end{aligned}\tag{2.19}$$

which, since $\|u_l(x,t)+g\varphi_{Bl}(x,t)\|^2 \geqslant 0$, implies the desired results.

Lemma 2.3 *Under the assumptions of lemma 2.1, we have*

$$\frac{d}{dt}\left(1+\|(u_l+g\varphi_{Bl})_x\|^2\right) \leqslant C_3\left(1+\|(u_l+g\varphi_{Bl})_x\|^2\right)^2,$$

where C_3 depending only on $a,b,c,d,m,U,g,\varphi_{B0l}$ and u_{0l}.

Proof: Using the equation (2.5), we can rewrite the equations (2.6) and (2.7),

$$\begin{aligned}
-id(u_{lt},-\Delta w_j) =& \left(-\frac{dg^2+1}{U}+a\right)(u_l,-\Delta w_j)+g\left[a+d(2\nu-2\mu)\right](\varphi_{Bl},-\Delta w_j)\\
&+\frac{c}{4m}(u_{lxx},-\Delta w_j)+\frac{g}{4m}(c-d)(\varphi_{Blxx},-\Delta w_j)\\
&-b(|u_l+g\varphi_{Bl}|^2(u_l+g\varphi_{Bl}),-\Delta w_j),
\end{aligned} \tag{2.20}$$

$$i(\varphi_{Blt},-\Delta w_j) = -\frac{g}{U}(u_l,-\Delta w_j)+(2\nu-2\mu)(\varphi_{Bl},-\Delta w_j)-\frac{1}{4m}(\varphi_{Blxx},-\Delta w_{jx}), \tag{2.21}$$

Combining (2.20) with (2.21), we get

$$\begin{aligned}
\int(u_l+g\varphi_{Bl})_t\Delta\overline{w_j}\mathrm{d}x =& \frac{ig}{dU}\int\varphi_{Bl}\Delta\overline{w_j}\mathrm{d}x+\left(\frac{ia}{d}-\frac{i}{dU}\right)\int(u_l+g\varphi_{Bl})\Delta\overline{w_j}\mathrm{d}x\\
&+\frac{ic}{4md}\int(u_l+g\varphi_{Bl})_{xx}\Delta\overline{w_j}\mathrm{d}x-\frac{ib}{d}\int|u_l\\
&+g\varphi_{Bl}|^2(u_l+g\varphi_{Bl})\Delta\overline{w_j}\mathrm{d}x.
\end{aligned} \tag{2.22}$$

Making the scalar product of $(\alpha_{jl}+g\beta_{jl})$ with (2.22), and β_{jl} with (2.21), respectively, summing up the resulting products for $j=1,2,\cdots,l$, and then taking the real part of the resulting equations, we get:

$$\begin{aligned}
&\frac{1}{2}\frac{\mathrm{d}}{\mathrm{d}t}\|(u_l+g\varphi_{Bl})_x\|^2\\
=&Re\left(-\frac{ig}{dU}\int\varphi_{Bl}(\overline{u_l+g\varphi_{Bl}})_{xx}\mathrm{d}x\right)+\left(\frac{ad_i}{|d|^2}-\frac{d_i}{|d|^2U}\right)\int|(u_l+g\varphi_{Bl})_x|^2\mathrm{d}x\\
&-\frac{cd_i}{4m|d|^2}\int|(u_l+g\varphi_{Bl})_{xx}|^2\mathrm{d}x-\frac{2bd_i}{|d|^2}\int|u_l+g\varphi_{Bl}|^2|(u_l+g\varphi_{Bl})_x|^2\mathrm{d}x\\
&-Re\left(\frac{ib}{d}\int(u_l+g\varphi_{Bl})^2\left((\overline{u_l+g\varphi_{Bl}})_x\right)^2\mathrm{d}x\right).
\end{aligned} \tag{2.23}$$

$$\frac{1}{2}\frac{\mathrm{d}}{\mathrm{d}t}\|\varphi_{Blx}\|^2 = Re\left(\frac{ig}{U}\int\varphi_{Bl}(\overline{u_l+g\varphi_{Bl}})_{xx}\mathrm{d}x\right). \tag{2.24}$$

From (2.23) and (2.24),

$$\begin{aligned}&\frac{1}{2}\frac{\mathrm{d}}{\mathrm{d}t}\left(\|(u_l+g\varphi_{Bl})_x\|^2+\|\varphi_{Blx}\|^2\right)\\ \leqslant&\left(\frac{|g|}{|d|U}+\frac{|g|}{U}\right)\int|\varphi_{Bl}||(u_l+g\varphi_{Bl})_{xx}|\mathrm{d}x+\left(\frac{ad_i}{|d|^2}-\frac{d_i}{|d|^2U}\right)\int|(u_l+g\varphi_{Bl})_x|^2\mathrm{d}x\\ &-\frac{cd_i}{4m|d|^2}\int|(u_l+g\varphi_{Bl})_{xx}|^2\mathrm{d}x+\left(\frac{|b|}{|d|}-\frac{2bd_i}{|d|^2}\right)\int|u_l+g\varphi_{Bl}|^2|(u_l+g\varphi_{Bl})_x|^2\mathrm{d}x.\end{aligned} \tag{2.25}$$

By Young's inequality,

$$\begin{aligned}&\left(\frac{|g|}{|d|U}+\frac{|g|}{U}\right)\int|\varphi_{Bl}||(u_l+g\varphi_{Bl})_{xx}|\mathrm{d}x\\ \leqslant&\,\varepsilon\|(u_l+g\varphi_{Bl})_{xx}\|^2+\frac{1}{\varepsilon}\left(\frac{|g|}{|d|U}+\frac{|g|}{U}\right)^2\|\varphi_{Bl}\|^2,\end{aligned} \tag{2.26}$$

and

$$\int|u_l+g\varphi_{Bl}|^2|(u_l+g\varphi_{Bl})_x|^2\mathrm{d}x\leqslant\|u_l+g\varphi_{Bl}\|_\infty^2\|(u_l+g\varphi_{Bl})_x\|^2. \tag{2.27}$$

Using Gagliardo-Nirenberg inequality:

$$\|f\|_p\leqslant C\|f\|_{H^k}^{\theta}\|f\|_q^{1-\theta},\quad \frac{1}{p}=\theta\left(\frac{1}{2}-k\right)+(1-\theta)\cdot\frac{1}{q}$$

to obtain Agmon's inequality for $p=\infty, k=1, q=2, \theta=\dfrac{1}{2}$:

$$\begin{aligned}\|u_l+g\varphi_{Bl}\|_\infty&\leqslant K\|u_l+g\varphi_{Bl}\|_{H^1}^{1/2}\|u_l+g\varphi_{Bl}\|_2^{1/2}\\ &=K\left(\|u_l+g\varphi_{Bl}\|^2+\|(u_l+g\varphi_{Bl})_x\|^2\right)^{1/4}\|u_l+g\varphi_{Bl}\|_2^{1/2}.\end{aligned} \tag{2.28}$$

From (2.28) and Young's inequality,

$$\begin{aligned}&\|u_l+g\varphi_{Bl}\|_\infty^2\|(u_l+g\varphi_{Bl})_x\|^2\\ \leqslant&\,K^2\left(\|u_l+g\varphi_{Bl}\|+\|(u_l+g\varphi_{Bl})_x\|\right)\|u_l+g\varphi_{Bl}\|\|(u_l+g\varphi_{Bl})_x\|^2\\ \leqslant&\,K_1\left(\|(u_l+g\varphi_{Bl})_x\|^2+\|(u_l+g\varphi_{Bl})_x\|\|(u_l+g\varphi_{Bl})_x\|^2\right)\\ \leqslant&\,K_2\left(\|(u_l+g\varphi_{Bl})_x\|^2+\|(u_l+g\varphi_{Bl})_x\|^4\right),\end{aligned} \tag{2.29}$$

where $K_2=K_2(u_{0l},\varphi_{B0l},d,U,a,b,g)$.

From (2.26), (2.29) and lemma 2.1, choosing $0<\varepsilon\leqslant\dfrac{cd_i}{4m|d|^2}$, (2.25) becomes

$$\frac{1}{2}\frac{\mathrm{d}}{\mathrm{d}t}\left(\|(u_l+g\varphi_{Bl})_x\|^2+\|\varphi_{Blx}\|^2\right)\leqslant\frac{1}{2}C_3\left(1+\|(u_l+g\varphi_{Bl})_x\|^2\right)^2, \tag{2.30}$$

where C_3 depending on $u_{0l}, \varphi_{B0l}, a, b, c, d, m, U, g$.

This completes the proof of lemma 2.3.

Lemma 2.4 *Under the assumptions of lemma* 2.1, *we have a priori uniform estimates*

$$\|u_l(x,t)+g\varphi_{Bl}(x,t)\|_{H^1_{per}} \leqslant C_5, \qquad \|\varphi_{Bl}(x,t)\|_{H^1_{per}} \leqslant C_6,$$

for constants C_5, C_6 *depending only on* $u_{0l}, \varphi_{B0l}, a, b, c, d, m, U, g, \tau$.

Proof fix $\widehat{\tau}=\min\left\{1, \frac{\tau}{2}, \sup\left\{t : \|u_l(x,t)+g\varphi_{Bl}(x,t)\|_{H^1_{per}} \leqslant 2\|u_{0l}+g\varphi_{B0l}\|_{H^1_{per}}\right.\right.$ and $\left.\left.\|\varphi_{Bl}(x,t)\|_{H^1_{per}} \leqslant 2\|\varphi_{B0l}\|_{H^1_{per}}\right\}\right\}$.

Notice that $\widehat{\tau}$ depends only on $u_{0l}, \varphi_{B0l}, \widehat{\tau} \in (0,1]$ and that for $t \in [0, \widehat{\tau}]$,

$$\|u_l+g\varphi_{Bl}\|_{H^1_{per}} \leqslant 2\|u_{0l}+g\varphi_{B0l}\|_{H^1_{per}} \quad \text{and} \quad \|\varphi_{Bl}\|_{H^1_{per}} \leqslant 2\|\varphi_{B0l}\|_{H^1_{per}}.$$

Now we only need to show that $\|u_l+g\varphi_{Bl}\|_{H^1_{per}}$ and $\|\varphi_{Bl}\|_{H^1_{per}}$ are also uniformly bounded on $t \geqslant \widehat{\tau}$.

From lemma 2.3, there have

$$\int_s^t \frac{d(1+\|(u_l(x,z)+g\varphi_{Bl}(x,z))_x\|^2)}{1+\|(u_l(x,z)+g\varphi_{Bl}(x,z))_x\|^2} \leqslant C_3 \int_s^t (1+\|(u_l(x,z)+g\varphi_{Bl}(x,z))_x\|^2)\mathrm{d}z$$

for $t \geqslant \widehat{\tau}$ and choose s such that $t-\widehat{\tau} \leqslant s \leqslant t$, then

$$\begin{aligned}&1+\|(u_l(x,t)+g\varphi_{Bl}(x,t))_x\|^2\\ \leqslant& (1+\|(u_l(x,s)+g\varphi_{Bl}(x,s))_x\|^2)\exp\left[C_3\int_s^t(1+\|(u_l(x,z)+g\varphi_{Bl}(x,z))_x\|^2)\mathrm{d}z\right].\end{aligned} \tag{2.31}$$

Using lemma 2.1 and lemma 2.2 imply that

$$\begin{aligned}&\int_s^t (1+\|(u_l(x,z)+g\varphi_{Bl}(x,z))_x\|^2)\mathrm{d}z \leqslant \int_{t-\widehat{\tau}}^t (1+\|(u_l(x,z)+g\varphi_{Bl}(x,z))_x\|^2)\mathrm{d}z\\ \leqslant& \frac{2m|d|^2}{cd_i}\left\{\|u_l(x,t-\widehat{\tau})+g\varphi_{Bl}(x,t-\widehat{\tau})\|^2\right.\\ &\left.+2C_1\left(\frac{|g|}{|d|U}+\frac{|ad_i|}{|d|^2}\right)(\|u_{0l}+g\varphi_{B0l}\|^2+\|\varphi_{B0l}\|^2)\left(e^{C_1t}-e^{C_1(t-\widehat{\tau})}\right)+\frac{cd_i}{2m|d|^2}\widehat{\tau}\right\}\\ \leqslant& \frac{2m|d|^2}{cd_i}\left\{(\|u_{0l}+g\varphi_{B0l}\|^2+\|\varphi_{B0l}\|^2)e^{C_1t}\right.\\ &\left.+2C_1\left(\frac{|g|}{|d|U}+\frac{|ad_i|}{|d|^2}\right)(\|u_{0l}+g\varphi_{B0l}\|^2+\|\varphi_{B0l}\|^2)\left(e^{C_1t}-e^{C_1(t-\widehat{\tau})}\right)+\frac{cd_i}{2m|d|^2}\widehat{\tau}\right\}\\ \equiv& C_4=C_4(u_{0l},\varphi_{B0l},a,b,c,d,m,U,g,\tau).\end{aligned} \tag{2.32}$$

Now (2.31) and (2.32) gives

$$1+\|(u_l(x,t)+g\varphi_{Bl}(x,t))_x\|^2\leqslant\left(1+\|(u_l(x,s)+g\varphi_{Bl}(x,s))_x\|^2\right)\exp(C_3C_4).\tag{2.33}$$

Integrating both sides of (2.33) with respect to s on $t-\widehat{\tau}\leqslant s\leqslant t$, we have

$$\widehat{\tau}\left(1+\|(u_l(x,t)+g\varphi_{Bl}(x,t))_x\|^2\right)\leqslant\int_{t-\widehat{\tau}}^{t}\left(1+\|(u_l(x,s)+g\varphi_{Bl}(x,s))_x\|^2\right)ds\exp(C_3C_4),$$

and using (2.32) once more, we obtain

$$1+\|(u_l(x,t)+g\varphi_{Bl}(x,t))_x\|^2\leqslant\frac{C_4\exp(C_3C_4)}{\widehat{\tau}}\quad\text{for}\quad t\geqslant\widehat{\tau},$$

which implies the uniform bound of $1+\|(u_l(x,t)+g\varphi_{Bl}(x,t))_x\|^2$. Combining this with lemma 2.1 we get that $\|u_l+g\varphi_{Bl}\|_{H^1_{per}}\leqslant C_5$.

Now we proceed to show the uniformly bounded of $\|\varphi_{Bm}\|_{H^1_{per}}$.

From (2.24), we get

$$\begin{aligned}\frac{1}{2}\frac{\mathrm{d}}{\mathrm{d}t}\|(\varphi_{Bl})_x\|^2&\leqslant\left|\frac{ig}{U}\right|\int|(u_l+g\varphi_{Bl})_x||(\varphi_{Bl})_x|\mathrm{d}x\\&\leqslant\frac{|g|}{2U}\|(u_l+g\varphi_{Bl})_x\|^2+\frac{|g|}{2U}\|(\varphi_{Bl})_x\|^2.\end{aligned}$$

Applying Gronwall inequality, we have

$$\|(\varphi_{Bl})_x\|^2\leqslant C_6'.\tag{2.34}$$

Together with the result of lemma 2.1, we complete the proof of lemma 2.4.

Lemma 2.5 *Under the assumptions of lemma* 2.1, *we have*

$$\|(u_l+g\varphi_{Bl})_t\|^2\leqslant C_9,\quad\text{and}\quad\|\varphi_{Blt}\|^2\leqslant C_9,$$

where C_9 is a constant independent of l.

Proof We differentiate the identity (2.9) with respect to t and multiply the result by $(\alpha_{jl}'(t)+g\beta_{jl}'(t))$, and summ up the products for $j=1,2,\cdots,l$ and then take the real part of the result equality, to obtain

$$\begin{aligned}&\frac{1}{2}\frac{\mathrm{d}}{\mathrm{d}t}\|(u_l+g\varphi_{Bl})_t\|^2\\&=Re\left(\frac{ig}{dU}\int\varphi_{Blt}(\overline{u_l+g\varphi_{Bl}})_t\mathrm{d}x\right)+\left(\frac{ad_i}{|d|^2}-\frac{d_i}{|d|^2U}\right)\|(u_l+g\varphi_{Bl})_t\|^2\\&\quad-\frac{cd_i}{4m|d|^2}\|(u_l+g\varphi_{Bl})_{xt}\|^2-\frac{2bd_i}{|d|^2}\int|u_l+g\varphi_{Bl}|^2|(u_l+g\varphi_{Bl})_t|^2\mathrm{d}x\\&\quad-Re\left(\frac{ib}{d}\int(u_l+g\varphi_{Bl})^2\left((\overline{u_l+g\varphi_{Bl}})_t\right)^2\mathrm{d}x\right).\end{aligned}\tag{2.35}$$

And then we differentiate (2.7) with respect to t and multiply the result by $\beta'_{jl}(t)$, by summing up the products for $j = 1, 2, \cdots, l$ and taking the real part of the result equality,

$$\frac{1}{2}\frac{\mathrm{d}}{\mathrm{d}t}\|\varphi_{Blt}\|^2 = Re\left(-\frac{ig}{U}\int \overline{\varphi}_{Blt}(u_l + g\varphi_{Bl})_t \mathrm{d}x\right). \tag{2.36}$$

From (2.35) and (2.36), and noting that $bd_i > 0, cd_i > 0, m > 0$, then by (2.28), lemma 2.1 and lemma 2.4, we get

$$\begin{aligned}
&\frac{1}{2}\frac{\mathrm{d}}{\mathrm{d}t}\left(\|(u_l + g\varphi_{Bl})_t\|^2 + \|\varphi_{Blt}\|^2\right)\\
&\leqslant C_7\left(\|(u_l + g\varphi_{Bl})_t\|^2 + \|\varphi_{Blt}\|^2\right) + \frac{|b|}{|d|}\|u_l + g\varphi_{Bl}\|_\infty^2\|(u_l + g\varphi_{Bl})_t\|^2\\
&\leqslant C_8\left(\|(u_l + g\varphi_{Bl})_t\|^2 + \|\varphi_{Blt}\|^2\right),
\end{aligned} \tag{2.37}$$

where $C_7 = \max\left\{\frac{|g|}{2|d|U} + \frac{|g|}{2U}, \frac{|g|}{2|d|U} + \frac{|g|}{2U} + \frac{ad_i}{|d|^2} - \frac{d_i}{|d|^2U}\right\}$.

Using Gronwall inequality, to give

$$\|(u_l + g\varphi_{Bl})_t\|^2 + \|\varphi_{Blt}\|^2 \leqslant C_9,$$

which implies the desired results.

3 Proof of the Main Result

In this section, we would complete the proof of the main result.

Lemma 3.1 *Assume that D is a bounded domain in $R_x^n \times R_t$, functions $f_l, f \in L^q(D)(1 < q < \infty)$, and*

$$\|f_l\|_{L^q(D)} \leqslant C, \quad f_l \to f \quad in \quad D \quad \text{a.e.}.$$

Then,

$$f_l \to f \quad \text{in} \quad L^q \quad weakly.$$

Proof Assume that a sequence N is a increasing sequence and

$$E_n = \{(x,t)|(x,t) \in D, |f_l(x,t) - f(x,t)| \leqslant 1, \quad \text{if} \quad l \geqslant n\}.$$

There exists a measurable set $\mathcal{E}_N$. The measurable set is increasing with respect to N and $\mathrm{mes}(\mathcal{E}_N) \to \mathrm{mes}(D)$, as $N \to \infty$.

Let $\Phi_n = \{\varphi | \varphi \in L^{q'}(D)\}, \mathrm{supp}\Phi_n = E_n, \Phi = \bigcup_{n\to\infty} \Phi_N$. Then Φ is dense in $L^{q'}(D)$, where $\frac{1}{q} + \frac{1}{q'} = 1$. Now, by Lebesgue theorem, for $\varphi \in \Phi$, we have

$$\int_D \varphi(f_l - f)dx \to 0, \quad \text{as} \quad l \to \infty. \tag{3.1}$$

In fact, for $\varphi \in \Phi_{N_0}$, choosing $l \geqslant N_0$, we get

$$|\varphi(f_l - f)| \leqslant \varphi, \qquad \text{and} \qquad |\varphi(f_l - f)| \to 0$$

almost everywhere. Which means the inequality (3.1) holds.

Noting that Φ is dense in $L^{q'}(D)$, then by (3.1) we get the desire result immediately.

This complete the proof of lemma 3.1. Then we proceed the proof of theorem 1.1.

Proof of Theorem 1.1 According to the conclusion of lemmas 2.1, 2.4 and 2.5, one can see that (2.6)–(2.8) admit a global solution on the interval $[0, T]$. Which also means that $u_l + g\varphi_{Bl}$ is bounded in $L^2(0, T; H^1_{per})$ and $(u_l + g\varphi_{Bl})_t$ is bounded in $L^2(0, T; L^2_{per})$. Thus, $u_l + g\varphi_{Bl}$ is in the bound set of $H^1(0, T; \Omega)$. Via Rellich theorem, there have

$$H^1(0, T; \Omega) \hookrightarrow L^2(\Omega) \quad \text{compact}. \tag{3.2}$$

Thus by a priori estimates of lemma 2.1, 2.4 and 2.5, we know that Galerkin approximate solutions $\{u_l(x, t), \varphi_{Bl}(x, t)\}$ have subsequence (also denoted by u_l and φ_{Bl}), such that

$$u_l(x, t) \to u(x, t) \quad \text{in} \quad L^\infty(0, T; H^1_{per}) \quad \text{weakly star}, \tag{3.3}$$

$$\varphi_{Bl}(x, t) \to \varphi_B(x, t) \quad \text{in} \quad L^\infty(0, T; H^1_{per}) \quad \text{weakly star}, \tag{3.4}$$

$$u_{lt}(x, t) \to u_t(x, t) \quad \text{in} \quad L^\infty(0, T; L^2_{per}) \quad \text{weakly star}, \tag{3.5}$$

$$\varphi_{Blt}(x, t) \to \varphi_{Bt}(x, t) \quad \text{in} \quad L^\infty(0, T; L^2_{per}) \quad \text{weakly star}, \tag{3.6}$$

$$u_l(x, t) + g\varphi_{Bl}(x, t) \to u(x, t) + g\varphi_B(x, t) \quad \text{in} \quad L^4(0, T; L^4_{per}) \quad \text{weakly}, \tag{3.7}$$

and from (3.2) we also have

$$u_l(x, t) + g\varphi_{Bl}(x, t) \to u(x, t) + g\varphi_B(x, t) \quad \text{in} \quad L^2(0, T; \Omega) \quad \text{strongly and a.e.}. \tag{3.8}$$

These imply that

$$|u_l + g\varphi_{Bl}|^2(u_l + g\varphi_{Bl}) \to P \quad \text{in} \quad L^\infty(0, T; L^{\frac{4}{3}}_{per}) \quad \text{weakly star}. \tag{3.9}$$

Then, we are in the present of showing that $P = |u + g\varphi_B|^2(u + g\varphi_B)$. By lemma 3.1, we can get that for $D = (0, T; \Omega)$, $f_l = |u_l + g\varphi_{Bl}|^2(u_l + g\varphi_{Bl})$ and $q = \frac{4}{3}$,

$$P = f = |u + g\varphi_B|^2(u + g\varphi_B).$$

Using (3.3)–(3.9), from the identity (2.6)–(2.7), let $l \to \infty$, we immediately obtain

$$\begin{aligned}
-id(u_t, w_j) &= \left(-\frac{dg^2+1}{U} + a\right)(u, w_j) + g\left[a + d(2\nu - 2\mu)\right](\varphi_B, w_j) - \frac{c}{4m}(u_x, w_{jx}) \\
&\quad - \frac{g}{4m}(c-d)(\varphi_{Bx}, w_{jx}) - b(|u + g\varphi_B|^2(u + g\varphi_B), w_j), \quad j = 1, 2, \cdots \\
i(\varphi_{Bt}, w_j) &= -\frac{g}{U}(u, w_j) + (2\nu - 2\mu)(\varphi_B, w_j) + \frac{1}{4m}(\varphi_{Bx}, w_{jx}), \qquad j = 1, 2, \cdots
\end{aligned}$$

In virtue of the density of $\{w_j\}_{j=1}^{\infty}$ in $H_0^1(\Omega) \cap L^\infty(\Omega)$, we get for $\forall \psi \in H^1(\Omega) \cap L^\infty(\Omega)$

$$\begin{aligned}
-id(u_t, \psi) &= \left(-\frac{dg^2+1}{U} + a\right)(u, \psi) + g\left[a + d(2\nu - 2\mu)\right](\varphi_B, \psi) - \frac{c}{4m}(u_x, \psi_x) \\
&\quad - \frac{g}{4m}(c-d)(\varphi_{Bx}, \psi_x) - b(|u + g\varphi_B|^2(u + g\varphi_B), \psi), \\
i(\varphi_{Bt}, \psi) &= -\frac{g}{U}(u, \psi) + (2\nu - 2\mu)(\varphi_B, \psi) + \frac{1}{4m}(\varphi_{Bx}, \psi_x) \\
(u(x+L, t), \psi) &= (u(x, t), \psi), \quad (\varphi_B(x+L, t), \psi) = (\varphi_B(x, t), \psi).
\end{aligned}$$

Now, it remains to verify that the vector-valued functions $u(x, t)$ and $\varphi_B(x, t)$ satisfy the initial condition.

From (2.3) and (2.4), we see that

$$u_l(x, 0) \to u_0(x) \quad \text{in} \quad L^2(R) \quad \text{weakly}, \tag{3.10}$$

$$\varphi_{Bl}(x, 0) \to \varphi_{B0}(x) \quad \text{in} \quad L^2(R) \quad \text{weakly}, \tag{3.11}$$

then for any $\psi \in L^2$, using (3.8), we get

$$\int u_l(x, 0)\psi \mathrm{d}x = \int u_l(x, t)\psi \mathrm{d}x - \int_0^t \int u_{lt}(x, \tau)\psi \mathrm{d}x\mathrm{d}\tau,$$

$$\int \varphi_{Bl}(x, 0)\psi \mathrm{d}x = \int \varphi_{Bl}(x, t)\psi \mathrm{d}x - \int_0^t \int \varphi_{Blt}(x, \tau)\psi \mathrm{d}x\mathrm{d}\tau.$$

Let $l \to \infty$, the above equalities give

$$\int u_0(x)\psi \mathrm{d}x = \int u\psi \mathrm{d}x - \int_0^t \int u_t(x, \tau)\psi \mathrm{d}x\mathrm{d}\tau = \int u(x, 0)\mathrm{d}x,$$

$$\int \varphi_{B0}\psi \mathrm{d}x = \int \varphi_B\psi \mathrm{d}x - \int_0^t \int \varphi_{Bt}(x,\tau)\psi \mathrm{d}x\mathrm{d}\tau = \int \varphi_B(x,0)\psi \mathrm{d}x.$$

Its mean that the identity (2.3) and (2.4) hold. Then the problem (2.1)–(2.4) admits at least one weak solution. Finally we would prove the uniqueness of the solutions.

Assume that (u, φ_B) and (u^*, φ_B^*) are the two different solutions of (2.1)–(2.4), and for $w = u - u^*, v = \varphi_B - \varphi_B^*$, we have

$$\begin{aligned}(w+gv)_t =& -\frac{i}{Ud}w + \frac{ia}{d}(w+gv) + \frac{ic}{4md}(w+gv)_{xx}\\ &- \frac{ib}{d}\left(|u+g\varphi_B|^2(u+g\varphi_B) - |u^*+g\varphi_B^*|^2(u^*+g\varphi_B^*)\right),\end{aligned} \tag{3.12}$$

$$v_t = \frac{ig}{U}w - i(2\nu - 2\mu)v + \frac{i}{4m}v_{xx}. \tag{3.13}$$

$$w(x,0) = 0, \quad v(x,0) = 0. \tag{3.14}$$

Multiplying the above (3.12) by $(w+gv)$, (3.13) by v and taking the real part,

$$\begin{aligned}&\frac{1}{2}\frac{\mathrm{d}}{\mathrm{d}t}\|w+gv\|^2\\ =& -Re\left(\frac{i}{Ud}\int w(\overline{w+gv})dx\right) + \frac{ad_i}{|d|^2}\|w+gv\|^2 - \frac{cd_i}{4m|d|^2}\|(w+gv)_x\|^2\\ &-Re\left(\frac{ib}{d}\left(|u+g\varphi_B|^2(u+g\varphi_B) - |u^*+g\varphi_B^*|^2(u^*+g\varphi_B^*)\right)(\overline{w+gv})\mathrm{d}x\right),\\ &\frac{1}{2}\frac{\mathrm{d}}{\mathrm{d}t}\|v\|^2 = Re\left(\frac{ig}{U}\int w\bar{v}\mathrm{d}x\right).\end{aligned}$$

Then, from (2.28) and lemma 2.1 and lemma 2.4, gives

$$\begin{aligned}&\frac{1}{2}\frac{\mathrm{d}}{\mathrm{d}t}\left(\|w+gv\|^2 + \|v\|^2\right)\\ \leqslant& \frac{|g|}{|d|U}\int |w||w+gv|\mathrm{d}x + \frac{|g|}{U}\int |w||w+gv|\mathrm{d}x + \left(\frac{ad_i}{|d|^2} - \frac{d_i}{|d|^2U}\right)\|w+gv\|^2\\ &- \frac{cd_i}{4m|d|^2}\|(w+gv)_x\|^2 + \frac{3b}{|d|}\int \max\left(|u+g\varphi_B|^2, |u^*+g\varphi_B^*|^2\right)|w+gv|^2\mathrm{d}x\\ \leqslant& C_{10}\left(\|w+gv\|^2 + \|v\|^2\right),\end{aligned}$$

for $C_{10} = \max\left\{\frac{|g|}{2|d|U} + \frac{|g|}{2U} + \frac{ad_i}{|d|^2} - \frac{d_i}{|d|^2U} + \frac{bK_3}{|d|}, \frac{|g|}{2|d|U} + \frac{|g|}{2U}\right\}$.

Noting that that $w(x,0) = 0$ and $v(x,0) = 0$, by Gronwall inequality, which means that $w + gv = 0$ and $v = 0$. This completes the proof of theorem 1.1.

Acknowledgements This work is supported by the National Natural Science Foundation of China (No: 11201415); Program for New Century Excellent Talents in Fujian Province University (No. JA14191).

References

[1] Aranson I, Kramer L. The would of the complex Ginzburg-Landau equation, Rev. Mod. Phys., 2002, 74: 99–143.

[2] Baranov M A, Petrov D S. Low-energy collective excitations in a superfluid trapped Fermi gas, Phy. Rev. A, 2000, 62: 041601(R).

[3] de Gennes P G. Superconductivity of Metals and Alloys (Addisom-Wesley, Reading, MA,1998); A.A. Abrikosov, Fundamentals of the Theory of Metals (Elsevier-Science Ltd., New York, 1988).

[4] De Palo S, Castelloni C, Dicastro C, Chakraverty B K. Effective action for superconductiors and BCS-Bose crossover, Phys. Rev. B, 1999, 60: 564–573.

[5] Drechsler M, Zwerger W. Crossover from BCS-superconductivity to Bose-condensation, Ann. Phys. 1992, 504(1)1: 15–23.

[6] Henry D. Geometric theory of semilinear parabolic equation, Berlin: Spring-Verlag, 1981.

[7] Holland M, Kokkelmans S J J M F, Chiofalo, M L. Walser, R. Resonance superfluidity in a quantum degenerate Fermi gas, Phys. Rev. Lett., 2001, 87(12): 120406.

[8] Machida M, Koyama T. Time-dependent Ginzburg-Landau theory for atomic Fermigases near the BCS-BEC crossover, Phy. Rev. A, 2006, 74: 033603.

[9] Ohashi Y, Griffin A. BCS-BEC crossover in a gas of Fermi atoms with a Feshbach resonance, Phys. Rev. Lett., 2002, 89: 130402.

[10] Pazy A. Semigroups of linear operators and applications to partial differential equation, Berlin: Spring-Verlag, 1983.

[11] de Melo C A R Sa, Randeria M, Engelbrecht J R. Crossover from BCS to Bose superconductivity: Transition temperature and time-dependent Ginzburg-Landau theory, Phys. Rev. Lett. 1993: 71: 3202–3205.

[12] Tempere J, Wouters M, Devreese J T. Path-integral mean-field description of the vortex state in the BEC-to-BCS crossover, Phys. Rev. A, 2005, 71: 033631.

[13] Timmermans E, Furuya K, Milonni P W, Kerman A K. Prospect of creating a composite Fermi-Bose superfluid, Phys. Lett. A, 2001, 285(3-4): 228–233.

[14] Wouters M, Tempere J, Devreese J T. Path integral formulation of the tunneling dynamics of a superfluid Fermi gas in an optical potential, Phys. Rev. A, 2004, 70: 013616.

Orbital Instability of Standing Waves for the Generalized 3D Nonlocal Nonlinear Schrödinger Equations*

Gan Zaihui (甘在会), Guo Boling (郭柏灵) and Jiang Xin (蒋芯)

Abstract The existence and orbital instability of standing waves for the generalized three dimensional nonlocal nonlinear Schrödinger equations is studied. By defining some suitable functionals and a constrained variational problem, we first establish the existence of standing waves, which relys on the inner structure of the equations under consideration to overcome the drawback that nonlocal terms violate the space-scale invariance. We then show the orbital instability of standing waves. The arguments depend upon the conservation laws of the mass and of the energy.

Keywords Nonlocal Nonlinear Schrödinger Equations; Standing waves; Orbital instability

1 Introduction

In this paper, we study the generalized three-dimensional nonlocal nonlinear Schrödinger equations:

$$\begin{aligned} &i\partial_t E_1 + \Delta E_1 + \left(|E_1|^2 + |E_2|^2 + |E_3|^2\right) E_1 \\ &\qquad + A_1(E_1, E_2, E_3) + A_2(E_1, E_2, E_3) = 0, \end{aligned} \tag{1.1}$$

$$\begin{aligned} &i\partial_t E_2 + \Delta E_2 + \left(|E_1|^2 + |E_2|^2 + |E_3|^2\right) E_2 \\ &\qquad + A_3(E_1, E_2, E_3) + A_4(E_1, E_2, E_3) = 0, \end{aligned} \tag{1.2}$$

$$\begin{aligned} &i\partial_t E_3 + \Delta E_3 + \left(|E_1|^2 + |E_2|^2 + |E_3|^2\right) E_3 \\ &\qquad + A_5(E_1, E_2, E_3) + A_6(E_1, E_2, E_3) = 0, \end{aligned} \tag{1.3}$$

along with the initial data

$$E_1(0, x) = E_{10}(x), E_2(0, x) = E_{20}(x), E_3(0, x) = E_{30}(x). \tag{1.4}$$

* Acta Math. Sci. Ser. B, 2015, 35B(5): 1163—1188. DOI:10.1016/S0252-9602(15)30047-3.

Here,

$$A_1(E_1,E_2,E_3) = -E_2\mathcal{F}^{-1}\left\{\frac{\eta}{|\xi|^2-\delta}\left[\xi_1\xi_3\mathcal{F}(E_2\overline{E_3}-\overline{E_2}E_3)\right.\right.$$
$$\left.\left.-(\xi_1^2+\xi_2^2)\mathcal{F}(E_1\overline{E_2}-\overline{E_1}E_2)+\xi_2\xi_3\mathcal{F}(\overline{E_1}E_3-E_1\overline{E_3})\right]\right\}, \qquad \text{(I-1)}$$

$$A_2(E_1,E_2,E_3) = E_3\mathcal{F}^{-1}\left\{\frac{\eta}{|\xi|^2-\delta}\left[\xi_2\xi_3\mathcal{F}(E_1\overline{E_2}-\overline{E_1}E_2)\right.\right.$$
$$\left.\left.-(\xi_1^2+\xi_3^2)\mathcal{F}(\overline{E_1}E_3-E_1\overline{E_3})+\xi_1\xi_2\mathcal{F}(E_2\overline{E_3}-\overline{E_2}E_3)\right]\right\}, \qquad \text{(I-2)}$$

$$A_3(E_1,E_2,E_3) = -E_3\mathcal{F}^{-1}\left\{\frac{\eta}{|\xi|^2-\delta}\left[\xi_1\xi_2\mathcal{F}(\overline{E_1}E_3-E_1\overline{E_3})\right.\right.$$
$$\left.\left.-(\xi_2^2+\xi_3^2)\mathcal{F}(E_2\overline{E_3}-\overline{E_2}E_3)+\xi_1\xi_3\mathcal{F}(E_1\overline{E_2}-\overline{E_1}E_2)\right]\right\}, \qquad \text{(I-3)}$$

$$A_4(E_1,E_2,E_3) = E_1\mathcal{F}^{-1}\left\{\frac{\eta}{|\xi|^2-\delta}\left[\xi_1\xi_3\mathcal{F}(E_2\overline{E_3}-\overline{E_2}E_3)\right.\right.$$
$$\left.\left.-(\xi_1^2+\xi_2^2)\mathcal{F}(E_1\overline{E_2}-\overline{E_1}E_2)+\xi_2\xi_3\mathcal{F}(\overline{E_1}E_3-E_1\overline{E_3})\right]\right\}, \qquad \text{(I-4)}$$

$$A_5(E_1,E_2,E_3) = -E_1\mathcal{F}^{-1}\left\{\frac{\eta}{|\xi|^2-\delta}\left[\xi_2\xi_3\mathcal{F}(E_1\overline{E_2}-\overline{E_1}E_2)\right.\right.$$
$$\left.\left.-(\xi_1^2+\xi_3^2)\mathcal{F}(\overline{E_1}E_3-E_1\overline{E_3})+\xi_1\xi_2\mathcal{F}(E_2\overline{E_3}-\overline{E_2}E_3)\right]\right\}, \qquad \text{(I-5)}$$

$$A_6(E_1,E_2,E_3) = E_2\mathcal{F}^{-1}\left\{\frac{\eta}{|\xi|^2-\delta}\left[\xi_1\xi_2\mathcal{F}(\overline{E_1}E_3-E_1\overline{E_3})\right.\right.$$
$$\left.\left.-(\xi_2^2+\xi_3^2)\mathcal{F}(E_2\overline{E_3}-\overline{E_2}E_3)+\xi_1\xi_3\mathcal{F}(E_1\overline{E_2}-\overline{E_1}E_2)\right]\right\}, \qquad \text{(I-6)}$$

$\mathcal{F}$ and $\mathcal{F}^{-1}$ denote the Fourier transform and the Fourier inverse transform, respectively ([14-17]), $\eta > 0$ and $\delta \leqslant 0$ are two constants, $(E_1,E_2,E_3)(t,x)$ are complex vector-valued functions from $\mathbb{R}^+\times\mathbb{R}^3$ into $\mathbb{C}^3$, $\overline{E}_i (i=1,2,3)$ denotes the complex conjugate of E_i. Due to rotational invariance of (1.1)–(1.3), let $\mathbf{E}=(E_1,E_2,E_3)$ and $\xi=(\xi_1,\xi_2,\xi_3)$, system (1.1)–(1.3) is equivalent to a vetor-valued nonlinear Schrödinger equations

$$i\mathbf{E}_t+\Delta\mathbf{E}+|\mathbf{E}|^2\mathbf{E}+i(\mathbf{E}\wedge\mathbf{B})=0, \qquad (M\text{-}S\text{-}1)$$
$$\mathbf{B}(\mathbf{E})=\mathcal{F}^{-1}\left[\frac{i\eta}{|\xi|^2-\delta}(\xi\wedge(\xi\wedge\mathcal{F}(\mathbf{E}\wedge\overline{\mathbf{E}})))\right], \qquad (M\text{-}S\text{-}2)$$

where $\wedge$ denotes the exterior product of vector-valued functions, and $\overline{\mathbf{E}}$ the complex conjugate of $\mathbf{E}$. (Indeed, a direct computation implies that equations (1.1)–(1.3) are the componential form of the equations (M-S-1)–(M-S-2). To understand the relationship of all the components E_1,E_2,E_3, we adopt the componential form (1.1)–(1.3) in the present paper.)

Equations (M-S-1)–(M-S-2) arise in the infinite ion acoustic speed limit of the self-generated magnetic field in a cold plasma, **E** denotes a slowly varying complex amplitude of the high-frequency electric field, and **B** the self-generated magnetic field [5, 13, 23, 24]. Due to the gauge invariance $A_j(e^{i\omega t}E_1, e^{i\omega t}E_2, e^{i\omega t}E_3) = e^{i\omega t}A_j(E_1, E_2, E_3)$, $j = 1, 2, 3, 4, 5, 6$, we can study the so-called standing wave solutions of the equations (1.1)–(1.3) in the form $E_i(t, x) = e^{i\omega t}u_i(x)$ $(i = 1, 2, 3)$ with the initial condition (1.4), where $\omega > 0$ is a real constant parameter called frequency and $u_i(x)(i = 1, 2, 3)$ is a complex-valued function. The search for standing waves of equations (1.1)–(1.3) leads to the following nonlinear elliptic equations (1.5)–(1.7):

$$\begin{aligned}
&-\omega u_1 + \Delta u_1 + (|u_1|^2 + |u_2|^2 + |u_3|^2)u_1 - u_2\mathcal{F}^{-1}\left[\frac{\eta\xi_1\xi_3}{|\xi|^2-\delta}\mathcal{F}(u_2\overline{u_3} - \overline{u_2}u_3)\right]\\
&+u_2\mathcal{F}^{-1}\left[\frac{\eta(\xi_1^2+\xi_2^2)}{|\xi|^2-\delta}\mathcal{F}(u_1\overline{u_2} - u_2\overline{u_1})\right] - u_2\mathcal{F}^{-1}\left[\frac{\eta\xi_2\xi_3}{|\xi|^2-\delta}\mathcal{F}(\overline{u_1}u_3 - u_1\overline{u_3})\right]\\
&+u_3\mathcal{F}^{-1}[\frac{\eta\xi_2\xi_3}{|\xi|^2-\delta}\mathcal{F}(u_1\overline{u_2} - \overline{u_1}u_2)] + u_3\mathcal{F}^{-1}\left[\frac{\eta(\xi_1^2+\xi_3^2)}{|\xi|^2-\delta}\mathcal{F}(u_1\overline{u_3} - \overline{u_1}u_3)\right]\\
&+u_3\mathcal{F}^{-1}\left[\frac{\eta\xi_1\xi_2}{|\xi|^2-\delta}\mathcal{F}(u_2\overline{u_3} - \overline{u_2}u_3)\right] = 0,
\end{aligned}\tag{1.5}$$

$$\begin{aligned}
&-\omega u_2 + \Delta u_2 + (|u_1|^2 + |u_2|^2 + |u_3|^2)u_2 - u_3\mathcal{F}^{-1}\left[\frac{\eta\xi_1\xi_2}{|\xi|^2-\delta}\mathcal{F}(\overline{u_1}u_3 - u_1\overline{u_3})\right]\\
&+u_3\mathcal{F}^{-1}\left[\frac{\eta(\xi_2^2+\xi_3^2)}{|\xi|^2-\delta}\mathcal{F}(u_2\overline{u_3} - \overline{u_2}u_3)\right] - u_3\mathcal{F}^{-1}\left[\frac{\eta\xi_1\xi_3}{|\xi|^2-\delta}\mathcal{F}(u_1\overline{u_2} - \overline{u_1}u_2)\right]\\
&+u_1\mathcal{F}^{-1}\left[\frac{\eta\xi_1\xi_3}{|\xi|^2-\delta}\mathcal{F}(u_2\overline{u_3} - \overline{u_2}u_3)\right] + u_1\mathcal{F}^{-1}\left[\frac{\eta(\xi_1^2+\xi_2^2)}{|\xi|^2-\delta}\mathcal{F}(\overline{u_1}u_2 - u_1\overline{u_2})\right]\\
&+u_1\mathcal{F}^{-1}\left[\frac{\eta\xi_2\xi_3}{|\xi|^2-\delta}\mathcal{F}(\overline{u_1}u_3 - u_1\overline{u_3})\right] = 0,
\end{aligned}\tag{1.6}$$

$$\begin{aligned}
&-\omega u_3 + \Delta u_3 + (|u_1|^2 + |u_2|^2 + |u_3|^2)u_3 - u_1\mathcal{F}^{-1}\left[\frac{\eta\xi_2\xi_3}{|\xi|^2-\delta}\mathcal{F}(u_1\overline{u_2} - \overline{u_1}u_2)\right]\\
&+u_1\mathcal{F}^{-1}\left[\frac{\eta(\xi_1^2+\xi_3^2)}{|\xi|^2-\delta}\mathcal{F}(\overline{u_1}u_3 - u_1\overline{u_3})\right] - u_1\mathcal{F}^{-1}\left[\frac{\eta\xi_1\xi_2}{|\xi|^2-\delta}\mathcal{F}(u_2\overline{u_3} - \overline{u_2}u_3)\right]\\
&+u_2\mathcal{F}^{-1}\left[\frac{\eta\xi_1\xi_2}{|\xi|^2-\delta}\mathcal{F}(\overline{u_1}u_3 - u_1\overline{u_3})\right] + u_2\mathcal{F}^{-1}\left[\frac{\eta(\xi_2^2+\xi_3^2)}{|\xi|^2-\delta}\mathcal{F}(\overline{u_2}u_3 - u_2\overline{u_3})\right]\\
&+u_2\mathcal{F}^{-1}\left[\frac{\eta\xi_1\xi_3}{|\xi|^2-\delta}\mathcal{F}(u_1\overline{u_2} - \overline{u_1}u_2)\right] = 0.
\end{aligned}\tag{1.7}$$

For the nonlinear Schrödinger equations without any nonlocal term, there have been many works on the stability results of their standing waves. Berestycki and

Cazenave [1], Grillakis [9], Jones [12], Shatah and Strauss [18], Weinstein [20] and Zhang [22] investigated the instability of solitons. On the other hand, Cazenave and Lions [4], Weinstein [21], Grillakis, Shatah and Strauss [10] studied the stability of the standing waves.

In the study of equations (1.1)–(1.3), we still concentrate on the existence and orbital instability of the standing waves. For the nonlocal nonlinear Schrödinger equations, to our best knowledge, there have been no any works on the existence and instability of the standing waves other than those in our former papers [6, 7, 8], where we studied the similar topic for a general Davey-Stewartson system [6] and a simplified version for the nonlocal nonlinear Schrödinger equations (1.1)–(1.3) [7, 8], in which $(E_1, E_2, E_3) = (E_1, E_2, 0)$, $(\xi_1, \xi_2, \xi_3) = (\xi_1, \xi_2, 0)$. To attain our goal in the present paper, the main difficulty is to deal with the nonlocal terms since the nonlocal terms may violate the space inner-scale invariance, and we are forced to make some additional arguments for them. Fortunately, by defining some suitable functionals and a constrained variational minimization problem, utilizing the monotonicity argument for some defined auxiliary functions to deal with these terms generated by the nonlocal effects, and applying the inner structure of the corresponding elliptic equations (1.5)–(1.7), we first show the existence of standing waves for the equations (1.1)–(1.3). In addition, we establish the orbital instability of the standing waves for the equations under consideration. The arguments of the result rely on the conservation of energy and mass as well as the construction of a suitable invariant manifold of solution flows. However, it should be pointed out that the uniqueness of these ground states for (1.5)–(1.7) is a much different and difficult problem, and we do not intend to discuss it in the present paper.

This paper arranges as follows. In section 2, we give some preliminaries and state the main results. The existence of standing waves with ground states will be established in Section 3. At the last section, we will show the orbital instability of standing waves.

For simplicity, we denote any positive constant by C throughout the present paper.

2 Preliminaries and main results

In this section, we first establish the conservation laws of total mass and total energy. Then we define some functionals, a set and a constrained variational minimization problem. At the end of this section we state the main results of this paper.

2.1 Conservation Laws of the Mass and of the Energy

According to the inner structure of equations (1.1)–(1.3), making some estimates on the nonlocal terms and using the standard contraction mapping theorem, we can establish the local well-posedness in the energy space $H^1(\mathbb{R}^3)\times H^1(\mathbb{R}^3)\times H^1(\mathbb{R}^3)$, the conservation laws of the total mass and of the total energy for the Cauchy problem (1.1)–(1.4).

Lemma 2.1 *The Cauchy problem* (1.1)–(1.4), for $\eta>0$, $\delta\leqslant 0$ and

$$(E_{10}(x),E_{20}(x),,E_{30}(x))\in H^1(\mathbb{R}^3)\times H^1(\mathbb{R}^3)\times H^1(\mathbb{R}^3),$$

has a unique solution

$$(E_1,E_2,E_3)\in C\left([0,T);H^1(\mathbb{R}^3)\times H^1(\mathbb{R}^3)\times H^1(\mathbb{R}^3)\right)$$

for some $T\in(0,+\infty)$ with $T=+\infty$ or $T<+\infty$ and

$$\lim_{t\to T}\left(\|E_1\|_{H^1(\mathbb{R}^3)}+\|E_2\|_{H^1(\mathbb{R}^3)}+\|E_3\|_{H^1(\mathbb{R}^3)}\right)=+\infty.$$

In addition, the total mass and total energy are conserved:

$$\int_{\mathbb{R}^3}(|E_1|^2+|E_2|^2+|E_3|^2)\mathrm{d}x=\int_{\mathbb{R}^3}(|E_{10}|^2+|E_{20}|^2+|E_{30}|^2)\mathrm{d}x, \tag{2.1}$$

$$\begin{aligned}
&\mathcal{H}(E_1,E_2,E_3)\\
=&\int_{\mathbb{R}^3}(|\nabla E_1|^2+|\nabla E_2|^2+|\nabla E_3|^2)\mathrm{d}x-\frac{1}{2}\int_{\mathbb{R}^3}(|E_1|^4+|E_2|^4+|E_3|^4)\mathrm{d}x\\
&-\int_{\mathbb{R}^3}(|E_1|^2|E_2|^2+|E_1|^2|E_3|^2+|E_2|^2|E_3|^2)\mathrm{d}x\\
&-\frac{1}{2}\int_{\mathbb{R}^3}\frac{\eta(\xi_1^2+\xi_2^2)}{|\xi|^2-\delta}(|\mathcal{F}(E_1\overline{E_2}-\overline{E_1}E_2)|^2\mathrm{d}\xi\\
&-\frac{1}{2}\int_{\mathbb{R}^3}\frac{\eta(\xi_1^2+\xi_3^2)}{|\xi|^2-\delta}(|\mathcal{F}(\overline{E_1}E_3-E_1\overline{E_3})|^2\mathrm{d}\xi\\
&-\frac{1}{2}\int_{\mathbb{R}^3}\frac{\eta(\xi_2^2+\xi_3^2)}{|\xi|^2-\delta}(|\mathcal{F}(E_2\overline{E_3}-\overline{E_2}E_3)|^2\mathrm{d}\xi\\
&+\mathrm{Re}\int_{\mathbb{R}^3}\frac{\eta\xi_1\xi_2}{|\xi|^2-\delta}\mathcal{F}(E_2\overline{E_3}-\overline{E_2}E_3)\overline{\mathcal{F}(\overline{E_1}E_3-E_1\overline{E_3})}\mathrm{d}\xi
\end{aligned}$$

$$
\begin{aligned}
&+\mathrm{Re}\int_{\mathbb{R}^3}\frac{\eta\xi_1\xi_3}{|\xi|^2-\delta}\mathcal{F}(E_1\overline{E_2}-\overline{E_1}E_2)\overline{\mathcal{F}(E_2\overline{E_3}-\overline{E_2}E_3)}\mathrm{d}\xi\\
&+\mathrm{Re}\int_{\mathbb{R}^3}\frac{\eta\xi_2\xi_3}{|\xi|^2-\delta}\mathcal{F}(\overline{E_1}E_3-E_1\overline{E_3})\overline{\mathcal{F}(E_1\overline{E_2}-\overline{E_1}E_2)}\mathrm{d}\xi\\
&=\mathcal{H}(E_{10},E_{20},E_{30}).
\end{aligned}\tag{2.2}
$$

To attain the conservation identities (2.1) and (2.2), besides using the standard arguments on the nonlinear Schrödinger equations without any nonlocal terms, we must give some extra discussions on the nonlocal terms, in which we will employ the Parseval identity, some properties of Fourier transform, suitable groupings and potential coupled arguments for these nonlocal terms. In Lemma 2.1 of [11], we established the detail of the proof for (2.1) and (2.2). □

2.2 Variational Structures

For $(u_1,u_2,u_3)\in H_r^1(\mathbb{R}^3)\times H_r^1(\mathbb{R}^3)\times H_r^1(\mathbb{R}^3)$ ($u_i(x)(i=1,2,3)$ is a complex-valued function), we define the following functionals:

$$
\begin{aligned}
S(u_1,u_2,u_3)=&\frac{1}{2}\int_{\mathbb{R}^3}(|\nabla u_1|^2+|\nabla u_2|^2+|\nabla u_3|^2)\mathrm{d}x\\
&+\frac{\omega}{2}\int_{\mathbb{R}^3}(|u_1|^2+|u_2|^2+|u_3|^2)\mathrm{d}x\\
&-\frac{1}{4}\int_{\mathbb{R}^3}(|u_1|^4+|u_2|^4+|u_3|^4)\mathrm{d}x\\
&-\frac{1}{2}\int_{\mathbb{R}^3}(|u_1|^2|u_2|^2+|u_1|^2|u_3|^2+|u_2|^2|u_3|^2)\mathrm{d}x\\
&-\frac{1}{4}\int_{\mathbb{R}^3}A(u_1,u_2,u_3)\mathrm{d}\xi+\frac{1}{2}Re\int_{\mathbb{R}^3}B(u_1,u_2,u_3)\mathrm{d}\xi,
\end{aligned}\tag{2.3}
$$

$$
\begin{aligned}
R(u_1,u_2,u_3)=&\int_{\mathbb{R}^3}(|\nabla u_1|^2+|\nabla u_2|^2+|\nabla u_3|^2)\mathrm{d}x\\
&-\frac{3}{4}\int_{\mathbb{R}^3}(|u_1|^4+|u_2|^4+|u_3|^4)\mathrm{d}x\\
&-\frac{3}{2}\int_{\mathbb{R}^3}(|u_1|^2|u_2|^2+|u_1|^2|u_3|^2+|u_2|^2|u_3|^2)\mathrm{d}x\\
&-\frac{3}{4}\int_{\mathbb{R}^3}A(u_1,u_2,u_3)\mathrm{d}\xi+\frac{1}{2}\delta\int_{\mathbb{R}^3}\frac{A(u_1,u_2,u_3)}{|\xi|^2-\delta}\mathrm{d}\xi\\
&+\frac{3}{2}Re\int_{\mathbb{R}^3}B(u_1,u_2,u_3)\mathrm{d}\xi-\delta Re\int_{\mathbb{R}^3}\frac{B(u_1,u_2,u_3)}{|\xi|^2-\delta}\mathrm{d}\xi,
\end{aligned}\tag{2.4}
$$

where

$$H_r^1(\mathbb{R}^3)\times H_r^1(\mathbb{R}^3)\times H_r^1(\mathbb{R}^3)$$

$$= \{(f_1(x), f_2(x), f_3(x)) \in H^1(\mathbb{R}^3) \times H^1(\mathbb{R}^3) \times H^1(\mathbb{R}^3),$$
$$f_i(x) = f_i(|x|) \ \text{ is a function of } |x| \text{ alone}, i = 1, 2, 3\},$$

$$\begin{aligned} A(u_1, u_2, u_3) =& \frac{\eta}{|\xi|^2 - \delta} \left[(\xi_1^2 + \xi_2^2)(|\mathcal{F}(u_1\overline{u_2} - \overline{u_1}u_2)|^2) \right. \\ &+ (\xi_1^2 + \xi_3^2)(|\mathcal{F}(u_1\overline{u_3} - \overline{u_1}u_3)|^2) \\ &\left. + (\xi_2^2 + \xi_3^2)(|\mathcal{F}(u_2\overline{u_3} - \overline{u_2}u_3)|^2) \right], \end{aligned} \tag{II-1}$$

$$\begin{aligned} B(u_1, u_2, u_3) =& \frac{\eta}{|\xi|^2 - \delta} \left[\xi_1\xi_2\mathcal{F}(\overline{u_2}u_3 - u_2\overline{u_3})\overline{\mathcal{F}(u_1\overline{u_3} - \overline{u_1}u_3)} \right. \\ &+ \xi_1\xi_3\mathcal{F}(u_2\overline{u_3} - \overline{u_2}u_3)\overline{\mathcal{F}(u_1\overline{u_2} - \overline{u_1}u_2)} \\ &\left. + \xi_2\xi_3\mathcal{F}(\overline{u_1}u_3 - u_1\overline{u_3})\overline{\mathcal{F}(u_1\overline{u_2} - \overline{u_1}u_2)} \right]. \end{aligned} \tag{II-2}$$

A natural attempt to find nontrivial solutions to (1.5)–(1.7) is to solve the constrained minimization problem

$$d := \inf_{(u_1,u_2,u_3)\in M} S(u_1, u_2, u_3), \tag{2.5}$$

where the set M is defined by

$$M = \{(u_1, u_2, u_3) \in H_r^1(\mathbb{R}^3) \times H_r^1(\mathbb{R}^3) \times H_r^1(\mathbb{R}^3) \setminus \{(0,0,0)\}, R(u_1, u_2, u_3) = 0\}. \tag{2.6}$$

From $(u_1, u_2, u_3) \in H_r^1(\mathbb{R}^3) \times H_r^1(\mathbb{R}^3) \times H_r^1(\mathbb{R}^3)$ $\eta > 0$, $\delta \leqslant 0$, the Sobolev's embedding theorem and the properties of Fourier transform, it follows that functionals $S(u_1, u_2, u_3)$ and $R(u_1, u_2, u_3)$ are both well defined.

Remark 2.1 *We note that for all* $\theta \geqslant 1$, $j, k, l, m = 1, 2, 3$,

$$\int_{\mathbb{R}^3} \frac{\eta\xi_l\xi_m}{(|\xi|^2 - \delta)^\theta} |\mathcal{F}(u_j\overline{v_k})|^2 \mathrm{d}\xi = \int_{\mathbb{R}^3} \frac{\eta\xi_l\xi_m}{(|\xi|^2 - \delta)^\theta} |\mathcal{F}(\overline{u_j}v_k)|^2 \mathrm{d}\xi.$$

We also note that if (u_1, u_2, u_3) *is a critical point of* $S(u_1, u_2, u_3)$ *and hence a solution of* (1.5)–(1.7), *then* $(E_1, E_2, E_3) = (e^{i\omega t}u_1, e^{i\omega t}u_2, e^{i\omega t}u_3)$ *is a standing wave solution of* (1.1)–(1.3).

2.3 Main results

Here, we state the main results of this paper.

Theorem 2.1 *For* $\eta > 0$ *and* $\delta \leqslant 0$, *there exists* $(Q_1, Q_2, Q_3) \in M$ *such that*

(1) $S(Q_1, Q_2, Q_3) = d = \inf_{(u_1,u_2,u_3)\in M} S(u_1, u_2, u_3)$;

(2) (Q_1, Q_2, Q_3) *is a ground state solution to* (1.5)–(1.7).

(3) (Q_1, Q_2, Q_3) *are functions of* $|x|$ *alone and decay exponentially at infinity.*

From the physical viewpoint, an important role is played by the ground state solution of (1.5)–(1.7). A solution (Q_1, Q_2, Q_3) to (1.5)–(1.7) is termed as a ground state if it has some minimal action among all solutions to (1.5)–(1.7). Here, the action of solution (u_1, u_2, u_3) is defined by $S(u_1, u_2, u_3)$.

Concerning the dynamics of the ground state solution (Q_1, Q_2, Q_3), we have the following orbital instability result. Here, we assume that the ground state solution (Q_1, Q_2, Q_3) of (1.5)–(1.7) is unique.

Theorem 2.2 *For $\eta > 0$ and $\delta \leqslant 0$, let $(Q_1, Q_2, Q_3) \in M$ be given by Theorem 2.1. For arbitrary $\varepsilon > 0$, there exists $(E_{10}, E_{20}, E_{30}) \in H_r^1(\mathbb{R}^3) \times H_r^1(\mathbb{R}^3) \times H_r^1(\mathbb{R}^3)$ with*

$$\|E_{10} - Q_1\|_{H_r^1(\mathbb{R}^3)} < \varepsilon, \qquad \|E_{20} - Q_2\|_{H_r^1(\mathbb{R}^3)} < \varepsilon, \qquad \|E_{30} - Q_3\|_{H_r^1(\mathbb{R}^3)} < \varepsilon \tag{2.7}$$

such that the solution (E_1, E_2, E_3) of the equations (1.1)–(1.3) with the initial data (1.4) has the following property: For some finite time $T < \infty$, (E_1, E_2, E_3) exists on $[0, T)$, $(E_1, E_2, E_3) \in C([0, T); H_r^1(\mathbb{R}^3) \times H_r^1(\mathbb{R}^3) \times H_r^1(\mathbb{R}^3))$ and

$$\lim_{t \to T} \left(\|E_1\|_{H_r^1(\mathbb{R}^3)} + \|E_2\|_{H_r^1(\mathbb{R}^3)} + \|E_3\|_{H_r^1(\mathbb{R}^3)}\right) = +\infty. \tag{2.8}$$

3 Existence of standing waves

In this section, we prove Theorem 2.1, which concerns the existence of minimal energy standing waves of the system (1.1)–(1.3). For that purpose, we first give some key Propositions and Lemmas.

Remark 3.1 *For $\eta > 0$, $\delta \leqslant 0$, $\theta = 1, 2$, $j, k, l, m = 1, 2, 3$, since*

$$\int_{\mathbb{R}^3} \frac{\eta \xi_l \xi_m}{(|\xi|^2 - \delta)^\theta} |\mathcal{F}(Q_j \overline{Q_k})|^2 \mathrm{d}\xi = \int_{\mathbb{R}^3} \frac{\eta \xi_l \xi_m}{(|\xi|^2 - \delta)^\theta} |\mathcal{F}(\overline{Q_j} Q_k)|^2 \mathrm{d}\xi, \tag{3.1}$$

$$\begin{aligned}
&\left|\mathrm{Re} \int_{\mathbb{R}^3} \frac{\eta \xi_l \xi_m}{(|\xi|^2 - \delta)^\theta} \mathcal{F}(Q_j \overline{Q_k}) \overline{\mathcal{F}(\overline{Q_j} Q_k)} \mathrm{d}\xi\right| \\
&\leqslant \frac{1}{4} \int_{\mathbb{R}^3} \frac{\eta(|\xi_l|^2 + |\xi_m|^2)}{(|\xi|^2 - \delta)^\theta} (|\mathcal{F}(Q_j \overline{Q_k})|^2 + |\mathcal{F}(\overline{Q_j} Q_k)|^2) \mathrm{d}\xi \\
&= \frac{1}{2} \int_{\mathbb{R}^3} \frac{\eta(|\xi_l|^2 + |\xi_m|^2)}{(|\xi|^2 - \delta)^\theta} |\mathcal{F}(Q_j \overline{Q_k})|^2 \mathrm{d}\xi,
\end{aligned} \tag{3.2}$$

by Theorem 2.1, (2.3), (2.4) and (2.5), one has

$$
\begin{aligned}
&S(Q_1,Q_2,Q_3)\\
=&\frac{1}{6}\int_{\mathbb{R}^3}(|\nabla Q_1|^2+|\nabla Q_2|^2+|\nabla Q_3|^2)\mathrm{d}x+\frac{w}{2}\int_{\mathbb{R}^3}(|Q_1|^2+|Q_2|^2+|Q_3|^2)\mathrm{d}x\\
&-\frac{1}{6}\delta\int_{\mathbb{R}^3}\frac{\eta(\xi_1^2+\xi_2^2)}{(|\xi|^2-\delta)^2}(|\mathcal{F}(Q_1\overline{Q_2}-\overline{Q_1}Q_2)|^2)\mathrm{d}\xi\\
&-\frac{1}{6}\delta\int_{\mathbb{R}^3}\frac{\eta(\xi_1^2+\xi_3^2)}{(|\xi|^2-\delta)^2}(|\mathcal{F}(Q_1\overline{Q_3}-\overline{Q_1}Q_3)|^2)\mathrm{d}\xi\\
&-\frac{1}{6}\delta\int_{\mathbb{R}^3}\frac{\eta(\xi_2^2+\xi_3^2)}{(|\xi|^2-\delta)^2}(|\mathcal{F}(Q_2\overline{Q_3}-\overline{Q_2}Q_3)|^2)\mathrm{d}\xi\\
&+\frac{1}{3}\delta\mathrm{Re}\int_{\mathbb{R}^3}\frac{\eta\xi_1\xi_2}{(|\xi|^2-\delta)^2}\mathcal{F}(\overline{Q_2}Q_3-Q_2\overline{Q_3})\overline{\mathcal{F}(Q_1\overline{Q_3}-\overline{Q_1}Q_3)}\mathrm{d}\xi\\
&+\frac{1}{3}\delta\mathrm{Re}\int_{\mathbb{R}^3}\frac{\eta\xi_1\xi_3}{(|\xi|^2-\delta)^2}\mathcal{F}(Q_2\overline{Q_3}-\overline{Q_2}Q_3)\overline{\mathcal{F}(Q_1\overline{Q_2}-\overline{Q_1}Q_2)}\mathrm{d}\xi\\
&+\frac{1}{3}\delta\mathrm{Re}\int_{\mathbb{R}^3}\frac{\eta\xi_2\xi_3}{(|\xi|^2-\delta)^2}\mathcal{F}(\overline{Q_1}Q_3-Q_1\overline{Q_3})\overline{\mathcal{F}(Q_1\overline{Q_2}-\overline{Q_1})Q_2}\mathrm{d}\xi>0. \qquad (3.3)
\end{aligned}
$$

□

We continue to give some key facts.

Lemma 3.1([19, 20]) *For $2<\sigma<6$, the embedding $H_r^1(\mathbb{R}^3)\hookrightarrow L_r^\sigma(\mathbb{R}^3)$ is compact, where*

$$H_r^1(\mathbb{R}^3)=\{f(x)\in H^1(\mathbb{R}^3):f(x)=f(|x|)\ is\ a\ function\ of\ |x|\ alone\}. \qquad \square$$

Proposition 3.1 *Let $\eta>0$ and $\delta\leqslant 0$. Then the non-trivial solution to* (1.5)–(1.7) *belongs to M.*

Proof Let (u_1,u_2,u_3) be a non-trivial solution to (1.5)–(1.7). Multiplying (1.5) by $\overline{u_1}$, (1.6) by $\overline{u_2}$, (1.7) by $\overline{u_3}$, then integrating with respect to x on $\mathbb{R}^3$, we obtain

$$
\begin{aligned}
&-\int_{\mathbb{R}^3}(|\nabla u_1|^2+|\nabla u_2|^2+|\nabla u_3|^2)\mathrm{d}x-\omega\int_{\mathbb{R}^3}(|u_1|^2+|u_2|^2+|u_3|^2)\mathrm{d}x\\
&+\int_{\mathbb{R}^3}(|u_1|^4+|u_2|^4+|u_3|^4)\mathrm{d}x+2\int_{\mathbb{R}^3}(|u_1|^2|u_2|^2+|u_1|^2|u_3|^2+|u_2|^2|u_3|^2)\mathrm{d}x\\
&+\int_{\mathbb{R}^3}A(u_1,u_2,u_3)\mathrm{d}\xi-2\mathrm{Re}\int_{\mathbb{R}^3}B(u_1,u_2,u_3)\mathrm{d}\xi=0. \qquad (3.4)
\end{aligned}
$$

We further attain the following identity:

$$
\begin{aligned}
&\frac{1}{2}\int_{\mathbb{R}^3}(|\nabla u_1|^2+|\nabla u_2|^2+|\nabla u_3|^2)\mathrm{d}x+\frac{3\omega}{2}\int_{\mathbb{R}^3}(|u_1|^2+|u_2|^2+|u_3|^2)\mathrm{d}x\\
&-\frac{3}{4}\int_{\mathbb{R}^3}(|u_1|^4+|u_2|^4+|u_3|^4)\mathrm{d}x-\frac{3}{2}\int_{\mathbb{R}^3}(|u_1|^2|u_2|^2+|u_1|^2|u_3|^2+|u_2|^2|u_3|^2)\mathrm{d}x
\end{aligned}
$$

$$
\begin{aligned}
&-\frac{3}{4}\int_{\mathbb{R}^3} A(u_1,u_2,u_3)\mathrm{d}\xi+\frac{3}{2}\mathrm{Re}\int_{\mathbb{R}^3} B(u_1,u_2,u_3)\mathrm{d}\xi\\
&-\frac{\delta}{2}\int_{\mathbb{R}^3}\frac{1}{|\xi|^2-\delta}A(u_1,u_2,u_3)\mathrm{d}\xi+\delta\mathrm{Re}\int_{\mathbb{R}^3}\frac{1}{|\xi|^2-\delta}B(u_1,u_2,u_3)\mathrm{d}\xi=0.
\end{aligned}\tag{3.5}
$$

Here, A and B are defined by (II-1) and (II-2) in Section 2, respectively. The identity (3.5) is obtained by multiplying (1.5) by $x\nabla\overline{u_1}$, (1.6) by $x\nabla\overline{u_2}$, (1.7) by $x\nabla\overline{u_3}$, then integrating with respect to x in $\mathbb{R}^3$ and taking the real parts for the resulting equations, finally using the following estimates:

$$
\begin{aligned}
&\mathrm{Re}\int_{\mathbb{R}^3}(-wu_1x\nabla\overline{u}_1-wu_2x\nabla\overline{u}_2-wu_3x\nabla\overline{u}_3)\mathrm{d}x\\
&=\frac{3w}{2}\int_{\mathbb{R}^3}(|u_1|^2+|u_2|^2+|u_3|^2)\mathrm{d}x,
\end{aligned}\tag{III-1}
$$

$$
\begin{aligned}
&\mathrm{Re}\int_{\mathbb{R}^3}(u_1x\nabla\overline{u}_1+u_2x\nabla\overline{u}_2+u_3x\nabla\overline{u}_3)\mathrm{d}x\\
&=\frac{1}{2}\int_{\mathbb{R}^3}(|u_1|^2+|u_2|^2+|u_3|^2)\mathrm{d}x,
\end{aligned}\tag{III-2}
$$

$$
\begin{aligned}
&\mathrm{Re}\int_{\mathbb{R}^3}(|u_1|^2+|u_2|^2+|u_3|^2)(u_1x\nabla\overline{u}_1+u_2x\nabla\overline{u}_2+u_3x\nabla\overline{u}_3)\mathrm{d}x\\
&=-\frac{3}{4}\int_{\mathbb{R}^3}(|u_1|^4+|u_2|^4+|u_3|^4)\mathrm{d}x\\
&\quad-\frac{3}{2}\int_{\mathbb{R}^3}(|u_1|^2|u_2|^2+|u_1|^2|u_3|^2+|u_2|^2|u_3|^2)\mathrm{d}x,
\end{aligned}\tag{III-3}
$$

$$
\begin{aligned}
&\mathrm{Re}\int_{\mathbb{R}^3}u_1x\nabla\overline{u}_2\mathcal{F}^{-1}[\frac{\eta(\xi_1^2+\xi_2^2)}{|\xi|^2-\delta}\mathcal{F}(u_1\overline{u}_2-\overline{u}_1u_2)]\mathrm{d}x\\
&-\mathrm{Re}\int_{\mathbb{R}^3}u_2x\nabla\overline{u}_1\mathcal{F}^{-1}[\frac{\eta(\xi_1^2+\xi_2^2)}{|\xi|^2-\delta}\mathcal{F}(u_1\overline{u}_2-\overline{u}_1u_2)]\mathrm{d}x\\
&=\frac{3}{4}\int_{\mathbb{R}^3}\frac{\eta(\xi_1^2+\xi_2^2)}{|\xi|^2-\delta}(|\mathcal{F}(u_1\overline{u}_2-\overline{u}_1u_2)|^2)\mathrm{d}\xi\\
&\quad+\frac{\delta}{2}\int_{\mathbb{R}^3}\frac{\eta(\xi_1^2+\xi_2^2)}{(|\xi|^2-\delta)^2}(|\mathcal{F}(u_1\overline{u}_2-\overline{u}_1u_2)|^2)\mathrm{d}\xi,
\end{aligned}\tag{III-4}
$$

$$
\begin{aligned}
&\mathrm{Re}\int_{\mathbb{R}^3}u_1x\nabla\overline{u}_3\mathcal{F}^{-1}[\frac{\eta(\xi_1^2+\xi_3^2)}{|\xi|^2-\delta}\mathcal{F}(u_1\overline{u}_3-\overline{u}_1u_3)]\mathrm{d}x\\
&-\mathrm{Re}\int_{\mathbb{R}^3}u_3x\nabla\overline{u}_1\mathcal{F}^{-1}[\frac{\eta(\xi_1^2+\xi_3^2)}{|\xi|^2-\delta}\mathcal{F}(u_1\overline{u}_3-\overline{u}_1u_3)]\mathrm{d}x
\end{aligned}
$$

$$
\begin{aligned}
&= \frac{3}{4}\int_{\mathbb{R}^3} \frac{\eta(\xi_1^2+\xi_3^2)}{|\xi|^2-\delta}(|\mathcal{F}(u_1\overline{u}_3-\overline{u}_1u_3)|^2)\mathrm{d}\xi \\
&\quad + \frac{\delta}{2}\int_{\mathbb{R}^3} \frac{\eta(\xi_1^2+\xi_3^2)}{(|\xi|^2-\delta)^2}(|\mathcal{F}(u_1\overline{u}_3-\overline{u}_1u_3)|^2)\mathrm{d}\xi, \qquad \text{(III-5)}
\end{aligned}
$$

$$
\begin{aligned}
&\mathrm{Re}\int_{\mathbb{R}^3} u_2x\nabla\overline{u}_3\mathcal{F}^{-1}[\frac{\eta(\xi_2^2+\xi_3^2)}{|\xi|^2-\delta}\mathcal{F}(u_2\overline{u}_3-\overline{u}_2u_3)]\mathrm{d}x \\
&- \mathrm{Re}\int_{\mathbb{R}^3} u_3x\nabla\overline{u}_2\mathcal{F}^{-1}[\frac{\eta(\xi_2^2+\xi_3^2)}{|\xi|^2-\delta}\mathcal{F}(u_2\overline{u}_3-\overline{u}_2u_3)]\mathrm{d}x \\
&= \frac{3}{4}\int_{\mathbb{R}^3} \frac{\eta(\xi_2^2+\xi_3^2)}{|\xi|^2-\delta}(|\mathcal{F}(u_2\overline{u}_3-\overline{u}_2u_3)|^2)\mathrm{d}\xi \\
&\quad + \frac{\delta}{2}\int_{\mathbb{R}^3} \frac{\eta(\xi_2^2+\xi_3^2)}{(|\xi|^2-\delta)^2}(|\mathcal{F}(u_2\overline{u}_3-\overline{u}_2u_3)|^2)\mathrm{d}\xi, \qquad \text{(III-6)}
\end{aligned}
$$

$$
\begin{aligned}
&\mathrm{Re}\int_{\mathbb{R}^3} u_2x\nabla\overline{u}_1\mathcal{F}^{-1}[\frac{\eta\xi_1\xi_3}{|\xi|^2-\delta}\mathcal{F}(u_2\overline{u}_3-u_3\overline{u}_2)]\mathrm{d}x \\
&+ \mathrm{Re}\int_{\mathbb{R}^3} u_3x\nabla\overline{u}_2\mathcal{F}^{-1}[\frac{\eta\xi_1\xi_3}{|\xi|^2-\delta}\mathcal{F}(u_1\overline{u}_2-\overline{u}_1u_2)]\mathrm{d}x \\
&- \mathrm{Re}\int_{\mathbb{R}^3} u_1x\nabla\overline{u}_2\mathcal{F}^{-1}[\frac{\eta\xi_1\xi_3}{|\xi|^2-\delta}\mathcal{F}(u_2\overline{u}_3-u_3\overline{u}_2)]\mathrm{d}x \\
&- \mathrm{Re}\int_{\mathbb{R}^3} u_2x\nabla\overline{u}_3\mathcal{F}^{-1}[\frac{\eta\xi_1\xi_3}{|\xi|^2-\delta}\mathcal{F}(u_1\overline{u}_2-\overline{u}_1u_2)]\mathrm{d}x \\
&= -\frac{3}{2}\mathrm{Re}\int_{\mathbb{R}^3} \frac{\eta\xi_1\xi_3}{|\xi|^2-\delta}\mathcal{F}(u_2\overline{u}_3-\overline{u}_2u_3)\overline{\mathcal{F}(u_1\overline{u}_2-\overline{u}_1u_2)}\mathrm{d}\xi \\
&\quad - \delta\mathrm{Re}\int_{\mathbb{R}^3} \frac{\eta\xi_1\xi_3}{(|\xi|^2-\delta)^2}\mathcal{F}(u_2\overline{u}_3-\overline{u}_2u_3)\overline{\mathcal{F}(u_1\overline{u}_2-\overline{u}_1u_2)}\mathrm{d}\xi, \qquad \text{(III-7)}
\end{aligned}
$$

$$
\begin{aligned}
&\mathrm{Re}\int_{\mathbb{R}^3} u_1x\nabla\overline{u}_2\mathcal{F}^{-1}[\frac{\eta\xi_2\xi_3}{|\xi|^2-\delta}\mathcal{F}(u_1\overline{u}_3-\overline{u}_1u_3)]\mathrm{d}x \\
&+ \mathrm{Re}\int_{\mathbb{R}^3} u_1x\nabla\overline{u}_3\mathcal{F}^{-1}[\frac{\eta\xi_2\xi_3}{|\xi|^2-\delta}\mathcal{F}(u_1\overline{u}_2-\overline{u}_1u_2)]\mathrm{d}x \\
&- \mathrm{Re}\int_{\mathbb{R}^3} u_3x\nabla\overline{u}_1\mathcal{F}^{-1}[\frac{\eta\xi_2\xi_3}{|\xi|^2-\delta}\mathcal{F}(u_1\overline{u}_2-\overline{u}_1u_2)]\mathrm{d}x \\
&- \mathrm{Re}\int_{\mathbb{R}^3} u_2x\nabla\overline{u}_1\mathcal{F}^{-1}[\frac{\eta\xi_2\xi_3}{|\xi|^2-\delta}\mathcal{F}(u_1\overline{u}_3-\overline{u}_1u_3)]\mathrm{d}x \\
&= -\frac{3}{2}\mathrm{Re}\int_{\mathbb{R}^3} \frac{\eta\xi_2\xi_3}{|\xi|^2-\delta}\mathcal{F}(u_1\overline{u}_3-\overline{u}_1u_3)\overline{\mathcal{F}(u_1\overline{u}_2-\overline{u}_1u_2)}\mathrm{d}\xi
\end{aligned}
$$

$$-\delta\mathrm{Re}\int_{\mathbb{R}^3}\frac{\eta\xi_2\xi_3}{(|\xi|^2-\delta)^2}\mathcal{F}(u_1\overline{u}_3-\overline{u}_1u_3)\overline{\mathcal{F}(u_1\overline{u}_2-\overline{u}_1u_2)}\mathrm{d}\xi, \quad \text{(III-8)}$$

$$\mathrm{Re}\int_{\mathbb{R}^3}u_3x\nabla\overline{u}_1\mathcal{F}^{-1}[\frac{\eta\xi_1\xi_2}{|\xi|^2-\delta}\mathcal{F}(u_2\overline{u}_3-\overline{u}_2u_3)]\mathrm{d}x$$

$$+\mathrm{Re}\int_{\mathbb{R}^3}u_3x\nabla\overline{u}_2\mathcal{F}^{-1}[\frac{\eta\xi_1\xi_2}{|\xi|^2-\delta}\mathcal{F}(u_1\overline{u}_3-\overline{u}_1u_3)]\mathrm{d}x$$

$$-\mathrm{Re}\int_{\mathbb{R}^3}u_1x\nabla\overline{u}_3\mathcal{F}^{-1}[\frac{\eta\xi_1\xi_2}{|\xi|^2-\delta}\mathcal{F}(u_2\overline{u}_3-\overline{u}_2u_3)]\mathrm{d}x$$

$$-\mathrm{Re}\int_{\mathbb{R}^3}u_2x\nabla\overline{u}_3\mathcal{F}^{-1}[\frac{\eta\xi_1\xi_2}{|\xi|^2-\delta}\mathcal{F}(u_1\overline{u}_3-\overline{u}_1u_3)]\mathrm{d}x$$

$$=-\frac{3}{2}\mathrm{Re}\int_{\mathbb{R}^3}\frac{\eta\xi_1\xi_2}{|\xi|^2-\delta}\mathcal{F}(u_2\overline{u}_3-\overline{u}_2u_3)\overline{\mathcal{F}(u_1\overline{u}_3-\overline{u}_1u_3)}\mathrm{d}\xi$$

$$-\delta\mathrm{Re}\int_{\mathbb{R}^3}\frac{\eta\xi_1\xi_2}{(|\xi|^2-\delta)^2}\mathcal{F}(u_2\overline{u}_3-\overline{u}_2u_3)\overline{\mathcal{F}(u_1\overline{u}_3-\overline{u}_1u_3)}\mathrm{d}\xi. \quad \text{(III-9)}$$

Combining (3.4) with (3.5), one easily verifies that $R(u_1,u_2,u_3)=0$, and hence $(u_1,u_2,u_3)\in M$. □

Proposition 3.2 *The functional S is bounded from below on M for $\eta>0$ and $\delta\leqslant 0$.*

Proof Let A and B are defined by (II-1) and (II-2) in Section 2, respectively. According to (2.3), (2.4) and (2.6), for $(u_1,u_2,u_3)\in M$ one gets

$$\begin{aligned}
&S(u_1,u_2,u_3)\\
&=\frac{1}{6}\int_{\mathbb{R}^3}(|\nabla u_1|^2+|\nabla u_2|^2+|\nabla u_3|^2)\mathrm{d}x\\
&+\frac{w}{2}\int_{\mathbb{R}^3}(|u_1|^2+|u_2|^2+|u_3|^2)\mathrm{d}x\\
&-\frac{\delta}{6}\int_{\mathbb{R}^3}\frac{A(u_1,u_2,u_3)}{|\xi|^2-\delta}\mathrm{d}\xi+\frac{\delta}{3}Re\int_{\mathbb{R}^3}\frac{B(u_1,u_2,u_3)}{|\xi|^2-\delta}\mathrm{d}\xi.
\end{aligned} \tag{3.6}$$

Noting that $\eta>0$, $\delta\leqslant 0$ and the inequality $2Re(ab)\leqslant a^2+b^2$, making some suitable rearrangements, we obtain

$$-\frac{\delta}{6}\int_{\mathbb{R}^3}\frac{A(u_1,u_2,u_3)}{|\xi|^2-\delta}\mathrm{d}\xi+\frac{\delta}{3}Re\int_{\mathbb{R}^3}\frac{B(u_1,u_2,u_3)}{|\xi|^2-\delta}\mathrm{d}\xi\geqslant 0. \tag{3.7}$$

(Indeed, recall that

$$A(u_1,u_2,u_3)=\frac{\eta}{|\xi|^2-\delta}\,[(\xi_1^2+\xi_2^2)(|\mathcal{F}(u_1\overline{u_2}-\overline{u_1}u_2)|^2)$$

$$+ (\xi_1^2+\xi_3^2)(|\mathcal{F}(u_1\overline{u_3}-\overline{u_1}u_3)|^2)$$
$$+(\xi_2^2+\xi_3^2)(|\mathcal{F}(u_2\overline{u_3}-\overline{u_2}u_3)|^2)\Big], \tag{II-1}$$

$$\begin{aligned}B(u_1,u_2,u_3)=&\frac{\eta}{|\xi|^2-\delta}\Big[\xi_1\xi_2\mathcal{F}(\overline{u_2}u_3-u_2\overline{u_3})\overline{\mathcal{F}(u_1\overline{u_3}-\overline{u_1}u_3)}\\&+\xi_1\xi_3\mathcal{F}(u_2\overline{u_3}-\overline{u_2}u_3)\overline{\mathcal{F}(u_1\overline{u_2}-\overline{u_1}u_2)}\\&+\xi_2\xi_3\mathcal{F}(\overline{u_1}u_3-u_1\overline{u_3})\overline{\mathcal{F}(u_1\overline{u_2}-\overline{u_1}u_2)}\Big].\end{aligned} \tag{II-2}$$

Since

$$\begin{aligned}&Re\xi_1\xi_2\mathcal{F}(\overline{u_2}u_3-u_2\overline{u_3})\overline{\mathcal{F}(u_1\overline{u_3}-\overline{u_1}u_3)}\\&\leqslant\frac{1}{2}\xi_1^2|\mathcal{F}(u_1\overline{u_3}-\overline{u_1}u_3)|^2+\frac{1}{2}\xi_2^2|\mathcal{F}(\overline{u_2}u_3-u_2\overline{u_3})|^2,\\&Re\xi_1\xi_3\mathcal{F}(u_2\overline{u_3}-\overline{u_2}u_3)\overline{\mathcal{F}(u_1\overline{u_2}-\overline{u_1}u_2)}\\&\leqslant\frac{1}{2}\xi_1^2|\mathcal{F}(u_1\overline{u_2}-\overline{u_1}u_2)|^2+\frac{1}{2}\xi_3^2|\mathcal{F}(u_2\overline{u_3}-\overline{u_2}u_3)|^2,\\&Re\xi_2\xi_3\mathcal{F}(\overline{u_1}u_3-u_1\overline{u_3})\overline{\mathcal{F}(u_1\overline{u_2}-\overline{u_1}u_2)}\\&\leqslant\frac{1}{2}\xi_2^2|\mathcal{F}(u_1\overline{u_2}-\overline{u_1}u_2)|^2+\frac{1}{2}\xi_3^2|\mathcal{F}(\overline{u_1}u_3-u_1\overline{u_3})|^2,\end{aligned}$$

through regrouping and applying some properties of Fourier transform, we conclude that for $\eta>0$, and $\delta\leqslant 0$,

$$\begin{aligned}&\frac{\delta}{3}Re\int_{\mathbb{R}^3}\frac{B(u_1,u_2,u_3)}{|\xi|^2-\delta}\mathrm{d}\xi\\&\geqslant\frac{\delta}{6}\int_{\mathbb{R}^3}\frac{1}{|\xi|^2-\delta}\cdot\frac{\eta}{|\xi|^2-\delta}\left[(\xi_1^2+\xi_2^2)|\mathcal{F}(u_1\overline{u_2}-\overline{u_1}u_2)|^2\right.\\&(\xi_1^2+\xi_3^2)|\mathcal{F}(u_1\overline{u_3}-\overline{u_1}u_3)|^2\\&\left.(\xi_2^2+\xi_3^2)|\mathcal{F}(u_2\overline{u_3}-\overline{u_2}u_3)|^2\right]\\&=\frac{\delta}{6}\int_{\mathbb{R}^3}\frac{A(u_1,u_2,u_3)}{|\xi|^2-\delta}\mathrm{d}\xi.\end{aligned}$$

Hence (3.7) is valid.)

Therefore, (3.6) and (3.7) imply that on M,

$$\begin{aligned}S(u_1,u_2,u_3)\geqslant&\frac{1}{6}\int_{\mathbb{R}^3}(|\nabla u_1|^2+|\nabla u_2|^2+|\nabla u_3|^2)\mathrm{d}x\\&+\frac{w}{2}\int_{\mathbb{R}^3}(|u_1|^2+|u_2|^2+|u_3|^2)\mathrm{d}x.\end{aligned} \tag{3.8}$$

This completes the proof of Proposition 3.2. □

Proposition 3.3 *For $\eta > 0$ and $\delta \leqslant 0$, let $(u_1, u_2, u_3) \in H_r^1(\mathbb{R}^3) \times H_r^1(\mathbb{R}^3) \times H_r^1(\mathbb{R}^3) \setminus \{(0,0,0)\}$. For $\lambda > 0$, we make the following scale transform:*

$$u_{1\lambda}(x) = \lambda^{\frac{3}{2}} u_1(\lambda x),\ u_{2\lambda}(x) = \lambda^{\frac{3}{2}} u_2(\lambda x),\ u_{3\lambda}(x) = \lambda^{\frac{3}{2}} u_3(\lambda x),$$

then there exists a unique $\mu > 0$ (relying on (u_1, u_2, u_3)) such that $R(u_{1\mu}, u_{2\mu}, u_{3\mu}) = 0$. Furthermore, the following three estimates will occur:

$$\begin{array}{ll} R(u_{1\mu}, u_{2\mu}, u_{3\mu}) > 0, & for\ \ \lambda \in (0, \mu); \\ R(u_{1\mu}, u_{2\mu}, u_{3\mu}) < 0, & for\ \ \lambda \in (\mu, \infty); \\ S(u_{1\mu}, u_{2\mu}, u_{3\mu}) \geqslant S(u_{1\lambda}, u_{2\lambda}, u_{3\lambda}), & for\ \ \forall \lambda > 0. \end{array}$$

Proof According to (2.3) and (2.4), $S(u_{1\lambda}, u_{2\lambda}, u_{3\lambda})$ and $R(u_{1\lambda}, u_{2\lambda}, u_{3\lambda})$ are of the following expressions:

$$\begin{aligned} & S(u_{1\lambda}, u_{2\lambda}, u_{3\lambda}) \\ = & \frac{1}{2}\lambda^2 \int_{\mathbb{R}^3} (|\nabla u_1|^2 + |\nabla u_2|^2 + |\nabla u_3|^2) \mathrm{d}x + \frac{w}{2} \int_{\mathbb{R}^3} (|u_1|^2 + |u_2|^2 + |u_3|^2) \mathrm{d}x \\ & - \frac{1}{4}\lambda^3 \int_{\mathbb{R}^3} (|u_1|^4 + |u_2|^4 + |u_3|^4) \mathrm{d}x \\ & - \frac{1}{2}\lambda^3 \int_{\mathbb{R}^3} (|u_1|^2|u_2|^2 + |u_1|^2|u_3|^2 + |u_2|^2|u_3|^2) dx \\ & - \frac{1}{4}\lambda^3 \int_{\mathbb{R}^3} \frac{\eta\lambda^2(\xi_1^2 + \xi_2^2)}{\lambda^2|\xi|^2 - \delta} (|\mathcal{F}(u_1\overline{u}_2 - \overline{u}_1 u_2)|^2) \mathrm{d}\xi \\ & - \frac{1}{4}\lambda^3 \int_{\mathbb{R}^3} \frac{\eta\lambda^2(\xi_3^2 + \xi_2^2)}{\lambda^2|\xi|^2 - \delta} (|\mathcal{F}(u_2\overline{u}_3 - \overline{u}_2 u_3)|^2) \mathrm{d}\xi \\ & - \frac{1}{4}\lambda^3 \int_{\mathbb{R}^3} \frac{\eta\lambda^2(\xi_1^2 + \xi_3^2)}{\lambda^2|\xi|^2 - \delta} (|\mathcal{F}(u_1\overline{u}_3 - \overline{u}_1 u_3)|^2) \mathrm{d}\xi \\ & + \frac{1}{2}\lambda^3 Re \int_{\mathbb{R}^3} \frac{\eta\lambda^2(\xi_1\xi_2)}{\lambda^2|\xi|^2 - \delta} \mathcal{F}(\overline{u_2}u_3 - u_2\overline{u_3}) \overline{\mathcal{F}(u_1\overline{u_3} - \overline{u_1}u_3)} \mathrm{d}\xi \\ & + \frac{1}{2}\lambda^3 Re \int_{\mathbb{R}^3} \frac{\eta\lambda^2(\xi_2\xi_3)}{\lambda^2|\xi|^2 - \delta} \mathcal{F}(\overline{u_1}u_3 - u_1\overline{u_3}) \overline{\mathcal{F}(u_1\overline{u_2} - \overline{u_1}u_2)} \mathrm{d}\xi \\ & + \frac{1}{2}\lambda^3 Re \int_{\mathbb{R}^3} \frac{\eta\lambda^2(\xi_1\xi_3)}{\lambda^2|\xi|^2 - \delta} |\mathcal{F}(u_2\overline{u_3} - \overline{u_2}u_3) \overline{\mathcal{F}(u_1\overline{u_2} - \overline{u_1}u_2)} \mathrm{d}\xi, \end{aligned} \tag{3.9}$$

$$
\begin{aligned}
&R(u_{1\lambda},u_{2\lambda},u_{3\lambda})\\
&=\lambda^2\left\{\int_{\mathbb{R}^3}(|\nabla u_1|^2+|\nabla u_2|^2+|\nabla u_3|^2)\mathrm{d}x\right.\\
&\quad-\frac{3}{4}\lambda\int_{\mathbb{R}^3}(|u_1|^4+|u_2|^4+|u_3|^4)\mathrm{d}x\\
&\quad-\frac{3}{2}\lambda\int_{\mathbb{R}^3}(|u_1|^2|u_2|^2+|u_1|^2|u_3|^2+|u_2|^2|u_3|^2)\mathrm{d}x
\end{aligned}
$$

$$
R_{1\lambda}(u_1,u_2,u_3):=\left\{\begin{aligned}
&-\frac{3}{4}\lambda\int_{\mathbb{R}^3}\frac{\eta\lambda^2(\xi_1^2+\xi_2^2)}{\lambda^2|\xi|^2-\delta}(|\mathcal{F}(u_1\overline{u}_2-\overline{u}_1u_2)|^2)\mathrm{d}\xi\\
&-\frac{3}{4}\lambda\int_{\mathbb{R}^3}\frac{\eta\lambda^2(\xi_3^2+\xi_2^2)}{\lambda^2|\xi|^2-\delta}(|\mathcal{F}(u_2\overline{u}_3-\overline{u}_2u_3)|^2)\mathrm{d}\xi\\
&-\frac{3}{4}\lambda\int_{\mathbb{R}^3}\frac{\eta\lambda^2(\xi_1^2+\xi_3^2)}{\lambda^2|\xi|^2-\delta}(|\mathcal{F}(u_1\overline{u}_3-\overline{u}_1u_3)|^2)\mathrm{d}\xi\\
&+\frac{\delta}{2}\lambda\int_{\mathbb{R}^3}\frac{\eta\lambda^2(\xi_1^2+\xi_2^2)}{(\lambda^2|\xi|^2-\delta)^2}(|\mathcal{F}(u_1\overline{u}_2-\overline{u}_1u_2)|^2)\mathrm{d}\xi\\
&+\frac{\delta}{2}\lambda\int_{\mathbb{R}^3}\frac{\eta\lambda^2(\xi_3^2+\xi_2^2)}{(\lambda^2|\xi|^2-\delta)^2}(|\mathcal{F}(u_2\overline{u}_3-\overline{u}_2u_3)|^2)\mathrm{d}\xi\\
&+\frac{\delta}{2}\lambda\int_{\mathbb{R}^3}\frac{\eta\lambda^2(\xi_1^2+\xi_3^2)}{(\lambda^2|\xi|^2-\delta)^2}(|\mathcal{F}(u_1\overline{u}_3-\overline{u}_1u_3)|^2)\mathrm{d}\xi\\
&+\frac{3}{2}\lambda^3\mathrm{Re}\int_{\mathbb{R}^3}\frac{\eta\lambda^2\xi_1\xi_2}{\lambda^2|\xi|^2-\delta}\mathcal{F}(\overline{u_2}u_3-u_2\overline{u_3})\overline{\mathcal{F}(u_1\overline{u_3}-\overline{u_1}u_3)}\mathrm{d}\xi\\
&+\frac{3}{2}\lambda\mathrm{Re}\int_{\mathbb{R}^3}\frac{\eta\lambda^2\xi_2\xi_3}{\lambda^2|\xi|^2-\delta}\mathcal{F}(\overline{u_1}u_3-u_1\overline{u_3})\overline{\mathcal{F}(u_1\overline{u_2}-\overline{u_1}u_2)}\mathrm{d}\xi\\
&+\frac{3}{2}\lambda\mathrm{Re}\int_{\mathbb{R}^3}\frac{\eta\lambda^2\xi_1\xi_3}{\lambda^2|\xi|^2-\delta}\mathcal{F}(u_2\overline{u_3}-\overline{u_2}u_3)\overline{\mathcal{F}(u_1\overline{u_2}-\overline{u_1}u_2)}\mathrm{d}\xi\\
&-\lambda\delta\mathrm{Re}\int_{\mathbb{R}^3}\frac{\eta\lambda^2\xi_1\xi_2}{(\lambda^2|\xi|^2-\delta)^2}\mathcal{F}(\overline{u_2}u_3-u_2\overline{u_3})\overline{\mathcal{F}(u_1\overline{u_3}-\overline{u_1}u_3)}\mathrm{d}\xi\\
&-\lambda\delta\mathrm{Re}\int_{\mathbb{R}^3}\frac{\eta\lambda^2\xi_2\xi_3^2}{(\lambda^2|\xi|^2-\delta)^2}\mathcal{F}(\overline{u_1}u_3-u_1\overline{u_3})\overline{\mathcal{F}(u_1\overline{u_2}-\overline{u_1}u_2)}\mathrm{d}\xi\\
&-\lambda\delta\mathrm{Re}\int_{\mathbb{R}^3}\frac{\eta\lambda^2\xi_1\xi_3}{(\lambda^2|\xi|^2-\delta)^2}\mathcal{F}(u_2\overline{u_3}-\overline{u_2}u_3)\overline{\mathcal{F}(u_1\overline{u_2}-\overline{u_1}u_2)}\mathrm{d}\xi\left.\right\}
\end{aligned}\right.
$$

$$
=\lambda^2R^*(u_{1\lambda},u_{2\lambda},u_{3\lambda}).\tag{3.10}
$$

Making a preliminary estimate, we can verify that

$$
R_{1\lambda}(u_1,u_2,u_3)\leqslant 0.\tag{3.11}
$$

((3.11) can be easily obtained by utilizing the method of verifying (3.7).)

First of all, we show that there exists $\mu > 0$ such that $R(u_{1\mu}, u_{2\mu}, u_{3\mu}) = 0$, and we divide the proof into two cases:

Case 1 $R(u_1, u_2, u_3) > 0$; **Case 2** $R(u_1, u_2, u_3) < 0$.

♣ If Case 1 occurs, and if there exists λ such that $R(u_{1\lambda}, u_{2\lambda}, u_{3\lambda}) = 0$, then $\lambda \in (1, \infty)$ according to (3.10) and (3.11). For $j, k = 1, 2, 3$, let $G(\lambda^2) = \frac{3}{4} \frac{\eta\lambda^2(\xi_j^2 + \xi_k^2)}{\lambda^2|\xi|^2 - \delta} - \frac{\delta}{2} \frac{\eta\lambda^2(\xi_j^2 + \xi_k^2)}{(\lambda^2|\xi|^2 - \delta)^2}$. Then $G'(\lambda^2) = \frac{(-\delta\eta\lambda^2|\xi|^2 + 5\delta^2\eta)(\xi_j^2 + \xi_k^2)}{4(\lambda^2|\xi|^2 - \delta)^3}$, which together with $\eta > 0$ and $\delta \leqslant 0$ leads to $G'(\lambda^2) \geqslant 0$. Thus, $G(\lambda^2)$ is an increasing function of λ^2 ($\lambda^2 \in (1, \infty)$). Thus there exists $\mu \in (1, \infty)$ such that $R(u_{1\mu}, u_{2\mu}, u_{3\mu}) = 0$.

♣ If Case 2 occurs, and if there exists λ such that $R(u_{1\lambda}, u_{2\lambda}, u_{3\lambda}) = 0$, then $\lambda \in (0, 1)$. Indeed, we consider functional $R^*(u_{1\lambda}, u_{2\lambda}, u_{3\lambda})$ defined by (3.10). Note that $R^*(u_{1\lambda}, u_{2\lambda}, u_{3\lambda}) \to R(u_1, u_2, u_3) < 0$ as $\lambda \to 1$, and $R^*(u_{1\lambda}, u_{2\lambda}, u_{3\lambda}) = \int_{\mathbb{R}^3} (|\nabla u_1|^2 + |\nabla u_2|^2 + |\nabla u_3|^2) dx > 0$ as $\lambda \to 0$, one can verifies that there exists $\mu \in (0, 1)$ such that $R^*(u_{1\mu}, u_{2\mu}, u_{3\mu}) = 0$. The latter implies that $\mu^2 R^*(u_{1\mu}, u_{2\mu}, u_{3\mu}) = R(u_{1\mu}, u_{2\mu}, u_{3\mu}) = 0$.

In both cases as above, there always exists $\mu > 0$ such that $R(u_{1\mu}, u_{2\mu}, u_{3\mu}) = 0$.

Furthermore, we can easily check that

$$\begin{aligned} &R(u_{1\lambda}, u_{2\lambda}, u_{3\lambda}) > 0 \quad for \quad \lambda \in (0, \mu), \\ &R(u_{1\lambda}, u_{2\lambda}, u_{3\lambda}) < 0 \quad for \quad \lambda \in (\mu, \infty). \end{aligned} \tag{3.12}$$

By a direct calculation, we achive for $j, k = 1, 2, 3$,

$$\frac{d\left(\eta\lambda^5(\xi_j^2 + \xi_k^2)/(\lambda^2|\xi|^2 - \delta)\right)}{d\lambda} = \frac{3\eta\lambda^4(\xi_j^2 + \xi_k^2)}{\lambda^2|\xi|^2 - \delta} - \frac{2\eta\delta\lambda^4(\xi_j^2 + \xi_k^2)}{(\lambda^2|\xi|^2 - \delta)^2},$$

$$\frac{d\left(\eta\lambda^5\xi_j\xi_k/(\lambda^2|\xi|^2 - \delta)\right)}{d\lambda} = \frac{3\eta\lambda^4\xi_j\xi_k}{\lambda^2|\xi|^2 - \delta} - \frac{2\eta\delta\lambda^4\xi_j\xi_k}{(\lambda^2|\xi|^2 - \delta)^2},$$

which together with (3.9) and (3.10) yield that

$$\frac{\mathrm{d}}{\mathrm{d}\lambda} S(u_{1\lambda}, u_{2\lambda}, u_{3\lambda}) = \lambda^{-1} R(u_{1\lambda}, u_{2\lambda}, u_{3\lambda}). \tag{3.13}$$

By $R(u_{1\mu}, u_{2\mu}, u_{3\mu}) = 0$, (3.12) and (3.13) imply that

$$S(u_{1\mu}, u_{2\mu}, u_{3\mu}) \geqslant S(u_{1\lambda}, u_{2\lambda}, u_{3\lambda}), \quad \forall \lambda > 0.$$

This completes the proof of Proposition 3.3. □

Now, we begin to prove Theorem 2.1.

Step 1 Proof of (1).

Let $\{(Q_{1n}, Q_{2n}, Q_{3n}), n \in \mathbf{N}\} \subset M$ be a minimizing sequence for (2.5). There then has

$$R(Q_{1n}, Q_{2n}, Q_{3n}) = 0$$

and

$$S(Q_{1n}, Q_{2n}, Q_{3n}) \to \inf_{(u_1,u_2,u_3)\in M} S(u_1, u_2, u_3) \ as\ n \to \infty. \tag{3.14}$$

In view of $\eta > 0$, $\delta \leqslant 0$ and Young's inequality, combining (3.6) with (3.14) yields that $\|Q_{1n}\|_{H_r^1(\mathbb{R}^3)}$, $\|Q_{2n}\|_{H_r^1(\mathbb{R}^3)}$, $\|Q_{3n}\|_{H_r^1(\mathbb{R}^3)}$ are all bounded for all $n \in \mathbb{N}$. Thus there exists a subsequence

$$\{(Q_{1nk}, Q_{2nk}, Q_{3nk}), k \in \mathbb{N}\} \subset \{(Q_{1n}, Q_{2n}, Q_{3n}), n \in \mathbb{N}\} \tag{3.15}$$

such that as $k \to \infty$,

$$Q_{1nk} \rightharpoonup Q_{1\infty}\ weakly\ in\ H_r^1(\mathbb{R}^3), \quad Q_{1nk} \to Q_{1\infty}\ a.e.\ in\ \mathbb{R}^3, \tag{3.16}$$

$$Q_{2nk} \rightharpoonup Q_{2\infty}\ weakly\ in\ H_r^1(\mathbb{R}^3), \quad Q_{2nk} \to Q_{2\infty}\ a.e.\ in\ \mathbb{R}^3, \tag{3.17}$$

and

$$Q_{3nk} \rightharpoonup Q_{3\infty}\ weakly\ in\ H_r^1(\mathbb{R}^3), \quad Q_{3nk} \to Q_{3\infty}\ a.e.\ in\ \mathbb{R}^3. \tag{3.18}$$

For simplicity, we still denote $\{(Q_{1nk}, Q_{2nk}, Q_{3nk}), k \in \mathbb{N}\}$ by $\{(Q_{1n}, Q_{2n}, Q_{3n}), n \in \mathbb{N}\}$. From Lemma 3.1, (3.16)–(3.18), it follows that

$$\begin{aligned} &Q_{1n} \to Q_{1\infty}\ strongly\ in\ L_r^4(\mathbb{R}^3),\\ &Q_{2n} \to Q_{2\infty}\ strongly\ in\ L_r^4(\mathbb{R}^3),\\ &Q_{3n} \to Q_{3\infty}\ strongly\ in\ L_r^4(\mathbb{R}^3). \end{aligned} \tag{3.19}$$

Since

$$\|Q_{1n}\|_{L_r^2(\mathbb{R}^3)} \leqslant C, \quad \|Q_{2n}\|_{L_r^2(\mathbb{R}^3)} \leqslant C, \|Q_{3n}\|_{L_r^2(\mathbb{R}^3)} \leqslant C,$$

the boundedness of $\{(Q_{1n}, Q_{2n}, Q_{3n}), n \in \mathbb{N}\}$ in $H_r^1(\mathbb{R}^3) \times H_r^1(\mathbb{R}^3) \times H_r^1(\mathbb{R}^3)$ and the Gagliardo-Nirenberg inequality

$$\|v\|^{p+1}_{L_r^{p+1}(\mathbb{R}^N)} \leqslant C\|\nabla v\|^{\frac{N}{2}(p-1)}_{L_r^2(\mathbb{R}^N)} \|v\|^{p+1-\frac{N}{2}(p-1)}_{L_r^2(\mathbb{R}^N)}, \quad v \in H_r^1(\mathbb{R}^N), \quad 1 \leqslant p < \frac{N+2}{N-2},$$

imply in particular that

$$\begin{aligned} &\int_{\mathbb{R}^3} |Q_{1n}|^4 \mathrm{d}x \leqslant C \left(\int_{\mathbb{R}^3} |\nabla Q_{1n}|^2 \mathrm{d}x\right)^{\frac{3}{2}},\\ &\int_{\mathbb{R}^3} |Q_{2n}|^4 \mathrm{d}x \leqslant C \left(\int_{\mathbb{R}^3} |\nabla Q_{2n}|^2 \mathrm{d}x\right)^{\frac{3}{2}},\\ &\int_{\mathbb{R}^3} |Q_{3n}|^4 \mathrm{d}x \leqslant C \left(\int_{\mathbb{R}^3} |\nabla Q_{3n}|^2 \mathrm{d}x\right)^{\frac{3}{2}}. \end{aligned} \tag{3.20}$$

Here and henceforth, $C>0$ denotes various positive constants. Via (3.20), $\eta>0$, $\delta\leqslant 0$ and

$$R(Q_{1n},Q_{2n},Q_{3n})=0,\quad \int_{\mathbb{R}^3}|Q_{jn}|^2|Q_{kn}|^2\mathrm{d}x\leqslant\frac{1}{2}\int_{\mathbb{R}^3}(|Q_{jn}|^4+|Q_{kn}|^4)\mathrm{d}x, \tag{3.21}$$

where $j,k=1,2,3$, we thus obtain

$$\begin{aligned}
&\int_{\mathbb{R}^3}(|\nabla Q_{1n}|^2+|\nabla Q_{2n}|^2+|\nabla Q_{3n}|^2)\mathrm{d}x\\
&\leqslant C\left(\int_{\mathbb{R}^3}|\nabla Q_{1n}|^2\mathrm{d}x\right)^{\frac{3}{2}}+\left(\int_{\mathbb{R}^3}|\nabla Q_{2n}|^2\mathrm{d}x\right)^{\frac{3}{2}}+\left(\int_{\mathbb{R}^3}|\nabla Q_{3n}|^2\mathrm{d}x\right)^{\frac{3}{2}}\\
&\leqslant C\left(\int_{\mathbb{R}^3}(|\nabla Q_{1n}|^2+|\nabla Q_{2n}|^2+|\nabla Q_{3n}|^2)\mathrm{d}x\right)^{\frac{3}{2}}.
\end{aligned}\tag{3.22}$$

(3.22) yields that $\|\nabla Q_{1n}\|^2_{L^2_r(\mathbb{R}^3)}+\|\nabla Q_{2n}\|^2_{L^2_r(\mathbb{R}^3)}+\|\nabla Q_{3n}\|^2_{L^2_r(\mathbb{R}^3)}$ is bounded away from 0. Furthermore, we claim that $(Q_{1\infty},Q_{2\infty},Q_{3\infty})\neq(0,0,0)$. Assume to the contrary that $(Q_{1\infty},Q_{2\infty},Q_{3\infty})\equiv(0,0,0)$, from (3.19), it follows that

$$\begin{aligned}
&Q_{1n}\to 0 \textit{ strongly in } L^4_r(\mathbb{R}^3),\\
&Q_{2n}\to 0 \textit{ strongly in } L^4_r(\mathbb{R}^3),\\
&Q_{3n}\to 0 \textit{ strongly in } L^4_r(\mathbb{R}^3).
\end{aligned}\tag{3.23}$$

Thus, (3.21) implies that as $n\to\infty$,

$$\begin{aligned}
&\int_{\mathbb{R}^3}(|Q_{1n}|^4+|Q_{2n}|^4+|Q_{3n}|^4)dx\to 0,\\
&\int_{\mathbb{R}^3}(|Q_{1n}|^2|Q_{2n}|^2+|Q_{1n}|^2|Q_{3n}|^2+|Q_{2n}|^2|Q_{3n}|^2)\mathrm{d}x\to 0,\\
&R_{1\lambda}(Q_{1n},Q_{2n},Q_{3n})|_{\lambda=1}\to 0,
\end{aligned}\tag{3.24}$$

where $R_{1\lambda}(Q_{1n},Q_{2n},Q_{3n})$ is defined by (3.10) with replacing (u_1,u_2,u_3) by (Q_{1n},Q_{2n},Q_{3n}). In view of $R(Q_{1n},Q_{2n},Q_{3n})=0$, one would then conclude that as $n\to\infty$,

$$\int_{\mathbb{R}^3}(|\nabla Q_{1n}|^2+|\nabla Q_{2n}|^2+|\nabla Q_{3n}|^2)\mathrm{d}x\to 0,$$

which contradicts to (3.22). Thus $(Q_{1\infty},Q_{2\infty},Q_{3\infty})\neq(0,0,0)$.

Let $Q_1=(Q_{1\infty})_\mu$, $Q_2=(Q_{2\infty})_\mu$, $Q_3=(Q_{3\infty})_\mu$ with $\mu>0$ uniquely determined by the condition $R(Q_1,Q_2,Q_3)=R[(Q_{1\infty})_\mu,(Q_{2\infty})_\mu,(Q_{3\infty})_\mu]=0$, where $(Q_{1\infty})_\mu,(Q_{2\infty})_\mu$ and $(Q_{3\infty})_\mu$ are defined by Proposition 3.3. Thus Lemma 3.1 yields that, as $n \to \infty$,

$$\begin{cases} (Q_{1n})_\mu \to Q_1, \quad (Q_{2n})_\mu \to Q_2, \quad (Q_{3n})_\mu \to Q_3, \ strongly\ in\ L_r^4(\mathbb{R}^3), \\ (Q_{1n})_\mu \to Q_1, \quad (Q_{2n})_\mu \to Q_2, \quad (Q_{3n})_\mu \to Q_3, \ weakly\ in\ H_r^1(\mathbb{R}^3), \end{cases} \tag{3.25}$$

whereas $R(Q_{1n}, Q_{2n}, Q_{3n}) = 0$ and Proposition 3.3 imply

$$S[(Q_{1n})_\mu, (Q_{2n})_\mu, (Q_{3n})_\mu] \leqslant S(Q_{1n}, Q_{2n}, Q_{3n}). \tag{3.26}$$

Hence, from (3.25) and (3.26) one concludes

$$\begin{aligned} S(Q_1, Q_2, Q_3) &\leqslant S[(Q_{1n})_\mu, (Q_{2n})_\mu, (Q_{3n})_\mu] \leqslant S(Q_{1n}, Q_{2n}, Q_{3n}) \\ &= \inf_{(u_1,u_2,u_3)\in M} S(u_1, u_2, u_3) \end{aligned} \tag{3.27}$$

which together with $R(Q_1, Q_2, Q_3) = 0$ yields that $(Q_1, Q_2, Q_3) \in M$. Therefore, (Q_1, Q_2, Q_3) solves the minimization problem

$$S(Q_1, Q_2, Q_3) = \min_{(u_1,u_2,u_3)\in M} S(u_1, u_2, u_3). \tag{3.28}$$

This completes the proof of (1) of Theorem 2.1 .

Step 2 Proofs of (2) and (3) of Theorem 2.1

We first prove (2). Since (Q_1, Q_2, Q_3) is a solution of the minimization problem (3.28), there exists a Lagrange multiplier Λ such that

$$\begin{aligned} &\delta'_{\overline{Q_1}}[S(Q_1, Q_2, Q_3) + \Lambda R(Q_1, Q_2, Q_3)] = 0, \\ &\delta'_{\overline{Q_2}}[S(Q_1, Q_2, Q_3) + \Lambda R(Q_1, Q_2, Q_3)] = 0, \\ &\delta'_{\overline{Q_3}}[S(Q_1, Q_2, Q_3) + \Lambda R(Q_1, Q_2, Q_3)] = 0. \end{aligned} \tag{3.29}$$

Here, $\delta'_{u_1} T(u_1, u_2, u_3)$ denotes the variation of $T(u_1, u_2, u_3)$ with respect to u_1. By the formula $\delta'_{u_1} T(u_1, u_2, u_3) = \frac{\partial}{\partial \zeta} T(u_1 + \zeta\delta' u_1, u_2, u_3)|_{\zeta=0}$, and by taking $\delta'\overline{Q_1} = \overline{Q_1}$, $\delta'\overline{Q_2} = \overline{Q_2}$ and $\delta'\overline{Q_3} = \overline{Q_3}$, we obtain from (3.29) that

$$B_1(Q_1, Q_2, Q_3) = 0, \quad B_2(Q_1, Q_2, Q_3) = 0, \quad B_3(Q_1, Q_2, Q_3) = 0, \tag{3.30}$$

where

$$\begin{aligned} B_1(Q_1, Q_2, Q_3) = &-(1 + 2\Lambda)\Delta Q_1 + \omega Q_1 - (1 + 3\Lambda)|Q_1|^2 Q_1 \\ &- (1 + 3\Lambda)Q_1|Q_2|^2 - (1 + 3\Lambda)Q_1|Q_3|^2 \\ &- (1 + 3\Lambda)Q_2\mathcal{F}^{-1}\left[\frac{\eta(\xi_1^2 + \xi_2^2)}{|\xi|^2 - \delta}\mathcal{F}(Q_1\overline{Q_2} - \overline{Q_1}Q_2)\right] \end{aligned}$$

$$
\begin{aligned}
&-(1+3\Lambda)Q_3\mathcal{F}^{-1}\left[\frac{\eta(\xi_1^2+\xi_3^2)}{|\xi|^2-\delta}\mathcal{F}(Q_1\overline{Q_3}-\overline{Q_1}Q_3)\right]\\
&+(1+3\Lambda)Q_2\mathcal{F}^{-1}\left[\frac{\eta\xi_1\xi_3}{|\xi|^2-\delta}\mathcal{F}(Q_2\overline{Q_3}-\overline{Q_2}Q_3)\right]\\
&+(1+3\Lambda)Q_2\mathcal{F}^{-1}\left[\frac{\eta\xi_2\xi_3}{|\xi|^2-\delta}\mathcal{F}(\overline{Q_1}Q_3-Q_1\overline{Q_3})\right]\\
&-(1+3\Lambda)Q_3\mathcal{F}^{-1}\left[\frac{\eta\xi_2\xi_3}{|\xi|^2-\delta}\mathcal{F}(Q_1\overline{Q_2}-\overline{Q_1}Q_2)\right]\\
&+(1+3\Lambda)Q_3\mathcal{F}^{-1}\left[\frac{\eta\xi_1\xi_2}{|\xi|^2-\delta}\mathcal{F}(\overline{Q_2}Q_3-Q_2\overline{Q_3})\right]\\
&+2\Lambda\delta Q_2\mathcal{F}^{-1}\left[\frac{\eta(\xi_1^2+\xi_2^2)}{(|\xi|^2-\delta)^2}\mathcal{F}(Q_1\overline{Q_2}-\overline{Q_1}Q_2)\right]\\
&+2\Lambda\delta Q_3\mathcal{F}^{-1}\left[\frac{\eta(\xi_1^2+\xi_3^2)}{(|\xi|^2-\delta)^2}\mathcal{F}(Q_1\overline{Q_3}-\overline{Q_1}Q_3)\right]\\
&-2\Lambda\delta Q_2\mathcal{F}^{-1}\left[\frac{\eta\xi_1\xi_3}{(|\xi|^2-\delta)^2}\mathcal{F}(Q_2\overline{Q_3}-\overline{Q_2}Q_3)\right]\\
&-2\Lambda\delta Q_2\mathcal{F}^{-1}\left[\frac{\eta\xi_2\xi_3}{(|\xi|^2-\delta)^2}\mathcal{F}(\overline{Q_1}Q_3-Q_1\overline{Q_3})\right]\\
&+2\Lambda\delta Q_3\mathcal{F}^{-1}\left[\frac{\eta\xi_2\xi_3}{(|\xi|^2-\delta)^2}\mathcal{F}(Q_1\overline{Q_2}-\overline{Q_1}Q_2)\right]\\
&-2\Lambda\delta Q_3\mathcal{F}^{-1}\left[\frac{\eta\xi_1\xi_2}{(|\xi|^2-\delta)^2}\mathcal{F}(\overline{Q_2}Q_3-Q_2\overline{Q_3})\right], \qquad (3.31)
\end{aligned}
$$

$$
\begin{aligned}
B_2(Q_1,Q_2,Q_3) &= -(1+2\Lambda)\Delta Q_2+\omega Q_2-(1+3\Lambda)|Q_2|^2Q_2\\
&-(1+3\Lambda)|Q_1|^2Q_2-(1+3\Lambda)|Q_3|^2Q_2\\
&-(1+3\Lambda)Q_1\mathcal{F}^{-1}\left[\frac{\eta(\xi_1^2+\xi_2^2)}{|\xi|^2-\delta}\mathcal{F}(\overline{Q_1}Q_2-Q_1\overline{Q_2})\right]\\
&-(1+3\Lambda)Q_3\mathcal{F}^{-1}\left[\frac{\eta(\xi_2^2+\xi_3^2)}{|\xi|^2-\delta}\mathcal{F}(Q_2\overline{Q_3}-\overline{Q_2}Q_3)\right]\\
&-(1+3\Lambda)Q_1\mathcal{F}^{-1}\left[\frac{\eta\xi_1\xi_3}{|\xi|^2-\delta}\mathcal{F}(Q_2\overline{Q_3}-\overline{Q_2}Q_3)\right]\\
&-(1+3\Lambda)Q_1\mathcal{F}^{-1}\left[\frac{\eta\xi_2\xi_3}{|\xi|^2-\delta}\mathcal{F}(\overline{Q_1}Q_3-Q_1\overline{Q_3})\right]\\
&-(1+3\Lambda)Q_3\mathcal{F}^{-1}\left[\frac{\eta\xi_1\xi_2}{|\xi|^2-\delta}\mathcal{F}(Q_1\overline{Q_3}-\overline{Q_1}Q_3)\right]\\
&+(1+3\Lambda)Q_3\mathcal{F}^{-1}\left[\frac{\eta\xi_1\xi_3}{|\xi|^2-\delta}\mathcal{F}(\overline{Q_1}Q_2-Q_1\overline{Q_2})\right]\\
&+2\Lambda\delta Q_1\mathcal{F}^{-1}\left[\frac{\eta(\xi_1^2+\xi_2^2)}{(|\xi|^2-\delta)^2}\mathcal{F}(\overline{Q_1}Q_2-Q_1\overline{Q_2})\right]
\end{aligned}
$$

$$
\begin{aligned}
&+2\Lambda\delta Q_3\mathcal{F}^{-1}\left[\frac{\eta(\xi_2^2+\xi_3^2)}{(|\xi|^2-\delta)^2}\mathcal{F}(Q_2\overline{Q_3}-\overline{Q_2}Q_3)\right]\\
&+2\Lambda\delta Q_1\mathcal{F}^{-1}\left[\frac{\eta\xi_1\xi_3}{(|\xi|^2-\delta)^2}\mathcal{F}(Q_2\overline{Q_3}-\overline{Q_2}Q_3)\right]\\
&+2\Lambda\delta Q_1\mathcal{F}^{-1}\left[\frac{\eta\xi_2\xi_3}{(|\xi|^2-\delta)^2}\mathcal{F}(\overline{Q_1}Q_3-Q_1\overline{Q_3})\right]\\
&+2\Lambda\delta Q_3\mathcal{F}^{-1}\left[\frac{\eta\xi_1\xi_2}{(|\xi|^2-\delta)^2}\mathcal{F}(Q_1\overline{Q_3}-\overline{Q_1}Q_3)\right]\\
&+2\Lambda\delta Q_3\mathcal{F}^{-1}\left[\frac{\eta\xi_1\xi_3}{(|\xi|^2-\delta)^2}\mathcal{F}(\overline{Q_1}Q_2-Q_1\overline{Q_2})\right],
\end{aligned}\tag{3.32}
$$

$$
\begin{aligned}
B_3(Q_1,Q_2,Q_3)=&-(1+2\Lambda)\Delta Q_3+\omega Q_3-(1+3\Lambda)|Q_3|^2Q_3\\
&-(1+3\Lambda)|Q_1|^2Q_3-(1+3\Lambda)|Q_2|^2Q_3\\
&-(1+3\Lambda)Q_1\mathcal{F}^{-1}\left[\frac{\eta(\xi_1^2+\xi_3^2)}{|\xi|^2-\delta}\mathcal{F}(\overline{Q_1}Q_3-Q_1\overline{Q_3})\right]\\
&-(1+3\Lambda)Q_2\mathcal{F}^{-1}\left[\frac{\eta(\xi_2^2+\xi_3^2)}{|\xi|^2-\delta}\mathcal{F}(\overline{Q_2}Q_3-Q_2\overline{Q_3})\right]\\
&-(1+3\Lambda)Q_1\mathcal{F}^{-1}\left[\frac{\eta\xi_1\xi_2}{|\xi|^2-\delta}\mathcal{F}(\overline{Q_2}Q_3-Q_2\overline{Q_3})\right]\\
&+(1+3\Lambda)Q_1\mathcal{F}^{-1}\left[\frac{\eta\xi_2\xi_3}{|\xi|^2-\delta}\mathcal{F}(Q_1\overline{Q_2}-\overline{Q_1}Q_2)\right]\\
&-(1+3\Lambda)Q_2\mathcal{F}^{-1}\left[\frac{\eta\xi_1\xi_3}{|\xi|^2-\delta}\mathcal{F}(\overline{Q_1}Q_2-Q_1\overline{Q_2})\right]\\
&+(1+3\Lambda)Q_2\mathcal{F}^{-1}\left[\frac{\eta\xi_1\xi_2}{|\xi|^2-\delta}\mathcal{F}(Q_1\overline{Q_3}-\overline{Q_1}Q_3)\right]\\
&+2\Lambda\delta Q_1\mathcal{F}^{-1}\left[\frac{\eta(\xi_1^2+\xi_3^2)}{(|\xi|^2-\delta)^2}\mathcal{F}(\overline{Q_1}Q_3-Q_1\overline{Q_3})\right]\\
&+2\Lambda\delta Q_2\mathcal{F}^{-1}\left[\frac{\eta(\xi_2^2+\xi_3^2)}{(|\xi|^2-\delta)^2}\mathcal{F}(\overline{Q_2}Q_3-Q_2\overline{Q_3})\right]\\
&+2\Lambda\delta Q_1\mathcal{F}^{-1}\left[\frac{\eta\xi_1\xi_2}{(|\xi|^2-\delta)^2}\mathcal{F}(\overline{Q_2}Q_3-Q_2\overline{Q_3})\right]\\
&-2\Lambda\delta Q_1\mathcal{F}^{-1}\left[\frac{\eta\xi_2\xi_3}{(|\xi|^2-\delta)^2}\mathcal{F}(Q_1\overline{Q_2}-\overline{Q_1}Q_2)\right]\\
&+2\Lambda\delta Q_2\mathcal{F}^{-1}\left[\frac{\eta\xi_1\xi_3}{(|\xi|^2-\delta)^2}\mathcal{F}(\overline{Q_1}Q_2-Q_1\overline{Q_2})\right]\\
&-2\Lambda\delta Q_2\mathcal{F}^{-1}\left[\frac{\eta\xi_1\xi_2}{(|\xi|^2-\delta)^2}\mathcal{F}(Q_1\overline{Q_3}-\overline{Q_1}Q_3)\right].
\end{aligned}\tag{3.33}
$$

Multiplying $B_1(Q_1,Q_2,Q_3)=0$ by $\overline{Q_1}$, $B_2(Q_1,Q_2,Q_3)=0$ by $\overline{Q_2}$ and $B_3(Q_1,Q_2,$

$Q_3) = 0$ by $\overline{Q_3}$, then integrating the resulting equations with respect to x on $\mathbb{R}^3$, we get

$$K_1(Q_1, Q_2, Q_3) + \Lambda K_2(Q_1, Q_2, Q_3) = 0, \tag{3.34}$$

where

$$\begin{aligned} K_1(Q_1, Q_2, Q_3) = & \int_{\mathbb{R}^3} (|\nabla Q_1|^2 + |\nabla Q_2|^2 + |\nabla Q_3|^2) \mathrm{d}x \\ & + \omega \int_{\mathbb{R}^3} (|Q_1|^2 + |Q_2|^2 + |Q_3|^2) \mathrm{d}x \\ & - \int_{\mathbb{R}^3} (|Q_1|^4 + |Q_2|^4 + |Q_3|^4) \mathrm{d}x \\ & - 2 \int_{\mathbb{R}^3} (|Q_1|^2 |Q_2|^2 + |Q_1|^2 |Q_3|^2 + |Q_2|^2 |Q_3|^2) \mathrm{d}x \\ & - \int_{\mathbb{R}^3} A(Q_1, Q_2, Q_3) \mathrm{d}\xi + 2Re \int_{\mathbb{R}^3} B(Q_1, Q_2, Q_3) \mathrm{d}\xi, \end{aligned} \tag{3.35}$$

$$\begin{aligned} K_2(Q_1, Q_2, Q_3) = & 2 \int_{\mathbb{R}^3} (|\nabla Q_1|^2 + |\nabla Q_2|^2 + |\nabla Q_3|^2) \mathrm{d}x \\ & - 3 \int_{\mathbb{R}^3} (|Q_1|^4 + |Q_2|^4 + |Q_3|^4) \mathrm{d}x \\ & - 6 \int_{\mathbb{R}^3} (|Q_1|^2 |Q_2|^2 + |Q_1|^2 |Q_3|^2 + |Q_2|^2 |Q_3|^2) \mathrm{d}x \\ & - 3 \int_{\mathbb{R}^3} A(Q_1, Q_2, Q_3) d\xi + 6 \mathrm{Re} \int_{\mathbb{R}^3} B(Q_1, Q_2, Q_3) \mathrm{d}\xi \\ & + 2\delta \int_{\mathbb{R}^3} \frac{A(Q_1, Q_2, Q_3)}{|\xi|^2 - \delta} \mathrm{d}\xi - 4\delta \mathrm{Re} \int_{\mathbb{R}^3} \frac{B(Q_1, Q_2, Q_3)}{|\xi|^2 - \delta} \mathrm{d}\xi, \end{aligned} \tag{3.36}$$

$$\begin{aligned} A(Q_1, Q_2, Q_3) = & \frac{\eta}{|\xi|^2 - \delta} [(\xi_1^2 + \xi_2^2)(|\mathcal{F}(Q_1 \overline{Q_2} - \overline{Q_1} Q_2)|^2) \\ & + (\xi_1^2 + \xi_3^2)(|\mathcal{F}(Q_1 \overline{Q_3} - \overline{Q_1} Q_3)|^2) \\ & + (\xi_2^2 + \xi_3^2)(|\mathcal{F}(Q_2 \overline{Q_3} - \overline{Q_2} Q_3)|^2)], \end{aligned} \tag{3.37}$$

$$\begin{aligned} B(Q_1, Q_2, Q_3) = & \frac{\eta}{|\xi|^2 - \delta} [\xi_1 \xi_2 \mathcal{F}(\overline{Q_2} Q_3 - Q_2 \overline{Q_3}) \overline{\mathcal{F}(Q_1 \overline{Q_3} - \overline{Q_1} Q_3)} \\ & + \xi_1 \xi_3 \mathcal{F}(Q_2 \overline{Q_3} - \overline{Q_2} Q_3) \overline{\mathcal{F}(Q_1 \overline{Q_2} - \overline{Q_1} Q_2)} \\ & + \xi_2 \xi_3 \mathcal{F}(\overline{Q_1} Q_3 - Q_1 \overline{Q_3}) \overline{\mathcal{F}(Q_1 \overline{Q_2} - \overline{Q_1} Q_2)}]. \end{aligned} \tag{3.38}$$

On the other hand, multiplying $B_1(Q_1, Q_2, Q_3) = 0$ by $x\nabla\overline{Q_1}$, $B_2(Q_1, Q_2, Q_3) = 0$ by $x\nabla\overline{Q_2}$, and $B_3(Q_1, Q_2, Q_3) = 0$ by $x\nabla\overline{Q_3}$, then integrating the resulting equations with respect to x on $\mathbb{R}^3$ and taking the real part, one obtains

$$K_3(Q_1,Q_2,Q_3)+\Lambda K_4(Q_1,Q_2,Q_3)=0, \tag{3.39}$$

where

$$\begin{aligned}K_3(Q_1,Q_2,Q_3)=&-\frac{1}{2}\int_{\mathbb{R}^3}(|\nabla Q_1|^2+|\nabla Q_2|^2+|\nabla Q_3|^2)\mathrm{d}x\\&-\frac{3\omega}{2}\int_{\mathbb{R}^3}(|Q_1|^2+|Q_2|^2+|Q_3|^2)\mathrm{d}x\\&+\frac{3}{4}\int_{\mathbb{R}^3}(|Q_1|^4+|Q_2|^4+|Q_3|^4)\mathrm{d}x\\&+\frac{3}{2}\int_{\mathbb{R}^3}(|Q_1|^2|Q_2|^2+|Q_1|^2|Q_3|^2+|Q_2|^2|Q_3|^2)\mathrm{d}x\\&+\frac{3}{4}\int_{\mathbb{R}^3}A(Q_1,Q_2,Q_3)\mathrm{d}\xi+\frac{\delta}{2}\int_{\mathbb{R}^3}\frac{A(Q_1,Q_2,Q_3)}{|\xi|^2-\delta}\mathrm{d}\xi\\&-\frac{3}{2}\mathrm{Re}\int_{\mathbb{R}^3}B(Q_1,Q_2,Q_3)\mathrm{d}\xi-\delta\mathrm{Re}\int_{\mathbb{R}^3}\frac{B(Q_1,Q_2,Q_3)}{|\xi|^2-\delta}\mathrm{d}\xi,\end{aligned} \tag{3.40}$$

$$\begin{aligned}K_4(Q_1,Q_2,Q_3)=&-\int_{\mathbb{R}^3}(|\nabla Q_1|^2+|\nabla Q_2|^2+|\nabla Q_3|^2)\mathrm{d}x\\&+\frac{9}{4}\int_{\mathbb{R}^3}(|Q_1|^4+|Q_2|^4+|Q_3|^4)\mathrm{d}x\\&+\frac{9}{2}\int_{\mathbb{R}^3}(|Q_1|^2|Q_2|^2+|Q_1|^2|Q_3|^2+|Q_2|^2|Q_3|^2)\mathrm{d}x\\&+\frac{9}{4}\int_{\mathbb{R}^3}A(Q_1,Q_2,Q_3)\mathrm{d}\xi-\frac{9}{2}\mathrm{Re}\int_{\mathbb{R}^3}B(Q_1,Q_2,Q_3)\mathrm{d}\xi\\&-\delta\int_{\mathbb{R}^3}\frac{A(Q_1,Q_2,Q_3)}{|\xi|^2-\delta}\mathrm{d}\xi-2\delta^2\int_{\mathbb{R}^3}\frac{A(Q_1,Q_2,Q_3)}{(|\xi|^2-\delta)^2}\mathrm{d}\xi\\&+2\delta\mathrm{Re}\int_{\mathbb{R}^3}\frac{B(Q_1,Q_2,Q_3)}{|\xi|^2-\delta}\mathrm{d}\xi+4\delta^2\mathrm{Re}\int_{\mathbb{R}^3}\frac{B(Q_1,Q_2,Q_3)}{(|\xi|^2-\delta)^2}\mathrm{d}\xi.\end{aligned} \tag{3.41}$$

Here, $A(Q_1,Q_2,Q_3)$ and $B(Q_1,Q_2,Q_3)$ are defined by (3.37) and (3.38), respectively. Thus by (3.34), we obtain

$$\frac{3}{2}K_1(Q_1,Q_2,Q_3)+\frac{3}{2}\Lambda K_2(Q_1,Q_2,Q_3)=0. \tag{3.42}$$

Noting that

$$\begin{aligned}&\frac{3}{2}K_1(Q_1,Q_2,Q_3)+K_3(Q_1,Q_2,Q_3)\\&=\int_{\mathbb{R}^3}(|\nabla Q_1|^2+|\nabla Q_2|^2+|\nabla Q_3|^2)\mathrm{d}x-\frac{3}{4}\int_{\mathbb{R}^3}(|Q_1|^4+|Q_2|^4+|Q_3|^4)\mathrm{d}x\\&\quad-\frac{3}{2}\int_{\mathbb{R}^3}(|Q_1|^2|Q_2|^2+|Q_1|^2|Q_3|^2+|Q_2|^2|Q_3|^2)\mathrm{d}x-\frac{3}{4}\int_{\mathbb{R}^3}A(Q_1,Q_2,Q_3)\mathrm{d}\xi\\&\quad+\frac{\delta}{2}\int_{\mathbb{R}^3}\frac{A(Q_1,Q_2,Q_3)}{|\xi|^2-\delta}\mathrm{d}\xi+\frac{3}{2}\mathrm{Re}\int_{\mathbb{R}^3}B(Q_1,Q_2,Q_3)\mathrm{d}\xi-\delta\mathrm{Re}\int_{\mathbb{R}^3}\frac{B(Q_1,Q_2,Q_3)}{|\xi|^2-\delta}\mathrm{d}\xi\\&=R(Q_1,Q_2,Q_3)=0,\end{aligned}\tag{3.43}$$

in view of (3.39), (3.42),(3.43), one can verify that

$$\frac{3}{2}\Lambda K_2(Q_1,Q_2,Q_3)+\Lambda K_4(Q_1,Q_2,Q_3)=0.\tag{3.44}$$

(3.44) is equivalent to

$$\Lambda\left[\frac{3}{2}K_2(Q_1,Q_2,Q_3)+K_4(Q_1,Q_2,Q_3)\right]=\Lambda K_5(Q_1,Q_2,Q_3)=0,\tag{3.45}$$

where

$$\begin{aligned}K_5(Q_1,Q_2,Q_3)&=3R(Q_1,Q_2,Q_3)-\int_{\mathbb{R}^3}(|\nabla Q_1|^2+|\nabla Q_2|^2+|\nabla Q_3|^2)\mathrm{d}x\\&\quad+\frac{\delta}{2}\int_{\mathbb{R}^3}\frac{A(Q_1,Q_2,Q_3)}{|\xi|^2-\delta}\mathrm{d}\xi-2\delta^2\int_{\mathbb{R}^3}\frac{A(Q_1,Q_2,Q_3)}{(|\xi|^2-\delta)^2}\mathrm{d}\xi\\&\quad-\delta\mathrm{Re}\int_{\mathbb{R}^3}\frac{B(Q_1,Q_2,Q_3)}{|\xi|^2-\delta}\mathrm{d}\xi+4\delta^2\mathrm{Re}\int_{\mathbb{R}^3}\frac{B(Q_1,Q_2,Q_3)}{(|\xi|^2-\delta)^2}\mathrm{d}\xi.\end{aligned}\tag{3.46}$$

Noting the expressions of $A(Q_1,Q_2,Q_3)$ and $B(Q_1,Q_2,Q_3)$ in (3.37) and (3.38), applying the Young's inequality, we have

$$\begin{aligned}&\frac{\delta}{2}\int_{\mathbb{R}^3}\frac{A(Q_1,Q_2,Q_3)}{|\xi|^2-\delta}\mathrm{d}\xi-2\delta^2\int_{\mathbb{R}^3}\frac{A(Q_1,Q_2,Q_3)}{(|\xi|^2-\delta)^2}\mathrm{d}\xi\\&-\delta\mathrm{Re}\int_{\mathbb{R}^3}\frac{B(Q_1,Q_2,Q_3)}{|\xi|^2-\delta}\mathrm{d}\xi+4\delta^2\mathrm{Re}\int_{\mathbb{R}^3}\frac{B(Q_1,Q_2,Q_3)}{(|\xi|^2-\delta)^2}\mathrm{d}\xi\leqslant 0.\end{aligned}$$

Then, in view of $(Q_1,Q_2,Q_3)\neq(0,0,0)$, $R(Q_1,Q_2,Q_3)=0$, $\eta>0$ and $\delta\leqslant 0$, there holds

$$K_5(Q_1,Q_2,Q_3)\leqslant-\int_{\mathbb{R}^3}(|\nabla Q_1|^2+|\nabla Q_2|^2+|\nabla Q_3|^2)\mathrm{d}x<0.\tag{3.47}$$

which implies that $K_5(Q_1,Q_2,Q_3)\neq 0$ and thus $\Lambda=0$ by (3.45). Hence, from (3.30), (3.31), (3.32) and (3.33), it follows that (Q_1,Q_2,Q_3) solves the following equations:

$$
\begin{aligned}
& -\omega Q_1+\Delta Q_1+(|Q_1|^2+|Q_2|^2+|Q_3|^2)Q_1 \\
& -Q_2\mathcal{F}^{-1}[\frac{\eta(\xi_1^2+\xi_2^2)}{|\xi|^2-\delta}\mathcal{F}(\overline{Q_1}Q_2-Q_1\overline{Q_2})]+Q_3\mathcal{F}^{-1}[\frac{\eta(\xi_1^2+\xi_3^2)}{|\xi|^2-\delta}\mathcal{F}(Q_1\overline{Q_3}-\overline{Q_1}Q_3)] \\
& -Q_2\mathcal{F}^{-1}\{\frac{\eta}{|\xi|^2-\delta}[\xi_1\xi_3\mathcal{F}(Q_2\overline{Q_3}-\overline{Q_2}Q_3)+\xi_2\xi_3\mathcal{F}(\overline{Q_1}Q_3-Q_1\overline{Q_3})]\} \\
& +Q_3\mathcal{F}^{-1}\{\frac{\eta}{|\xi|^2-\delta}[\xi_2\xi_3\mathcal{F}(Q_1\overline{Q_2}-\overline{Q_1}Q_2)+\xi_1\xi_2\mathcal{F}(Q_2\overline{Q_3}-\overline{Q_2}Q_3)]\} \\
= {} & 0,
\end{aligned}
\tag{3.48}
$$

$$
\begin{aligned}
& -\omega Q_2+\Delta Q_2+(|Q_1|^2+|Q_2|^2+|Q_3|^2)Q_2 \\
& -Q_1\mathcal{F}^{-1}Q_1[\frac{\eta(\xi_1^2+\xi_2^2)}{|\xi|^2-\delta}\mathcal{F}(Q_1\overline{Q_2}-\overline{Q_1}Q_2)]+Q_3\mathcal{F}^{-1}[\frac{\eta(\xi_2^2+\xi_3^2)}{|\xi|^2-\delta}\mathcal{F}(Q_2\overline{Q_3}-\overline{Q_2}Q_3)] \\
& -Q_3\mathcal{F}^{-1}\{\frac{\eta}{|\xi|^2-\delta}[\xi_1\xi_2\mathcal{F}(\overline{Q_1}Q_3-Q_1\overline{Q_3})+\xi_1\xi_3\mathcal{F}(Q_1\overline{Q_2}-\overline{Q_1}Q_2)]\} \\
& +Q_1\mathcal{F}^{-1}\{\frac{\eta}{|\xi|^2-\delta}[\xi_1\xi_3\mathcal{F}(Q_2\overline{Q_3}-\overline{Q_2}Q_3)+\xi_2\xi_3\mathcal{F}(\overline{Q_1}Q_3-Q_1\overline{Q_3})]\} \\
= {} & 0,
\end{aligned}
\tag{3.49}
$$

$$
\begin{aligned}
& -\omega Q_3+\Delta Q_3+(|Q_1|^2+|Q_2|^2+|Q_3|^2)Q_3 \\
& -Q_1\mathcal{F}^{-1}[\frac{\eta(\xi_3^2+\xi_1^2)}{|\xi|^2-\delta}\mathcal{F}(Q_1\overline{Q_3}-\overline{Q_1}Q_3)]+Q_2\mathcal{F}^{-1}[\frac{\eta(\xi_2^2+\xi_3^2)}{|\xi|^2-\delta}\mathcal{F}(\overline{Q_2}Q_3-Q_2\overline{Q_3})] \\
& -Q_1\mathcal{F}^{-1}\{\frac{\eta}{|\xi|^2-\delta}[\xi_2\xi_3\mathcal{F}(Q_1\overline{Q_2}-\overline{Q_1}Q_2)+\xi_1\xi_2\mathcal{F}(Q_2\overline{Q_3}-\overline{Q_2}Q_3)]\} \\
& +Q_2\mathcal{F}^{-1}\{\frac{\eta}{|\xi|^2-\delta}[\xi_1\xi_2\mathcal{F}(\overline{Q_1}Q_3-Q_1\overline{Q_3})+\xi_1\xi_3\mathcal{F}(Q_1\overline{Q_2}-\overline{Q_1}Q_2)]\} \\
= {} & 0.
\end{aligned}
\tag{3.50}
$$

That is, (Q_1, Q_2, Q_3) solves the equations (1.5)–(1.7). As (1.5)–(1.7) are the Euler-Lagrange equations of the functional $S(Q_1, Q_2, Q_3)$, applying Proposition 3.1, we conclude (Q_1, Q_2, Q_3) is a ground state solution of (1.5)–(1.7). Furthermore, it is obvious that (Q_1, Q_2, Q_3) are functions of $|x|$ alone. Motivated by the works [2,3], we can obtain that (Q_1, Q_2, Q_3) has exponential decay at infinity, which will be shown in the Appendix A for convenience.

This completes the proof of Theorem 2.1. □

4 Orbital instability of standing waves in $\mathbb{R}^3$

In this section, we show the instability of standing waves of (1.1)–(1.3) in $\mathbb{R}^3$ obtained in Theorem 2.1 (Theorem 2.2). We first give a key proposition to show Theorem 2.2.

Proposition 4.1 *Let* $\delta \leqslant 0$, $\eta > 0$ *and* $u_{1\lambda}(x) = \lambda^{\frac{3}{2}} u_1(\lambda x)$, $u_{2\lambda}(x) = \lambda^{\frac{3}{2}} u_2(\lambda x)$, $u_{3\lambda}(x) = \lambda^{\frac{3}{2}} u_3(\lambda x)$ *for* $\lambda > 0$. *Suppose that* $(u_1, u_2, u_3) \in H_r^1(\mathbb{R}^3) \times H_r^1(\mathbb{R}^3) \times H_r^1(\mathbb{R}^3) \backslash \{(0,0,0)\}$ *and* $(u_1, u_2, u_3) \in K$, *where*

$$\begin{aligned} K &= \{(u_1, u_2, u_3) \in H_r^1(\mathbb{R}^3) \times H_r^1(\mathbb{R}^3) \times H_r^1(\mathbb{R}^3), \\ &\quad R(u_1, u_2, u_3) < 0, S(u_1, u_2, u_3) < S(Q_1, Q_2, Q_3)\}. \end{aligned} \tag{4.1}$$

Then there exists $0 < \mu < 1$ *such that* $R(u_{1\mu}, u_{2\mu} u_{3\mu}) = 0$ *and*

$$S(u_1, u_2, u_3) - S(u_{1\mu}, u_{2\mu} u_{3\mu}) \geqslant \frac{1}{2} R(u_1, u_2, u_3). \tag{4.2}$$

Here, $S(u_1, u_2, u_3)$ *and* $R(u_1, u_2, u_3)$ *are defined by* (2.3) *and* (2.4), *respectively.*

Proof By a direct calculation, we can easily show the estimate (4.2). □

Now, we begin to show Theorem 2.2.

Proof of Theorem 2.2 Let $(E_1, E_2, E_3) \in H_r^1(\mathbb{R}^3) \times H_r^1(\mathbb{R}^3) \times H_r^1(\mathbb{R}^3)$ be a solution to the equations (1.1)–(1.3) with (1.4) on $[0, T)$. By the conservation laws of the total mass and of the total energy (2.1) and (2.2), we get

$$S(E_1(t), E_2(t), E_3(t)) = S(E_{10}, E_{20}, E_{30}), t \in [0, T). \tag{4.3}$$

Let

$$J(t) = \int_{\mathbb{R}^3} |x|^2 (|E_1|^2 + |E_2|^2 + |E_3|^2) \mathrm{d}x. \tag{4.4}$$

By a direct calculation, one achieves

$$\frac{\mathrm{d}^2}{\mathrm{d}t^2} J(t) = 8R(E_1, E_2, E_3). \tag{4.5}$$

We further need to show, for some initial data, that the right-hand side of (4.5) is strictly negative (that is, $R(E_1, E_2, E_3) < 0$). One first notices that

$$S(Q_1, Q_2, Q_3) = \min_{(u_1, u_2, u_3) \in M} S(u_1, u_2, u_3) > 0. \tag{4.6}$$

Let $(E_{10}, E_{20}, E_{30}) \in K$ such that

$$(|x| E_{10}, |x| E_{20}, |x| E_{30}) \in L_r^2(\mathbb{R}^3) \times L_r^2(\mathbb{R}^3) \times L_r^2(\mathbb{R}^3). \tag{4.7}$$

We shall see later that such (E_{10}, E_{20}, E_{30}) exists. We claim that there is a finite time T such that

$$\lim_{t \to T} (\|E_1\|_{H_r^1(\mathbb{R}^3)} + \|E_2\|_{H_r^1(\mathbb{R}^3)} + \|E_3\|_{H_r^1(\mathbb{R}^3)}) = +\infty. \tag{4.8}$$

Indeed, for such $(E_{10}, E_{20}, E_{30}) \in K$, one has

$$S(E_1, E_2, E_3) = S(E_{10}, E_{20}, E_{30}) < S(Q_1, Q_2, Q_3), t \in [0, T), \tag{4.9}$$

and

$$R(E_1, E_2, E_3) < 0, t \in [0, T). \tag{4.10}$$

The latter is true, for otherwise, by continuity, there would exist a $t_1 > 0$ such that $0 < t_1 < T$, and

$$R(E_1(t_1), E_2(t_1), E_3(t_1)) = 0, \tag{4.11}$$

which implies that $(E_1(t_1), E_2(t_1), E_3(t_1)) \in M$. This contradicts Theorem 2.1 and (4.9).

Next, for a fixed $t \in [0, T)$, $(E_1, E_2, E_3) = (E_1(t), E_2(t), E_3(t))$, and let $0 < \mu < 1$ be such that

$$R(E_{1\mu}, E_{2\mu}, E_{3\mu}) = 0, (E_{1\mu}(x), E_{2\mu}(x), E_{3\mu}(x)) = (\mu^{\frac{3}{2}} E_1(\mu x), \mu^{\frac{3}{2}} E_2(\mu x), \mu^{\frac{3}{2}} E_3(\mu x))$$

(see Proposition 3.3). Since

$$S(E_{1\mu}, E_{2\mu}, E_{3\mu}) \geqslant S(Q_1, Q_2, Q_3), S(E_1, E_2, E_3) = S(E_{10}, E_{20}, E_{30}), \tag{4.12}$$

in view of Proposition 4.1, we have

$$\begin{aligned} R(E_1, E_2, E_3) &\leqslant 2[S(E_1, E_2, E_3) - S(E_{1\mu}, E_{2\mu}, E_{3\mu})] \\ &\leqslant 2[S(E_{10}, E_{20}, E_{30}) - S(Q_1, Q_2, Q_3)] \\ &=: \varphi < 0. \end{aligned} \tag{4.13}$$

(4.5) and (4.13) then yield that

$$\frac{\mathrm{d}^2}{\mathrm{d}t^2} J(t) \leqslant 8\varphi < 0, \tag{4.14}$$

which implies that T must be finite and that

$$\lim_{t \to T} (\|E_1\|_{H_r^1(\mathbb{R}^3)} + \|E_2\|_{H_r^1(\mathbb{R}^3)} + \|E_3\|_{H_r^1(\mathbb{R}^3)}) = +\infty.$$

In order to complete the proof of Theorem 2.2, we need to show $(E_{10}, E_{20}, E_{30}) \in K$ with (4.7). Let

$$E_{10}(x) = \lambda^{\frac{3}{2}} Q_1(\lambda x), E_{20}(x) = \lambda^{\frac{3}{2}} Q_2(\lambda x), E_{30}(x) = \lambda^{\frac{3}{2}} Q_3(\lambda x), \lambda > 1. \tag{4.15}$$

By Proposition 3.3, the functions $E_{10}(x), E_{20}(x), E_{30}(x)$ verify

$$R(E_{10}, E_{20}, E_{30}) < 0, \ S(E_{10}, E_{20}, E_{30}) < S(Q_1, Q_2, Q_3), \ \lambda > 1. \tag{4.16}$$

In addition, by Theorem 2.1, one sees that $(Q_1(x), Q_2(x), Q_3(x))$ have exponential decays at infinity, and hence

$$(|x|E_{10}, |x|E_{20}, |x|E_{30}) \in L_r^2(\mathbb{R}^3) \times L_r^2(\mathbb{R}^3) \times L_r^2(\mathbb{R}^3).$$

As $\lambda \to 1$,

$$\|E_{10} - Q_1\|_{H_r^1(\mathbb{R}^3)}, \|E_{20} - Q_2\|_{H_r^1(\mathbb{R}^3)}, \|E_{30} - Q_3\|_{H_r^1(\mathbb{R}^3)}$$

can be made arbitrarily small. We thus complete the proof of Theorem 2.2. □

References

[1] Berestycki H, Cazenave T. Instabilité des états stationnaires dans les équations de Schrödinger et de Klein-Gordon non linéairees. C. R. Acad. Sci. Paris, 1981, **293**: 489–492.

[2] Berestycki H, Lions P L. Nonlinear scalar field equations, I. Existence of a ground state. Arch. Rat. Mech. Anal., 1983, **82**: 313–345.

[3] Berestycki H, Lions P L. Nonlinear scalar field equations, II. Existence of infinitely many solutions. Arch. Rat. Mech. Anal., 1983, **82**: 347–375.

[4] Cazenave T, Lions P L. Orbital stability of standing waves for some nonlinear Schrödinger equations. Commun. Math. Phys., 1982, **85**: 549–561.

[5] Dendy R O. Plasma Dynamics. Oxford University Press, 1990.

[6] Gan Z H, Zhang J. Sharp threshold of global existence and instability of stangding wave for a Davey-Stewartson System. Commun. Math. Phys., 2008, **283**: 93–125.

[7] Gan Z H, Zhang J. Blow-up, global existence and standing waves for the magnetic nonlinear Schrödinger equations. Discrete and Continuous Dynamical Systems, 2012, **32**(3): 827–846.

[8] Gan Z H, Zhang J. Nonlocal nonlinear Schrödinger equations in $\mathbb{R}^3$. Arch. Rational Mech. Anal., 2013, **209**: 1–39.

[9] Grillakis M. Linearized instability for nonlinear Schrödinger and Klein-Gordon equations . Comm. Pure Appl. Math., 1988, **XLI**: 747–774.

[10] Grillakis M, Shatah J, Strauss W. Stability theory of solitary waves in the presence of symmetry, I*. J. Funct. Anal., 1987, **74**: 160–197.

[11] Jiang X, Gan Z H. Collapse for the generalized three-dimensional nonlocal nonlinear Schrödinger equations. Advanced Nonlinear Studies, 2014, **14**: 777–790.

[12] Jones C. An instability mechanism for radically symmetric standing waves of a nonlinear Schrödinger equation. J. Differential Equations, 1988, **71**: 34–62.

[13] Kono M, Skoric M M, Ter Haar D. Spontaneous excitation of magnetic fields and collapse dynamics in a Langmuir plasma. J. Plasma Phys., 1981, **26**: 123–146.

[14] Laurey C. The Cauchy problem for a generalized Zakharov system. Diffe. Integral Equ., 1995, **8**(1): 105–130.

[15] Miao C X. Harmonic Analysis and Applications to Partial Differential Equations. Monographs on Modern Pure Mathematics No. 89. Second Edition. Beijing: Science Press, 2004.

[16] Miao C X. The Modern Method of Nonlinear Wave Equations. Lectures in Contemporary Mathematics, No. 2. Beijing: Science Press, 2005.

[17] Miao C X, Zhang B. Harmonic Analysis Method of Partial Differential Equations. Monographs on Modern Pure Mathematics, No. 117. Second Edition. Science Press, Beijing, 2008.

[18] Shatah J, Strauss W. Instability of nonlinear bound states. Commun. Math. Phys., 1985, **100**: 173–190.

[19] Strauss W A. Existence of solitary waves in high dimensions. Commun. Math. Phys., 1977, **55**: 149–162.

[20] Weinstein M I. Nonlinear Schrödinger equations and sharp interpolation estimates. Commun. Math. Phys., 1983, **87**: 567–576.

[21] Weinstein M I. Lyapunov stability of ground states of nonlinear dispersive evolution equations. Commun. Pure Appl. Math., 1986, **39**: 51–68.

[22] Zhang J. Sharp threshold for blowup and global existence in nonlinear Schrödinger equations under a harmonic potential. Commun. in PDE, 2005, **30**: 1429-1443.

[23] Zakharov V E. The collapse of Langmuir waves. Soviet Phys., JETP, 1972,**35**: 908–914.

[24] Zakharov V E, Musher S L, Rubenchik A M. Hamiltonian approach to the description of nonlinear plasma phenomena. Physics Reports, 1985, **129**(5): 285–366.

Appendix

In this Appendix, we will show the exponential decay at infinity of the solution (Q_1, Q_2, Q_3) to the equations (3.48)–(3.50).

Consider the equations (3.48)–(3.50):

$$\begin{aligned}
&-\omega Q_1+\Delta Q_1+(|Q_1|^2+|Q_2|^2+|Q_3|^2)Q_1\\
&-Q_2\mathcal{F}^{-1}\left[\frac{\eta(\xi_1^2+\xi_2^2)}{|\xi|^2-\delta}\mathcal{F}(\overline{Q_1}Q_2-Q_1\overline{Q_2})\right]+Q_3\mathcal{F}^{-1}\left[\frac{\eta(\xi_1^2+\xi_3^2)}{|\xi|^2-\delta}\mathcal{F}(Q_1\overline{Q_3}-\overline{Q_1}Q_3)\right]\\
&-Q_2\mathcal{F}^{-1}\left\{\frac{\eta}{|\xi|^2-\delta}[\xi_1\xi_3\mathcal{F}(Q_2\overline{Q_3}-\overline{Q_2}Q_3)+\xi_2\xi_3\mathcal{F}(\overline{Q_1}Q_3-Q_1\overline{Q_3})]\right\}\\
&+Q_3\mathcal{F}^{-1}\left\{\frac{\eta}{|\xi|^2-\delta}[\xi_2\xi_3\mathcal{F}(Q_1\overline{Q_2}-\overline{Q_1}Q_2)+\xi_1\xi_2\mathcal{F}(Q_2\overline{Q_3}-\overline{Q_2}Q_3)]\right\}\\
=&\,0, \qquad (3.48)
\end{aligned}$$

$$-\omega Q_2+\Delta Q_2+(|Q_1|^2+|Q_2|^2+|Q_3|^2)Q_2$$

$$
\begin{aligned}
&-Q_1\mathcal{F}^{-1}Q_1\left[\frac{\eta(\xi_1^2+\xi_2^2)}{|\xi|^2-\delta}\mathcal{F}(Q_1\overline{Q_2}-\overline{Q_1}Q_2)\right]+Q_3\mathcal{F}^{-1}\left[\frac{\eta(\xi_2^2+\xi_3^2)}{|\xi|^2-\delta}\mathcal{F}(Q_2\overline{Q_3}-\overline{Q_2}Q_3)\right]\\
&-Q_3\mathcal{F}^{-1}\left\{\frac{\eta}{|\xi|^2-\delta}[\xi_1\xi_2\mathcal{F}(\overline{Q_1}Q_3-Q_1\overline{Q_3})+\xi_1\xi_3\mathcal{F}(Q_1\overline{Q_2}-\overline{Q_1}Q_2)]\right\}\\
&+Q_1\mathcal{F}^{-1}\left\{\frac{\eta}{|\xi|^2-\delta}[\xi_1\xi_3\mathcal{F}(Q_2\overline{Q_3}-\overline{Q_2}Q_3)+\xi_2\xi_3\mathcal{F}(\overline{Q_1}Q_3-Q_1\overline{Q_3})]\right\}\\
=&\,0, \qquad (3.49)
\end{aligned}
$$

$$
\begin{aligned}
&-\omega Q_3+\Delta Q_3+(|Q_1|^2+|Q_2|^2+|Q_3|^2)Q_3\\
&-Q_1\mathcal{F}^{-1}\left[\frac{\eta(\xi_3^2+\xi_1^2)}{|\xi|^2-\delta}\mathcal{F}(Q_1\overline{Q_3}-\overline{Q_1}Q_3)\right]+Q_2\mathcal{F}^{-1}\left[\frac{\eta(\xi_2^2+\xi_3^2)}{|\xi|^2-\delta}\mathcal{F}(\overline{Q_2}Q_3-Q_2\overline{Q_3})\right]\\
&-Q_1\mathcal{F}^{-1}\left\{\frac{\eta}{|\xi|^2-\delta}[\xi_2\xi_3\mathcal{F}(Q_1\overline{Q_2}-\overline{Q_1}Q_2)+\xi_1\xi_2\mathcal{F}(Q_2\overline{Q_3}-\overline{Q_2}Q_3)]\right\}\\
&+Q_2\mathcal{F}^{-1}\left\{\frac{\eta}{|\xi|^2-\delta}[\xi_1\xi_2\mathcal{F}(\overline{Q_1}Q_3-Q_1\overline{Q_3})+\xi_1\xi_3\mathcal{F}(Q_1\overline{Q_2}-\overline{Q_1}Q_2)]\right\}\\
=&\,0. \qquad (3.50)
\end{aligned}
$$

Let

$$
\begin{aligned}
&g_1(Q_1,Q_2,Q_3)\\
=&-\omega Q_1+(|Q_1|^2+|Q_2|^2+|Q_3|^2)Q_1\\
&-Q_2\mathcal{F}^{-1}\left[\frac{\eta(\xi_1^2+\xi_2^2)}{|\xi|^2-\delta}\mathcal{F}(\overline{Q_1}Q_2-Q_1\overline{Q_2})\right]\\
&+Q_3\mathcal{F}^{-1}\left[\frac{\eta(\xi_1^2+\xi_3^2)}{|\xi|^2-\delta}\mathcal{F}(Q_1\overline{Q_3}-\overline{Q_1}Q_3)\right]\\
&-Q_2\mathcal{F}^{-1}\left\{\frac{\eta}{|\xi|^2-\delta}[\xi_1\xi_3\mathcal{F}(Q_2\overline{Q_3}-\overline{Q_2}Q_3)+\xi_2\xi_3\mathcal{F}(\overline{Q_1}Q_3-Q_1\overline{Q_3})]\right\}\\
&+Q_3\mathcal{F}^{-1}\left\{\frac{\eta}{|\xi|^2-\delta}[\xi_2\xi_3\mathcal{F}(Q_1\overline{Q_2}-\overline{Q_1}Q_2)+\xi_1\xi_2\mathcal{F}(Q_2\overline{Q_3}-\overline{Q_2}Q_3)]\right\}, \qquad (A\text{-}1)
\end{aligned}
$$

$$
\begin{aligned}
&g_2(Q_1,Q_2,Q_3)\\
=&-\omega Q_2+(|Q_1|^2+|Q_2|^2+|Q_3|^2)Q_2\\
&-Q_1\mathcal{F}^{-1}Q_1\left[\frac{\eta(\xi_1^2+\xi_2^2)}{|\xi|^2-\delta}\mathcal{F}(Q_1\overline{Q_2}-\overline{Q_1}Q_2)\right]\\
&+Q_3\mathcal{F}^{-1}\left[\frac{\eta(\xi_2^2+\xi_3^2)}{|\xi|^2-\delta}\mathcal{F}(Q_2\overline{Q_3}-\overline{Q_2}Q_3)\right]\\
&-Q_3\mathcal{F}^{-1}\left\{\frac{\eta}{|\xi|^2-\delta}[\xi_1\xi_2\mathcal{F}(\overline{Q_1}Q_3-Q_1\overline{Q_3})+\xi_1\xi_3\mathcal{F}(Q_1\overline{Q_2}-\overline{Q_1}Q_2)]\right\}
\end{aligned}
$$

$$+Q_1\mathcal{F}^{-1}\left\{\frac{\eta}{|\xi|^2-\delta}[\xi_1\xi_3\mathcal{F}(Q_2\overline{Q_3}-\overline{Q_2}Q_3)+\xi_2\xi_3\mathcal{F}(\overline{Q_1}Q_3-Q_1\overline{Q_3})]\right\}. \tag{A-2}$$

$$\begin{aligned}
&g_3(Q_1,Q_2,Q_3)\\
=&-\omega Q_3+(|Q_1|^2+|Q_2|^2+|Q_3|^2)Q_3\\
&-Q_1\mathcal{F}^{-1}\left[\frac{\eta(\xi_3^2+\xi_1^2)}{|\xi|^2-\delta}\mathcal{F}(Q_1\overline{Q_3}-\overline{Q_1}Q_3)\right]\\
&+Q_2\mathcal{F}^{-1}\left[\frac{\eta(\xi_2^2+\xi_3^2)}{|\xi|^2-\delta}\mathcal{F}(\overline{Q_2}Q_3-Q_2\overline{Q_3})\right]\\
&-Q_1\mathcal{F}^{-1}\left\{\frac{\eta}{|\xi|^2-\delta}[\xi_2\xi_3\mathcal{F}(Q_1\overline{Q_2}-\overline{Q_1}Q_2)+\xi_1\xi_2\mathcal{F}(Q_2\overline{Q_3}-\overline{Q_2}Q_3)]\right\}\\
&+Q_2\mathcal{F}^{-1}\left\{\frac{\eta}{|\xi|^2-\delta}[\xi_1\xi_2\mathcal{F}(\overline{Q_1}Q_3-Q_1\overline{Q_3})+\xi_1\xi_3\mathcal{F}(Q_1\overline{Q_2}-\overline{Q_1}Q_2)]\right\}.
\end{aligned}\tag{A-3}$$

Then equations (3.48)–(3.50) reduce to the following system:

$$-\Delta Q_1=g_1(Q_1,Q_2,Q_3), \tag{A-4}$$

$$-\Delta Q_2=g_2(Q_1,Q_2,Q_3), \tag{A-5}$$

$$-\Delta Q_3=g_3(Q_1,Q_2,Q_3). \tag{A-6}$$

By the expressions of $g_i(Q_1,Q_2,Q_3)$ $(i=1,2,3)$, for $\eta>0$ and $\delta\leqslant 0$, we conclude the following properties:

$$\begin{aligned}
-\infty&<\underline{\lim}_{(Q_1,Q_2,Q_3)\to(0^+,0^+,0^+)}\quad g_i(Q_1,Q_2,Q_3)/Q_i\\
&\leqslant\overline{\lim}_{(Q_1,Q_2,Q_3)\to(0^+,0^+,0^+)}\quad g_i(Q_1,Q_2,Q_3)/Q_i\\
&=-\omega<0,
\end{aligned}\tag{A-7}$$

$$-\infty\leqslant\overline{\lim}_{(Q_1,Q_2,Q_3)\to(+\infty,+\infty,+\infty)}\quad g_i(Q_1,Q_2,Q_3)/Q_i^5\leqslant 0. \tag{A-8}$$

Furthermore, motivated by Lemma 1 and Radial lemma A.II in [2,3], we can establish the following two lemmas.

Lemma A1 Under the properties (A-7)–(A-8), if (Q_1,Q_2,Q_3) is a spherically symmetric solution of (A-4)–(A-6) then $(Q_1,Q_2,Q_3)\in C^2(\mathbb{R}^3)\times C^2(\mathbb{R}^3)\times C^2(\mathbb{R}^3)$.

Lemma A2 Let $N=3$. The radial function $Q_i\in H^1(\mathbb{R}^3)$ $(i=1,2,3)$ is almost everywhere equal to a function $U_i(x)$, continuous for $x\neq 0$ and such that

$$|U_i(x)|\leqslant C_3|x|^{-1}\|Q_i\|_{H^1(\mathbb{R}^3)}\quad for\quad |x|\geqslant\alpha_3, \tag{A-9}$$

where C_3 and α_3 are two constants depend only on the dimension N (N=3).
We now begin to prove the exponential decay at infinity of the solution (Q_1,Q_2,Q_3) to (A-4)–(A-6). That is, we need to show the following proposition.

Proposition A1 Under the properties (A-7)–(A-8), if (Q_1,Q_2,Q_3) is a spherically symmetric solution of (A-4)–(A-6) then

$$|D^\alpha Q_i(x)| \leqslant Ce^{-\beta|x|}, \quad x\in\mathbb{R}^3 \tag{A-10}$$

for some C, $\beta>0$ and for $|\alpha|\leqslant 2$.

Proof By Lemma A1 $Q_i(i=1,2,3)$ is of class $C^2(\mathbb{R}^3)$, and it satisfies the equations below:

$$-\frac{\partial^2 Q_i}{\partial r^2}-\frac{2}{r}\frac{\partial Q_i}{\partial r}=g_i(Q_1,Q_2,Q_3), \tag{A-11}$$

where $i=1,2,3$. Let $P_i=rQ_i$ $(i=1,2,3)$, then P_i satisfies

$$\frac{\partial^2 P_i}{\partial r^2}=\left[\frac{-g_i(Q_1,Q_2,Q_3)}{Q_i}\right]P_i.$$

For r large enough, say $r\geqslant r_0$, one gets $-g_i(Q_1,Q_2,Q_3)/Q_i\geqslant\omega/\varepsilon$ *for any* $\varepsilon\geqslant 1$.

(Indeed, Lemma A2 yields that $Q_i(r)\to 0$ as $r\to+\infty$).

Next, let $W_i=P_i^2$ $(i=1,2,3)$, then W_i solves

$$\frac{1}{2}\frac{\partial^2 W_i}{\partial r^2}=\left(\frac{\partial P_i}{\partial r}\right)^2+(-g_i(Q_1,Q_2,Q_3)/Q_i)\,W_i.$$

Thus for $r\geqslant r_0$ one has $\dfrac{\partial^2 W_i}{\partial r^2}\geqslant\dfrac{2\omega}{\varepsilon}W_i$, and $W_i\geqslant 0$.
Further, let

$$Z_i=e^{-\sqrt{2\omega/\varepsilon}\,r}\left(\frac{\partial W_i}{\partial r}+\sqrt{2\omega/\varepsilon}W_i\right). \tag{A-12}$$

Direct calculation yields that

$$\frac{\partial Z_i}{\partial r}=e^{-\sqrt{2\omega/\varepsilon}\,r}\left(\frac{\partial^2 W_i}{\partial r^2}-\frac{2\omega}{\varepsilon}W_i\right)\geqslant 0. \tag{A-13}$$

This implies that Z_i is nondecreasing on $(r_0,+\infty)$.

We now claim that

$$Z_i(r)\leqslant 0 \quad for \quad r\geqslant r_1>r_0. \tag{A-14}$$

(Otherwise, if there exists $r_1>r_0$ such that $Z_i(r_1)>0$, then $Z_i(r)\geqslant Z_i(r_1)>0$ for all $r\geqslant r_1$. In view of (A-12),

$$\frac{\partial W_i}{\partial r}+\sqrt{2\omega/\varepsilon}W_i\geqslant Z_i(r_1)e^{\sqrt{2\omega/\varepsilon}\,r}, \tag{A-15}$$

which implies that $\frac{\partial W_i}{\partial r}+\sqrt{2\omega/\varepsilon}W_i$ is not integrable on $(r_1,+\infty)$. But P_i^2 and $P_i\frac{\partial P_i}{\partial r}$ are integrable near ∞ ($P_i=rQ_i, Q_i\in H^1(\mathbb{R}^3)$), so that $\frac{\partial W_i}{\partial r}$, and W_i are also integrable ($W_i=P_i^2$), a contradiction).

(A-14) then implies that

$$\frac{\partial(e^{\sqrt{2\omega/\varepsilon}r}W_i)}{\partial r}=e^{2\sqrt{2\omega/\varepsilon}r}Z_i\leqslant 0 \quad for \quad r\geqslant r_1. \tag{A-16}$$

Hence $W_i(r)\leqslant Ce^{-\sqrt{2\omega/\varepsilon}r}$ and in turn

$$|Q_i(r)|\leqslant Cr^{-1}e^{(-\sqrt{2\omega/\varepsilon}/2)r} \quad for \quad r\geqslant r_1, \tag{A-17}$$

for certain positive constants C, r_1, $\omega>0$ and $\varepsilon\geqslant 1$.

Next, we show the exponential decay of $\frac{\partial Q_i}{\partial r}$ $(i=1,2,3)$ at infinity.

Note that $\frac{\partial Q_i}{\partial r}$ satisfies

$$\frac{\partial\left(r^2\frac{\partial Q_i}{\partial r}\right)}{\partial r}=-r^2g_i(Q_1,Q_2,Q_3). \tag{A-18}$$

Applying (A-7) and the exponential decay of Q_i, it is easily verified that for r large enough, say $r\geqslant r_0$, one has

$$\omega_1|Q_i|\leqslant|g_i(Q_1,Q_2,Q_3)|\leqslant\omega_2|Q_i|, \tag{A-19}$$

where $\omega_2\geqslant\omega_1\geqslant 0$. Hence integrating (A-18) on (r,R), using (A-17) and letting $r,R\to+\infty$ shows that $r^2\frac{\partial Q_i}{\partial r}$ has a limit as $r\to+\infty$. This limit can only be zero by (A-17).

Now, integrating (A-18) on $(r,+\infty)$ then yields that $\frac{\partial Q_i}{\partial r}$ has exponential decay.

Finally, the exponential decay of $\frac{\partial^2 Q_i}{\partial r^2}$ (and thus $|D^\alpha Q_i(x)|$ for $|\alpha|\leqslant 2$) follows immediately from equations (A-11).

The proof of the exponential decay of Q_i at infinity is completed. □

Existence of the Global Smooth Solution to a Fractional Nonlinear Schrödinger System in Atomic Bose-Einstein Condensates*

Guo Boling (郭柏灵) and Li Qiaoxin (李巧欣)

Abstract In this paper, the fractional nonlinear Schrödinger equations for atomic Bose-Einstein condensates are studied. By using the Galërkin method and *a priori* estimates, the existence and uniqueness of global smooth solution are obtained.

Keywords Fractional Schrödinger equations; the Galërkin method; *a priori* estimates; global smooth solution

1 Introduction

In this paper we consider the following fractional nonlinear coupled Schrödinger system [17]

$$\begin{cases} \mathrm{i}\hbar u_t = \left(\dfrac{\hbar^2}{2M}(-\Delta)^\alpha + \lambda_u|u|^2 + \lambda|v|^2\right) u + \sqrt{2}\beta\bar{u}v, \\ \mathrm{i}\hbar v_t = \left(\dfrac{\hbar^2}{4M}(-\Delta)^\alpha + \varepsilon + \lambda_v|v|^2 + \lambda|u|^2\right) v + \dfrac{\beta}{\sqrt{2}}u^2, \end{cases} \tag{1.1}$$

with the initial condition and periodic boundary condition

$$u(x,0) = u_0(x), \qquad v(x,0) = v_0(x), \qquad x \in \Omega, \tag{1.2}$$

$$u(x+2L,t) = u(x,t), \qquad v(x+2L,t) = v(x,t), \quad x \in \Omega, \quad t \geqslant 0, \tag{1.3}$$

where $\Delta = \dfrac{\partial^2}{\partial x^2}$, $\dfrac{1}{2} < \alpha < 1$, $\mathrm{i} = \sqrt{-1}$, $L > 0$, $\Omega = (-L, L)$, $\hbar$ is Planck constant, $M > 0$ is the mass of a single atom, $\lambda_u, \lambda_v, \lambda$ represent the strengths of the atom-atom, molecule-molecule and atom-molecule interactions, respectively and ε, β are any real constants.

* J. Appl. Anal. Comp., 2015, 5(4): 793 – 808. DOI: 10.11948_2015060.

Nonlinear Schrödinger equations have been used to analyze several physical situations, and have attracted the attention of researchers, especially in optics and hydrodynamics. In optics, systems of coupled nonlinear equations can be used to describe the propagation of light along birefringent optical fibers [11].

Fractional differential equations are extensively used in modeling phenomena in various fields of science and engineering [3]. Rida etal. [15] have studied nonlinear Schrödinger equation of fractional order. The investigation of the exact solutions of nonlinear evolution equations play an important role in the study of nonlinear physical phenomena. Then many authors have considered the fractional nonlinear Schrödinger equation. In 2008, Guo boling, Han yongqian and Xin jie [9] proved the existence and uniqueness of the global smooth solution to the period boundary value problem of fractional nonlinear Schrödinger equation by using energy method. In 2011, Jiaqian Hu, Jie Xin , Hong Lu [11] considered a class of systems of fractional nonlinear Schrödinger equations. They proved the existence and uniqueness of the global solution to the periodic boundary value problem by using the Galërkin method. Further discussion can be found in Refs [2, 10, 12, 13, 16].

As far as we know, the fractional nonlinear Schrödinger system (1.1) has not yet been fully studied. In this paper, we prove the existence and uniqueness of the global solution to the periodic boundary value problem for a class of system of fractional nonlinear Schrödinger equations by using the Galërkin method.

Before starting the main results, we review the notations and the calculus inequalities used in this paper.

To simplify the notation in this paper, we shall denote by $\int U(x)\,dx$ the integration $\int_\Omega U(x)\,dx$, C is a generic constant and may assume different values in different formulates. And denote $L^p = L^p(\Omega)$ be the Banach space endowed with the norm $\|\cdot\|_{L^p}$, when $p = 2$, $L^2(\Omega)$ denote the Hilbert space with the usual scalar product $(\cdot,\cdot)$. Here, (u, v) denotes the integral $\int uv\,dx$ as usual.

The Fourier transform $\hat{f} = \mathcal{F}(f)$ of a tempered distribution $f(x)$ on $\mathbf{R}^d$ is defined as

$$\mathcal{F}(f)(\xi) = \hat{f}(\xi) := \frac{1}{(2\pi)^d}\int_{\mathbf{R}^d} f(x)e^{-ix\cdot\xi}\,dx,$$

where $\xi = (\xi_1, \xi_2, \cdots, \xi_n)$. For $\forall\alpha \in \mathbf{R}$, $(-\Delta)^{\frac{\alpha}{2}} f$ can be defined as

$$\mathcal{F}((-\Delta)^{\frac{\alpha}{2}} f)(\xi) = |\xi|^\alpha \hat{f}(\xi) = \frac{1}{(2\pi)^d}\int_{\mathbf{R}^d} |\xi|^\alpha f(x)e^{-ix\cdot\xi}\,dx.$$

Using the Fourier inverse transform, $(-\Delta)^{\frac{\alpha}{2}} f$ can be denoted as

$$(-\Delta)^{\frac{\alpha}{2}} f = \frac{1}{(2\pi)^d} \int_{\mathbf{R}^d} |\xi|^\alpha \hat{f}(\xi) e^{ix\cdot\xi}\, \mathrm{d}\xi.$$

Then the Sobolev space H^α is

$$H^\alpha = H^\alpha(\mathbf{R}^d) = \left\{ f \in S'(\mathbf{R}^d : \hat{f}) \text{ is a function and } \ \|f\|_{H^\alpha}^2 < \infty \right\},$$

where

$$\|f\|_{H^\alpha}^2 := \int_{\mathbf{R}^d} (1+|\xi|^2)^\alpha |\hat{f}(\xi)|^2 \mathrm{d}\xi < \infty.$$

Under this definition, it is clear to see that H^α is a Banach space. And if φ and ϕ belong to H^α, combining the Parseval's identity we conclude the following equation

$$\int_{\mathbf{R}^d} (-\Delta)^\alpha \varphi \cdot \phi \, \mathrm{d}x = \int_{\mathbf{R}^d} (-\Delta)^{\alpha_1} \varphi \cdot (-\Delta)^{\alpha_2} \phi \, \mathrm{d}x,$$

where α_1 and α_2 are nonnegative and $\alpha_1 + \alpha_2 = \alpha$.

These concepts can be easily generalized to the periodic case, and we make no explicit distinctions about the notations for the two situations

The following auxiliary lemmas will be needed.

Lemma 1.1 (Gagliardo-Nirenberg inequality) *Assuming* $u \in L^q(\mathbf{R})$, $\partial_x^m u \in L^r(\mathbf{R})$, $1 \leqslant q, r \leqslant \infty$. *Let* p *and* α *satisfy*

$$\frac{1}{p} = j + \theta\left(\frac{1}{r} - m\right) + (1-\theta)\frac{1}{q}; \quad \frac{j}{m} \leqslant \theta \leqslant 1.$$

Then

$$\|\partial_x^j u\|_p \leqslant C(p,m,j,q,r) \|\partial_x^m u\|_r^\theta \|u\|_q^{1-\theta}. \tag{1.4}$$

In particular, as $m = \alpha, j = 0, p = 4, r = 2, q = 2$, *we have*

$$\|u\|_4^4 \leqslant C \|(-\Delta)^{\frac{\alpha}{2}} u\|_2^{\frac{1}{\theta}} \|u\|_2^{4-\frac{1}{\theta}}. \tag{1.5}$$

Lemma 1.2 (The Gronwall inequality) *Let* c *be a constant, and* $b(t), u(t)$ *be non-negative continuous functions in the interval* $[0,T]$ *satisfying*

$$u(t) \leqslant c + \int_0^t b(\tau) u(\tau)\, \mathrm{d}\tau, \qquad t \in [0,T].$$

Then $u(t)$ *satisfies the estimate*

$$u(t) \leqslant c \exp\left(\int_0^t b(\tau)\, \mathrm{d}\tau \right), \qquad \textit{for} \quad t \in [0,T]. \tag{1.6}$$

Theorem 1.1 *Let $u_0(x) \in H^{\alpha}_{per}(\Omega), v_0(x) \in H^{\alpha}_{per}(\Omega)$, and $0 < \alpha < 1$. Then for $\forall T > 0$, the system (1.1)–(1.3) has a global weak solution*

$$(u,v) \in L^{\infty}\left([0,T); H^{\alpha}_{per}(\Omega)\right)^2, \qquad (u_t,v_t) \in L^{\infty}\left([0,T); H^{-\alpha}_{per}(\Omega)\right)^2. \tag{1.7}$$

Theorem 1.2 *Let $u_0(x) \in H^{4\alpha}_{per}(\Omega), v_0(x) \in H^{4\alpha}_{per}(\Omega)$, and $\frac{1}{2} < \alpha < 1$. Then for $\forall T > 0$, the system (1.1)–(1.3) has a uniquely global smooth solution*

$$(u,v) \in L^{\infty}\left([0,T); H^{4\alpha}_{per}(\Omega)\right)^2, \qquad (u_t,v_t) \in L^{\infty}\left([0,T); H^{2\alpha}_{per}(\Omega)\right)^2. \tag{1.8}$$

Theorem 1.3 *Let $u_0(x) \in H^{4\alpha}_{per}(\mathbf{R}), v_0(x) \in H^{4\alpha}_{per}(\mathbf{R})$, and $\frac{1}{2} < \alpha < 1$. Then the system (1.1)–(1.3) has a uniquely global smooth solution*

$$(u,v) \in L^{\infty}\left([0,\infty); H^{4\alpha}_{per}(\mathbf{R})\right)^2, \qquad (u_t,v_t) \in L^{\infty}\left([0,\infty); H^{2\alpha}_{per}(\mathbf{R})\right)^2. \tag{1.9}$$

2 A priori estimates

In this section, we give the demonstration of *a priori* estimates that guarantee the existence of the global smooth solution of the system (1.1)–(1.3).

Lemma 2.1 *Let $u_0(x) \in L^2(\Omega), v_0(x) \in L^2(\Omega)$ and (u,v) be a solution of the system (1.1) with initial data (u_0,v_0), then we have the identity*

$$\|u(x,t)\|_2^2 + 2\|v(x,t)\|_2^2 \equiv \|u_0(x)\|_2^2 + 2\|v_0(x)\|_2^2. \tag{2.10}$$

Proof Taking the inner product for the first equation of the system (1.1) with $\overline{u}$ and the second equation with $\overline{v}$, respectively, and integrating the resulting equations with respect to x on Ω, and then taking the imaginary part of the resulting equations, we obtain

$$\begin{cases} \dfrac{\hbar}{2}\dfrac{\mathrm{d}}{\mathrm{d}t}\|u\|_2^2 = \sqrt{2}\beta \mathrm{Im} \int (\overline{u})^2 v \,\mathrm{d}x, \\ \dfrac{\hbar}{2}\dfrac{\mathrm{d}}{\mathrm{d}t}\|v\|_2^2 = \dfrac{\beta}{\sqrt{2}} \mathrm{Im} \int u^2 \overline{v} \,\mathrm{d}x. \end{cases} \tag{2.11}$$

Multiplying the second equation of the system (2.11) by 2 and then sum up the first equation, it follows that

$$\frac{\hbar}{2}\frac{\mathrm{d}}{\mathrm{d}t}\|u\|_2^2 + \hbar\frac{\mathrm{d}}{\mathrm{d}t}\|v\|_2^2 = 0,$$

which implies the identity (2.10).

This completes the proof of Lemma 2.1.

Lemma 2.2 *Let* $u_0 \in H^{\alpha}_{per}(\Omega), v_0 \in H^{\alpha}_{per}(\Omega)$, $0 < \alpha < 1$, *then for the solution* (u, v) *of the system* (1.1), *we can get*

$$\sup_{0\leqslant t\leqslant\infty} (\|(-\Delta)^{\frac{\alpha}{2}} u\|_2^2 + \|(-\Delta)^{\frac{\alpha}{2}} v\|_2^2) \leqslant C, \tag{2.12}$$

where C *is a constant depending only on* $\|u_0\|_{H^{\alpha}_{per}}, \|v_0\|_{H^{\alpha}_{per}}$.

Proof The inner product is taken to the first equation of the system (1.1) with $\overline{u}_t$ and the second equation with $\overline{v}_t$, and then integrating and taking the real part of the resulting equations, we get

$$\begin{cases} 0 = \dfrac{\hbar^2}{4M}\dfrac{\mathrm{d}}{\mathrm{d}t}\int |(-\Delta)^{\frac{\alpha}{2}} u|^2 \,\mathrm{d}x + \dfrac{\lambda_u}{4}\dfrac{\mathrm{d}}{\mathrm{d}t}\int |u|^4 \,\mathrm{d}x + \lambda \mathrm{Re}\int |v|^2 u\overline{u}_t \,\mathrm{d}x \\ \quad +\sqrt{2}\beta \mathrm{Re}\int \overline{u}v\overline{u}_t \,\mathrm{d}x, \\ 0 = \dfrac{\hbar^2}{8M}\dfrac{\mathrm{d}}{\mathrm{d}t}\int |(-\Delta)^{\frac{\alpha}{2}} v|^2 \,\mathrm{d}x + \dfrac{\varepsilon}{2}\dfrac{\mathrm{d}}{\mathrm{d}t}\int |v|^2 \,\mathrm{d}x + \dfrac{\lambda_v}{4}\dfrac{\mathrm{d}}{\mathrm{d}t}\int |v|^4 \,\mathrm{d}x + \lambda \mathrm{Re}\int |u|^2 v\overline{v}_t \,\mathrm{d}x \\ \quad +\dfrac{\beta}{\sqrt{2}}\mathrm{Re}\int u^2\overline{v}_t \,\mathrm{d}x. \end{cases} \tag{2.13}$$

Summing up the two equations of the system (2.13), we have

$$\begin{aligned} &\frac{\hbar^2}{4M}\frac{\mathrm{d}}{\mathrm{d}t}\left(\int |(-\Delta)^{\frac{\alpha}{2}} u|^2 \,\mathrm{d}x + \frac{1}{2}\int |(-\Delta)^{\frac{\alpha}{2}} v|^2 \,\mathrm{d}x\right) \\ &+\frac{1}{4}\frac{\mathrm{d}}{\mathrm{d}t}\left(\lambda_u \int |u|^4 \mathrm{d}x + \lambda_v \int |v|^4 \,\mathrm{d}x\right) \\ &+\frac{\varepsilon}{2}\frac{\mathrm{d}}{\mathrm{d}t}\int |v|^2 \,\mathrm{d}x + \frac{\lambda}{2}\frac{\mathrm{d}}{\mathrm{d}t}\int |u|^2|v|^2 \,\mathrm{d}x + \frac{\beta}{\sqrt{2}}\mathrm{Re}\frac{\mathrm{d}}{\mathrm{d}t}\int u^2\overline{v} \,\mathrm{d}x = 0. \end{aligned}$$

Let

$$\mathrm{I} := \frac{\hbar^2}{4M}\left(\int |(-\Delta)^{\frac{\alpha}{2}} u|^2 \,\mathrm{d}x + \frac{1}{2}\int |(-\Delta)^{\frac{\alpha}{2}} v|^2 \,\mathrm{d}x\right)$$

$$\mathrm{II} := \frac{1}{4}\left(\lambda_u \int |u|^4 \,\mathrm{d}x + \lambda_v \int |v|^4 \,\mathrm{d}x\right),$$

$$\mathrm{III} := \frac{\lambda}{2}\int |u|^2|v|^2 \,\mathrm{d}x, \qquad \mathrm{IV} := \frac{\varepsilon}{2}\int |v|^2 \,\mathrm{d}x, \qquad \mathrm{V} := \frac{\beta}{\sqrt{2}}\mathrm{Re}\int u^2\overline{v} \,\mathrm{d}x.$$

Then

$$E(t) = \mathrm{I} + \mathrm{II} + \mathrm{III} + \mathrm{IV} + \mathrm{V} \equiv E(0). \tag{2.14}$$

Applying Lemma 1.1 and the Young inequality, we have

$$\|u\|_4^4 \leqslant C\|(-\Delta)^{\frac{\alpha}{2}} u\|_2^{\frac{1}{\theta}} \|u\|_2^{4-\frac{1}{\theta}} \leqslant \delta\|(-\Delta)^{\frac{\alpha}{2}} u\|_2^2 + C_1\|u\|_2^{\frac{2(4\theta-1)}{2\theta-1}}, \tag{2.15}$$

$$\|v\|_4^4 \leqslant C\|(-\Delta)^{\frac{\alpha}{2}} v\|_2^{\frac{1}{\theta}} \|v\|_2^{4-\frac{1}{\theta}} \leqslant \delta\|(-\Delta)^{\frac{\alpha}{2}} v\|_2^2 + C_1\|v\|_2^{\frac{2(4\theta-1)}{2\theta-1}}. \tag{2.16}$$

Then we can bound the term II by

$$|\mathrm{II}| \leqslant \delta \left(\|(-\Delta)^{\frac{\alpha}{2}} u\|_2^2 + \|(-\Delta)^{\frac{\alpha}{2}} u\|_2^2\right) + C\left(\|u\|_2^{\frac{2(4\theta-1)}{2\theta-1}} + \|v\|_2^{\frac{2(4\theta-1)}{2\theta-1}}\right). \tag{2.17}$$

For the term III, using the Hölder's inequality

$$\frac{\lambda}{2}\int |u|^2|v|^2 dx \leqslant \frac{\lambda}{4}\left(\|u\|_4^4 + \|v\|_4^4\right). \tag{2.18}$$

Combining the inequalities (2.15) and (2.16), the term III can be bounded by

$$|\mathrm{III}| \leqslant \delta \left(\|(-\Delta)^{\frac{\alpha}{2}} u\|_2^2 + \|(-\Delta)^{\frac{\alpha}{2}} u\|_2^2\right) + C\left(\|u\|_2^{\frac{2(4\theta-1)}{2\theta-1}} + \|v\|_2^{\frac{2(4\theta-1)}{2\theta-1}}\right). \tag{2.19}$$

The term

$$\mathrm{V} = \frac{\beta}{\sqrt{2}}\mathrm{Re}\int u^2\overline{v}\,\mathrm{d}x \leqslant \frac{\beta}{\sqrt{2}}\int |u|^2|v|\,\mathrm{d}x \leqslant C\|v\|_2\|u\|_4^2. \tag{2.20}$$

By applying the inequality (2.15) and Lemma 2.1 to yield

$$|\mathrm{V}| \leqslant \delta\|(-\Delta)^{\frac{\alpha}{2}} u\|_2^2 + C\|u\|_2^{\frac{2(4\theta-1)}{2\theta-1}}. \tag{2.21}$$

Using the estimates of the term II, III and V, we deduce

$$|\mathrm{II}| + |\mathrm{III}| + |\mathrm{V}| \leqslant \delta \left(\|(-\Delta)^{\frac{\alpha}{2}} u\|_2^2 + \|(-\Delta)^{\frac{\alpha}{2}} u\|_2^2\right) + C\left(\|u\|_2^{\frac{2(4\theta-1)}{2\theta-1}} + \|v\|_2^{\frac{2(4\theta-1)}{2\theta-1}}\right). \tag{2.22}$$

In view of Lemma 2.1, it follows that

$$\mathrm{IV} = \frac{\varepsilon}{2}\|v\|_2^2 \leqslant C. \tag{2.23}$$

Combining the estimates (2.22) and (2.23)

$$\mathrm{I} - \delta\left(\|(-\Delta)^{\frac{\alpha}{2}} u\|_2^2 + \|(-\Delta)^{\frac{\alpha}{2}} v\|_2^2\right) \leqslant C,$$

a.e.

$$\|(-\Delta)^{\frac{\alpha}{2}} u\|_2^2 + \|(-\Delta)^{\frac{\alpha}{2}} v\|_2^2 \leqslant C,$$

where C is a constant depending only on $\|u_0\|_{H^\alpha_{per}}, \|v_0\|_{H^\alpha_{per}}$.

This completes the proof of Lemma 2.2.

Lemma 2.3 *Let T be any positive number, $u_0 \in H^{2\alpha}_{per}(\Omega), v_0 \in H^{2\alpha}_{per}(\Omega)$, for $\frac{1}{2} < \alpha < 1$. Then the solution (u,v) of the system (1.1) satisfies the following estimate*

$$\sup_{0\leqslant t\leqslant T}(\|(-\Delta)^\alpha u\|_2^2 + \|(-\Delta)^\alpha v\|_2^2) \leqslant C, \qquad \forall T>0, \tag{2.24}$$

where the constant C depends only on T and $\|u_0\|_{H^{2\alpha}_{per}}, \|v_0\|_{H^{2\alpha}_{per}}$.

Proof Differentiate (1.1) with respect to t, multiply the first equation of the system (1.1) by $\overline{u}_t$, the second equation by $\overline{v}_t$, and then integrate with respect to x , take the imaginary part to get

$$\frac{\hbar}{2}\frac{\mathrm{d}}{\mathrm{d}t}\|u_t\|_2^2 = Im\left(\frac{\mathrm{d}}{\mathrm{d}t}(\lambda_u|u|^2u),\overline{u}_t\right) + \mathrm{Im}\left(\frac{\mathrm{d}}{\mathrm{d}t}(\lambda|v|^2u),\overline{u}_t\right) + \mathrm{Im}\left(\frac{\mathrm{d}}{\mathrm{d}t}(\sqrt{2}\beta\overline{u}v),\overline{u}_t\right),$$

$$\frac{\hbar}{2}\frac{\mathrm{d}}{\mathrm{d}t}\|v_t\|_2^2 = \mathrm{Im}\left(\frac{\mathrm{d}}{\mathrm{d}t}(\lambda|u|^2v),\overline{v}_t\right) + \mathrm{Im}\left(\frac{\mathrm{d}}{\mathrm{d}t}(\lambda_v|v|^2v),\overline{v}_t\right) + \mathrm{Im}\left(\frac{\mathrm{d}}{\mathrm{d}t}(\frac{\beta}{\sqrt{2}}u^2),\overline{v}_t\right).$$

But

$$\mathrm{Im}\left(\frac{\mathrm{d}}{\mathrm{d}t}(\lambda_u|u|^2u),\overline{u}_t\right) = \mathrm{Im}\int\frac{\mathrm{d}}{\mathrm{d}t}(\lambda_u|u|^2u)\overline{u}_t\,\mathrm{d}x = \lambda_u\mathrm{Im}\int u^2\overline{u}_t^2\,\mathrm{d}x,$$

$$\mathrm{Im}\left(\frac{\mathrm{d}}{\mathrm{d}t}(\lambda|v|^2u),\overline{u}_t\right) = \mathrm{Im}\int\frac{\mathrm{d}}{\mathrm{d}t}(\lambda|v|^2u)\overline{u}_t\,\mathrm{d}x = \lambda\mathrm{Im}\int|v|_t^2u\overline{u}_t\,\mathrm{d}x,$$

$$\begin{aligned}\mathrm{Im}\left(\frac{\mathrm{d}}{\mathrm{d}t}(\sqrt{2}\beta\overline{u}v),\overline{u}_t\right) &= \mathrm{Im}\int\frac{\mathrm{d}}{\mathrm{d}t}(\sqrt{2}\beta\overline{u}v)\overline{u}_t\,\mathrm{d}x\\ &= \sqrt{2}\beta\mathrm{Im}\int\overline{u}_t^2v\,\mathrm{d}x + \sqrt{2}\beta\mathrm{Im}\int\overline{u}v_t\overline{u}_t\,\mathrm{d}x.\end{aligned}$$

Similarly

$$\mathrm{Im}\left(\frac{\mathrm{d}}{\mathrm{d}t}(\lambda_v|v|^2v),\overline{v}_t\right) = \mathrm{Im}\int\frac{\mathrm{d}}{\mathrm{d}t}(\lambda_v|v|^2v)\overline{v}_t\,\mathrm{d}x = \lambda_v\mathrm{Im}\int v^2\overline{v}_t^2\,\mathrm{d}x,$$

$$\mathrm{Im}\left(\frac{\mathrm{d}}{\mathrm{d}t}(\lambda|u|^2v),\overline{v}_t\right) = \mathrm{Im}\int\frac{\mathrm{d}}{\mathrm{d}t}(\lambda|u|^2v)\overline{v}_t\mathrm{d}x = \lambda\mathrm{Im}\int|u|_t^2v\overline{v}_t\,\mathrm{d}x,$$

$$\mathrm{Im}\left(\frac{\mathrm{d}}{\mathrm{d}t}\left(\frac{\beta}{\sqrt{2}}u^2\right),\overline{v}_t\right) = \mathrm{Im}\int\frac{\mathrm{d}}{\mathrm{d}t}\left(\frac{\beta}{\sqrt{2}}u^2\right)\overline{v}_t\,\mathrm{d}x = \sqrt{2}\beta\mathrm{Im}\int uu_t\overline{v}_t\,\mathrm{d}x.$$

Therefore

$$\begin{aligned}\frac{\hbar}{2}\frac{\mathrm{d}}{\mathrm{d}t}\|u_t\|_2^2 =& \lambda_u\mathrm{Im}\int u^2\overline{u}_t^2\,\mathrm{d}x + \lambda\mathrm{Im}\int|v|_t^2u\overline{u}_t\,\mathrm{d}x + \sqrt{2}\beta\mathrm{Im}\int\overline{u}_t^2v\,\mathrm{d}x\\ &+ \sqrt{2}\beta\mathrm{Im}\int\overline{u}v_t\overline{u}_t\,\mathrm{d}x,\end{aligned}$$

$$\frac{\hbar}{2}\frac{\mathrm{d}}{\mathrm{d}t}\|v_t\|_2^2=\lambda_v\mathrm{Im}\int v^2\overline{v}_t^2\,\mathrm{d}x+\lambda\mathrm{Im}\int|u|_t^2v\overline{v}_t\,\mathrm{d}x+\sqrt{2}\beta\mathrm{Im}\int uu_t\overline{v}_t\,\mathrm{d}x.$$

Integrating the above two equality from 0 to t, we have

$$\begin{aligned}&\frac{\hbar}{2}(\|u_t\|_2^2+\|v_t\|_2^2)\\
=&\ \mathrm{Im}\int_0^t\int\lambda_uu^2\overline{u}_t^2\,\mathrm{d}x\mathrm{d}s+\int_0^t\mathrm{Im}\int\lambda_vv^2\overline{v}_t^2\,\mathrm{d}x\mathrm{d}s+2\int_0^t\mathrm{Im}\int\lambda uv\overline{u}_t\overline{v}_t\,\mathrm{d}x\mathrm{d}s\\
&+\int_0^t\mathrm{Im}\int\sqrt{2}\beta(\overline{u}_t^2v+\overline{u}v_t\overline{u}_t+uu_t\overline{v}_t)\,\mathrm{d}x\mathrm{d}s+\frac{\hbar}{2}(\|u_t(x,0)\|_2^2+\|v_t(x,0)\|_2^2)\\
\leqslant&\,C_1\left(\int_0^t\int|u|^2|u_t|^2\,\mathrm{d}x\mathrm{d}s+\int_0^t\int|v|^2|v_t|^2\,\mathrm{d}x\mathrm{d}s+\int_0^t\int|u||v||u_t||v_t|\,\mathrm{d}x\mathrm{d}s\right)\\
&+C_2\left(\int_0^t\int|u_t|^2|v|\,\mathrm{d}x\mathrm{d}s+2\int_0^t\int|u||u_t||v_t|\,\mathrm{d}x\mathrm{d}s\right)\\
&+\frac{\hbar}{2}(\|u_t(x,0)\|_2^2+\|v_t(x,0)\|_2^2).\end{aligned}$$

Applying Sobolev embedding inequality $\|u\|_\infty\leqslant C\|u\|_{H^\alpha}\leqslant C_1,\quad\left(\alpha>\frac{1}{2}\right)$, we have

$$\begin{aligned}&\frac{\hbar}{2}(\|u_t\|_2^2+\|v_t\|_2^2)\\
\leqslant&\,C\left(\int_0^t\|u\|_\infty^2\|u_t\|_2^2\,\mathrm{d}s+\int_0^t\|v\|_\infty^2\|v_t\|_2^2\,\mathrm{d}s+\int_0^t\|u\|_\infty\|v\|_\infty\|u_t\|_2\|v_t\|_2\,\mathrm{d}s\right)\\
&+C_1\left(\int_0^t\|v\|_\infty\|u_t\|_2^2\,\mathrm{d}s+2\int_0^t\|u\|_\infty\|u_t\|_2\|v_t\|_2\,\mathrm{d}s\right)\\
&+\frac{\hbar}{2}\left(\|u_t(x,0)\|_2^2+\|v_t(x,0)\|_2^2\right)\\
\leqslant&\,C\left(\int_0^t\|u_t\|_2^2\,\mathrm{d}s+\int_0^t\|v_t\|_2^2\,\mathrm{d}s\right)+\frac{\hbar}{2}\left(\|u_t(x,0)\|_2^2+\|v_t(x,0)\|_2^2\right)\end{aligned}$$

In term of Gronwall inequality, we deduce that

$$\|u_t\|_2^2+\|v_t\|_2^2\leqslant C.\tag{2.25}$$

Applying the system (1.1), we obtain

$$\left\|\frac{\hbar^2}{2M}(-\Delta)^\alpha u\right\|_2^2\leqslant\|\hbar u_t\|_2^2+\|\lambda_u|u|^2u\|_2^2+\|\lambda|v|^2u\|_2^2+\|\sqrt{2}\beta\overline{u}v\|_2^2,$$

$$\left\|\frac{\hbar^2}{4M}(-\Delta)^\alpha v\right\|_2^2\leqslant\|\hbar v_t\|_2^2+\|\varepsilon v\|_2^2+\|\lambda_v|v|^2v\|_2^2+\|\lambda|u|^2v\|_2^2+\left\|\frac{\beta}{\sqrt{2}}u^2\right\|_2^2.$$

Using the inequality (2.25),

$$\begin{aligned}
\|(-\Delta)^{\alpha}u\|_2^2 &\leqslant C + C_1\|u\|_\infty^4\|u\|_2^2 + C_2\|v\|_\infty^4\|u\|_2^2 + C_3\|u\|_\infty\|v\|_\infty\|u\|_2^2\|v\|_2^2 \leqslant C_4,\\
\|(-\Delta)^{\alpha}v\|_2^2 &\leqslant C + C_1\|v\|_\infty^4\|v\|_2^2 + C_2\|u\|_\infty^4\|v\|_2^2 + C_3\|u\|_\infty^2\|u\|_2^2 \leqslant C_4,
\end{aligned}$$

a.e.

$$\|(-\Delta)^{\alpha}u\|_2^2 + \|(-\Delta)^{\alpha}v\|_2^2 \leqslant C,$$

where the constant C depends only on T and $\|u_0\|_{H^{2\alpha}_{per}}, \|v_0\|_{H^{2\alpha}_{per}}$.

This completes the proof of Lemma 2.3.

Lemma 2.4 *Let* $\frac{1}{2} < \alpha < 1$, $u_0 \in H^{3\alpha}_{per}(\Omega), v_0 \in H^{3\alpha}_{per}(\Omega)$, *and* (u, v) *be the solution of the system* (1.1). *Then*

$$\sup_{0\leqslant t\leqslant\infty} (\|(-\Delta)^{\frac{\alpha}{2}}u_t\|_2^2 + \|(-\Delta)^{\frac{\alpha}{2}}v_t\|_2^2) \leqslant C, \tag{2.26}$$

where the constant C *depends only on* $\|u_0\|_{H^{3\alpha}_{per}}, \|v_0\|_{H^{3\alpha}_{per}}$.

Proof Differentiate (1.1) with respect to t two times, multiply the first equation of the system (1.1) by $\overline{u}_t$, the second equation by $\overline{v}_t$, and then integrate with respect to x , take the imaginary part to get

$$\frac{\hbar}{2}\|u_{tt}\|_2^2 = \mathrm{Im}\left(\frac{\mathrm{d}^2}{\mathrm{d}t^2}(\lambda_u|u|^2u), \overline{u}_{tt}\right) + \mathrm{Im}\left(\frac{\mathrm{d}^2}{\mathrm{d}t^2}(\lambda|v|^2u), \overline{u}_{tt}\right) + \mathrm{Im}\left(\frac{\mathrm{d}^2}{\mathrm{d}t^2}(\sqrt{2}\beta\overline{u}v), \overline{u}_{tt}\right),$$

$$\frac{\hbar}{2}\|v_{tt}\|_2^2 = \mathrm{Im}\left(\frac{\mathrm{d}^2}{\mathrm{d}t^2}(\lambda_v|v|^2v), \overline{v}_{tt}\right) + \mathrm{Im}\left(\frac{\mathrm{d}^2}{\mathrm{d}t^2}(\lambda|u|^2v), \overline{v}_{tt}\right) + \mathrm{Im}\left(\frac{\mathrm{d}^2}{\mathrm{d}t^2}\left(\frac{\beta}{\sqrt{2}}u^2\right), \overline{v}_{tt}\right).$$

But applying Sobolev embedding theorem and the Young inequality, we get

$$\mathrm{Im}\left(\frac{\mathrm{d}^2}{\mathrm{d}t^2}(\lambda_u|u|^2u), \overline{u}_{tt}\right) = \mathrm{Im}\left(\lambda_u|u|_{tt}^2u + 2\lambda_u|u|_t^2u_t, \overline{u}_{tt}\right) \leqslant C\|u_{tt}\|_2^2 + \delta\|u_t\|_4^4,$$

$$\begin{aligned}
\mathrm{Im}\left(\frac{d^2}{dt^2}(\lambda|v|^2u), \overline{u}_{tt}\right) &= \mathrm{Im}\left(\lambda|v|_{tt}^2u + 2\lambda|v|_t^2u_t, \overline{u}_{tt}\right)\\
&\leqslant C(\|u_{tt}\|_2^2 + \|v_{tt}\|_2^2) + \delta(\|u_t\|_4^4 + \|v_t\|_4^4),
\end{aligned}$$

$$\begin{aligned}
\mathrm{Im}\left(\frac{d^2}{dt^2}(\sqrt{2}\beta\overline{u}v), \overline{u}_{tt}\right) &= \mathrm{Im}(\sqrt{2}\beta(\overline{u}_{tt}v + \overline{u}_tv_t + 2\overline{u}_tv_t), \overline{u}_{tt})\\
&\leqslant C(\|u_{tt}\|_2^2 + \|v_{tt}\|_2^2) + \delta(\|u_t\|_4^4 + \|v_t\|_4^4),
\end{aligned}$$

and

$$\mathrm{Im}\left(\frac{\mathrm{d}^2}{\mathrm{d}t^2}(\lambda_v|v|^2v), \overline{v}_{tt}\right) = \mathrm{Im}\left(\lambda_v|v|_{tt}^2v + 2\lambda_v|v|_t^2v_t, \overline{v}_{tt}\right) \leqslant C\|v_{tt}\|_2^2 + \delta\|v_t\|_4^4,$$

$$\operatorname{Im}\left(\frac{d^2}{dt^2}(\lambda|u|^2v),\overline{v}_{tt}\right)=\operatorname{Im}\left(\lambda|u|_{tt}^2v+2\lambda|u|_t^2v_t,\overline{v}_{tt}\right)$$
$$\leqslant C(\|u_{tt}\|_2^2+\|v_{tt}\|_2^2)+\delta(\|u_t\|_4^4+\|v_t\|_4^4),$$

$$\operatorname{Im}\left(\frac{\mathrm{d}^2}{\mathrm{d}t^2}(\frac{\beta}{\sqrt{2}}u^2),\overline{v}_{tt}\right)=\operatorname{Im}(\sqrt{2}\beta(u_{tt}u+u_t^2),\overline{v}_{tt})\leqslant C\|u_t\|_4^2\|v_{tt}\|_2+\|u_{tt}\|_2\|v_{tt}\|_2.$$

Taking the above inequality to obtain

$$\frac{\mathrm{d}}{\mathrm{d}t}(\|u_{tt}\|_2^2+\|v_{tt}\|_2^2)\leqslant C(\|u_{tt}\|_2^2+\|v_{tt}\|_2^2)+\delta(\|u_t\|_4^4+\|v_t\|_4^4). \tag{2.27}$$

Let $\theta=\dfrac{1}{8\alpha}<\dfrac{1}{4}$. Then using Gagliardo–Nirenberg inequality and the inequality (2.25), we have

$$\|u_t\|_4\leqslant C\|u_t\|^{1-\theta}\|(-\Delta)^{\frac{\alpha}{2}}u_t\|^\theta\leqslant C_1\|(-\Delta)^{\frac{\alpha}{2}}u_t\|^\theta, \tag{2.28}$$

$$\|v_t\|_4\leqslant C\|v_t\|^{1-\theta}\|(-\Delta)^{\frac{\alpha}{2}}v_t\|^\theta\leqslant C_1\|(-\Delta)^{\frac{\alpha}{2}}v_t\|^\theta. \tag{2.29}$$

Combining the inequalities (2.27), (2.28) and (2.29), we get

$$\frac{\mathrm{d}}{\mathrm{d}t}(\|u_{tt}\|_2^2+\|v_{tt}\|_2^2)\leqslant C(\|u_{tt}\|_2^2+\|v_{tt}\|_2^2)+\delta(\|(-\Delta)^{\frac{\alpha}{2}}u_t\|^2+\|(-\Delta)^{\frac{\alpha}{2}}v_t\|^2) \tag{2.30}$$

Differentiate (1.1) with respect to t, multiply the first equation of the system (1.1) by $\overline{u}_t$, the second equation by $\overline{v}_t$, and then integrate with respect to x , take the real part to get

$$-\operatorname{Re}\int i\hbar u_{tt}\overline{u}_t\,\mathrm{d}x+\frac{\hbar^2}{4M}\|(-\Delta)^{\frac{\alpha}{2}}u_t\|^2+\operatorname{Re}\int\frac{\mathrm{d}}{\mathrm{d}t}(\lambda_u|u|^2u)\overline{u}_t\,\mathrm{d}x$$
$$+\operatorname{Re}\int\frac{\mathrm{d}}{\mathrm{d}t}(\lambda|v|^2u)\overline{u}_t\,\mathrm{d}x+\operatorname{Re}\int\frac{\mathrm{d}}{\mathrm{d}t}(\sqrt{2}\beta\overline{u}v)\overline{u}_t\,\mathrm{d}x=0,$$
$$-\operatorname{Re}\int i\hbar v_{tt}\overline{v}_t\,\mathrm{d}x+\frac{\hbar^2}{8M}\|(-\Delta)^{\frac{\alpha}{2}}v_t\|^2+\frac{\varepsilon}{2}\frac{\mathrm{d}}{\mathrm{d}t}\|v_t\|_2^2+\operatorname{Re}\int\frac{\mathrm{d}}{\mathrm{d}t}(\lambda_v|v|^2v)\overline{v}_t\,\mathrm{d}x$$
$$+\operatorname{Re}\int\frac{\mathrm{d}}{\mathrm{d}t}(\lambda|u|^2v)\overline{v}_t\,\mathrm{d}x+\operatorname{Re}\int\frac{\mathrm{d}}{\mathrm{d}t}\left(\frac{\beta}{\sqrt{2}}u^2\right)\overline{v}_t\,\mathrm{d}x=0.$$

But applying the inequality (2.25), we have

$$\operatorname{Re}\int\frac{\mathrm{d}}{\mathrm{d}t}(\lambda_u|u|^2u)\overline{u}_t\,\mathrm{d}x+\operatorname{Re}\int\frac{\mathrm{d}}{\mathrm{d}t}(\lambda|v|^2u)\overline{u}_t\,\mathrm{d}x+\operatorname{Re}\int\frac{\mathrm{d}}{\mathrm{d}t}(\sqrt{2}\beta\overline{u}v)\overline{u}_t\,\mathrm{d}x\leqslant C,$$

$$\frac{\varepsilon}{2}\frac{\mathrm{d}}{\mathrm{d}t}\|v_t\|_2^2+\operatorname{Re}\int\frac{\mathrm{d}}{\mathrm{d}t}(\lambda_v|v|^2v)\overline{v}_t\,\mathrm{d}x+\operatorname{Re}\int\frac{\mathrm{d}}{\mathrm{d}t}(\lambda|u|^2v)\overline{v}_t\,\mathrm{d}x$$
$$+\operatorname{Re}\int\frac{\mathrm{d}}{\mathrm{d}t}\left(\frac{\beta}{\sqrt{2}}u^2\right)\overline{v}_t\,\mathrm{d}x\leqslant C.$$

Therefore

$$(\|(-\Delta)^{\frac{\alpha}{2}}u_t\|^2+\|(-\Delta)^{\frac{\alpha}{2}}v_t\|^2)\leqslant C(\|u_{tt}\|_2^2+\|v_{tt}\|_2^2)+C_1. \tag{2.31}$$

Using (2.30) and the above inequality, we get

$$\frac{\mathrm{d}}{\mathrm{d}t}(\|u_{tt}\|_2^2+\|v_{tt}\|_2^2)\leqslant C(\|u_{tt}\|_2^2+\|v_{tt}\|_2^2)+C_1.$$

By Gronwall inequality to obtain

$$(\|u_{tt}\|_2^2+\|v_{tt}\|_2^2)\leqslant C. \tag{2.32}$$

Then combing the inequalities (2.31) and (2.32), the below inequality is true

$$(\|(-\Delta)^{\frac{\alpha}{2}}u_t\|^2+\|(-\Delta)^{\frac{\alpha}{2}}v_t\|^2)\leqslant C,$$

where C is a constant depending only on $\|u_0\|_{H^{3\alpha}_{per}}, \|v_0\|_{H^{3\alpha}_{per}}$.

This completes the proof of Lemma 2.4.

Lemma 2.5 *Let* $\frac{1}{2}<\alpha<1$, $u_0\in H^{4\alpha}_{per}(\Omega), v_0\in H^{4\alpha}_{per}(\Omega)$, *the solution* (u,v) *of the system* (1.1) *satisfies the following estimate*

$$\sup_{0\leqslant t<\infty}(\|(-\Delta)^{2\alpha}u\|+\|(-\Delta)^{2\alpha}v\|)\leqslant C,$$

where the constant C *depends only on* $\|u_0\|_{H^{4\alpha}_{per}}, \|v_0\|_{H^{4\alpha}_{per}}$.

Proof Using the system (1.1), we have

$$\begin{aligned}\|(-\Delta)^{2\alpha}u\|_2^2\leqslant &C_1\|(-\Delta)^{\alpha}u_t\|_2^2+C_2\|(-\Delta)^{\alpha}|u|^2u\|+C_3\|(-\Delta)^{\alpha}|v|^2u\|\\&+C_4\|(-\Delta)^{\alpha}\overline{u}v\|,\end{aligned}$$

$$\begin{aligned}\|(-\Delta)^{2\alpha}v\|_2^2\leqslant &C\|(-\Delta)^{\alpha}v_t\|_2^2+C_1\|(-\Delta)^{\alpha}v\|_2^2+C_2\|(-\Delta)^{\alpha}|v|^2v\|\\&+C_3\|(-\Delta)^{\alpha}|u|^2v\|+C_4\|(-\Delta)^{\alpha}u^2\|.\end{aligned}$$

For $\frac{1}{2}<\alpha<1$,

$$\|(-\Delta)^{2\alpha}u\|_2^2\leqslant C_1\|(-\Delta)^{\alpha}u_t\|_2^2+C_2\|\Delta(|u|^2u)\|+C_3\|\Delta(|v|^2u)\|+C_4\|\Delta(\overline{u}v)\|,$$

$$\begin{aligned}\|(-\Delta)^{2\alpha}v\|_2^2\leqslant &C\|(-\Delta)^{\alpha}v_t\|_2^2+C_1\|(-\Delta)^{\alpha}v\|_2^2+C_2\|\Delta(|v|^2v)\|\\&+C_3\|\Delta(|u|^2v)\|+C_4\|\Delta u^2\|.\end{aligned}$$

By Lemma 2.5 and the simple computation, we have

$$\left(\|(-\Delta)^{2\alpha}u\|_2^2+\|(-\Delta)^{2\alpha}v\|_2^2\right)\leqslant C+C_1\left(\|\Delta u\|+\|\Delta v\|\right)+C_2\left(\|\nabla u\|_4^2+\|\nabla v\|_4^2\right). \tag{2.33}$$

Let $\theta=\dfrac{2}{4\alpha}<1$. Using Gagliardo–Nirenberg inequality to have

$$C\|\Delta u\|\leqslant C_1\|(-\Delta)^{2\alpha}u\|^{\theta}\|u\|^{1-\theta}\leqslant\frac{1}{4}\|(-\Delta)^{2\alpha}u\|+C_2, \tag{2.34}$$

$$C\|\Delta v\|\leqslant C_1\|(-\Delta)^{2\alpha}v\|^{\theta}\|v\|^{1-\theta}\leqslant\frac{1}{4}\|(-\Delta)^{2\alpha}v\|+C_2. \tag{2.35}$$

Define $\gamma=\dfrac{1}{16\alpha-4}<\dfrac{1}{4}$. By Gagliardo–Nirenberg inequality, we have

$$\begin{aligned}C\|\nabla u\|_4^2&\leqslant C_1\|(-\Delta)^{2\alpha}u\|^{2\gamma}\|\nabla u\|^{2(1-\gamma)}\\&\leqslant C\|(-\Delta)^{2\alpha}u\|^{2\gamma}\|(-\Delta)^{\alpha}u\|^{2(1-\gamma)}\leqslant\frac{1}{4}\|(-\Delta)^{2\alpha}u\|+C_1,\end{aligned} \tag{2.36}$$

$$\begin{aligned}C\|\nabla v\|_4^2&\leqslant C_1\|(-\Delta)^{2\alpha}v\|^{2\gamma}\|\nabla v\|^{2(1-\gamma)}\\&\leqslant C\|(-\Delta)^{2\alpha}v\|^{2\gamma}\|(-\Delta)^{\alpha}v\|^{2(1-\gamma)}\leqslant\frac{1}{4}\|(-\Delta)^{2\alpha}v\|+C_1.\end{aligned} \tag{2.37}$$

Combining the inequalities (2.33)–(2.37), we have

$$\left(\|(-\Delta)^{2\alpha}u\|+\|(-\Delta)^{2\alpha}v\|\right)\leqslant C,$$

where C is a constant depending only on $\|u_0\|_{H^{4\alpha}_{per}}, \|v_0\|_{H^{4\alpha}_{per}}$.

This completes the proof of Lemma 2.5.

3 Proof of the main results

In this section, we prove the existence of weak solution to the problem (1.1)–(1.3) by using Galerkin-Fourier method. We need the following lemmas.

Lemma 3.1 *Let B_0, B and B_1 be three Banach spaces. Assume that $B_0\subset B\subset B_1$ and B_i, $i=0,1$ are reflective. Suppose also that B_0 is compactly embedded in B. Let*

$$W=\left\{v|v\in L^{p_0}(0,T;B_0), v'=\frac{\mathrm{d}v}{\mathrm{d}t}\in L^{p_1}(0,T;B_1)\right\}$$

where T is finite and $1<p_i<\infty, i=0,1$. W is equipped with the norm

$$\|v\|_{L^{p_0}(0,T;B_0)}+\|v'\|_{L^{p_1}(0,T;B_1)}.$$

Then W is compactly embedded in $L^{p_0}(0,T;B)$.

Lemma 3.2 *Suppose that Q is a bounded domain in $\mathbf{R}_x^n \times \mathbf{R}_t, g_\mu, g \in L^q(Q)(1 < q < \infty)$ and $\|g_\mu\|_{L^q(Q)} \leqslant C$. Furthermore, suppose that*

$$g_\mu \to g \text{ a.e. in } Q.$$

Then

$$g_\mu \rightharpoonup g \text{ weakly in } L^q(Q).$$

Lemma 3.3 *X is a Banach space. Suppose that $g \in L^p(0,T;X), \dfrac{\partial g}{\partial x} \in L^p(0,T;X)$ $(1 \leqslant p \leqslant \infty)$. Then $g \in C([0,T],X)$ (after possibly being redefined on a set of measure zero).*

In the following, we prove the existence of weak solution to the problem (1.1)–(1.3).

Proof of Theorem 1.1 We prove theorem 1.1. by the following three steps.

Step 1 Constructing the approximate solutions by the Galerkin-Fourier method.

Let $\{\omega_j(x)\}(j = 1, 2, \cdots)$ be a complete orthonormal basis of eigenfunctions for the periodic boundary problem $-\Delta u = \lambda u$ in Ω For every integer m, we are looking for an approximate solution of the system (1.1) of the form

$$u_m(t) = \sum_{j=1}^m \xi_{jm}(t)\omega_j, \quad v_m(t) = \sum_{j=1}^m \mu_{jm}(t)\omega_j,$$

where ξ_{jm}, μ_{jm} satisfy the following nonlinear equations

$$\begin{cases} \left(-i\hbar u_{mt} + \left(\dfrac{\hbar^2}{2M}(-\Delta)^\alpha + \lambda_u|u_m|^2 + \lambda|v_m|^2\right) u_m + \sqrt{2}\beta\overline{u}_m v_m, \omega_j\right) = 0, \\ \left(-i\hbar v_{mt} + \left(\dfrac{\hbar^2}{4M}(-\Delta)^\alpha + \varepsilon + \lambda_v|v_m|^2 + \lambda|u_m|^2\right) v_m + \dfrac{\beta}{\sqrt{2}}u_m^2, \omega_j\right) = 0, \end{cases} \tag{3.38}$$

the nonlinear equations (3.38) satisfy the following initial-value conditions

$$\begin{cases} u_m(0) = u_{0m},\ u_{0m} = \displaystyle\sum_{i=1}^m f_{im}\omega_i \to u_0 \ \ in \ \ H_{per}^\alpha(\Omega) \ as \ m \to \infty, \\ v_m(0) = v_{0m},\ v_{0m} = \displaystyle\sum_{i=1}^m g_{im}\omega_i \to v_0 \ \ in \ \ H_{per}^\alpha(\Omega) \ \ as \ \ m \to \infty. \end{cases} \tag{3.39}$$

Then (3.38) becomes the system of nonlinear ODE subject to the initial condition (3.39). According to standard existence theory for nonlinear ordinary differential equations, there exists a unique solution of (3.38) and (3.39) for a.e. $0 \leqslant t \leqslant t_m$. By a priori estimates we obtain that $t_m = T$.

Step 2 A priori estimates.

As the proof of Lemmas 2.1 and 2.2, we have

$$(u_m, v_m) \in L^\infty(0,T; H^\alpha_{per}(\Omega))^2. \tag{3.40}$$

For $\forall(\varphi, \phi) \in H^\alpha_{per}(\Omega) \times H^\alpha_{per}(\Omega)$, we have

$$\begin{cases} \left(-i\hbar u_{mt} + \left(\frac{\hbar^2}{2M}(-\Delta)^\alpha + \lambda_u|u_m|^2 + \lambda|v_m|^2\right) u_m + \sqrt{2}\beta \overline{u}_m v_m, \varphi\right) = 0, \\ \left(-i\hbar v_{mt} + \left(\frac{\hbar^2}{4M}(-\Delta)^\alpha + \varepsilon + \lambda_v|v_m|^2 + \lambda|u_m|^2\right) v_m + \frac{\beta}{\sqrt{2}}u_m^2, \phi\right) = 0. \end{cases} \tag{3.41}$$

So

$$\begin{aligned} &|(u_{mt}, \varphi)| \\ \leqslant & C_1|((-\Delta)^\alpha u_m, \varphi)| + C_2\left|(|u_m|^2 u_m, \varphi)\right| + C_3\left|(|v_m|^2 u_m, \varphi)\right| + C_4\left|(\overline{u}_m v_m, \varphi)\right| \\ \leqslant & C_1\|D^\alpha u_m\|\|D^\alpha \varphi\| + C_2\|u_m\|_4^3\|\varphi\|_4 + C_3\|v_m\|_4^2\|u_m\|_4\|\varphi\|_4 + C_4\|u_m\|_3\|v_m\|_3\|\varphi\|_3 \end{aligned} \tag{3.42}$$

$$\begin{aligned} &|(v_{mt}, \phi)| \\ \leqslant & C_1|((-\Delta)^\alpha v_m, \phi)| + \varepsilon\left|(v_m, \phi)\right| + C_2\left|(|v_m|^2 v_m, \phi)\right| + C_3|(|u_m|^2 v_m, \psi)| \\ & + C_4|(u_m^2, \psi)| \\ \leqslant & C_1\|D^\alpha v_m\|\|D^\alpha \phi\| + \varepsilon\|v_m\|_2\|\phi\|_2 + C_2\|v_m\|_4^3\|\phi\|_4 + C_3\|u_m\|_4^2\|v_m\|_4\|\phi\|_4 \\ & + C_4\|u_m\|_4^2\|\phi\|_2 \end{aligned} \tag{3.43}$$

Using the Sobolev embedding theorem, we have

$$\|\varphi\|_4 \leqslant C\|D^\alpha \varphi\| + C_1, \quad \|\varphi\|_3 \leqslant C_2\|D^\alpha \varphi\| + C_3, \quad \|\phi\|_4 \leqslant C_4\|D^\alpha \phi\| + C_5.$$

So by (3.42) and (3.43), we get

$$|(u_{mt}, \varphi)| \leqslant C\|D^\alpha \varphi\| + C_1, \quad |(v_{mt}, \phi)| \leqslant C_2\|D^\alpha \phi\| + C_3, \quad \forall \varphi, \phi \in H^\alpha_{per}(\Omega).$$

Therefore

$$(u_{mt}, v_{mt}) \in L^\infty(0,T; H^{-\alpha}_{per}(\Omega))^2. \tag{3.44}$$

Step 3 Passaging to the limit.

By applying (3.40) and (3.44), we deduce that there exists a subsequence u_μ from u_m, v_k from v_m such that

$$u_\mu \rightharpoonup u \text{ *-weakly in } L^\infty(0,T;H^\alpha_{per}(\Omega)),\quad u_{\mu t} \rightharpoonup u_t \text{ *-weakly in } L^\infty(0,T;H^{-\alpha}_{per}(\Omega)). \tag{3.45}$$

$$v_k \rightharpoonup v \text{ *-weakly in } L^\infty(0,T;H^\alpha_{per}(\Omega)),\quad v_{kt} \rightharpoonup v_t \text{ *-weakly in } L^\infty(0,T;H^{-\alpha}_{per}(\Omega)). \tag{3.46}$$

By (3.40), we have

$$(u_m, v_m) \text{ is bounded in } L^2(0,T;H^\alpha_{per}(\Omega))^2. \tag{3.47}$$

By (3.44), we have

$$(u_{mt}, v_{mt}) \text{ is bounded in } L^2(0,T;H^{-\alpha}_{per}(\Omega))^2. \tag{3.48}$$

Define

$$W = \left\{v | v \in L^2(0,T;H^\alpha_{per}(\Omega)), v_t \in L^2(0,T;H^{-\alpha}_{per}(\Omega))\right\}$$

We equip W with the norm:

$$\|v\|_W = \|v\|_{L^2(0,T;H^\alpha_{per}(\Omega))} + \|v_t\|_{L^2(0,T;H^{-\alpha}_{per}(\Omega))}.$$

Since $H^\alpha_{per}(\Omega)$ is compactly embedded in $L^2(\Omega)$ for $\frac{1}{2} < \alpha < 1$, by Lemma 3.1 we have that W is compactly embedded in $L^2(0,T;L^2(\Omega))$. By (3.47) and (3.48), $u_m \in W$. Then, there exists the subsequence u_μ, v_k (not rebelled) which satisfies

$$u_\mu \to u,\quad v_k \to v \text{ strongly in } L^2(0,T;L^2(\Omega)) \text{ and a. e.} \tag{3.49}$$

By using (3.40), (3.49) and Lemma 3.2, we have

$$|u_\mu|^2 u_\mu \rightharpoonup |u|^2 u \text{ *-weakly in } L^\infty(0,T;L^{\frac{4}{3}}(\Omega)), \tag{3.50}$$

$$|v_k|^2 v_k \rightharpoonup |v|^2 v \text{ *-weakly in } L^\infty(0,T;L^{\frac{4}{3}}(\Omega)). \tag{3.51}$$

Fixing j, we get

$$\begin{cases} \left(-i\hbar u_{mt} + \left(\dfrac{\hbar^2}{2M}(-\Delta)^\alpha + \lambda_u|u_m|^2 + \lambda|v_m|^2\right)u_m + \sqrt{2}\beta\overline{u}_m v_m, \omega_j\right) = 0, \\ \left(-i\hbar v_{mt} + \left(\dfrac{\hbar^2}{4M}(-\Delta)^\alpha + \varepsilon + \lambda_v|v_m|^2 + \lambda|u_m|^2\right)v_m + \dfrac{\beta}{\sqrt{2}}u_m^2, \omega_j\right) = 0, \end{cases} \tag{3.52}$$

By applying (3.45), (3.46), (3.50) and (3.51), we deduce that there exists a subsequence u_μ from u_m, v_k from v_m such that

$$((-\Delta)^{\frac{\alpha}{2}}u_\mu,\omega_j)\rightharpoonup((-\Delta)^{\frac{\alpha}{2}}u,\omega_j)\ \text{*-weakly in } L^\infty(0,T),$$

$$(u_{\mu t},\omega_j)\rightharpoonup(u_t,\omega_j)\ \text{*-weakly in } L^\infty(0,T),$$

$$((\lambda_u|u_\mu|^2+\lambda|v_\mu|^2)u_\mu,\omega_j)\rightharpoonup((\lambda_u|u|^2+\lambda|v|^2)u,\omega_j)\ \text{*-weakly in } L^\infty(0,T),$$

$$(\overline{u}_\mu v_\mu,\omega_j)\rightharpoonup(\overline{u}v,\omega_j)\ \text{*-weakly in } L^\infty(0,T),$$

$$((-\Delta)^{\frac{\alpha}{2}}v_\mu,\omega_j)\rightharpoonup((-\Delta)^{\frac{\alpha}{2}}v,\omega_j)\ \text{*-weakly in } L^\infty(0,T),$$

$$(v_{\mu t},\omega_j)\rightharpoonup(v_t,\omega_j)\ \text{*-weakly in } L^\infty(0,T),$$

$$(v_\mu,\omega_j)\rightharpoonup(v,\omega_j)\ \text{*-weakly in } L^\infty(0,T),$$

$$((\lambda_v|v_\mu|^2+\lambda|u_\mu|^2)v_\mu,\omega_j)\rightharpoonup((\lambda_v|v|^2+\lambda|u|^2)v,\omega_j)\ \text{*-weakly in } L^\infty(0,T),$$

$$(u_\mu^2,\omega_j)\rightharpoonup(u^2,\omega_j)\ \text{*-weakly in } L^\infty(0,T),$$

Then from (3.52), we have

$$\begin{cases}\left(-i\hbar u_t+\left(\dfrac{\hbar^2}{2M}(-\Delta)^\alpha+\lambda_u|u|^2+\lambda|v|^2\right)u+\sqrt{2}\beta\overline{u}v,\omega_j\right)=0,\\ \left(-i\hbar v_t+\left(\dfrac{\hbar^2}{4M}(-\Delta)^\alpha+\varepsilon+\lambda_v|v|^2+\lambda|u|^2\right)v+\dfrac{\beta}{\sqrt{2}}u^2,\omega_j\right)=0,\end{cases}\tag{3.53}$$

the above equalities hold for any fixed j. By the density of the basis $\omega_j,(j\in Z)$, we have:

$$\begin{cases}\left(-i\hbar u_t+\left(\dfrac{\hbar^2}{2M}(-\Delta)^\alpha+\lambda_u|u|^2+\lambda|v|^2\right)u+\sqrt{2}\beta\overline{u}v,h\right)=0,\quad\forall h\in H^\alpha_{per}(\Omega),\\ \left(-i\hbar v_t+\left(\dfrac{\hbar^2}{4M}(-\Delta)^\alpha+\varepsilon+\lambda_v|v|^2+\lambda|u|^2\right)v+\dfrac{\beta}{\sqrt{2}}u^2,g\right)=0,\quad\forall g\in H^\alpha_{per}(\Omega)\end{cases}\tag{3.54}$$

Hence (u,v) satisfies the system (1.1). By (3.40), (3.44) and Lemma 3.3, we obtain that

$$u_\mu\in C(0,T;H^{-\alpha}_{per}(\Omega))\qquad v_k\in C(0,T;H^{-\alpha}_{per}(\Omega)).$$

Then $u_\mu(0)\rightharpoonup u(0)$ weakly in $H^{-\alpha}_{per}(\Omega)$, $v_k(0)\rightharpoonup v(0)$ weakly in $H^{-\alpha}_{per}(\Omega)$.

But from (3.39), we have

$u_\mu(0)\to u_0$ weakly in $H^\alpha_{per}(\Omega)$, $v_k(0)\to v_0$ weakly in $H^\alpha_{per}(\Omega)$.

Therefore, $u(0)=u_0,\ \ v(0)=v_0$.

Theorem 1.4 (in Ref. [3]) generalizes the result of the global existence of weak solution to the nonlinear Schrödinger equations. So that Theorem 1.1 is complete.

By the *a priori* estimates from Lemma 2.1 to Lemma 2.5 and Theorem 1.1, there exists a global smooth solution (u,v) for the system (1.1)–(1.3) such that

$$(u,v)\in L^\infty\left([0,T];H^{4\alpha}_{per}(\Omega)\right)^2,\qquad (u_t,v_t)\in L^\infty\left([0,T];H^{2\alpha}_{per}(\Omega)\right)^2.$$

Finally, we prove the uniqueness of the solution to the system (1.1)–(1.3) in the following.

Let $(u_1,v_1),(u_2,v_2)$ be two solutions which satisfy the system (1.1)–(1.3), then $(s=u_1-u_2, m=v_1-v_2)$ satisfies

$$\begin{cases} i\hbar s_t = \dfrac{\hbar^2}{2M}(-\Delta)^\alpha s+\lambda_u(|u_1|^2u_1-|u_2|^2u_2)+\lambda(|v_1|^2u_1-|v_2|^2u_2)\\ \qquad +\sqrt{2}\beta(\overline{u}_1v_1-\overline{u}_2v_2),\\ i\hbar m_t = \dfrac{\hbar^2}{4M}(-\Delta)^\alpha m+\varepsilon m+\lambda_v(|v_1|^2v_1-|v_2|^2v_2)+\lambda(|u_1|^2v_1-|u_2|^2v_2)\\ \qquad +\dfrac{\beta}{\sqrt{2}}(u_1^2-u_2^2), \end{cases} \tag{3.55}$$

with the initial condition

$$s(0)=0,\qquad m(0)=0.$$

Taking the inner product of the first equation of the system (3.55) with $\overline{s}$ and the second equation with $\overline{m}$, considering the imaginary part of the resulting equations, we obtain:

$$\begin{aligned}\frac{\hbar}{2}\frac{\mathrm{d}}{\mathrm{d}t}\|s\|_2^2 =&\lambda_u\mathrm{Im}\int(|u_1|^2u_1-|u_2|^2u_2)\overline{s}\,\mathrm{d}x+\lambda\mathrm{Im}\int(|v_1|^2u_1-|v_2|^2u_2)\overline{s}\,\mathrm{d}x\\ &+\sqrt{2}\beta\mathrm{Im}\int(\overline{u}_1v_1-\overline{u}_2v_2)\overline{s}\,\mathrm{d}x\\ \frac{\hbar}{2}\frac{\mathrm{d}}{\mathrm{d}t}\|m\|_2^2 =&\lambda_v\mathrm{Im}\int(|v_1|^2v_1-|v_2|^2v_2)\overline{m}\,\mathrm{d}x+\lambda\mathrm{Im}\int(|u_1|^2v_1-|u_2|^2v_2)\overline{m}\,\mathrm{d}x\\ &+\frac{\beta}{\sqrt{2}}\mathrm{Im}\int(u_1^2-u_2^2)\overline{m}\,\mathrm{d}x.\end{aligned}$$

But

$$\begin{aligned}&\lambda_u\mathrm{Im}\int(|u_1|^2u_1-|u_2|^2u_2)\overline{s}\,\mathrm{d}x\\ \leqslant& C\int|(|u_1|^2s\overline{s}+(|u_1|^2-|u_2|^2)u_2\overline{s})|\,\mathrm{d}x\end{aligned}$$

$$
\begin{aligned}
&\leqslant C\|u_1\|_\infty^2\|s\|_2^2 + C\int(|u_1|^2 - |u_2|^2)u_2\overline{s}\,\mathrm{d}x\\
&\leqslant C\|s\|_2^2.
\end{aligned}
$$

And

$$
\begin{aligned}
&\lambda\mathrm{Im}\int(|v_1|^2u_1 - |v_2|^2u_2)\overline{s}\,\mathrm{d}x\\
\leqslant& C\int|(|v_1|^2s\overline{s} + (|v_1|^2 - |v_2|^2)u_2\overline{s})|\,\mathrm{d}x\\
\leqslant& C\|v_1\|_\infty^2\|s\|_2^2 + C\|u_2\|_\infty\|(|v_1|^2 - |v_2|^2)\|_2\|s\|_2\\
\leqslant& C(\|s\|_2^2 + \|m\|_2^2),
\end{aligned}
$$

$$
\begin{aligned}
&\int(\overline{u}_1v_1 - \overline{u}_2v_2)\overline{s}\,\mathrm{d}x\\
=&\int(\overline{u}_1v_1 - \overline{u}_1v_2 + \overline{u}_1v_2 - \overline{u}_2v_2)\overline{s}\,\mathrm{d}x\\
\leqslant& C_1\int(m\overline{s} + |s|^2)\,\mathrm{d}x \leqslant C_2(\|m\|_2^2 + \|s\|_2^2),
\end{aligned}
$$

$$
\begin{aligned}
&\int(u_1^2 - u_2^2)\overline{m}\,\mathrm{d}x\\
=&\int\left((u_1^2 - u_1u_2) + (u_1u_2 - u_2^2)\right)\overline{m}\,\mathrm{d}x\\
\leqslant& C_1\int s\overline{m}\,\mathrm{d}x \leqslant C_2(\|s\|_2^2 + \|m\|_2^2).
\end{aligned}
$$

By the above inequalities, one can easily obtain

$$
\frac{\mathrm{d}}{\mathrm{d}t}(\|s\|_2^2 + \|m\|_2^2) \leqslant C(\|s\|_2^2 + \|m\|_2^2).
$$

Applying the Gronwall inequality, we get $s = 0, m = 0$. Thus the uniqueness is obtained.

So we complete Theorem 1.2.

Remark 3.1 *All the above estimates are unconcerned with the period L and only depend on the norm of initial data. Therefore, by using the a priori estimates of the solution to the system* (1.1)–(1.3) *for L, as in Ref.* [18], *we derive the global smooth solution as $L \to \infty$. So that, Theorem* 1.3 *is obtained.*

References

[1] Adhikari S, Muruganandam P. Bose-Einstein condensation dynamics from the numerical solution of the Gross-Pitaevskii equation [J]. J. Phys. B, 2002, 35.

[2] Alexandru D, Fabio P. Nonlinear fractional Schrödinger equations in one dimension [J]. J. Funct. Anal., 2014, 266: 139–176.

[3] Alireza K, AliK G,Dumitru B. On nonlinear fractional Klein-Gordon equation [J]. Signal Processing, 2011, 91: 446–451.

[4] Ginibre J, Velo G. The global Cauchy problem for the nonlinear Schrödinger equation revisited [J]. Ann Inst H Poincaré Anal Non Linéaire, 1985, 2: 309–327.

[5] Guo B, Huo Z. Global Well-Posedness for the Fractional Nonlinear Schrödinger Equation [J]. Comm. Partial Differential Equations , 2010, 36: 247–255.

[6] Guo B. The global solution for some systems of nonlinear Schrödinger equations [J]. Proc of D-1 Symposium, 1980, 3: 1227–1246.

[7] Guo B. The initial and periodic value problem of one class nonlinear Schrödinger equations describing excitons in molecular crystals [J]. Acta Math. Sci, 1982, 2: 269–276.

[8] Guo B. The initial value problems and periodic boundary value problem of one class of higher order multi-dimensional nonlinear Schrödinger equations [J]. Chinese Science Bulletin, 1982, 6: 324–327.

[9] Guo B, Han Y, Xin J. Existence of the global smooth solution to the period boundary value problem of fractional nonlinear Schrödinger equation [J]. Applied Mathematics and Computation, 2008, 204: 468–477.

[10] Guo B, Zeng M. Solutions for the fractional Landau-Lifshitz equation [J]. J. Math. Anal. Appl, 2009, 361: 131–138.

[11] Hu J, Xin J, Lu H. The global solution for a class of systems of fractional nonlinear Schrödinger equations with periodic boundary condition, Computers and Mathematics with Applications, 2011, 62: 1510–1521.

[12] Pu X, Guo B. Existence and decay of solutions to the two-dimensional fractional quasigeostrophic equation [J]. J. Math. Phys., 2010, 51: 1–15.

[13] Pu X, Guo B, Zhang J. Global weak solutions to the 1-d fractional Landau-Lifshitz equation [J]. Discret. Contin. Dyn. Syst.Ser. B, 2010, 14: 199–207.

[14] Pu X, Guo B. The fractional Landau-Lifshitz-Gilbert equation and the heat flow of harmonic maps [J]. Calculus of Variations, 2011, 42: 1–19.

[15] Rida S, El-Sherbiny H, Arafa A. On the solution of the fractional nonlinear Schrödinger equation [J]. Phys. Lett. A, 2008, 372.

[16] Shang X, Zhang J. Ground states for fractional Schrödinger equations with critical growth [J]. Nonlinearity, 2014, 27: 187–207.

[17] Timmermans E, Tommasini P, Hussein M, Kerman A. Feshbach resonances in atomic Bose-Einstein condensates [J]. Physics Reports, 1999, 315: 199–230.

[18] Zhou Y, Guo B. Periodic boundary problem and initial value problem for the generalized Korteweg-de Vries systems of higher order [J]. Acta Math. Sci, 1984, 27: 154–176.

Diffusion Limit of 3D Primitive Equations of the Large-scale Ocean under Fast Oscillating Random Force*

Guo Boling (郭柏灵), Huang Daiwen (黄代文) and Wang Wei (王伟)

Abstract The three-dimensional (3D) viscous primitive equations describing the large-scale oceanic motions under fast oscillating random perturbation are studied. Under some assumptions on the random force, the solution to the initial boundary value problem (IBVP) of the 3D random primitive equations converges in distribution to that of IBVP of the limiting equations, which are the 3D stochastic primitive equations describing the large-scale oceanic motions under a white in time noise forcing. This also implies the convergence of the stationary solution of the 3D random primitive equations.

Keywords Random primitive equations; stationary solution; martingale; statistical solution

1 Introduction

The important 3D viscous primitive equations of the large-scale ocean in a Cartesian coordinate system, are written as the following system on a cylindrical domain

$$\frac{\partial v}{\partial t}+(v\cdot\nabla)v+\Phi(v)\frac{\partial v}{\partial z}+fk\times v+\nabla p_b-\int_{-1}^{z}\nabla T\mathrm{d}z'$$
$$-\Delta v-\frac{\partial^2 v}{\partial z^2}=\Psi_1\,, \tag{1.1}$$

$$\frac{\partial T}{\partial t}+(v\cdot\nabla)T+\Phi(v)\frac{\partial T}{\partial z}-\Delta T-\frac{\partial^2 T}{\partial z^2}=\Psi_2\,, \tag{1.2}$$

$$\int_{-1}^{0}\nabla\cdot v\,\mathrm{d}z=0. \tag{1.3}$$

* J. Diff. Equ., 2015, 259: 2388 − 2407.

with boundary value conditions

$$\frac{\partial v}{\partial z}=0, \quad \frac{\partial T}{\partial z}=-\alpha_u T \qquad \text{on} \quad M\times\{0\}=\Gamma_u\,, \tag{1.4}$$

$$\frac{\partial v}{\partial z}=0\,, \quad \frac{\partial T}{\partial z}=0 \qquad \text{on} \quad M\times\{-1\}=\Gamma_b\,, \tag{1.5}$$

$$v\cdot \boldsymbol{n}=0, \quad \frac{\partial v}{\partial \boldsymbol{n}}\times \boldsymbol{n}=0, \quad \frac{\partial T}{\partial \boldsymbol{n}}=0 \quad \text{on } \partial M\times[-1,0]=\Gamma_l\,, \tag{1.6}$$

and the initial value conditions

$$u|_{t=t_0}=(v|_{t=t_0}\,, T|_{t=t_0})=u_{t_0}=(v_{t_0}\,, T_{t_0})\,, \tag{1.7}$$

where the unknown functions are v, p_b, T, $v=(v^{(1)},v^{(2)})$ the horizontal velocity, p_b the pressure, T temperature, $\Phi(v)(t,x,y,z)=-\int_{-1}^{z}\nabla\cdot v(t,x,y,z')\,dz'$ vertical velocity, $f=f_0(\beta+y)$ the Coriolis parameter, k vertical unit vector, Ψ_1 a given forcing field, Ψ_2 a given heat source, $\nabla=(\partial_x,\ \partial_y)$, $\Delta=\partial_x^2+\partial_y^2$, α_u a positive constant, $\vec{n}$ the norm vector to Γ_l and M a smooth bounded domain in $\mathbb{R}^2$. For more details for (1.1)–(1.7), see [3,23] and references therein.

In the past two decades, there are several research works about the well-posedness of the above 3D deterministic primitive equations of the large-scale ocean. In [18], Lions, Temam and Wang obtained the global existence of weak solutions for the primitive equations. In [15], Guillén-González etc. obtained the global existence of strong solutions to the primitive equations with small initial data. Moreover, they proved the local existence of strong solutions to the equations. In [3], Cao and Titi developed a beautiful approach to proving that L^6-norm of the horizontal velocity is uniformly in t bounded, and obtained the global well-posedness for the 3D viscous primitive equations.

In study of the primitive equations of the large-scale ocean or atmosphere, taking the stochastic external factors into account is reasonable and necessary. There are many works about mathematical study of some stochastic climate models, see, e.g., [7,8,9,21,22]. [9] is one of the first works on a 3D stochastic quasi-geostrophic model. Guo and Huang in [14] considered the global well-posedness and long-time dynamics for the 3D stochastic primitive equations of the large-scale ocean under a white in time noise forcing.

In realistic model, random fluctuation always exits. We consider the following 3D primitive equations with fast oscillating random force

$$\frac{\partial v^\epsilon}{\partial t}+(v^\epsilon\cdot\nabla)v^\epsilon+\Phi(v^\epsilon)\frac{\partial v^\epsilon}{\partial z}+fk\times v^\epsilon+\nabla p_b^\epsilon-\int_{-1}^{z}\nabla T^\epsilon \mathrm{d}z'-\Delta v^\epsilon$$
$$-\frac{\partial^2 v^\epsilon}{\partial z^2}=\frac{1}{\sqrt{\epsilon}}\eta_1^\epsilon, \tag{1.8}$$

$$\frac{\partial T^\epsilon}{\partial t}+(v^\epsilon\cdot\nabla)T^\epsilon+\Phi(v^\epsilon)\frac{\partial T^\epsilon}{\partial z}-\Delta T^\epsilon-\frac{\partial^2 T^\epsilon}{\partial z^2}=\Psi+\frac{1}{\sqrt{\epsilon}}\eta_2^\epsilon, \tag{1.9}$$

$$\int_{-1}^{0}\nabla\cdot v^\epsilon\,\mathrm{d}z=0. \tag{1.10}$$

with boundary value conditions

$$\frac{\partial v^\epsilon}{\partial z}=0,\quad \frac{\partial T^\epsilon}{\partial z}=-\alpha_u T^\epsilon,\quad \text{on}\quad \Gamma_u, \tag{1.11}$$

$$\frac{\partial v^\epsilon}{\partial z}=0,\quad \frac{\partial T^\epsilon}{\partial z}=0,\quad \text{on}\quad \Gamma_b, \tag{1.12}$$

$$v^\epsilon\cdot\boldsymbol{n}=0,\quad \frac{\partial v^\epsilon}{\partial \boldsymbol{n}}\times\boldsymbol{n}=0,\quad \frac{\partial T}{\partial \boldsymbol{n}}=0,\quad \text{on}\quad \Gamma_l, \tag{1.13}$$

and initial value conditions

$$u^\epsilon|_{t=t_0}=(v^\epsilon|_{t=t_0},T^\epsilon|_{t=t_0})=u_{t_0}^\epsilon=(v_{t_0}^\epsilon,T_{t_0}^\epsilon), \tag{1.14}$$

where $(v_0^\epsilon,T_0^\epsilon)\in V$ is $\mathcal{F}_0^0$-measurable random variable. Here η_1^ϵ and η_2^ϵ are random forces which are stationary processes satisfying some assumptions given in subsection, and Ψ is a given heat source defined on $\Omega=M\times(-1,0)$.

One reason of considering such fast oscillating random force in the primitive equation is that white noise is an idealistic model. On the other hand the random model (1.8)–(1.10) converges in some sense to the white noise driven primitive equations as $\epsilon\to 0$ which implies that the random model (1.8)–(1.10) is more appropriate to describe some physical phenomena if stochastic primitive equations do.

There are lots of models of such random forces η_1 and η_2. Here, for simplicity, we assume that these random forces are some Gaussian processes solving linear stochastic differential equation (see (2.4)). Classical result [1, Chapter 10, e.g.] shows that

$$\frac{1}{\sqrt{\epsilon}}\int_0^t\eta(s/\epsilon)\,\mathrm{d}s\quad\text{converges in distribution to}\quad W\quad\text{as}\quad \epsilon\to 0$$

for some scalar Wiener process W provided that process η has some mixing property. Then for small $\epsilon>0$, the limit of the above 3D random primitive equations is expected to be 3D primitive equations driven by white noise. In fact by a weak convergence method, we show that the limit of 3D random primitive equations is the limiting model, which is the 3D stochastic primitive equations driven by white in time noise (see Theorem 4.1) as $\epsilon \to 0$. To show the limit of stationary solution, we work on the statistical solution on $[0, \infty)$. The limit of the statistical solution of the 3D random primitive equations is shown to be that of the 3D primitive equations driven by a white noise, which implies the limit of stationary solution of the 3D random primitive equations is that of the 3D stochastic primitive equations. One difficulty here is the singularity caused by the randomly fast oscillation. For this we define a martingale to replace this fast oscillation term, which eliminates the ϵ^{-1} term. This method is applied in both energy estimates for solutions and passing limit of $\epsilon \to 0$, see the proofs of Lemma 4.1 and Theorem 4.1.

The above limit approach is also called a diffusion limit which has been applied to study the asymptotic behavior of stochastic Burgers type equations with stochastic advection [26]. There are also some works on diffusion approximation for some random PDEs with different approachs [5, 16, e.g.].

The paper is organized as follows. In section 2, the 3D random primitive equations are introduced. Our working spaces and a new formulation of the initial boundary value problem for the primitive equations with fast oscillating random force are given in this section. We obtain the global well-posedness to 3D primitive equations of the large-scale ocean under fast oscillating random force in section 3. We prove main results of our paper in section 4.

2 New formulation for the 3D random primitive equations

2.1 New formulation for the 3D random primitive equations

Before formulating a new formulation for the 3D random primitive equations, notations for some function spaces, functionals and operators are given.

Let $L^p(\Omega)$ be the usual Lebesgue space with the norm $|\cdot|_p$, $1 \leqslant p \leqslant \infty$. $H^m(\Omega)$ is the usual Sobolev space(m is a positive integer) with the norm

$$\|h\|_m = \left[\int_\Omega \left(\sum_{1\leqslant k\leqslant m}\ \sum_{i_j=1,2,3;j=1,\cdots,k} |\nabla_{i_1}\cdots\nabla_{i_k} h|^2 + |h|^2\right)\right]^{\frac{1}{2}},$$

where $\nabla_1 = \dfrac{\partial}{\partial x}$, $\nabla_2 = \dfrac{\partial}{\partial y}$ and $\nabla_3 = \dfrac{\partial}{\partial z}$. $\int_\Omega \cdot \mathrm{d}\Omega$ and $\int_M \cdot \mathrm{d}M$ are denoted by $\int_\Omega \cdot$ and

$\int_M \cdot$ respectively.

Define our working spaces for (1.8)–(1.14) as

$$\mathcal{V}_1 := \left\{ v \in (C^\infty(\Omega))^2; \frac{\partial v}{\partial z}|_{\Gamma_u,\Gamma_b} = 0, v\cdot \boldsymbol{n}|_{\Gamma_l} = 0, \frac{\partial v}{\partial \boldsymbol{n}} \times \boldsymbol{n}|_{\Gamma_l} = 0, \int_{-1}^0 \nabla\cdot v dz = 0 \right\},$$

$$\mathcal{V}_2 := \left\{ T \in C^\infty(\Omega); \frac{\partial T}{\partial z}|_{\Gamma_u} = -\alpha_u T, \frac{\partial T}{\partial z}|_{\Gamma_b} = 0, \frac{\partial T}{\partial \boldsymbol{n}}|_{\Gamma_l} = 0 \right\},$$

$$V_1 = \text{the closure of } \mathcal{V}_1 \text{ with respect to the norm } \|\cdot\|_1,$$

$$V_2 = \text{the closure of } \mathcal{V}_2 \text{ with respect to the norm } \|\cdot\|_1,$$

$$H_1 = \text{the closure of } \mathcal{V}_1 \text{ with respect to the norm } |\cdot|_2,$$

$$V = V_1 \times V_2, \qquad H = H_1 \times L^2(\Omega).$$

The inner products and norms on V, H are given by

$$\langle u, u_1\rangle_V = \langle v, v_1\rangle_{V_1} + \langle T, T_1\rangle_{V_2},$$

$$\langle u, u_1\rangle = \langle v^{(1)}, (v_1)^{(1)}\rangle + \langle v^{(2)}, (v_1)^{(2)}\rangle + \langle T, T_1\rangle,$$

$$\|u\| = \langle u, u\rangle_V^{\frac{1}{2}} = \langle v, v\rangle_{V_1}^{\frac{1}{2}} + \langle T, T\rangle_{V_2}^{\frac{1}{2}} = \|v\| + \|T\|, \quad |u|_2 = \langle u, u\rangle^{\frac{1}{2}},$$

where $u = (v, T)$, $u_1 = (v_1, T_1) \in V$, and $\langle \cdot,\cdot\rangle$ denotes the inner product in $L^2(\Omega)$.

Then, we define the functionals $a : V\times V \to \mathbb{R}$, $a_1 : V_1\times V_1 \to \mathbb{R}$, $a_2 : V_2\times V_2 \to \mathbb{R}$, and their corresponding linear operators $A : V \to V'$, $A_1 : V_1 \to V_1'$, $A_2 : V_2 \to V_2'$ by

$$a(u, u_1) = \langle Au, u_1\rangle = a_1(v, v_1) + a_2(T, T_1),$$

where

$$a_1(v, v_1) = \langle A_1 v, v_1\rangle = \int_\Omega \left(\nabla v \cdot \nabla v_1 + \frac{\partial v}{\partial z}\cdot\frac{\partial v_1}{\partial z} \right),$$

$$a_2(T, T_1) = \langle A_2 T, T_1\rangle = \int_\Omega \left(\nabla T\cdot\nabla T_1 + \frac{\partial T}{\partial z}\frac{\partial T_1}{\partial z} \right) + \alpha_u \int_{\Gamma_u} TT_1.$$

According to Lemma 3.1 in [14], we know that a and A have the following properties. a is coercive and continuous, and $A : V \to V'$ is isomorphism. Moreover,

$$a(u, u_1) \leqslant c\|v\|\|v_1\| + c\|T\|\|T_1\| \leqslant c\|u\|\|u_1\|,$$

$$a(u, u) \geqslant c\|v\|^2 + c\|T\|^2 \geqslant c\|u\|^2.$$

The isomorphism $A : V \to V'$ can be extended to a self-adjoint unbounded linear operator on H with a compact inverse $A^{-1} : H \to H$ and with the domain of definition of the operator $D(A) = V \cap [(H^2(\Omega))^2 \times H^2(\Omega)]$. Denote by $0 < \lambda_1 \leqslant \lambda_2 \leqslant \cdots$ the eigenvalues of A and by e_1, $e_2 \cdots$ the corresponding complete orthonormal system of eigenvectors.

We define a nonlinear operator $N = (N_1, N_2) : V \times V \to V'$ by

$$\langle N(u_1, u_1), u_2\rangle_H = \langle N_1(v_1, v_1), v_2\rangle + \langle N_2(v_1, T_1), T_2\rangle_{H_2},$$

where

$$\langle N_1(v, v_1), v_2\rangle = \int_\Omega \left((v \cdot \nabla)v_1 + \Phi(v)\frac{\partial v_1}{\partial z}\right) \cdot v_2,$$

$$\langle N_2(v, T_1), T_2\rangle = \int_\Omega \left((v_1 \cdot \nabla)T_1 + \Phi(v_1)\frac{\partial T_1}{\partial z}\right) T_2,$$

Related to the linear terms, we define a bilinear functional $l : V \to \mathbb{R}$ and its corresponding operator $L : V \to V$ by

$$l(u, u_1) = \langle Lu, u_1\rangle_H = \int_\Omega f(k \times v) \cdot v_1 - \int_\Omega \left(\int_{-1}^z \nabla T \mathrm{d}z'\right) T_1.$$

Now, we rewrite (1.8)–(1.14) as the following abstract stochastic evolution equations

$$\frac{\partial v^\epsilon}{\partial t} + N_1(v^\epsilon, v^\epsilon) + Lu^\epsilon + A_1 v^\epsilon = \frac{1}{\sqrt{\epsilon}}\eta_1^\epsilon, \tag{2.1}$$

$$\frac{\partial T^\epsilon}{\partial t} + N_2(v^\epsilon, T^\epsilon) + A_2 T^\epsilon = \Psi + \frac{1}{\sqrt{\epsilon}}\eta_2^\epsilon, \tag{2.2}$$

$$u^\epsilon(0) = (v^\epsilon(0), T^\epsilon(0)) = (v_0^\epsilon, T_0^\epsilon). \tag{2.3}$$

For the above random system we give the following definition.

Definition 2.1 *For any $T > t_0$, a process $u^\epsilon(t, \omega) = (v^\epsilon, T^\epsilon)$ is called a strong solution to (2.1)–(2.3) in $[t_0, T]$, if, for $\mathbb{P}$-a.e. $\omega \in \boldsymbol{\Omega}$, u^ϵ satisfies*

$$\langle v^\epsilon(t), \varphi_1\rangle - \int_{t_0}^t [\langle N_1(v^\epsilon, \varphi_1), v\rangle - \langle Lu^\epsilon, \varphi\rangle_H] + \int_{t_0}^t \langle v^\epsilon, A_1\varphi_1\rangle$$

$$= \langle v_{t_0}^\epsilon, \varphi_1\rangle + \frac{1}{\sqrt{\epsilon}}\int_{t_0}^t \langle \eta_1^\epsilon(s, \omega), \varphi_1\rangle,$$

$$\langle T^\epsilon(t), \varphi_2\rangle - \int_{t_0}^t [\langle N_2(v^\epsilon, \varphi_2), T^\epsilon\rangle - \langle T^\epsilon, A_2\varphi_2\rangle]$$

$$= \langle T_{t_0}^\epsilon, \varphi_2\rangle + \int_{t_0}^t \langle \Psi, \varphi_2\rangle + \frac{1}{\sqrt{\epsilon}}\int_{t_0}^t \langle \eta_2^\epsilon(s, \omega), \varphi_2\rangle,$$

for all $t \in [t_0, T]$ and $\varphi = (\varphi_1, \varphi_2) \in D(A_1) \times D(A_2)$, moreover $u^\epsilon \in L^\infty(t_0, T; V) \cap L^2(t_0, T; (H^2(\Omega))^2)$ and is progressively measurable in these topologies.

2.2 A model for $(\eta_1^\epsilon, \eta_2^\epsilon)$

To detail the random model (1.8)–(1.14), we assume that the stationary process $\eta^\epsilon(t) = (\eta_1^\epsilon(t), \eta_2^\epsilon(t))$ solves the following linear stochastic system

$$\epsilon d\frac{\eta^\epsilon(t)}{\sqrt{\epsilon}} = -\frac{\eta^\epsilon(t)}{\sqrt{\epsilon}}\mathrm{d}t + \mathrm{d}W(t), \tag{2.4}$$

where $W(t) = (W_1(t), W_2(t))$ is H-valued Wiener process with covariance operator $\mathbb{Q} = (Q_1, Q_2)$. Here we assume that $\dot{W}_i$, $i = 1, 2$, have the following form

$$\frac{\mathrm{d}W_i(t)}{\mathrm{d}t} = \sqrt{Q_i}\frac{\mathrm{d}\tilde{W}_i(t)}{\mathrm{d}t},$$

where $\tilde{W}_i(t)$ is a cylindrical Wiener process in H_i defined on a complete probability space $(\boldsymbol{\Omega}, \mathcal{F}, \mathbb{P})$ with expectation denoted by $\mathbb{E}$, and $\sqrt{Q_i}$ is a linear operator.

Assumption (H$_1$) Q_i, $i = 1, 2$, satisfy

$$\mathrm{Tr}A^3 Q_i < +\infty.$$

Remark 2.1 *An example for W is a two-sided in time finite-dimensional Brownian motion with the form*

$$W = \sum_{i=1}^{m} \delta_i \beta_i(t, \omega) e_i.$$

In the above formula, $\beta_1, \cdots, \beta_m$ are independent standard one-dimensional Brownian motions on a complete probability space $(\boldsymbol{\Omega}, \mathcal{F}, P)$, and δ_i are real coefficients.

Remark 2.2 *An another example for W is a two-sided in time infinite dimensional Brownian motion with the form*

$$W(t) = \sum_{i=1}^{+\infty} \mu_i \beta_i(t, \omega) e_i.$$

Here $\beta_1, \beta_2, \cdots$ is a sequence of independent standard one-dimensional Brownian motions on a complete probability space $(\boldsymbol{\Omega}, \mathcal{F}, P)$ and the coefficients μ_i satisfy $\sum_{i=1}^{+\infty} \lambda_i^3 \mu_i^2 < +\infty$.

By the stationarity and the assumption $(\mathbf{H}_1)$, a simple application of Itô formula yields

$$\mathbb{E}\|\eta^\epsilon(t)\|_3^2 = \mathrm{Tr}(A^3\mathbb{Q}) < \infty\,. \tag{2.5}$$

Moreover, the system (2.4) is strong mixing. To describe, this we define

$$\mathcal{F}_{s/\epsilon}^{t/\epsilon} = \sigma\{\eta^\epsilon(\tau) : s \leqslant \tau \leqslant t\}$$

and

$$\phi(t/\epsilon) = \sup_{s\geqslant 0}\ \sup_{A\in\mathcal{F}_0^{s/\epsilon},B\in\mathcal{F}_{s/\epsilon+t/\epsilon}^{\infty}} |\mathbb{P}(AB) - \mathbb{P}(A)\mathbb{P}(B)|\,.$$

Then

$$\int_0^\infty \phi^k(t)\,\mathrm{d}t < \infty$$

for any $k > 0$. In fact by the exponentially stable of the linear system (2.4), we have for $t > 0$

$$\phi(t/\epsilon) < Ce^{-t/\epsilon}$$

for some constant $C > 0$. Moreover for $s \geqslant t$

$$\mathbb{E}[\eta^\epsilon(s)|\mathcal{F}_0^{t/\epsilon}] = \eta^\epsilon(t)e^{-(s-t)/\epsilon}, \tag{2.6}$$

$$\mathbb{E}\eta_i^\epsilon(x,t)\eta_i^\epsilon(y,s) = \frac{1}{2}q_i(x,y)\exp\left(-\frac{|t-s|}{\epsilon}\right), \quad i = 1, 2, \tag{2.7}$$

where $q_i(x, y)$ satisfy $q_i(x, y) = q_i(y, x)$ and

$$Q_i f(x) = \int_\Omega q_i(x,y)f(y)\,\mathrm{d}y\,, \quad i = 1, 2\,.$$

Notice that (v^ϵ, T^ϵ) is not Markovian. So in the following we consider the Markov process $(v^\epsilon, T^\epsilon, \eta_1^\epsilon, \eta_2^\epsilon)$.

3 The global well-posedness and existence of stationary measure

Let $U^\epsilon = (u^\epsilon, \eta^\epsilon)$, with $u^\epsilon = (v^\epsilon, T^\epsilon)$. To show the global well-posedness to (2.1)–(2.3), we introduce process $\mathcal{Z}^\epsilon$ solving

$$\dot{\mathcal{Z}}^\epsilon = -A\mathcal{Z}^\epsilon + \frac{1}{\sqrt{\epsilon}}\eta^\epsilon\,, \quad \mathcal{Z}^\epsilon(0) = 0.$$

One can see that

$$\mathcal{Z}^\epsilon(t) = \frac{1}{\sqrt{\epsilon}}\int_0^t e^{-A(t-s)}\eta^\epsilon(s)\,\mathrm{d}s.$$

Then we have the following estimates.

Lemma 3.1 *For any integer $k \geqslant 0$, there is a constant $c > 0$, such that for any $t > 0$,*

$$\mathbb{E}\|\mathcal{Z}^\epsilon(t)\|_k^2 \leqslant c\mathrm{Tr}A^k\mathbb{Q}.$$

We need some estimates on $Z_\alpha^\epsilon(t)$. First

$$\mathbb{E}\|\mathcal{Z}^\epsilon(t)\|_0^2 = \frac{1}{\epsilon}\mathbb{E}\int_0^t\int_0^t \left\langle e^{-A(t-s)}\eta^\epsilon(s), e^{-A(t-r)}\eta^\epsilon(r)\right\rangle \mathrm{d}r\,\mathrm{d}s,$$

then by (2.6)–(2.7) we have for any $t > 0$

$$\mathbb{E}\|\mathcal{Z}^\epsilon(t)\|_0^2 \leqslant c\left[\|q_1\|_{L^2}^2 + \|q_2\|_{L^2}^2\right]$$

for some constant $c > 0$. Similarly by the regularity of η^ϵ, for $k > 0$,

$$\mathbb{E}\|\mathcal{Z}^\epsilon(t)\|_k^2 \leqslant c\left[\|q_1\|_{H^k}^2 + \|q_2\|_{H^k}^2\right].$$

Now define

$$w^\epsilon = v^\epsilon - \mathcal{Z}^{\epsilon,1}, \quad \theta^\epsilon = T^\epsilon - \mathcal{Z}^{\epsilon,2}, \quad \text{and} \quad \tilde{u}^\epsilon = u^\epsilon - \mathcal{Z}^\epsilon,$$

where $\mathcal{Z}^\epsilon = (\mathcal{Z}^{\epsilon,1}, \mathcal{Z}^{\epsilon,2})$ Then, to obtain the global existence of strong solutions to the system (2.1)–(2.3), we just consider that of the following system

$$\frac{\partial w^\epsilon}{\partial t} + N_1(w^\epsilon + \mathcal{Z}^{\epsilon,1}, w^\epsilon + \mathcal{Z}^{\epsilon,1}) + L(\tilde{u}^\epsilon + \mathcal{Z}^\epsilon) + A_1 w^\epsilon = 0, \tag{3.1}$$

$$\frac{\partial \theta^\epsilon}{\partial t} + N_2(w^\epsilon + \mathcal{Z}^{\epsilon,1}, \theta^\epsilon + \mathcal{Z}^{\epsilon,2}) + A_2\theta^\epsilon = \Psi, \tag{3.2}$$

$$\tilde{u}^\epsilon(0) = (w^\epsilon(0), \theta^\epsilon(0)) = u_0^\epsilon = (v_0^\epsilon, T_0^\epsilon). \tag{3.3}$$

Theorem 3.1 [Global well-posedness of (2.1)–(2.3)] *Assume that $\Psi \in V_2$ and $(\mathbf{H}_1)$ hold, then*

(1) *For any initial data $u_0^\epsilon \in V$, there exists globally a unique strong solution u^ϵ to* (2.1)–(2.3), *i.e., for any $\mathcal{T} > 0$,*

$$u^\epsilon \in C(0, \mathcal{T}; V) \cap L^2(0, \mathcal{T}; (H^2(\Omega))^2)$$

for $\mathbb{P}$ a.e. $\omega \in \mathbf{\Omega}$.

(2) *The process U^ϵ is a Markov process in $\mathbb{V} := V \times H^1 \times H^1$.*

Remark 3.1 *Result (1) is proved by Lemma* 3.1 *and the method for* 3*D stochastic primitive equations* [14]. *Result* (2) *can be justified by a standard argument* [24, Theorem 9.8].

Proposition 3.1 *Assume $\Psi \in V_2$ and $B_\rho = \{u; \|u\| \leqslant \rho, u \in V\}$. Then there exist $r_0(\omega, \|\Psi\|_1)$ and $t(\omega,\rho) \leqslant -1$ such that for any $t_0 \leqslant t(\omega,\rho)$, $u^\epsilon_{t_0} \in B_\rho$,*

$$\|u^\epsilon(0,\omega,u^\epsilon_{t_0})\| \leqslant r_0(\omega),$$

where $u^\epsilon(t,\omega,u^\epsilon_{t_0})$ is the strong solution of (2.1)–(2.3) with initial data $u^\epsilon(t_0) = u^\epsilon_{t_0}$. Moreover, the family of random variables $\{u^\epsilon(0,\omega,u^\epsilon_{t_0}) : -\infty < t_0 < t(\omega,\rho)\}$ is tight in V.

Remark 3.2 *The estimate in the above proposition can be followed by a similar discussion for 3D stochastic primitive equations* [14] *and the tight result can be derived by the similar method* [13].

Denote by P^ϵ_t, $t \geqslant 0$, the associated Markov semigroup to U^ϵ. Let $P^{\epsilon*}_t$ be the dual semigroup which is defined, in space $Pr(\mathbb{V})$ consisting of probability measures on $\mathbb{V}$, as

$$\int_{\mathbb{V}} \varphi \mathrm{d}P^{\epsilon*}_t \mu = \int_{\mathbb{V}} P^\epsilon_t \varphi \mathrm{d}\mu$$

for all $\varphi \in C_b(\mathbb{V})$ and $\mu \in Pr(\mathbb{V})$. Then a probability measure $\mu \in Pr(\mathbb{V})$ is called a stationary one if

$$P^{\epsilon*}_t \mu = \mu.$$

Then we have the following result.

Theorem 3.2 [The existence of stationary measure] *Under the assumption* $(\mathbf{H}_1)$, *the system of 3D random primitive equations (2.1)–(2.3) coupled with (2.4) has at least a stationary measure.*

Proof By the Proposition 3.1, the distribution of U^ϵ in space $\mathbb{V}$ is tight, then by the Kryloff–Bogoliubov procedure [25] one can construct one stationary measure. □

Let $\mathfrak{y}^{*\epsilon}$ be a stationary measure of 3D random primitive equations (2.1)–(2.3) coupled with (2.4). Then $U^{*\epsilon} = (v^{*\epsilon}, T^{*\epsilon}, \eta^\epsilon)$, the solution with initial value distributes as $\mathfrak{y}^{*\epsilon}$, is a stationary solution to 3D random primitive equations (2.1)–(2.3) coupled with (2.4). Then we call $(v^{*\epsilon}, T^{*\epsilon})$ a stationary solution to 3D random primitive equations (2.1)–(2.3).

4 The diffusion limit for the 3D random primitive equations

Notice that the distributions of η^ϵ_1 and η^ϵ_2 are independent of ϵ. We just consider the limit of (v^ϵ, T^ϵ). We have showed the tightness of the distributions of

$\{(v^\epsilon, T^\epsilon)\}_{0<\epsilon\leqslant 1}$ in space $C([0,\infty);V)$. Further the following result determine the limit of (v^ϵ, T^ϵ) in the sense of distribution. For this we first give the following limiting equation, 3D stochastic primitive equations

$$\frac{\partial v}{\partial t}+(v\cdot\nabla)v+\Phi(v)\frac{\partial v}{\partial z}+fk\times v+\nabla p_b-\int_{-1}^{z}\nabla T\mathrm{d}z'$$

$$-\Delta v-\frac{\partial^2 v}{\partial z^2}=\dot{W}_1\,, \tag{4.1}$$

$$\frac{\partial T}{\partial t}+(v\cdot\nabla)T+\Phi(v)\frac{\partial T}{\partial z}-\Delta T-\frac{\partial^2 T}{\partial z^2}=\Psi+\dot{W}_2\,, \tag{4.2}$$

$$\int_{-1}^{0}\nabla\cdot v\,\mathrm{d}z=0. \tag{4.3}$$

with boundary value conditions (1.4)–(1.6) and initial value

$$u(0)=(v_0,T_0)\,. \tag{4.4}$$

Remark 4.1 *The above equations are called as 3D stochastic primitive equations of the large-scale ocean which are considered in the article [14]. There are two main reasons for considering the stochastic model. Firstly, it is impossible to accurately predict long-term oceanic motions in deterministic frameworks since the space scale and time scale of the forecast target are bounded in the deterministic oceanic forecasting. To predict long-term oceanic motions more objectively, it is suitable to apply some stochastic modes. Secondly, it is suitable and useful to consider the sochastic primitive equations of the large-scale ocean with a white in time noise since the primitive oceanic equations are usually used to understand the mechanism of long-term oceanic motions.*

We need some estimates on the solution.

Lemma 4.1 *Under the assumption, for any $T>0$ the following estimate holds*

$$\sup_{0\leqslant t\leqslant T}\mathbb{E}\|u^\epsilon(t)\|+\mathbb{E}\int_0^T\|\nabla u^\epsilon(s)\|^2\,\mathrm{d}s\leqslant C_T$$

for some constant $C_T>0$.

Proof The difficulty is the existence of singular terms. To overcome this, we introduce the following processes

$$
\begin{aligned}
M_t^{1,\epsilon} = & \epsilon[\langle \eta_1^\epsilon(t), v^\epsilon(t)\rangle - \langle \eta_1^\epsilon(0), v^\epsilon(0)\rangle] + \frac{1}{\sqrt{\epsilon}}\int_0^t \langle \eta_1^\epsilon(s), v^\epsilon(s)\rangle \mathrm{d}s \\
& + \epsilon \int_0^t \langle \eta_1^\epsilon(s), -A_1 v^\epsilon(s) - Lu^\epsilon(s) - N_1(v^\epsilon(s), v^\epsilon(s))\rangle \,\mathrm{d}s \\
& - \sqrt{\epsilon}\int_0^t \langle \eta_1^\epsilon(s), \eta_1^\epsilon(s)\rangle \,\mathrm{d}s
\end{aligned}
$$

and

$$
\begin{aligned}
M_t^{2,\epsilon} = & \epsilon[\langle \eta_2^\epsilon(t), T^\epsilon(t)\rangle - \langle \eta_2^\epsilon(0), T^\epsilon(0)\rangle] + \frac{1}{\sqrt{\epsilon}}\int_0^t \langle \eta_2^\epsilon(s), T^\epsilon(s)\rangle \mathrm{d}s \\
& + \epsilon \int_0^t \langle \eta_2^\epsilon(s), -A_2 T^\epsilon(s) - N_2(v^\epsilon(s), T^\epsilon(s)) + \Psi\rangle \,\mathrm{d}s \\
& - \sqrt{\epsilon}\int_0^t \langle \eta_2^\epsilon(s), \eta_2^\epsilon(s)\rangle \,\mathrm{d}s\,.
\end{aligned}
$$

For fixed $\epsilon > 0$, by direct calculation, we have

$$
\sup_{0\leqslant t\leqslant \mathcal{T}} \mathbb{E}\|u^\epsilon(t)\| + \int_0^{\mathcal{T}} \|\nabla u^\epsilon(t)\|^2 \,\mathrm{d}t \leqslant C_{\epsilon,\mathcal{T}},
$$

which implies that, by a direct verification [6, Lemma 2], $M_t^{1,\epsilon}$ and $M_t^{2,\epsilon}$ are square integrable martingale with respect to $\mathcal{F}_0^{t/\epsilon}$. Multiplying equation (2.1) with v^ϵ on both sides in H_1 yields

$$
\begin{aligned}
& d\left[\frac{1}{2}\|v^\epsilon(t)\|^2 + \epsilon\langle \eta_1^\epsilon(t), v^\epsilon(t)\rangle\right] \\
= & \langle -A_1 v^\epsilon(t) - Lu^\epsilon(t) - N_1(v^\epsilon, v^\epsilon), v^\epsilon(t)\rangle \,\mathrm{d}t \\
& - \epsilon\langle -A_1 v^\epsilon(t) - Lu^\epsilon(t) - N_1(v^\epsilon, v^\epsilon), \eta_1^\epsilon\rangle \,\mathrm{d}t \\
& - \sqrt{\epsilon}\|\eta_1^\epsilon\|^2 \,\mathrm{d}t + \mathrm{d}M_t^{1,\epsilon}
\end{aligned}
$$

and

$$
\begin{aligned}
& d\left[\frac{1}{2}\|T^\epsilon(t)\|^2 + \epsilon\langle \eta_2^\epsilon(t), T^\epsilon(t)\rangle\right] \\
= & \langle -A_2 T^\epsilon(t) - N_2(v^\epsilon, T^\epsilon) + \Psi, T^\epsilon(t)\rangle \,\mathrm{d}t \\
& - \epsilon\langle -A_2 T^\epsilon(t) - N_2(v^\epsilon, T^\epsilon), \eta_2^\epsilon\rangle \,\mathrm{d}t \\
& - \sqrt{\epsilon}\|\eta_2^\epsilon\|^2 \,\mathrm{d}t + \mathrm{d}M_t^{2,\epsilon}\,.
\end{aligned}
$$

Then for $\epsilon > 0$ small enough, by the martingale property of $M_t^{1,\epsilon}$, $M_t^{2,\epsilon}$ and Gronwall lemma, we have the estimate of this lemma. □

Theorem 4.1 *Assume $(v_0^\epsilon, T_0^\epsilon) \in V$ converges in distribution to (v_0, T_0) as $\epsilon \to 0$. The solution (v^ϵ, T^ϵ) of 3D random primitive equations (1.8)–(1.14) converges in distribution, as $\epsilon \to 0$, to the solution of 3D stochastic primitive equations (4.1)–(4.4) in space $C([0,\infty); V)$.*

Remark 4.2 *We apply a weak convergence method developed by Kushner [17] to prove this result. Such method is also applied to study the stochastic self-similarity in stochastic Burgers equation [26].*

Proof Denote by (v, T) one limit point in the sense of distribution of (v^ϵ, T^ϵ) as $\epsilon \to 0$ in space $C([0,\infty); V)$. For simplicity we assume (v^ϵ, T^ϵ) converges in distribution to (v, T) as $\epsilon \to 0$ in $C([0,\infty); V)$. Notice that convergence in distribution is not enough to pass limit $\epsilon \to 0$, by Skorohod theorem we can construct a new probability space and new random variables without changing distributions in $C([0,\infty); V)$, of which for simplicity we do not change the notations, such that $(v_0^\epsilon, T_0^\epsilon)$ converges almost surely to (v_0, T_0) in V and (v^ϵ, T^ϵ) converges almost surely to (v, T) in $C([0,\infty); V)$.

In the following, we shall prove that (v, T) is a solution to (4.1)–(4.4). Now for any $\varphi = (\varphi_1, \varphi_2) \in C_0^\infty$ and C^3-differentiable compactly supported real valued function F, we consider processes

$$\{F(\langle (v^\epsilon, T^\epsilon), \varphi \rangle)\}_{0<\epsilon \leqslant 1}. \tag{4.5}$$

We derive from equations (1.8)–(1.14)

$$\begin{aligned} &F(\langle (v^\epsilon(t), T^\epsilon(t)), \varphi \rangle) - F(\langle (v_0^\epsilon, T_0^\epsilon), \varphi \rangle) \\ =& \int_0^t F'(\langle (v^\epsilon(s), T^\epsilon(s)), \varphi \rangle) \left[\langle G_1^\epsilon(s), \varphi_1 \rangle + \langle G_2^\epsilon(s), \varphi_2 \rangle \right] \mathrm{d}s \\ &+ \frac{1}{\sqrt{\epsilon}} \int_0^t F'(\langle (v^\epsilon(s), T^\epsilon(s)), \varphi \rangle) \langle \eta_1^\epsilon(s), \varphi_1 \rangle \, \mathrm{d}s \\ &+ \frac{1}{\sqrt{\epsilon}} \int_0^t F'(\langle (v^\epsilon(s), T^\epsilon(s)), \varphi \rangle) \langle \eta_2^\epsilon(s), \varphi_2 \rangle \, \mathrm{d}s, \end{aligned} \tag{4.6}$$

where

$$G_1^\epsilon = \Delta v^\epsilon - (v^\epsilon \cdot \nabla) v^\epsilon - \Phi(v^\epsilon) \frac{\partial v^\epsilon}{\partial z} - fk \times v^\epsilon - \nabla p_b^\epsilon + \int_{-1}^z \nabla T^\epsilon \mathrm{d}z'$$

and

$$G_2^\epsilon = \Delta T^\epsilon - (v^\epsilon \cdot \nabla) T^\epsilon - \Phi(v^\epsilon) \frac{\partial T^\epsilon}{\partial z} + \Psi.$$

To treat the singular terms in (4.6) we introduce

$$F_1^\epsilon(t) := \frac{1}{\sqrt{\epsilon}} \mathbb{E} \left[\int_t^\infty F'(\langle (v^\epsilon(t), T^\epsilon(t)), \varphi \rangle) \langle \eta_1^\epsilon(s), \varphi_1 \rangle \, \mathrm{d}s \middle| \mathcal{F}_0^{t/\epsilon} \right]$$

and

$$F_2^\epsilon(t) := \frac{1}{\sqrt{\epsilon}}\mathbb{E}\left[\int_t^\infty F'(\langle(v^\epsilon(t),T^\epsilon(t)),\varphi\rangle)\langle\eta_2^\epsilon(s),\varphi_2\rangle\,\mathrm{d}s\Big|\mathcal{F}_0^{t/\epsilon}\right].$$

Then by the property (2.6) of η^ϵ we have

$$F_1^\epsilon(t) = \sqrt{\epsilon}F'(\langle(v^\epsilon(t),T^\epsilon(t)),\varphi\rangle)\langle\eta_1^\epsilon(t),\varphi_1\rangle$$

and

$$F_2^\epsilon(t) = \sqrt{\epsilon}F'(\langle(v^\epsilon(t),T^\epsilon(t)),\varphi\rangle)\langle\eta_2^\epsilon(t),\varphi_2\rangle.$$

Moreover

$$\mathbb{E}|F_i^\epsilon(t)| \leqslant C\sqrt{\epsilon}, \quad i=1,2$$

for some constant $C>0$.

Next we construct a martingale depends on ϵ and pass the limit $\epsilon\to 0$ in this martingale. For this we introduce the pseudo-differential operator A^ϵ defined by

$$A^\epsilon X(t) = \mathbb{P}-\lim_{\delta\to 0}\frac{1}{\delta}\mathbb{E}\left[X(t+\delta)-X(t)\mid\mathcal{F}_0^{t/\epsilon}\right] \tag{4.7}$$

for any $\mathcal{F}_0^{t/\epsilon}$ measurable function X with $\sup_t\mathbb{E}|X(t)|<\infty$. Then Ethier and Kurtz's proposition [10, Proposition 2.7.6] yields that

$$X(t)-\int_0^t A^\epsilon X(s)\,\mathrm{d}s$$

is a martingale with respect to $\mathcal{F}_0^{t/\epsilon}$. Now define (Y^ϵ,Z^ϵ) as

$$Y^\epsilon(t) = F(\langle(v^\epsilon(t),T^\epsilon(t)),\varphi\rangle)+F_1^\epsilon(t)+F_2^\epsilon(t), \quad Z^\epsilon(\tau)=A^\epsilon Y^\epsilon(t).$$

Then

$$\begin{aligned}
&Z^\epsilon(t)\\
=&F'(\langle(v^\epsilon(t),T^\epsilon(t)),\varphi\rangle)\left[\langle G_1^\epsilon(t),\varphi_1\rangle+\langle G_2^\epsilon(t),\varphi_2\rangle\right]\\
&+\sqrt{\epsilon}F''(\langle(v^\epsilon(t),T^\epsilon(t)),\varphi\rangle)\langle\eta_1^\epsilon(t),\varphi_1\rangle\left[\langle G_1^\epsilon(t),\varphi_1\rangle+\langle G_2^\epsilon(t),\varphi_2\rangle\right]\\
&+\sqrt{\epsilon}F''(\langle(v^\epsilon(t),T^\epsilon(t)),\varphi\rangle)\langle\eta_2^\epsilon(t),\varphi_2\rangle\left[\langle G_1^\epsilon(t),\varphi_1\rangle+\langle G_2^\epsilon(t),\varphi_2\rangle\right]\\
&+F''(\langle(v^\epsilon(t),T^\epsilon(t)),\varphi\rangle)\langle\eta_1^\epsilon(t),\varphi_1\rangle^2+F''(\langle(v^\epsilon(t),T^\epsilon(t)),\varphi\rangle)\langle\eta_2^\epsilon(t),\varphi_2\rangle^2\\
&+2F''(\langle(v^\epsilon(t),T^\epsilon(t)),\varphi\rangle)\langle\eta_1^\epsilon(t),\varphi_1\rangle\langle\eta_2^\epsilon(t),\varphi_2\rangle.
\end{aligned} \tag{4.8}$$

In fact first by equaton (4.6) and the definition of A^ϵ, we have

$$
\begin{aligned}
&A^\epsilon F(\langle(v^\epsilon(t),T^\epsilon(t)),\varphi\rangle)\\
&=F'(\langle(v^\epsilon(t),T^\epsilon(t)),\varphi\rangle)\left[\langle G_1^\epsilon(t),\varphi_1\rangle+\langle G_2^\epsilon(t),\varphi_2\rangle\right]\\
&\quad+\frac{1}{\sqrt{\epsilon}}F'(\langle(v^\epsilon(t),T^\epsilon(t)),\varphi\rangle)\langle\eta_1^\epsilon(t),\varphi_1\rangle\\
&\quad+\frac{1}{\sqrt{\epsilon}}F'(\langle(v^\epsilon(t),T^\epsilon(t)),\varphi\rangle)\langle\eta_2^\epsilon(t),\varphi_2\rangle\,.
\end{aligned}
$$

Now by the construction of η^ϵ and F_1^ϵ,

$$
\begin{aligned}
&\mathbb{E}\big[F_1^\epsilon(t+\delta)|\mathcal{F}_0^{t/\epsilon}\big]\\
&=\sqrt{\epsilon}\mathbb{E}\big\{\big[F'(\langle(v^\epsilon(t+\delta),T^\epsilon(t+\delta)),\varphi\rangle)\\
&\quad-F'(\langle(v^\epsilon(t),T^\epsilon(t)),\varphi\rangle)\big]\langle\eta_1^\epsilon(t+\delta),\varphi_1\rangle\big|\mathcal{F}_0^{t/\epsilon}\big\}\\
&\quad+\sqrt{\epsilon}F'(\langle(v^\epsilon(t),T^\epsilon(t)),\varphi\rangle)\langle\eta_1^\epsilon(t),\varphi_1\rangle e^{-\delta/\epsilon}\,.
\end{aligned}
$$

Then

$$
\begin{aligned}
A^\epsilon F_1^\epsilon(t)=&\sqrt{\epsilon}F''(\langle(v^\epsilon(t),T^\epsilon(t)),\varphi\rangle)\langle\eta_1^\epsilon(t),\varphi_1\rangle\big[\langle G_1^\epsilon(t),\varphi_1\rangle+\langle G_2^\epsilon(t),\varphi_2\rangle\big]\\
&+F''(\langle(v^\epsilon(t),T^\epsilon(t)),\varphi\rangle)\langle\eta_1^\epsilon(t),\varphi_1\rangle^2\\
&+F''(\langle(v^\epsilon(t),T^\epsilon(t)),\varphi\rangle)\langle\eta_1^\epsilon(t),\varphi_1\rangle\langle\eta_2^\epsilon(t),\varphi_2\rangle\\
&-\frac{1}{\sqrt{\epsilon}}F'(\langle(v^\epsilon(t),T^\epsilon(t)),\varphi\rangle)\langle\eta_1^\epsilon(t),\varphi_1\rangle\,.
\end{aligned}
$$

Similarly

$$
\begin{aligned}
A^\epsilon F_2^\epsilon(t)=&\sqrt{\epsilon}F''(\langle(v^\epsilon(t),T^\epsilon(t)),\varphi\rangle)\langle\eta_2^\epsilon(t),\varphi_1\rangle\big[\langle G_1^\epsilon(t),\varphi_1\rangle+\langle G_2^\epsilon(t),\varphi_2\rangle\big]\\
&+F''(\langle(v^\epsilon(t),T^\epsilon(t)),\varphi\rangle)\langle\eta_2^\epsilon(t),\varphi_2\rangle^2\\
&+F''(\langle(v^\epsilon(t),T^\epsilon(t)),\varphi\rangle)\langle\eta_1^\epsilon(t),\varphi_1\rangle\langle\eta_2^\epsilon(t),\varphi_2\rangle\\
&-\frac{1}{\sqrt{\epsilon}}F'(\langle(v^\epsilon(t),T^\epsilon(t)),\varphi\rangle)\langle\eta_2^\epsilon(t),\varphi_2\rangle
\end{aligned}
$$

which shows (4.8).

Now denote by $Z_i^\epsilon(t)$, $i=1,\cdots,5$, the five terms on the righthand side of $Z^\epsilon(t)$, then

$$
\mathbb{E}\big[|Z_2^\epsilon(t)|+|Z_3^\epsilon(t)|\big]=\mathcal{O}(\sqrt{\epsilon}),\quad \epsilon\to 0\,.
$$

To pass the limit $\epsilon\to 0$, we further need more processes. Define

$$
F_3^\epsilon(t)=F''(\langle(v^\epsilon(t),T^\epsilon(t)),\varphi\rangle)\int_t^\infty\mathbb{E}\left[\langle\eta_1^\epsilon(s),\varphi_1\rangle^2-\frac{1}{2}\langle Q_1\varphi_1,\varphi_1\rangle\big|\mathcal{F}_0^{t/\epsilon}\right]\mathrm{d}s,
$$

$$F_4^\epsilon(t) = F''(\langle(v^\epsilon(t), T^\epsilon(t)), \varphi\rangle) \int_t^\infty \mathbb{E}\left[\langle\eta_2^\epsilon(s), \varphi_2\rangle^2 - \frac{1}{2}\langle Q_2\varphi_2, \varphi_2\rangle \middle| \mathcal{F}_0^{t/\epsilon}\right] \mathrm{d}s$$

and

$$F_5^\epsilon(t) = F''(\langle(v^\epsilon(t), T^\epsilon(t)), \varphi\rangle) \int_t^\infty \mathbb{E}\left[\langle\eta_1^\epsilon(s), \varphi_1\rangle\langle\eta_2^\epsilon(s), \varphi_2\rangle \middle| \mathcal{F}_0^{t/\epsilon}\right] \mathrm{d}s\,.$$

Then by the property of η^ϵ we have

$$F_3^\epsilon(t) = \frac{\epsilon}{2} F''(\langle(v^\epsilon(t), T^\epsilon(t)), \varphi\rangle) \left[\langle\eta_1^\epsilon(t), \varphi_1\rangle^2 - \frac{1}{2}\langle Q_1\varphi_1, \varphi_1\rangle\right],$$

$$F_4^\epsilon(t) = \frac{\epsilon}{2} F''(\langle(v^\epsilon(t), T^\epsilon(t)), \varphi\rangle) \left[\langle\eta_2^\epsilon(t), \varphi_2\rangle^2 - \frac{1}{2}\langle Q_2\varphi_2, \varphi_2\rangle\right]$$

and

$$F_5^\epsilon(t) = \frac{\epsilon}{2} F''(\langle(v^\epsilon(t), T^\epsilon(t)), \varphi\rangle) \left[\langle\eta_1^\epsilon(t), \varphi_1\rangle\langle\eta_2^\epsilon(t), \varphi_2\rangle\right].$$

Further by direct calculation and estimates on (v^ϵ, T^ϵ) we have

$$\sup_{t\geqslant 0} \mathbb{E} F_i^\epsilon(t) = \mathcal{O}(\epsilon)\,, \quad i = 3, 4, 5\,, \quad \text{as} \quad \epsilon \to 0\,.$$

Moreover

$$A^\epsilon F_3^\epsilon(t) = F''(\langle(v^\epsilon(t), T^\epsilon(t)), \varphi\rangle) \left[\frac{1}{2}\langle Q_1\varphi_1, \varphi_1\rangle - \langle\eta_1^\epsilon(t), \varphi_1\rangle^2\right] + R_3^\epsilon(t)\,,$$

and

$$A^\epsilon F_4^\epsilon(t) = F''(\langle(v^\epsilon(t), T^\epsilon(t)), \varphi\rangle) \left[\frac{1}{2}\langle Q_2\varphi_2, \varphi_2\rangle - \langle\eta_2^\epsilon(t), \varphi_2\rangle^2\right] + R_4^\epsilon(t)$$

with

$$\sup_{t\geqslant 0} \mathbb{E}|R_3^\epsilon(t)| = \mathcal{O}(\epsilon) \quad \text{and} \quad \sup_{t\geqslant 0} \mathbb{E}|R_4^\epsilon(t)| = \mathcal{O}(\epsilon)$$

and

$$\sup_{t\geqslant 0} \mathbb{E}|A^\epsilon F_5^\epsilon(t)| = \mathcal{O}(\epsilon) \quad \text{as} \quad \epsilon \to 0\,.$$

Now we have the following $\mathcal{F}_0^{t/\epsilon}$ martingale

$$\begin{aligned}
&\mathcal{M}^\epsilon(t) \\
=\,& F(\langle(v^\epsilon(t), T^\epsilon(t)), \varphi\rangle) - F(\langle(v_0^\epsilon, T_0^\epsilon), \varphi\rangle) + F_1^\epsilon(t) + F_2^\epsilon(t) + F_3^\epsilon(t) + F_4^\epsilon(t) \\
&+ F_5^\epsilon(t) - \int_0^t F'(\langle(v^\epsilon(s), T^\epsilon(s)), \varphi\rangle)\left[\langle G_1^\epsilon(s), \varphi_1\rangle + \langle G_2^\epsilon(s), \varphi_2\rangle\right] \mathrm{d}s \\
&- \frac{1}{2}\int_0^t F''(\langle(v^\epsilon(s), T^\epsilon(s)), \varphi\rangle)\langle Q_1\phi_1, \phi_1\rangle \, \mathrm{d}s \\
&- \frac{1}{2}\int_0^t F''(\langle(v^\epsilon(s), T^\epsilon(s)), \varphi\rangle)\langle Q_2\phi_2, \phi_2\rangle \, \mathrm{d}s + R^\epsilon(t)
\end{aligned}$$

where

$$R^\epsilon(t) = \int_0^t [Z_2^\epsilon(s) + Z_3^\epsilon(s) + R_3^\epsilon(s) + R_4^\epsilon(s) + A^\epsilon F_5^\epsilon(s)]\, \mathrm{d}s$$

with $\mathbb{E}|R^\epsilon(t)| = \mathcal{O}(\epsilon)$ as $\epsilon \to 0$. Now passing to the limit $\epsilon \to 0$ in $\mathcal{M}^\epsilon(t)$ shows the distribution of the limit (v, T) solves the following martingale problem

$$\begin{aligned}\mathcal{M}(t) = & F(\langle\langle(v(t), T(t)), \varphi\rangle\rangle) - F(\langle\langle(v_0, T_0), \varphi\rangle\rangle)\\ & - \int_0^t F'(\langle\langle(v(s), T(s)), \varphi\rangle\rangle)\big[\langle G_1(s), \varphi_1\rangle + \langle G_2(s), \varphi_2\rangle\big]\, \mathrm{d}s\\ & - \frac{1}{2}\int_0^t F''(\langle\langle(v(s), T(s)), \varphi\rangle\rangle)\langle Q_1\phi_1, \phi_1\rangle\, \mathrm{d}s\\ & - \frac{1}{2}\int_0^t F''(\langle\langle(v(s), T(s)), \varphi\rangle\rangle)\langle Q_2\phi_2, \phi_2\rangle\, \mathrm{d}s\end{aligned}$$

which is equivalent to the martingale solution to the 3D stochastic primitive equations (4.1)–(4.4)([20]). By the global well-posedness of equations (4.1)–(4.4), we complete the proof. □

By Theorem 4.1 we have the following result on convergence of stationary solution of the 3D random primitive equations. Denote by $\mathfrak{P}^{*\epsilon} = \mathcal{D}(\hat{v}^{*\epsilon}, \hat{T}^{*\epsilon}, \hat{\eta}_1^\epsilon, \hat{\eta}_2^\epsilon)$, a stationary statistical solution (see Appendix A) to the system of random primitive equations (1.8)–(1.14) coupled with (2.4). Let $\mathbb{P}^{*\epsilon} = \mathcal{D}(\hat{v}^{*\epsilon}, \hat{T}^{*\epsilon})$, then we have

Corollary 4.1 *For $\epsilon \to 0$, there is sequence $\epsilon_n \to 0$, as $n \to \infty$, such that*

$$\mathbb{P}^{*\epsilon_n} \to \mathbb{P}^* \quad \textit{weakly as} \quad n \to \infty$$

where $\mathbb{P}^*$ is a probability measure on $C([0,\infty); V)$, which is a stationary statistical solution to 3D stochastic primitive equations (4.1)–(4.4).

Proof By the tightness of $\{\mathbb{P}^{*\epsilon}\}_\epsilon = \{\mathcal{D}(\hat{v}^{*\epsilon}, \hat{T}^{*\epsilon})\}_\epsilon$ in $C((0,\infty]; V)$, there is a sequence $\epsilon_n \to 0$, such that $\mathcal{D}(\hat{v}^{*\epsilon_n}(0), \hat{T}^{*\epsilon_n}(0))$ converges weakly to $\mathcal{D}(\hat{v}^*(0), \hat{T}^*(0))$ for some random variable $(\hat{v}^*(0), \hat{T}^*(0))$. Denote by $(\hat{v}^*, \hat{T}^*)$ the solution to 3D stochastic primitive equations (4.1)–(4.4) with initial data distributes as $(\hat{v}^*(0), \hat{T}^*(0))$. Then by Theorem 4.1,

$$\mathbb{P}^{*\epsilon_n} \to \mathbb{P} \quad \text{weakly as} \quad n \to \infty.$$

Here $\mathbb{P} = \mathcal{D}(\hat{v}^*, \hat{T}^*)$, by the stationary property of $\mathbb{P}^{*\epsilon_n}$, is a statistical stationary solution to 3D stochastic primitive equations (4.1)–(4.4). □

The above result shows that for $\epsilon \to 0$ and any stationary solution $(v^{*\epsilon}, T^{*\epsilon})$ to 3D random primitive equations, there is a subsequence $\epsilon_n \to 0$, as $n \to \infty$, such that $(v^{*\epsilon_n}, T^{*\epsilon_n})$ converges in distribution to a stationary solution (v^*, T^*) to 3D stochastic primitive equations.

A Statistical solution

We give an introduction of statistical solution of 3D random primitive equations (1.8)–(1.14) coupled with (2.4).

Statistical solution was introduced to study universal properties of turbulent flows [11, 12, 27, e.g.]. We say the system of 3D random primitive equations (1.8)–(1.14) coupled with (2.4) has a statistical solution in space $C([0,\infty);\mathbb{V})$ if there is a probability measure $\mathfrak{P}^\epsilon$ supported on $C([0,\infty);\mathbb{V})$, and there are processes $(\hat{v}^\epsilon, \hat{T}^\epsilon, \hat{\eta}_1^\epsilon, \hat{\eta}_2^\epsilon) \in C([0,\infty);\mathbb{V})$, $\hat{W} = (\hat{W}_1, \hat{W}_2)$ defined on a new probability space, such that

(1) $\mathcal{D}(\hat{v}^\epsilon, \hat{T}^\epsilon, \hat{\eta}_1^\epsilon, \hat{\eta}_2^\epsilon) = \mathfrak{P}^\epsilon$;

(2) $\hat{W}_1$ and $\hat{W}_2$ are Wiener processes distribute same as W_1 and W_2 respectively;

(3) $\mathcal{D}(\hat{v}^\epsilon(0), \hat{T}^\epsilon(0)) = \mathcal{D}(v_0^\epsilon, T_0^\epsilon)$, $\mathcal{D}(\hat{\eta}_1^\epsilon, \hat{\eta}_2^\epsilon) = \mathcal{D}(\eta_1^\epsilon, \eta_2^\epsilon)$ and $(\hat{v}^\epsilon(0), \hat{T}^\epsilon(0))$ are independent from $\hat{W}_1$ and $\hat{W}_2$;

(4) The process $(\hat{v}^\epsilon, \hat{T}^\epsilon)$ is a weak solution of 3D random primitive equations (1.8)–(1.14) with η_1^ϵ, η_2^ϵ replaced by $\hat{\eta}_1^\epsilon$, $\hat{\eta}_2^\epsilon$ respectively. Here $\hat{\eta}^\epsilon = (\hat{\eta}_1^\epsilon, \hat{\eta}_\epsilon^2)$ is stationary process solving (2.4) with W replaced by $\hat{W}$.

The above definition of statistical solutions are also used in [4] to study stochastic 3D Navier–Stokes equations.

A stationary statistical solution is a statistical solution, a Borel measure $\mathfrak{P}^\epsilon$, which is invariant under the following translation on $C([0,\infty);\mathbb{V})$

$$(v(\cdot), T(\cdot), \eta_1(\cdot), \eta_2(\cdot)) \mapsto (v(\cdot+t), T(\cdot+t), \eta_1(\cdot+t), \eta_2(\cdot+t)), \quad t \geqslant 0$$

for $(v, T, \eta_1, \eta_2) \in C([0,\infty);\mathbb{V})$. For a statistical solution of the system of random 3D primitive equations (1.8)–(1.14) coupled with (2.4), we denote by $\mathfrak{P}_t^\epsilon = \mathcal{D}(\hat{v}^\epsilon(\cdot + t), \hat{T}^\epsilon(\cdot+t), \hat{\eta}_1^\epsilon(\cdot+t), \hat{\eta}_2^\epsilon(\cdot+t))$, which is also a statistical solution of the random 3D primitive equations (1.8)–(1.14) coupled with (2.4). For a stationary statistical solution $\mathfrak{P}^*$ we have

$$\mathfrak{P}_t^* = \mathfrak{P}^*, \quad t \geqslant 0.$$

The following result shows that the relation between the stationary measure and stationary statistical solution.

Lemma A.1 *The 3D random primitive equations* (1.8)–(1.14) *coupled with* (2.4) *has a stationary measure supported on* $\mathbb{V}$, *then there is a stationary statistical solution in* $C([0,\infty);\mathbb{V})$.

Proof The proof is direct by the following observation [4]: Let $\mathfrak{P}^{*\epsilon}=\mathcal{D}(\hat{v}^{*\epsilon},\hat{T}^{*\epsilon},\hat{\eta}_1^{*\epsilon},\hat{\eta}_2^{*\epsilon})$ be a stationary statistical solution to the 3D random primitive equations coupled with (2.4), then

$$\mathfrak{y}^{*\epsilon}=\mathcal{D}(\hat{v}^{*\epsilon}(0),\hat{T}^{*\epsilon}(0),\hat{\eta}_1^{*\epsilon}(0),\hat{\eta}_2^{*\epsilon}(0))$$

is a stationary measure for the Markov process defined by the 3D random primitive equations coupled with (2.4); Conversely, assume $\mathfrak{y}^{*\epsilon}$ is a stationary measure of the random 3D primitive equations coupled with (2.4), let $(v^{*\epsilon},T^{*\epsilon},\eta_1^{\epsilon},\eta_2^{\epsilon})$ be a solution of the 3D random primitive equations coupled with (2.4) with $\mathcal{D}(v^{*\epsilon}(0),T^{*\epsilon}(0),\eta_1^{\epsilon}(0),\eta_2^{\epsilon}(0))=\mathfrak{y}^{*\epsilon}$, then $\mathfrak{P}^{*\epsilon}=\mathcal{D}(v^{*\epsilon},T^{*\epsilon},\eta_1^{\epsilon},\eta_2^{\epsilon})$ is a stationary statistical solution of the 3D random primitive equations coupled with (2.4). □

For stochastic 3D primitive equation (4.1)–(4.4), a statistical solution in space $C([0,\infty;V)$ is a probability measure $\mathbb{P}$ supported on $C([0,\infty;V)$ and there are processes $(\hat{v},\hat{T})\in C([0,\infty;V)$, $\hat{W}=(\hat{W}_1,\hat{W}_2)$ defined on a new probability space such that

(i) $\mathcal{D}(\hat{v},\hat{T})=\mathbb{P}$;

(ii) $\hat{W}_1$ and $\hat{W}_2$ are Wiener processes distribute same as W_1 and W_2 respectively;

(iii) $\mathcal{D}(\hat{v}(0),\hat{T}(0))=\mathcal{D}(v_0,T_0)$ and $(\hat{v}(0),\hat{T}(0))$ are independent from $\hat{W}_1$ and $\hat{W}_2$;

(iv) The process $(\hat{v},\hat{T})$ is a weak solution of stochastic 3D primitive equations (4.1)–(4.4) with W replaced by $\hat{W}$.

Notice the above definition of statistical solution is in fact a solution to a martingale problem [20, Chapter V].

Similarly we also have stationary statistical solution and relation in Lemma A.1 holds.

Acknowledgements The authors would like to express my heartful thanks to the referee for the valuable comments and suggestions. This work was supported by 973 Program (grant No. 2013CB834100), and National Natural Science Foundation of China (91130005,11271052).

References

[1] Arnold L. Stochastic Differential Equations: Theory and Applications, Wiley, New York–London–Sydney, 1974.

[2] Billingsley P. Convergence of Probability Measures 2nd, John Wiley & Sons, New York, 1999.

[3] Cao C & Titi E S. Global well-posedness of the three-dimensional viscous primitive equations of large-scale ocean and atmosphere dynamics, Ann. of Math., 2007, **166**: 245–267.

[4] Chueshov I & Kuksin S. Stochastic 3D Navier–Stokes equations in a thin domain and its α-approximation, Physica D, 2008, **10–12**: 1352–1367.

[5] Debussche A & Vovelle J. Diffusion limit for a stochastic kinetic problem, preprinted, 2011.

[6] Diop M A, Iftimie B, Pardoux E, Piatnitski A L. Singular homogenization with stationary in time and periodic in space coefficients, J. Funct. Anal., 2006, **231**: 1–46.

[7] Duan J, Gao H and Schmalfuss B. Stochastic dynamics of a coupled atmosphereocean model, Stoch. and Dynam., 2002, **2(3)**: 357–380.

[8] Duan J, Kloeden P E and Schmalfuss B. Exponential stability of the quasigeostrophic equation under random perturbations, Prog. in Probability, 2001, **49**: 241–256.

[9] Duan J & Schmalfuss B. The 3D quasi-geostrophic fluid dynamics under random forcing on boundary, Comm. Math. Sci., 2003, **1**: 133–151.

[10] Ethier S N & Kurtz T G. Markov Processes: Characterization and Convergence, John Wiley & Sons, 1986.

[11] Foias C. Statistical study of Navier–Stokes equations I., Rend. Sem. Mat. Univ. Padova, 1972, **48**: 219–348.

[12] Foias C. Statistical study of Navier–Stokes equations II., Rend. Sem. Mat. Univ. Padova, 1973, **49**: 9–123.

[13] Gao H & Sun C. Random attractor for the 3D viscous stochastic primitive equations with additive noise, Stoch. Dyn., 2009, **9**: 293–313.

[14] Guo B & Huang D. 3D stochastic primitive equations of the large-scale ocean: global well-posedness and attractors, Comm. Math. Phys., 2009, **286**: 697–723.

[15] Guillén-González F, Masmoudi N & Rodríguez-Bellido M A. Anisotropic estimates and strong solutions for the primitive equations, Diff. Int. Equ., 2001, **14**: 1381–1408.

[16] Kifer Y. L^2-diffusion approximation for slow motion in averaging, Stoch. and Dynam., 2003, **3**: 213–246.

[17] Kushner H. Approximation and Weak Convergence Methods for Random Processes with Applications to Stochastic Systems Theory, MIT Press, 1984.

[18] Lions J L, Temam R and Wang S. On the equations of the large scale ocean, Nonlinearity, 1992, **5**: 1007–1053.

[19] Maslowski B. On probability distributions of solutions of semilinear stochastic evolu-

tion equations, Stoc. Stoc. Rep., 1993, **45**: 265–289.

[20] Metivier M. Stochastic Partial Differential Equations in Infinite Dimensional Spaces, Scuola Normale Superiore, Pisa, 1988.

[21] Majda A & Eijnden E V. A mathematical framework for stochastic climate models, Comm. Pure Appl. Math., 2001, **54**: 891–974.

[22] Müller P. Stochastic forcing of quasi-geostrophic eddies, Stochastic Modelling in Physical Oceanography, edited by R. J. Adler, P. Müller and B. Rozovskii, Birkhäuser, Basel, 1996.

[23] Pedlosky J. Geophysical Fluid Dynamics, 2nd Edition, Springer- Verlag, Berlin/New York, 1987.

[24] Da Prato G & Zabczyk J. Stochastic Equations in Infinite Dimensions, Cambridge University Press, 1992.

[25] Da Prato G & Zabczyk J. Ergodicitiy for Infinite Dimensions, Cambridge University Press, 1996.

[26] Wang W & Roberts A J. Diffusion approximation for self-similarity of stochastic advection in Burgers' equation, Comm. Math. Phys., to appear, 2014. http://dx.doi.org/10.1007/s00220-014-2117-7.

[27] Vishik M I, Fursikov A V. Mathematical problems of statistical hydromechanics, Kluwer Academic Publishers, Dordrecht, 1988.

[28] Watanabe H. Averaging and fluctuations for parabolic equations with rapidly oscillating random coefficients, Probab. The. Rel. Fields, 1988, **77**: 359–378.

Uniqueness of Weak Solutions to the 3D Ginzburg-Landau Superconductivity Model*

Fan Jishan (樊继山), Gao Hongjun (高洪俊) and Guo Boling (郭柏灵)

Abstract We prove the uniqueness of weak solutions of the 3D time-dependent Ginzburg-Landau model for superconductivity with L^3 initial data in the case of Coulomb and type I Lorentz gauges.

Keywords 3D Ginzburg-Landau Superconductivity model; uniqueness; weak solutions

1 Introduction

We consider the uniqueness problem for the Ginzburg-Landau model in superconductivity:

$$\eta\psi_t + \mathrm{i}\eta k\phi\psi + \left(\frac{\mathrm{i}}{k}\nabla + A\right)^2 \psi + (|\psi|^2 - 1)\psi = 0, \tag{1.1}$$

$$A_t + \nabla\phi + \mathrm{curl}^2 A + Re\left\{\left(\frac{\mathrm{i}}{k}\nabla\psi + \psi A\right)\bar{\psi}\right\} = \mathrm{curl}H. \tag{1.2}$$

in $Q_T := (0,T)\times\Omega \subset \mathbb{R}\times\mathbb{R}^3$, with boundary and initial conditions

$$\nabla\psi\cdot\nu = 0, A\cdot\nu = 0, \mathrm{curl}A\times\nu = H\times\nu \text{ on } (0,T)\times\partial\Omega. \tag{1.3}$$

$$(\psi, A)|_{t=0} = (\psi_0, A_0) \text{ in } \Omega \subseteq \mathbb{R}^3. \tag{1.4}$$

Here, the unknowns ψ, A, and ϕ are $\mathbb{C}$-valued, $\mathbb{R}^d$-valued, and $\mathbb{R}$-valued functions, respectively, and they stand for the order parameter, the magnetic potential, and the electric potential, respectively. $H := H(t,x)$ is the applied magnetic field, η and k are Ginzburg-Landau positive constants, and $\mathrm{i} = \sqrt{-1}$. Ω is a simply connected and

* Intern. Math. Res. Notices, 2013, 2015(5): 1239 – 1246. DOI:10.1093/imrn/rnt253.

bounded domain with smooth boundary $\partial\Omega$ and ν is the outward normal to $\partial\Omega$. $\bar{\psi}$ denotes the complex conjugate of ψ, $\mathbf{Re}\psi := (\psi+\bar{\psi})/2$, $|\psi|^2 := \psi\bar{\psi}$ is the density of superconducting carriers. T is any given positive constant. It is well known that the Ginzburg-Landau equations are gauge invariant, that is, if (ψ, A, ϕ) is a solution of (1.1)–(1.4), then for any real-valued smooth function χ, $(\psi e^{ik\chi}, A+\nabla\chi, \phi-\chi_t)$ is also a solution of (1.1)–(1.4). So in order to obtain the well-posedness of the problem, we need to impose the gauge condition. One usually has four types of the gauge condition:

(1) Coulomb gauge: divA=0 in Ω and $\int_\Omega \phi \mathrm{d}x = 0$.

(2) Lorentz gauge: $\phi = -\mathrm{div}A$ in Ω.

(3) Type I Lorentz gauge: $\phi_t = -\mathrm{div}A$ in Ω.

(4) Type II Temporal gauge: $\phi = 0$ in Ω.

For the initial data $\psi_0 \in H^1(\Omega), |\psi_0| \leqslant 1$ in Ω and $A_0 \in H^1(\Omega)$, Chen et al. [3, 5], Du [6], Fan and Ozawa [9], and Tang [14] proved the existence and uniqueness of global strong solutions to (1.1)–(1.4) in the case of the Coulomb and Lorentz as well as temporal gauges. For the initial data $\psi_0 \in H^1(\Omega)$, $A_0 \in H^1(\Omega)$, Tang and Wang [15] obtained the existence and uniqueness of global strong solutions. Fan and Jiang [8] showed the existence of global weak solutions when $\psi_0, A_0 \in L^2(\Omega)$. Fan and Gao [7], Fan and Ozawa [11,10] proved some uniqueness criteria of weak solutions. Zaouch [17] proved the existence of time-periodic solutions when the applied magnetic field H is time periodic. Phillips and Shin [13], Chen and Hoffmann [4] studied the well-posedness of classical solutions to the nonisothermal models for superconductivity. Wang [16] proved the uniqueness of weak spatially periodic solutions under the type I Lorentz gauge when $\psi_0 \in L^4, A_0 \in L^3$, while Zhan [18] established a uniqueness result for a simplified model to (1.1), (1.2) when $\psi_0, A_0 \in L^3$.

Recently, Fan and Gao [7] proved the following conditional uniqueness result.

Theorem 1.1 ([7]) *Let $(\psi_0, A_0) \in L^2$. $H \in L^2(Q_T)$. Assume that*

$$(\psi, A) \in L^r(0,T;L^p(\Omega)) \quad \text{with } \frac{2}{r}+\frac{3}{p}=1, 3<p\leqslant\infty. \tag{1.5}$$

Then there exists at most one weak solution (ψ, A) to the problem (1.1)–(1.4) *satisfying $(\psi, A) \in V_2(Q_T) := L^\infty(0,T;L^2)\cap L^2(0,T;H^1)$ with the type I Lorentz and Coulomb gauges.*

The condition (1.5) comes from scaling for (1.1) and (1.2). Let us be more precise: If $(\psi(t,x),A(t,x))$ is a solution of (1.1)–(1.2) associated with the initial value $(\psi_0(x),A_0(x))$ without linear lower order term $-\psi$ and H, then

$$(\lambda\psi(\lambda^2 t,\lambda(x-x_0)),\lambda A(\lambda^2 t,\lambda(x-x_0)))$$

is also a solution associated with

$$(\lambda\psi_0(\lambda(x-x_0)),\lambda A_0(\lambda(x-x_0)))$$

for any $\lambda>0$ and $x_0\in\mathbb{R}^3$. A Banach space B of distributions on $\mathbb{R}\times\mathbb{R}^3$ is a critical space if its norm verifies for any λ and any $u\in B$,

$$\|u\|_B=\|\lambda u(\lambda^2\cdot,\lambda(\cdot-x_0))\|_B.$$

If we choose B as $L^r(0,\infty;L^p(\mathbb{R}^3))$, then (r,p) should satisfy

$$\frac{2}{r}+\frac{3}{p}=1.$$

The aim of this paper is to completely solve the uniqueness of the weak solutions to the problem (1.1)–(1.4). For simplicity, we will take the applied magnetic field $H\equiv 0$. We will prove the following theorem.

Theorem 1.2 *Let $\psi_0,A_0\in L^3$ and $H\equiv 0$ in Q_T. Then the problem* (1.1)–(1.4) *has a unique weak solution (ψ,A) satisfying*

$$\psi,A\in L^2(Q_T)\cap V_2(Q_T) \tag{1.6}$$

for any $T>0$ with the choice of type I *Lorentz or Coulomb gauges.*

Remark 1.1 *Here, the weak solution ψ,A to* (1.1)–(1.4) *under the Coulomb gauge means that $\psi,A,|\psi|^{\frac{3}{2}},A^{\frac{3}{2}}\in V_2(Q_T),\nabla\phi\in L^{5/3}(Q_T)\cap L^2(0,T;L^{3/2})$ and the Equations* (1.1)–(1.2) *holds weakly.*

We will use Theorem 1.1 *to prove our Theorem* 1.2*. We only need to establish a priori estimates* (1.6)*. For simplicity, we only prove the case of Coulomb gauge. In our proof, we will use the following lemmas.*

Lemma 1.1 ([1,12]) *Let Ω be a smooth and bounded open set in $\mathbb{R}^3$. Then there exists $C>0$ such that*

$$\|f\|_{L^p(\Omega)}\leqslant C\|f\|_{L^p(\Omega)}^{1-\frac{1}{p}}\|f\|_{W^{1,p}(\Omega)}^{\frac{1}{p}} \tag{1.7}$$

for any $1<p<\infty$ and $f:\Omega\to\mathbb{R}^3$ be in $W^{1,p}(\Omega)$.

Lemma 1.2 ([2]) *Let Ω be a regular bounded domain in $\mathbb{R}^3$, let $f:\Omega\to\mathbb{R}^3$ be a smooth enough vector field, and let $1<p<\infty$. Then, the following identity holds true:*

$$\begin{aligned}-\int_\Omega \Delta f\cdot f|f|^{p-2}\mathrm{d}x=&\frac{1}{2}\int_\Omega |f|^{p-2}|\nabla f|^2\mathrm{d}x+\frac{4(p-2)}{p^2}\int_\Omega |\nabla|f|^{p/2}|^2\mathrm{d}x\\&-\int_{\partial\Omega}|f|^{p-2}(\nu\cdot\nabla)f\cdot f\mathrm{d}S.\end{aligned}\tag{1.8}$$

Lemma 1.3([8]) *Let $\psi, A\in V_2(Q_T)\cap L^5(Q_T)\cap L^\infty(0,T;L^3)$ and $\left|\frac{i}{k}\nabla\psi+\psi A\right||\psi|^{\frac{1}{2}}\in L^2(Q_T)$, then $\nabla\phi\in L^{5/3}(Q_T)\cap L^2(0,T;L^{3/2})$ satisfies*

$$-\Delta\phi=\operatorname{div} Re\left\{\left(\frac{\mathrm{i}}{k}\nabla\psi+\psi A\right)\bar\psi\right\}\ \text{in } Q_T,\tag{1.9}$$

$$\frac{\partial\phi}{\partial\nu}=0 \text{ on } (0,T)\times\partial\Omega.\tag{1.10}$$

2 Proof of Theorem 1.2

In this section, we use Theorem 1.1 to prove Theorem 1.2. We only need to prove the a priori estimate (1.6) with the choice of Coulomb gauge.

Multiplying (1.1) by $\bar\psi$, integrating by parts, and then taking real part, we see that

$$\frac{\eta}{2}\frac{\mathrm{d}}{\mathrm{d}t}\int|\psi|^2\mathrm{d}x+\int\left|\frac{\mathrm{i}}{k}\nabla\psi+\psi A\right|^2\mathrm{d}x+\int(|\psi|^2-1)^2\mathrm{d}x+\int|\psi|^2\mathrm{d}x=|\Omega|.$$

Integrating the above equation in t, we obtain

$$\sup_{0\leqslant t\leqslant T}\int|\psi|^2\mathrm{d}x+\int_0^T\int\left|\frac{\mathrm{i}}{k}\nabla\psi+\psi A\right|^2\mathrm{d}x\mathrm{d}t\leqslant C.\tag{2.1}$$

Multiplying (1.1) by $|\psi|\psi$, integrating by parts, and then taking the real part, we find that

$$\frac{1}{3}\frac{\mathrm{d}}{\mathrm{d}t}\int|\psi|^3\mathrm{d}x+\int\left|\frac{\mathrm{i}}{k}\nabla\psi+\psi A\right|^2|\psi|\mathrm{d}x+\int|\psi|^5\mathrm{d}x\leqslant\int|\psi|^3\mathrm{d}x.$$

Applying the Gronwall inequality to the above inequality, we obtain

$$\sup_{0\leqslant t\leqslant T}\int|\psi|^3\mathrm{d}x+\int_0^T\int\left|\frac{\mathrm{i}}{k}\nabla\psi+\psi A\right|^2|\psi|\mathrm{d}x\mathrm{d}s+\int_0^T\int|\psi|^5\mathrm{d}x\mathrm{d}s\leqslant C.\tag{2.2}$$

Testing (1.2) by A and using (2.2), we obtain

$$\begin{aligned}
&\frac{1}{2}\frac{\mathrm{d}}{\mathrm{d}t}\int |A|^2 dx + \int |\mathrm{curl}A|^2 \mathrm{d}x - \int H\mathrm{curl}A\mathrm{d}x \\
&\leqslant \int \left|\frac{\mathrm{i}}{k}\nabla\psi + \psi A\right| |\psi||A|\mathrm{d}x \\
&= \int \left|\frac{\mathrm{i}}{k}\nabla\psi + \psi A\right| |\psi|^{1/2}\cdot|\psi|^{1/2}\cdot|A|\mathrm{d}x \\
&\leqslant \left\|\left|\frac{\mathrm{i}}{k}\nabla\psi + \psi A\right| |\psi|^{1/2}\right\|_{L^2} \||\psi|^{1/2}\|_{L^6}\|A\|_{L^3} \\
&\leqslant C\left\|\left|\frac{\mathrm{i}}{k}\nabla\psi + \psi A\right| |\psi|^{1/2}\right\|_{L^2} (\|A\|_{L^2} + \|\mathrm{curl}A\|_{L^2}) \\
&\leqslant \frac{1}{2}\int |\mathrm{curl}A|^2 dx + C\left\|\left|\frac{\mathrm{i}}{k}\nabla\psi + \psi A\right| |\psi|^{1/2}\right\|_{L^2}^2 + C\|A\|_{L^2}^2,
\end{aligned}$$

which gives

$$A \in V_2(Q_T). \tag{2.3}$$

Since

$$\int_0^T \int |\psi A|^2 \mathrm{d}x\mathrm{d}t \leqslant \|\psi\|_{L^\infty}^2 \int_0^T \|A\|_{L^6}^2 \mathrm{d}t \leqslant C,$$

it follows from (2.1) that

$$\psi \in V_2(Q_T). \tag{2.4}$$

Testing (1.2) by $|A|A$ and letting $u := |A|^{3/2}$, using (1.3), (1.7), (1.8) and the vector identities

$$(\nu\cdot\nabla)A\cdot A = (A\cdot\nabla)A\cdot\nu + (\mathrm{curl}A\times\nu)\cdot A,$$

and

$$(A\cdot\nabla)A\cdot\nu = -(A\cdot\nabla)\nu\cdot A, \quad (A\cdot\nu = 0 \text{ on } (0,T)\times\partial\Omega), \tag{2.5}$$

we obtain

$$\begin{aligned}
&\frac{\mathrm{d}}{\mathrm{d}t}\int u^2 dx + C_0\int |\nabla u|^2 \mathrm{d}x \\
&\leqslant \int \left|\frac{\mathrm{i}}{k}\nabla\psi + \psi A\right| |\psi| u^{4/3}\mathrm{d}x + \int |\nabla\phi|\cdot u^{4/3}\mathrm{d}x + C\int_{\partial\Omega} u^2 \mathrm{d}x \\
&\leqslant C\int \left|\frac{\mathrm{i}}{k}\nabla\psi + \psi A\right| |\psi|^{1/2}\cdot|\psi|^{1/2}\cdot u^{4/3}\mathrm{d}x
\end{aligned}$$

$$
\begin{aligned}
&+\int|\nabla\phi|\cdot u^{4/3}dx+C\|u\|_{L^2(\Omega)}\|u\|_{H^1(\Omega)}\\
\leqslant & C\left\|\left|\frac{\mathrm{i}}{k}\nabla\psi+\psi A\right||\psi|^{1/2}\right\|_{L^2}\||\psi|^{1/2}\|_{L^6}\|u^{4/3}\|_{L^3}+\frac{C_0}{4}\|\nabla u\|^2_{L^2(\Omega)}+C\|u\|^2_{L^2(\Omega)}\\
\leqslant & C\left\|\left|\frac{\mathrm{i}}{k}\nabla\psi+\psi A\right||\psi|^{1/2}\right\|_{L^2}\|u\|^{4/3}_{L^4}+\frac{C_0}{4}\|\nabla u\|^2_{L^2(\Omega)}+C\|u\|^2_{L^2(\Omega)}\\
\leqslant & C\left\|\left|\frac{\mathrm{i}}{k}\nabla\psi+\psi A\right||\psi|^{1/2}\right\|_{L^2}(\|u\|^{1/3}_{L^2}\|\nabla u\|_{L^2}+\|u\|^{4/3}_{L^2})\\
&+\frac{C_0}{4}\|\nabla u\|^2_{L^2(\Omega)}+C\|u\|^2_{L^2(\Omega)}\\
\leqslant & \frac{C_0}{2}\|\nabla u\|^2_{L^2}+C\left(\left\|\left|\frac{\mathrm{i}}{k}\nabla\psi+\psi A\right||\psi|^{1/2}\right\|^2_{L^2}+1\right)(\|u\|^2_{L^2}+1),
\end{aligned}
$$

which gives

$$
A\in L^\infty(0,T;L^3)\cap L^5(Q_T),\ \int_0^T\int|\nabla|A|^{3/2}|^2\mathrm{d}x\mathrm{d}t\leqslant C. \tag{2.6}
$$

Here, we have used

$$
\begin{aligned}
\int|\nabla\phi|\cdot u^{4/3}\mathrm{d}x\leqslant&\|\nabla\phi\|_{L^{3/2}}\|u^{4/3}\|_{L^3}\\
\leqslant& C\left\|\left(\frac{\mathrm{i}}{k}\nabla\psi+\psi A\right)\psi\right\|_{L^{3/2}}\|u^{4/3}\|_{L^3}\\
\leqslant& C\left\|\left|\frac{\mathrm{i}}{k}\nabla\psi+\psi A\right||\psi|^{1/2}\right\|_{L^2}\||\psi|^{1/2}\|_{L^6}\|u\|^{4/3}_{L^4}
\end{aligned}
$$

and the Gagliardo-Nirenberg inequality

$$
\|u\|_{L^4}\leqslant C\|u\|^{1/4}_{L^2}\|\nabla u\|^{3/4}_{L^2}+C\|u\|_{L^2}.
$$

Finally, if follows from (2.2) and (2.6) that

$$
\psi,A\in L^5(0,T;L^5)\text{ with }\frac{2}{5}+\frac{3}{5}=1.
$$

Using Theorem 1.1, this completes the proof.

Acknowledgements The authors would like to thank the referee for some nice suggestions which improved the presentation of the paper. This paper was supported by NSFC (No. 11171154, 11171158) of China, “333” and Qing Lan Project of Jiangsu Province and the NSF of the Jiangsu Higher Education Committee of China (11KJA110001).

References

[1] Adams R A and Fournier J J F. *Sobolev Spaces*, 2nd ed, Pure and Applied Mathematics(Amsterdam) 140. Amsterdam: Elsevier/Academic Press, 2003.

[2] Beirão da Veiga H and Crispo F. *Sharp inviscid limit results under Navier type boundary conditions: An L^p theory*, Journal of Mathematical Fluid Mechanics, 2010, 12(3): 397–411.

[3] Chen Z M, Elliott C and Tang Q. *Justification of a two-dimensional evolutinary Ginzburg-Landau superconductivity model*, RAIRO Modelisation Mathematique et Analyse Numerique, 1998, 32(1): 25–50.

[4] Chen Z M and Hoffmann K H. *Global classical solutions to a nonisothermal dynamical Ginzburg-Landau model in superconductivity*, Numerical Functional Analysis and Optimization, 1997, 18(9-10): 901–920.

[5] Chen Z M, Hoffmann K H and Liang J. *On a nonstationary Ginzburg-Landau superconductivity model*, Mathematical Methods in the Applied Sciences, 1993, 16(12): 855–875.

[6] Du Q. *Global existence and uniqueness of solutions of the time dependent Ginzburg-Landau model for superconductivity*, Applicable Analysis, 1994, 52(1-2): 1–17.

[7] Fan J and Gao H. *Uniqueness of weak solutions in critical spaces of the 3-D time-dependent Ginzburg-Landau equations for superconductivity*, Mathematische Nachrichten, 2010, 283(8): 1134–1143.

[8] Fan J and Jiang S. *Global existence of weak solutions of a time-dependent 3-D Ginzburg-Landau model for superconductivity*, Applied Mathematics Letters, 2003, 16(3): 435–440.

[9] Fan J and Ozawa T. *Global strong solutions of the time-dependent Ginzburg-Landau model for superconductivity with a new gauge*, International Journal of Mathematical Analysis, 2012, 6(3): 1679–1684.

[10] Fan J and Ozawa T. *Uniqueness of weak solutions to the Ginzburg-Landau model for superconductivity*, International Journal of Mathematical Analysis, 2012, 6(21-24): 1095–1104.

[11] Fan J and Ozawa T. *Uniqueness of weak solutions to the Ginzburg-Landau model for superconductivity*, Zeitschrift für Angewandte Mathematik und Physik, 2012, 63(3): 453–459.

[12] Lunardi A. *Interpolation Theory*, 2nd ed. Lecture Notes, Scuola Normale Superiore di Pisa(New Series). Pisa: Edizioni dekka Normale, 2009.

[13] Phillips D and Shin E. *On the analysis of a non-isothermal model for superconductivity*, European Journal of Applied Mathematics, 2004, 15(2): 147–179.

[14] Tang Q. *On an evolutionary system of Ginzburg-Landau equations with foxed total magnetic flux*, Communications in Partiak Differential Equations, 1995, 20(1-2): 1–36.

[15] Tang Q and Wang S. *Time dependent Ginzburg-Landau equations of superconductivity*, Physica D, 1995, 88(3-4): 139–166.

[16] Wang B. *Uniqueness of solutions for the Ginzburg-Landau model of superconductivity in three spatial dimensions*, Journal of Mathematical Analysis and Applications, 2002, 266(1): 1–20.

[17] Zaouch F. *Time-periodic solutions of the time-dependent Ginzburg-Landau equations of superconductivity*, Zeitschrift für Angewandte Mathematik und Physik, 2003, 54(6): 905–918.

[18] Zhan M. *Well-posedness of phase-lock equations of superconductivity*, Applied Mathematics Letters, 2005, 18(11): 1210–1215.

High-Order Soliton Solution of Landau-Lifshitz Equation*

Bian Dongfen (边东芬), Guo Boling (郭柏灵) and Ling Liming (凌黎明)

Abstract Landau-Lifshitz equation is analyzed via the inverse scattering method. First, we give the well-posedness theory for Landau-Lifshitz equation with the frame of inverse scattering method. The generalized Darboux transformation is rigorous considered in the frame of inverse scattering transformation. Finally, we give the high-order soliton solution formula of Landau-Lifshitz equation and vortex filament equation.

Keywords Landau-Lifshitz equation; Darboux transformation; well-pasedness; in verse scattering method

1 Introduction

The Landau-Lifshitz (L-L) equation [37]

$$\vec{S}_t = \vec{S} \times \vec{S}_{xx}, \quad \vec{S}(x,t) = (S^x, S^y, S^z)^{\mathrm{T}} \in \mathbb{R}^3, \quad \vec{S} \cdot \vec{S} = 1 \tag{1.1}$$

describes nonlinear spin waves in an isotropic ferromagnet, where the symbols ^{T} and $\times$ mean the transpose and vector product respectively, $\vec{S}(x,t)$ is magnetization vector. Setting $\vec{S} = \vec{\gamma}_x$ and integrating (1.1) with respect to x, we can obtain another relative physical model—vortex filament equation (VFE) or localization induction equation

$$\vec{\gamma}_t = \vec{\gamma}_x \times \vec{\gamma}_{xx}, \ \vec{\gamma} = (\gamma^x, \gamma^y, \gamma^z)^{\mathrm{T}}, \tag{1.2}$$

which is the simplest model of dynamics of Eulerian vortex filament, where space vector $\vec{\gamma}(x,t)$ represent the vortex filament, x is the arclength parameter, t is time. The model (1.2) was firstly derived by Da Rios, a student of Levi-Civita, in 1906 [42],

* Stud. Appl. Math., 134: 181-214. Doi: 10.1111/sapm.12051.

and rediscovered by Arms and Hama in 1965 [4]. The model (1.2) also can be used to describe the flow of superfluids [44], to investigate the turbulent fluid [8, 50] and high-temperature superconductors [12].

It is well known that the inverse scattering method [1, 5, 25] is a powerful method to solve the cauchy problem of nonlinear integrable partial differential equation. In the past forty years, the inverse scattering method had made great development in the field of mathematical physics. Initially the inverse scattering transformation utilizes Marchenko integral equation to reconstruct the potential function [25]. Afterwards Shabat used Riemann-Hilbert problem (RHP) to reconstruct the inverse scattering method [47]. In the last century 90s the RHP method had made important progress. For instance, the Deift-Zhou method [15, 17, 18] and initial-boundary problem [22, 23].

In the case of KdV equation, the poles of discrete spectrum must be simple, since the Lax operator is self-adjoint. However, to the focusing nonlinear Schrödinger (NLS) equation, the corresponding Lax operator is no longer self-adjoint. Thus it allows high order pole, which corresponds to the high order soliton. The scattering data is demanded for simple pole in the classical paper of Beals and Coifman [5]. Several years later, this restraint was removed by [45] and [54] respectively. However, they didn't give the exact soliton formula. The exact high-order soliton solution for NLS equation was given in [24] by the dressing method. The general soliton formula for NLS type equation had been constructed by Shchesnovich and Yang [49, 48]. And the high order transmission coefficient by the Marchenko equation method was considered by Cohen and Kappeler [11]. Recently Aktosun, Demontis and van der Mee consider the high order soliton solution of NLS equation with inverse scattering method by GLM equation [3]. The exact second order soliton solution of L-L equation was obtained by bilinear method in 1990 [6]. Besides the high-order pole, another interesting problem is the infinite pole and infinite soliton. To the best of our knowledge, the concept of infinite soliton was provided by Zhou firstly [54]. The explicit infinite soliton solution for KdV equation was rigorous established by Gesztesy et.al [26]. The infinite soliton solution of NLS equation is obtained by Kamvissis [35]. Besides the high order pole and infinite pole, the spectral singularity is also an obstacles to the inverse scattering method. This problem was first solved by Zhou [54] via the- deformed RHP.

As well as the inverse scattering method, the Darboux-Bäcklund transformation is another powerful method to derive the multi-soliton and other interesting physical solution. There are several methods to derive the Darboux transformation. For instance, state space method [40,43,27], the loop group method [52] and gauge transformation [28,39]. The relation between different versions of Darboux-Bäcklund transformation had been indicated by Ciéliński [9]. Generally speaking, the Darboux transformation is merely a way to obtain the soliton solution in soliton theory. However, it has other utilization also. Deift and Trubowitz combined Darboux transformation with inverse scattering method for the Schrödinger spectral problem [14]. In this work, we would like to inherit their idea. The Darboux transformation can be used to deal with the initial boundary problem either [23,13]. Besides, the Darboux transformation can be used for the analysis of orbitally stability property of soliton as well [41].

Finally, we recalled some results of L-L equation (1.1). In 1977, Takhtajan used inverse scattering method to derive the 2-soliton solution and infinite sets of constants for the first time [51]. The gauge equivalence between NLS equation and L-L equation was obtained by Zakharov and Takhtajan [56] in the frame of inverse scattering transformation. Indeed, this gauge transformation is another version of Hasimoto transformation essentially. The generalized Hasimoto transformation was rigorously considered with tools of differential geometry in [10]. Recently, Calini, Keith and Lafortune considered the spectral stability property for soliton and periodical solution of vortex filament equation [7].

In this work, firstly we prove the global well-posedness for L-L equation with initial data in space $H^{2,1}(\mathbb{R})$ without discrete scattering data via RHP method. Secondly, we handle generalized Darboux transformation [29, 30] in a rigorous way with the frame of inverse scattering method. Via to this method, the global solution and the general soliton solution formula of L-L equation is obtained. What need alludes is, for the evolution of discrete scattering data, we use the evolution of eigenfunction replace with the proportionality coefficient. With this way, we can readily deal with the evolution of high order spectrum.

This paper is organized as following. In section 2, we give the scattering and inverse scattering analysis for L-L equation. To establish the well-posedness theory, we combine the gauge transformation and inverse scattering method. In section 3, we give the Darboux transformation in the frame of inverse scattering. In section 4, the explicit general soliton formula of L-L equation is constructed. Final section includes some discussions and remarks.

2 The scattering, inverse scattering, and well-posedness theory

It is well known that the KdV, MKdV, sine-Gordan and NLS can be obtained from the AKNS hierarchy [1] by some symmetry reduction. The symmetry reduction is called the reality condition [52], which is also the solvable condition for the RHP. Thus in this section, we first recall the symmetry condition. Then we give the scattering, inverse scattering analysis and gauge transformation theory to L-L equation.

The focusing NLS

$$\mathrm{i}q_t + q_{xx} + 2|q|^2 q = 0 \tag{2.3}$$

is the second flow of $su(2)$ (the fixed-point set of the involution $\sigma(y) = -y^\dagger$, where superscript "†" represents hermite conjugation) hierarchy, and turns out to be a compatibility condition for the following linear system

$$\begin{aligned} &\Phi_x = U(\lambda)\Phi, \quad U(\lambda) \equiv -\mathrm{i}\lambda\sigma_3 + Q, \\ &\Phi_t = V(\lambda)\Phi, \quad V(\lambda) \equiv -2\mathrm{i}\lambda^2\sigma_3 + 2\lambda Q - \mathrm{i}(Q^2 + Q_x)\sigma_3, \end{aligned} \tag{2.4}$$

here

$$Q = \begin{pmatrix} 0 & q(x,t) \\ -q^*(x,t) & 0 \end{pmatrix},$$

superscript "∗"" represents complex conjugation and σ_3 is standard Pauli matrix. It is readily see that the matrix $U(\lambda)$ and $V(\lambda)$ possess the reality relation $U^\dagger(\lambda^*) = -U(\lambda)$ and $V^\dagger(\lambda^*) = -V(\lambda)$.

The L-L equation (1.1) can be rewritten as

$$S_t = \frac{\mathrm{i}}{2}[S, S_{xx}], \quad S \in AO(2), \quad [A,B] \equiv AB - BA, \tag{2.5}$$

where

$$AO(2) \equiv \{S | S^2 = I, S = S^\dagger, \text{and } \mathrm{tr}S = 0\}, \quad S = \begin{pmatrix} S^z & S^- \\ S^+ & -S^z \end{pmatrix}, \quad S^\pm = S^x \mp \mathrm{i}S^y,$$

is also located in the $su(2)$ hierarchy. Equation (2.5) can be rewritten as the compatibility condition for the following system

$$\begin{aligned} &\Psi_x = -\mathrm{i}\lambda S\Psi, \\ &\Psi_t = W(\lambda)\Psi, \quad W(\lambda) \equiv -2\mathrm{i}\lambda^2 S + \lambda S S_x. \end{aligned} \tag{2.6}$$

It is readily to verify the reality condition $W^\dagger(\lambda^*) = -W(\lambda)$. The coefficient matrix of system (2.4) and (2.6) possess the same reality condition. Besides this, a gauge transformation can be related between these two linear systems, this fact was found by Zakharov and Takhtajan [56].

The aim in this section is to solve the cauchy problem of (2.5) with initial data

$$S(x,0)=S_0(x), \quad |S_{0,x}|\in H^{1,1}(\mathbb{R}) \tag{2.7}$$

where $|\cdot|$ stands the matrix or vector norm $|A|=(\mathrm{tr}A^\dagger A)^{1/2}$, $H^{1,1}(\mathbb{R})$ is the weighted Sobolev space

$$H^{1,1}(\mathbb{R})=\{f|f,f_x,xf\in L^2(\mathbb{R})\}$$

and boundary condition

$$\lim_{|x|\to\infty} S=\sigma_3. \tag{2.8}$$

Notation: We denote $e^{\mathrm{ad}\sigma_3}\cdot \equiv e^{\sigma_3}\cdot e^{-\sigma_3}$.

2.1 Scattering problem for spectral problem

The spectral problem for L-L equation (2.5) is the first equation of (2.6). If we directly analysis the spectral problem (2.6), similar reason as the derivative NLS equation [38], it is not convenience to analyse the asymptotical behavior of analytical solution. Thus we use the gauge transformation. First, we establish the following lemma:

Lemma 2.1 *If $S\in AO(2)$, then S can be decomposed into $S=g\sigma_3 g^\dagger$ uniquely, where g satisfies $g^\dagger g=I$, $g_x^\dagger g+\sigma_3 g_x^\dagger g\sigma_3=0$, $\lim_{x\to-\infty} g(x)=I$.*

Proof We use the linear algebra method to construct the matrix g directly. Because matrix S is an unitary matrix, it can be diagonalizable. Simple algebra, we can see that the eigenvalue of S is ±1. Then S can be decomposed into

$$S=g_0\exp(\mathrm{i}\theta\sigma_3)\sigma_3\exp(-\mathrm{i}\theta\sigma_3)g_0^\dagger, \tag{2.9}$$

where θ is an undetermined real function and

$$g_0=\begin{pmatrix}\sqrt{\dfrac{1+S^z}{2}} & -\dfrac{S^-}{\sqrt{2(1+S^z)}}\\ \dfrac{S^+}{\sqrt{2(1+S^z)}} & \sqrt{\dfrac{1+S^z}{2}}\end{pmatrix}.$$

In order to satisfy the condition $g_x^\dagger g+\sigma_3 g_x^\dagger g\sigma_3=0$, we can adjust the function θ. Directly calculating, we have

$$\begin{aligned}&\left[e^{-\mathrm{i}\theta\sigma_3}g_0^\dagger\right]_x g_0 e^{\mathrm{i}\theta\sigma_3}\\ =&e^{-\mathrm{i}\theta\mathrm{ad}\sigma_3}(g_{0,x}^\dagger g_0)-\mathrm{i}\theta_x\sigma_3,\end{aligned}$$

and

$$g_{0,x}^{\dagger}g_0 = \begin{pmatrix} \dfrac{S_x^z S^z + S_x^- S^+}{2(1+S^z)} & \dfrac{1}{2}\left(S_x^- - \dfrac{S^- S_x^z}{1+S^z}\right) \\ -\dfrac{1}{2}\left(S_x^+ - \dfrac{S^+ S_x^z}{1+S^z}\right) & \dfrac{S_x^z S^z + S_x^+ S^-}{2(1+S^z)} \end{pmatrix}.$$

If we demand

$$\mathrm{i}\theta_x = \frac{S_x^- S^+ + S_x^z S^z}{2(1+S^z)}, \quad \text{i.e. } \theta_x = \frac{S^x S_x^y - S_x^x S^y}{2(1+S^z)},$$

then g satisfies the condition $g_x^{\dagger}g + \sigma_3 g_x^{\dagger} g \sigma_3 = 0$. Finally, to satisfy the condition $\lim\limits_{x\to-\infty} g(x) = I$, we take

$$\theta(x) = \int_{-\infty}^{x} \frac{S^x S_x^y - S_x^x S^y}{2(1+S^z)} \mathrm{d}s.$$

This completes the proof.

Furthermore, we have

$$g_x^{\dagger} g \equiv Q, \quad Q = \begin{pmatrix} 0 & q \\ -q^* & 0 \end{pmatrix},$$

$$q = \frac{1}{2}\left(S_x^- - \frac{S^- S_x^z}{1+S^z}\right) \exp\left(\mathrm{i}\int_{-\infty}^{x} \frac{S_x^x S^y - S^x S_x^y}{1+S^z} \mathrm{d}s\right),$$

and $4|q|^2 = (S_x^x)^2 + (S_x^y)^2 + (S_x^z)^2$. Via the relation $S = g\sigma_3 g^{\dagger}$, we have $S_x = g[\sigma_3, Q]g^{\dagger}$, and $S_{xx} = g([\sigma_3, Q_x] + [[\sigma_3, Q], Q])g^{\dagger}$. It follows that $4(|q_x|^2 + 4|q|^4) = (S_{xx}^x)^2 + (S_{xx}^y)^2 + (S_{xx}^z)^2$. Then ones obtain $q(x) \in H^{1,1}(\mathbb{R})$.

In addition, because $\lim\limits_{|x|\to\infty} S = \sigma_3$, we have

$$\lim_{x\to+\infty} g^{\dagger} = g_{\infty} = \mathrm{diag}(a^*(0), a(0)), \ a(0) = e^{\mathrm{i}\theta(+\infty)}.$$

As a byproduct, we can obtain a conservation law. Indeed, we can see that $\lim\limits_{|x|\to\infty} g_0 = I$. It follows that $\lim\limits_{x\to+\infty} g = \exp[\mathrm{i}\theta(+\infty)\sigma_3]$, i.e.

$$\int_{-\infty}^{+\infty} \frac{S_x^x S^y - S^x S_x^y}{1+S^z} \mathrm{d}s = 2\arg(a(0)). \tag{2.10}$$

Via the gauge transformation $\Phi = g^{\dagger}\Psi$, we have

$$\Phi_x = (-\mathrm{i}\lambda\sigma_3 + Q)\Phi, \tag{2.11}$$

which is a standard AKNS spectral problem. Thus it is convenient to make the scattering analysis. To write spectral problem (2.11) as the integral equation, we make the following transformation $\Phi^{\pm}(x,t) = m^{(\pm)}(x,t)e^{-\mathrm{i}\lambda x\sigma_3}$. Associated with asymptotical behavior, we have

$$m^{(\pm)}(x;\lambda) = I + \int_{\pm\infty}^{x} e^{-\mathrm{i}(x-y)\lambda \mathrm{ad}\sigma_3} Q(y) m^{(\pm)} \mathrm{d}y \equiv I + K_{Q,\lambda,\pm} m^{(\pm)}.$$

The properties of above Jost solutions can be summarized as following:

Proposition 2.1 ([2]) Suppose $Q \in L^1(\mathbb{R})$, then $(m_1^{(-)}, m_2^{(+)})$ is analytic in the upper half plane $\{\lambda \in \mathbb{C}|\text{Im}(\lambda) > 0\}$, and $(m_1^{(+)}, m_2^{(-)})$ is analytic in the lower half plane $\{\lambda \in \mathbb{C}|\text{Im}(\lambda) < 0\}$. And they are all continuous on the real line.

Proof Firstly we prove

$$m_1^{(-)}(x;\lambda) = \begin{pmatrix}1\\0\end{pmatrix} + \int_{-\infty}^{x} \begin{pmatrix}1 & 0\\ 0 & e^{2\mathrm{i}(x-y)\lambda}\end{pmatrix} \begin{pmatrix}0 & q(y)\\ -q^*(y) & 0\end{pmatrix} m_1^{(-)}(y;\lambda)\mathrm{d}y \tag{2.12}$$

have a unique analytic solution in the upper half plane. It is readily to obtain that the estimation from (2.12),

$$|m_1^{(-)}(x;\lambda)| \leqslant 1 + \int_{-\infty}^{x} |Q||m_1^{(-)}(y;\lambda)|\mathrm{d}y. \tag{2.13}$$

To prove the solvability of (2.12), we iterate the series as following:

$$m_1^{(-)}(x;\lambda) = g_0 + \sum_{n=1}^{+\infty} g_n(x;\lambda) \tag{2.14}$$

where

$$g_0 = \begin{pmatrix}1\\0\end{pmatrix}, \quad g_{k+1} = \int_{-\infty}^{x} \begin{pmatrix}0 & q(y)\\ -q^*(y)e^{2\mathrm{i}\lambda(x-y)} & 0\end{pmatrix} g_k(y)\mathrm{d}y.$$

We can see that

$$|g_1(x;\lambda)| \leqslant \int_{-\infty}^{x} |Q(y)|\mathrm{d}y,$$

it follows that

$$|g_k(x;\lambda)| \leqslant \frac{1}{k!}\left(\int_{-\infty}^{x} |Q(y)|\mathrm{d}y\right)^k$$

From the above estimate, the series (2.14) converges uniformly in the upper half plane, thus the solution $m_1^{(-)}$ is analytical in the upper half plane and can be continuous extended to the real line. In addition, we have estimation

$$|m_1^{(-)}(x;\lambda)| \leqslant \exp\left(\int_{-\infty}^{x} |Q(y)|\mathrm{d}y\right).$$

Via the inequality (2.13) and the Growall inequality, the uniqueness is proved.

We have the parallel results for $m_1^{(+)}$, $m_2^{(\pm)}$. This completes the proof.

Corollary 2.1 *If $|S_x| \in L^1(\mathbb{R})$, then the Jost solution $\Psi^{\pm}$ for spectral problem (2.6) can be obtained as $\Psi^- = gm^{(-)}\exp(-\mathrm{i}\lambda x\sigma_3)$, $\Psi^+ = gm^{(+)}g_\infty\exp(-\mathrm{i}\lambda x\sigma_3)$. Let $n^{(-)} = gm^{(-)}$ and $n^{(+)} = gm^{(+)}g_\infty$, then $n^{(\pm)}$ possess the analytic and continuity property as $m^{(\pm)}$. Finally, we have $g^\dagger = m^{(-)}(x,t;\lambda = 0)$.*

Proof The first two arguments are direct results from above propositions. The last argument is valid for the existence and uniqueness of ODE.

In the following, we analysis the scattering matrix. By the Abel formula, we have $\det(m^{(\pm)}) = \det(n^{(\pm)}) = 1$. Thus we can define a matrix function $A(\lambda)$ for real λ with $\det(A(\lambda)) = 1$ and

$$m^{(+)} = m^{(-)}e^{-\mathrm{i}\lambda x \mathrm{ad}\sigma_3}A(\lambda), \qquad A(\lambda) = \begin{pmatrix} a(\lambda) & -b^*(\lambda) \\ b(\lambda) & a^*(\lambda) \end{pmatrix}, \tag{2.15}$$

where

$$\begin{aligned} a(\lambda) &= \det(m_1^{(+)}, m_2^{(-)}) = 1 - \int_{\mathbb{R}} q(y)m_{21}^{(+)}\mathrm{d}y = 1 - \int_{\mathbb{R}} q^*(y)m_{12}^{(-)}\mathrm{d}y, \\ b(\lambda) &= e^{-2\mathrm{i}x\lambda}\det(m_1^{(-)}, m_1^{(+)}) = \int_{\mathbb{R}} q^*(y)e^{-2\mathrm{i}\lambda y}m_{11}^{(+)}\mathrm{d}y = \int_{\mathbb{R}} q^*(y)e^{-2\mathrm{i}\lambda y}m_{11}^{(-)}\mathrm{d}y. \end{aligned}$$

It follows that $A(0) = g_\infty^{-1}$ and

$$n^{(+)} = n^{(-)}e^{-\mathrm{i}\lambda x \mathrm{ad}\sigma_3}A_1(\lambda), \tag{2.16}$$

where $n^{(+)} = gm^{(+)}g_\infty$, $n^{(-)} = gm^{(-)}$ and $A_1(\lambda) \equiv A(\lambda)g_\infty$. In summary, we describe the above process with the following arrow diagram

$$(\Psi, S(x,0)) \xrightarrow{\ g\ } (\Phi, Q(x,0)) \xrightarrow{\text{scattering}} \left(A(\lambda), A(0) = g_\infty^{-1}\right).$$

According to the above propositions, we can obtain that $A(\lambda) - 1 \in H^k(\mathrm{d}\lambda)$ [55]. It follows that we can define a solution m normalized as $x \to +\infty$:

$$m = \begin{cases} m_+ = \left(m_1^{(-)}, m_2^{(+)}\right)\begin{pmatrix} (a^*(\lambda^*))^{-1} & 0 \\ 0 & 1 \end{pmatrix}, & \mathrm{Im}(\lambda) > 0, \\ m_- = \left(m_1^{(+)}, m_2^{(-)}\right)\begin{pmatrix} 1 & 0 \\ 0 & a^{-1} \end{pmatrix}, & \mathrm{Im}(\lambda) < 0. \end{cases} \tag{2.17}$$

Then we could have the following decomposition:

$$m_+ = m_- e^{-\mathrm{i}\lambda x \mathrm{ad}\sigma_3}v, \quad \lambda \in \mathbb{R} \tag{2.18}$$

where $m_\pm = m^{(+)}e^{-\mathrm{i}\lambda x \mathrm{ad}\sigma_3}v_\pm$, $v \equiv v_-^{-1}v_+$,

$$\begin{aligned} v_+ &= \begin{pmatrix} 1 & 0 \\ r(\lambda) & 1 \end{pmatrix}, \quad v_- = \begin{pmatrix} 1 & -r^*(\lambda) \\ 0 & 1 \end{pmatrix}, \\ v &= \begin{pmatrix} 1 + |r(\lambda)|^2 & r^*(\lambda) \\ r(\lambda) & 1 \end{pmatrix}, \quad r = -\frac{b(\lambda)}{a^*(\lambda)}. \end{aligned}$$

To complete the RHP, we need the boundary condition [5]

$$m \to I \text{ as } \lambda \to \infty. \tag{2.19}$$

Thus equations (2.17), (2.18) and (2.19) construct the normalized RHP (see [19]) with the constraint $r(0) = 0$. The similar manner, we define $\widetilde{m}_+ = m_+[a^*(\lambda)]^{\sigma_3}$, and $\widetilde{m}_- = m_-[a(\lambda)]^{-\sigma_3}$. The solution $\widetilde{m}$ is normalized as $x \to -\infty$. And $\widetilde{m}_+ = \widetilde{m}_- e^{-\mathrm{i}\lambda x\sigma_3}\widetilde{v}(\lambda)$, $\widetilde{v}(\lambda) = a^{\sigma_3}v[a^*]^{\sigma_3}$. Accordingly, define

$$n = \begin{cases} n_+ = (n_1^{(-)}, n_2^{(+)})\mathrm{diag}(1/a_1^*(\lambda), 1), & \mathrm{Im}(\lambda) > 0, \\ n_- = (n_1^{(-)}, n_2^{(+)})\mathrm{diag}(1, 1/a_1(\lambda)), & \mathrm{Im}(\lambda) < 0, \end{cases} \qquad a_1(\lambda) = \frac{a(\lambda)}{a(0)}.$$

Then we have the RH problem for $n_\pm$

$$n_+ = n_- e^{-\mathrm{i}\lambda x \mathrm{ad}\sigma_3} v_1, \quad n_\pm = g m_\pm g_\infty \tag{2.20}$$

where

$$v_1 = \begin{pmatrix} 1 + |r_1| & r_1^* \\ r_1 & 1 \end{pmatrix}, \quad r_1 = -\frac{a_1^*(\lambda)}{b_1(\lambda)}, \quad b_1(\lambda) = \frac{b(\lambda)}{a(0)}.$$

However, in this case, when $\lambda \to \infty$, $n \to g g_\infty$. Thus this RHP is not a normalized RHP (see [19]). For convenience, we merely consider the normalized RHP (2.17), (2.18) and (2.19).

When m has no spectral singularities, the scattering data can be represent as

$$\{m, e^{-\mathrm{i}\lambda x \mathrm{ad}\sigma_3} v(\lambda), \lambda \in \mathbb{R}; \quad e^{-\mathrm{i}\lambda x \mathrm{ad}\sigma_3} v_{\lambda'} \in V_{\lambda'}, \lambda' \in P\} \tag{2.21}$$

where $\{v_{\lambda'}, \lambda \in P\}$ is the discrete part of the scattering data [55]. In the next section, we would like to deal with the discrete spectrum by generalized Darboux transformation. Thus ones can set $P = \emptyset$. In this way, it is convenient to consider argument contour. The Zhou's method [54, 55] deals with the poles by adding small circle centered at the poles. The spectral singularity is solved by reconstructing a new RHP on $\Gamma = \mathbb{R} \cup S_\infty$. However, when the poles is located in the inside of S_∞. It is not convenient to define the new RHP. If we deal with poles or high order poles by the generalized Darboux transformation, that problem will avoid automatically.

To describe the general case, we first consider the following equation

$$(m_1^{(+)}, m_2^{(-)}) = I + \int_{x_0}^{x} e^{-\mathrm{i}(x-y)\lambda \mathrm{ad}\sigma_3} Q(y)(m_1^{(+)}(y), m_2^{(-)}(y))\mathrm{d}y, \tag{2.22}$$

where $x_0=-\infty$ for the $(1,2)$ and $(2,2)$ entries, $x_0=+\infty$ for the $(1,1)$ and $(2,1)$ entries. For the entry $(2,2)$ of (2.22), using (2.15) we can obtain

$$(m_1^{(+)},m_2^{(-)})=\begin{pmatrix}1 & 0\\ 0 & a(\lambda)\end{pmatrix}+\int_{x_0'}^{x}e^{-\mathrm{i}(x-y)\lambda\mathrm{ad}\sigma_3}Q(y)(m_1^{(+)}(y),m_2^{(-)}(y))\mathrm{d}y,$$

where $x_0'=+\infty$ for the $(1,1)$, $(2,1)$ and $(2,2)$ entries, $x_0'=-\infty$ for the $(1,2)$ entry. It follows that

$$\begin{aligned}&(m_1^{(+)},m_2^{(-)})\begin{pmatrix}1 & 0\\ 0 & a(\lambda)^{-1}\end{pmatrix}\\ =&I+\int_{x_0'}^{x}e^{-\mathrm{i}(x-y)\lambda\mathrm{ad}\sigma_3}Q(y)(m_1^{(+)}(y),m_2^{(-)}(y))\begin{pmatrix}1 & 0\\ 0 & a(\lambda)^{-1}\end{pmatrix}\mathrm{d}y.\end{aligned}$$

Similar, we can obtain

$$\begin{aligned}&(m_1^{(-)},m_2^{(+)})\begin{pmatrix}a^*(\lambda^*)^{-1} & 0\\ 0 & 1\end{pmatrix}\\ =&I+\int_{-x_0'}^{x}e^{-\mathrm{i}(x-y)\lambda\mathrm{ad}\sigma_3}Q(y)(m_1^{(+)}(y),m_2^{(-)}(y))\begin{pmatrix}a^*(\lambda^*)^{-1} & 0\\ 0 & 1\end{pmatrix}\mathrm{d}y.\end{aligned}$$

Finally, we have

$$m(x,\lambda)=I+\int_{-x_0'\mathrm{sgn}(\mathrm{Im}\lambda)}^{x}e^{-\mathrm{i}(x-y)\lambda\mathrm{ad}\sigma_3}Q(y)m(y,\lambda)\mathrm{d}y,\tag{2.23}$$

which is the Fredholm integral equation. Hence by the analytical Fredholm theorem, it induces that directly solutions m is meromorphic in $\mathbb{C}\backslash\mathbb{R}$.

The following results were similar as [55], thus here we merely give the main step and results.

Let $x_0\in\mathbb{R}$ be such that $|q|_{L^1([x_0,+\infty))}<1$. Using (2.23), we have a bounded solution $m^{(0)}$ normalized as $x\to+\infty$ for the potential $Q\chi_{(x_0,+\infty)}$. This solution does not have poles and spectral singularities. On the other hand define a solution

$$m^{(1)}=I-\int_{x}^{x_0}e^{-\mathrm{i}(x-y)\lambda\mathrm{ad}\sigma_3}Q(y)m^{(1)}(y,\lambda)\mathrm{d}y,$$

and another solution for $Q(x)$

$$m^{(2)}(x,\lambda)=m^{(1)}(x,\lambda)e^{-\mathrm{i}(x-x_0)\lambda\mathrm{ad}\sigma_3}m^{(0)}(x_0,\lambda).$$

This solution is consistence with $m^{(0)}$ at $x=x_0$, since the existence and uniqueness property of ODE. It follows that $m^{(2)}$ is normalized as $x\to+\infty$. Since $m^{(1)}$ is entire in λ and $m^{(0)}(x,\cdot) \in \mathbf{A}H^k(\mathbb{C}\backslash\mathbb{R})$, then $m^{(2)}(x,\cdot) - I \in \mathbf{A}H^k(\mathbb{C}\backslash(\mathbb{R}\cup S_{R,r}))$, where $\mathbf{A}H^k(\Omega)$ denotes the space of functions analytic on Ω with H^k boundary values, $S_{R,r} = \{|\lambda| = R, |\lambda| = r\}$ for some $R > r > 0$.

Since a approaches 1 as $\lambda \to \infty$ and $a(0)$ as $\lambda \to 0$, they have no zero near $\lambda = \infty$ and $\lambda = 0$. Hence we use m near $\lambda = \infty$ and $\lambda = 0$, and $m^{(2)}$ elsewhere. Set $\Gamma = \mathbb{R}\cup S_{R,r}$, where $\Omega_+ = \Omega_1 \cup \Omega_4$ and $\Omega_- = \Omega_2 \cup \Omega_3$,

$$\Omega_1 = \{\lambda|\mathrm{Im}(\lambda) > 0, |\lambda| > R, \text{or } |\lambda| < r\}, \quad \Omega_4 = \{\lambda|\mathrm{Im}(\lambda) < 0, r < |\lambda| < R\}$$

and

$$\Omega_2 = \{\lambda|\mathrm{Im}(\lambda) < 0, |\lambda| > R, \text{or } |\lambda| < r\}, \quad \Omega_3 = \{\lambda|\mathrm{Im}(\lambda) > 0, r < |\lambda| < R\}.$$

Define $\mathbf{m} = m$ on $\Omega_1\cup\Omega_2$, $\mathbf{m} = m^{(2)}$ on $\Omega_3\cup\Omega_4$. It follows that $e^{-\mathrm{i}x\lambda\mathrm{ad}\sigma_3}\mathbf{v} = \mathbf{m}_-^{-1}\mathbf{m}_+$. Then we have the following theorem which can be established as the work of [55]

Theorem 2.1([55],Zhou)

(C1) The matrix $\mathbf{v}$ admits a triangular factorization $\mathbf{v} = \mathbf{v}_-^{-1}\mathbf{v}_+$, where $\mathbf{v}_\pm - I \in H^k(\partial\Omega_\pm)$, $\mathbf{v}_+|_{\partial\Omega_1} - I$ ($\mathbf{v}_+|_{\partial\Omega_4} - I$) is strictly lower (upper) triangular, and $\mathbf{v}_-|_{\partial\Omega_1} - I$ ($\mathbf{v}_-|_{\partial\Omega_3} - I$) is strictly upper (lower) triangular.

(C2) There exists an auxiliary scattering matrix s such that $s_-^{-1}\mathbf{v}s_+ = \tilde{\mathbf{v}}_-^{-1}\tilde{\mathbf{v}}_+$ for some invertible matrices $\tilde{\mathbf{v}}_\pm \in I + H^k(\partial\Omega_\pm)$ with $\tilde{v}_\pm$ having opposite triangularities of $\mathbf{v}_\pm$.

(C3) The RH problem $(e^{-\mathrm{i}x\lambda\mathrm{ad}\sigma_3}\mathbf{v}, \Gamma)$ is solvable for all $x \in \Gamma$.

Since the symmetry property $Q^\dagger = -Q$, then

$$m(x;\lambda)m^\dagger(x;\lambda^*) = I, \qquad m^{(0)}(x;\lambda)m^{(0)\dagger}(x;\lambda^*) = I$$

it follows that

$$\mathbf{m}(x;\lambda)\mathbf{m}^\dagger(x;\lambda^*) = I.$$

Using this and the fact that the contour Γ is Schwarz-reflection-invariant with the orientation, we have the symmetry condition of $\mathbf{v}$ is

$$\mathbf{v}(\lambda) = \mathbf{v}^\dagger(\lambda^*). \tag{2.24}$$

This symmetry condition keep the the solvability of RHP [53].

Therefore we have established the scattering map

$$\mathbf{S}: S_{0,x} \mapsto \mathbf{v}(\lambda), \qquad H^{1,1} \to H_0^{1,1} \equiv H^{1,1} \cap \{\mathbf{v}(0) = I\}. \tag{2.25}$$

Following [19, 55], ones can establish the following theorem. Because the proof is similar as [19, 55], we omit the explicit proof.

Theorem 2.2 *If $S_{0,x} \in H^{1,1}$, then $\mathbf{v}_\pm - I \in H_0^{1,1}$.*

2.2 Inverse scattering

Suppose the scattering data is given, we can resolve the potential function $Q(x)$. For conveniently, denote $w_x = e^{-\mathrm{i}\lambda x \mathrm{ad}\sigma_3} w$. Indeed, the RHP $(\mathbf{v}_x, \Gamma = \mathbb{R} \cup S_{R,r})$ is equivalent to the integral equation problem

$$\mu = I + C_{\mathbf{v}_{x\pm}}\mu, \qquad C_{\mathbf{v}_{x\pm}}\mu = C_\Gamma^+ \mu(\mathbf{v}_{x+} - I) + C_\Gamma^- \mu(I - \mathbf{v}_{x-}),$$

where $\mathbf{v}_{x\pm} = e^{-\mathrm{i}\lambda x \mathrm{ad}\sigma_3}\mathbf{v}_\pm$,

$$C_\Gamma f = \frac{1}{2\pi}\int_\Gamma \frac{f(\zeta)}{\zeta - \lambda}\mathrm{d}\zeta,$$

$\lambda \notin \Gamma$, $\mu = m^{(+)}$. The symmetry condition for RHP $(\mathbf{v}_x, \Gamma = \mathbb{R} \cup S_{R,r})$ guarantee the existence and uniqueness of RHP [53].

Once this integral equation is solved, $\mathbf{m}$ can be constructed through

$$\mathbf{m} = I + C_\Gamma \mu e^{-\mathrm{i}x\lambda \mathrm{ad}\sigma_3}(\mathbf{v}_+ - \mathbf{v}_-) \tag{2.26}$$

and

$$Q = \mathrm{iad}\sigma_3 \mathbf{m}_{\infty,1} = -\frac{\mathrm{ad}\sigma_3}{\pi}\int_\Gamma \mu(\mathbf{v}_{x+} - \mathbf{v}_{x-})\mathrm{d}\lambda$$

where we denote $\mathbf{m} = I + \mathbf{m}_{\infty,1}/\lambda + o(1/\lambda)$ as $\lambda \to \infty$. Simple calculation, we have the RHP

$$M_+ = M_- e^{-\mathrm{i}\lambda x \sigma_3}\mathbf{v}(\lambda), \quad M_\pm = m_{\pm,x} + \mathrm{i}\lambda[\sigma_3, m_\pm] - Qm_\pm.$$

Together with $M_\pm \in \partial C(L^2)$ [19], we have $M_\pm = 0$.

To prove the well-posedness of L-L equation (2.5), we construct the gauge transformation

$$g(x) = m^{(-)}(x; \lambda = 0)^\dagger = \mathrm{diag}(a(0), a^*(0)) m^{(+)}(x; \lambda = 0)^\dagger. \tag{2.27}$$

It is readily see that $g(x)$ is an Hermite matrix, i.e. $gg^\dagger = I$. And the boundary condition is

$$\lim_{x \to -\infty} g(x) = I, \text{ and } \lim_{x \to +\infty} g(x) = \mathrm{diag}(a(0), a^*(0)). \tag{2.28}$$

Proposition 2.2 ([56],Zakhrov-Taktajan) If $q(x)$ belongs to $H^{1,1}(\mathbb{R})$ and satisfies the boundary conditions $\lim\limits_{|x|\to\infty} q(x) = 0$ and scattering data restraint $r(0) = 0$, $g(x)$ satisfies equation (2.27) and (2.28), then the function $S(x) = g(x)\sigma_3 g^\dagger(x)$ satisfies the boundary condition (2.8), and $|S_x(x)| \in H^{1,1}(\mathbb{R})$, $\Psi^\pm$ satisfy the spectral problem $\Psi_x = -\mathrm{i}\lambda S\Psi$.

If we expand n in the neighborhood of 0, i.e.

$$n = gmv^{-1}(0) = I + n_1\lambda + o(\lambda^2)$$

then we can resolve

$$S = \sigma_3 + \mathrm{i}n_{1,x}. \tag{2.29}$$

Via this resolvent formula, ones can obtain a compact formula. Indeed, $mv^{-1}(0)$ satisfy the equation

$$(mv^{-1}(0))_x = -\mathrm{i}\lambda[\sigma_3, mv^{-1}(0)] + Qmv^{-1}(0)$$

Then we can expand $mv^{-1}(0)$ in the neighborhood of 0:

$$mv^{-1}(0) = g^{-1} + \lambda m_1(x,t) + o(\lambda^2)$$

where $m_{1,x} = -\mathrm{i}[\sigma_3, g^{-1}] + Qm_1$. Together with $n_1 = gm_1$, we have $S = g\sigma_3 g^\dagger$.

Similar as [19,55], together with the fact $4|q|^2 = \vec{S}_x^2$ and $4(|q_x|^2 + 4|q|^4) = \vec{S}_{xx}^2$ and Sobolev embedding $H^1(\mathbb{R}) \hookrightarrow L^\infty(\mathbb{R})$, we can establish the following theorems:

Theorem 2.3([55],Zhou) *Under the conditions*

- $r \in H^k(\Gamma)$, $r(0) = 0$
- $W_\Gamma(1 + |r|^2) = 0$, W *stands for the winding-number constraint,*

and RHP $(\mathbf{v}_x, \Gamma)$ *is solvable for all* $\lambda \in \Gamma$, *then we have* $|S_x| \in L^2((1 + x^2)\mathrm{d}x)$.

Theorem 2.4 *In addition, if* $\mathbf{v}_\pm - I, \tilde{\mathbf{v}}_\pm - I \in H_0^{1,1}$, *then* $|S_x| \in H^{1,1}$.

Thus the above theorems establish the local existence and uniqueness theorem for L-L equation (2.5) in $H^{1,1}$ without discrete scattering data. If the initial data $|S_x(x,0)| \in H^{1,1}$ without discrete scattering data, then the solution $S(x,t)$ is existence and uniqueness in the local part of $t = 0$.

2.3 Time evolution and global well-posedness without discrete scattering data

Up to now, we can prove L-L equation (2.5) is global existence and uniqueness in the space $H^{1,1}(\mathbb{R})$. To obtain the time evolution for scattering data, we use the time evolution part of Lax pair (2.4) or (2.6). However, the gauge transformation between two linear systems had been established in reference [56]. Thus, we merely need to analysis the one of them. We still analysis the time evolution part of Lax pair (2.4). We know NLS equation (2.3) is equivalent with the following compatibility condition

$$U_t - V_x + [U, V] = 0.$$

Differential spectral problem (2.4) with t, together with compatibility condition, we have

$$(\Phi_t^{\pm} - V\Phi^{\pm})_x = U(\lambda)(\Phi_t^{\pm} - V\Phi^{\pm}).$$

For arbitrary $t \in [0, \infty)$, by asymptotical analysis we can obtain

$$m_t^{(\pm)} = -2\mathrm{i}\lambda^2[\sigma_3, m^{(\pm)}] + [2\lambda Q - \mathrm{i}(Q^2 + Q_x)\sigma_3]m^{(\pm)}. \tag{2.30}$$

Proposition 2.3 *The evolution of the continuous scattering data is given by the following equation*

$$A_t = -2\mathrm{i}\lambda^2[\sigma_3, A].$$

Proof Suppose we have

$$m^{(+)}e^{-\mathrm{i}\lambda x\sigma_3} = m^{(-)}e^{-\mathrm{i}\lambda x\sigma_3}A(\lambda).$$

By the Lebesgue dominated convergence theorem, it follows that

$$A(\lambda) = \lim_{x\to-\infty} e^{\mathrm{i}\lambda x \mathrm{ad}\sigma_3}m^{(+)}.$$

It is readily to see that

$$e^{\mathrm{i}\lambda x\mathrm{ad}\sigma_3}m_t^{(+)} = -2\mathrm{i}\lambda^2[\sigma_3, e^{\mathrm{i}\lambda x\mathrm{ad}\sigma_3}m^{(+)}] + e^{\mathrm{i}\lambda x\mathrm{ad}\sigma_3}[(2\lambda Q - \mathrm{i}(Q^2 + Q_x)\sigma_3)m^{(+)}]$$

Taking the limit $x \to -\infty$ both sides, we obtain

$$A_t = -2\mathrm{i}\lambda^2[\sigma_3, A],$$

which completes the proof.

Thus the RHP (2.18) becomes

$$m_+ = m_- e^{-\mathrm{i}\lambda(x+2\lambda t)\sigma_3} v(\lambda). \tag{2.31}$$

And $r(\lambda,t) = r(\lambda,0)e^{4\mathrm{i}\lambda^2 t} \in H_0^{1,1}(\mathbb{R})$. Thus scattering data persists the solvability property. It follows that the global existence and uniqueness of L-L equation (2.5) without discrete scattering data is proved.

In the following, we consider the discrete scattering data evolution. Firstly, we rewrite equation (2.30) with the following equations

$$\begin{aligned}(\Phi_1^+ e^{-2\mathrm{i}\lambda^2 t})_t &= V(\lambda)(\Phi_1^+ e^{-2\mathrm{i}\lambda^2 t}),\\ (\Phi_2^- e^{2\mathrm{i}\lambda^2 t})_t &= V(\lambda)(\Phi_2^- e^{2\mathrm{i}\lambda^2 t}).\end{aligned} \tag{2.32}$$

We know the discrete spectrum λ_i corresponds L^2 eigenfunction

$$\Phi_1^+(x,0;\lambda_i) = \gamma_i \Phi_2^-(x,0;\lambda_i),\ \ \lambda_i \in \mathbb{C}_-. \tag{2.33}$$

It follows that

$$\Phi_1^+(x,t;\lambda_i)e^{-2\mathrm{i}\lambda_i^2 t} = \gamma_i \Phi_2^-(x,t;\lambda_i)e^{2\mathrm{i}\lambda_i^2 t}.$$

If the discrete spectrum is multiple algebraic spectrum, we have

$$\begin{aligned}\frac{1}{j!}\left(\frac{\mathrm{d}}{\mathrm{d}\lambda}\right)^j [\Phi_1^+(x,0;\lambda_i)] =& \frac{\gamma_i}{j!}\left(\frac{d}{d\lambda}\right)^j [\Phi_2^-(x,0;\lambda_i)]\\ &+\sum_{k=1}^{j}\frac{\beta_{i,k}}{(j-k)!}\left(\frac{\mathrm{d}}{\mathrm{d}\lambda}\right)^{j-k}[\Phi_2^-(x,0;\lambda_i)],\ j=1,2,\cdots,r_i.\end{aligned} \tag{2.34}$$

Firstly, we can obtain the following equation

$$\frac{1}{j!}\left[\frac{\mathrm{d}^j}{\mathrm{d}\lambda^j}(\Phi_1^+ e^{-2\mathrm{i}\lambda^2 t})\right]_t = \frac{1}{j!}\sum_{l=0}^{j} C_j^l \left(\frac{\mathrm{d}^l}{\mathrm{d}\lambda^l}V(\lambda)\right)\left(\frac{\mathrm{d}^{j-l}}{\mathrm{d}\lambda^{j-l}}(\Phi_1^+ e^{-2\mathrm{i}\lambda^2 t})\right), \tag{2.35}$$

where $C_j^l = \frac{j!}{l!(j-l)!}$. On the other hand, we have

$$\frac{\gamma_i}{j!}\left[\frac{\mathrm{d}^j}{\mathrm{d}\lambda^j}(\Phi_2^- e^{2\mathrm{i}\lambda^2 t})\right]_t = \frac{\gamma_i}{j!}\sum_{l=0}^{j} C_j^l \left(\frac{\mathrm{d}^l}{\mathrm{d}\lambda^l}V(\lambda)\right)\left(\frac{\mathrm{d}^{j-l}}{\mathrm{d}\lambda^{j-l}}(\Phi_2^- e^{2\mathrm{i}\lambda^2 t})\right),$$

$$\frac{\beta_{i,k}}{(j-k)!}\left[\frac{\mathrm{d}^{j-k}}{\mathrm{d}\lambda^{j-k}}(\Phi_2^- e^{2\mathrm{i}\lambda^2 t})\right]_t = \frac{\beta_{i,k}}{(j-k)!}\sum_{l=0}^{j-k} C_{j-k}^l \left(\frac{\mathrm{d}^l}{\mathrm{d}\lambda^l}V(\lambda)\right)\left(\frac{\mathrm{d}^{j-k-l}}{\mathrm{d}\lambda^{j-k-l}}(\Phi_2^- e^{2\mathrm{i}\lambda^2 t})\right),$$

$j=1,2,\cdots,r_i$, it follows that

$$\begin{aligned}
&\left[\frac{\gamma_i}{j!}\frac{\mathrm{d}^j}{\mathrm{d}\lambda^j}(\Phi_2^- e^{2\mathrm{i}\lambda^2 t})+\sum_{k=0}^{j}\frac{\beta_{i,k}}{(j-k)!}\frac{\mathrm{d}^{j-k}}{\mathrm{d}\lambda^{j-k}}(\Phi_2^- e^{2\mathrm{i}\lambda^2 t})\right]_t\\
=&V(\lambda)\left[\frac{\gamma_i}{j!}\frac{\mathrm{d}^j}{\mathrm{d}\lambda^j}(\Phi_2^- e^{2\mathrm{i}\lambda^2 t})+\sum_{k=1}^{j}\frac{\beta_{i,k}}{(j-k)!}\frac{\mathrm{d}^{j-k}}{\mathrm{d}\lambda^{j-k}}(\Phi_2^- e^{2\mathrm{i}\lambda^2 t})\right]\\
&+\sum_{l=0}^{j}\frac{1}{l!}\left(\frac{\mathrm{d}^l}{\mathrm{d}\lambda^l}V(\lambda)\right)\left[\frac{\gamma_i}{(j-l)!}\frac{\mathrm{d}^{j-l}}{\mathrm{d}\lambda^{j-l}}(\Phi_2^- e^{2\mathrm{i}\lambda^2 t})\right.\\
&\left.+\sum_{k=1}^{j-l}\frac{\beta_{i,k}}{(j-k-l)!}\frac{\mathrm{d}^{j-l-k}}{\mathrm{d}\lambda^{j-k-l}}(\Phi_2^- e^{2\mathrm{i}\lambda^2 t})\right].
\end{aligned}\tag{2.36}$$

By mathematical induction and existence and uniqueness of ordinary differential equation, we can obtain the time evolution relation

$$\left[\frac{1}{j!}\frac{\mathrm{d}^j}{\mathrm{d}\lambda^j}(\Phi_1^+ e^{-2\mathrm{i}\lambda^2 t})\right]=\left[\frac{\gamma_i}{j!}\frac{\mathrm{d}^j}{\mathrm{d}\lambda^j}(\Phi_2^- e^{2\mathrm{i}\lambda^2 t})+\sum_{k=0}^{j}\frac{\beta_{i,k}}{(j-k)!}\frac{\mathrm{d}^{j-k}}{\mathrm{d}\lambda^{j-k}}(\Phi_2^- e^{2\mathrm{i}\lambda^2 t})\right].\tag{2.37}$$

3 The discrete spectrum and Darboux transformation

In this section, we use the Darboux transformation method to delete or add the discrete spectrum of L-L spectral problem (2.6). In order to derive the Darboux transformation for L-L equation (2.5), we first give the Darboux transformation of NLS (2.3).

3.1 Darboux transformation of NLS

The Darboux transformation for NLS is well known for us, we can readily establish the following theorem:

Theorem 3.1([9,29,52]) *Assume we have N distinct parameters λ_1, λ_2, $\cdots$, $\lambda_N\in\mathbb{C}_-$ and the corresponding special solution matrices $|y_1\rangle$, $|y_2\rangle$, $\cdots$, $|y_N\rangle$, then the Darboux matrix can be represented as*

$$T_N=I-\left[|y_1\rangle,\quad |y_2\rangle,\quad \cdots,\quad |y_N\rangle\right]M^{-1}(\lambda-S)^{-1}\begin{bmatrix}\langle y_1|\\ \langle y_2|\\ \vdots\\ \langle y_N|\end{bmatrix}$$

where $\mathbb{C}_-$ represents the lower half complex plane

$$M=\left(\frac{\langle y_i|y_j\rangle}{\lambda_j-\lambda_i^*}\right)_{1\leqslant i\leqslant N,1\leqslant j\leqslant N},$$

$$S=\operatorname{diag}(\lambda_1,\lambda_2,\cdots,\lambda_N),$$

and $|y_i\rangle=\left(m_1^{(+)}(\lambda_i),m_2^{(-)}(\lambda_i)\right)e^{-\mathrm{i}\lambda_i x\sigma_3}C_i$, $C_i=(1,-\gamma_i)^T$ is a non-zero column vector.

Lemma 3.1 *The matrix*

$$M=\left(\frac{\langle y_i|y_j\rangle}{\lambda_j-\lambda_i^*}\right)_{N\times N}$$

is a non-singular.

Proof Similar as [32].

Indeed the essence of Darboux transformation is a kind of special gauge transformation. A important step is to find the seed solution for original spectral problem. Suppose we have a fundamental solution $\Phi(\lambda)=(\Phi_1(\lambda),\Psi_1(\lambda))$ of a spectral problem (2.11), the high order Darboux transformation can be construct as following arrow diagram

$$(\Phi_1,\Psi_1)\xrightarrow[\Phi_1(\lambda_1)\in\mathrm{Ker}(T_0[1])]{T_0[1]}\left(\Phi_1^{[1]},\Psi_1^{[1]}\right)\xrightarrow[\Phi_1[1](\lambda_1)\in\mathrm{Ker}(T_1[1])]{T_1[1]}\cdots$$

where $\Psi_1^{[1]}=T_0[1]\Psi_1$, $\Phi_1^{[1]}=[(T_0[1]\Phi_1)_\lambda+\beta_1T_0[1]\Psi_1]\,|_{\lambda=\lambda_1}$, β_1 is a complex constants. We can see that the parameters β_1 is not convenient to calculate the exact solution. Indeed, we can absorb the parameter β_1 into Φ_1. We need the following lemma:

Lemma 3.2 *Assume Φ_1 is a seed solution for (2.11) at $\lambda=\lambda_1$, and T is the Darboux transformation by Φ_1, Ψ_1 is another linear dependent solution with Φ_1, then $T\Psi_1$ is uniquely determined module a nozero constant.*

Proof The Darboux matrix is

$$T=I+\frac{\lambda_1^*-\lambda_1}{\lambda-\lambda_1^*}\frac{\Phi_1\Phi_1^\dagger}{\Phi_1^\dagger\Phi_1},$$

direct calculating, it follows that

$$T\Psi_1=\frac{\det(\Psi_1,\Phi_1)}{\Phi_1^\dagger\Phi_1}\begin{pmatrix}0&1\\-1&0\end{pmatrix}\Phi_1^*.$$

However, by the Abel formula, $\det(\Psi_1,\Phi_1)_x=0$. This completes the proof.

By above lemma, we can see that the new seed function $\Phi_1[1]$ does not depend with exact form of Ψ_1. Thus we can choose function Ψ_1 arbitrary. Thus, $\Phi_1[1]$ can be rewritten as

$$\Phi_1^{[1]}=\lim_{\xi\to 0}\frac{T_1[1](\lambda_1+\xi)(\Phi_1(\lambda_1+\xi)+\xi\beta_1\Psi_1(\lambda_1+\xi))}{\xi}.$$

Generally, we can obtain

$$\Phi_1^{[N-1]}=\lim_{\xi\to 0}\frac{T_{N-1}[1]\cdots T_1[1](\lambda_1+\xi)\left(\Phi_1(\lambda_1+\xi)+\sum_{i=1}^{N-1}\xi^i\beta_i\Psi_1(\lambda_1+\xi)\right)}{\xi^{N-1}}.$$

Remark 3.1 *In reference [29, 30, 32], we use the relation*

$$\left[\exp\left(\sum_{i=1}^{N-1}\delta_i\xi^i\right)\right]_{[N-1]}=1+\sum_{i=1}^{N-1}\xi^i\beta_i$$

where the symbol $_{[N-1]}$ *represents the taylor expansion truncate from* ξ^{N-1}. *And* δ_i *can be determined by* β_i *through elementary Schur polynomial. When the spectral is branch spectral, ones need to make small modification the above polynomial [29,30,32].*

Theorem 3.2 *Generalized Darboux matrix*

$$T_N=\prod_{i=1}^{s}T[i],\ N=\sum_{i=1}^{s}r_i,$$

where

$$\begin{aligned}T[i]&=T_{r_i}[i]T_{r_i-1}[i]\cdots T_0[i],\ T_j[i]=\left(I+\frac{\lambda_i^*-\lambda_i}{\lambda-\lambda_i^*}P_i^{(j)}\right),\ j=1,2,\cdots,r_i,\\ T_0[i]&=T[0]=I,\ P_i^{(j)}=\frac{|y_{i,j}\rangle\langle y_{i,j}|}{\langle y_{i,j}|y_{i,j}\rangle},\\ |y_{i,j}\rangle&=\lim_{\xi\to 0}\frac{(T_{j-1}[i]\cdots T_0[i])(\lambda_i+\xi)\prod_{m=1}^{i-1}T[m](\lambda_i+\xi)}{\xi^{j-1}}\left(|y_i\rangle-\sum_{k=1}^{j-1}\xi^k\beta_{i,k}|x_i\rangle\right).\end{aligned}$$

and $\beta_{i,0}=0$, $|y_i\rangle=\Phi_1^+(\lambda_i+\xi)-\gamma_i\Phi_2^-(\lambda_i+\xi)$, $|x_i\rangle=\Phi_2^-(\lambda_i+\xi)$. *The function* $\Phi_1^+[N](\lambda_i)$ *is* $L^2(\mathbb{R})$ *eigenfunction for spectral problem* $L\Phi=\lambda\Phi$, *where* $L=\mathrm{i}\sigma_3(\partial_x-Q[N])$, $\Phi_1^+[N]=T_N\Phi_1^+$, *and*

$$Q[N]=Q+\mathrm{i}\left[\sigma_3,\sum_{i=1}^{s}\sum_{j=1}^{r_i}(\lambda_i^*-\lambda_i)P_i^{(j)}\right].$$

And the eigenfunctions satisfy the following relation

$$\frac{1}{j!}\frac{\mathrm{d}^j}{\mathrm{d}\lambda^j}(\Phi_1^+[N])|_{\lambda=\lambda_i}=\frac{\gamma_i}{j!}\frac{\mathrm{d}^j}{\mathrm{d}\lambda^j}(\Phi_2^-[N])|_{\lambda=\lambda_i}+\sum_{k=0}^{j}\frac{\beta_{i,k}}{(j-k)!}\frac{\mathrm{d}^{j-k}}{\mathrm{d}\lambda^{j-k}}(\Phi_2^-[N])|_{\lambda=\lambda_i},$$

$$j=1,2,\cdots,r_i,$$

where $\Phi_2^-[N]=T_N\Phi_2^-$. *By above relations, its imply that* $\frac{d^j}{d\lambda^j}(\Phi_1^+[N])|_{\lambda=\lambda_i}$ *are the generalized eigenfunction and belong to space* $L^2(\mathbb{R})$.

Proof The generalized Darboux transformation is constructed in reference [29]. In the following, we derive the properties of eigenfunction. Firstly, we expand the following function

$$T_N(\lambda_i+\xi)\left(|y_i\rangle-\sum_{k=0}^{j-1}\xi^k\beta_{i,k}|x_i\rangle\right)=\sum_{k=0}^{+\infty}Q_k\xi^k,$$

where

$$Q_k=\frac{1}{k!}\frac{\mathrm{d}^k}{\mathrm{d}\lambda^k}(\Phi_1^+[N])|_{\lambda=\lambda_i}-\frac{\gamma_i}{k!}\frac{\mathrm{d}^k}{\mathrm{d}\lambda^k}(\Phi_2^-[N])|_{\lambda=\lambda_i}-\sum_{l=0}^{k}\frac{\beta_{i,l}}{(k-l)!}\frac{\mathrm{d}^{k-l}}{\mathrm{d}\lambda^{k-l}}(\Phi_2^-[N])|_{\lambda=\lambda_i},$$

and $k=0,1,\cdots,r_i-1$. By the construction of generalized Darboux transformation, we can obtain $Q_k=0$.

Since $\Phi_1^+[N](\lambda_i)\to 0$ exponentially as $x\to+\infty$ and $\Phi_2^-[N](\lambda_i)\to 0$ exponentially as $x\to-\infty$, we can deduce that $\Phi_1^+[N](\lambda_i)\in L^2(\mathbb{R})$. This completes the proof.

Theorem 3.3([49],Lemma 4) *The above Darboux matrix can be represented as*

$$T_N=I-\begin{bmatrix}Y_1, & Y_2, & \cdots, & Y_s\end{bmatrix}M^{-1}D\begin{bmatrix}Y_1^\dagger\\ Y_2^\dagger\\ \vdots\\ Y_s^\dagger\end{bmatrix},$$

where

$$Y_i=\begin{bmatrix}|z_i\rangle, & |z_i\rangle^{(1)}, & \cdots & \frac{1}{(r_i-1)!}|z_i\rangle^{(r_i-1)}\end{bmatrix}_{\xi=0},\quad D=\mathrm{diag}(D_1,D_2,\cdots,D_s),$$

$$D_i=\begin{bmatrix}\frac{1}{\lambda-\lambda_i^*} & 0 & \cdots & 0\\ \frac{1}{(\lambda-\lambda_i^*)^2} & \frac{1}{\lambda-\lambda_i^*} & \cdots & 0\\ \vdots & \vdots & & \vdots\\ \frac{1}{(\lambda-\lambda_i^*)^{r_i}} & \frac{1}{(\lambda-\lambda_i^*)^{r_i-1}} & \cdots & \frac{1}{\lambda-\lambda_i^*}\end{bmatrix},\quad M=\begin{bmatrix}M^{[11]} & M^{[12]} & \cdots & M^{[1s]}\\ M^{[21]} & M^{[22]} & \cdots & M^{[2s]}\\ \vdots & \vdots & & \vdots\\ M^{[s1]} & M^{[s2]} & \cdots & M^{[ss]}\end{bmatrix}.$$

and symbol $^{(i)}$ means the derivative with respect to ξ,

$$\begin{aligned}
|z_i(\xi)\rangle &= |y_i(\lambda_i+\xi)\rangle + \sum_{k=1}^{r_i-1} \xi^k \beta_{i,k} |x_i(\lambda_i+\xi)\rangle, \\
M^{[ij]} &= \left(M^{[ij]}_{m,n}\right)_{r_i\times r_j}, \\
M^{[ij]}_{m,n} &= \frac{1}{(m-1)!(n-1)!} \frac{\partial^{n-1}}{\partial \xi^{n-1}} \frac{\partial^{m-1}}{\partial(\xi^*)^{m-1}} \frac{\langle z_i|z_j\rangle}{\lambda_j-\lambda_i^*+\xi-\xi^*}.
\end{aligned}$$

Proof Directly calculating, we can obtain

$$(T_N-I)_{lk} = -\frac{\det(M_1)}{\det(M)}, \quad M_1 = \begin{bmatrix} M & Y_k^\dagger \\ Y_l & 0 \end{bmatrix}$$

where Y_l means the l-th row of $\left[Y_1, Y_2, \cdots, Y_s\right]$. Taking the limits with respect to $\xi \to 0$ from above formula, we can obtain the results.

The above theorem we obtained through generalized Darboux transformation is consistent with the Lemma 4 in reference [49]. In the following, we consider the relation between Darboux transformation and scattering data.

Proposition 3.1 *The Darboux matrix T_N transform the scattering data $\{a(\lambda), b(\lambda)\}$ into*

$$\{\widetilde{a(\lambda)}, \widetilde{b(\lambda)}; \lambda_i, \gamma(\lambda_i), \beta_{i,k}\},$$

where

$$\begin{aligned}
\widetilde{a(\lambda)} &= a(\lambda) \prod_{i=1}^{s} \left(\frac{\lambda-\lambda_i}{\lambda-\lambda_i^*}\right)^{r_i}, \\
\widetilde{b(\lambda_i)} &= b(\lambda_i).
\end{aligned} \tag{3.38}$$

Proof Direct calculating, we obtain

$$\begin{aligned}
a(\lambda) &= \det(m_1^{(+)}, m_2^{(-)}), \\
b(\lambda) &= \det(m_1^{(-)} e^{-\mathrm{i}\lambda x}, m_1^{(+)} e^{-\mathrm{i}\lambda x}).
\end{aligned}$$

It follows that the Darboux transformation

$$T_N(m_1^{(+)}, m_2^{(-)}) = (\widetilde{m_1}^{(+)}, \widetilde{m_2}^{(-)})$$

give the first equation of (3.38). By symmetry relation, we have

$$m_2^{(+)} = \begin{bmatrix} 0 & 1 \\ -1 & 0 \end{bmatrix} (m_1^{(+)}(\lambda^*))^*, \; m_1^{(-)} = -\begin{bmatrix} 0 & 1 \\ -1 & 0 \end{bmatrix} (m_2^{(-)}(\lambda^*))^*.$$

It follows that

$$\widehat{T}_N(m_1^{(-)}, m_2^{(+)}) = (\widetilde{m_1}^{(-)}, \widetilde{m_2}^{(+)}), \quad \widehat{T}_N = \begin{bmatrix} 0 & -1 \\ 1 & 0 \end{bmatrix} T_N^*(\lambda^*) \begin{bmatrix} 0 & 1 \\ -1 & 0 \end{bmatrix}.$$

By theorem (3.2), we know that T_N is determined by spectral parameters λ_i, γ_i, $\beta_{i,k}$. Furthermore, we have

$$T_N \to \begin{pmatrix} 1 & 0 \\ 0 & \prod_{i=1}^{s} \left(\dfrac{\lambda - \lambda_i}{\lambda - \lambda_i^*}\right)^{r_i} \end{pmatrix}, \quad \widehat{T}_N \to \begin{pmatrix} \prod_{i=1}^{s} \left(\dfrac{\lambda - \lambda_i^*}{\lambda - \lambda_i}\right)^{r_i} & 0 \\ 0 & 1 \end{pmatrix}, \quad x \to +\infty,$$

$$T_N \to \begin{pmatrix} \prod_{i=1}^{s} \left(\dfrac{\lambda - \lambda_i}{\lambda - \lambda_i^*}\right)^{r_i} & 0 \\ 0 & 1 \end{pmatrix}, \quad \widehat{T}_N \to \begin{pmatrix} 1 & 0 \\ 0 & \prod_{i=1}^{s} \left(\dfrac{\lambda - \lambda_i^*}{\lambda - \lambda_i}\right)^{r_i} \end{pmatrix}, \quad x \to -\infty.$$

It follows that

$$\widetilde{b(\lambda)} = \lim_{x\to+\infty} \det(\widehat{T}_N m_1^{(-)} e^{-\mathrm{i}\lambda x}, T_N m_1^{(+)} e^{-\mathrm{i}\lambda x}) = b(\lambda).$$

The above proposition can be considered as adding the zeros of the scattering data $a(\lambda)$. The inverse process is the delete zeros of scattering data $a(\lambda)$, which can be established as references [20, 31].

3.2 Darboux transformation of L-L equation

Since the Darboux transformation for NLS is constructed by above in detail. On the other hand, as we know, the Darboux transformation is a special gauge transformation. Based on these ideas, we could construct the Darboux transformation for L-L equation by combining the above two gauge transformation.

In order to give the Darboux transformation with a linear factional transformation or a simple element $L_-(\mathrm{GL}(2,\mathbb{C}))$, we use the loop group representation [52]. If matrix functions $\Phi^\pm$ satisfy

$$\begin{cases} \Phi_x^\pm = (-\mathrm{i}\lambda\sigma_3 + Q)\Phi^\pm, \\ \lim\limits_{x\to\pm\infty} \Phi^\pm = \exp(-\mathrm{i}\lambda\sigma_3 x), \end{cases}$$

then such $\Phi^\pm$ will be called the trivialization of potential function Q at $\pm\infty$. Similarly, if matrix functions $\Psi^\pm$ satisfy

$$\begin{cases} \Psi_x^\pm = -\mathrm{i}\lambda S\Psi^\pm, \\ \lim\limits_{x\to\pm\infty} \Psi^\pm = \exp(-\mathrm{i}\lambda\sigma_3 x), \end{cases}$$

then such $\Psi^\pm$ will be called the trivialization of function S at $\pm\infty$.

Theorem 3.4 *Let S satisfies the boundary condition (2.8), and $\Psi^{\pm}$ are the trivialization of S at $\pm\infty$ respectively, and π is the projection of $\mathbb{C}^2$. For each $x \in \mathbb{R}$, set*

$$\Psi_1 = (\Psi_1^+(\lambda_1), \Psi_2^-(\lambda_1))\,(a(0), -\gamma(\lambda_1))^T, \quad \lambda_1 \in \mathbb{C}_-,$$

$$\widehat{T} = I + \frac{\zeta_1^* - \zeta_1}{\zeta - \zeta_1^*}\pi,\ \pi = \frac{\Psi_1\Psi_1^\dagger}{\Psi_1^\dagger\Psi_1},\ \zeta = \lambda^{-1}.$$

Then

$$\widehat{S} = D_1(S + 2\mathrm{Im}(\zeta_1)\pi_x)D_1^{-1},\ D_1 = \mathrm{diag}\left(\frac{\lambda_1}{\lambda_1^*}, 1\right) \tag{3.39}$$

is the global solution for L-L equation defined on $\mathbb{R}^2$, and

$$\widehat{\Psi}^+ = D_1(\widehat{T}\Psi_1^+, \sigma\widehat{T}^*(\lambda^*)\sigma^{-1}\Psi_2^+)D_1^{-1},$$
$$\widehat{\Psi}^- = D_1(\sigma\widehat{T}^*(\lambda^*)\sigma^{-1}\Psi_1^-, \widehat{T}\Psi_2^-)D_1^{-1}$$

are the trivialization of $\widehat{S}$ at $+\infty$ and $-\infty$ respectively, where

$$\sigma = \begin{pmatrix} 0 & -1 \\ 1 & 0 \end{pmatrix}.$$

Proof Suppose the analytical matrix $\Phi_- = (\Phi_1^+, \Phi_2^-)$, $\Phi_+ = (\Phi_1^-, \Phi_2^+)$. The elementary Darboux transformation for spectral problem (2.11) is

$$T_- = I + \frac{\lambda_1^* - \lambda_1}{\lambda - \lambda_1^*}\frac{\Phi_1\Phi_1^\dagger}{\Phi_1^\dagger\Phi_1}, \quad T_+ = \sigma T_-^*(\lambda^*)\sigma^{-1}, \quad \Phi_1 = (\Phi_1^+(\lambda_1), \Phi_2^-(\lambda_1))(1, -\gamma(\lambda_1))^T.$$

By above Darboux transformation, we can obtain $\widehat{\Phi}_- = (\widehat{\Phi}_1^+, \widehat{\Phi}_2^-) = T_-\Phi_-$, $\widehat{\Phi}_+ = (\widehat{\Phi}_1^-, \widehat{\Phi}_2^+) = T_+\Phi_+$. It follows that $\widehat{\Phi}^+ = (\widehat{\Phi}_1^+, \widehat{\Phi}_2^+)$ and $\widehat{\Phi}^- = (\widehat{\Phi}_1^-, \widehat{\Phi}_2^-)$ are the trivialization of

$$\widehat{Q} = Q + \mathrm{i}(\lambda_1^* - \lambda_1)\left[\sigma_3, \frac{\Phi_1\Phi_1^\dagger}{\Phi_1^\dagger\Phi_1}\right],$$

at $\pm\infty$ respectively.

On the other hand, by gauge transformation, we obtain that

$$\Psi^- = g\Phi^-, \quad \Psi^+ = g\Phi^+\mathrm{diag}(a^*(0), a(0)),$$

are the trivialization of S at $\pm\infty$ respectively, where $g = [\Phi^-]^\dagger|_{\lambda=0}$. And

$$\widehat{\Psi}^- = \widehat{g}\widehat{\Phi}^-, \quad \widehat{\Psi}^+ = \widehat{g}\widehat{\Phi}^+\mathrm{diag}(\widehat{a}^*(0), \widehat{a}(0)),$$

are the trivialization of $\widehat{S}$ at $\pm\infty$ respectively, where $\widehat{g} = [\widehat{\Phi}^{-}]^{\dagger}|_{\lambda=0}$ and $\widehat{a}(0) = \dfrac{\lambda_1}{\lambda_1^*}a(0)$, the function $\widehat{S}$ we will be given in the following. In order to obtain the relation between $\widehat{S}$ and S, we use the following analytical function

$$\Psi_- = (\Psi_1^+, \Psi_2^-), \quad \widehat{\Psi}_- = (\widehat{\Psi}_1^+, \widehat{\Psi}_2^-).$$

It follows that

$$\begin{aligned}\widehat{g} =& \mathrm{diag}\left(a(0)\frac{\lambda_1}{\lambda_1^*}, 1\right)\left[\Phi_1^+|_{\lambda=0}, \Phi_2^-|_{\lambda=0}\right]^{\dagger}(T_-^{\dagger}|_{\lambda=0})\\ =& D_1 g(T_-^{\dagger}|_{\lambda=0}).\end{aligned}$$

Together with above equation, we have

$$\begin{aligned}\widehat{\Psi}_- =& \widehat{g}\left[\widehat{\Phi}_1^+(\lambda)\widehat{a}^*(0), \widehat{\Phi}_2^-(\lambda)\right]\\ =& D_1 g\left(T_-^{\dagger}|_{\lambda=0}\right)T_- g^{\dagger}\Psi_- D_1^{-1}\\ =& D_1\widehat{T}\Psi_- D_1^{-1},\end{aligned}$$

where

$$\begin{aligned}\Psi_1 =& g(\Phi_1^+, \Phi_2^-)(1, -\gamma(\lambda_1))^T\\ =& (\Psi_1^+(\lambda_1), \Psi_2^-(\lambda_1))\,(a(0), -\gamma(\lambda_1))^T.\end{aligned}$$

Then we can obtain

$$\widehat{S} = D_1\left(S + \mathrm{i}(\zeta_1^* - \zeta_1)\pi_x\right)D_1^{-1}.$$

Then $\widehat{\Psi}^{\pm}$ are the trivialization of $\widehat{S}$ at $\pm\infty$ respectively. Finally, the estimation

$$|\pi| \leqslant 1$$

implies that the solutions are global for $(x, t) \in \mathbb{R}^2$.

We define the matrix $\widehat{T}$ as the Darboux matrix of L-L equation (2.5). In the following, we consider the N-fold Darboux transformation for L-L equation (2.5). We give the following theorem

Lemma 3.3 *The N-fold Darboux transformation for L-L equation* (2.5) *can be represented as*

$$\widehat{T}_N = I - \left[|y_1\rangle, \ |y_2\rangle, \ \cdots, \ |y_N\rangle\right] M^{-1}(\zeta - D)^{-1}\begin{bmatrix}\langle y_1|\\ \langle y_2|\\ \vdots\\ \langle y_N|\end{bmatrix},$$

where $|y_i\rangle = (\Psi_1^+(\lambda_i), \Psi_2^-(\lambda_i)))(a(0), -\gamma_i)^T s$ *are special solutions of Lax pair* (2.6) *at* $\lambda = \lambda_i$, $\gamma_i \in \mathbb{C}$,

$$M = \left(\frac{\langle y_i | y_j \rangle}{\zeta_j - \zeta_i^*} \right)_{1 \leqslant i \leqslant N, 1 \leqslant j \leqslant N},$$

and

$$D = \operatorname{diag}(\zeta_1^*, \zeta_2^*, \cdots, \zeta_N^*).$$

Proof The N-fold Darboux transformation can be constructed by N times iteration of Darboux transformation, i.e.

$$\widehat{T}_N = \widehat{T}[N]\widehat{T}[N-1] \cdots \widehat{T}[1]$$

where

$$\widehat{T}[i] = I + \frac{\zeta_i^* - \zeta_i}{\zeta - \zeta_i^*} \frac{\Psi_i[i-1]\Psi_i[i-1]^\dagger}{\Psi_i[i-1]^\dagger \Psi_i[i-1]},$$
$$\Psi_i[i-1] = T[i-1]T[i-2] \cdots T[1] |y_i\rangle|_{\zeta=\zeta_i}.$$

Since the residue of $\widehat{T}_N$, we can write the above Darboux transformation $\widehat{T}_N$ with the following linear fractional transformation

$$\widehat{T}_N = I + \sum_{i=1}^{N} \frac{P_i}{\zeta - \zeta_i^*}$$

where P_is are 2×2 matrices with rank equals 1. So we can suppose $P_i = |x_i\rangle\langle y_i|$. Because P_is are uniquely determined by the iteration. Thus if $\langle y_i|$s are determined, then $|x_i\rangle$s are uniquely determined.

On the other hand, we know

$$\widehat{T}^{-1} = \widehat{T}^\dagger(\zeta^*) = I + \sum_{i=1}^{N} \frac{P_i^\dagger}{\zeta - \zeta_i}.$$

By the residue relation of $\widehat{T}_N \widehat{T}_N^{-1} = I$, we have

$$|y_j\rangle + \sum_{i=1}^{N} |x_i\rangle \frac{\langle y_j | y_i \rangle}{\zeta_j - \zeta_i^*} = 0, \quad i, j = 1, 2, \cdots, N.$$

In addition, since $\operatorname{Rank}(\widehat{T}_N(\zeta_i)) = 1$, so we can suppose

$$\operatorname{Ker}(\widehat{T}_N(\zeta_i)) = |y_i\rangle = (\Psi_1^+(\lambda_i), \Psi_2^-(\lambda_i)))(a(0), -\gamma_i)^T.$$

By simple linear algebra, we can obtain the N-fold Darboux transformation for L-L equation. This completes the proof.

In the following, we consider the high order algebraic poles for the scattering problem. Similar as the above section, we can obtain the following theorem:

Theorem 3.5 *The generalized Darboux matrix for L-L equation* (2.5) *can be represented as*

$$\widehat{T}_N = I - YM^{-1}DY^{\dagger},$$

where

$$Y = \left[Y_1, \quad Y_2, \quad \cdots, \quad Y_s\right],$$

$$Y_i = \left[|z_i\rangle, \quad |z_i\rangle^{(1)}, \quad \cdots, \quad \frac{1}{(r_i-1)!}|z_i\rangle^{(r_i-1)}\right]|_{\xi=0}, \quad D = \mathrm{diag}(D_1, D_2, \cdots, D_s),$$

$$D_i = \begin{bmatrix} \dfrac{-\lambda\lambda_i^*}{\lambda-\lambda_i^*} & 0 & \cdots & 0 \\ \dfrac{-\lambda^2}{(\lambda-\lambda_i^*)^2} & \dfrac{-\lambda\lambda_i^*}{\lambda-\lambda_i^*} & \cdots & 0 \\ \vdots & \vdots & & \vdots \\ \dfrac{-\lambda^2}{(\lambda-\lambda_i^*)^{r_i}} & \dfrac{-\lambda^2}{(\lambda-\lambda_i^*)^{r_i-1}} & \cdots & \dfrac{-\lambda\lambda_i^*}{\lambda-\lambda_i^*} \end{bmatrix}, \quad M = \begin{bmatrix} M^{[11]} & M^{[12]} & \cdots & M^{[1s]} \\ M^{[21]} & M^{[22]} & \cdots & M^{[2s]} \\ \vdots & \vdots & & \vdots \\ M^{[s1]} & M^{[s2]} & \cdots & M^{[ss]} \end{bmatrix},$$

and

$$|z_i(\xi)\rangle = |y_i(\lambda_i+\xi)\rangle + \sum_{k=1}^{r_i-1} \xi^k \beta_{i,k} |x_i(\lambda_i+\xi)\rangle,$$

$$M^{[ij]} = \left(M^{[ij]}_{m,n}\right)_{r_i\times r_j},$$

$$M^{[ij]}_{m,n} = \frac{1}{(m-1)!(n-1)!}\frac{\partial^{n-1}}{\partial\xi^{n-1}}\frac{\partial^{m-1}}{\partial(\xi^*)^{m-1}}\frac{\langle z_i(\xi^*)|z_j(\xi)\rangle}{(\lambda_j+\xi)^{-1}-(\lambda_i^*+\xi^*)^{-1}},$$

and symbol $^{(i)}$ *means the derivative with respect to* ξ, *the transformations between the field functions are*

$$(S[N])_{kl} = S_{kl} + \mathrm{i}\left(\frac{A_{kl}}{\det(M)}\right)_x, \quad A_{kl} = \det\begin{bmatrix} M & Y[l]^{\dagger} \\ Y[k] & 0 \end{bmatrix}, \tag{3.40}$$

where $|x_i\rangle$ *is the linear dependent solution with* $|y_i\rangle$, $Y[i]$ *denotes the i-th row of the matrix* Y *and the subscript* $_{kl}$ *represents the k-th row and l-th column element.*

By simple linear algebra, we can obtain the following compact soliton formula

$$(S[N])_{kl} = S_{kl} + \mathrm{i}\left(\frac{\det(M_{kl})}{\det(M)}\right)_x, \quad M_{kl} = M - Y[l]^{\dagger}Y[k]. \tag{3.41}$$

Remark 3.2 *Integrating the above expression,*

$$\int (S[N])_{kl}\mathrm{d}x = \int S_{kl}\mathrm{d}x + \mathrm{i}\left(\frac{\det(M_{kl})}{\det(M)} - 1\right), \quad M_{kl} = M - Y[l]^\dagger Y[k]. \tag{3.42}$$

it follows that the soliton formula for VFE (1.2)

$$\begin{aligned} \gamma^x[N] =& \mathrm{Re}\left(\int (S[N])_{12}\mathrm{d}x\right), \\ \gamma^y[N] =& \mathrm{Im}\left(\int (S[N])_{12}\mathrm{d}x\right), \\ \gamma^z[N] =& \int (S[N])_{11}\mathrm{d}x. \end{aligned}$$

Finally, we give the transformation between (Ψ_-, S) and $(\Psi_-[N], \widehat{S}[N])$

$$\begin{aligned} \Psi_- &\to \Psi_-[N] = D_N\widehat{T}_N\Psi_-D_N^{-1}, \\ S &\to \widehat{S}[N] = D_N S[N] D_N^{-1}, \end{aligned} \tag{3.43}$$

where (Ψ_-, S) represents wave function and potential function without discrete scattering data, $(\Psi_-[N], \widehat{S}[N])$ represents wave function and potential function possess discrete scattering data and

$$D_N = \mathrm{diag}\left(\prod_{i=1}^{s}\left(\frac{\lambda_i}{\lambda_i^*}\right)^{r_i}, 1\right).$$

Since D_N is a trivial gauge transformation, we omit it in the process of obtaining exact solution.

And the Darboux matrix $\widehat{T}_N$ is determined uniqueness by the parameters λ_i, γ_i and $\beta_{i,k}$ and is nonsingular for $(x,t) \in \mathbb{R}^2$. By the transformation (3.43), if S is global existence and uniqueness, it follows that $D_N S[N] D_N^{-1}$ is global existence and uniqueness. Thus the global existence and uniqueness of L-L equation (2.5) is proved.

Theorem 3.6 *If*

$$S_0 \in \left\{S | S_x \in H^{1,1}, S \in AO(2), \lim_{|x|\to\infty} S = \sigma_3\right\},$$

then the solutions of L-L equation (2.5) *are global existence and uniqueness.*

4 High order soliton solution

In this subsection, we consider the high order soliton solution for L-L equation (2.5). The mixed rational and exponential function solution (or high order soliton) is obtained. Besides, we give the explicit expression for high order soliton solution of L-L equation (2.5) and VFE (1.2).

For the classical integrable L-L equation (1.1), the single soliton or the N-soliton and the interaction of N-soliton have been studied in detail by the Riemann-Hilbert method [20]. In our case, we derive the Darboux transformation of L-L equation (2.5) by the gauge transformation. With the Darboux transformation, we obtain a simple generalized soliton solution formula for L-L equation (2.5).

Single soliton can be generated by Darboux transformation from the vacuum solution. In this case, there is no reflection coefficient. Then the RHP (2.26) can be solved evidently, i.e. $\mathbf{m} = I$. Then we have $Q = 0$ and $S = \sigma_3$. To obtain the pure soliton solution, we use the Darboux transformation to yield the discrete spectrum. The vector functions Ψ_1^+ and Ψ_2^- can be represented as

$$\Psi_1^+ = \begin{pmatrix} e^{-\mathrm{i}\lambda x} \\ 0 \end{pmatrix}, \quad \Psi_2^- = \begin{pmatrix} 0 \\ e^{\mathrm{i}\lambda x} \end{pmatrix}$$

and $a(0) = 1$, $\gamma(t;\lambda_1) = -c_1 e^{4\mathrm{i}\lambda_1^2 t}$. Then the standard single soliton for L-L equation (2.5) can be obtain by formula (3.39), i.e.

$$\begin{pmatrix} S^z & S^- \\ S^+ & -S^z \end{pmatrix}, \quad S^+ = (S^-)^* \tag{4.44}$$

where

$$\begin{aligned} S^z &= 1 - \frac{2b^2}{a^2+b^2}\mathrm{sech}^2(A), \quad A = 2b(x + 4at + x_0) \\ S^- &= \frac{2b}{a^2+b^2} e^{\mathrm{i}B}\mathrm{sech}(A)[b\tanh(A) + \mathrm{i}a], \\ B &= -2ax + 4[b^2 - a^2]t - \varphi_1. \end{aligned}$$

with defintion

$$a = \mathrm{Re}(\lambda_1), \quad b = \mathrm{Im}(\lambda_1), \quad x_0 = -\frac{\ln|c_1|}{2\mathrm{Im}(\lambda_1)}, \quad \varphi_0 = \arg(c_1).$$

It follows that the single soliton of VFE (1.2) are

$$\begin{aligned} \gamma^x &= \frac{-b}{a^2+b^2}\mathrm{sech}(A)\cos(B), \\ \gamma^y &= \frac{-b}{a^2+b^2}\mathrm{sech}(A)\sin(B), \\ \gamma^z &= x - \frac{b}{a^2+b^2}(1 + \tanh(A)). \end{aligned}$$

To obtain the high order soliton solution, we firstly give the following lemma

Lemma 4.1 *If $A(\xi)$ possesses the series expansions*

$$A(\xi)=\sum_{n=0}^{+\infty}\gamma_n\xi^n,$$

then we have the following series expansions

$$\frac{A(\xi)\overline{A(\xi)}}{\bar{\lambda_1}-\lambda_1+(\bar{\xi}-\xi)}=\frac{1}{\bar{\lambda_1}-\lambda_1}\sum_{n=0}^{\infty}\sum_{m=0}^{n}\left(\sum_{j=0,i\leqslant j,}^{n}\sum_{i=0,j-i\leqslant n-m}^{m}\frac{(-1)^{n-j-m+i}C_{n-j}^{m-i}\gamma_i\bar{\gamma}_{j-i}}{(\bar{\lambda_1}-\lambda_1)^{n-j}}\right)\xi^m\bar{\xi}^{n-m}. \tag{4.45}$$

Proof Indeed this lemma can be proved by directly calculating. To be convenience for reading, we give the details of calculating:

$$\begin{aligned}
&\frac{A(\xi)\overline{A(\xi)}}{\bar{\lambda_1}-\lambda_1+(\bar{\xi}-\xi)}\\
=&\frac{1}{\bar{\lambda_1}-\lambda_1}\left(\sum_{n=0}^{+\infty}\gamma_n\xi^n\right)\left(\sum_{n=0}^{+\infty}\bar{\gamma}_n\bar{\xi}^n\right)\sum_{k=0}^{+\infty}\left(\frac{\xi-\bar{\xi}}{\bar{\lambda_1}-\lambda_1}\right)^k\\
=&\frac{1}{\bar{\lambda_1}-\lambda_1}\sum_{n=0}^{+\infty}\left(\sum_{k=0}^{n}\gamma_k\bar{\gamma}_{n-k}\xi^k\bar{\xi}^{n-k}\right)\sum_{k=0}^{\infty}\left(\frac{\xi-\bar{\xi}}{\bar{\lambda_1}-\lambda_1}\right)^k\\
=&\frac{1}{\bar{\lambda_1}-\lambda_1}\sum_{n=0}^{+\infty}\sum_{j=0}^{n}\left(\frac{1}{(\bar{\lambda_1}-\lambda_1)^{n-j}}\sum_{l=0}^{j}\gamma_l\bar{\gamma}_{j-l}\xi^l\bar{\xi}^{j-l}\sum_{k=0}^{n-j}(-1)^{n-j-k}C_{n-j}^{k}\xi^k\bar{\xi}^{n-j-k}\right)\\
=&\frac{1}{\bar{\lambda_1}-\lambda_1}\sum_{n=0}^{+\infty}\sum_{j=0}^{n}\\
&\left(\frac{1}{(\bar{\lambda_1}-\lambda_1)^{n-j}}\sum_{m=0}^{n}\left(\sum_{i=0,i\leqslant j,j-i\leqslant n-m}^{m}(-1)^{n-j-(m-i)}C_{n-j}^{m-i}\gamma_i\bar{\gamma}_{j-i}\xi^m\bar{\xi}^{n-m}\right)\right)\\
=&\frac{1}{\bar{\lambda_1}-\lambda_1}\sum_{n=0}^{+\infty}\sum_{m=0}^{n}\left(\sum_{j=0,i\leqslant j,}^{n}\sum_{i=0,j-i\leqslant n-m}^{m}\frac{(-1)^{n-j-(m-i)}C_{n-j}^{m-i}\gamma_i\bar{\gamma}_{j-i}}{(\bar{\lambda_1}-\lambda_1)^{n-j}}\right)\xi^m\bar{\xi}^{n-m}.
\end{aligned}$$

Lemma 4.2 *The expansion*

$$B(\xi)=(\lambda_1+\xi)e^{-\mathrm{i}(\lambda_1+\xi)(x+2(\lambda_1+\xi)t)}=\sum_{i=0}^{\infty}\beta_i\xi^i,\quad \beta_i=\lambda_1\widehat{\beta}_i+\widehat{\beta_{i-1}},$$

$$C(\xi)=(\lambda_1+\xi)e^{\mathrm{i}(\lambda_1+\xi)(x+2(\lambda_1+\xi)t)}\sum_{i=0}^{\infty}\alpha_i\xi^i=\sum_{i=0}^{\infty}\delta_i\xi^i,\ \delta_i=\sum_{k=0}^{i}\alpha_k(\lambda_1\widetilde{\beta}_i+\widetilde{\beta_{i-1}})$$

where

$$\widehat{\beta_i}=\begin{cases} e^{-\mathrm{i}\lambda_1(x+2\lambda_1 t)}\sum\limits_{j=0}^{n}\dfrac{\alpha^{2j}}{(2j)!}\dfrac{(-1)^{n-j}\beta^{n-j}}{(n-j)!}, & k=2n,\quad n\geqslant 0,\\ e^{-\mathrm{i}\lambda_1(x+2\lambda_1 t)}\sum\limits_{j=0}^{n}\dfrac{\alpha^{2j+1}}{(2j+1)!}\dfrac{(-1)^{n-j+1}\beta^{n-j}}{(n-j)!}, & k=2n+1,\end{cases}$$

$$\widetilde{\beta_i}=\begin{cases} e^{-\mathrm{i}\lambda_1(x+2\lambda_1 t)}\sum\limits_{j=0}^{n}\dfrac{\alpha^{2j}}{(2j)!}\dfrac{\beta^{n-j}}{(n-j)!}, & k=2n,\quad n\geqslant 0,\\ e^{-\mathrm{i}\lambda_1(x+2\lambda_1 t)}\sum\limits_{j=0}^{n}\dfrac{\alpha^{2j+1}}{(2j+1)!}\dfrac{\beta^{n-j}}{(n-j)!}, & k=2n+1,\end{cases}$$

$\widehat{\beta_{-1}}=0$, $\widetilde{\beta_{-1}}=0$, $\alpha=\mathrm{i}(x+4\lambda_1 t)$ *and* $\beta=2\mathrm{i}t$.

Proof Simple algebra, we have

$$(\lambda_1+\xi)e^{-\mathrm{i}(\lambda_1+\xi)(x+2(\lambda_1+\xi)t)}=e^{-\mathrm{i}\lambda_1(x+2\lambda_1 t)}(\lambda_1+\xi)\left(\cosh(\alpha\xi)+\sinh(\alpha\xi)\right)e^{\beta\xi^2}.$$

It follows that we can obtain the expansion.

By above two lemmas, we have expansion

$$\begin{aligned}&\frac{B(\xi)\overline{B(\xi)}+C(\xi)\overline{C(\xi)}}{\bar{\lambda_1}-\lambda_1+(\bar{\xi}-\xi)}\\ =&\frac{1}{\bar{\lambda_1}-\lambda_1}\sum_{n=0}^{\infty}\sum_{m=0}^{n}A_{m,n-m}\xi^m\bar{\xi}^{n-m}.\end{aligned}\tag{4.46}$$

where

$$A_{m,n-m}=\left(\sum_{j=0,i\leqslant j,}^{n}\ \sum_{i=0,j-i\leqslant n-m}^{m}\frac{(-1)^{n-j-m+i}C_{n-j}^{m-i}(\beta_i\bar{\beta}_{j-i}+\delta_i\bar{\delta}_{j-i})}{(\bar{\lambda_1}-\lambda_1)^{n-j}}\right)$$

and

$$\begin{aligned}M_{k,m}&=\frac{\partial^{m+k}}{\partial\xi^m\partial\bar{\xi}^k}\left(\frac{B(\xi)\overline{B(\xi)}+C(\xi)\overline{C(\xi)}}{\bar{\lambda_1}-\lambda_1+(\bar{\xi}-\xi)}\right)_{\xi=0}\\ &=\frac{1}{\bar{\lambda_1}-\lambda_1}\sum_{j=0,i\leqslant j,}^{m+k}\ \sum_{i=0,j-i\leqslant k}^{m}(-1)^{k+i-j}C_{m+k-j}^{m-i}\frac{\beta_i\bar{\beta}_{j-i}+\delta_i\bar{\delta}_{j-i}}{(\bar{\lambda_1}-\lambda_1)^{m+k-j}}.\end{aligned}$$

Then we can obtain the following theorem:

Theorem 4.1 *The N-th order soliton solution of L-L equation* (2.5) *and VFE* (1.2) *can be represented as*

$$S^z[N]=1+\mathrm{i}\left(\frac{\det(M^z)}{\det(M)}\right)_x,\quad M^z=(M_{k,m}-\bar{\beta}_k\beta_m)_{1\leqslant k,m\leqslant N},\ M=(M_{k,m})_{1\leqslant k,m\leqslant N},$$
$$S^-[N]=\mathrm{i}\left(\frac{\det(M^-)}{\det(M)}\right)_x,\quad M^-=(M_{k,m}-\bar{\delta}_k\beta_m)_{1\leqslant k,m\leqslant N}. \tag{4.47}$$

and

$$\begin{aligned}\gamma^x[N]&=\mathrm{Re}\left(\mathrm{i}\frac{\det(M^-)}{\det(M)}-\mathrm{i}\right),\\ \gamma^y[N]&=\mathrm{Im}\left(\mathrm{i}\frac{\det(M^-)}{\det(M)}-\mathrm{i}\right),\\ \gamma^z[N]&=x+\mathrm{i}\left(\frac{\det(M^z)}{\det(M)}-1\right).\end{aligned} \tag{4.48}$$

respectively.

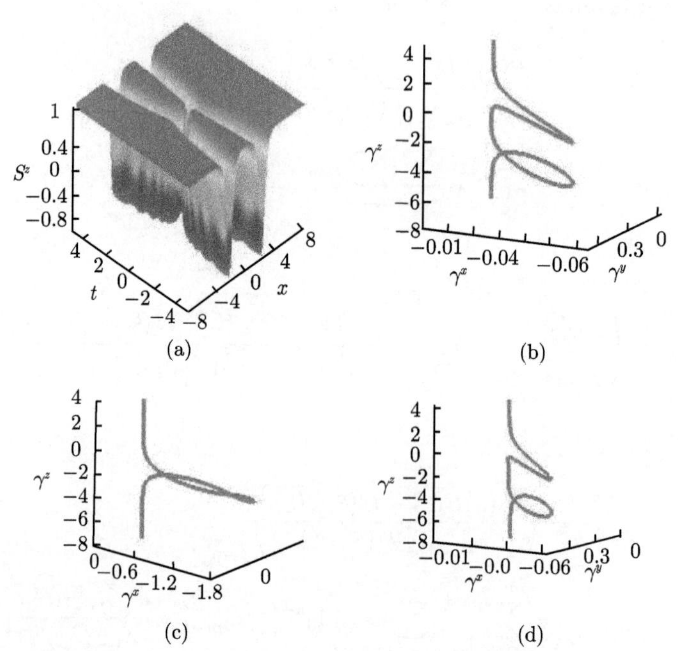

Figure 1 Second order soliton for L-L equation S^z and VFE γ^x, γ^y and γ^z. Parameters (a) $\lambda_1=\mathrm{i}$, $\gamma_1=1$, $\beta_{1,1}=0$. (b) $t=-4\pi$, (c) $t=0$, (d) $t=4\pi$.

By above formula, we can readily obtain the high order soliton solution for VFE (1.2) equation. Specially, we take parameters $\lambda_1 = bi$, $\gamma = 1$ and $\beta_{1,1} = c + id$, then we can obtain the second order soliton solution for L-L equation (1.1) and VFE (1.2):

$$\gamma^x = \frac{-4\left[\cosh(2bx) + b(2x+d)\sinh(2bx)\right]\cos(4b^2t) - 4b(8bt-c)\cosh(2bx)\sin(4b^2t)}{2b^2(d+2x)^2 + 2b^2(8bt-c)^2 + \cosh(4bx) + 1},$$

$$\gamma^y = \frac{-4\left[\cosh(2bx) + b(2x+d)\sinh(2bx)\right]\sin(4b^2t) + 4b(8bt-c)\cosh(2bx)\cos(4b^2t)}{2b^2(d+2x)^2 + 2b^2(8bt-c)^2 + \cosh(4bx) + 1},$$

$$\gamma^z = x - \frac{2}{b}\left(1 + \frac{\sinh(4bx) - 2b(d+2x)}{2b^2(d+2x)^2 + 2b^2(8bt-c)^2 + \cosh(4bx) + 1}\right),$$

$S^x = \gamma^x_x$, $S^y = \gamma^y_x$, $S^z = \gamma^z_x$.

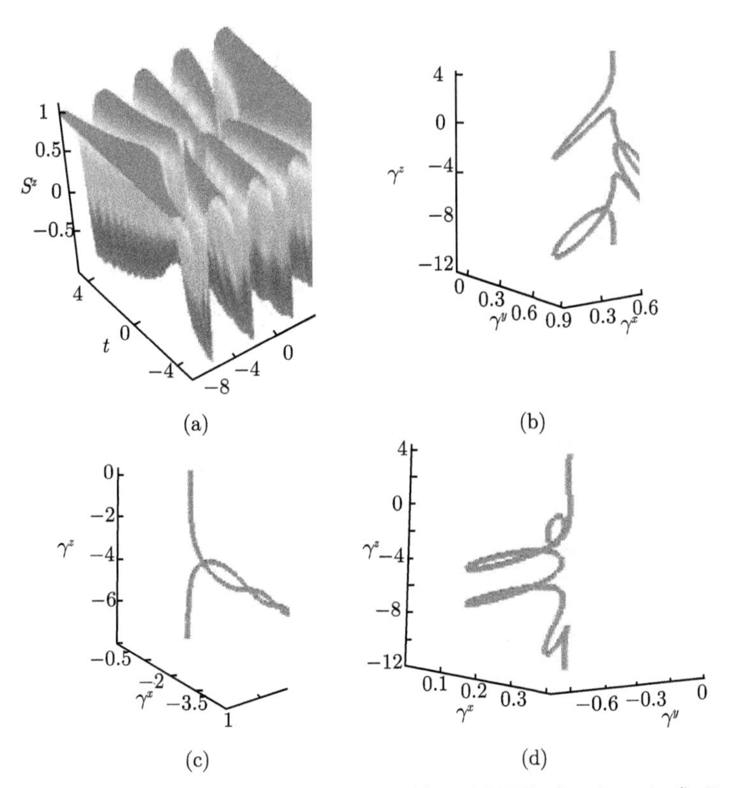

Figure 2 Forth order soliton for L-L equation S^z and VFE γ^x, γ^y and γ^z. Parameters (a) $\lambda_1 = \mathrm{i}$, $\gamma_1 = 1$, $\beta_{1,1} = 0$, $\beta_{1,2} = 0$ and $\beta_{1,3} = 0$. (b) $t = -4\pi$, (c) $t = 0$, (d) $t = 4\pi$.

Finally, we give the dynamics of L-L equation (1.1) and VFE (1.2) by plotting picture. The second order soliton solution and forth order solution are showed with special parameters (Fig. 1 and Fig. 2). It is seen that high order soliton possesses the similar structure as multi-soliton solution. The difference is that the velocity of high order soliton is no longer a constant.

5 Conclusion and discussion

In conclusion, we analysis the L-L equation by inverse scattering method and generalized Darboux transformation. The generalized Darboux transformation [29,30] is general version for the Darboux transformation in [9, 52]. These results are self-contained. Besides, we remark that the general soliton solution for VFE (1.2) can readily obtain by our formula (3.40). We intend to research the long time asymptotics of high order soliton in a subsequent publication.

Acknowledgements This work is supported by National Natural Science Foundation of China (Contact No. 11271052).

References

[1] Ablowitz M, Kaup D, Newell A, Segur H. The inverse scattering transform-Fourier analysis for nonlinear problems [J]. Stud. Appl. Math., 1974, 53: 249–315.

[2] Ablowitz M, Prinari B, Trubatch A. Discrete and Continuous Nonlinear Schrödinger Systems [M]. UK: Cambridge University Press, 2004.

[3] Aktosun T, Demontis F, van der Mee C. Exact solutions to the focusing nonlinear Schrödinger equation [J]. Inverse Problems, 2007, 23: 2171.

[4] R. J. Arms and F. R. Hama, Localized-induction concept on a curved vortex and motion of an elliptic vortex ring [J]. Phys. Fluids, 1965, 8/4: 553.

[5] Beals R, Coifman R R. Scattering and inverse scattering for first order systems [J]. Comm. Pure Appl. Math., 1984, 37: 39–90.

[6] Bezmaternih G V. Exact solutions of the sine-gordon and landau-lifshitz equations: rational-exponential solutions [J]. Phys. Lett. A, 1990, 146: 492–495.

[7] Calini A, Keith S F, Lafortune S. Squared eigenfunctions and linear stability properties of closed vortex filaments [J]. Nonlinearity, 2011, 24: 3555–3583.

[8] Chorin A J. Equilibrium statistics of a vortex filament with applications [J]. Commun. Math. Phys., 1991, 141: 619–631.

[9] Ciéliński J. Algebraic construction of the Darboux matrix revisited [J]. J. Phys. A: Math. Theor., 2009, 42: 404003.

[10] Chang N, Shatah J, Uhlenbeck K. Schröinger Maps [J]. Comm. Pure Appl. Math., 2000, 53: 590–602.

[11] Cohen A, Kappeler T. Solutions to the cubic Schödinger equation by the inverse scattering method [J]. SIAM J. Math. Anal., 1992, 23: 900–922.

[12] Crabtree G W, Nelson D R. Vortex physics in high-temperature superconductors [J]. Physics Today, 1997, 50: 38–45.

[13] Deift P, Park J. Long-time asymptotics for solutions of the NLS equation with a delta potential and even initial data [J]. IMRN, 2011, 24: 5505–5624.

[14] Deift P, Trubowitz E. Inverse Scattering on the Line [J]. Comm. Pure Appl. Math., 1979, 32: 121–251.

[15] Deift P, Zhou X. A steepest descent method for oscillatory Riemann Hilbert problems. Asymptotics for the MKdV equation [J]. Ann. of Math., 1993, 137: 295–368.

[16] Deift P, Zhou X. Long-time asymptotics for integrable systems. High order theory [J]. Commun. Math. Phys., 1994, 165: 175–191.

[17] Deift P, Zhou X. Perturbation theory for infinite-dimensional integrable systems on the line. A case study [J]. Acta Math., 2002, 188: 163–262.

[18] Deift P, Zhou X. A priori L^p estimates for solutions of Riemann-Hilbert problems [J]. IMRN., 2002, 40: 2121–2154.

[19] Deift P, Zhou X. Long-time asymptotics for solutions of the NLS equation with initial data in a weighted Sobolev space [J]. Comm. Pure Appl. Math., 2003, 56: 1029–1077.

[20] Faddeev L D, Takhtajan L A. Hamiltonian Methods in the Theory of Solitons [M]. Berlin-New York: Springer-Verlag, 1987.

[21] Fordy A P, Kulish P P. Nonlinear Schrödinger equations and simple Lie algebras [J]. Commun. Math. Phys., 1983, 89: 427–443.

[22] Fokas A S. Integrable nonlinear evolution equations on the half-line [J]. Commun. Math. Phys., 2002, 230: 1–39.

[23] Fokas A S, Its A R. The linearization of the initial-boundary value problem of the nonlinear Schödinger equation [J]. SIAM J. Math. Anal., 1996, 27: 738–764.

[24] Gagnon L, Stièvenart N. N-soliton interaction in optical fibers: the multiple-pole case [J]. Opt. Lett., 1986, 11: 19.

[25] Gardner C S, Greene J M, Kruskal M D, Miura R M. Method for solving the Kortcmeg-deVries equation [J]. Phys. Rev. Lett., 1967, 19: 1095–1097.

[26] Gesztesy F, Karwowski W, Zhao Z. Limits of soliton solutions [J]. Duke Math. J., 1992, 68: 101–150.

[27] Gohberg I, Kaashoek M A, Sakhnovich A. Pseudo-canonical systems with rational Weyl functions: explicit formulas and applications [J]. J. Diff. Equa., 1998, 146: 375–398.

[28] Gu C H, Hu H S, Zhou Z X. Darboux Transformation in Soliton Theory, and its Geometric Applications [M]. Shanghai: Shanghai Science and Technology Publishers, 2005.

[29] Guo B, Ling L, Liu Q P. Nonlinear Schrödinger equation: Generalized Darboux transformation and rogue wave solutions [J]. Phys. Rev. E, 2012, 85: 026607.

[30] Guo B,Ling L, Liu Q P. High-Order Solutions and Generalized Darboux Transformations of Derivative Nonlinear Schrödinger Equations [J]. Stud. Appl. Math., 2013, 130: 317–344.

[31] Guo B, Ling L. Riemann-Hilbert approach and N-soliton formula for coupled derivative Schrödinger equation [J]. J. Math Phys., 2012, 53: 073506.

[32] Guo B, Ling L. Landau-Lifshitz equation: Riemann-Hilbert method and rogue wave [J]. 2013, Preprint.

[33] Krüger H, Teschl G. Long-time asymptotics of the Toda lattice for decaying initial data revisited [J]. Rev. Math. Phys., 2009, 21: 61–109.

[34] Grunert K, Teschl G. Long-Time asymptotics for the Korteweg C de Vries equation via nonlinear steepest descent [J]. Math. Phys. Anal. Geom., 2009, 12: 287–324.

[35] Kamvissis S. Focusing nonlinear Schrödinger equation with infinitely many solitons [J]. J. Math. Phys., 1995, 36: 4175.

[36] Kamvissis S. Long Time Behavior for the Focusing Nonlinear Schroedinger Equation with Real Spectral Singularities [J]. Commun. Math. Phys., 1996, 180: 325–341.

[37] Landau L, Lifshits E. On the theory of the dispersion of magnetic permeability in ferromagnetic bodies [J]. Phys. Zeitsch. der Sow., 1935, 8: 153–169.

[38] Lee J. Analytic properties of zakharov-shabat inverse scattering problem with a polynomial [D]. Dissertation, Yale University, 1983.

[39] Matveev V B, Salle M. Darboux transformation and solitons [M]. Berlin: Springer-Verlag, 1991.

[40] Mennichen R, Sakhnovich A, Tretter C. Direct and inverse spectral problem for a system of differential equations depending rationally on the spectral parameter [J]. Duke Math. J., 2001, 109: 413–449.

[41] Mizumachi T, Pelinovsky D. Bäcklund Transformation and L^2-stability of NLS Solitons [J]. IMRN, 2012, 9: 2034–2067.

[42] Ricca R L. Rediscovery of the Da Rios equations [J]. Nature, 1991, 352: 561.

[43] Sakhnovich A. Iterated Bäcklund-Darboux transform for canonical system [J]. J. Func. Anal., 1997, 144: 359–370.

[44] Samuels D C. Vortex filament methods for superfluids Quantized Vortex Dynamics and Superfluid Turbulence [M]. (Lecture Notes in Physics vol 571) (Berlin: Springer) pp 97-113, 2001.

[45] Sattinger D, Zurkowski V. Gauge theory of Bäcklund transformations. II [J]. Physica D., 1987, 26: 225–250.

[46] Schuur P C. Asymptotic analysis of soliton problems. An inverse scattering approach [M]. Lecture Notes in Mathematics, 1232. Berlin: Springer-Verlag, 1986.

[47] Shabat A. One dimensional perturbations of a differential operator and the inverse scattering problem [M]. Problems in Mechanics and Mathematical Physics, 279–296, Nauka, Moscow, 1976.

[48] Shchesnovich V S, Yang J. Higher-order solitons in the N-wave system [J]. Stud. Appl. Math., 2003, 110: 297.

[49] Shchesnovich V S, Yang J. General soliton matrices in the Riemann-Hilbert problem for integrable nonlinear equations [J]. J. Math. Phys., 2003, 44: 4604.

[50] She Z S, Jackson E, Orszag S A. Intermittent vortex structures in homogeneous isotropic turbulence [J]. Nature, 1990, 344: 226–228.

[51] Takhtajan L A. Integration of the continous Heisenberg spin chain through the inverse scattering method [J]. Phys. Lett. A, 1977, 64: 235–237.

[52] Terng C L, Uhlenbeck K. Bäcklund transformations and loop group actions [J]. Comm. Pure Appl. Math., 2000, 53: 1–75.

[53] Zhou X. The Riemann-Hilbert problem and inverse scattering [J]. SIAM J. Math. Anal., 1989, 20: 966–986.

[54] Zhou X. Direct and inverse scattering transforms with arbitrary spectral singularities [J]. Comm. Pure Appl. Math., 1990, 42: 895–938.

[55] Zhou X. L^2-Sobolev space bijectivity of the scattering and inverse scattering transforms [J]. Comm. Pure Appl. Math., 1998, 51: 697–731.

[56] Zakharov V E, Takhtajan L A. Equivalence of the nonlinear Schödinger equation and the equation of a Heisenberg ferromagnet [J]. Theor. Math. Phys., 1979, 38: 17–23.

Random Attractor and Stationary Measure for Stochastic Long-short Wave Equations*

Li Donglong (李栋龙), Guo Yanfeng (郭艳凤) and Guo Boling (郭柏灵)

Abstract Asymptotic behaviors of stochastic long–short equations driven by random force, which is smooth enough in space and white noise in time, are mainly considered. The existence and uniqueness of solutions for stochastic long–short equations are obtained via Galerkin approximation by the stopping time and Borel–Cantelli Lemma on the basis of *a priori* estimates in the sense of expectation. A global random attractor and the existence of a stationary measure are investigated by Birkhoff ergodic theorem and Chebyshev inequality.

Keywords Stochastic long-short equations; Existence and uniqueness; Global random attractor; Stationary measure

1 Introduction and main results

Long–short wave equations describe the resonance interaction between the long wave and the short wave, which are first derived by Djordjevic and Redekop in [5]. The long–short wave equations can be written as

$$\mathrm{i}u_t + u_{xx} - nu = 0, \tag{1.1}$$

$$n_t + (|u|^2)_x = 0, \tag{1.2}$$

where u is the envelope of short wave and complex, n is amplitude of long wave and real. The physical significance of this system is represented by the dispersion of the short wave balanced by nonlinear interaction of the long wave with short wave. This system also appears in analysis of internal wave in [12] and in plasma physics, which describes the resonance of high–frequency electron plasma oscillation and the associated low–frequency ion density perturbation as in [17]. Due to its important

* Commun. Math. Sci. , 2015, 13(2): 539 – 555.

physical and mathematical properties, the long–short equations have drawn greatly attention of many mathematicians and physicists and have been quite extensively studied in theory. The existence of global solution for long–short wave equations and generalized long–short wave equations are obtained by Guo in [8] and [9], respectively. The orbital stability of solitary waves and the existence of a global attractor have been studied in [6,10,11,15,22].

It is well known that stochastic partial differential equations(SPDEs) play an important role in understanding the dynamics of many interesting phenomena. Recently, the importance of taking random effects into account in modeling, analyzing, simulating and predicting complex phenomena has been widely recognized in geophysical and climate dynamics, materials science, chemistry, biology and other areas, see [7,13]. SPDEs are more realistic mathematical models for complex systems under random environmental effects or errors of measurement. The existence and uniqueness of the solution and attractors for SPDEs have been studied by many authors, see [1,2,20,21].

As is well known, there are mainly two sources of random effects coming from the random environmental effects and the errors of measurement. Hence, Random effects now have a broad rang of applications. In particular, in 1960 Kalman and in 1961 Kalman and Bucy proved what is now known as the Kalman–Bucy filter which is an example of a recent mathematical discovery and gives some random procedure. It now has a wide application[18]. Therefore, when the random environmental effects or the errors of measurement are included into the model of long–short wave equations, the stochastic long–short wave equations can be obtained. So far, there are few papers considering it. Hence, it is significant to study the stochastic long–short wave equations.

In this paper, we consider the following stochastic long–short equations in $\mathbf{R}^1$ on a regular domain $D=[0,L]$ in Itô sense

$$n_t+(|u|^2)_x+\delta n=f+\dot{W}_1, \tag{1.3}$$

$$\mathrm{i}u_t+u_{xx}-nu+\mathrm{i}\alpha u=g+\dot{W}_2, \tag{1.4}$$

where W_1 and W_2 are independent $L^2(D)$ value Wiener processes which come from the random environmental effects or the errors of measurement and will be detailed in the next section.

There are two points being worthy of note in this paper. Firstly, it is well known that it is difficult to find some technical and tricky issues to treat for the higher dimensional noises. We consider only one dimensional noise term that is a natural choice for some particular application. For the higher dimensional noises, the investigation will be further considered in the future. Secondly, in process of the study, the differences to the deterministic literature are to use the Itô formula and the some stochastic inequalities and obtain the estimates in the meaning of the expectation. Therefore, we need more skills to deal with the difficulties coming from the coupled nonlinear terms in the stochastic literature than in the deterministic literature for the groups of equations.

In Section 3, some techniques and Itô formula are used for the coupled nonlinear terms in the meaning of the expectation. The maximal estimates for stochastic integral are important, see Lemma 2.2. Random dynamical system of (1.3)–(1.4) is investigated in Section 4. The existence and uniqueness of the solutions for the stochastic long–short equations via the standard Galerkin approximation method by the stopping time and Borel–Cantelli Lemma on the basis of *a priori* estimates in different investigated spaces V_1 and V_2 are given. Usually, Galerkin method for SPDEs leads to martingale solutions instead of weak solutions. However, in this approach we transform (1.3)–(1.4) into partial differential equations with random coefficient. Then by *a priori* estimates obtained in Section 3, we construct full probability measure set $\tilde{\Omega}$. So the classic Galerkin approximation can be applied to obtain a unique solution for (1.3)–(1.4). This method has been applied to the stochastic porous medium equation in [14]. Now we give the main results of this paper. First, the existence and uniqueness of solutions for (1.3)–(1.4) is given as follows.

Theorem 1.1 *If* $(n_0, u_0) \in V_1$, $q_1 \in H^1(D), q_2 \in H^2(D), f \in H^1(D)$ *and* $g \in H^1(D)$. *Then there exists a unique solution* $(n, u) \in L^\infty(\mathbf{R}^+; V_1)$ *almost surely for equations* (1.3)–(1.4). *Moreover* (n, u) *is continuous from* $\mathbf{R}^+$ *to* V_1.

Here, q_1 and q_2 are coefficients of standard complex and real valued Wiener process, which are the parts of the Wiener process terms in (1.3)–(1.4) and are given in detail in Section 2. Similar to the Theorem 1.1, using the same method, we can obtain the corresponding results in V_2 as follows.

Theorem 1.2 *If* $(n_0, u_0) \in V_2$, $q_1 \in H^2(D), q_2 \in H^3(D), f, g \in H^2(D)$. *Then there exists a unique solution* $(n, u) \in L^\infty(\mathbf{R}^+; V_2)$ *almost surely for equations* (1.3)–(1.4). *Moreover* (n, u) *is continuous from* $\mathbf{R}^+$ *to* V_2.

In fact, by Theorem 1.1 and 1.2, we can define a continuous random dynamical system in V_1 and V_2, respectively. Then a random attractor endowed with weak topology can be constructed for the continuous random dynamical system in V_1 and V_2, respectively.

Theorem 1.3 *If* $(n_0, u_0) \in V_1$, $q_1 \in H^1(D)$, $q_2 \in H^2(D)$, *and* $f, g \in H^1(D)$. *Then* (1.3)–(1.4) *have a global random weak attractor* $\mathcal{A}(\omega)$ *which is a random tempered compact set in* V_1 *endowed with weak topology.*

Theorem 1.4 *If* $(n_0, u_0) \in V_2$, $q_1 \in H^2(D)$, $q_2 \in H^3(D)$, *and* $f, g \in H^2(D)$. *Then* (1.3)–(1.4) *have a global random weak attractor* $\mathcal{A}(\omega)$ *which is a random tempered compact set in* V_2 *endowed with weak topology.*

Although the random attractor is constructed in weak topology, we can prove the existence of a stationary measure in some phase spaces endowed with strong topology. But we cannot prove the existence of a random weak attractor in V_2 endowed with usual topology for the existence of noise term. By random dynamics theory, the existence of a random attractor with usual topology yields the existence of one stationary measure. However, the random attractor is obtained in weak topology of V_2 which does not yield the existence of a stationary measure in V_2. But in the following we can construct a stationary measure in V_2.

Theorem 1.5 *If* $(n_0, u_0) \in V_2$, $q_1 \in H^2(D), q_2 \in H^4(D)$, *then* (1.3)–(1.4) *have one stationary measure on* V_1 *and* V_2.

The spaces V_1 and V_2 can be seen below. The paper is organized as follows. In Section 2, we give some functional sets for the problem and some conditions. In Section 3, we establish a series of *a* time uniform *priori* estimates in different energy spaces which are the key step for solving the problem. The Section 4 is devoted to proofs of Theorems 1.1 and 1.2. In Section 5, The proofs of Theorems 1.3, 1.4 and 1.5 will be given. In the paper the letters c and C are generic positive constants independent of T which may change their values from term to term. In addition, c_T and C_T are generic positive constants dependent on T which also may change the values from term to term.

2 Preliminaries

We denote the bounded set by $D = [0, L]$ in $\mathbf{R}^1$. Consider the stochastic long–short equations (1.3)–(1.4) on D with Dirichlet boundary condition $u(0,t) = u(L,t) = 0, n(0,t) = n(L,t) = 0$ and initial condition $u(x,0) = u_0(x), n(x,0) = n_0(x), x \in D$, here $\delta > 0, \alpha > 0$. Now we introduce some functional spaces. The scalar product on $L^2(D)$ will be denoted by $(u,v) = \int_D u(x)\overline{v(x)}\,\mathrm{d}x$, and the norm by $\|u\|^2 = (u,u)^{\frac{1}{2}} = \left(\int_D |u|^2\,\mathrm{d}x\right)^{\frac{1}{2}}$. And that the general p–norm of $L^p(D)(p \geqslant 1)$ is denoted by $\|u\|_{L^p} = \left(\int_D |u|^p\,\mathrm{d}x\right)^{\frac{1}{p}}$. The norm in $H^s(D)$ (s is a nonnegative integer in normal Sobolev spaces) is denoted by $\|u\|_{H^s} = \left(\sum_{0\leqslant|\alpha|\leqslant s}\|D^\alpha u\|^2\right)^{\frac{1}{2}}$. The space $H_0^s(D)$ is the set of $H_0^s(D) = \{u : \|u\|_{H^s(D)} < \infty, u(0,t) = u(L,t) = 0\}$. It is well known that the norm of u in $H_0^s(D)$ that is $\|u\|_{H_0^s}$, is equivalent of the norm of u in $H^s(D)$ in bounded domain D. We define $H^{-s}(D)$ is the dual space of $H^s(D)$. We know that the Poincaré inequality $\|u\| \leqslant \lambda_1^{-\frac{1}{2}}\|\nabla u\|$ can be used for bounded domain D with $\lambda_1 = \dfrac{4\pi^2}{L^2}$.

Here we give a complete probability space $(\Omega, \mathcal{F}, \{\mathcal{F}_t\}_{t\geqslant 0}, \mathbf{P})$. The expectation operator with respect to $\mathbf{P}$ is denoted by $\mathbf{E}$. Stochastic terms $W_1(t)$ and $W_2(t)$ are defined on $(\Omega, \mathcal{F}, \{\mathcal{F}_t\}_{t\geqslant 0}, \mathbf{P})$ by

$$W_1(t) = q_1(x)\omega_1(t), W_2(t) = q_2(x)\omega_2(t),$$

where $\omega_1(t)$ is a standard real valued Wiener process, $\omega_2(t)$ is a standard complex valued Wiener process which is independent of $\omega_1(t)$, and $q_1(x)$, $q_2(x)$ are sufficiently smooth functions in some sense.

We also define the different product spaces for the solution (n,u) of the (1.3)–(1.4)

$$\begin{aligned} V_0 &= L^2(D) \times H_0^1(D), \\ V_1 &= H_0^1(D) \times (H^2(D) \cap H_0^1(D)), \\ V_2 &= (H^2(D) \cap H_0^1(D)) \times \{\varphi \in H^3(D) \cap H_0^1(D), \varphi_{xx} \in H_0^1(D)\}. \end{aligned}$$

Endow each $V_i (i = 0,1,2)$ with the usual norm and we have $V_2 \subset V_1 \subset V_0$ with compact embeddings.

Let $(\mathcal{X}, \|\cdot\|_{\mathcal{X}}) \subset (\mathcal{Y}, \|\cdot\|_{\mathcal{Y}}) \subset (\mathcal{Z}, \|\cdot\|_{\mathcal{Z}})$ be three Banach reflective spaces and $\mathcal{X} \subset \mathcal{Y}$ with compact and dense embedding. Define Banach space

$$\mathcal{G} = \{v : v \in L^2(0,T;\mathcal{X}), \frac{\mathrm{d}v}{\mathrm{d}t} \in L^2(0,T;\mathcal{Z})\}$$

with norm

$$\|v\|_{\mathcal{G}}^2 = \int_0^T \|v\|_{\mathcal{X}}^2\,\mathrm{d}s + \int_0^T \|\frac{\mathrm{d}v}{\mathrm{d}t}\|_{\mathcal{Z}}^2\,\mathrm{d}s, v \in \mathcal{G}.$$

We have the following lemma about compactness result [16].

Lemma 2.1 *If K is bounded in $\mathcal{G}$, then K is precompact in $L^2(0,T;\mathcal{Y})$.*

Another lemma is needed for some maximal estimates on stochastic integral. Assume U and H are separable Hilbert spaces and W is a Q–Wiener process on U_0 with $U_0 = Q^{\frac{1}{2}}U$. Let $L_2^0 = L_2^0(U_0, H)$ be the space of Hilbert–Schmidt operators from U_0 to H. For such operator we have the following results in [3].

Lemma 2.2 *For any $r \geqslant 1$ and any L_2^0–valued predictable process $\Phi(t), t \in [0,T]$, we have*

$$\boldsymbol{E}\int_0^t \Phi(s)\,\mathrm{d}W(s) = 0$$

and

$$\begin{aligned}\boldsymbol{E}\left(\sup_{s\in[0,t]}\left|\int_0^s \Phi(\sigma)\,\mathrm{d}W(\sigma)\right|^{2r}\right) &\leqslant c_r \sup_{s\in[0,t]} \boldsymbol{E}\left(\left|\int_0^s \Phi(\sigma)\,\mathrm{d}W(\sigma)\right|^{2r}\right)\\ &\leqslant C_r \boldsymbol{E}\left(\int_0^t \|\Phi(s)\|_{L_2^0}^2\,\mathrm{d}s\right)^r, \end{aligned} \tag{2.1}$$

where c_r and C_r are some positive constants dependent on r.

3 Time uniform *a* priori estimates

3.1 A priori estimates in V_0

Lemma 3.1 *Assume that $u_0, g, q_2 \in L^2(D)$. Then for any $T > 0$ and $p \geqslant 1$, we have*

$$u \in L^{2p}(\Omega; L^\infty(0,T; L^2(D))) \cap L^\infty(0,\infty; L^{2p}(\Omega; L^2(D))). \tag{3.1}$$

Proof Applying Itô formula to $\|u\|^2$, one gets

$$\frac{\mathrm{d}}{\mathrm{d}t}\|u\|^2 = -2\alpha\|u\|^2 + 2\mathrm{Im}\int_D g\bar{u}\,\mathrm{d}x + 2\mathrm{Im}\int_D \bar{u}\dot{W}_2\,\mathrm{d}x + \|q_2\|^2. \tag{3.2}$$

From (3.2), using Hölder and Young's inequality, we can obtain

$$\frac{\mathrm{d}}{\mathrm{d}t}\|u\|^2 + \alpha\|u\|^2 \leqslant \frac{1}{\alpha}\|g\|^2 + \|q_2\|^2 + 2\mathrm{Im}\int_D \bar{u}\dot{W}_2\,\mathrm{d}x. \tag{3.3}$$

Then multiplying $\mathrm{e}^{\alpha t}$ and integrating from 0 to t on the both sides of (3.3) and taking the expectation, we can get

$$\mathbf{E}\|u\|^2 \leqslant \mathrm{e}^{-\alpha t}\mathbf{E}\|u_0\|^2 + \frac{1}{\alpha^2}\|g\|^2 + \frac{1}{\alpha}\|q_2\|^2 \leqslant C, \qquad t > 0, \tag{3.4}$$

where C is independent of T.

On the other hand, integrating from 0 to t and taking the supremum and the expectation on the both sides of (3.3), we can obtain

$$\begin{aligned}\mathbf{E}\sup_{0\leqslant t\leqslant T}\|u\|^2 \leqslant &\mathbf{E}\|u_0\|^2+\left(\frac{1}{\alpha}\|g\|^2+\|q_2\|^2\right)T\\&+\mathbf{E}\sup_{0\leqslant t\leqslant T}|\int_0^t \mathrm{Im}\int_D \bar{u}\dot{W}_2\,\mathrm{d}x\mathrm{d}s|^2+1.\end{aligned}\tag{3.5}$$

By Lemma 2.2, for some positive constant C, we have

$$\mathbf{E}\sup_{0\leqslant t\leqslant T}|\int_0^t \mathrm{Im}\int_D \bar{u}\dot{W}_2\,\mathrm{d}x\mathrm{d}s|^2\leqslant C\|q_2\|^2\mathbf{E}\int_0^T\|u\|^2\,\mathrm{d}s.\tag{3.6}$$

So by (3.4), for any $T>0$, there exists a positive constant C_T such that

$$\begin{aligned}\mathbf{E}\sup_{0\leqslant t\leqslant T}\|u\|^2 &\leqslant \mathbf{E}\|u_0\|^2+\left(\frac{1}{\alpha}\|g\|^2+\|q_2\|^2\right)T+C\|q_2\|^2\mathbf{E}\int_0^T\|u\|^2\,\mathrm{d}s+1\\&\leqslant C_T(\mathbf{E}\|u_0\|^4+\|g\|^4+\|q_2\|^4+1).\end{aligned}\tag{3.7}$$

By the above estimates we can further give an estimate of $\|u(t)\|_0^{2p}$ for any $p\geqslant 1$. Now applying Itô formula and Hölder inequality, we have

$$\frac{\mathrm{d}}{\mathrm{d}t}\|u\|^{2p}\leqslant -\frac{\alpha p}{2}\|u\|^{2p}+c(\|g\|^{2p}+\|q_2\|^{2p})+2p\|u\|^{2(p-1)}\mathrm{Im}\int_D\bar{u}\dot{W}_2\,\mathrm{d}x.\tag{3.8}$$

Multiplying $\mathrm{e}^{\frac{\alpha p}{2}t}$ and integrating from 0 to t and taking the expectation on the both sides of (3.8) yields

$$\mathbf{E}\|u\|^{2p}\leqslant \mathrm{e}^{-\frac{\alpha p}{2}t}\mathbf{E}\|u_0\|^{2p}+c(\|g\|^{2p}+\|q_2\|^{2p})\leqslant C,\qquad t>0,\tag{3.9}$$

where C is independent of T. On the other hand, from (3.8) we have

$$\frac{\mathrm{d}}{\mathrm{d}t}\|u\|^{2p}\leqslant c(\|g\|^{2p}+\|q_2\|^{2p})+2p\|u\|^{2(p-1)}\mathrm{Im}\int_D\bar{u}\dot{W}_2\,\mathrm{d}x.\tag{3.10}$$

Integrating from 0 to t for above inequality and taking the supremum and the expectation, we can obtain

$$\begin{aligned}\mathbf{E}\sup_{0\leqslant t\leqslant T}\|u(t)\|^{2p}\leqslant &\mathbf{E}\|u(0)\|^{2p}+c(\|g\|^{2p}+\|q_2\|^{2p})T+1\\&+\mathbf{E}\sup_{0\leqslant t\leqslant T}\left|\int_0^t p\|u\|^{2(p-1)}\mathrm{Im}\int_D\bar{u}\dot{W}_2\,\mathrm{d}x\mathrm{d}s\right|^2.\end{aligned}\tag{3.11}$$

By Lemma 2.2, for some positive constant C, we obtain

$$\mathbf{E}\sup_{0\leqslant t\leqslant T}\left|\int_0^t p\|u\|^{2(p-1)}\mathrm{Im}\int_D \bar{u}\dot{W}_2\,\mathrm{d}x\mathrm{d}s\right|^2 \leqslant Cp^2\|q_2\|^2\mathbf{E}\int_0^T \|u\|^{4p-2}\,\mathrm{d}s, \tag{3.12}$$

where $2p-1\geqslant 1$. Then inserting (3.12) into (3.11) and by (3.9), for any $T>0$ we have

$$\mathbf{E}\sup_{0\leqslant t\leqslant T}\|u\|^{2p}\leqslant C_T(u_0,g,q_2) \tag{3.13}$$

where $C_T(u_0,g,q_2)$ is a positive constant depending on u_0,g,q_2 and T.

Lemma 3.2 *Assume that* $(n_0,u_0)\in V_0$, $q_1\in L^2(D), q_2\in H^1(D), f,g\in L^2(D)$. *Then for any* $T>0$ *and* $p\geqslant 1$, *we have* $(n,u)\in L^{2p}(\Omega;L^\infty(0,T;V_0))\cap L^\infty(0,\infty;L^{2p}(\Omega;V_0))$.

Proof Applying Itô formula to $\|n\|^2$, since $n_t=-(|u|^2)_x-\delta n+f+\dot{W}_1$, we can obtain

$$\frac{\mathrm{d}}{\mathrm{d}t}\|n\|^2-\|q_1\|^2+2\int_D n(|u|^2)_x\,\mathrm{d}x+2\delta\|n\|^2-2\int_D fn\,\mathrm{d}x-2\int_D n\dot{W}_1\,\mathrm{d}x=0. \tag{3.14}$$

Noticing that

$$\frac{\mathrm{d}}{\mathrm{d}t}\int_D \mathrm{i}(u_x\bar{u}-u\bar{u}_x)\mathrm{d}x=2\int_D \mathrm{i}(u_x\bar{u}_t-\bar{u}_xu_t)\mathrm{d}x+2\mathrm{Im}\int_D q_2\bar{q}_{2x}\,\mathrm{d}x,$$

from (3.14), we can obtain

$$\begin{aligned}&\frac{\mathrm{d}}{\mathrm{d}t}\left(\|n\|^2+\int_D \mathrm{i}(u_x\bar{u}-u\bar{u}_x)\,\mathrm{d}x\right)+2\delta\|n\|^2\\ =&-2\int_D \mathrm{i}\alpha(u_x\bar{u}-\bar{u}_xu)\,\mathrm{d}x+2\int_D fn\,\mathrm{d}x+4\mathrm{Re}\int_D u\bar{g}_x\,\mathrm{d}x\\ &+2\mathrm{Im}\int_D q_2\bar{q}_{2x}\,\mathrm{d}x+\|q_1\|^2+4\mathrm{Re}\int_D u\bar{\dot{W}}_{2x}\,\mathrm{d}x+2\int_D n\dot{W}_1\,\mathrm{d}x.\end{aligned} \tag{3.15}$$

Applying Itô formula to $\|u_x\|^2$, and taking the inner for (1.4) with $u_t+\alpha u$, we can obtain

$$\begin{aligned}&\frac{\mathrm{d}}{\mathrm{d}t}\|u_x\|^2+2\alpha\|u_x\|^2+2\mathrm{Re}\int_D nu\bar{u}_t\,\mathrm{d}x+2\alpha\int_D n|u|^2\,\mathrm{d}x\\ &+2\mathrm{Re}\int_D g\bar{u}_t\,\mathrm{d}x+2\alpha\mathrm{Re}\int_D g\bar{u}\,\mathrm{d}x+2\mathrm{Re}\int_D(\bar{u}_t+\alpha\bar{u})\dot{W}_2\,\mathrm{d}x=\|q_{2x}\|^2.\end{aligned} \tag{3.16}$$

Notice that

$$\frac{\mathrm{d}}{\mathrm{d}t}\int_D n|u|^2\,\mathrm{d}x = \int_D n_t|u|^2\,\mathrm{d}x + 2\mathrm{Re}\int_D nu_t\bar{u}\,\mathrm{d}x + 2\int_D q_1\mathrm{Im}(q_2\bar{u})\,\mathrm{d}x.$$

In addition, since $n_t = -((|u|^2)_x + \delta n - f - \dot{W}_1)$ and

$$-2\mathrm{Re}\int_D (\bar{u}_t + \alpha\bar{u})\dot{W}_2\,\mathrm{d}x = 2\mathrm{Im}\int_D \bar{u}_x\dot{W}_{2x}\,\mathrm{d}x + 2\mathrm{Im}\int_D (n\bar{u} + \bar{g})\dot{W}_2\,\mathrm{d}x,$$

from (3.16), we can obtain

$$\begin{aligned}
&\frac{\mathrm{d}}{\mathrm{d}t}\left(\|u_x\|^2 + \int_D n|u|^2\,\mathrm{d}x + 2\mathrm{Re}\int_D g\bar{u}\,\mathrm{d}x\right) + 2\alpha\|u_x\|^2\\
=&-\delta\int_D n|u|^2\,\mathrm{d}x + \int_D f|u|^2\,\mathrm{d}x - 2\alpha\int_D n|u|^2\,\mathrm{d}x - 2\alpha\mathrm{Re}\int_D g\bar{u}\,\mathrm{d}x + 2\int_D q_1\mathrm{Im}(q_2\bar{u})\,\mathrm{d}x\\
&+\|q_{2x}\|^2 + 2\mathrm{Im}\int_D \bar{u}_x\dot{W}_{2x}\,\mathrm{d}x + 2\mathrm{Im}\int_D (n\bar{u} + \bar{g})\dot{W}_2\,\mathrm{d}x + \int_D |u|^2\dot{W}_1\,\mathrm{d}x.
\end{aligned}\tag{3.17}$$

Then we can estimate each term using Hölder inequality, Gagliardo–Nirenberg inequality and Young's inequality. Now let $H_0(t) = \|u_x\|^2 + \|n\|^2 + \int_D n|u|^2\,\mathrm{d}x + 2\mathrm{Re}\int_D g\bar{u}\,\mathrm{d}x + \int_D \mathrm{i}(u_x\bar{u} - u\bar{u}_x)\,\mathrm{d}x$ and take $\eta = \min\{\alpha, \delta\}$. From the sum of (3.15) and (3.17), it can be inferred that

$$\begin{aligned}
\frac{\mathrm{d}}{\mathrm{d}t}H_0(t) + \eta H_0(t) \leqslant{}& c(f, g, q_1, q_2) + c\|u\|^6 + \int_D |u|^2\dot{W}_1\,\mathrm{d}x + 2\int_D n\dot{W}_1\,\mathrm{d}x\\
&+ 4\mathrm{Re}\int_D u\dot{\bar{W}}_{2x}\,\mathrm{d}x + 2\mathrm{Im}\int_D \bar{u}_x\dot{W}_{2x}\,\mathrm{d}x + 2\mathrm{Im}\int_D (n\bar{u} + \bar{g})\dot{W}_2\,\mathrm{d}x.
\end{aligned}\tag{3.18}$$

Multiplied by $\mathrm{e}^{\eta t}$ and integrating from 0 to t and taking expectation on the both sides of (3.18), then yields

$$\mathbf{E}H_0(t) \leqslant \mathrm{e}^{-\eta t}\mathbf{E}H_0(0) + c(f, g, q_1, q_2) + c\mathbf{E}\int_0^t e^{-\eta(t-s)}\|u\|^6\,\mathrm{d}s. \tag{3.19}$$

By (3.9), we can estimate (3.19) and obtain

$$\mathbf{E}H_0(t) \leqslant \mathrm{e}^{-\beta t}\mathbf{E}H_0(0) + c(f, g, q_1, q_2, u_0) \leqslant c(f, g, q_1, q_2, u_0) \leqslant C, \quad t > 0, \tag{3.20}$$

where C is independent of T.

Since

$$H_0(t) \geqslant \frac{1}{2}(\|u_x\|^2 + \|n\|^2) - c(\|u\|^2 + \|g\|^2 + \|u\|^6), \tag{3.21}$$

for any $t > 0$, we can obtain

$$\mathbf{E}(\|u_x\|^2 + \|n\|^2) \leqslant c\mathbf{E}(\|u\|^2 + \|g\|^2 + \|u\|^6) + c\mathbf{E}H_0(t) \leqslant C, \tag{3.22}$$

where C is independent of T.

Further, we estimate $H_0^p(t)$ for $p \geqslant 1$. Now applying Itô formula to $H_0^p(t)$, we have

$$\begin{aligned}\frac{\mathrm{d}}{\mathrm{d}t}H_0^p(t) \leqslant& -\frac{\eta p}{2}H_0^p(t) + c(\|u\|^{6p} + c) + pH_0^{p-1}(t)\left(\int_D |u|^2\dot{W}_1\,\mathrm{d}x + 2\int_D n\dot{W}_1\,\mathrm{d}x\right)\\ &+ pH_0^{p-1}(t)\left(4\mathrm{Re}\int_D u\bar{\dot{W}}_{2x}\,\mathrm{d}x + 2\mathrm{Im}\int_D \bar{u}_x\dot{W}_{2x}\,\mathrm{d}x + 2\mathrm{Im}\int_D (n\bar{u}+\bar{g})\dot{W}_2\,\mathrm{d}x\right).\end{aligned} \tag{3.23}$$

Multiplying $\mathrm{e}^{\frac{\eta p}{2}t}$ and integrating from 0 to t and taking expectation on the both sides of (3.23), then yields

$$\mathbf{E}(H_0^p(t)) \leqslant \mathrm{e}^{-\frac{\eta p}{2}t}\mathbf{E}H_0^p(0) + c + c\mathbf{E}\int_0^t e^{-\frac{\eta p}{2}(t-s)}\|u\|^{6p}\,\mathrm{d}s. \tag{3.24}$$

According to (3.9) and (3.24) we can obtain

$$\mathbf{E}(H_0^p(t)) \leqslant \mathrm{e}^{-\frac{\eta p}{2}t}\mathbf{E}H_0^p(0) + c \leqslant C, \quad t > 0, \tag{3.25}$$

where C is independent of T. Therefore, by (3.21), we have

$$\mathbf{E}(\|u_x\|^{2p} + \|n\|^{2p}) \leqslant C, \qquad t > 0. \tag{3.26}$$

On the one hand, integrating from 0 to t on the both sides of (3.18), one deduces

$$\begin{aligned}H_0(t) \leqslant& H_0(0) + c(f,g,q_1,q_2)t + c\int_0^t \|u\|^6\,\mathrm{d}s\\ &+ \int_0^t\left(\int_D |u|^2\dot{W}_1\,\mathrm{d}x + 2\int_D n\dot{W}_1\,\mathrm{d}x\right)\mathrm{d}s\\ &+ \int_0^t\left(4\mathrm{Re}\int_D u\bar{\dot{W}}_{2x}\,\mathrm{d}x + 2\mathrm{Im}\int_D \bar{u}_x\dot{W}_{2x}\,\mathrm{d}x + 2\mathrm{Im}\int_D (n\bar{u}+\bar{g})\dot{W}_2\,\mathrm{d}x\right)\mathrm{d}s.\end{aligned} \tag{3.27}$$

Now taking the supremum and expectation on the both sides of (3.27), then yields

$$
\begin{aligned}
&\mathbf{E}\sup_{0\leqslant t\leqslant T} H_0(t)\leqslant \mathbf{E}H_0(0)+c(f,g,q_1,q_2)T+c+c\mathbf{E}\sup_{0\leqslant t\leqslant T}\int_0^t\|u\|^6\,\mathrm{d}s\\
&+\mathbf{E}\sup_{0\leqslant t\leqslant T}\Big|\int_0^t\int_D|u|^2\dot{W}_1\,\mathrm{d}x\mathrm{d}s\Big|^2+\mathbf{E}\sup_{0\leqslant t\leqslant T}\Big|\int_0^t\int_D n\dot{W}_1\,\mathrm{d}x\mathrm{d}s\Big|^2\\
&+\mathbf{E}\sup_{0\leqslant t\leqslant T}\Big|\int_0^t\mathrm{Re}\int_D u\bar{\dot{W}}_{2x}\,\mathrm{d}x\mathrm{d}s\Big|^2+\mathbf{E}\sup_{0\leqslant t\leqslant T}\Big|\int_0^t\mathrm{Im}\int_D \bar{u}_x\dot{W}_{2x}\,\mathrm{d}x\mathrm{d}s\Big|^2\\
&+\mathbf{E}\sup_{0\leqslant t\leqslant T}\Big|\int_0^t\mathrm{Im}\int_D(n\bar{u}+\bar{g})\dot{W}_2\,\mathrm{d}x\mathrm{d}s\Big|^2.
\end{aligned}\tag{3.28}
$$

After estimating each term of the right hand side of (3.28), we can obtain

$$
\mathbf{E}\sup_{0\leqslant t\leqslant T} H_0(t)\leqslant C_T(E_0,f,g,q_1,q_2).\tag{3.29}
$$

On the other hand, for $H_0^p(t)(p\geqslant 1)$, integrating from 0 to t, and taking the supremum and the expectation on the both sides of (3.23), we can obtain

$$
\begin{aligned}
&\mathbf{E}\sup_{0\leqslant t\leqslant T} H_0^p(t)\leqslant \mathbf{E}H_0^p(0)+c(q_1,q_2)T+c\mathbf{E}\sup_{0\leqslant t\leqslant T}\int_0^t\|u\|^{6p}\,\mathrm{d}s+c\\
&+p^2\mathbf{E}\sup_{0\leqslant t\leqslant T}\Big|\int_0^t H_0^{p-1}(s)\int_D|u|^2\dot{W}_1\,\mathrm{d}x\mathrm{d}s\Big|^2\\
&+p^2\mathbf{E}\sup_{0\leqslant t\leqslant T}\Big|\int_0^t 2H_0^{p-1}(s)\int_D n\dot{W}_1\,\mathrm{d}x\mathrm{d}s\Big|^2\\
&+p^2\mathbf{E}\sup_{0\leqslant t\leqslant T}\Big|\int_0^t H_0^{p-1}(s)4\mathrm{Re}\int_D u\bar{\dot{W}}_{2x}\,\mathrm{d}x\mathrm{d}s\Big|^2\\
&+p^2\mathbf{E}\sup_{0\leqslant t\leqslant T}\Big|\int_0^t 2H_0^{p-1}(s)\mathrm{Im}\int_D \bar{u}_x\dot{W}_{2x}\,\mathrm{d}x\mathrm{d}s\Big|^2\\
&+p^2\mathbf{E}\sup_{0\leqslant t\leqslant T}\Big|\int_0^t 2H_0^{p-1}(s)\mathrm{Im}\int_D(n\bar{u}+\bar{g})\dot{W}_2\,\mathrm{d}x\mathrm{d}s\Big|^2.
\end{aligned}\tag{3.30}
$$

Now we estimate each term of (3.30). For the third term on the right hand of (3.30),

$$
c\mathbf{E}\sup_{0\leqslant t\leqslant T}\int_0^t\|u\|^{6p}\,\mathrm{d}s\leqslant c\mathbf{E}\int_0^T\|u\|^{6p}\,\mathrm{d}s\leqslant C_T.
$$

For the fifth and sixth terms on the right hand of (3.30), we have

$$
\begin{aligned}
&p^2\mathbf{E}\sup_{0\leqslant t\leqslant T}\Big|\int_0^t H_0^{p-1}(s)\int_D|u|^2\dot{W}_1\,\mathrm{d}x\mathrm{d}s\Big|^2+p^2\mathbf{E}\sup_{0\leqslant t\leqslant T}\Big|\int_0^t 2H_0^{p-1}(s)\int_D n\dot{W}_1\,\mathrm{d}x\mathrm{d}s\Big|^2\\
&\leqslant c\|q_1\|^2\mathbf{E}\int_0^T(H_0^{2p}(s)+\|u\|^{2p}+\|u\|^{6p}+\|g\|^{2p})\,\mathrm{d}s\leqslant C_T.
\end{aligned}
$$

For the seventh and eighth terms on the right hand of (3.30), using the similar method, we can estimate

$$
\begin{aligned}
&p^2\left|\int_0^t H_0^{p-1}(s)4\mathrm{Re}\int_D u\bar{\dot{W}}_{2x}\,\mathrm{d}x\mathrm{d}s\right|^2+p^2\left|\int_0^t 2H_0^{p-1}(s)\mathrm{Im}\int_D \bar{u}_x\dot{W}_{2x}\,\mathrm{d}x\mathrm{d}s\right|^2\\
&\leqslant c\|q_{2x}\|^2\mathbf{E}\int_0^T(H_0^{2p}(s)+\|u\|^{2p}+\|u\|^{6p}+\|g\|^{2p})\,\mathrm{d}s\leqslant C_T.
\end{aligned}
$$

For the last term on the right hand of (3.30), we can estimate

$$
\begin{aligned}
&p^2\mathbf{E}\sup_{0\leqslant t\leqslant T}\left|\int_0^t 2H_0^{p-1}(s)\mathrm{Im}\int_D(n\bar{u}+\bar{g})\dot{W}_2\,\mathrm{d}x\mathrm{d}s\right|^2\\
&\leqslant c\|q_{2x}\|^2\mathbf{E}\int_0^T(H_0^{2p}(s)+\|u\|^{2p}\\
&+\|u\|^{6p}+\|g\|^{2p})\,\mathrm{d}s+c\|q_2\|^2\mathbf{E}\int_0^T(H_0^{2p}(s)+\|g\|^{2p})\,\mathrm{d}s\leqslant C_T.
\end{aligned}
$$

Then according to the above estimates, from (3.30) we can obtain

$$
\mathbf{E}\sup_{0\leqslant t\leqslant T}H_0^p(t)\leqslant \mathbf{E}H_0^p(0)+C_T(q_1,q_2,E_0,n_1,n_0)\leqslant C_T(q_1,q_2,E_0,n_1,n_0). \tag{3.31}
$$

Moreover, from (3.21) it is inferred that for $p\geqslant 1$

$$
\begin{aligned}
&\mathbf{E}\sup_{0\leqslant t\leqslant T}(\|u_x\|^{2p}+\|n\|^{2p})\\
&\leqslant c\mathbf{E}\sup_{0\leqslant t\leqslant T}H_0^p(t)+c\mathbf{E}\sup_{0\leqslant t\leqslant T}(\|u\|^{2p}+\|u\|^{6p}+\|g\|^{2p})\leqslant C_T.
\end{aligned} \tag{3.32}
$$

Then we can obtain

$$
\mathbf{E}\sup_{0\leqslant t\leqslant T}\|(n,u)\|_{V_0}^{2p}\leqslant C_T(f,g,q_1,q_2,u_0,n_0). \tag{3.33}
$$

Then the proof is complete.

3.2 A priori estimates in V_1

Lemma 3.3 *Assume that $(n_0,u_0)\in V_1$, $q_1\in H^1(D)$, $q_2\in H^2(D)$, $f,g\in H^1(D)$. Then for any $T>0$ and $p\geqslant 1$, we have $(n,u)\in L^{2p}(\Omega;L^\infty(0,T;V_1))\cap L^\infty(0,\infty;L^{2p}(\Omega;V_1))$.*

Proof Applying Itô formula to $\|u_{xx}\|^2$, we have

$$
\frac{\mathrm{d}}{\mathrm{d}t}\|u_{xx}\|^2=2\mathrm{Re}\int_D\bar{u}_{txx}u_{xx}\,\mathrm{d}x+\|q_{2xx}\|^2. \tag{3.34}
$$

Since $u_t = \mathrm{i}(u_{xx} - nu + \mathrm{i}\alpha u - g - \dot{W}_2)$ and

$$\frac{\mathrm{d}}{\mathrm{d}t}\mathrm{Re}\int_D nu\bar{u}_{xx}\,\mathrm{d}x$$
$$= \mathrm{Re}\int_D (n_t u\bar{u}_{xx} + nu_t\bar{u}_{xx} + nu\bar{u}_{xxt} + q_1\bar{q}_{2xx} + \mathrm{i}(q_1\bar{q}_{2xx}u - q_1q_2\bar{u}_{xx}))\,\mathrm{d}x, \quad (3.35)$$

we can obtain

$$\frac{\mathrm{d}}{\mathrm{d}t}\left(\|u_x\|^2 - 2\mathrm{Re}\int_D nu\bar{u}_{xx}\,\mathrm{d}x - 2\mathrm{Re}\int_D g\bar{u}_{xx}\,\mathrm{d}x\right) + 2\alpha\|u_{xx}\|^2 = \|q_{2xx}\|^2 + 2\|q_{2x}\|^2$$
$$+ 2\mathrm{Re}\int_D \mathrm{i}(q_1q_2\bar{u}_{xx} - q_1\bar{q}_{2xx}u)\,\mathrm{d}x$$
$$+ 2\alpha\mathrm{Re}\int_D nu\bar{u}_{xx}\,\mathrm{d}x + 2\alpha\mathrm{Re}\int_D g\bar{u}_{xx}\,\mathrm{d}x$$
$$- 2\mathrm{Re}\int_D n_tu\bar{u}_{xx}\,\mathrm{d}x - 2\mathrm{Re}\int_D nu_t\bar{u}_{xx}\,\mathrm{d}x + 2\mathrm{Re}\int_D \dot{W}_2(\bar{u}_{xxt} + \alpha\bar{u}_{xx})\,\mathrm{d}x. \quad (3.36)$$

Similar to the former, we can estimate the each term on the right hand side of (3.36) using Hölder inequality, Gagliardo–Nirenberg inequality and Young's inequality. Then from (3.36) we can obtain

$$\frac{\mathrm{d}}{\mathrm{d}t}\left(\|u_x\|^2 - 2\mathrm{Re}\int_D nu\bar{u}_{xx}\,\mathrm{d}x - 2\mathrm{Re}\int_D g\bar{u}_{xx}\,\mathrm{d}x\right) + 2\alpha\|u_{xx}\|^2$$
$$\leqslant \frac{\alpha}{2}\|u_{xx}\|^2 + \frac{\delta}{2}\|n_x\|^2 + c(\|q_1\|_{H^1}^4 + \|q_2\|_{H^2}^4 + \|f\|^4 + \|g\|^8 + 1 + \|u\|_{H^1}^8 + \|n\|^{12})$$
$$+ 2\mathrm{Re}\int_D \dot{W}_2(\bar{u}_{xxt} + \alpha\bar{u}_{xx})\,\mathrm{d}x - 2\mathrm{Re}\int_D u\bar{u}_{xx}\dot{W}_1\,\mathrm{d}x + 2\mathrm{Re}\int_D \mathrm{i}n\bar{u}_{xx}\dot{W}_2\,\mathrm{d}x. \quad (3.37)$$

In addition, applying Itô formula to $\|n_x\|^2$, we have

$$\frac{\mathrm{d}}{\mathrm{d}t}\|n_x\|^2 = 2\int_D n_xn_{xt}\,\mathrm{d}x + \|q_{1x}\|^2. \quad (3.38)$$

Since $n_t = -(|u|^2)_x - \delta n + f + \dot{W}_1$ and

$$\frac{\mathrm{d}}{\mathrm{d}t}\mathrm{i}\int_D (u_x\bar{u}_{xx} - \bar{u}_xu_{xx})\,\mathrm{d}x = 2\mathrm{i}\int_D (u_{xt}\bar{u}_{xx} - \bar{u}_{xt}u_{xx})\,\mathrm{d}x + 2\mathrm{i}\mathrm{Re}\int_D q_{2x}\bar{q}_{2xx}\,\mathrm{d}x,$$

we can obtain

$$\frac{\mathrm{d}}{\mathrm{d}t}(\|n_x\|^2 + \mathrm{i}\int_D (u_x\bar{u}_{xx} - \bar{u}_xu_{xx})\,\mathrm{d}x) + 2\delta\|n_x\|^2 \leqslant \frac{\alpha}{2}\|u_{xx}\|^2 + \frac{\delta}{2}\|n_x\|^2$$
$$+ c(\|g_x\|^2 + \|f_x\|^2 + \|q_2\|_{H^2}^2 + \|q_{1x}\|^2)$$
$$+ 2\int_D n_x\dot{W}_{1x}\,\mathrm{d}x + 4\mathrm{Re}\int_D \bar{u}_{xx}\dot{W}_{2x}\,\mathrm{d}x. \quad (3.39)$$

So from (3.37) and (3.39), taking $\eta = \min\{\alpha, \delta\}$ and letting

$$H_1(t) = \|u_x\|^2 + \|n_x\|^2 - 2\mathrm{Re}\int_D n u \bar{u}_{xx}\,\mathrm{d}x - 2\mathrm{Re}\int_D g\bar{u}_{xx}\,\mathrm{d}x + \mathrm{i}\int_D (u_x\bar{u}_{xx} - \bar{u}_x u_{xx})\,\mathrm{d}x,$$

we can obtain

$$\begin{aligned}
&\frac{\mathrm{d}}{\mathrm{d}t}H_1(t) + \eta H_1(t)\\
&\leqslant c(\|u\|_{H^1}^8 + \|n\|^{12} + \|q_1\|_{H^1}^4 + \|q_2\|_{H^2}^4 + \|f\|_{H^1}^4 + \|g\|_{H^1}^8 + 1)\\
&+ 2\mathrm{Re}\int_D \dot{W}_2(\bar{u}_{xxt} + \alpha\bar{u}_{xx})\,\mathrm{d}x - 2\mathrm{Re}\int_D u\bar{u}_{xx}\dot{W}_1\,\mathrm{d}x + 2\mathrm{Re}\int_D \mathrm{i}n\bar{u}_{xx}\dot{W}_2\,\mathrm{d}x\\
&+ 2\int_D n_x\dot{W}_{1x}\,\mathrm{d}x + 4\mathrm{Re}\int_D \bar{u}_{xx}\dot{W}_{2x}\,\mathrm{d}x.
\end{aligned}\tag{3.40}$$

Multiplied by $\mathrm{e}^{\eta t}$ and integrating from 0 to t and taking expectation on the both sides of (3.40), then by (3.22), we can obtain

$$\mathbf{E}H_1(t) \leqslant \mathrm{e}^{-\beta t}\mathbf{E}H_1(0) + c(f,g,q_1,q_2,u_0) \leqslant c(f,g,q_1,q_2,u_0) \leqslant C, \quad t > 0, \tag{3.41}$$

where C is independent of T.

Since

$$H_1(t) \geqslant \frac{1}{2}(\|u_{xx}\|^2 + \|n_x\|^2) - c(\|u_x\|^8 + \|n\|^4 + 1 + \|g\|^2), \tag{3.42}$$

for any $t > 0$, we can obtain

$$\mathbf{E}(\|u_{xx}\|^2 + \|n_x\|^2) \leqslant c\mathbf{E}(\|u_x\|^8 + \|n\|^4 + 1 + \|g\|^2) + c\mathbf{E}H_1(t) \leqslant C, \tag{3.43}$$

where C is independent of T.

Further, we estimate $H_1^p(t)$ for $p \geqslant 1$. Similarly to the former, applying Itô formula to $H_1^p(t)$, and taking expectation, one gets

$$\mathbf{E}H_1^p(t)) \leqslant \mathrm{e}^{-\frac{\eta p}{2}t}\mathbf{E}H_1^p(0) + c + c\mathbf{E}\int_0^t \mathrm{e}^{-\frac{\eta p}{2}(t-s)}(\|u_x\|^{8p} + \|n\|^{12p} + 1)\,\mathrm{d}s. \tag{3.44}$$

According to (3.22) we can obtain

$$\mathbf{E}H_0^p(t)) \leqslant \mathrm{e}^{-\frac{\eta p}{2}t}\mathbf{E}H_0^p(0) + c \leqslant C, \quad t > 0, \tag{3.45}$$

where C is independent of T. Therefore by (3.42) for any $t > 0$ we have

$$\mathbf{E}(\|u_{xx}\|^{2p} + \|n_x\|^{2p}) \leqslant C. \tag{3.46}$$

After integrating from 0 to t and taking the supremum and expectation on the both sides of (3.40), Similarly to the estimates in the Lemma 3.2 for the each term, we can deduce

$$\mathbf{E}\sup_{0\leqslant t\leqslant T} H_1(t) \leqslant C_T(u_0, f, g, q_1, q_2). \tag{3.47}$$

For $H_0^p(t)(p \geqslant 1)$, integrating from 0 to t and taking the supremum and the expectation, and estimating each term, we can obtain

$$\mathbf{E}\sup_{0\leqslant t\leqslant T} H_1^p(t) \leqslant \mathbf{E}H_1^p(0) + C_T(f, g, q_1, q_2, u_0, n_0) \leqslant C_T(f, g, q_1, q_2, u_0, n_0). \tag{3.48}$$

Moreover, from (3.42) it is inferred that for $p \geqslant 1$

$$\mathbf{E}\sup_{0\leqslant t\leqslant T} (\|u_{xx}\|^{2p} + \|n_x\|^{2p}) \leqslant C_T(f, g, q_1, q_2, u_0, n_0). \tag{3.49}$$

Then we can obtain

$$\mathbf{E}\sup_{0\leqslant t\leqslant T} \|(n, u)\|_{V_1}^{2p} \leqslant C_T(f, g, q_1, q_2, u_0, n_0). \tag{3.50}$$

Then the proof is complete.

3.3 A priori estimates in V_2

Similarly to the Subsection 3.1 and 3.2, using the same method and idea, we can obtain *a priori* estimates in V_2. For the convenience we only give the idea of the proof. Applying Itô formula to $\|n_{xx}\|^2$ and $\|u_{xxx}\|^2$ respectively, and using (1.3)–(1.4), we can obtain some inequalities by Hölder inequality, Young's inequality and Gagliardo–Nirenberg inequality together. Then taking the supremum and the expectation for corresponding inequalities and using Gronwall–type estimates to $\|n_{xx}\|^2$ and $\|u_{xxx}\|^2$, it is easy to obtain the results needed for the proof of the following theorem.

Lemma 3.4 *Assume that* $(n_0, u_0) \in V_2$, $q_1 \in H^2(D)$, $q_2 \in H^3(D)$, $f, g \in H^2(D)$. *Then for any* $T > 0$ *and* $p \geqslant 1$, *we have* $(n, u) \in L^{2p}(\Omega; L^\infty(0, T; V_2)) \cap L^\infty(0, \infty; L^{2p}(\Omega; V_2))$.

4 Proofs of Theorem 1.1 and Theorem 1.2

By the above *a priori* estimates, we prove the existence and uniqueness of solution for stochastic equations (1.3)–(1.4) in spaces V_1 as Theorem 1.1.

The proof of the Theorem 1.1.

Proof First we know that $(n_0, u_0) \in V_1$. Consider $\{e_i(x)\}_{i=1}^{\infty}$ an orthonormal basis of eigenvectors of the Laplace operator on D, which is an orthonormal basis of $L^2(D)$. Let P^k be the projection from $L^2(D)$ onto the space spanned by $\{e_i : i = 1, 2, \cdots, k\}$. Then the approximation solution (n^k, u^k) solves the approximation problem

$$n_t^k + P^k(|u^k|^2)_x + \delta n^k = f^k + \dot{W_1}^k, \tag{4.1}$$

$$\mathrm{i}u_t^k + u_{xx}^k - P^k(n^k u^k) + \mathrm{i}\alpha u^k = \dot{W_2}^k, \tag{4.2}$$

where P^k is the projector onto the first k vectors e_i, $\dot{W_1}^k = P^k \dot{W}_1$, $\dot{W_2}^k = P^k \dot{W}_2$, and P^k commutes with the operator Δ. Now we will treat the above equations pathwise by introducing the following random processes solving

$$\eta_t^k + \delta\eta^k = \dot{W_1}^k, \tag{4.3}$$

$$\mathrm{i}\xi_t^k + \xi_{xx}^k + \mathrm{i}\alpha\xi^k = \dot{W_2}^k, \tag{4.4}$$

with Dirichlet boundary conditions $\eta(0,t) = \eta(L,t) = 0, \xi(0,t) = \xi(L,t) = 0$, and initial conditions $\eta_t(x,0) = 0, \eta(x,0) = 0, \xi(x,0) = 0, x \in D$, here $\delta > 0, \alpha > 0$.

Following the same methods as Section 3, for any $T > 0$ and almost all $\omega \in \Omega$, we have

$$\eta \in C(0,T; H_0^1(D)), \quad \xi \in C(0,T; H^2(D) \cap H_0^1(D)). \tag{4.5}$$

And there is the following estimate

$$\mathbf{E}(\|\eta_x\|^2 + \|\xi_{xx}\|^2) \leqslant C, \tag{4.6}$$

for a positive constant C independent of T. And for any $T > 0$,

$$\mathbf{E} \sup_{0 \leqslant t \leqslant T} (\|\eta_x\|^2 + \|\xi_{xx}\|^2) \leqslant C_T \tag{4.7}$$

holds for a positive constant C_T dependent of T. Now let $R^{k,M} = (N^{k,M}, U^{k,M})$ be the solution of the following equations

$$N_t^{k,M} + \chi_M(\|R^{k,M}\|_{V_1}) P^k(|u^{k,M}|^2)_x + \delta N^{k,M} = f^k, \tag{4.8}$$

$$\mathrm{i}U_t^{k,M} + U_{xx}^{k,M} - \chi_M(\|R^{k,M}\|_{V_1}) P^k(n^{k,M} u^{k,M}) + \mathrm{i}\alpha U^{k,M} = g^k, \tag{4.9}$$

$$N^{k,M}(x,0) = P^k n_0, U^{k,M}(x,0) = P^k u_0.$$

Here $n^{k,M} = N^{k,M}+P^k\eta$, $u^{k,M} = U^{k,M}+P^k\xi$, and $\chi_M \in C_0^\infty(\mathbb{R})$ such that $\chi_M(r) = 1$ for $|r| \leqslant M$ and $\chi_M(r) = 0$ for $|r| \geqslant 2M$. Notice that (4.8)–(4.9) are random differential equations with Lipschitz nonlinearity in finite dimension. Then for almost all $\omega \in \Omega$ we have a unique solution $(N^{k,M}, U^{k,M})$ for (4.8)–(4.9). Define the stopping time by

$$\tau_M = \inf\{t > 0 : \|R\|_{V_1}^{k,M} \geqslant M\}$$

if the set $\{\|R\|_{V_1}^{k,M} \geqslant M\}$ is nonempty, otherwise $\tau_M = \infty$. Since τ_M is increasing in M, let $\tau_\infty = \lim_{M\to\infty} \tau_M$ almost surely. And for $t < \tau_M$, we have

$$(N^{k,M}, U^{k,M}) + (P^k\eta, P^k\xi)$$

satisfing (4.1)–(4.2). By the estimates given in Subsection 3.2 and (4.6)–(4.7), for any $t \geqslant 0$ we have

$$\mathbf{E}\|(N^{k,M}, U^{k,M})\|_{V_1}^2 \leqslant C(f, g, q_1, q_2, n_0, u_0) \tag{4.10}$$

where the positive constant $C(f, g, q_1, q_2, n_0, u_0)$ is independent of T and M. And for $T > 0$ we have

$$\mathbf{E} \sup_{0\leqslant t\leqslant T\wedge\tau_M} \|(N^{k,M}, U^{k,M})\|_{V_1}^2 \leqslant C_T(f, g, q_1, q_2, n_0, u_0) \tag{4.11}$$

with the positive constant $C(f, g, q_1, q_2, n_0, u_0)$, which is dependent on T but independent of M. Here $T \wedge \tau_M = \min\{T, \tau_M\}$. On the other hand we have

$$\begin{aligned}&\mathbf{E}\|(N^{k,M}(T\wedge\tau_M), U^{k,M}(T\wedge\tau_M))\|_{V_1}^2\\ \geqslant&\mathbf{E}[I(\tau_M \leqslant T)|(N^{k,M}(T\wedge\tau_M), U^{k,M}(T\wedge\tau_M))|_{V_1}^2] \geqslant M^2\mathbf{P}(\tau_M \leqslant T),\end{aligned}$$

where $I(\tau_M \leqslant T) = 1$ for $\tau_M \leqslant T$ and $I(\tau_M \leqslant T) = 0$ for $\tau_M > T$. Then according to (4.10), we have

$$\mathbf{P}(\tau_M \leqslant T) \leqslant \frac{1}{M^2} C(f, g, q_1, q_2, n_0, u_0).$$

According to the above estimate and Borel–Cantelli lemma, for any $T > 0$ we have

$$\mathbf{P}(\tau_\infty > T) = 1.$$

So we know that

$$(N^k, U^k) = \lim_{M\to\infty} (N^{k,M}, U^{k,M})$$

satisfies the following random differential equations

$$N_t^k + P^k(|u^{k,M}|^2)_x + \delta N^k = f^k, \tag{4.12}$$

$$\mathrm{i}U_t^k + U_{xx}^k - P^k(n^k u^k) + \mathrm{i}\alpha U^k = g^k, \tag{4.13}$$

with initial conditions $N^k(0) = P^k n_0, U^k(0) = P^k u_0$. Then (N^k, U^k) satisfies the estimates (4.10) and (4.11), and for any $t \geqslant 0$ we get $(n^k, u^k) = (N^k, U^k) + (P^k\eta, P^k\xi)$ is the unique global solution of (4.1)–(4.2).

Now we will consider (4.12)–(4.13) for fixed ω. Firstly, by (4.11) for any $T > 0$ we have

$$\mathbf{P}\left(\bigcap_{L=1}^{\infty}\bigcup_{l=1}^{\infty}\cap_{k=l}^{\infty}\left\{\sup_{0\leqslant t\leqslant T}\|(N^k, U^k)\|_{V_1}^2 \geqslant L\right\}\right) = 0.$$

We let

$$\tilde{\Omega} = \bigcap_{L=1}^{\infty}\bigcup_{l=1}^{\infty}\cap_{k=l}^{\infty}\left\{\sup_{0\leqslant t\leqslant T}\|(N^k, U^k)\|_{V_1}^2 \leqslant L\right\}.$$

Then $\mathbf{P}(\Omega\backslash\tilde{\Omega}) = 0$. Therefore, for any fixed $\omega \in \tilde{\Omega}$, there is a $r(\omega)$ with $0 < r(\omega) < \infty$ such that(ref [14])

$$\sup_{0\leqslant t\leqslant T}\|(N^k, U^k)\|_{V_1}^2 \leqslant r(\omega). \tag{4.14}$$

Then we can extract a subsequence still denoted by (N^k, U^k), such that for any $T > 0$ N^k converges to N weakly star in $L^\infty(0, T; H_0^1(D))$ and U^k converges to U weakly star in $L^\infty(0, T; H^2(D) \cap H_0^1(D))$. These convergence are sufficient to pass the limit $k \to \infty$ in linear terms, but we need a strong convergence of U^k for nonlinear terms. In fact, by (4.13) and the estimate (4.14), we know $U_t^k \in L^\infty(0, T; L^2(D))$. Then we can further extract a subsequence still denoted by U^k such that U^k converges to U strongly in $L^\infty(0, T; H_0^1(D))$. Then by a standard procedure we can pass the limit $k \to \infty$ for the nonlinear term. So we can show that $(N, U) \in L^\infty(0, T; V_1)$ be a weak solution of

$$N_t + (|u|^2)_x + \delta N = f, \tag{4.15}$$

$$\mathrm{i}U_t + U_{xx} - nu + \mathrm{i}\alpha U = g, \tag{4.16}$$

with initial conditions $N(0) = n_0(x), U(0) = u_0(x), x \in D$.

Then $(n, u) = (N, U) + (\eta, \xi)$ is a solution of (1.3)–(1.4) and satisfies the estimates given in Subsection 3.2. In the following we prove the continuity of the solution. For any $\omega \in \tilde{\Omega}$, $(|u|^2)_x \in L^\infty(0,T;H_0^1(D))$, we can obtain $N_t = -(|u|^2)_x - \delta N + f \in H_0^1(D)$. Then by Lemma 3.2 in [19], we can see that for almost all $\omega \in \Omega$ there is $N \in C(0,T;H_0^1(D))$. Using similar methods and noticing $U_t \in L^\infty(0,T;L^2(D))$ almost surely, according to [16] we can obtain $U \in C(0,T;H^2(D) \cap H_0^1(D))$. Then by definition of N and U we have $(n,u) \in C(0,T;V_1)$ almost surely. So the solution (n,u) is continuous from $[0,T]$ to V_1 almost surely.

As the noise is additive, we can follow the same approach as [11]. So the solution (n,u) is unique in $L^\infty(0,T;V_1)$ almost surely. The proof is complete.

The proof of Theorem 1.2 is similar to Theorem 1.1. It is omitted.

5 Proofs of Theorem 1.3–Theorem 1.5

Now we study the asymptotic behavior of the solution for (1.3)–(1.4). We will construct a random attractor for stochastic long–short wave equations in V_1 equipped with the weak topology. Some basic concepts related to random attractors for random dynamical systems are referred to [1,2,20,21].

The following existence result for a random attractor for a continuous RDS can be found in [1,2]. It gives a sufficient condition for the existence of a random attractor.

Theorem 5.1(see [1,2]) *Suppose Φ is a RDS on a Polish space (E,d) and there exists a random compact set $K(\omega)$ absorbing every bounded deterministic set $D \subset E$. Then the set*

$$\mathcal{A}(\omega) = \bigcap_{\tau \geqslant 0} \overline{\bigcup_{t \geqslant \tau} \Phi(t, \theta_{-t}\omega, K(\theta_{-t}\omega))},$$

is a global random attractor for RDS Φ.

Next, using Theorem 5.1, we consider the random attractors for the stochastic long–short wave equations in V_1 and V_2 on the basis of *a priori* estimates in Section 3.

The proof of Theorem 1.3

Proof According to the former analysis, we can consider the properties of solution (N,U) of the system (4.15)–(4.16) instead of (n,u) of stochastic long–short wave equations. For any (n_0,u_0) and any $T > 0$, the system (4.15)–(4.16) has a unique solution $(N,U) \in C(0,T;V_1)$ for almost all $\omega \in \Omega$. Noticing the system (4.15)–(4.16) with coefficients driven by θ_t, then (N,U) defines a random dynamical system on V_1. Therefore, $(n,u) = (N+\eta, U+\xi)$ also defines a continuous random dynamical system on V_1, which is denoted by $\Phi(t,\omega)$. And $\Phi(t,\omega)$ is weakly continuous almost surely on V_1.

Denoted the ball center at 0 with radius r in V_1 by $B(0,r)$. Then by the estimates given in Section 3, there is a random variable $R(\omega)$ such that for any $r>0$, $(n,u)\in B(0,r)$. So there exists a random time $t_r(\omega)>0$, such that for all $t>t_r(\omega)$ and almost all $\omega\in\Omega$ satisfies

$$\|\Phi(t,\theta_{-t}\omega)(n_0,u_0)\|_{V_1}\leqslant R(\omega).$$

Let

$$\mathcal{A}(\omega)=\cap_{\tau\geqslant 0}\overline{\bigcup_{t\geqslant\tau}\Phi(t,\theta_{-t}\omega)B(0,R(\omega))}^{V_1^w},$$

where the closure is taken with respect to the weak topology of V_1. Now we show that random attractor $\mathcal{A}(\omega)$ is tempered. By the estimates obtained in Section 3, we have

$$\mathbf{E}\sup_{0\leqslant t\leqslant 1}R^2(\theta_t\omega)<\infty.$$

Then by Birkhoff ergodic theorem[4]

$$\lim_{s\to\pm\infty}\frac{\sup_{t\in[0,1]}R^2(\theta_{t+s}\omega)}{s}=0$$

on a $\theta--$invariant subset of Ω with full probability measure, i.e., $R(\omega)$ is tempered. Then we know $\mathcal{A}(\omega)$ is tempered. The proof is completed.

Similarly to Theorem 1.3, using the same methods and ideas of proof, we can prove Theorem 1.4. It is that to say, there is a random attractor for the stochastic long–short wave equations in V_2.

Next the proof of Theorem 1.5 is given.

The proof of Theorem 1.5

Proof If $(n_0,u_0)\in V_2$, by the results given in Section 3 and Section 4, (1.3)–(1.4) has a unique solution (n,u) with $(n(0),u(0))=(n_0,u_0)$, and for any $t>0$ satisfies

$$\mathbf{E}(\|n_{xx}\|^2+\|u_{xxx}\|^2)\leqslant C,\tag{5.1}$$

for a positive constant $C>0$ which is independent of $t>0$.

Now let μ_t be the distribution of (n_t,n,E) for $t\geqslant 0$. Following the classical Bogolyubov–Krylov argument[3], we define

$$\bar{\mu}_t=\frac{1}{t}\int_0^t\mu_s\,\mathrm{d}s$$

as

$$\bar{\mu}_t(\Gamma) = \frac{1}{t}\int_0^t \mu_s(\Gamma)\,\mathrm{d}s$$

for any Borel set Γ of V_1, that is $\Gamma \in \mathscr{B}(V_1)$. By (5.1) we have

$$\int_{V_1} \|(n,u)\|_{V_2}^2 \bar{\mu}_t(\mathrm{d}v) = \frac{1}{t}\int_0^t \mathbf{E}\|(n(s),u(s))\|_{V_2}^2\,\mathrm{d}s \leqslant C.$$

By Chebyshev inequality and V_2 is compact embedding into V_1, $\{\bar{\mu}_t\}_{t\geqslant 0}$ is tight in V_1. Then there is a sequence $\{\bar{\mu}_{t_k}\}$ with $t_k \to \infty$ as $k \to \infty$ and a probability measure μ on V_1 such that $\{\bar{\mu}_{t_k}\} \to \mu$ weakly as $k \to \infty$. Therefore, μ is a stationary measure for stochastic long–short wave equations on V_1 by using the standard arguments as in [4]. Moreover, by (5.1) μ is in fact support on V_2, namely, μ is a stationary measure for stochastic long–short wave equations on V_2. The proof is completed.

Acknowledgements The work is supported by NNSFC(No. 11301097), CSC, GXNSF Grant(No. 2013GXNSFAA019001), Guangxi Education Institution Scientific Research Item(No. 2013YB170), Guangxi University of Science and Technology Grant(No. 03081587, 03081588).

References

[1] Crauel H, Flandoli F. Attractors for random dynamical systems[J]. Probab. Theory Rel., 1994, 100: 365–393.

[2] Crauel H, Debussche A, Flandoli F. Random attractors[J]. J. Dyn. Differ. Equ., 1997, 9: 307–341.

[3] Da Prato G, Zabczyk J. Stochastic equations in infinite dimensional system[M]. Cambridge University Press, 1992.

[4] Da Prato G, Zabczyk J. Ergodicity for Infinite Dimensional System[M]. Cambridge University Press, 1996.

[5] Djordjevic V D, Redekopp L G. On two–dimensional packets of capillary–gravity waves[J]. J. Fluid. Mech., 1977, 79: 703–714.

[6] Du X Y, Guo B L. The global attractor for LS type equation in $\mathbb{R}^1$[J]. Acta Math. Appl. Sin., 2005, 28: 723–734.

[7] E W, Li X, Vanden–Eijnden E. Some recent progress in multiscale modeling, Multiscale modeling and simulation[M]. Lect. Notes in Computer Science Engineering, Springer, Berlin, 2004.

[8] Guo B L. The global solution for one class of the system of long–short nonlinear wave interaction[J]. J. Math. Res Exposition, 1987, 1: 69–76.

[9] Guo B L. The periodic initial value problems and initial value problems for one class of generalized long–short type equations[J]. J. Engineering Math., 1991, 8: 47–53.

[10] Guo B L, Chen L. Orbital stability of solitary waves of the long–short wave resonance equations[J]. Math. Mechods Allo. Sci., 1998, 21: 883–894.

[11] Guo B L, Wang B X. Attractors for the Long–short wave equations[J]. J. Partial Diff. Equ., 1998, 11(4): 361–383.

[12] Grimshaw R H J. The modulation of an internal gravity–wave packet and the resonance with the mean motion[J]. Studies in Appl. Math., 1977, 56: 241–266.

[13] Imkeller P, Monahan A H. Conceptual stochastic climate models[J]. Stoch. Dynam., 2002, 2: 311–326.

[14] Kim J U. On the stochastic porous medium equation[J]. J. Diff. Equ., 2006, 220: 163–194.

[15] Li Y S. Long time behavior for the weakly dampe driven long–wave–short–wave resonance equations[J]. J. Diff. Equ., 2006, 223(2): 261–289.

[16] Lions J L, Macenes E. Problemes aux Limites Non Homogenes et Applications[M]. Dunod, Paris, 1968.

[17] Nicholson D R, Goldman M V. Damped nonlinear Schrödinger equation[J]. Phys. Fluids, 1976, 19: 1621–1635.

[18] Øksendal B. Stochastic Differential Equations, An Introduction with Applications[M]. Springer-Verlag Berlin Heidelberg, 2006.

[19] Temam R. Infinite-dimensional Dynamical Systems in Mechanics and Physics[M]. Springer-Verlag, New York, 1997.

[20] Wang B. Random attractors for the stochastic Benjamin-Bona-Mahony equation on unbounded domains [J]. J. Diff. Equ., 2009, 246: 2506–2537.

[21] Wang B. Random attractors for the stochastic FitzHugh C Nagumo system on unbounded domains [J]. Nonlinear Anal. TMA., 2009, 71(7–8): 2811–2828.

[22] Zhang R F. Existence of global attractor for LS type equation [J]. J. Math. Res. Exposition, 2006, 26: 708–714.

Global Well-Posedness of Stochastic Burgers Systerm*

Guo Boling (郭柏灵), Han Yongqian (韩永前) and Zhou Guoli (周国立)

Abstract In this paper stochastic Burgers system in Itö form is considered. The global well-posedness is proved in H^1 framework. The proof relies on energy estimates about velocity. We use maximum principle of deterministic parabolic equations to overcome the difficulties arising from higher order norms. This is the first result concerning stochastic Burgers system in Itö form.

Keywords S2D Stochastic Burgers Equation; Wiener Noise; Global Solution

1 Introduction

The paper is concerned with the 2-dimensional Burgers equation in a bounded domain with Wiener noise as the body forces like this:

$$\begin{cases} du = (\nu\Delta u + (u \cdot \nabla)u)dt + dW, & t \in (0,T],\ x \in D \subset \mathbb{R}^2, \\ u(t,x) = 0, & t \in [0,T],\ x \in \partial D, \\ u(0,x) = u_0(x), & x \in D, \end{cases} \tag{1.1}$$

where D is a regular bounded open domain of $\mathbb{R}^2$, $u(t,x) = (u^1(t,x), u^2(t,x)) \in \mathbb{R}^2$, $\nu > 0$ is viscid coefficient, Δ denotes the Laplace operator, ∇ represents the gradient operator, W stands for the Wiener process taking values in $L^2(D;\mathbb{R}^2)$ and is defined on a complete probability space $(\Omega, \mathcal{F}, P)$, with normal filtration $\mathcal{F}_t = \sigma\{W(s) : s \leqslant t\}, t \in [0,T]$. Burgers equation has received an extensive amount of attention since the studies by Burgers in the 1940s(and it has been considered even earlier by Beteman [1] and Forsyth [10]). But it is well known that the Burgers' equation is not a good model for turbulence, since it does not perform any chaos. Even if a force is added to equation, all solutions will converge to a unique stationary solution as

* Commun.Math.Sci., 2015, 13(1): 153–169. Doi: 10.4310/CMS.2015.v13.n1.a8.

time goes to infinity. However if the force is a random one, the result is completely different.

So several authors have indeed suggested to use the stochastic Burgers' equation to model turbulence, see [3], [4], [13], [14]. The stochastic equation has also been proposed in [16] to study the dynamics of interfaces.

One dimensional stochastic Burgers equation has been fairly well studied. Bertini et al. [2] solved the equation with additive space-time white noise by an adaptation of the Hopf-cole transformation. Da Prato et al. [6] studied the equation via a different approach based on semigroup property for the heat equation on a bounded interval. The more general equation with multiplicative noise was considered by Da Prato and Debussche [5]. With a similar method Gyöngy and Nualart [12] extended the Burgers equation from bounded interval to real line. A large deviation principle for the solution was obtained by Mathieu Gourcy [11]. Concerning the Ergodicity, an important paper E. Weinan etc [9] proved that there exists a unique stationary distribution for the solutions of the random inviscid Burgers equation, and typical solutions are piecewise smooth with a finite number of jump discontinuities corresponding to shocks. For model with Lévy jumps, Dong and Xu [7] proved the global existence and uniqueness of the strong, weak and mild solutions. When the noise is fractal, Guolian Wang etc [21] got the global well-posedness.

Concerning multidimensional Burgers equation, there are few works. Kiselev, Ladyzhenskaya [15] proved the existence and uniqueness of a global solution to the deterministic Burgers equation on a bounded domain $\mathcal{O}$ in the class of functions $L^\infty(0,T;L^\infty(\mathcal{O}))\cap L^2(0,T;H_0^{1,2}(\mathcal{O}))$. When the limit $\nu\to 0$ and the initial condition is zero, Ton [19] proved convergence of solutions on small time interval. In this article, we consider the two dimensional stochastic Burgers equation with the viscid coefficient $\nu=1$. Using classical fixed point theorem for contractions, we obtain local mild solution v. In order to prove global well-posedness, we try to do priori estimates in L^2. But this will produce $\|v\|_{L^4}^4$ which can not be dominated by the dissipative term $\|\Delta v\|_{L^2}^2$. However, if the noise of the stochastic equation acts only in one coordinate, we can make a change to the two dimensional stochastic Burgers equation such that we can use Maximum Principle to get the estimates uniform in time and space. Using these uniform estimates, we successively do priori estimates and prove the global well-posedness.

The remaining of this paper is organized as follows. Some preliminaries are presented in Section 2, the local existence is presented in section 3, and the last section is for the global existence. As usual, constants C may change from one line to the next, unless, we give a special declaration ; we denote by $C(a)$ a constant which depends on some parameter a.

2 Preliminaries on the burgers system

For $p \geqslant 1$, let $L^p(D;\mathbb{R}^2)$ be the vector valued L^p-space in which the norm is denoted by$\|\cdot\|_{L^p}$. In particularly, when $p=\infty$, $L^p(D;\mathbb{R}^2)$ denote the collection of vector valued functions which are essentially bounded on D . And we denote the norm of $L^\infty(D;\mathbb{R}^2)$ by $\|\cdot\|_{L_x^\infty}$.

Let $C^\infty(D;\mathbb{R}^2)$ be the set of all smooth functions from D to $\mathbb{R}^2$, and denote it's subset with compact supports by $C_0^\infty(D;\mathbb{R}^2)$. Let $\mathbb{H}^\alpha$ be the closure of $C_0^\infty(D;\mathbb{R}^2)$ in $[H^\alpha(D)]^2$, for all real α. And we denote by $\|\cdot\|_{\mathbb{H}^\alpha}$ the norm in $\mathbb{H}^\alpha$. Obviously, when $\alpha=0$, $\mathbb{H}^\alpha=L^2(D;\mathbb{R}^2)$, and we denote by $\langle.,.\rangle$ the inner product in $L^2(D;\mathbb{R}^2)$. We define the bilinear operator $B(u,v):\mathbb{H}^1\times\mathbb{H}^1\to\mathbb{H}^{-1}$ as

$$\langle B(u,v),z\rangle=\int_D z(x)\cdot(u(x)\cdot\nabla)v(x)\mathrm{d}x$$

for all $z\in\mathbb{H}^1$. Then (1.1) is equivalent to the abstract equation

$$\mathrm{d}u(t)+[Au(t)+B(u(t),u(t))]\mathrm{d}t=\mathrm{d}W(t). \tag{2.2}$$

W is the Q Wiener process having the representative:

$$W(t)=\sum_{n=1}^{\infty}\sqrt{\lambda_n}e_n\beta_n(t),\quad t\in[0,T],$$

in which $\sum_{n=1}^{\infty}\lambda_n<\infty$ and $\{\beta_n\}_{n\in\mathbb{N}}$ is a sequence of mutually independent 1-dimensional Brownian motions in the probability space $(\Omega,\mathcal{F},P)$ adapted to the filtration $\{\mathcal{F}_t\}_{t\geqslant0}$. It can be derived from [8] that the solution to the linear problem

$$\begin{aligned}
&\mathrm{d}u=\Delta u\mathrm{d}t+\mathrm{d}W,\quad \text{on } [0,T]\times D,\\
&u(t,x)=0,\ t\in[0,T],\quad x\in\partial D,\\
&u(0,x)=u_0(x),\quad x\in D,
\end{aligned}$$

is unique, and when $u_0=0$, it has the form of

$$W_A(t)=\int_0^t e^{(t-s)A}\mathrm{d}W(s).$$

By Theorem 5.20 in [8], we know W_A is Gaussian process taking values in $L^2(D;\mathbb{R}^2)$, and the process has a version $W_A(t,x),(t,x)\in[0,T]\times D$, which is, a.s. for $w\in\Omega$, α- Hölder continuous with respect to (t,x). Let

$$v(t)=u(t)-W_A(t),\ t\geqslant 0.$$

Then u is a mild solution, which is defined below, to (1.1) if and only if v solves the following evolution equation:

$$\begin{cases}\dfrac{\partial v}{dt}+Av+B(v+W_A,v+W_A)=0, & t\in(0,T],x\in D,\\ v(t,x)=0, & t\in(0,T],x\in\partial D,\\ v(0,x)=u_0(x), & x\in D.\end{cases}\tag{2.3}$$

Definition 2.1 *We say a $(\mathcal{F}(t))_{t\geqslant 0}$ adapted process $(v(t))_{t\in[0,T]}$ is a mild solution to* (2.3), *if $(v(t))_{t\in[0,T]}\in C([0,T];\mathbb{H}^1)$ P-a.e. and it satisfies*

$$v(t)\ =\ e^{-tA}u_0+\int_0^t e^{-(t-s)A}B(v+W_A,v+W_A)\mathrm{d}s,\ \ t\in[0,T].$$

Equivalently, $(u(t))_{t\in[0,T]}$ is a mild solution to (1.1), if it is a $(\mathcal{F}(t))_{t\geqslant 0}$ adapted process which belongs to $C([0,T];\mathbb{H}^1)$ P-a.s. and satisfies

$$u(t)\ =\ e^{-tA}u_0+\int_0^t e^{-(t-s)A}B(u,u)\mathrm{d}s+\int_0^t e^{-(t-s)A}\mathrm{d}W(s),\ \ t\in[0,T].$$

From now on, we will study the equation of the form (2.3) to get the existence and uniqueness of the solution a.s.$w\in\Omega$.

3 Local existence in time

In this section, we will use the classical fixed point theorem for contractions to prove the local existence in time of the mild solution to (2.3).

Theorem 3.1 *Let $v_0=(v_0^1,v_0^2)\in\mathbb{R}^2, v_0\in\mathbb{H}^1$, and v_0^i be adapted to $\mathcal{F}_0, i=1,2$. And we assume $\sum_{i=1}^{\infty}\lambda_n\alpha_n^2<\infty$. Then, for P-a.e. $w\in\Omega$, there exists $T^*(w)>0$ and a unique mild solution v, in sense of definition* 2.1, *to* (2.3) *on* $[0,T^*(w)]$.

Proof For arbitrary constant $T>0$ and $j\in\mathbb{N}$, we define

$$W_A^j(t)=\sum_{n=1}^{j}\sqrt{\lambda_n}\int_0^t e^{-A(t-s)}e_n d\beta_n(s),\ \ t\in[0,T].$$

Obviously,

$$W_A^j(w) \in C([0,T];\mathbb{H}^3),\ \ P-a.e.\ w \in \Omega.$$

For $k \in \mathbb{N}$ and $k > j$, by Burkholder-Davis-Gundy inequalities, we have

$$\begin{aligned} & E \sup_{t\in[0,T]} \|A^{\frac{3}{2}}(W_A^j - W_A^k)\|_{L^2}^2 \\ \leqslant & \sum_{n=j+1}^{k} \lambda_n \alpha_n^3 \int_0^T e^{-2\alpha_n s} ds \\ = & \sum_{n=j+1}^{k} \lambda_n \alpha_n^2 \to 0,\quad as\ j \to \infty. \end{aligned}$$

Therefore

$$W_A(w) \in C([0,T];\mathbb{H}^3),\ \ P-a.e.\ w \in \Omega.$$

We let $(\mathcal{F}(t))_{t\geqslant 0}$ adapted process $v \in C([0,T];\mathbb{H}^1)$ and define

$$\mathcal{L}(v) := e^{-tA}v_0 + \int_0^t e^{-(t-s)A}[(v+W_A)\cdot\nabla](v+W_A)ds,\quad t \in [0,T].$$

We will show that $\mathcal{L}$ is a contraction mapping in

$$\begin{aligned} B_R^{T^*} = \{v \in C([0,T^*];\mathbb{H}^1) :\ & \sup_{t\in[0,T^*]} \|v(t)\|_{\mathbb{H}^1} \\ & + \sup_{t\in[0,T^*]} t^{\frac{7}{12}}\|v(t)\|_{\mathbb{H}^2} \leqslant R, \|v_0\|_{\mathbb{H}^1} \leqslant \frac{R}{3}\}, \end{aligned}$$

where

$$R = 3\Big(\sup_{t\in[0,T]} \|W_A\|_{\mathbb{H}^3} + \|v_0\|_{\mathbb{H}^1}\Big)$$

and T^* is chosen sufficiently small. We will see that the value of R and T^* depend on $w \in \Omega$. Choose $v \in B_R^{T^*}$, and set $u = v + W_A$. Then

$$\begin{aligned} \|\mathcal{L}(v)\|_{\mathbb{H}^1} & \leqslant \|e^{-tA}v_0\|_{\mathbb{H}^1} + \int_0^t \|e^{-(t-s)A}(u\cdot\nabla u)ds\|_{\mathbb{H}^1} ds \\ & \leqslant \|v_0\|_{\mathbb{H}^1} + \int_0^t (t-s)^{-\frac{1}{2}}\|u\cdot\nabla u\|_{L^2} ds \\ & \leqslant \|v_0\|_{\mathbb{H}^1} + \int_0^t (t-s)^{-\frac{1}{2}}\|u\|_{L_x^\infty}\|\nabla u\|_{L^2} ds. \end{aligned}$$

By Gagliardo-Nirenberg interpolation inequalities(see [17]), we have

$$\|u\|_{L_x^\infty} \leqslant C\|u\|_{L^2}^{\frac{1}{2}}\|u\|_{\mathbb{H}^2}^{\frac{1}{2}},$$

where C is a positive constant which does not depend on $t \in [0,T]$. So,

$$\begin{aligned}\|\mathcal{L}(v)\|_{\mathbb{H}^1} &\leqslant \|v_0\|_{\mathbb{H}^1} + C\int_0^t (t-s)^{-\frac{1}{2}}\|u\|_{L^2}^{\frac{1}{2}} \cdot \|u\|_{\mathbb{H}^2}^{\frac{1}{2}}\|\nabla u\|_{L^2}\mathrm{d}s \\ &\leqslant \|v_0\|_{\mathbb{H}^1} + C\int_0^t (t-s)^{-\frac{1}{2}}s^{-\frac{7}{24}}\|u\|_{\mathbb{H}^1}^{\frac{3}{2}}(s^{\frac{7}{12}}\|u\|_{\mathbb{H}^2})^{\frac{1}{2}}\mathrm{d}s \\ &\leqslant \frac{R}{3} + CR^2\int_0^t (t-s)^{-\frac{1}{2}}s^{-\frac{7}{24}}\mathrm{d}s.\end{aligned}$$

Denote $\frac{s}{t} = u$. Then we have

$$\begin{aligned}\|\mathcal{L}(v)\|_{\mathbb{H}^1} &\leqslant \frac{R}{3} + CR^2t^{1-\frac{1}{2}-\frac{7}{24}}\int_0^1 (1-u)^{-\frac{1}{2}}u^{-\frac{7}{24}}\mathrm{d}u \\ &\leqslant \frac{R}{3} + CR^2t^{\frac{5}{24}}.\end{aligned} \tag{3.4}$$

For $t \leqslant T^*$,

$$\begin{aligned}t^{\frac{7}{12}}\|\mathcal{L}(v)\|_{\mathbb{H}^2} &\leqslant t^{\frac{7}{12}}\|e^{-At}v_0\|_{\mathbb{H}^2} + t^{\frac{7}{12}}\int_0^t \|e^{-A(t-s)}u\cdot\nabla u\|_{\mathbb{H}^2}\mathrm{d}s \\ &\leqslant t^{\frac{1}{12}}\|v_0\|_{\mathbb{H}^1} + t^{\frac{7}{12}}\int_0^t (t-s)^{-\frac{1}{2}}\|u\cdot\nabla u\|_{\mathbb{H}^1}\mathrm{d}s \\ &\leqslant t^{\frac{1}{12}}\|v_0\|_{\mathbb{H}^1} + t^{\frac{7}{12}}\int_0^t (t-s)^{-\frac{1}{2}}(\|\nabla u\|_{L^4}^2 + \|u\|_{L_x^\infty}\|u\|_{\mathbb{H}^2})\mathrm{d}s.\end{aligned}$$

By Gagliardo–Nirenberg interpolation inequalities, we have

$$\|u\|_{L^4} \leqslant C\|u\|_{L^2}^{\frac{1}{2}}\|u\|_{\mathbb{H}^1}^{\frac{1}{2}},$$

where C is a positive constant which does not depend on t. Therefore, we have

$$\begin{aligned}t^{\frac{7}{12}}\|\mathcal{L}(v)\|_{\mathbb{H}^2} &\leqslant t^{\frac{1}{12}}\|v_0\|_{\mathbb{H}^1} + Ct^{\frac{7}{12}}\int_0^t (t-s)^{-\frac{1}{2}} \\ &\quad\times(\|u\|_{\mathbb{H}^1}\|u\|_{\mathbb{H}^2} + \|u\|_{L^2}^{\frac{1}{2}}\|u\|_{\mathbb{H}^2}^{\frac{3}{2}})\mathrm{d}s \\ &\leqslant t^{\frac{1}{12}}R + CRt^{\frac{7}{12}}\int_0^t (t-s)^{-\frac{1}{2}}s^{-\frac{7}{12}}s^{\frac{7}{12}}\|v\|_{\mathbb{H}^2}\mathrm{d}s\end{aligned}$$

$$+CR^2t^{\frac{7}{12}}\int_0^t(t-s)^{-\frac{1}{2}}\mathrm{d}s$$
$$+CR^{\frac{1}{2}}t^{\frac{7}{12}}\int_0^t(t-s)^{-\frac{1}{2}}s^{-\frac{7}{8}}(s^{\frac{7}{12}}\|v\|_{\mathbb{H}^2})^{\frac{3}{2}}\mathrm{d}s$$
$$+CR^2t^{\frac{7}{12}}\int_0^t(t-s)^{-\frac{1}{2}}\mathrm{d}s.$$

After elementary calculation, we obtain

$$t^{\frac{7}{12}}\|\mathcal{L}(v)\|_{\mathbb{H}^2}\leqslant t^{\frac{1}{12}}R+CR^2(t^{\frac{5}{24}}+t^{\frac{1}{2}}+t^{\frac{13}{12}}). \tag{3.5}$$

By (3.4) and (3.5), we have

$$\begin{aligned}&\|\mathcal{L}(v)\|_{\mathbb{H}^1}+t^{\frac{7}{12}}\|\mathcal{L}(v)\|_{\mathbb{H}^2}\\ \leqslant&\frac{R}{3}+C(R+R^2)(t^{\frac{1}{12}}+t^{\frac{5}{24}}+t^{\frac{1}{2}}+t^{\frac{13}{12}}).\end{aligned} \tag{3.6}$$

For v_1 and $v_2\in B_R^{T^*}$, we denote

$$u_1=v_1+W_A,\quad u_2=v_2+W_A.$$

Then, we have

$$\mathcal{L}(v_1)-\mathcal{L}(v_2)=\int_0^t(u_1\cdot\nabla u_1-u_2\cdot\nabla u_2)\mathrm{d}s.$$

So,

$$\begin{aligned}&\|\mathcal{L}(v_1)-\mathcal{L}(v_2)\|_{\mathbb{H}^1}\\ \leqslant&\int_0^t\|u_1\cdot\nabla u_1-u_2\cdot\nabla u_2\|_{\mathbb{H}^1}\mathrm{d}s\\ \leqslant&\int_0^t(t-s)^{-\frac{1}{2}}\|u_1\cdot\nabla u_1-u_1\cdot\nabla u_2\|_{L^2}\mathrm{d}s\\ &+\int_0^t(t-s)^{-\frac{1}{2}}\|u_1\cdot\nabla u_2-u_2\cdot\nabla u_2\|_{L^2}\mathrm{d}s\\ \leqslant&\int_0^t(t-s)^{-\frac{1}{2}}\|u_1\|_{L_x^\infty}\|u_1-u_2\|_{\mathbb{H}^1}\mathrm{d}s\\ &+\int_0^t(t-s)^{-\frac{1}{2}}\|u_1-u_2\|_{L^4}\|\nabla u_2\|_{L^4}\mathrm{d}s.\end{aligned}$$

By Gagliardo–Nirenberg interpolation inequality and Sobolev embedding theorem, we have

$$\begin{aligned}
&\|\mathcal{L}(v_1)-\mathcal{L}(v_2)\|_{\mathbb{H}^1}\\
\leqslant & C\int_0^t (t-s)^{-\frac{1}{2}}\|u_1\|_{L^2}^{\frac{1}{2}}\|u_1\|_{\mathbb{H}^2}^{\frac{1}{2}}\|v_1-v_2\|_{\mathbb{H}^1}\mathrm{d}s\\
&+C\int_0^t (t-s)^{-\frac{1}{2}}\|u_2\|_{\mathbb{H}^1}^{\frac{1}{2}}\|u_2\|_{\mathbb{H}^2}^{\frac{1}{2}}\|v_1-v_2\|_{\mathbb{H}^1}\mathrm{d}s\\
=: & I_1+I_2.
\end{aligned}$$

For I_1,

$$\begin{aligned}
I_1\leqslant & C\int_0^t (t-s)^{-\frac{1}{2}}R^{\frac{1}{2}}[R^{\frac{1}{2}}+s^{-\frac{7}{24}}(s^{\frac{7}{12}}\|v_1\|_{\mathbb{H}^2})^{\frac{1}{2}}]\|v_1-v_2\|_{\mathbb{H}^1}\mathrm{d}s\\
\leqslant & C\int_0^t (t-s)^{-\frac{1}{2}}R\|v_1-v_2\|_{\mathbb{H}^1}\mathrm{d}s\\
&+C\int_0^t (t-s)^{-\frac{1}{2}}s^{-\frac{7}{24}}R\|v_1-v_2\|_{\mathbb{H}^1}\mathrm{d}s\\
\leqslant & CR(t^{\frac{1}{2}}+t^{\frac{5}{24}})\sup_{t\in[0,T^*]}\|v_1-v_2\|_{\mathbb{H}^1}.
\end{aligned}$$

Analogously to derive I_1, we have

$$I_2\leqslant CR(t^{\frac{1}{2}}+t^{\frac{5}{24}})\sup_{t\in[0,T^*]}\|v_1-v_2\|_{\mathbb{H}^1}.$$

So, by the estimates of I_1 and I_2, we have

$$\|\mathcal{L}(v_1)-\mathcal{L}(v_2)\|_{\mathbb{H}^1}\leqslant CR(t^{\frac{1}{2}}+t^{\frac{5}{24}})\sup_{t\in[0,T^*]}\|v_1-v_2\|_{\mathbb{H}^1}. \tag{3.7}$$

Next, we consider

$$\begin{aligned}
&t^{\frac{7}{12}}\|\mathcal{L}(v_1)-\mathcal{L}(v_2)\|_{\mathbb{H}^2}\\
\leqslant & t^{\frac{7}{12}}\int_0^t \|e^{-(t-s)A}(u_1\cdot\nabla u_1-u_2\cdot\nabla u_2)\|_{\mathbb{H}^2}\mathrm{d}s\\
\leqslant & t^{\frac{7}{12}}\int_0^t (t-s)^{-\frac{1}{2}}\|(u_1-u_2)\nabla u_1\|_{\mathbb{H}^1}\mathrm{d}s\\
&+t^{\frac{7}{12}}\int_0^t (t-s)^{-\frac{1}{2}}\|u_2\nabla(u_1-u_2)\|_{\mathbb{H}^1}\mathrm{d}s\\
=: & I_3+I_4.
\end{aligned}$$

For I_3, by elementary calculation, we have that

$$\begin{aligned}
I_3 \leqslant & t^{\frac{7}{12}} \int_0^t (t-s)^{-\frac{1}{2}} \|u_1 - u_2\|_{L_x^\infty} \|u_1\|_{\mathbb{H}^2} \mathrm{d}s \\
& + t^{\frac{7}{12}} \int_0^t (t-s)^{-\frac{1}{2}} \|\nabla u_1\|_{L^4} \|\nabla(u_1 - u_2)\|_{L^4} \mathrm{d}s \\
\leqslant & C t^{\frac{7}{12}} \int_0^t (t-s)^{-\frac{1}{2}} \|v_1 - v_2\|_{L^2}^{\frac{1}{2}} \|v_1 - v_2\|_{\mathbb{H}^2}^{\frac{1}{2}} \|u_1\|_{\mathbb{H}^2} \mathrm{d}s \\
& + t^{\frac{7}{12}} \int_0^t (t-s)^{-\frac{1}{2}} \|\nabla u_1\|_{L^2}^{\frac{1}{2}} \|\nabla u_1\|_{\mathbb{H}^1}^{\frac{1}{2}} \\
& \times \|\nabla(v_1 - v_2)\|_{L^2}^{\frac{1}{2}} \|\nabla(v_1 - v_2)\|_{\mathbb{H}^1}^{\frac{1}{2}} \mathrm{d}s \\
\leqslant & C t^{\frac{7}{12}} \int_0^t (t-s)^{-\frac{1}{2}} s^{-\frac{7}{8}} \|v_1 - v_2\|_{\mathbb{H}^1}^{\frac{1}{2}} \\
& \times (s^{\frac{7}{12}} \|v_1 - v_2\|_{\mathbb{H}^2})^{\frac{1}{2}} (s^{\frac{7}{12}} \|u_1\|_{\mathbb{H}^2}) \mathrm{d}s \\
& + t^{\frac{7}{12}} \int_0^t (t-s)^{-\frac{1}{2}} s^{-\frac{7}{12}} \|u_1\|_{\mathbb{H}^1}^{\frac{1}{2}} \\
& \times (s^{\frac{7}{12}} \|u_1\|_{\mathbb{H}^2})^{\frac{1}{2}} \|v_1 - v_2\|_{\mathbb{H}^1}^{\frac{1}{2}} (s^{\frac{7}{12}} \|v_1 - v_2\|_{\mathbb{H}^2})^{\frac{1}{2}} \mathrm{d}s \\
\leqslant & C R t^{\frac{7}{12}} \int_0^t (t-s)^{-\frac{1}{2}} s^{-\frac{7}{8}} \mathrm{d}s \Big(\sup_{t \in [0,T^*]} \|v_1 - v_2\|_{\mathbb{H}^1} \\
& + \sup_{t \in [0,T^*]} t^{\frac{7}{12}} \|v_1 - v_2\|_{\mathbb{H}^2} \Big) \\
& + C R t^{\frac{7}{12}} \int_0^t (t-s)^{-\frac{1}{2}} s^{-\frac{7}{12}} \mathrm{d}s \Big(\sup_{t \in [0,T^*]} \|v_1 - v_2\|_{\mathbb{H}^1} \\
& + \sup_{t \in [0,T^*]} t^{\frac{7}{12}} \|v_1 - v_2\|_{\mathbb{H}^2} \Big) \\
\leqslant & C R (t^{\frac{5}{24}} + t^{\frac{1}{2}}) \Big(\sup_{t \in [0,T^*]} \|v_1 - v_2\|_{\mathbb{H}^1} \\
& + \sup_{t \in [0,T^*]} t^{\frac{7}{12}} \|v_1 - v_2\|_{\mathbb{H}^2} \Big),
\end{aligned}$$

where the second inequality follows by interpolation inequalities and the third inequality follows by Sobolev embedding theorem. Analogously to I_3, we have

$$I_4 \leqslant C R t^{\frac{5}{24}} \sup_{t \in [0,T^*]} t^{\frac{7}{12}} \|v_1 - v_2\|_{\mathbb{H}^2}.$$

So, by the estimate of I_3 and I_4, we have that

$$\begin{aligned}
& t^{\frac{7}{12}} \| \mathcal{L}(v_1) - \mathcal{L}(v_2) \|_{\mathbb{H}^2} \\
& \leqslant C R (t^{\frac{5}{24}} + t^{\frac{1}{2}}) \Big(\sup_{t \in [0,T^*]} \|v_1 - v_2\|_{\mathbb{H}^1} + \sup_{t \in [0,T^*]} t^{\frac{7}{12}} \|v_1 - v_2\|_{\mathbb{H}^2} \Big).
\end{aligned} \tag{3.8}$$

By (3.7) and (3.8), we have

$$\sup_{t\in[0,T^*]}\|\mathcal{L}(v_1)-\mathcal{L}(v_2)\|_{\mathbb{H}^1}+\sup_{t\in[0,T^*]}t^{\frac{7}{12}}\|\mathcal{L}(v_1)-\mathcal{L}(v_2)\|_{\mathbb{H}^2}$$
$$\leqslant CR(t^{\frac{5}{24}}+t^{\frac{1}{2}})(\sup_{t\in[0,T^*]}\|v_1-v_2\|_{\mathbb{H}^1}+\sup_{t\in[0,T^*]}t^{\frac{7}{12}}\|v_1-v_2\|_{\mathbb{H}^2}). \quad (3.9)$$

By (3.6) and (3.9), when T^* is small enough, we can get

$$\sup_{t\in[0,T^*]}\|\mathcal{L}(v)\|_{\mathbb{H}^1}+\sup_{t\in[0,T^*]}t^{\frac{7}{12}}\|\mathcal{L}(v)\|_{\mathbb{H}^2}\leqslant R \quad (3.10)$$

and

$$2CR(t^{\frac{5}{24}}+t^{\frac{1}{2}})\leqslant 1,\ \forall t\in[0,T^*], \quad (3.11)$$

where the constant C is as in (3.9). By interpolation inequalities and elementary calculations, we have that

$$\begin{aligned}\|u\nabla u\|_{\mathbb{H}^1}&\leqslant\|\nabla u\|_{L^4}^2+\|u\|_{L^\infty}\|u\|_{\mathbb{H}^2}\\&\leqslant C\|\nabla u\|_{L^2}\|u\|_{\mathbb{H}^2}+\|u\|_{L^2}^{\frac{1}{2}}\|u\|_{\mathbb{H}^2}^{\frac{3}{2}}\\&\leqslant CRt^{-\frac{7}{12}}(t^{\frac{7}{12}}\|v\|_{\mathbb{H}^2}+t^{\frac{7}{12}}\|W_A\|_{\mathbb{H}^2})\\&\quad+CR^{\frac{1}{2}}t^{-\frac{7}{8}}(t^{\frac{7}{12}}\|v\|_{\mathbb{H}^2}+t^{\frac{7}{12}}\|W_A\|_{\mathbb{H}^2})^{\frac{3}{2}}\\&\leqslant CR^2(t^{-\frac{7}{12}}+t^{-\frac{7}{8}}).\end{aligned}$$

Since $u=v+W_A$, by dominated theorem, it is easy to check that

$$\int_0^t e^{-(t-s)A}[(v+W_A)\cdot\nabla](v+W_A)\mathrm{d}s\in C([0,T^*];\mathbb{H}^1),\quad t\in[0,T^*],\quad P-\text{a.s.}.$$

So for $v\in\mathcal{B}_R^{T^*}$, it is easy to see

$$\mathcal{L}(v)\in C([0,T^*];\mathbb{H}^1),\quad t\in[0,T^*],\quad P-\text{a.s.}. \quad (3.12)$$

By (3.6) and (3.9) – (3.12) , we can see that $\mathcal{L}$ maps $\mathcal{B}_R^{T^*}$ into itself and is a strict contraction in $\mathcal{B}_R^{T^*}$. Hence, $\mathcal{L}$ has a unique fixed point in $\mathcal{B}_R^{T^*}$, which is a solution to (2.3) on $[0,T^*(w)]$.

Remark 3.1 *An example of the noise satisfying condition of Theorem* 3.1 *is*

$$\mathrm{d}W(t)=\sum_{n=1}^{\infty}\sqrt{\lambda_n}e_n\mathrm{d}\beta_n(t),$$

where $\{\beta_n\}$ *is a sequence of independent* $1-$*dimensional Brownian motion, and* $\{\lambda_n\}$ *satisfies*

$$\lambda_n = n^{-(3+2\theta)}, \ \alpha_n = n,$$

where $\theta > 0$ *and* $n \in \mathbb{N}$. *It is so because the eigenvalues* α_n *of the operator* A, *in* $2-$*dimensional space, behave like* n *(cf. [18]).*

Remark 3.2 *Another example of stochastic noise satisfying Theorem* 3.1 *is*

$$A^{-\gamma} L dW(t)$$

where $W(t) = \sum_{n=1}^{\infty} e_n d\beta_n(t)$, L *is an isomorphism in* $L^2(D;\mathbb{R}^2)$ *and* $\gamma > 2$.

4 Global existence

In Theorem 3.1, the result is valid a.s. for $w \in \Omega$; in particular T^* depend on w. In this section, we will prove, if the noise acts only in one coordinate, the solution exists in space $C([0,T];\mathbb{H}^1)$ for arbitrary constant $T > 0$. So, let $e_k = (\bar{e}_k, 0) \in \mathbb{R}^2, k = 1, 2...$, $(\bar{e}_k)_{k\in\mathbb{N}}$ is a complete orthonormal system on $L^2(D;\mathbb{R}^1)$ which is the usual Lebesgue spaces of real-valued functions on D. We still denote by $(\alpha_n)_{n\in\mathbb{N}}$ the eigenvalues of A, and by $(\bar{e}_n)_{n\in\mathbb{N}}$ the corresponding eigenvectors. Then for $t \in [0,T], x \in D$,

$$W(t,x) = \sum_{n=1}^{\infty} \sqrt{\lambda_n} e_n \beta_n(t) = (\sum_{n=1}^{\infty} \sqrt{\lambda_n} \bar{e}_n \beta_n(t), 0) \in \mathbb{R}^2, \text{ a.s..} \tag{4.13}$$

Therefore

$$\begin{aligned} W_A(t,x) &= \int_0^t e^{-(t-s)A} dW = \sum_{n=1}^{\infty} \sqrt{\lambda_n} \int_0^t e^{-(t-s)A} e_n d\beta_n(s) \\ &= (\sum_{n=1}^{\infty} \sqrt{\lambda_n} \int_0^t e^{-(t-s)A} \bar{e}_n d\beta_n(s), 0) \in \mathbb{R}^2, \text{ a.s..} \end{aligned}$$

In the proof, we will use some real valued spaces. For $p \in [1,\infty]$, we denote by $|\cdot|_{L^p}$ the norm in $L^p(D;\mathbb{R}^1)$ which is the usual Lebesgue spaces of real-valued functions on D. When $p = 2$, we still let $\langle\cdot,\cdot\rangle$ be the inner product in $L^2(D;\mathbb{R}^1)$. Let $C_0^\infty(D;\mathbb{R}^1)$ be the set of all smooth functions from D to $\mathbb{R}^1$ with compact supports contained in D. For $\alpha \in \mathbb{R}$, we denote by $|\cdot|_{H^\alpha} = |A^{\frac{\alpha}{2}} \cdot|_{L^2}$ the norm in Hilbert spaces H^α which is closure of $C_0^\infty(D;\mathbb{R}^1)$ under norm $|\cdot|_{H^\alpha}$.

Theorem 4.1 *Under conditions of Theorem* 3.1, *we consider problem* (2.3) *with noise in form of* (4.13) *and, in addition, assume the initial condition of problem* (2.3) *satisfies* $\|v_0\|_{L_x^\infty} < \infty$. *Then there exists a unique solution* $(v(t))_{t\in[0,T]}$ *to problem* (2.3) *in sense of Definition* 3.1, *for arbitrary* $T > 0$. *Moreover,*

$$\sup_{t\in[0,T]} \|v\|_{\mathbb{H}^1} \leqslant C(T, \|v_0\|_{L_x^\infty}, \|W_A\|_{L_t^\infty L_x^\infty}, \|\nabla W_A\|_{L_t^\infty L_x^\infty})$$

where $\|\cdot\|_{L_t^\infty L_x^\infty} := \sup\limits_{(t,x)\in[0,T]\times D} |\cdot|$.

Proof Let $\{v_n^0\}_{n\geqslant 1}$ be a sequence of vectors in $C_0^\infty(D;\mathbb{R}^2)$ such that

$$v_n^0 \to v_0, \ as \ n\to\infty \tag{4.14}$$

in $L^\infty(D;\mathbb{R}^2)\cap\mathbb{H}^1$. As $W_A \in C([0,T];\mathbb{H}^3)$ a.s., we can choose a sequence of regular process $\{W_A^n(t,x)\}_{n\geqslant 1} = \{(W_{A,1}^n(t,x),0)\}_{n\geqslant 1}, t\in[0,T], x\in D$ such that

$$W_A^n(t) \to W_A(t), \ as \ n\to\infty \tag{4.15}$$

in $C([0,T];\mathbb{H}^3)$ a.s.. Then, by (4.15), we have

$$\sup_{\{n\geqslant 1\}} \|W_A^n\|_{L_t^\infty L_x^\infty} < \infty$$

and

$$\sup_{\{n\geqslant 1\}} \|A^{\frac{1}{2}} W_A^n\|_{L_t^\infty L_x^\infty} < \infty.$$

By Theorem 3.1, there exists positive random variable T_n^* such that, for $t\in[0,T_n^*]$, v_n is the solution of the following equation

$$v_n(t) = e^{tA}v_n^0 + \int_0^t e^{(t-s)A}[(v_n + W_A^n)\cdot\nabla](v_n + W_A^n)\mathrm{d}s.$$

Let $T_{\max}$ be maximal existence time of solution v_n. Obviously, $T_{\max} \leqslant T$ a.s.. In the following, we will prove

$$T_{\max} = T, \ \text{a.s.}.$$

For $t\in[0,T_{\max})$, v_n is regular such that

$$\frac{\partial v_n}{\partial t} + Av_n + B(v_n + W_A^n, v_n + W_A^n) = 0. \tag{4.16}$$

Let

$$\bar{v}_n = v_n e^{-\int_0^t (1+\|\nabla W_A^n\|_{L_x^\infty})\mathrm{d}s} - \|W_A^n\|_{L_t^\infty L_x^\infty}\|\nabla W_A^n\|_{L_t^\infty L_x^\infty} I, \tag{4.17}$$

where $I=(1,1)$. Substituting (4.17) into (4.16), we have

$$\begin{aligned}&\Delta\bar{v}_n-(v_n+W_A^n)\nabla\bar{v}_n-\bar{v}_n(\nabla W_A^n+\|\nabla W_A^n\|_{L_x^\infty}+1)-\frac{\mathrm{d}\bar{v}_n}{\mathrm{d}t}\\=&2\|W_A^n\|_{L_t^\infty L_x^\infty}\|\nabla W_A^n\|_{L_t^\infty L_x^\infty}I(1+\|\nabla W_A^n\|_{L_x^\infty}+\nabla W_A^n)\\&+2W_A^n\nabla W_A^n e^{-\int_0^t(1+\|\nabla W_A^n\|_{L_x^\infty})\mathrm{d}s}>0.\end{aligned}\tag{4.18}$$

Denote $v_n:=(v_n^1,v_n^2),\bar{v}_n:=(\bar{v}_n^1,\bar{v}_n^2)$. To simplify the notations, we set

$$\partial_i=\frac{\partial}{\partial x_i},\quad i=1,2.$$

Then by (4.18), we get

$$\begin{aligned}&\Delta\bar{v}_n^2-[(v_n^1+W_{A,1}^n)\partial_1+(v_n^2+W_{A,2}^n)\partial_2]\bar{v}_n^2\\&-(\|\nabla W_A^n\|_{L_x^\infty}+1)\bar{v}_n^2-\frac{\mathrm{d}\bar{v}_n^2}{\mathrm{d}t}>0.\end{aligned}$$

By the Maximum Principle of parabolic equations(see Theorem 7, p.174, [20]), we obtain

$$\max_{(t,x)\in[0,T_{\max})\times D}\bar{v}_n^2(t,x)\leqslant\max_{x\in D}v_n^0(x),\ \text{a.s.}.\tag{4.19}$$

If we denote

$$\hat{v}_n=v_ne^{-\int_0^t(1+\|\nabla W_A^n\|_{L_x^\infty})\mathrm{d}s}+\|W_A^n\|_{L_t^\infty L_x^\infty}\|\nabla W_A^n\|_{L_t^\infty L_x^\infty}I,\tag{4.20}$$

where I is the vector in (4.17). Substituting (4.20) into (4.16), we get

$$\Delta\hat{v}_n-(v_n+W_A^n)\nabla\hat{v}_n-\hat{v}_n(\nabla W_A^n+\|\nabla W_A^n\|_{L_x^\infty}+1)-\frac{\mathrm{d}\hat{v}_n}{\mathrm{d}t}<0.$$

We denote $\hat{v}_n=(\hat{v}_n^1,\hat{v}_n^2)\in\mathbb{R}^2$. By the Minimum Principle of parabolic equations(see Theorem 7, p.174, [20]), we have

$$\min_{(t,x)\in[0,T_{\max})\times D}\hat{v}_n^2(t,x)\geqslant\min_{x\in D}v_n^0(x),\text{a.s.}.\tag{4.21}$$

By (4.19) and (4.21), we can conclude that

$$\begin{aligned}&\sup_{t\in[0,T_{\max})}\|v_n^2\|_{L_x^\infty}\\&\leqslant(\|v_n^0\|_{L_x^\infty}+\|W_A^n\|_{L_t^\infty L_x^\infty}\|\nabla W_A^n\|_{L_t^\infty L_x^\infty})e^{\int_0^T(1+\|\nabla W_A^n\|_{L_x^\infty})\mathrm{d}s},\ \text{a.s.}.\end{aligned}\tag{4.22}$$

In the following, we will estimate $\sup\limits_{t\in[0,T_{\max})}\|v_n^1\|_{L_x^\infty}$. Let

$$\begin{aligned}\tilde{v}_n^1 =& v_n^1 e^{-\int_0^t (1+\|\nabla W_A^n\|_{L_x^\infty})\mathrm{d}s} - (\|W_A^n\|_{L_t^\infty L_x^\infty} \\ &+ \sup_{t\in[0,T_n^*]}\|v_n^2\|_{L_x^\infty})\|\nabla W_A^n\|_{L_t^\infty L_x^\infty}.\end{aligned} \tag{4.23}$$

By (4.16), we have

$$\begin{aligned}&\frac{\partial v_n^1}{\partial t} - \Delta v_n^1 + [(W_{A,1}^n + v_n^1)\partial_1 + (W_{A,2}^n + v_n^2)\partial_2]v_n^1 \\ &+ v_n^1\partial_1 W_{A,1}^n = -v_n^2\partial_2 W_{A,1}^n - W_{A,1}^n\partial_1 W_{A,1}^n - W_{A,2}^n\partial_2 W_{A,1}^n.\end{aligned} \tag{4.24}$$

Substituting (4.23) into (4.24), we can get

$$\begin{aligned}&\Delta\tilde{v}_n^1 - [(W_{A,1}^n + v_n^1)\partial_1 + (W_{A,2}^n + v_n^2)\partial_2]\tilde{v}_n^1 \\ &- \tilde{v}_n^1(1 + \|\nabla W_A^n\|_{L_x^\infty} + \partial_1 W_{A,1}^n e^{-\int_0^t (1+\|\nabla W_A^n\|_{L_x^\infty})\mathrm{d}s}) - \frac{\partial \tilde{v}_n^1}{\partial t} > 0.\end{aligned}$$

By Maximum Principle for parabolic equations, we have

$$\max_{(t,x)\in[0,T_{\max})\times D} \tilde{v}_n^1 \leqslant \max_{x\in D} v_n^0, \quad \text{a.s..}$$

Let

$$\begin{aligned}\check{v}_n^1 =& v_n^1 e^{-\int_0^t (1+\|\nabla W_A^n\|_{L_x^\infty})\mathrm{d}s} + (\|W_A^n\|_{L_t^\infty L_x^\infty} \\ &+ \sup_{t\in[0,T_n^*]}\|v_n^2\|_{L_x^\infty})\|\nabla W_A^n\|_{L_t^\infty L_x^\infty}.\end{aligned} \tag{4.25}$$

Substituting (4.25) into (4.24), we can get

$$\begin{aligned}&\Delta\check{v}_n^1 - [(W_{A,1}^n + v_n^1)\partial_1 + (W_{A,2}^n + v_n^2)\partial_2]\check{v}_n^1 \\ &- \tilde{v}_n^1(1 + \|\nabla W_A^n\|_{L_x^\infty} + \partial_1 W_{A,1}^n e^{-\int_0^t (1+\|\nabla W_A^n\|_{L_x^\infty})\mathrm{d}s}) - \frac{\partial \check{v}_n^1}{\partial t} < 0.\end{aligned}$$

By Minimum Principle for parabolic equations, we have

$$\min_{(t,x)\in[0,T_{\max})\times D} \check{v}_n^1 \geqslant \min_{x\in D} v_n^0, \quad \text{a.s..}$$

Therefore, we conclude that

$$\sup_{t\in[0,T_{\max})}\|v_n^1\|_{L_x^\infty} \leqslant C(T, \|v_n^0\|_{L_x^\infty}, \|W_A^n\|_{L_t^\infty L_x^\infty}, \|\nabla W_A^n\|_{L_t^\infty L_x^\infty}).$$

So far, we proved

$$\sup_{t\in[0,T_{\max})}\|v_n\|_{L_x^\infty}\leqslant C(T,\|v_n^0\|_{L_x^\infty},\|W_A^n\|_{L_t^\infty L_x^\infty},\|\nabla W_A^n\|_{L_t^\infty L_x^\infty}). \tag{4.26}$$

Taking inner product with respect to v_n in (4.16), we have

$$\langle\frac{\partial v_n}{\partial t},v_n\rangle+\langle Av_n,v_n\rangle+\langle B(v_n+W_A^n,v_n+W_A^n),v_n\rangle=0. \tag{4.27}$$

First we calculate the third term on the left hand side of (4.27) .

$$\begin{aligned}
&\langle B(v_n+W_A^n,v_n+W_A^n),v_n\rangle\\
=&\langle(v_n^1+W_{A,1}^n)\partial_1(v_n^1+W_{A,1}^n),v_n^1\rangle\\
&+\langle(v_n^2+W_{A,2}^n)\partial_2(v_n^1+W_{A,1}^n),v_n^1\rangle\\
&+\langle(v_n^1+W_{A,1}^n)\partial_1(v_n^2+W_{A,2}^n),v_n^2\rangle\\
&+\langle(v_n^2+W_{A,2}^n)\partial_2(v_n^2+W_{A,2}^n),v_n^2\rangle\\
=&J_1+J_2+J_3+J_4.
\end{aligned} \tag{4.28}$$

For J_1, we have

$$\begin{aligned}
J_1=&\langle v_n^1\partial_1v_n^1,v_n^1\rangle+\langle W_{A,1}^n\partial_1v_n^1,v_n^1\rangle\\
&+\langle v_n^1\partial_1W_{A,1}^n,v_n^1\rangle+\langle W_{A,1}^n\partial_1W_{A,1}^n,v_n^1\rangle.
\end{aligned}$$

In the next, we estimate the four terms of J_1 respectively. For the first term

$$\langle v_n^1\partial_1v_n^1,v_n^1\rangle=\int_D(v_n^1)^2\partial_1v_n^1\mathrm{d}x=\int_D\partial_1[\frac{(v_n^1)^3}{3}]\mathrm{d}x=0.$$

For the seconde term, by (4.15), we have

$$\langle W_{A,1}^n\partial_1v_n^1,v_n^1\rangle\leqslant C|v_n^1|_{L^2}^2+\varepsilon|v_n^1|_{H^1}^2.$$

Similarly, for the third term, we have

$$|\langle v_n^1\partial_1W_{A,1}^n,v_n^1\rangle|=|\int_D(v_n^1)^2\partial_1W_{A,1}^n\mathrm{d}x|\leqslant C|v_n^1|_{L^2}^2.$$

For the last term, we have

$$|\langle W_{A,1}^n\partial_1W_{A,1}^n,v_n^1\rangle|\leqslant C|\int_D\partial_1v_n^1dx|\leqslant C+C|v_n^1|_{L^2}^2.$$

Therefore, for J_1, we have

$$J_1 \leqslant C(1+\|v_n\|_{L^2}^2)+\varepsilon\|v_n\|_{\mathbb{H}^1}^2.$$

Similarly,

$$J_4 \leqslant C(1+\|v_n\|_{L^2}^2)+\varepsilon\|v_n\|_{\mathbb{H}^1}^2.$$

For J_3,

$$\begin{aligned} J_3 = & \langle v_n^1\partial_1 v_n^2, v_n^2\rangle + \langle v_n^1\partial_1 W_{A,2}^n, v_n^2\rangle \\ & +\langle W_{A,1}^n\partial_1 v_n^2, v_n^2\rangle + \langle W_{A,1}^n\partial_1 W_{A,2}^n, v_n^2\rangle. \end{aligned}$$

For the first term of J_3, we have

$$\begin{aligned} |\langle v_n^1\partial_1 v_n^2, v_n^2\rangle| = & \frac{1}{2}\left|\int_D v_n^1\partial_1(v_n^2)^2 \mathrm{d}x\right| = \frac{1}{2}\left|\int_D \partial_1 v_n^1\cdot(v_n^2)^2\mathrm{d}x\right| \\ \leqslant & \frac{1}{2}|v_n^2|_{L^4}^2\cdot|v_n^1|_{H^1} \leqslant C|v_n^2|_{L^4}^4+\varepsilon|v_n^1|_{H^1}^2. \end{aligned}$$

For the seconde term of J_3, we have

$$|\langle v_n^1\partial_1 W_{A,2}^n, v_n^2\rangle| \leqslant C\|v_n\|_{L^2}^2.$$

Analogously, for the third term of J_3, we have

$$|\langle W_{A,1}^n\partial_1 v_n^2, v_n^2\rangle| \leqslant C\|v_n\|_{L^2}^2+\varepsilon\|v_n\|_{\mathbb{H}^1}^2.$$

For the last term of J_3, we have

$$|\langle W_{A,1}^n\partial_x W_{A,2}^n, v_n^2\rangle| \leqslant C+C\|v_n\|_{L^2}^2.$$

Therefore, for J_3, we get

$$J_3 \leqslant C\|v_n^2\|_{L^4}^4+\varepsilon\|v_n\|_{\mathbb{H}^1}^2+C\|v_n\|_{L^2}^2+C.$$

Analogously, for J_2, we obtain

$$J_2 \leqslant C\|v_n^1\|_{L^4}^4+\varepsilon\|v_n\|_{\mathbb{H}^1}^2+C\|v_n\|_{L^2}^2+C.$$

By (4.28) and the estimates of J_1, J_2, J_3 and J_4, we have

$$\langle B(v_n+W_A^n, v_n+W_A^n), v_n\rangle \leqslant C(1+\|v_n\|_{L^2}^2)+4\varepsilon\|v_n\|_{\mathbb{H}^1}^2+C\|v_n\|_{L^4}^4.$$

Therefore by (4.27) , we get

$$\frac{\partial}{\partial t}\|v_n\|_H^2+\|v_n\|_{H^1}^2\leqslant C(1+\|v_n\|_{L^2}^2)+4\varepsilon\|v_n\|_{\mathbb{H}^1}^2+C\|v_n\|_{L^4}^4. \tag{4.29}$$

For $t\in[0,T_{\max})$, integrating over $[0,t]$ on both sides of (4.29), we have

$$\|v_n(t)\|_{L^2}^2+\int_0^t\|v_n(s)\|_{\mathbb{H}^1}^2\mathrm{d}s\leqslant\|v_n(0)\|_{L^2}^2+Ct+C\int_0^t\|v_n(s)\|_{L^4}^4\mathrm{d}s. \tag{4.30}$$

Since, for $t\in[0,T_{\max})$,

$$\|v_n(0)\|_{L^2}^2\leqslant C\|v_n(0)\|_{L_x^\infty}^2$$

and

$$\|v_n(t)\|_{L^4}^4\leqslant C\|v_n(t)\|_{L_x^\infty}^4$$

where $C>0$. Thus, for all $t\in[0,T_{\max})$, by (4.26) and (4.30), we have

$$\|v_n(t)\|_{L^2}^2\leqslant C(T,\|v_n^0\|_{L_x^\infty},\|W_A^n\|_{L_t^\infty L_x^\infty},\|\nabla W_A^n\|_{L_t^\infty L_x^\infty}) \tag{4.31}$$

and

$$\int_0^t\|v_n(s)\|_{\mathbb{H}^1}^2\mathrm{d}s\leqslant C(T,\|v_n^0\|_{L_x^\infty},\|W_A^n\|_{L_t^\infty L_x^\infty},\|\nabla W_A^n\|_{L_t^\infty L_x^\infty}). \tag{4.32}$$

Multiply (4.16) by Av_n and integrating over D, we find

$$\langle\frac{\partial v_n}{\partial t},Av_n\rangle+\langle Av_n,Av_n\rangle=\langle B(v_n+W_A^n,v_n+W_A^n),Av_n\rangle,$$

which is equivalent to

$$\frac{1}{2}\frac{\partial}{\partial t}\|v_n\|_{\mathbb{H}^1}^2+\|v_n\|_{\mathbb{H}^2}^2=\langle B(v_n+W_A^n,v_n+W_A^n),Av_n\rangle. \tag{4.33}$$

For the term on the right hand side of (4.33),

$$\begin{aligned}
&\langle B(v_n+W_A^n,v_n+W_A^n),Av_n\rangle\\
=&\langle v_n^1+W_{A,1}^n\partial_1(v_n^1+W_{A,1}^n),Av_n^1\rangle\\
&+\langle v_n^2+W_{A,2}^n\partial_2(v_n^1+W_{A,1}^n),Av_n^1\rangle\\
&+\langle v_n^1+W_{A,1}^n\partial_1(v_n^2+W_{A,2}^n),Av_n^2\rangle\\
&+\langle v_n^2+W_{A,2}^n\partial_2(v_n^2+W_{A,2}^n),Av_n^2\rangle\\
=&K_1+K_2+K_3+K_4.
\end{aligned} \tag{4.34}$$

For K_1, we have

$$K_1 = \langle v_n^1 \partial_1 v_n^1, Av_n^1 \rangle + \langle v_n^1 \partial_1 W_{A,1}^n, Av_n^1 \rangle$$
$$+\langle W_{A,1}^n \partial_1 v_n^1, Av_n^1 \rangle + \langle W_{A,1}^n \partial_1 W_{A,1}^n, Av_n^1 \rangle$$
$$= l_1 + l_2 + l_3 + l_4. \tag{4.35}$$

For l_1, we have

$$l_1 \leqslant \varepsilon |v_n^1|_{H^2}^2 + C|v_n^1|_{L^4}^2 \cdot |\partial_1 v_n^1|_{L^4}^2.$$

By interpolation inequality, there exists some $C > 0$, such that

$$|v_n^1|_{L^4} \leqslant C|v_n^1|_H^{\frac{1}{2}} |v_n^1|_{H^1}^{\frac{1}{2}}, |\partial_1 v_n^1|_{L^4} \leqslant C|\partial_1 v_n^1|_{L^2}^{\frac{1}{2}} |\partial_1 v_n^1|_{H^1}^{\frac{1}{2}} = C|v_n^1|_{H^1}^{\frac{1}{2}} |v_n^1|_{H^2}^{\frac{1}{2}}.$$

Then

$$l_1 \leqslant \varepsilon |v_n^1|_{H^2}^2 + C|v_n^1|_{L^2} \cdot |v_n^1|_{H^1}^2 \cdot |v_n^1|_{H^2}$$
$$\leqslant 2\varepsilon |v_n^1|_{H^2}^2 + C|v_n^1|_{H^1}^4$$

where the last inequality follows from (4.31). For l_2, we have

$$l_2 \leqslant \varepsilon |v_n^1|_{H^2}^2 + C\int_D (v_n^1)^2 (\partial_1 W_{A,1}^n)^2 \mathrm{d}x$$
$$\leqslant \varepsilon |v_n^1|_{H^2}^2 + C|v_n^1|_{L^2}^2.$$

For l_3, we have

$$l_3 \leqslant C\int_D |\partial_1 v_n^1 \cdot Av_n^1| \mathrm{d}x \leqslant \varepsilon |v_n^1|_{H^2}^2 + C|v_n^1|_{H^1}^2.$$

For l_4, we have

$$l_4 \leqslant C + \varepsilon |v_n^1|_{H^2}^2.$$

By the estimates of $l_1 - l_4$, (4.31) and (4.35), we have

$$K_1 \leqslant 5\varepsilon |v_n^1|_{H^2}^2 + C|v_n^1|_{H^1}^4 + C|v_n^1|_{H^1}^2$$
$$+C(T, \|v_n^0\|_{L_x^\infty}, \|W_A^n\|_{L_t^\infty L_x^\infty}, \|\nabla W_A^n\|_{L_t^\infty L_x^\infty}).$$

Similarly, for K_4 we have

$$K_4 \leqslant 5\varepsilon |v_n^2|_{H^2}^2 + C|v_n^2|_{H^1}^4 + C|v_n^2|_{H^1}^2$$
$$+C(T, \|v_n^0\|_{L_x^\infty}, \|W_A^n\|_{L_t^\infty L_x^\infty}, \|\nabla W_A^n\|_{L_t^\infty L_x^\infty}).$$

For K_2, we have

$$K_2=\langle v_n^2\partial_2 v_n^1, Av_n^1\rangle+\langle W_{A,2}^n\partial_2 v_n^1, Av_n^1\rangle$$
$$+\langle v_n^2\partial_2 W_{A,1}^n, Av_n^1\rangle+\langle W_{A,2}^n\partial_2 W_{A,1}^n, Av_n^1\rangle.$$

For the first term of K_2, by interpolation inequality and (4.31), we have

$$\begin{aligned}\langle v_n^2\partial_2 v_n^1, Av_n^1\rangle &\leqslant \varepsilon|v_n^1|_{H^2}^2+C|v_n^2|_{L^4}^2|\partial_1 v_n^1|_{L^4}^2\\ &\leqslant \varepsilon|v_n^1|_{H^2}^2+C|v_n^2|_{L^2}|v_n^2|_{H^1}|v_n^1|_{H^1}|v_n^1|_{H^2}\\ &\leqslant 2\varepsilon|v_n^1|_{H^2}^2+C\|v_n\|_{\mathbb{H}^1}^4.\end{aligned}$$

For the second term of K_2, we have

$$\langle W_{A,2}^n\partial_2 v_n^1, Av_n^1\rangle\leqslant C\int_D|\partial_2 v_n^1|\cdot|Av_n^1|\mathrm{d}x\leqslant \varepsilon|v_n^1|_{H^2}^2+C|v_n^1|_{H^1}^2.$$

For the third term of K_2, we have

$$\begin{aligned}\langle v_n^2\partial_2 W_{A,1}^n, Av_n^1\rangle &\leqslant C\int_D|v_n^2|\cdot|Av_n^1|\mathrm{d}x\\ &\leqslant \varepsilon|v_n^1|_{H^2}^2+C(T,\|v_n^0\|_{L_x^\infty},\|W_A^n\|_{L_t^\infty L_x^\infty},\|\nabla W_A^n\|_{L_t^\infty L_x^\infty}).\end{aligned}$$

For the last term of K_2, we have

$$\langle W_{A,2}^n\partial_2 W_{A,1}^n, Av_n^1\rangle\leqslant \varepsilon|v_n^1|_{H^2}^2+C.$$

Therefore, we get

$$K_2\ \leqslant\ 5\varepsilon|v_n^1|_{H^2}^2+C\|v_n\|_{\mathbb{H}^1}^4+C|v_n^1|_{H^1}^2+C(T,\|v_n^0\|_{L_x^\infty},\|W_A^n\|_{L_t^\infty L_x^\infty},\|\nabla W_A^n\|_{L_t^\infty L_x^\infty}).$$

Analogously to K_2, we can derive

$$\begin{aligned}K_3\leqslant&\, 5\varepsilon\|v_n^2\|_{H^2}^2+C\|v_n\|_{\mathbb{H}^1}^4+C|v_n^2|_{H^1}^2\\ &+C(T,\|v_n^0\|_{L_x^\infty},\|W_A^n\|_{L_t^\infty L_x^\infty},\|\nabla W_A^n\|_{L_t^\infty L_x^\infty}).\end{aligned}$$

By the estimates of K_1-K_4, we obtain that

$$\begin{aligned}&\langle B(v_n+W_A^n, v_n+W_A^n), Av_n\rangle\\ &\leqslant 10\varepsilon\|v_n\|_{\mathbb{H}^2}^2+C(T,\|v_n^0\|_{L_x^\infty},\|W_A^n\|_{L_t^\infty L_x^\infty},\|\nabla W_A^n\|_{L_t^\infty L_x^\infty})(\|v_n\|_{\mathbb{H}^1}^4+1).\end{aligned}$$

So by (4.33), we get

$$\begin{aligned}&\frac{1}{2}\frac{\partial}{\partial t}\|v_n\|_{\mathbb{H}^1}^2+\|v_n\|_{\mathbb{H}^2}^2\\ &\leqslant 10\varepsilon\|v_n\|_{\mathbb{H}^2}^2+C(T,v_n^0,W_A^n,W_A^n)(\|v_n\|_{\mathbb{H}^1}^4+1).\end{aligned}\tag{4.36}$$

By (4.32), (4.36) and Gronwall inequality, we get

$$\sup_{t\in[0,T_{\max})}\|v_n(t)\|_{\mathbb{H}^1}^2\leqslant C(T,\|v_n^0\|_{L_x^\infty},\|W_A^n\|_{L_t^\infty L_x^\infty},\|\nabla W_A^n\|_{L_t^\infty L_x^\infty}).$$

By (4.36), we can also get that for all $t\in[0,T_{\max})$

$$\int_0^t\|v_n(s)\|_{\mathbb{H}^2}^2\mathrm{d}s\leqslant C(T,\|v_n^0\|_{L_x^\infty},\|W_A^n\|_{L_t^\infty L_x^\infty},\|\nabla W_A^n\|_{L_t^\infty L_x^\infty}). \tag{4.37}$$

For simplicity, we write

$$C(T,v_n^0,W_A^n):=C(T,\|v_n^0\|_{L_x^\infty},\|W_A^n\|_{L_t^\infty L_x^\infty},\|\nabla W_A^n\|_{L_t^\infty L_x^\infty}).$$

Since v_n is mild solution to (4.16), we have, for $t\in[0,T_{\max})$

$$v_n(t)=e^{-tA}v_n^0+\int_0^t e^{-(t-s)A}B(v_n+W_A^n,v_n+W_A^n)\mathrm{d}s.$$

Let $u_n=v_n+W_A^n$. Similarly to derive (3.5) we have

$$\begin{aligned}
&t^{\frac{7}{12}}\|v_n\|_{\mathbb{H}^2}\\
\leqslant&t^{\frac{1}{12}}\|v_n^0\|_{\mathbb{H}^1}+t^{\frac{7}{12}}\int_0^t(t-s)^{-\frac12}\|u_n\nabla u_n\|_{\mathbb{H}^1}\mathrm{d}s\\
\leqslant&t^{\frac{1}{12}}\|v_n^0\|_{\mathbb{H}^1}+Ct^{\frac{7}{12}}\int_0^t(t-s)^{-\frac12}\\
&\times(\|u_n\|_{\mathbb{H}^1}\|u_n\|_{\mathbb{H}^2}+\|u_n\|_{L^2}^{\frac12}\|u_n\|_{\mathbb{H}^2}^{\frac32})\mathrm{d}s\\
\leqslant&C(T,v_n^0,W_A^n)[t^{\frac{1}{12}}+t^{\frac{7}{12}}\int_0^t(t-s)^{-\frac12}s^{-\frac{7}{12}}s^{\frac{7}{12}}\|v_n\|_{\mathbb{H}^2}\mathrm{d}s]\\
&+C(T,v_n^0,W_A^n)t^{\frac{7}{12}}\int_0^t(t-s)^{-\frac12}\mathrm{d}s\\
&+C(T,v_n^0,W_A^n)t^{\frac{7}{12}}\int_0^t(t-s)^{-\frac12}s^{-\frac78}(s^{\frac{7}{12}}\|v_n\|_{\mathbb{H}^2})^{\frac32}\mathrm{d}s\\
&+C(T,v_n^0,W_A^n)t^{\frac{7}{12}}\int_0^t(t-s)^{-\frac12}\mathrm{d}s.
\end{aligned}$$

By Gronwall inequality, we have

$$\begin{aligned}
&t^{\frac{7}{12}}\quad\|v_n\|_{\mathbb{H}^2}\\
&\leqslant C(T,v_n^0,W_A^n)(t^{\frac{1}{12}}+t^{\frac{13}{12}})\\
&\quad\times e^{C(T,v_n^0,W_A^n)t^{\frac{7}{12}}\int_0^t[(t-s)^{-\frac12}s^{-\frac{7}{12}}+(t-s)^{-\frac12}s^{-\frac{7}{12}}\|v_n\|_{\mathbb{H}^2}^{\frac12}]\mathrm{d}s}\\
&\leqslant C(T,v_n^0,W_A^n)(t^{\frac{1}{12}}+t^{\frac{13}{12}})e^{C(T,v_n^0,W_A^n)[t^{\frac12}+t^{\frac14}(\int_0^t\|v_n\|_{\mathbb{H}^2}^2\mathrm{d}s)^{\frac14}]}\\
&\leqslant C(T,v_n^0,W_A^n),
\end{aligned}$$

By Gronwall inequality, we have

$$
\begin{aligned}
t^{\frac{7}{12}} \quad & \|v_n\|_{\mathbb{H}^2} \\
& \leqslant C(T, v_n^0, W_A^n)(t^{\frac{1}{12}} + t^{\frac{13}{12}}) \\
& \quad \times e^{C(T,v_n^0,W_A^n)t^{\frac{7}{12}} \int_0^t [(t-s)^{-\frac{1}{2}} s^{-\frac{7}{12}} + (t-s)^{-\frac{1}{2}} s^{-\frac{7}{12}} \|v_n\|_{\mathbb{H}^2}^{\frac{1}{2}}] \mathrm{d}s} \\
& \leqslant C(T, v_n^0, W_A^n)(t^{\frac{1}{12}} + t^{\frac{13}{12}}) e^{C(T,v_n^0,W_A^n)[t^{\frac{1}{2}} + t^{\frac{1}{4}} (\int_0^t \|v_n\|_{\mathbb{H}^2}^2 \mathrm{d}s)^{\frac{1}{4}}]} \\
& \leqslant C(T, v_n^0, W_A^n),
\end{aligned}
$$

where the seconde inequality follows by Hölder inequality and the last inequality follows by (4.37). Therefore, we can repeat the proof of Theorem 3.1 on $[0, T_n^*], [T_n^*, 2T_n^*], \ldots$ for v_n. If there exists $\Omega_1 \in \mathcal{F}$ satisfying $P(\Omega_1) > 0$ and $T_{\max}(w) < T$, for $w \in \Omega_1$. Then for any $w \in \Omega_1$, there exists $m \in \mathbb{N}$ such that $T_{\max} \in ((m-1)T_n^*, mT_n^*)$. This is a contradiction with the definition of $T_{\max}$. Therefore, $T_{\max} = T$,a.s. and $v_n \in C([0,T]; \mathbb{H}^1)$ satisfying

$$
\begin{aligned}
& \sup_{t \in [0,T]} \|v_n(t)\|_{\mathbb{H}^1} + \sup_{t \in [0,T]} t^{\frac{7}{12}} \|v_n(t)\|_{\mathbb{H}^2} \\
& \leqslant C(T, \|v_n^0\|_{L_x^\infty}, \|W_A^n\|_{L_t^\infty L_x^\infty}, \|\nabla W_A^n\|_{L_t^\infty L_x^\infty}).
\end{aligned}
$$

Obviously the space

$$
\mathcal{B}^T := \{v \in C([0,T]; \mathbb{H}^1); \ \sup_{t \in [0,T]} \|v\|_{\mathbb{H}^1} + \sup_{t \in [0,T]} t^{\frac{7}{12}} \|v\|_{\mathbb{H}^2} < \infty\}
$$

is complete. Since $(v_n)_{n \in \mathbb{N}}$ is bounded in $\mathcal{B}^T$, it is weekly star convergent in this space to a function $\tilde{v}$ which satisfies

$$
\sup_{t \in [0,T]} \|\tilde{v}(t)\|_{\mathbb{H}^1} + \sup_{t \in [0,T]} t^{\frac{7}{12}} \|\tilde{v}(t)\|_{\mathbb{H}^2} \leqslant C(T, \|v_0\|_{L_x^\infty}, \|W_A\|_{L_t^\infty L_x^\infty}, \|\nabla W_A\|_{L_t^\infty L_x^\infty}).
$$

Let us define the mapping $\mathcal{L}_n$ in the same way as $\mathcal{L}$; it is easy to check that $\mathcal{L}_n$ is a strict contraction uniformly in n on $B_{r(w)}^{t(w)}$ where

$$
r(w) = 3(\sup_{n \in \mathbb{N}} \sup_{t \in [0,T]} \|W_A^n\|_{\mathbb{H}^3} + \sup_{t \in [0,T]} \|\tilde{v}(t)\|_{\mathbb{H}^1}^2)
$$

and

$$
2Cr(w)[(t(w))^{\frac{5}{24}} + (t(w))^{\frac{1}{2}}] \leqslant 1,
$$

where the constant C is as in (3.11). Then by a standard arguments, we can prove that

$$
v_n \to v
$$

in $B^{t(w)}_{r(w)}$, implying

$$v=\tilde{v} \quad \text{on } [0,t(w)]$$

and

$$\|v(t(w))\|_{\mathbb{H}^1}+(t(w))^{\frac{7}{12}}\|v(t(w))\|_{\mathbb{H}^2}\leqslant \sup_{s\in[0,T]}\|\tilde{v}(s)\|_{\mathbb{H}^1}+\sup_{s\in[0,T]}s^{\frac{7}{12}}\|\tilde{v}(s)\|_{\mathbb{H}^2}.$$

Thus we can construct a solution on $[t(w),2t(w)]$ starting from $v(t(w))$. We get the unique global solution on $[0,T]$ by reiterating this argument.

References

[1] Beteman H. Some recent researches of the motion of fluid, Monthly Weather Rev. 1915, 43: 163–170.

[2] Bertini L, Cancrini N, Jona-Lasinio G. The stochastic Burgers equation, Comm. Math. Phys., 1994, 165: 211–232.

[3] Chambers D H, Adrian R J, Moin P, Stewart D S, Sung H J. Karhunen-Loéve expansion of Burgers' model of turbulence, Phys. Fluids 1988, 31: 2573–2582.

[4] Choi H, Teman R, Moin P, Kim J. Feedback control for unsteady flow and its application to Burgers equation, Center for Turbulence Research, Stanford University, CTR Manuscript 131. J. Fluid Mechanics 1993, 509–543.

[5] Da Prato G, Debussche A. Stochastic Cahn-Hilliard equation, Nonlinear Anal., 1996, 26: 241–263.

[6] Da Prato G. Arnaud Debussche, Roger Teman, Stochastic Burgers' equation, NoDEA, 1994, 1: 389–402.

[7] Dong Z, Xu T G. One-dimensional stochastic Burgers equation driven by Lévy processes, Journal of Functional Analysis, 2007, 243: 631–678.

[8] Da Prato G, Zabczyk J. Stochastic equations in infinite dimensions, in Encyclopedia of Mathematics and its Application, Cambridge University Press, Cambridge, 1992.

[9] Weinan E, Khanin K, Mazel A. Ya. Sinai, Invariant measures for Burgers equation with stochastic forcing, Ann. of Math., 2000, 151: 877–900.

[10] Forsyth A R. Theory of Differential Equations, Vol. 6, Cambridge University Press, 1906.

[11] Mathieu Gourcy, Large deviation principle of occupation measure for stochastic Burgers equation, Ann. I. H. Poincaré-PR 2007, 43: 441–459.

[12] Gyöngy I, Nualart D. On the stochastic Burgers equation in the real line, Ann. Probab., 1999, 27: 782–802.

[13] Dah Teng Jeng, Forced Model Equation for Turbulence, The Physics of Fluids 1969, 12: 2006–2010.

[14] Hosokawa I, Yamamoto K, Turbulence in the randomly forced one dimensional Burgers flow, J. Stat. Phys., 1975, 245.

[15] Kiselev A, Ladyzhenskaya O A. On the existence and uniqueness of the solution of the nonstationary problem for a viscous, incompressible fluid. (Russian) Izv. Akad. Nauk SSSR. Ser. Mat. 1957, 21: 655–680.

[16] Kardar M, Parisi M, Zhang J C. Dynamical scaling of growing interfaces, Phys. Rev. Lett., 1986, 55: 889.

[17] Nirenberg L. On elliptic partial differential equations, Ann. Sc. Norm. Sup. Pisa 13 (1959), pp. 116–162.

[18] Temam R. Infinite Dimensional Dynamics Systems in Mechanics and Physics. Berlin, Heidelberg, New York: Springer, 1988.

[19] Ton, Bui An, Non-stationary Burgers flows with vanishing viscosity in bounded domains of $\mathbb{R}^3$. Math. Z. 1975, 145: 69–79.

[20] Protter M H, Weinberger H F. Maximum principles in differential equations, Prentice-Hall, Inc., Englewood Cliffs, N.J. 1967 x+261 pp.

[21] Wang G L. Ming Zeng, Boling Guo, Stochastic Burgers equation driven by fractional Brownian motion, J. Math. Anal. Appl., 2010, 371: 210–222.